Conversion factors

1 ft	=	0.305 m
1 mi	=	1.61 km
1 mph	=	0.447 m/s
1 y	=	3.16×10^7 s
1 u	=	1.66×10^{-27} kg
1 lb	=	4.45 N
1 dyn	=	10^{-5} N
1 erg	=	10^{-7} J
1 ft-lb	=	1.36 J
1 eV	=	1.60×10^{-19} J
1 cal	=	4.18 J
1 Btu	=	1.06×10^3 J
1 kWh	=	3.60×10^6 J
1 atm	=	1.01×10^5 Pa
1 cmHg	=	1.33×10^3 Pa
1 psi	=	6.90×10^3 Pa
1 u·c²	=	932 MeV

Values are to three significant figures.

See Appendix D for further information.

PHYSICS

Richard T. Weidner *Rutgers University*

in collaboration with
Michael E. Browne *University of Idaho*

Allyn and Bacon, Inc.
Boston • London • Sydney • Toronto

Developmental Editor: Jane Dahl

Production Editor: Mary Beth Finch

Library of Congress Cataloging in Publication Data

Weidner, Richard T.
   Physics.

   Includes index.
   1. Physics.   I. Browne, Michael E.   II. Title.
QC21.2.W425   1985b   530   84-18571
ISBN 0-205-08078-2 (complete version)
ISBN 0-205-08547-4 (international version)

Printed in the United States of America.

10 9 8 7 6 5 4 3   89 88 87 86

Credits

The cover art is a special-effects photograph created by Michael Freeman, one of a series appearing in the May, 1983, issue of *Smithsonian,* as part of the article by James Trefil titled, ''How the Universe Began.''

   The picture shows events coming about three minutes after the ''Big Bang,'' the event that marked the creation of the Universe about 15 billion years ago. Free quarks are no longer present; they have aggregated in groups of three to form protons and neutrons. Neutrons and protons, each symbolized by a sphere, have started to form the nuclei of low-mass atoms. The temperature of the Universe is still so high, as symbolized by the red clouds of radiation in the foreground, that atoms cannot form. If an electron, symbolized by a yellow streak and star, should happen to get close to a nucleus, it is knocked away by collisions. This is happening to the electrons in the upper left. A mass concentration, perhaps the seed of a future galaxy, is shown as a blue cloud.

Figure 2-16: © BEELDRECHT, Amsterdam / V.A.G.A. New York, Collection Haags Gemeentemusem—The Hague, 1981. Reproduction rights arranged courtesy of the Vorpal Galleries: New York City, San Francisco, Laguna Beach, California. **Page 65:** (top) Russ Kinne/Photo

(Continued on page A-30.)

Contents

19   Thermal Properties of an Ideal Gas, Macroscopic View   401

20   Thermal Properties of an Ideal Gas, Microscopic View   419

21   Thermal Properties of Solids and Liquids   448

22   The Second Law of Thermodynamics and Heat Engines   469

23   Point Electric Charges   491

Preface

- *To concentrate on basic principles* and skip much of the rest   *The aim*
- *To be persuasive*, with crystalline clarity on the fundamentals, no cutting of corners, no superficial explanations, lots of worked examples that illustrate and apply basic ideas, a large variety and number of problems and questions for students to work
- *To be informal*, with a conversational tone and interesting asides
- And then, *to stop*.

That's what I've tried to do in this calculus-based introductory physics textbook for students of engineering and science.

If you're going to do more than just name topics and give a superficial   *Coverage* once-over to each, if the book is to be small enough to be carried around easily, and if students taking introductory calculus-based physics are to have at least some time for other courses and interests, you simply cannot cover all of the topics that, in some ideal world, would be "nice" to include. Something's got to give.

I've concentrated on the fundamental topics, the topics that every budding engineer or scientist simply has to know, and treated these basic ideas in much more detail than is typical in introductory textbooks. Why, for example, is that factor $\cos \theta$ in the definition of work? What do heat and work mean at the microscopic level? These are the sorts of questions that I've not brushed over. And dealing thoroughly with the most basic concerns pays off, in my opinion, in at least two ways—students find the physics to be easier, and they even come to like it.

But which are the truly basic topics? Before this work was begun, I had the benefit of a survey conducted by the publisher and the opinions of several hundred college instructors. Of course, they did not agree in detail. But on some central questions they reached substantial convergence: treat fewer topics thoroughly, and don't attempt to be encyclopedic. Or, don't cover a lot, uncover a little. Although the final choices reflect my own predilections, the topics included here are those that most physics instructors said were most important. This meant dropping entirely some really nice items; probably every instructor will find one or more favorites missing. As a scan of the table of contents will show, this is pretty much straight physics, with a strong emphasis on classical physics. There is little on the history of physics, little on experimental details, little on trendy items. The panels are intended to add a touch of spice to what is basically a meat-and-potatoes diet.

Departures from convention

The sequence of topics is pretty much canonical. Here are specific instances in which the text departs from tradition:

• Chapter 1 is short (barely 4 pages). It does just three things — tells the student to be on the look-out for the "Message of Physics" (that things hang together, and the basic laws of physics are simple); cautions the student that this is "textbook physics" with all of the limitations that implies; and, in the fashion of a preface for the student, tells "What's Where".

• Chapter 2, mostly on the properties of vectors, also includes the statics of a particle to illustrate vector properties applied to forces. This item may, of course, be treated at a later stage.

• Rotation progresses from the simple to the more sophisticated as follows: first equilibrium of a rigid body (Chapter 12), then rotational dynamics (Chapter 13), and finally angular momentum (Chapter 14).

• The first chapter on electrostatics deals with point charges (Chapter 23) and the next one with continuous distributions of electric charges, including Gauss's law (Chapter 24).

• Topics in modern physics are condensed into three (longish) chapters — special relativity (Chapter 40), Quantum Physics (Chapter 41), and Atomic Structure (Chapter 42). This is what most instructors surveyed said that they wanted.

Antecedents

This book is in some respects a lineal descendant of earlier texts of the same publisher that I co-authored with Robert L. Sells. But the changes relative to these earlier works are substantial enough as to make this effectively a new book.

Optional sections

Instructors may wish to skip sections marked *optional*. Some are on detailed proofs (the parallel-axis theorem, for example) that are good to have for the record but may not get much class attention. The results of an optional section are not required in later sections, and they are not included in chapter summaries nor in the problem and question sets.

Use of mathematics

Very little calculus is needed in the first several chapters, and the calculus that does appear there is pretty simple. Elementary calculus is a co-requisite. The dot and cross products are introduced where they are first needed for the physics (work and torque, respectively). Unit vectors, introduced in separate sections, are not used much in problems and can easily be skipped.

Panels

The *panels* are short items that appear on a colored background and are sprinkled throughout the text. They are stories about physics and physicists, asides, extras, independent of the text development. Their style is informal, even breezy, but their subject matter is serious. The chapter location, title, and (in parentheses) the topics are these:

Chapter 2 "Ballpark Figure" (using approximations)
Chapter 4 "Scaling" (strength of materials and scaling dimensions)

Chapter 5 "The Hell with Friction, Use More Grease." (idealization in physics)

Chapter 6 "I Was in the Prime of My Life for Invention . . ." (biographical sketch, Newton)

Chapter 7 "That Professor Goddard" (biographical sketch, Goddard)

Chapter 10 "I Understand the Physics, but I Just Can't Do the Problems." (strategy for solving problems)

Chapter 14 "Show Work" (the role of mathematics in physics)

Chapter 15 "But Why Does it Fall?" (gravitational fields and waves)

Chapter 15 "Happy Birthday, Dear Albert!" (principle of equivalence)

Chapter 16 "The Sicilian" (biographical sketch, Archimedes)

Chapter 19 "The End of the Caloric Theory" (biographical sketch, Benjamin Thompson)

Chapter 20 "Only a Theory" (meaning and role of theory in physics)

Chapter 23 "Famous Physicist, Founding Father" (biographical sketch, Benjamin Franklin)

Chapter 26 "Knowing the Connections" (how physics connects disparate items)

Chapter 30 "And the Beat Goes On" (electromagnetism viewed from different levels of sophistication)

Chapter 31 "How Does Physics Advance?" (strategies for progress in physics)

Chapter 34 "The Odd Couple" (biographical sketches of Faraday and Maxwell, and the Maxwell demon)

Chapter 35 "Everything Was Big but the Particles" (1984 Nobel Prize in Physics)

Chapter 36 "Using a Point Source" (geodesy with radiointerferometry)

Chapter 38 "Hey, Phenomenal!" (biographical sketch, Thomas Young)

Chapter 40 "The Italian Navigator Has Just Landed" (biographical sketch, Enrico Fermi)

Chapter 41 "Particles, Fields" (the electromagnetic interaction from the point of view of quantum field theory)

Chapter 42 "Aha, That Did It!" (biographical sketch, Wolfgang Pauli)

Surely every physics instructor has had the following experience. You're fed up with working yet one more inclined-plane problem, you're afraid that the engineering students are getting a distorted picture of what's really important in physics, so you launch into a discussion of, say, the complementary roles of theory and experiment in physics. After just a few minutes have gone by, and you silently congratulate yourself on the apt phrase you've used, the telling point you've made, someone near the back of the room asks in an exasperated tone, "Is this going to be on the test?" It is hard to present these crucial ideas to intensely goal-oriented freshman engineering students by a frontal attack. What the panels attempt is to do this by guile.

Examples Of worked examples there are many, on the average about one per section. They stress insight, not computation. Skipped steps are only in simple algebra, never in the explanation. I should think that a typical student can read and understand the examples strictly on his own.

Units It's SI all the way. Well, almost. English and cgs units and their relation to SI units are given, usually in footnotes. In a very few instances English units are used in problems where to do otherwise would seem highly contrived (whoever, for example, heard of the power of an automobile engine given in kW, rather than hp?) In thermal physics I've eschewed calories.

Chapter summaries

The summaries at the chapter ends all have the same format with entries organized under four main headings: *Definitions, Units, Fundamental Principles,* and *Important Results.* Deliberately telegraphic, the summaries are intended simply to remind students, of the most essential items, not tell them everything they must know.

Problems and questions

There are many Problems (P) and Questions (Q) at the chapter ends. As defined here, a problem requires some computation, algebraic or numerical; a question does not. Multiple-choice items are most often marked Q. Each P or Q is identified by level of difficulty:

Easy $\equiv$ ·   Medium $\equiv$:   Hard $\equiv$:

Of course, people may disagree on the level of difficulty, but I am pretty confident that no one will think that an item marked easy should have been called hard, or conversely. A student who has the fundamental ideas straight should be able readily to work easy Ps and Qs.

Problems and Questions are also identified by the section number and title. Answers to odd-numbered Ps and Qs are given in the back of the book. So-called *supplementary problems* are found at the very ends of some chapters; not identified by section or difficulty, these problems involve some interesting quirk or applied aspect.

Professor Michael E. Browne of the University of Idaho is identified as collaborator in this work because he produced a large fraction of the problems and questions.

Acknowledgments

I am especially grateful for the comments and criticisms on some portions of the manuscript by my long-time partner and friend, Professor Robert L. Sells of State University College (SUNY), Geneseo. By mutual agreement, Duke is not co-author in this work. His special style and insights have influenced me over many years.

The following professors of physics have reviewed at least portions of the manuscript and made especially valuable comments and suggestions:

Paul A. Bender, Washington State University
George H. Bowen, Iowa State University
Jack Brennan, University of Central Florida
Keith H. Brown, California State Polytechnic University, Pomona
Roger W. Clapp, University of South Florida
Roger Creel, University of Akron
John E. Crew, Illinois State University
Harriet H. Forster, University of Southern California
Simon George, California State University, Long Beach
George Goedecke, New Mexico State University
George W. Greenlees, University of Minnesota
Alvin W. Jenkins, Jr., North Carolina State University
Mohan Kalelkar, Rutgers University

Clement J. Kevane, Arizona State University
Brij M. Khorana, Rose-Hulman Institute of Technology
Sung K. Kim, Macalester College
David Markowitz, University of Connecticut
W. F. Parks, University of Missouri, Rolla
Philip C. Peters, University of Washington
R. L. Place, Ball State University
Marllin Simon, Auburn University
F. B. Stumpf, Ohio University
George A. Williams, University of Utah
John S. Zetts, University of Pittsburgh at Johnstown
E. J. Zimmerman, University of Nebraska, Lincoln
Earl Zwicker, Illinois Institute of Technology

These physicists have made valuable contributions in checking answers and solutions to problems and questions and in other aspects of manuscript preparation:

Professor Roger W. Clapp, Jr., University of South Florida, Tampa
Professor A. Douglas Davis, Eastern Illinois University
Professor William T. Franz, Randolph-Macon College
David Klesch, Michigan State University
Dr. Arthur E. Walters, freelance consultant
Jake Zwart, University of Waterloo, Ontario

I also want to thank members of the staff at the Air Force Academy for their careful reviews of the first and second printings of this text.

I have had a happy association over many years with Allyn and Bacon, Inc. and I am particularly appreciative of the special efforts of the following people in helping to see this work reach completion:

John Gilman, Vice President
Gary Folven, Editor-in-Chief
James M. Smith, Science Editor
Jane Dahl, Developmental Editor
Mary Beth Finch, Production Editor

Above all I am grateful to the students who were not satisfied with facile, stock answers and who kept pressing me for simpler, more persuasive explanations.

The manuscript was effectively typed by Allegra L. Cushing and Cynthia Sells.

I accept responsibility for all errors, and I should like to be informed of those residual infelicities that escaped earlier detection.

R. T. W.

Some Preliminaries 1

We'll skip much of the usual preliminaries — what the subject is about, why it is good for you — and get right into it promptly. For one thing, you probably have already studied some physics and know some of this. More importantly, we want you to believe that physics is fundamental and powerful and immensely practical and beautiful, not because you are simply told that it is, but because you see this from first-hand experience and inner conviction.

1-1   The Message

Albert Einstein said it well:

"The most incomprehensible thing about the universe is that it *is* comprehensible."*

That's the miracle. That's the message. The universe *does* make sense. Things hang together, and they do so in an extraordinarily simple and elegant way.

The notion that the physical universe behaves in a coherent fashion (that there are "laws of nature") is, of course, the fundamental assumption of all of science. But it is especially the message of physics, the fundamental experimental science.

* On page 18 in *Albert Einstein: Creator and Rebel,* by Banesh Hoffman (Viking: New York, 1972).

Be on the look-out for the message of physics. It is all too easy to lose sight of the big picture with too much attention to such admittedly important concerns as "How do you do problem 12-3?" or "What will be covered in the test next week?" So many new items are introduced so rapidly in the typical introductory college physics course that it is easy to be overwhelmed by details. But the message of physics — that the fundamental ideas are few and simple — can also be an important learning aid: if some topic appears to be messy and highly complicated, it is very likely that you have not grasped the basic ideas.

1-2   Textbook Physics

The physics you will read about here is, we hope, clear, straight-forward, persuasive, and digestible. It is not all of physics — not by a long shot — but only those fundamental ideas that every student of science or engineering will have to know. The arguments are simplified, many complications and details go unmentioned, the physics is clean and lean. In short, it is *textbook* physics, and it has these shortcomings:

• Relatively little is said about the historical development of physics, a fascinating and important story in itself. You'll hear little about who did what, when, and where. More importantly, you'll hear little about the failures, the sidetracks, the false starts, the misconceptions and confusions. The physics presented here will be unrealistically simple.

• Physics is first of all an *experimental* science. It deals with what happens both in the real world and, especially, under controlled conditions in the laboratory. To learn about physics only through the simplifications and omissions of a textbook is to miss an essential ingredient. To see what physics is really like, you should experience it in the laboratory — you should hear electrically charged objects crackle; you should see white light dispersed into a spectrum; you should see that the so-called "laws of physics" are in fact *laws* only because they seem to work every time; and you should experience directly the frustrations of trying to do a "clean" experiment, one in which outside influences have presumably been eliminated. The demonstration experiments your instructor performs in lecture sessions can also help to inject a note of realism into your study of physics.

1-3   What's Where

• Each chapter opens with a *table of contents* by section, a preview of what's to come.

• The *Summary* at the end of each chapter reviews the essential content briefly and always in the same format: *Definitions, Units, Fundamental Principles,* and *Important Results.*

The *Definitions* give the vocabulary we must all agree to use when we talk physics. You cannot argue about a definition; it simply must be known precisely. Having a feel for or knowing the gist of a new term is not good enough.

Physics builds: a definition introduced in week w depends on a definition introduced in week $w - 1$, which in turn depends on a definition introduced in week $w - 2$, and so on. It is always useful to ask yourself why the definition is made, rather than merely memorizing it.

Next come newly introduced *Units*, never more than one or two per chapter.

Then come the *Fundamental Principles*. One of the virtues of physics is that there are relatively few of these. Some chapters may have none at all. Don't give inordinate attention to the equation that may summarize a fundamental principle; the equation itself is not the important thing, the idea is.

Important Results are listed at the end. Think of this and all other parts of the *Summary*, not as giving everything that you really need to know, but merely as a reminder of the most crucial items.

• *Problems* (P) and *Questions* (Q) at the end of each chapter are distinguished by the fact that a problem requires some computation, algebraic or numerical, whereas a question does not. The *P*s and *Q*s are identified by the section number and title in which they fall. They are also identified by level of difficulty:

$$\text{Easy} \equiv \cdot \quad \text{Medium} \equiv : \quad \text{Hard} \equiv \vdots$$

Try working easy problems and questions, even if your instructor does not assign them. If you have trouble with easy items, you probably have a serious misconception of the related subject matter and should study it again. Some chapters also have Supplementary Problems at the very end. Unclassified as to difficulty, these problems have an interesting quirk or wrinkle and are often on especially applied matters. Answers to selected problems and questions are given in the back of the book.

• *Optional sections* are not required for later material and usually deal with secondary matter. They are excluded from the chapter summaries and problem and question sets.

• As for *units*, this textbook uses the SI system almost exclusively. (SI is the official abbreviation in all languages for *Systeme Internationale*.) This system of units used to be called the *metric* system, because the meter was the fundamental unit for length. Recently, however, the speed of light has been defined to be the fundamental speed and is assigned a numerical value in meters per second. Together with the fundamental unit for time (defined in terms of an atomic clock), the speed of light now yields the standard for length. For the basic units of the SI system see Appendix A.

• Undoubtedly you know already that the prefix "k" in a unit means the factor 10^3, as in km = kilometer = 10^3 meter. Other official prefixes for *factors of ten* with which you should become familiar are found in Appendix B.

• Physical *constants* that are commonly needed are in Appendix C. Here you will also find astronomical data. Some representative values for physical parameters, such as density or thermal conductivity, appear in tables throughout the book; they can be located with the index.

• *Conversion factors* between various types of units are in Appendix D. (The inside front and back covers also have the most commonly needed constants and conversion factors.)

- Appendix E lists *references* for further or related reading on topics in the book.
- Sprinkled throughout the text are *panels.* Each is printed on a colored background. These are extras—stories about physics or physicists, helpful hints, asides. Although they are written in a lighter tone, each has a serious purpose.

Vectors and Force Equilibrium

2

2-1 Displacement as a Vector
2-2 Vector Components
2-3 Vector Addition by Components
2-4 Vectors in Three Dimensions
2-5 Unit Vectors and Vector Addition
2-6 Force Vectors and Equilibrium
     Summary

Some quantities in physics—a volume of 22.4 l, a mass of 2.0 kg, a time interval of 50 min—have magnitude only. Such quantities are called *scalars.* Other physical quantities have both magnitude and direction and are called *vectors.* Here we consider first the basic properties of displacement vectors. Then we apply the vector properties of forces to a particle in equilibrium.

2-1 Displacement as a Vector

Figure 2-1 shows the path on a map for a trip made by a car beginning at Pittsburgh (point *a* on the map), continuing to New York City (point *b*), and ending at Boston (point *c*).

Displacement is defined as a *change* in the *position of a point,* here that of the car. The displacement of New York (*b*) relative to Pittsburgh (*a*) is represented by a *displacement vector,* the directed straight line labeled **A**. The length of **A** tells how far New York is from Pittsburgh according to the scale on the map; the direction of **A** tells in what direction one must travel along a straight line to get from Pittsburgh to New York. The displacement vector **A** does not give the distance traveled by the car, but rather the final location *b* relative to the starting point *a*. Indeed, if the car had taken a different route between the cities (shown as dashed path in Figure 2-1), the displacement for the trip would still be **A**. In short, any displacement vector always shows

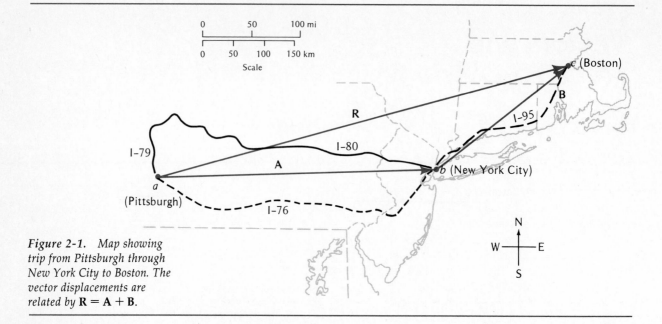

Figure 2-1. *Map showing trip from Pittsburgh through New York City to Boston. The vector displacements are related by* **R = A + B**.

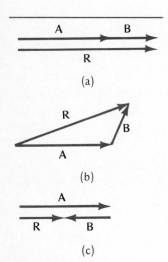

Figure 2-2. *With vector* **A** *fixed in magnitude and direction and vector* **B** *fixed in magnitude only, the resultant vector* **R = A + B** *has its maximum magnitude (a) when* **A** *and* **B** *are aligned, and its minimum magnitude (b) when* **A** *and* **B** *are anti-aligned.*

- Distance from the starting point.
- Direction.

We represent the vector a to b by the boldface symbol **A**. Because it is difficult to write boldface symbols on paper or on the blackboard, a vector quantity, such as **A**, is commonly distinguished from a scalar quantity by an arrow above the symbol, $\vec{A}$, or by a wavy underscore, $\underset{\sim}{A}$, the printer's symbol for boldface type. The magnitude of the vector **A** is a scalar quantity and is symbolized by A in lightface type or by $|\mathbf{A}|$. Thus, in Figure 2-1, if b is 450 km distant from a, $A = 450$ km.

In the second leg of the trip, the car's displacement from New York to Boston is **B**. Vector **R** is the displacement from Pittsburgh to Boston; it is the single displacement representing the location of Boston relative to Pittsburgh, a straight line "as the crow flies." It is clear from geometry that displacement **A** together with displacement **B** is the same as a single displacement **R**; going from a to b to c gives the same result as going directly from a to c. The overall, or *resultant*, displacement **R** can then be written

$$\mathbf{A} + \mathbf{B} = \mathbf{R} \qquad (2\text{-}1)$$

This vector equation says that the resultant vector **R** is the *vector sum* of **A** and **B**. It expresses in symbols what is shown in the geometry of Figure 2-1. The signs $+$ and $=$ imply "addition" and "equality" in the sense shown in the figure. Note that the scalar equation $A + B = R$ is not true unless **A** and **B** happen to point in the same direction. In fact, if **A** is fixed in both magnitude and direction whereas **B** is fixed in magnitude only, the *magnitude* of **R** can have any value ranging from $A + B$ (with **A** and **B** aligned) to $A - B$ (with **A** and **B** in opposite directions). See Figure 2-2.

We restate the definition of a vector. It is a quantity, such as displacement, that has magnitude and direction and that follows the law of vector addition

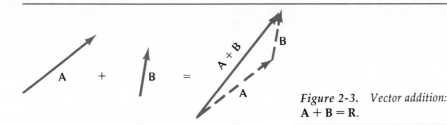

Figure 2-3. *Vector addition:* $\mathbf{A} + \mathbf{B} = \mathbf{R}$.

illustrated in Figure 2-3 and symbolized by (2-1). But, a quantity with magnitude and direction is not a vector unless it obeys the law of vector addition.

Now we consider certain basic operations in vector algebra: the equality of two vectors, the negative of a vector, the commutative and associative laws of vector addition, the subtraction of vectors, and the multiplication of a vector by a scalar. Some of these operations have counterparts in ordinary algebra, and it is useful to recall first these seemingly obvious (yet fundamental) operations as they apply to ordinary numbers. The negative of 3 is such that $3 + (-3) = 0$. The commutative law holds because $3 + 2 = 2 + 3$. The associative law holds because $(2 + 3) + 4 = 2 + (3 + 4)$. Multiplication is simply iterative addition, as in $3 \times 2 = 2 + 2 + 2$. Subtraction of 2 from 5 is simply the addition of -2 to 5, that is, $5 - 2 = 5 + (-2)$.

Equality of Two Vectors   Two vectors are equal when they have the same magnitude and point in the same direction. The equal vectors may, however, be at different locations. Since displacement means *change* in position, two displacements of say 10 km north correspond to the *same* displacement, even though the starting locations may be different. All vectors in Figure 2-4 show the same displacement. This property is important because we can then shift a vector from one location to another in a diagram without changing the vector in any way, again so long as its magnitude and direction are not changed.

Negative of a Vector   If $\mathbf{A}$ is a displacement from a to b, then $-\mathbf{A}$ is a displacement from b to a. As Figure 2-5 shows, vectors $\mathbf{A}$ and $-\mathbf{A}$ have the same magnitude but opposite directions. To find the negative of a vector, we merely reverse its arrow.

Commutative Law in Vector Addition   To arrive at the vector sum of $\mathbf{A}$ and $\mathbf{B}$, we can add (vectorially) either $\mathbf{A}$ to $\mathbf{B}$ or $\mathbf{B}$ to $\mathbf{A}$. See Figure 2-6. Vectors

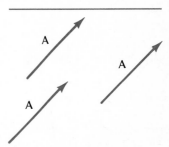

Figure 2-4.   *All vectors* $\mathbf{A}$ *with the same magnitude and direction are equal.*

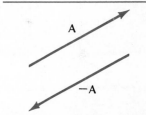

Figure 2-5.   *Vector* $-\mathbf{A}$, *the negative of vector* $\mathbf{A}$, *is equal in magnitude to* $\mathbf{A}$ *but points in the opposite direction.*

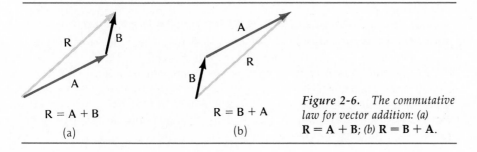

$\mathbf{R} = \mathbf{A} + \mathbf{B}$

(a)

$\mathbf{R} = \mathbf{B} + \mathbf{A}$

(b)

Figure 2-6.   *The commutative law for vector addition: (a)* $\mathbf{R} = \mathbf{A} + \mathbf{B}$; *(b)* $\mathbf{R} = \mathbf{B} + \mathbf{A}$.

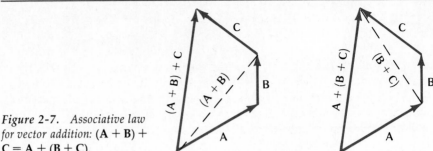

Figure 2-7. Associative law for vector addition: $(\mathbf{A} + \mathbf{B}) + \mathbf{C} = \mathbf{A} + (\mathbf{B} + \mathbf{C})$.

$\mathbf{A} + \mathbf{B}$ and $\mathbf{B} + \mathbf{A}$ are the same because they are parallel and have the same magnitude. Their order in the addition does not matter. Therefore,

$$\mathbf{R} = \mathbf{A} + \mathbf{B} = \mathbf{B} + \mathbf{A}$$

Associative Law in Vector Addition Suppose that we are to find the vector sum of $\mathbf{A}$, $\mathbf{B}$, and $\mathbf{C}$. Then we can either add $\mathbf{C}$ to the vector sum $\mathbf{A} + \mathbf{B}$ or add $\mathbf{A}$ to the vector sum $\mathbf{B} + \mathbf{C}$. As Figure 2-7 shows,

$$(\mathbf{A} + \mathbf{B}) + \mathbf{C} = \mathbf{A} + (\mathbf{B} + \mathbf{C})$$

In fact, for any number of vectors, we can merely draw them in head-to-tail fashion; the sum of all vectors is then just a vector from the tail of the first to the head of the last vector.

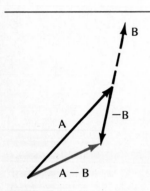

Figure 2-8. Subtraction of vectors: $\mathbf{A} - \mathbf{B} = \mathbf{A} + (-\mathbf{B})$.

Subtraction of Vectors If we know $\mathbf{A}$ and $\mathbf{B}$, how do we find $\mathbf{A} - \mathbf{B}$? We simply add $-\mathbf{B}$ (vector $\mathbf{B}$ with direction reversed) to $\mathbf{A}$:

$$\mathbf{A} - \mathbf{B} = \mathbf{A} + (-\mathbf{B})$$

Figure 2-8 shows the vector difference $\mathbf{A} - \mathbf{B}$. Compare this with the vector sum $\mathbf{A} + \mathbf{B}$ (Figure 2-3).

Change in Vector We shall often need to find the change in a vector. Suppose $\mathbf{R}_i$ gives the displacement of a particle at some initial time. At some later (final) time, displacement of this particle is $\mathbf{R}_f$. What is the change in the displacement, represented by $\Delta\mathbf{R}$? The delta symbol Δ means "the change in," so that $\Delta\mathbf{R}$ means "the change in displacement $\mathbf{R}$." The change in any quantity can always be described as follows; it is what must be added to the initial value to get the final value. For vectors, we have

$$\mathbf{R}_f = \mathbf{R}_i + \Delta\mathbf{R}$$

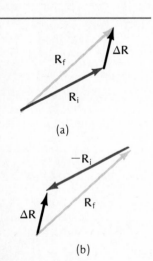

Figure 2-9. Change $\Delta\mathbf{R}$ from vector $\mathbf{R}_i$ to $\mathbf{R}_f$. (a) $\mathbf{R}_f = \mathbf{R}_i + \Delta\mathbf{R}$; (b) $\Delta\mathbf{R} = \mathbf{R}_f - \mathbf{R}_i$.

See Figure 2-9(a).

The change $\Delta\mathbf{R}$ can also be described equivalently as the final vector $\mathbf{R}_f$ minus (in the vector sense) the initial vector $\mathbf{R}_i$:

$$\Delta\mathbf{R} = \mathbf{R}_f - \mathbf{R}_i$$

as shown in Figure 2-9(b). Figure 2-1 shows a vector change. The car's displacement relative to point a (Pittsburgh) changes from $\mathbf{A}$ to $\mathbf{R}$ as the car goes from city b (New York) to city c (Boston); the change in the car's displacement (between New York and Boston) is $\mathbf{B}$.

Multiplication of a Vector by a Scalar   From the addition property of vectors (2-1), we can see that the vector 2**A** is simply the sum of **A** and **A** (2**A** = **A** + **A**), since two successive and identical displacements, **A** and **A**, yield a resultant with the same direction as **A** but twice its magnitude.

More generally, if we multiply vector **A** by scalar *s*, the result is another vector with the direction of **A** but magnitude *sA*. If *s* is negative, *s***A** has a direction opposite that of **A**. See Figure 2-10. Of course, if quantities *s* and **A** both happen to have units, then vector *s***A** is expressed in units indicated for such a product. For example, if **A** = 10 m/s north and *t* = 2 s, then *t***A** = (10 m/s north)(2 s) = 20 m north.

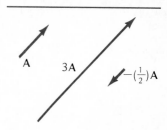

Figure 2-10.   *Multiplication of vector* **A** *by scalars.*

2-2   Vector Components

When we add two vectors, such as **A** and **B**, we replace the separate vectors by a single equivalent resultant vector, **R** = **A** + **B**. We can do this in reverse, replacing a single vector by any two (or more) vectors whose sum gives us back the original vector. Any two nonparallel vectors (in a plane) whose sum yields vector **A** are said to be the *vector components* of **A**. It is usually preferable to replace a vector **A** by its *rectangular* components or resolve it into them; these would constitute two vectors at right angles to one another (along *x* and *y* axes), such as **A**$_x$ and **A**$_y$ in Figure 2-11. Then,

$$\mathbf{A} = \mathbf{A}_x + \mathbf{A}_y$$

From the geometry of Figure 2-11, we see that the magnitudes of components A_x and A_y are related to the magnitude of **A** by

$$A_x = A \cos \theta$$
$$A_y = A \sin \theta \tag{2-2}$$

Following the usual convention, the angle θ is measured counterclockwise from the positive *x* axis. (Angle θ can range from 0° to 360°, or from 0 to 2π radians.) The rectangular components themselves are *vectors*, but we can specify them as *algebraic* quantities. For example, $A_x = -10$ m means that this *x* component has a magnitude of 10 m toward the negative *x* axis.

Equation (2-2) gives the components in terms of the magnitude *A* and direction θ of vector **A**. Conversely, the magnitude and direction of **A** can be expressed in terms of its components. Again, from the geometry of Figure 2-11, we have

$$A = \sqrt{A_x^2 + A_y^2} \tag{2-3}$$

and

$$\tan \theta = \frac{A_y}{A_x}$$

In short, it takes *two* numbers to specify vector **A** in a plane: either *A* and θ or A_x and A_y. Converting from *A* and θ to A_x and A_y is equivalent to converting from polar to rectangular coordinates, and conversely.

We shall have many occasions to find the component of a vector along some particular directed line. Consider, for example, the component A_p of **A** along the line *p* in Figure 2-12(a). The rectangular component of any vector

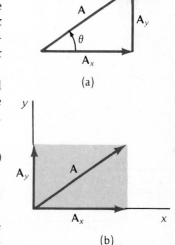

Figure 2-11.   *Vector* **A** *with rectangular components* **A**$_x$ *and* **A**$_y$.

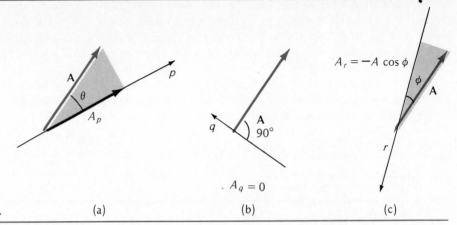

Figure 2-12. *Components of vector* **A** *along three directed lines: (a)* $A_p = A \cos \theta$; *(b)* $A_q = 0$; *and (c)* $A_r = -A \cos \phi$.

(a)   (b)   (c)

along a particular line is the orthogonal projection of the vector along that line; the component is, so to speak, the shadow of the vector on the line, with the vector imagined illuminated at right angles to the line. We see that

$$A_p = A \cos \theta$$

The component is the magnitude of **A** multiplied by the cosine of the angle θ between **A** and the positive end of the directed line. In parts (b) and (c) of Figure 2-12 components are found for the same vector **A**, but now along different directions. Line q is perpendicular to **A**, so that the component $A_q = 0$. The component A_r of **A** along line r is $-A \cos \phi$, a negative quantity, because the angle between the directions of **A** and r is more than $90°$.

2-3   Vector Addition by Components

Rectangular components are particularly useful for finding the vector sum of several vectors. Consider, for example, two vectors **A** and **B** and their sum **R** in Figure 2-13. From the geometry of Figure 2-13, we see that the component of

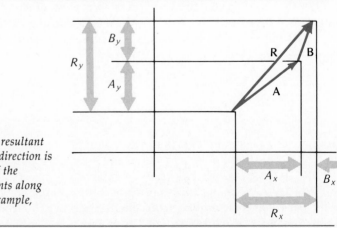

Figure 2-13. *The resultant component in some direction is the algebraic sum of the individual components along that direction, for example,* $R_x = A_x + B_x$.

the vector **R** along any line equals the sum of the components of the vectors **A** and **B** along that line. Along the x axis, $R_x = A_x + B_x$, and along the y axis, $R_y = A_y + B_y$. The magnitude and direction of the resultant vector are then*

$$R = \sqrt{R_x^2 + R_y^2} \quad \text{and} \quad \tan \theta = \frac{R_y}{R_x}$$

Here is the general procedure for finding vector sums using rectangular components:

- Choose conveniently oriented rectangular coordinate axes (so that as many vectors as possible lie along x or y).
- Resolve each vector into its x and y components.
- Sum (algebraically) the x components of the individual vectors to find the x component of the resultant; similarly for the y components.
- Find the magnitude and direction of the resultant from its x and y components.
- To detect gross errors in computation, sketch the vectors roughly in head-to-tail fashion to find the approximate resultant by geometrical construction, and check this result against what was obtained analytically.

Example 2-1. Using the component method, find the resultant vector **R** for the sum of the three vectors **A**, **B**, and **C** shown in Figure 2-7.
We follow the procedure above.

Step 1
Choose the $+y$ axis along the direction of **B**, and the $+x$ axis to the right. The three vectors in Figure 2-7 are also shown in Figure 2-14, where the tails of all vectors have been placed at the common origin of our chosen x, y axes.

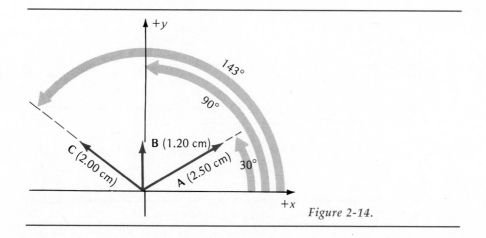

Figure 2-14.

---

* Note that for any particular value of (R_y/R_x), angle θ has two possible values. For example, if R_y and R_x are both positive, vector **R** is in the first quadrant and θ lies between $0°$ and $90°$. If R_y and R_x are both negative, then while R_y/R_x is still positive, vector **R** is in the third quadrant and θ lies between $180°$ and $270°$. In using a hand calculator to find θ, one must be aware of the double-valuing—the calculator always gives the *smaller* value of θ.

Step 2

VECTOR	x COMPONENT (cm)	y COMPONENT (cm)
A	$A_x = (2.50 \text{ cm}) \cos 30° = 2.17$	$A_y = (2.50 \text{ cm}) \sin 30° = 1.25$
B	$B_x = 0.00$	$B_y = 1.20$
C	$C_x = -1.60$	$C_y = 1.20$

Step 3

$$R_x = A_x + B_x + C_x = (2.17 + 0.00 - 1.60) \text{ cm} = 0.57 \text{ cm}$$

$$R_y = A_y + B_y + C_y = (1.25 + 1.20 + 1.20) \text{ cm} = 3.65 \text{ cm}$$

Step 4

$$R = \sqrt{R_x^2 + R_y^2} = \sqrt{(0.57)^2 + (3.65)^2} \text{ cm} = 3.7 \text{ cm}$$

$$\tan \theta = \frac{R_y}{R_x} = \frac{3.65 \text{ cm}}{0.57 \text{ cm}} = 6.4$$

$$\theta = 81°$$

Step 5

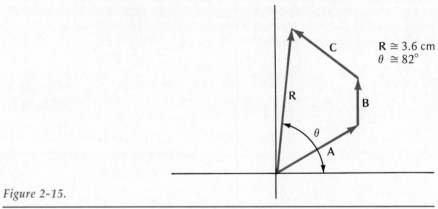

$$R \cong 3.6 \text{ cm}$$
$$\theta \cong 82°$$

Figure 2-15.

2-4   Vectors in Three Dimensions

All vectors we have considered thus far have been in a single plane. The vector properties hold equally well in three-dimensional space. Adding vectors in three dimensions is shown in Figure 2-16.

In Figure 2-17 the three components $\mathbf{A}_x$, $\mathbf{A}_y$, and $\mathbf{A}_z$ of a vector $\mathbf{A}$ are shown along the x, y, z coordinate axes. It follows that

$$\mathbf{A} = \mathbf{A}_x + \mathbf{A}_y + \mathbf{A}_z$$

with

$$A = \sqrt{A_x^2 + A_y^2 + A_z^2}$$

where A_z is the component magnitude of $\mathbf{A}$ along the $+z$ axis, and similarly for A_x and A_y.

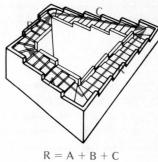

$$R = A + B + C$$

Figure 2-16.   *M. C. Escher's* Ascending-Descending *(with vectors added). Here* **R** = **A** + **B** + **C**.

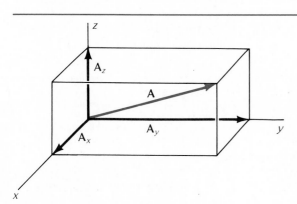

Figure 2-17.   *Vector* **A** *and its three rectangular components,* A_x, A_y, *and* A_z.

The Ballpark Figure

Approximate Calculations

You've forgotten your pocket calculator. Or its battery is run down. Or maybe you're discussing a problem with friends at dinner. In any event, you have no easy way of making computations to, say, eight significant figures. You'll have to work with pencil and the back of an envelope. And for the problem at hand, you'll settle for a ballpark figure—an answer with the right power of ten, or order of magnitude. Actually, with just a little care an approximate calculation will, in fact, not be off by more than about a factor of 2 or 3.

To do approximate calculations easily, you must first relax. Learn to round off numbers to easy ones. Thus, take 92 to be 100, and 0.0326 to be 3×10^{-2}. For example, what is π^2, roughly? Pi is 3. Three squared is nine. Nine is roughly ten. The result: $\pi^2 \simeq 10$.

Another example: estimate the volume of a spherical water droplet having a diameter of 0.21 mm.

The volume V of a sphere is given by

$$V = \frac{4}{3} \pi R^3$$
$$= \frac{4}{3} \pi \left(\frac{D^3}{8} \right) = \frac{\pi D^3}{6}.$$

Taking $\pi \simeq 3$ and $D \simeq 0.2$ mm gives

$$V \simeq \frac{3}{6} (2 \times 10^{-1} \text{ mm})^3 = 4 \times 10^{-3} \text{ mm}^3$$

This rough estimate is within 20 percent of the more precise calculation 4.8×10^{-3} mm³.

Metric Units

Except for everyday usage in the United States and the United Kingdom, the world is effectively metricized. You should be able to give ballpark figures to common physical quantities in metric units. The exact conversion factors can always be looked up when required. Having a good feel for the approximate conversion figure is far more valuable. Here are some basic metric-English conversion factors that everyone should know. They are off by no more than 10 percent.

$$
\begin{aligned}
1 \text{ m} &\simeq 3 \text{ ft} \\
25 \text{ mm} &\simeq \text{ diameter of a quarter} \\
&\qquad (25¢) \simeq 1 \text{ inch} \\
100 \text{ km} &\simeq 60 \text{ mi} \\
1 \text{ liter} &\simeq 1 \text{ quart} \\
1 \text{ newton} &\simeq \text{ weight of small apple} \\
1 \text{ kg} &\simeq 2 \text{ lb} \\
20° \text{ Celsius} &\simeq 70° \text{ Fahrenheit}
\end{aligned}
$$

To add vectors in three dimensions, it is usually simplest first to resolve each vector into its *three* rectangular components.

To a vector equation like

$$\mathbf{R} = \mathbf{A} + \mathbf{B} + \mathbf{C}$$

there correspond three component algebraic equations:

$$R_x = A_x + B_x + C_x$$
$$R_y = A_y + B_y + C_y$$
$$R_z = A_z + B_z + C_z$$

Herein lie the economy, generality, and elegance of the vector algebra. A single vector equation replaces three component equations. Moreover, the vector equation expresses a relation that is independent of the particular choice of coordinate axes.

Temperature differ-ence of 1 Celsius degree	$\simeq$	temperature differ-ence of 2 Fahrenheit degrees
1 cc water	$\simeq$	1 ml water $\simeq$ 1 gm water
Size of atom	$\simeq$	10^{-10} m = 0.1 nm
Avogadro's number (number of atoms in one mole)	$\simeq$	number of grains of sand in a cube 1 mile on an edge

Mathematical Approximations

When mathematical approximations can be used, they can save a lot of work. Here are some of the most useful.

Binomial Expansion

$$(1 + x)^n = 1 + nx + \frac{n(n-1)}{2!}x^2$$
$$+ \frac{n(n-1)(n-2)}{3!}x^3 + \cdots$$

If $x \ll 1$,

$$(1 + x)^n \simeq 1 + nx$$

Example: $\dfrac{1}{0.97} = ?$

$$\frac{1}{0.97} = \frac{1}{1 - 0.03}$$
$$= (1 - 0.03)^{-1} = 1 + (-1)(-0.03) = 1.03$$

Trigonometric Expansions

When angle θ is measured in *radians*,

$$\sin \theta = \theta - \frac{\theta^3}{3!} + \frac{\theta^5}{5!} - \cdots$$

$$\cos \theta = 1 - \frac{\theta^2}{2!} + \frac{\theta^4}{4!} - \cdots$$

For a *small* angle θ,

$$\sin \theta \simeq \theta \quad \text{and} \quad \cos \theta \simeq 1$$

Exponential Expansion

$$e^x = 1 + x + \frac{x^2}{2!} + \frac{x^3}{3!} + \cdots$$

where $e = 2.71828.$ . . is the base of the natural logarithms.

Physics and Precision

Are physicists obsessed with measuring all quantities with extraordinarily high precision? Not so. The precision must be appropriate to the circumstances.

When a physicist cooks a meal, measuring the amount of salt by such a vague unit as "a pinch" is perfectly all right. But when a physicist is devising an atomic clock as the standard for measuring time intervals, the precision of 1 part in 10^{12} (less than 1 second in 10 millennia) is hardly excessive.

2-5 Unit Vectors and Vector Addition

In vector algebra, it is often useful to represent a vector by its magnitude and a unit vector. Any vector $\mathbf{A}$, for example, can always be written

$$\mathbf{A} = A\mathbf{u}$$

where A is the magnitude of $\mathbf{A}$ and $\mathbf{u}$ is a dimensionless vector, of magnitude 1, pointing in the same direction as $\mathbf{A}$. That is, $\mathbf{u}$ is a *unit* vector.

A unit vector may point in any direction. For a rectangular coordinate system with $x, y,$ and z axes (Figure 2-17) the unit vectors along these respective positive axes are usually designated $\mathbf{i}, \mathbf{j},$ and $\mathbf{k}$. (See also Figure 2-18.)

Any three-dimensional vector $\mathbf{A}$ can then be represented in either of the following ways:

- As the vector sum of its three rectangular vector components,

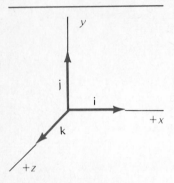

Figure 2-18. *Unit vectors* **i**, **j**, **k** *along x, y, z positive axes, respectively.*

$$\mathbf{A} = \mathbf{A}_x + \mathbf{A}_y + \mathbf{A}_z$$

- In terms of unit vectors,

$$\mathbf{A} = A_x\mathbf{i} + A_y\mathbf{j} + A_z\mathbf{k}$$

Here A_x, A_y, and A_z, the respective components (with dimensions) along the directions of these unit vectors, are algebraic quantities that may take both positive and negative values.

The sum of two or more vectors in unit-vector notation is

$$\mathbf{A} + \mathbf{B} = (A_x\mathbf{i} + A_y\mathbf{j} + A_z\mathbf{k}) + (B_x\mathbf{i} + B_y\mathbf{j} + B_z\mathbf{k})$$

$$= (A_x + B_x)\mathbf{i} + (A_y + B_y)\mathbf{j} + (A_z + B_z)\mathbf{k}$$

Example 2-2. Use unit vectors to find the vector sum of a displacement **A** of 2.0 m at 30° and a displacement **B** of 2.5 m at 133° counterclockwise. See Figure 2-19.

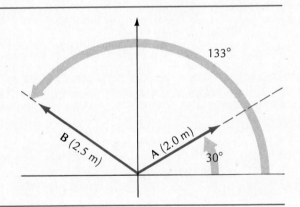

Figure 2-19.

$$A_x = (2.0 \text{ m}) \cos 30° = 1.7 \text{ m} \qquad A_y = (2.0 \text{ m}) \sin 30° = 1.0 \text{ m}$$

$$B_x = (2.5 \text{ m}) \cos 133° = -1.7 \text{ m} \qquad B_y = (2.5 \text{ m}) \sin 133° = 1.8 \text{ m}$$

The vector sum is

$$\mathbf{A} + \mathbf{B} = (A_x + B_x)\mathbf{i} + (A_y + B_y)\mathbf{j}$$

$$= (1.7 - 1.7)\mathbf{i} + (1.0 + 1.8)\mathbf{j}$$

$$\mathbf{A} + \mathbf{B} = 0\mathbf{i} + (2.8 \text{ m})\mathbf{j}$$

The resultant vector $\mathbf{R} = \mathbf{A} + \mathbf{B}$ is therefore 2.8 m along the $+y$ axis.

2-6 Force Vectors and Equilibrium

Force is one important example of a physical quantity with *vector* properties — magnitude and direction. When two or more forces act simultaneously on an object, we can replace the individual forces by a single force equal to their vector sum. We add forces as vectors to arrive at a single equivalent resultant force. We can also replace a single force by its components.

A precise definition of force will be given later (Section 5-3), as will evidence for its vector properties (Section 5-4). Now we are content to use our intuitive notion of force as a push or a pull, as in muscular exertion. Among the simple forces we consider here are:

- Weight of an object (the downward force on the object arising from gravitational attraction of the earth).
- Force on an object due to a stretched or compressed spring.
- Tension in a cord when its two ends are pulled.
- Force of one object touching or pressing against another object.

In the SI unit system (Système Internationale), the unit for force is the *newton* (named for the founder) and abbreviated N.

$$1 \text{ N} \simeq \tfrac{1}{5} \text{ lb} \simeq \text{weight of a 0.1-kg object (such as a small apple)}$$

More precisely, $1 \text{ N} = 0.2248$ lb. Since the direction of a force must always be specified, the weight of a 100-gm object is represented by a force vector with a magnitude of approximately 1 N in the down direction (toward the center of the earth).

A simple procedure for measuring and comparing forces is shown in Figure 2-20. Here an ordinary spring has its upper end fixed to a ceiling; various objects may be suspended from its lower end. Suppose that a standard weight of 1 N is attached to the spring and after the weight has come to rest, it just happens to have stretched the spring by 1 cm [Figure 2-20(b)]. Now suppose that the standard weight is replaced by another weight, and the elongation of the spring turns out to be 3 cm [Figure 2-20(c)]. Then by definition, the size of the second weight attached is three times the standard weight, or 3 N. In other words, the elongation of the spring is directly proportional to the force applied to it. This behavior, a linear relation between applied force and elongation, is followed by any springs for which the change in length is small compared with the relaxed length.

Here we are concerned with forces acting on a *particle*. For our purposes a particle is simply an object so small that we need not be concerned with its internal structure. Even an object of finite size can be regarded as equivalent to a particle if all forces on it pass through a single point.

In all situations in this chapter we consider the particle to be at rest and to *remain at rest*. Such a particle is said to be in *static equilibrium*. What is the relation among the forces acting on a particle in static equilibrium? We can

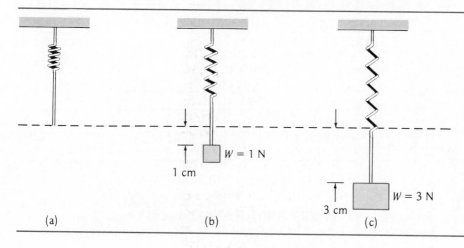

(a)   (b)   (c)

Figure 2-20. *The magnitude of the force produced by a stretched spring is directly proportional to the amount of stretch.*

readily accept the notion that if no forces at all act on a particle, the undisturbed particle will, if initially at rest, remain at rest. (Actually, a particle would be truly free of all forces only if it were located in interstellar space with no other objects nearby.) But—and this is the essential point—*no force* at all is equivalent to *zero resultant force*. Said differently, a particle remains in equilibrium if the vector sum of all forces *acting on it is zero*.*

We can express the condition for equilibrium of a particle in symbols. The resultant force **R** on a particle is the vector sum of the individual forces **F**$_1$, **F**$_2$, **F**$_3$, acting on the particle. For equilibrium,

$$\mathbf{R} = \mathbf{F}_1 + \mathbf{F}_2 + \mathbf{F}_3 + \cdots = 0 \qquad (2\text{-}4)$$

If the resultant force **R** is zero, then so too are its x and y components, R_x and R_y,

$$R_x = F_{1x} + F_{2x} + F_{3x} + \cdots = 0$$

and $\qquad\qquad\qquad\qquad\qquad\qquad\qquad\qquad\qquad\qquad\qquad\qquad (2\text{-}5)$

$$R_y = F_{1y} + F_{2y} + F_{3y} + \cdots = 0$$

In (2-5) we have used a result already proved; the x component of the resultant vector is merely the algebraic sum of the x components of the individual vectors, and similarly for y. (We do not write an equation for z components since in our examples, all forces will lie in the single, xy plane.)

The following examples illustrate these ideas. We shall identify a variety of forces. Each force is specified completely only if we give:

- Magnitude of the force (in appropriate units).
- Direction of the force.
- Object on which the force acts.

Example 2-3.   A block having a weight of 100 N (~22 lb) remains at rest on a horizontal surface [Figure 2-21(a)]. What is the upward force of the surface on the block?

The only forces acting on the block are its weight **W** downward and the *normal force* **N** upward [Figure 2-21(b)]. The normal force is so named because it is normal (perpendicular) to the surface. Of course, the block presses down upon the surface and there-

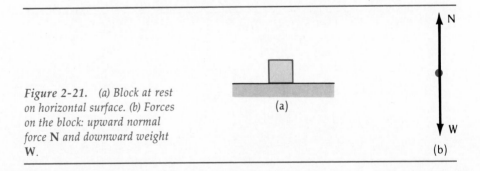

*Figure 2-21.   (a) Block at rest on horizontal surface. (b) Forces on the block: upward normal force **N** and downward weight **W**.*

(a)

(b)

---

* If a particle remains at rest, the resultant force on it is zero. If the resultant force on a particle is zero, the particle will remain at rest *if initially at rest,* or the particle may move in a straight line at constant speed. Here we deal only with particles at rest.

fore exerts a downward force on it. However, this downward force of block on surface *does not act on the object whose equilibrium we are studying*—the *block;* accordingly, *we exclude it.*

With the block in equilibrium, we have

$$\mathbf{N} + \mathbf{W} = 0 \quad \text{or} \quad \mathbf{N} = -\mathbf{W}$$

Since $\mathbf{W} = 100$ N down,

$$\mathbf{N} = -\mathbf{W} = 100 \text{ N up}$$

Example 2-4.   The 100-N block is now supported by a cord attached to a ceiling (Figure 2-22). What is the tension force $\mathbf{T}$ in the cord?

The cord replaces the horizontal surface of Example 2-3, and it produces an upward force $\mathbf{T}$ on the block. The forces acting on *the block* are its weight $\mathbf{W}$ and the tension $\mathbf{T}$ [Figure 2-22(b)].

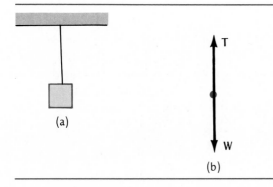

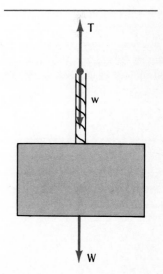

Figure 2-22. *(a) Block suspended from cord. (b) Forces on block: upward tension $\mathbf{T}$ and downward weight $\mathbf{W}$.*

For equilibrium,

$$\mathbf{T} + \mathbf{W} = 0$$

$$\mathbf{T} = -\mathbf{W} = -100 \text{ N down} = 100 \text{ N up}$$

A tension exists, in fact, at all points along the cord, not merely at the point of attachment to the block.

By definition, the magnitude of the tension is the force one side applies on the adjoining side at any imaginary cross section of the cord. For a weightless cord, or one whose weight is small compared with all other acting forces, the tension has the same magnitude at all points along the cord. See Figure 2-23. Here $\mathbf{T} + \mathbf{W} + \mathbf{w} = 0$, where $\mathbf{w}$ is the weight of the cord section. Since $\mathbf{w} \simeq 0$, $\mathbf{T} \simeq -\mathbf{W}$, and each piece of cord produces a force of 100 N on the adjoining piece.

If a perfectly flexible cord with negligible weight is in contact with a frictionless

Figure 2-23. *For a weightless cord ($\mathbf{w} \simeq 0$), the tension $\mathbf{T}$ is nearly equal in magnitude to the suspended weight $\mathbf{W}$.*

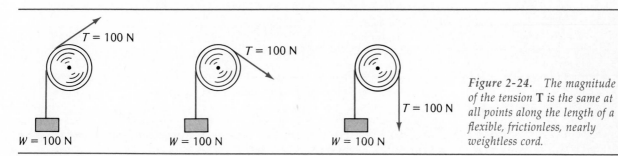

Figure 2-24. *The magnitude of the tension $\mathbf{T}$ is the same at all points along the length of a flexible, frictionless, nearly weightless cord.*

surface, the *magnitude* of the tension is again unchanged throughout its length. The direction of the tension force changes; it is always tangent to the cord. In Figure 2-24, where a weightless cord is in contact with a frictionless pulley and the 100-N weight is stationary, the magnitude of tension **T** is 100 N.

Example 2-5.   A weight of 100 N is supported as shown in Figure 2-25(a). Three cords of negligible weight are joined at a knot. What are the magnitudes of the three tensions T_1, T_2, and T_3?

Our first decision: What particle (or equivalent particle) in equilibrium do we *choose*? After all, every object shown in the figure remains at rest and is in equilibrium. Suppose we choose the 100-N block. Tension T_2 acts on it, but not tensions T_1 and T_3, so we have no way of determining them. Similarly, the piece of wall to which the cord with tension T_1 is attached is in equilibrium, but T_2 and T_3 do not act on this piece. We choose the knot, because all three forces act on it.

Figure 2-25(b) shows the knot (isolated from all else in a so-called free-body diagram) and vectors representing the forces on the knot. The weight of the knot is negligible.

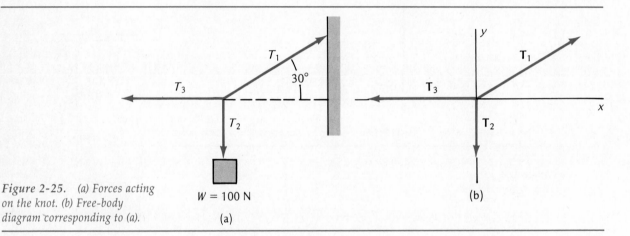

Figure 2-25.   *(a) Forces acting on the knot. (b) Free-body diagram corresponding to (a).*

From Example 2-5, we know that T_2 is down and has a magnitude of 100 N. Although lengths have been drawn for vectors T_1 and T_3 in Figure 2-25, we really do not yet know their actual magnitudes. In fact, that is the problem we're solving.

The x and y axes have been superimposed on the diagram, since it will be useful to replace any vector not along x or y by its rectangular components. For T_1 the components are

$$T_{1x} = T_1 \cos 30° \quad \text{and} \quad T_{1y} = T_1 \sin 30°$$

The tension T_3 is along the negative x axis (and therefore has a zero y component); T_2 is along the negative y axis (and has a zero x component);

Since the knot remains at rest and the resultant force on it is zero, we know that the resultant force component along x is zero:

$$R_x = T_1 \cos 30° - T_3 = 0$$

Likewise, the resultant force component along y is zero:

$$R_y = T_1 \sin 30° - 100 \text{ N} = 0$$

We have finished the physics. The rest is algebra. With two equations, we can solve for the two unknowns (T_1 and T_3). The second equation, solved for T_1, yields

$$T_1 = \frac{100 \text{ N}}{\sin 30°} = \frac{100 \text{ N}}{0.500} = 200 \text{ N}$$

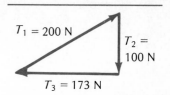

Figure 2-26.

and the first equation then gives

$$T_3 = T_1 \cos 30° = (200 \text{ N})(0.866) = 173 \text{ N}$$

Note that the tension $T_1 = 200$ N in the oblique cord exceeds the magnitude of the block's weight of 100 N.

Figure 2-26 shows the force vectors for this example once more, now arranged in head-to-tail fashion. We see that the vectors form a closed triangle. In other words, it is impossible to draw a vector (the resultant) from the tail of the first to the head of the last vector. The resultant force vector is zero. More generally, whenever a particle is in equilibrium, the vectors representing the forces acting on the particle form a *closed* polygon. A resultant force vector can never be drawn; it is always zero.

The three force vectors as portrayed in Figure 2-26 are the *same* vectors shown originally in Figure 2-25. In Figure 2-25, the tail of each force vector was located at the chosen particle simply for convenience. But we recall that we can shift any vector around on a sheet without changing the vector, so long as we do not change its magnitude and direction.

Example 2-6.   A 100-N block is at rest on a rough inclined plane that makes an angle of 37° with the horizontal. Find the magnitude of (a) the normal force **N** of the surface on the block and (b) the force of friction **f** along the surface.

A sketch of the physical situation is shown in Figure 2-27.

Next, we choose the object whose equilibrium we shall examine; that is the block.

Figure 2-28 shows all the forces on the block: the downward 100-N weight, the normal force **N** perpendicular to the surface (which keeps the block from sinking into the surface), and the friction force **f** acting upward along the surface (which keeps the block from sliding down the incline).

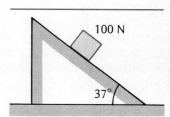

Figure 2-27.

Here it is advantageous to orient the x and y axes as shown in the figure, parallel and perpendicular to the incline.

With this orientation of axes, two of the three forces (**f** and **N**) lie entirely along x or y. The 100-N weight **W** now makes an angle of 37° with the negative y axis, a result from geometry.*

We first resolve all forces into x and y components (Figure 2-29). For the weight **W** they are

$$W_x = W \sin 37°$$

$$W_y = -W \cos 37°$$

(here the *cosine* does *not* go with the x component, since the 37° angle is measured from the y axis).

Any vector is equivalent to the vector sum of its components. Therefore, the weight **W** of the block, vertically downward, is equivalent in its effect on the block to two simultaneous forces: a force of $W \sin 37°$ down the incline and a force of $W \cos 37°$ perpendicular to the surface.

With the block remaining at rest on the incline, the resultant force along x is zero,

$$R_x = W \sin 37° - f = 0$$

and likewise for the y component,

$$R_y = N - W \cos 37° = 0$$

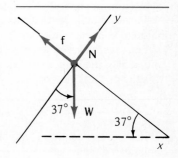

Figure 2-28.

* Two sets of mutually perpendicular lines — here (a) horizontal and vertical and (b) x and y — make equal angles with corresponding sides.

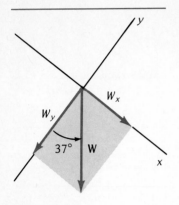

Figure 2-29.

The algebraic signs for the force components tell whether the component is right or left, up or down.

Again the physics is done, and the rest is algebra. We have deliberately retained symbols to the end (the weight was given as W, not as 100 N), instead of plugging in numerical values. This is almost always advisable.

With W taken to be 100 N, the first equation yields

$$f = W \sin 37° = (100 \text{ N})(0.60) = 60 \text{ N}$$

and the second equation

$$N = W \cos 37° = (100 \text{ N})(0.80) = 80 \text{ N}$$

With an inclined surface, the normal force (80 N) is *less* in magnitude than the weight (100 N).

In these examples, we have followed a procedure that applies to all problems for a particle in equilibrium:

- Draw a sketch of the situation, no matter how simple.
- Choose the object (or particle) whose equilibrium will be studied.
- Draw separately the particle alone and vectors representing all forces acting on the chosen particle. For problems considered here, the only possible forces are the particle's weight and those produced by adjoining objects with which it is in contact. Do not include forces the chosen particle exerts on nearby objects; these are not forces *on the chosen particle.* If the magnitude or the direction of a force is not known, choose a symbol.
- Draw x and y axes to make the resolution of force vectors into components as simple as possible. Find the x and y components of all forces with proper attention to plus and minus signs.
- Write down two equations, one for the sum of x-force components, the other for the sum of y-force components, with each sum set equal to zero.
- Solve for the unknowns and check whether the magnitudes and directions make sense.

Summary

Definitions

Scalar: a quantity that has magnitude only.

Vector: a quantity that has magnitude and direction and follows the laws for adding displacement vectors.

Equal vectors: vectors that have the same magnitude and point in the same direction.

Change $\Delta\mathbf{R}$ going from displacement vector $\mathbf{R}_i$ to vector $\mathbf{R}_f$:

$$\Delta\mathbf{R} = \mathbf{R}_f - \mathbf{R}_i$$

Resultant vector for a perpendicular x-y coordinate system in two dimensions:

$$\mathbf{R} = \mathbf{A} + \mathbf{B} \qquad \begin{cases} R_x = A_x + B_x \\ R_y = A_y + B_y \end{cases} \qquad (2\text{-}1)$$

Relation between rectangular coordinates and polar coordinates in two dimensions:

$$A_x = A \cos \theta$$
$$A_y = A \sin \theta \qquad (2\text{-}2)$$
$$A = \sqrt{A_x^2 + A_y^2} \qquad (2\text{-}3)$$

$$\tan \theta = \frac{A_y}{A_x}$$

Fundamental Principles

The resultant force on a particle is the vector sum of the individual forces:

$$\mathbf{R} = \mathbf{F}_1 + \mathbf{F}_2 + \cdots \qquad \begin{cases} R_x = F_{1x} + F_{2x} + \cdots \\ R_y = F_{1y} + F_{2y} + \cdots \end{cases}$$

If the particle is in static equilibrium, the resultant force is zero:

$$\mathbf{R} = 0 \quad \begin{cases} R_x = F_{1x} + F_{2x} + \cdots = 0 \\ R_y = F_{1y} + F_{2y} + \cdots = 0 \end{cases} \quad \text{(2-4), (2-5)}$$

Important Results

The negative $-\mathbf{V}$ of a vector has the same magnitude as $\mathbf{V}$ but points in the opposite direction.

The commutative and associative laws hold for vector addition.

The product of a scalar quantity s and a vector $\mathbf{V}$ is a vector pointing in the same direction as $\mathbf{V}$ and having magnitude sV. The units of this product consist of the product of the units of s and $\mathbf{V}$.

A *unit vector* is dimensionless and has magnitude 1. Unit vectors respectively along the x, y, and z axes are written $\mathbf{i}$, $\mathbf{j}$, $\mathbf{k}$.

Problems and Questions

P ≡ problem   Q ≡ question
· ≡ easy   : ≡ medium   : ≡ hard

Section 2-1   Displacement as a Vector

· **2-1 Q** Which of the following are vector quantities?
(A) Number of students in a class
(B) Velocity of a thrown baseball
(C) Weight of a car
(D) Energy content of a piece of coal
(E) Distance from New York to Milwaukee
(F) Displacement from Billings to Denver
(G) Time duration of the Mesozoic era

· **2-2 Q** When two vectors **A** and **B** are added, their resultant will have a magnitude
(A) $A + B$
(B) $\sqrt{A^2 + B^2}$
(C) $\frac{1}{2}(A + B)$
(D) zero.
(E) that cannot be determined without more information.

· **2-3 Q** How many angles must be given to specify the direction of a vector?

· **2-4 Q** If $\mathbf{A} + \mathbf{B} = \mathbf{C}$, is it possible that more than one vector **B** exists which will satisfy this equation for given vectors **A** and **C**?

· **2-5 Q** Which of the following is a true statement?
(A) It is possible for a vector to have zero magnitude even if one of its components is not zero.
(B) A road sign in the shape of an arrow is an example of a vector quantity.
(C) Several vectors may be added graphically by placing them together head–tail–tail–head–head–tail–tail . . . etc.
(D) When two vectors of magnitudes 2 N and 3 N are added they yield a vector of magnitude 5 N, but the direction of this resultant vector cannot be determined without more information.
(E) If three vectors add up to zero, they must all lie in the same plane.

: **2-6 Q** A vector is unchanged if
(A) it is rotated through an arbitrary angle θ.

(B) it is rotated through 180°, but not through other angles.
(C) it is multiplied by a dimensionless factor.
(D) it is slid parallel to itself.
(E) a unit vector is added to it.

· **2-7 P** Two people push horizontally on a packing crate with forces of 100 N and 200 N. What is the total force exerted on the crate? (Estimate graphically.)

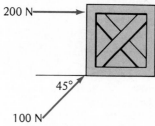

Figure 2-30.   *Problem 2-7.*

: **2-8 P** Many substances occur in a cubic crystal structure with a particular kind of atom at the center of a cube face. Suppose that such a cube of side a is positioned in the first quadrant with one corner at the origin and with cube edges along the xyz axes. Write the position vectors for atoms at the center of each cube face.

· **2-9 P** A canoeist wants to stay in radio contact with the ranger at park headquarters where she starts her trip. She plans to travel 34 km southeast and then 22 km approximately due east. Estimate graphically how far from headquarters she will then be. This information will help her know how powerful her transmitter must be.

· **2-10 P** Keller's Gulch is 30 km northeast from a woodsman's camp. He starts out from camp one morning at 7:00 A.M. and hikes 10 km due east to Camp 9. He gets there at 9:00 A.M. After having a cup of coffee and a bite to eat he sets out for Keller's Gulch. (a) Assuming he always hikes at the same rate, how long will it take him to go from Camp 9 to Keller's Gulch? (b) In what compass direction, given with respect to north, should he head?

: **2-11 P** An airplane flies 332 km on a compass bearing of 95° from Boise to Idaho Falls and then 440 km on a bearing of 8° from Idaho Falls to Great Falls, Montana. How far and on what bearing must the plane fly to return directly to Boise from Great Falls? (Compass bearings are determined by measuring clockwise on a map, starting with due north as zero.)

Section 2-2 Vector Components

· 2-12 P A vector has components $A_x = 6$ and $A_y = 8$. What angle does the vector make with the x axis?

· 2-13 P A gravitational force of 60 N acts vertically downward on an object resting on a plane inclined at 30° above horizontal. What is the component of this force parallel to the plane?

· 2-14 P An electric field vector has components $E_x = 200$ volts/meter and $E_y = 150$ volts/meter. What is the magnitude of the electric field?

· 2-15 P In Washington, D.C., the earth's magnetic field vector points down toward the earth at an angle of 71° below horizontal. What are (a) the vertical and (b) the horizontal components of the field if its magnitude is 0.57×10^{-4} Tesla?

: 2-16 P Consider a coordinate system $x'y'$ which is rotated counterclockwise about the z axis through an angle θ from another coordinate system xy. If a vector has components A_x and A_y with respect to the xy system, what are its components with respect to the $x'y'$ system? Show that the magnitude of **A** is the same in both cases, i.e.:

$$\sqrt{A_x^2 + A_y^2} = \sqrt{(A_x')^2 + (A_y')^2}$$

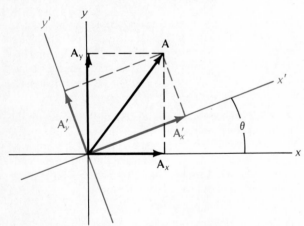

Figure 2-31. Problem 2-16.

Section 2-3 Vector Addition by Components

· 2-17 Q The result of adding two vectors of magnitudes 4 quork and 6 quork will be a resultant vector whose magnitude is

(A) 2 quorks
(B) 4 quorks

(C) 6 quorks
(D) 8.94 quorks
(E) 10 quorks
(F) not possible to determine without more information.

· 2-18 P A hawser (a big rope) used to moor a ship passes around a piling (a big pole sticking up out of the water). It is tied at one end, and the other end is attached to the ship, which exerts a steady force of 5000 N on the hawser. However, a force of 5000 N exerted directly against the piling at the point where the hawser passes around it will cause the piling to break. What is the maximum value the angle θ can have if the piling is not to break?

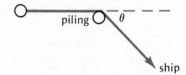

Figure 2-32. Problem 2-18.

· 2-19 P Graphically find the magnitude and direction of adding the following vectors. The components of each vector are given in the form (A_x, A_y).

$$\mathbf{R} = (2, 2) + (-4, 0) + (8, -1) + (-1, -6)$$

· 2-20 P Find (a) $\mathbf{A} + \mathbf{B}$ and (b) $\mathbf{A} - \mathbf{B}$ given $\mathbf{A} = 40$ m directed east and $\mathbf{B} = 50$ m directed 37° south of east.

: 2-21 P An orthodontist attaches two metal bands to a tooth he wishes to move, as shown in Figure 2-33. The tension in band A is adjusted to 100 N. If the maximum force which can safely be exerted on the tooth is 200 N, what is the maximum tension which can be provided by band B?

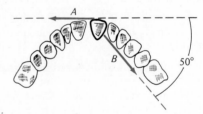

Figure 2-33. Problem 2-21.

: 2-22 P Vectors **A** and **B** have magnitudes 5 and 2, respectively. How must they be oriented to give resultants of magnitudes (a) 3, (b) 4, (c) 5.4, (d) 7?

: 2-23 P Compass bearings are given in degrees measured clockwise from north. Thus, due east is a bearing of 90°, south is a bearing of 180°, west is a bearing of 270°, etc. A Coast Guard radar operator picks up a signal from a disabled yacht, the Golden Vanity out of Vancouver, at a distance of 400 km on a bearing of 315°. He locates the Lucky Lassy, a commercial salmon fishing boat, at a distance of 100 km from his transmission on a bearing of 30°. He radios the fishing boat to request that she sail to help the Golden Vanity. How far from the disabled yacht is the Lucky Lassy, and on what bearing should she sail to reach the yacht?

: **2-24 P** Two vectors **A** and **B** intersect at an angle θ. By combining their components show that the magnitude of their resultant is $C = \sqrt{A^2 + B^2 + 2\,AB\cos\theta}$, where A and B are the magnitudes of the individual vectors. This is called the law of cosines.

Section 2-4   Vectors in Three Dimensions

: **2-25 P** If $\mathbf{A} + \mathbf{B} = \mathbf{C}$, determine the components and magnitude of **C** given $\mathbf{A}_x = 2$, $\mathbf{A}_y = 3$, $\mathbf{A}_z = -1$ and $\mathbf{B}_x = 3$, $\mathbf{B}_y = 1$, $\mathbf{B}_z = 9$. Are **A** and **B** perpendicular to each other?

: **2-26 P** A plane surface may be represented by a vector directed perpendicular to the surface with magnitude equal to the area of the surface. Write the six vectors representing the surfaces of a cube of side a with its edges aligned along the x, y, and z axes.

: **2-27 P** A plane surface may be represented by a vector pointing outward perpendicular to the surface with magnitude equal to the area of the surface. Consider a cube of edge b with its edges aligned along the x, y, and z axes as shown in Figure 2-34. Write a vector representing the

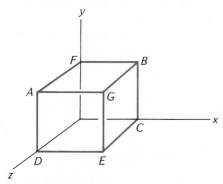

Figure 2-34.   Problem 2-27.

planes formed by joining each of the following sets of points on the cube: *ABCD* and *ABE*.

: **2-28 P** Write a vector with magnitude 1 directed perpendicular to the plane passing through the points $(0,0,0)$, $(1,0,0)$ and $(0,3,-4)$

: **2-29 P** Given that the components of **A** are $A_x = -10$, $A_y = 4$, $A_z = 6$, and the components of **B** are $B_x = 5$, $B_y = -2$, $B_z = 2$, what is the magnitude of $\mathbf{A} + \mathbf{B}$?

Section 2-6   Force Vectors and Equilibrium

· **2-30 P** Three persons push horizontally on a packing crate, exerting the forces shown in Figure 2-35. What is the net force exerted?

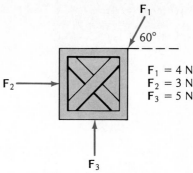

$F_1 = 4$ N
$F_2 = 3$ N
$F_3 = 5$ N

Figure 2-35.   Problem 2-30.

· **2-31 Q** Suppose that a particle is subject to a force with $\mathbf{F}_x = -2$ N and $\mathbf{F}_y = +2$ N. Since -2 N $+ 2$ N $= 0$, is it true that the particle is in static equilibrium?

· **2-32 Q** A 10 kg fish is suspended from two spring scales connected together, as shown in Figure 2-36. The scales are of negligible weight. In this case
(A) each scale will read 5 kg.
(B) each scale will read 10 kg.
(C) the top scale will read 10 kg, the bottom one zero.
(D) the bottom scale will read 10 kg, the top one zero.
(E) each scale will show a reading between zero and 10 kg such that the sum of the two readings is 10 kg, but without more information we cannot determine exact values.

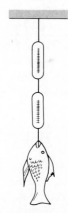

Figure 2-36.   Question 2-32.

· **2-33 Q** Which of the following groups of force vectors could be in static equilibrium?
(A) 5 N, 6 N, 7 N
(B) 8 N, 8 N, 8 N
(C) 20 N, 13 N, 7 N
(D) 6 N, 24 N, 16 N
(E) 2 N, 4 N, 7 N, 15 N

: **2-34 Q** Two tug-of-war teams are holding each other to a standstill. Each exerts a force of 3000 N on the rope. What is the tension in the rope?

· **2-35 P** What force is exerted on the patient's leg by the traction device shown in Figure 2-37?

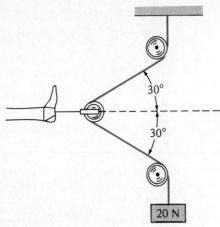

Figure 2-37.   Question 2-35.

· **2-36 P** What force must be exerted with the block and tackle in Figure 2-38 to lift a 2000-N load? Friction is negligible.

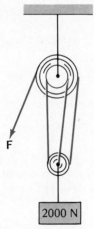

Figure 2-38.   Problem 2-36.

: **2-37 P** A hot air balloon experiences an upward buoyant force of 1200 N. It is moored to the ground by a rope, which is inclined at 20° to the vertical because a stiff wind is blowing the balloon sideways. What is the tension in the rope?

: **2-38 P** In an effort to pull a pickup truck out of a ditch, a farmer ties a steel cable tightly between the truck and a tree.

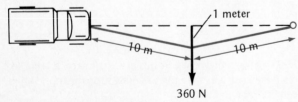

Figure 2-39.   Problem 2-38.

The length of the cable is 20 meters, and by pulling sideways at its midpoint with a force of 360 N she deflects it a distance of 1 meter. What force is she able to exert on the truck in this way?

: **2-39 P** A painter weighing 800 N stands on a platform that weighs 400 N. He supports himself as shown in Figure 2-40. What is the tension in the rope he is holding?

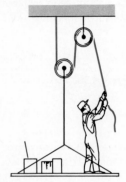

Figure 2-40.   Problem 2-39.

: **2-40 P** A skier weighing 600 N is in equilibrium while waiting for a stalled ski tow to start up. If friction is negligible, what is the tension in the cable holding him?

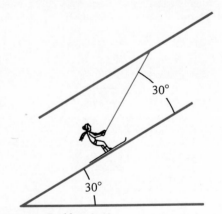

Figure 2-41.   Problem 2-40.

: **2-41 P** A person weighing 650 N is moved from one ship

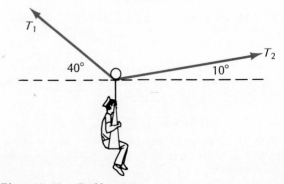

Figure 2-42.   Problem 2-41.

to another in the bosun's chair shown in Figure 2-42. The pulley catches on a knot in the line and the person is stopped. What then is the tension in the rope on each side?

: **2-42 P** A packing crate of weight 400 N is placed on a loading ramp inclined at 30° above horizontal. If a frictional force of 100 N acts on the crate, what additional horizontal force must be applied to keep the crate from sliding down the ramp?

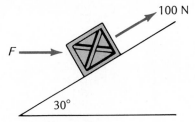

Figure 2-43.   Problem 2-42.

: **2-43 P** A sign weighing 20 N is suspended from a light strut and a cable. The strut is pivoted at the wall to which it is attached. What is (a) the tension in the cable and (b) the force exerted by the strut? (c) Is the strut in compression or tension?

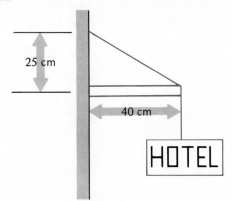

Figure 2-44.   Problem 2-43.

: **2-44 P** A chain rests without friction on a right triangle of sides a and b. What is the ratio of the length on side a to the length on side b in equilibrium?

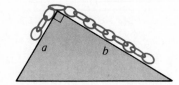

Figure 2-45.   Problem 2-44.

: **2-45 P** Determine the weights W_1 and W_2 and the tensions T_1, T_2 and T_3 in the system in Figure 2-46.

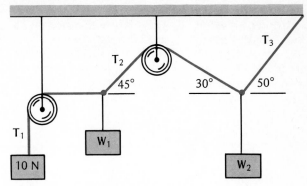

Figure 2-46.   Problem 2-45.

: **2-46 P** What is the tension in each cord in Figure 2-47?

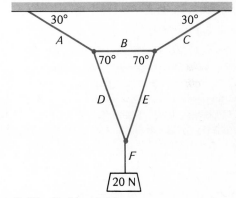

Figure 2-47.   Problem 2-46.

: **2-47 P** The force exerted by a spring is proportional to the distances which it has been stretched, i.e. $\mathbf{F} = ks$. k is called the *spring constant*. If a weight W is attached to two identical springs of unstretched length L, what is the value of y in equilibrium?

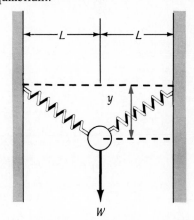

Figure 2-48.   Problem 2-47.

: **2-48 P** When two surfaces are pressed together with a force $\mathbf{F}_N$ perpendicular to the interface of the surfaces (called the *normal force*) they exert a frictional force $\mathbf{F}_F$ on each other. Frequently $F_F = \mu F_N$, where the constant of proportionality μ is called the *coefficient of friction.*

Suppose that a metal wedge of negligible weight is driven into an axe handle. What is the minimum coefficient of friction between the wedge and the wood so that the wedge will not slide out after it is driven in? Express the result in terms of the wedge angle, θ.

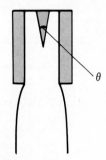

Figure 2-49.   Problem 2-48.

: **2-49 P** A bridge truss is constructed as sketched here. The members are joined by pins and can exert forces only along their length. If the weight of each member is negligible, what force is exerted by each member when a load *W* is applied at the midpoint of the truss? State whether each member is in tension or compression. (All members are of equal length.)

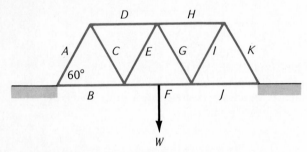

Figure 2-50.   Problem 2-49.

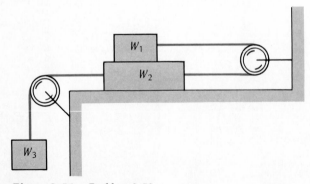

Figure 2-51.   Problem 2-50.

: **2-50 P** Two blocks of weights $W_1 = 3$ N and $W_2 = 6$ N are connected to a third block of weight W_3 as shown in Figure 2-51. The frictional force acting on each block surface is equal to $\frac{1}{2}$ N, where N is the normal force on that surface. What is the maximum weight W_3 for which the system will still be in equilibrium?

: **2-51 P** Two small objects of weights W_1 and W_2 are joined by a rod of negligible weight. They rest in equilibrium on two frictionless planes, as shown in Figure 2-52. What is the angle θ? $W_1 = 20$ N, $W_2 = 10$ N

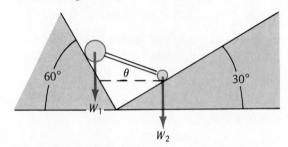

Figure 2-52.   Problem 2-51.

: **2-52 P** Four large metal spheres, each of radius R and weight W, are to be stacked so that one rests on top of the other three, with their centers positioned at the vertices of a regular tetrahedron. A cable is passed around the lower spheres to hold them together. What is the tension in the cable?

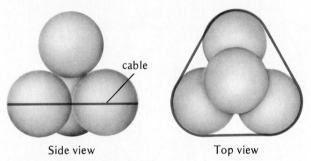

Side view   Top view
Figure 2-53.   Problem 2-52.

Supplementary Problem
2-53 Q Prove the following:

And hence no force, however great,
Can stretch a cord, however fine,
Into a horizontal line
That shall be absolutely straight.

This example of accidental verse appeared on page 44 of the 1819 edition of *An Elementary Treatise on Mechanics* by William Whewell, Master of Trinity College, Cambridge University. When the rhyme was pointed out to Whewell, he was considerably annoyed and changed this sentence to eliminate the verse in the next edition. Whewell composed intentional poetry, but only this verse has survived.

Motion along a Line

3

Dynamics deals with the motions of objects as they are related to such physical concepts as mass, force, momentum, and energy, and with the principles unifying these concepts. Kinematics, on the other hand, has a more modest program. Combining the ideas of geometry and time, it simply describes motion without giving attention to its causes. We shall explore the kinematics of a particle moving along a straight line.

3-1   Meaning of a Particle

What is a particle? It is not necessarily a tiny, hard sphere, as commonly imagined; rather we take a particle to be simply an object that is small enough for its size to be unimportant for the scale of our observations. Moreover, we are not concerned with its internal structure. A star can be considered a particle when viewed as one part of a galaxy, but an atom is too large to be considered a particle when we examine its component parts. It is all a matter of scale. The particle is the physical counterpart of the point in mathematics; it has not only the property of precise localizability but also such physical attributes as mass and electric charge.

3-2 Constant Velocity

In this and following sections we shall use three equivalent ways of defining and describing kinematical quantities:

- Verbal definition.
- Mathematical definition in terms of symbols.
- Graphical definition.

Graphical definitions are especially useful; from a plot of displacement or velocity as a function of time, one can see at a glance the important qualitative features of the motion.

Our concern now is only with rectilinear motion, motion along a straight line, here the x axis. The particle's location along the x axis is given by the coordinate x, with some appropriate unit of length. Locations to the right or left of the origin are distinguished by the algebraic sign of x.

The coordinate x then specifies the *vector displacement* relative to the origin. See Figure 3-1. The two possible directions of the vector displacement relative to the origin—toward the right, or toward more positive values of x; and toward the left, or toward more negative values of x—are given by the algebraic sign of x. (The absolute value $|x|$ gives *distance* from the origin.) Other vector quantities derived from the displacement (such as velocity and acceleration) have their directions also indicated by their algebraic signs. In short, for the simple and special case of one-dimensional motion along a single *straight* line, the two possible directions of vectors—right or left—are given by the corresponding algebraic signs (+ or −).

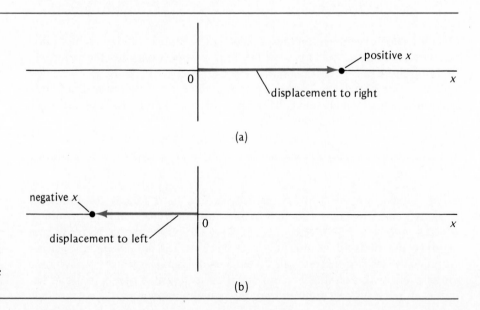

positive x

0 x

displacement to right

(a)

negative x

0 x

displacement to left

(b)

Figure 3-1. Displacements along the x axis: (a) positive x; (b) negative x.

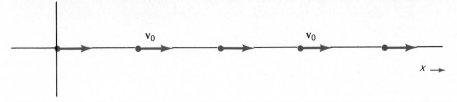

Figure 3-2. *Motion at constant velocity* $\mathbf{v}_0$ *along the x axis.*

The simplest type of rectilinear motion is motion with *constant velocity,* that motion in which a particle makes equal displacements in equal time intervals. A particle starting from the origin at constant velocity v_0 has a displacement x after a time t given by

$$v_0 = x/t$$

Subscript zero indicates that the velocity at the start (time $t = 0$) is the *same* as the velocity at any later time.

Appropriate units for velocity are distance divided by time; in the SI system, velocity units are meters per second, or m/s. Positive velocity means motion to the right; negative velocity, motion to the left. Zero velocity is the state of rest. Figure 3-2 shows the location and velocity vectors v_0 at different equal time intervals for an object moving along the x axis at constant velocity; here the tail of the velocity vector is at the location of the object in motion.

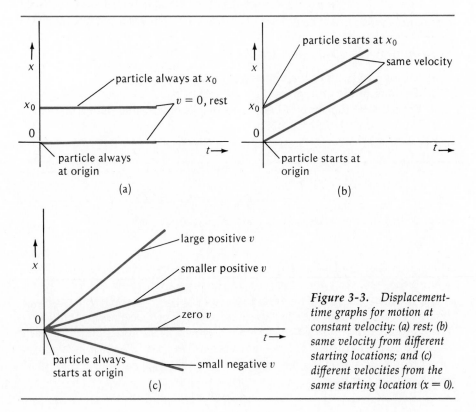

Figure 3-3. *Displacement-time graphs for motion at constant velocity: (a) rest; (b) same velocity from different starting locations; and (c) different velocities from the same starting location (x = 0).*

A graphical representation of the motion, in which one plots displacement of a particle as a function of time, is particularly helpful. Figure 3-3 shows several examples for constant velocity.

In Figure 3-3(a), two particles are at rest; for each, the coordinate x does not change with time and the slope of the x-t graph is zero. The two particles remain at rest, however, at two different locations: one at the origin, and the other at $x = x_0$.

In Figure 3-3(b), two particles have the same constant velocity but start from different locations. Both particles cover equal distances (in the same direction) in equal time intervals. Since velocity is the change in displacement divided by the corresponding time interval, the *slope of a straight x-t line* is equal to the constant velocity. For a particle starting from x_0 at time $t = 0$, its displacement at any later time t relative to its starting displacement is $(x - x_0)$. Then, from the definition of constant velocity as displacement $(x - x_0)$ divided by elapsed time, we have

$$v_0 = \frac{x - x_0}{t}$$

or
$$x = x_0 + v_0 t \tag{3-1}$$

This is a general relation giving the displacement x for any later time t in terms of the initial displacement x_0 and the particle's *constant* velocity v_0.

Figure 3-3(c) shows the motion of four particles all starting from the origin $(x_0 = 0)$. The small positive slope corresponds to low velocity to the right. Larger positive slope corresponds to higher velocity to the right. The horizontal line corresponds to rest $(v_0 = 0)$, and the downward-sloping line with negative slope corresponds to a velocity to the left (motion toward increasingly negative values of x).

In short, for *constant* velocity:

- The sign of v_0 gives the direction of motion.
- The magnitude of v_0 gives the time rate of covering distance.
- The slope of the straight line in the displacement-time plot equals v_0.
- The value of the displacement x at any time t is given by $x = x_0 + v_0 t$.

3-3 Average Velocity

We represent the change in the displacement of the particle by the symbol Δx. For a particle first at x_i and later x_f, the displacement Δx is

$$\Delta x = x_f - x_i$$

where x_i is the "initial" and x_f the "final" location. Clearly the sign of Δx denotes the sense in which the particle shifts position along the axis. When $x_f > x_i$, then Δx is positive and the displacement is to the right; when $x_f < x_i$, then Δx is negative and the displacement is to the left. Displacement must be distinguished from the distance traveled, which is the total length of path traversed by the particle and therefore necessarily positive. A particle making a round trip to its starting point undergoes zero displacement, even though it travels a finite distance.

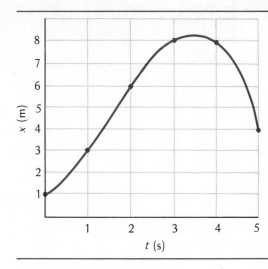

Figure 3-4. *Displacement x of a particle plotted as a function of time t.*

Consider Figure 3-4, in which the coordinate x of some particle in motion is plotted as a function of time t at which the particle was at each position. (Data of this sort might, for example, be derived from a motion picture in which each frame shows one event in the history of the particle, or from a stroboscopic photograph with multiple images giving the position of a particle at equal time intervals.) Since the x-t line is not straight, this is clearly not motion at constant velocity.

The *average velocity* $\bar{v}$ is defined as the particle's displacement divided by the elapsed time for this displacement. In symbols,

$$\bar{v} = \frac{\Delta x}{\Delta t} \tag{3-2}$$

where $\Delta t = t_f - t_i$ and $\Delta x = x_f - x_i$.

To compute the average velocity for the data in Figure 3-4, we apply this definition for a number of time intervals. More specifically, we use (3-2), with x and t values taken from Figure 3-4, to compute the average velocity for several different time intervals.

t_i(s)	t_f(s)	Δt(s)	x_i(m)	x_f(m)	Δx(m)	$\bar{v}_{if}$ (m/s)
1	2	1	3	6	3	$\bar{v}_{12} = 3$
2	4	2	6	8	2	$\bar{v}_{24} = 1$
3	4	1	8	8	0	$\bar{v}_{34} = 0$
4	5	1	8	4	-4	$\bar{v}_{45} = -4$

For the data from Figure 3-4, the value of average velocity clearly depends on the time interval chosen; $\bar{v}_{12}$ is positive, $\bar{v}_{24}$ has a smaller positive magnitude (a smaller average velocity over this interval), $\bar{v}_{34}$ is zero (no displacement over this interval), and $\bar{v}_{45}$ is negative (motion toward the origin). Only if the displacement-time graph is a single straight line will the average velocity have the same magnitude for all possible pairs of points; only then will we have constant velocity, or uniform velocity.

3-4   Instantaneous Velocity

The slope of any line is defined as the change in the ordinate (here x) divided by the corresponding change in the abscissa (here t). For a displacement-time line, the ratio $\Delta x/\Delta t$ is the average velocity over the chosen interval Δt. A horizontal line, one with zero slope, indicates rest; a positive slope implies motion to the right; and a negative slope, motion to the left.

See Figure 3-5, where the motion portrayed is of an object that moves slowly at first, picks up speed, and then slows down. Actually, to characterize the velocity in this way, we need the concept of instantaneous velocity, the velocity at a particular instant, which is really the average velocity computed over an infinitesimally small time interval. This is what is read directly, for example, on an automobile's speedometer.

Figure 3-6 shows how instantaneous velocity is computed. Time interval Δt is chosen to be so small that the actual curve cannot be distinguished from a straight line drawn between the two end points. Then a line drawn tangent to the displacement-time curve will coincide with the actual curve over a small region at the point of contact; the *instantaneous velocity* is the *slope* of the

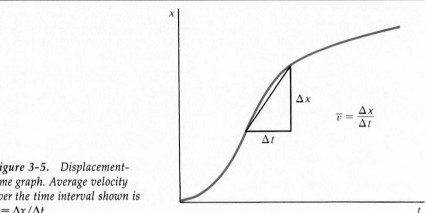

$$\overline{v} = \frac{\Delta x}{\Delta t}$$

Figure 3-5.   *Displacement-time graph. Average velocity over the time interval shown is* $\overline{v} = \Delta x/\Delta t.$

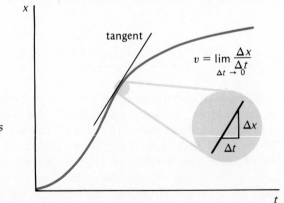

$$v = \lim_{\Delta t \to 0} \frac{\Delta x}{\Delta t}$$

Figure 3-6.   *The instantaneous velocity is the tangent of the displacement-time graph, which is the limit of average velocity for a vanishingly small time interval.*

x-versus-t curve. In formal terms, the instantaneous velocity v at time t is the limit, as the time interval Δt approaches zero duration, of the displacement Δx divided by the corresponding time interval Δt, the interval Δt being centered at time t. In symbols,

$$\text{Instantaneous velocity} = v = \lim_{\Delta t \to 0} \frac{\Delta x}{\Delta t} = \frac{dx}{dt} \tag{3-3}$$

In the notation of the calculus we have, $v = dx/dt$, instantaneous velocity is the derivative of x with respect to t, or the time rate of change of displacement.* Hereafter, the symbol v (called simply the velocity) will represent the instantaneous velocity.

By instantaneous speed is meant the *magnitude* of the instantaneous velocity; speed is always positive. To specify velocity we must give the direction of motion in addition to the speed.

Example 3-1.   The displacement of an object is given as a function of time by the relation $x = (4 \text{ m/s}^3)t^3$. Compute the average velocity for the following conditions: (a) $\Delta t = 2$ s, and centered at $t = 3$ s; and (b) $\Delta t = 0.2$ s, and again centered at $t = 3$ s.

(a) $x_i = (4 \text{ m/s}^3)(2\text{s})^3 = 32$ m   and $x_f = (4 \text{ m/s}^3)(4\text{s})^3 = 256$ m

$$\bar{v} = \frac{\Delta x}{\Delta t} = \frac{256 \text{ m} - 32 \text{ m}}{2 \text{ s}} = 112 \text{ m/s}$$

(b) $x_i = (4 \text{ m/s}^3)(2.9 \text{ s})^3 = 97.556$ m   and $x_f = (4 \text{ m/s}^3)(3.1 \text{ s})^3 = 119.164$ m

$$\bar{v} = \frac{\Delta x}{\Delta t} = \frac{119.164 \text{ m} - 97.556 \text{ m}}{0.2 \text{ s}} = 108.04 \text{ m/s}$$

Taking a still smaller Δt, we should find that the average velocity approaches closely 108.0000 . . . m/s; that is, the instantaneous velocity at $t = 3$ s is $v = 108$ m/s exactly.

The same result follows from taking the time derivative of $x = (4 \text{ m/s}^3)t^3$ to find the relation for the instantaneous velocity:

$$v = \frac{dx}{dt} = (12 \text{ m/s}^3)t^2 = (12 \text{ m/s}^3)(3 \text{ s})^2 = 108 \text{ m/s}$$

3-5   Displacement as Area Under the Velocity-Time Graph

Figure 3-7(a) gives the instantaneous velocity as a function of time. It was derived from the displacement-time curve of Figure 3-6 by the procedure of taking slopes.

As we have seen, a velocity-time curve can be derived from a record of displacement versus time. We can do this in reverse; we can find a displacement-time curve if we are first given the velocity-time curve. Consider Figure 3-7(b). Here numerous narrow rectangles, each of width Δt, are fitted under the curve. Over a small time interval Δt, the velocity can be regarded as essentially constant. The corresponding displacement Δx is $v \, \Delta t$, which is, as

---

* Although we give some definitions in this chapter in the language of calculus, we shall not require the mathematical apparatus of the calculus in the next several chapters.

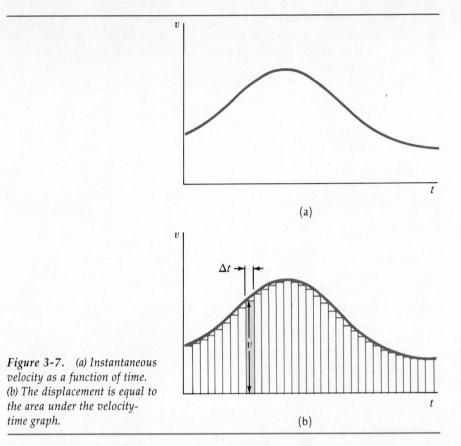

Figure 3-7. *(a) Instantaneous velocity as a function of time. (b) The displacement is equal to the area under the velocity-time graph.*

we see from the figure, the area of one shaded rectangle. Then if the displacement $x - x_0$ (from the start at $t = 0$ to the time t) is to be found, we add the areas of all elementary rectangles. This amounts to saying that the net displacement at time t equals the entire area under the velocity-time curve up to t. If the velocity is negative and the velocity-time curve lies below the time axis, the contribution of the area "under" the curve is negative. In the language of integral calculus,

$$x_f - x_i = \lim_{\Delta t \to 0} \Sigma v \, \Delta t = \int_{t_i}^{t_f} v \, dt$$

3-6   Acceleration

Instantaneous velocity is the time rate of change of displacement. Velocity tells us how displacement changes with time. Another kinematic quantity, acceleration, tells us how the velocity changes with time. *Acceleration* is defined as the time rate of change of velocity. The average acceleration $\bar{a}$ is the velocity change Δv divided by the corresponding finite elapsed time interval Δt:

$$\bar{a} = \frac{\Delta v}{\Delta t} \qquad (3\text{-}4)$$

Graphically, average acceleration for a chosen time interval is the slope of the

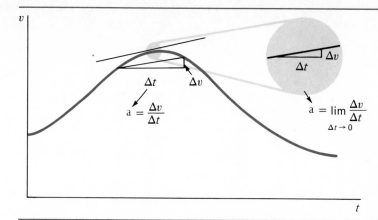

Figure 3-8. *Instantaneous acceleration is the tangent of the velocity-time graph, which is the limit of the average acceleration for a vanishingly small time interval.*

straight line drawn between two points on a velocity-time curve. Instantaneous acceleration *a* is the average acceleration over a vanishingly small time interval (see Figure 3-8). Equivalently, *a* is the slope of the velocity-time curve. In symbols,

$$\text{Instantaneous acceleration} = a = \lim_{\Delta t \to 0} \frac{\Delta v}{\Delta t} = \frac{dv}{dt} \qquad (3\text{-}5)$$

Hereafter *acceleration* will mean "instantaneous acceleration."

Just as area under a velocity-time curve is net displacement, so too the area under an acceleration-time curve is change in velocity.

Acceleration has the units of velocity divided by time. Its dimensions are length/time^2, for example, mi/(h)(s) or m/(s)(s). It is customary procedure to write m/(s)(s) as m/s^2. Thus an object having a constant acceleration of, say, $+10$ m/s^2 has its velocity increasing by $+10$ m/s (to the right) during each 1-s interval.

In summary, instantaneous *velocity* can be defined verbally, mathematically, and graphically as:

- Time rate of change of displacement.
- $v = \lim_{\Delta t \to 0} \dfrac{\Delta x}{\Delta t} = \dfrac{dx}{dt}$
- Slope of the *x-t* graph.

And *acceleration* can be defined similarly:

- Time rate of change of velocity.
- $a = \lim_{\Delta t \to 0} \dfrac{\Delta v}{\Delta t} = \dfrac{dv}{dt}$
- Slope of the *v-t* graph.

Net displacement is area under the *v-t* graph.*

---

* The procedure of defining the velocity as the time rate of change of the displacement, and the acceleration as the time rate of change of velocity, could be extended to yield a quantity giving the time rate of change of acceleration (a measure of "jerkiness" of the motion); but this is not usually necessary. The concepts of velocity and acceleration are sufficient to describe motion adequately. The reasons for this lie in the physics, or dynamics, of the motion, not in the mathematics or kinematics of the motion—as we shall see later.

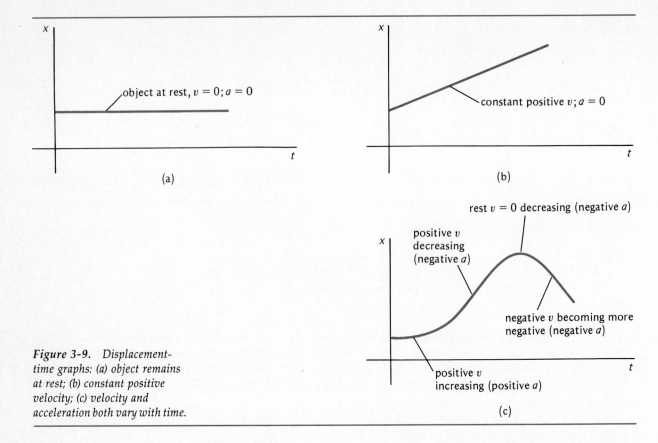

Figure 3-9. *Displacement-time graphs: (a) object remains at rest; (b) constant positive velocity; (c) velocity and acceleration both vary with time.*

You should now be able to interpret the velocity and acceleration at a glance in parts (a), (b), and (c) of Figure 3-9. In each, displacement is plotted against time.

Example 3-2. Figure 3-10, a velocity-time graph, shows the performance characteristics of a Porsche 924 Turbo sports car. The car accelerates from rest through the five forward gears; the horizontal notches in the curve correspond to the short time intervals during which gears are shifted and the car moves at nearly constant speed.

(a) What is the average acceleration of the car while in first gear (from rest to 13 m/s in 2.3 s)?

(b) What is the average acceleration for fifth gear (45 m/s at 23.2 s, and 54 m/s at 48.9 s)?

(c) Suppose that the car's acceleration were *constant* from rest to the speed of 54 m/s (120 mph), achieved in 48.9 s. Would the car have covered more distance than in the actual performance shown in Figure 3-10, or less?

(a) For first gear, the average acceleration is

$$\bar{a}_{1st} = \frac{\Delta v}{\Delta t} = \frac{(13 \text{ m/s} - 0)}{(2.3 \text{ s} - 0)} = 5.7 \text{ m/s}^2$$

(The acceleration initially is more than half that of a freely falling object, $g \approx 10 \text{ m/s}^2$.)

(b) For fifth gear, the average acceleration is

$$\bar{a}_{5th} = \frac{\Delta v}{\Delta t} = \frac{(54 \text{ m/s} - 45 \text{ m/s})}{(48.9 \text{ s} - 23.2 \text{ s})} = 0.4 \text{ m/s}^2$$

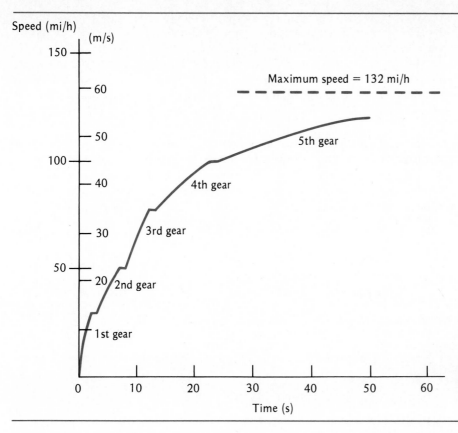

Speed (mi/h)

Figure 3-10. *Velocity-time graph for a Porsche 924 Turbo sports car driven for maximum acceleration. The acceleration (the slope) decreases as the speed increases. Notches in the curve correspond to gear shifts.*

This is a much smaller acceleration than that for first gear. The Porsche has maximum acceleration at the start and progressively smaller acceleration as its speed increases; this reflects that the slope of the *v-t* curve (the instantaneous acceleration) decreases with time.

(c) The displacement generally equals the area under the *v-t* curve. Clearly, then, a straight line from the origin to the point 54 m/s and 48.9 s, which corresponds to constant acceleration, would have an area under it that is less than that shown in Figure 3-10. Under constant acceleration the car does not travel as far.

3-7 Constant Acceleration

A simple, special but important type of rectilinear motion is that in which acceleration is constant. If a particle's acceleration is constantly zero, the particle has constant velocity; it either moves with unchanging speed in a straight line or remains at rest.

Suppose that we know the velocity and displacement for some starting time of a particle traveling with known constant acceleration. Then it is easy to compute the particle's velocity and displacement for any future time. A simple example shows how this works.

Suppose that a particle has an initial velocity of 10 m/s as it starts from the origin and that its acceleration is a constant 2 m/s². What is the particle's velocity after 3 s? What is its displacement after these 3 s?

With velocity changing at the constant rate of 2 m/s for each 1 s interval, the total velocity change over 3 s is simply $(2 \text{ m/s}^2)(3 \text{ s}) = 6$ m/s. Therefore, the velocity 3 s after the start is $10 \text{ m/s} + 6 \text{ m/s} = 16$ m/s.

The displacement of a particle is always its average velocity multiplied by the elapsed time interval. Here the velocity changes at a steady rate from 10 m/s to 16 m/s, so that the average velocity over the 3 s time interval is $(10 \text{ m/s} + 16 \text{ m/s})/2 = 13$ m/s. Then the displacement is $(13 \text{ m/s})(3 \text{ s}) = 39$ m.

To generalize from this specific numerical example to *any* rectilinear motion at constant acceleration is simple. We merely substitute general algebraic symbols for what had been specific numerical quantities.

Over a time interval t, the velocity change is $\Delta v = at$, so that the final velocity v at time t is

$$v = v_0 + \Delta v$$
$$v = v_0 + at \tag{3-6}$$

where v_0 is the initial velocity at $t = 0$.

For *constant* acceleration, the average velocity $\bar{v}$ over the time t is*

$$\bar{v} = \frac{v_0 + v}{2} \tag{3-7}$$

The final displacement x relative to the initial displacement x_0 at $t = 0$ can be written

$$x - x_0 = \bar{v}t$$

or by (3-7),

$$x - x_0 = \frac{v_0 + v}{2} t \tag{3-8}$$

We can replace v in (3-8) by its value from (3-6), to arrive at

$$x - x_0 = \frac{v_0 + (v_0 + at)}{2} t$$
$$x - x_0 = v_0 t + \tfrac{1}{2}at^2 \tag{3-9}$$

Note that if a particle's acceleration is zero, so that its velocity is constant, (3-9) reduces to (3-1).

Equations (3-6), (3-8), and (3-9) all contain variable t. We can easily arrive at a fourth kinematic relation by eliminating time t between (3-6) and (3-8). Solving for t in (3-6), we have

$$t = \frac{v - v_0}{a}$$

so that (3-8) can be written

$$x - x_0 = \frac{v_0 + v}{2} t = \frac{v_0 + v}{2} \frac{v - v_0}{a}$$
$$x - x_0 = \frac{v^2 - v_0^2}{2a} \tag{3-10}$$

---

* The validity of this relation for average velocity is reexamined in Section 3-8.

Table 3-1. *Kinematic equations for constant acceleration*

EQUATION		VARIABLE NOT APPEARING
$v = v_0 + at$	(3-6)	x
$x - x_0 = v_0 t + \frac{1}{2}at^2$	(3-9)	v
$x - x_0 = \frac{1}{2}(v_0 + v)t$	(3-8)	a
$x - x_0 = \dfrac{(v^2 - v_0^2)}{2a}$	(3-10)	t

The four kinematic equations for constant acceleration are displayed in Table 3-1. Each equation is identified also by the variable (x, v, a, or t) that *does not* appear in it.

These equations merely express in general terms the algebraic and logical consequence of the definitions of velocity and constant acceleration. The first two, (3-6) and (3-9), can be considered the primary relations; they yield, respectively, the final velocity v and final displacement x in terms of their initial values, v_0 and x_0, the acceleration a, and the elapsed time t. Equations (3-6) and (3-9) permit the future course (and past history) of a particle to be projected, and these two equations alone always suffice to solve kinematic problems for constant acceleration. It is sometimes more convenient to use the last two equations, (3-8) and (3-10).

3-8 Graphs and Constant Acceleration

Figure 3-11 is a velocity-time graph for constant a. The slope, the acceleration a, is constant; the line is straight. Clearly, the final velocity v is the sum of two parts — the initial velocity v_0 and the velocity change at:

$$v = v_0 + at \qquad (3\text{-}6)$$

From the velocity-time graph of Figure 3-11, we can derive the relation for the displacement x as a function of time t. Recall that the area under any velocity-time curve equals the change in displacement, $x - x_0$. The figure under the straight line in Figure 3-11 is a trapezoid; it consists of a rectangle of width t and height v_0 and a right triangle of sides t and at. Thus, the displacement change $x - x_0$ is the sum of $v_0 t$ and $\frac{1}{2}(at)(t)$, or

$$x - x_0 = v_0 t + \tfrac{1}{2}at^2 \qquad (3\text{-}9)$$

the same result as that derived from algebraic considerations.

Now compare the velocity-time graphs in (a) and (b) of Figure 3-12. The shaded area under the v-t curve is the same for both: in Figure 3-12(a), the area under the trapezoid; and in Figure 3-12b), the same area under the rectangle. Both areas equal the displacement change, $x - x_0$. The horizontal line in Figure 3-12(b) has the value $(v_0 + v)/2$; it is in fact the average velocity $\bar{v}$. Therefore, the area under the curve in Figure 3-12(b) is the width t of the rectangle multiplied by its height $\bar{v} = (v_0 + v)/2$, or

$$x - x_0 = \frac{(v_0 + v)}{2}\, t \qquad (3\text{-}8)$$

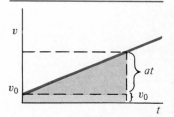

Figure 3-11. *Velocity-time graph for constant acceleration.*

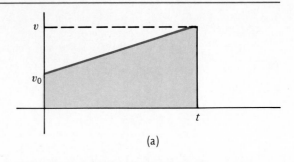

(a)

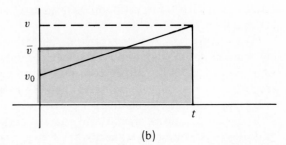

Figure 3-12. *Velocity-time graphs with identical areas under the curves: (a) constant acceleration; and (b) equivalent constant velocity $\bar{v}$.*

(b)

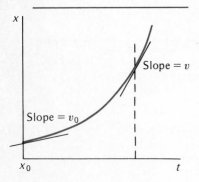

Figure 3-13. *Displacement-time graph for constant acceleration.*

The average velocity equals the average of the initial and final velocities only for *constant* acceleration, or for a velocity-time curve that is a *straight* line.

A displacement-time graph for constant acceleration is shown in Figure 3-13. This is a plot of (3-9):

$$x = x_0 + v_0 t + \tfrac{1}{2}at^2$$

The x-t curve is a parabola. The slope increases uniformly from v_0 at $t = 0$ to v at time t, corresponding to constant acceleration. The three terms in (3-9) have a simple interpretation: x_0 gives the particle's displacement at the starting time; $v_0 t$ is what the displacement would be if the velocity were constantly v_0; and $\tfrac{1}{2}at^2$ is the additional displacement resulting from acceleration.

3-9   Calculus and Constant Acceleration (Optional)

The kinematic relations for constant acceleration are easily verified by the differential calculus:

$$x = x_0 + v_0 t + \tfrac{1}{2}at^2$$

$$v = \frac{dx}{dt} = v_0 + at$$

$$a = \frac{dv}{dt} = a$$

Conversely, by integrating we can obtain the velocity-time and displacement-time formulas directly from the definitions:

$$a = \frac{dv}{dt}$$

or
$$dv = a\, dt$$

We integrate velocity from the initial value v_0 to the final value v, the corresponding limits on time being zero and t:

$$\int_{v_0}^{v} dv = a \int_{0}^{t} dt$$

$$v - v_0 = at$$

$$v = v_0 + at \qquad\qquad (3\text{-}6)$$

We integrate once more to find x:

$$\frac{dx}{dt} = v_0 + at$$

or
$$dx = v_0\, dt + at\, dt$$

The displacement goes from x_0 at $t = 0$ to x at time t. Thus, the definite integrals are

$$\int_{x_0}^{x} dx = v_0 \int_{0}^{t} dt + a \int_{0}^{t} t\, dt$$

$$x - x_0 = v_0 t + \tfrac{1}{2}at^2 \qquad\qquad (3\text{-}9)$$

Note that an essential part of the analysis is the assumption that v_0 and a are constant.

Example 3-3.   A preliminary engineering study (technical feasibility, costs, engineering design) has been made for a rapid-transit coast-to-coast transport system.* So-called Planetrans would travel in evacuated tubes (about 0.1 percent of atmospheric pressure) several hundred feet below the earth's surface. The trains would be floated by magnetic repulsion (using superconductors) and driven by electromagnetic propulsion. With air resistance minimized and friction with rails eliminated, the trains would have a

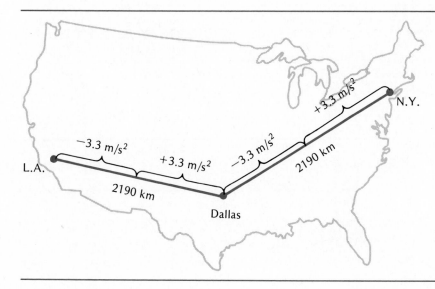

Figure 3-14.   Rapid-transit coast to coast transport system (New York → Dallas → Los Angeles). Cars in evacuated underground tubes would travel at constant acceleration of 3.3 m/s².

lift-to-drag ratio 60 times that for a supersonic aircraft. Therefore, the Planetrans could travel at constant and relatively large accelerations.

The coast-to-coast trip would be broken into two legs: New York to Dallas, a distance of 2190 km (1360 mi), and Dallas to Los Angeles, also 2190 km. See Figure 3-14. In the first leg, the Planetran would have a constant acceleration of 3.3 m/s² (approximately one-third the acceleration of a freely falling object) up to the midway point. Then there would be constant deceleration, also of 3.3 m/s² magnitude, for the remainder of the first leg, with the Planetran arriving at Dallas at rest. The second leg, Dallas to Los Angeles, would be similar: constant acceleration in the first half and deceleration in the second half.

(a) What would be the total travel time from New York to Dallas (or Dallas to Los Angeles)? (b) What would Planetran's maximum speed be?

For the first half of the New York–Dallas leg, we are given

$$v_0 = 0 \quad \text{(Planetran starts from rest at N.Y.)}$$

$$a = +3.3 \text{ m/s}^2$$

$$x_0 = 0 \quad \text{(Choose origin at N.Y.)}$$

$$x = \tfrac{1}{2}(2190 \text{ km}) = 1.09 \times 10^6 \text{ m}$$

We are to find t.

Equation (3-9) involves only these quantities (see Table 3-1), and we have

$$x = x_0 + v_0 + \tfrac{1}{2}at^2 = \tfrac{1}{2}at^2$$

Solving for the unknown t gives

$$t = \sqrt{\frac{2x}{a}}$$

$$= \sqrt{\frac{2(1.09 \times 10^6 \text{ m})}{3.30 \text{ m/s}^2}}$$

$$= 813 \text{ s} = 13.5 \text{ min}$$

The second half of the first leg is just like the first in distance traveled and magnitude of acceleration. Therefore, it too would take 13.5 min. The time for the N.Y.–Dallas or Dallas–L.A. trip would then be 27 min. An express trip from coast to coast (with no appreciable time spent in Dallas) would take only 54 min, compared with 6 h by commercial aircraft.

(b) The maximum speed would be achieved at the midpoint in each leg of the trip, after 13.5 min of constant acceleration of 3.3 m/s², starting from rest. Here we are given

$$v_0 = 0$$

$$a = 3.3 \text{ m/s}^2$$

$$t = 13.5 \times 60 \text{ s}$$

$$v = ?$$

From (3-6),

$$v = v_0 + at$$

$$= 0 + (3.30 \text{ m/s}^2)(13.5 \times 60 \text{ s})$$

$$= 2673 \text{ m/s} = 2.7 \text{ km/s}$$

$$= (2.7 \times 10^3 \text{ m/s}) \frac{(2.24 \text{ mi/h})}{1 \text{ m/s}} = 6048 \text{ mi/h}$$

The Planetran's maximum speed would be about one-third the speed of a satellite orbiting the earth close to its surface and three times the speed of a supersonic commercial aircraft.

* R. M. Salter, "N.Y. to L.A. in 54 Minutes by Subway," in *Science and the Future Yearbook*. Chicago: Encyclopedia Britannica Educational Corp., 1980.

Note two points, illustrated in the example and applicable to all problems in kinematics—indeed, to all problems in physics:

• We solve algebraically for the unknown *before* substituting numerical values for symbols.
• The units are carried along with the numbers and treated algebraically. One necessary (but not sufficient) test of the correctness of the solution is whether the dimensions are appropriate. Any formula in physics must be dimensionally consistent; that is, all its terms must have the same dimensions. For example, the terms x, x_0, $v_0 t$, and $\frac{1}{2}at^2$ in the equation $x = x_0 + v_0 t + \frac{1}{2}at^2$ must have the same units; for example, all the lengths must be meters. No answer is complete if it does not specify the units.

Example 3-4. A motorist traveling at a constant speed of 50 km/h ($\approx$ 30 mi/h) runs through a red traffic light. A police car hiding behind a billboard at the intersection gives chase at the instant the motorist passes the light. Assume that the police car can move at a constant acceleration of 2.8 m/s². (a) How far from the traffic light does the police car catch up with the motorist? (b) How long does it take? (c) What is the police car's speed as it overtakes the motorist?

(a), (b) We use the subscripts m for the motorist and p for the police car. Question (a) asks for the displacement x when $x_p = x_m$, and question (b) asks for the time t when $x_p = x_m$. It is useful, therefore, to write relations for the motorist's displacement x_m and the police car's displacement x_p, both as functions of time t.

For the motorist,

$$x_{0m} = 0$$

$$v_{0m} = \frac{50 \text{ km}}{\text{h}} \frac{1 \text{ h}}{3600 \text{ s}} \frac{10^3 \text{ m}}{1 \text{ km}} = 14 \text{ m/s}$$

$$a_m = 0$$

Displacement zero ($x_{0m} = 0$) has been chosen to correspond to the position of the traffic light. The displacement x_m of the motorist can be written as a function of time, using (3-9):

$$x = x_0 + v_0 t + \frac{1}{2}at^2 \tag{3-9}$$

$$x_m = 0 + (14 \text{ m/s})t + 0$$

For the police car,

$$x_{0p} = 0$$

$$v_{0p} = 0$$

$$a_p = 2.8 \text{ m/s}^2$$

and (3-9) gives

$$x_p = 0 + 0 + \frac{1}{2}(2.8 \text{ m/s}^2)t^2$$

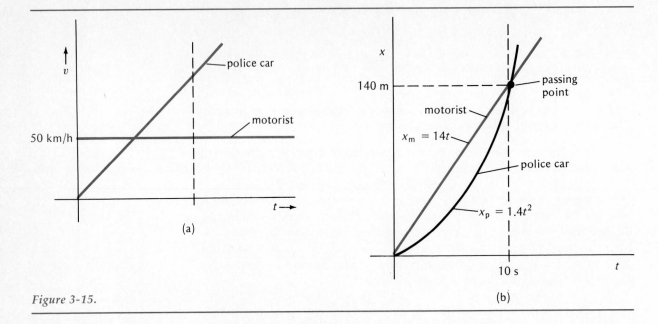

Figure 3-15.

(b)

When the vehicles pass,

$$x_m = x_p$$

$$14t = 1.4t^2$$

$$t = 0 \text{ s}, 10 \text{ s}$$

The two vehicles are at the same location at *two* times: $t = 0$, when both are at the traffic light, and $t = 10$ s, as the police car overtakes the motorist.

At $t = 10$ s, the displacements are

$$x_m = (14 \text{ m/s})t = (14 \text{ m/s})(10 \text{ s}) = 140 \text{ m}$$

or $\qquad\qquad x_p = \tfrac{1}{2}(2.8 \text{ m/s}^2)t^2 = (1.4 \text{ m/s}^2)(10 \text{ s})^2 = 140 \text{ m}$

It is useful to look at a plot of both x_m and x_p as a function of time t, as shown in Figure 3-15(b). For the motorist, x_m versus t is a straight line (constant velocity); for the police car, x_p versus t is a parabola (constant acceleration). The intersection of the two lines corresponds to the passing point.

(c) When the police car passes the motorist, its velocity is

$$v_p = v_{0p} + a_p t$$
$$= 0 + (2.8 \text{ m/s}^2)(10 \text{ s})$$
$$= 28 \text{ m/s} \approx 60 \text{ mi/h}$$

or just twice the speed of the motorist; this is also shown in Figure 3-15(a). We can see this result on another basis; if both vehicles are to have the same displacement over the same time interval, their average velocities must be the same. For the motorist, $\bar{v}_m = 14$ m/s; for the police car at constant acceleration,

$$\bar{v}_p = \frac{v_{0p} + v_p}{2} = \frac{0 + 28 \text{ m/s}}{2} = 14 \text{ m/s}$$

Example 3-5.   An electron enters the region between two parallel electrically charged metal plates initially with a velocity of 4.0×10^6 m/s to the right. See Figure

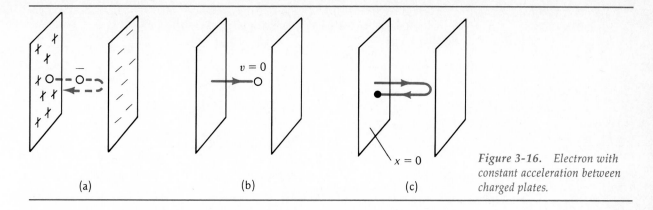

Figure 3-16. Electron with constant acceleration between charged plates.

(a)   (b)   (c)

3-16(a). The negatively charged electron is attracted by positive charges on the left plate and repelled by negative charges on the right plate. While the electron is between the plates, it has a constant acceleration of 8.0×10^{14} m/s^2 to the left. As the electron travels to the right, it slows down, is brought to rest momentarily, and then accelerates back to the left plate. (a) How far from the left plate does the electron travel before turning around? (b) How long does the round trip from the left plate and then back take?

(a) Choosing the left plate to correspond to $x = 0$, we have

$$x_0 = 0$$

$$v_0 = 4.0 \times 10^6 \text{ m/s}$$

$$a = -8.0 \times 10^{14} \text{ m/s}^2$$

The electron's velocity is positive as the electron travels to the right, negative as it travels left, and *zero* at the turning point. See Figure 3-16(b). We then can write

$$v = 0 \qquad \text{for} \qquad x = ?$$

Since x_0, v_0, a, and v are given and x is asked for, we use (3-10) of Table 3-1:

$$v^2 = v_0^2 + 2a(x - x_0)$$

We have, using the previous information,

$$x = -\frac{v_0^2}{2a}$$

$$= -\frac{(4.0 \times 10^6 \text{ m/s})^2}{2(-8.0 \times 10^{14} \text{ m/s}^2)}$$

$$= 1.0 \times 10^{-2} \text{ m} = 1.0 \text{ cm}$$

(b) When the electron is back again at the starting point, its displacement x is zero [Figure 3-16(c)], so that the time t for the round trip is described by

$$t = ? \qquad \text{for } x = 0$$

Again using (3-9), with $x_0 = 0$, yields

$$x = v_0 t + \tfrac{1}{2}at^2$$

$$0 = (v_0 + \tfrac{1}{2}at)t$$

$$t = 0 \qquad \text{or} \qquad t = \frac{-2v_0}{a}$$

There are, of course, *two* solutions to the quadratic equation; this means that there are two times when the electron was at the left plate ($x = 0$): the starting time, $t = 0$, and the return time,*

$$t = \frac{2v_0}{a}$$

$$= \frac{2(4.0 \times 10^6 \text{ m/s})}{(-8.0 \times 10^{14} \text{ m/s}^2)}$$

$$= 1.0 \times 10^{-8} \text{ s} = 10 \text{ ns}$$

The values for speed and acceleration here, although extraordinarily high for ordinary objects, are typical for an electron. A consequence is that travel times are correspondingly short.

* The answer 10 nanoseconds uses the prefix definition $1 \text{ n} = 1 \text{ nano} \equiv 1 \times 10^{-9}$. See Appendix B for a comprehensive listing of abbreviations of multiples of 10.

3-10   Freely Falling Bodies

Close to the earth's surface, all freely falling bodies have a nearly constant magnitude of acceleration toward the center of the earth. By a "freely falling" body is meant not only some object dropped from rest, but also a thrown object, in the vicinity of the earth's surface. Any object truly falls free only when it moves through a vacuum, but for relatively low speeds and for smooth, compact, dense objects, the resistance of the air is negligible and their motion closely approximates free fall. On the other hand, descending parachutes, flying birds, and falling leaves do not qualify as freely falling objects. Unless we state otherwise, we ignore the effects of air resistance or take it to be negligible.

The acceleration of a freely falling object does not depend on the object's mass or weight. Experiment shows that an apple, an atom, an orbiting satellite, or any other object near the earth's surface falls with constant acceleration of 9.8 m/s^2, or 32.2 ft/s^2. This acceleration is so important in physics that it is designated by the special symbol g and referred to as the *acceleration due to gravity* (not as *gravity* and especially not as the *force of gravity*). How g is related to the general phenomenon of gravitation will be dealt with later; here we treat only falling-body kinematics.

Constant acceleration for any object in free flight implies that its velocity is constantly changing. The instantaneous velocity may be zero, so that the object is momentarily at rest but the falling object's acceleration is still g. Consider an object thrown upward; at the highest point its velocity is zero as the object stops moving upward (positive v) and starts moving downward (negative v). The painting *Castle in the Pyrenees* by René Magritte (Figure 3-17) shows a huge boulder floating, defying gravity. That's one way of looking at it. But here is another interpretation: somehow the boulder was thrown upward, and the picture is a snapshot of the boulder at its highest point when it is momentarily at rest.

We must qualify some of our assertions. Strictly, g is not constant; its value differs at various places on the earth's surface, although by no more than 0.4 percent. Moreover, g decreases with increasing distance from the earth's center. (As an extreme case, the moon "falls" toward earth, but with an acceler-

Figure 3-17.   The Castle of
the Pyrenees, *1959, Réné
Magritte. Photo © G. D. Hackett.*

ation of only 2.7×10^{-3} m/s².) For altitudes of up to 40 mi above sea level, g is
constant within 2 percent.

Falling-body motion is one of the very few situations in physics approxi-
mating motion at constant acceleration. (Another example is a charged particle
between parallel, charged plates.)

Example 3-6.   An object is thrown vertically upward at 30 m/s. (a) What is its
displacement after 4.0 s? (b) What is its velocity after 4.0 s? (c) What is the maximum
height it attains? (d) How long is the object in flight? (e) What is its velocity on striking
the ground?

The kinematic relations for constant acceleration apply here. Since the direction of
motion is along the vertical, we label the displacement y, rather than x, with the positive
y direction taken to be vertically upward. Then $a = -g$. We put $y_0 = 0$. In this example,
we round off the value of g to 9.8 m/s² and use no more than two significant figures in
the computed results.

(a) We know that $v_0 = +30$ m/s and $a = -9.8$ m/s²; we are to find y for $t = 4.0$ s.
Equation (3-9), $y = y_0 + v_0 t + \frac{1}{2}at^2$, becomes

$$y = 0 + (30 \text{ m/s})(4.0 \text{ s}) + \tfrac{1}{2}(-9.8 \text{ m/s}^2)(4.0 \text{ s})^2 = +42 \text{ m}$$

(b) What is v after 4 s? Equation (3-6), $v = v_0 + at$, gives

$$v = 30 \text{ m/s} + (-9.8 \text{ m/s}^2)(4.0 \text{ s}) = -9 \text{ m/s}$$

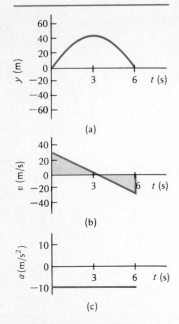

(a)

(b)

(c)

Figure 3-18. (a) Displacement-time, (b) velocity-time, and (c) acceleration-time graphs for a projectile thrown vertically upward.

Four seconds after the object is thrown, it is 42 m above the ground and traveling downward with an instantaneous speed of 9 m/s.

(c) At maximum height, where the object is changing direction of motion, its instantaneous velocity is zero (its acceleration is still g!). Therefore, we want to know what y is when $v = 0$. From (3-10), $v^2 = v_0^2 + 2a(y - y_0)$, we have

$$y = \frac{v^2 - v_0^2}{2a} = \frac{0 - (30 \text{ m/s})^2}{2(-9.8 \text{ m/s}^2)} = +46 \text{ m}$$

See Figure 3-18.

(d) The displacement y is zero when the object strikes the ground; we wish to find the corresponding time of flight. Using (3-9), $y = y_0 + v_0 t + \frac{1}{2}at^2$, to find t for $y = 0$, we have

$$0 = (v_0 + \tfrac{1}{2}at)t$$

$$t = 0 \quad \text{or} \quad -\frac{2v_0}{a} = -\frac{2(30 \text{ m/s})}{-9.8 \text{ m/s}^2}$$

$$t = 0 \quad \text{or} \quad t = 6.1 \text{ s}$$

The time of flight is 6.1 s.

(e) The average velocity $\bar{v}$ of the entire motion is zero, since the overall displacement y is zero. Therefore, (3-7), $\bar{v} = (v_0 + v)/2$, gives

$$v = 2\bar{v} - v_0 = 0 - 30 \text{ m/s} = -30 \text{ m/s}$$

[Alternatively, we can find v for $t = 6.1$ s by using (3-7).] The object strikes the ground with the same speed as that with which it was thrown. The entire motion is

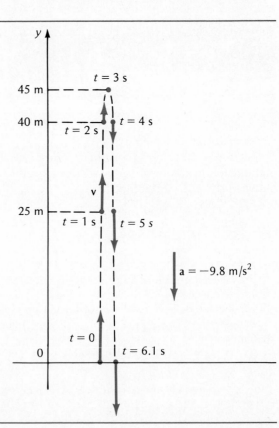

Figure 3-19. Velocity vectors for a projectile thrown vertically upward. (The vectors for the downward motion are shown shifted to the right for clarity.)

symmetrical in time about the midtime. Stated differently, it is the same motion with time running backward; one cannot distinguish between motion pictures of a thrown object, one with the film run forward and the other backward.

The displacement-time, velocity-time, and acceleration-time curves in Figure 3-18 display all the information we have computed here (and more). Note that the displacement is zero at 6.1 s, as indicated not only on the displacement-time curve but also by the net area, shaded, on the velocity-time graph.

The displacement and velocity of the thrown object are also shown in Figure 3-19. Here velocity vectors are located with their tails at the corresponding positions of the object.

Summary

Definitions

Average velocity: change in displacement (Δx) divided by the corresponding time interval (Δt)

$$\bar{v} \equiv \frac{\Delta x}{\Delta t} \qquad (3\text{-}2)$$

Velocity (instantaneous): time rate of change of displacement

$$v \equiv \lim_{\Delta t \to 0} \frac{\Delta x}{\Delta t} = \frac{dx}{dt} = \text{slope of } x\text{-}t \text{ graph} \qquad (3\text{-}3)$$

Acceleration (instantaneous): time rate of change of velocity

$$a \equiv \lim \frac{\Delta v}{\Delta t} = \frac{dv}{dt} = \text{slope of } v\text{-}t \text{ graph} \qquad (3\text{-}5)$$

Units

Displacement: meter = m

Velocity: meter/second = m/s

Acceleration: (meter/second)/second = m/s^2

Important Results

For constant acceleration **a**, the kinematic variables are related by

$$v = v_0 + at \qquad (3\text{-}6)$$

$$x - x_0 = v_0 t + \tfrac{1}{2}at^2 \qquad (3\text{-}9)$$

$$x - x_0 = \tfrac{1}{2}(v_0 + v)t \qquad (3\text{-}8)$$

$$x - x_0 = \frac{v^2 - v_0^2}{2a} \qquad (3\text{-}10)$$

Freely falling objects near the earth's surface have a constant downward acceleration, $g = 9.8$ m/s^2.

Problems and Questions

Section 3-2   Constant Velocity

· **3-1 P** How long will it take to drive the 90 miles from Sacramento to San Francisco at the legal speed limit of 55 miles per hour? How much time would you save by driving at 70 mi/h?

· **3-2 P** You are driving at 40 km/h when a pedestrian catches your eye. You glance away for 2 seconds. How far did you travel in this time toward the person crossing the street at the intersection ahead?

· **3-3 P** A world-class sprinter can run 100 yards in 9.4 seconds. How long would it take him to run 100 meters?

· **3-4 P** Light travels through empty space at 3×10^8 m/s. A *light year* is the distance light travels in one year. (a) How many meters in a light year? (b) The distance from the earth to the sun is about 1.49×10^8 km. How long does it take light to travel this distance? (c) The average radius of the

earth's orbit is called one *astronomical unit* (1 AU). How many AU are there in 1 light year?

: **3-5 P** On the last lap of a relay race the East German runner has a 10 meter lead when the American anchor man gets the baton. The American can do the remaining 400 meters in 44.0 seconds. The German can run the final 400 meters of a relay in 45.5 seconds. Who will win the race between the two, and by how many meters?

: **3-6 P** A quarterback can throw a football with an average velocity of 20 meters per second. A wide receiver breaks across the middle, running 7 meters per second on a line perpendicular to the flight of the ball. If he is 20 meters downfield from the quarterback when he catches the ball, how much did the quarterback have to lead him?

: **3-7 P** In a 1500 meter race run on a 300 meter oval track the eventual winner laps the last runner with one lap to go.

The winner finishes the race in a time of 4 minutes 10 seconds. Assuming both runners run at their average speed during the final lap, what is the time for the race of the slowest runner?

: **3-8 P** A boy was walking across a field with his dog one day when he saw a friend approaching. The dog was excited and dashed to the friend, then back to his master, and then to the friend, and so on, never stopping. How far would you estimate the dog ran if his speed was 30 km/hr and each boy walked at 4 km/hr, starting 400 meters apart?

Section 3-3   Average Velocity
· **3-9 P** A motorist travels 60 km at 30 km/hr and 60 km at 60 km/hr. What is her average speed for the trip?

· **3-10 P** A motorist wishes to average 60 km/hr for a trip. She finds that halfway through the trip she has averaged only 40 km/hr. What average speed would she have to maintain for the remainder of the trip in order to achieve an overall average speed of 60 km/hr?

· **3-11 P** A person making a 200 km trip drives the first 100 km at a steady 40 km/hr. How fast must he drive the remaining 100 km if he is to average 50 km/hr for the total trip?

· **3-12 P** A world-class sprinter can run 100 meters in 10.2 seconds. What is his average speed in miles per hour?

: **3-13 P** A motorist drives north on Highway 95 at 70 km/hr for 1 hour. She spends 20 minutes cutting down a Christmas tree and then returns home driving 90 km/hr. For the entire trip, what are (a) the average speed, (b) the average velocity, and (c) the total distance travelled? (d) If the stopping time had been 30 minutes, how fast would the motorist have had to drive on the return trip to achieve the same average speed for the entire trip?

Section 3-4   Instantaneous Velocity
: **3-14 Q** The position of a bug crawling on a meter stick is sketched here as a function of time. Thus we see that the bug

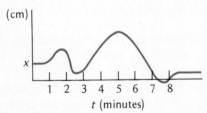

Figure 3-20.   Question 3-14.

(A) stopped four times.
(B) was going fastest near $t = 2$ minutes.
(C) was going fastest near $t = 5$ minutes.
(D) returned only once to its starting point.
(E) turned around five times.

: **3-15 Q** Speed is to velocity as
(A) centimeters are to meters.
(B) velocity is to acceleration.
(C) distance is to displacement.
(D) a vector is to a scalar.
(E) distance is to time.

Section 3-6   Acceleration
· **3-16 P** Suppose that a rocket ship could accelerate at a rate of 10 m/s² (about one g). How many years would it take the ship to reach a speed of one tenth the speed of light if it started from rest? The speed of light is 3×10^8 m/s and 1 year $= 3 \times 10^7$ seconds (approximately).

· **3-17 Q** If a particle moving along a straight line has constant acceleration
(A) it is always moving.
(B) its velocity is also constant.
(C) a plot of its position as a function of time will be a parabola only in those cases where it does not change direction.
(D) a plot of its position as a function of time will be a straight line.
(E) it may be speeding up or slowing down.

· **3-18 P** A racquetball strikes a wall with a speed of 30 m/s and rebounds with a speed of 26 m/s. The collision takes 20 ms (20×10^{-3} s). What is the direction and magnitude of the average acceleration?

Section 3-7   Constant Acceleration
· **3-19 Q** If an object moving in one dimension experiences constant non-zero acceleration,
(A) its speed is constant.
(B) it will go faster and faster without stopping.
(C) its velocity is always changing.
(D) it will move twice as far in two seconds as it moves in one second.
(E) a graph of v vs. t will be a parabola.

· **3-20 Q** When you throw a ball straight up in the air,
(A) its velocity is constant.
(B) it is always moving.
(C) it is always accelerating.
(D) the time required to rise a height h is half the time required to rise a height 2h.
(E) it moves equal distances in equal times.

: **3-21 Q** Which of the following is an accurate statement?
(A) If velocity and acceleration have opposite signs, the object is slowing down.
(B) If position and velocity have opposite signs, the object is slowing down.
(C) If an object is not moving it must have zero acceleration at that instant.
(D) It is not possible for an object to have both negative velocity *and* negative acceleration.

(E) If an electron is accelerated from rest with constant acceleration, it will move twice as far in the second millisecond as it moved in the first millisecond.

⋮ **3-22 Q** Two cars run a fixed-distance time trial as follows. The Ferrari accelerates at a constant rate a for the first half of the distance and then at a rate $2a$ for the second half. The Lotus accelerates at $2a$ for the first half and a for the second half. Who will have the better time, or will they have equal times?

· **3-23 P** An easy and fairly accurate way to measure the depth of an old mine is to drop a rock down it and time the fall until you hear a clunk at the bottom. Do this by counting the seconds ("One Mississippi, two Mississippis . . ."). How deep is a four-Mississippi mine?

· **3-24 P** An automobile travelling 20 m/s slams on the brakes and decelerates uniformly at 2 m/s². (a) How long does it take to stop? (b) How far does it travel while decelerating? Repeat the calculation for an initial velocity of 40 m/s.

· **3-25 P** A 1 kg rock and a 2 kg rock are dropped from rest from a tower 80 meters high. If the 2 kg rock is dropped 1 second after the 1 kg rock, how far above the ground will it be when the first rock hits?

· **3-26 P** In an accident, a passenger's head hits the dashboard of a car with a speed of 4 m/s and comes to rest in 10 cms. What deceleration does she experience?

· **3-27 P** An electron with an initial velocity of 2×10^4 m/s enters a region of space where it is accelerated at 2×10^{15} m/s² for a distance of 1 cm. This is the process that occurs in an electron gun, such as is used in a TV tube. (a) What is the electron's velocity when it leaves the accelerating region? (b) What velocity would be achieved if the accelerated electron were directed into a second 1 cm long accelerating region?

⋮ **3-28 P** The driver of a car travelling 80 km/hr sees a truck coming at him at 60 km/hr. Both drivers hit the brakes simultaneously. The truck decelerates at 2 m/s² and the car decelerates at 4 m/s². What must be the minimum distance separating them at the time they hit the brakes if they are to avoid a collision?

⋮ **3-29 P** If the average driver's reaction time is 0.1 s, how long should a traffic light stay yellow (allowing no extra safety factor) for cars travelling (a) 30 km/hr, (b) 60 km/hr, and (c) 90 km/hr to stop as it turns red? What are the stopping distances in each case? (See Problem 3-28.)

⋮ **3-30 P** A car stopped at a traffic light accelerates at 2 m/s² just as soon as the light turns green. Three seconds after the light turns green a truck travelling at a constant speed of 12 m/s passes the car. (a) Draw a neat graph (not necessarily with exact numerical values, but schematically correct) that shows the positions of the two vehicles as a function of time, with the traffic light at $x = 0$ turning green

at $t = 0$. (b) How long after the truck passed the car will the car overtake the truck? (c) How fast will the car then be going? (d) How far from the traffic light will the car catch up with the truck? (Be careful about where your distances are measured from.)

⋮ **3-31 P** A free-falling skydiver reaches a terminal velocity of 65 m/s before opening his parachute. His parachute can provide a deceleration of 30 m/s². A parachutist cannot hit the ground at a speed greater than 5 m/s without risking injury. What is the minimum elevation at which the skydiver can open his parachute?

⋮ **3-32 P** A bicyclist and a stock car racer run the following race: The bicycle is given a running start with a speed of 40 km/hr, which he maintains throughout the 100 meter race. The Chevy starts from rest at the moment the bike crosses the start line and accelerates throughout the race at 1.65 m/s². (a) Who will win the race and (b) where will the loser be when the winner crosses the finish line?

· **3-33 P** A model rocket engine provides a constant upward acceleration of 10 m/s² for 5 seconds. After burnout the rocket decelerates at a rate of about 10 m/s² because of gravity. (a) What is the maximum velocity achieved? (b) How high will the rocket go? (c) How long will the rocket be in the air?

⋮ **3-34 P** A Datsun 280Z is speeding at 120 km/hr in an 80 km/hr speed zone. As the Datsun passes a parked highway patrol car, the officer accelerates the patrol car at a uniform rate, reaching 60 km/hr in 10 seconds. The officer continues to accelerate at this rate until reaching the car's top speed of 150 km/hr, and continues at this top speed until overtaking the speeder.

(a) How long did the chase take? (Hint: Break the problem into two parts: the accelerating part and the constant speed part.) (b) How far did the highway patrol officer travel to overtake the speeder? (c) Draw a careful graph plotting the positions of the Datsun and of the parked police car as a function of time. Take $x = 0$ as the position of the parked police car and $t = 0$ when the police car starts.

⋮ **3-35 P** A. J. Foyt pulls out of a pit stop at Indianapolis accelerating at a constant 3 m/s². Five seconds later Bobby Unser whizzes past at 190 mph. If each continues in this fashion, (a) how long after leaving the pit will Foyt catch Unser? (b) How far will he have travelled from the point where Unser passed him?

⋮ **3-36 P** A motorist travelling at speed v_0 rounds a curve and sees a slow moving truck on the roadway ahead. The truck is moving at a constant speed $v_T < v_0$. If the driver's reaction time is t_R and she can decelerate at a rate a while braking, what is the minimum distance she can be from the truck when she first sees it if she is to avoid a collision?

Section 3-8   Graphs and Constant Acceleration

· **3-37 Q** Consider a particle moving in one dimension

whose motion is represented by the graph shown in Figure 3-21. (a) How many times did the particle return to its starting point? (b) At what time was the velocity a maximum? (c) Was the acceleration positive, negative, or zero at time t_1?

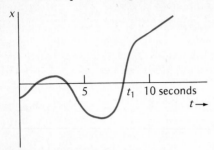

Figure 3-21.   Question 3-37.

· **3-38 Q** The position of a motorcycle as a function of time is plotted in Figure 3-22. Tell at one second intervals, starting with $t = 0$, if the velocity is positive or negative, and if the acceleration is positive or negative.

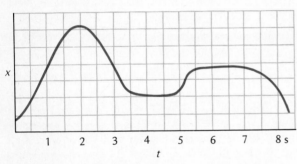

Figure 3-22.   Question 3-38.

: **3-39 Q** The variation of velocity as a function of time is shown in Figure 3-23 for a moving object. From this we see that
(A) the object returned to its starting point.
(B) the object did not turn around.
(C) the object had zero acceleration.
(D) the speed of the object steadily increased.
(E) the object never stopped.

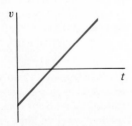

Figure 3-23.   Question 3-39.

: **3-40 Q** Figure 3-24 shows a schematic plot of distance vs. time for two moving objects. Which of the following descriptions best matches the graph shown?
(A) A ball is thrown into the air and a moment later an-

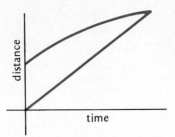

Figure 3-24.   Question 3-40.

other ball is thrown, but with higher initial velocity than the first.
(B) While accelerating in an attempt to pass, Gordon Johncock plows into the back of A. J. Foyt on the back straightaway at the Indy 500.
(C) Leading the Indianapolis 500 with two laps to go, Johnny Rutherford throws a wheel bearing and pulls in for a pit stop as Bobby Unser zooms by and goes on to win the race.
(D) A rock and a crumpled piece of paper are dropped from rest at $t = 0$ in the presence of air (*i.e.*, friction is present).
(E) A speeder races past a parked police car. A moment later the police car starts up and takes off after the speeder with maximum acceleration.

· **3-41 P** The position of a particle as a function of time is measured and the data below are obtained. Plot these data and draw a smooth curve through it. Then plot corresponding graphs of velocity vs. time, and acceleration vs. time, for the particle.

t (s)	0	1	2	3	4	5	6	7	8	9	10	15
x (cm)	0	1	3	7	15	31	39	43	45	46	46.5	46.5

: **3-42 P** For the graph of x vs. t shown in Figure 3-25, estimate the instantaneous velocity at (a) $t = 3$ s, (b) $t = 9$ s and (c) $t = 10$ s. (d) What is the average velocity during the interval $t = 0$ to $t = 9$ s?

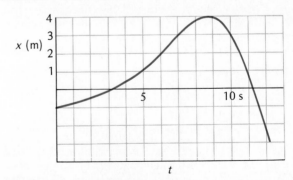

Figure 3-25.   Problem 3-42.

: **3-43 P** Draw a graph of x vs. t and of acceleration vs. t for the particle whose velocity is graphed in Figure 3-26.

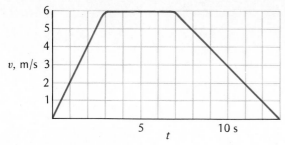

Figure 3-26.   Problem 3-43.

: **3-44 P** An elevator starts from rest at the ground floor of a building and accelerates as indicated in Figure 3-27. Where is it at $t = 18$ seconds?

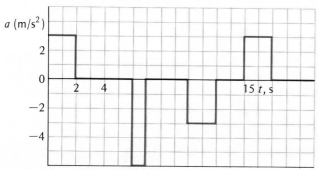

Figure 3-27.   Problem 3-44.

Section 3-10   Freely Falling Objects

· **3-45 Q** When a ball is thrown straight up with velocity v_0 its position as a function of time is given by

$$y = v_0 t - \tfrac{1}{2} g t^2$$

This quadratic equation has two roots for t for a given value of y. What does this mean?

(A) One root corresponds to throwing the ball upward, the other to throwing the ball downward.

(B) One root gives the time the ball reaches a given elevation on the way up; the other gives the time the ball passes that point on the way down.

(C) One of the roots has no physical significance and must be discarded.

(D) Although there are in principle two roots, in fact they always turn out to be equal, *i.e.*, they are not distinct roots.

· **3-46 Q** Two packages are dropped from an airplane. The first is 20 meters below the plane when the second is pushed out. If we neglect air resistance effects, we deduce that when the first package hits the ground

(A) the second will hit simultaneously.

(B) the second package will have hit the ground already.

(C) the second package will be less than 20 meters from the ground.

(D) the second package will be 20 meters above the ground.

(E) the second package will be more than 20 meters above the ground.

(F) the distance of the second package above the ground will depend on the relative weights of the two packages.

: **3-47 Q** If the separation between two falling objects is observed to increase steadily, which of the following is a valid observation?

(A) Air friction is not negligible.

(B) The objects may have been released simultaneously at different heights and are subject to negligible air friction.

(C) The objects may have been released at different times and are subject to air friction.

(D) The object that is ahead of the other must have been released at a lower elevation than was the trailing object.

(E) One of the objects must have had a greater initial velocity than the other.

· **3-48 P** A bullet from a 30.06 rifle has a muzzle velocity of approximately 900 m/s. How high would such a bullet go if fired straight up, neglecting friction?

· **3-49 P** A car travelling 90 km/hr crashes into a solid brick wall. From what height would the car have to be dropped to experience a similar impact?

· **3-50 P** Here is a simple way to measure your reaction time. Have a friend hold a meter stick vertical, next to, but not touching, your outstretched hand. Let him drop it unexpectedly. When you see it drop, clasp it. Note the position of the stick before and after catching it. It is then easy to calculate the time of fall if you know the acceleration due to gravity. A typical value is a drop of 13 cm. To what reaction time does this correspond? You might try this to see if you get the same results through stimuli other than visual, such as tactile or aural.

: **3-51 P** A boy drops a rock from a high bridge, and 1 second later he throws downward a second rock with a velocity of 10 m/s. (a) How long after the second rock is thrown will it catch up with the first rock? (b) How far will it have fallen then?

: **3-52 P** A hot air balloon is released from the ground and ascends with a constant acceleration of 0.5 m/s². After 10 seconds a sandbag is released overboard and falls to the ground. How long does it take the sandbag to reach the ground?

: **3-53 P** From the top of a cliff 30 meters above his friend, a boy drops a watermelon at the same instant his friend shoots an arrow upward with an initial velocity of 20 m/s. If the arrow were to hit the watermelon, how high above the ground would they be?

4

Motion in a Plane

The kinematics treated in Chapter 3 was special—motion along a single straight line. The most general motion of a particle takes place in three-dimensional space. Since most situations we shall encounter in our present study involve a particle moving in a single plane, we concentrate on kinematics in two dimensions. The generalization to three dimensions is then an obvious extension.

Our first concern is the definition, meaning, and interrelation of displacement, velocity, and acceleration, all as *vector* quantities. We then consider in some detail two special but important examples of two-dimensional kinematics:

- Motion in which the acceleration is constant in both magnitude and direction, illustrated by the motion of a projectile
- Motion in a circle at constant speed, illustrated by a satellite in circular orbit about the earth.

Finally, we consider how the velocity of a particle measured in one reference frame is related to the velocity of the same particle as measured in some second, moving reference frame.

4-1   Velocity and Acceleration Vectors

Two numbers (two coordinates) are required to specify the location of a particle in a plane. If we choose x and y axes in this plane, then the particle's location can be given by its rectangular coordinates, x and y. We can also specify a particle's location in a plane by polar coordinates, r and θ. See Figure 4-1.

The two sets of coordinates are related as follows:

$$x = r \cos \theta$$

$$y = r \sin \theta$$

and

$$r = \sqrt{x^2 + y^2}$$

$$\tan \theta = \frac{y}{x}$$

A particle's displacement from the origin can then be given in either of two equivalent ways:

• By the magnitude r of the *radius vector* **r**, together with its direction at angle θ relative to the positive x axis (θ is taken as positive for counterclockwise rotation).

• By the rectangular components (or coordinates), x and y, of displacement vector **r**.

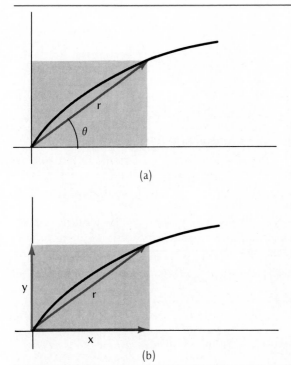

(a)

(b)

Figure 4-1. *Vector displacement* **r** *in terms of (a) magnitude r and direction θ and (b) x and y components.*

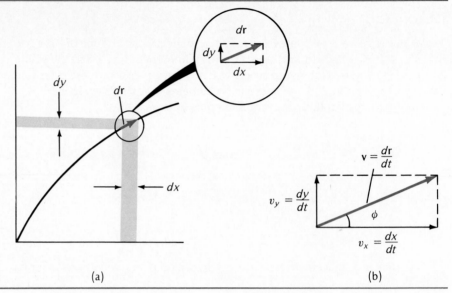

Figure 4-2. *(a) Incremental displacement vector d**r** over time interval dt has rectangular components dx and dy. (b) Velocity vector **v** = d**r**/dt has rectangular components v_x and v_y.*

(a)   (b)

Suppose that a particle undergoes a change in position over a very short time interval dt, as in Figure 4-2(a). The x and y coordinates change respectively by infinitesimal amounts dx and dy. The vector displacement over dt is $d\mathbf{r}$; its x and y components are dx and dy. The x and y components of the particle's velocity are

$$v_x = \frac{dx}{dt}$$
$$v_y = \frac{dy}{dt} \tag{4-1}$$

given in exactly the same fashion as in Chapter 3.* Each velocity component, v_x and v_y, is the time derivative, or the time rate of change, of the respective displacement component.

The *instantaneous velocity vector* **v** is defined as the *time rate of change of the displacement vector* **r**. In symbols,

$$\mathbf{v} \equiv \frac{d\mathbf{r}}{dt} \tag{4-2}$$

where $d\mathbf{r}$ is the incremental vector displacement, with components dx and dy, over the vanishingly small time interval dt. Expressed in terms of the rectangular components, v_x and v_y, the magnitude of the velocity **v**, which is called the speed, is

$$v = \sqrt{v_x^2 + v_y^2}$$

---

* Here and hereafter we use the conventional definition, (3-3), of the derivative:

$$v_x \equiv \lim_{\Delta t \to 0} \frac{\Delta x}{\Delta t} \equiv \frac{dx}{dt}$$

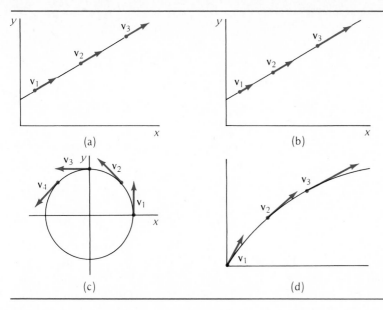

Figure 4-3. *Instantaneous velocity vectors superimposed on the path for (a) constant velocity, (b) velocity changing in magnitude only, (c) velocity changing in direction only, and (d) velocity changing in both magnitude and direction.*

The velocity direction is given by

$$\tan \phi = \frac{v_y}{v_x}$$

where ϕ is the angle between $\mathbf{v}$ and the x axis.* See Figure 4-2(b).

The instantaneous velocity vector $\mathbf{v} = d\mathbf{r}/dt$ has the same direction as that of the small vector displacement $d\mathbf{r}$ over time dt. (To find velocity $\mathbf{v}$, we merely multiply vector $d\mathbf{r}$ by the scalar $1/dt$, so that $d\mathbf{r}$ and $\mathbf{v}$ have the *same* direction.) The vanishingly small displacement vector $d\mathbf{r}$ coincides with the particle's path; therefore, the instantaneous *velocity vector* $\mathbf{v}$ is always *tangent to the particle's path*. See Figure 4-3, where several velocity vectors are superimposed on a particle's path. The tail of each velocity vector is placed at the corresponding location of the particle. The direction of $\mathbf{v}$ shows where the particle is heading; the magnitude of $\mathbf{v}$ tells how fast it is moving.

In Figure 4-3(a), the particle's velocity is constant; the particle moves along a straight line at constant speed. In Figure 4-3(b), the velocity changes in magnitude only. In Figure 4-3(c), the velocity changes in direction only as the particle moves in a circle at constant speed. In Figure 4-3(d), the velocity changes in both magnitude and direction; here the particle changes both direction of motion and speed.

The instantaneous velocity vector is the time rate of change of the displacement vector, $\mathbf{v} = d\mathbf{r}/dt$. Likewise, the *instantaneous acceleration vector* $\mathbf{a}$ is defined as the *time rate of change of velocity*,[†]

---

* Angle ϕ between $\mathbf{v}$ and the x axis is not in general the same as angle θ between $\mathbf{r}$ and the x axis. Compare figures 4-1(a) and 4-2(b).

† Instantaneous acceleration $\mathbf{a}$ is the limit, as the time interval Δt approaches zero, of average acceleration $\bar{\mathbf{a}} = \Delta \mathbf{v}/\Delta t$:

$$\mathbf{a} = \lim_{\Delta t \to 0} \frac{\Delta \mathbf{v}}{\Delta t} = \frac{d\mathbf{v}}{dt}$$

Figure 4-4. *(a) Initial velocity* **v**$_i$ *and "final" velocity* **v**$_f$ *superimposed on a particle's path. (b) The velocity change d**v** found from the vector diagram for the vector difference,* d**v** = **v**$_f$ − **v**$_i$. *(c) The velocity change found from the vector diagram for the vector* sum, **v**$_f$ = **v**$_i$ + d**v**.

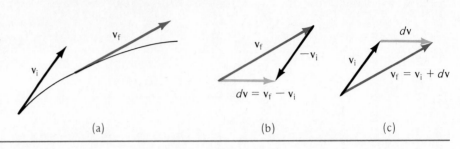

(a)　　　　　　(b)　　　　　　(c)

$$\mathbf{a} = \frac{d\mathbf{v}}{dt} \qquad (4\text{-}3)$$

Vector d**v** is the change in velocity occurring over an infinitesimal time interval, dt. Figure 4-4(a) shows a particle with velocity **v**$_i$ initially and a velocity **v**$_f$ a short time later; the velocity vectors are superimposed on the particle's path. Figures 4-4(b) and (c) show the *same* velocity vectors, now arranged so that velocity change d**v** can readily be found; d**v** can be regarded in two equivalent ways:

- As the difference between the "final" and the initial velocity: d**v** = **v**$_f$ − **v**$_i$ [Figure 4-4(b)].
- As the vector that must be added to the initial velocity to yield the final velocity: **v**$_f$ = **v**$_i$ + d**v** [Figure 4-4(c)].

By definition, d**v** = **a** dt; therefore, the direction of the acceleration vector **a** is the same as the direction of the velocity change d**v**. Note especially that a particle is accelerated whenever its velocity changes, either in *magnitude* or in *direction*. In short, **a** = 0 only for a particle with constant velocity; this means that a particle must move along a straight line at constant speed or else it must remain at rest if its acceleration is to remain zero.

Any vector can be replaced by its rectangular components, so the instantaneous acceleration vector **a** can be replaced by its x and y acceleration components, a_x and a_y. See Figure 4-5(a). The component a_x tells us at what rate the x velocity component v_x changes with time, and similarly for v_y.

An acceleration vector **a** can also be resolved into a different set of perpen-

Figure 4-5. *Acceleration components: (a)* a_x *and* a_y, *respectively along the* x *and* y *axes; and (b)* $a_\parallel$ *and* $a_\perp$, *respectively parallel with and perpendicular to the particle's velocity.*

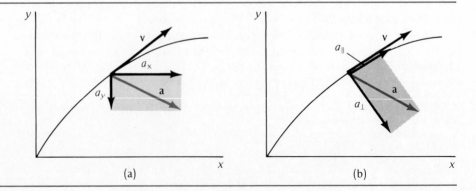

(a)　　　　　　(b)

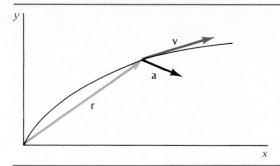

Figure 4-6. *Displacement vector* **r**, *velocity vector* **v**, *and acceleration vector* **a**.

dicular components — a component ($a_{\parallel}$) parallel to the particle's velocity and a component ($a_{\perp}$) perpendicular to **v**, as shown in Figure 4-5(b). In this representation, the acceleration components have a particularly simple meaning (proved in detail in Section 4-4):

- Acceleration component $a_{\parallel}$, *parallel* to the velocity, gives the time rate at which velocity changes in *magnitude.*
- Acceleration component $a_{\perp}$, *perpendicular* to the velocity, gives the time rate at which velocity changes in *direction.*

In summary, displacement **r**, velocity **v**, and acceleration **a** are all *vector* quantities:

- Vector **r** shows where the particle is *located.*
- Vector **v** points in the direction in which the particle is *headed* and gives its *speed.*
- Vector **a** shows the *direction of velocity change* and the *rate* at which this change is occurring.

If a particle moves along a single straight line, these three quantities are necessarily along that line. But for motion in two (or three) dimensions, vectors **r**, **v**, and **a** generally have different directions. See Figure 4-6. Moreover, both the magnitude and direction of **r**, **v**, and **a** may change with time.

The motion of a particle can be described by giving the direction and magnitude of vectors **r**, **v**, and **a** at each instant. Another, equivalent way of describing motion is the component representation. For motion in a plane, one gives the rectangular x and y components of **r**, **v**, and **a**. Then we can regard the motion in the plane as the superposition of two simultaneous motions along two straight lines — motion along the x axis with components x, v_x, and a_x together with motion along the y axis, with components, y, v_y, and a_y.

Example 4-1.   A particle moves in the curved path shown in Figure 4-7(a). Its instantaneous velocity at one point is v_i of 4.0 m/s at an angle $\theta = 10°$ above the x axis; one-tenth of a second later its velocity v_f still has the magnitude 4.0 m/s but now the velocity direction is at an angle of $10°$ below the x axis. What is the particle's average acceleration over the 0.10-s time interval?

To find the average acceleration, $\bar{\mathbf{a}} = \Delta \mathbf{v}/\Delta t$, we must first find the change in velocity $\Delta \mathbf{v}$. The initial and final velocity vectors are shown in Figure 4-7(b) with their tails at a common point. The vector construction shows that the velocity change, $\Delta \mathbf{v} = \mathbf{v}_f - \mathbf{v}_i$, is downward. From the geometry of the triangle in Figure 4-7(b), we see

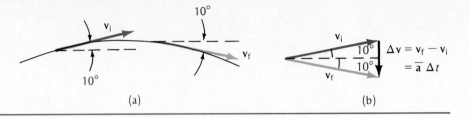

Figure 4-7. *(a) Initial and final velocity vectors,* $\mathbf{v}_i$ *and* $\mathbf{v}_f$. *(b) Vectors* $\mathbf{v}_i$ *and* $\mathbf{v}_f$ *relocated to find the change in velocity* $\Delta\mathbf{v} = \mathbf{v}_f - \mathbf{v}_i$.

that the magnitude is $\Delta v = 2v \sin \theta$, where v now represents the magnitude of either $\mathbf{v}_i$ or $\mathbf{v}_f$.

Applying the definition, we find that the magnitude of the downward average acceleration $\bar{a}$ is

$$\bar{a} = \frac{\Delta v}{\Delta t} = \frac{2v \sin \theta}{\Delta t} = \frac{2(4.0 \text{ m/s}) \sin 10°}{0.10 \text{ s}} = 14 \text{ m/s}^2$$

4-2 Motion at Constant Acceleration

Constant acceleration means that a particle's acceleration *vector* **a** is constant in both direction and magnitude. This in turn means that each of the x and y components of the constant acceleration vector **a** is separately constant. We have then *constant* acceleration a_x along x, and also *constant* acceleration a_y (generally not of the same magnitude as a_x) along y.

Before developing the general mathematical relations for constant acceleration, we first consider a simple example of this type of motion. Suppose that an object is projected horizontally and falls under the influence of gravity, as shown in Figure 4-8.

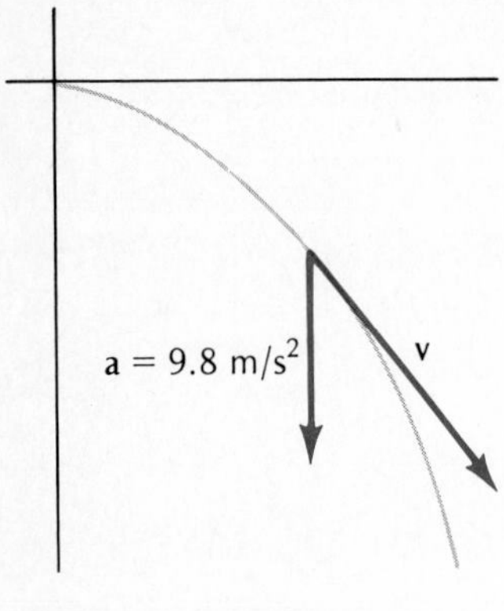

We choose the positive y axis to be vertically up, so that the constant acceleration lies along the negative y axis. The x axis lies along the horizontal, and here we choose the positive direction of the x axis to coincide with the direction of the object's initial velocity.

More specifically, suppose that an object is projected from the origin at an initial velocity of 30 m/s along the positive x axis. We wish to find its location and velocity at future times, $t = 1$ s, 2 s, 3 s, ...

The circumstances of the motion can be described as follows:

Horizontal (x axis): $a_x = 0$

$v_x = \text{constant} = 30 \text{ m/s}$

Vertical (y axis): $a_y = -g \approx -9.8 \text{ m/s}^2$ for $v_{0y} = 0$

Figure 4-8. *Projectile thrown horizontally. Its acceleration* **a** *is constant.*

In short, the object coasts with a *constant horizontal velocity* component of 30 m/s, and it falls with *constant vertical acceleration*.

Table 4-1 gives values for x, y, v_x, and v_y for several times t. Let us see how these entries are arrived at. First, the horizontal velocity component v_x is always 30 m/s. This means that the x displacement increases by 30 m over each 1-s interval, so that $x = 0$, 30 m, 60 m, 90 m, . . . at $t = 0$, 1 s, 2 s, ... The vertical velocity component v_y, initially zero, increases in the downward direction by 9.8 m/s over each 1-s interval. Along y we have, in effect, an

Table 4-1.

t (s)	v_x (m/s)	v_y (m/s)	x (m)	y (m)
0	30	0	0	0
1	30	−9.8	30	−4.9
2	30	−19.6	60	−19.6
3	30	−29.4	90	−44.1
4	30	−39.2	120	−78.4

object dropped from rest ($v_{0y} = 0$) and falling with constant acceleration $a_y = -9.8$ m/s². The general equation, (3-9),

$$y - y_0 = v_{0y}t + \tfrac{1}{2}a_y t^2$$

then reduces, for $v_{0y} = 0$, to

$$y = \tfrac{1}{2}(-9.8 \text{ m/s}^2)t^2$$

$$= 0, -4.9 \text{ m}, -19.6 \text{ m}, -44.1 \text{ m}, -78.4 \text{ m}$$

$$\text{at } t = 0 \text{ s}, \quad 1 \text{ s}, \quad 2 \text{ s}, \quad 3 \text{ s}, \quad 4 \text{ s}.$$

The results listed in Table 4-1 are plotted in Figure 4-9. The path is a parabola; this is proved for general projectile motion in Section 4-3. The downward vertical velocity component increases at a constant rate with time; the horizontal velocity component does not change. At any instant, the velocity **v** of the object has the magnitude $v = \sqrt{v_x^2 + v_y^2}$. Velocity vector **v** is always tangent to the path.

Now consider Figure 4-10. It shows the locations at equal time intervals of two objects: one dropped from rest, the other projected horizontally (as in Figure 4-9) at the same instant. Along the horizontal, each ball has a constant

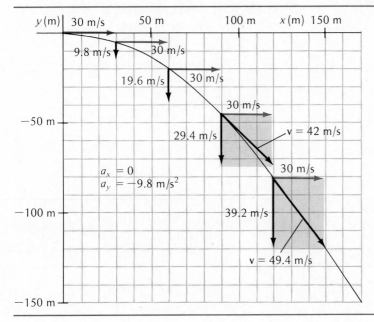

Figure 4-9. Velocity vector components superposed on the parabolic path of a projectile.

Scaling

Suppose that you're an engineer who has just finished designing a bridge for a 100-meter span. You've carefully taken into account how strong the materials must be to guarantee a bridge that is safe under all foreseeable circumstances. Now comes a request to design a bridge for a 400-m span. Can you simply use the drawings and specifications for the 100-m bridge with every dimension increased by a factor of 4?

As we shall see with easy arguments below, the answer is *no.* If you simply scaled up the dimensions by a factor of 4, the bridge would be seriously weak.

To see how the strength of a solid depends on its dimensions, we first consider a simpler situation—breaking a string.

Suppose that a certain kind of string rips into two pieces when its tension is more than 10 N. Then the force required to break two such strands side by side would be 20 N. To tear a rope with 100 such strands would require 100 × 10 N, or 1000 N, all told. Clearly, the breaking strength of the rope is directly proportional to the number of strands in it.

We can think of a solid under tension as analogous in some ways to a rope. A single strand now corresponds to a line of atoms along the direction in which the solid is stressed. It is, after all, the force of one atom on a neighboring atom that is responsible basically for the strength of the solid. Since the number of "atomic strands" is proportional to the cross-sectional area of the solid, the strength of a particular solid, we see, is directly proportional to its cross-sectional area, the area at right angles to directions of the stressing forces.

"Strength," or what is sometimes called breaking strength, can mean the force applied at the ends that actually rips the solid apart. Or in a less extreme case, by *strength* we can mean the force required to change

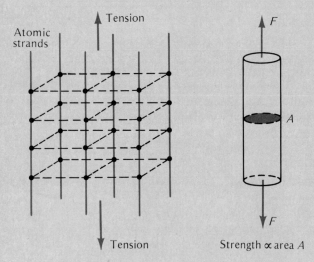

Atomic strands
Tension
Tension

F
A
F
Strength ∝ area A

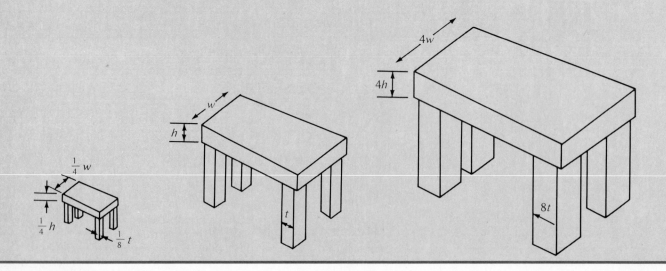

$\frac{1}{4} w$
$\frac{1}{4} h$
$\frac{1}{8} t$

w
h
t

$4w$
$4h$
$8t$

the length by, say, 1 part in 10³. Whatever the particular criterion we choose, however, the strength of a solid of a given material is proportional to its area of cross section. We are all familiar with this result—when a body builder makes his arm muscles stronger by ''pumping iron,'' the cross-sectional area of his biceps increases.

Let d stand for some characteristic dimension of a solid, for example, the length of a cylinder or its diameter. Then, for any given shape, the *volume* is proportional to the *cube* of dimension d (L^3 for a cube,

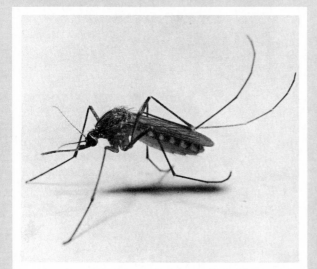

$4\pi r^3/3$ for a sphere, $\pi R^2 h$ for a cylinder). If every dimension of any solid object is doubled, then regardless of the shape, the object's volume increases by a factor 8. On the other hand, the *strength* of a solid member is proportional to cross-sectional area, or proportional to the *square* of d:

$$\text{Volume} \propto d^3$$

$$\text{Strength} \propto d^2$$

The fact that volume (and therefore also mass and weight of the object) varies directly with d^3, whereas strength varies with d^2, rules out a simple scaling of dimensions while maintaining constant strength.

To return to the original question on bridge design, we consider the simpler case of designing legs strong enough for an ordinary table of large size. Suppose that we start with a properly designed table and wish to construct, using the same materials, another table whose top has dimensions four times larger.

The volume of the new table top is larger by a factor $4^3 = 64$. Therefore, the weight of the table top is also greater by a factor of 64. This in turn means that each of the legs should be stronger by a factor of 64. Now, if the transverse dimension of each leg were to increase by the same factor 4, each leg's strength would increase by a factor of only $4^2 = 16$, not the factor 64 required. So the transverse leg dimension must be increased by the factor $(64)^{1/2} = 8$. (For the same reasons, the center of the table top would be more vulnerable to sagging.)

The reverse effect is seen in scaling down the size of the table top. With each dimension reduced by a factor 4, the top's weight is down by a factor 64, and the transverse dimension of each leg can be reduced by a factor 8. The big table has fat legs; the small table has skinny legs.

The very same effect is seen in the animal world, where transverse dimension of limbs are a measure of muscle strength; elephants are, as animals go, relatively fat and mosquitoes are skinny.*

*For further discussion of scaling, see J. B. S. Haldane, ''On Being The Right Size,'' *World of Mathematics,* vol. 2, ed. James R. Newman (New York: Simon & Schuster, 1956).

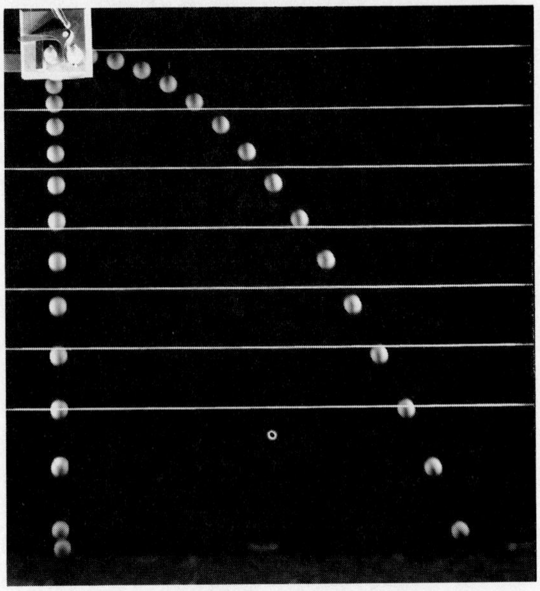

Figure 4-10. *Multiflash photograph of object dropped from rest and a second object simultaneously projected horizontally. The vertical motions of the two objects are identical. Both objects also have constant horizontal velocity components: for the dropped object, the horizontal velocity component is zero; for the projected object, it is not.*

velocity; one is zero and one is not. The motions along the vertical are identical. If air resistance is neglected, the two balls would be found to strike a horizontal floor *simultaneously*, quite apart from the initial speed of the ball thrown horizontally.

Now we put on record the general equations that give the displacement components (x and y) and velocity components (v_x and v_y) for any particle with constant acceleration. For motion along x at a constant acceleration a_x, we have from (3-9) and (3-6)

$$x - x_0 = v_{0x}t + \tfrac{1}{2}a_x t^2$$
$$v_x = v_{0x} + a_x t \tag{4-4}$$

To get corresponding equations along y, we merely replace every x in the previous equations by y.

$$y - y_0 = v_{0y}t + \tfrac{1}{2}a_y t^2$$
$$v_y = v_{0y} + a_y t \tag{4-5}$$

The time t is the *same* variable for both sets of equations; it links motion along x to that along y.

The specific example worked at the beginning of this section corresponds to the following values of the parameters:

$$x_0 = 0 \qquad v_{0x} = 30 \text{ m/s} \qquad a_x = 0$$
$$y_0 = 0 \qquad v_{0y} = 0 \qquad a_y = -9.81 \text{ m/s}^2$$

In this instance, the equations above reduce to

$$x = (30 \text{ m/s})t$$
$$v_x = 30 \text{ m/s}$$

$$y = (-4.9 \text{ m/s}^2)t^2$$

$$v_y = (-9.8 \text{ m/s})t$$

and the entries in Table 4-1 can be directly computed from these relations.

Here is a procedure that always applies to problems involving constant acceleration:

- Choose the orientation of the x and y axes and the location of the system's origin. One axis should always be along the direction of the constant acceleration.
- Write down the values of any given parameters (for example, x_0, y_0, v_{0x}, v_{0y}, a_x, a_y).
- Write equations for x, y, v_x, v_y, all as functions of time t. Refer to (4-4) and (4-5).
- Solve for the unknowns.

Algebraic equations (4-4) and (4-5) give the x and y *components* for displacement and velocity. They may be combined into *vector* relations for displacement and velocity. The vector displacement $\mathbf{r}$ has components x and y; the vector velocity $\mathbf{v}$ has components v_x and v_y; and the constant acceleration $\mathbf{a}$ has components a_x and a_y. The equivalent vector relations for the displacement and velocity as functions of time are then

$$\mathbf{r} - \mathbf{r}_0 = \mathbf{v}_0 t + \tfrac{1}{2}\mathbf{a}t^2$$

$$\mathbf{v} = \mathbf{v}_0 + \mathbf{a}t \tag{4-6}$$

Their meaning is illustrated in Figure 4-11 (which shows the particular situation for $t = 4$ s in Table 4-1 and Figure 4-8). As Figure 4-11(a) shows, the resultant displacement $\mathbf{r}$ is the vector sum of the displacement $\mathbf{v}_0 t$ the object would have made had there been no acceleration, and the additional displacement $\tfrac{1}{2}\mathbf{a}t^2$ arising from the constant acceleration. Similarly, we see in Figure 4-11(b) that the final velocity $\mathbf{v}$ is the vector sum of the initial velocity $\mathbf{v}_0$ and the velocity change $\mathbf{a}t$ arising from constant acceleration. Figure 4-11 shows that once again $\mathbf{r}$, $\mathbf{v}$, and $\mathbf{a}$ generally do not have the same direction.

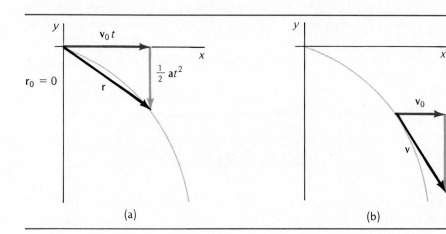

Figure 4-11. (a) Displacement vectors, $\mathbf{r} = \mathbf{v}_0 t + \tfrac{1}{2}\mathbf{a}t^2$, for constant acceleration; (b) velocity vectors, $\mathbf{v} = \mathbf{v}_0 + \mathbf{a}t$, for constant acceleration.

4-3   Projectile Motion

The most familiar example of motion in a plane at constant acceleration is that of a projectile. By *projectile* we mean here any small object thrown into the air and traveling distances small compared with the earth's radius. Then, if air resistance can be neglected, the object has a constant acceleration **g** vertically downward.

We saw one simple example of projectile motion—an object thrown horizontally—in Section 4-2. Here we consider more general features of projectile motion.

Our basic strategy is the same; we separate the motion into:

- Motion along the *horizontal* (*x* axis) at *constant velocity.*
- Motion along the *vertical* (*y* axis) at *constant acceleration.*

The motions along *x* and *y* are independent of one another but linked by the common time *t*.

Consider Figure 4-12, which represents a projectile being fired from the origin with an initial velocity $\mathbf{v}_0$ at angle θ above the horizontal. Its rise, fall, and return once again to the *x* axis are shown. We wish to find the time of flight *T*, the horizontal range *R*, the maximum height *H*, and the shape of the path. The initial velocity components are

$$v_{0x} = v_0 \cos \theta$$

$$v_{0y} = v_0 \sin \theta$$

and the acceleration components are

$$a_x = 0$$

$$a_y = -g$$

(The negative sign preceding *g* implies that in what follows, *g* will stand for the *magnitude* of the gravitational acceleration.)

The essential features of the motion are now on record; the object coasts at constant velocity along *x*, and it has a constant acceleration parallel to the *y* direction.

Using the specific values above for the velocity and acceleration components in the general relations (4-4) and (4-5) for the *x* and *y* velocity and displacement components, we get

$$v_x = v_0 \cos \theta \tag{4-7a}$$

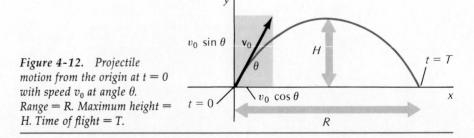

Figure 4-12. *Projectile motion from the origin at $t = 0$ with speed v_0 at angle θ. Range = R. Maximum height = H. Time of flight = T.*

$$v_y = v_0 \sin \theta - gt \tag{4-7b}$$

$$x = (v_0 \cos \theta)t \tag{4-7c}$$

$$y = (v_0 \sin \theta)t - \tfrac{1}{2}gt^2 \tag{4-7d}$$

When we put $y = 0$ in (4-7d), we get

$$0 = (v_0 \sin \theta)t - \tfrac{1}{2}gt^2 = t[(v_0 \sin \theta) - \tfrac{1}{2}gt]$$

The two roots of this quadratic equation are

$$t = 0 \quad \text{and} \quad t = \frac{2v_0 \sin \theta}{g}$$

What is the flight time T? It is the time elapsing from the firing to the time the projectile is once again at the x axis. In symbols,

When $y = 0$, $t = T$ (as well as when $t = 0$)

Therefore, the flight time is

$$T = \frac{2v_0 \sin \theta}{g} \tag{4-8}$$

The horizontal range R corresponds to the distance traveled along x during the time T. In symbols,

$$x = R \quad \text{when } t = T$$

Substituting the value of T from (4-8) in (4-7c) yields

$$R = \frac{2v_0^2 \sin \theta \cos \theta}{g} \tag{4-9a}$$

This relation can be put into simpler form by using the trigonometric identity $2 \sin \theta \cos \theta \equiv \sin 2\theta$. The range can then be written as

$$\text{Range} = R = \frac{v_0^2 \sin 2\theta}{g} \tag{4-9b}$$

We see, for a fixed projection angle θ, that R is proportional to v_0^2; doubling the projection speed quadruples the range. Further, for a given v_0, the range is proportional to $\sin 2\theta$. But $\sin 2\theta$ has its maximum value for $2\theta = 90°$, or $\theta = 45°$. Therefore, the range is a maximum (in the absence of air resistance) for projection at $45°$; $R_{\max} = v_0^2/g$. See Figure 4-13.

Another consequence of the range relation (4-9a) is that for a given v_0 the range is the same for any two angles whose sum is $90°$, that is, for complementary angles θ and $90° - \theta$. If we substitute $90° - \theta$ in (4-9a), we have

$$R = \frac{v_0^2[2 \sin (90° - \theta) \cos (90° - \theta)]}{g}$$

By the identities $\sin (90° - \theta) = \cos \theta$ and $\cos (90° - \theta) = \sin \theta$, the above equation becomes

$$R = \frac{v_0^2(2 \cos \theta \sin \theta)}{g}$$

which is just (4-9a).

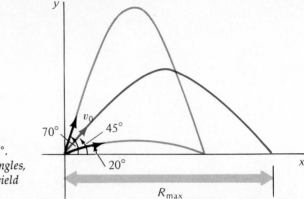

Figure 4-13. Maximum projectile range for $\theta = 45°$. Any two complementary angles, for example 20° and 70°, yield the same range.

Thus, in Figure 4-13, the range for $\theta = 20°$ is the same as the range for the complementary angle, $\theta = 70°$. Although two complementary projection angles yield the same range when air resistance is neglected (unrealistically), the two paths differ in time of flight and in maximum height.

How do we characterize the maximum height H? For $y = H$, we have $v_y = 0$; that is, the projectile stops going upward ($v_y > 0$) and starts going downward ($v_y < 0$) at maximum height. In symbols,

$$v_y = 0 \qquad \text{for } y = H$$

From (4-7b), the time at which $v_y = 0$ is given by

$$t = \frac{v_0 \sin \theta}{g}$$

Substituting this value in (4-7d) yields

$$\text{Maximum height} = H = \frac{v_0^2 \sin^2 \theta}{2g} \tag{4-10}$$

Finally, let us find the equation for the projectile path. In (4-7c) and (4-7d), x and y are given as functions of the common time t. If we eliminate the time by

Figure 4-14. Projectiles trace out parabolic paths.

solving for t in the first equation and we then substitute this value in the second one, we get

$$y = (\tan \theta)x - \left(\frac{g}{2v_o^2 \cos^2 \theta}\right)x^2 \qquad (4\text{-}11)$$

This equation shows that y is a quadratic function of x; that is, y plotted against x is a *parabola* whose symmetry axis is parallel to the direction of the constant acceleration. See Figure 4-14.

The equations for the flight time, range, maximum height, and path are special relations (they hold only for projection *from the origin* and return *to the x axis*). These specialized equations need not be memorized; they can always be looked up when needed. The *procedures* used to derive them *are* important; they apply to any problem in projectile motion.

Example 4-2.   A projectile is fired from the edge of a cliff 80 m above a plane surface. Its initial velocity is 38 m/s at an angle of 30° above the horizontal. (a) How far from the base of the cliff does it land? (b) What is its velocity when it hits the ground? See Figure 4-15.

As usual, we choose the launch point as the location of the origin of the x and y axes and neglect air resistance. The initial velocity components are

$$v_x = (38 \text{ m/s}) \cos 30° = 33 \text{ m/s}$$

$$v_y = (38 \text{ m/s}) \sin 30° = 19 \text{ m/s}$$

Therefore, the equations (4-7) for the x and y components of the projectile's velocity and displacement become

$$v_x = 33$$

$$v_y = 19 - 9.8t$$

$$x = 33t$$

$$y = 19t - 4.9t^2$$

where we have used $a_y = -g = -9.8 \text{ m/s}^2$, and dropped units for simplicity.

(a) The horizontal distance D from the base of the cliff to the place where the projectile lands may be described as follows:

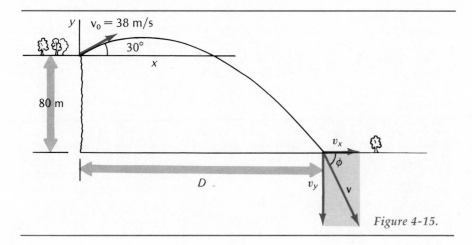

Figure 4-15.

$$x = D \qquad \text{when } y = -80$$

To find D, we first find the time t for which $y = -80$. Then we substitute this value for t in the equation for x to find D. We have now solved the problem conceptually; the rest is simple algebra.

When $y = -80$, the equation for y above becomes $-80 = 19t - 4.9t^2$, which may be written (check it!) as

$$t^2 - 3.9t - 16.3 = 0$$

We use the standard form for the roots of a quadratic equation to get t:

$$t = \frac{-b \pm \sqrt{b^2 - 4ac}}{2a} = \frac{3.9 \pm \sqrt{(3.9)^2 - 4(1)(-16.3)}}{2}$$

$$= +6.4 \text{ s}, -2.5 \text{ s}$$

We want the positive root, $+6.4$ s, which is the flight time. (What meaning can be attached to $t = -2.5$ s, a second time for which $y = -80$ m?)

Finally, using the relation above for x, we have

$$D = x = 33t = (33 \text{ m/s})(6.4 \text{ s})$$

$$= 211 \text{ m}$$

(b) The projectile hits the ground at $t = 6.4$ s. The velocity components at this time are

$$v_x = 33 \text{ m/s}$$

$$v_y = 19 - 9.8t = 19 - 9.8(6.4) = -44 \text{ m/s}$$

The magnitude of the projectile's velocity vector $\mathbf{v}$ is then

$$v = \sqrt{v_x^2 + v_y^2} = \sqrt{33^2 + (-44)^2} = 55 \text{ m/s}$$

The direction downward from the positive x axis is given by ϕ, where

$$\tan \phi = 44/33$$

$$\phi = 53°$$

The projectile will penetrate the ground at $53°$ below the horizontal.

Example 4-3. A demonstration experiment often used in physics lectures on projectile motion is the "monkey-and-hunter" experiment. See Figure 4-16(a). A steel ball (the monkey) is initially held at rest by an electromagnet. A tube with another ball within it (the "dart" in the hunter's gun) is aimed directly at the suspended ball, or monkey. Then the ball is blown out or otherwise ejected from the tube. As the ball leaves the tube, a switch is opened, which cuts the electric current to the electromagnet. The result is that the suspended ball is released from rest at exactly the time when the dart leaves the tube. Show that the dart always hits the monkey.

First, suppose that there were no gravity. Then the monkey would remain motionless; the dart would travel out along a straight line and would surely hit the monkey. Now, *with gravity*, both monkey and dart fall at the same acceleration $\mathbf{g}$ over the same time t. The monkey falls a vertical distance $\frac{1}{2}gt^2$ from rest; during this *same* time, the dart "falls" the *same* vertical distance $\frac{1}{2}gt^2$ but from its straight-line motion (along the direction of its initial velocity $\mathbf{v}_0$). See Figure 4-16(b), a graphical representation of (4-6). The monkey and dart reach the same place at the same time, so that they always collide, quite apart from the magnitude of $\mathbf{v}_0$ or even the magnitude of $\mathbf{g}$.

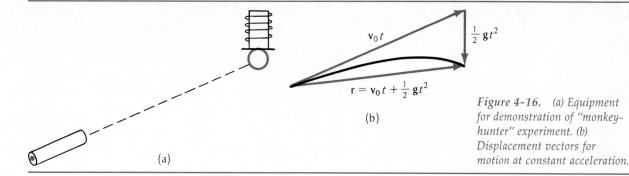

(b)

(a)

Figure 4-16. *(a) Equipment for demonstration of "monkey-hunter" experiment. (b) Displacement vectors for motion at constant acceleration.*

4-4   Uniform Circular Motion

A particle moving in a *circular arc at constant speed* is said to be in *uniform circular motion.* It covers equal distances along the circumference in equal times. Although the velocity is constant in magnitude, it continually changes direction. Consequently, there is an acceleration. We wish to find its magnitude and direction.

It is convenient to locate a particle traveling in a circle by radius vector **r** with its origin at the center of the circle. Then, with the particle speed constant, the radius vector sweeps through equal angles in equal times. Consider the infinitesimally small displacement $d\mathbf{r}$ occurring in a very small time interval dt. The vector displacement $d\mathbf{r}$ coincides with the path of the particle for a short segment of circular arc; see Figure 4-17. The radius vectors $\mathbf{r}_i$ and $\mathbf{r}_f$, both of magnitude r, give the initial and final displacements over the small interval dt. As the figure shows,

$$\mathbf{r}_f - \mathbf{r}_i = d\mathbf{r} = \mathbf{v}\,dt$$

where **v** is the velocity over the interval dt.

The velocity vector is always at right angles to the radius vector. As **r** rotates, so does **v**. See Figure 4-18(a), where velocity vectors, all of the same magnitude, are shown with their tails on the circumference of the circle. Figure 4-18(b) shows the *same* velocity vectors but now with their tails at a common point.

The initial and final velocities $\mathbf{v}_i$ and $\mathbf{v}_f$, both of magnitude v, differ only in direction. Their vector difference $d\mathbf{v}$ is given by

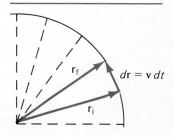

Figure 4-17. *Infinitesimal displacement d**r** over time interval dt for a particle in uniform circular motion.*

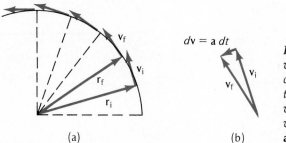

(a)

(b)

Figure 4-18. *(a) Velocity vectors $\mathbf{v}_i$ and $\mathbf{v}_f$ for uniform circular motion superposed on the path. (b) The same velocity vectors arranged to find the velocity change $d\mathbf{v} = \mathbf{v}_f - \mathbf{v}_i = \mathbf{a}\,dt$.*

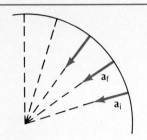

Figure 4-19. *For uniform circular motion, the acceleration is radially inward and of constant magnitude.*

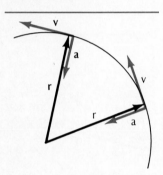

Figure 4-20. *Instantaneous radius vector* **r**, *velocity vector* **v**, *and acceleration vector* **a** *for a particle in uniform circular motion.*

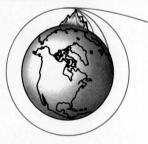

Figure 4-21. *Object projected horizontally from earth at various initial speeds. The circular path corresponds to that of an earth satellite.*

$$\mathbf{v}_f - \mathbf{v}_i = d\mathbf{v} = \mathbf{a}\, dt$$

where **a** is the instantaneous acceleration in the infinitesimal time interval dt. Now, vectors $\mathbf{r}_i$, $\mathbf{r}_f$, and $d\mathbf{r}$ form an isosceles triangle in Figure 4-17 that is similar to the triangle formed by the corresponding velocity vectors $\mathbf{v}_i$, $\mathbf{v}_f$, and $d\mathbf{v}$ in Figure 4-18(b). This is so because the displacement vector sweeps through the *same* angle as the velocity vector does during the time dt. Consequently, the magnitudes of corresponding sides are in the same ratio:

$$dr/r = dv/v$$

or

$$v\,(dt)/r = a\,(dt)/v$$

Thus,

$$a = \frac{v^2}{r} \tag{4-12}$$

The acceleration of a particle moving at constant speed v in a circle of radius r has a constant magnitude of v^2/r. Now for the direction of the acceleration.

Vector $d\mathbf{r} = \mathbf{v}\,dt$ is perpendicular to **r**, and vector $d\mathbf{v} = \mathbf{a}\,dt$ is perpendicular to **v**; see figures 4-18 and 4-19. Therefore, acceleration **a** has a direction always opposite that of the displacement **r**; that is, **a** is directed toward the center of the circle. *Uniform circular* motion is, then, characterized by an acceleration that is:

- Constant in magnitude.
- Continually changing in direction.
- Always radially inward.* See figures 4-19 and 4-20. The term *radial acceleration*, or sometimes *centripetal acceleration*, is used to designate such an acceleration.

Example 4-4.   An object is projected horizontally from the peak of an imaginary high mountain extending well above most of the earth's atmosphere. See Figure 4-21. Its path is a parabola and it strikes the earth near the mountain's base.

Suppose that the object is thrown with a higher speed. Then it may travel a distance comparable to the earth's radius before striking the earth; and the object's acceleration is not constant in direction and its path is not a parabola. If the initial speed is enormous, the object flies out into space, essentially undeviated from a straight line. There is one particular speed at which an object will always fall the same distance toward the earth as that by which it curves away from its straight-line motion; this is to say, there is one particular speed for which the object will travel in a circle. The object becomes an earth satellite.*

Find the speed and the period (the time for one complete revolution) of a satellite in circular orbit 161 km (~ 100 mi) above the earth's surface and outside its atmosphere. The satellite's orbital radius is 6.53×10^6 m; at this distance from the earth's center, $g = 9.32$ m/s^2.

The radial acceleration has the magnitude 9.32 m/s^2, and we have from (4-12)

$$a = \frac{v^2}{r}$$

---

* *Nonuniform* circular motion is treated in Section 13-1.

Thus

$$v = \sqrt{ar}$$
$$= \sqrt{(9.32 \text{ m/s}^2)(6.53 \times 10^6 \text{ m})}$$
$$= 7.8 \times 10^3 \text{ m/s}$$

The time T for one revolution is the circumference $2\pi r$ divided by the speed v:

$$T = \frac{2\pi r}{v} = \frac{2\pi(6.53 \times 10^6 \text{ m})}{7.8 \times 10^3 \text{ m/s}}$$
$$= 5.3 \times 10^3 \text{ s} = 88 \text{ min}$$

We are now in a position to understand the simpler kinematical aspects of satellite launching and recovery. After the liftoff rockets have stopped firing, the rocket follows a nearly parabolic path. At its highest point, when the velocity is parallel to the earth's surface, a booster rocket is fired to increase the speed of the released satellite to the value required for orbital motion. If the satellite is well above the earth's atmosphere, so that atmospheric drag is indeed negligible, the speed remains constant and the satellite orbits indefinitely; on the other hand, if the drag of the atmosphere is appreciable, the satellite spirals inward.

* The argument developed here was first put forth by Sir Isaac Newton (1642–1727).

4-5   Other Relations for Circular Motion

It is useful to express radial acceleration also in terms of two quantities that characterize any particle moving in a circle: the *period T*, defined as the *time required for one complete orbit;* and the *angular speed ω*, defined as the *time rate at which the radius vector sweeps through angles.*

In Figure 4-22, the radius vector r rotates and undergoes a small angular displacement $d\theta$; the corresponding distance ds along the circumference of the circle of radius r is given by

$$ds = r\, d\theta$$

This relation is merely the definition of angle $d\theta$ in *radian measure.** Dividing both sides of this equation by the short time interval dt in which the angle changes by $d\theta$ and the particle advances a distance ds along the circular arc, we have

$$\frac{ds}{dt} = r\frac{d\theta}{dt}$$

But $ds/dt = v$, the speed of the particle around the circle. The angular speed at which the radius vector rotates is, by definition,

$$\omega \equiv \frac{d\theta}{dt} \qquad (4\text{-}13)$$

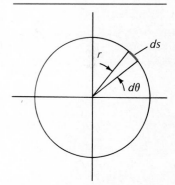

Figure 4-22. *Radius vector **r** undergoes an infinitesimal angular displacement $d\theta$. The corresponding distance along the circumference is $ds = r\, d\theta$.*

* For angle θ in radians, the distance s along the corresponding circular arc can be written $s = r\theta$. Then, for one complete turn around the circle, s is the circumference $2\pi r$; therefore, $\theta = s/r = 2\pi r/r = 2\pi$ radians. Radian measure is related to degree measure by 2π rad $= 360°$.

Angular speed is the *time rate of change of angular displacement.* (By convention, ω is taken to be positive for counterclockwise rotation.) Then the relation above becomes

$$v = r\omega \qquad (4\text{-}14)$$

The speed v of a particle along a circular arc is merely the radius of its path multiplied by the angular speed. Equation (4-14) applies only if ω is expressed in units of *radians per unit time.** Equation (4-14) then says that for a given angular speed ω, the speed of any particle in a circle is proportional to the distance r of the particle from the rotation axis.

Using (4-14) in (4-12) for the centripetal acceleration, we get

$$a = \frac{v^2}{r} = \frac{(\omega r)^2}{r} = \omega^2 r \qquad (4\text{-}15)$$

In this alternative form, we see that the magnitude of the radial acceleration is directly proportional to r and to the square of ω

The period T, the time for one complete loop, is easily related to v and ω. The particle travels a distance of $2\pi r$ around the circle in a time T, so that

$$v = \frac{2\pi r}{T}$$

Using this relation in (4-14), we then get

$$\omega = \frac{2\pi}{T} \qquad (4\text{-}16)$$

which says merely that the radius vector makes one rotation and turns through an angle of 2π rad in time T.

Using (4-16) in (4-15) then gives

$$a = \omega^2 r = \left(\frac{2\pi}{T}\right)^2 r$$

$$a = \frac{4\pi^2}{T^2} r \qquad (4\text{-}17)$$

We now have three equivalent ways of expressing the centripetal acceleration:

$$\mathbf{a} = -\omega^2 \mathbf{r} = -\frac{4\pi^2}{T^2} \mathbf{r} = -\left(\frac{v}{r}\right)^2 \mathbf{r} \qquad (4\text{-}18)$$

In these *vector* relations for the radial acceleration, the minus sign indicates that the acceleration is *radially inward,* opposite the direction of radius vector $\mathbf{r}$. All three forms contain the radius $\mathbf{r}$; they differ according to whether the angular speed ω, period T, or linear speed v appears.

Example 4-5.   What is the acceleration of a point on the earth's equator arising from the daily rotation of the earth? (Earth's radius $= 6.4 \times 10^6$ m.)

---

* If ω is given, for example, as 60 rpm, we must convert to radians per second to use (4-14):
60 rpm $= 60$ rot/60 s $= 1$ rot/s $= 2\pi$ rad/s.

Since the period of rotation is known (1 day), it is most straightforward to use (4-17):

$$a = \frac{4\pi^2}{T^2} r$$

$$= \frac{4\pi^2(6.4 \times 10^6 \text{ m})}{(24 \times 60 \times 60 \text{ s})^2}$$

$$= 3.4 \times 10^{-2} \text{ m/s}^2 = 0.3 \text{ percent of } g$$

4-6   Relative Velocities

Any velocity **v** is known completely only if we can specify:

- Object in motion.
- Direction and magnitude of **v**.
- Reference frame relative to which **v** is measured.

Here we shall see how the velocity of an object measured by an observer at rest in one reference frame is related to the velocity of the same object measured by a different observer at rest in another reference frame.

Consider two reference frames in relative motion. Let A represent the earth as one reference frame and B some other reference frame, say, a truck moving relative to the earth. (Of course, A and B may represent any two reference frames; we identify them here as the earth and a truck merely to avoid an abstract argument.) The velocity of **B** relative to **A** is represented by $\mathbf{v}_{BA}$, where the first subscript B then denotes the object (or reference frame) in motion and the second subscript A is the reference frame relative to which the velocity is measured. Velocity vectors $\mathbf{v}_{BA}$ and $\mathbf{v}_{AB}$ are shown in Figure 4-23. Here, and generally,

$$\mathbf{v}_{BA} = -\mathbf{v}_{AB}$$

Suppose, for example, that the truck is moving east at 10 m/s relative to earth. Then, relative to the truck, the earth is moving *in the opposite direction at the same speed,* 10 m/s to the west. To change velocity vector $\mathbf{v}_{BA}$ into vector $\mathbf{v}_{AB}$, we merely reverse the arrow.

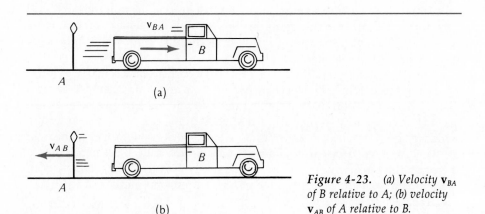

Figure 4-23.   (a) Velocity $\mathbf{v}_{BA}$ of B relative to A; (b) velocity $\mathbf{v}_{AB}$ of A relative to B.

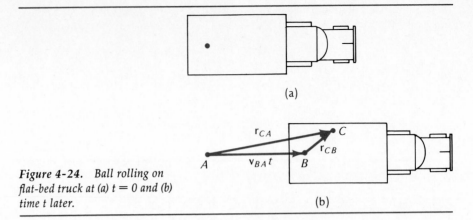

Figure 4-24.   *Ball rolling on flat-bed truck at (a) t = 0 and (b) time t later.*

Now suppose that a small ball *C* rolls along the smooth, flat floor of the truck at constant velocity while the truck moves at constant velocity relative to the earth. We wish to find how the object's velocity $\mathbf{v}_{CA}$ relative to earth is related to $\mathbf{v}_{CB}$, its velocity relative to the truck. Figure 4-24(a) shows things at the start, and Figure 4-24(b) at time *t* later. From Figure 4-24(b), we see that the ball's displacement $\mathbf{r}_{CB}$ relative to its starting point on the truck and its displacement $\mathbf{r}_{CA}$ relative to its starting point on earth are related by

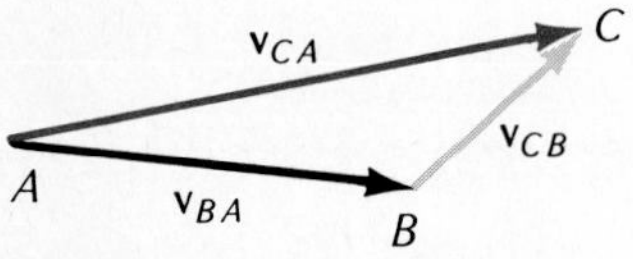

$$\mathbf{r}_{CA} = \mathbf{r}_{CB} + \mathbf{v}_{BA}t$$

where $\mathbf{v}_{BA}t$ is the displacement of *B* relative to *A* occurring in time *t*. This vector relation is nothing more than the basic equation for adding displacement vectors. We divide both sides of this equation by *t* and recognize that by definition $\mathbf{r}_{CA}/t = \mathbf{v}_{CA}$ and $\mathbf{r}_{CB}/t = \mathbf{v}_{CB}$. Then we get

$$\mathbf{v}_{CA} = \mathbf{v}_{CB} + \mathbf{v}_{BA} \qquad (4\text{-}19)$$

This is the basic relation for relative velocities.

Figure 4-25.   *Vector diagram for relative velocities:* $\mathbf{v}_{CA} = \mathbf{v}_{CB} + \mathbf{v}_{BA}$.

The subscripts give us a useful mnemonic for keeping straight the meanings of the terms and their signs. As we know, the first subscript gives the object in motion and the second subscript the reference frame relative to which it is measured. Now we see that the "outside" subscripts on the right side of (4-19) correspond to the pair of subscripts on the left, while the "inner" subscripts are the same. This is also shown in Figure 4-25, where a *single* symbol appears at each vertex of the velocity vector diagram.

A simple conclusion follows from (4-19): If a particle has a constant velocity in one reference frame, then its velocity relative to any other reference frame is also constant (but different), provided the two reference frames have a constant relative velocity. Moreover, a reference frame can always be found in which the body's velocity is zero. Suppose that a hockey puck slides to the right at 20 m/s relative to the earth. Then, with respect to a truck traveling 20 m/s to the right, the puck *is* (not merely "appears to be") at rest. Thus, the states of motion with constant velocity and of rest differ from one another only through the arbitrary choice of a reference frame.

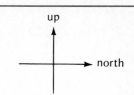

Example 4-6. Raindrops are falling vertically at 5.0 m/s and a man is walking north at 2.0 m/s, both relative to the earth. What is the velocity of a raindrop relative to the man?

The velocity vectors are shown in Figure 4-26, with the subscripts assigned the following meanings: d = drop, e = earth, and m = man. To have a consistent vector triangle, we have drawn $\mathbf{v}_{em}$, rather than $\mathbf{v}_{me}$. We are given that $\mathbf{v}_{me}$, of man-relative-to-earth, is 2.0 m/s *north*; therefore, velocity $\mathbf{v}_{em}$ of earth-relative-to-man is 2.0 m/s *south*. A single subscript appears at each vertex in the velocity-vector triangle, and the three velocities are related by

$$\mathbf{v}_{dm} = \mathbf{v}_{de} + \mathbf{v}_{em}$$

The magnitude of $\mathbf{v}_{dm}$, the velocity of drop-relative-to-man, is given by

$$v_{dm}^2 = v_{de}^2 + v_{em}^2$$
$$v_{dm} = \sqrt{(5.0)^2 + (2.0)^2} \text{ m/s} = 5.4 \text{ m/s}$$

Angle ϕ for the direction of $\mathbf{v}_{dm}$ relative to the vertical is given by

$$\tan \phi = \frac{v_{em}}{v_{de}} = \frac{2.0}{5.0} = 0.4$$
$$\phi = 22°$$

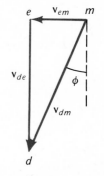

Figure 4-26.

If the man carries an umbrella, he gets maximum protection by tipping it 22° from the vertical toward the north.

Summary

Definitions

Velocity $\mathbf{v}$: time rate of change of displacement vector $\mathbf{r}$,

$$\mathbf{v} \equiv \frac{d\mathbf{r}}{dt} \qquad (4\text{-}2)$$

with rectangular components

$$v_x \equiv \frac{dx}{dt} \quad \text{and} \quad v_y \equiv \frac{dy}{dt} \qquad (4\text{-}1)$$

Acceleration $\mathbf{a}$: time rate of change of velocity $\mathbf{v}$,

$$\mathbf{a} \equiv \frac{d\mathbf{v}}{dt} \qquad (4\text{-}3)$$

with rectangular components

$$a_x \equiv \frac{dv_x}{dt} \quad \text{and} \quad a_y \equiv \frac{dv_y}{dt}$$

The acceleration can also be resolved into rectangular components $a_\parallel$ and $a_\perp$, respectively parallel and perpendicular to the velocity.

Angular speed ω: time rate of change of angular displacement θ:

$$\omega \equiv \frac{d\theta}{dt} \qquad (4\text{-}13)$$

Time for one complete rotation: the period T; for constant angular speed in a circle,

$$\omega = \frac{2\pi}{T} \qquad (4\text{-}16)$$

Centripetal acceleration: toward the center of a circle.

Units

Angular displacement θ: radians
Angular speed ω: rad/s = s^{-1}

Important Results

Motion in a plane can be regarded as the superposition of independent motions along x and y.

For constant acceleration $\mathbf{a}$, displacement:

$$x - x_0 = v_{0x}t + \tfrac{1}{2}a_x t^2 \qquad (4\text{-}4)$$
$$y - y_0 = v_{0y}t + \tfrac{1}{2}a_y t^2 \qquad (4\text{-}5)$$

Equivalently,

$$\mathbf{r} - \mathbf{r}_0 = \mathbf{v}_0 t + \tfrac{1}{2}\mathbf{a}t^2 \qquad (4\text{-}6)$$

For constant acceleration $\mathbf{a}$, velocity:

$$v_x = v_{0x} + a_x t \qquad (4\text{-}4)$$
$$v_y = v_{0y} + a_y t \qquad (4\text{-}5)$$

Equivalently,

$$\mathbf{v} = \mathbf{v}_0 + \mathbf{a}t \qquad (4\text{-}6)$$

For uniform circular motion at radius r, linear speed v, angular speed ω, and period T, the radially inward, centripetal acceleration is

$$a = \frac{v^2}{r} = \omega^2 r = \frac{4\pi^2 r}{T^2} \qquad (4\text{-}18)$$

The relation between linear speed v and angular speed ω is

$$v = r\omega \qquad (4\text{-}14)$$

The basic rule for combining relative velocities is

$$\mathbf{v}_{CA} = \mathbf{v}_{CB} + \mathbf{v}_{BA} \qquad (4\text{-}19)$$

where the first subscript denotes the object (or reference frame) in motion and the second subscript denotes the object (or reference frame) relative to which the motion is measured.

Problems and Questions

Section 4-1   Velocity and Acceleration Vectors
· **4-1 Q** If a particle is accelerating it is
(A) either speeding up or slowing down.
(B) turning.
(C) moving.
(D) subject to the force of gravity.
(E) None of the above is correct.

· **4-2 Q** Which one of the following situations is impossible?
(A) An object has velocity directed east and acceleration directed west.
(B) A body has velocity directed east and acceleration directed east.
(C) A body has zero velocity but acceleration not zero.
(D) A body has constant acceleration and changing velocity.
(E) A body has constant velocity and changing acceleration.

· **4-3 Q** Which of the following is a true statement concerning general motion in three dimensions?
(A) A moving object does not have zero acceleration.
(B) An object can be moving in one direction and accelerating in the opposite direction.
(C) Velocity and speed have the same meaning.
(D) The distance an object moves in time t is directly proportional to t.
(E) An object moving with constant speed is said to have zero acceleration.

Section 4-2   Motion at Constant Acceleration
: **4-4 P** A Camaro car races a BMW motorcycle under the following conditions: The car crosses the START line with a speed of 42 m/s (about 94 MPH) and maintains this constant speed throughout the race. The motorcycle is given a 120 meter head start, but it starts from rest. The motorcycle maintains a constant acceleration of 6 m/s² throughout the race. (a) At what distance from the starting point will the car catch the motorcycle? (b) At what distance from the START line will the motorcycle catch the car? (c) Draw a careful

sketch, approximately to scale, showing the positions of the two vehicles as a function of time and indicate the times at which they draw abreast.

: **4-5 P** The x and y coordinates of a particle are given as a function of time t by the relations:

$$x = 10 - 5t + 4t^2 \text{ and } y = 5t - 2t^2.$$

(a) Find the relations for the x and y components of the particle's velocity and acceleration. (b) Sketch the particle's path and draw velocity and acceleration at several locations along the path.

Section 4-3   Projectile Motion
· **4-6 Q** A key to understanding projectile motion is the recognition that
(A) gravity always causes the velocity of a particle to increase.
(B) any increase in the y-component of velocity is compensated for by a decrease in the x-component of velocity.
(C) horizontal motion is independent of vertical motion.
(D) acceleration is zero at the highest point in the trajectory.

: **4-7 Q** A girl throws a ball into the air with initial speed of 10 m/s. If we neglect friction effects,
(A) the ball will always be moving when it is in the air.
(B) the time the ball is in the air is independent of the angle at which the ball is thrown.
(C) the height to which the ball rises depends on the ball's weight.
(D) the ball is accelerating all the time it is in the air.
(E) the sum of the horizontal distance traveled plus the vertical height reached will be independent of the angle at which the ball is thrown.

: **4-8 Q** A rifle's gunsight is so adjusted that you aim the gunsight directly at a target 200 yards distant at the same level in order to hit the target. When you use this rifle to shoot at downhill targets 200 yards away, you should aim

(A) right at the target.

(B) high.

(C) low.

(D) high for some angles, low for others.

: 4-9 Q Figure 4-27 shows several trajectories followed by a ball thrown by a baseball outfielder. Assuming that air effects are negligible, which ball stayed in the air longest?

(A) *A*

(B) *B*

(C) *C*

(D) All were in the air for the same time.

(E) Cannot be determined without knowing initial velocities.

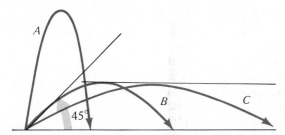

Figure 4-27.   Question 4-9.

: 4-10 Q If the effects of air resistance are included, the problem of analyzing the motion of a projectile becomes much more difficult. Some important qualitative features can be deduced without complicated mathematics simply from knowing that the resistive force increases with increasing velocity. For example, we can deduce that

(A) the time while ascending will still be equal to the time while descending.

(B) the time while descending will be greater than the time while ascending.

(C) maximum horizontal range will occur for a launching angle greater than 45° above the horizontal.

(D) the projectile will hit the ground with a greater speed than that with which it was launched.

: 4-11 Q Two bullets are fired horizontally simultaneously from the same elevation over a flat field. Which will hit the ground first if air friction is negligible?

(A) The faster one.

(B) The slower one.

(C) The heavier one.

(D) The lighter one.

(E) Both will hit the ground simultaneously, independent of bullet weight or initial speed.

: 4-12 P How high should a rifle be aimed if it is to hit a target 200 meters away and along the horizontal with a bullet whose muzzle velocity is 800 m/s?

· 4-13 P An arrow is shot at an elevation of 30° with an initial velocity of 40 m/s. What is its minimum velocity while in the air?

: 4-14 P A U.S. Forest Service plane swings low to drop some supplies to firefighters on the ground below. If the plane is traveling horizontally with a speed of 50 m/s at an elevation of 120 meters when a bundle is pushed out of the plane, how far ahead of the spot where it was pushed will the bundle land?

: 4-15 P A diver jumps downward with a speed of 4 m/s at an angle 30° below horizontal. With what speed will he hit the water if he jumps from a platform 6 meters high?

: 4-16 P A child throws a ball from the balcony of a building with a speed of 10 m/s. How fast will the ball be moving at a point 10 meters below where it was released? (Will you need to know the angle at which the ball was thrown? Call the angle θ and see what happens.)

: 4-17 P A golf ball leaves the club with a velocity of 40 m/s at an angle of elevation of 50°. What is its position and velocity after 4 seconds? Neglect air resistance.

: 4-18 P An electron with a horizontal velocity of 5×10^6 m/s is projected into the region between two parallel metal plates where it experiences a uniform vertical acceleration of 6×10^{13} m/s². The plates have a horizontal length of 3 cm. A screen is positioned 24 cm from where the electron first enters the accelerating region. (*a*) What is the electron's speed and direction as it leaves the region between the plates? (*b*) What is the vertical displacement of the electron at the screen? (*c*) What is the time of flight from the moment the electron enters the region between the plates until it strikes the screen?

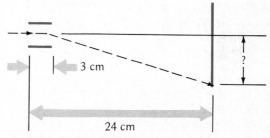

Figure 4-28.   Problem 4-18.

: 4-19 P At what *two* angles of elevation can a gun be fired with a muzzle velocity of 160 m/s if it is to have a horizontal range of 1200 meters? What is the time of flight in each case?

· 4-20 P Galileo once observed that projectiles launched at angles of elevation of $45° + \theta$ and $45° - \theta$ have equal horizontal ranges. Prove or disprove this assertion.

: 4-21 P A punter can kick a football 60 meters so that it is in the air for 3 seconds. Suppose air friction to be negligible and assume that the ball was kicked at an angle of elevation of 50°. What is the maximum height the ball could reach?

: 4-22 P An artillery shell of mass 2 kg is fired at an elevation of 20° over a level field. What is its range if its initial velocity is 400 m/s?

: 4-23 P During World War I, the Germans shelled Paris with "Big Bertha," a cannon located 120 km from the city. The cannon was fired at an elevation of 70° above horizontal. Neglecting (unrealistically!) frictional effects, what muzzle velocity would be needed?

: 4-24 P A spawning salmon wishes to jump up a waterfall 0.8 meter high. If its maximum swimming speed is 5 m/s, what is the greatest distance from the base of the waterfall from which it can begin to leap and still reach the top of the falls?

: 4-25 P A ball rolls off the top of a stairway with a horizontal velocity of 1.5 m/s. The steps are 20 cm tall and 24 cm wide. (a) Which step, measured down from the top, will the ball hit first? (b) How far from the front edge will it hit?

: 4-26 P A batter hits a hard drive to right field; the ball just barely clears a 7 meter high wall some 100 meters away. The ball was struck 1 meter above home plate and left the bat at an elevation of 30°. What was its initial velocity?

: 4-27 P Suppose you threw a superball from the roof of a building onto a playground below in such a way that the ball takes the longest possible bounce (greatest range between successive collisions with the horizontal pavement). Assume the ball rebounds with the same speed with which it lands, and that the ball is thrown with a speed of 20 m/s from an elevation of 10 meters. At what angle above or below horizontal should the ball be thrown? (You may have to use numerical methods to solve the resulting equation.)

: 4-28 P Lake Chabot golf course has a short par 3 hole where one hits from an elevated tee. The green is 30 meters below the tee and 100 meters away by horizontal measure, and is circular with a 15 meter diameter. If a shot is hit with a speed of 30 m/s directed at 60° above horizontal, will it land on the green? At what angle with the vertical will it land?

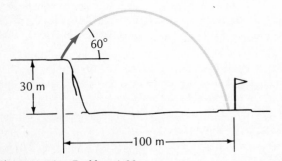

Figure 4-29.   Problem 4-28.

: 4-29 P What angle does a ramp make with the horizontal if a motorcyclist leaves the ramp with speed v_0 and just makes it to an elevated platform a height h above the launching ramp and a horizontal distance d away? Neglect the size of the motorcycle and air effects.

: 4-30 P At what angle of elevation must a projectile be fired in order that its maximum height equal its horizontal range?

Section 4-4   Uniform Circular Motion

· 4-31 P (a) What is the angular speed of a point on the rim of a bicycle wheel of outer diameter 95 cm when the bike is moving 6 m/s? (b) What is the angular speed of a point 12 cm from the axis?

· 4-32 Q An automobile speedometer is calibrated so that it gives an accurate speed reading when the car is equipped with regular highway tires. (Basically, the speedometer works by measuring the number of turns of an auto wheel per unit time.) What would happen if the tires were replaced with thick-tread snow tires?
(A) This would cause the speedometer to give a reading that is too low.
(B) This would cause the speedometer to give a reading that is too high.
(C) This would have no effect on the accuracy of the readings, since the speedometer is calibrated for given angular velocity of the wheel, not for linear velocity.
(D) This would have no effect on the accuracy of the readings since the larger tire radius would result in a reduced angular velocity and hence no change in speed.
(E) This would have no effect on the accuracy of the readings, since a speedometer measures auto speed, no matter what kind of tires are used.

· 4-33 P Express the following angles in degrees: (a) $\pi/4$ radians (b) $\pi/2$ radians (c) 0.3 radians (d) 1.5 radians (e) 0.5 radians

· 4-34 P Express the following angles in radians: (a) 30° (b) 45° (c) 90° (d) 180° (e) $\frac{1}{3}$ revolution

· 4-35 P (a) What is the angular speed of the second hand on a watch? (b) Through what angle, in radians and in degrees, does it turn in 2 minutes and 12 seconds?

· 4-36 P A jet pilot flying at 300 m/s pulls up out of a dive by turning upward in a circle of radius 1800 meters. What is his acceleration, expressed as a multiple of g?

· 4-37 P An ultracentrifuge used to separate and study biological macromolecules has an effective radius of 4 cm and rotates at 120,000 rpm. What is the centripetal acceleration of this device, expressed as a multiple of g?

: 4-38 P In the Octupus carnival ride, four people are allowed to sit in seats mounted on the end of a tentacle,

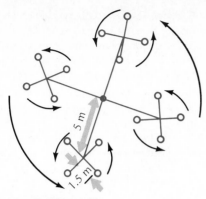

Figure 4-30.   *Problem 4-38.*

which whirls around in a horizontal plane, as is shown in Figure 4-30. The seats make 0.6 revolution per second about a vertical axis at the same time the tentacles rotate at 0.3 revolution per second. What maximum acceleration, expressed as a multiple of g, does a rider in this contraption experience?

: **4-39 P** A motorcycle of mass 200 kg has two wheels whose axles are separated by 1.5 meters. The outer diameter of each tire is 50 cm. When the motorcycle is moving straight ahead at a speed of 12 m/s, what is the speed with respect to the ground of the point at the top of the front tire?

: **4-40 P** The path of a particle is described by $x = r \cos \omega t$ and $y = r \sin \omega t$; r and ω are constants. (*a*) Show that the path of the particle is a circle of radius r which the particle traces out with angular speed ω. (*b*) Calculate a_x and a_y and thereby show that the vector acceleration has magnitude $r\omega^2$ and is directed toward the center of rotation.

Section 4-5   Other Relations for Circular Motion

· **4-41 P** A flywheel rotates at 1800 revolutions per minute. Its radius is 10 cm. (*a*) What is its angular speed? (*b*) Through what angle, expressed in degrees and in radians, does the wheel turn in 8 ms? (*c*) What is the tangential speed of a point on the perimeter? (*d*) What is the period of the rotation?

: **4-42 P** You have probably noticed in movies that sometimes the wheels of the stagecoach seem to be stationary. The wheel has a 1 meter diameter and 8 spokes. The camera takes 16 frames per second. What is the minimum stagecoach speed for this phenomenon to occur?

· **4-43 P** The moon circles the earth once every 27.3 days at an average distance of 3.8×10^8 meters. (*a*) What is the moon's angular speed? (*b*) What is the moon's frequency of revolution? (*c*) What is the moon's centripetal acceleration? (The moon to earth distance is 3.84×10^5 km.)

: **4-44 P** Calculate the following for points A, B, and C, where A is at the earth's north pole, B is at 45° north lati-

tude, and C is at the equator: (*a*) Frequency of rotation. (*b*) Period. (*c*) Angular speed. (*d*) Tangential speed. (*e*) Centripetal acceleration, expressed as a multiple of g. (The radius of the earth is 6.37×10^3 km.)

Section 4-6   Relative Velocities

· **4-45 P** A person who can walk up a stalled escalator in 90 seconds is carried up in 60 seconds on the same escalator when it is moving. How long would it take to walk up the moving escalator?

· **4-46 P** Two cars, one driving north-south and the other east-west, are approaching an intersection at speeds of 60 km/hr and 100 km/hr. At what speed are they approaching each other?

: **4-47 P** A pilot heads her plane due north according to her compass bearing. Her air speed is 300 km/hr, and she is flying in a wind of 80 km/hr blowing from the southeast. What is her ground speed?

: **4-48 P** A fishing boat can travel 6 m/s in still water. The captain wishes to sail due north in the presence of an ocean current of 2 m/s flowing to the northeast. Draw a careful velocity vector diagram, to scale, showing the angle θ, with respect to north, at which the boat should steer. Also, estimate graphically (not using equations) the speed of the boat with respect to the earth in this case.

: **4-49 P** A fisherman has a boat that will travel through still water at 10 km/hr. He wishes to travel from the Ferry Building in San Francisco due north to Sausalito, a distance of 6 km. He must travel against an ocean current of 5 km/hr directed toward the southeast. (*a*) Draw a careful vector diagram which shows the angle θ in which the boat must head. (*b*) Determine how long the trip will take.

: **4-50 P** A pilot wishes to fly 580 km due south from Reno to Ventura. The plane's air speed is 400 km/hr, and a wind of 80 km/hr is blowing toward the northwest. (*a*) In what direction should the pilot steer the plane? (*b*) How long will the trip take?

: **4-51 Q** Suppose that you have to run from your house to your car in a rainstorm. What is the best strategy for avoiding the raindrops? Should you run as fast as possible, or is there some optimum speed that depends on how fast the raindrops are falling? Does it matter how fast you run?

: **4-52 P** A pilot flies an old biplane from Denver to Durango, a distance of 380 km. His air speed is 200 km/hr, and he makes the trip in 2 hours. Suppose that the wind speed and direction were steady during the trip. What was the wind direction for minimum air velocity and the magnitude of this minimum air velocity?

: **4-53 P** A swimmer who can swim 1.8 m/s in still water plans to swim across a river which is 200 meters wide and flowing 1 m/s. (a) In what direction should she swim in order to land directly opposite her starting point? (b) How long will it take her to cross the river? (c) If she heads directly across the river how far downstream will she land? (d) How long will the route in part (c) require? (e) In what direction should the swimmer head in order to cross the river in the shortest possible time?

: **4-54 P** An aircraft carrier sails at 20 knots on a course 45° west of north in the Gulf Stream, which is flowing 45° east of north at 7 knots. A helicopter pilot flying at 40 knots air speed due east is trying to rendevous with the carrier. The task is complicated by a 30 knot wind blowing due south. (a) How fast is the carrier moving with respect to the helicopter? (b) In what direction?

Supplementary Problems

4-55 P A Porsche 944 has (a) an average acceleration of 0.42 g in going from rest to 50 mph. How long does this take? (b) Its average acceleration in covering a quarter mile from rest is 0.31 g. How long does this take?

4-56 P The Mazda RX-7 GS has the following performance characteristics: (a) from 0 to 30 mph in 4.3 s, (b) from 0 to 60 mph in 12.5 s, (c) speed of 78 mph at the end of a quarter mile, (d) passing from 45 to 65 mph in 8.0 s, (e) braking from 60 mph with no wheels locked to rest in 185 ft. Compute the acceleration in m/s², assumed to be constant, for each item.

4-57 P Sorin Munteanu of the Circus Willy Hagenbeck can juggle seven clubs at one time. Each club reaches a maximum height close to 3.0 meters. (a) What is the time interval between successive tossings of a club? (b) To what height would Sorin have to toss each club if he were to juggle eight clubs with the same time interval between successive tossings?

4-58 P Nolan Ryan is well-known for his fast ball; one pitched baseball has been clocked at 100.9 mph (August 20, 1974). Suppose that Ryan throws a baseball at this speed horizontally. Assuming that there are no complications—no spin, no curving, no air resistance—how far vertically will the ball drop as it travels the roughly 56 ft from the pitcher's extended hand to home plate?

4-59 P The high-speed French TGV train (*Train à Grande Vitesse*) can achieve an *average* speed of 260 km/h (~156 mph) over trips several hours long. The train's maximum speed is 400 km/h (~250 mph). All curves (on flat roadbed) are 3.7 km or more in radius. (a) What is the maximum radial acceleration when the train rounds a curve at maximum speed? (b) To bring a TGV to rest from 250 km/h takes a little more than 3 km of track. What is the deceleration, assuming it to be constant?

4-60 P The men's world record for the "free-fall" distance for delayed opening in parachute jumping was set by J. K. Kittinger of the U.S.A. in Tularosa, New Mexico, in 1960. Kittinger descended 25.7 km as a skydiver traveling at a terminal speed of about 60 m/s before opening his parachute. The time to acquire the terminal speed was about 10 s. (a) How long was Kittinger in flight? (b) What would his final speed have been in the absence of air resistance?

4-61 P The cornering ability of a Volkswagen Rabbit GTI is characterized in terms of a maximum lateral acceleration of "0.82 g on the skid pad". Find the minimum turn radius if the car is traveling at 60 mph.

4-62 P According to the *Guinness Book of Records* the longest measured basketball shot is the 89-feet, 3-inch field goal made by Les Henson in 1980 as the last shot in the championship game between Virginia Tech and Florida State. The final buzzer sounded while the ball was in the air. Virginia Tech won, 79–77. (A basketball court is only 94 feet long. The 89 feet 3 inches in the record shot is measured horizontally from Henson's back foot to the back of the rim.) Make reasonable assumptions and show that the basketball must have been thrown at approximately 60 km/h and was in flight for about 2.3 s.

Newton's Laws I

5

Everything we have considered to this point has been a prelude to Newton's laws of motion; nearly everything in classical mechanics that follows is an application of Newton's laws. They are the foundation of classical mechanics. When we know the forces acting on a particle, we can use these laws to predict its future motion in detail. Conversely, when we study the motion of a particle in detail, we can deduce the net force acting on it. In short, with Newton's laws we relate the forces that influence an object to its motion.

You may have already encountered Newton's laws in an earlier course in physics. In modern language, they may be stated as follows:

• First Law: When a particle is subject to no net force, the particle has a constant velocity; an isolated particle either coasts at constant speed in a straight line or remains at rest. In symbols, if $\Sigma \mathbf{F} = 0$, then $\mathbf{v} =$ constant.

• Second Law: When a particle is subject to a resultant force, the particle is accelerated. The vector sum of the forces acting on a particle, $\Sigma \mathbf{F}$, is equal to the particle's mass m multiplied by its acceleration $\mathbf{a}$: $\Sigma \mathbf{F} = m\mathbf{a}$.

• Third Law: Every force involves an interaction between two particles. The force of the first particle on the second is equal in magnitude to the force of the second particle on the first but opposite in direction: $\mathbf{F}_{1 \text{ on } 2} = -\mathbf{F}_{2 \text{ on } 1}$. Equivalently, the action and reaction forces are equal and opposite.

It is extremely important that we be clear on exactly what these fundamental principles say and what the terms *mass* and *force* mean. We must get straight what is a definition and what, on the other hand, is a general result of observation (a law of physics). We cannot say, for example, that mass is *defined* by the relation $\mathbf{F} = m\mathbf{a}$ and, at the same time, that $\mathbf{F} = m\mathbf{a}$ is a *law* of physics.

The purpose of this chapter is to examine the basis for the three laws of motion of Sir Isaac Newton. Chapter 6 will also illustrate their wide applicability.

We shall consider each of the three laws separately. These fundamental propositions are, however, not independent of one another; to some degree each depends on the other two and is also clarified by them. In their totality they are the postulational basis of classical mechanics.

5-1   The First Law

A hockey puck slides along a horizontal concrete pavement and soon comes to rest. If projected along a polished floor, it coasts farther before coming to rest. On a smooth surface of ice, the puck goes even farther. If the puck is projected on an "air hockey" table, so that it floats on streams of air emerging upward through many small holes in the surface, it coasts so freely that it is hard to measure any slowing. In short, as the influence of friction is progressively reduced, we find that the hockey puck more nearly approaches motion in a straight line at constant speed. See Figure 5-1. We can well imagine that if we were to take the hockey puck, or any other object, into interstellar space, where it would be far from any other object, and then project it there, it would move in a straight line at constant speed forever (or at least until it came close to another object). It follows also as a special case that if an object were placed at rest (given an initial velocity of "zero") while effectively isolated from all other objects, it would remain at rest indefinitely.

No one has done such an experiment. No one can observe an object completely free from external influence. To say that a completely isolated particle —one subject to no force—has constant velocity amounts to extrapolating the results of actual observation to an unrealizable ideal situation. Yet this is exactly what is done in formulating Newton's first law of motion. How can an idealization, approximated but never realized in any laboratory, be regarded as a *law* of physics?

The answer is simply that it meets the requirement for any law of physics; *it always works.* The first law of motion—that an isolated particle has constant

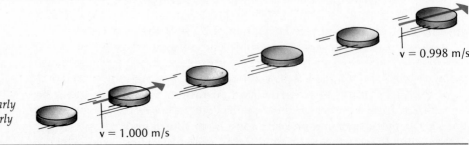

Figure 5-1. *An object nearly free of friction coasts at nearly constant velocity.*

v = 1.000 m/s

v = 0.998 m/s

velocity — works in the sense that whenever a particle is found to depart from motion in a straight line at constant speed or from the state of rest, this *departure* can always be attributed to and accounted for in detail by *forces* acting on the particle.

Newton's first law "answers" a fundamental question, What does it take to keep an object moving? Just as nothing is required to keep an object at rest, *nothing* is required to keep an object in uniform motion. Recognizing this fundamental fact — that motion is just as "natural" as rest, that motion requires no explanation — is perhaps the most profound result in all physics. Its formulation, first by Galileo Galilei as the so-called *law of inertia,* and later by Newton as his first law of motion, marked the beginning of physics as a science in the modern sense.

The law of inertia is restricted to certain special reference frames (Section 4-6). More specifically, a reference frame in which the law of *inertia* holds is known as an inertial reference frame, or simply an *inertial frame.* When an isolated object described by Newton's first law is said to have constant velocity, either zero or nonzero, we mean that the velocity is constant relative to some observer in an *inertial* reference frame.

Not all reference frames are inertial frames. Actually, what is meant by an inertial frame is defined by Newton's first law. An airplane accelerating on the runway at takeoff or a rotating merry-go-round are examples of reference frames that do not qualify as inertial reference frames; an undisturbed object in these reference frames does not maintain a constant velocity.

The earth's surface is an approximate inertial frame. It is approximate because the earth rotates daily about its axis (radial acceleration of 3.4×10^{-2} m/s^2 at the equator); it revolves annually about the sun (5.9×10^{-3} m/s^2); and the sun revolves about the center of our galaxy (10^{-10} m/s^2). The most nearly perfect inertial frame is one at rest with respect to the fixed stars, those distant stars that astronomers find at the same relative locations night after night.

Suppose a particle has a constant velocity relative to one reference frame. This particle has a different but still *constant* velocity when it is observed in a second reference frame moving at constant velocity relative to the first. Suppose, for example, that an observer moves relative to the earth at the same speed and in the same direction as a freely coasting puck. This observer says that the puck is *at rest.* This same traveling observer would say that a puck at rest relative to the earth is moving with constant velocity. The states of uniform motion and of rest are equivalent.

We shall hereafter assume, unless specifically indicating otherwise, that all observations are from the point of view of an inertial frame.

5-2   Mass

Mass Defined   An ordinary plastic hockey puck coasts along a smooth surface. We catch the puck and bring it to rest. Now suppose that a second puck with the same dimensions as the first, but made of lead, coasts along the smooth surface at the same speed as the first. To catch the lead puck and bring it to rest in the same time requires far more effort, a larger force from the hand. The lead puck has the greater "inertia"; it resists a change in its velocity to a greater degree than the ordinary plastic puck does. It is harder to change the velocity of the puck with the greater inertia.

In like fashion, with a plastic puck and a lead puck of the same dimensions both initially at rest, a greater effort (that is, a greater force) is required to put the lead puck in motion than to put the plastic puck in motion at the same speed in the same time interval. Again the lead puck has the greater inertia; it offers more resistance to a change in its initial state of rest.

All these observations are qualitative, and we are familiar with still other examples of the inertia of objects in motion or at rest. For example, when a car in motion is suddenly braked, a passenger moving with the car lurches forward, thereby maintaining, or trying to maintain, his inertia in the forward direction. Or when the car is suddenly accelerated, the passenger feels the back of his seat pressing hard against him; the passenger again resists a change in his velocity.

To formulate physical laws, we must have a *quantitative* measure of inertia, a quantitative measure of an object's *mass*. We shall use the simple term *mass* for what is more properly denoted the *inertial mass*.

What, then, do we mean by the mass of an object? "The amount of material in the object," as customarily given in a dictionary definition? If so, then what is meant by "amount of material"? The mass? Clearly, a strictly verbal definition leads us in circles and simply won't do.

We must have an *operational definition* for mass. That is, we must specify precisely the operations, or laboratory procedures, that could be used at least in principle by anyone, anywhere, to define and measure the physical quantity. These procedures must allow us to assign a number (together with a mass unit) to any object.

We remark in passing that the mass and the weight of an object are not the same thing, so that weighing an object (placing an object on a balance, or hanging it from a spring) does not measure its mass directly. (How weight and mass are related is treated in Section 5-7.)

The simplest conceivable arrangement of objects is that in which we have a single object in empty space. We know what happens to it — by Newton's first law, its velocity is constant. The next simplest arrangement is that with just two objects, again isolated from the rest of the universe, but somehow interacting. Suppose that two objects, labeled 1 and 2, interact through a very light spring. See Figure 5-2(a), in which the two objects shown are pressed by an observer against the two ends of the spring to compress it; then the objects are released simultaneously from rest. The objects fly off in opposite directions as the spring expands back to its equilibrium configuration. Their respective final velocities are $\mathbf{v}_1$ and $\mathbf{v}_2$. Then, *by definition*, the ratio of their respective masses m_1/m_2 is

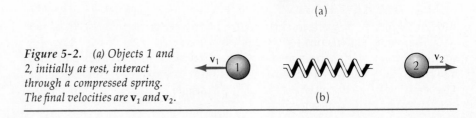

Figure 5-2. (a) Objects 1 and 2, initially at rest, interact through a compressed spring. The final velocities are $\mathbf{v}_1$ and $\mathbf{v}_2$.

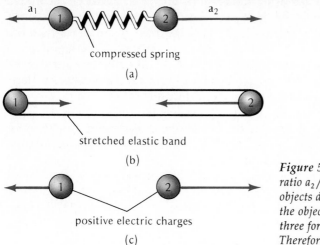

(a)

compressed spring

(b)

stretched elastic band

(c)

positive electric charges

Figure 5-3. *The acceleration ratio a_2/a_1 for a given pair of objects does not depend on how the objects interact. For all three forces $\mathbf{a}_2 = -2\mathbf{a}_1$. Therefore, $m_1 = 2m_2$.*

equal to the inverse ratio of their velocities:

$$\frac{m_1}{m_2} \equiv -\frac{\mathbf{v}_2}{\mathbf{v}_1} \qquad (5\text{-}1)$$

(The minus sign simply reflects that the two velocity directions are opposite.)

Therefore, if the two objects happen to emerge with the same speed, then the physical property they have in common — whatever differences there may be in their size, shape, color, composition, temperature, or any other property — is *mass.* Further, if object 1 has half the speed of object 2 (if $\mathbf{v}_1 = -\frac{1}{2}\mathbf{v}_2$), so that object 1 responds more sluggishly, then object 1 has twice the mass of object 2 ($m_1 = 2m_2$).

Actually, (5-1), which defines the mass ratio in terms of the ratio of the emerging *final* velocities, is more general. Observation shows that the velocity ratio is the *same constant* at *every instant* while the spring is expanding; that is, the mass of an object does not depend on its velocity.*

We may rewrite (5-1) as

$$m_1\mathbf{v}_1 = -m_2\mathbf{v}_2$$

and recognize that m_1 and m_2 are constants. Then, when we take the derivative with respect to time of both sides, we have

$$m_1\mathbf{a}_1 = -m_2\mathbf{a}_2 \qquad (5\text{-}2)$$

(The accelerations are written as vector quantities to emphasize that their directions are opposite.)

Equation (5-2) is a still more general *definition* of the mass ratio, now given in terms of the ratio of accelerations. Not only is a_2/a_1 the same number for any instant with objects interacting by, for example, an expanding spring. Experiment also shows that the acceleration ratio has the same numerical value for the same two objects interacting by *any* means — for example, pulled towards one another by a stretched elastic band or repelling one another because each carries a positive electric charge. See Figure 5-3. This means that the accelera-

---

* Strictly, the velocities must be small compared with the speed of light, 3×10^8 m/s.

tion ratio a_2/a_1 is characteristic *of the objects,* not of their interaction mode.*
Note that the procedure we have described for comparing and measuring
masses does not involve any explicit knowledge of what is meant by a force.

Mass Units   In the International System of Units (SI), the standard of mass,
to which ultimately all mass measurements are related, is a platinum-iridium
cylinder carefully stored at the International Bureau of Weights and Measures
near Paris. The mass of this cylinder is by definition exactly one kilogram
(1 kg).† Replicas of the standard kilogram have been made, and the one residing under double bell jars in the U.S. Bureau of Standards is the standard of
mass for the United States.

The familiar unit the pound appears as a unit of weight, not mass, in the
U.S. Customary or English System of Units. Roughly, 1 lb corresponds to
454 gm‡; strictly, the pound is defined as the weight of a 0.45359237-kg mass
at a location where $g = 32.17398$ ft/s². Comprehensive conversion factors for
all physical quantities are given in Appendix D.

Because the masses of atoms are small, a special system of units is used for
measuring them. The basic unit is called the *atomic mass unit* (u). By definition,
the mass of an electrically neutral atom of carbon isotope 12 is exactly 12 u.
Measurements show that 1 u $= 1.66057 \times 10^{-27}$ kg.

Scalar Additivity of Mass   Masses add as scalar quantities; the mass of a
body is the sum of the masses of its parts. This assertion is not obvious a priori;
it must be, and is, confirmed by experiment. (After all, physical volumes do not
always add as scalars: 1.0 m³ of water mixed with 1.0 m³ of alcohol does not
yield 2.0 m³ of liquid.) The scalar additivity of mass is verified by noting that
the mass of a composite object does not depend on how its constituent parts are
attached. The mass of the parts, added together, equals the mass of the whole.
For example, the mass of a water molecule is, within experimental error, the
sum of the masses of two hydrogen atoms and one oxygen atom. Thus, scalar
additivity of masses holds extremely well even in the atomic domain. There are
departures in the nuclear domain, which are accounted for, however, by the
relativistic mass-energy equivalence.

---

* By having each of the objects 1 and 2 interact in turn with a third object 3, one may confirm that
the assignment of masses is unambiguous and unique. With 3 and 1 interacting by any means, the
acceleration ratio has the constant value a_3/a_1; with 3 and 2 interacting, the ratio has the constant
value a_3/a_2. Then it is an *experimental fact* that the three acceleration ratios are related by

$$\frac{a_3/a_1}{a_3/a_2} = \frac{a_2}{a_1}$$

which is equivalent, by (5-2), to

$$\frac{m_1/m_3}{m_2/m_3} = \frac{m_1}{m_2}$$

The last equation shows that the masses of 1 and 2 can both be expressed in terms of a standard
mass, here that of object 3.
† The standard unit for mass was originally chosen to have a simple relation to the volume of
water. One cubic meter of water has a mass of approximately 10^3 kg, or 1 cm³ has a mass of 1 gm.
‡ The official abbreviation for gram in the SI unit system is "g." To avoid confusion with the
symbol g representing the acceleration due to gravity, we shall always abbreviate *gram* as "gm."

The Law of Mass Conservation Suppose that we build a completely leak-proof container through which no particles can pass, in or out. Whatever the nature of the contents of the container and despite any chemical or other changes that take place within, the total mass of this isolated system is constant. This is the law of conservation of mass, the first of several fundamental conservation laws of physics: The total mass of an isolated system is constant.

5-3   Force Defined

Roughly speaking, a force is a push or a pull. That is how we considered forces on the static equilibrium of a particle (Section 2-6) and so far in this chapter. But we must have a precise, quantitative definition of force.

Newton's first law gives us a qualitative criterion for recognizing when a force acts.* We can express this in several equivalent ways:

- A force causes a particle to depart from motion at constant velocity.
- A force speeds up an object, slows it down, and changes its direction of motion.
- A force accelerates an object.

Clearly, we must associate the net force acting on a particle with the particle's acceleration.

An ordinary helical spring can be used to measure the magnitude of a force; we can, at least tentatively, take the change in the length of a stretched spring to be directly proportional to the magnitude of the force applied to the spring (or of the force the spring applies to objects to which it is attached).

Consider Figure 5-4(a), in which some object on a frictionless horizontal surface is pulled to the right by a stretched spring whose *length* is maintained *constant*. Suppose we record the position of the object as a function of time (with flashing lights and a camera, for example), so that its changing velocity and its acceleration can be computed. Experiment shows that the *acceleration is constant*. We repeat the experiment with the spring stretched by a different but still constant amount [Figure 5-4(b)]. The acceleration is again constant; furthermore, the acceleration is proportional to the amount of stretch [Figure 5-4(c)]. These observations can be summarized as follows:

- Constant stretch produces constant **a**.
- Stretch $\propto$ **a**.

If, now, the force **F** produced by the stretched spring is taken to be proportional to the amount of stretch, we have

$$\mathbf{F} \propto \mathbf{a} \qquad \text{for constant } m$$

Note that this proportionality relation also implies that the *direction of force* is the *same as* the direction of *acceleration*. For the situation in Figure 5-4, both are to the right.

Now suppose that the first object affected by the spring force in Figure 5-4 is replaced by another one of different mass. The results of observation are these. For a constant force **F** (a constant spring stretch), the magnitude of the

---

* By a *force* here is meant, more strictly, a *net, resultant,* or *unbalanced* force.

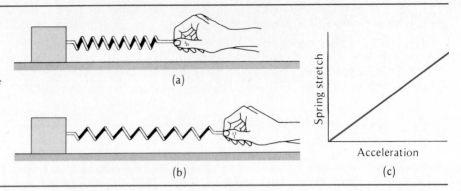

Figure 5-4. (a) A stretched spring of constant length gives an object a constant acceleration. (b) Increasing the stretch of the spring increases the magnitude of the object's acceleration. (c) By observation, spring stretch is directly proportional to acceleration for a given object.

acceleration is inversely proportional to the mass of the accelerated object,

$$a \propto \frac{1}{m} \quad \text{for constant } \mathbf{F}$$

or equivalently,

$$ma \propto \text{constant} \quad \text{for constant } \mathbf{F}$$

This implies that more massive objects (those with greater inertia) accelerate more slowly than less massive objects when both are subject to the same force.

The two proportionality relations can be combined into a single equation *defining* force:

$$\mathbf{F} \equiv m\mathbf{a} \tag{5-3}$$

The appropriateness of this definition must be tested by further considerations. It incorporates the important result that the quantity *ma*, which has the *same value* for any of numerous different objects affected by the *same* force (the same spring stretched the same amount), is a property not of any one object but *of the force.*

Equation (5-3) agrees with our intuitive sense of what constitutes a large or a small push or pull. A large force applied to a particular object accelerates it more rapidly than a small force does. A given force applied to two different objects accelerates the less massive object more rapidly than the more massive object. Further, we find that with force defined as the product of an object's mass and acceleration, fundamental forces in physics have a particularly simple mathematical form. For example, both the gravitational force between point masses and the electric force between point charges are inversely proportional to the square of the distance separating the point objects. Then the weight of any one object near the earth has a constant magnitude. Figure 5-5(a) shows a weight hung from a spring; the stretched spring produces a constant force up and the weight a constant force down. In Figure 5-5(b) we have a weight attached to a cord passing over a frictionless pulley, and the spring is fastened to an object on a horizontal frictionless surface. Again the constant weight stretches the spring by a *constant* amount, and the *constant* force of the spring gives the object a *constant* acceleration.

Acceleration is measured in meters per second per second (m/s²), mass in kilograms (kg). Therefore, a 1-kg object with an acceleration of 1 m/s² is subject to a force whose magnitude is

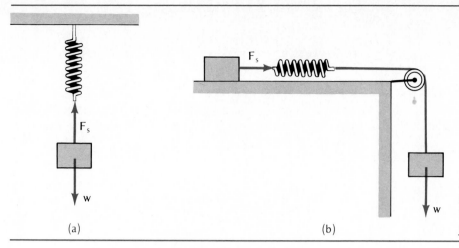

Figure 5-5. *(a) Weight* **w** *balances spring force* **F**$_s$. *(b) Weight produces a constant force* **F**$_s$ *on the spring.*

$$F = ma = (1 \text{ kg})(1 \text{ m/s}^2)$$

$$= 1 \text{ kg} \cdot \text{m/s}^2$$

The name *newton* (abbreviated N) is given to the combination of units $1 \text{ kg} \cdot \text{m/s}^2$; the newton is the SI unit for force:

$$1 \text{ N} \equiv 1 \text{ kg} \cdot \text{m/s}^2$$

Example 5-1.   A 2.0-kg object sliding east at 3.0 m/s on a horizontal frictionless surface is subject to a single constant force of 8.0 N to the east for 1.0 s. What is the object's velocity at the end of this 1.0-s interval?

While the 2.0-kg object is under the influence of a resultant force of 8.0 N, its velocity changes at the rate

$$\mathbf{a} = \frac{\mathbf{F}}{m} = \frac{8.0 \text{ N east}}{2.0 \text{ kg}} = 4.0 \text{ m/s}^2 \text{ east}$$

Over the 1.0-s interval, the object's velocity changes by the amount

$$\Delta \mathbf{v} = \mathbf{a}\Delta t = (4.0 \text{ m/s}^2 \text{ east})(1.0 \text{ s}) = 4.0 \text{ m/s east}$$

Therefore, the final velocity is

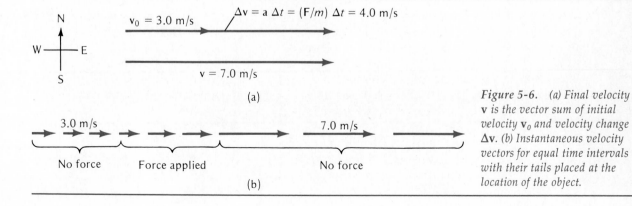

Figure 5-6.   *(a) Final velocity* **v** *is the vector sum of initial velocity* **v**$_0$ *and velocity change* Δ**v**. *(b) Instantaneous velocity vectors for equal time intervals with their tails placed at the location of the object.*

$$\mathbf{v} = \mathbf{v}_0 + \Delta\mathbf{v} = (3.0 \text{ m/s east}) + (4.0 \text{ m/s east})$$

$$= 7.0 \text{ m/s east}$$

In this simple example, with the force applied parallel to the initial velocity, the object changes its speed but not the direction of its velocity. See Figure 5-6.

Example 5-2.   The circumstances are like those in Example 5-1. A 2.0-kg object is initially sliding east at 3.0 m/s, and it is subject to a single constant force of 8.0 N for 1.0 s. Now, however, the force is in the direction *north*. What is the object's velocity at the end of the 1.0-s interval during which the external force is applied?

The object's constant acceleration is now

$$\mathbf{a} = \frac{\mathbf{F}}{m} = \frac{8.0 \text{ N north}}{2.0 \text{ kg}} = 4.0 \text{ m/s}^2 \text{ north}$$

and over the 1.0-s interval, its velocity changes by the amount

$$\Delta\mathbf{v} = \mathbf{a} \, \Delta t = (4.0 \text{ m/s}^2 \text{ north})(1.0 \text{ s}) = 4.0 \text{ m/s north}$$

The object's initial velocity was 3.0 m/s east; under the influence of the external force, it acquired a velocity component of 4.0 m/s north. See Figure 5-7. Its final velocity is then just the vector sum of the velocity components, or 5.0 m/s in the direction 53° north of east, where tan 53° = (4.0 m/s)/(3.0 m/s).

As this example emphasizes, the direction of the applied force is also the direction of the object's acceleration, or of the *change* in its velocity. Since the force is here applied in the direction north, the velocity component of 3.0 m/s to the east is unaffected.

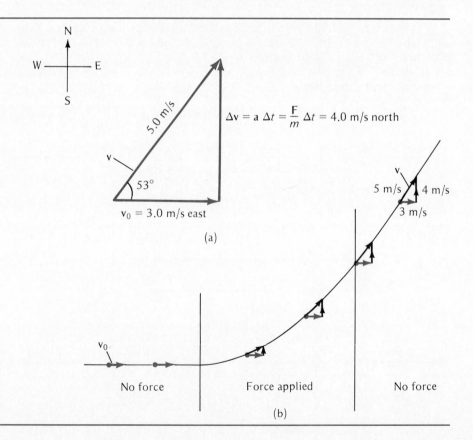

Figure 5-7.   (a) Final velocity **v** is the vector sum of initial velocity **v**₀ and velocity change Δ**v**. (b) Instantaneous velocity vectors for equal time intervals with their tails placed at the location of the object.

5-4   Newton's Second Law

If $\mathbf{F} = m\mathbf{a}$ is a definition of force, what then is Newton's second law? It is primarily a statement about the accelerating effects of *two or more forces applied simultaneously* to the same object.

Suppose that an object of mass m interacts with several other objects that produce respective individual forces $\mathbf{F}_1$, $\mathbf{F}_2$, $\mathbf{F}_3$, . . . on it. The result of experiments is that simultaneous forces produce an acceleration $\mathbf{a}$ given by

$$\mathbf{a} = \frac{\Sigma \mathbf{F}}{m} \tag{5-4}$$

where

$$\Sigma \mathbf{F} \equiv \mathbf{F}_1 + \mathbf{F}_2 + \mathbf{F}_3 + \cdots$$

Here, $\Sigma \mathbf{F}$ is the *vector sum* of the individual forces acting on the chosen object; $\Sigma \mathbf{F}$ is also referred to as the *resultant* force, the *net* force, or the *unbalanced* force. This is Newton's *second law of motion:* the resultant $\Sigma \mathbf{F}$ of all forces acting on a particle of mass m gives the particle an acceleration $\mathbf{a}$, where

$$\Sigma \mathbf{F} = m\mathbf{a} \tag{5-5}$$

Equation (5-5) is verified by experiment. Consider a specific case. Suppose that a particle of mass m is first accelerated east by force $\mathbf{F}_1$ alone (produced, for example, by a spring stretched a constant amount); the acceleration is found to be 1 m/s² east. See Figure 5-8(a). Then a second spring is attached to the same particle; it produces a constant force $\mathbf{F}_2$ north, and the acceleration is now found to be 2 m/s² north. See Figure 5-8(b). Now suppose that *both* springs are attached to the particle and stretched the same amounts and in the same directions as before. (The particle is subject now to both $\mathbf{F}_1$ and $\mathbf{F}_2$.) What is the observed acceleration? It is $\mathbf{a} = \sqrt{5}$ m/s² at 63° north of east, as shown in Figure 5-8(c). But this acceleration is just the *vector sum* of separate accelerations $\mathbf{a}_1$ and $\mathbf{a}_2$:

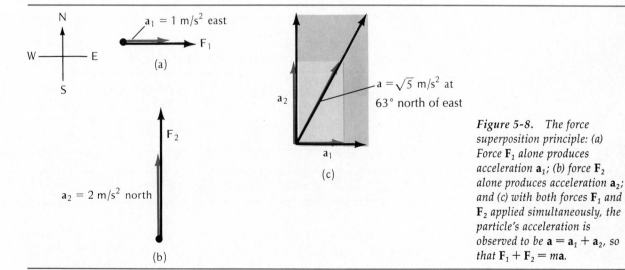

Figure 5-8. *The force superposition principle: (a) Force $\mathbf{F}_1$ alone produces acceleration $\mathbf{a}_1$; (b) force $\mathbf{F}_2$ alone produces acceleration $\mathbf{a}_2$; and (c) with both forces $\mathbf{F}_1$ and $\mathbf{F}_2$ applied simultaneously, the particle's acceleration is observed to be $\mathbf{a} = \mathbf{a}_1 + \mathbf{a}_2$, so that $\mathbf{F}_1 + \mathbf{F}_2 = m\mathbf{a}$.*

$$\mathbf{a} = \mathbf{a}_1 + \mathbf{a}_2$$

Using the *definitions* $\mathbf{F}_1 = m\mathbf{a}_1$ and $\mathbf{F}_2 = m\mathbf{a}_2$, we have

$$\mathbf{a} = \frac{\mathbf{F}_1}{m} + \frac{\mathbf{F}_2}{m}$$

or

$$\mathbf{F}_1 + \mathbf{F}_2 = m\mathbf{a}$$

$$\Sigma\mathbf{F} = m\mathbf{a}$$

The adding of individual forces as vectors to arrive at a single equivalent resultant force vector is sometimes referred to as the *superposition principle for forces.* We now see that the definition of force as $m\mathbf{a}$ is further justified, because only with this definition do simultaneous forces add as vectors. One important application of the force superposition principle is that any force vector can be replaced by its components.

Here are implications of Newton's second law.

• It is a *vector* relation. Simultaneous forces acting on the same object must be added *as vectors* to arrive at the single equivalent resultant force $\Sigma\mathbf{F}$ on the object. The direction of the acceleration $\mathbf{a}$ is the *same* as the direction of the resultant force.

• The *change in velocity* is controlled by the *net force.* This resultant force does not necessarily point in the direction in which a particle is moving (the particle's velocity does that); the net force gives the direction of the velocity change, or acceleration.

• The *vector equation* $\Sigma\mathbf{F} = m\mathbf{a}$ is equivalent to three component algebraic equations,

$$\Sigma F_x = ma_x$$

$$\Sigma F_y = ma_y \qquad\qquad (5\text{-}6)$$

$$\Sigma F_z = ma_z$$

where ΣF_x is the sum of the x components of the forces on the particle and a_x is the x component of its acceleration; similarly for y and z.

• *Acceleration* is the link between the forces acting on an object *and its motion,* the link between dynamics and kinematics. See Figure 5-9(a). If we know the forces acting on a particle, we can, by means of the second law, deduce its acceleration. Knowing the particle's acceleration, we can then find at each instant the change in its velocity; this, in turn, allows us to find the particle's new velocity and location at any future time at which it is still under the influence of known forces. In short, when we know the forces acting, we can predict the motion.

We can also run the procedure in reverse. Knowing the motion enables us to know the resultant force. If we know where a particle is at each instant, we can find its acceleration and therefore, with $\Sigma\mathbf{F} = m\mathbf{a}$, the resultant force on it. See figure 5-9(b).

• If the vector sum of several simultaneous forces on a particle is zero, these forces are equivalent to no force at all. Newton's first law may be regarded in one way as a special case of the second law; when $\Sigma\mathbf{F} = 0$, then $\mathbf{a} = 0$ and

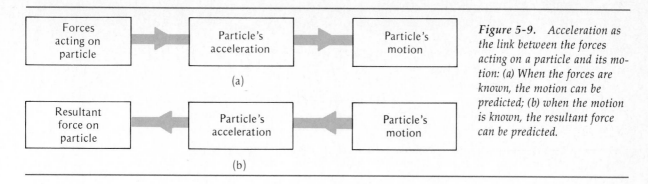

(a)

(b)

Figure 5-9. Acceleration as the link between the forces acting on a particle and its motion: (a) When the forces are known, the motion can be predicted; (b) when the motion is known, the resultant force can be predicted.

$\mathbf{v}$ = constant. Static equilibrium (Section 2-6) is the still more special case in which the particle remains at rest: $\mathbf{v} = 0$ = constant when $\Sigma \mathbf{F} = 0$.

5-5 Newton's Third Law

In Section 5-2, on the definition of mass, we considered two objects, 1 and 2, that interacted in isolation from other objects (Figures 5-2 and 5-3). There we found that the results of observation, together with the definition of mass, led to the relation

$$m_1 \mathbf{a}_1 = -m_2 \mathbf{a}_2 \tag{5-2}$$

where object 1 has a mass m_1 and acceleration $\mathbf{a}_1$, and similarly for object 2. The minus sign implies that the two objects accelerate in *opposite* directions.

We can interpret this relation in a new light through the definition of force. The force of object 2 on object 1 is defined as

$$\mathbf{F}_{2 \text{ on } 1} \equiv m_1 \mathbf{a}_1$$

Similarly, the force of 1 on 2 is

$$\mathbf{F}_{1 \text{ on } 2} \equiv m_2 \mathbf{a}_2$$

Substituting these results in the equation above yields

$$\mathbf{F}_{2 \text{ on } 1} = -\mathbf{F}_{1 \text{ on } 2} \tag{5-7}$$

This is Newton's *third law of motion:* Whenever two objects interact, the force of the second on the first ($\mathbf{F}_{2 \text{ on } 1}$) is *equal in magnitude* but *opposite in direction* to the force of the first on the second ($\mathbf{F}_{1 \text{ on } 2}$). See Figure 5-10. The two forces are often referred to as the *action* and *reaction* forces. Which of the two is called the action force and which the reaction force is strictly arbitrary. We should not think of the reaction force as somehow a "response" to the action force; both act simultaneously.

An interaction always involves a *pair* of objects. Likewise, forces occur in *pairs* only. There is no such thing as a single, isolated force. Indeed, to describe a force completely we must give, besides its magnitude and direction, the object on which the force acts, and the second object producing this force. Thus, each force is identified with *two* objects.

For example, when a hammer strikes a nail, the force of hammer on nail is

Figure 5-10. Action-reaction forces, $\mathbf{F}_{2 \text{ on } 1} = -\mathbf{F}_{1 \text{ on } 2}$.

"The Hell with Friction, Use More Grease"

Strings that have no mass and cannot be stretched. Surfaces that are perfectly frictionless. Massless springs. These are just a few of the many idealizations that keep showing up in examples and problems in physics. It's unreal. Why not deal with things as they really are?

First, simplification and idealization make analysis easier. We can thereby concentrate on the most important effects, ignoring secondary influences, at least at the start. For example, what is the motion of an object dropped from rest? We have *constant* acceleration as a *first approximation*. Of course, there is resistance for any object falling through air, but we choose to examine first the principal effect—gravity—and after this is well understood, then turn to more subtle influences. Galileo was not deterred when his critics complained that he studied the motion of falling lead balls but not falling leaves. Indeed, making appropriate idealizations is a central strategy in both experimental and theoretical physics. The experimentalist tries hard in the laboratory to isolate and study just one phenomenon at a time, and much effort is directed toward eliminating secondary disturbing influences. The theorist can do it even more easily; by the mere stroke of a pen he can decide that friction *is* eliminated, that strings *are* massless and stretchless and adjust the equations accordingly.

Models are important in advancing physics. Invariably any model implies simplification and idealization. In trying to understand the behavior of a gas from a molecular point of view, for example, molecules can be imagined to be tiny billiard balls. No molecule is actually a tiny hard sphere, but it can be so imagined for the purpose of studying the most important gross properties of a gas. When it comes, however, to knowing what happens when a gas condenses into a liquid, then the detailed properties of a molecule—the details of its force on another molecule, its atomic constituents, even the constituents of the atom—become important.

In short, when a simplification can be made, by all means make it. It takes practice and experience to develop a sense of when it's appropriate. The first concern of a physicist is whether the physics is *right*, but the next most important feature is that it be *simple*.

precisely equal in magnitude to the force of nail on hammer. When a ball crashes through a glass window, the force of ball on window is exactly equal to the force of window on ball. And so it goes. In each instance, we identify the reaction force simply by reversing the nouns.

The two forces of an action-reaction pair act on *different* objects. Consequently, the action and reaction forces cannot by themselves place any object in equilibrium. Furthermore, just because two forces happen to be equal and opposite does not necessarily mean that they constitute an action-reaction pair. We have an action-reaction pair only if they satisfy the *A-on-B* and *B-on-A* requirement.

Example 5-3. A 5.0-gm bullet moving at 300 m/s penetrates a wood post a distance of 5.0 cm before the bullet is brought to rest. Assuming the force of the bullet on the post to be constant, determine its magnitude.

From the information given, we cannot *directly* compute the force of bullet *on post*. We should have to know all other forces acting on the post, and we don't know these forces. But we can compute the average force of *post on bullet*. From Newton's third law, the magnitude of the force of post on bullet is the same as the force of bullet on post. With

$$v_0 = 300 \text{ m/s}$$

$$v = 0$$

$$x_0 = 0$$

$$x = 5.0 \text{ cm} = 0.050 \text{ m}$$

we compute the bullet's average acceleration directly from the kinematic relation, (3-10):

$$a = \frac{v^2 - v_0^2}{2(x - x_0)} = \frac{0 - (300 \text{ m/s})^2}{2(0.050 \text{ m})} = -9.0 \times 10^5 \text{ m/s}^2$$

The force on the bullet is therefore

$$F_{\text{post on bullet}} = ma = (5.0 \times 10^{-3} \text{ kg})(-9.0 \times 10^5 \text{ m/s}^2) = -4.5 \times 10^3 \text{ N}$$

From Newton's third law, then,

$$F_{\text{bullet on post}} = -F_{\text{post on bullet}} = +4.5 \times 10^3 \text{ N}$$

5-6   Weight

Any object in free flight over short distances near the earth's surface has a constant acceleration **g** toward the earth's center. This implies that any object is acted on by a constant force arising from its interaction with the entire earth. The phenomenon of an object's being attracted by the earth is called *gravitation*, and the force on the object due to its gravitational interaction with the earth is called its *weight*.

Applying Newton's second law to a freely falling object of mass m, weight **w**, and acceleration **g**, we have (Figure 5-11)

$$\Sigma \mathbf{F} = m\mathbf{a}$$

$$\mathbf{w} = m\mathbf{g} \tag{5-8}$$

The weight magnitude is simply the mass multiplied by g, the acceleration due to gravity. The direction of the weight is "down," which is to say, toward the earth's center. The acceleration of a freely falling object is, from another point of view, the vector $\mathbf{g} = \mathbf{w}/m$ that gives the gravitational force per unit mass.

The so-called standard acceleration due to gravity, close to the measured value at latitudes of 45° near sea level, is

$$\text{Standard } g = 9.80665 \text{ m/s}^2 = 32.17398 \text{ ft/s}^2$$

Therefore, the weight of a 1-kg object is about 9.81 N. A small apple weighs about 1 N.

In nonscientific usage, the terms *mass* and *weight* are often taken as interchangeable. Mass and weight are, however, different quantities.

Weight is a vector; it is a measure of the earth's pull on an object (or of the object's pull on the earth). The weight magnitude for a particular object is not constant but depends on the magnitude of g; although the magnitude of **g** is constant at one location near the earth, it varies with altitude and latitude and is, in fact, zero for objects far from the earth in interstellar space. Thus the weight of a 1-kg mass is 9.81 N at a location where g is 9.81 m/s²; at another location on earth, where g is 9.79 m/s², its weight is 9.79 N. An object of any mass is truly weightless when in interstellar space, far from other objects.

On the other hand, mass is a scalar. Mass is an intrinsic property of an

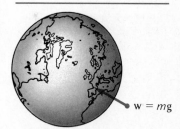

$$\mathbf{w} = m\mathbf{g}$$

Figure 5-11.   Newton's second law applied to a freely falling object with mass m and weight **w**.

object, not dependent on its interaction with other objects. A net force of 1 N must act on a mass of 1 kg to give it an acceleration of 1 m/s², whether the object is at the earth's surface or center or at any other location.

Suppose we compare the weights at the same location of two objects with masses m_1 and m_2. Both have the same acceleration **g** when falling; therefore,

$$\mathbf{w}_1 = m_1\mathbf{g} \qquad \mathbf{w}_2 = m_2\mathbf{g}$$

and

$$\frac{w_1}{w_2} = \frac{m_1}{m_2} \tag{5-9}$$

At any location, the mass ratio of two objects is the same as their weight ratio. Thus, an ordinary beam balance, which compares the weights of two bodies, also indicates, when it is in balance, that the masses are equal.

An object's weight does not depend on its state of motion. This result is simple but important, and we can see that it is true when we consider an object tossed up in the air. The object's acceleration is always **g**, whether it is going up or down or is momentarily at rest at the zenith. It follows that the gravitational force $\mathbf{w} = m\mathbf{g}$ is likewise independent of the object's velocity.

Consider, for example, a 1-kg block at rest on a horizontal surface. The block is in equilibrium; there is no net force on it. The earth exerts the force **w** downward; there must be an upward force N, equal in magnitude to the weight, as shown in Figure 5-12(a). Upward normal force **N** arises from the block's interaction with the surface. In magnitude, $N = w$ for this situation. The normal force and weight are equal and opposite forces, but they do not make up an action-reaction pair. Both forces act on the same object, the block, and put it in equilibrium.

Now consider the reaction forces. They are shown in Figure 5-12(b). By weight **w** we mean the force of earth on block; the reaction force is then the force **w'** of block on earth. The normal force **N** is the force of surface on block; the reaction force is then the force **N'** of block on surface. The *only* forces relevant in describing the state of motion, or rest, of the block are the forces that act *on it* — its weight and the normal force. The reaction forces **w'** and **N'** do not act on the block, and they are irrelevant to its motion.

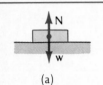

Figure 5-12. (a) Forces on a block at rest on a horizontal surface. (b) The reaction forces to the forces shown in part (a): $\mathbf{N'} = -\mathbf{N}$ and $\mathbf{w'} = -\mathbf{w}$.

Example 5-4.   A 10-gm coin is resting on the outstretched palm of a woman's hand. (a) She *lowers* her hand with an acceleration of 12 m/s². What is the resultant force on the coin and the contact force of her hand on the coin? (b) Suppose that the woman *raises* her hand with an acceleration of 12 m/s². What is now the resultant force on the coin and the force of her hand on the coin?

(a) The woman's hand accelerates downward at a rate that exceeds the acceleration of a freely falling object, so that the coin loses contact with her hand. See Figure 5-13(a). The hand produces no force on the coin, and the resultant force on the coin is simply its weight,

$$\mathbf{w} = m\mathbf{g} = (10 \times 10^{-3}\ \text{kg})(9.8\ \text{m/s}^2) = 9.8 \times 10^{-2}\ \text{N down}$$

(b) With the hand accelerating upward, the coin's acceleration is also 12 m/s² upward, so that the resultant force on it is

$$\Sigma \mathbf{F} = m\mathbf{a} = (10 \times 10^{-3}\ \text{kg})(12\ \text{m/s}^2) = 0.12\ \text{N up}$$

The coin is subject to two forces: its weight $m\mathbf{g}$ down and the larger force of the

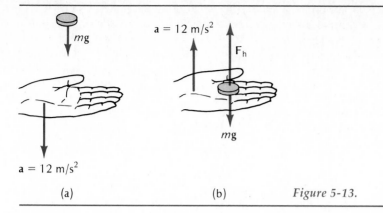

(a)   (b)   *Figure 5-13.*

hand $\mathbf{F}_h$ up. See Figure 5-13(b). The resultant force is then

$$\Sigma\mathbf{F} = \mathbf{F}_h - m\mathbf{g}$$

Here algebraic signs indicate the force directions:

$$\mathbf{F}_h = \Sigma\mathbf{F} + m\mathbf{g} = 0.12 \text{ N} + 0.098 \text{ N} = 0.22 \text{ N up}$$

Example 5-5.   A 70-kg man jumps down from a height of 1.0 m above floor level. (a) What is his speed when his feet first touch the floor? (b) What is the force of the floor on his feet if his body comes to rest during a time interval of 6.00 milliseconds?

(a) The man's motion consists of two parts: first, his falling freely at constant downward acceleration $\mathbf{g}$; secondly, his coming to rest after his feet first touch the floor.

In the first part of the motion we take the man's initial velocity v_0 as he jumps to be zero. His velocity v after falling through a vertical displacement $y = -1.0$ m is easily found through the kinematical relation

$$y = \frac{v^2 - v_0^2}{2a}$$

$$v = -\sqrt{2gy} = -\sqrt{2(-9.8 \text{ m/s}^2)(-1 \text{ m})} = -4.4 \text{ m/s}$$

(The negative sign corresponds to a downward velocity.)

(b) In the second part of the motion, where the man's body comes to rest, the *initial* velocity is now $v_0 = -4.4$ m/s and the final velocity $v = 0$. For simplicity, we assume that the acceleration is constant over the 6.0×10^{-3}-second time interval. Then we find the acceleration to be

$$a = \frac{v - v_0}{t} = \frac{0 - (-4.4 \text{ m/s})}{(6.0 \times 10^{-3} \text{ s})} = 7.4 \times 10^2 \text{ m/s}^2$$

Note that the man's acceleration is *upward* as his downward motion is arrested. The resultant force on the man must also be upward. Individual forces acting on the man are his downward weight $m\mathbf{g}$ and the upward force $\mathbf{F}_f$ of the floor on his feet. See Figure 5-14. Newton's second law gives

$$\Sigma\mathbf{F} = m\mathbf{a}$$

$$F_f - mg = ma$$

$$F_f = m(g + a)$$

$$= 70 \text{ kg}(9.8 \text{ m/s}^2 + 7.4 \times 10^2 \text{ m/s}^2)$$

$$= 5.2 \times 10^4 \text{ N}$$

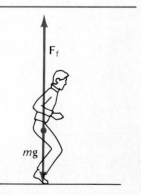

Figure 5-14.   Forces on man jumping onto floor: upward force of floor $\mathbf{F}_f$ and the man's weight $m\mathbf{g}$.

The man would actually break one or both of his legs if he held them straight. In a typical human, the femur bone of the upper leg fractures when a force of about 2×10^4 N (≈ 4400 lb) is applied to it. In this example we have computed a total force of 5.2×10^4 N on both legs, or a force of 2.6×10^4 N on each. The man's landing is relatively hard because his body comes to rest in a relatively short time interval. This would surely happen if he struck the floor with his legs out straight. But people do not ordinarily land on hard surfaces with outstretched legs. The knees are bent on first touching the floor; then the legs are straightened. The flexed legs act as a spring to lengthen the time interval over which the body (strictly, the torso) comes to rest, thereby reducing the body's acceleration and producing a relatively soft landing.

5-7   Procedures for Applying Newton's Laws: Constant Acceleration

Solving problems for Newton's law involves far more than blindly following a set routine. But the following procedure can help in organizing your work. (It is like what was suggested in Section 2-4 for problems in static equilibrium of a particle.)

- Draw a simple, clear diagram.
- Choose the object whose motion is to be analyzed. It is best to draw this object as a particle, isolated from its surroundings, on a separate diagram. This drawing is referred to as a force diagram, or a *free-body diagram.*
- Draw vectors representing all forces acting on the chosen object. Such forces would include the object's weight and forces produced by adjoining objects with which the chosen object is in contact. Unless it is stated specifically to the contrary, we assume frictional or resistive forces to be negligible compared with other forces. We shall later encounter, in addition to contact forces, such noncontact forces as gravitational, electric, and magnetic forces. (Remember that when we apply Newton's second law, we are concerned only with external forces acting *on the chosen object.* It is true that the chosen body exerts forces on its surroundings; such forces are irrelevant, however, since they do not act on the chosen object.)
- On the force diagram, indicate the magnitude and direction of each force; choose symbols to represent unknown quantities.
- Choose appropriate axes for finding components of the forces. For motion at constant acceleration, the x or y axis should be chosen in the direction of the acceleration. Replace each force vector by its components along the chosen axes.
- Use the component form of Newton's second law, (5-6), to solve algebraically for the unknowns.
- Recall that *acceleration* is the link between force and motion. The acceleration may not be asked for explicitly, but it is the quantity that allows the motion to be predicted given the forces acting, or the acting forces to be computed given the motion.

The examples of motion at constant acceleration that follow illustrate these procedures. The simplest example of an object with constant acceleration is a freely falling object near the earth's surface. It is subject to *constant* force $m\mathbf{g}$; it has a *constant* acceleration $\mathbf{g}$; and it traces a parabolic path (Section 4-2).

Example 5-6.   A 2.0-kg block is on a perfectly smooth ramp that makes an angle of 30° with the horizontal. (a) What is the block's acceleration down the ramp and the force of the ramp on the block? (b) What force applied upward along the ramp would allow the block to move with constant velocity?

(a) In this problem, as in most problems, it is advisable first to use symbols for all quantities, and only after the problem has been solved, to substitute numerical quantities. In short, use numerical quantities only when the problem can be carried no further and you are forced to do so. In so doing, you solve, in effect, all problems of that particular type. Moreover, certain quantities may cancel! Here the block's mass is simply m and the angle of the ramp is θ. See Figure 5-15(a).

The forces acting on the block are shown by force vectors in Figure 5-15(b); they are the weight $m\mathbf{g}$ and the normal force $\mathbf{N}$ of the incline on the block. Because the surface is taken to be perfectly smooth, there can be no frictional force parallel to the incline.

The block travels and accelerates along the incline, and therefore we choose the x axis to be parallel to the incline and the y axis perpendicular to the incline. Then $\mathbf{N}$ is along y. The weight $m\mathbf{g}$, which acts vertically downward, is replaced by its x- and y-force components: $mg \sin \theta$ along the positive x axis and $mg \cos \theta$ along the negative y axis. (The earth pulls vertically downward on the block; its pull is altogether equivalent to the simultaneous action of the force components of the weight.)

Applying Newton's second law, (5-6), to the x- and y-force components in turn, we have

$$\Sigma F_x = ma_x$$

$$mg \sin \theta = ma_x$$

$$\Sigma F_y = ma_y$$

$$N - mg \cos \theta = 0$$

We have set a_y equal to zero, because the block slides along the incline and does not accelerate along y. Note further that the force component $mg \cos \theta$ carries a negative algebraic sign, since it points along the negative y direction.

The acceleration and normal force are then, from the equations above,

$$a_x = g \sin \theta = (9.8 \text{ m/s}^2) \sin 30° = 4.9 \text{ m/s}^2$$

$$N = mg \cos \theta = (2.0 \text{ kg})(9.8 \text{ m/s}^2)(0.87) = 17 \text{ N}$$

We see that the block's acceleration is constant and does not depend on its mass. We see further that the normal force of incline on block is not equal to the block's weight.

Imagine that the incline's angle θ is a variable. We can then get useful results for two extreme cases. First, suppose that $\theta = 0$ (the incline is horizontal). Then, $a = g \sin \theta =$

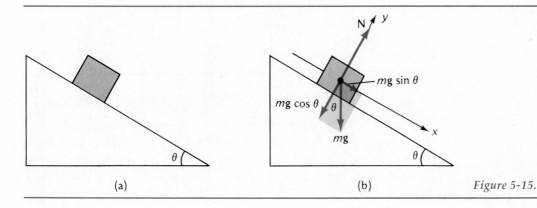

(a)   (b)   *Figure 5-15.*

0. Now the block is on a smooth horizontal surface; it has zero acceleration and a constant horizontal velocity. Furthermore, $N = mg \cos \theta = mg$.

Now suppose that $\theta = 90°$, so that the incline becomes effectively a smooth vertical wall. Then, $a = g \sin 90° = g$. The block now falls freely vertically downward (the x axis now points downward) with an acceleration g. For this situation the normal force is $N = mg \cos 90° = 0$.

For any intermediate fixed angle ($0 < \theta < 90°$), $a = g \sin \theta$; the block accelerates at a constant rate, which is, however, less than g. Motion along a smooth incline is, in effect, slow motion of a freely falling object. Galileo recognized this as a way of "diluting" gravity. He proved that a freely falling object has a constant acceleration by observing that an object sliding down an incline has a more easily measured constant acceleration.

(b) The resultant force accelerating the block down the incline has a magnitude $mg \sin \theta$. If the block is to have a constant velocity, an additional force **F** must be applied upward along the incline with the magnitude

$$\mathbf{F}_u = mg \sin \theta = (2.0 \text{ kg})(9.8 \text{ m/s}^2)(0.50) = 9.8 \text{ N}$$

With this additional force, the block is placed in equilibrium (see Example 2-4). If the block is initially at rest, it remains at rest. If the block is initially in motion, it continues moving up or down the incline at constant velocity.

Example 5-7. A 50-kg woman stands on a scale in an elevator. (See Figure 5-16a). The elevator is originally moving downward at constant speed; then it comes to rest with a constant acceleration of 2.0 m/s^2. What does the scale read while the elevator is coming to rest?

The scale shows the force of *woman on scale*. We know nothing about the properties of the scale itself, so we cannot compute directly the forces on it. But we know, from Newton's third law, that the magnitude of the force of *scale on woman* is the same as that of woman on scale. The object we choose for applying Newton's second law is *the woman*, and in Figure 5-16 we see vectors representing the forces on her: her weight $m\mathbf{g}$ down and force of the scale $\mathbf{F}_s$ up.

We must be careful to get the *direction* of the acceleration right. There are a couple of ways of doing it:

• The woman is traveling downward initially and later she is at rest. Therefore the woman's downward motion is arrested by a resultant force that must be *up*. The

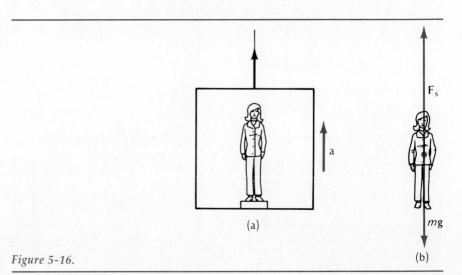

(a)

(b)

Figure 5-16.

direction of the acceleration is the same as the direction of the resultant force, so the acceleration is *up*.

• The velocity *decreases* in the *downward* direction, so that the acceleration is up. To see this in more detail, take "up" to be positive. The initial velocity v_{0y} is then *negative* and the final velocity v_y is zero. Therefore, we have for acceleration, $a_y = (v_y - v_{0y})/t = -v_{0y}/t$, a *positive* (or upward) quantity because of the negative value for v_{0y}.

Applying Newton's second law to the forces in Figure 5-16(b) and with upward quantities taken as positive, we have

$$\Sigma F_y = ma_y$$

$$F_s - mg = ma_y$$

$$F_s = m(g + a_y)$$

$$= mg\left(1 + \frac{a_y}{g}\right)$$

$$= mg\left(1 + \frac{2.0 \text{ m/s}^2}{9.8 \text{ m/s}^2}\right)$$

$$= 1.2mg$$

The scale reads a force that is 20 percent higher than the woman's ordinary weight.

When the elevator moves downward with *constant* speed, its acceleration a_y is zero. Then, with $a_y = 0$ in the equation above, we have $F_s = mg$. With the elevator moving down (or up) at *constant* velocity, the scale reads the woman's weight mg.

We can readily see what the scale would read if the woman's acceleration had been *downward* instead of upward. Downward acceleration would correspond, for example, to the elevator's traveling upward initially and then coming to rest, or we could have the elevator starting from rest and traveling downward. The forces are as before. The only change is in the direction (or sign) of the acceleration a_y. For downward acceleration, the relation above becomes

$$F_s = mg\left(1 - \frac{a_y}{g}\right)$$

Now the scale reading is less than the person's ordinary weight. If the elevator should fall freely (if its cable were cut, for example), the downward acceleration a_y would be equal in magnitude to g, and we find that $F_s = 0$. The scale would read *zero* weight because the elevator, the woman, and the scale would all fall at the same acceleration.

Example 5-8.   A 4.0-kg block is on a smooth horizontal table. This block is connected to a second block with a mass of 1.0 kg by an essentially massless, flexible cord that passes over a frictionless pulley. The 1.0-kg block is initially 1.0 m above the floor. [See Figure 5-17(a).] The blocks are released from rest with the string taut. (a) With what speed does the 1.0-kg block hit the floor? (b) What is the tension in the string during the fall?

To find the block's speed we must first find its acceleration, and we do this by applying Newton's second law.

Here we have two *coupled* masses. Their motions are not independent. With the string taut, both objects have the same magnitude of acceleration. The 4.0-kg mass accelerates to the right; the 1.0-kg mass accelerates down. With two objects, we must apply Newton's second law twice; once to each of the two objects in turn.

First consider the forces on the 4.0-kg block (labelled mass m_4), as shown in Figure 5-17b. The normal force **N** is cancelled by weight $m_4\mathbf{g}$, and the resultant force on the object is **T**, the force of the string toward the right on this block. This force is equal to the *tension* in the string, the force of any piece of string on an adjoining piece of string, as

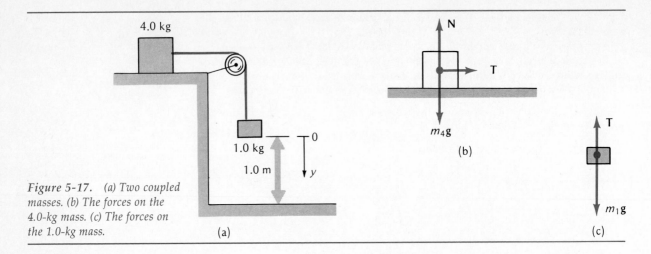

Figure 5-17. *(a) Two coupled masses. (b) The forces on the 4.0-kg mass. (c) The forces on the 1.0-kg mass.*

will be shown in Example 5-9. Newton's second law then gives

$$\Sigma F_x = ma_x$$

$$T = m_4a$$

where a is the acceleration of m_4 to the right.

Now we consider the motion of m_1. Its motion is not independent of m_4. This system of two coupled objects is subject to a *constraint:* the string connecting them has a *constant* length. Therefore, the magnitude of m_1's acceleration is the same as that of m_4. Furthermore, when m_4 moves to the right, m_1 must move down. This implies that the direction of *positive* acceleration for m_1 must be *down,* if m_4's positive acceleration is to the right.

The forces on m_1 are shown in Figure 5-17(c). They are its weight $m_1\mathbf{g}$ down, and the string tension $\mathbf{T}$ up. Newton's second law yields, with downward as positive,

$$\Sigma F_y = ma_y$$

$$m_1g - T = m_1a$$

Eliminating T from the two equations yields

$$m_1g = (m_1 + m_4)a$$

The magnitude of the acceleration can now be computed.

$$a = \frac{m_1}{m_1 + m_4}g = \left(\frac{1 \text{ kg}}{1 \text{ kg} + 4 \text{ kg}}\right)(9.8 \text{ m/s}^2) = 2.0 \text{ m/s}^2$$

The tension is given by

$$T = m_4a = \frac{m_4m_1}{m_1 + m_4}g = \frac{(4)(1) \text{ kg}}{1 + 4}(9.8 \text{ m/s}^2) = 7.8 \text{ N}$$

Finally, the speed v_y with which m_1 hits the floor is found from kinematic relations for constant acceleration:

$$y - y_0 = \frac{v_y^2 - v_{0y}^2}{2a_y}$$

From Figure 5-17(a), we see that $y - y_0 = 1.0$ m.

$$v_y = \sqrt{2a_y(y - y_0)} = \sqrt{2(2.0 \text{ m/s}^2)(1.0 \text{ m})} = 2.0 \text{ m/s}$$

Example 5-9. What is the nature of the tension forces exerted by an ideal rope, one that is perfectly flexible, inextensible, and massless? No such rope (or cord, string, or wire) exists. But in many situations a real rope closely approximates the ideal.

We prove that the tension in the rope is the same at all points along the length of the rope, whether the rope is accelerated or in equilibrium.

We concentrate first on the short segment of cord shown in Figure 5-18(a). The forces on it from the adjoining cords at the two ends are labeled **T** and **T'**; its weight is $m\mathbf{g}$. Applying Newton's second law to this effective particle, we have

$$\Sigma\mathbf{F} = m\mathbf{a}$$

$$\mathbf{T} + \mathbf{T'} + m\mathbf{g} = m\mathbf{a}$$

where **a** is the segment's acceleration. But if the segment's mass m is neglibly small, we may discard both the $m\mathbf{g}$ and $m\mathbf{a}$ terms, and the above equation becomes

$$\mathbf{T} = -\mathbf{T'}$$

or in magnitude

$$T = T'$$

The *tension T* has the same magnitude at both ends of the small segment. Moreover, the tension, defined as the force exerted by *any* cord segment on an adjoining segment, is the same at all points along the cord, provided the cord's mass and weight are negligible. Consider Figure 5-18(b), where the force of the small segment on the portion above it is again T. By Newton's third law, the force at the imaginary cut of the short segment on the cord immediately above it is equal in magnitude to the force of the upper cord on the segment.

When a flexible rope passes over a light pulley, the magnitude of the tension is again the same at all points along the rope, and the pulley changes the direction of the force without affecting its magnitude. Hereafter, unless stated otherwise, it will be assumed that any rope is ideal and therefore the tension has the same magnitude at all points along the rope.

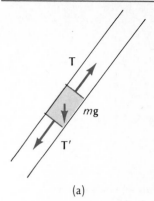

(a)

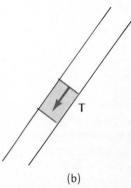

(b)

Figure 5-18. (a) Forces on a small segment of cord. (b) The short segment produces a force **T** on an adjoining cord.

5-8 Force and Mass Units in Other Systems (Optional)

The SI unit system is also referred to as the mks system, since the *m*eter, the *k*ilogram, and the *s*econd are the fundamental units for length, mass, and time. Another related metric system is the cgs system, in which the fundamental units are the *c*entimeter, the *g*ram, and the *s*econd. The cgs unit for force is the *dyne*, defined as

$$1 \text{ dyn} \equiv (1 \text{ gm})(1 \text{ cm/s}^2) = (10^{-3} \text{ kg})(10^{-2} \text{ m/s}^2) = 10^{-5} \text{ N}$$

In the English Gravitational System (also referred to sometimes as the English Engineering System or the U.S. Customary System), the *pound* is regarded as a unit for *weight*, not mass. Since the mass of an object with a weight w is $m = w/g$, we may write the mass of, say, a 64-lb object as $m = (64 \text{ lb})/(32 \text{ ft/s}^2) = 2 \text{ lb}/(\text{ft/s}^2)$. The mass of an object in the English unit system can then be expressed in units of $\text{lb}/(\text{ft/s}^2)$; the odious name *slug* is sometimes given to this combination of units (it is a measure of sluggishness), but the use of this term is not required. Comprehensive tables of conversion factors for various systems of units are given in Appendix D.

Example 5-10.   A 2-lb object is subject to a single external force of 10 lb. What is its acceleration?

A 2-lb object has a *weight* of 2 lb. Its mass is then $m = w/g = 2$ lb$/g$. With a resultant force of 10 lb on an object whose mass is 2 lb$/g$, Newton's second law gives

$$F = ma$$

$$10 \text{ lb} = \left(\frac{2 \text{ lb}}{g}\right) a$$

$$a = 5g$$

Summary

Definitions

Inertial frame: a reference frame in which an isolated particle has constant velocity (a reference frame in which Newton's first law is valid). Newton's laws of motion apply only to inertial reference frames.

Mass: with two objects interacting only with one another, the ratio of the masses of the two objects is defined as the inverse ratio of their respective accelerations

$$\frac{m_1}{m_2} = -\frac{a_2}{a_1} \qquad (5\text{-}2)$$

Force: the force $\mathbf{F}$ on a particle of mass m with acceleration $\mathbf{a}$ is

$$\mathbf{F} = m\mathbf{a} \qquad (5\text{-}3)$$

Weight: the gravitation force $\mathbf{w}$ on a particle of mass m at a location for which the acceleration due to gravity is g is

$$\mathbf{w} = m\mathbf{g} \qquad (5\text{-}7)$$

Units

Mass: kilogram (kg)
Force: newton (1 N = 1 kg·m/s²)

Fundamental Principles

The conservation of mass law states that the total mass of a completely isolated system is constant.

Newton's laws of motion are as follows:

1. An isolated particle has a constant velocity:

   $$\mathbf{v} = \text{constant} \qquad \text{when } \Sigma\mathbf{F} = 0$$

2. The vector sum of all forces acting on a particle equals its mass multiplied by its acceleration:

   $$\Sigma\mathbf{F} = m\mathbf{a} \qquad (5\text{-}5)$$

3. The action and reaction forces are equal and opposite:

   $$\mathbf{F}_{2 \text{ on } 1} = -\mathbf{F}_{1 \text{ on } 2} \qquad (5\text{-}7)$$

Problems and Questions

Section 5-1   The First Law

· **5-1 Q** The force required to keep a rocket ship moving at constant velocity in free space, far from all stellar objects, is
(A) equal to its weight.
(B) equal to the force required to stop it.
(C) dependent on how fast the rocket is moving.
(D) dependent on the particular inertial reference frame to which the force is referred.
(E) zero.

: **5-2 Q** Suppose that the force of gravity here on earth could suddenly be switched off. What would happen to a person standing on the equator? Some possible paths she might follow are shown in Figure 5-19. Looking down from the North Pole,
(A) she would fly off along path *A*.

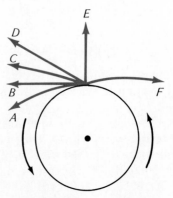

Figure 5-19.   Question 5-2.

(B) she would fly off along path *B*.
(C) she would fly off along path *C*.
(D) she would fly off along path *D*.
(E) she would fly off along path *E*.
(F) she would fly off along path *F*.
(G) she would not fly off at all, but would instead stay put.

· **5-3 Q** Consider a situation in which your car, when coasting down a straight downhill road, acquires a constant velocity. In this situation
(A) gravity is the only force acting on the car.
(B) friction is the only force acting on the car.
(C) the net force acting on the car is zero.
(D) there must be some net force acting on the car since it is moving.
(E) we see that this could not possibly correspond to a situation in real life, since in fact a car will continue to gain speed as it goes downhill.

Section 5-2   Mass

· **5-4 Q** The concept of mass (or inertia) is most closely associated with which of the following?
(A) The way in which units are defined in the metric system.
(B) The ratio of the accelerations for two interacting objects.
(C) The relation between the force on an object and how fast it is moving.
(D) The observation that forces always occur in pairs.
(E) The fact that there is such a thing as gravitational force.

· **5-5 P** Objects *A* and *B* are attached to the ends of a stretched spring. When the spring is released, *A* has an acceleration 1.0 m/s² and *B* an acceleration of 2.0 m/s². What is the ratio of mass *A* to mass *B*?

Section 5-4   Newton's Second Law

· **5-6 Q** The force acting on a particle, the velocity, and the acceleration of the particle, are all vectors. What is the relative orientation of these vectors?
(A) **f** and **v** are always parallel.
(B) **f** and **a** are always parallel.
(C) **a** and **v** are always parallel.
(D) **a**, **v**, and **f** are always parallel.
(E) None of these three vectors need be parallel, since they are all independent.

· **5-7 Q** In which of the following situations can we *not* deduce that a force is acting on a particle?
(A) The particle is moving.
(B) The particle is going in a circle at constant speed.
(C) The particle is changing speed.
(D) The particle is accelerating.
(E) The particle is turning.

· **5-8 P** What average force must be exerted on a car of mass 1200 kg, initially at rest, in order to give it a velocity of 20 m/s after 10 seconds?

· **5-9 P** An electron (mass 9.11×10^{-31} kg) is subject to an accelerating force of 1.6×10^{-18} N. (*a*) If its initial velocity is 2×10^5 m/s, what is its velocity 0.2 μs later? (*b*) How far does it move during this time period? (*c*) What is its acceleration?

: **5-10 P** In an automobile accident a child is struck by a car. You are asked to aid in the investigation. The driver pleads that he was barely moving when the child ran out of a playground into his path. The police provide you the following information:

   Victim: Female, 8 years old, 30 kg mass, height 120 cm.
   Vehicle: Ford sedan, mass 1500 kg.

   Victim struck front grill and did not bounce off. Grill was dented 10 cm. Tests on a similar vehicle show this would require a force of 6×10^4 Newtons. Skid marks were not sufficiently well defined to determine stopping distance. (Accident occurred on a gravel road.)
   What would be your best estimate of the speed of the vehicle when it struck the child?

: **5-11 P** A drag racer starts from rest and accelerates with constant acceleration for 5 seconds. At this time he is moving 40 m/s and puts out a drogue parachute to slow himself. The mass of the car plus driver is 1000 kg, and the parachute exerts a drag force of 3000 N. (*a*) What was the acceleration of the dragster? (*b*) How far did he travel while accelerating? (*c*) How far did he travel from starting point to stopping point?

: **5-12 P** A locomotive exerts a force **F** on three coupled cars of masses *m*, 2*m* and 3*m*. Determine the tension in the coupling between each pair of cars.

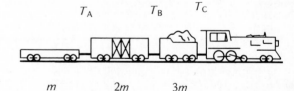

Figure 5-20.   Problem 5-12.

: **5-13 P** A girl of mass 40 kg and a boy of mass 20 kg hold the ends of a light rope while standing still on ice skates. The girl starts pulling on the rope with a steady force of 10 N (pulling the rope in steadily as she does so). If no friction acts on them as they slide together, and if they are initially 8 meters apart, how long will it take them to come together?

: **5-14 P** A "snow dragon" is a "crack-the-whip" activity in which each skier holds the poles of the person in front of him. Then, in close snowplow formation, the group takes off down the mountain. To see why skiers are easily lost from the chain, suppose masses 3*m*, 2*m*, and *m* are rigidly

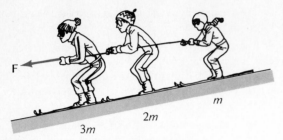

Figure 5-21.   Problem 5-14.

connected and coasting along at constant velocity. If you give a big jerk and apply force **F** to the largest mass, what force will be exerted on the smallest mass?

Section 5-5   Newton's Third Law

· **5-15 Q** An adult of mass 60 kg and a child of mass 40 kg stand on a frictionless surface (such as an ice rink). They hold hands. When the adult pulls on the child with a force of 120 Newtons,
(A) the child will automatically exert a force of 120 N on the adult also.
(B) the child will exert a force of 180 N on the adult.
(C) the child will exert a force of 80 N on the adult.
(D) the force exerted on the adult will depend on whether or not the child digs his skates into the ice and resists the adult's pull.
(E) the child will not exert any force on the adult as long as no friction acts.

: **5-16 Q** Suppose that in an attempt to get your motorcycle started, you push on it with a force F while running alongside it. In this situation:
(A) the force exerted on you by the motorcycle may be zero.
(B) the force exerted on you by the motorcycle will depend on whether or not the motorcycle is moving with constant velocity.
(C) the motorcycle will exert a force F back on you independent of how fast the motorcycle is going.
(D) the force exerted on you by the motorcycle will depend on how heavy it is.
(E) it is not possible for the force F to exceed your weight.

: **5-17 Q** A man pushes against a rigid, immovable wall. Which of the following is the most accurate statement concerning this situation?
(A) The man can never exert a force on the wall which exceeds his weight.
(B) If the man pushes on the wall with a force of 200 N, we can be sure that the wall is pushing back with a force of exactly 200 N on him.
(C) Since the wall cannot move, it cannot exert any force on the man.
(D) The man cannot be in equilibrium, since he is exerting a net force on the wall.

(E) The friction force on the man's feet is directed to the left.

· **5-18 Q** Suppose you hold a lead brick in your hand and give it a whack with a hammer. Which of the following is the most accurate statement concerning what happens?
(A) The force exerted by the brick on your hand is equal to the force exerted by the hammer on the brick.
(B) The force exerted by the brick on your hand is much less than that exerted by the hammer on the brick, mainly because the brick has a much greater surface area than does the face of the hammer.
(C) The force exerted by the brick on your hand is much less than the force exerted by the hammer on the brick, mainly because the brick has a much greater mass than does your hand.
(D) If the hammer face had the same surface area as the brick, it would not matter to you whether the hammer hit your hand first or the brick first, as long as you hold the brick securely in your hand.

Section 5-6   Weight

· **5-19 Q** Which of the following is an accurate statement?
(A) Mass and weight are the same thing expressed in different units.
(B) If an object has no weight, it must also have no mass.
(C) If the weight of an object varies, so must the mass.
(D) Mass and inertia are different concepts.
(E) Weight is always proportional to mass.

· **5-20 P** A big king salmon can weigh 200 N. How many pounds is this? What is the mass of such a fish?

· **5-21 P** A certain physics professor weighs 200 lb. What is his weight in Newtons and his mass in kilograms?

· **5-22 P** How much does 2 kilograms of hamburger weigh (a) in pounds? (b) in Newtons?

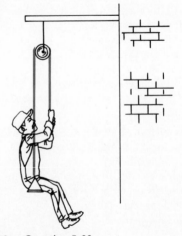

Figure 5-22.   Question 5-23.

: **5-23 Q** A painter working on the side of a building pulls himself up to where he wants to work by pulling on the rope suspending him, as shown in Figure 5-22. In order to free his hands to work, he then ties the end of the rope to the building. As soon as he does so, the rope breaks. Why might this have happened?

: **5-24 P** A light string is attached to a 160-gm ball of 2 cm radius. The string passes over a pulley 40 cm above the surface on which the sphere rests. A constant force of 2 N is applied to the string. The sphere is originally at rest at a horizontal distance of 100 cm from the pulley. How far will the ball have moved horizontally when it loses contact with the surface?

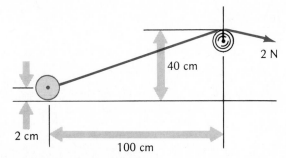

Figure 5-23.   Problem 5-24.

: **5-25 Q** Can a person pull herself up on a platform by pulling on a rope as in Figure 5-24? If she and the platform together weigh 600 N, what force would she have to exert?

Figure 5-24.   Question 5-25.

· **5-26 P** When the brakes are slammed on in a small car, a braking force equal to $\frac{1}{4}$ the weight of the car is exerted. *(a)* In what distance will such a car come to stop when traveling 90 km/hr initially? *(b)* How long will it take to come to rest?

· **5-27 Q** A piece of string is cut into two pieces and used to suspend an object as shown in Figure 5-25. If you pull down on the lower string with enough force:
(A) both strings will break at the same time.
(B) the lower string will break first.

Figure 5-25.   Question 5-27.

(C) which string breaks first depends on whether you pull slowly or give a sudden jerk.

· **5-28 Q** A mass m is suspended from a cord attached to the ceiling of an elevator. The elevator has acceleration a upward. The tension in the cord is:
(A) ma
(B) mg
(C) $ma - mg$
(D) $mg - ma$
(E) $mg + ma$

· **5-29 Q** Consider what happens when you jump up in the air. Which of the following is the most accurate statement?
(A) It is the upward force exerted by the ground that pushes you up, but this force can never exceed your weight.
(B) You are able to spring up because the earth exerts a force upward on you which is larger than the downward force you exert on the earth.
(C) Since the ground is stationary, it cannot exert the upward force necessary to propel you into the air. Instead, it is the internal forces of your muscles acting on your body itself which propels the body into the air.
(D) When you push down on the earth with a force greater than your weight, the earth will push back with the same force and thus propel you into the air.
(E) When you jump up the earth exerts a force F_1 on you and you exert a force F_2 on the earth. You go up because $F_1 > F_2$, and this is so because F_1 is to F_2 as the earth's mass is to your mass.

· **5-30 Q** A woman sits in a chair. The reaction force to the woman's weight is:
(A) the force of the woman on the chair.
(B) the force of the chair on the woman.
(C) the force of the ground on the chair.
(D) the weight of the chair.
(E) the force of the woman on the earth.

Section 5-7   Newton's Laws, Constant Acceleration

· **5-31 Q** A ball rolls down a smooth hill to a flat surface, so that
(A) its speed increases and its acceleration decreases.
(B) its speed decreases and its acceleration increases.
(C) both speed and acceleration increase.
(D) both speed and acceleration remain constant.
(E) both speed and acceleration decrease.

· **5-32 Q** On a frictionless plane inclined at angle θ above horizontal, there rests a block of mass m. The magnitude of the force of the block on the plane is:
(A) zero.
(B) mg.
(C) $mg \cos \theta$.
(D) $mg \sin \theta$.
(E) $mg \tan \theta$.

: **5-33 P** A toy rocket of mass 50 gm can accelerate horizontally with an acceleration of 29.8 m/s². The rocket engine operates for 3 seconds before burnout, and provides constant thrust. Assuming negligible air friction, (a) How high will it go when fired straight up? (b) How long will it stay in the air?

: **5-34 P** (a) If you stood on a bathroom scale in an elevator accelerating upward with an acceleration of $\frac{1}{3}g$, what would the scale read (expressed as a multiple of your normal weight, W)? (b) What would the scale read if the elevator were accelerating downward with $\frac{1}{2}g$?

: **5-35 P** A space vehicle traveling 50 m/s fires its retro-rockets to slow down preparatory to making a landing on the moon. The vehicle decelerates at 1.0 m/s². What would a 150 lb astronaut apparently weigh during the process? (The acceleration due to gravity on the moon's surface is $\frac{1}{6}$ that on earth.)

: **5-36 P** Bob Beamon made a fantastic long jump of more than 29 feet in the Mexico City Olympic Games of 1968. A successful long jumper must combine great sprint speed with powerful jumping legs. A world-class male long jumper can run the hundred meters in about 10.0 seconds. Assuming that his foot is in contact with the take-off board for the time it takes him to make one stride (2 meters, say) and that a 70 kg jumper can exert a downward force of 2100 N during his take-off, how far would you estimate he could jump, neglecting air resistance? (Assume his horizontal motion is unaffected by his springing up from the take-off board.)

: **5-37 P** A sign of weight W is attached to a wall by two light rods of negligible weight. The rods are pivoted at the wall and at the point where they join together. Determine the magnitude and direction of the force exerted on each of the two pivots P_1 and P_2.

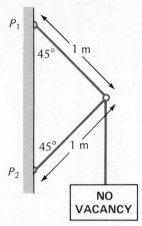

Figure 5-26.   Problem 5-37.

: **5-38 P** A bead is released from rest to slide without friction down one of two straight wires, AB or AC, which join points of a circle as shown in Figure 5-27. Show that the time to slide down is the same in both cases, independent of where on the circle point C is located. AB is a diameter of the circle.

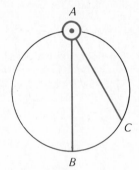

Figure 5-27.   Problem 5-38.

: **5-39 P** The device shown in Figure 5-28 is intended to carry ore up out of an old mine. The idea was to fill an oil drum with water from a diverted stream, so that when there was enough water in the drum, it would pull the ore car up to the top. There it would be stopped, the water dumped out of the oil drum, and the cart sent back down

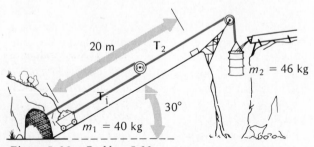

Figure 5-28.   Problem 5-39.

again. Some technical problems developed, but the idea is theoretically workable. Suppose friction is negligible. If only the ore car and the oil drum have appreciable mass, determine the tensions in the two ropes and how long it would take to pull the car to the top.

⋮ 5-40 P A 2 kg block and a 4 kg block are connected by a light string as shown in Figure 5-29. Friction is negligible. What is (a) the acceleration of each block and (b) the tension in the string?

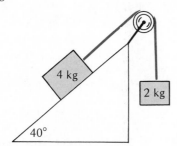

Figure 5-29. Problem 5-40.

⋮ 5-41 P What should mass m be if the 8 kg mass is to accelerate up the inclined plane in Figure 5-30 with an acceleration of 0.5 m/s²? Friction is negligible, as is the mass of the pully.

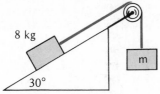

Figure 5-30. Problem 5-41.

⋮ 5-42 P Three blocks are connected as shown in Figure 5-31. Friction is negligible. Determine (a) the acceleration of the system and the tensions (b) T_a and (c) T_b.

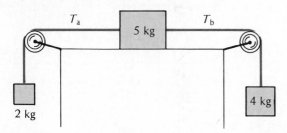

Figure 5-31. Problem 5-42.

⋮ 5-43 P In Air Force tests designed to investigate the ability of humans to survive large accelerations, a test pilot is launched along a horizontal track in a rocket sled. High-speed cameras positioned along the track record the person's response to the motion. In such experiments it is useful to have a record of the instantaneous acceleration, and

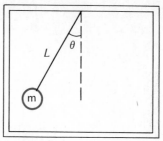

Figure 5-32. Problem 5-43.

this can readily be obtained from the photographs if a small acceleration meter is mounted on the sled. A simple form of accelerometer consists of a mass m attached to a thread of length L and hung as a simple pendulum. (See Figure 5-32.) When the sled is accelerating, the mass will not hang straight down; the string will be inclined at an angle θ to the vertical. In one experiment it was found that $\theta = 60°$ for a mass of 50 grams attached to a thread of length 12 cm. What was the acceleration of the sled under these circumstances?

⋮ 5-44 P A falling parachutist experiences a frictional drag force cv^2, where c is a constant that depends on the size and shape of the person and on the density of air. For a person of mass 70 kg, c typically has the value of 0.23 kg/m. (a) Determine the terminal velocity (i.e., the final velocity) reached by a person if the parachute does not open. (b) If the parachute did not open and the parachutist landed in some snow or foliage, he would still have a finite chance of surviving even if he had reached terminal velocity. If he were to land flat on his back, he would have a 50% chance of survival if the force stopping him did not exceed 1.2×10^5 N. In what minimum distance would he have to stop if the force acting on him was not to exceed this value?

⋮ 5-45 P In the apparatus in Figure 5-33, $m_1 = 0.5$ kg and $m_2 = 0.3$ kg. Neglecting friction and the mass of the pulleys, calculate (a) the tension in the cord and (b) the acceleration of mass m_1.

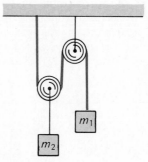

Figure 5-33. Problem 5-45.

⋮ 5-46 P A frictional force equal to half the weight of the 4 kg block acts in the system shown in Figure 5-34. Deter-

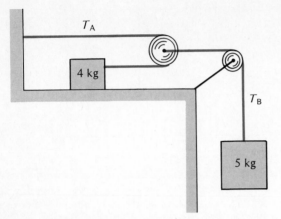

Figure 5-34.   Problem 5-46.

mine the tension in each string and the acceleration of each block.

: **5-47 P** A mass of 100 gm can slide down the frictionless post mounted on the 600 gm cart shown in Figure 5-35. The pulleys are frictionless and of negligible mass. If the small mass is released from rest when the cart is stationary, how long will it take the cart to move 1 meter?

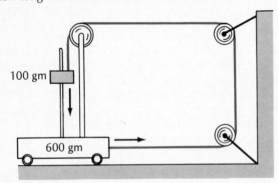

Figure 5-35.   Problem 5-47.

: **5-48 P** A small block of mass m rests on a larger wedge-shaped block of mass M. No friction is present. (See Figure 5-36.) *(a)* What horizontal force must be exerted on M if the small block is not to move with respect to the large block? *(b)* What horizontal force must be exerted on m if the small block is not to move with respect to the large block? *(c)* If the

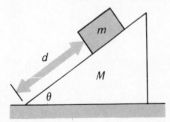

Figure 5-36.   Problem 5-48.

small block is released from rest a distance d from the bottom of the wedge, how long will it take to slide down?

: **5-49 P** Blocks of masses 1 kg and 3 kg are connected by a light string that passes over a frictionless pulley of negligible mass. Originally the two masses are at rest on a table with the string taut. An upward force **F** is then applied to the pulley. *(a)* For what range of forces **F** will the string remain taut and the 3 kg mass remain on the table? *(b)* If **F** = 50 N, what is the tension in the string and the acceleration of each mass? *(c)* If **F** = 80 N, what is the tension in the string and the acceleration of each mass?

: **5-50 P** Suppose that you had a rocket with two engines, each of which could provide a constant thrust (force) for a time T. (Real engines are not like this. Their thrust depends on rocket velocity.) For simplicity suppose the mass of fuel exhausted is small compared to the rocket's total mass. How could you achieve the greatest rocket velocity?
(A) Fire both engines simultaneously.
(B) Fire first one engine, then when it had burned out fire the second.
(C) The velocity achieved will not depend on whether or not the engines are fired simultaneously.
(D) This cannot be answered without specifying whether the rocket is to be fired vertically or horizontally.

: **5-51 P** A limp rope of mass 200 gm and length 80 cm lies on a frictionless table. It is slid so that half its length hangs over the edge of the table, and it is then released from rest. What is the velocity of the rope just as it leaves the table?

Supplementary Problems

5-52 P Here's an old question relating to Newton's third law of motion: A horse pulls forward on a cart with a certain force. The cart pulls back on the horse with, according to Newton's third law, exactly the same magnitude of force. How can the cart and the horse ever move?

5-53 P According to L.X. Finegold, *Physics Today* **34,** 15 (July 1981), the ideal weight for a man or woman is very closely specified by W/H^2. where W is the person's mass and H the height. The range of ratios for maximum longevity is between 19 and 24 kg/m² for women and between 20 and 25 kg/m² for men. Try computing W/H^2 for your own weight and height.

5-54 P At lift-off, the three main engines and the two booster rockets of the 4.5×10^6 lb space shuttle, Columbia, produced 6.4×10^6 lb of thrust vertically downward. *(a)* What was the acceleration at lift-off? *(b)* Assuming the acceleration to remain constant, how long did it take for Columbia to rise through its own 184-ft height? *(c)* What was its velocity then?

Newton's Laws II

Having developed Newton's three laws of motion and applied them to simple situations, we can now apply these fundamental principles of classical mechanics to a greater variety of problems. Among the new items to be examined are resistive and frictional forces, forces arising for objects moving in circles, and centrifugal force (and other types of fictitious, or inertial, forces).

6-1   Friction

Force from either friction or resistance opposes an object's motion; it is always opposite to the object's velocity. Friction can arise when one object slides on a second object, and a resistive force comes into play when an object moves through a gas or liquid (Section 6-4). Friction can also be present even when two objects are at rest with respect to each other.

Kinetic Friction   An exact analysis of friction is very complicated because it involves the interaction between protuberances on two surfaces in contact, and ultimately, the forces between atoms and molecules on the surfaces. (For metals in contact, the two objects may actually become welded together at points of contact; the frictional force then depends on making and breaking welds.)

Kinetic friction applies to two surfaces in relative motion. Consider a block sliding along a rough surface, as in Figure 6-1. With the block's velocity to the right (relative to surface), the frictional force of the surface on the block is to the left, opposing the motion. By the same token, the frictional force of block on

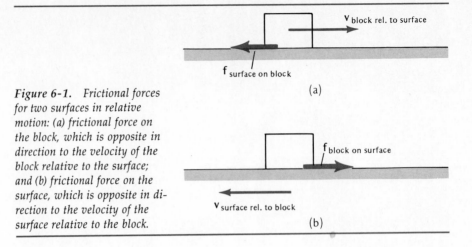

Figure 6-1. *Frictional forces for two surfaces in relative motion: (a) frictional force on the block, which is opposite in direction to the velocity of the block relative to the surface; and (b) frictional force on the surface, which is opposite in direction to the velocity of the surface relative to the block.*

surface is to the right, as in Figure 6-1(b); it is again opposite the velocity, this time the velocity to the left of the surface relative to the block. The two frictional forces are, by Newton's third law, equal in magnitude and opposite in direction, and they act on different objects.

By experiment it is found that for a given pair of surfaces (particular kinds of materials on the two surfaces) the kinetic-frictional force f_k is:

- *Proportional* to the *normal* force N pressing the two surfaces together.
- *Independent* of the *speed* of one surface relative to the other.
- *Independent* of the *area* in contact for a particular object on a particular surface.

These rules are very rough and approximate.*

We introduce the proportionality constant μ_k, the *coefficient of kinetic friction,* so that the magnitude of the kinetic-frictional force f_k can be written

$$f_k = \mu_k N \tag{6-1}$$

Since it is defined as the ratio of two forces, μ_k is a pure number for a given pair of surfaces. The numerical value of μ_k depends on the particular surfaces in contact (their composition, their relative roughness). The rougher the surface the larger the μ_k. Values of μ_k are usually, but not necessarily, less than 1.00. See Table 6-1 for some values.

   Example 6-1. A block is projected at 2.0 m/s to slide on a fairly smooth horizontal surface. The block comes to rest in a distance of 4.0 m. What is the coefficient of kinetic friction between the block and the surface?
   The negative acceleration of the block can be related, through kinematic relations, to the information given on initial and final speeds and distance traveled; the acceleration is also related, through Newton's second law, to the forces on the block. These forces are shown in Figure 6-2: **N** up, mg down, and $\mathbf{f}_k$ opposite in direction to the block velocity **v**.
   Along the vertical we have

---

* They are, frankly, assumed to hold generally just to allow problems involving friction to be solved by simple methods.

$$\Sigma F_y = ma_y$$

$$N - mg = 0$$

or

$$N = mg$$

and along the horizontal

$$\Sigma F_x = ma_x$$

$$-f_k = ma_x$$

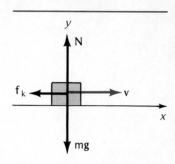

Figure 6-2. Forces f_k, N, and mg on a block sliding on a rough surface. Note that the frictional force $\mathbf{f}_k$ is opposite to the block's velocity $\mathbf{v}$.

The minus sign corresponds to the frictional force opposing the motion. In general $f_k = \mu_k N$, and here, with $N = mg$,

$$f_k = \mu_k(mg)$$

But we have already found that $-f_k = ma_x$. Therefore,

$$a_x = -\mu_k g$$

a constant *negative* acceleration (the block is slowed to rest). Using the kinematic relation for constant acceleration

$$x - x_0 = \frac{v^2 - v_0^2}{2a_x}$$

in the equation above, we have

$$\mu_k = -\frac{a_x}{g} = -\frac{v^2 - v_0^2}{2(x - x_0)g}$$

We are given that $v_0 = 2.0$ m/s, $v = 0$, and $x - x_0 = 4.0$ m, so our final result is

$$\mu_k = -\frac{0 - (2.0 \text{ m/s})^2}{2(4.0 \text{ m})(9.8 \text{ m/s}^2)} = 0.051$$

The small kinetic coefficient of friction indicates that the surfaces in contact were relatively smooth.

Static Friction A frictional force may act on each of two surfaces in contact even if there is no relative motion between them. Then it is a force of *static friction.* Consider a block at rest on a rough horizontal surface, as shown in Figure 6-3. With no horizontal force applied to the block, the static-frictional force $\mathbf{f}_s$ is zero. Now suppose that a small force is applied to the block. The block does *not* move. This means that the block is also subject to a static-fric-

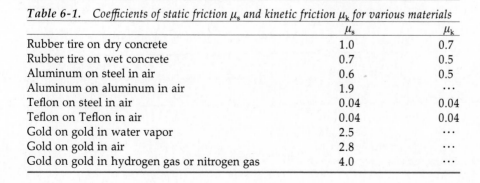

Table 6-1. Coefficients of static friction μ_s and kinetic friction μ_k for various materials

	μ_s	μ_k
Rubber tire on dry concrete	1.0	0.7
Rubber tire on wet concrete	0.7	0.5
Aluminum on steel in air	0.6	0.5
Aluminum on aluminum in air	1.9	...
Teflon on steel in air	0.04	0.04
Teflon on Teflon in air	0.04	0.04
Gold on gold in water vapor	2.5	...
Gold on gold in air	2.8	...
Gold on gold in hydrogen gas or nitrogen gas	4.0	...

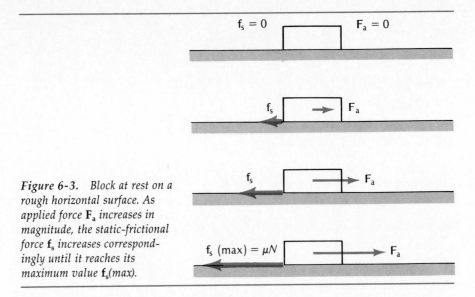

Figure 6-3. Block at rest on a rough horizontal surface. As applied force $\mathbf{F}_a$ increases in magnitude, the static-frictional force $\mathbf{f}_s$ increases correspondingly until it reaches its maximum value $\mathbf{f}_s$(max).

tional force $\mathbf{f}_s$ opposite in direction and equal in magnitude to the applied force $\mathbf{F}_a$. (The forces $\mathbf{f}_s$ and $\mathbf{F}_a$ are not an action-reaction pair of forces; they act on the same object and, in fact, keep it in equilibrium.) If the magnitude of $\mathbf{F}_a$ increases and the block still does not move, $\mathbf{f}_s$ must increase correspondingly. Finally the block does move. The maximum value of the static force of friction is proportional, for a given pair of surfaces, to the normal force N pressing the two surfaces together. We may write

$$f_s(\text{max}) = \mu_s N$$

where μ_s is the coefficient of static friction, a dimensionless number characteristic of the two surfaces in contact. See Table 6-2. Again, the rougher the surface, the larger the μ_s. Since f_s may assume any value up to its maximum, we can write more generally

$$f_s \le \mu_s N \qquad \text{with } f_s(\text{max}) = \mu_s N \qquad (6\text{-}2)$$

Typically, $\mu_s > \mu_k$ for a given pair of surfaces; it takes a larger applied force to set an object in motion than to keep it in motion at a constant speed. See Table 6-1.

Example 6-2.   A block sits on a rough incline whose angle θ with the horizontal can be varied. A trial shows that the block does not begin sliding down the incline until

Table 6-2.   Coefficients of static friction μ_s for steel on steel

CONDITION	μ_s
Degassed in vacuum at high temperature	(Welds on contact)
Grease-free in vacuum	0.78
Grease-free in air	0.39
Clean and coated with light mineral oil	0.23
Clean and coated with castor oil	0.15
Clean and coated with stearic acid	0.005 to 0.013

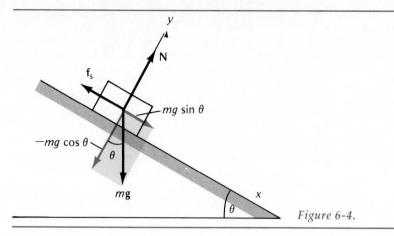

Figure 6-4.

$\theta = 30°$. What is the coefficient of static friction between the block and the surface of the incline?

Until the block begins sliding, it remains in static equilibrium. See Figure 6-4. Therefore, the resultant force on block is zero, and so are the x and y components of the resultant force.

The forces are the ones that arise for a smooth inclined surface (Example 5-6, Figure 5-15). There is, however, an additional static-frictional force $\mathbf{f}_s$ opposing the block's motion; it is up the incline, as shown in Figure 6-4. As before, the x and y axes are chosen parallel and perpendicular, respectively, to the incline. The weight of the block is replaced by its x component $mg \sin \theta$ and its y component $-mg \cos \theta$.

Newton's second law applied in turn to the x and y components of the forces gives

$$\Sigma F_x = ma_x$$

$$-f_s(\text{max}) + mg \sin \theta = 0$$

since $a_x = 0$, and

$$\Sigma F_y = ma_y$$

$$N - mg \cos \theta = 0$$

since $a_y = 0$ also.

Solving for $f_s(\text{max})$ and N, we have

$$f_s(\text{max}) = mg \sin \theta$$

and

$$N = mg \cos \theta$$

Angle θ corresponds to the block's just starting to slide, so the static-frictional force has its maximum value and

$$\mu_s = \frac{f_s(\text{max})}{N} = \frac{mg \sin \theta}{mg \cos \theta}$$

$$= \tan \theta = \tan 30° = 0.58$$

Note that the block's mass did not enter into the computation.

Example 6-3. A sports car with *front-wheel drive* can accelerate from rest in first gear at 5.7 m/s². (See Example 3-2 and Figure 3-10.) We suppose that 65 percent of the

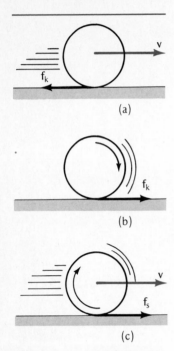

(a)

(b)

(c)

Figure 6-5. *(a) Brakes locked and the car skids; kinetic friction acts between the wheel and roadway. (b) Wheels spin and car is motionless; kinetic friction acts between wheel and roadway. (c) Wheels roll without slipping; static-frictional force acts on the wheels from the roadway in the* same *direction as vehicle velocity* **v**.

car's weight is on the front wheels. What is the minimum coefficient of *static* friction between tires and roadway to produce this acceleration?

We must be clear on when the frictional force between tires and roadway is static and when it is kinetic. Suppose first that an automobile is in motion with its wheels locked; then the tires skid on the surface and the opposing frictional force is *kinetic* since the tires are in motion relative to the roadway at the point of contact. See Figure 6-5(a). We also have a *kinetic*-frictional force when the wheels spin but the car remains at rest. See Figure 6-5(b). Here again the tire is in motion relative to the roadway at the point of contact.

When the wheels roll without slipping, a *static*-frictional force acts on the tires *in the direction* of the car's acceleration. See Figure 6-5(c). The point of contact between tires and roadway advances at the same velocity as the car, and a tire is always momentarily *at rest* relative to the roadway at the point of contact.

To verify this, suppose that the car's velocity relative to the roadway is v. Then, relative to an observer in the car, the velocity of the roadway is $-v$; but this observer looking down at a rotating, rolling tire sees that the point of the tire in contact with the roadway also moves back at each instant with a velocity $-v$. Both roadway and point of contact on the tire have the *same* velocity $-v$ relative to an observer traveling with the car; therefore, relative to an observer at rest on the ground, these two objects also have the *same velocity*, but now its value is *zero*.

The maximum static-frictional force always exceeds the kinetic-frictional force. Therefore, the frictional force on the car tires is greater when the wheels are rolling than when they are spinning or locked.

Consider now the forces on the sports car. See Figure 6-6. The force accelerating the car is the static-frictional force $\mathbf{f}_s$. It acts at the front wheels in the direction of the car's forward acceleration. The other forces on the two front wheels are 65 percent of the car's weight ($0.65mg$) and the upward normal force N on the front wheels.

From Newton's second law,

$$\Sigma F_x = ma_x$$

$$f_s = ma_x$$

$$\Sigma F_y = ma_y$$

$$N - 0.65mg = 0$$

For the car to accelerate at the maximum rate without slipping the tires, the static-frictional force has its maximum value

$$f_s(\text{max}) = \mu_s N = \mu_s(0.65mg)$$

Figure 6-6.

with $f_s(\text{max}) = ma_x$, from above, we have

$$\mu_s(0.65mg) = ma_x$$

or

$$\mu_s = \frac{a_x}{0.65g} = \frac{5.7 \text{ m/s}^2}{0.65(9.8 \text{ m/s}^2)} = 0.89$$

The required minimum static coefficient of friction between tire rubber and roadway is 0.89. For such a tire, it takes almost as much horizontal force to start dragging the tire along the roadway as to lift it.

6-2   Uniform Circular Motion

Whenever a particle is accelerated, there is a resultant force on the particle acting in the direction of its acceleration. If the resultant force is parallel to the particle's velocity, its speed changes but not its direction of motion. If the resultant force is perpendicular to a particle's velocity, the direction of motion changes but not the speed. The particle is again accelerated.

See Figure 6-7, where resultant force **F** is perpendicular to a particle's initial velocity $\mathbf{v}_0$. The direction of the particle's acceleration $\Delta\mathbf{v}/\Delta t$ over the short time interval Δt is the same as the direction for **F**. Therefore, the velocity change $\Delta\mathbf{v}$ is perpendicular to $\mathbf{v}_0$. At a time Δt later, the velocity, $\mathbf{v} = \mathbf{v}_0 + \Delta\mathbf{v}$, has the *same magnitude* as $\mathbf{v}_0$ but a *different direction*. So if a particle is to continue to move with constant speed in a circle (uniform circular motion), there must be a resultant radial force F_r of constant magnitude perpendicular to the particle's velocity and always pointing toward the center of the circle. See Figure 6-8(a). If the particle's speed is changing, there is also a force component F_t tangent to the particle's velocity. See Figure 6-8(b).

The kinematics of uniform circular motion was treated in Section 4-4. We have a particle moving at constant speed v in a circle of radius r. The radius vector **r** from the center of the circle to the location of the particle then sweeps through angles at angular speed ω (in radians per unit time), and the time for one complete rotation is the period T. Under these circumstances, the particle has a radially inward (centripetal) acceleration $\mathbf{a}_r$, whose magnitude is written in three different forms:

$$a_r = \frac{v^2}{r} = \omega^2 r = \frac{4\pi^2 r}{T^2} \tag{4-18}$$

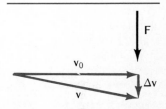

Figure 6-7.   Force **F** applied perpendicular to a particle's velocity $\mathbf{v}_0$ changes the velocity direction but not the speed. $\mathbf{v} = \mathbf{v}_0 + \Delta\mathbf{v}$.

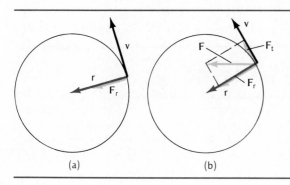

(a) (b)

Figure 6-8. (a) Radially inward force **F**$_r$ perpendicular to **v** produces uniform circular motion; (b) with a force component **F**$_t$ tangent to the particle velocity **v**, the velocity changes magnitude as well as direction.

"I Was in the Prime of My Life for Invention. . . ."

That's how Isaac Newton (1642–1727) later described the year and a half at age 23 that he spent at his home, close to the legendary Robin Hood's Sherwood Forest; he had to leave Cambridge University because it was shut down by the "Great Plague," an epidemic that killed nearly 20 percent of England's population. To say that he was in his "prime . . . for invention" was one of the great understatements of all time. During that brief period, Newton laid the foundations of several of his most important scientific discoveries: gravity, the laws of mechanics, and the nature of light. It was a creative burst almost unparalleled in the history of science.* In mathematics, Newton was responsible for calculus, introducing the general binomial expansion, and polar coordinates; he also made significant advances in infinite series, algebra, number theory, classical and analytical geometry, finite differences, the classification of curves, mathematical approximations, and probability.

Alexander Pope said it well:

Nature and Nature's laws lay hid in night;
God said, Let Newton be! and all was light.

*Albert Einstein, at age 26 in 1905, while working as a patent examiner in Berne, Switzerland, wrote fundamental papers on the special theory of relativity, the quantum theory of the photoelectric effect, Brownian motion, and statistical mechanics.

How did Newton view his own work? He wrote: "I do not know what I may appear to the world, but to myself I seem to have been only a boy playing on the sea-shore, and diverting myself in now and then finding a smoother pebble or a prettier shell than ordinary, whilst the great ocean of truth lay all undiscovered before me." Or, consider what he once wrote to Robert Hooke (Hooke's law, $F = -kx$) about his indebtedness to his scientific predecessors: "If I have seen further, it is by standing on the shoulders of Giants."

Actually, Newton spent years fighting Hooke over questions of priority. Hooke claimed that he first had the idea of an inverse-square force; he also disputed Newton's theory of colors. Newton simply could not accept criticism of any sort; the most trifling hint of disagreement could set him into a rage. In addition to Hooke, Newton carried on running battles with a number of prominent contemporaries, including the astronomer Edmond Halley (Halley's comet), the philosopher John Locke, the diarist Samuel Pepys, and the Astronomer Royal John Flamsteed. But Newton's most celebrated adversary was Gottfried Wilhelm von Leibnitz, the German mathematician and philosopher. It is now agreed that Newton and Leibnitz each invented calculus independently, but Newton fought Leibnitz for 25 years, even after Leibnitz had died, on who deserved credit for calculus.

Then, from Newton's second law, the resultant radially inward force must have the constant magnitude ΣF_r, where

$$\Sigma F_r = ma_r$$

$$\Sigma F_r = \frac{mv^2}{r} = m\omega^2 r = m\left(\frac{4\pi^2 r}{T^2}\right) \tag{6-3}$$

for a particle moving in a circle at *constant* speed. (If the speed of a particle moving in a circular arc changes, then in addition to a radial component of the resultant force, there must also be a force component tangent to the circle and parallel (or antiparallel) to the particle's velocity.

As always, it is essential that we apply Newton's second law from the point of view of an observer in an *inertial frame*. This deserves special emphasis when one deals with objects moving in circles. Suppose, for example, that we have a man in an automobile traveling in a circle. Then we must view his motion as an observer fixed in the inertial frame of the earth (or one moving with constant velocity relative to it). We cannot properly apply Newton's

Newton was the quintessential theoretical physicist. His arguments were elegant and compelling, his highly geometrical mathematics was neat. But Newton was also an experimenter. At age 18, he studied the refraction of light by prisms in his dormitory room in college. He observed the interference phenomenon we now call Newton's rings (but did not interpret the effect correctly). He invented the reflecting telescope. He studied diffraction fringes, and yet steadfastly defended a particle theory of light.

After years of polishing, Newton at age 44 published his grand opus, *Philosophiae Naturalis Principia Mathematica* (mathematical principles of natural philosophy, an earlier name for what we now call physics); the *Principia* set forth in elaborate detail the foundations and applications of mechanics and universal gravitation. His other great work, *Optiks,* was published still later, nearly 30 years after it had been written.

Then Newton effectively quit physics. He turned to interests that had also occupied him for many years. One was alchemy, the precurser of chemistry, one of whose aims was that of converting lead to gold. Newton had a direct but bizarre way of doing qualitative analysis in his alchemy experiments—he simply tasted the reagents. Another special enthusiasm was the interpretation of Biblical prophesy, especially in the books of *Daniel* and *Revelation.* The writings in these areas equal in length those in physics.

Newton was appointed warden, and later master, of The Mint. The pay was good; his obligations were minimal. But Newton took the job seriously. He pursued counterfeiters relentlessly and saw at least 19 of them hanged. To cut down on the chiselling of gold coins, he invented the milled edge.

In middle age, Newton became increasingly eccentric, reclusive, combative, even irrational. He had a couple of serious nervous breakdowns. His associates wondered—the question dared hardly be whispered about one of the towering intellects of all time—whether the great master might actually be losing his mind.

There is good reason to believe that Newton actually suffered from mercury poisoning, a common affliction of hat makers, who used mercury to cure the felt for hats ("mad as a hatter"). The evidence comes from the recent analysis of just four hairs from Newton's head, in which the concentration of mercury was found to be nearly 40 times normal, quite enough to produce maddening effects.*

Newton was the most celebrated person of his age. Lionized by his contemporaries, knighted by the queen (a first for a scientist), he ended up with a good spot in Westminster Abbey. Whatever the quirks in his temperament and personality, he had—through his extraordinary accomplishments—indeed seen the vision glorious.

*A summary of the work by P. E. Spargo and C. A. Pounds is given in *Science,* vol. 213, 18 September, 1981.

---

second law from the point of view of the rider, as is perhaps psychologically more appealing. Such an observer is in an accelerated, and therefore noninertial, frame; the law of inertia does not hold in such a reference frame.

In Section 6-3, we consider what happens when an observer in a noninertial, rotating reference frame tries, nevertheless, to apply Newton's second law. As shown there, a strictly fictitious centrifugal force must then be introduced. We cannot emphasize too strongly that when Newton's laws are applied in an inertial frame, as we shall do in all examples in this section, a *centrifugal force simply does not exist.*

The resultant force acting on a body in uniform circular motion is sometimes referred to as the *centripetal force.* This designation can be mischievous, since the centripetal force is not a distinctive type of force. An object may move in a circle under the influence of its weight, the tension of a cord, a friction force, a normal force, or a combination of forces. If the resultant of all forces acting on the object produces circular motion, this resultant is *the* centripetal force.

A general procedure was suggested in Section 5-7 for solving problems involving Newton's second law. In dealing with an object moving in a circle, these additional considerations apply.

- Be sure to identify clearly the circle in which the particle moves and the location of its center. The resultant force on the particle must always point towards this center.
- Resolve forces into components that are (a) in the direction of and (b) perpendicular to the line joining the particle to the center of the circle.

Example 6-4.   A small mass of 1.0 kg is attached to the lower end of a string 1.0 m long whose upper end is fixed. The mass moves in a horizontal circle while the string maintains a constant angle of 30° with respect to the vertical. (The string sweeps out a cone with an angle of 30°, and the arrangement is known as a *conical pendulum.*) What is the period of the motion and the tension of the string?

The mass moves in a horizontal circle with a radius of $L \sin \theta$. See Figure 6-9. The only forces on the mass are its weight mg and the tension $\mathbf{T}$ of the string. (There is *no* outward force!) We resolve the two forces into horizontal and vertical components. We take the horizontal component to be positive when it is directed radially inward, toward the center of the circle and in the direction of acceleration.

Applying Newton's second law to the vertical components gives

$$\Sigma F_y = ma_y$$

$$T \cos \theta - mg = 0$$

where we have used the fact that the acceleration a_y along the vertical is zero.

The tension then is

$$T = \frac{mg}{\cos \theta} = \frac{(1.0 \text{ kg})(9.8 \text{ m/s}^2)}{0.87} = 11 \text{ N}$$

Applying Newton's second law now to the horizontal force components gives

$$\Sigma F_r = ma_r$$

$$T \sin \theta = ma_r$$

We can write the acceleration as

$$a_r = \frac{4\pi^2 r}{P^2} = \frac{4\pi^2 L \sin \theta}{P^2}$$

Here we use P to represent the period (to avoid confusion with tension T). The radius r of the circle has been identified as $L \sin \theta$. Using this result in the equation above it gives

$$T \sin \theta = m \frac{4\pi^2 L \sin \theta}{P^2}$$

We use the earlier equation to eliminate T, and then solve for P:

$$P = 2\pi \sqrt{\frac{L \cos \theta}{g}} = 2\pi \sqrt{\frac{(1.0 \text{ m})(0.87)}{9.8 \text{ m/s}^2}} = 1.9 \text{ s}$$

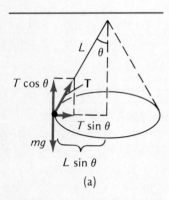

$L \sin \theta$

(a)

Figure 6-9.   Forces on a particle of a conical pendulum.

Example 6-5.   An automobile rounds a curve of radius R on a flat roadway. The coefficient of static friction between the car's tires and the roadway is μ_s. (a) What maximum speed can the car attain and still avoid slipping? (b) What happens when the car reaches an icy stretch of roadway?

(a) Figure 6-10(a) shows the forces acting on the car: its weight $m\mathbf{g}$, the upward

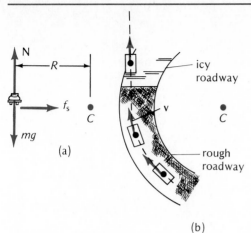

(a)

icy roadway

C

rough roadway

(b)   *Figure 6-10.*

normal force **N**, and a static-frictional force $\mathbf{f}_s$ pointing to the center C of the circle. We can be sure that the radially inward frictional force $\mathbf{f}_s$ is *static* friction because the car maintains's a *constant* distance R from the circle's center.

Applying Newton's second law in turn to the radial and vertical forces gives

$$\Sigma F_r = ma_r$$

$$f_s = \frac{mv^2}{R}$$

$$\Sigma F_y = ma_y$$

$$N - mg = 0$$

The static-frictional force has a maximum value

$$f_s(\text{max}) = \mu_s N = \mu_s mg$$

Using this result in the first (radial) equation, we have for the maximum speed v_{max} without slipping

$$\mu_s mg = \frac{mv_{\text{max}}^2}{R}$$

$$v_{\text{max}} = \sqrt{\mu_s gR}$$

(b) See Figure 6-10(b), where the car's motion is viewed from above. On a perfectly smooth road surface there can be no static-frictional force accelerating the car in a circle. The resultant force in the plane of the roadway is zero, so that the car flies off the circular path along a tangent.

Example 6-6.   Identify the forces acting on the chosen object for the following circumstances: (a) an automobile traveling at constant speed on a banked curved roadway (without friction); (b) a small object at rest relative to the rough horizontal surface of a rotating platform; (c) a car at the highest point in a loop-the-loop; (d) an automobile at the crest of a hill; (e) a satellite orbiting the earth; (f) a person in a spinning cylinder on an amusement-park ride; and (g) an astronaut experiencing artificial gravity in a spinning doughnut-shaped spacecraft.

The forces, identified by their customary symbols, are shown in Figure 6-11. In each instance, the resultant force acts toward the center of the circle in which the object travels.

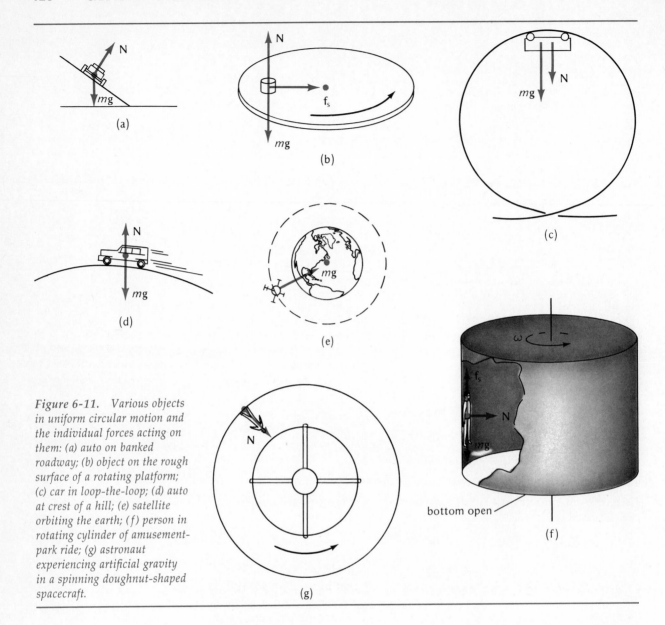

Figure 6-11. Various objects in uniform circular motion and the individual forces acting on them: (a) auto on banked roadway; (b) object on the rough surface of a rotating platform; (c) car in loop-the-loop; (d) auto at crest of a hill; (e) satellite orbiting the earth; (f) person in rotating cylinder of amusement-park ride; (g) astronaut experiencing artificial gravity in a spinning doughnut-shaped spacecraft.

6-3   Centrifugal Force and Other Inertial Forces (Optional)

To apply Newton's second law in its customary formulation, an observer must be in an *inertial frame*, a reference frame in which the law of inertia, or Newton's first law of motion, holds. An observer can easily make a preliminary test of whether he is indeed in an inertial frame; he sees whether an isolated object maintains constant velocity. If it does, he can use Newton's second law with impunity. In this section we see what happens if an observer in an accelerated, and therefore noninertial, frame tries to apply Newton's second law.

We first consider a rotating reference frame, a platform or a merry-go-round rotating uniformly at the constant angular velocity ω relative to an inertial frame. Suppose that a disk of mass m is attached by a spring to a point on the rotating platform. The disk also rotates at angular velocity ω. How does an inertial observer I (who is standing outside the rotating platform) describe the motion of the disk?

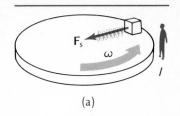

(a)

He sees the disk executing circular motion at constant speed under the action of the spring force $\mathbf{F_s}$ toward the center of the circle. See Figure 6-12(a).

How do things look to a noninertial observer *non-I*, one who sits on the rotating platform and rotates with it? Suppose that this observer applies, or tries to apply, Newton's second law to the disk's motion. *Non-I* sees the disk at rest; for him, it is in equilibrium. *Non-I* sees the spring stretched and so he knows that an inward force $\mathbf{F_s}$ acts on the disk. There must then, says *non-I*, be another force $\mathbf{F_i}$ of the same magnitude acting on the disk in the radially *outward* direction. See Figure 6-12(b). *The noninertial observer invents the inertial force $\mathbf{F_i}$ to preserve Newton's second law in his accelerating reference frame.* The inertial force, sometimes called a pseudo force, is fictitious in the sense that it does not have its origins in the interaction of the disk with other objects. It arises simply because *non-I* insists on applying Newton's laws under circumstances in which they are really not valid.

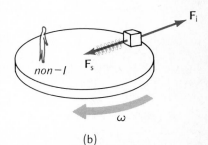

(b)

Figure 6-12. (a) As observed by inertial observer I, the disk is subject only to an inward spring force $\mathbf{F_s}$; (b) as observed by noninertial observer non-I on the rotating platform, the disk is subject to an inward spring force $\mathbf{F_s}$ and an outward inertial force $\mathbf{F_i}$.

Each of the two observers applies Newton's second law to the motion, as he sees it, of the same disk:

$$\Sigma \mathbf{F} = m\mathbf{a}$$

Inertial observer: $\mathbf{F_s} = -m\omega^2\mathbf{r}$

(The negative sign simply indicates that the acceleration is *inward*, in a direction opposite that of radius vector $\mathbf{r}$.)

Noninertial observer: $\mathbf{F_s} + \mathbf{F_i} = 0$

The spring force $\mathbf{F_s}$ has the same magnitude (same length) for I and *non-I*. Therefore, from the equations above

$$\mathbf{F_i} = m\omega^2\mathbf{r} \tag{6-4}$$

The *radially outward inertial force* has a magnitude $m(v^2/r) = m(\omega^2 r) = m(4\pi^2 r/T^2)$, and it exists only for an observer at rest in a rotating reference frame. It is called the *centrifugal force*. A rotating observer sees an object at rest in this rotating frame subject to a centrifugal force, in addition to other "real" forces that may act on the object.

A centrifugal force arises only in a rotating reference frame. It is the fictitious force that exists, for example, for an occupant in an automobile rounding a curve; the centrifugal force pushes him outward from the center of the circle. The centrifugal force also exists for an astronaut circling the earth in a spaceship. From *his* point of view, every object in the spaceship is subject to two forces: the inward pull of the earth and an equal centrifugal force away from the earth. The net force on any object is, in *his* view, zero. Unattached objects float *relative to the astronaut*, who may conclude that all objects in the spaceship *are weightless*. On the other hand, an inertial observer says that the objects in the spaceship *appear* to be weightless because all such objects have the *same* acceleration.

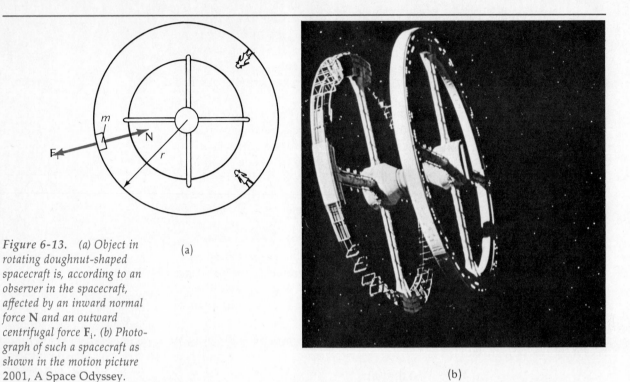

(a)

(b)

*Figure 6-13.   (a) Object in rotating doughnut-shaped spacecraft is, according to an observer in the spacecraft, affected by an inward normal force **N** and an outward centrifugal force $\mathbf{F}_i$. (b) Photograph of such a spacecraft as shown in the motion picture 2001, A Space Odyssey.*

Example 6-7.   Artificial gravity equivalent to the gravity on the earth's surface ($g = 9.8$ m/s²) is to be created by spinning a doughnut-shaped spacecraft (with an outer radius of 10 m) about its symmetry axis. What is the required period of rotation?

We examine the motion of an object of mass m sitting on the "floor" (the outer circumference) from the point of view of an observer fixed to the rotating doughnut. See Figure 6-13(a). The object is in equilibrium under the action of two forces: the radially inward normal force **N** of the floor on the object and the radially outward centrifugal force, whose magnitude may be written $m(4\pi^2 r/T^2)$, where T is the period of rotation. See (6-4). If gravity on the earth's surface is to be simulated, the normal force on the object must be equal in magnitude to the object's weight mg. Therefore,

$$N = mg = m\,\frac{4\pi^2 r}{T^2}$$

$$T = 2\pi\sqrt{\frac{r}{g}} = 2\pi\sqrt{\frac{10 \text{ m}}{9.8 \text{ m/s}^2}} = 6.3 \text{ s}$$

What is the behavior of an apple dropped from rest in the spacecraft? From the point of view of an observer in the spacecraft, when the apple is released from rest, it is subject to an outward force of magnitude mg and "falls" outward. It is interesting to examine the motion of the dropped apple from the point of view of an observer in an inertial frame. The inertial observer sees the spacecraft turning and the apple subject to a single inward force *before* it is released. Once released, the apple is *free* of all forces; it flies off at a tangent from the circle in which it was initially traveling, and continues motion in a straight line at constant speed. See Figure 6-14(a). The apple hits the floor again, essentially at the point "beneath" its release point because of the rotation of the

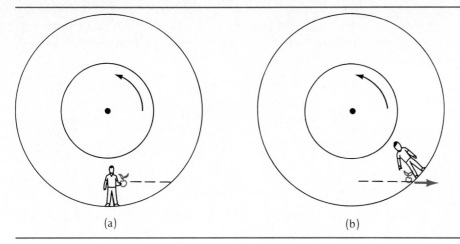

Figure 6-14. As viewed by an inertial observer, an apple is released from rest in the rotating spacecraft. The apple (a) flies off at a tangent in a straight line and (b) hits the "floor" below the point of release.

spacecraft. See Figure 6-14(b). The noninertial observer sees the apple subject to a centrifugal force and falling; the inertial observer sees the apple in free flight.

A rotating reference frame is one special kind of accelerated reference frame. A reference frame that accelerates in a straight line relative to an inertial frame is also a noninertial frame, and a noninertial observer *non-I* in such a reference frame must also invoke a fictitious inertial force to apply Newton's second law of motion.

Suppose that our disk attached to a spring is now on a smooth platform on a truck that accelerates to the right, as shown in Figure 6-15(a). An inertial observer *I* sees the disk in motion with an acceleration $\mathbf{a}_i$ to the right under the influence of the spring force $\mathbf{F}_s$. A noninertial observer *non-I* traveling with the

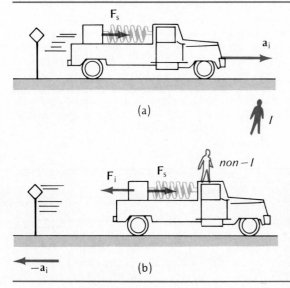

Figure 6-15. Disk attached to a stretched spring on the platform of an accelerated reference frame (truck) as seen by (a) inertial observer I and (b) noninertial observer non-I.

truck sees the disk at rest; *non-I* says that the disk, besides being acted on by the real spring force $\mathbf{F}_s$ to the right, must be acted on also by an inertial force $\mathbf{F}_i$ of the same magnitude to the left. See Figure 6-15(b). Each observer applies Newton's second law to the same disk as follows:

$$\Sigma\mathbf{F} = m\mathbf{a}$$

Inertial observer: $\mathbf{F}_s = m\mathbf{a}_i$

Noninertial observer: $\mathbf{F}_s + \mathbf{F}_i = 0$

Solving for $\mathbf{F}_i$ in the two equations, we see that the inertial force $\mathbf{F}_i$ must be given by

$$\mathbf{F}_i = -m\mathbf{a}_i$$

This inertial force, like the centrifugal force, has a direction opposite the object's acceleration in the inertial frame.

If a noninertial observer is to apply Newton's second law, he must include, in addition to any real forces that act on the body, the inertial force $\mathbf{F}_i = -m\mathbf{a}_i$, where m is the body's mass and $\mathbf{a}_i$ is the acceleration of the noninertial reference system relative to an inertial frame. This allows us to write Newton's second law in its most general form, applicable to any observer whether inertial or noninertial:

$$\Sigma\mathbf{F} + \mathbf{F}_i = m\mathbf{a} \qquad \text{where } \mathbf{F}_i = -m\mathbf{a}_i \qquad (6\text{-}5)$$

Example 6-8.   A small mass m is attached to the lower end of a string whose upper end is attached to the ceiling of a train traveling with an acceleration $\mathbf{a}_i$ relative to the earth. (a) What is the angle between the string and the vertical when the mass is at rest with respect to the train? (b) What is the effective acceleration due to gravity $\mathbf{g}$ in the accelerating train?

We examine the forces acting on m as a noninertial observer traveling with the train. The mass is then in equilibrium under the action of *three* forces: its weight $m\mathbf{g}$, the tension $\mathbf{T}$, and an inertial force $m\mathbf{a}_i$ toward the rear of the train. See Figure 6-16(a).

The horizontal and vertical components separately add to zero.

$$T \sin\theta = ma_i$$

$$T \cos\theta = mg$$

$$\therefore \tan\theta = \frac{a_i}{g}$$

(b) If we think of the mass at the end of the string as a plumb bob, then the direction of the string gives the direction "down." See Figure 6-16(b). The effective weight $m\mathbf{g}'$ now balances the tension $\mathbf{T}$. Comparing Figures 6-16(a) and (b), we see that

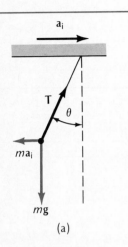

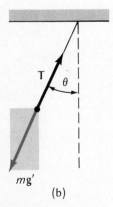

Figure 6-16.   *(a) Forces acting on a plumb bob in an accelerated reference frame. (b) Effective weight $m\mathbf{g}'$ in the accelerated reference frame.*

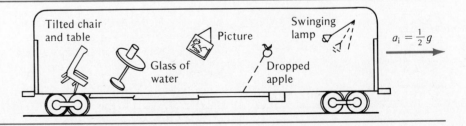

Figure 6-17.   *In a train with constant acceleration $(= g/2)$, the effective direction for "down" is at an angle relative to the vertical.*

$$g' = \sqrt{g^2 + a_i^2}$$

For an observer in the accelerating train, the direction "down" is the direction of **g'**. See Figure 6-17.

6-4   Resistive Force Proportional to Speed (Optional)

A smooth, dense, streamlined object that slips easily through a gas or a liquid without disturbing the viscous fluid appreciably is subject to a resistive, or damping, force directly proportional to its velocity—a *linear* resistive force. An example is a billiard ball falling through molasses. Other objects such as a sky-diver or a shuttlecock falling through air produce turbulence in the fluid, and the magnitude of the resistive force is proportional to the square of the speed.

It is most convenient, for reasons soon to be evident, to write the linear resistive force in the form

$$\mathbf{F_R} = -km\mathbf{v} \qquad (6\text{-}6)$$

where m is the mass of the object moving through the fluid, **v** is its velocity, and k is a constant. The minus sign implies that the resistive force is opposite in direction to the object's velocity.

We want to study the motion of an object dropped from rest and acted on by a linear velocity-dependent resistive force. In particular, we want to find how the velocity and acceleration of such an object varies with time. We can easily arrive at the principal qualitative features of the motion.

Immediately after the object is released from rest, the resistive force is zero ($F_R = 0$ when $v = 0$). Then the object, under the influence of its weight alone, has downward acceleration g. See Figure 6-18(a). As the falling object acquires a downward velocity, the upward resistive force increases in magnitude. The net force down is then smaller than the object's weight, and its acceleration is less than g; therefore, the downward velocity increases at a lower rate. See Figure 6-18(b). Finally, at a still higher speed, the upward resistive force becomes equal in magnitude to the object's weight. Then the resultant force is zero, the acceleration is zero, and the object thereafter falls downward at a *constant terminal velocity* $\mathbf{v_T}$. In summary, the velocity grows from zero to a final constant value v_T, while the acceleration drops from g to zero. See Figure 6-19.

The terminal velocity v_T can easily be related to the other parameters; we simply recognize that when $v = v_T$, the downward weight mg is balanced by the upward resistive force of magnitude kmv_T:

$$mg = kmv_T$$

or

$$v_T = \frac{g}{k} \qquad (6\text{-}7)$$

In applying Newton's second law to get an analytical solution of the motion, it is most convenient to take downward quantities as positive. We then have

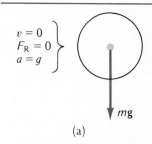

(a)

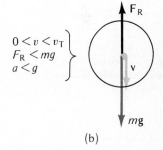

(b)

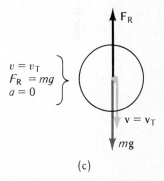

(c)

Figure 6-18.   Object dropped from rest and also subject to an upward linear resistive force F_R. (a) Immediately after release from rest, $v = 0$ so that $F_R = 0$ and $a = g$. (b) Later, with $F_R < mg$, we have $a < g$. (c) Still later, with $F_R = mg$, the object falls at the constant terminal velocity v_T.

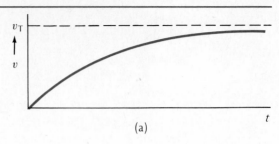

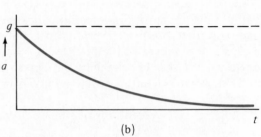

Figure 6-19. *Object dropped from rest under the influence of gravity and a linear resistive force. (a) Velocity-time graph: v grows from 0 to the terminal velocity v_T. (b) Acceleration-time graph: a drops from g to 0.*

$$\Sigma F_y = ma_y$$

$$mg - kmv = m\frac{dv}{dt} \tag{6-8}$$

where $a_y = dv/dt$. Removing the common factor m from all terms in the equation above, using (6-7), and rearranging, we have

$$g\left(1 - \frac{v}{v_T}\right) = \frac{dv}{dt} \tag{6-9}$$

Separating the variables v and t on the two sides of the equation gives

$$\int_0^t g\, dt = \int_0^v \frac{dv}{1 - v/v_T}$$

The time variable is integrated from an initial value zero to any later time t, and the velocity from its initial zero value to the later value v.

If you carry out the integration in the equation above, you will get for the velocity as a function of time

$$v = v_T(1 - e^{-kt}) \tag{6-10}$$

Check that (6-10) is indeed a solution of (6-9) by substituting relations for v and dv/dt in (6-9).

Using (6-10) in (6-8), we get for the acceleration dv/dt

$$\frac{dv}{dt} = g - kv = g - kv_T(1 - e^{-kt})$$

By (6-7), this reduces to

$$a_y = \frac{dv}{dt} = ge^{-kt} \tag{6-11}$$

Figure 6-19(a) is a plot of (6-10), and Figure 6-19(b) of (6-11). The velocity grows exponentially from zero to the final value v_T; the acceleration decays exponentially from g to zero. The rapidity of the growth and decay is determined by the factor e^{-kt}; when k is large, the terminal velocity is achieved in a short time. But we see from (6-6) that a large k implies a large resistive force for a given mass and velocity.

Summary

Important Results

Force component parallel to object's velocity **v** changes magnitude of **v**; perpendicular force component changes direction of **v**.

The kinetic-frictional force between the surfaces of two objects moving relative to one another is proportional to normal force

$$f_k = \mu_k N \qquad (6\text{-}1)$$

The static-frictional force between two objects at rest with respect to one another is always less than the maximum static-frictional force that is proportional to normal force:

$$f_s \leq f_s(\text{max}) = \mu_s N \qquad (6\text{-}2)$$

When an object rolls without slipping, a *static*-frictional force acts at the point of contact between the object and the surface.

Problems and Questions

Section 6-1   Friction

· **6-1 Q** Can a friction coefficient exceed 1.0? Give an example.

· **6-2 Q** You wish to push a block along a rough floor. Why is it easier to move the block by pushing on its back side than by pressing down and pushing on the top of the block?

· **6-3 Q** A block slides up a rough ramp, comes to rest momentarily at the top, and then slides down. Why is the magnitude of the block's acceleration up the ramp greater than that down the ramp?

: **6-4 Q** Why are motorists traveling on slippery roads advised to avoid skidding by applying the brakes *slowly*?

: **6-5 Q** Why should a motorist starting from rest on an icy roadway accelerate gently, and perhaps start in second or third gear?

: **6-6 Q** How could you find, at least approximately, the coefficient of kinetic friction between your shoes and a fairly smooth floor?

: **6-7 Q** A crate is on the bed of a flat-bed truck. (*a*) The driver starts the truck from rest suddenly and stops it gradually. What happens to the crate? (*b*) The driver starts the truck from rest gradually and stops it suddenly. What happens to the crate? (The crate cannot tip.)

: **6-8 Q** Crawling typically consists of pushing limbs outward *slowly* and retracting them *rapidly*. Why?

· **6-9 P** A crate of 50 kg mass rests on a rough floor. The coefficient of static friction between the crate and the floor is 0.40. What is the force of *static* friction on the crate when a person pushes horizontally on the crate with a force of (*a*) 50 N? (*b*) 100 N? (*c*) 150 N? (*d*) 200 N?

· **6-10 P** A crate of 50 kg mass is sliding east at a speed of 2.0 m/s on a rough floor. The coefficient of kinetic friction between the crate and the floor is 0.30. What is the magnitude and direction of the force of kinetic friction on the floor?

· **6-11 P** As inclines go, a grade of 3.5% (a rise of 3.5 m for every 100 m along the horizontal) is relatively steep for a railroad. What is the minimum coefficient of static friction between a train's wheels and the track if the train is to climb a 3.5% grade at constant speed?

· **6-12 P** See Figure 6-20. If $m_2 = 0$, m_1 will, once started, slide down the incline at constant speed. What value of m_2

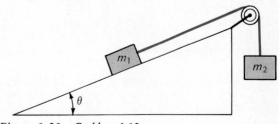

Figure 6-20.   Problem 6-12.

will allow m_1 to slide *up* the incline at constant speed? (The pulley is massless and frictionless; the cord is massless and inextensible.)

· **6-13 P** A block is projected at an initial speed of 2.0 m/s to slide on a rough floor. The coefficient of kinetic friction between the floor and the block is 0.30. How far does the block slide before coming to rest?

: **6-14 P** In order to keep a 5.0-kg block from sliding down a rough wall, a person must press it horizontally against the wall with a force of at least 140 N. What is the coefficient of static friction between the block and wall?

: **6-15 P** A block is projected with an initial speed v_0 up a rough plane inclined at angle θ relative to the horizontal. The block travels a distance d up the incline and comes to rest. What is (*a*) the kinetic coefficient of friction between the block and the surface of the incline, and (*b*) the minimum value of the static coefficient of friction?

: **6-16 P** See Figure 6-21. A rope is tied to a 40-kg box on a rough horizontal surface. The rope maintains a constant angle of 30° with the horizontal as the man pulls on the rope. It takes a force of 120 N to start the box sliding and a force of 100 N to keep it sliding at constant speed. What are (*a*) the static and (*b*) the kinetic coefficients of friction between the box and the surface?

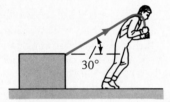

Figure 6-21.   *Problem 6-16.*

: **6-17 P** See Figure 6-22, where the 4.0-kg block is on a rough horizontal surface and is connected to mass m, which hangs over a massless, frictionless pulley. The static and kinetic coefficients of friction between the block and the surface are 0.25 and 0.20, respectively. What value of m will (*a*) start the block moving and (*b*) keep it moving at constant speed?

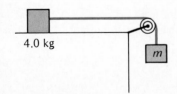

Figure 6-22.   *Problem 6-17.*

: **6-18 P** A book slides a distance of 3.0 m along the floor in a hallway before coming to rest. If the book is projected at twice the initial speed, how far will it slide before coming to rest?

: **6-19 P** A small block is placed on a spherical surface, as shown in Figure 6-23. The coefficient of static friction between the block and the surface is 0.5. What is the minimum angle θ at which the block will start sliding?

Figure 6-23.   *Problem 6-19.*

: **6-20 P** A block is projected initially and slides along a horizontal surface. The block then slides up a ramp in the form of a circular arc. See Figure 6-24. The coefficients of kinetic and static friction between the block and the ramp are μ_k and μ_s, respectively. What is the maximum angle θ to which the block can go and remain at rest there?

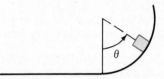

Figure 6-24.   *Problem 6-20.*

: **6-21 P** See Figure 6-25, where m_1 sits on m_2, which in turn sits on a horizontal surface. The coefficient of static friction between blocks m_1 and m_2, and between m_2 and the surface, is μ_s. The coefficient of kinetic friction between any two surfaces is μ_k. (*a*) What is the maximum acceleration that m_2 can have without the block m_1 slipping? (*b*) What maximum horizontal force can be applied to m_2 without m_1 slipping?

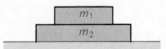

Figure 6-25.   *Problem 6-21.*

: **6-22 P** A package of mass m is being dragged across a rough floor at constant speed by a force **F** applied at an angle θ with respect to the horizontal. The coefficient of kinetic friction between the package and the floor is μ_k. At what angle θ is the force magnitude at its minimum?

Section 6-2   Uniform Circular Motion
· **6-23 Q** Goliath is west of David, as shown in Figure 6-26. With his slingshot, David whirls a stone in a circle overhead in a counterclockwise direction (as viewed from above). At what compass point (N, S, E, or W) does David release the stone in order to slay Goliath?

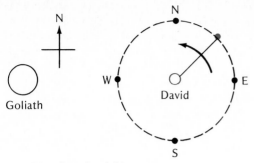

Figure 6-26.   Question 6-23.

· **6-24 Q** An automatic washing machine is in the spin-dry cycle. What is the path of water droplets as they leave through the small holes in the spinning cylindrical shell?

· **6-25 Q** A particle moves along a twisting, curving path. Its speed changes from point to point. Show that the resultant force on the particle at any point must point toward the concave, rather than the convex, side of its path.

· **6-26 Q** Which is more effective in improving the performance of a centrifuge, doubling the radius or doubling the number of rotations per unit time?

: **6-27 Q** A plumb line points to the center of the earth when the cord and plumb bob are at the equator or at either pole, but not for any other location. Why? (For the purposes of this problem, assume the earth to be perfectly spherical.)

: **6-28 Q** Why does the earth bulge at the equator?

: **6-29 Q** A merry-go-round is on a pond of ice. A girl sits at the outer edge of the rotating merry-go-round holding a hockey puck. Identify the forces acting on the puck as observed by an observer at rest on the pond. The girl releases the puck and it slides on the ice. What is the path of the puck immediately after its release as seen by *(a)* an observer at rest on the pond, and *(b)* the girl?

· **6-30 Q** A blindfolded person rides in an automobile. How does she tell when the car speeds up, slows down, turns left, turns right?

: **6-31 Q** You are riding in a high-speed train which is rounding a banked curve at the proper design speed. *(a)* How is a plumb bob suspended from the ceiling of the train oriented? *(b)* How would it be oriented if the train were traveling too slowly? *(c)* Too fast?

: **6-32 Q** Explain why ruts in dirt roads are found mostly at the bottoms of hills rather than at the tops.

· **6-33 P** A 50-gram particle makes 10 turns per second in a radius of 20 cm. What is the resultant force on the particle?

· **6-34 P** A particle of mass m travels in a horizontal circle of radius r at constant speed v. What is the average result-

ant force on the particle over the time: *(a)* for one complete loop, *(b)* for a half loop, and *(c)* for 1/1000 of a loop?

· **6-35 P** A particle traveling in a circle triples its speed and doubles its radius. By what factor does the resultant force on the particle increase?

· **6-36 P** See Figure 6-27. Mass m is connected to M by a massless cord passing through a small hole in the plate. The mass m slides at speed v on a perfectly smooth horizontal plate in a circle of radius r. What is the appropriate value for mass M?

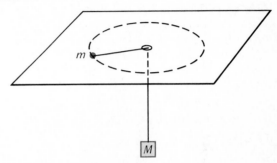

Figure 6-27.   Problem 6-36.

: **6-37 P** A 1.0-kg object is attached to the end of a 1.0-m cord and moves in a horizontal circle as a conical pendulum (see Figure 6-9). The cord breaks when the tension exceeds 30 N. *(a)* What is the period of the conical pendulum at which the string breaks? *(b)* What is the angle between the string and the vertical when this happens?

: **6-38 P** A highway engineer wishes to design a properly banked roadway for a turn of 100 m radius for vehicles traveling at 25 m/s (55 mph). [See Figure 6-11(a).] At what angle is the roadway to be banked relative to the horizontal?

: **6-39 P** A car in an amusement park travels in a circle in a vertical plane around a loop-the-loop, as shown in Figure 6-11(c). The circle's diameter is 10 m. What must the car's minimum speed be at the top of the loop if it is not to lose contact with the track?

: **6-40 P** A coin is near the edge of a 12-inch, 33 rpm phonograph record. See Figure 6-11(b). What is the minimum coefficient of static friction if the coin is not to slide off?

: **6-41 P** A 1000-kg automobile is at the crest of a hill whose radius of curvature in a vertical plane is 40 m. See Figure 6-11(d). What is the force of the car on the roadway at the crest if the car's speed is *(a)* 15 m/s? *(b)* 30 m/s?

: **6-42 P** Figure 6-11(f) shows an amusement park ride consisting of a spinning cylindrical shell open at the bottom. The inner surface of the shell is typically a wire mesh, for which the static coefficient of friction can easily be as large as 0.5. Choose reasonable values for the cylinder's radius and its spin rate and show that the ride is not as dangerous as it might at first seem to be.

: **6-43 P** On a rainy day the coefficient of static friction between a car's tires and a certain flat but curved road surface is reduced to one quarter of its usual value. Show that the maximum safe speed for the car rounding a curve is half of its usual value.

: **6-44 P** The photograph in Figure 6-28 shows a cyclist rounding a turn of 15 m radius. The bicycle is leaning 20° from the vertical. (a) What is the cyclist's speed? (b) What is the minimum coefficient of static friction between the bicycle's tires and the roadway?

Figure 6-28.   Problem 6-44.

: **6-45 P** When a person stands on a scale at the equator, the force the scale registers (the person's apparent weight) is less than the gravitational force of the earth on the person. By what percentage is the scale reading less than the person's true weight?

: **6-46 P** A physics student is traveling in a train rounding a curve on a flat roadbed. She estimates the radius of the curve to be about 100 m. She sees that a strap hanging from the ceiling makes an angle of about 25° with the vertical. What will the student estimate the train's speed to be?

: **6-47 P** You are standing with your arm straight down and your palm facing backward. Your fingers are turned up and hold a tumbler containing water. You believe that you can swing the tumbler forward, overhead, and around in a vertical circle in about one second. Are you likely to get wet?

: **6-48 P** Suppose that the earth's rotation rate could be speeded up so that its period would be the same as that of an earth satellite orbiting close to the earth's surface, rather than 24 hours. What would be the behavior of unattached objects at (a) the equator and (b) higher latitudes?

: **6-49 P** Astronauts are trained to experience the state of "weightlessness" in planes that fly around the top of a circle in a vertical plane. What are the forces acting on an astronaut trainee as seen by an observer (a) on earth? (b) In the plane? (c) What is the correct radius of curvature for a plane flying at a speed of 500 km/h?

: **6-50 P** A 60-kg woman is riding in a Ferris wheel at a distance of 20 m from its center. The wheel makes 6.0 turns in a minute. What is the magnitude of the force of the seat on the woman when she is (a) at the top, (b) at the bottom, and (c) halfway up in the circle?

: **6-51 P** Two small blocks of equal mass m attached to two strings of equal length r, as shown in Figure 6-29, are set into rotation on a horizontal frictionless plane at constant angular speed ω. Find the ratio of the tension in the inner string to that in the outer string.

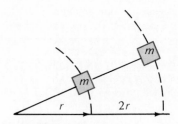

Figure 6-29.   Problem 6-51.

: **6-52 P** A particle is thrown from the origin along the x-axis with initial speed v_0. Its acceleration is **g** in the negative y direction. (a) Show that the equation for the particle's path can be written as $y = -(1/2R)x^2$, where R is the initial radius of curvature of the path. (b) Sketch the parabolic path with the initial circular path superimposed.

: **6-53 P** A pitcher throws a baseball at 40 m/s (≈ 100 mi/h) toward home plate, a distance of 20 m away. (See Problem 6-52.) (a) Calculate the initial radius of curvature of the pitched baseball. (b) Compare this radius with the 20 m from pitcher to catcher and satisfy yourself that the baseball is moving essentially at constant speed along a circular arc.

: **6-54 P** A binary consists of two stars, each traveling in a circle about the same center. A certain binary consists of two stars of the same mass m separated by a constant distance d as each travels in the same circle. The resultant force on each star is F. What is the period of revolution for either star?

Momentum

7

A rifle recoils when a bullet is fired. A rocket reacts in a forward direction when particles leave in the exhaust. A speeding truck runs into a car and the two travel together at a lesser speed.

These are all simple examples of a basic principle—the conservation law for momentum—according to which the total momentum of any isolated system of particles is constant. The momentum concept is introduced in physics simply because under appropriate circumstances, the vector quantity called momentum is conserved.

7-1 Momentum Defined

The resultant force $\mathbf{F}$ acting on a particle of mass m is

$$\mathbf{F} = m\mathbf{a}$$

where $\mathbf{a}$ is the particle's acceleration. Since acceleration $\mathbf{a}$ is the time rate of change of velocity, $\mathbf{a} = d\mathbf{v}/dt$, we can also write Newton's law as

$$\mathbf{F} = m\,\frac{d\mathbf{v}}{dt} = \frac{d}{dt}\,(m\mathbf{v}) \tag{7-1}$$

In the last step we have taken a particle's mass to be constant.

Equation (7-1) says that the time-rate of change of the quantity $m\mathbf{v}$ is equal to the resultant force on the particle.

The vector quantity $m\mathbf{v}$ is defined as the *momentum* of the particle, typically

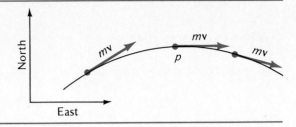

Figure 7-1. *A particle's momentum* **p** = *m***v** *is tangent to its path.*

represented by **p**:

$$\text{Momentum} = \mathbf{p} \equiv m\mathbf{v} \tag{7-2}$$

Then Newton's second law can be written

$$\mathbf{F} = \frac{d\mathbf{p}}{dt} \tag{7-3}$$

Strictly, *m***v** is a particle's *linear* momentum—its momentum along a line.* A particle may also have *angular* momentum (Chapter 14), momentum related to rotation. Hereafter when we use the term *momentum,* the linear momentum will always be implied.

From the definition, we see that a particle's momentum is a vector quantity, parallel to the particle's velocity. Momentum is always tangent to the particle's path and equal in magnitude to the particle's mass multiplied by its speed. See Figure 7-1. For example, a 1-kg particle moving east at 1 m/s has a momentum of 1 kg·m/s east (point *p* in Figure 7-1).

Units for momentum are those of mass multiplied by velocity: kilogram meters per second (kg·m/s). From (7-3) we see that equivalent momentum units are newton seconds (N·s).

7-2 Momentum Conservation

Momentum is an important physical quantity for one primary reason; the momentum-conservation law. In this section we restrict our consideration to *two* particles; a general proof of the momentum-conservation law, applicable to any number of particles, is given in Section 7-4.

---

* The relation **p** = *m***v** for linear momentum applies when the particle's velocity is much less than the speed of light, $c = 3.0 \times 10^8$ m/s. For high speeds, the correct relation for the conserved vector quantity **p** of an object is

$$\mathbf{p} = \frac{m_0 \mathbf{v}}{\sqrt{1 - (v/c)^2}}$$

where m_0 denotes the so-called *rest mass,* the mass of the object at zero speed. The quantity **p** is called the *relativistic momentum.* For an isolated system of particles, the vector sum of the relativistic momenta of the particles in the system is conserved.

The relation for relativistic momentum **p** of a particle, given above, reduces to the classical momentum relation, $\mathbf{p} = m_0\mathbf{v}$, for speeds much less than that of light, that is, $v/c \ll 1$. For example, with $v = 30{,}000$ km/s $= 3 \times 10^7$ m/s, we have $v/c = 1/10$, and the denominator differs from 1 by a mere 0.5 percent. On the other hand, as the particle speed approaches c, the denominator becomes appreciably smaller than 1, and the relativistic momentum approaches infinity.

Consider the two particles, 1 and 2, in Figure 7-2. They interact but are isolated from the rest of the universe; no external forces act on the system of two particles. Therefore, the only force on particle 1 arises from particle 2, and conversely.

Newton's second law, applied to particle 1, gives

$$\mathbf{F}_{2\ on\ 1} = \frac{d}{dt}(m_1\mathbf{v}_1)$$

where m_1 and $\mathbf{v}_1$ are the mass and velocity of particle 1.

Similarly, for particle 2,

$$\mathbf{F}_{1\ on\ 2} = \frac{d}{dt}(m_2\mathbf{v}_2)$$

Now we add the two equations to get

$$\mathbf{F}_{1\ on\ 2} + \mathbf{F}_{2\ on\ 1} = \frac{d}{dt}(m_1\mathbf{v}_1 + m_2\mathbf{v}_2)$$

Newton's third law requires that the two forces be equal and opposite: $\mathbf{F}_{1\ on\ 2} = -\mathbf{F}_{2\ on\ 1}$. Therefore, the left side of the equation above is zero, and we are left with

$$0 = \frac{d}{dt}(m_1\mathbf{v}_1 + m_2\mathbf{v}_2)$$

The time derivative of the quantity $(m_1\mathbf{v}_1 + m_2\mathbf{v}_2)$ is zero. This vector quantity must then be constant with time:

$$m_1\mathbf{v}_1 + m_2\mathbf{v}_2 = \text{constant (magnitude and direction)}$$

The momentum of particle 1 is $\mathbf{p}_1 = m_1\mathbf{v}_1$, and the momentum of particle 2 is $\mathbf{p}_2 = m_2\mathbf{v}_2$. The above relation then says that the *total vector momentum* of the two particles is *constant*. We have proved that if particles interact but are isolated from external influence and therefore have no resultant external force acting on them, the total (vector) momentum of the particles does not change.* This is the *conservation of (linear) momentum law*. In symbols,

$$\Sigma\mathbf{p} = \text{total momentum} = \text{constant} \qquad \text{if } \Sigma\mathbf{F}_{ext} = 0 \qquad (7\text{-}4)$$

Note the crucial role played by Newton's third law of motion in the derivation above. With no net external force, the only forces acting between the two particles are internal equal and opposite action-reaction forces, which add to zero. Just as Newton's laws of motion agree with observation (for all particle speeds much less than that of light), so too the momentum-conservation principle has met every experimental test.

So long as interacting particles are isolated from the rest of the world, their total momentum, a vector quantity, is unchanged. The individual particles may change speed and direction of motion. A force may act on each particle from the other particle. Particles may undergo a "hard" collision (say two billiard balls with a short interaction time and a rapid momentum change), or they may have a "soft" collision (say two tennis balls with a long interaction time and

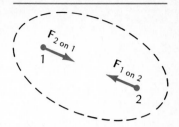

Figure 7-2. An isolated system of particles 1 and 2 interacting by internal forces, $\mathbf{F}_{2\ on\ 1}$ and $\mathbf{F}_{1\ on\ 2}$.

* The relation for momentum conservation is the same as the equation (5-1) defining mass.

slow momentum change). The objects may separate after a collision, or stick together. They may touch or merely come close together (like two repelling magnets). In every such collision or interaction, the total vector momentum *in* equals the total vector momentum *out;* indeed, the total vector momentum is the same at *every instant.* The momentum lost by one object is always compensated exactly by the momentum gained by the other object.

Example 7-1. A 1.0-kg block moving initially to the right at 12 m/s collides with and sticks to a 2.0-kg block initially at rest. See Figure 7-3(a). What is the velocity of the composite object after the collision?

The two objects compose an isolated system, and therefore their total momentum is constant.

First let's settle the direction in which the composite object moves. Before the collision, the momentum of the 1.0-kg object was to the right (the 2.0-kg object then had no momentum). Therefore, the direction of the total momentum after the collision must also be to the right.

The composite object of 3.0 kg traveling at speed v is nothing more than both the 1.0 kg and 2.0 kg, each traveling together at speed v.

Applying momentum conservation

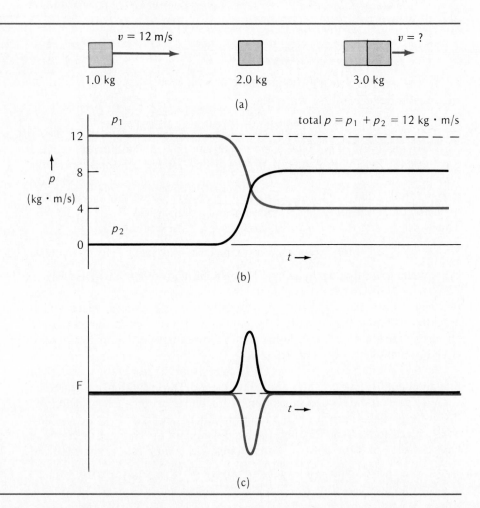

Figure 7-3. (a) Head-on collision. (b) Momentum-time graphs. (c) Force-time graphs.

Momentum before = momentum after

we have

$$(1.0 \text{ kg})(12 \text{ m/s}) + (2.0 \text{ kg})(0) = (1.0 \text{ kg} + 2.0 \text{ kg})v$$

or

$$v = 4.0 \text{ m/s}$$

It is informative to see how the momentum of each of the two blocks changes with time. This is sketched in Figure 7-3(b). The momentum of the 1-kg block was constant while it coasted before the collision; its momentum dropped during the collision; and after the collision, the 1-kg part of the composite object had a constant but smaller momentum. On the other hand, the momentum of the 2-kg mass increased during the collision, and by the same amount. Some momentum was transferred from the 1-kg to the 2-kg object, but the total momentum of both objects — before, at each instant during, and after the collision — was the same amount, 12 kg·m/s to the right.

Now look at the instantaneous force on each of the two objects, shown as a function of time in Figure 7-3(c). No force acts on either block, except when they interact. Then the 1-kg object is slowed (the force $\mathbf{F}_{2 \text{ on } 1}$ is to the left, or negative) while the 2-kg object is speeded up from rest (the force $\mathbf{F}_{1 \text{ on } 2}$ is to the right, or positive).

It is easy to see that the two force curves have exactly the same shape, one curve being the other flipped upside down. This is nothing more than Newton's third law:

$$\mathbf{F}_{1 \text{ on } 2} = -\mathbf{F}_{2 \text{ on } 1}$$

A force acts on either block only while its momentum is changing (compare parts (b) and (c) of Figure 7-3). This follows immediately from (7-2): $\mathbf{F} = d(m\mathbf{v})/dt = d\mathbf{p}/dt$. In graphical terms, the force on an object is just the slope, dp/dt, of the object's momentum plotted as a function of time.

Example 7-2. An object initially at rest explodes into two pieces with masses m_1 and m_2, respectively. How are the velocities of the two pieces related?

An explosion may be regarded as a collision in which the two interacting objects are initially at rest with respect to one another. Here the total momentum is initially zero, and it must remain zero:

$$\mathbf{p}_1 + \mathbf{p}_2 = 0$$

$$m_1 v_1 + m_2 v_2 = 0$$

$$\frac{m_1}{m_2} = -\frac{v_2}{v_1}$$

The speeds are in the inverse ratio of the masses, just as in Section 5-2 on the definition of mass. The minus sign implies that the two pieces fly off in opposite directions. For a fired rifle, the bullet's speed exceeds the rifle recoil speed by the same factor that the rifle's mass exceeds the mass of the bullet. If the two masses happen to be equal, $m_1 = m_2$, then the velocities are equal in magnitude as well as opposite, $v_1 = -v_2$. This may be used as a test for equality of masses: If two pieces fly off with equal speeds, their masses must be equal. Similarly, if two pieces fly together at equal speeds and stick together to form a composite object at rest, we can infer that the two pieces have equal masses.

Example 7-3. A 1.0-kg object (object 1) with a velocity of 12.0 m/s to the right strikes a second object (object 2) of 2.0 kg originally at rest. See Figure 7-4. Object 1 is deflected in the collision from its original direction through an angle of 30°; its speed after the collision is 11.2 m/s. What is (a) the angle θ of object 2's velocity after the

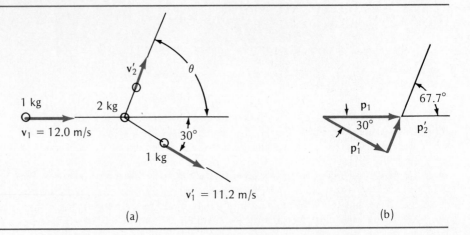

Figure 7-4.   (a) Non-head-on
collision. (b) The corresponding
momentum vectors.

(a)

(b)

collision (with respect to the original direction of object 1)? (b) the speed v_2' of 2 after the collision?

No net external force acts on the colliding objects, and momentum conservation requires that

$$\mathbf{p}_1 = \mathbf{p}_1' + \mathbf{p}_2'$$

See Figure 7-4, where $\mathbf{p}_1$ and $\mathbf{p}_1'$ are the momenta of 1 before and after the collision, respectively, and $\mathbf{p}_2'$ is the momentum of 2 after the collision. The system's total vector momentum is constant throughout the collision. Therefore, the momentum components of 1 and 2 along any given direction are also constant. Along the original direction of 1, we have

$$p_1 = p_1' \cos 30° + p_2' \cos \theta$$

(7-5)

$$(1.0 \text{ kg})(12.0 \text{ m/s}) = (1.0)(11.2 \text{ m/s}) \cos 30° + (2.0 \text{ kg})(v_2') \cos \theta$$

Equation (7-5) simply says that the total linear momentum is unchanged along the forward direction; that is, object 2 acquired a forward momentum component just equal to the momentum component lost by 1.

Now consider the momentum components along a perpendicular direction:

$$0 = p_1' \sin 30° - p_2' \sin \theta$$

(7-6)

$$0 = (1.0 \text{ kg})(11.2 \text{ m/s}) \sin 30° - (2.0 \text{ kg})(v_2') \sin \theta$$

Before the collision, there is no component of momentum along this transverse direction; after the collision, the momentum components of 1 and 2 are just balanced. (There is no component of momentum, before or after the collision, in a direction perpendicular to the plane of vectors $\mathbf{p}_1'$ and $\mathbf{p}_2'$.)

We have two equations, (7-5) and (7-6), and two unknowns, v_2' and θ. We can eliminate v_2' from the two equations by solving for v_2' in (7-6) and substituting in (7-5). The result, after a little algebra, is (check it!)

$$\theta = 67.7°$$

Using $\theta = 67.7°$ in (7-5) or (7-6) gives

$$v_2' = 3.03 \text{ m/s}$$

Example 7-4.   The Saturn V rocket, used in the Apollo missions to the moon, was the largest of its type, with an overall height of 110 m (twice that of the Statue of

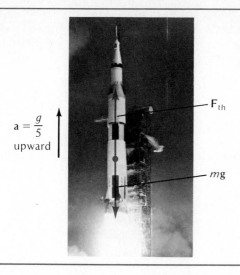

$a = \dfrac{g}{5}$
upward

$\mathbf{F}_{th}$

$m\mathbf{g}$

Figure 7-5.   Photo of Saturn V rocket at takeoff. The rocket's weight is mg; the upward thrust is $\mathbf{F}_{th}$.

Liberty) and a mass at liftoff of 2.8×10^6 kg ($\sim$3000 tons). See Figure 7-5. A motion picture at liftoff shows that it takes 10.5 s for the rocket to ascend a distance equal to its height. What are (a) Saturn V's acceleration and (b) the thrust of the first-stage rocket (the force of the exhaust gases on the rocket) during the first 10.5 s? (c) The exhaust particles have a speed of about 2.0 km/s (relative to the rocket). What is the rate at which the first-stage rocket ejects mass? (d) How much mass is ejected during the first 10.5 s?

(a) We are given

$$y_0 = 0,\ v_0 = 0,\ y = 110\ \text{m},\ t = 10.5\ \text{s}$$
$$a = ?$$

Taking acceleration a to be constant, we have:

$$y - y_0 = v_0 t + \tfrac{1}{2} a t^2$$

Therefore,

$$a = \frac{2y}{t^2} = \frac{2(110\ \text{m})}{(10.5)^2} = 2.00\ \text{m/s}^2$$

The upward acceleration is about $\tfrac{1}{5}g$.

(b) Figure 7-5 shows the forces acting on the Saturn V; its weight is $m\mathbf{g}$ down and the thrust is $\mathbf{F}_{th}$ upward. Obviously, the thrust must exceed the weight for liftoff. More specifically, from Newton's second law

$$\Sigma F_y = ma_y$$

$$F_{th} - mg = ma$$

$$F_{th} = m(g + a) = (2.8 \times 10^6\ \text{kg})(9.8 + 2.0)\ \text{m/s}^2 = 3.3 \times 10^7\ \text{N}$$

(c) To find the mass ejection rate, we first need a general relation for thrust. For simplicity, suppose that a rocket is in empty space and initially at rest. Its momentum is zero. The *total* momentum of rocket and exhaust particles remains zero by the momentum-conservation principle. As exhaust particles acquire momentum to the rear on leaving the rocket, the rocket acquires equal momentum in the forward direction. The magnitude of the force F_{ex} of *rocket on exhaust particles* is, by Newton's third law, the same as the force (the thrust) F_{th} of *exhaust particles on rocket*. That is, $|F_{th}| = F_{ex}$. But the force F_{ex} on the exhaust particles is, from $F = dp/dt$, just the rate at which the momentum of these particles changes. While the particles are inside the rocket, their speed is zero; after leaving the rocket, they have an exhaust velocity v_{ex}. If a very small mass dm is ejected in time dt, then the rate at which the momentum of exhaust particle changes is

$$F = \frac{dp}{dt}$$

$$F_{th} = v_{ex} \frac{dm}{dt} \tag{7-7}$$

Generally the thrust is the product of the exhaust velocity v_{ex} and the rate dm/dt at which mass is ejected. (The velocity v_{ex} is essentially fixed by the type of chemical reaction taking place in the rocket, so that a large thrust F_{th} requires a high rate dm/dt of ejecting mass.)

For the data on the Saturn V rocket,

$$F_{th} = 3.3 \times 10^7 \text{ N} \quad \text{and} \quad v_{ex} = 2.0 \times 10^3 \text{ m/s}$$

Equation (7-7) then yields

$$\frac{dm}{dt} = \frac{F_{th}}{v_{ex}} = \frac{3.3 \times 10^7 \text{ N}}{2.0 \times 10^3 \text{ m/s}} = 1.7 \times 10^4 \text{ kg/s}$$

(d) Over the first 10.5 s, the total mass ejected is

$$\text{Mass} = \left(\frac{dm}{dt}\right)(10.5 \text{ s}) = (1.7 \times 10^4 \text{ kg/s})(10.5 \text{ s}) = 1.8 \times 10^5 \text{ kg}$$

The Saturn V loses 1.8×10^5 kg (roughly the mass of the Statue of Liberty), or about 6 percent of its initial mass, merely in climbing its own height. Small wonder that spectators are urged to stand back.

7-3   Impulse and Momentum

Newton's second law, expressed in terms of momentum, is

$$\mathbf{F} = \frac{d\mathbf{p}}{dt} \tag{7-3}$$

The incremental momentum change $d\mathbf{p}$ over the short time interval dt is

$$d\mathbf{p} = \mathbf{F} \, dt$$

where $\mathbf{F}$ is the instantaneous force. Now let us integrate the above relation from some initial state i to a final state f:

$$\int_{\mathbf{p}_i}^{\mathbf{p}_f} d\mathbf{p} = \int_{t_i}^{t_i} \mathbf{F} \, dt$$

$$\mathbf{p}_f - \mathbf{p}_i = \Delta\mathbf{p} = \int_{t_1}^{t_2} \mathbf{F} \, dt \tag{7-8}$$

where $\mathbf{p}_f$ is the final and $\mathbf{p}_i$ the initial momentum, and $\Delta\mathbf{p}$ is the net momentum change.

The integral of force over time $\int \mathbf{F} \, dt$ is known as the *impulse* of the force. As (7-8) shows, impulse equals change in momentum. This is the *impulse-momentum theorem.*

Force is the derivative of momentum with respect to time. Momentum change is the integral of force over time. Force is a measure of the rate of flow of

That Professor Goddard

"That Professor Goddard . . . does not know the relation of action and reaction, and of the need to have something better than a vacuum against which to react—to say that would be absurd. Of course, he only seems to lack the knowledge ladled out daily in high schools." (*New York Times,* January 13, 1920, editorial comment on page 12)

True, the anonymous editorial writer for the prestigious *New York Times* did not come right out and say that Robert H. Goddard (1882–1945), the American rocket pioneer, did not have his basic physics straight, but that was the snide implication. It was the sort of abuse and ridicule that Goddard, a professor of physics at Clark University, had to endure over many years of persistent and lonely research on the fundamentals of rocketry. The particular occasion for the comments quoted above was the front-page article, "Believes Rocket Can Reach Moon," that appeared in the *Times* the day before. The Smithsonian Institution had put up some money for Goddard's research and it authorized a press release telling of a new type of high-efficiency multiple-charge rocket that might rise above the earth's atmosphere and ". . . possibly [go] even as far as the moon itself."

The *New York Times* continued its criticism of Goddard by saying that Jules Verne, the French science-fiction writer, "deliberately seemed to make the same mistake that Professor Goddard seems to make. That was one of Verne's few scientific slips, or else a deliberate step aside from scientific accuracy, pardonable enough in him as a romancer, but its like is not so easily explained when made by a savant who isn't writing a novel of adventure."

Well, does a rocket actually work in a vacuum? The physics of this is perfectly clear: a rocket need not push on air; it pushes on the exhaust particles, just as the exhaust particles push with equal force on the rocket. Nevertheless, Goddard showed by direct experiment that a rocket can indeed work in a vacuum. It was one of Goddard's minor accomplishments. His imagination was seized at age six by H. G. Wells' *War of the Worlds,* and he devoted his life to rockets. He produced the first two-stage solid-fuel rocket in 1914. A special occasion, in 1926 on his Aunt Effie's farm in Auburn, Massachusetts, was the lift-off of a liquid-propelled (gasoline and liquid oxygen) rocket engine. All told, Goddard held 214 patents in rocketry.

Actually, Goddard was not alone. Two other investigators independently arrived at many of the same results. Konstantin E. Tsiolkovsky, an obscure schoolteacher in a remote Russian village, published an article entitled, "Investigations of Space by Means of Rockets" in 1903; and in 1923 the German scientist Hermann Oberth, published "Die Rakete zu den Planetenräumen."

Recognition finally came to Goddard. The National Aeronautics and Space Administration named its principal research establishment for Goddard, and the U. S. government acknowledged its debt to Goddard by posthumously settling $1 million with his heirs for its use of his patents.

momentum between interacting objects. Impulse is a measure of the net amount of momentum transferred, as shown schematically in Figure 7-6.

When a force changes with time, it is often useful to speak of the equivalent time-average force $\overline{\mathbf{F}}$, defined by

$$\text{Impulse} = \overline{\mathbf{F}}\,\Delta t \equiv \int_{t_i}^{t_f} \mathbf{F}\,dt \tag{7-9}$$

where $\Delta t = t_f - t_i$. In words, a constant average force $\overline{\mathbf{F}}$ has the same impulse and produces the same momentum change as a time-varying force $\mathbf{F}$ when the forces are related as in (7-9).

The term *impulse* used colloquially usually implies a force that acts only very briefly, for example, the force of a hammer on a nail. But the formal definition of *impulse* includes any force, however long the period during which it acts.

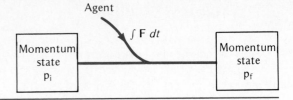

Figure 7-6. Schematic representation of the transfer of momentum to an object by the impulse of an agent.

Impulse has the same units as momentum: kg·m/s or N·s.

It is useful to examine the graphical relations among momentum, time, force, and impulse. Figure 7-7(a) shows the momentum of an object along a line plotted as a function of time. The momentum is *constant* before and after the interaction period Δt; during Δt, the momentum increases at a constant rate. The average force on the object $\overline{F} = \Delta p / \Delta t$ is constant over the time Δt, and otherwise zero, as shown in Figure 7-7(b). In graphical terms, force is the slope of the momentum-time graph, whereas impulse is the area under a force-time graph.

A more realistic situation is portrayed in the momentum-time and corresponding force-time graphs of Figure 7-8. Here the momentum changes smoothly, and the force rises continuously to a peak and falls back to zero. The instantaneous force $F = dp/dt$ is plotted, as derived from the slope of the

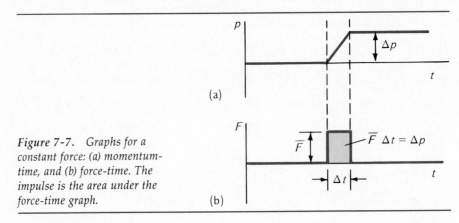

Figure 7-7. Graphs for a constant force: (a) momentum-time, and (b) force-time. The impulse is the area under the force-time graph.

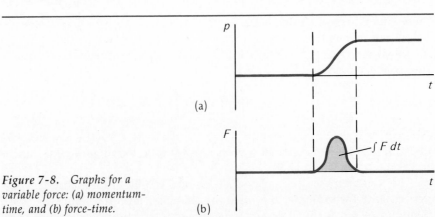

Figure 7-8. Graphs for a variable force: (a) momentum-time, and (b) force-time.

momentum-time graph. Again impulse is the area under the force-time graph; the area of a slender rectangle is $\mathbf{F}\,dt$, and the entire area under the curve $\int \mathbf{F}\,dt$.

Example 7-5. A fast ball thrown by a baseball pitcher reaches the batter at a speed of 40 m/s (~90 mi/h). The baseball leaves the bat at the same speed, having been in contact with it for one millisecond (1 ms). The baseball's mass is 0.16 kg. What is the average force of bat on ball during the time this force acts?

First, we don't worry about the baseball's weight, since it is far smaller than the force of the bat on the ball. As the ball comes in contact with the bat, the force on the ball rises to a peak and then drops back to zero, as shown in Figure 7-9. The impulse on the baseball is $\int \mathbf{F}\,dt$, the area under the force-time curve. The equivalent constant average force $\bar{F}$ acting over a time interval Δt has the same impulse (same area under the force-time curve). From (7-8) and (7-9), we have

$$\int F\,dt = \bar{F}\,\Delta t = p_f - p_i = mv_f - mv_i$$

where v_i and v_f are the baseball's initial and final speeds. If we take the direction from batter to pitcher to be positive, then $v_i = -40$ m/s and $v_f = +40$ m/s. The relation above gives

$$\bar{F} = \frac{mv_f - mv_i}{\Delta t}$$

$$= \frac{(0.16 \text{ kg})[(+40 \text{ m/s}) - (-40 \text{ m/s})]}{1.0 \times 10^{-3} \text{ s}}$$

$$= 1.3 \times 10^4 \text{ N}$$

The maximum instantaneous force on the baseball is still larger than 1.3×10^4 N (~1.5 tons).

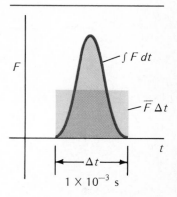

Figure 7-9. The area under the force-time graph is $\int F\,dt = \bar{F}\,\Delta t$.

7-4   General Proof of Momentum Conservation (Optional)

We have shown that the total momentum of a system consisting of *two* particles is constant when the particles are isolated from external influence (Section 7-2). The proof here of the momentum-conservation principle is more general. It applies for three or more particles.

Consider a system of three particles, 1, 2, and 3, as shown in Figure 7-10(a). The term *system* is intentionally vague. The particles that a system comprises

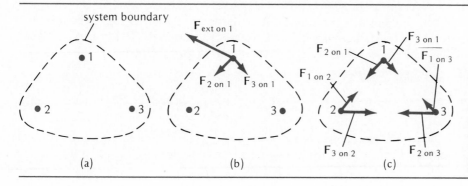

Figure 7-10. (a) System of three particles. (b) Internal and external forces on particle 1. (c) The internal forces only on particles within the system.

can be of any sort. We don't know, nor do we need to know, precisely what force one particle exerts on another particle. What matters here is only that particles 1, 2, and 3 are inside the boundary of a well-defined system.

The particles may interact. The force of 2 on 1 is $\mathbf{F}_{2 \text{ on } 1}$; the force of 1 on 2 is $\mathbf{F}_{1 \text{ on } 2}$ [see Figure 7-10(c)]. And so on for other combinations of particles. Such forces are *internal forces,* forces between particles *inside* the system. The vectors for internal forces lie within the system's boundary.

Particles 1, 2, and 3 may also be acted on by forces arising from objects outside the chosen system. For example, the external force on particle 1 is denoted $\mathbf{F}_{\text{ext on } 1}$ [Figure 7-10(b)]. Forces of this sort, which penetrate the boundary, are called *external forces.*

The resultant force on any one particle then consists of two parts:

- Internal forces from all other particles within the system.
- External forces from objects outside the system.

More specifically, the resultant force on particle 1 consists of the external force $\mathbf{F}_{\text{ext on } 1}$ and the internal forces $\mathbf{F}_{2 \text{ on } 1}$ and $\mathbf{F}_{3 \text{ on } 1}$.

From Newton's second law, in the form of (7-1), we can then write for the resultant force on particle 1

$$\mathbf{F}_{\text{ext on } 1} + \mathbf{F}_{2 \text{ on } 1} + \mathbf{F}_{3 \text{ on } 1} = \frac{d}{dt}(m_1\mathbf{v}_1)$$

where m_1 and $\mathbf{v}_1$ are this particle's mass and instantaneous velocity.

In like fashion, Newton's second law applied in turn to particles 2 and 3 yields

$$\mathbf{F}_{\text{ext on } 2} + \mathbf{F}_{1 \text{ on } 2} + \mathbf{F}_{3 \text{ on } 2} = \frac{d}{dt}(m_2\mathbf{v}_2)$$

and

$$\mathbf{F}_{\text{ext on } 3} + \mathbf{F}_{1 \text{ on } 3} + \mathbf{F}_{2 \text{ on } 3} = \frac{d}{dt}(m_3\mathbf{v}_3)$$

We add the three previous equations, and use parentheses to group terms on the left:

$$(\mathbf{F}_{\text{ext on } 1} + \mathbf{F}_{\text{ext on } 2} + \mathbf{F}_{\text{ext on } 3}) + (\mathbf{F}_{1 \text{ on } 2} + \mathbf{F}_{2 \text{ on } 1}) + (\mathbf{F}_{1 \text{ on } 3} + \mathbf{F}_{3 \text{ on } 1})$$
$$+ (\mathbf{F}_{2 \text{ on } 3} + \mathbf{F}_{3 \text{ on } 2}) = \frac{d}{dt}(m_1\mathbf{v}_1 + m_2\mathbf{v}_2 + m_3\mathbf{v}_3)$$

Newton's third law applied to particles 1 and 2 requires that $\mathbf{F}_{2 \text{ on } 1} = -\mathbf{F}_{1 \text{ on } 2}$; therefore, the second parenthesis on the left is zero. Similarly, the third and fourth parentheses are also zero. The internal forces drop out because they come in equal and opposite pairs. See Figure 7-10(c).

We are left then with

$$\mathbf{F}_{\text{ext on } 1} + \mathbf{F}_{\text{ext on } 2} + \mathbf{F}_{\text{ext on } 3} = \frac{d}{dt}(m_1\mathbf{v}_1 + m_2\mathbf{v}_2 + m_3\mathbf{v}_3) \qquad (7\text{-}10)$$

Now the resultant external force $\Sigma\mathbf{F}_{\text{ext}}$ on the chosen system of particles is just the vector sum of the individual external forces on 1, 2, and 3 separately:

$$\Sigma \mathbf{F}_{ext} = \mathbf{F}_{ext\ on\ 1} + \mathbf{F}_{ext\ on\ 2} + \mathbf{F}_{ext\ on\ 3}$$

so that (7-10) can be written*

$$\Sigma \mathbf{F}_{ext} = \frac{d}{dt}(m_1\mathbf{v}_1 + m_2\mathbf{v}_2 + m_3\mathbf{v}_3)$$

$$\Sigma \mathbf{F}_{ext} = \frac{d}{dt}(\Sigma m\mathbf{v}) \qquad\qquad (7\text{-}11)$$

Equation 7-11 is a fundamental result. Important consequences follow from it.

- *Momentum-conservation law*

Suppose that the system is isolated, and therefore the resultant external force on the system is zero.

$$\frac{d}{dt}(\Sigma m\mathbf{v}) = 0 \qquad \text{for } \Sigma\mathbf{F}_{ext} = 0$$

or

$$\text{Total momentum of isolated system} = \Sigma(m\mathbf{v}) = \text{constant} \qquad (7\text{-}12)$$

This completes the general proof of the momentum-conservation law.

- *External forces and the momentum of the system*

If we denote the momentum of each particle of the system by $\mathbf{p}_1 = m_1\mathbf{v}_1$, $\mathbf{p}_2 = m_2\mathbf{v}_2$, . . . , then

$$\Sigma m\mathbf{v} = \Sigma\mathbf{p} = \mathbf{P} \qquad\qquad (7\text{-}13)$$

where $\mathbf{P}$ is the system's total momentum, the vector sum of the momenta of the individual particles.

Equation 7-11 can then be written

$$\Sigma\mathbf{F}_{ext} = \frac{d\mathbf{P}}{dt} \qquad\qquad (7\text{-}14)$$

This relation says that the time rate of change of the system's total momentum $d\mathbf{P}/dt$ is controlled solely by the *external* forces on the system. It is analogous to (7-3) for a single particle, $\mathbf{F} = d\mathbf{p}/dt$. We must know *all* the forces acting on a single particle to find how the particle's momentum changes with time. But for any system of particles that are interacting by any means, the *external forces alone* determine how the system's total momentum changes with time. For this reason, any ordinary object, which consists of many interacting particles, can be considered a single mass point.

- *External forces and center of mass*

The total momentum of the system $\mathbf{P}$ can be written as the product of a mass M and a velocity $\mathbf{V}$:

$$\mathbf{P} \equiv M\mathbf{V} \qquad\qquad (7\text{-}15)$$

Since $\mathbf{P}$ is the momentum of the entire system, it is natural to take M to be the system's total mass:

---

* Although our system has consisted of only three particles, it is clear that the result applies for any number of particles.

$$M = m_1 + m_2 + m_3 + \cdots = \Sigma m \tag{7-16}$$

Then (7-14) can be written

$$\Sigma \mathbf{F}_{ext} = \frac{d}{dt}(M\mathbf{V}) = M\frac{d\mathbf{V}}{dt} = M\mathbf{A} \tag{7-17}$$

with

$$\mathbf{A} = \frac{d\mathbf{V}}{dt}$$

Equation 7-17, $\Sigma \mathbf{F}_{ext} = M\mathbf{A}$, even though it applies for *any system of particles,* is just like Newton's second law applied to a *single particle,* $\mathbf{F} = m\mathbf{a}$. The relation $\Sigma \mathbf{F}_{ext} = M\mathbf{A}$ says, in effect, that an entire system of particles may be thought to be like a *single* particle of mass M (equal to the mass of the whole system) having an acceleration $\mathbf{A}$. The equivalent single particle of mass M has velocity $\mathbf{V}$ and instantaneous acceleration $\mathbf{A}$ that are given, through (7-13), (7-15), and (7-17) by

$$M\mathbf{V} = \mathbf{P} = \Sigma m\mathbf{v}$$

$$\mathbf{V} = \frac{\Sigma m\mathbf{v}}{M} \tag{7-18}$$

and

$$M\mathbf{A} = \frac{d\mathbf{P}}{dt} = \frac{d}{dt}(\Sigma m\mathbf{v}) = \Sigma m\frac{d\mathbf{v}}{dt} = \Sigma m\mathbf{a}$$

$$\mathbf{A} = \frac{\Sigma m\mathbf{a}}{M} \tag{7-19}$$

The single equivalent particle, whose velocity and acceleration are given by (7-18) and (7-19), is located at the system's *center of mass.* (It is the subject of Chapter 8.)

Summary

Definitions

Momentum (linear) of a particle:

$$\mathbf{p} \equiv m\mathbf{v} \tag{7-2}$$

Impulse of force:

$$\int_{t_i}^{t_f} \mathbf{F}\, dt = \overline{\mathbf{F}}\, \Delta t = \mathbf{p}_f - \mathbf{p}_i \tag{7-8}$$

Units

Momentum (linear): $kg \cdot m/s = N \cdot s$
Impulse: $N \cdot s = kg \cdot m/s$

Fundamental Principles

The momentum (linear) conservation law states that the total (vector) momentum of an isolated system is constant:

$$\Sigma \mathbf{p} = \text{total momentum} = \text{constant} \quad \text{if } \Sigma \mathbf{F}_{ext} = 0 \tag{7-4}$$

Important Results

For any system of particles, each with a momentum $m\mathbf{v}$,

$$\Sigma \mathbf{F}_{ext} = \frac{d}{dt}(\Sigma m\mathbf{v}) \tag{7-11}$$

where $\Sigma \mathbf{F}_{ext}$ is the vector sum of the external forces on the system.

Problems and Questions

Section 7-1   Momentum Defined

· **7-1 Q** A race car travels in a circle at constant speed. At the instant when the car is headed east, its momentum is
(A) zero.
(B) south.
(C) north.
(D) somewhat south of east.
(E) changing.

· **7-2 Q** Which of the following units for momentum is *wrong?*
(A) kg·m/s
(B) N·s
(C) (J·kg)$^{\frac{1}{2}}$
(D) J·s
(E) All of the above are correct.

· **7-3 P** What momentum has a 3-kg object acquired after it has fallen freely from rest for 3 seconds?

· **7-4 P** A 50-gm rock is thrown with a speed of 10 m/s. How fast would a 75-gm rock have to be thrown to have the same momentum?

Section 7-2   Momentum Conservation

: **7-5 Q** Show that the definition of mass given in Chapter 5 is equivalent to momentum conservation.

: **7-6 Q** A 1.0-kg object is fired from ground level at an initial speed of 5.0 m/s, at an angle 37° from the horizontal. Over the time when the object is in flight above ground level, the *maximum* magnitude of the object's momentum is
(A) 3.0 kg·m/s.
(B) 4.0 kg·m/s.
(C) 5.0 kg·m/s.
(D) 9.0 kg·m/s.
(E) impossible to determine.

· **7-7 Q** Newton's third law is intimately associated with — in fact, can be generally derived from—
(A) the law of inertia.
(B) conservation of linear momentum.
(C) the superposition principle for forces.
(D) conservation of mass.
(E) the definition of equilibrium.

· **7-8 Q** A ball of mass m is dropped from rest at a height h above a floor. The ball makes four bounces with the floor and then is at rest. What is the overall change in the ball's momentum?

· **7-9 Q** A golf ball is released from rest. It hits the ground and bounces back up. Is the momentum of the ball conserved in this process? If not, how would you enlarge the "system" so that momentum is conserved, or is this possible?

: **7-10 P** A 60-lb girl and a 50-lb boy face each other on frictionless roller skates. The girl pushes the boy, who moves away at a speed of 4 ft/s. The girl's speed is
(A) 2.1 ft/s.
(B) 3.3 ft/s.
(C) 4.0 ft/s.
(D) 4.8 ft/s.
(E) 4.2 ft/s.

: **7-11 Q** The frictional force (called the *viscous drag force*) acting on an object in a fluid is proportional to the velocity of the object. Knowing this, can you figure out how a person who has lost her oars while rowing on a lake could get to shore simply by moving around in the boat? Explain your reasoning.

: **7-12 Q** An artillery shell, subject to negligible friction, explodes in midair. Discuss the motion of the fragments in relation to the path the shell would have followed had it not exploded.

: **7-13 Q** A ball of mass m is dropped from rest from height h above a floor. The ball bounces back to height h'. What is the change in the momentum of the earth as a consequence of the ball's having bounced from it?

: **7-14 Q** Rockets are typically made in several stages, with each successive stage being jettisoned when its fuel is used up. What is the reason for doing this? Explain in terms of momentum concepts.

: **7-15 P** Two particles of equal mass, one moving east at 30 m/s, the other heading north at 40 m/s, collide and stick together. The velocity of the resulting composite is
(A) 50 m/s, 53° north of east.
(B) 70 m/s, 37° west of south.
(C) 35 m/s, 37° north of east.
(D) 25 m/s, 53° north of east.
(E) impossible to determine by the information given.

: **7-16 Q** A rubber bullet and a metal bullet, each of the same mass, speed, and size, are to be fired at a block of wood, hitting it in the same spot. Which is more likely to knock the block over?
(A) The rubber bullet.
(B) The metal bullet.
(C) Both have equal probability of knocking over the block.
(D) This cannot be answered without knowing whether the mass of the wooden block is greater than or less than the mass of a bullet.

: **7-17 Q** In which of the following cases is the linear momentum of the object(s) most nearly conserved?
(A) A *ball* falling freely in vacuum.
(B) An *automobile* making a turn at constant speed.
(C) A *rubber ball* as it bounces from the floor.

(D) *Two pool balls* as they collide at right angles, making a loud click.

(E) A *golf ball* when it is struck by a club.

: 7-18 P A freight car with an empty mass of 2000 kg rolls without friction under a hopper that drops sand into the car at a steady rate. If the empty car approaches the hopper with a speed of 3 m/s, and 5000 kg of sand are loaded into the car, determine the final speed of the car.

· 7-19 P A series of four identical freight cars, each of mass 20 000 kg, are rolling slowly along a track at 5 m/s. They are separated by 100 meters between each car. The front car bumps into a fifth stationary car of the same mass and couples to it. Succeeding cars then catch up with the ones ahead and couple to them. What is the final speed of the string of five cars, assuming negligible friction?

· 7-20 P Two air-track gliders moving toward each other collide. Initially, one has a mass of 100 gm and a speed of 45 cm/s, and the other has a mass of 200 gm and a speed of 30 cm/s. After the collision the smaller glider is knocked back with a velocity of 55 cm/s. What is the velocity of the 200-gm mass after the collision?

: 7-21 P When a neutron of mass m collides head-on with a deuteron of mass $2m$ at rest in a reactor, the deuteron is given a velocity $2v_0/3$, where v_0 is the initial velocity of the neutron. With what velocity does the neutron recoil?

: 7-22 P A 1200-kg Toyota traveling west at 30 km/hr collides with a 1000-kg Honda traveling north at 40 km/hr. They stick together in the collision. In what direction and with what speed are they moving just after the collision?

: 7-23 P A mass initially at rest explodes and splits into three parts. A 3-gm fragment moves along the y axis at 40 m/s. A 2-gm fragment moves at 3 m/s along the x axis. Determine the momentum (magnitude and direction) of the third fragment.

: 7-24 P In a pool shot a cue ball traveling with speed v_0 strikes a stationary ball of the same size and mass. The cue ball is deflected 30° to the left, and the other ball moves off on a line at 60° to the right. What is the final speed of each ball in terms of v_0?

: 7-25 P A 2000-kg car moving east at 4.0 m/s collides with and sticks to a 600-kg car moving south at 10.0 m/s. What is the direction of motion and speed of the two cars just after the collision?

: 7-26 P A 200-gm air track glider has mounted on it a toy gun of mass 50 gm, which fires a 10-gm rubber dart. The velocity of the dart as observed in the lab is 4 m/s. What is the recoil velocity of the glider and gun if they are at rest when the dart is fired?

: 7-27 P An artillery piece of mass 60 kg fires a shell of mass 10 kg with a muzzle velocity (with respect to the cannon) of 500 m/s. What is the recoil velocity of the cannon with respect to the earth?

: 7-28 P Mary is skating 4 m/s when her little brother, going 3 m/s, runs into her from the side as shown in Figure 7-11. They grab on to each other to keep from falling. In what direction, and with what speed, would they be moving just after the collision if Mary's weight is 400 N and her brother's weight is 250 N?

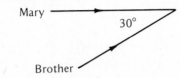

Figure 7-11. *Problem 7-28.*

: 7-29 P A 10-gm bullet traveling 600 m/s horizontally strikes a 600-gm wood block that is at rest on a frictionless surface. The bullet passes through the block and emerges with speed of 200 m/s. Sketch momentum versus time graphs for the bullet and for the block, and determine the speed acquired by the block.

: 7-30 P A 40-kg boy stands on a log floating in a pond. The log has a mass of 400 kg and is 20 m long. It experiences no friction while floating. The boy runs from one end to the other with an average speed of 3 m/s (with respect to the log). (a) How far does the log move when he does this? (b) What is its average speed?

: 7-31 P A large space ship of total mass M (including all fuel and cargo) is coasting through the solar system subject to no external force. Its constant velocity (with respect to the solar system) is v_0, directed from the sun toward Polaris. It now sends out a small rocket of mass $\frac{1}{4}M$ with velocity $\frac{1}{2}v_0$ with respect to the large ship. The small rocket is fired along a line perpendicular to the direction of travel of the large ship. What is the direction and velocity of the large ship just after firing?

: 7-32 P A speeding Volkswagen of mass 1100 kg crashes into a stationary Ford sedan of mass 1500 kg. The two cars stick together in a crumpled mass of metal and slide a distance of 6.2 meters before coming to rest. Tests show that such sliding cars experience a frictional force of one fifth their weight for the pavement conditions under which the accident occurred. How fast was the VW going when it hit the other car?

: 7-33 P A 0.12-kg glider moving 53° south of east on a smooth horizontal surface collides and sticks to an identical glider moving directly northward. The resulting composite object of total mass 0.24-kg then moves directly eastward at 3.0 m/s. See Figure 7-12.

(a) What is the speed of the first glider as it initially moves 53° south of east? (b) What is the speed of the second glider as it initially moves directly northward? (c) Is the linear momentum of the second glider, which initially

Before Collision   After Collision

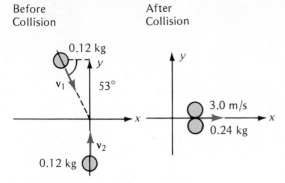

Figure 7-12.   *Problem 7-33.*

moves northward, the same before and after the collision? Is this result consistent with the general law for momentum conservation?

: 7-34 P Show that Newton's laws can be derived from the principle of momentum conservation.

: 7-35 P A proton (atomic mass 1) bounces off, at the same speed, a far more massive lead nucleus (atomic mass 207) at an angle of 45° from its initial direction of travel. In what direction with respect to the proton's original line of motion does the lead nucleus move?

Section 7-3   Impulse and Momentum

· 7-36 Q In simple terms, impulse is a measure of
(A) the force exerted on something.
(B) the change of momentum of something.
(C) an object's momentum.
(D) acceleration.
(E) how suddenly a force is applied.

· 7-37 Q Turbine blades are often designed so that they reverse the direction of the water striking them. Why is this done?

· 7-38 Q Impulse is
(A) measured in the same units as momentum.
(B) the time rate of change of momentum.
(C) force per unit time.
(D) the area under a graph of acceleration versus distance.
(E) the area under a graph of momentum versus time.

: 7-39 Q Which of the following statements is most accurate?
(A) When a ball is thrown against a wall, the force exerted by the ball on the wall depends only on the mass and velocity of the ball, not on whether or not the ball bounces back from the wall.
(B) When a ball is thrown against a wall it exerts a force on the wall. Thus we can imagine that a moving ball carries with it a certain amount of force.
(C) Suppose you wish to calculate the force exerted against a wall when a ball undergoes an elastic collision with it. In this case all you need to know is the kinetic energy of the ball, assuming the motion of the ball is perpendicular to the wall.
(D) Suppose that a ball moving perpendicular to a wall undergoes an elastic collision with the wall. You can determine the force exerted on the wall provided you know only the momentum of the ball.
(E) Suppose two balls have the same size and mass, but one is a hardball and the other is a softball. If they are each thrown against a wall on the same trajectory with the same speed, the hardball will exert a greater force on the wall because its collision time is shorter.

: 7-40 Q Consider two different automobile crashes. In the first, you are driving 30 mph and crash head on into an identical car also going 30 mph. In the second, you are driving 30 mph and crash head on into a stationary brick wall. In neither case does your car bounce off the thing it hits, and the collision time is the same in both cases. Which of these two situations would result in the greatest impact force?
(A) Hitting the other car.
(B) Hitting the brick wall.
(C) The force would be the same in both cases.
(D) We cannot answer this question without more information.
(E) None of the above is true.

: 7-41 Q A ball of mass 0.1 kg traveling 20 m/s strikes a concrete wall and bounces back with a velocity of 18 m/s. Which of the following additional pieces of information would most easily enable us to determine the average force the ball exerted on the wall when it struck?
(A) the ball was deformed by 3 mm.
(B) The ball lost 3.8 joules of energy.
(C) The ball was in contact with the wall for 0.01 seconds.
(D) The ball acquired a velocity of 10^{-6} m/s.
(E) 3.8 kg·m/s of momentum was transferred to the wall.

· 7-42 P Wheat is poured onto a conveyor belt at a steady rate of 20 kg/s. What force must be exerted to keep the belt moving at a steady speed of 2 m/s?

· 7-43 P High-speed photographs reveal that a golf club imparts a speed of 60 m/s to a ball of mass 55 gm in a collision which lasts 0.2 ms. What is the average force applied to the ball in this process?

: 7-44 P A hose squirts a horizontal stream of water against a building. The water strikes the building and runs down the wall. If the water flow is 4 l/s through a circular nozzle of 1 cm diameter, what average force is exerted against the building?

: 7-45 P A machine gun mounted on a stationary platform fires 480 bullets per minute. Each bullet has mass 10 gm and velocity 500 m/s. What average force is exerted on the platform?

: 7-46 P How many tennis balls, each of mass 100 gm and speed 10 m/s, must strike a wall horizontally each second in order to exert an average force of 500 N on the wall? Assume that each ball collides elastically with the wall.

: 7-47 P A flexible chain of length L and mass M is lifted vertically with constant velocity by applying a force F to one end. Derive an expression for F and plot the force as a function of time, starting while the chain is still on the table and continuing until it is completely off the table.

: 7-48 P An empty freight car of mass 2000 kg rolls at 3 m/s under a hopper, which loads sand into the car at a steady rate of 800 kg/s. If the car is 9 meters long and the sand is loaded uniformly over the length, what average force must be exerted during loading to keep the car moving at a constant speed?

: 7-49 P A fire hose directs a stream of water against a wall, delivering 500 l/s with a velocity of 40 m/s. The water strikes the wall at an angle of 45° and spreads out over the surface. What normal force is exerted on the wall?

: 7-50 P A 4-kg bowling ball is dropped from a height of 2 meters onto a concrete floor. It bounces back up to a height of 1.8 meters. What was the impulse applied to the floor?

: 7-51 P A baseball of mass 160 gm is moving 25 m/s when it is struck by the bat. It travels out toward the shortstop at 50 m/s. High speed photographs show that the collision with the bat lasted 1.2 ms. (a) What was the impulse of the collision? (b) What average force was exerted by the bat?

: 7-52 P Raindrops (2 mm diameter) typically fall with a terminal velocity of 6.5 m/s, whereas hailstones (2 cm diameter) fall at 20 m/s. What is the average force on 1 m² of a flat roof when struck by (a) a rain storm that delivers 1 inch of water per hour? (b) a hailstorm that delivers the same amount of water as in (a)?

: 7-53 P The force exerted on a 160-gm ball as a function of time is shown in Figure 7-13. If the ball is initially at rest, estimate graphically its final velocity.

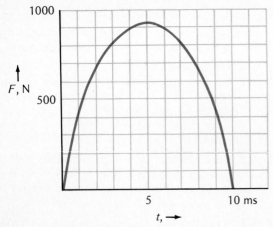

Figure 7-13.   Problem 7-53.

: 7-54 P A rocket has an initial mass of 20 000 kg, including fuel. At burnout, its mass is only 8000 kg. Exhaust gases leave the rocket at 160 kg/sec with a velocity of 2000 m/s relative to the rocket. The rocket is fired straight up from rest. (a) What is the initial thrust? (b) What is the thrust just before burnout? (c) If no external forces whatever (including gravity) act upon the rocket, what maximum speed could it reach?

: 7-55 P The acceleration of the 4-kg head of a dummy when hit by a heavyweight prizefighter has been measured, and the data obtained plotted in Figure 7-14. Graphically estimate the total impulse delivered and the resulting final velocity, assuming the dummy head was initially at rest and unconnected to anything else.

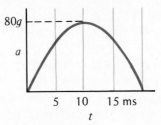

Figure 7-14.   Problem 7-55.

: 7-56 P A bullet of mass 12 gm and velocity 900 m/s is fired horizontally into a 1-kg block mounted on a frictionless air track. The block is initially at rest, and the bullet penetrates 20 cm into the block. Calculate (a) the velocity imparted to the block, (b) the average force on the block during impact, (c) the duration of the collision (the time from when the bullet struck the block until it came to rest in the block).

: 7-57 P A water jet squirts a stream of water at a constant rate R (kg/s) against the end of an air glider cart of mass M, initially at rest. Show that (a) the velocity of the cart after time t is $v(t) = v_0(1 - e^{-Rt/M})$; (b) in time t the cart has moved a distance $x(t) = v_0 t - \dfrac{Mv_0}{R}(1 - e^{-Rt/M})$.

Figure 7-15.   Problem 7-57.

: 7-58 P A beaker of mass m_0 is placed on a scale and water is poured into it from a height h at a rate of R kg/second. What is the reading of the scale as a function of time?

: 7-59 P A rocket is launched straight up from the earth's surface from rest. Its total initial mass, including fuel, is 15 000 kg. Exhaust gases are ejected at a constant rate with velocity 2500 m/s relative to the rocket. At burnout the rocket has used 10 000 kg of fuel and reached a speed of 2160 m/s. (a) What is the burn time of the engine? (Integration required). (b) What is the acceleration just before burnout?

Center of Mass

As we have seen in Chapter 7, an isolated system of particles behaves, in one respect, just like a single isolated particle. Both have constant momentum.

This implies that we can effectively replace a system of particles by a single equivalent particle. We do so through the concept of *center of mass*, and it is pertinent to examine the meaning of the location, velocity, and acceleration of a system's center of mass, and the special significance of a reference frame in which the center of mass is at rest.

8-1   Center of Mass Defined

Suppose that we have two mass points lying along the x axis, m_1 at x_1 and m_2 at x_2. Then, it is useful to define the x coordinate of their center of mass (abbreviated CM) as

$$X \equiv \frac{m_1 x_1 + m_2 x_2}{m_1 + m_2} \tag{8-1}$$

Equation (8-1) shows that each coordinate is, so to speak, "weighted" according to the mass at its location.

Example 8-1.   A 1-kg particle is at $x_1 = 0$ and a 2-kg particle is at $x_2 = 6$ m. Where is the center of mass of the two particles?
From (8-1),

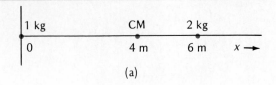

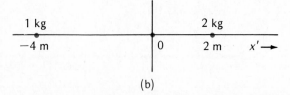

Figure 8-1. *Center of mass of two particles: (a) the origin of the x axis is at the left particle; and (b) the origin of the x′ axis is at CM.*

$$X = \frac{(1 \text{ kg})(0) + (2 \text{ kg})(6 \text{ m})}{(1 + 2) \text{ kg}} = 4 \text{ m}$$

In Figure 8-1(a), we see that the CM is 4 m from the 1-kg particle and 2 m from the 2-kg particle.

Let's locate the CM again, this time choosing the origin to be at $x = 4$ m, the location we have just computed for X.

With the new coordinate origin, the 1-kg particle is at $x_1' = -4$ m, and the 2-kg at $x_2' = 2$ m. (The new coordinates are identified by primes.) Applying (8-1) again, we have

$$X' = \frac{m_1 x_1' + m_2 x_2'}{m_1 + m_2}$$

$$X' = \frac{(1 \text{ kg})(-4 \text{ m}) + (2 \text{ kg})(2 \text{ m})}{(1 + 2) \text{ kg}} = 0$$

See Figure 8-1(b). The CM is at the *same* location relative to the two particles. Once again it is along the line joining the particles, and once again it is 4 m from the 1-kg mass and 2 m from the 2-kg mass. The respective distances of the particles from the CM is in the inverse ratio of their masses: $m_1/m_2 = 1 \text{ kg}/2 \text{ kg} = \frac{1}{2}$; and $x_1'/x_2' = (-4 \text{ m})/(2 \text{ m}) = -2$. The minus sign simply reflects that m_1 is to the left and m_2 to the right of the origin. More generally, when the CM is at the origin and $X' = 0$, we have, from (8-1),

$$\frac{x_1'}{x_2'} = -\frac{m_2}{m_1} \tag{8-2}$$

It is easy to generalize to any number of particles. Mass m_1 has vector displacement $\mathbf{r_1}$ relative to the origin, and its x and y coordinates are x_1 and y_1. (We suppose that all particles lie in the xy plane with zero z coordinates.) Other particles 2, 3, . . . are located in the same way. See Figure 8-2.

The vector displacement $\mathbf{R}$ of the CM is, by analogy with (8-1), given by

$$\mathbf{R} = \frac{m_1 \mathbf{r_1} + m_2 \mathbf{r_2} + m_3 \mathbf{r_3} + \cdots}{m_1 + m_2 + m_3 + \cdots}$$

which we can write more compactly as

$$\mathbf{R} = \frac{\Sigma m \mathbf{r}}{\Sigma m} \tag{8-3}$$

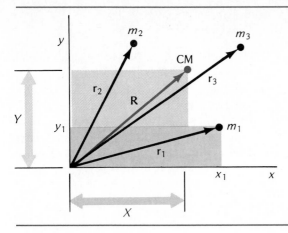

Figure 8-2. *Particle 1 has mass m_1 and a displacement* **r_1** *relative to the origin with rectangular components x_1 and y_1. Similarly for particles 2, 3,. . . .*

where the summation signs imply that we are to add contributions from all particles.

We can also write vector equation (8-3) in terms of its rectangular components. The x coordinate of the CM is designated X, and the y coordinate Y:

$$X = \frac{\Sigma mx}{\Sigma m} \quad \text{and} \quad Y = \frac{\Sigma my}{\Sigma m} \tag{8-4}$$

An analogous relation would give Z for the z coordinate.

8-2   Locating the Center of Mass

Finding the CM for a collection of mass points simply involves applying (8-3) or its equivalent, (8-4). For a continuous mass distribution—a solid rigid object, for example—we must, in effect, imagine the object to be subdivided into a collection of tiny objects, each small enough to be considered a mass point. Then we can again use the general relation for mass points.

Symmetrical Objects   It is easy to see that for a uniform symmetrical solid object, the CM is located on the geometrical line, or center of symmetry, of the object. Consider the uniform right circular cylinder shown in Figure 8-3. The coordinate origin is located at the cylinder's center of symmetry (halfway between the ends on the central line), which we shall show is also the location of the CM. First consider equal tiny volumes at the symmetrical locations, 1 and 2. With a *uniform* cylinder—one with the same density, or mass per unit volume, throughout—equal volumes contain equal masses. Mass at location 1 has a y coordinate y_1; the same volume of mass at location 2 has a y coordinate y_2. But $y_2 = -y_1$, with the two points symmetrically located above and below the midway, and for this pair of points $my_1 + my_2 = 0$. Indeed, the sum is zero for *any* pair of mass points symmetrically located along the y axis. The y coordinate of the CM is zero, $Y = 0$. By the same token, pairs of symmetrically located points along the x axis, such as 3 and 4 in Figure 8-3, yield $X = 0$. In short, the CM for every uniform symmetrical solid *is* at the center or on the line of symmetry—at the center of a sphere, a spherical shell, or a ring, for exam-

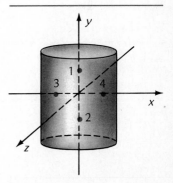

Figure 8-3. *Solid uniform right circular cylinder. Particles 1 and 2 are located symmetrically relative to the x axis; particles 3 and 4 are located symmetrically relative to the y axis.*

ple. Note that the CM need not be located *within* the material of a solid object; this certainly is the case for a spherical shell, a ring, or a doughnut.

Continuous Mass Distribution   To find the center of mass for a continuous distribution of mass, we replace the summation Σmx by the integral $\int x\,dm$, where dm is a small element of mass, small enough so that all atomic particles within dm can be taken as having the same coordinate x, but large enough so that the actually discrete atomic particles are equivalent to a continuous mass distribution. If the mass per unit volume, or density, is ρ, the mass element dm may be written $dm = \rho\,dv$, where dv represents the volume element containing mass dm of density ρ. Therefore, the center-of-mass coordinates become

$$X = \frac{\Sigma mx}{\Sigma m} = \frac{\int x\,dm}{\int dm} = \frac{\int \rho x\,dv}{\int \rho\,dv}$$

For constant density ρ throughout the continuous mass distribution,

$$X = \frac{\int x\,dv}{\int dv} \qquad (8\text{-}5a)$$

where the limits of the integrals correspond to the physical boundary of the object.

The corresponding relation for the Y coordinate is

$$Y = \frac{\int y\,dv}{\int dv} \qquad (8\text{-}5b)$$

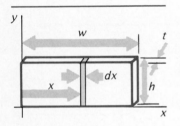

Figure 8-4.   *Center of mass of a rectangular plate. An element of mass is contained in the vertical strip of width dx and thickness t at displacement x.*

Example 8-2.   Find the center of mass of a uniform rectangular plate. The width is w, the height h, and the thickness t. The coordinate origin is chosen to be at the lower left corner of a rectangle, as shown in Figure 8-4. To find the x component of the center of mass, we imagine the rectangle to be divided into thin vertical sections, each of width dx. All mass points within any one section have the same x coordinate. We then sum the contributions from the vertical sections from $x = 0$ to $x = w$. The volume element is $dv = ht\,dx$, and (8-5a) gives

$$X = \frac{\int x\,dv}{\int dv} = \frac{\int_0^w x(ht\,dx)}{hwt} = \frac{\int_0^w x\,dx}{w} = \frac{w^2/2}{w} = \frac{w}{2}$$

The CM is halfway between the ends. Similarly, it is halfway up. In short, the center of mass is the center of symmetry of the rectangle.

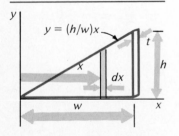

Figure 8-5.   *Center of mass of a uniform right triangular plate.*

Example 8-3.   Find the center of mass of a uniform right-triangular plate, height h and width w.

See Figure 8-5, where the volume element dv has width dx, height y, and thickness t. The hypotenuse of the triangle is the line $y = (h/w)x$. Therefore, $dv = yt\,dx = (h/w)tx\,dx$, and (8-5) gives

$$X = \frac{\int x\,dv}{\int dv} = \frac{\int_0^w (h/w)tx^2\,dx}{hwt/2} = \frac{2}{3}w$$

In similar fashion, $Y = h - \frac{2}{3}h = \frac{1}{3}h$, again one-third the distance from the square corner.

Reduction to Mass Points Suppose that the shape of an object is too complicated to allow for integration to locate the CM. It may be possible to subdivide the object into parts, each of whose CM location is known. In effect, an extended object is then reduced to a collection of mass points. Consider Figure 8-6, for example. There a right circular cylinder is placed against a rectangular sheet. The CM of the cylinder is at its center of symmetry (CM$_c$ in Figure 8-6), and the CM of the rectangular sheet at its center of symmetry, CM$_s$. The two objects may be replaced by point masses at their respective centers of symmetry. The CM of the composite object then lies along the line joining CM$_c$ and CM$_s$, and the distances to the individual CM's are in the inverse ratio to their respective masses, following (8-2).

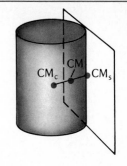

Figure 8-6. Uniform right circular cylinder with CM$_c$. Uniform rectangular sheet with CM$_s$. The CM of the entire system lies on the line joining CM$_c$ and CM$_s$.

Example 8-4. A uniform square (4 × 4) plate has one square (2 × 2) missing from a corner. See Figure 8-7(a). Where is the plate's center of mass?

Imagine the plate to be subdivided into a 2 × 4 rectangle and a 2 × 2 square, as shown in Figure 8-7(b). The CM of the rectangle is at its center of symmetry, with x and y coordinates (1, 2); the CM of the square has coordinates (3, 1). Since the plate is uniform, the mass of the rectangle is proportional to its area and can be given as 8k, where k is the mass per unit area. The mass of the square is 4k.

The L-shaped plate has now been reduced to two equivalent particles: a mass of 8k at (1, 2) and a mass of 4k at (3, 1). The x component of the CM of these two particles is given by

$$X = \frac{\Sigma mx}{\Sigma m} = \frac{(8k)(1) + (4k)(3)}{8k + 4k} = 1.67$$

It is easy to see, even without computation, that Y is also 1.67. Consider Figure 8-7(c), in which the dashed line at 45° now subdivides the plate into two symmetrical halves. Clearly, the CM of the plate must lie on this 45° line. Therefore,

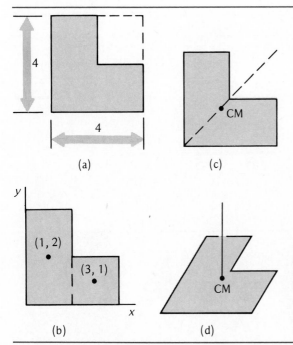

(a)

(c)

(b)

(d)

Figure 8-7. (a) Uniform square with a missing square corner. (b) CM of rectangle and CM of square. (c) CM lies along symmetry axis. (d) The plate is in equilibrium when suspended from its CM.

$$X = 1.67 \quad \text{and} \quad Y = 1.67$$

If the plate were suspended by a cord attached at the CM, it would remain in equilibrium, as shown in Figure 8-7(d). (The detailed proof is given in Chapter 12.)

Irregular Shapes You are to locate the CM of an irregularly shaped solid not necessarily made of material of the same density throughout. Any of the mathematical procedures given previously are too tedious to be practicable. Here is an experimental procedure. Suspend the object from any point 1 and let it come to rest; then the object's CM will lie on the vertical line L_1 passing through 1. See Figure 8-8(a). Now suspend the object from a second point 2, and again the CM lies somewhere on the new vertical line L_2. The object's CM must lie on L_1 and on L_2. The *intersection* of L_1 and L_2 satisfies this condition; it is the location of the CM.

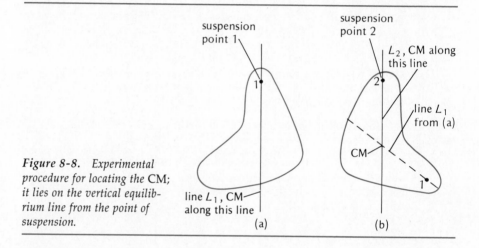

Figure 8-8. *Experimental procedure for locating the* CM; *it lies on the vertical equilibrium line from the point of suspension.*

Figure 8-9. *Mobile by Alexander Calder.*

This procedure depends on the fact that for an object small compared with the earth, the *center of mass* has the *same location* as the *center of gravity.* By *center of mass,* we mean that location at which we can imagine (in the sense discussed in this chapter) all the object's *mass* to be concentrated. By *center of gravity* is meant the location at which all the object's *weight* can be imagined to be located. (A formal definition of the center of gravity and the proof that it has the same location as the center of mass for objects small compared with the earth is given in Section 12-7.) Finally, with an object replaced by a point mass — really a point weight — we see that for equilibrium, this point must lie beneath the suspension point, just as a single mass point on a massless cord (a pendulum) lies along the vertical. See Figure 8-9.

8-3 Velocity of the Center of Mass

Thus far, we have been concerned with the location of the CM at some one instant. But the individual particles of a system need not be at rest. If the system consists of a solid object, it may move through space. If the system consists merely of unattached particles, they too may move through space; moreover, the individual particles need not execute the same motion. So the location of the CM may change with time. Here we are concerned with the velocity $\mathbf{V}$ of the CM of the system and how it is related to the velocities of individual particles.

We start with (8-3),

$$\mathbf{R} = \frac{\Sigma\, m\mathbf{r}}{\Sigma\, m}$$

where $\mathbf{r}_1, \mathbf{r}_2, \ldots$ are the displacements relative to the same origin of particles with respective masses $m_1, m_2, \ldots$. We take the derivative with respect to time of both sides of this equation. The individual particle velocities are $\mathbf{v}_1 = d\mathbf{r}_1/dt$, $\mathbf{v}_2 = d\mathbf{r}_2/dt$, ... , so that the CM velocity $\mathbf{V} = d\mathbf{R}/dt$ is

$$\mathbf{V} = \frac{\Sigma\, m\mathbf{v}}{\Sigma m} \tag{8-6}$$

The equivalent relations of the x and y velocity components are

$$V_x = \frac{\Sigma m v_x}{\Sigma m} \quad \text{and} \quad V_y = \frac{\Sigma m v_y}{\Sigma m} \tag{8-7}$$

The sum Σm is just the mass M of the entire system of particles, $\Sigma m = M$, so that (8-6) can be rewritten as

$$\text{Total momentum of system} = \Sigma m\mathbf{v} = M\mathbf{V} \tag{8-8}$$

Here we have recognized $\Sigma m\mathbf{v}$ to be just the total momentum of the system. From (8-8), we see that this total momentum equals the momentum of a single equivalent particle with mass M and velocity $\mathbf{V}$. We can, so far as momentum is concerned, replace an entire system of particles, possibly flying in all sorts of directions, by a single mass point M at the center of mass.

We know from the momentum-conservation principle that if a system of particles is isolated from external influence, the system's momentum is constant:

$$\Sigma m\mathbf{v} = \text{constant} \quad \text{if } \Sigma \mathbf{F}_{\text{ext}} = 0$$

From (8-8), we see that that now means also

$$MV = \text{constant} \qquad \text{if } \Sigma \mathbf{F}_{ext} = 0 \qquad (8\text{-}9)$$

For an isolated system, the CM has a *constant* velocity. Actually, this simple result should not surprise us. We know that an isolated particle is like an isolated system of particles in one respect — both have constant momentum.

Example 8-5. A 4-kg object initially at rest explodes into three pieces. A 1-kg piece flies east at 5.0 m/s, and a 2-kg piece flies 60° south of west at 4.0 m/s. What is the velocity of the system's center of mass?

The momentum of the system is initially zero, and it remains zero. Therefore, quite apart from any of the details of what happens after the explosion, we know that the center-of-mass velocity must be initially *zero*, and must remain so.

Example 8-6. A 1-kg block travels to the right at 12 m/s and collides with a 2-kg block initially at rest. See Figure 8-10. The two blocks stick together after the collision. Find the velocity of the CM before and after the collision.

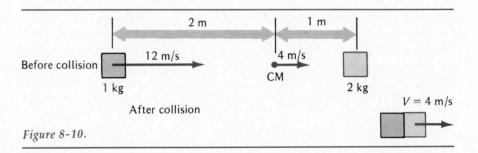

Figure 8-10.

By applying momentum conservation, we found in Example 7-1 that these two objects travel to the right at 4 m/s after the collision. Now reexamine this simple collision to find the location and velocity of the CM before and after the collision.

From (8-7), the CM velocity is

$$V_x = \frac{\Sigma m v_x}{\Sigma m} = \frac{(1 \text{ kg})(12 \text{ m/s}) + (2 \text{ kg})(0)}{(1 + 2) \text{ kg}} = 4 \text{ m/s}$$

Here we added the contribution of the two blocks *before* the collision.

Let's do it again, this time using the velocities *after* the collision. We have in effect a single 3-kg composite object moving at 4 m/s:

$$V_x = \frac{\Sigma m v_x}{\Sigma m} = \frac{(3 \text{ kg})(4 \text{ m/s})}{3 \text{ kg}} = 4 \text{ m/s}$$

The result is the same. The two objects are isolated from external forces, and the CM necessarily travels at *constant* velocity. Before the collision, the CM lies between the two objects and the distances from the CM to the 1-kg and 2-kg objects are always in a 2-to-1 ratio. After the collision, the CM travels with the single composite object.

Example 8-7. Reexamine the collision of Example 7-3 to find the CM velocity.

The masses and velocities before collision are exactly the same as in Example 8-6, so that once again the CM velocity is 4 m/s to the right, *before, during, and after the collision.* Again the distances of the two particles from the CM are in the ratio 2 to 1, as shown in Figure 8-11.

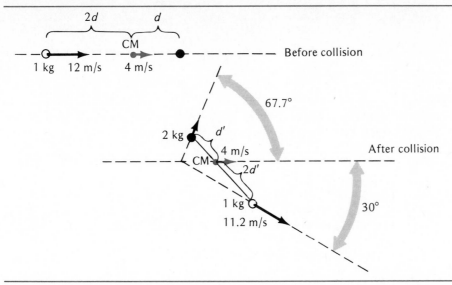

Figure 8-11.

8-4   Center-of-Mass Reference Frame

An isolated system's center of mass has a constant velocity in any inertial frame. But one particular inertial frame is of special interest—the *center-of-mass reference frame*—because in it an isolated system's *total momentum* is not only constant but *zero* and the *center-of-mass velocity* not only constant but *zero*. That the center-of-mass frame is indeed the zero-momentum frame follows immediately from (8-8):

$$M\mathbf{V} = \Sigma m\mathbf{v}$$

$$\Sigma m\mathbf{v} = 0 \qquad \text{if } \mathbf{V} = 0$$

In words, if the center of mass is at rest, the system's total momentum $\Sigma m\mathbf{v}$ must be zero.

Example 8-8.   Let us reexamine the collisions appearing earlier in Figures 8-10 and 8-11 both from the point of view of an observer in the laboratory reference frame (unprimed coordinates, velocities) and now also from the point of view of an observer riding with the CM reference frame (primed coordinates, velocities). For both collisions, the CM velocity is 4 m/s to the right with respect to the laboratory. Therefore, relative

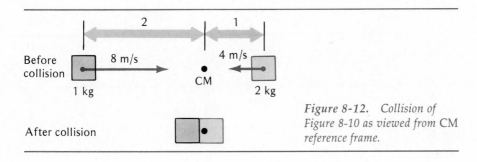

Figure 8-12.   Collision of Figure 8-10 as viewed from CM reference frame.

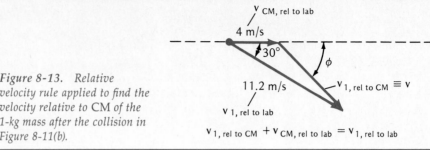

Figure 8-13. Relative velocity rule applied to find the velocity relative to CM of the 1-kg mass after the collision in Figure 8-11(b).

to the CM reference frame, the 1-kg object has a velocity of $v_1' = 12$ m/s $- 4$ m/s $= 8$ m/s, and the 2-kg object a velocity of $v_2' = 0 - 4$ m/s $= -4$ m/s. See Figure 8-12. [We have used the rule for adding relative velocities, (4-19).] The two objects approach the CM in opposite directions. Their distances to the CM are again always in a 2-to-1 ratio. We now see that their velocities are also in a 2-to-1 ratio. This ensures that the system's total momentum is zero: $\Sigma m\mathbf{v}' = (1 \text{ kg})(8 \text{ m/s}) + (2 \text{ kg})(-4 \text{ m/s}) = 0$. After the collision, the composite object is still at rest in the CM reference frame, and the total momentum is still zero.

   In Figure 8-11(b), we have the 1-kg object moving at 11.2 m/s at 30° off the forward direction. In what direction does this object travel relative to the CM? We apply the rule for relative velocity vectors (Section 4-6), shown in Figure 8-13.

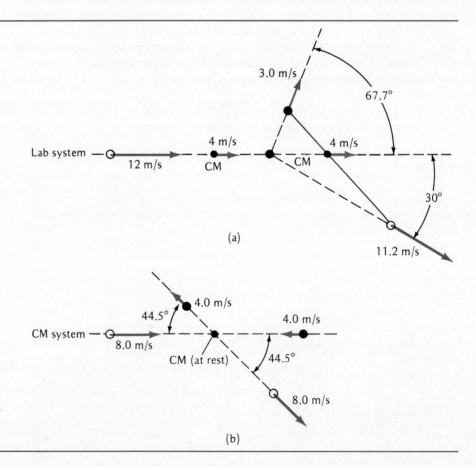

Figure 8-14. Collision of Figure 8-11: (a) laboratory reference frame, and (b) CM reference frame.

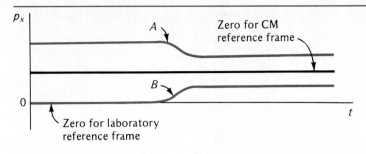

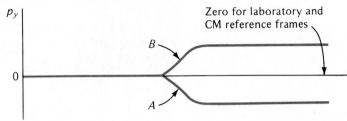

Figure 8-15. Momentum-time graphs for the x and y components for the collision shown in Figure 8-14. Note that the momenta for the laboratory and CM reference frames differ only in the zero for momentum.

For the y' velocity components, we have

$$v' \sin \phi' = (11.2 \text{ m/s}) \sin 30°$$

and for the x' components,

$$(4 \text{ m/s}) + v' \cos \phi' = (11.2 \text{ m/s}) \cos 30°$$

Solving these equations simultaneously yields (check it!)

$$\phi' = 44.5° \quad \text{and} \quad v' = 8.0 \text{ m/s}$$

There is no need to compute separately the direction of the 2-kg mass relative to the CM. Since the total momentum is zero in the CM reference frame, the 2-kg object moves in a direction opposite that for the 1-kg object and at half the speed. Here again the speeds and distances are in a 2-to-1 ratio. See Figure 8-14.

Consider the momentum-time graphs for the laboratory and center-of-mass reference frames, as shown in Figure 8-15. The x and y components of the momenta are plotted separately. The x-momentum components of the two particles are different in the two reference frames. In fact, the two reference frames differ simply in the choice of the zero of momentum. Only in the center-of-mass reference frame is the system's total momentum zero. Because the center of mass moves along the x axis, the y-momentum components are the same for the laboratory and center-of-mass reference frames.

8-5   Acceleration of the Center of Mass

When a system is isolated from forces, the velocity of its CM is constant. But when the particles are subject to external forces, the CM may undergo an acceleration. How is the CM acceleration, $\mathbf{A} = d\mathbf{V}/dt$, related to the accelerations of the particles, $\mathbf{a}_1 = d\mathbf{v}_1/dt$, $\mathbf{a}_2 = d\mathbf{v}_2/dt$, . . . , and especially to the forces on the particles?

When we take the derivative with respect to time of both sides of (8-6), $\mathbf{V} = \Sigma m\mathbf{v}/\Sigma m$, we get

$$\mathbf{A} = \frac{\Sigma m\mathbf{a}}{\Sigma m} \tag{8-10}$$

Figure 8-16. *Exploding fireworks rockets.*

The relation for the CM acceleration is similar to (8-3) for its displacement and (8-6) for its velocity.

A proof completed in Section 7-4 gives the true significance of (8-10). There it was shown, in (7-17) and (7-19), that

$$\Sigma \mathbf{F}_{ext} = M\mathbf{A} = \Sigma m\mathbf{a} \qquad (8\text{-}11)$$

where $\Sigma \mathbf{F}_{ext}$ represents the resultant *external* force on the system of particles. (The *internal* forces — between any one particle in the system and any other particle also in the system — don't enter the picture at all; they come in equal and opposite pairs and therefore add to *zero.*)

Equation (8-11) then says that we can think of an entire system of particles replaced by a single equivalent particle at the system's CM. Its mass $M = \Sigma m$ is equal to that of the system, and its acceleration $\mathbf{A}$ is controlled by the external forces alone. Moreover, this simple behavior in no way depends on how the particles within the system act on one another.

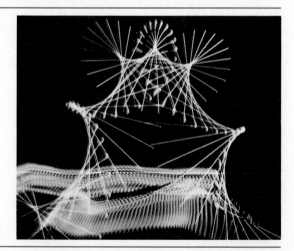

Figure 8-17. *Photograph showing that the CM of a thrown object traces out a parabolic path.*

For example, a fireworks rocket explodes into numerous flares. See Figure 8-16. While the unexploded projectile was in free flight traveling upward, it followed a parabolic path, which was also the motion of its CM. After the explosion into many fragments, the system's CM *continues in the same parabolic path,* since the only external force on the rocket, before or after the explosion, is the constant gravitational force downward.

Figure 8-17 shows the motion of a rigid body thrown into the air. Its center of mass traces out a parabolic path, irrespective of the motions of the individual parts of the body.

Summary

Definitions

Center-of-mass (CM) displacement **R**, velocity **V**, and acceleration **A** in terms of the corresponding properties of the particles in the system:

$$\mathbf{R} = \frac{\Sigma m\mathbf{r}}{\Sigma m} \qquad (8\text{-}3)$$

$$\mathbf{V} = \frac{\Sigma m\mathbf{v}}{\Sigma m} \qquad (8\text{-}6)$$

$$\mathbf{A} = \frac{\Sigma m\mathbf{a}}{\Sigma m} \qquad (8\text{-}10)$$

Important Results

Computing the location of the CM always amounts to reducing the system to a collection of particles.

A system of particles may, in effect, be replaced by a single equivalent particle at the CM with a mass $M = \Sigma m$ equal to that of the entire system and with velocity **V** and acceleration **A** whose significance is contained in the relations:

$$\text{Total momentum of system} = \Sigma m\mathbf{v} = M\mathbf{V} = \text{constant}$$
$$\text{when } \Sigma\mathbf{F}_{\text{ext}} = 0 \qquad (8\text{-}8)$$

and

$$\Sigma\mathbf{F}_{\text{ext}} = \Sigma m\mathbf{a} = M\mathbf{A} \qquad \text{when } \Sigma\mathbf{F}_{\text{ext}} \neq 0 \quad (8\text{-}11)$$

For the inertial reference frame attached to an isolated system's CM:

$$\Sigma m\mathbf{v} = 0 \qquad \text{and} \qquad \mathbf{V} = 0$$

The center-of-mass frame is the zero-momentum frame.

Problems and Questions

Section 8-1   Center of Mass Defined

· **8-1 Q** The center of mass of an object
(A) always lies within the object.
(B) is located halfway between the ends of the object.
(C) may lie outside of the object.
(D) has no meaning for an inhomogeneous material.
(E) has meaning only for symmetric objects.

: **8-2 Q** Why are trees, plants, and animals symmetrical?

: **8-3 Q** Two women sit at opposite ends of a canoe at rest in the water. The woman on the left throws a ball toward the one on the right. While the ball is in flight the center of mass of the system (canoe plus women plus ball)
(A) is stationary with respect to the shore, but moves to the right with respect to the canoe.
(B) is stationary with respect to the shore, but moves to the left with respect to the canoe.

(C) is stationary with respect to the canoe, but moves to the right with respect to the shore.
(D) is stationary with respect to the canoe, but moves to the left with respect to the shore.
(E) is stationary with respect to the canoe as well as the shore.

Section 8-2   Locating the Center of Mass

: **8-4 Q** A uniform sheet of cardboard has the shape shown in Figure 8-18(a). As was shown in Example 8-4, the coordinates of this sheet's center of mass are (1.67, 1.67) relative to the origin at the lower left corner. Then the cardboard is folded over, as shown in part (b). The center of mass of the new configuration now has coordinates (x, y) where
(A) $x = 1.67; y = 1.67$
(B) $x < 1.67; y > 1.67$

(C) $x > 1.67; y < 1.67$
(D) $x < 1.67; y < 1.67$
(E) $x > 1.67; y > 1.67$

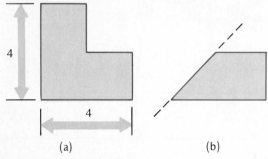

(a)   (b)

Figure 8-18.   Question 8-4.

: **8-5 Q** Two people of mass 60 kg each sit on opposite ends of a stationary sled. Assume that the sled slides without friction on the ice, that there is no air resistance, and that the two people remain stationary with respect to the sled. The person on the left holds a ball which he throws directly toward the person on the right. After the ball is thrown, the center of mass of the system (the two people plus the sled plus the ball)
(A) is stationary with respect to the sled, but moves to the right with respect to the ice.
(B) is stationary with respect to the sled, but moves to the left with respect to the ice.
(C) is stationary with respect to the ice but moves to the right with respect to the sled.
(D) is stationary with respect to the ice but moves to the left with respect to the sled.
(E) none of the other answers.

: **8-6 Q** See the statement in Problem 8-5. The person on the right end of the sled misses catching the ball entirely. The ball rolls up on land, where it finally comes to rest. We know, then, that the horizontal momentum of the entire system (the two people plus the sled plus the ball) does not have a constant value
(A) before the ball is thrown
(B) during the throwing process
(C) while the ball is in flight
(D) during the time the ball is being brought to rest on the bank
(E) after the ball has come to rest on the bank.

: **8-7 P** Asteroids A and B have masses m_A and m_B. They are initially at rest with separation d. Their mutual gravitational attraction then draws them together. How far from A's original position will they collide?
(A) $\frac{1}{2}d$
(B) $m_A d / m_B$
(C) $m_B d / m_A$
(D) $m_A d / (m_A + m_B)$
(E) $m_B d / (m_A + m_B)$

· **8-8 P** Masses 2 kg and 4 kg are placed at $x = 2\ m$ and $x = 5\ m$ on the x-axis. Locate the center of mass of the system.

· **8-9 P** Determine the location of the center of mass of the earth-moon system. (Earth mass $= 5.98 \times 10^{24}$ kg; moon mass $= 7.36 \times 10^{22}$ kg; earth-moon separation distance $= 3.84 \times 10^8$ m)

· **8-10 P** From a uniform plate is cut a "T" shape, as shown in Figure 8-19. Locate its center of mass.

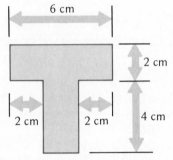

Figure 8-19.   Problem 8-10.

: **8-11 P** Locate the center of mass of a uniform rectangular plate of dimension 10 cm $\times$ 20 cm in which a circular hole of diameter 10 cm has been cut.

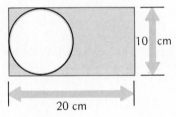

Figure 8-20.   Problem 8-11.

: **8-12 P** Locate the center of mass of a uniform sphere of radius R in which there is a spherical void of diameter R, as shown in Figure 8-21.

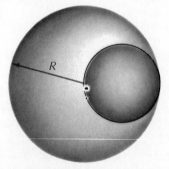

Figure 8-21.   Problem 8-12.

: **8-13 P** What should be the mass of the fish to balance the mobile in Figure 8-22?

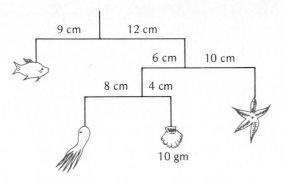

Figure 8-22. Problem 8-13.

: 8-14 P Determine the center of mass of a uniform semi-circular plate of radius R.

: 8-15 P Three identical masses are placed at the vertices of an equilateral triangle of side a. Locate the center of mass.

: 8-16 P Find the location of the center of mass of the thin uniform sheet shown shaded in Figure 8-23.

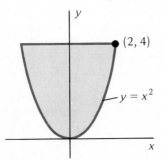

Figure 8-23. Problem 8-16.

: 8-17 P Three atoms each of mass M are placed at the three corners of an equilateral triangle that is the base of a regular tetrahedron. A fourth atom of mass $2M$ is placed at the upper corner of the tetrahedron. Locate the center of mass of the molecule.

: 8-18 P A uniform wire is bent into a semicircle of radius R. Locate the center of mass of this C-shaped wire.

· 8-19 P Particles are placed as follows: 1 kg at (0, 0); 2 kg at (2, 0); 3 kg at (2, 2). Locate the center of mass.

: 8-20 P Show that the center of mass of a hemisphere of uniform density and radius R is along the symmetry axis at a distance $3R/8$ from the flat end.

: 8-21 P Two hemispheres of radius R, one with mass M_1 and the other with mass M_2, are joined together to form a sphere. Locate the center of mass. (See also Problem 8-20.)

: 8-22 P Find the center of mass of a uniform plate of mass M which is bounded by the x-axis and the curve $4x^2 + y^2 = 4$, with $y > 0$.

: 8-23 P Find the center of mass of a uniform plate which is bounded by the x-axis, the y-axis, and the curve $x = 9 - y^2$.

: 8-24 P Locate the center of mass of a solid right circular cone of uniform density. The radius of the base is R and the height is h.

Section 8-3   Velocity of the Center of Mass
: 8-25 P Three particles move in the x-y plane. Determine (a) the coordinates of the center of mass and (b) the x and y components of the velocity of the center of mass.

	(x, y) meters	(v_x, v_y) m/s
$m_1 = 2$ kg	(0, 2)	(−2, 0)
$m_2 = 1$ kg	(3, 2)	(1, −1)
$m_3 = 3$ kg	(4, 0)	(−1, 1)

· 8-26 Q A system consists of two particles with $m_1 = 1$ kg and $m_2 = 2$ kg. The velocity of this system's CM
(A) is zero when the CM is at the origin.
(B) is three times the speed of m_1.
(C) is to the east when m_2 moves east.
(D) is the same as the velocity of m_1 when both particles have the same speed.
(E) is equal to the system's total momentum divided by its total mass.

: 8-27 P An object explodes into two pieces of equal mass. One piece goes east at speed v, while the other goes west at speed $3v$. What was the velocity of the center of mass of the two pieces before the explosion?
(A) Zero
(B) v west
(C) $2v$ west
(D) $4v$ west
(E) $2v$ east

: 8-28 P A 2-kg object initially traveling 4.0 m/s explodes in flight into two pieces, each of mass 1 kg. One piece travels at 8.0 m/s in the same direction as the 2-kg object was moving initially. What is the speed of the center of mass of the two pieces after the explosion?

: 8-29 P A 2.0-kg object moves in the positive x direction at 3 m/s. A 4.0-kg object is at rest at the origin. A 6.0-kg object moves in the negative x direction at 7.0 m/s. What is the x component of the velocity of the center of mass of these objects?

: 8-30 P A point mass of 1 gm is placed at the top of the

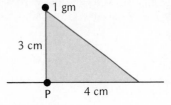

Figure 8-24. Problem 8-30.

wedge in Figure 8-24. The wedge has mass 4 gm. Assuming negligible friction, where, with respect to fixed point P on the horizontal floor, will the small mass strike the floor if the point mass and wedge are released from rest?

Section 8-4   Center-of-Mass Reference Frame

· **8-31 Q** An important reason for using the center-of-mass reference frame in analyzing collisions is that
(A) the total velocity of the system is zero before and after the collision.
(B) the total momentum of the system is zero before and after the collision.
(C) the velocity of each particle is zero in this frame.
(D) all of the mass of the system is localized at the CM.
(E) one of the particles involved in the collision is always at rest in this frame.

· **8-32 Q** In the center-of-mass reference frame
(A) all particles have constant velocity.
(B) the net force acting on the system is zero.
(C) the center of mass always has zero velocity.
(D) it is possible that at one instant all particles could have positive velocity components in the x direction.
(E) the total momentum of the system has a constant, non-zero value.

⋮ **8-33 Q** Consider two particles of masses m_1 and m_2, which may or may not be moving with respect to the laboratory.
(A) If the center of mass of the particles is moving, the CM reference frame is not an inertial coordinate system.
(B) The velocity vectors of the particles need not be anti-parallel in the CM frame.
(C) The momentum of the particles in the CM frame need not be zero.
(D) The speed of the CM is less than or equal to the speed of the fastest particle.
(E) The velocity of the CM in the CM frame need not be zero.

⋮ **8-34 P** A particle of 2 gm mass, with velocity 5 m/s, collides with an identical stationary particle. The incoming particle is deflected by 30° from its initial line of motion, and has a velocity of 4 m/s after the collision.
(a) Draw the momentum vectors for the particles in the center-of-mass frame before and after the collision, and in so doing determine the direction of motion of the second particle. Draw in the momentum vectors before and after the collision in the lab frame as well. (b) Draw a diagram showing the velocity vectors of the particles before and after the collision in the lab frame and in the CM frame.

⋮ **8-35 P** A billiard ball traveling at speed v strikes an identical stationary ball. After the collision each ball moves with speed $v/\sqrt{2}$, and the incoming ball has been deflected by 45° from its initial line of motion.
(a) What is the velocity of the system's center of mass? (b) What are the velocities of the two balls before the collision in the center-of-mass frame? (c) What are the velocities of the two balls after the collision in the center-of-mass frame? (d) What are the momenta of the two balls in the center-of-mass frame before and after the collision?

⋮ **8-36 P** An object moving north at 20 m/s explodes in flight into two pieces of equal mass. One travels 80 m/s 60° west of north. (a) What is the velocity of the second piece as measured in the laboratory? (b) What is the velocity of the system's center of mass? (c) What is the velocity of the first piece in the CM frame? (d) What is the velocity of the second piece in the CM frame?

⋮ **8-37 P** A particle of mass 1 gm with velocity 4 m/s to the east strikes a second 1-gm particle at rest. After collision, one particle travels in the direction 30° north of east; the struck particle moves in the direction 60° south of east. Draw four diagrams, showing the velocity vectors in the center-of-mass frame before and after the collision, and the momentum vectors in the center-of-mass frame before and after the collision.

Section 8-5   Acceleration of the Center of Mass

· **8-38 P** A particle of mass 3 kg is acted on by a force of 4 N directed along the x-axis and a second particle of mass 2 kg is acted on by a force of 6 N directed along the y-axis. Determine the acceleration of the system's center of mass.

· **8-39 Q** The net external force on a system of two objects is zero. It is then necessarily true that
(A) both objects are at rest.
(B) the objects have the same speed, but are moving in opposite directions.
(C) neither object is accelerating.
(D) the center of mass of the objects has a constant velocity.
(E) the objects have momenta of equal magnitude but opposite direction.

⋮ **8-40 Q** A uniform rod of length L is tipped over 45° from the vertical and released from rest while its bottom end rests on a frictionless plane. Describe qualitatively its motion while falling. Where does it land with respect to its original position?

Work and Kinetic Energy

9

9-1 Kinetic Energy Defined
9-2 Work Defined and the Work-Energy Theorem
9-3 Properties of Work
9-4 The Dot (or Scalar) Product
9-5 Unit Vectors and the Dot Product (Optional)
9-6 Work Done By Gravity
9-7 Conservative Force
9-8 Work Done by a Spring
9-9 Power
   Summary

Energy is a fundamental concept of physics. Here we are concerned specifically with the work done by a force and its relation to the change in the kinetic energy of a particle. We shall examine in some detail the work done by the constant force of gravity and by a compressed or a stretched spring. We shall find that both forces are conservative forces, a circumstance that will lead us to the energy-conservation principle (Chapter 10).

9-1 Kinetic Energy Defined

A simple but special situation introduces us to the concept of kinetic energy. Subsequently we shall see how the change in a particle's kinetic energy is

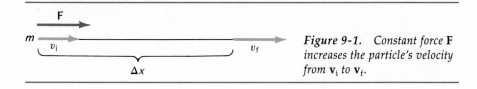

Figure 9-1. *Constant force* **F** *increases the particle's velocity from* $\mathbf{v}_i$ *to* $\mathbf{v}_f$.

related, in general, to the forces acting on the particle and the work done by these forces.

A constant force **F** acts on a particle of mass m along the direction of its displacement $\Delta\mathbf{x}$ along a straight line. See Figure 9-1. As a consequence, the particle's speed goes from an initial value v_i to a final speed v_f.

Newton's second law requires that

$$F = ma$$

where the constant acceleration a is related to Δx, v_i, and v_f through the kinematic equation

$$2a\,\Delta x = v_f^2 - v_i^2 \tag{3-10}$$

If we eliminate the acceleration a from the two previous equations, we arrive at a relation that can be written in the form

$$F\,\Delta x = \tfrac{1}{2}mv_f^2 - \tfrac{1}{2}mv_i^2 \tag{9-1}$$

In words, the constant force F multiplied by the displacement Δx over which it acts equals the change in the quantity $\tfrac{1}{2}mv^2$. We shall return in later sections to the left side. (It is the work $F\,\Delta x$ done by force F over the displacement Δx.) Here we concentrate on the right side and the quantity $\tfrac{1}{2}mv^2$, which is defined as the *kinetic energy* of a particle of mass m in motion at speed v:*

$$\text{Kinetic energy} = K = \tfrac{1}{2}mv^2 \tag{9-2}$$

A particle in motion has kinetic energy. Because m and v^2 are both positive, the kinetic energy is a scalar and always positive. Kinetic energy is, we shall see, as significant a property of a particle as its vector momentum $\mathbf{p} = m\mathbf{v}$. A

---

* The relativistic kinetic energy K of a particle of rest mass m_0 and speed v is given by

$$K = m_0 c^2 \left[\frac{1}{\sqrt{1 - (v/c)^2}} - 1 \right]$$

where c is the speed of light. This relation looks quite different from the classical relation $\tfrac{1}{2}m_0 v^2$ applicable for low speeds. Actually, the relativistic relation reduces, as it must, to the classical relation for kinetic energy for low particle speeds, or $v/c \ll 1$. Consider the term with the radical, which can be expanded using the binomial theorem:

$$\frac{1}{\sqrt{1 - (v/c)^2}} = \left[1 - (v/c)^2 \right]^{-\frac{1}{2}} = 1 + \frac{1}{2}\left(\frac{v}{c}\right)^2 + \frac{3}{8}\left(\frac{v}{c}\right)^4 + \cdots$$

For low speeds the term with $(v/c)^4$ can be discarded, so that the low speed kinetic energy becomes

$$K = m_0 c^2 \left[1 + \frac{1}{2}\left(\frac{v}{c}\right)^2 + \cdots - 1 \right] = \frac{1}{2}m_0 v^2$$

At the high speed extreme $v \to c$, the kinetic energy becomes infinite.

The relation for relativistic kinetic energy can be written in simpler form by introducing the following definitions:

$$\text{Total relativistic energy of a particle} = E = mc^2 \equiv \frac{m_0 c^2}{\sqrt{1 - (v/c)^2}}$$

where the so-called relativistic mass m is $m_0 / \sqrt{1 - (v/c)^2}$. When the particle is at rest ($v = 0$) the relation immediately above becomes

$$\text{Rest energy} = E_0 = m_0 c^2$$

The general relation for relativistic kinetic energy can then be written simply as

$$K = E - E_0$$

particle's momentum changes if the particle changes direction of motion or speed; the kinetic energy of a particle changes only if its speed changes. Further, doubling the speed of a particle doubles also the magnitude of its momentum, but a twofold increase in speed corresponds to kinetic energy increased by the factor 4.

For a system of two or more particles in motion, the system's total kinetic energy can never be zero. Consider, for example, two particles of equal mass m moving in opposite directions with equal speeds v. Their total momentum is zero: $m\mathbf{v} - m\mathbf{v} = 0$. Their total kinetic energy is not: $\frac{1}{2}mv^2 + \frac{1}{2}mv^2 = 2(\frac{1}{2}mv^2)$.

The kinetic energy of a particle may also be expressed in terms of the magnitude of its linear momentum, $p = mv$.

$$K = \frac{1}{2}\,mv^2 = \frac{(mv)^2}{2m}$$

$$K = \frac{p^2}{2m} \tag{9-3}$$

Kinetic energy, from its definition, has units of mass multiplied by the square of speed ($\text{kg}\cdot\text{m}^2/\text{s}^2$), or from (9-1), force multiplied by distance ($\text{N}\cdot\text{m}$). For convenience, kinetic energy is measured in the SI system in joules (abbreviated J), where by definition*

$$1 \text{ joule} = 1 \text{ J} = 1 \text{ kg}\cdot\text{m}^2/\text{s}^2 = 1 \text{ N}\cdot\text{m}$$

It follows from (9-2) that a 2-kg particle moving at 1 m/s has a kinetic energy of $K = 1$ J.

The most common energy unit for atomic or subatomic particles is the *electron volt* (abbreviated eV), which is defined as

$$1 \text{ electron volt} = 1 \text{ eV} = 1.60 \times 10^{-19} \text{ J}$$

(The reasons for the name and the numerical value are given in Section 25-4.)

Example 9-1. A 2-kg mass is dropped from rest. What is its kinetic energy (a) 0.1 s later? (b) 0.2 s later?

(a) The dropped object's acceleration is approximately 10 m/s², so that its speed after 0.1 s is about 1 m/s. Therefore, its kinetic energy is

$$K = \tfrac{1}{2}mv^2 = \tfrac{1}{2}(2 \text{ kg})(1 \text{ m/s})^2 = 1 \text{ J}$$

(b) After 0.2 s, the speed is approximately 2 m/s, and the object's kinetic energy is $\frac{1}{2}(2 \text{ kg})(2 \text{ m/s})^2 = 4$ J.

9-2 Work Defined and the Work-Energy Theorem

Look again at the circumstances shown in Figure 9-1 that led us to define a particle's kinetic energy as $K = \frac{1}{2}mv^2$. They were very special—the particle subject to a *constant* resultant force acting *along the direction* of the particle's *straight-line* displacement. We then found that

$$F\,\Delta x = \text{change in } K \tag{9-1}$$

---

* In the cgs system kinetic energy is expressed in ergs, where $1 \text{ erg} = 1 \text{ dyn}\cdot\text{cm} = 1 \text{ g}\cdot\text{cm}^2/\text{s}^2 = 10^{-7}$ J. The energy unit in the U.S. Customary System is the foot-pound, where $1 \text{ ft}\cdot\text{lb} = 1.356$ J. See Appendix D for other energy units and their conversion factors.

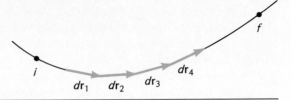

Figure 9-2. A general curved path from initial point i to final point f considered as a succession of infinitesimal straight-line displacements.

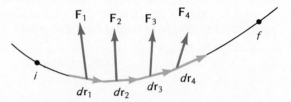

Figure 9-3. Over displacement $d\mathbf{r}_1$ the force is $\mathbf{F}_1$, and similarly for the other infinitesimal displacements.

The change in kinetic energy was equal to the product $F\,\Delta x$ of the resultant force F and displacement Δx.

But what if the resultant force on the particle varies from one location to another, both in magnitude and direction, and the particle's path is curved? How is kinetic energy change related *in general* to the resultant force on the particle and its displacement? What we are really asking is, How, if at all, must we modify the left side of (9-1) if the right side is still to be just the change in kinetic energy?[*]

Our strategy is as follows.

• Imagine any general curved path to be subdivided into a succession of infinitesimal straight-line displacements, $d\mathbf{r}_1$, $d\mathbf{r}_2$, $d\mathbf{r}_3$, . . . , each one coinciding with the path. See Figure 9-2.

• Take the force to have constant magnitude and direction over each vanishingly small displacement ($\mathbf{F}_1$ over $d\mathbf{r}_1$, $\mathbf{F}_2$ over $d\mathbf{r}_2$, $\mathbf{F}_3$ over $d\mathbf{r}_3$, . . .), as shown in Figure 9-3.

• Replace the effectively constant force $\mathbf{F}$ over each infinitesimal displacement by its force components parallel and perpendicular to $d\mathbf{r}$, and consider separately the effect on the particle's kinetic energy of each force component.

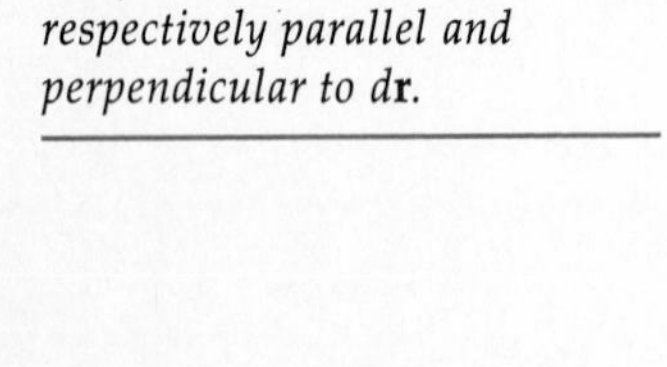

As Figure 9-4 shows, the angle between $\mathbf{F}$ and $d\mathbf{r}$ is θ. We know that the parallel component $F\cos\theta$ of the resultant force changes the particle's speed but not its direction of motion, while the perpendicular force component, $F\sin\theta$, changes the direction of motion but not the speed. In other words, only the *parallel* force component $F\cos\theta$, *along* the direction of the *displacement* $d\mathbf{r}$ can change the particle's kinetic energy by an infinitesimal amount dK. The general relation corresponding to the special relation (9-1) is therefore

$$F\cos\theta\,dr = dK \qquad (9\text{-}4)$$

The quantity appearing on the left side of this equation is given the special name *work*. Indeed, the work dW done by force $\mathbf{F}$ on the particle as the particle

Figure 9-4. Force $\mathbf{F}$ and its components $F\cos\theta$ and $F\sin\theta$, respectively parallel and perpendicular to $d\mathbf{r}$.

[*] Essentially, our question is, How should the work done by the resultant force be defined so that work done always equals kinetic energy change?

undergoes a displacement dr is defined, for the reasons given previously, as

$$dW \equiv F \cos \theta \, dr \qquad \text{(9-5)}$$

where θ is the angle between vectors $\mathbf{F}$ and $d\mathbf{r}$.

What is the total work done on a particle as it undergoes a finite displacement from some initial state i to a final (really just a later) state f? Equation (9-5) applies to each infinitesimal segment where F and θ have their respective local values at each segment. To get the total work done $W_{i \to f}$, we simply add up the contributions from all the segments, $W_{i \to f} = \int_i^f F \cos \theta \, dr$.

What happens when we add up the kinetic energy changes, $dK_1 + dK_2 + dK_3 + \cdots$, over the entire path? The kinetic energy with which the particle exits segment 1 is the same as that with which it enters segment 2, and so on for all the other adjacent segments. Therefore, the overall change in kinetic energy $dK_1 + dK_2 + dK_3 + \cdots$ is simply the kinetic energy K_f with which the particle exits the last segment less the kinetic energy K_i with which it enters the first segment. Adding up the contributions to work and kinetic energy over the entire path, we have, from (9-4) and (9-5),

$$\int_i^f dW = \int_i^f F \cos \theta \, dr = \int_i^f dK$$

$$W_{i \to f} \text{ (by resultant force)} = \int_i^f F \cos \theta \, dr = K_f - K_i \qquad \text{(9-6)}$$

This is the general form of the fundamental *work-energy theorem*; the *work done by the resultant force on a particle equals the change in the particle's kinetic energy.*

9-3   Properties of Work

Here are important properties of work that follow directly from its definition and its relation to kinetic energy:

- Work has the same units as kinetic energy, joules.
- Work is a *scalar* quantity that carries an *algebraic* sign.

If $\theta < 90°$, there is a *net* force component along the direction of a particle's displacement, so that the work done is *positive* and the particle's kinetic energy *increases.*

If $\theta > 90°$, there is a *net force component opposite* the direction of a particle's displacement, so that the force does *negative* work and the particle's kinetic energy *decreases.*

If $\theta = 90°$, *no* work is done on a particle and its kinetic energy remains *constant.*

- Work is done by a force *only if the particle undergoes a displacement.* The word *work* as used in physics has a more restricted meaning than the "effort" or "exertion" referred to in everyday speech. The athlete shown in Figure 9-5 does *no* work as she holds herself *stationary* at the highest point in the handstand.
- Work is done by a force only if there is a *force component along* the line of the object's *displacement.* No work is done, for example, by the earth's force of gravity on a satellite orbiting the earth in a circle at constant speed. More generally, any particle in uniform circular motion has zero work done on it by the resultant force, and its kinetic energy remains constant.

Figure 9-5. Olympic gold medalist Mary Lou Retton does no work while stationary at the top of the handstand.

- The work done by the resultant force **F**, where $\mathbf{F} = \mathbf{F}_1 + \mathbf{F}_2 + \mathbf{F}_3 + \cdots$, equals the total work done by the individual forces.

We prove this as follows. With $\mathbf{F} = \mathbf{F}_1 + \mathbf{F}_2 + \mathbf{F}_3 + \cdots$, we have for some displacement $d\mathbf{r}$

$$F \cos \theta \, dr = F_1 \cos \theta_1 \, dr + F_2 \cos \theta_2 \, dr + F_3 \cos \theta_3 \, dr + \cdots$$

where θ_1 is the angle between $\mathbf{F}_1$ and $d\mathbf{r}$, and likewise for the other forces, $\mathbf{F}_2, \mathbf{F}_3, \ldots$. Then

$$\int F \cos \theta \, dr = \int F_1 \cos \theta_1 \, dr + \int F_2 \cos \theta_2 \, dr + \int F_3 \cos \theta_3 \, dr + \cdots$$

This equation then implies that

$W_{i \to f}$ (by resultant of forces $\mathbf{F}_1, \mathbf{F}_2, \mathbf{F}_3, \ldots$) =

$$W_{i \to f} \text{ (by 1)} + W_{i \to f} \text{ (by 2)} + W_{i \to f} \text{ (by 3)} + \cdots \quad \text{(9-7)}$$

In many instances, this result facilitates the computation of work. Instead of finding the work done by the resultant force, we can find the total work done by all the individual forces.

- The work done by any force is simply the sum of the work done by its rectangular components:

$$\int_i^f F \cos \theta \, dr = \int_i^f F_x \, dx + \int_i^f F_y \, dy + \int_i^f F_z \, dz \quad \text{(9-8)}$$

This result follows immediately from (9-7) and the fact that any vector is merely the vector sum of its rectangular components.

- To describe completely work done, one must specify not only the magnitude (with the appropriate algebraic sign) but also the particular force doing the work and the object on which work is done.

- For a force whose direction and magnitude at each location does not change (this is not true for a friction force), the following statement is true:

If the *route* the particle follows from i to f is *reversed*, the *work done changes sign:*

$$\int_i^f F \cos \theta \, dr = - \int_f^i F \cos \theta \, dr$$

$$\text{(9-9)}$$

or

$$W_{i \to f} = -W_{f \to i}$$

This follows immediately from the fact that reversing the path means reversing the direction of $d\mathbf{r}$ at each point.

- In a plot of *force F_x against displacement x*, the *area under the curve* is equal in magnitude to the *work* done by F_x. See Figure 9-6. Over some small displacement dx, the force F_x may be taken to be constant, and the work done by F_x over dx is $F_x \, dx$. But $F_x \, dx$ is also the area of the slender rectangle in the graph. The total work done from i to f is $W_{i \to f} = \int_i^f F_x \, dx$, which is just the area, from i to f, under the $F_x - x$ curve.

- The kinetic energy of an object in motion may be interpreted as the ability of the object to do work in coming to rest. The work done by the arresting body A on the particle p is $W_{A \text{ on } p} = K_f - K_i = -K_i$ since $K_f = 0$ with the particle brought to rest. But $W_{p \text{ on } A} = -W_{A \text{ on } p} = -(-K_i) = K_i$.

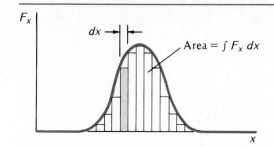

F_x

dx

Area = $\int F_x\, dx$

x

Figure 9-6.   *Work done is area under the force-displacement graph.*

Example 9-2.   A particle of mass m falls freely from rest and traverses a vertical distance y. (a) What is the work done by the resultant force on m? (b) What is the particle's final speed?

(a) The particle is subject to a single constant force $m\mathbf{g}$, its weight, in the downward direction. Its displacement is also y downward. Therefore the total work done is

$$W = mgy$$

(b) Since the work done by the resultant force equals the change in kinetic energy, initially zero, we have

$$W_{i\to f} = K_f - K_i$$

$$mgy = \tfrac{1}{2}mv^2 - 0$$

or

$$v = (2gy)^{1/2}$$

a result we could have arrived at simply from the kinematics of falling objects.

Example 9-3.   A 2-kg object is raised 1.5 m vertically upward by a man who holds the object in his hand. The object travels at constant speed. What is the work done by (a) the resultant force on the object? (b) the man? and (c) the gravitational force? See Figure 9-7.

(a) If the object travels at constant speed, the resultant force on it must be zero. The downward weight $m\mathbf{g}$ is just balanced by a force of equal magnitude from the man's hand. Since the resultant force is zero, it does *zero* work, and the object's kinetic energy does not change.

(b) The man's hand produces a force of magnitude $mg = (2.0\text{ kg})(9.8\text{ m/s}^2) = 19.6\text{ N}$ on the object. This force is upward, and the object travels 1.5 m upward. Therefore, with the force and displacement along the same direction (upward), the work done by the man's hand is $F\,\Delta x = (19.6\text{ N})(1.5\text{ m}) = 29.4\text{ J}$.

(c) The weight of the object is downward and of magnitude 19.6 N, whereas the object is displaced upward. Therefore, the work done by this force is negative ($\theta = 180°$) and is $F\cos\theta\,\Delta x = -(19.6\text{ N})(1.5\text{ m}) = -29.4\text{ J}$.

Note that the results from the three parts are consistent with the general theorem: The work by the resultant force equals the total work done by the individual forces, whose sum is the resultant force. Here we have

Work by resultant force = work by hand + work by gravitational force

$$0 = \quad 29.4\text{ J} \quad + \quad (-29.4\text{ J})$$

Example 9-4.   A package is dragged along a rough floor at constant speed by a rope making a fixed angle of 30° with the horizontal and exerting a tension of 100 N. The package moves 2.00 m horizontally. See Figure 9-8. What is the work done on the package by each of the following forces: (a) tension $\mathbf{T}$ in the rope; (b) weight $m\mathbf{g}$ of the package; (c) normal force $\mathbf{N}$; (d) friction force $\mathbf{f_k}$ with floor; (e) resultant force on the package?

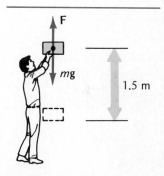

F

mg

1.5 m

Figure 9-7.

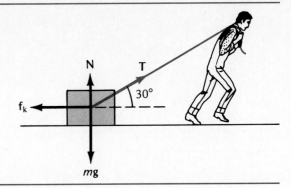

Figure 9-8.

(a) The work W_T done by the tension is

$$W_T = F \cos \theta \, d = (100 \text{ N})(0.866)(2.00 \text{ m}) = 173 \text{ J}$$

(b) The force $m\mathbf{g}$ is at right angles to the horizontal displacement. Therefore, no work is done by the weight: $W_{mg} = 0$.

(c) The normal force $\mathbf{N}$ is also at right angles to the displacement and it too does no work: $W_N = 0$.

(d) The kinetic friction force f_k to the left is opposite the displacement; $\mathbf{f}_k$ does negative work. Since the block moves at constant speed, it is in equilibrium. Therefore, $\mathbf{f}_k$ is equal in magnitude to the horizontal component of tension $\mathbf{T}$ and opposite in direction:

$$\mathbf{f}_k = -T \cos \theta = -(100 \text{ N})(0.866) = -86.6 \text{ N}$$

And the work done by $\mathbf{f}_k$ is

$$W_f = f_k \cos \theta \, d = (-86.6 \text{ N})(2.00 \text{ m}) = -173 \text{ J}$$

(e) The block is in equilibrium. The resultant force F_r on it is zero, and this force does zero work: $W_r = 0$.

It is interesting to compare the work W_r done by the resultant force with the total work done by all the individual forces (9-7).

First, the resultant force is related to the separate forces by

$$\mathbf{F}_r = \mathbf{T} + m\mathbf{g} + \mathbf{N} + \mathbf{f}_k = 0$$

The sum of the work is, from the values computed above,

$$W_T + W_{mg} + W_N + W_f = +173 \text{ J} + 0 + 0 + -173 \text{ J} = 0$$

which is also the work done by the resultant force, $W_r = 0$.

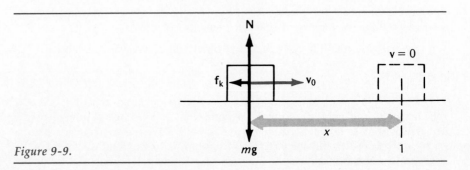

Figure 9-9.

Example 9-5.   A block is projected at 2.0 m/s to slide on a fairly smooth horizontal surface. How far does it travel before coming to rest if the coefficient of kinetic friction between the block and the relatively smooth surface is 0.050? See Figure 9-9. (This is exactly Example 6-1, solved earlier with Newton's second law.)

Both the weight $m\mathbf{g}$ and $\mathbf{N}$ are perpendicular to the block's horizontal displacement; each of these forces does zero work. The frictional force has magnitude $f_k = \mu_k N = \mu_k mg$, and it does negative work since its direction is opposite the block's velocity. The block comes to rest after coasting a distance x. The work-energy theorem then gives

$$W_{i \to f} = K_f - K_i$$

$$-(\mu_k mg)x = 0 - \tfrac{1}{2}mv_i^2$$

$$x = \frac{v_i^2}{2\mu_k g} = \frac{(2.0 \text{ m/s})^2}{2(0.050)(9.8 \text{ m/s}^2)} = 4.1 \text{ m}$$

Note that the block's mass was not required.

9-4   The Dot (or Scalar) Product

The rule for adding vectors was given in Section 2-1. Another important operation among vector quantities is the multiplication of two vectors. Here we consider the scalar product of two vectors, so named because the *product* itself is a *scalar* quantity. (In Section 12-2, we shall give the properties of the vector, or cross, product of two vector quantities.)

Consider two vectors $\mathbf{A}$ and $\mathbf{B}$ with angle θ between them. The *dot product* of $\mathbf{A}$ and $\mathbf{B}$ (also called the scalar product) is symbolized by $\mathbf{A} \cdot \mathbf{B}$ and read "*A* dot *B*." By definition $\mathbf{A} \cdot \mathbf{B}$ is

$$\mathbf{A} \cdot \mathbf{B} = AB \cos \theta \qquad (9\text{-}10)$$

That is, the dot product of $\mathbf{A}$ and $\mathbf{B}$ is a scalar quantity, the product of the magnitudes of $\mathbf{A}$ and $\mathbf{B}$ multiplied by the cosine of θ. Another way of saying the same thing is that $\mathbf{A} \cdot \mathbf{B}$ is the magnitude of $\mathbf{A}$ multiplied by the component of $\mathbf{B}$ along the direction of $\mathbf{A}$, as in Figure 9-10; or it is B multiplied by the component of $\mathbf{A}$ along $\mathbf{B}$. The dimensions of $\mathbf{A} \cdot \mathbf{B}$ are the product of the dimensions of $\mathbf{A}$ and $\mathbf{B}$.

When $\mathbf{A}$ and $\mathbf{B}$ are aligned ($\theta = 0°$), then $\mathbf{A} \cdot \mathbf{B} = AB$. When $\mathbf{A}$ and $\mathbf{B}$ are antialigned ($\theta = 180°$), then $\mathbf{A} \cdot \mathbf{B} = -AB$. When $\mathbf{A}$ and $\mathbf{B}$ are oriented at right angles to one another ($\theta = 90°$), then $\mathbf{A} \cdot \mathbf{B} = 0$. See Figure 9-10.

The *commutative law* for the dot product follows directly from the definition:

$$\mathbf{A} \cdot \mathbf{B} = AB \cos \theta = BA \cos \theta = \mathbf{B} \cdot \mathbf{A}$$

We can prove the *distributive law* by referring to Figure 9-11. If $\mathbf{C} = \mathbf{A} + \mathbf{B}$, then

$$\mathbf{D} \cdot \mathbf{C} = \mathbf{D} \cdot (\mathbf{A} + \mathbf{B}) = \mathbf{D} \cdot \mathbf{A} + \mathbf{D} \cdot \mathbf{B}$$

We see that the component of vector $\mathbf{C}$ along the direction of $\mathbf{D}$ is the sum of the components of vectors $\mathbf{A}$ and $\mathbf{B}$ along $\mathbf{D}$.

Consider vectors $\mathbf{A}$ and $\mathbf{B}$, each written as a *vector* sum of its rectangular vector components:

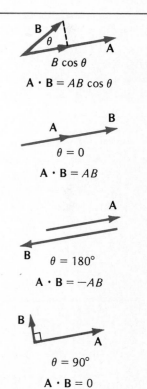

$$\mathbf{A} \cdot \mathbf{B} = AB \cos \theta$$

$$\theta = 0$$

$$\mathbf{A} \cdot \mathbf{B} = AB$$

$$\theta = 180°$$

$$\mathbf{A} \cdot \mathbf{B} = -AB$$

$$\theta = 90°$$

$$\mathbf{A} \cdot \mathbf{B} = 0$$

Figure 9-10. The dot product $\mathbf{A} \cdot \mathbf{B}$ of vectors $\mathbf{A}$ and $\mathbf{B}$ for several values of the angle θ between them.

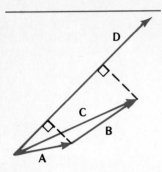

Figure 9-11. *Graphical proof of the distributive law for dot products:* $\mathbf{D} \cdot \mathbf{C} = \mathbf{D} \cdot (\mathbf{A} + \mathbf{B}) = \mathbf{D} \cdot \mathbf{A} + \mathbf{D} \cdot \mathbf{B}$.

$$\mathbf{A} = \mathbf{A}_x + \mathbf{A}_y + \mathbf{A}_z \qquad \text{and} \qquad \mathbf{B} = \mathbf{B}_x + \mathbf{B}_y + \mathbf{B}_z$$

($\mathbf{A}_x$ is the vector component of $\mathbf{A}$ along x; A_x is the algebraic value of $\mathbf{A}_x$.) The coordinate axes are mutually perpendicular, $\mathbf{A}_x \cdot \mathbf{B}_x = A_x B_x$. On the other hand, $\mathbf{A}_x \cdot \mathbf{B}_y = 0$, and similarly for other combinations. It follows then that

$$\mathbf{A} \cdot \mathbf{B} = \mathbf{A}_x \cdot \mathbf{B}_x + \mathbf{A}_y \cdot \mathbf{B}_y + \mathbf{A}_z \cdot \mathbf{B}_z$$

or

$$\mathbf{A} \cdot \mathbf{B} = A_x B_x + A_y B_y + A_z B_z \tag{9-11}$$

As a special case, when $\mathbf{B} = \mathbf{A}$,

$$\mathbf{A} \cdot \mathbf{A} = A_x^2 + A_y^2 + A_z^2$$

or

$$\mathbf{A} \cdot \mathbf{A} = A^2 \tag{9-12}$$

Example 9-6 (Optional).   The derivation of the general work-energy theorem (9-6) was complete. The properties of the dot product may, however, be exploited to derive the work-energy theorem in particularly elegant fashion.

First we recognize that the resultant force $\mathbf{F}$ may be written $\mathbf{F} = m\, d\mathbf{v}/dt$ and the infinitesimal displacement $d\mathbf{r}$ as $d\mathbf{r} = \mathbf{v}\, dt$. Then the work done by $\mathbf{F}$ becomes

$$\int F \cos \theta \, dr = \int \mathbf{F} \cdot d\mathbf{r} = \int \left(m \frac{d\mathbf{v}}{dt} \right) \cdot (\mathbf{v}\, dt) = m \int \mathbf{v} \cdot d\mathbf{v}$$

Now the square of the particle's speed v^2 may be written, from (9-12), as $\mathbf{v} \cdot \mathbf{v}$. Taking the differential of v^2 gives

$$d(v^2) = d(\mathbf{v} \cdot \mathbf{v}) = d\mathbf{v} \cdot \mathbf{v} + \mathbf{v} \cdot d\mathbf{v} = 2\mathbf{v} \cdot d\mathbf{v}$$

Substituting this result in the preceding equation yields

$$\int_i^f \mathbf{F} \cdot d\mathbf{r} = \frac{m}{2} \int_i^f d(v^2)$$

or

$$\int_i^f \mathbf{F} \cdot d\mathbf{r} = \tfrac{1}{2} m v_f^2 - \tfrac{1}{2} m v_i^2$$

9-5   Unit Vectors and the Dot Product (Optional)

Using unit vectors (Section 2-5) $\mathbf{i}$, $\mathbf{j}$, and $\mathbf{k}$, we can write

$$\mathbf{A} = A_x \mathbf{i} + A_y \mathbf{j} + A_z \mathbf{k} \qquad \text{and} \qquad \mathbf{B} = A_B \mathbf{i} + B_y \mathbf{j} + B_z \mathbf{k}$$

Because $\mathbf{i}$ is parallel to $\mathbf{i}$ and each has unit magnitude, and likewise for $\mathbf{j}$ and $\mathbf{k}$,

$$\mathbf{i} \cdot \mathbf{i} = 1 \qquad \mathbf{j} \cdot \mathbf{j} = 1 \qquad \mathbf{k} \cdot \mathbf{k} = 1 \tag{A}$$

On the other hand, the three unit vectors are mutually perpendicular, so that

$$\mathbf{i} \cdot \mathbf{j} = 0 \qquad \mathbf{j} \cdot \mathbf{k} = 0 \qquad \mathbf{k} \cdot \mathbf{i} = 0 \tag{B}$$

It then follows that

$$\mathbf{A} \cdot \mathbf{B} = (A_x \mathbf{i} + A_y \mathbf{j} + A_z \mathbf{k}) \cdot (B_x \mathbf{i} + B_y \mathbf{j} + B_z \mathbf{k})$$

$$\mathbf{A} \cdot \mathbf{B} = A_x B_x + A_y B_y + A_z B_z \qquad\qquad \text{(C)}$$

as also given in (9-11).

Example 9-7. Find the angle θ between vectors $\mathbf{A}$ and $\mathbf{B}$, where

$$\mathbf{A} = -2\mathbf{i} + 2\mathbf{j} \qquad \text{and} \qquad \mathbf{B} = 2\sqrt{3}\mathbf{i} + 2\mathbf{j}$$

Vectors $\mathbf{A}$ and $\mathbf{B}$ are shown in Figure 9-12.

As given

$$A_x = -2 \qquad B_x = 2\sqrt{3}$$
$$A_y = 2 \qquad B_y = 2$$

Therefore, combining (9-10) ($\mathbf{A} \cdot \mathbf{B} = AB \cos \theta$) with (C) above, we have

$$\cos \theta = \frac{A_x B_x + A_y B_y}{\sqrt{A_x^2 + A_y^2}\sqrt{B_x^2 + B_y^2}} = \frac{(-2)(2\sqrt{3}) + (2)(2)}{\sqrt{(-2)^2 + (2)^2}\sqrt{(2\sqrt{3})^2 + (2)^2}} = -0.259$$
$$\theta = 105°$$

This result follows also since A makes an angle of $135°$ with the positive x axis, whereas the angle for B is $30°$.

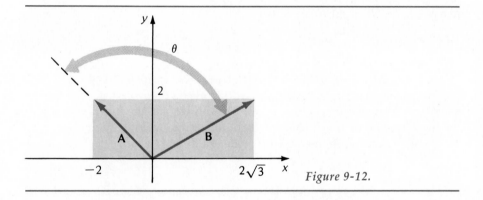

Figure 9-12.

9-6 Work Done by Gravity

We wish to find the work done by the constant gravitational force on an object of mass m. We imagine that m is displaced in a number of different paths, but in each of these paths the object is displaced downward by y.

See Figure 9-13, in which a particle is shown to start at the upper edge of a ramp of angle θ at an elevation y above ground level.

Path: vertically downward In Figure 9-13(a), the gravitational force $m\mathbf{g}$ is downward and the displacement $\mathbf{y}$ is also downward. The work done by gravity is positive and equal to

$$W_{i \to f} = mgy$$

(The mass might have dropped from rest at its starting point and acquired kinetic energy in falling. Or it might be carried downward at any speed. Under

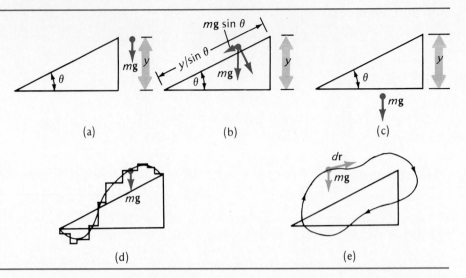

Figure 9-13. *Work done by gravity on a particle of mass m that starts at the upper edge of the ramp of angle θ at an elevation y above ground level and undergoes various displacements: (a) vertically downward; (b) along the incline over a distance y/sin θ; (c) vertically down then horizontally; (d) any path to the lower edge of the ramp; and (e) any closed path.*

any circumstance, the *work done by the weight only* for a vertical downward displacement is *mgy*.)

Path: down along the ramp [Figure 9-13(b)] The displacement down the ramp is $y/\sin \theta$. The component of the weight parallel to the incline is $mg \sin \theta$. Just as before, the total work done is

$$W = \left(\frac{y}{\sin \theta}\right)(mg \sin \theta) = mgy$$

Path: vertically down then horizontally to base of ramp [Figure 9-13(c)] For the downward displacement, the work done by the weight is, as we have seen, *mgy*. When the object travels along the horizontal, the force (down) and the displacement (to the left) are at right angles; no work is done by weight along this horizontal segment. The total work done is again*

$$W = mgy$$

Path: any from upper to lower end of the ramp [Figure 9-13(d)] The arbitrary curved path can always be broken into equivalent small vertical and horizontal segments. Along any horizontal segment, no work is done. Along a *downward* vertical segment, the work done is *positive* in the amount *mg dy*; along an *upward* vertical segment, the work done is *negative* and equal to −*mg dy*. Overall, the net work done by gravity once again is

$$W = mgy \text{ for } any \text{ route with vertical displacement } y \text{ downward} \qquad (9\text{-}13)$$

---

* The result follows also from (9-8):

$$\int_i^f F \cos \theta \, dr = \int_i^f F_x \, dx + \int_i^f F_y \, dy + \int_i^f F_z \, dz$$

With $F_x = 0$, $F_y = mg$, $F_z = 0$, and y taken as positive downward, we have

$$W_{i \to f} = \int_0^y mg \, dy = mgy$$

Path: upward displacement   If the object goes from the base of the ramp to its upper edge, the net work, again by *any* route is

$$W_{up} = -mgy$$

The result follows also from (9-9); reversing the path reverses the sign of the work done.

Path: any closed loop   [Figure 9-13(e)] Now the object is imagined to move by any path that brings it back to its starting point; it goes around a closed loop. It is clear that the net work done by gravity is always *zero;* positive work is done in a downward displacement and an equal amount of negative work is done in an upward displacement of the same magnitude. No work is done for any horizontal displacement.

Example 9-8.   A block slides down a perfectly smooth surface; the initial speed is v_i. What is its speed at a point a distance y vertically beneath the starting point?

As Figure 9-14 shows, the forces acting on the particle are its weight $m\mathbf{g}$ and the normal force $\mathbf{N}$ of the surface on the block. (That the surface is perfectly smooth is reflected in the fact that the force of the surface is, in fact, perpendicular to the surface, with no component along the surface.) The work done by the resultant force on the block between the two end points is the change in the block's kinetic energy:

$$\int_i^f \mathbf{F} \cdot d\mathbf{r} = \tfrac{1}{2}mv_f^2 - \tfrac{1}{2}mv_i^2$$

But here $\mathbf{F} = m\mathbf{g} + \mathbf{N}$, and we can compute the work done by the resultant force by adding the work by $m\mathbf{g}$ alone to work done by $\mathbf{N}$ alone. Since the block slides along the plane, the normal force $\mathbf{N}$ is always at right angles to the displacement $d\mathbf{r}$, so that $\mathbf{N} \cdot d\mathbf{r} = 0$. This leaves the work done by $m\mathbf{g}$, which we know is just mgy.

Therefore, the total work done by the resultant force is mgy, and the work-energy theorem yields

$$W_{i \rightarrow f} = K_f - K_i$$
$$mgy = \tfrac{1}{2}mv_f^2 - \tfrac{1}{2}mv_i^2$$

or a final speed,

$$v_f = \sqrt{v_i^2 + 2gy}$$

Note that the final speed is independent not only of the mass m but also of any details concerning the shape of the surface leading from the starting point to the ending point. It follows that all objects starting from rest and sliding on frictionless surfaces connecting two points with the same vertical separation reach the same final speed. The various paths may differ in the total time required to traverse the particular path, but the final speeds are the same.

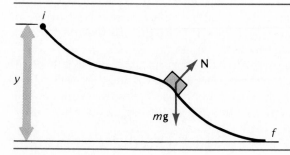

Figure 9-14.   *Block sliding on a smooth surface from i to f is subject to two forces: normal force* $\mathbf{N}$ *and weight* $m\mathbf{g}$.

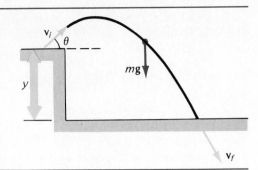

Figure 9-15. Object thrown with initial velocity v_i at angle θ from elevation y.

Example 9-9. An object is thrown at an initial speed v_0 at an angle θ relative to the horizontal. With what speed does it arrive at a point a vertical distance y down from the starting point? See Figure 9-15.

The only force acting on the object in flight is its weight $m\mathbf{g}$, and as we have seen, the work done by $m\mathbf{g}$ over a displacement having a downward vertical component is mgy. Therefore, the work-energy theorem yields

$$mgy = \tfrac{1}{2}mv_f^2 - \tfrac{1}{2}mv_i^2$$

or

$$v_f^2 = v_i^2 + 2gy$$

Projection angle θ does not enter into the final result. An object of any mass tossed in any direction, upward or downward from the horizontal, will change its speed by the same amount when traversing a vertical distance y.

One case is of special interest; that with $y = 0$. Then the result above gives $v_f = v_i$, and the total work done by the gravitational force is zero. For example, an object sliding on a perfectly smooth roller coaster, as in Figure 9-16, will return to its starting point with precisely the same kinetic energy as that with which it started the complete cycle, and the particle's kinetic energy at all points at the same horizontal level will be the same.

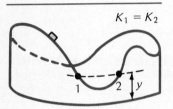

Figure 9-16. Object sliding on a frictionless roller coaster has the same kinetic energy at all points on the same horizontal level.

9-7   Conservative Force

A *conservative force* is a force for which *the total work done around a closed loop*, a path that returns the particle to its starting point, is *zero*. In symbols, we signify the work done around a closed loop by

$$\oint \mathbf{F} \cdot d\mathbf{r} = 0 \tag{9-14}$$

where the little circle on the integral sign is to remind us that the loop is indeed closed. The term *conservative* is used to denote such a force, because the energy-conservation principle applies to conservative forces (as we shall see in Chapter 10). In mathematical language, (9-14) says that the *line integral* of a conservative force around a closed loop is zero.

Figure 9-17 shows several paths all leading from initial point i to final point f. For *any* round trip—for example, from i to f by route A and then back from f to i by route B—the net amount of work done is zero. Positive work done in one segment of the path is balanced exactly by negative work done in another segment, so that overall no net work is done.

An equivalent way of describing a conservative force is this: the work done

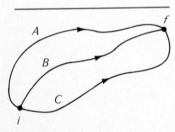

Figure 9-17. Three paths—A, B, and C—from i to f.

in going by *any* path from *i* to *f* is *independent of the route taken between the end points*. In Figure 9-17, any path leading from *i* to *f* involves the same amount of work done:

$$W_{i \to f} \text{ (by route } A) = W_{i \to f} \text{ (by route } B) = W_{i \to f} \text{ (by route } C) \qquad (9\text{-}15)$$

But as shown in (9-9), reversing the direction in which a path is traversed merely changes the sign of the work done:

$$W_{i \to f} \text{ (by route } A) = -W_{f \to i} \text{ (by route } A)$$

From (9-15), we have

$$W_{i \to f} \text{ (by route } A) = W_{i \to f} \text{ (any other route)}$$

so that

$$W_{i \to f} \text{ (any route)} = -W_{f \to i} \text{ (by route } A)$$

or

$$W_{i \to f} + W_{f \to i} = 0$$

which is equivalent to

$$\oint \mathbf{F} \cdot d\mathbf{r} = \int_i^f \mathbf{F} \cdot d\mathbf{r} + \int_f^i \mathbf{F} \cdot d\mathbf{r} = 0$$

We have already identified one conservative force: the constant gravitational force on a particle. It has the essential property of a conservative force; work around a closed loop is zero, or equivalently, work is independent of the path. We shall encounter other conservative forces: the general gravitational force (Chapter 15), the Coulomb electric force (Chapter 23), and the force produced by a compressed or a stretched spring (Section 9-8).

9-8   Work Done by a Spring

If an ordinary helical spring is not stretched or compressed too much, it will return to its undeformed, or equilibrium, configuration when the deforming forces are removed. The force F_s produced by the spring along its length is, for small deformations, given by

$$F_s = -kx \qquad (9\text{-}16)$$

where x is the displacement of one end relative to the other and k is a constant.* See Figure 9-18(a).

The minus sign appears in (9-16) because the spring's *force* is always *opposite* in direction to the *displacement*. When $x = 0$ and the spring is relaxed, $F_s = 0$ [Figure 9-18(b)]. When the spring is stretched [Figure 9-18(c)] and x is positive, the force produced by the spring is in the opposite direction, or negative. Likewise, when the spring is compressed and x is negative, as in Figure 9-18(d), the force F_s is positive. In every instance the spring applies a force to an object attached to it that tends to restore this object and the spring to the equilibrium configuration. The spring is said to exert a restoring force. The

* This relation is known as Hooke's law after its discoverer, Robert Hooke (1635–1703).

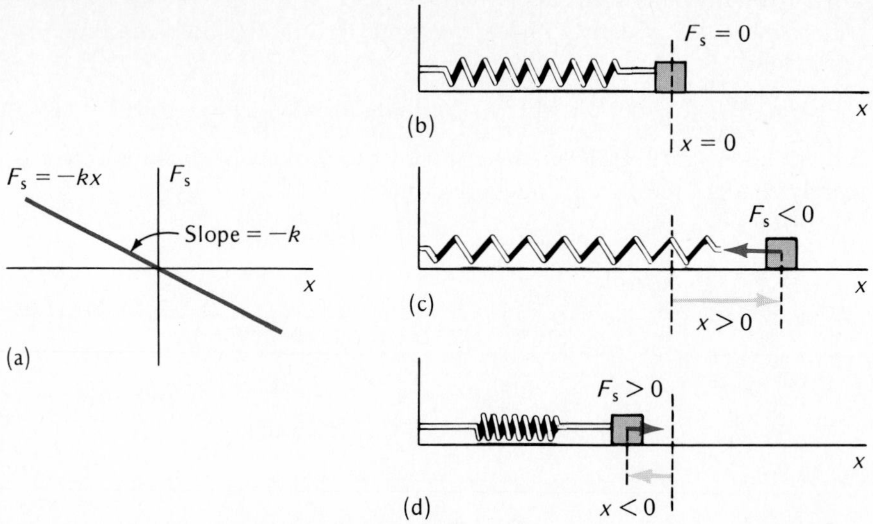

Figure 9-18. *Force characteristics of a spring. (a) Spring force F_s as a function of displacement x. (b) The spring in its relaxed state. (c) The spring stretched by a displacement x to the right with spring force F_s to the left. (d) The spring compressed, with displacement x to the left and spring force F_s to the right.*

$F_s = 0$

(b)

$x = 0$

$F_s = -kx$

Slope $= -k$

(a)

$F_s < 0$

(c)

$x > 0$

$F_s > 0$

(d)

$x < 0$

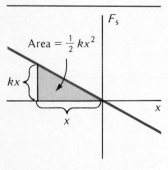

Figure 9-19. *The work done by a spring expanding from $-x$ to 0 is equal to the area $\frac{1}{2}kx^2$ under the force-displacement graph.*

Area $= \frac{1}{2} kx^2$

spring constant, or *force constant*, k is a measure of a spring's stiffness; the larger the k, the stiffer the spring.

For simplicity we shall take the mass of a spring to be zero, or to be more realistic, small compared with other masses.

In Chapter 11 we shall consider in detail the motion of an object attached to a spring. Here we are concerned with the work done by a stretched or compressed spring, an example of work done by a force that varies with displacement.

Suppose that a spring is initially compressed, as in Figure 9-18(d); it applies a force F_s to the right to a block pressed against the spring. We wish to find the work done *by an expanding spring* as its displacement goes from $-x$ to zero. Note that the force and displacement are both to the right, so that the angle θ between the resultant force **F** and the displacement $d\mathbf{r}$ is zero. Applying the general definition of work, we then have

$$W_{i \to f} = \int_i^f \mathbf{F} \cdot d\mathbf{r} = \int_{-x}^0 F_s \, dx = \int_{-x}^0 (-kx) \, dx$$

$$W = \tfrac{1}{2}kx^2 \tag{9-17}$$

We may arrive at the same result graphically by considering the area under the F_s-x graph of Figure 9-19. The area of the triangle is half the height kx multiplied by the base x, or $\frac{1}{2}kx^2$.

It is easy to see that the force produced by a spring obeying the relation $F_x = -kx$ is a conservative force. Suppose that a spring is stretched by some small amount dx; the spring does work

$$dW_{\text{stretched}} = F_s \, dx = -kx \, dx$$

Now we imagine that the spring is "unstretched" by the same amount, so that the displacement in the opposite direction is now $-dx$. The work done is

$$dW_{\text{unstretched}} = F_s \, (-dx) = -kx \, (-dx) = +kx \, dx$$

Over the round trip we have

$$dW_{\text{stretched}} + dW_{\text{unstretched}} = 0$$

The net work done by the spring for this infinitesimal round trip *is* zero. But any other round trip is effectively a succession of infinitesimal round trips, so that in general the work done by a spring around a closed loop is zero and

$$\oint \mathbf{F_s} \cdot d\mathbf{r} = 0$$

Example 9-10.   A force of 50 N must be applied to a certain spring to compress it 8.0 cm. This spring has one end fixed to the wall, and the other end free. A 1.0-kg block slides along a horizontal frictionless surface at 1.0 m/s and strikes the spring. See Figure 9-20. (a) What is the spring's force constant? (b) What is the maximum compression of the spring? (c) At what speed does the block slide after losing contact with the spring?

(a) The force constant is

$$k = \frac{F}{x} = \frac{50 \text{ N}}{8.0 \times 10^{-2} \text{ m}} = 6.25 \times 10^2 \text{ N/m}$$

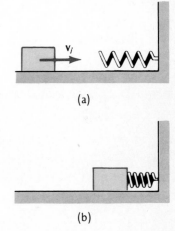

(a)

(b)

Figure 9-20.

(b) The block initially slides to the right with constant kinetic $K_i = \frac{1}{2}mv_i^2$. When the spring is compressed by the maximum amount, and the block is momentarily at rest, its kinetic energy is $K_f = 0$. The spring does *negative* work $-\frac{1}{2}kx^2$ on the block in bringing it to rest, where x is the maximum compression of the spring. (The block does *positive* work $\frac{1}{2}kx^2$ on the spring in compressing it.) From the work-energy theorem applied to the block,

$$W_{i \to f} = K_f - K_i$$
$$-\tfrac{1}{2}kx^2 = 0 - \tfrac{1}{2}mv_i^2$$
$$x = v_i \sqrt{\frac{m}{k}} = 1.0 \text{ m/s} \sqrt{\frac{1.0 \text{ kg}}{6.25 \times 10^2 \text{ N/m}}}$$
$$= 4.0 \text{ cm}$$

(c) When the block loses contact with the spring, the spring is back again in its undeformed state. As the spring was being compressed, it did negative work on the block in bringing it to rest; as the spring expanded back to its undeformed state, it did positive work in the same amount on the block. The net work done by the spring is zero all told. Therefore, the block must emerge with the same kinetic energy and travel at the same speed of 1.0 m/s (but in the opposite direction) as initially.

9-9   Power

A backhoe can lift a large pile of sand in one fell swoop. The same work can be done by an army of ants lifting the grains one at a time. The backhoe is more powerful than the ants; it does the same work in less time.

  Power is the *rate at which work is done,* or more generally, the *rate at which energy is transferred.* If work $W_{i \to f}$ is done in a time interval Δt, the average power $\overline{P}$ is

$$\overline{P} = \frac{W_{i \to f}}{\Delta t}$$

The instantaneous power P is the time rate of doing work:

$$P = \frac{dW}{dt} \qquad (9\text{-}18)$$

Work may be written $dW = \mathbf{F} \cdot d\mathbf{r} = \mathbf{F} \cdot \mathbf{v} \, dt$, so that instantaneous power may also be written

$$P = \frac{dW}{dt} = \frac{\mathbf{F} \cdot \mathbf{v} \, dt}{dt}$$

$$P = \mathbf{F} \cdot \mathbf{v} \qquad (9\text{-}19)$$

In words, the instantaneous power transferred to a particle is the dot product of the force on the particle and the particle's velocity.

Appropriate units for power are joules per second.* This combination of units is termed the *watt,* abbreviated W. Therefore,

$$1 \text{ watt} = 1 \text{ W} \equiv 1 \text{ J/s} = 1 \text{ kg} \cdot \text{m}^2/\text{s}^3$$

An *energy* unit related to the kilowatt is the kilowatt hour (kW·h), where

$$1 \text{ kW} \cdot \text{h} = (10^3 \text{ J/s})(3600 \text{ s}) = 3.60 \times 10^6 \text{ J}$$

Example 9-11. A sports car is driven at its maximum speed of 54 m/s (~120 mph) on a flat roadway. The power output delivered by the engine to the wheels is 110 kW (~150 hp). What is the resultant resistive force on the car?

Since the car travels at constant speed, the resultant force on it is zero. This means that the forward force of the roadway on the car's tires is equal in magnitude to the backward force of air resistance and friction on the car. The car's kinetic energy does not change; therefore, energy delivered to the wheels must equal energy dissipated by friction and air resistance. From (9-19), the magnitude of the total resistive force is

$$F = \frac{P}{v} = \frac{110 \times 10^3 \text{ W}}{54 \text{ m/s}} = 2.0 \times 10^3 \text{ N}$$

---

* The unit for power in the English system is the horsepower, abbreviated hp; 1 hp = 550 ft·lb/s = 0.7457 kW.

Summary

Definitions

Kinetic energy:

$$K = \tfrac{1}{2}mv^2 \qquad (9\text{-}2)$$

Work done by force $\mathbf{F}$ over a displacement $d\mathbf{r}$ (at angle θ relative to $\mathbf{F}$):

$$dW = F \cos \theta \, dr = \mathbf{F} \cdot d\mathbf{r} \qquad (9\text{-}4)$$

Dot product of vectors $\mathbf{A}$ and $\mathbf{B}$:

$$\mathbf{A} \cdot \mathbf{B} = AB \cos \theta \qquad (9\text{-}10)$$

with angle θ between them.

Conservative force: a force for which the total work done around *any* closed loop is zero:

$$\oint \mathbf{F} \cdot d\mathbf{r} = 0 \qquad (9\text{-}14)$$

Power: the rate of doing work $= P = dW/dt = \mathbf{F} \cdot \mathbf{v}$

$$(9\text{-}18, 9\text{-}19)$$

Units

Work, kinetic energy: 1 Joule $= 1$ J $= 1$ N·m
Power: 1 watt $= 1$ W $= 1$ J/s

Fundamental Principles

The work-energy theorem states that the work $W_{i \to f}$ done by the resultant force on a particle from some initial state i to a final state f equals the change in the particle's kinetic energy:

$$W_{i \to f} \text{ (by resultant force)} = K_f - K_i$$

$$\int_i^f \mathbf{F} \cdot d\mathbf{r} = \tfrac{1}{2}mv_f^2 - \tfrac{1}{2}mv_i^2 \qquad (9\text{-}6)$$

Important Results

The work done by the resultant force on a particle equals the sum of the work done by the individual forces:

$$W_{i \to f} \text{ (by } \mathbf{F}) = W_{i \to f} \text{ (by 1)} + W_{i \to f} \text{ (by 2)} + \cdots \qquad (9\text{-}7)$$

The work done by a force F_x plotted as a function of displacement x is equal in magnitude to the area under the F-x curve.

The force F_s produced by a spring with force constant k for a small displacement x is

$$F_s = -kx \qquad (9\text{-}16)$$

The work done by the constant force of gravity on a particle of mass m as the particle undergoes a vertically downward displacement y is

$$W = mgy \qquad (9\text{-}13)$$

The work done by the spring expanding from $-x$ to 0 is

$$W = \tfrac{1}{2}kx^2 \qquad (9\text{-}17)$$

The constant gravitational force and the displacement-dependent force of a spring are both conservative forces.

Problems and Questions

Section 9-1   Kinetic Energy Defined

· **9-1 Q** Suppose that in distant outer space, where no friction or other forces act, an amount of energy E is needed to accelerate a rocket ship from rest to a speed v. What additional energy would be required to further increase the speed from v to $2v$?
(A) Zero additional energy would be needed.
(B) $\tfrac{1}{2}E$
(C) E
(D) $2E$
(E) $3E$
(F) $4E$
(G) The added energy required cannot be determined without more information.

· **9-2 Q** A particle travels in a circle at constant speed. We know then that
(A) the particle's velocity is constant.
(B) the particle's acceleration is constant.
(C) the particle's momentum is constant.
(D) the particle's kinetic energy is constant.
(E) None of the preceding statements is true.

· **9-3 P** When an electron is accelerated from rest through a potential difference of 25,000 volts in a cathode ray tube, it acquires a kinetic energy of 25,000 eV. What is the speed of such an electron? (Electron mass $= 9.11 \times 10^{-31}$ kg)

: **9-4 P** A 2.0-kg object is pulled straight up by a cord. If the object initially moves upward at 0.80 m/s and the tension in the cord is 26 N, what is the kinetic energy of the object after the cord has pulled it up a distance 1.0 m above its initial position?

· **9-5 P** Compare the energy needed to increase an object's speed from 1 m/s to 2 m/s to that needed to increase the speed from 2 m/s to 3 m/s.

: **9-6 P** A 3-kg object initially moving east at 10 m/s explodes in flight into a 1-kg piece and a 2-kg piece. The 1-kg piece moves northeast at 15 m/s. What is (a) the speed, and (b) the direction of the 2-kg piece? (c) How much kinetic energy is released in the explosion?

: **9-7 P** The wind may be a useful source of electric energy in the future for such small isolated users as ranchers, forest rangers, and people who live in heavily mountainous regions. Modern windmills have an overall efficiency of 40%; that is, 40% of the energy in the wind hitting the blades can be converted to electric energy. What blade area would be needed to generate 5 kW in a 50 km/hr wind? One cubic meter of air has a mass of 1.2 kg.

Section 9-2   Work Defined and the Work-Energy Theorem

· **9-8 Q** A particle is subject continuously to a single resultant force of constant magnitude. It necessarily follows that
(A) the particle's kinetic energy is changing.
(B) the particle's momentum vector changes.
(C) the particle's momentum vector changes both in magnitude and direction.
(D) the particle has work done on it.
(E) the particle's speed changes.

: **9-9 P** A force center is at the origin. The magnitude of the repulsive force acting on a particle at the displacement x is given by $F = A/x^2$. What is the work done by this repulsive force in displacing the particle from (a) $x = 1$ to $x = 3$ and (b) $x = 3$ to $x = 1$?

: **9-10 P** In an automobile accident, a pedestrian was struck and injured by a car. The motorist insisted he was traveling at a legal speed when the pedestrian stepped out in front of him. He slammed on his brakes and his car skidded into the person. Police measured the skid marks and found they extended 24 meters. In tests with a similar car they found it would skid 12 meters with brakes locked from a speed of 50 km/hr. Using this information, what speed would you estimate the car was traveling before the accident?

: 9-11 P Ten bricks, each of mass 1 kg, are to be stacked up on a floor to make a pile 50 cm high. The ten bricks originally lie on the floor. How much work is done in stacking up the bricks?

: 9-12 P A woman picks up a 1.0-kg mass from the floor, raises it 1.5 meters and then throws it with a speed of 4.0 m/s. How much work has she done, all told?

: 9-13 P A jet airliner crashed shortly after take-off and smashed into a concrete parking garage. In their investigation of the accident, Federal Aviation Administration inspectors concluded that it was important to determine the speed of the plane just before impact. Knowing this, they could decide whether the plane had lost power or the engines were still driving it forward when it hit the building. They found a nacelle (a metal engine cover) of mass 40 kg which had separated from the fuselage just before the crash. It was crushed from its normal length of 3.2 meters to 1.7 meters. Laboratory tests established that this deformation would be caused by an average force of 3.2×10^5 N. From this information, estimate the speed of the plane just before impact, assuming the plane and the nacelle had the same speed. Treat the nacelle as if all of its mass were concentrated at its center.

: 9-14 P A Ford sedan with a mass of 1500 kg is able to come to a stop in a distance of 60 meters on a level, dry pavement if the brakes are applied when traveling at a speed of 20 m/s. (a) What is the minimum coefficient of static friction between the tires and the pavement? (b) Suppose the car is going down a 6% grade (one that drops 6 meters for every 100 meters traveled along the roadway). If the brakes are again applied when traveling 20 m/s, in what distance could the car stop? (c) What would be the magnitude of the deceleration?

: 9-15 P Police investigators determined that a car skidded for 50 meters before hitting a train broadside. Tests with a similar vehicle showed that the effective coefficient of friction between the tire and the roadway was 0.7. Using the work-energy principle, estimate the minimum speed of the car just before the brakes were applied.

: 9-16 P A particle of mass m is accelerated from rest for a time t by a constant force **F**. An observer in the laboratory could use the work-energy theorem to calculate the particle's final speed.

Now consider a second observer moving at constant speed u parallel to the direction of motion of the particle. Both the stationary observer and the moving observer agree that the force of magnitude **F** acts, but they get different values for the kinetic energy gained by the particle. Show that the work-energy theorem still applies for each observer. (Each observer sees the force acting over a *different* distance).

: 9-17 P A person falling through air experiences a drag force which may be written av^2, where a is a constant that depends on the size and shape of the person and on the density and viscosity of air. A typical value of a for a person of mass 70 kg is about 0.24 kg/m.

(a) What terminal velocity will such a person reach when falling? (b) If on impact with the ground the pressure on your body exceeds 3.5×10^5 N/m² (about 50 lb/in²) you have about a 50% chance of surviving the crash. Suppose that when falling at the terminal velocity calculated in (a) you make full body contact with the ground. Assume that the area of your body in contact with the ground is 0.34 m². What is the minimum distance in which you can stop and still have a 50% chance of survival? (You may be surprised at the result. People have survived falls from great heights if they landed in snow or foliage.)

Section 9-3 Properties of Work

· 9-18 Q A planet is moving at constant speed in a circular orbit of radius R around a star. Gravitational force F acts on the planet in the direction of the star. The work done by the force on the planet in one complete revolution is
(A) zero.
(B) $-RF$
(C) $2RF$
(D) $-R^2F$
(E) $2\pi rF$

· 9-19 Q A man clutches his hands together and uses his arm muscles to try to separate his hands. Since they are tightly clenched, nothing moves. Does the man do any work? Does he get tired? Why?

· 9-20 Q An elevator descends at constant speed. What is the sign of the work done on the elevator by (a) the resultant force on it? (b) the cable from which the elevator car is suspended? (c) the weight of the elevator car?

: 9-21 Q A man on roller skates stands next to a vertical wall. When the man pushes on the wall he gains kinetic energy, but the man's force on the wall does not act over a displacement. Is there a contradiction here? (Hint: Consider carefully the system under consideration, the forces that act on it, and the motion of the center-of-mass system.)

· 9-22 P A tractor exerts a force of 2000 N on a farm implement used in preparing a field for planting. If the hitch attached to the tractor is inclined at 30° above horizontal, how much work does the tractor do in pulling the implement 400 meters across a field?

: 9-23 P An elevator of mass m is lowered with acceleration $\frac{1}{4}g$ over a vertical distance h. How much work is done on the elevator by the cable supporting it?

: 9-24 P Figure 9-21 shows a screw jack of the type used to lift buildings. The screw has 4 threads per centimeter, and the twisting force is applied at a point 30 cm from the axis of the screw. Assuming no friction, what force must be applied to lift an object of mass 1000 kg?

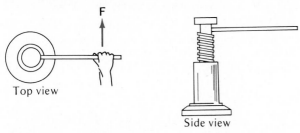

Figure 9-21.   *Problem 9-24.*

: 9-25 P In the machine in Figure 9-22 the large gears have 32 teeth and the small gears have 8 teeth. The radius of the large gears is $5r$, where r is the radius of the inner gear. What force **F** must be exerted to lift a mass of 100 kg? Neglect friction.

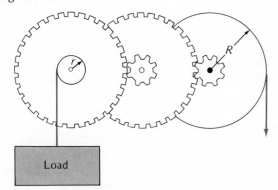

Figure 9-22.   *Problem 9-25.*

: 9-26 P A winch is used to drag shipping crates across a warehouse floor, as in Figure 9-23. A cable passes over a pulley a height h above the floor. Consider a crate of mass m for which the coefficient of kinetic friction with the floor is μ_k. (a) If the crate is dragged at a constant speed, derive an expression for the tension in the cable as a function of x, the distance horizontally from the crate to the pulley. (b) How

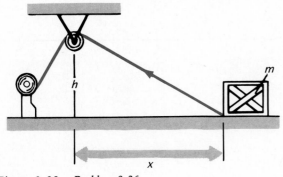

Figure 9-23.   *Problem 9-26.*

much work is done in dragging a 200-kg crate from $x_1 = 6$ m to $x_2 = 3$ m if $h = 3$ m and $\mu_k = 0.6$?

Section 9-4   The Dot (or Scalar) Product

· 9-27 Q If the dot (scalar) product of two vectors is zero, then

(A) one of the vectors must have zero magnitude.
(B) they must be constant vectors.
(C) the vectors are parallel.
(D) it is possible the vectors are antiparallel.
(E) the two vectors must be perpendicular.

· 9-28 P Vector **A** has components $A_x = 6$ and $A_y = 2$. Vector **B** has components $B_x = -3$ and $B_y = 9$. What is **A·B**?

: 9-29 P Vectors **A** and **B** are two sides of a triangle. The third side is **A + B**. The angle between **A** and **B** is θ. Derive the law of cosines by computing $(\mathbf{A} + \mathbf{B}) \cdot (\mathbf{A} + \mathbf{B})$.

: 9-30 P Vector **A** has components $A_x = 2$, $A_y = 3$, and $A_z = 4$. Vector **B** has components $B_x = 5$, $B_y = 6$, and $B_z = 7$. What is the angle θ between **A** and **B**?

Section 9-6   Work Done by Gravity

: 9-31 P A boy throws a ball with a speed of 10 m/s at an angle 30° above horizontal. He is on top of a building, and the ball strikes a playground surface. This surface is 8 m below roof level. How fast is it then going? The boy throws the ball with his hand a height of 1.0 m above roof level.

· 9-32 P A woman raises a 1.0-kg object vertically upward at constant speed through a distance of 0.5 m. (a) What is the work done by the woman? (b) What is the work done by the force of gravity on the object? (c) What is the work done by the resultant force on the object?

: 9-33 P A mass m is attached to a string of length L and hung as a pendulum. The mass is pulled to the side until the string makes an angle θ with the vertical, and then released. Show in detail, using integration, that the work done by the force of gravity acting on the mass as it swings down to its lowest point is $W = mgL(1 - \cos \theta)$

Section 9-7   Conservative Force

· 9-34 Q Which of the following best characterizes a "conservative" force?

(A) A conservative force is one that conserves momentum.
(B) The direction in which a force acts can be changed, but the force itself cannot be created or destroyed if it is conservative.
(C) If you do work against a conservative force the energy used is not "lost," but can be regained later.
(D) A conservative force changes a system's potential energy, but a nonconservative force changes the system's kinetic energy.
(E) The magnitude and direction of the force depends only on the location.

· **9-35 Q** A conservative force is one that

(A) cannot be created or destroyed.

(B) is the same as a "reaction" force.

(C) does no net work.

(D) does work on a particle that is independent of the particular path the particle takes between two points.

(E) does equal amounts of positive and negative work in any process.

Section 9-8   Work Done by a Spring

· **9-36 Q** When a mass m is attached to a spring it experiences a force $F = -kx$. The minus sign in this equation indicates that

(A) F is a conservative force.

(B) F is a restoring force; it tends to move m back to its equilibrium position.

(C) F is always opposite to the direction of motion.

(D) F is always negative.

(E) F is a damping force.

: **9-37 P** A spring with force constant k is compressed an amount x and two identical objects, each of mass m, are attached to the ends of the spring. Then the objects are released from rest. What velocity will each acquire?

: **9-38 P** A 1-kg block rests on a rough horizontal surface ($\mu_k = 0.40$) and is held in contact with one end of a spring ($k = 400$ N/m) which is compressed 20 cm. The other end of the spring is fixed permanently. If the block is released, how far will it travel over the surface from its initial position until it comes to rest?

: **9-39 P** Two springs of spring constants k_1 and k_2 and of equal unstretched length may be connected to form a composite spring in the two ways drawn in Figure 9-24. By considering the work done to stretch them a given amount, deduce the effective force constant for each combination.

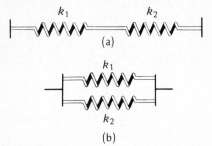

Figure 9-24.   Problem 9-39.

Section 9-9   Power

· **9-40 P** A 4.0-kg object is thrown vertically upward at a speed of 30 m/s. What is the rate at which the force of gravity does work on the object one second after it has been thrown?

· **9-41 P** A 50-kg woman runs up a flight of stairs in 10 seconds. There are 15 steps, each one 20 cm high and 20 cm deep. What is the rate at which the woman is doing work?

: **9-42 P** A 1000-kg automobile is accelerated from rest to 20 m/s in 40 seconds. (*a*) How much energy is required to do this? (*b*) What average power level is needed? (*c*) If the acceleration is constant, is the power also constant?

· **9-43 P** When running at 5.5 mi/hr (about 11 minutes per mile) a typical metabolic rate (i.e., rate of using energy) for a 70-kg person is 580 kcal/hr. At this rate, how far would such a person have to run to lose one pound (455 gm) of fatty tissue? One gram of fat when "burned" yields 9 kcal.

: **9-44 P** The world land speed record was set October 23, 1970, on the Bonneville salt flats in Utah by G. Gabelich, U.S.A. The rocket-propelled "Blue Flame" racer he drove reached a speed of 1002 km/hr using an engine that provided 13,000 lb of thrust. What effective power (in horsepower) did the engine deliver?

· **9-45 P** The force required to tow an ore barge is proportional to the speed. If it requires 80 hp to tow a 50,000-lb barge at 3 km/hr, what power is required to tow it at 9 km/hr?

: **9-46 P** In 1979 the Gossamer Albatross made the world's longest controlled, human-powered flight when Bryan Allen pedaled it 22 miles across the English Channel in 2.6 hours. He delivered power to the pedals at a rate of 0.33 hp. Against what average drag force was he working?

: **9-47 P** The aerodynamic drag force acting on a car moving at high speeds is proportional to v^2, where v is the speed of the car. Assuming that this is the only frictional force acting on a car, suppose that you are cruising at 30 mph with a power output of 10 kW from your car's engine. If you step on the accelerator and increase the car's speed by 10% to 33 mph, what power output from the engine is required?

: **9-48 P** In towing a trailer of mass 4000 kg at 36 km/hr a truck provides 50 kW of power. What force does the truck exert on the trailer?

: **9-49 P** Suppose that you wish to design a T-bar ski tow with the following parameters: Pull N skiers per hour up a

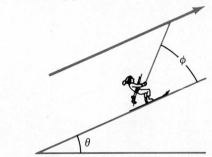

Figure 9-25.   Problem 9-49.

slope D meters long with a spacing of d meters between each pair of skiers (two people ride on each T-bar). Coefficient of kinetic friction between snow and skis is μ_k. Efficiency of drive machinery is e. Weight of average skier is W. What average power is required for such a ski tow?

: **9-50 P** A typical car has a drive-train thermal efficiency of about 12%. This is the fraction of the energy in the fuel that is finally delivered to the wheels. One gallon of gasoline yields about 1.8×10^8 joules of energy. For a car of mass (including driver) of 1000 kg, (a) calculate the energy needed to accelerate the car from rest to 55 mph (neglecting frictional resistance due to the motion). How much gasoline would this use? (b) Suppose that the power output of the engine is 20 hp while cruising at 55 mph. How much gasoline is required to drive at this speed for one mile?

: **9-51 P** A person at rest has a basal metabolic rate of 120 watts. Her metabolic rate increases to 480 watts when she is riding her bike on a level road at 6 m/s (13.5 mph). Twenty percent of this added power is used to do work.

(a) What is the average frictional force acting on her? (b) If she bends low over the handlebars she can reduce the frictional force to 9 N. If she continues to pedal with a constant power output, what will her speed be now?

: **9-52 P** The power delivered to the drive train of an automobile for an engine of given efficiency determines the gasoline mileage of the car. About 10% of this power is lost because of friction in the transmission and ancillary gearing, wheel bearings, and brakes. The remainder drives the car forward against friction arising from air resistance and the flexing of tires. The tire-rolling resistance is typically 1% of the car's weight, and the aerodynamic drag force is given approximately by the relation

$$F_D = \tfrac{1}{2}C_D\rho Av^2$$

where C_D = a dimensionless aerodynamic drag coefficient
A = frontal area of the car (in m², here 2.0 m²)
ρ = air density, about 1.25 kg/m³ at STP
v = velocity, m/s
Modern experimental car designs can reduce a car's curb weight (car plus 300-lb load) to 2200 lb. C_D can be reduced from about 0.5 for older cars to about 0.4 for contemporary cars. Under these conditions, what engine power, in hp, is required to drive a car at 55 mi/hr up a 5% grade (5 m rise for 100 m on road)?

: **9-53 P** Chevrolet advertises that one of its new models (of mass 1200 kg) can cruise at 22 m/s (about 50 mph) at a power level of 9 kW (about 12 hp).

(a) What is the average frictional force acting on the car? (b) If the frictional force were independent of speed (not true, but a simplifying first approximation), how far would the car coast on level ground if the engine were turned off when traveling 22 m/s? (c) If the car gets 40 miles per gallon on level ground, what mileage would you expect it to get at the same power level when going up a 2% grade (2 meter rise for each 100 meters along the roadway)?

: **9-54 P** A 70-kg hiker climbs 1.2 km in elevation from a base camp at Two Pan Camp on the Lostine River to the peak of Eagle Cap, a distance of 15 km. He makes the trip in 4 hours. The efficiency of his muscles is 15%.

(a) How much mechanical work against gravity (in both joules and kilocalories) does the hiker do on this trip? (b) At what rate is he doing work? (c) Taking into account the efficiency of his muscles, how much energy is needed for the trip? (d) A doughnut yields about 140 kcal when eaten. How many doughnuts would a hiker have to eat to get the energy needed for this hike? (e) One gram of fatty tissue yields 9 kcal of energy when used up. How much weight (in fat) could one expect to lose on a hike like this?

Supplementary Problems

9-55 P Steve McPeak is celebrated for his high-wire stunts. He climbed from Rio de Janeiro, Brazil, to the top of Sugar Loaf Mountain—a total distance of 2400 ft. The average inclination of the 1.7-inch cable was 25°, and the 675-ft vertical ascent took 65 minutes. The pole McPeak carried weighed 35 pounds; his own weight was 155 pounds. (a) What was the total work done by McPeak? (b) What was his average power? (c) The most steeply inclined wire McPeak climbed was at an angle of 37° at Santa Cruz Amusement Park. What was the minimum static coefficient of friction between McPeak's shoes and the wire?

9-56 P The world's strongest woman, Jan Todd, set the record for the two-handed dead lift at 2.134 kN (479.7 lb) in 1981. The weights, initially sitting on the floor, were raised upward a distance of 41 cm in a time interval of about $\tfrac{1}{5}$ s. What was Jan's average power output while lifting the weights?

9-57 P A Porsche 924 Turbo sports car requires only 16.5 hp to cruise at 55 mph. Its top speed is 132 mph, and the engine then produces 154 hp at 5750 rpm. Calculate the total resistive force at the two speeds. Is the resistive force linear in the speed, quadratic with speed?

9-58 P According to published test results, the Mazda RX-7 sports car requires just 13 hp to cruise at 50 mph; half of this power is used to overcome areodynamic drag and the rest is dissipated in other frictional losses. Find the aerodynamic drag force at (a) 50 mph and (b) the rated top speed of 118 mph. (c) What is the power required to overcome aerodynamic drag at the top speed? (Assume that the drag force is given by the equation $F_D = -(\tfrac{1}{2})C_D\rho Av^2$, where C_D is the dimensionless drag coefficient, ρ is the air density, A is the area of the car projected normal to its direction of travel, and v is its speed relative to air.)

10 Potential Energy and Energy Conservation

The term *energy* may conjure up in the imagination all sorts of mysterious, almost mystical, meanings and associations. But *energy,* as used in physics, has an ordinary origin. The idea of kinetic energy and the work-energy theorem come directly from Newton's laws. The extension of these ideas, especially the concept of potential energy and the energy-conservation principle that follows from it, are now our primary concerns.

The concept of potential energy ensures that a system's total energy is conserved. When we use the energy conservation principles, it becomes particularly easy to analyze and solve problems that could also be solved using Newton's laws of motion.

10-1   Potential Energy Defined

You have $20 in bills in your pocket and $30 in your bank account. How much money do you have all told? It is, of course, $50; there is $20 in circulation ("kinetic" money, so to speak) and $30 that can become money in circulation simply by being withdrawn from your account. While on deposit, the $30 is just potential money ("potentially kinetic" money). The total amount you have does not change if you deposit cash or withdraw some of your credit. If the bank is sound (a "conservative" bank), you can be sure that you can always get back money you temporarily relinquish by depositing it (converting kinetic money into potential money).

This homely example has a close analogy in the concepts of kinetic energy, potential energy, and the energy-conservation principle. Suppose a particle loses kinetic energy but you are sure that you can get back all the "lost" kinetic energy. Then, if the kinetic-energy debt is sure to be repaid, you can think of the missing kinetic energy as potentially kinetic energy, or *potential energy* for short. In fact, the potential energy for a system of particles is defined as that quantity U, whose addition to the particle's kinetic energy K ensures that the total is constant. If we label the system's total energy E, we can then write

$$E = K + U = \text{constant} \qquad \text{(10-1a)}$$

If K_1 and U_1 are the kinetic and potential energies of the system at one time and K_2 and U_2 their respective values at some second time, then the constancy of E implies that

$$K_1 + U_1 = K_2 + U_2 \qquad \text{(10-1b)}$$

Using still another form, we so define the system's potential energy that for any change ΔK in the system's kinetic energy, there is a corresponding change ΔU in the potential energy, so that

$$\Delta K + \Delta U = 0$$
$$\Delta U = -\Delta K \qquad \text{(10-1c)}$$

Thus, when K decreases (increases), U must increase (decrease) by the same amount to ensure that $\Delta E = 0$.

In summary, we *invent* potential energy *so that* a system's total energy is *constant*. Equations 10-1a, b, c are alternative expressions of the *law of energy conservation*.

What does it take to be sure that any kinetic energy lost can be recovered completely? As we saw in Section 9-5, if the net work done on a particle is zero when the particle is brought back to its starting position by any route, then the net change in the particle's kinetic energy is zero. Over a complete cycle, any kinetic energy "lost" *is* returned. Therefore, potential energy can be defined and the energy-conservation principle applied if the interaction force is a *conservative force*, symbolized by

$$\oint \mathbf{F} \cdot d\mathbf{r} = 0 \qquad \text{(10-2)}$$

The gravitational force (Section 9-6) and the spring force (Section 9-8) are both conservative forces. Therefore, we can associate a potential energy function with each of these interactions (to be given in Sections 10-2 and 10-3, respectively).

How is the change in potential energy ΔU related to ΔW, the work done by the interaction force? From the work-energy theorem (Section 9-2), we have

$$\Delta W = \Delta K$$

and potential energy is defined (10-1c) so that

$$\Delta U = -\Delta K$$

Eliminating ΔK from these two equations gives

$$\Delta U = -\Delta W \tag{10-3}$$

In words, the change in the potential energy of a system of particles is equal to the negative of the work done by the interaction force. Equation (10-3) can be written in more detail as

$$U_f - U_i = -\int_i^f \mathbf{F} \cdot d\mathbf{r} \tag{10-4}$$

where i and f are the initial and final states.

10-2   Gravitational Potential Energy

Here we consider the potential energy for a system consisting of a mass m interacting with the earth, of effectively infinite mass. We suppose that mass m undergoes displacements that are small compared with the earth's radius. Then the acceleration due to gravity has constant magnitude g and constant direction, vertically downward. (The general relation for gravitational potential energy, applicable for all separation distances, is given in Section 15-5.)

Suppose that mass m is thrown vertically upward from the earth's surface. It rises to its maximum height and then falls back to earth. To describe the motion in terms of work and kinetic energy, we say the object loses kinetic energy on the way up because the gravitational force $m\mathbf{g}$ on it does negative work (force down, displacement up). On the way down the object gains kinetic energy because the gravitational force now does positive work (force down, displacement down). The net work done over this round trip, and any other, is *zero*; the object hits the earth with the *same* kinetic energy it had initially. The gravitational force $m\mathbf{g}$ *is* a conservative force and a potential energy *may* be associated with it.

Equation 10-3 relates the change in potential energy ΔU to the work done ΔW by the interaction force:

$$\Delta U = -\Delta W$$

As an object of mass m rises from ground level (taken to be $y = 0$) to an elevation y, the work ΔW done by gravity on this object is

$$\Delta W = -mgy$$

where the minus sign indicates that with an *upward* displacement y and *downward* force mg, the work done is negative.

Combining the two equations above, we have

$$\Delta U_g = mgy$$

Since it is only the potential-energy *change* that equals the kinetic-energy *change*, we can choose any point as the zero for U. Here it is simplest to take the zero for gravitational potential energy U_g to correspond to $y = 0$. Then

$$U_g = mgy \tag{10-5}$$

It must be emphasized that (10-5) applies only when $\mathbf{g}$ can be considered constant.

We now have two equivalent ways of describing the energetics of an object acted on by gravitational force:

- The *work-energy theorem:* We attribute a kinetic energy change to the work done by external forces acting on the object (without invoking the idea of potential energy).
- The *energy-conservation principle:* We attribute a kinetic energy change to a corresponding change in the system's potential energy (without involving the work done by forces).

The potential-energy concept came from the work-energy theorem. The work-energy theorem in turn came from Newton's laws, so that the energy-conservation principle, too, has its origin in Newton's laws. Anything that can be done with energy conservation can also be done with Newton's laws. But the energy-conservation principle has special advantages in analyzing many situations:

- We deal with *scalar* quantities (K, U, and E) rather than vector quantities. Once the potential-energy function U is known, we need not concern ourselves with forces and their vector components.
- We need not be concerned in detail with what goes on between some initial state and some final state, as is required in applying Newton's laws. In the energy-conservation principle, we deal only with the *end points*, because the total energy at one point is the same as at any other point.

On the other hand, not all possible questions can be answered by energy conservation alone. For example, with energy conservation we can find a particle's future kinetic energy and speed, but we cannot find directly the time of flight or the direction of its final velocity. The examples following illustrate these ideas.

Example 10-1.   An object of mass m is thrown with an initial speed v_1 at angle θ relative to a horizontal plane at an elevation y above the earth's surface. With what speed v_2 does the object strike the earth?

This is just Example 9-9, worked earlier by applying the work-energy theorem. Now we apply energy conservation. We take 1 and 2 to designate the initial and final states, as shown in Figure 10-1. From energy conservation,

$$K_1 + U_1 = K_2 + U_2$$

$$\tfrac{1}{2}mv_1^2 + mgy = \tfrac{1}{2}mv_2^2 + 0$$

or

$$v_2 = (v_1^2 + 2gy)^{1/2}$$

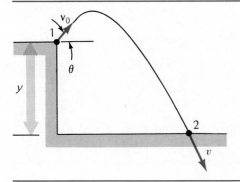

Figure 10-1.   An object projected with speed v_1 at point 1 has speed v at point 2, a vertical distance y below 1.

Note again that the final speed is independent of the object's mass and of the projection angle θ. The great utility of the energy-conservation principle is that we need not be concerned with what happens in detail at all the intermediate stages but can instead deal only with the end points. Energy conservation does not, however, give us all details of the motion. In this example, for instance, we cannot find the time of flight or the direction of the final velocity from energy considerations alone.

Example 10-2.   Suppose that the object in the previous example, instead of being projected at speed v_1 to fly through the air and strike the ground a vertical distance y below, is now sliding at initial speed v_1 on a smooth surface, as in Figure 10-2. We wish to find its speed v_2 at points a distance y vertically beneath the starting point.

The only significant difference from the previous example is that we now have, in addition to the object's weight, a normal force N, which constrains the object to remain on the surface. But the normal force is always at right angles to the object's displacement along the surface. It does no work, and no potential energy can be associated with it. Therefore, exactly the same analysis applies here, and we have at once that $v_2 = (v_1^2 + 2gy)^{1/2}$ at points 2, 2', and 2" in Figure 10-2.

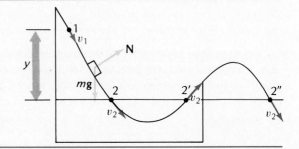

Figure 10-2.   An object sliding on a frictionless surface has speed v_1 at point 1. Its speed at points 2, 2', and 2", all a vertical distance y below point 1 is v_2.

Example 10-3.   A small object is attached to the end of a taut cord of negligible mass as shown in Figure 10-3. At point 1 its speed is v_1, and it swings downward in a vertical circle. What is the object's speed at point 2, a distance y below point 1?

Again we have essentially the same problem as in the two previous examples. The string's tension T is always at right angles to the particle's displacement along the circular arc, so that it does no work, and a potential energy cannot be associated with this constraint force. And again the gravitational force (weight) is accounted for by the gravitational potential energy. The speed is, once more, $v_2 = (v_1^2 + 2gy)^{1/2}$.

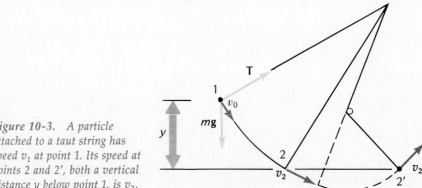

Figure 10-3.   A particle attached to a taut string has speed v_1 at point 1. Its speed at points 2 and 2', both a vertical distance y below point 1, is v_2.

Notice in the figure that the string, after swinging downward, is interrupted by a post, so that a shortened segment of string now turns about it. The force of this post does no work on the system—it does not act over a displacement—and has no influence on the system's total energy. Consequently, the particle at the string's end will also have the same speed and kinetic energy for all points along a horizontal line.

Example 10-4. A 4.0-kg block (m_4) is on a smooth horizontal table. This block is connected to a second block with a mass of 1.0 kg (m_1) by an essentially massless, flexible taut cord that passes over a frictionless pulley. See Figure 10-4. The 1.0-kg block is 1.0 m above the floor. The two blocks are released from rest. With what speed does the 1.0-kg block hit the floor?

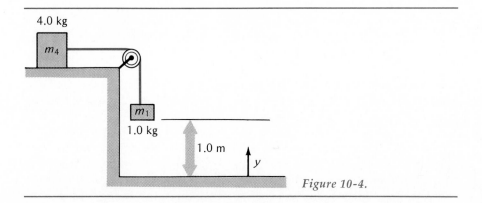

Figure 10-4.

This is just Example 5-8, solved earlier by applying Newton's second law. Now we can solve for the speed far more simply by applying the energy-conservation principle.

We denote the initial state (when blocks m_1 and m_4 are released from rest) by i; the final state (when the block m_1 hits the floor) by f. The origin for the y axis is chosen to be at floor level. Then the energy-conservation principle implies

$$K_i + U_i = K_f + U_f$$

where K is the total kinetic energy of the two blocks and U is the gravitational potential energy of m_1. (We need not include the gravitational potential energy of m_4 since it does not change.) We have, then,

$$(0 + 0) + (m_1 g y) = (\tfrac{1}{2}m_1 v_f^2 + \tfrac{1}{2}m_4 v_f^2) + 0$$

where v_f is the speed of both m_1 and m_4 at the instant when m_1 hits the floor. Note that we have effectively worked this problem in *one line.*

Solving for v_f gives

$$v_f = \sqrt{\frac{2m_1 g y}{m_1 + m_4}}$$

$$= \sqrt{\frac{2(1.0 \text{ kg})(9.8 \text{ m/s}^2)(1.0 \text{ m})}{(5.0 \text{ kg})}}$$

$$= 2.0 \text{ m/s}$$

Example 10-5. An apple of mass m is dropped from rest at height y above the earth, of mass M. How does the earth compare with the apple in kinetic energy when the two collide?

The apple is subject to a force $m\mathbf{g}$ downward and it acquires a final kinetic energy

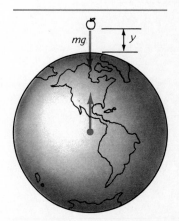

*Figure 10-5. Apple with weight m**g** is dropped from rest at a distance y above the earth's surface. The downward momentum of the apple is always equal in magnitude to the upward momentum of the earth.*

$K_m = \frac{1}{2}mv^2 = p^2/2m$, where its momentum is $p = mv$. The earth is subject to an upward force of the same magnitude (Newton's third law), and it acquires a final kinetic energy $K_M = \frac{1}{2}MV^2 = P^2/2M$, where the earth's final momentum is $P = MV$. See Figure 10-5. Actually, in a reference frame in which the center of mass of the system (m and M) is at rest, we see the apple falling down and the earth falling up. Then the system's total momentum is zero: $\mathbf{p} + \mathbf{P} = 0$. In magnitude, $p = P$. The ratio of kinetic energies is, then,

$$\frac{K_M}{K_m} = \frac{(P^2/2M)}{(p^2/2m)} = \frac{m}{M}$$

We take $m = 0.3$ kg and use $M = 6 \times 10^{24}$ kg. Then $K_M/K_m = 5 \times 10^{-26}$. The apple and the earth have exactly the same momentum magnitude, but the earth has a trivial kinetic energy compared with the apple.

10-3   Spring Potential Energy

Our system is now a block of mass m and a spring of stiffness k attached to a second object of effectively infinite mass. See Figure 10-6. We wish to find the potential energy function U_s associated with the compressed or stretched spring. Suppose that mass m with initial kinetic energy K_i approaches the undeformed spring, as in Figure 10-6. We know that the spring does negative work on the block as it is brought to rest and the spring is being compressed; the spring does an equal amount of positive work on the block as the spring expands back to its equilibrium configuration. Since the spring force is conservative, the net work done by this spring over a cycle that brings it back to its initial configuration is zero;

Figure 10-6.

$$\oint \mathbf{F_s} \cdot d\mathbf{r} = 0$$

There is then no net change in the block's kinetic energy; it leaves the spring (again undeformed) with the same kinetic energy as that with which it approached.

When a spring is compressed by an amount x from its equilibrium configuration, the spring does negative work in the amount*

$$W_s = -\tfrac{1}{2}kx^2$$

as shown in Section 9-6.

The change ΔU_s in the spring's potential energy is, then, from (10-3),

$$\Delta U_s = -\Delta W_s = -(-\tfrac{1}{2}kx^2)$$

$$= \tfrac{1}{2}kx^2$$

We take the spring's potential energy to be zero when it is relaxed ($U_s = 0$ when $x = 0$ and $F_s = 0$). Therefore,

$$U_s = \tfrac{1}{2}kx^2 \tag{10-6}$$

Note that either a compression ($x < 0$) or an extension ($x > 0$) of a spring produces an increase in its potential energy.

---

* The spring does negative work, $-\tfrac{1}{2}kx^2$, on the block striking it. The block, or a person, compressing the spring does positive work, $+\tfrac{1}{2}kx^2$, on the spring.

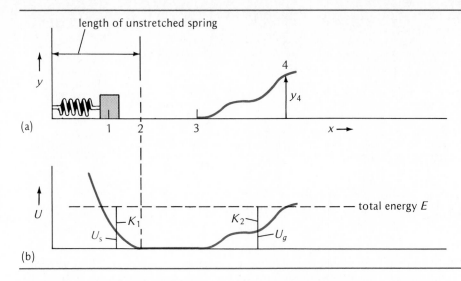

(a)

(b)

Figure 10-7. *(a) A block is released at point 1 from a compressed spring, slides freely from point 2 to 3, and slides up the incline to point 4.*
(b) Potential energy plotted as a function of position.

Example 10-6. A 1.0-kg block is pressed against, but not attached to, a light spring having a stiffness constant of 1.0×10^3 N/m. When the spring has been compressed 7.0 cm, the block is released. The block slides along a frictionless surface and up an incline, as shown in Figure 10-7(a). What maximum height y does the block achieve?

After the spring has been compressed and released, the total energy content of the system remains constant. First, potential energy of the spring $\frac{1}{2}kx_1^2$ is transformed into kinetic energy $\frac{1}{2}mv_2^2$ of the sliding block; then at point 4, the block's kinetic energy becomes gravitational potential energy mgy_4 as the block comes to rest at the highest point:

$$\tfrac{1}{2}kx_1^2 = \tfrac{1}{2}mv_2^2 = mgy$$

$$y = \frac{kx_1^2}{2mg} = \frac{(1.0 \times 10^3 \text{ N/m})(7.0 \times 10^{-2} \text{ m})^2}{2(1.0 \text{ kg})(9.8 \text{ m/s}^2)} = 0.25 \text{ m}$$

This example illustrates the usefulness of the potential-energy concept and the conservation principle following from it. We merely equate the total energy of the system at the end points, taking no concern for such details of the motion as the shape of the incline or the velocity at intermediate points. Figure 10-7(b) shows how the total energy remains constant while it changes from one form to another.

10-4 Properties of Potential Energy

• *Potential energy can be defined only for a conservative force and can be associated only with such a force.* A force on a particle is conservative if the net work is zero as a particle goes around a loop and returns to its initial position (Section 9-7):

$$\text{Conservative force:} \quad \oint \mathbf{F} \cdot d\mathbf{r} = 0$$

Said differently, positive work done by the interaction force along one part of the path must be compensated exactly by an equal amount of negative work elsewhere.

A *nonconservative force* is one for which the work done around a closed path is *not* zero. Potential energy cannot be associated with such a force:

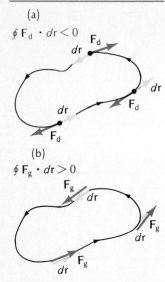

Figure 10-8. *Nonconservative forces: (a) An energy-dissipative force $\mathbf{F_d}$ is opposite to the displacement and $\oint \mathbf{F_d} \cdot d\mathbf{r} < 0$ around a closed loop; (b) an energy-generative force $\mathbf{F_g}$ is along the displacement and $\oint \mathbf{F_g} \cdot d\mathbf{r} > 0$ around a closed loop.*

$$\text{Nonconservative force:} \quad \oint \mathbf{F} \cdot d\mathbf{r} \neq 0$$

Suppose some person pushes an object around a closed path on a rough horizontal surface. The force of friction is always opposite in direction to the object's displacement; the friction force always does *negative* work. See Figure 10-8. For any *energy-dissipative force,* such as friction,

$$\oint \mathbf{F_d} \cdot d\mathbf{r} < 0$$

Now consider the work done by the hand of the person pushing the object. The force of the hand must, for each point in the path, be along the displacement so as to balance out the frictional force. Therefore the hand force always does *positive* work. Around a closed path, the net work done by such an *energy-generative* force is positive:

$$\oint \mathbf{F_g} \cdot d\mathbf{r} > 0$$

Considering now both the friction and hand forces, we see that the object returns to its starting location with no net change in kinetic energy, since the negative work done by the frictional (energy-dissipative) force is just balanced by the positive work done by the hand (energy-generative) force.

• *Potential energy is a property of a system of interacting bodies as a whole.* One cannot speak of the potential energy of a single object. For example, when a ball falls toward the earth, it is perhaps natural to speak of the ball as losing potential energy; after all, it is the ball rather than the earth that gains most of the kinetic energy. However, it is the earth-ball *system* that gains kinetic energy as the two objects approach one another, and it is the *system* that loses potential energy.

• Our concern is always with differences in potential energy; therefore, the *choice of the zero of potential energy is arbitrary.* If the force associated with the potential energy is *constant*—for example, the gravitational force on an object close to the earth—we may choose any convenient horizontal level (usually the lowest, or ground, level) as the zero for gravitational potential energy. On the other hand, if the force varies with displacement—for example, a spring—it is customary to choose the potential energy to be zero at that displacement for which the force is zero (both $U_s = \frac{1}{2}kx^2$ and $F_s = -kx$ are zero at $x = 0$). There is also, in fact, an arbitrariness in the choice of the zero for kinetic energy—it is the choice of reference frame. An object always has zero kinetic energy relative to the reference frame in which it is at rest.

Both kinetic and potential energies are scalar quantities, but whereas kinetic energy must always be positive, potential energy may be either positive or negative.

• While kinetic energy is energy of motion, *potential energy is energy of position,* or more properly, *the energy associated with the relative separation of interacting particles.* A spring has potential energy when it is deformed by being stretched or compressed, and it retains this potential energy so long as it remains deformed. For example, if we clamp a compressed spring, the spring's potential energy is locked in. No matter how long a time elapses, the spring has the potential for doing work, and if later we release the clamp, this stored energy is then released. When a spring is deformed, the atoms that it comprises must change their separation distances. Any potential energy is, in fact, related ultimately to the relative separation of particles.

• *The interaction force may be deduced from the potential-energy function.*
First, recall that we already have the general relation that allows us to find U,
given $\mathbf{F}$:

$$U_f - U_i = -\int_i^f \mathbf{F} \cdot d\mathbf{r} \qquad (10\text{-}4)$$

Now we wish to find the inverse relation, one for finding $\mathbf{F}$, given U.

Consider a particle with kinetic energy $K = \frac{1}{2}mv^2$ interacting with an effectively infinitely massive second object. The system's total energy E is

$$E = K + U = \tfrac{1}{2}mv^2 + U(x)$$

For simplicity, we suppose that the particle is confined to motion along the x
axis so that U depends only on x.

Since energy is conserved, E does not change with time. Taking the derivative with respect to time of the preceding equation, we then have

$$\frac{dE}{dt} = 0 = mv\,\frac{dv}{dt} + \frac{d}{dt}\,U(x) \qquad (10\text{-}7)$$

$U(x)$ is a function of x only, not of time, so we compute dU/dt by using the chain
rule for derivatives:

$$\frac{d}{dt}\,U(x) = \frac{dU}{dx}\frac{dx}{dt} = \frac{dU}{dx}\,v$$

We recognize that the particle's velocity v is dx/dt and that its acceleration a is
dv/dt. Using the result, we can then write

$$0 = ma + \frac{dU}{dx}$$

But the force F_x on the particle equals ma, so that we have

$$F_x = -\frac{dU}{dx} \qquad (10\text{-}8)$$

In words, the force in a given direction is the negative derivative of the potential-energy function with respect to the coordinate in that direction.

Consider, for example, the gravitational potential energy $U_g = mgy$. Applying (10-8), we have

$$F_y = -\frac{dU}{dy} = -\frac{d}{dy}\,(mgy) = -mg$$

The interaction force is simply the downward weight, $-mg$.

Now consider a spring with potential energy $U_s = \frac{1}{2}kx^2$. From (10-8),

$$F_x = -\frac{dU}{dx} = -\frac{d}{dx}\left(\frac{1}{2}\,kx^2\right) = -kx$$

The force is that of a spring, $F_x = -kx$.

In general, the potential energy may depend on all three coordinates,
$U = U(x, y, z)$. Then the force components are given by*

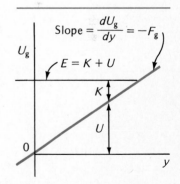

Figure 10-9. *Gravitational
potential energy U_g plotted as a
function of vertical displacement y for a constant gravitational force F_g. The force is the
negative slope of the graph of
U_g against y: $F_g = -dU_g/dy$.*

---

* In computing the *partial derivative* $\partial U/\partial x$ of U with respect to x, one takes variables y and z to be
constant.

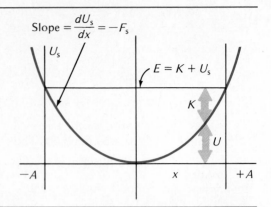

Figure 10-10. *Plot of a spring's potential energy U_s as a function of its displacement x. The curve is a parabola. The object's motion is restricted to displacements between x = ±A, for which the kinetic energy K = E − U_s is positive.*

$$F_x = -\frac{\partial U}{\partial x} \qquad F_y = -\frac{\partial U}{\partial y} \qquad F_z = -\frac{\partial U}{\partial z} \qquad (10\text{-}9)$$

• *A graph of potential energy as a function of displacement is useful in analyzing the energetics of motion.*

Consider a plot of gravitational potential energy $U_g = mgy$ against the displacement y, as shown in Figure 10-9. The plot is a straight line with positive slope mg. A horizontal line may be used to represent the system's *constant* total energy, E, where $E = K = U_g$. The vertical segment between the y axis and the line represents potential energy U_g; the vertical segment between the sloped line and the horizontal line for constant E represents kinetic energy K. Each of K and U may change with time; their sum E is always the same.

The potential-energy curve for a spring, $U_s = \frac{1}{2}kx^2$, is a parabola, as shown in Figure 10-10. Again, the slope of the curve equals the negative of the force: $dU_s/dx = kx = -F_s$. The system's total energy $E = K + U$ is again represented by a horizontal line. The particle's motion is restricted to those values of x for which the kinetic energy is positive. Therefore, the motion is *bound;* an oscillating particle attached to a spring is confined by the *potential well* to the region between $x = -A$ and $x = +A$, where A is the maximum displacement from the equilibrium position. At displacement $x = \pm A$, kinetic energy is a minimum ($K = 0$) and the potential energy for the spring is therefore a maximum ($U_s = \frac{1}{2}kA^2$). Outside the region, for which $-A < x < A$, the kinetic energy would have to be negative—an impossibility.

Suppose that a particle is displaced from the equilibrium position at $x = 0$. Then a restoring force acts to return it to the lowest point in the potential-energy valley. This is in contrast to the *unbound* motion of the falling object in Figure 10-9; there the kinetic energy is positive for *all* points lower than the highest point, at which the particle has zero kinetic energy and maximum potential energy.

10-5   The Energy Conservation Law

Systems with Conservative Forces   When an isolated system consists of particles that interact by strictly conservative forces, the total energy E of the system is constant. By definition,

$$E = K + U = \text{constant}$$

where K is the sum of the kinetic energies of the particles $a, b, c, \ldots$, in the system $K = K_a + K_b + \cdots$, and U is the sum of the potential energies between each pair of interacting particles, $U = U_{ab} + U_{bc} + U_{ac} + \cdots$. The particles of the system may lose kinetic energy; if they do, the potential energy then increases to keep the total energy constant. The "initial" energy E_i of a conservative system is exactly equal to the "final" energy E_f at any later time:

$$\text{Conservative system:} \quad E_i = E_f$$

Equivalently, the system's energy does not change with time: $dE/dt = 0$.

If this system is no longer isolated from its surroundings and an external agent does work $W_{i \to f}$ on the system, then

$$W_{i \to f} = E_f - E_i$$

where the energy content of the system has been increased by the work $W_{i \to f}$ done on it. Conversely, if the system does work on its surroundings, the energy content of the system decreases.

Suppose, for example, that a man raises a block of weight mg vertically from y_1 to y_2. The man interposes himself, so to speak, between the earth and the block, pushing on each to increase their separation. See Figure 10-11. The man, as the external agent, does work on the system, thereby increasing its total energy content.

Suppose the man raises the block at constant velocity. Since the resultant force on the block must then be zero, he must apply a force of magnitude mg to the block (and also to the earth). Neither the kinetic energy of the block nor the kinetic energy of the earth changes. Therefore, the potential energy of the earth-block system must then have increased by $mg(y_2 - y_1)$. But $mg(y_2 - y_1)$ is also the work done by the man. Here, and in general, the work done by an external agent in changing the separation distance between interacting objects

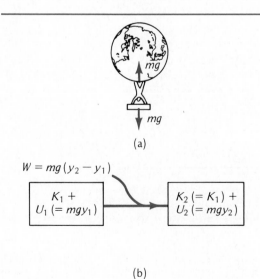

(a)

$$W = mg(y_2 - y_1)$$

$K_1 +$ $U_1 (= mgy_1)$

$$\longrightarrow$$

$K_2 (= K_1) +$ $U_2 (= mgy_2)$

(b)

Figure 10-11. *(a) An external agent increases the separation distance between a block and the earth by applying a force of magnitude mg to both. (b) The change in the potential energy of the earth-block system equals the work $W = mg(y_2 - y_1)$ done by the external agent separating the objects at constant velocity.*

"I Understand the Physics, but I Just Can't Do the Problems."

Ask any physics professor who has been teaching physics for at least a few years, and he or she will tell you that the remark above is the most common one by students who are having trouble with the introductory physics course. Why so much attention to problems? It is by far the best way your instructor has—in fact, the best way *you* have—to tell whether you know the basic physics. You can't do the problems unless you really understand the physics.

It is easy enough to tell how not to do physics problems: Your instructor's remarks in class were so reasonable and the way he or she worked problems in class made them seem so easy you're certain that there is no need to waste time making sure that you have the definitions straight, have understood clearly the basic ideas, and have checked the worked examples in the textbook. Instead, you plunge ahead, read the problem statement rapidly, find out what quantities are given and which are to be found, search for a handy-dandy formula that fits, and plug in numbers. It (almost) never works.

What does work? Just as the answer to "What does it take to get to Carnegie Hall?" is "Practice! Practice!" so too experience in doing problems is the best teacher. But what do experienced experts do when they solve problems? Are they successful because they have memorized all the many specialized formulas and can cough up just the right relation to fit each problem exactly? Not at all! It's just the reverse. An expert organizes knowledge into large, connected, coherent chunks and, at first, skips the details. The expert concentrates first on the big picture and general principles, and then moves by successive refinements to the particulars.

Here is a general strategy that works for solving physics problems:*

• *Describe* the problem. Read the problem statement slowly, even a second or third time, to make sure that you are clear on what is given, what circumstances apply, what simplifications can be made,

*See R. G. Fuller, *Physics Today* **35,** 43 (Sept. 1982) and F. Reif, J. H. Larkin, and G. C. Brackett, *Am. J. Physics* **44,** 212 (1976).

at constant velocity equals the change in the system's potential energy. As another example, a person pulling on the two ends of a spring to elongate it by x does work $\frac{1}{2}kx^2$, and the spring's potential energy increases by $\frac{1}{2}kx^2$.

Systems with Nonconservative Forces It is impossible to construct any large-scale system of objects in which all frictional and other nonconservative forces are absent. Perpetual-motion machines are impossible. Suppose an isolated system has nonconservative internal forces. When particles in such a system lose kinetic energy, the potential energy of the system does *not* increase in the same amount, and conversely. Consequently, the initial total energy of an isolated system with nonconservative forces always *exceeds* the final energy. The total energy *decreases* with time in a nonconservative system:

Nonconservative system: $E_i > E_f$

Energy is not conserved, if by *energy* is meant the kinetic energy of *large-scale* objects one can see and the potential energy one can identify with a *discernible* change in their relative separation. Thus, if energy is still to be conserved in a system with nonconservative forces, a new form, or perhaps several new forms, of energy must be identified, so that the inequality above can be replaced by an equation such as

$$E_i = E_f + E_{\text{non-m}} \tag{10-10}$$

where $E_{\text{non-m}}$ represents the sum of all forms of what might be called *nonme-*

what can be neglected, and what is asked for. Draw a clear diagram, no matter how simple the circumstances. Avoid sketching an angle that comes out close to 45°; you can draw incorrect conclusions from apparent symmetries that do not apply.

• *Plan* a solution. This is always the toughest part. Ask yourself what fundamental principles and relationships might apply. Are there symmetries that can simplify the analysis? To what other problems worked before is this one analogous? Don't, before you see what basic ideas apply, idly go through algebraic manipulations.

• *Implement* the solution. Only after you have identified the principal ideas that apply, should you write them down in algebraic form and solve for the unknowns. This part of the solution may be strictly mathematics since the physics has already been done.

• *Check* the result.

How can you tell if your answer is right? (Don't necessarily count on the answer in the back of the book. It has been checked, but after all, the author of this, or any other physics textbook, is not infallible.)

The following steps are useful not only for testing the solution to a problem but, more generally, for testing your understanding of any theoretical result in physics. These steps can never guarantee that a result is right, but they will always illuminate the meaning of a result and tell you when the result is wrong in some fundamental respect:

• *Check dimensions.* Every valid equation in physics must be dimensionally consistent; the dimensions (and finally the units) on one side must be the same as on the other.

• *Take the limits* of variable quantities. Imagine each quantity to be a variable and let that variable approach zero and infinity. What happens, for example, when a mass approches zero and infinity? Do the results make sense?

• *Sketch the functional dependence* to understand the mathematical character of the result.

Only after the preceding steps have been completed, *try substituting numerical values.* Number crunching comes *last.* Do the results seem reasonable? Do they make sense? Can you compare them with other results?

chanical energy—that is, energy different from the kinetic and potential energies of large-scale bodies.

One of the greatest discoveries in classical physics was that the relation shown in (10-10) agreed with experiment. As we shall see in later chapters, physicists found that nonmechanical forms of energy exist in that they can be measured in meaningful ways, their amounts being such as to make the *total energy content* of an isolated system constant. When the nonconservative force is friction, $E_{\text{non-m}}$ is heat, or thermal energy (mostly). But still other forms of energy are recognized: the energy associated with electromagnetic radiation, atomic energy, and nuclear energy, among others.

Experiments in all branches of physics are consistent with the *general conservation of energy law: the total energy content of an isolated system is constant.* By an *isolated system* is now meant a collection of objects on which no work is done and into (or out of) which neither thermal energy nor radiation flows.

Within the system, energy may be converted from one form to another, but the total amount is unchanged. Together with the conservation laws of mass and of linear momentum, the energy-conservation law ranks as one of the truly fundamental principles of physics. (In the theory of special relativity, the separate conservation laws of mass and energy are combined into a single conservation law, the law of mass-energy, which is thus even more simple and general.) Physicists' confidence in the universality of energy conservation has always been vindicated by experiment.

Recognizing that various forms of energy can be delineated, that energy appears as thermal energy, or *radiation*, as well as macroscopic kinetic energy and potential energy, was a remarkable discovery. But even more remarkable perhaps has been the realization that so-called nonmechanical forms of energy are, after all, just kinetic and potential energy on a submicroscopic scale. Thermal energy is, in fact, kinetic and potential energy associated with the disordered motion of atoms or molecules. Energy of electromagnetic radiation is the kinetic energy of particlelike photons traveling at the speed of light. Chemical energy is traceable to the kinetic energy of subatomic particles and the electric potential energy of their interaction. Nuclear energy is the kinetic energy of the nuclear constituents and the potential energy associated with their interaction. Macroscopically, many forms of energy must be delineated; submicroscopically — at the level of the smallest particles known to physicists — one needs only kinetic and potential energy.

Example 10-7.   Starting from rest, a sled slides down a hill, as shown in Figure 10-12. At point 2 its speed is 9.0 m/s and its elevation is 18 m below the starting point. What fraction of the sled's initial potential energy has been dissipated in going from 1 to 2?

From (10-10), the dissipated energy is

$$E_{\text{non-m}} = E_i - E_f$$
$$= (mgy_1 + \tfrac{1}{2}mv_1^2) - (mgy_2 + \tfrac{1}{2}mv_2^2)$$
$$= mgy_1 - \tfrac{1}{2}mv_2^2$$

Therefore, the fraction dissipated is

$$\frac{E_{\text{non-m}}}{E_i} = \frac{mgy_1 - \tfrac{1}{2}mv_2^2}{mgy_1}$$
$$= 1 - \frac{v_2^2}{2gy_1}$$
$$= 1 - \frac{(9.0 \text{ m/s})^2}{2(9.8 \text{ m/s}^2)(18 \text{ m})} = 0.77$$

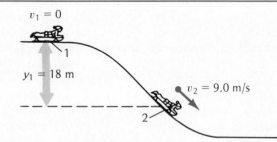

Figure 10-12.

Example 10-8.   Two masses, m and $2m$, are attached to the ends of a massless spring with force constant k. The spring is stretched an amount A, and the masses are then released from rest. See Figure 10-13. What is the kinetic energy (in terms of k and A) of each of the two masses when the spring is relaxed?

The potential energy of the initially stretched spring $U_s = \tfrac{1}{2}kx^2 = \tfrac{1}{2}kA^2$ becomes the sum of the kinetic energies of the two masses. Therefore, energy conservation gives

$$K_i + U_i = K_f + U_f$$

$$(0 + 0) + \tfrac{1}{2}kA^2 = (\tfrac{1}{2}mv_1^2 + \tfrac{1}{2}(2m)v_2^2) + 0$$

where v_1 is the speed of m and v_2 the speed of $2m$.

Since the system is isolated from external forces, its total momentum remains zero:

$$0 = mv_1 + (2m)v_2$$

or

$$v_2 = -\tfrac{1}{2}v_1$$

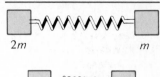

Figure 10-13.

Mass $2m$ moves in the direction opposite that of m and at half the speed of m. Substituting $v_2 = -\tfrac{1}{2}v_1$ in the energy equation and solving for $\tfrac{1}{2}mv_1^2$ yields

$$K_m = \tfrac{1}{2}mv_1^2 = \tfrac{2}{3}(\tfrac{1}{2}kA^2)$$

Mass m has two-thirds of the initial spring potential energy.

10-6   Collisions

A collision can be described as follows. One object with constant initial velocity approaches a second object with a different initial velocity (including zero). The two objects interact; each produces a force on the other. When the collision is over, each of the objects again has a constant velocity, but generally not the same as its initial one. See Figure 10-14. The interacting objects may actually touch in a collision, as billiard balls do. But a collision occurs even if the interacting objects do not touch, as when a comet encounters a planet but does not hit it.

We have already examined collisions from the point of view of the momentum-conservation principle (Section 7-2); if the colliding objects interact only with one another and are isolated from the rest of the world, the total momentum of the system does not change — total (vector) momentum before collision equals total momentum after collision.

Now we have another physical quantity, energy, to characterize colliding objects. Collisions between large-scale objects may be classified according to whether the total *kinetic* energy of both objects before the collision equals or exceeds their total kinetic energy after the collision. The terms *elastic* and *inelastic* are used to characterize these two possibilities:

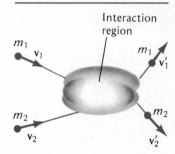

Figure 10-14. A generalized collision. Particles m_1 and m_2 enter the interaction region with velocities $\mathbf{v}_1$ and $\mathbf{v}_2$ and emerge with velocities $\mathbf{v}_1'$ and $\mathbf{v}_2'$.

Elastic collision: Total K before = total K after

Inelastic collision: Total K before > total K after

In an elastic collision, some of the initial kinetic energy is converted temporarily, during the collision, to potential energy; after the elastic collision is over, the potential energy has been entirely converted back to kinetic energy. The colliding objects must interact by a conservative force. Although the interacting objects are deformed during the collision, they spring back elastically to precisely their original shapes after the collision. For example, billiard balls make nearly elastic collisions; and each of the balls looks the same after as it did before.

In an inelastic collision, on the other hand, the objects do not spring back to their original shapes. For example, when two pieces of putty make an inelastic collision, the pieces do not look the same before and after. A nonconservative force acts, and at least some of the original kinetic energy is dissipated. The energy is not "lost"; it is converted, at least partly, to essentially invisible kinetic energy and potential energy on a microscopic scale. Even in an inelastic collision, the total energy of the two colliding objects is the same before and after collision. The macroscopically visible kinetic energy *decreases*, however.

A *completely inelastic* collision is one in which the two colliding objects are *at rest with respect to one another* after the collision. Momentum and total energy are each separately conserved when a collision is observed in any inertial reference frame. But almost every collision is more simply analyzed when it is examined from the point of view of an observer at rest with the system's center of mass. Recall that, by definition, the system's total momentum is zero at every instant in the center-of-mass reference frame (Section 8-4). Thus, in the CM, or zero-momentum, reference frame, the two interacting objects always move in opposite directions with the same momentum magnitude.

The examples on collisions that follow illustrate these ideas:

- Example 10-9: Head-on elastic collision; properties of the moderator in a nuclear reactor.
- Example 10-10: Head-on completely inelastic collision; ballistic pendulum; colliding beams.

Example 10-9.   *Head-on Elastic Collision.*   Particle m_1 hits m_2 (initially at rest) in a head-on elastic collision. What fraction of m_1's initial kinetic energy is transferred to m_2?

First, some matters of definition and nomenclature:

A *head-on* collision is one in which both particles move along a *single straight line* before and after the collision.

- Subscript 1 will identify the incident particle and 2 the struck particle.
- Velocities before the collision are shown without primes; velocities after the collision are shown with primes.
- A velocity measured relative to the laboratory is denoted by **v**; a velocity measured relative to the system's center of mass is denoted by **u**.
- The velocity of the particles' center of mass is denoted by V. From (8-6),

$$V = \frac{m_1 v_1 + m_2 v_2}{m_1 + m_2} = \frac{m_1 v_1}{m_1 + m_2}$$

since $v_2 = 0$.

See Figure 10-15(a) for the situation, as seen in the laboratory, before the collision takes place.

We are to find the fraction F_{trans} of m_1's initial kinetic energy, $K_1 = \frac{1}{2} m_1 v_1^2$, transferred to m_2. Particle m_2 has kinetic energy of $K_2' = \frac{1}{2} m_2 v_2'^2$ after the collision. Therefore,

$$F_{\text{trans}} = \text{fraction of } K_1 \text{ transferred}$$

$$= \frac{K_2'}{K_1} = \frac{\frac{1}{2} m_2 v_2'^2}{\frac{1}{2} m_1 v_1^2} = \frac{m_2 v_2'^2}{m_1 v_1^2}$$

To find F_{trans} we must first find v_2'. We can do this most easily by transforming our point of view of the collision, becoming an observer traveling with the system's center of mass. See Figure 10-15(b).

Comparing Figures 10-15(a) and (b), we see that the CM has velocity V relative to the lab; V is also the velocity of the CM relative to m_2 before the collision. Therefore, before the collision the velocity of m_2 relative to the CM is $-V$. The collision is perfectly elastic, so that the kinetic energy and speed of m_2 are the same relative to the CM before

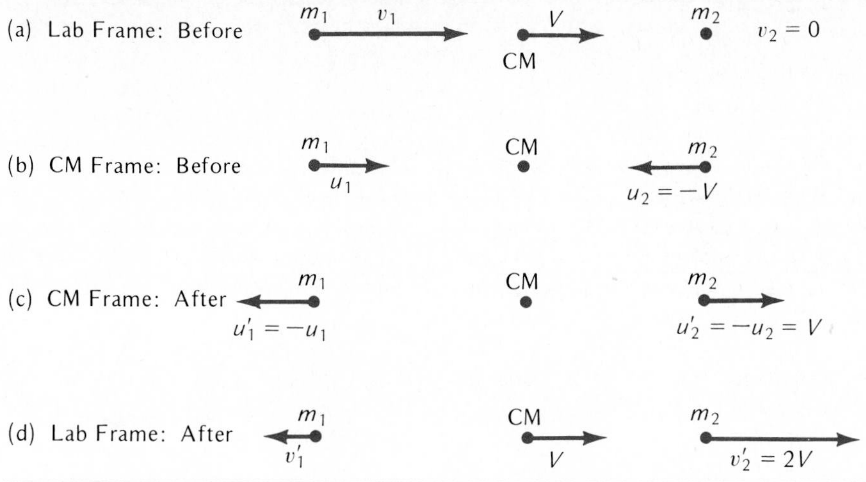

(a) Lab Frame: Before

(b) CM Frame: Before

(c) CM Frame: After

(d) Lab Frame: After

Figure 10-15. *Collision from four points of view: (a) lab frame before collision; (b) CM frame before; (c) CM after; and (d) lab frame after collision.*

and after the collision. After the collision, m_2's velocity in the CM frame is $u_2' = V$ [Figure 10-15(c)]. In other words, m_2 approaches and recedes from the CM at the *same* speed (m_1 also approaches and recedes from the CM with the same speed, but it generally has a different speed from that of m_2). We arrive then at this general result: *The speed of one object measured relative to the other object is unchanged in an elastic collision.*

The speed v_2' of m_2 in the lab after the collision is easily found by comparing Figures 10-15(c) and (d). To switch from the CM back to the lab frame, we merely add a velocity $+V$ to all particles shown in (c) to arrive at (d). We see then that

$$v_2' = 2V$$

By the relation for V above, this becomes

$$v_2' = 2V = \frac{2m_1 v_1}{m_1 + m_2}$$

Substituting this value for v_2' in the relation for F_{trans}, we have

$$F_{trans} = \frac{m_2 v_2'^2}{m_1 v_1^2} = \frac{m_2}{m_1 v_1^2} \left(\frac{2m_1 v_1}{m_1 + m_2} \right)^2 = \frac{4m_1 m_2}{(m_1 + m_2)^2}$$

or

$$F_{trans} = \frac{K_2'}{K_1} = \frac{4 \left(\dfrac{m_1}{m_2} \right)}{\left(1 + \dfrac{m_1}{m_2} \right)^2} \tag{10-11}$$

Consider a special case, one with $m_1/m_2 = 1$, for example, a billiard ball hitting head-on a second billiard ball at rest. Then, (10-11) shows that $F_{trans} = 1$. *All* the kinetic energy is transferred to the struck object. The colliding objects trade velocities, as well as kinetic energies and momenta. Even if the masses are unequal, but differ by no more than an order of magnitude, an appreciable fraction of the incident particle's kinetic energy is transferred to the struck particle. Take $m_1/m_2 = 10$, for example; then (10-11) gives $F_{trans} = 0.33$, or a 33 percent kinetic-energy transfer. The inverse mass ratio $m_1/m_2 = 1/10$ gives the same result.

Moderator. The function of the *moderator* in a nuclear reactor is to reduce the kinetic energy of neutrons by having them collide with other particles of comparable mass. In a typical thermal nuclear reactor using uranium as fuel, the uranium is distributed as lumps or rods in ordinary water. The water acts as moderator as well as heat transfer material. A neutron released in the fission of a uranium nucleus has relatively high kinetic energy. But a neutron is most likely to be captured by another uranium nucleus and produce still another nuclear fission reaction only if the neutron's kinetic energy is relatively small. Therefore, a large fraction of the high-energy neutrons produced in the fission of uranium nuclei must be slowed down before being captured by other uranium nuclei; the water molecules of the moderator do this.

A high-energy neutron hitting a hydrogen atom in a water molecule head-on can lose all its kinetic energy, since the neutron and hydrogen atom have essentially the same mass. Less kinetic energy is lost when the neutron collides with an oxygen atom or when it makes a non-head-on collision.

Example 10-10   *Head-On Completely Inelastic Collision.*   Particle m_1 hits m_2 (initially at rest) in a completely inelastic collision. The two particles stick together and travel off as a single composite object after the collision. What fraction of the initial kinetic energy is dissipated in this collision?

See Figure 10-16, where the sequence is as follows: part (a) is before collision in the lab, part (b) before collision in the CM frame, part (c) after collision in the CM, and part (d) after collision back in the lab. Note that the particles are *at rest* after collision in the CM frame; this always characterizes a completely inelastic collision. We use here the same notation as for Example 10-9.

To find the fraction of kinetic energy dissipated in the collision, we must compare the kinetic energy of the composite object, $\frac{1}{2}(m_1 + m_2)V^2$, with the kinetic energy of the incident particle, $\frac{1}{2}m_1v_1^2$. Note that the composite particle has the velocity $\mathbf{V}$, that of the CM relative to the lab.

$$V = \frac{m_1 v_1}{m_1 + m_2} \text{ with } v_2 = 0 \tag{10-12}$$

The final kinetic energy K' is

$$K' = \frac{1}{2}(m_1 + m_2)V^2 = \frac{1}{2}(m_1 + m_2)\left(\frac{m_1 v_1}{m_1 + m_2}\right)^2 = \left(\frac{m_1}{m_1 + m_2}\right)\left(\frac{1}{2}m_1 v_1^2\right)$$

The initial kinetic energy is $K = \frac{1}{2}m_1 v_1^2$. Therefore, the ratio of final to initial kinetic energy is

(a) Lab Frame: before collision m_1 v_1 CM V m_2 $v_2 = 0$

(b) CM Frame: before collision m_1 u_1 CM $V = 0$ m_2 u_2

(c) CM Frame: after collision   CM $m_1 + m_2$

(d) Lab Frame: after collision $m_1 + m_2$ V

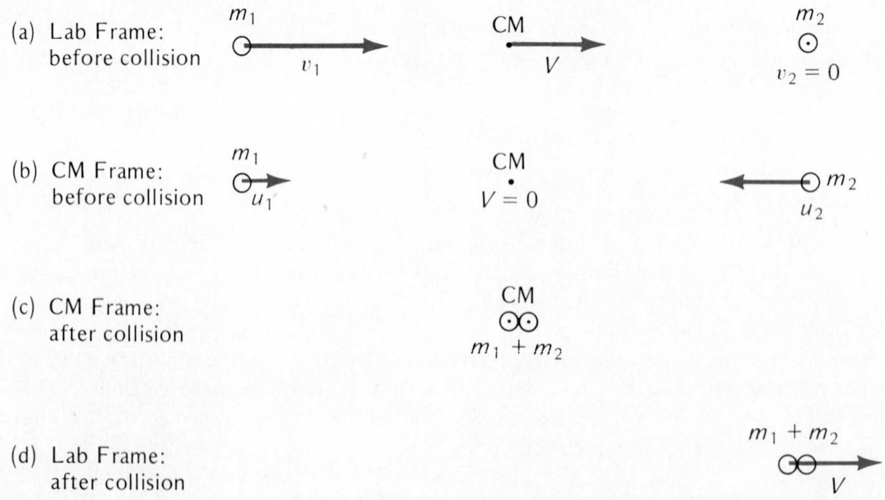

Figure 10-16.   *Completely inelastic collision: (a) lab frame before collision; (b) CM frame before; (c) CM frame after; and (d) lab frame after.*

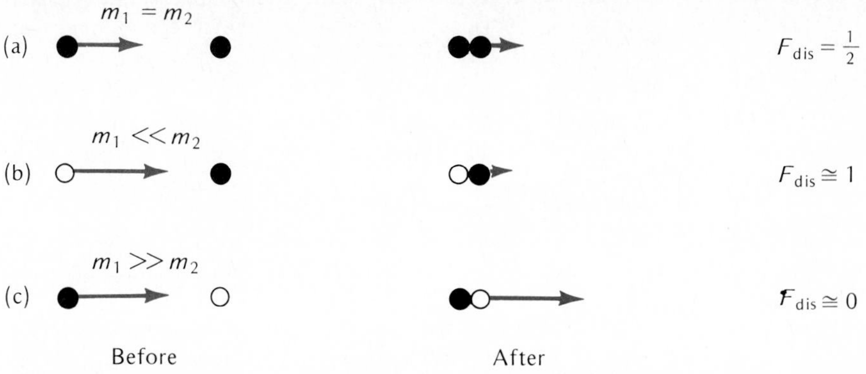

$m_1 = m_2$

(a) ●——→ ●   ●●→ $F_{dis} = \frac{1}{2}$

$m_1 \ll m_2$

(b) ○——→ ●   ○●→ $F_{dis} \cong 1$

$m_1 \gg m_2$

(c) ●——→ ○   ●○——→ $F_{dis} \cong 0$

Before   After

Figure 10-17. The fraction F_{dis} of the initial kinetic energy dissipated in a completely inelastic collision depends on the mass ratio: (a) with $m_1 = m_2$, $F_{dis} = \frac{1}{2}$; (b) with $m_1 \ll m_2$, $F_{dis} \cong 1$; and (c) with $m_1 \gg m_2$, $F_{dis} \cong 0$.

$$F_{rem} = \frac{K'}{K} = \frac{m_1}{m_1 + m_2} \tag{10-13}$$

Equation (10-13) gives F_{rem}, the fraction of kinetic energy remaining after the collision. The rest of the initial kinetic energy is dissipated; therefore F_{dis}, the fraction of initial kinetic energy dissipated in the completely inelastic collision, is

$$F_{dis} = 1 - F_{rem} = 1 - \frac{m_1}{m_1 + m_2} = \frac{m_2}{m_1 + m_2} \tag{10-14}$$

To see the implications of (10-13) and (10-14), consider first a simple, special case: equal masses, $m_1 = m_2$. See Figure 10-17(a). Then (10-13) and (10-14) show that both F_{rem} and F_{dis} are equal to one-half; half the initial kinetic energy is dissipated and the other half remains with the composite object.

Now consider the other extremes. First, suppose that $m_1 \ll m_2$ (for example, a Ping-Pong ball in motion hitting and sticking to a billiard ball at rest). Almost all its energy is dissipated ($F_{dis} \cong 1$); hardly any survives ($F_{rem} \cong 0$). Consider the other extreme, with $m_1 \gg m_2$ (for example, a billiard ball hits and sticks to a Ping-Pong ball initially at rest). Now almost no energy is dissipated ($F_{dis} \cong 0$), and most of it remains ($F_{rem} \cong 1$); this result corresponds to the circumstance that the billiard ball is slowed hardly at all in colliding with and picking up a Ping-Pong ball.

Ballistic Pendulum.   The ballistic pendulum is a device often used in the laboratory of an introductory physics course to illustrate how the speed of a high-speed bullet can be measured indirectly with relatively high precision. More important, the ballistic pendulum illustrates how the conservation principles of momentum and energy apply to a completely inelastic collision.

The essential parts of the device are shown in Figure 10-18. A bullet of mass m_1 hits and comes to rest in a pendulum block of mass m_2. As (10-12) shows, the velocity V of block plus bullet just after collision is

$$V = \frac{m_1 v_1}{m_1 + m_2}$$

Since $m_1 \ll m_2$, the speed V is far less than the bullet speed v_1 and therefore more readily measured.

Actually, V need not be measured directly. After being hit by the bullet, the block swings upward through the vertical distance y in coming to rest. Applying energy conservation to the motion (1) immediately after the bullet is captured and (2) when the block is at its highest point, we have

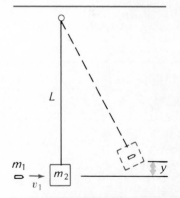

Figure 10-18. A ballistic pendulum. A bullet of mass m_1 and velocity v_1 strikes and becomes imbedded in a block of mass m_2 attached to the end of a light rod of length L. The block rises a vertical distance y in coming to rest.

$$K_1 + U_1 = K_2 + U_2$$

$$\tfrac{1}{2}(m_1 + m_2)V^2 + 0 = 0 + (m_1 + m_2)gy$$

or

$$V = \sqrt{2gy}$$

This equation, together with the preceding equation, allows the bullet speed to be computed from the masses, the measured height y, and the constant g. Note that the time interval during which the bullet is decelerated and the block acquires the speed V is very short indeed. The block is acted on by an impulsive force for a time interval much less than the time required for the block and rod to swing upward and come to rest at a height y above the lowest point. Thus, the complete motion consists of two parts: (1) the short time interval during which the bullet is brought to rest in the block and the block acquires the speed V without rising appreciably, and (2) the much longer time interval during which the block swings to its highest point. During interval 1, momentum is conserved (no *external* force on the system of bullet and block), but kinetic energy is not conserved (a nonconservative *internal* force acts); during interval 2, mechanical energy is conserved (the block is subject only to conservative forces), but momentum is not conserved (the bullet-block system is now subject to an unbalanced external force).

Colliding Particle Beams.   A block collides with a second, identical block initially at rest and sticks to it. How much of the initial kinetic energy is dissipated in the collision?

We know [Figure 10-17(a) and (10-14)] that with equal masses, one of which is at rest, *only half* the initial kinetic energy is dissipated. Not all the initial kinetic energy can be dissipated; if this were to happen, the composite object would have to be at rest in the laboratory after the collision. But then the composite would have no momentum, in contradiction to the momentum-conservation principle.

There is only one way to dissipate all the initial kinetic energy; both of the two colliding objects must be in motion initially in opposite directions, so that the system's total momentum is zero. Then, the composite object can be at rest after the collision. Said differently, the system's center of mass must be at rest in the laboratory, so that the center-of-mass reference frame and the laboratory reference frame are the same frame.

Suppose a block with 1 J of kinetic energy collides with a second, identical block initially at rest. Only 0.5 J is dissipated. On the other hand, if two identical blocks (each with 1 J initially) collide head-on, all the initial kinetic energy of 2 J can be dissipated. See Figure 10-19.

This effect is exploited in the high-energy accelerators used in present-day research on the properties of the fundamental particles. In a typical accelerator, such particles as protons are accelerated from rest to high speed and then hurled at target particles. If the target also consists of protons (hydrogen atoms) and these target particles are at rest, then, as we have seen, no more than half the kinetic energy of the incident proton can be dissipated as it collides with another proton initially at rest. Actually, in the collision between such particles as protons, the energy is not really "dissipated"; instead, the

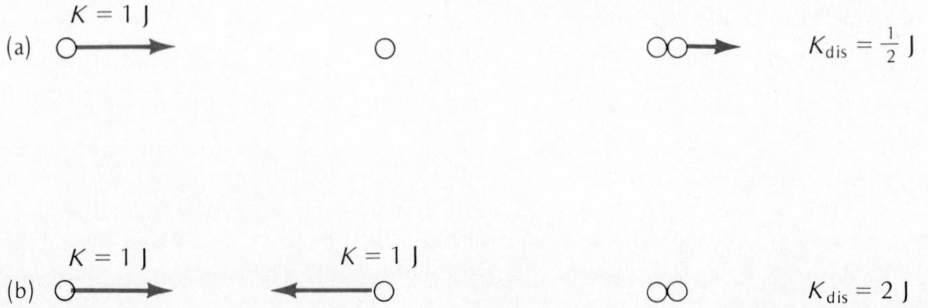

Figure 10-19. *Completely inelastic collision between two particles of the same mass. (a) With one particle initially at rest, only half of the initial kinetic energy is dissipated. (b) With the two particles approaching one another at equal kinetic energies, all the initial kinetic energy is dissipated.*

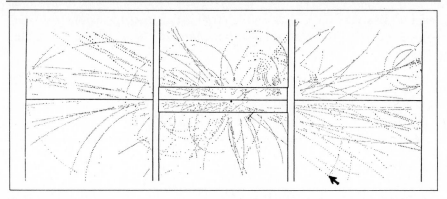

Figure 10-20. *Tracks produced by particles created in the head-on collision of a proton having kinetic energy of 270 GeV (270 × 10⁹ electron volts) with an antiproton of the same kinetic energy. The protons and antiprotons were accelerated in the Super Proton Synchrotron at CERN (European Laboratory for Particle Physics) near Geneva, Switzerland. The arrow identifies a track made by an electron; a neutrino — massless, chargeless, invisible — went in the opposite direction. This and similar observations in 1982 were compelling evidence for the existence of the so-called W particle, the charged intermediate vector boson that mediates the electroweak force and was long hypothesized to exist from theoretical considerations.*

energy "lost" in kinetic energy may appear as the energy required to create still other particles.

In a colliding-beam experiment, on the other hand, a proton collides with a particle of the same mass moving in the opposite direction with the same kinetic energy (Figure 10-20), and all the initial kinetic energy can be dissipated. It is at least *four* times the energy available when one proton hits another at rest. See Figure 10-20.

10-7 Energy and the Center-of-Mass Reference Frame (Optional)

We now prove a general proposition relating three kinetic energies for any system of particles:

- $K_{\text{rel lab}}$, the total kinetic energy of all particles as measured relative to any *laboratory reference frame.*

- $K_{\text{rel CM}}$, the total kinetic energy of the same particles now as measured relative to the system's *center-of-mass (CM) reference frame.*

- K_{CM}, the kinetic energy $\frac{1}{2}MV^2$ associated with the system's center of mass — we shall call it the *kinetic energy of the CM* — where M is the total mass of all particles in the system and V is the velocity of the system's CM relative to the laboratory.

The result to be proved is

$$K_{\text{rel lab}} = K_{\text{CM}} + K_{\text{rel CM}} \tag{10-15}$$

We consider a simple system with only two particles; the generalization to systems with more particles is an obvious extension. The particles have masses

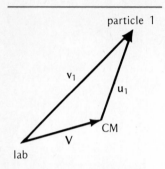

Figure 10-21. Velocity of particle 1 relative to the lab (v_1) and to CM (u_1). Velocity of CM relative to lab is V.

m_1 and m_2 and their respective velocities are $\mathbf{v}_1$ and $\mathbf{v}_2$, both as measured relative to the laboratory. The velocities of the two particles measured relative to the system's CM are denoted $\mathbf{u}_1$ and $\mathbf{u}_2$. As Figure 10-21 shows, $\mathbf{v}_1$ is related to $\mathbf{u}_1$ and to the velocity $\mathbf{V}$ of the CM relative to the laboratory by

$$\mathbf{v}_1 = \mathbf{u}_1 + \mathbf{V} \tag{10-16a}$$

after the rule for relative velocities [(4-19) and Figure 4-25].

In like fashion,

$$\mathbf{v}_2 = \mathbf{u}_2 + \mathbf{V} \tag{10-16b}$$

The CM velocity is

$$\mathbf{V} = \frac{m_1\mathbf{v}_1 + m_2\mathbf{v}_2}{m_1 + m_2} \tag{8-6}$$

By definition, a system's total momentum is zero in the CM reference frame (Section 8-4), so that

$$m_1\mathbf{u}_1 + m_2\mathbf{u}_2 = 0 \tag{10-17}$$

Now the system's total kinetic energy as measured in the laboratory is

$$K_{\text{rel lab}} = \tfrac{1}{2}m_1 v_1^2 + \tfrac{1}{2}m_2 v_2^2$$

But we recognize that v_1^2 can also be written $\mathbf{v}_1 \cdot \mathbf{v}_1 = (\mathbf{u}_1 + \mathbf{V}) \cdot (\mathbf{u}_1 + \mathbf{V})$ [using (10-16a)]. Similarly, $v_2^2 = (\mathbf{u}_2 + \mathbf{V}) \cdot (\mathbf{u}_2 + \mathbf{V})$. Using these relations in the equation above, we get

$$K_{\text{rel lab}} = \tfrac{1}{2}m_1(\mathbf{u}_1 + \mathbf{V}) \cdot (\mathbf{u}_1 + \mathbf{V}) + \tfrac{1}{2}m_2(\mathbf{u}_2 + \mathbf{V}) \cdot (\mathbf{u}_2 + \mathbf{V})$$
$$= \tfrac{1}{2}m_1(u_1^2 + V^2 + 2\mathbf{u}_1 \cdot \mathbf{V}) + \tfrac{1}{2}m_2(u_2^2 + V^2 + 2\mathbf{u}_2 \cdot \mathbf{V})$$

Regrouping terms in this equation, we get

$$K_{\text{rel lab}} = (\tfrac{1}{2}m_1 u_1^2 + \tfrac{1}{2}m_2 u_2^2) + \tfrac{1}{2}(m_1 + m_2)V^2 + \mathbf{V} \cdot (m_1\mathbf{u}_1 + m_2\mathbf{u}_2) \tag{10-18}$$

The first term in this equation is the system's kinetic energy $K_{\text{rel CM}}$ measured relative to the CM. The second term (10-14) represents what we call the kinetic energy of the CM. The third term in (10-18) is zero because of (10-17).

Equation (10-18) can then be written

$$K_{\text{rel lab}} = K_{\text{rel CM}} + K_{\text{CM}} \tag{10-15}$$

In summary, the total kinetic energy of any system of particles can be considered the sum of two parts:

- Kinetic energy relative to the center of mass, $K_{\text{rel CM}}$.
- Kinetic energy of the center of mass, K_{CM}.

Suppose an observer is at rest in the CM reference frame. Then for this CM observer, $V = 0$ and $K_{\text{CM}} = \tfrac{1}{2}MV^2 = 0$. In the CM reference frame, the system's kinetic energy consists of $K_{\text{rel CM}}$ and has its minimum value. Recall that in the CM frame, any system's total momentum also has its minimum value (zero).

The kinetic energy of any system of particles depends on the reference frame relative to which it is measured. What about potential energy? As we have seen, the potential energy of any pair of particles depends at most on their separation distance, but not on the particle velocities. Therefore, the potential energy of any system has the same value in all reference frames. This in turn means that any system's total mechanical energy (kinetic plus potential) has its minimum value in the CM reference frame.*

Summary

Definitions
Potential energy of a system of two or more particles: the quantity U that, when added to the kinetic energy K of the particles, ensures that the system's total energy E is constant:

$$K + U = E = \text{constant} \qquad (10\text{-}1a)$$

Equivalently,

$$K_1 + U_1 = K_2 + U_2 \qquad (10\text{-}1b)$$

$$\Delta U = -\Delta K \qquad (10\text{-}1c)$$

Elastic collision: a collision in which the macroscopic kinetic energy out of the collision equals macroscopic kinetic energy entered into it.

Inelastic collision: a collision in which the macroscopic kinetic energy out of the collision is less than macroscopic kinetic energy entered into it.

Fundamental Principles
The energy-conservation principle states that the total energy of an isolated system is constant. If the system has nonconservative forces, the total macroscopic kinetic energy will decrease with a corresponding increase in nonmechanical forms of energy.

Important Results
Gravitational potential energy U_g of a particle of mass m at a vertical displacement y interacting with the earth:

$$U_g = mgy \qquad (10\text{-}5)$$

Potential energy U_s of a spring of stiffness k stretched or compressed through a distance x:

$$U_s = \tfrac{1}{2}kx^2 \qquad (10\text{-}6)$$

Potential energy function derived from known conservative force:

$$U_f - U_i = -\int_i^f \mathbf{F} \cdot d\mathbf{r} \qquad (10\text{-}4)$$

Force component derived from known potential energy function:

$$F_x = -\frac{\partial U}{\partial x} \qquad (10\text{-}9)$$

Problems and Questions

Section 10-1   Potential Energy Defined
· **10-1 Q** Two blocks of *unequal* masses m_1 and m_2 are attached to the ends of a massless string passing over a frictionless pulley. The blocks are released from rest. After they are released,
(A) the system's potential energy decreases while the kinetic energy increases.
(B) the system's potential energy stays constant while the kinetic energy changes.
(C) the system's potential energy changes while the kinetic energy stays constant.
(D) the system's potential and kinetic energies both stay constant.
(E) the system's potential energy increases while the kinetic energy decreases.

· **10-2 Q** A person initially standing on a hard floor jumps into the air. Describe this process in terms of work, kinetic energy, and potential energy.

: **10-3 P** Masses m_1 and m_2 are attached to the ends of a very light rod of length L. Through the center of the rod passes a horizontal pin that allows the rod to rotate in a

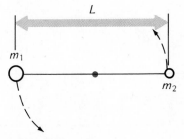

Figure 10-22.   Problem 10-3.

---

* In thermodynamics, the total energy of the system in the CM frame is termed the *internal energy* of the system.

vertical plane. The rod is placed so it is horizontal and is released from rest. What will be the maximum speed reached by each mass? See Figure 10-22.

· **10-4 P** Niagra Falls drops 53 meters and has a flow of 4.2×10^5 m³/min. About 20% of this water is used to generate power. Assuming 100% generating efficiency, what would be the output power of such a hydroelectric plant?

· **10-5 P** How fast can a hoist driven by a 2 kW motor lift a load of 600 N if friction is negligible?

⠸ **10-6 P** A 2-kg block moving 3 m/s initially slides up a frictionless plane inclined at 30° above horizontal. How far along the plane will the block slide before stopping?

⠸ **10-7 P** A toy car is released from rest as shown in Figure 10-23. Friction is negligible. The mass of the car is 50 gm and the radius of the loop-the-loop is 20 cm. From what minimum height h can the car be released without leaving the track?

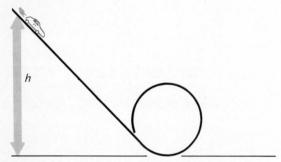

Figure 10-23. *Problem 10-7.*

⠸ **10-8 P** In an amusement park ride called the Orbiting Rocket, the rider is strapped into the cockpit of a small "rocket" attached to the end of a rod that can rotate in a vertical plane. See Figure 10-24. The rocket is whipped around and around. Frequently a rocket stalls at the top of the loop, only to fall from rest to make yet another loop. Suppose that a person weighing 600 N is stalled at the top and then whooshes down. What is the maximum force the seat will exert on her?

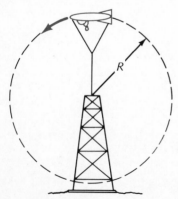

Figure 10-24. *Problem 10-8.*

⠸ **10-9 P** A world-class pole vaulter can clear the bar at about 18 feet. Typically his center of mass is 1 meter above the ground, and by arching his body he is able to clear the bar while his center of mass does not go appreciably higher than the bar. Assuming the pull he exerts with his arms in going over the bar is insignificant, estimate the speed with which he must take off to clear a bar 18 ft high.

⠸ **10-10 P** A mass m is attached to a string of length L and suspended as a pendulum. A peg is placed a distance D below the point of suspension. The mass is pulled to the side until the thread makes an angle θ with the vertical and is released from rest. See Figure 10-25. What is the minimum value of θ for which the thread will not go slack when the mass circles around the peg?

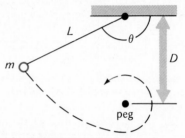

Figure 10-25. *Problem 10-10.*

⠸ **10-11 P** Tidal power is a potentially useful source of energy. The largest installation in use is at the mouth of the Rance River in Brittany, where the difference between high and low tides is about 35 feet, and there are two high tides daily. The area behind the dam is approximately 9 square miles.

(a) What average power could be generated there? (Assume 100% efficiency.) (b) What is the ratio of this power output to that of Hoover Dam, which can produce about 1 GW (10^9 W)?

⠸ **10-12 P** A 2-kg ball is thrown down at a speed of 5 m/s from a window sill 10 meters above the ground. The ball bounces back and reaches a height of 10 meters. (a) What was the kinetic energy of the ball immediately after leaving the ground? (b) How much energy was lost in the collision with the ground?

⠸ **10-13 P** A particle of mass m starts from rest on the track leading into a circular loop-the-loop, at height of $4R$ above the bottom of the loop. The surface is frictionless. When the particle is at the top of the loop-the-loop (radius R), the force of the loop on the particle has the magnitude
(A) mg
(B) $2mg$
(C) $3mg$
(D) $4mg$
(E) $5mg$

⠸ **10-14 P** The American Eagle roller coaster in Gurnee, Illinois, climbs up 39 meters, then drops 45 meters into an underground tunnel. What is the maximum speed it can achieve if it barely makes it to the height at which it begins the descent? Express your answer in m/s and in mi/h.

· **10-15 P** A ball is dropped from rest from a height h. Its speed drops by a factor of two when it hits the ground. What is the maximum height, in terms of h, that it will reach on the rebound?

: **10-16 P** A typical small car plus driver has a mass of 1100 kg. Such a car can cruise at 55 mi/h at a power level of 15 hp from the engine. By what factor would the gas mileage be reduced when going up a 3% grade as compared to traveling on level ground? (A 3% grade rises 3 meters for every 100 meters driven along the roadway.)

: **10-17 P** A block starts from rest at a height R above a horizontal surface, slides down a frictionless curved surface, and then travels a distance L along the rough horizontal surface before coming to rest. The coefficient of kinetic friction between the block and the horizontal surface is:
(A) $\sqrt{2gR/L}$
(B) $2R/L$
(C) $R/2L$
(D) R/L
(E) $L/\sqrt{2gR}$

: **10-18 P** The Tidal Wave, a monster roller coaster in the Midwest, reaches speeds of 55 mph. (a) Through what vertical drop must it fall to reach this speed? (b) It also whirls the riders through a vertical loop of 23 meters diameter. From what height would it have to fall from rest to make such a loop without leaving the track?

Section 10-3 Spring Potential Energy

· **10-19 P** A toymaker wishes to design a spring-loaded gun that will propel a 40-gm rubber dart with a velocity of 5 m/s when the spring is compressed 4 cm. What force constant should the spring have?

Figure 10-26. *Problem 10-20.*

: **10-20 P** Starting from rest, a 1-kg mass slides down a frictionless incline of 30° from an elevation of 0.50 m above ground level. The mass then slides along a frictionless horizontal surface and strikes an initially relaxed spring with a force constant $k = 9.0 \times 10^4$ N/m. See Figure 10-26. (a) What is the mass's speed along the horizontal surface? (b) By how much is the spring compressed when the mass comes momentarily to rest? (c) By how much is the spring compressed when the mass's speed has been reduced to half of that before hitting the spring?

: **10-21 P** A 5-kg block resting on a frictionless horizontal surface is attached to two identical springs, each of unstretched length 0.2 m with force constant of 80 N/m. The block is released from rest at point A, at which point each spring has length 0.4 m. See Figure 10-27. What is the speed of the block as it passes point B?

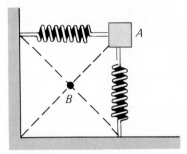

Figure 10-27. *Problem 10-21.*

: **10-22 P** An object of weight 120 N is hung from a spring scale, as sketched in Figure 10-28. The object is supported from below so the scale reads zero. When the support is removed, what will be the maximum momentary reading of the scale?

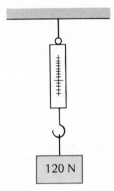

120 N

Figure 10-28. *Problem 10-22.*

: **10-23 P** The cushion on a pool table acts like a spring with force constant of 10^5 N/m. What is the maximum compression of the cushion when it is struck by a billiard ball of mass 100 gm moving 1.0 m/s?

: **10-24 P** A 0.5-kg sphere is dropped from rest from 3.5 meters above the top of a spring that has a relaxed length of 2 meters. The sphere lands in a cup on the top of the spring and stays there, continuing downward as the spring is compressed.

(a) What is the kinetic energy of the sphere when it hits the spring? (b) If the spring constant is 160 N/m, how far does the sphere compress the spring?

: **10-25 P** A spring with spring constant k and unstretched length L is hung from the ceiling. A mass M is attached to the lower end. Then the mass is lifted until it is a distance $(L - a)$ below the ceiling (i.e., it is lifted up a distance a). If the mass is now released from rest, what maximum distance below the ceiling will it reach?

: **10-26 P** A block of mass m is attached to the end of a spring of force constant k; the other end of the spring is fixed. The block rests on a horizontal surface with coefficients of static and kinetic friction μ_s and μ_k. The spring is compressed a distance x_1 and the block is released from rest. (a) How far does the mass slide before it first stops? (b) With what minimum value of μ_s will the mass slide to a stop without changing direction?

Section 10-5   The Energy Conservation Law

· **10-27 Q** The essence of the meaning of energy conservation is:
(A) Energy cannot be created or destroyed. It can only be changed from one form into another.
(B) The kinetic energy of a system does not change in time.
(C) When the kinetic energy of a system increases, the potential energy increases by the same amount.
(D) Since energy is the only true measure of real wealth, it is vital that it not be wasted.
(E) Any natural system will tend to save or "conserve" its energy, dissipating as little as possible as heat due to friction.

· **10-28 Q** A man pumps repeatedly on an automobile jack, and the automobile rises several centimeters at the point where the jack holds the car. (a) How does the work done by the man on the jack compare with the work done by the jack on the car? (b) How does the total displacement of the man's hand compare with the total displacement of the jack? (c) How does the force applied by the man compare with the force applied by the jack?

: **10-29 Q** A roller coaster car passes point X with speed v_0, passes over a hill at point Y and travels on to point Z. See Figure 10-29. Assuming friction is negligible, the speed of the car at point Z will be
(A) v_0.
(B) dependent on the difference in height of the two hills, $h_1 - h_2$.
(C) $v_0 + \sqrt{2gh_1}$.
(D) independent of the height h_2, as long as $h_2 < h_1$.
(E) equal to $\sqrt{2gh_2}$.

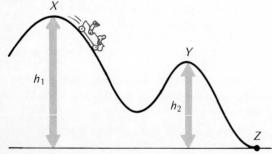

Figure 10-29.   *Question 10-29.*

: **10-30 Q** A boy stands on a cliff and throws stones into a lake below. Each stone is thrown with the same initial

speed, but the stones are of different sizes. One of mass M_A is thrown on trajectory A. One of mass M_B is thrown on trajectory B, and the largest stone, of mass $M_C = M_A + M_B$, is thrown on trajectory C. See Figure 10-30. Neglecting friction, which stone will hit the water with the greatest speed?
(A) The one on path A.
(B) The one on path B.
(C) The one on path C.
(D) The lightest one.
(E) All three will hit with the same speed.
(F) This question cannot be answered without knowing the relative values of M_A and M_B.

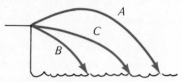

Figure 10-30.   *Question 10-30.*

: **10-31 P** A universal hoist is used for purposes such as removing the engine from a car. For the hoist sketched in Figure 10-31, the geared pulley wheels have teeth that engage the links of the chain. The two top pulleys are welded together. Determine the force that must be exerted to lift a 2000 N load with this device, assuming 50% efficiency. (Use the energy conservation principle, recognizing that 50% of the input energy is dissipated against friction and 50% does useful work in lifting the load.)

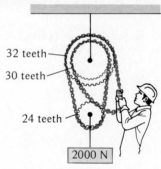

32 teeth
30 teeth
24 teeth
2000 N

Figure 10-31.   *Problem 10-31.*

: **10-32 P** In the Matterhorn ride at Disneyland, a roller coaster car races down a steep hill and is slowed by plowing

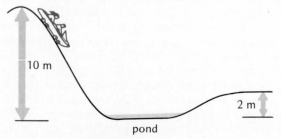

10 m
2 m
pond

Figure 10-32.   *Problem 10-32.*

through a pond of water at the bottom. Suppose that such a car comes over a hill at 10 m/s, drops 10 meters to the pond, and then after leaving the pond rises 2 meters to the landing area, where it comes to a stop. See Figure 10-32. What fraction of the car's kinetic energy must be lost in the pond if this system is to work?

: **10-33 P** A block is projected at 8 m/s up a plane inclined at 30°. It slides 4 meters up the plane and then slides back down again. If the force of friction is independent of speed, how fast will the block be moving when it returns to its starting point?

: **10-34 P** A block, moving 4 m/s at the base of a plane inclined at 20° above horizontal, slides up the plane and then back down. It returns to its starting point with a speed of 3 m/s. (a) How far up along the plane did it slide? (b) What was the coefficient of kinetic friction on the plane?

: **10-35 P** Consider a mass m attached to a cord of length L. Calculate the maximum tension in the cord if (a) the mass is pulled to the side until the cord makes an angle θ with the vertical and released from rest; (b) the mass is pulled to the side until the cord makes an angle θ with the vertical, and is then given a sideways velocity which causes it to move in a horizontal circle with constant velocity and with the string making a constant angle θ with the vertical (it moves as a conical pendulum).

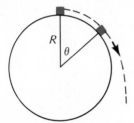

Figure 10-33. *Problem 10-36.*

: **10-36 P** A small mass is released from rest at the top of a frictionless sphere. At what value of θ will it leave the surface?

Section 10-7   Collisions

· **10-37 Q** When two objects collide, we can be certain that the collision was inelastic if
(A) after the collision the paths of the particles are not collinear.
(B) after the collision the paths of the particles are collinear.
(C) the particles do not have equal momenta after the collision.
(D) the particles stick together.
(E) the kinetic energies of the two particles are not equal to their respective values before the collision.

: **10-38 P** A particle collides elastically with a second identical particle initially at rest in a non-head-on collision. Show that the angle between the particles' velocity vectors after the collision is always 90°. (Hint: express momentum conservation in a momentum vector diagram.)

: **10-39 Q** Two billiard balls in motion with the same velocity collide head-on with a line of identical billiard balls at rest. Why do just two billiard balls emerge from the far end of the line with the same velocity, while the initially moving billiard balls are brought to rest?

: **10-40 Q** A particle of mass m is to lose at least one-quarter of its initial kinetic energy in a head-on elastic collision with a second particle initially at rest. What is (a) the minimum and (b) the maximum mass of the struck particle?

: **10-41 P** The radioactive nucleus radium-226 decays into an alpha particle (helium-4) and a radon-222 nucleus, releasing 4.87 MeV of energy, which appears as kinetic energy of the decay products. What are the kinetic energies of the alpha particle and of the radon nucleus? The masses of these two particles are in the ratio $4:222$.

: **10-42 P** An object is to explode into two parts, the ratio of whose kinetic energies is three-to-one. What is the required mass ratio?

: **10-43 P** An impressive lecture demonstration is as follows: A marble of mass m is placed on top of a large super-ball of mass M. The two are dropped from rest from a height h. They collide elastically with the floor and with each other. To what maximum height will the marble rise? Show that in the limit $M \gg m$, the marble rises to a height $9h$. (Hint: Imagine a small separation between the objects so that the small one is still going down when it collides with the large one going up.)

: **10-44 P** A particle of mass m and speed v makes a head-on elastic collision with a particle of mass $3m$ and speed v. Calculate the velocities of each particle after the collision.

· **10-45 P** Two identical plastic balls are attached to strings of length L. Each is pulled to the side and released from rest simultaneously in such a way that they collide. See Figure 10-34. After colliding, each swings back up only to a height $\frac{1}{2}L$ below the point of suspension. By what factor was the speed of each reduced in the collision? This factor, called the *coefficient of restitution,* gives a measure of how elastic a collision is.

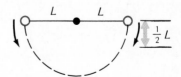

Figure 10-34. *Problem 10-45.*

11 Simple Harmonic Motion

The simplest type of repetitive motion along a straight line is simple harmonic motion. It is the motion executed by any object acted on by a linear restoring force. Many situations illustrate this behavior: an oscillating spring, a swinging pendulum, the vibrating atoms in a solid, the alternating current in an electric circuit, the oscillating electric or magnetic fields in an electromagnetic wave, the pressure variations in a sound wave. Indeed, *any particle undergoing small oscillations about a point of stable equilibrium executes simple harmonic motion.*

11-1 Definitions

We shall define simple harmonic motion (abbreviated SHM) in two different ways:

- The kinematics: what the motion itself is.
- The dynamics: what produces SHM.

How the kinematics is related to the dynamics is treated in Section 11-2.

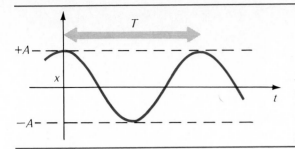

Figure 11-1. *Displacement-time graph for SHM. Displacement x varies sinusoidally with time t. The amplitude is A; the period is T.*

Kinematics   A particle executes SHM when its *displacement varies sinusoidally with time;* that is, the displacement x varies with time as either the sine or the cosine. One simple example is

$$x = A \cos \omega_0 t \qquad (11\text{-}1)$$

See Figure 11-1.

Amplitude A is the particle's *maximum displacement from its equilibrium position,* which is taken to be at $x = 0$. The motion of the particle is bounded; it oscillates between $x = +A$ and $x = -A$.

Period T is the *time for one complete oscillation,* as shown in Figure 11-1.

Frequency f is the number of oscillations per unit time. Clearly, if an oscillating particle has a period $T = 0.10$ s, the number of oscillations per unit time is $f = 10$ s^{-1}. More generally,

$$T = \frac{1}{f} \qquad (11\text{-}2)$$

The official SI unit for frequency is the *Hertz* (abbreviated Hz), where 1 Hz $= 1$ s^{-1}.

Angular frequency ω_0 (also called radial frequency) is the *number of radians per unit time.* (The subscript zero is to emphasize that the oscillations take place in the absence of damping: *zero* damping force.) From (11-1), we see that one oscillation has been completed when $\omega_0 t$ goes from its initial value of zero to the value 2π radians as time t goes from zero to T. Therefore,

$$\omega_0 T = 2\pi$$

or
$$T = \frac{1}{f} = \frac{2\pi}{\omega_0} \qquad (11\text{-}3)$$

Equation 11-1 can then be written in the alternative forms:

$$x = A \cos \omega_0 t = A \cos \frac{2\pi}{T} t = A \cos 2\pi f t \qquad (11\text{-}4)$$

Simple harmonic motion can also be defined as the *projection along a diameter of uniform circular motion.* Consider Figure 11-2, where a particle moves counterclockwise in a circle of radius A with an angular velocity of magnitude ω_0. The angle θ, measured in radians, which the radius vector makes with the x axis, is $\theta = \omega_0 t$, where $\theta = 0$ at $t = 0$. Therefore, the x component of the particle's circular motion is

$$x = A \cos \theta = A \cos (\omega_0 t) \qquad (11\text{-}1)$$

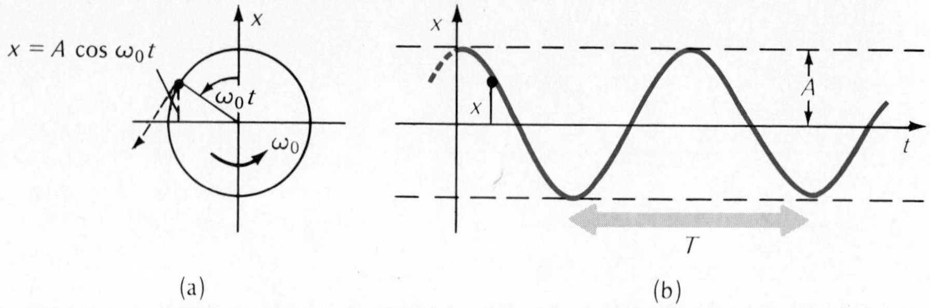

$x = A \cos \omega_0 t$

(a)

(b)

Figure 11-2. *(a) Representation of uniform circular motion (at angular velocity ω_0 around a circle of radius A) corresponding to SHM. (b) Displacement-time graph with $x = A \cos (\omega_0 t)$.*

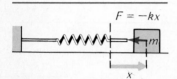

$F = -kx$

Figure 11-3. *A simple harmonic oscillator. Mass m attached to a spring of force constant k is subject to restoring force $F = -kx$, where x is the displacement of the mass from its equilibrium position.*

Dynamics   Simple harmonic motion takes place if a particle is subject to a *linear restoring force*, for example, the force F_s exerted by a spring stretched or compressed by a small amount x from its equilibrium configuration,

$$F_s = -kx \qquad (11\text{-}5)$$

The spring force is *linear* because a plot of F_x against x is, for small deformations, a *straight* line. See Figure 9-18(a). The spring force is a *restoring* force because its direction is *opposite* to the direction of the displacement [the minus sign in (11-5)]; the deformed spring always pushes or pulls a particle attached to one end towards the equilibrium position. See Figure 11-3.

The spring force is conservative, so that an alternative way of describing a linear restoring force is this: *the potential energy varies as the square of the displacement from the equilibrium position.* For a spring, the potential energy is

$$U_s = \tfrac{1}{2}kx^2 \qquad (10\text{-}6),\ (11\text{-}6)$$

If the force on the particle is of the form given in (11-5), then the potential energy of the system must be of the form given by (11-6). Actually, the spring serves as a prototype of the situation arising in many areas of physics. For a particle oscillating at one end of the spring, constant k is the stiffness of the spring. For other types of SHM, k represents the equivalent force constant.

11-2   Dynamics of Simple Harmonic Motion

If a particle is subject to a linear restoring force, its displacement will vary sinusoidally with time. In other words, the kinematical and dynamical definitions of simple harmonic motion are equivalent.

Suppose that a particle of mass m moves along the x axis while attached to the end of a spring of stiffness k. Newton's second law gives

$$\Sigma F_x = ma_x$$

$$-kx = m\frac{d^2x}{dt^2}$$

where d^2x/dt^2 is the particle's acceleration a_x along x.

Our problem: find x as a function of time t. First we rearrange the preceeding equation:

$$\frac{d^2x}{dt^2} + \frac{k}{m}x = 0$$

We then set the ratio k/m equal to ω_0^2:

$$\omega_0^2 \equiv \frac{k}{m} \tag{11-7}$$

so that the equation can be written as

$$\frac{d^2x}{dt^2} + \omega_0^2 x = 0 \tag{11-8}$$

Any physical situation that leads to a linear second-order differential equation of motion of the form of (11-8) implies SHM.

It is easy to show that the general solution of (11-8) is

$$x = A\cos(\omega_0 t + \delta) \tag{11-9}$$

where ω_0, A, and δ are all constants.* [Equation 11-1 is just a simple form of (11-9), with $\delta = 0$.]

We take the first and second time derivations of (11-9) to find the particle's velocity $v = dx/dt$ and acceleration $a = d^2x/dt^2$:

$$v = \frac{dx}{dt} = -\omega_0 A\sin(\omega_0 t + \delta) \tag{11-10}$$

and

$$a = \frac{d^2x}{dt^2} = \frac{dv}{dt} = -\omega_0^2 A\cos(\omega_0 t + \delta) \tag{11-11}$$

Comparing (11-11) and (11-8), we see that (11-9) *is* indeed a solution.

The displacement x, velocity dx/dt, and acceleration d^2x/dt^2 all vary sinusoidally with time. See Figure 11-4.

The *phase constant* δ in (11-9) specifies the phase of the oscillation at the starting time, $t = 0$. Suppose that $\delta = 0$; then x varies with time as $\cos \omega t$ [(11-9) reduces to (11-1)]. On the other hand, if $\delta = -90° = -\pi/2$, we find that x varies with time as $\cos(\omega_0 t - \pi/2) = \sin \omega_0 t$. The phase constant is determined in general by the displacement x_0 and velocity v_0 at the starting time $t = 0$. From (11-9), we have $x_0 = A\cos\delta$, and from (11-10), we have $v_0 = -\omega_0 A\sin\delta$. Dividing the second relation by the first one then gives, with some rearranging, the general relation for the phase constant,

$$\tan\delta = -\frac{v_0}{\omega_0 x_0} \tag{11-12}$$

General features of SHM can be seen in Figure 11-4:

• The displacement ranges between $-A$ and $+A$ with the equilibrium position at $x = 0$.

---

* The solution of (11-9) may be written in different but equivalent mathematical forms as $x = B\sin(\omega_0 t + \epsilon)$ or $x = C\cos\omega_0 t + D\sin\omega_0 t$, where B, C, D, and ϵ are all constants.

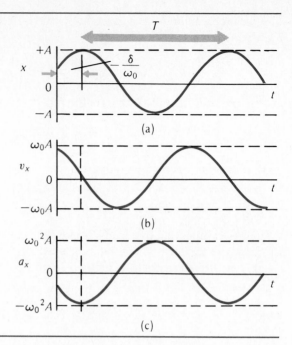

Figure 11-4. *(a) Displacement, (b) velocity, and (c) acceleration for a particle in SHM plotted as a function of time.*

• The velocity also varies sinusoidally and it is $90°\,(\pi/2$ rad) out of phase with respect to the displacement x. The particle's speed has its maximum value $\omega_0 A$ at the equilibrium position (at $x = 0$); the speed is zero at the amplitude positions, $x = \pm A$.

• The acceleration also varies sinusoidally and it is $180°\,(\pi$ rad) out of phase with respect to the displacement x. The particle's acceleration has its maximum value $\omega_0^2 A$ when the particle is at the amplitude position (where the force is also a maximum, $F_s = -kA$). At the equilibrium position, $x = 0$, the acceleration is zero and so is the force. Note that the acceleration varies as the *square* of ω_0.

How is the oscillation period T and frequency f related to k and m, the physical properties of the oscillator? From (11-3) and (11-7), we have

$$T = \frac{1}{f} = \frac{2\pi}{\omega_0} = 2\pi \sqrt{\frac{m}{k}} \tag{11-13}$$

This relation agrees with our intuition; increasing the mass m or using a less stiff spring (smaller k) will increase the time T for one oscillation. We can say that the particle's inertia (m) makes it overshoot the equilibrium position while the spring restoring force (measured by k) binds the particle to the equilibrium position.

Equally important is what the period T and frequency f in (11-13) do *not* depend on — the amplitude A. The time for one oscillation does not depend on how far the particle travels from equilibrium. This independence of T and f on oscillation amplitude holds only if the restoring force is strictly linear. For a large enough deformation, any spring force departs from strict proportionality to the amount of stretch; oscillatory motion is then not simple harmonic, and the period and frequency then depend on the maximum displacement from equilibrium (see Section 11-6).

Example 11-1.   A 2.0-kg block rests on a horizontal frictionless surface, attached to the right end of a spring whose left end is fixed. See Figure 11-3. The block is displaced 5.0 cm to the right from its equilibrium position and held motionless at this position by an external force of 10.0 N. (a) What is the spring's force constant? (b) The block is then released. What is the period of the block's oscillations? (c) What is the force on the block, at the time $t = \pi/15$ s?
   (a) From (11-5),

$$k = -\frac{F}{x} = -\frac{-10.0 \text{ N}}{0.050 \text{ m}} = 200 \text{ N/m}$$

(b) The period is given by (11-13):

$$T = 2\pi \sqrt{\frac{m}{k}} = 2\pi \sqrt{\frac{2.0 \text{ kg}}{200 \text{ N/m}}} = \frac{\pi}{5} \text{ s}$$

(c) The force of the spring on the block at any time t is given by

$$F = -kx = -kA \cos(\omega t_0 + \delta) = -kA \cos\left(\frac{2\pi t}{T} + \delta\right)$$

The motion is started from rest at the amplitude position $A = 0.050$ m; therefore $\delta = 0$. At time $t = \pi/15$ s, the equation above becomes

$$F = -(200 \text{ N/m})(0.050 \text{ m}) \cos\left[\frac{2\pi(\pi/15 \text{ s})}{\pi/5 \text{ s}}\right]$$

$$= -(10.0 \text{ N}) \cos\frac{2\pi}{3} = -(10.0 \text{ N}) \cos 120°$$

$$= +5.0 \text{ N}$$

At time $t = \pi/15$ s, the 5.0-N force of the spring on the block is to the right. The spring is compressed at this instant ($x < 0$).

Example 11-2.   At time $t = 0$, a particle oscillating in SHM is at $x_0 = +\frac{1}{2}A$ and heading along the positive x axis. What is the phase constant δ in the general relation $x = A \cos(\omega_0 t + \delta)$?
   Substituting $x = +\frac{1}{2}A$ at $t = 0$ in the general equation for the displacement yields

$$+\tfrac{1}{2}A = A \cos \delta$$

$$\cos \delta = +\tfrac{1}{2}$$

$$\therefore \delta = \pm 60°$$

The *two* smallest possible values for δ are $+60°$ and $-60°$. Which is it? We must use the fact that the particle is, at $t = 0$, heading along the *positive* x axis. That is, $v = dx/dt > 0$ at $t = 0$.
   The velocity is given by

$$v = -A\omega_0 \sin(\omega_0 t + \delta)$$

At $t = 0$,

$$v_0 = -A\omega_0 \sin \delta$$

To have v_0 positive, we must have δ itself negative. Therefore, δ must be $-60°$.

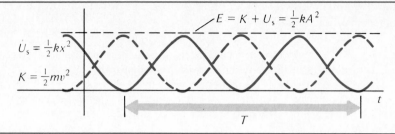

Figure 11-5. *Variation with time of kinetic energy K and elastic potential energy U_s for a simple harmonic oscillator. The total energy $E = K + U_s$ is constant.*

11-3   Energetics of Simple Harmonic Motion

An oscillating particle's kinetic energy K is, from (11-10),

$$K = \tfrac{1}{2}mv^2 = \tfrac{1}{2}m\omega_0^2A^2 \sin^2(\omega_0t + \delta) = \tfrac{1}{2}kA^2 \sin^2(\omega_0t + \delta)$$

We used (11-7) in the last step.

The potential energy of the spring is, from (11-6) and (11-9),

$$U_s = \tfrac{1}{2}kx^2 = \tfrac{1}{2}kA^2 \cos^2(\omega_0t + \delta)$$

Therefore the total energy E of the oscillating system is

$$E = K + U = \tfrac{1}{2}kA^2[\sin^2(\omega_0t + \delta) + \cos^2(\omega_0t + \delta)]$$

$$E = \tfrac{1}{2}kA^2 \tag{11-14}$$

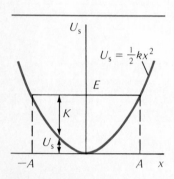

Figure 11-6. *Elastic potential energy U_s as a function of displacement x for a simple harmonic oscillator. The motion is restricted to the region between the amplitude positions, $x = \pm A$, for which the kinetic energy is positive.*

The energy of a simple harmonic oscillator is proportional to the *square of the amplitude.*

The total energy of an undamped simple harmonic oscillator is constant. When dissipative forces are absent, the oscillations persist indefinitely with undiminished amplitude. Both the kinetic and the potential energies vary sinusoidally with time, as shown in Figure 11-5, but their sum E is constant. The oscillating mass has a maximum kinetic energy when it passes through the equilibrium position ($x = 0$), and the spring has a maximum potential energy when the body is momentarily at rest at the amplitude position ($x = \pm A$).

Figure 11-5 shows the potential energy U_s as a function of time. It is also useful, since simple harmonic motion is repetitive, to examine the potential energy as a function of the displacement. Figure 11-6 shows U_s as a function of x: $U_s = \tfrac{1}{2}kx^2$, a parabola, following (11-6).

Example 11-3.   A particle attached to a spring of stiffness k undergoes simple harmonic motion with an amplitude A. What is the particle's kinetic energy (in terms of k and A) at the instant when its displacement is $\tfrac{1}{2}A$?

For $x = \tfrac{1}{2}A$, the spring's potential energy is

$$U_s = \frac{1}{2}kx^2 = \frac{1}{2}k\left(\frac{A}{2}\right)^2 = \frac{1}{8}kA^2$$

The total energy of the oscillator is $E = \tfrac{1}{2}kA^2$. Therefore, the particle's kinetic energy at this time is

$$K = E - U = \tfrac{1}{2}kA^2 - \tfrac{1}{8}kA^2$$

$$= \tfrac{3}{8}kA^2$$

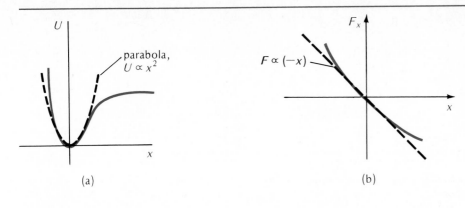

Figure 11-7. *(a) Potential energy as a function of displacement for a particle in stable equilibrium. The solid line is the typical potential energy for an atom bound in a molecule; the dashed line, which approximates the actual potential-energy curve near its minimum, is parabolic. (b) The force-displacement graph corresponding to part (a). Note that the force is strictly linear only near $x = 0$.*

At the midway point, the oscillator's energy is one-quarter potential, $\frac{1}{4}(\frac{1}{2}kA^2)$, and three-quarters kinetic, $\frac{3}{4}(\frac{1}{2}kA^2)$.

11-4   Small Oscillations

It is easy to show that ordinarily a particle undergoing small oscillations about a point of stable equilibrium executes SHM. Suppose a particle has a stable equilibrium position. Then the particle must be subject to a force back toward the equilibrium position when displaced away from equilibrium. Said differently, if the particle moves away from equilibrium, it is subject to a force that slows it and reduces its kinetic energy. Therefore, the system's potential energy must increase to maintain a constant total energy. We know then that the potential energy U must *rise* for either a positive or a negative displacement x from equilibrium, as shown in Figure 11-7(a). In the vicinity of the potential-energy minimum (the equilibrium position), the potential energy can be approximated by a parabola with a vertical symmetry axis, and we can write:*

$$U \propto x^2 \qquad \text{for small } x$$

Equivalently, the force $F_x = -dU/dx$ varies linearly with x near the point of stable equilibrium. See Figure 11-7(b). But a linear restoring force is implied, as we have seen, when the potential energy varies directly with the square of the displacement from equilibrium. A particle displaced a small distance from $x = 0$ undergoes SHM.

---

* We can write the potential energy U close to the minimum as an expansion in the displacement x:

$$U(\text{near } x = 0) = a_0 + a_1 x + a_2 x^2 + a_3 x^3 + \cdots$$

where $a_0, a_1, a_2, \ldots$ are constants. Since the zero for potential energy is arbitrary, and we may choose $U = 0$ when $x = 0$, the constant $a_0 = 0$. If a_1 were not zero, its contribution to U would make U rise for positive x and fall for negative x, in contradiction to the fact that U must rise in both directions. Thus $a_1 = 0$ for an equilibrium point. For sufficiently small x, then, the leading term is $a_2 x^2$, and $U \approx a_2 x^2$. For stable equilibrium, we must have $a_2 > 0$.

11-5   The Simple Pendulum

Consider a particle of mass m attached to the lower end of a massless cord of length L, whose upper end is fixed. See Figure 11-8. This system constitutes a *simple pendulum.**

The angle of the cord with the vertical is θ. We assume in what follows that the angle θ is always small, so that the particle, when released from the side with the string taut, undergoes oscillations of small amplitude. The forces on the particle are the cord tension **T** and the particle's weight $m\mathbf{g}$.

The particle moves in a circular arc of radius L. The resultant force along the direction of motion is of magnitude $mg \sin \theta$, in a direction opposite the angular displacement θ. If θ is small enough, the particle's linear displacement along the circular arc can be closely approximated by a horizontal displacement x from the equilibrium position, where $x = \theta L$. Thus, for small θ,

$$\frac{x}{L} \approx \sin \theta \approx \theta$$

so that in applying Newton's second law, we have

$$\Sigma F_x = ma_x$$

$$-mg \sin \theta = m \frac{d^2x}{dt^2}$$

$$-mg \frac{x}{L} \approx m \frac{d^2x}{dt^2}$$

or

$$\frac{d^2x}{dt^2} + \frac{g}{L} x = 0$$

The equation of motion above has exactly the same form as (11-8), the general equation for simple harmonic motion. Here,

$$\omega_0^2 = \frac{g}{L}$$

Therefore, *for small amplitude*, the period of a simple pendulum is

$$T = \frac{1}{f} = 2\pi \left(\frac{L}{g}\right)^{1/2} \tag{11-15}$$

where we have used (11-3). The period depends only on the length L and the acceleration g. For small amplitudes, the period is independent of the amplitude (whether of x or of θ) of the oscillation.† The period is also independent of

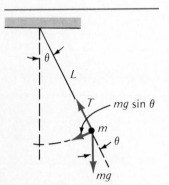

Figure 11-8. A simple pendulum. The resultant force on the particle along the circular arc is mg sin θ.

---

* In the more general type of pendulum, known as a compound pendulum, or a physical pendulum, the mass is not concentrated at a single point. An example is a rigid object undergoing oscillations about a fixed pivot. The physical pendulum is discussed in Section 13-6.

† The general relation giving the period for an arbitrary angular amplitude θ_m (in radians) is

$$T = 2\pi \sqrt{\frac{l}{g}} \left(1 + \frac{1}{4} \sin^2 \frac{\theta_m}{2} + \frac{9}{64} \sin^4 \frac{\theta_m}{2} + \cdots \right)$$

With $\theta_m = 30°$, $T = 2\pi \sqrt{l/g} \,(1.017)$; thus, even for this large angle of swing, the actual period is greater than that given by the simple relation in (11-15) by less than 2 percent.

the particle's mass; all simple pendulums of the same length oscillate at the same rate at the same location.

A pendulum is a suitable timing device, since it is isochronous; that is, its period is independent (nearly) of its amplitude.* A pendulum provides a simple and precise basis for measuring g. One simply measures the period of the pendulum of known length and applies (11-15).

Example 11-4.   A particle is released from rest and slides on the smooth surface of a hemispherical bowl with radius of curvature R. See Figure 11-9(a). (a) Show that if the particle is released not far from the bottom of the bowl, it executes simple harmonic motion. (b) Find the period of oscillation from energy considerations.

The center of curvature of the bowl is denoted C; any point on the surface is a distance R from C. The location of the particle is given by angle θ measured relative to the vertical.

(a) The forces on the particle are shown in Figure 11-9(b); they are its weight $m\mathbf{g}$ and the normal force $\mathbf{N}$. The situation here is just like that for the simple pendulum (Figure 11-8), with the normal force $\mathbf{N}$ replacing the tension $\mathbf{T}$. Furthermore, the radius R replaces the pendulum's constant length L. Therefore, for small θ, the particle's period T of oscillation is given by the pendulum relation (11-15) with R replacing L:

$$T = 2\pi \sqrt{\frac{R}{g}}$$

(b) Our strategy in finding the oscillation period from energy considerations will be to write the potential energy in the form

$$U = \tfrac{1}{2}kx^2$$

in order to find the equivalent "force constant" k. Then the period can be found immediately through the following relations:

$$T = \frac{2\pi}{\omega_0} \qquad\qquad \textbf{(11-3)}$$

where

$$\omega_0^2 = \frac{k}{m} \qquad\qquad \textbf{(11-7)}$$

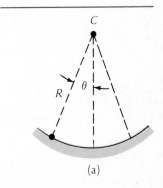

(a)

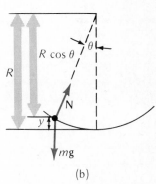

(b)

Figure 11-9.   Particle sliding in a hemispherical bowl.

Relative to the equilibrium position—the low point in its path—the gravitational potential energy of the suspended mass m is

$$U = mgy = mgR(1 - \cos\theta)$$

See Figure 11-9(b).

The cosine of any angle θ (in radians) can be written as a series expansion:

$$\cos\theta = 1 - \frac{\theta^2}{2!} + \frac{\theta^4}{4!} - \cdots$$

If θ is small, we can discard all but the first two terms in the above approximation, and the general relation for the potential energy above becomes

$$U = \tfrac{1}{2}mgR\theta^2$$

The particle's displacement along the circular arc can be closely approximated by x,

---

* The first recorded observation that the period of a pendulum depended on its length but not the amplitude of oscillation was made by Galileo at age 17 while in a cathedral. He timed the swinging lamp, using his pulse as a clock.

its horizontal displacement from equilibrium, where $x = R\theta$, so that the equation above may be written

$$U = \frac{1}{2} mgR \left(\frac{x}{R}\right)^2 = \frac{1}{2} \left(\frac{mg}{R}\right) x^2$$

The standard form for the potential energy of a simple harmonic oscillator is

$$U = \tfrac{1}{2}kx^2 \tag{11-6}$$

Comparing these equations, we discover that the effective value of the "spring constant" for the simple pendulum is

$$k = \frac{mg}{R}$$

Using the relations for T and ω_0 given above, we have finally

$$\omega_0^2 = \frac{k}{m} = \frac{mg/R}{m} = \frac{g}{R}$$

$$T = \frac{2\pi}{\omega_0} = 2\pi \left(\frac{R}{g}\right)^{1/2}$$

Example 11-5 (Optional). Show that the period T of a simple pendulum of length L and mass m must, from considerations of the dimensions of these quantities alone (dimensional analysis), have the functional dependence $T \propto (L/g)^{1/2}$, or equivalently, $T = b(L/g)^{1/2}$, where b is a dimensionless constant.

The basic dimensions of any quantity appearing in mechanics are time (T), length (L), and mass (M). All other mechanical quantities can be expressed by these dimensions alone. For example, the dimensions of momentum (mv) are MLT^{-1}, and the dimensions of kinetic energy ($\tfrac{1}{2}mv^2$) are ML^2T^{-2}.

The quantities that might a priori be imagined to determine the pendulum's period T are m, L, and g.

We assume the period T to vary as

$$T = m^a L^b g^c$$

where the exponents are constants to be determined and the dimensions of these quantities are as shown in the table.

QUANTITY	DIMENSIONS
T	T
m	M
L	L
g	LT^{-2}

Substituting the dimensions in the equation above, we have

$$(T)^1 = (M)^a (L)^b (LT^{-2})^c$$

Since the dimensions of M, L, and T are independent, we may equate the powers for each dimension:

$$T: \quad 1 = -2c$$

$$M: \quad 0 = a$$

$$L: \quad 0 = b + c$$

$$\therefore \quad a = 0, \, b = \tfrac{1}{2}, \, c = -\tfrac{1}{2}$$

The period has the functional dependence

$$T \propto L^{1/2}g^{-1/2} = \sqrt{\frac{L}{g}}$$

The proportionality above may be written as an equation:

$$T = b \left(\frac{L}{g}\right)^{1/2}$$

The value of the dimensionless constant b [2π from our detailed dynamical analysis, (11-15)] cannot be determined from dimensional analysis. The power of this procedure is that without concern with details or even with physical principles, in some instances we can arrive at the required functional dependence simply by considering the dimensions of physical quantities.

11-6   Periodic Motion and the Fourier Theorem (Optional)

Simple harmonic motion is one important but special example of periodic motion. By *periodic motion* is meant repetitive motion. The *period T* is defined as generally the smallest time interval that allows representing the motion for all times. If the motion is known for any interval of duration T, then other times can be determined merely by shifting the displacement-time curve to the right or left an integral number of periods. The motion illustrated in Figure 11-10, though periodic, is clearly not SHM.

A general mathematical theorem known as the *Fourier theorem* implies the following important property. Any periodic motion can be regarded as the superposition, or summation, of a number of SHM's which may differ in amplitude, frequency, and phase. More specifically,

$$x(t) = \frac{A_0}{2} + \sum_{n=1}^{\infty} [A_n \cos (n\omega_0 t) + B_n \sin (n\omega_0 t)] \qquad (11\text{-}16)$$

where the A's and B's are the Fourier coefficients.*

Consider, for example, the *square wave*, shown in Figure 11-11(a); the displacement makes abrupt jumps between $x = -\pi$ to $x = +\pi$ with a period of

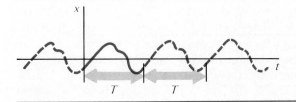

Figure 11-10. Displacement-time graph for one example of periodic motion with period T.

---

* Many hand-held programmable calculators compute approximate values for the A and B Fourier coefficients in (11-16) for any periodic function by specifying values of x associated with corresponding values of t over one period.

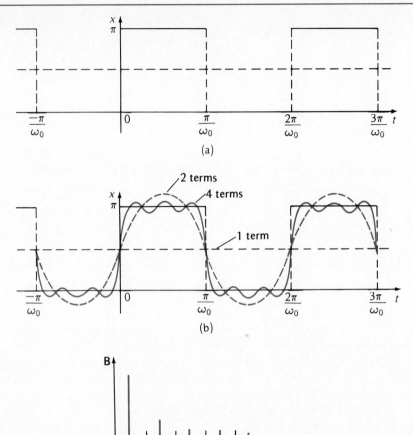

Figure 11-11. *(a) Periodic square wave. (b) Superposition of harmonic oscillations to approximate the square wave. (c) The corresponding frequency spectrum with amplitude plotted as a function of angular frequency.*

$2\pi/\omega_0$. This special type of periodic motion may be approximated as closely as one wishes by a sum of sines of different frequency and amplitude. More specifically, the Fourier expansion, (11-16), gives for the square-wave oscillation:

$$\text{Square wave:} \quad x = \frac{\pi}{2} + \frac{2}{1}\sin\omega_0 t + \frac{2}{3}\sin 3\omega_0 t + \frac{2}{5}\sin 5\omega_0 t + \cdots$$

Note that for this particular periodic oscillation, sines alone appear in the equation and only odd integral multiples (ω_0, $3\omega_0$, $5\omega_0$, . . .) of the fundamental angular frequency ω_0. The coefficient of each sine term gives the amplitude of that harmonic. These amplitudes are plotted as a function of the frequencies of the harmonics in Figure 11-11(c), which is known as the *frequency spectrum.* An important consequence of the equivalence of any periodic oscillation and simple harmonic oscillations is this: to learn how any system responds to a square-wave oscillation, one may see how it responds separately to each of the harmonic oscillators.

Another common example of a periodic oscillation is the *sawtooth wave,*

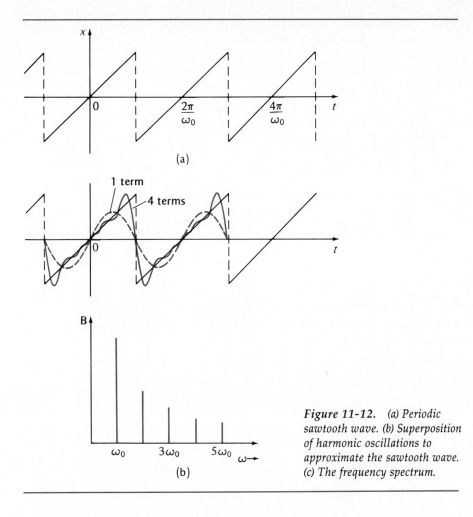

Figure 11-12. *(a) Periodic sawtooth wave. (b) Superposition of harmonic oscillations to approximate the sawtooth wave. (c) The frequency spectrum.*

shown in Figure 11-12(a).* The equivalent harmonic summation is

Sawtooth wave:

$$x = \frac{2}{1} \sin \omega_0 t - \frac{2}{2} \sin 2\omega_0 t + \frac{2}{3} \sin 3\omega_0 t - \frac{2}{4} \sin 4\omega_0 t + \cdots$$

Simple harmonic motion has a *single* frequency for any amplitude. If an oscillation is not sinusoidal but periodic, with a period $T = 2\pi/\omega_0$, then a frequency *spectrum*, with integral multiples, or *harmonics*, of the fundamental frequency ω_0, represents the motion.

If the restoring force for an oscillator is strictly linear, we get sinusoidal motion at a single frequency; we have *simple harmonic* motion. On the other hand, if the restoring force is nonlinear (for example, $F_s = -kx^3$), then the oscillation is nonsinusoidal and necessarily has a spectrum of frequencies.

---

* A sawtooth oscillation takes place on the face of a cathode-ray oscilloscope, or television picture tube. The spot produced by the electron beam moves horizontally to the right at constant speed to the edge of the tube, then jumps back abruptly to the left and begins the next sweep.

Therefore, a nonlinear device can be used to generate oscillations at integral multiples of the fundamental frequency.

11-7 Damped Oscillations (Optional)

With a particle subject to a linear restoring force only, the system's energy remains constant, and the oscillatory motion persists indefinitely with undiminished amplitude. But dissipative forces are always present, and this ideal situation cannot be achieved. We have, besides the restoring force, a force that opposes the motion. This dissipative force always does negative work. As a consequence, the system's total energy decreases with time, as shown in Figure 11-13(a). For small dissipative forces, the body oscillates with continuously decreasing amplitude; the motion is *damped oscillations*. See Figure 11-13(b).

We suppose that an oscillator is subject to a resistive force proportional to the particle's speed (as is sometimes the case for an object immersed in a liquid). We write the resistive force $\mathbf{F_r}$ as (Section 6-6)

$$\mathbf{F_r} = -r\mathbf{v}$$

where $\mathbf{v}$ is the body's velocity and r is a constant. The forces acting on the object are shown in Figure 11-14.

Newton's second law then gives

$$\Sigma F_x = ma_x$$

$$-kx - rv_x = ma_x$$

$$-kx - r\frac{dx}{dt} = m\frac{d^2x}{dt^2} \tag{11-17}$$

The solution to this second-order differential equation is, for small r,

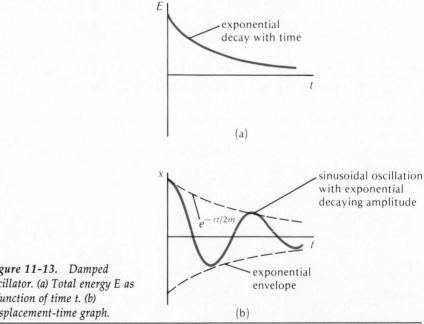

(a)

(b)

Figure 11-13. Damped oscillator. (a) Total energy E as a function of time t. (b) Displacement-time graph.

$$x = Ae^{-(r/2m)t} \cos{(\omega t + \delta)} \qquad \text{(11-18)}$$

where

$$\omega = \sqrt{\frac{k}{m} - \left(\frac{r}{2m}\right)^2} \qquad \text{(11-19)}$$

This can be verified by substituting the solution, (11-18), in (11-17). Figure 11-13(b) is a plot of (11-18). Figure 11-13(a) gives the total energy $E = \frac{1}{2}mv_x^2 + \frac{1}{2}kx^2$ as a function of time t.

The displacement x varies sinusoidally with time but the amplitude $Ae^{-(r/2m)t}$ decreases exponentially with time. The larger the resistive constant r, the more rapid the damping.

The angular frequency ω of the damped oscillations, given by (11-19) is less than the natural angular frequency $\omega_0 = \sqrt{k/m}$ in the absence of damping ($r = 0$). This is to be expected, since any resistive force slows the motion. It must be emphasized that (11-18) and (11-19) apply only for a linear resistive force (proportional to the body's speed) and when the damping is relatively small, that is, $(r/2m)^2$ smaller than k/m.

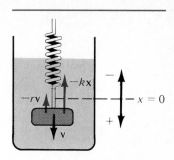

Figure 11-14.   A physical situation giving rise to damped oscillations: object attached to a spring and immersed in a liquid. The object is acted on by two forces — restoring force $-kx$ and resistive force $-rv$, where v is the object's velocity.

11-8   Forced Oscillations and Resonance (Optional)

When a damped oscillator is set into motion it loses energy and eventually comes to rest. How can the oscillator's energy be maintained constant? Work must be done on the system, so that the energy fed into the system compensates for the energy dissipated. Of course, energy is fed to an oscillator only when *positive* work is done on it by some external agent; the agent must push in the direction the oscillator moves. For example, a person can keep such a damped oscillator as a playground swing oscillating with constant amplitude by pushing it (at the right time) once each cycle. The frequency of the pushes will equal the natural frequency of the swing, and positive work will be done on the swing each time the person pushes.

Consider, as an example of forced oscillations, a damped oscillator driven continuously by an external force varying sinusoidally with time at an angular frequency ω_e. A physical situation illustrating this behavior is shown in Figure 11-15; it is the damped oscillator of Figure 11-14 with the added driving force $F(t)$. We suppose that the external sinusoidal force has a constant force amplitude, so that the driving force is written as $F(t) = F_0 \cos{\omega_e t}$, where F_0 is constant.

Applied to the mass m in Figure 11-15, Newton's second law is written

$$\Sigma F_x = ma_x$$

$$-kx - rv + F_0 \cos{\omega_e t} = ma_x$$

or

$$-kx - r\frac{dx}{dt} + F_0 \cos{\omega_e t} = m\frac{d^2x}{dt^2} \qquad \text{(11-20)}$$

The solution of this equation and the object's motion (which this solution represents) consists of two parts:

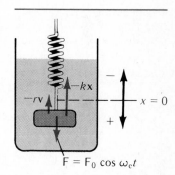

$$F = F_0 \cos{\omega_e t}$$

Figure 11-15.   Forced oscillation of a damped oscillator. In addition to restoring force $-kx$ and resistive force $-rv$, the object is acted on by an external driving force, $F = F_0 \cos{\omega_0 t}$.

- The *transient* motion that ensues immediately after the driving force is first applied but quickly damps out, given by (11-18).
- The *steady-state* solution, which can be shown to be

$$x = A_\omega \cos(\omega_e t + \delta)$$

$$= \frac{(F_0/m) \cos(\omega_e t + \delta)}{\sqrt{(\omega_e^2 - \omega_0^2)^2 + (r\omega_e/m)^2}} \tag{11-21}$$

where $\omega_0 = \sqrt{k/m}$ is the oscillator's frequency in the absence of damping ($r = 0$). The phase angle δ depends on the parameters F_0, m, r, ω, and ω_e.*

The driven mass undergoes oscillations *at the frequency ω_e of the external driving force.* The oscillation amplitude A_ω, given by the term preceding $\cos(\omega_e t + \delta)$ in (11-21), strongly depends of the frequency ω_e of the external force. Indeed, A_ω is a maximum when $\omega_e = \omega_0$ (assuming r to be relatively small). That is, the mass undergoes oscillations of the greatest amplitude when the oscillator is driven at its natural frequency of oscillation ($\omega_e = \omega_0$), as we should expect from the qualitative arguments given above. The oscillator is said to be in *resonance* with the driving agent when the oscillator's natural frequency equals that of the driving force.

The amplitude A_ω of (11-21) is shown as a function of the driving frequency ω_e in Figure 11-16. Two curves are shown. The sharp curve corresponds to a small value of r, and the broader curve (with its maximum at a slightly lower frequency than ω_0) corresponds to larger damping. The *resonance is sharp* when the *damping is small*; then the oscillator responds substantially to the driving force only over the narrow band in which the natural and driving frequencies are nearly equal. On the other hand, if the *damping is large*, the *resonance is broad*; then the oscillator is influenced less by the external driving force but responds over a broader band of frequencies.

The forced oscillator has a constant displacement amplitude A_ω for a given frequency ω_e. That is to say, the total energy of the oscillating mass and spring remains constant with time. Since energy is dissipated in the system continuously, through damping, it follows that the driving force feeds energy into the system at the same rate. The resonance frequency represents that frequency at which energy is most readily transferred from the external agent to the oscillating system.

Resonance phenomena appear in many areas of physics: in resonating mechanical devices, in electric circuits, in atomic and molecular structure. In fact, resonance occurs whenever an oscillator is acted on by a second driving oscillator at the same frequency.

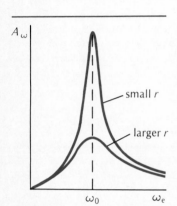

Figure 11-16. Amplitude of forced oscillations as a function of the frequency of the driving force. Resonance occurs when the driving frequency matches the oscillator's natural frequency. The resonance peak is narrow for small damping, broad for large damping.

* See, for example, K. R. Symon, *Mechanics*, 3d ed. (Reading, Mass.: Addison-Wesley Publishing Co., 1971), Section 2.10.

Summary

Definitions

Simple Harmonic Motion (SHM):

- Kinematics: displacement varies sinusoidally with time,

$$x = A \cos(\omega_0 t + \delta) \tag{11-9}$$

where

$$\omega_0 = 2\pi f = \frac{2\pi}{T} \tag{11-3}$$

- Dynamics: linear restoring force; equivalently, the system's potential energy varies with the square of displacement,

$$F_s = -kx \tag{11-5}$$

and

$$U_s = \tfrac{1}{2}kx^2 \qquad (11\text{-}6)$$

Amplitude A: maximum displacement.
Period T: time for one oscillation.
Frequency f: number of oscillations per unit time.
Angular frequency ω_0: radians per unit time.
Phase constant δ: argument of $\cos(\omega_0 t + \delta)$ at $t = 0$.
Damped oscillation: oscillator subject to resistive force; amplitude decays with time.
Forced oscillation: oscillator subject to external, oscillating force.
Resonance: external oscillating force at same frequency as oscillator alone. At resonance the amplitude is a maximum and power is transferred to the oscillator at maximum rate.

Units
Frequency: 1 Hertz = 1 Hz = 1 s^{-1}

Important Results
Total energy of oscillator with amplitude A is:

$$E = \tfrac{1}{2}kA^2 \qquad (11\text{-}14)$$

The period (for small A) of an oscillator with mass m attached to a spring of stiffness k is:

$$T = \frac{1}{f} = 2\pi \sqrt{\frac{m}{k}} \qquad (11\text{-}13)$$

The period (for small amplitude) of simple pendulum of length L is:

$$T = \frac{1}{f} = 2\pi \sqrt{\frac{L}{g}} \qquad (11\text{-}15)$$

Any particle undergoing oscillations with small amplitude about a point of stable equilibrium undergoes SHM.

The period of any SHM oscillator is independent of the oscillation amplitude.

Problems and Questions

Section 11-1   Definitions
· **11-1 P** A ball of mass m is dropped on a steel plate from a height h. The subsequent collisions are perfectly elastic. The ball bounces up and down with a frequency of
(A) $2\sqrt{g/2h}$
(B) $(\tfrac{1}{4})\sqrt{g/2h}$
(C) $(\tfrac{1}{2})\sqrt{2h/g}$
(D) $\sqrt{2h/g}$
(E) $(\tfrac{1}{2})\sqrt{2gh}$

: **11-2 P** Show that the superposition of two SHMs along the x and y axes will result in an elliptical path, when the separate motions have the same frequency, different amplitudes, and a phase difference of $90°$.

: **11-3 P** (a) Show that uniform circular motion can be represented by the superposition of two SHMs with the same frequency and amplitude and with a phase difference of $90°$, when the two SHMs oscillate along two mutually perpendicular lines. (b) Show that SHM along a line can be represented by superimposing two uniform circular motions having the same radius and angular frequency but rotating in opposite senses.

Section 11-2   Dynamics of Simple Harmonic Motion
· **11-4 Q** Shown here are the displacements of two oscillating systems, X and Y. With respect to their phase differences, we can say that
(A) X and Y are in phase.
(B) X leads Y by $90°$.
(C) Y leads X by $90°$.
(D) X and Y are $180°$ out of phase.
(E) X and Y differ in phase by $\pi/4$ radians.

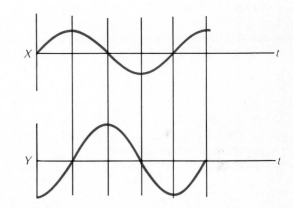

Figure 11-17.   Question 11-4.

: **11-5 Q** The phase relationships between displacement, velocity, and acceleration play an important role in many physical problems. For an undamped SHM,
(A) displacement, velocity, and acceleration are always in phase.
(B) the phase difference between displacement and velocity oscillates at the natural frequency.
(C) displacement and acceleration are always $180°$ out of phase.
(D) the phase differences depend on the amplitude of the motion.
(E) acceleration and velocity are always $180°$ out of phase.

: **11-6 P** A particle travels a total distance of 2.0 mm in 0.10 s as it completes one complete oscillation in SHM. (a) What is the particle's average speed? (b) What is its maximum speed? (c) What is its maximum acceleration?

: 11-7 Q When a mass is hung from a spring, it stretches 4 cm. If the mass is then displaced from equilibrium, it will oscillate with a period
(A) that cannot be determined without knowing the value of the mass.
(B) that cannot be determined without knowing the spring constant.
(C) that cannot be determined without knowing both the spring constant and the mass.
(D) that cannot be determined without knowing the mass, the spring constant, and the amplitude.
(E) that *can* be computed from the information given.

· 11-8 P The displacement of a particle is described by $x = 4 \sin(3t)$. What are the (a) amplitude, (b) frequency, (c) angular frequency, (d) period, and (e) maximum velocity?

· 11-9 P When a mass of 100 gm is attached to a spring attached to the ceiling, the weight of the mass stretches the spring downward 4 cm. The mass is then lifted 2 cm and released with a downward velocity of 3 cm/s at $t = 0$. Write expressions for the displacement, velocity, and acceleration as a function of time. (Take the equilibrium position to correspond to $y = 0$ with the positive axis upward.)

: 11-10 P A coin is placed on a platform that is set into oscillation vertically with an amplitude of 0.50 mm. What is the maximum oscillation frequency of the platform if the coin is not to lose contact with it?

: 11-11 P A mass is suspended from a spring and found to oscillate with frequency f. If the spring is cut in half and the same mass is attached to the shortened spring, with what frequency will the mass now oscillate?

: 11-12 P When a mass m is attached to one end of a horizontal spring (the other end of which is fixed) it oscillates with frequency f. If two particles, each of mass m, are attached to the two ends of this spring, with what frequency will they oscillate?

: 11-13 P A mass m is attached to a spring with force constant k. If the spring is hung from the ceiling, show that the effect of gravity is only to shift the equilibrium position of the mass downward. Show that when the spring oscillates it still undergoes SHM, and that the frequency is unaffected by the presence of the gravitational force.

: 11-14 P A 1-kg block rests on top of a 2-kg block that is undergoing SHM on a frictionless table. The frequency of the motion is 2 Hz. The coefficient of static friction between the two blocks is 0.5. What is the maximum allowable am-

plitude of oscillation if the top block is not to slide off the bottom one?

: 11-15 P When a mass m is hung from a spring it oscillates with frequency 60 Hz. When a 2-kg mass is added to m the combination oscillates at 40 Hz. Determine (a) the value of m and (b) the value of the force constant of the spring.

: 11-16 P A block is on a horizontal plane surface that is moving horizontally with simple harmonic motion of frequency 2 Hz. The coefficient of static friction between the block and the plane is 0.5. If the block is not to slip along the surface, what is the maximum possible amplitude for the SHM?

: 11-17 P A hummingbird's wings oscillate with a frequency of 70 Hz. What are (a) the maximum velocity, and (b) the maximum acceleration for a point near the midpoint of the wing. Choose reasonable values for the dimensions of the wing.

: 11-18 P What is the maximum speed of a particle that undergoes simple harmonic motion with a period of 0.5 s and an amplitude of 2 cm?

: 11-19 P A 400-kg mass is attached to two springs, both of length 20 cm, and with spring constants k_1 and k_2. (a) Determine the value of x for equilibrium if $k_1 = k_2 = 800$ N/m. (b) Determine the natural frequency of the system. (c) Repeat the preceding calculation for $k_1 = 600$ N/m, $k_2 = 1200$ N/m.

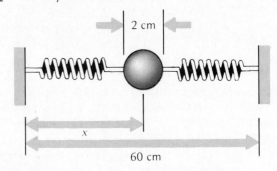

Figure 11-19.   Problem 11-19.

: 11-20 P A 300-gram ball is to be catapulted into the air with the device shown here. The piston has a mass of

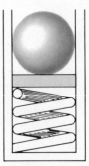

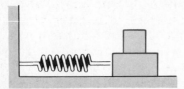

Figure 11-18.   Problem 11-14.

Figure 11-20.   Problem 11-20.

100 gm and the spring a force constant of 2000 N/m. The piston is pushed down 6 cm below the equilibrium point of piston and ball and released.

(a) Determine how far above the equilibrium position the piston is when the ball loses contact with it. (b) How long after the time of release does this happen? (c) How fast is the ball then moving?

⋮ **11-21 P** A 2-kg block rests on top of a spring but is not attached to it. The block is depressed 1 cm below its equilibrium position (this is the maximum initial displacement for which the block never loses contact with the spring) and is released. Determine (a) the force constant of the spring, and (b) the displacement of the block from equilibrium as a function of time.

⋮ **11-22 P** Suppose that a mass m is attached to a long uniform spring of length L and observed to oscillate at a frequency f_0. Now the spring is cut into two pieces, of lengths xL and $(1 - x)L$. Mass m is divided into two pieces in this same ratio, with $m_1 = xm$ and $m_2 = (1 - x)m$. The larger mass is attached to the shorter spring and the smaller mass to the longer spring. Calculate the frequency of oscillation for each of the two springs.

⋮ **11-23 P** When a 100-gm mass is hung from a spring it stretches the spring 5 cm. With the 100-gm mass at rest in equilibrium, a second 100-gm mass is attached to the first and the combination is released from rest. The 200 gm now oscillates up and down on the spring. Write an equation that gives the displacement of the 200-gm mass with respect its equilibrium position.

⋮ **11-24 P** Before rockets are launched, the payloads are rigidly attached to "shake tables" to simulate the accelerations experienced during launch. Suppose a table is capable of undergoing simple harmonic motion at a frequency of 100 Hz. What is the amplitude of motion required if the maximum acceleration to be experienced during launch is $8g = 80\text{m/s}^2$?

Section 11-3 Energetics of Simple Harmonic Motion

· **11-25 Q** Which of the following is an accurate statement concerning SHM?
(A) If the amplitude is doubled, the maximum stored energy is also doubled.
(B) If the frequency is doubled, the amplitude is doubled.
(C) If the frequency is doubled, the maximum velocity is also doubled.
(D) If the amplitude is doubled, the acceleration is unchanged.
(E) The potential energy is constant.

· **11-26 Q** Two identical masses are attached to two different springs, displaced the same distance from equilibrium, and let go. If one spring has a force constant twice as great as the other, which of the following statements is true?

(A) The two masses will oscillate at different frequencies but have the same total energy.
(B) The two masses will oscillate at different frequencies but will have the same maximum velocity.
(C) The two masses will oscillate at different frequencies but will have the same maximum displacement.
(D) The two masses will oscillate at the same frequency but have different maximum velocities.
(E) The two masses will oscillate at the same frequency but have different maximum displacements.

· **11-27 Q** A particle of mass m is attached to a spring with a force constant k. The particle executes simple harmonic motion with an amplitude A. When the particle is at a distance $\frac{1}{2}A$ from its equilibrium position, its *potential* energy is equal to
(A) $(\frac{1}{2})kA^2$
(B) $(\frac{1}{8})kA^2$
(C) $(\frac{3}{4})kA^2$
(D) $(\frac{3}{8})kA^2$
(E) cannot be given without knowing the speed of the particle.

· **11-28 Q** For undamped simple harmonic motion, which of the following statements is *not* true?
(A) The energy alternates continuously between kinetic and potential energy.
(B) The speed reaches a maximum value each time the displacement goes through zero.
(C) The total energy is proportional to the square of the amplitude.
(D) The kinetic energy varies sinusoidally with time.
(E) The restoring force is proportional to the square of the displacement away from the equilibrium position.

⋮ **11-29 Q** A mass attached to a spring executes simple harmonic motion with an amplitude of 4.0 cm. When the mass is 2.0 cm from its equilibrium position,
(A) half of the oscillator's energy is potential and the other half kinetic.
(B) three-quarters of the oscillator's energy is potential and one-fourth kinetic.
(C) three-quarters of the oscillator's energy is kinetic and one-fourth potential.
(D) questions concerning the relative amounts of kinetic and potential energy cannot be answered without also knowing the particle's speed and mass.
(E) questions concerning the relative amounts of kinetic and potential energy cannot be answered without also knowing the particle's mass and the spring's stiffness constant.

· **11-30 Q** A particle attached to a spring oscillates in simple harmonic motion with an amplitude A. The stiffness of the spring is k. Over a time interval equal to one period of oscillation, the net work done by the spring on the particle is
(A) $\frac{1}{2}kA^2$

(B) 0
(C) $\frac{1}{4}kA^2$
(D) $2kA^2$
(E) $4kA^2$

· **11-31 P** A 1-kg object is suspended from spring with stiffness constant $k = 1.0 \times 10^4$ N/m. The 1-kg object is then pulled downward a distance of 6.0 cm from its equilibrium position and released from rest. What are (a) the period of oscillation, (b) the maximum kinetic energy of the object, (c) the resultant force on the object when it is at the highest point in its oscillation?

: **11-32 P** A block of mass M slides along the frictionless track shown here. The block is released from rest at a distance R above the bottom of the track, and subsequently hits a spring of spring constant k. The maximum compression of the spring is
(A) $\sqrt{2Mg/k}$
(B) MgR/k
(C) $2\sqrt{MgR/k}$
(D) $\sqrt{2MgR/k}$
(E) $2Mg/k$

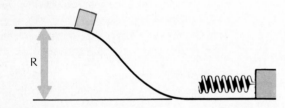

Figure 11-21.   Problem 11-32.

· **11-33 P** A particle undergoes SHM with a frequency f. Show that the kinetic energy and the potential energy each oscillate with frequency $2f$.

· **11-34 P** Show that the velocity of a particle undergoing SHM can be written as a function of its amplitude and displacement as $v = \omega \sqrt{A^2 - x^2}$

: **11-35 P** A 2-kg mass is held at the top of a massless vertical spring. When the mass is released from rest it compresses the spring 20 cm. What is the spring constant of the spring?

: **11-36 P** A spring with an unstretched length of 30 cm is hung vertically. Then two masses, of 100 gm and 200 gm, are attached to the lower end, and they cause the spring to stretch 4 cm. The masses are now pulled down an additional 6 cm and released from rest. The larger mass happens to fall off when the masses are at their lowest point. Over what range of distances from the top end of the spring will the smaller mass now vibrate?

: **11-37 P** Suppose that an 800-kg jeep is dropped by parachute from an aircraft and hits the ground with a speed of 4 m/s. Suppose that it lands on all four tires, and that each wheel is mounted on a spring suspension with a force constant of 10^5 N/m. Neglect the damping (shock absorbers) in real car springs. By how much will the springs be compressed when the jeep lands, and what would be the amplitude of the subsequent SHM?

Section 11-4   Small Oscillations

: **11-38 P** Two protons are positioned at $x = +a$ and $x = -a$ in molecule. An electron is positioned at the origin and is in equlibrium under the forces exerted on it by the protons. Suppose that the electron is displaced a small distance y along the y axis. Each proton will exert on the electron an attractive force with the magnitude $F = ke^2/r^2$ where k is a constant of nature and e is the magnitude of the electric charge on the proton and on the electron, directed as shown in Figure 11-22. Show that the electron will undergo SHM along the y axis, and determine the frequency of oscillation.

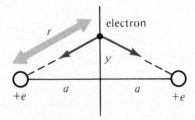

Figure 11-22.   Problem 11-38.

: **11-39 P** 20 gm (20 cm³) of water are poured into a U-shaped tube of 1 cm² cross-sectional area. Show that if the surface of the water on one side of the tube is depressed momentarily and released the water will oscillate in SHM. Determine the frequency of oscillation.

Section 11-5   The Simple Pendulum

· **11-40 P** The gravitational acceleration at the surface of the moon is one sixth the gravitational acceleration at the earth's surface. A long simple pendulum has a period of 10 s on the earth. The same pendulum would then have a period T on the moon, where T is
(A) impossible to determine from the information given.
(B) 10 s.
(C) 24.5 s.
(D) 60 s.
(E) 4.1 s.

: **11-41 Q** A mass of 1 kg swings on the end of a cord of length 0.5 m. At time $t = 0$, the mass is released from rest at a position 5 cm to the right of the vertical. We may write the subsequent x coordinates of the mass as a function of time t as $x = A \cos(\omega_0 t + \delta)$. In this case (with the positive x axis to the right),
(A) $A = 5$ cm and $\delta = 0$
(B) $A = 5$ cm and $\delta = -\pi/2$

(C) $A = 10$ cm and $\delta = 0$
(D) $A = 0$ cm and $\delta = -\pi/2$
(E) none of the above is correct.

: **11-42 P** A simple pendulum of length L and mass m is suspended in a car that accelerates at a constant rate a. What is the frequency of the pendulum for small oscillations?

: **11-43 Q** A pendulum bob is raised to a height h and released from rest. At what height will it attain half of its maximum speed?
(A) $\frac{1}{4}h$
(B) $\frac{1}{2}h$
(C) $3h/4$
(D) $0.707h$

: **11-44 P** A rise in temperature causes the pendulum in a grandfather clock to increase its length by 1 part in 10^4. How many seconds will the clock gain or lose in a day because of this?

: **11-45 P** A simple pendulum consists of a 50-gm ball attached to a string 80 cm long. The ball is pulled to the side until the string makes an angle of $10°$ with the vertical, and is released from rest at time $t = 0$. Write an equation for the angle the string makes with the vertical, as a function of time.

· **11-46 P** In analyzing the simple pendulum we make the small angle approximation $\sin \theta = \theta$. (a) Which is larger, $\sin \theta$ or θ? (b) evaluate the ratio $\theta/\sin \theta$ for $\theta = 1°, 5°, 10°, 15°, 20°, 30°, 45°, 60°$.

: **11-47 P** A 100-gm mass is attached to a string 1.2 m long and suspended as a simple pendulum. At a point 1 meter below the point of suspension is placed a peg, which the string hits when the pendulum swings down. If the mass is pulled a small distance to one side and released, what will be the period of its motion?

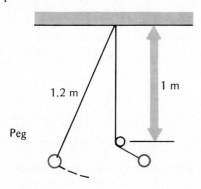

1.2 m

1 m

Peg

Figure 11-23. Problem 11-47.

: **11-48 Q** A plastic bottle full of water is attached to a string and suspended as a pendulum. There is a small hole in the bottom of the bottle, and as it swings back and forth the water slowly leaks out until the bottle is empty. Carefully describe in qualitative terms the motion of the bottle during this process.

Supplementary Problems

11-49 P A rocket engine can be characterized by its thrust. Another useful parameter is a rocket's *specific impulse*, which is defined as the thrust of the rocket divided by the weight of exhaust particles discharged per unit time. (a) Show that the specific impulse is equal to v_e/g, where v_e is the exhaust velocity. (b) Show that specific impulse has units of time. (The limit of the specific impulse for a rocket with ordinary chemical fuel is about 430 s. A rocket with nuclear fuel would have a specific impulse of 700 s to 1,000 s. It is estimated that an ion rocket, with an electrostatic field to accelerate dust particles or electrically charged ions, would have a specific impulse of 5,000 s to 10^5 s.)

11-50 P A conventional 10-speed ("Phase II") racing bicycle is 95 percent efficient on level ground at the racing speed of about 30 mph. An experienced cyclist expends energy at the rate of about 0.2 to 0.3 horsepower. (a) What is the total equivalent resistive force on the bike and cyclist? (b) A "Phase III" bike has a streamlined enclosure for most of the bike and rider(s); such a bicycle can achieve a much higher speed. For example, the two-seater Vector Tandem set a world speed record of 62.93 mph (*The New York Times Magazine*, August 10, 1980). Why is it noteworthy that with twice as many drivers you can achieve twice the speed? (Consider these facts: a rider in moderately good shape cruising at 10 mph expends one-third of his/her energy pushing away air; at full speed downhill, with a speed approaching 20 mph, that figure rises to more than half the energy; and at a racing speed of 30 mph, more than 90 percent of the cyclist's energy is used to push air around the bike.)

11-51 P "The Beast," a roller coaster in Kings Island in Cincinnati, Ohio, is the fastest in the world; it reaches a speed of 64.77 mph after a 141-ft drop. A roller coaster in Panama City, Florida, goes 60 mph after a drop of 80 ft. Is either of the two claims plausible?

11-52 P The "Vomit Comet" is the name given by NASA to a modified 707 aircraft that is used to simulate zero gravity in the training of astronauts. The plane, flying at 500 mph at an altitude of 24,000 ft, flies in a parabolic path in vertical plane in the fashion of a projectile. Any unattached objects inside the aircraft also fall freely and appear to float weightless. What is the required radius of curvature near the peak in the parabolic path?

12 Equilibrium of a Rigid Body

The equilibrium of a rigid body is one simple aspect of rotation. In treating rotation in this chapter and the two following, we shall progress from the simple and specific, the equilibrium of a rigid body, to the more complex. That means that we shall go on to the more general case of a rigid body that accelerates in its rotational motion (Chapter 13), and finally the most general case — any collection of particles, not merely a rigid body — and the special role of angular momentum. The propositions we assume as reasonable in studying the equilibrium of a rigid body will subsequently be rigorously proved.

12-1   Definitions: Rigid Body, Translation, Rotation, Vector Representation of Rotation

A perfectly *rigid body* is an ideal object in which the *distance between any pair of points* in the body is always *the same*. A truly rigid body could not be deformed in any way. No such thing exists in the real world; every solid object is deformed to some degree by forces applied to it. Even a solid sitting at rest on a surface is deformed by its own weight downward and an upward normal force. For simplicity we assume in what follows that all solids are perfectly rigid.

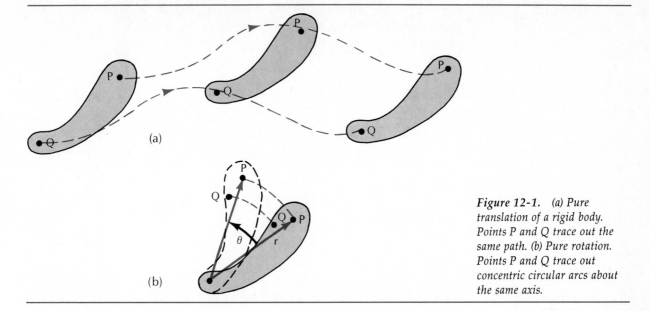

Figure 12-1. *(a) Pure translation of a rigid body. Points P and Q trace out the same path. (b) Pure rotation. Points P and Q trace out concentric circular arcs about the same axis.*

Any change in the position of a rigid body can be regarded as the superposition, or combination, of two simple motions:

• *Pure translation*, in which every particle in the object has the same displacement and therefore travels the *same trajectory* from its starting point. See Figure 12-1(a).

• *Pure rotation*, in which every particle travels in a circle about the *same axis*, as shown in Figure 12-1(b). The rotation axis is the straight line that contains the centers of all the circular paths. Another equivalent of defining pure rotation is this: if radius vector **r** perpendicular to the rotation axis to some particle turns through angle θ, then so do the corresponding radius vectors for all other particles.

The center of mass of a rigid body is highly important in the body's motion, and it is often most convenient (although not necessary) to describe any change in position of a rigid body in terms of:

• The translational motion of the center of mass.
• The rotational motion about the center of mass.

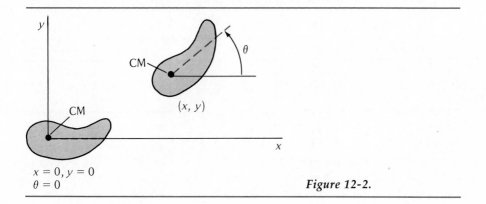

Figure 12-2.

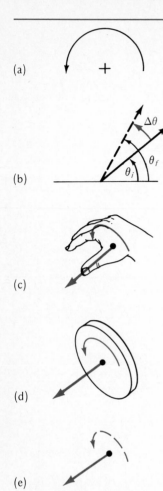

(a)

(b)

(c)

(d)

(e)

Figure 12-3. *Equivalent ways of showing counterclockwise rotation: (a) a curl; (b) positive $\Delta\theta$; (c) right hand with curled fingers and extended thumb; (d) curl sense and associated vector; (e) vector alone.*

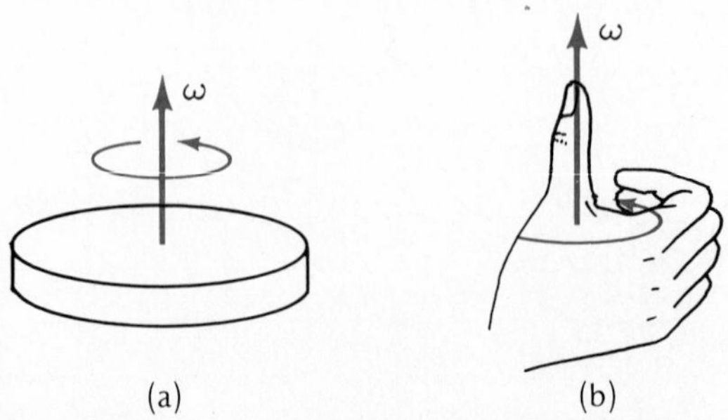

In this chapter and the next, the translational motion of a rigid body will be in a single plane (containing the x and y axes); the rotational motion will be about a single axis (the z axis). See Figure 12-2. It then takes three coordinates to specify the location of a rigid body:

- The x and y coordinates of its center of mass.
- The polar angle θ of some line through the center of mass relative to the $+x$ axis.

How can we describe the sense of rotation of an object? One obvious way is to use the terms *counterclockwise* and *clockwise*. See Figure 12-3(a). It is customary to call counterclockwise rotation positive (angle θ increases) and clockwise rotation negative (θ decreases). See Figure 12-3(b).

Another natural way of showing the sense of rotation is with curled fingers — to be specific, those of the right hand. See Figure 12-3(c). Now notice something special about the curled right-hand fingers; the sense in which they are curled is related uniquely to the direction in which the right thumb points. As Figure 12-3(c) shows, with the right fingers curled counterclockwise, the right thumb must come *out* of the paper; with the right fingers curled *clockwise*, the right thumb points *into* the paper. In fact, the right thumb alone can give the curl sense.

We can think of our right hand as a useful device for keeping straight that *counterclockwise* is associated with direction *out* and *clockwise* with direction *in*. Then the same association between rotation sense and direction in space can be shown, *without* a right hand, as in Figure 12-3(d). Finally, as in Figure 12-3(e), we see that a single direction in space can show rotation sense. [The dashed lines for curl sense appear in Figure 12-3(e) merely as a reminder.] Indeed, using a direction in space to represent rotation sense is, as we shall see, the simplest representation.

We have then several equivalent ways of describing, let's say, *counterclockwise rotation*:

- A counterclockwise curl.
- Positive rotation.
- Polar angle θ increasing.
- Right-hand fingers curled counterclockwise in the plane of the paper.
- Right thumb pointing out of the paper.
- Simply an arrow pointing out of the paper.

(For clockwise rotation all items are, of course, just reversed.)

The fact that a rotation sense can be indicated by an arrow suggests that

Figure 12-4. *(a) Angular velocity vector ω for a rotating object. (b) Right-hand rule correlating curl sense and direction of ω.*

(a)

(b)

physical rotation quantities may be vector quantities. Actually, many rotational quantities *are* vectors. The most familiar is angular velocity ω, as shown in Figure 12-4.

The right-hand fingers and thumb again correlate rotation sense with the direction of ω. Indeed, through its magnitude and direction, a single angular velocity vector ω gives all the following information.

- The orientation in space of the line containing the ω vector is also the orientation of the rotation axis.
  - The direction of ω gives the rotation sense about the rotation axis.
  - The magnitude of ω is the time rate of rotation, $d\theta/dt$.*

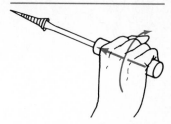

Figure 12-5. A right hand turns a screwdriver that turns a right-hand screw. The torque direction is the direction of advance of a right-hand screw.

Another vector quantity with a rotation sense is *torque*. We postpone a detailed definition of torque to Section 12-4, and for the moment, take torque to mean, roughly, the turning effect of one or more forces. A right-hand rule relates the rotation sense of a torque with a direction in space of a vector representing torque. Consider Figure 12-5, where a torque is applied by a right hand turning a screwdriver, which in turn turns an ordinary (right-hand) screw. The torque direction is the same as the direction of advance of the right-hand screw.

12-2   The Cross (or Vector) Product

Several fundamental quantities in rotational dynamics—especially torque and the angular momentum of a particle—are defined in terms of a cross product of two vectors. First we set forth the general mathematical properties of the cross product. Then we define the torque of a force in terms of a cross product (Section 12-4).

The *cross product* of vectors **A** and **B** is another vector **C**, symbolized by

$$C = A \times B \qquad (12\text{-}1)$$

The cross product **C** is also called the *vector product* of vectors **A** and **B** because **C** is itself a vector.† The quantity $A \times B$ is read "A cross B."

The direction and magnitude of cross product $C = A \times B$ are defined as follows.

- The vector **C** for the cross product, $C = A \times B$, is always oriented *perpendicular* to the plane containing the vectors **A** and **B**. See Figure 12-6(a).
- The *direction* of cross product **C** is that of the outstretched right thumb when the first of the two vectors **A** is imagined turned by the right-hand fingers, through the smaller angle θ between **A** and **B**, to align with the second vector **B**. See Figure 12-6(b), where vectors **A** and **B** are drawn from a common point.
- The *magnitude* of $A \times B$ is given by

$$C \equiv AB \sin \theta \qquad (12\text{-}2)$$

where θ is the magnitude of the smaller angle between **A** and **B**.

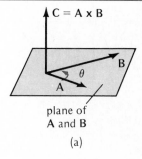

plane of
A and B

(a)

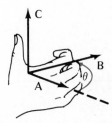

(b)

Figure 12-6. (a) Relative orientation of the vectors for $C = A \times B$. (b) The right-hand rule for the cross product.

---

* Curiously, although angular velocity is a vector, an angular displacement is not unless the displacement is infinitesimally small. The proof is given in Section 12-6.

† Recall that the dot product (or scalar product) of two vectors is itself a *scalar* quantity (Section 9-4).

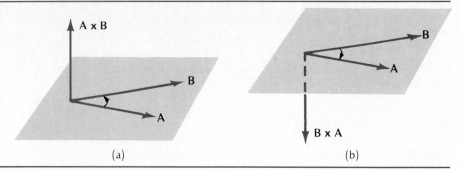

Figure 12-7. The direction of
A × **B** *in (a) is opposite to the
direction of* **B** × **A** *in (b).*

(a)

(b)

We emphasize that the direction and magnitude of the cross product are *defined* to have the characteristics shown in Figure 12-6 and (12-2). The reason? These very definitions turn out to endow the cross product with a powerful importance; a cross product says succinctly what would require a much longer story in words.

Suppose we switch the order of the two vectors from **A** × **B** to **B** × **A**. Then, as Figure 12-7 shows, the direction of the cross product is reversed. Therefore,

$$\mathbf{A} \times \mathbf{B} = -\mathbf{B} \times \mathbf{A}$$

The cross product does not obey the commutative law.

The magnitude $AB \sin \theta$ of the cross product can be interpreted by reference to Figure 12-8, where we see in part (a) the components ($B_\perp$ and $B_\parallel$) of vector **B** along and perpendicular to vector **A**, and in part (b) just the reverse, the components ($A_\perp$ and $A_\parallel$) of **A** along and perpendicular to **B**. From the geometry of the figures, it is clear that

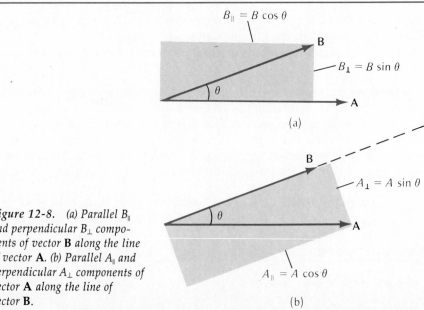

Figure 12-8. (a) Parallel $B_\parallel$
and perpendicular $B_\perp$ components of vector **B** along the line
of vector **A**. (b) Parallel $A_\parallel$ and
perpendicular $A_\perp$ components of
vector **A** along the line of
vector **B**.

(a)

(b)

$$B_\perp = B \sin \theta \text{ and } A_\perp = A \sin \theta$$

Thus, another way of writing the magnitude of the cross product is this:

$$C = AB \sin \theta = AB_\perp = A_\perp B \qquad (12\text{-}3)$$

The magnitude of a cross product is simply the magnitude of one vector multiplied by the perpendicular component of the other vector.

Suppose vectors **A** and **B** are parallel or antiparallel. The angle θ between them is either $0°$ or $180°$. Then $\mathbf{A} \times \mathbf{B} = 0$. The cross product of *parallel* (or antiparallel) vectors is *zero*.

Although the commutative law does not hold for cross products, the distributive law does apply:

$$\mathbf{A} \times (\mathbf{D} + \mathbf{F}) = (\mathbf{A} \times \mathbf{D}) + (\mathbf{A} \times \mathbf{F})$$

The proof is omitted here.

The cross and the dot product (Section 9-4) must be carefully distinguished.

- The cross product of two vectors is a vector; the dot product is a scalar.
- The magnitude of the cross product is $AB \sin \theta$; the magnitude of the dot product is $AB \cos \theta$.
- The cross product of two vectors is zero when the vectors are parallel; the dot product of two vectors is zero when the vectors are perpendicular.
- The cross product does not obey the commutative law; the dot product does.

12-3 Unit Vectors and the Cross Product (Optional)

For unit vectors **i**, **j**, and **k** along the respective x, y, and z directions of a right-handed set of coordinate axes (see Section 2-5), we have

$$\mathbf{i} \times \mathbf{i} = 0 \qquad \mathbf{i} \times \mathbf{j} = \mathbf{k}$$
$$\mathbf{j} \times \mathbf{j} = 0 \qquad \mathbf{j} \times \mathbf{k} = \mathbf{i}$$
$$\mathbf{k} \times \mathbf{k} = 0 \qquad \mathbf{k} \times \mathbf{i} = \mathbf{j}$$

The cross product of two vectors can easily be expressed in terms of unit vectors. For example, if $\mathbf{A} = 2\mathbf{i}$ and $\mathbf{B} = -2\mathbf{i} + 4\mathbf{j}$ then

$$\mathbf{A} \times \mathbf{B} = (2\mathbf{i}) \times (-2\mathbf{i} + 4\mathbf{j})$$
$$= -4(\mathbf{i} \times \mathbf{i}) + 8(\mathbf{i} \times \mathbf{j})$$
$$= 0 + 8\mathbf{k}$$

Note that in this example we see again that the plane of **A** and **B** (the **ij** plane) is perpendicular to the direction of $\mathbf{A} \times \mathbf{B}$ (the **k** direction).

12-4 Torque

If a rigid body is subject to two or more external forces and yet does not turn, then the turning effect of one force must somehow be offset by the opposite

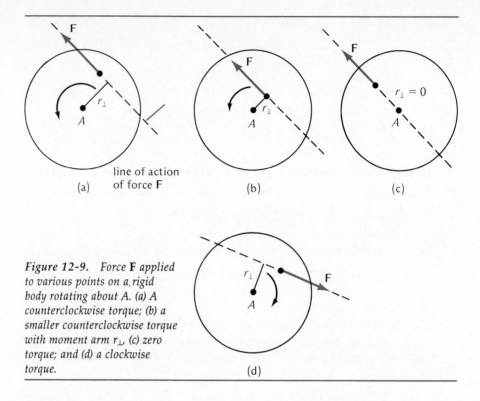

Figure 12-9. Force **F** applied to various points on a rigid body rotating about A. (a) A counterclockwise torque; (b) a smaller counterclockwise torque with moment arm $r_\perp$, (c) zero torque; and (d) a clockwise torque.

turning effect of one or more other forces. Roughly speaking, torque is a measure of the turning effect of a force.

Before defining torque, we set forth some items of usage.

• The arrow representing a force vector will be located on the diagram with its tail at that point on the rigid body at which the force is applied.
• By the *line of action* of a force is meant the infinite straight line containing the force vector.
• The *rotation axis* relative to which such things as torque are measured must always be specified.

Consider Figure 12-9, in which a force **F** is applied at various points to a rigid body rotating about a fixed axis at A. Clearly, in Figures 12-9(a) and (b) force **F** acting alone will cause the object to rotate in the counterclockwise sense. The sense or direction of such a torque τ can then be described alternatively as:

• Positive.
• Counterclockwise.
• Out of the paper, from the right-hand rule for rotation sense.

The magnitude of torque vector τ is defined as

$$\tau \equiv r_\perp F \qquad (12\text{-}4)$$

where $r_\perp$, called the *moment arm* of the force, is the perpendicular distance between the line of action force **F** and rotation axis A. Appropriate SI units for torque are meter newtons (m·N.)

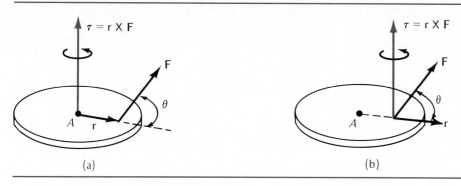

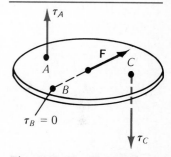

Figure 12-10. *(a) The torque $\tau = \mathbf{r} \times \mathbf{F}$ of force $\mathbf{F}$ applied at the point given by radius vector $\mathbf{r}$ relative to an origin at A. (b) The same vectors; now their tails are at the point of application of force $\mathbf{F}$.*

Note that the torque magnitude depends both on the magnitude of **F** and on $r_\perp$. In Figure 12-9(b) the force **F** has the same direction and magnitude as in Figure 12-9(a), but **F** is applied at a different point and $r_\perp$ is smaller. Consequently, the torque is smaller for Figure 12-9(b) than for Figure 12-9(a). In Figure 12-9(c) the line of action of the force passes through point A. Therefore, $r_\perp = 0$ and the torque $\tau = r_\perp F$ is also zero. In fact, the torque of any force that points through the rotation axis is always zero.

Now consider the torque in Figure 12-9(d). Here the sense or direction of the torque can be described as negative, clockwise, or into the paper. The magnitude is again $r_\perp F$.

Both the magnitude and the direction of torque are incorporated in the more general definition of torque τ expressed as the cross product of radius vector **r** and force **F**:

$$\tau \equiv \mathbf{r} \times \mathbf{F} \qquad (12\text{-}5)$$

Vector **r** has its tail at the axis A and extends to the point of application of force **F**, as shown in Figure 12-10(a). The same vectors are shown in Figure 12-10(b); in this figure the tails of **r** and **F** are at the same point. The angle between **r** and **F** is θ. The direction and the magnitude of the torque vector τ are these:

- Turn **r** (or imagine **r** to turn) through the smaller angle θ with the right-hand curled fingers until **r** lines up with **F**. Then the right thumb gives the direction of τ.
- From (12-3), the torque magnitude τ is given by

$$\tau = rF \sin \theta = r_\perp F = rF_\perp \qquad (12\text{-}6)$$

It cannot be emphasized too strongly that both the magnitude and the direction of a torque depend on the point (or axis) relative to which torque is measured. See Figure 12-11. To specify a torque completely, one must give:

- Its direction (or rotation sense).
- Its magnitude.
- The axis relative to which τ is measured.

Both torque and work are physical quantities in which distance is multiplied by force. But their meanings are quite different. Torque is the *cross* product of **r** and **F**; work is the *dot* product of **r** and **F**. Torque units are given as meter newtons (m·N), work units as newton meters (N·m = J).

Figure 12-11. *Torque of same force **F** relative to three different points A, B, and C as axes. τ_A is up; τ_B is zero; τ_C is down.*

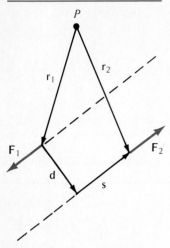

Figure 12-12. A pair of equal and opposite forces, $\mathbf{F}_1$ and $\mathbf{F}_2$, separated by distance d.

Example 12-1. A Couple (Optional). Two forces equal in magnitude and opposite in direction, but not collinear, constitute a *couple*. The magnitude of either force is F, and their respective lines of action are separated by a distance d. Let's show that the torque of a couple has the magnitude Fd for *any* reference point and a direction perpendicular to the plane containing the two forces.

The forces $\mathbf{F}_1$ and $\mathbf{F}_2$ are shown in Figure 12-12. The displacements from arbitrary point P to the respective points of application of the forces are $\mathbf{r}_1$ and $\mathbf{r}_2$. Vector $\mathbf{d}$ is perpendicular to $\mathbf{F}_1$ and $\mathbf{F}_2$; its magnitude is the perpendicular distance between the lines of action of the two forces. The resultant torque relative to P is

$$\tau = \mathbf{r}_1 \times \mathbf{F}_1 + \mathbf{r}_2 \times \mathbf{F}_2$$

But Figure 12-12 shows that

$$\mathbf{r}_2 = \mathbf{r}_1 + \mathbf{d} + \mathbf{s}$$

so that

$$\tau = \mathbf{r}_1 \times \mathbf{F}_1 + (\mathbf{r}_1 + \mathbf{d} + \mathbf{s})(-\mathbf{F}_1)$$

where we used the fact that $\mathbf{F}_2 = -\mathbf{F}_1$.

Since vector $\mathbf{s}$ is parallel to $\mathbf{F}_1$, then $\mathbf{s} \times \mathbf{F}_1 = 0$, and since vector $\mathbf{d}$ is perpendicular to $\mathbf{F}_1$, then $|\mathbf{d} \times \mathbf{F}_1| = Fd$. The relation above reduces to

$$|\tau| = Fd$$

The torque magnitude is simply the magnitude of either force F multiplied by separation distance d for any reference point. A couple represents what might be called a pure torque; it can rotate an object but cannot accelerate it translationally. The only way to cancel one couple is with another couple of the same magnitude in the opposite direction.

12-5 Center of Gravity

The *center of mass* of an object is that point at which all the *mass* may be imagined to be concentrated (Section 8-5). The *center of gravity*, on the other hand, is that point at which all the object's *weight* $M\mathbf{g}$ may be imagined to be concentrated. More specifically, the center of gravity is the place at which the torque produced by a single force $M\mathbf{g}$ acting there is the same as the sum of the gravitational torques from all individual parts of the object.

It is proved in Section 12-7 that, if the gravitational acceleration $\mathbf{g}$ has the same magnitude and direction at the locations for all particles of a rigid body, the center of gravity as defined above is *at the object's center of mass.* For all practical purposes, the center of mass and the center of gravity coincide for any rigid body whose size is small compared with the earth.

Example 12-2. A uniform rod of length L and mass M is suspended from one end. See Figure 12-13. What is the resultant gravitational torque on the rod relative to the pivot point A when the rod makes an angle θ with the vertical?

The rod's center of gravity (or center of mass) is at its geometrical center, $\frac{1}{2}L$ from either end. We imagine the rod's weight $M\mathbf{g}$ to act at this point. The moment arm about the pivot point A has the magnitude $r_\perp = (L/2) \sin \theta$, so that the magnitude of the torque is

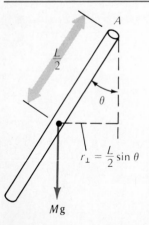

Figure 12-13.

$$\tau = r_\perp F = \frac{L}{2} \sin \theta \, Mg$$

The direction of τ is, for the counterclockwise gravitational torque shown in Figure 12-13, out of the paper.

12-6 Conditions for Rotational Equilibrium

What does it take to put a rigid body in equilibrium?

First, there is *translational equilibrium*; the object cannot accelerate along any line. We know that for any system of particles, the resultant force on it equals

$$\Sigma \mathbf{F}_{ext} = M\mathbf{A}$$

where M is the system's mass and $\mathbf{A}$ is the acceleration of its center of mass. Therefore, for translational equilibrium, with $\mathbf{A} = 0$, the resultant of all forces acting on the body must be zero:

$$\Sigma \mathbf{F}_{ext} = 0 \qquad\qquad (12\text{-}7a)$$

Equivalently, the x and y force components must separately add to zero:

$$\Sigma F_x = 0 \qquad \Sigma F_y = 0 \qquad\qquad (12\text{-}7b)$$

These are just the relations we used earlier (Section 2-6) to treat the equilibrium of a single particle.

For a rigid body there is also *rotational equilibrium*; the body cannot accelerate about a rotation axis. The condition to be met is, reasonably enough, that the resultant torque on the body be zero:

$$\Sigma \tau = 0 \qquad\qquad (12\text{-}8a)$$

If all forces lie in the xy plane, their torques are along the z axis. Then we can equivalently write

$$\Sigma \tau_z = 0 \qquad\qquad (12\text{-}8b)$$

The formal proof for (12-8) is given in Chapter 14. Torques may be measured relative to an axis through the rigid body's center of mass. But if the rigid body is in equilibrium, *any* axis will do. After all, if the rigid body does not rotate about its center of mass, it does not rotate about any other point. Equation (12-8) applies to any rotation axis. The proof is given in Section 12-7.

Our examples will be primarily with rigid bodies at rest, but it must be emphasized that:

- If the *resultant force* is zero, the center of mass has *constant linear velocity* (including the special case $\mathbf{V} = 0$).
- If the *resultant torque* is zero, the rigid body rotates about the center of mass with *constant angular velocity* (including the special case $\omega = 0$).

In Section 2-6 we gave a procedure that is useful in analyzing the static equilibrium of a single particle. The static equilibrium of a rigid body involves a similar routine:

- Choose the rigid body whose equilibrium is being studied.
- Draw a neat sketch.
- Draw arrows for vectors representing the forces acting on the chosen

object. It is helpful to locate the tail of each force vector at the place where the force is applied. The weight of the object is represented by a single force vector at the center of mass (or center of gravity). It is best to use symbols until the problem has been solved algebraically and only then insert numerical values.

- Draw x and y axes and resolve each force into its equivalent x and y components.
- Write down equations corresponding to $\Sigma F_x = 0$ and $\Sigma F_y = 0$.
- Choose the location for the rotation axis. The axis need not be at the obvious location (at a hinge, for example); put the axis at a place that will make the computation of torques easy. Obviously, if the axis is on the line of action of some force, the torque of that force must be zero, since $r_\perp = 0$ for it.
- Write down the equation corresponding to $\Sigma \tau_z = 0$.
- Solve for the unknowns. (With three equations, there can be at most three unknowns).

Example 12-3. A uniform beam, weighing 100 N and making an angle of 30° with the horizontal, is hinged at its lower end; a horizontal wire is attached to its upper end, as shown in Figure 12-14. What is the tension in the wire and the direction and magnitude of the force of the hinge on the beam?

The three external forces on the beam are shown in Figure 12-14(a). They are the weight **w** of the beam acting at its center of gravity, a distance $L/2$ from either end; the force of the hinge on the beam, replaced for convenience by its vertical and horizontal components $\mathbf{F}_y$ and $\mathbf{F}_x$; and the tension in the wire, **T**. Translational equilibrium requires that

$$\Sigma F_x = 0 \qquad \Sigma F_y = 0$$

$$F_x - T = 0 \qquad F_y - w = 0$$

A convenient choice of axis of rotation is one passing through the hinge. This choice is made, not primarily because the hinge represents a natural point of rotation, but rather because the forces $\mathbf{F}_y$ and $\mathbf{F}_x$ have lines of action through this point, thereby making their torques zero relative to this axis. For this axis, **w** produces a clockwise (negative) torque with moment arm $(L/2) \cos 30°$; the force **T** has a counterclockwise (positive) torque with moment arm $L \sin 30°$. Rotational equilibrium then requires that

$$\Sigma \tau_z = 0$$

$$-\frac{L}{2} \cos 30°(w) + L \sin 30°(T) = 0$$

Solving simultaneously for F_x, F_y, and T in the equations given earlier and setting $w = 100$ N yields

$$F_x = 87 \text{ N} \qquad F_y = 100 \text{ N} \qquad T = 87 \text{ N}$$

Notice that the length L of the beam does *not* enter.

We can readily compute the magnitude F and direction θ with respect to the horizontal of the force of the hinge on the beam:

$$F = \sqrt{F_x^2 + F_y^2} = \sqrt{87^2 + 100^2} \text{ N} = 132 \text{ N}$$

$$\tan \theta = \frac{F_y}{F_x} = \frac{100}{86.6} = 1.15$$

$$\theta = 49°$$

The force of the hinge on the beam is not along the length of the beam. This is also

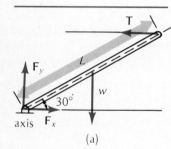

(a)

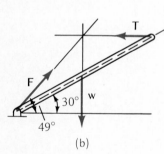

(b)

*Figure 12-14. External forces acting on a uniform beam. The force **F** on the hinge at the left end can be replaced by its vertical and horizontal components F_y and F_x. (b) The three forces on the beam interact at a common point. Note that the force on the beam's left end is not along the direction of the beam.*

obvious from Figure 12-14(b). We see that the lines of action of the three forces on the beam intersect at a single point. Indeed, this result applies whenever an object acted on by three forces is in equilibrium. Here is the proof. Clearly, we can always choose an axis of rotation passing through the point at which the lines of action of two forces intersect; the torques of these forces are then zero about this axis. But if the resultant torque is to be zero, the torque of the third force must be zero relative to the chosen axis. Hence, its line of action must also pass through the intersection point.

Example 12-4.   A billboard painter is standing on a uniform, horizontal platform supported by cords at both ends. See Figure 12-15. The painter, with weight **W**, stands one-quarter of the way from the left end. The platform has weight $\frac{1}{2}W$. The left cord is under tension **T**$_1$ and makes an angle of 60° with the horizontal. The tension T_r in the right cord makes an angle θ with the horizontal. What are T_1, T_r, and θ?

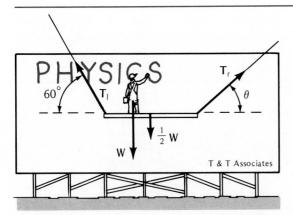

Figure 12-15.

The object whose equilibrium we study is the platform. The forces on it are shown in Figure 12-15. We choose the platform's right end as axis for computing torques in order to eliminate the torque from unknown tension **T**$_r$ at unknown angle θ. The three conditions for equilibrium can then be written in turn as:

$$\Sigma F_x = 0$$

$$-T_1 \cos 60° + T_r \cos \theta = 0 \qquad \text{(A)}$$

$$\Sigma F_y = 0$$

$$T_1 \sin 60° + T_r \sin \theta - W - \tfrac{1}{2}W = 0 \qquad \text{(B)}$$

$$\Sigma \tau_2 = 0$$
$$\text{(right end)}$$

$$(T_1 \sin 60°)L - W(\tfrac{3}{4}L) - (\tfrac{1}{2}W)(\tfrac{1}{2}L) = 0 \qquad \text{(C)}$$

Solving (A), (B), and (C) for the unknowns, we get

$$T_1 = 1.15\ W \qquad T_r = 0.76W \qquad \theta = 41°$$

By the way, if the painter in Figure 12-15 were to step either left or right, the platform would no longer be horizontal, and the tensions would change magnitude and direction.

12-7   Three Proofs (Optional)

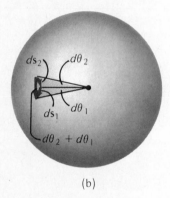

Angular Velocity Is a Vector   Angular velocity, defined as $\omega = d\theta/dt$, is taken to be a *vector* quantity, as in Figure 12-4. A direction and a magnitude can be assigned to ω. But to establish that ω is truly a vector quantity, we must test whether angular velocity obeys the rules of vector addition. Specifically, we must check whether the commutative law, which requires that $\omega_1 + \omega_2 = \omega_2 + \omega_1$, holds for two different angular velocities.

Since angular velocity is the product of a scalar, $1/dt$, and a vector, $d\theta$, we prove that $d\theta$ is a vector by showing that two infinitesimal angular displacements, $d\theta_1$ and $d\theta_2$, obey the commutative law in vector addition:

$$d\theta_1 + d\theta_2 = d\theta_2 + d\theta_1$$

The proof is apparent from an examination of Figures 12-16(a) and (b). Here we see the arcs ds_1 and ds_2 traced out on a spherical surface by a radius vector $\mathbf{r}$ undergoing, in succession, two small angular displacements. The magnitudes of $d\theta_1$ and $d\theta_2$ are proportional to the corresponding circular arcs $ds_1 = r\, d\theta_1$ and $ds_2 = r\, d\theta_2$, respectively. When the angular displacements are very small, these arcs are approximately straight lines on an essentially flat surface. Figure 12-16(a) shows $d\theta_1 + d\theta_2$; Figure 12-16(b) shows $d\theta_2 + d\theta_1$. The resultant angular displacement is the same for the two cases because the resultant linear displacements are the same. *Infinitesimal* angular displacements obey the commutative law.

Finite angular displacements, on the other hand, do not commute. Figures 12-17(a) and (b), in which a die undergoes two successive angular displacements of 90°, indicate that for *finite* angular displacements

$$\theta_1 + \theta_2 \neq \theta_2 + \theta_1$$

Thus angular velocity ω is a vector because $d\theta$ *is* a vector and $1/dt$ is a scalar.

Center of Gravity and Center of Mass   Consider the extended object shown in Figure 12-18 in a uniform gravitational field. Particle of mass m_i at displacement $\mathbf{r}_i$ is subject to a gravitational torque $\tau_i = \mathbf{r}_i \times m_i\mathbf{g}$. The resultant gravitational torque for all particles of the rigid body is then $\Sigma \mathbf{r}_i \times m_i\mathbf{g}$. This

Figure 12-16.   Geometrical demonstration of the vector addition of infinitesimal angular displacements. The angles $d\theta_1$ and $d\theta_2$ are represented by the arcs ds_1 and ds_2 on a sphere. For infinitesimal angular displacements, the arcs obey the commutative law for vector addition, $ds_1 + ds_2 = ds_2 + ds_1$. Therefore, $d\theta_1 + d\theta_2 = d\theta_2 + d\theta_1$.

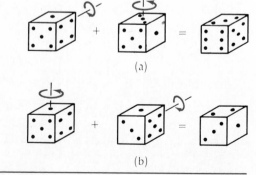

Figure 12-17.   Finite angular displacements—in this case, two 90° displacements of a die—do not obey the commutative law: (a) $\theta_1 + \theta_2$. (b) $\theta_2 + \theta_1$. Note that $\theta_1 + \theta_2 \neq \theta_2 + \theta_1$, and therefore θ_1 and θ_2 are not vectors.

sum must equal, by the definition of center of gravity, the torque produced by weight $M\mathbf{g}$ alone acting at the location $\mathbf{R}_{CG}$ of the center of gravity. Therefore,

$$\mathbf{R}_{CG} \times M\mathbf{g} = \Sigma \mathbf{r}_i \times m_i\mathbf{g}$$

Shifting the location of M and m_i in this equation gives us

$$M\mathbf{R}_{CG} \times \mathbf{g} = \Sigma m_i\mathbf{r}_i \times \mathbf{g}$$

Since $\mathbf{g}$ has the same constant value for all particles, this relation simplifies to

$$M\mathbf{R}_{CG} = \Sigma m_i\mathbf{r}_i$$

But the location $\mathbf{R}_{CM}$ of the center of mass is defined as

$$M\mathbf{R}_{CM} \equiv \Sigma m_i\mathbf{r}_i \qquad (8\text{-}3)$$

so that

$$\mathbf{R}_{CG} = \mathbf{R}_{CM}$$

The center of gravity of a rigid body is at the location of its center of mass if the acceleration due to gravity has the same value for all particles within the rigid body.

Independence of Choice of Axis When a rigid body is in equilibrium, the resultant torque computed relative to the body's center of mass must be zero. Here it is proved that the resultant torque is zero relative to *any* choice of axis of rotation for a rigid body in equilibrium.

Figure 12-19(a) shows a rigid body in equilibrium under the action of three coplanar forces, $\mathbf{F}_1$, $\mathbf{F}_2$, and $\mathbf{F}_3$. The points of application of these forces relative to an axis of rotation through the center of mass are the radius vectors $\mathbf{r}_1$, $\mathbf{r}_2$, and $\mathbf{r}_3$.

The sum of external forces is zero:

$$\mathbf{F}_1 + \mathbf{F}_2 + \mathbf{F}_3 = 0$$

The sum of the external torques about the center of mass is zero:

$$(\mathbf{r}_1 \times \mathbf{F}_1) + (\mathbf{r}_2 \times \mathbf{F}_2) + (\mathbf{r}_3 \times \mathbf{F}_3) = 0$$

Now we compute the torques about a second axis. From Figure 12-19(b) we see that the vector $\mathbf{d}_1$ from the new origin to the point of application of $\mathbf{F}_1$ is given by

$$\mathbf{d}_1 = \mathbf{r}_1 + \mathbf{R}$$

where vector $\mathbf{R}$ represents the location of the center of mass relative to the new origin. Similar relations give $\mathbf{d}_2$ and $\mathbf{d}_3$.

Substituting these equations in the equation above leads to

$$(\mathbf{d}_1 - \mathbf{R}) \times \mathbf{F}_1 + (\mathbf{d}_2 - \mathbf{R}) \times \mathbf{F}_2 + (\mathbf{d}_3 - \mathbf{R}) \times \mathbf{F}_3 = 0$$

Regrouping these terms yields

$$[(\mathbf{d}_1 \times \mathbf{F}_1) + (\mathbf{d}_2 \times \mathbf{F}_2) + (\mathbf{d}_3 \times \mathbf{F}_3)] - \mathbf{R} \times (\mathbf{F}_1 + \mathbf{F}_2 + \mathbf{F}_3) = 0$$

The left-hand term represents the resultant torque about the arbitrarily chosen, new axis of rotation. Within the parentheses of the right-hand term is the sum of the external forces, which is zero by the first equation. The left-hand bracket

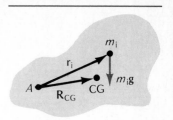

Figure 12-18.

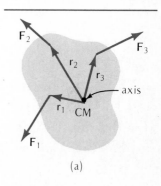

(a)

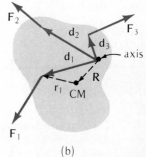

(b)

Figure 12-19. (a) A rigid body in equilibrium under the action of the three forces $\mathbf{F}_1$, $\mathbf{F}_2$, and $\mathbf{F}_3$ applied at the radius vectors $\mathbf{r}_1$, $\mathbf{r}_2$, and $\mathbf{r}_3$, respectively; the origins of the latter are at the object's center of mass. (b) The same object and forces as in (a) except that now the radius vectors $\mathbf{d}_1$, $\mathbf{d}_2$, and $\mathbf{d}_3$ give the points of application of the forces relative to a second rotation axis. The displacement of the center of mass relative to the new origin is $\mathbf{R}$.

must then also be zero. Therefore, when a rigid body is in equilibrium,

$$(\mathbf{d}_1 \times \mathbf{F}_1) + (\mathbf{d}_2 \times \mathbf{F}_2) + (\mathbf{d}_3 \times \mathbf{F}_3) = 0$$

We have proved that for a rigid body in equilibrium, the rotation axis can be chosen *anywhere*.

Summary

Definitions

Right-hand rule for rotation (for example, angular velocity vector ω): curled fingers give rotation sense; outstretched thumb gives associated direction in space.

Cross product $\mathbf{C}$ of vectors $\mathbf{A}$ and $\mathbf{B}$:

$$\mathbf{C} \equiv \mathbf{A} \times \mathbf{B} \qquad (12\text{-}1)$$

Direction of $\mathbf{C}$: right-hand rule.

Magnitude of $\mathbf{C}$:

$$C = AB \sin\theta = AB_\perp = A_\perp B \quad (12\text{-}2), (12\text{-}3)$$

where θ is the smaller angle between $\mathbf{A}$ and $\mathbf{B}$.

Torque of force $\mathbf{F}$ relative to the origin of radius vector $\mathbf{r}$:

$$\tau = \mathbf{r} \times \mathbf{F} \qquad (12\text{-}5)$$

$$\tau = rF \sin\theta = Fr_\perp = rF_\perp \quad (12\text{-}4), (12\text{-}6)$$

The moment arm is $r_\perp$.

Center of gravity: (a) point relative to which the sum of the gravitational torques on the object is zero, or (b) the point at which the object's weight may be imagined to be concentrated. For objects of ordinary size, the center of gravity coincides with the center of mass.

Units

Torque: $m \cdot N$

Important Results

Conditions for the equilibrium of a rigid body:

Translation $\Sigma\mathbf{F} = 0$

Rotation $\Sigma\tau = 0$ about *any* axis

When the forces lie entirely in the xy plane, then $\Sigma F_x = 0$, $\Sigma F_y = 0$, $\Sigma\tau_z = 0$.

Problems and Questions

Section 12-1   Definitions

· **12-1 Q** When a rigid body rotates, we can be sure that every point on the object has the same
(A) velocity.
(B) angular velocity.
(C) centripetal acceleration.
(D) speed.
(E) tangential acceleration.

· **12-2 Q** If while riding your bike forward you could look down and see the angular velocity vector of the wheels, it would be
(A) pointing straight ahead.
(B) pointing straight back.
(C) pointing to your left.
(D) pointing to your right.
(E) rotating and hence not pointing in any one direction.

Section 12-3   The Cross (or Vector) Product

· **12-3 Q** (a) What do you get when you cross a downhill skier with a mountain climber? (Answer: You can't cross a downhill skier with a mountain climber. A mountain

climber is a scaler.) (b) What do you get when you cross an apple with an orange? [Answer: (Apple)(Orange) sin θ.]

· **12-4 Q** If $\mathbf{A}$ and $\mathbf{B}$ form two sides of a parallelogram, show that the magnitude of $\mathbf{A} \times \mathbf{B}$ is equal to the area of the parallelogram.

⋮ **12-5 P** Vectors $\mathbf{A}$, $\mathbf{B}$, and $\mathbf{C}$ are three sides of a parallelepiped. (a) Show that the volume of the parallelepiped is $\mathbf{A} \cdot (\mathbf{B} \times \mathbf{C})$. (b) Using the result of part (a), prove that $\mathbf{A} \cdot (\mathbf{B} \times \mathbf{C}) = \mathbf{B} \cdot (\mathbf{C} \times \mathbf{A}) = \mathbf{C} \cdot (\mathbf{A} \times \mathbf{B})$

Section 12-4   Torque

· **12-6 Q** The mechanical advantage of a machine is defined as the ratio of the "input" force to the "output" force. The force that can be exerted by the biceps muscle can be measured by means of the apparatus shown here. A spring scale measures the force. For the two situations shown in Figure 12-20,
(A) the mechanical advantage of the biceps is greatest in part (a).
(B) the mechanical advantage of the biceps muscle is greatest in part (b).

(C) the mechanical advantage of the biceps is the same in parts (a) and (b).
(D) the mechanical advantage of the biceps depends on the force exerted on the spring.
(E) None of the above statements is true.

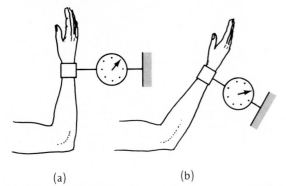

(a)   (b)

Figure 12-20.   Question 12-6.

· **12-7 P** A logger tries to dislodge a jammed log by pulling on it with a chain attached to a bulldozer. If the bulldozer exerts a force of 3000 N, what is the magnitude and direction of the torque about point P when the chain is pulled in each of the directions indicated in Figure 12-21?

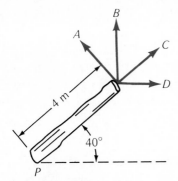

Figure 12-21.   Problem 12-7.

: **12-8 Q** A spool on which some thread has been wrapped is placed on a rough surface, and the thread is pulled gently in the direction shown in Figure 12-22. If F_N = normal force exerted by the floor, F_F = friction force, and mg = spool weight, we can deduce that
(A) $T \sin \theta \leq F_F$
(B) $T \cos \theta \leq F_F$

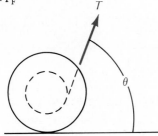

Figure 12-22.   Question 12-8.

(C) $F_N = mg$
(D) $T \sin \theta = mg$
(E) $F_F + mg = F_N$

: **12-9 Q** A billiard ball sits on a table with a rough sandpaper surface, so that there is a high coefficient of friction. Then a cue delivers a blow to the top of the ball (see Figure 12-23). During the time the cue is in contact with the ball, it is true that
(A) there is a net force to the left on the ball, but no resultant torque about the ball's center.
(B) there is a resultant torque about the ball's center, but the resultant force on the ball is zero.
(C) there is a net force on the ball to the left and a resultant torque about the ball's center to the left.
(D) there is a net force to the left and a resultant torque about the ball's center out of the paper.
(E) there is a net force to the left and a resultant torque about the ball's center into the paper.

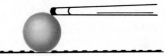

Figure 12-23.   Question 12-9.

: **12-10 P** A very large load can be controlled by snubbing a line around a post, as sketched in Figure 12-24. This technique is used widely in ships, in logging, and in belaying in mountaineering. If the coefficient of friction between the rope and the post is 0.5, what force **F** must be exerted to support a load $\mathbf{F}_L = 4000$ N? Assume the rope is wrapped four times around the post.

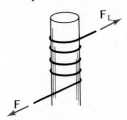

Figure 12-24.   Problem 12-10.

: **12-11 P** Two identical masses are joined by a light rod of

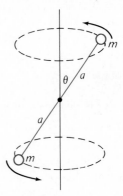

Figure 12-25.   Problem 12-11.

length 2*a*. The rod rotates with constant angular velocity *ω* about an axis that passes through the midpoint of the rod and makes an angle *θ* with the rod. What is the torque about the midpoint of the rod?

Section 12-5   Center of Gravity

· **12-12 Q** Why are people, trees, and animals approximately symmetric?

: **12-13 Q** Which of the following is an accurate statement?
(A) The center of gravity of a solid body necessarily lies within the body.
(B) A system consists of two objects whose centers of gravity are at points P_1 and P_2. The center of mass of the system thus lies halfway between P_1 and P_2 on the line joining them.
(C) It is possible for a high jumper to pass over a bar while his center of gravity passes below the bar.
(D) The center of gravity of the earth-moon system is closer to the moon than to the earth.
(E) At the moment an shell explodes in midair, the center of gravity of the shell begins to fall straight down, like a point mass falling under the influence of gravity.

: **12-14 Q** A teeter-totter made of a heavy plank is unbalanced when no one is on it. Suppose two children climb on and balance themselves. Each now moves halfway in to the fulcrum. What will happen?
(A) The short end will rise.
(B) The long end will rise.
(C) The teeter totter will remain in balance.
(D) The end with the lighter child will rise.
(E) The end with the heavier child will rise.

: **12-15 P** Suppose that a meter stick of mass 100 grams rests horizontally on your two index fingers. Your right index finger is under the 5 cm mark and your left index finger is under the 80 cm mark. The coefficient of static friction is 0.5, and the coefficient of kinetic friction is 0.4, between the stick and your fingers.

(*a*) What upward force are you exerting with each finger? (*b*) Suppose you now move your hands together. How far will the stick slide on your hand (or hands) before it first stops? Try this for yourself to see how it works.

Section 12-6   Conditions for Rotational Equilibrium

: **12-16 Q** Two people, Elaine and Robert, are carrying a log as shown in Figure 12-26. If someone were to come along and cut a piece off the end of the log at point *P*, the load supported by Robert would
(A) not change.
(B) decrease.
(C) increase.
(D) increase or decrease, depending on the log's weight.

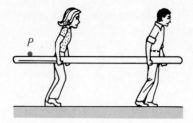

Figure 12-26.   Question 12-16.

: **12-17 Q** Each of the objects shown in Figure 12-27 has equal uniform thickness perpendicular to the plane of the paper. Each is of homogeneous composition, but they are of different materials. Assuming that no slipping occurs, which object will be most stable as the incline angle *θ* of the plane is increased?
(A) A
(B) B
(C) C
(D) D
(E) E
(F) F

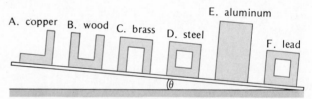

Figure 12-27.   Question 12-17.

: **12-18 P** Meter sticks *A* and *B*, each of mass 100 gm, are glued rigidly together at their 100 cm marks, with an angle of 120° between them. See Figure 12-28. The two meter sticks lie in a vertical plane. A pivot at the junction allows them to rotate in this vertical plane. What point mass must be attached to stick *B* at the 0 cm mark to maintain rotational equilibrium with *A* horizontal?

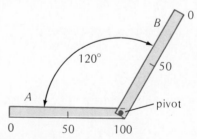

Figure 12-28.   Problem 12-18.

· **12-19 P** The location of a person's center of gravity can be determined by weighing him on two scales, as shown in

Figure 12-29. What is the distance x from the person's feet to his center of gravity for the scale reading shown?

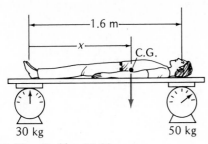

Figure 12-29.   Problem 12-19.

· **12-20 P** A woman is seated on a chair with her legs straight down in the normal position. Show that she cannot stand up unless she first leans her torso forward.

· **12-21 P** What force must be exerted by the biceps muscle to support the 50 N weight shown in Figure 12-30? The mass of the arm is 2 kg and its center of mass is 20 cm from the elbow pivot.

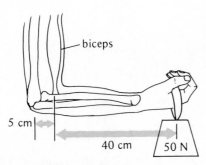

Figure 12-30.   Problem 12-21.

· **12-22 P** A uniform plank of mass 40 kg and length 8 m projects 3 m out over a cliff. How far out on the plank can a boy of mass 20 kg walk before the plank starts to tip? See Figure 12-31.

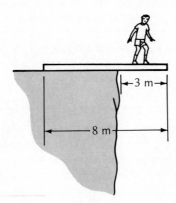

Figure 12-31.   Problem 12-22.

· **12-23 P** Two people, A and B, are carrying a uniform log

6 meters long. The log weighs 600 N. What force must each exert if they are positioned as shown in Figure 12-32?

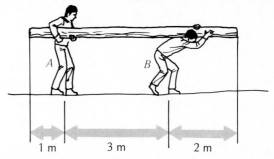

Figure 12-32.   Problem 12-23.

· **12-24 P** What minimum force must be exerted on the crank handle of the windlass shown in Figure 12-33 in order to lift a load of 100 kg? The length of the crank is 30 cm and the diameter of the shaft is 6 cm.

Figure 12-33.   Problem 12-24.

: **12-25 P** (a) A refrigerator packing crate measures 80 cm × 80 cm × 180 cm, and its center of gravity is at its geometrical center. It is placed on a loading ramp as shown in Figure 12-34(a). Assuming it does not slide, what is the maximum angle of incline θ the ramp can have without the crate tipping over?

(b) A workman tries to move the crate on a handtruck tipped back 30° from vertical. The crate's mass is 150 kg and the man holds the handtruck handles 1 meter from the wheels. See Figure 12-34(b). What is the minimum force he can exert and still support the crate? In what direction must he exert this force?

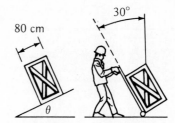

Figure 12-34.   Problem 12-25.

: **12-26 P** Bricks of length L and thickness t are stacked up with their sides aligned as shown in Figure 12-35. Each brick extends out over the brick below by a distance L/n, where n is an integer. How many bricks can be stacked up in this way, counting the bottom brick?

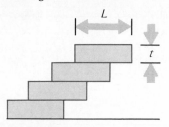

Figure 12-35.   Problem 12-26.

: **12-27 P** A mountain climber of 600 N weight is rapelling on a cliff, as shown in Figure 12-36. What are (*a*) the tension in his rope and (*b*) the direction and magnitude of the force exerted on his boots by the wall? His center of gravity is 1.2 m from the wall, and his shoulders are pivoted 1.5 m from the wall.

Figure 12-36.   Problem 12-27.

: **12-28 P** A woman tries to roll a barrel of oil up over a curb by pushing on it horizontally with a force **F** applied in line with the center of the barrel. See Figure 12-37. What force must she exert if the barrel has a mass of 220 kg and a radius of 0.3 m, and the curb is 10 cm high?

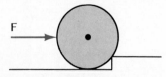

Figure 12-37.   Problem 12-28.

: **12-29 P** A ladder of mass 20 kg leans against a wall, as shown in Figure 12-38. The coefficient of static friction between the ladder and the wall and between the ladder and the floor is 0.4. What is the minimum value of the angle

θ if the ladder is not to slip when a 50 kg person climbs all the way to the top of the ladder?

Figure 12-38.   Problem 12-29.

: **12-30 P** A block of limestone weighs 1600 N. Its base is 30 cm × 40 cm, and it is 50 cm tall. What is the least force a stonemason can exert on it in order to tip it over? In what direction and where on the block should the force be applied?

: **12-31 P** The gear mechanism shown in Figure 12-39 is used to transfer a torque from shaft A to shaft B. The large gear has a radius of 6 cm and the small gear has a radius of 1.5 cm. What is the ratio of the torque applied to shaft A to that applied to shaft B?

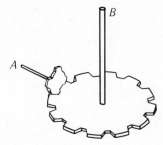

Figure 12-39.   Problem 12-31.

: **12-32 P** The center of gravity of a sports car is 50 cm above the ground and the wheelbase (measured to the outside of the tires) is 1.50 m. At what maximum speed can such a car take a flat (unbanked) curve of 20 m radius without tipping over?

: **12-33 P** Four identical bricks, each 24 cm long and 5 cm thick, are to be stacked one on top of the other so that the top brick projects as far as possible beyond the bottom brick. How far can it project beyond the bottom brick?

: **12-34 P** In the construction of a bridge over the Snake River, structures are cantilevered out from each bank, eventually being joined at the midpoint of the bridge. Suppose the structure under construction is represented by the model sketched in Figure 12-40. The uniform bridge of mass 8,000 kg is effectively pivoted at point P and supported by a cable. The cable will support a mass of

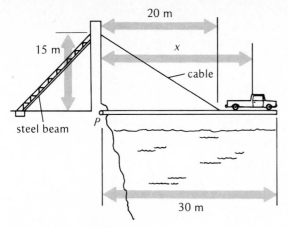

Figure 12-40.   *Problem 12-34.*

14,000 kg without breaking. How far out on the bridge would it be safe to drive a 2,000-kg truck?

: 12-35 P By pressing down with a force of 40 N, the person shown in Figure 12-41 is able to hold a uniform board weighing 20 N. His hands are separated by 30 cm. He weighs 600 N. What is the length of the board?

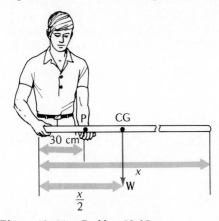

Figure 12-41.   *Problem 12-35.*

: 12-36 P A boom and cable on the deck of a merchant ship are used to support a load. What are (a) the tensions in the cables and (b) the compressional force exerted by the

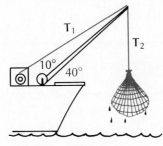

Figure 12-42.   *Problem 12-36.*

boom when a load of 10,000 N is supported? The weight of the boom is negligible. (Consider the cable in Figure 12-42 to be fixed to the end of the boom, so that the tensions in the two segments of cable are not necessarily the same.)

: 12-37 P Three power lines, running parallel to the ground, are attached to a power pole supported by a guy wire. See Figure 12-43. Determine the tension in the guy wire in terms of the parameters shown. The tension in each power line is T_p.

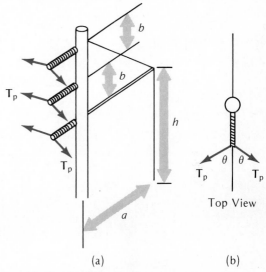

(a)   (b)

Figure 12-43.   *Problem 12-37.*

: 12-38 P A shipping container filled with wheat has a total weight W. In an attempt to tip it, a rope is attached as shown in Figure 12-44 and pulled horizontally. With what minimum force must the rope be pulled in order just to tip the crate? (Assume no sliding occurs.)

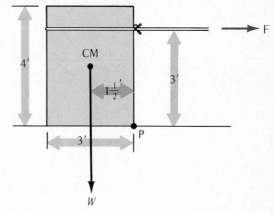

Figure 12-44.   *Problem 12-38.*

: 12-39 P A 50°–60°–70° triangle is cut from a uniform

sheet of wood and placed on edge on an inclined plane such that the edge is along the line of steepest descent, as shown in Figure 12-45. What is the angle θ of the most steeply inclined plane on which the triangle can be placed without tipping over?

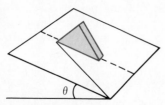

Figure 12-45.   *Problem 12-39.*

12-40 P A thin walled tube of weight W and inner diameter $3R$ stands upright on the table. Two identical balls, each of radius R, are dropped into the tube. What is the maximum weight of a ball if the tube is not to tip over?

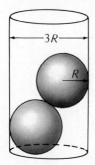

Figure 12-46.   *Problem 12-40.*

12-41 P A rod of length L is attached to a string of length L, which is fastened to a rough wall. If the rod is propped against the wall as shown in Figure 12-47, it will remain in place only if θ is greater than some critical angle. What is the critical angle if the coefficient of static friction between the wall and the rod is μ_s?

12-42 P A billiard ball of weight W and radius R sits on a table where the coefficient of kinetic friction is μ. The ball is

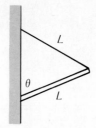

Figure 12-47.   *Problem 12-41.*

struck with a pool cue held parallel to the table, with a force F. See Figure 12-48. At what height above the table must the force be applied if the ball is to slide without rolling?

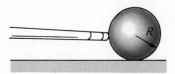

Figure 12-48.   *Problem 12-42.*

12-43 P A person pushes horizontally on the edge of a square table with constant force F. The table weighs 60 N and is 70 cm high; its center of gravity is 60 cm above the floor. The legs are at the corners of a square of side 1 meter.

(a) If the coefficient of static friction between the legs and the floor is 0.3, with what force must the person push to start moving the table? (b) What weight is supported by each leg?

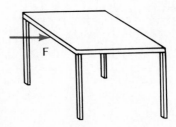

Figure 12-49.   *Problem 12-43.*

Rotational Dynamics

13

In describing rotational motion and dynamics, we exploit an enormously helpful fact—*corresponding translational and rotational relations are very closely analogous.* We do not have to learn a whole set of totally new relations for rotation. Knowing the translational relation allows the rotational relation to be written at once. Corresponding to $\mathbf{F} = m\mathbf{a}$, for example, is the rotational relation $\tau = I\alpha$. Torque τ is the rotational analog of force $\mathbf{F}$, angular acceleration α is the rotational analog of linear acceleration $\mathbf{a}$, and the quantity I is the rotational analog of inertial mass m.

13-1   Rotational Kinematics

For simplicity, all rotations to be considered in this chapter will be about axes perpendicular to the plane of the paper. Counterclockwise angular displacements are positive. Then it is sufficient to consider the angular velocity as an algebraic quantity, without concern for its vector properties (Section 12-1).

All particles of a perfectly rigid body undergoing rotation have the same angular velocity ω. By definition

$$\omega \equiv \frac{d\theta}{dt} \qquad \text{(4-13), (13-1)}$$

Angular velocity ω is the time rate of change of angular displacement θ. Positive ω represents counterclockwise rotation. Appropriate units for ω are radians per second ($\text{rad/s} = \text{s}^{-1}$).

If the rate of rotation of a rotating object changes, the object undergoes an angular acceleration. The instantaneous angular acceleration α is defined as the time rate of change of angular velocity:

$$\alpha \equiv \frac{d\omega}{dt} \tag{13-2}$$

Positive angular acceleration implies that the object's rotational velocity is increasing in the counterclockwise sense (or equivalently, decreasing in the clockwise sense). For $\alpha = 0$, the body rotates at constant ω. Angular acceleration is expressed as radians per second squared (rad/s^2), so that its fundamental units are s^{-2}.

Equations (13-1) and (13-2), which relate angular velocity ω and angular acceleration α to angular displacement θ and time t, are exactly analogous to the relations between the linear displacement x, linear velocity v, linear acceleration a, and time t. Therefore, to arrive at the kinematic equations describing rotational motion at a constant angular acceleration is easy. We simply replace the linear quantity by the corresponding angular quantity. More specifically, we replace.

$$x \qquad \text{by} \qquad \theta$$

$$v = \frac{dx}{dt} \qquad \text{by} \qquad \omega = \frac{d\theta}{dt}$$

$$a = \frac{dv}{dt} \qquad \text{by} \qquad \alpha = \frac{d\omega}{dt}$$

in the kinematic equations for constant linear acceleration, as shown in the accompanying chart.

CONSTANT LINEAR ACCELERATION a		CONSTANT ANGULAR ACCELERATION α	
$v = v_0 + at$	(3-6)	$\omega = \omega_0 + \alpha t$	(13-3)
$x - x_0 = v_0 t + \frac{1}{2}at^2$	(3-9)	$\theta - \theta_0 = \omega_0 t + \frac{1}{2}\alpha t^2$	(13-4)
$x - x_0 = \dfrac{(v_0 + v)t}{2}$	(3-8)	$\theta - \theta_0 = \dfrac{(\omega_0 + \omega)t}{2}$	(13-5)
$x - x_0 = \dfrac{v^2 - v_0^2}{2a}$	(3-10)	$\theta - \theta_0 = \dfrac{\omega^2 - \omega_0^2}{2\alpha}$	(13-6)

Here the values of angular displacement and angular velocity at time $t = 0$ are designated θ_0 and ω_0, respectively.

At any instant, all particles of a rigid body have the same angular displacement θ (relative to that at $t = 0$), the same ω, and the same α. But the linear speeds and accelerations of various particles in a rotating rigid body (each particle in a circular arc) depend on how far each particle is from the rotation axis.

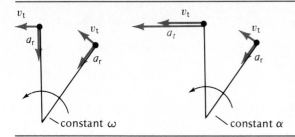

Figure 13-1. *(a) Motion at constant angular velocity ω. The magnitudes of tangential speed v_t and radial acceleration a_r are constant. (b) Constant angular acceleration, so that v_t and tangential acceleration a_t increase in magnitude.*

As we have already seen (Section 4-5), a particle a distance r from the rotation axis has a tangential speed v_t around a circular arc of

$$v_t = r\omega \qquad\qquad \text{(4-14), (13-7)}$$

If a circling particle's angular speed ω changes with time, so does its tangential linear speed v_t, and we can find the particle's tangential acceleration a_t simply by taking the time derivative of (13-7) with r constant:

$$\frac{dv_t}{dt} = r\,\frac{d\omega}{dt}$$

or

$$a_t = r\alpha \qquad\qquad \text{(13-8)}$$

Every particle traveling in a circle has an acceleration, even when its linear speed v_t and angular speed ω are constant. [See Figure 13-1(a).] This *radially inward*, or *centripetal*, acceleration a_r has the magnitude

$$a_r = \frac{v_t^2}{r} = \omega^2 r \qquad\qquad \text{(4-15)}$$

If a particle's linear speed around the circle is changing, that is, if there is an angular acceleration [see Figure 13-1(b)], then there will be a *tangential* acceleration component as well as the ever-present radial acceleration component.

Example 13-1.   A bicycle wheel spins initially about a fixed axis at 60 rpm. It is brought to rest in 10 s. (a) What is the wheel's angular acceleration, assumed to be constant? (b) How many turns does the wheel make in coming to rest?

(a) We have, from (13-3),

$$\omega = \omega_0 + \alpha t$$

Here we have $\omega_0 = 60$ rpm $= (60 \text{ rot/min})(2\pi \text{ rad/rot})(1 \text{ min/60 s}) = 2\pi \text{ s}^{-1}$, and $\omega = 0$ at $t = 10$ s.

$$\alpha = -\frac{\omega_0}{t} = -\frac{2\pi \text{ s}^{-1}}{10 \text{ s}} = -\frac{1}{5}\pi \text{ s}^{-2}$$

The minus sign for α, the opposite of the sign for ω_0, indicates that the spin rate is decreasing.

(b) To find the total angular displacement θ, we use (13-5), taking θ_0 to be zero:

$$\theta - \theta_0 = \frac{(\omega_0 + \omega)t}{2} = \frac{1}{2}(2\pi \text{ s}^{-1})(10 \text{ s})\left(\frac{1 \text{ turn}}{2\pi \text{ rad}}\right) = 5.0 \text{ turns}$$

13-2 Rotational Kinetic Energy, Work, and Power

Recall how the dynamics of a single particle was developed. We first considered Newton's laws (Chapter 5 and 6) and from them derived the work-energy theorem (Chapter 9). For rotational dynamics, it is easier to use the reverse sequence. Therefore, we shall first consider kinetic energy and work, as applied to a rotating rigid body, and then later derive Newton's second law for rotation from the work-energy relation.

First, some restrictions. The orientation of the rotation axis is fixed. The direction of torques applied to the rigid body is parallel (or antiparallel) to this rotation axis. More specifically, the rotation axis is perpendicular to the plane of the paper and the external forces produce counterclockwise or clockwise torques about this axis.

In Figure 13-2, we see a rigid body rotating about a fixed axis at instantaneous angular velocity ω. The ith particle has mass m_i and speed v_i and travels in a circular arc of radius r_i. The kinetic energy of this single particle is $K_i = \frac{1}{2}m_i v_i^2$. The total kinetic energy K for all particles in the rigid body is

$$K = \Sigma K_i = \Sigma \tfrac{1}{2} m_i v_i^2$$

Now we recognize that all particles in a rigid body have the same angular velocity ω. When we use (13-7), $v_i = \omega r_i$ in the relation above, we get

$$K = \Sigma \tfrac{1}{2} m_i (\omega r_i)^2 = \tfrac{1}{2}(\Sigma m_i r_i^2)\omega^2$$

The quantity $\Sigma m_i r_i^2$ appears repeatedly in rotational dynamics. It is called the object's *moment of inertia I*. By definition,

$$I \equiv \Sigma m_i r_i^2 \tag{13-9}$$

Then the kinetic energy of a rotating rigid body can be written

$$K = \tfrac{1}{2} I \omega^2 \tag{13-10}$$

This relation for the kinetic energy of a spinning object is analogous to the one for the kinetic energy of a single particle, $\frac{1}{2}mv^2$. Angular velocity ω replaces linear velocity v and moment of inertia I replaces mass m.* (This is another relation between analogous translational and rotational quantities; a comprehensive listing is given in Table 13-1.)

Clearly, a rigid body's moment of inertia, which is analogous to the mass of a single particle, is central to rotational dynamics. Basic properties of the moment of inertia are these:

- I is a property of a rigid body that depends not merely on the object's total mass but on how the mass is distributed relative to the rotation axis. To the degree that the mass is far from the axis (large r_i), moment of inertia I, or $\Sigma m_i r_i^2$, is large.
- I depends on the choice of the axis of rotation. Since the moment of inertia $m_i r_i^2$ of each individual mass particle of the rigid body depends on the

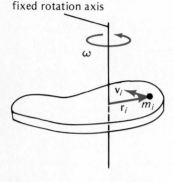

Figure 13-2. Representative particle i in a rotating rigid body.

fixed rotation axis

ω

v_i

r_i m_i

---

* It should not be thought that rotational kinetic energy written $\frac{1}{2}I\omega^2$ is somehow different from kinetic energy written as $\frac{1}{2}mv^2$. It's the same old thing. As the derivation above shows, we write rotational kinetic energy as $\frac{1}{2}I\omega^2$ rather than as $\Sigma\frac{1}{2}m_i v_i^2$, to exploit the fact that all particles have the same ω.

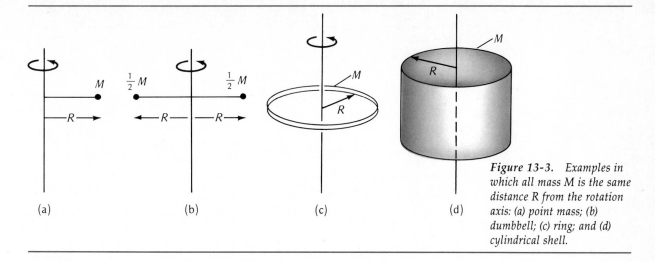

Figure 13-3.   Examples in which all mass M is the same distance R from the rotation axis: (a) point mass; (b) dumbbell; (c) ring; and (d) cylindrical shell.

choice of the axis, $I = \Sigma m_i r_i^2$ also depends on the axis. Therefore, one must specify the rotation axis (its location and orientation) relative to which I is measured.

The simplest mass distribution for a rotating object is that in which all mass is at the same distance R from the rotation axis. Examples (see Figure 13-3) are (a) a point mass m rotating in a circle of radius R about a perpendicular axis; (b) a dumbbell with two masses, each of $\frac{1}{2}M$, at a distance R from the midpoint, rotating about a perpendicular axis; (c) a ring of radius R rotating about its symmetry axis; and (d) a cylindrical shell of radius R, also rotating about the symmetry axis. For all four shapes, $r_i = R$ for every m_i. Therefore, $I = \Sigma m_i r_i^2 = \Sigma m_i R^2 = MR^2$, where M is the object's total mass.

Here we relate the rotational kinetic energy to the work done on a rigid body.* Figure 13-4 shows external force $\mathbf{F}$ applied at a distance r from the axis of rotation of a rigid body. ($\mathbf{F}$ lies in the plane of the paper; the direction of the torque $\boldsymbol{\tau} = \mathbf{r} \times \mathbf{F}$ is along the rotation axis.) We replace $\mathbf{F}$ by its components, $\mathbf{F}_\perp$ and $\mathbf{F}_\parallel$, which are respectively perpendicular and parallel to radius vector $\mathbf{r}$. The point on the rigid body at which force $\mathbf{F}$ is applied moves a distance ds along a circular arc of radius r as the body turns through an angle $d\theta$, where $ds = r\, d\theta$. Force component $\mathbf{F}_\parallel$ is always *perpendicular to the displacement* $d\mathbf{s}$; this component $\mathbf{F}_\parallel$ does no work. The force component $\mathbf{F}_\perp$ does work dW in the amount

$$dW = \mathbf{F} \cdot d\mathbf{s} = F_\perp\, ds = F_\perp r\, d\theta$$

But the torque $\boldsymbol{\tau}$ has magnitude $\tau = rF_\perp$. Therefore, the work done by the torque can be written

$$dW = \tau_{\text{ext}}\, d\theta \tag{13-11}$$

If several external torques act simultaneously on the body, then τ_{ext} represents

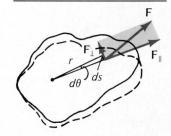

Figure 13-4.   A rigid body undergoing an angular displcement $d\theta$ under the influence of force $\mathbf{F}$ applied at $\mathbf{r}$. The perpendicular and parallel components of $\mathbf{F}$ are $\mathbf{F}_\perp$ and $\mathbf{F}_\parallel$.

* A reasonable guess for the work dW done by torque τ as it turns a rigid body through an angle $d\theta$ would be $dW = \tau\, d\theta$. This follows simply from $dW = F\, ds$ and the fact that torque τ is the rotational analog of force F and angular displacement $d\theta$ is the rotational analog of linear displacement ds. As we shall see, the guess is correct.

their sum. The work done is positive when the object rotates in the same sense as the resultant torque and the rotational kinetic energy increases. On the other hand, for a braking torque, the work done is negative and rotational kinetic energy decreases.

The general work-energy theorem (Section 9-2) applied to the work done by a resultant torque on a rigid body is*

$$W_{i \to f} = K_f - K_i$$

$$\int_i^f \tau_{\text{ext}} \, d\theta = \tfrac{1}{2} I \omega_f^2 - \tfrac{1}{2} I \omega_i^2 \tag{13-12}$$

Since the work dW done by torque τ is $dW = \tau \, d\theta$, the rate at which the torque does work, dW/dt, or power is†

$$\frac{dW}{dt} = \tau \frac{d\theta}{dt}$$

$$P = \tau \omega \tag{13-13}$$

Equation 13-13 gives the instantaneous rate at which energy is transferred to a rigid body rotating at angular speed ω by the instantaneous resultant torque τ. For example, an engine producing a torque of 120 m·N at an angular speed of $6000 \text{ rpm} = 200\pi \text{ rad/s}$ delivers energy at the rate $P = \tau \omega = (120 \text{ m·N})(200\pi \text{ rad/s}) = 75 \text{ kW} = 100 \text{ hp}$.

The general energy conservation principle applies to rotation; if the forces within a system are conservative, the total of kinetic energy (written in any form) and potential energy is constant.

Example 13-2.   See Figure 13-5. A wheel of radius R has nearly all its mass M at its outer edge. The wheel rotates about a fixed horizontal axis. Mass m is attached to the end of a rope that is wrapped around and attached to the wheel. Mass m is released from rest and drops a distance y before hitting a floor. (a) With what speed does m hit the floor (in terms of m, M, g, and y)? (b) What is the angular velocity of the wheel at the instant of impact?

(a) The typical idealized simplifying assumptions are made: no friction at the axle; a massless, perfectly flexible, stretchless cord; no slipping of the cord on the wheel; all the wheel's mass at its circumference. In applying the energy conservation principle, we designate the initial state by 1 and the final one (when m hits the floor) by 2. Then we have

$$K_1 + U_1 = K_2 + U_2$$

$$(0 + 0) + (mgy) = (\tfrac{1}{2} m v^2 + \tfrac{1}{2} I \omega^2) + 0$$

This relation says simply that as the system's gravitational potential energy decreases, the total kinetic energy of mass m and the rotating wheel increases by the same amount. The speed v of mass m is the same as the speed ωR of any point on the wheel relative to its center:

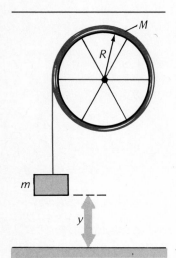

Figure 13-5.

---

* The work done by torque τ may be expressed in more general mathematical form using the dot product, as $\int \boldsymbol{\tau} \cdot d\boldsymbol{\theta}$, where the infinitesimal angular displacement $d\boldsymbol{\theta}$ is taken to be a *vector* whose direction is associated with the rotation sense in the same fashion as in the right-hand rule for angular velocity $\boldsymbol{\omega}$.

† A more general form of (13-13), using the dot product, is

$$P = \boldsymbol{\tau} \cdot \boldsymbol{\omega}$$

$$v = \omega R$$

Further, here we have

$$I = MR^2$$

The equation above can then be rewritten as

$$mgy = \frac{1}{2} mv^2 + \frac{1}{2} (MR^2) \left(\frac{v}{R} \right)^2$$

so that

$$v = \sqrt{\frac{2mgy}{(m + M)}}$$

(b) The wheel's angular speed is

$$\omega = \frac{v}{R} = \frac{1}{R} \sqrt{\frac{2mgy}{m + M}}$$

13-3 Moment-of-Inertia Computations

By definition, an object's moment of inertia I is

$$I \equiv \Sigma m_i r_i^2 \qquad \text{(13-9)}$$

The value for I is the sum of contributions from each particle of mass m_i at a distance r_i from the rigid body's rotation axis. Clearly, for a single object, the value of I depends on choice of axis.

For a continuous distribution of mass in a solid, we imagine the mass to be subdivided into a collection of effectively point masses dm, so that when we sum over all contributions, we have

$$I = \int r^2 \, dm$$

The mass per unit volume of the solid (density) is ρ. So we can write $dm = \rho \, dV$, where dV is the volume element containing mass dm. The relation for moment of inertia can then be written

$$I = \int r^2 \, dm = \int r^2 \rho \, dV = \rho \int r^2 \, dV \qquad \text{(13-14)}$$

The last step here, in which ρ was moved outside the integral sign, can be done only when the density has the same constant value throughout.

A term closely related to moment of inertia is *radius of gyration k*. By definition, k is that distance from the rotation axis at which all the object's mass M would have to be concentrated to yield the same value for I as the actual mass distribution. Clearly, with $I = \Sigma m_i r_i^2$ in general, the radius of gyration k is defined by

$$I \equiv Mk^2 \qquad \text{(13-15)}$$

A useful general theorem relates the moment of inertia I_{CM} of any object of mass M about an axis passing through its center of mass with the moment of inertia I about any other axis parallel to the first and a distance p away from it.

See Figure 13-6. The so-called *parallel-axis theorem* (proved in detail in Example 13-6) is this:

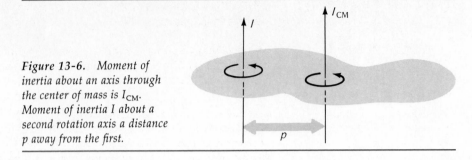

Figure 13-6. *Moment of inertia about an axis through the center of mass is I_{CM}. Moment of inertia I about a second rotation axis a distance p away from the first.*

$$I = I_{CM} + Mp^2 \tag{13-16}$$

Two consequences of (13-16) are these:

- If the moment of inertia I_{CM} is known — it is often relatively easy to calculate I_{CM} for a geometrically symmetrical mass distribution — then, using (13-16), it is very easy to calculate the moment of inertia I about a parallel axis.
- When a rigid body rotates about an axis passing through its center of mass, its moment of inertia has the minimum value when compared with the

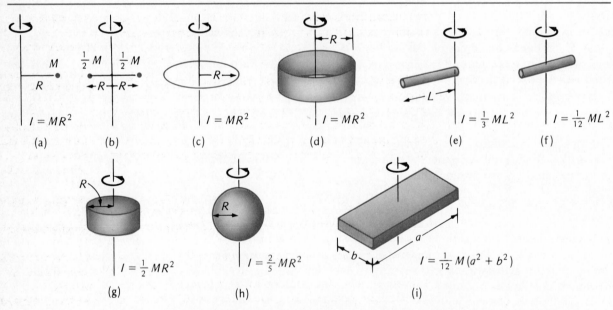

Figure 13-7. *Moments of inertia for several simple shapes, all of constant density: (a) a point particle; (b) a dumbbell about an axis through its center and perpendicular to the rod; (c) a circular ring; (d) a cylindrical shell; (e) a thin rod about an axis through one end and perpendicular to the rod axis; (f) a thin rod about an axis through its center and perpendicular to the rod axis; (g) a right circular cylinder through the symmetry axis; (h) a sphere through a diametrical axis; and (i) a rectangular plate with sides a and b about an axis perpendicular to the plate and passing through its center.*

moments of inertia about other, parallel axes. [In (13-16), when $p = 0$, then $I = I_{CM}$.]

The moments of inertia for several simple shapes are given in Figure 13-7.

Example 13-3. A dumbbell has two 1.0-kg point masses at the ends of a massless rod 1.0 m long. (a) What is the rod's moment of inertia relative to an axis perpendicular to the rod and passing through its center [Figure 13-8(a)]? (b) perpendicular to the rod and passing through one end [Figure 13-8(b)]? (c) What is the radius of gyration for part (b)?

(a) From the definition of moment of inertia, we have

$$I = \Sigma m_i r_i^2$$
$$= (1 \text{ kg})(0.5 \text{ m})^2 + (1 \text{ kg})(0.5 \text{ m})^2$$
$$I_{CM} = 0.50 \text{ kg} \cdot \text{m}^2$$

The subscript CM is used since this moment of inertia is computed relative to an axis through the center of mass.

(b) With one 1-kg mass at the axis, the second mass alone contributes to I:

$$I = \Sigma m_i r_i^2$$
$$= (1 \text{ kg})(1 \text{ m})^2 = 1 \text{ kg} \cdot \text{m}^2$$

This result can also be arrived at by using the parallel-axis theorem (13-16), together with the results of part (a):

$$I = I_{CM} + Mp^2$$
$$= 0.5 \text{ kg} \cdot \text{m}^2 + (2 \text{ kg})(0.5 \text{ m})^2$$
$$= 1.0 \text{ kg} \cdot \text{m}^2$$

Here we have used the fact that the two parallel axes are separated by a distance $p = 0.5$ m.

(c) By definition, the radius of gyration k is related to the moment of inertia by

$$I = Mk^2 \qquad\qquad \textbf{(13-15)}$$

or

$$k = \sqrt{\frac{I}{M}} = \sqrt{\frac{1 \text{ kg} \cdot \text{m}^2}{2 \text{ kg}}} = 0.71 \text{ m}$$

This result means that in Figure 13-8(b), if the entire mass of 2 kg were placed a distance $k = 0.71$ m from the rotation axis, the moment of inertia would be the same as for the

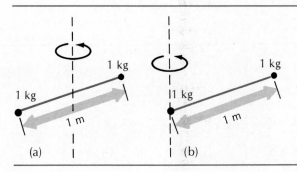

(a)   (b)   *Figure 13-8.*

actual mass distribution. Although effectively the *gyroradius* (radius of gyration) is the distance from the rotation axis at which we can imagine all of a body's mass to have been concentrated for rotational motion, it is not equal to the actual distance of the body's center of mass from the axis.

Example 13-4. What is the moment of inertia of a thin uniform rod of length L and mass M with respect to a rotation axis perpendicular to the rod and passing (a) through the rod's center? (b) one end of the rod? See Figure 13-9.

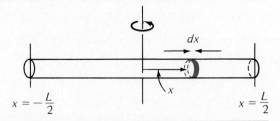

Figure 13-9. Computation of the moment of inertia of a thin rod of length L about a perpendicular axis through its center. The thin slice of thickness dx is located at x.

$$x = -\frac{L}{2} \qquad\qquad x = \frac{L}{2}$$

Our strategy here is to imagine the rod subdivided into thin slices stretched along the rod's length. Then the contribution to I from the thin slice of thickness dx at a distance x from the axis is $x^2\, dm$.

If the rod has a uniform density along its length, the mass of the thin slice is $dm = M(dx/L)$. Then the moment of inertia can be written

$$I_{CM} = \int r^2\, dm = \frac{M}{L} \int_{-L/2}^{+L/2} x^2\, dx = \frac{M}{L}\frac{L^3}{12} = \frac{1}{12}ML^2$$

Note that the integration extended from the left end of the rod ($x = -L/2$) to the right ($x = +L/2$.)

(b) To find I_{end} about one end, we simply choose different limits for the definite integral above. Now the rod extends from $x = 0$ to $x = L$, so that

$$I_{end} = \frac{M}{L}\int_0^L x^2\, dx = \frac{1}{3}ML^2$$

This result is consistent with the parallel-axis theorem:

$$I_{end} = I_{CM} + Mp^2$$
$$= \frac{1}{12}ML^2 + M\left(\frac{L}{2}\right)^2 = \frac{1}{3}ML^2$$

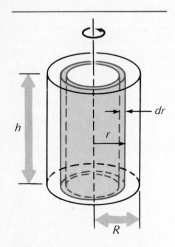

Figure 13-10. Computation of the moment of inertia of a right circular cylinder of radius R and height h, about an axis of rotation through the cylinder's symmetry axis. A thin cylindrical shell of radius r and thickness dr is shown.

Example 13-5. What is the moment of inertia of a uniform right circular cylinder of mass M and radius R relative to its geometrical symmetry axis?

Here we imagine the solid cylinder to consist of thin cylindrical shells each of radius r and thickness dr. See Figure 13-10. If the mass of a cylindrical shell is dm, its moment of inertia is $r^2\, dm$, since all the shell's mass is at the *same* distance r from the rotation axis. But $dm = \rho\, dV$, where the shell's volume dV is the product of its height h, circumference $2\pi r$, and thickness dr:

$$dm = \rho\, dV = \rho h(2\pi r)\, dr$$

Therefore, summing contributions from $r = 0$ to $r = R$, we have

$$I = \int r^2\, dm = \rho \int_0^R 2\pi h r^3\, dr = 2\pi\rho h\left(\frac{R^4}{4}\right)$$

The cylinder's density ρ can be written, in terms of its mass M and volume $V = \pi R^2 h$, as

$$\rho = \frac{M}{\pi R^2 h}$$

so that the relation above becomes

$$I_{CM} = \tfrac{1}{2} M R^2$$

Example 13-6. Proof of the Parallel-Axis Theorem (Optional).   We wish to relate the moment of inertia I_{CM} of a rigid body, relative to an axis passing through the center of mass, to the moment of inertia I of the same body relative to another axis, parallel to the first and a distance p from it.

First consider a particle i of mass m_i having a displacement $\mathbf{R}_i$ relative to an arbitrarily chosen axis of rotation A. The displacement of this same particle relative to the CM is $\mathbf{r}_i$; the displacement of CM relative to the axis A is $\mathbf{p}$ (see Figure 13-11); therefore,

$$\mathbf{R}_i = \mathbf{r}_i + \mathbf{p}$$

The contribution of the ith particle to the moment of inertia measured with respect to an axis through A is $m_i R_i^2$, which we may write equivalently, using the dot product, as

$$m_i R_i^2 = m_i \mathbf{R}_i \cdot \mathbf{R}_i = m_i (\mathbf{r}_i + \mathbf{p}) \cdot (\mathbf{r}_i + \mathbf{p})$$
$$= m_i (\mathbf{r}_i \cdot \mathbf{r}_i + \mathbf{p} \cdot \mathbf{p} + 2\mathbf{r}_i \cdot \mathbf{p})$$
$$= m_i (r_i^2 + p^2 + 2\mathbf{r}_i \cdot \mathbf{p})$$

If we now add the contributions to the moment of inertia about A of all the particles in the rigid body, we have

$$I = \Sigma m_i R_i^2 = \Sigma m_i (r_i^2 + p^2 + 2\mathbf{r}_i \cdot \mathbf{p})$$

which we may write as

$$I = \Sigma m_i r_i^2 + (\Sigma m_i) p^2 + 2(\Sigma m_i \mathbf{r}_i) \cdot \mathbf{p}$$

The three terms in this equation are interpreted as follows: The first term is the moment of inertia $I_{CM} = \Sigma m_i r_i^2$ of the body relative to an axis parallel to A and passing through the CM. The second term is the body's total mass $M = \Sigma m_i$ multiplied by the square of the distance separating the two axes: $(\Sigma m_i) p^2 = Mp^2$. The third term contains the vector $\Sigma m_i \mathbf{r}_i$, which gives, by definition (8-3), the location of CM relative to the chosen origin. But the CM is at the origin of each one of the displacements $\mathbf{r}_i$. Therefore, $\Sigma m_i \mathbf{r}_i = 0$ and the third term is zero. The equation for I then reduces to

$$I = I_{CM} + Mp^2 \qquad (13\text{-}16)$$

which is the parallel-axis theorem.

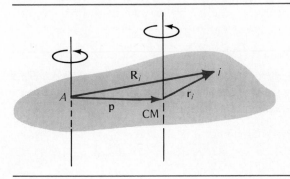

Figure 13-11.   *A particle i of a rigid body has a displacement* $\mathbf{r}_i$ *relative to the body's center of mass, and a displacement* $\mathbf{R}_i$ *relative to point A. The displacement of the center of mass relative to A is* $\mathbf{p}$.

13-4   Energy of a Rolling Object

Suppose that a symmetrical object rolls on a flat surface without slipping, like a tire rolling on a roadway. For a symmetrical object the horizontal rotation axis is also the object's geometrical symmetry axis. Therefore, the object rotates about an axis through its center of mass.

Rolling is the superposition of two motions:

- *Translation of the center of mass* at velocity V along a line.
- *Rotation about the center of mass* at angular velocity ω See Figures 13-12(a) and (b).

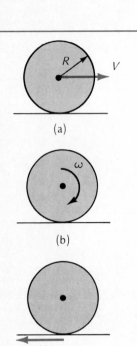

Linear velocity V and angular velocity ω are not independent of one another when a rigid body rolls without slipping. To see this, we note in Figure 13-12(a) first that relative to an observer on ground, the center advances to the right at speed V. Therefore, relative to an observer who travels with the rolling object and stays at its center, the roadway moves left at speed V, as in Figure 13-12(c). Moreover, relative to the wheel observer, the point on the wheel touching the road must also travel left at the *same* speed V, as in Figure 13-12(d). But a point on the wheel's circumference has the speed ωR, where R is the wheel's radius, so that

$$V = \omega R \qquad (13\text{-}17)$$

This relation represents a constraint between V and ω. It links translation and rotation for rolling without slipping.

As we have seen, rolling on a flat surface is the superposition of two motions: translation of the object's center of mass and rotation about its center of mass. Similarly, the total kinetic energy of a rolling object is the sum of two contributions:

- Translational kinetic energy $\frac{1}{2}MV^2$, where M is the mass of the object and V the speed of its center of mass.
- Rotational kinetic energy $\frac{1}{2}I\omega^2$, where I is the object's moment of inertia relative to a horizontal rotation axis through its center of mass and ω is the angular velocity.

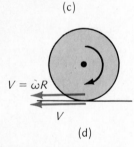

The total kinetic energy can then be written

$$K = \tfrac{1}{2}MV^2 + \tfrac{1}{2}I\omega^2 \qquad (13\text{-}18)$$

The validity of (13-18) was proved in Section 10-7, where it was shown that the total kinetic energy measured in the laboratory can be written, for any system of particles, as

$$K_{\text{rel lab}} = \tfrac{1}{2}MV^2 + K_{\text{rel CM}} \qquad (10\text{-}11)$$

The term $\frac{1}{2}MV^2$ is often called the kinetic energy of the center of mass. For a rolling object, the kinetic energy $K_{\text{rel CM}}$ measured relative to the center of mass is simply the rotational kinetic energy of a spinning rigid body, $\frac{1}{2}I\omega^2$.

Figure 13-12.   A symmetrical object of radius R rolls without slipping on a horizontal surface. (a) The center advances at speed V. (b) Angular velocity about the rotation axis is ω. (c) Relative to an observer traveling with the rolling wheel, the roadway has backward speed V. (d) The point of the wheel touching the roadway has the same speed $V = \omega R$ backward as the point it touches on the roadway, as observed from a reference frame traveling with the wheel.

Example 13-7.   A billiard ball of mass M rolls along a surface at speed V. The ball's radius is not known. What is the ball's total kinetic energy?

The moment of inertia of a uniform sphere about a diametrical axis is [Figure 13-7(h)]

$$I = \tfrac{2}{5}MR^2$$

where R is the sphere's radius.

Since the ball rolls without slipping, its angular velocity is, from (13-17),

$$\omega = \frac{V}{R}$$

Therefore,

$$
\begin{aligned}
K &= \frac{1}{2}MV^2 + \frac{1}{2}I\omega^2 \\
&= \frac{1}{2}MV^2 + \frac{1}{2}\left(\frac{2}{5}MR^2\right)\left(\frac{V}{R}\right)^2 \\
&= \frac{7}{10}MV^2
\end{aligned}
$$

Note that all uniform spheres with the same M and V have the same total kinetic energy, independent of radius R.

The translational kinetic energy divided by the total kinetic energy is $(\tfrac{1}{2}MV^2)/(\tfrac{7}{10}MV^2) = 71$ percent. The other 29 percent is rotational kinetic energy.

13-5 Newton's Law for Rotation

Any change in the position of a rigid body is a combination of its translation through space and rotation about an axis (Section 12-1).

The *translational motion* of the object's center of mass is controlled by the external forces acting on it:

$$\Sigma \mathbf{F}_{\text{ext}} = M\mathbf{A} \qquad (8\text{-}11)$$

where M is the total mass and $\mathbf{A}$ the acceleration of the center of mass. In effect, the object is replaced by an equivalent particle at the center of mass.

The *rotational motion* of a rigid body is controlled by the external torques on it. Since τ is the rotational analog of $\mathbf{F}$, and I the rotational analog of m, and α the rotational analog of $\mathbf{a}$, we guess that the rotational form of Newton's second law of motion is

$$\Sigma \tau_{\text{ext}} = I\alpha \qquad (13\text{-}19)$$

It is easy to prove this rigorously. We start with the work-energy relation (13-12):

$$\int_i^f \tau_{\text{ext}}\, d\theta = \frac{1}{2}I\omega_f^2 - \frac{1}{2}I\omega_i^2$$

Applied to an infinitesimal angular displacement $d\theta$, the work done is $\tau_{\text{ext}}\, d\theta$, and the infinitesimal change in kinetic energy is $d(\tfrac{1}{2}I\omega^2)$:

$$\tau_{\text{ext}}\, d\theta = d(\tfrac{1}{2}I\omega^2)$$

Now if we divide this equation by the short time interval dt, to take the time derivative, we get

Table 13-1

QUANTITY (ROTATIONAL ANALOG)	LINEAR OR TRANSLATIONAL	ANGULAR OR ROTATIONAL
Displacement	x	θ
Velocity	$v = dx/dt$	$\omega = d\theta/dt$
Acceleration	$a = dv/dt$	$\alpha = d\omega/dt$
Mass (moment of inertia)	m	$I = \Sigma m_i r_i^2$
Force (torque)	$\mathbf{F}$	$\tau = \mathbf{r} \times \mathbf{F}$
Newton's second law	$\mathbf{F} = m\mathbf{a}$	$\tau = I\alpha$
Work	$\mathbf{F} \cdot d\mathbf{s}$	$\tau \cdot d\theta$
Power	$\mathbf{F} \cdot \mathbf{v}$	$\tau \cdot \omega$
Kinetic energy	$\frac{1}{2}mv^2$	$\frac{1}{2}I\omega^2$

$$\tau_{\text{ext}} \frac{d\theta}{dt} = \frac{d}{dt}\left(\frac{1}{2}I\omega^2\right) = \left(\frac{1}{2}I\right)(2\omega)\frac{d\omega}{dt}$$

$$= I\omega\frac{d\omega}{dt}$$

But $d\theta/dt = \omega$ and $d\omega/dt = \alpha$, so that the preceding relation reduces to*

$$\Sigma\tau_{\text{ext}} = I\alpha$$

Clearly, if $\Sigma\tau_{\text{ext}} = 0$, then $\alpha = 0$; so that a rigid body is in rotational equilibrium if the resultant external torque on it is zero.

Table 13-1 shows analogous translational and rotational physical quantities.

Example 13-8. A rope is wrapped around the circumference of a uniform circular disk of radius R and mass m. The other end of the rope is attached to the ceiling. The disk descends with the rope unwinding from it. See Figure 13-13. What is (a) the acceleration of the center of the disk? (b) the tension in the rope?

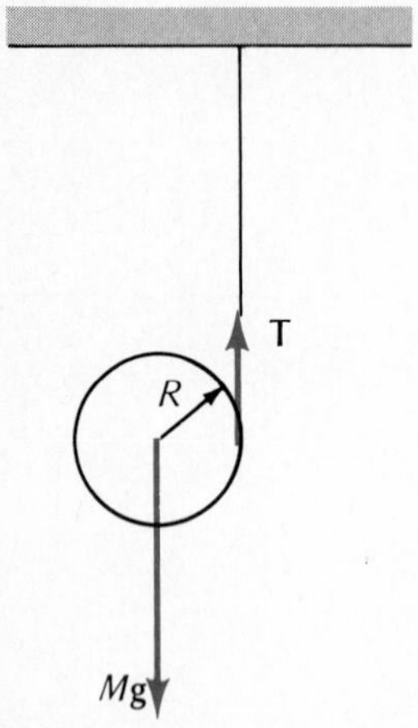

The forces on the disk are its weight $m\mathbf{g}$ and the upward rope tension $\mathbf{T}$. Choosing the downward direction as positive, we have for Newton's second law applied to the translational motion:

$$\Sigma\mathbf{F}_{\text{ext}} = m\mathbf{a}$$

$$mg - T = ma$$

Where a is the acceleration of the disk's center of mass.

We may apply Newton's second law for rotational motion to the accelerating disk with the rotation axis at the object's center of mass.[†]

$$\Sigma\tau_{\text{ext}} = I\alpha$$

The cylindrical disk's moment of inertia relative to its symmetry axis is $I = \frac{1}{2}mR^2$, and α is the angular acceleration of the spinning disk. Now at the point where the rope leaves the cylinder, its acceleration relative to the disk's center is $a = R\alpha$; this acceleration is the same as the acceleration of the disk's center relative to the ceiling. The only torque

Figure 13-13.

---

* Both τ_{ext} and α can be written as vectors only if the special circumstances assumed in this chapter apply: the directions of the torques and rotation axis are the same. A more general form for Newton's second law of motion for torques, applicable to *any* system of particles (not merely a rigid body rotating about a symmetry axis), is developed in Chapter 14.

relative to the center arises from the tension, and it has the magnitude RT. (The weight has no torque since its moment arm is zero.) Substituting these quantities in the equation above, we have

$$RT = \frac{1}{2}\, mR^2\, \frac{a}{R}$$

or

$$T = \frac{1}{2}\, ma$$

But from the previous equation we had

$$mg - T = ma$$

so that $a = \frac{2}{3}g$ and $T = \frac{1}{3}mg$. Note that disk's radius R cancels out.

† Newton's laws can be applied to an accelerating, and therefore noninertial, reference frame only if an inertial force acting at the center of mass is included (Section 6-3). But if the rotation axis also passes through the center of mass, the moment arm for the inertial force is zero, as is the torque due to the inertial force.

Example 13-9.   A billiard ball of mass m and radius R initially at rest on a rough inclined plane of angle θ rolls down the incline. (a) What is the acceleration of the ball's center of mass? (b) What is the minimum angle of the incline if the ball is to roll without slipping? (c) We supposed above that the ball rolls down the incline. If the ball were rolled up the incline, what would be the direction of the static-frictional force?

(a) Figure 13-14 shows the forces on the ball: its weight $m\mathbf{g}$, the normal force $\mathbf{N}$, and the frictional force $\mathbf{f_s}$. The tail of each force vector is located at the spot at which the force is applied to the ball.

The frictional force $\mathbf{f_s}$ is *static*; the ball rolls without slipping and there is no relative motion between the ball and incline at the point of contact. Figure 13-14 shows $\mathbf{f_s}$ as up the incline, the natural choice perhaps, since a frictional force usually opposes motion. We shall first have to check, however, whether this is, in fact, the right direction for $\mathbf{f_s}$.

Let us choose down the incline as the direction for positive linear acceleration a of the ball's center. As the ball rolls down the incline, it rotates about its center in the clockwise sense. So, to have consistent signs for the linear acceleration a of the ball's center down the incline, and the angular acceleration α of its rotation about its center, we must choose positive angular acceleration α to correspond to clockwise rotation. The resultant torque computed relative to the ball's center must then also be clockwise, and static-frictional force $\mathbf{f_s}$ must, for the conditions given previously, remain up the incline.

Applying Newton's second law to the translational motion down the incline gives

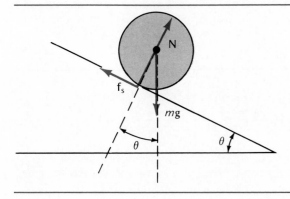

Figure 13-14.

$$\Sigma \mathbf{F}_{\text{ext}} = M\mathbf{A} \qquad \text{(8-11)}$$

$$mg \sin \theta - f_s = ma$$

Now we apply Newton's second law for rotation about the center of mass. The ball's moment of inertia is $I_{CM} = \frac{2}{5}mR^2$, and rolling without slipping implies that $\alpha = a/R$. We see in Figure 13-14 that both N and mg pass through ball's center, and therefore neither of these two forces produces a torque. The only torque is that from f_s; with no static-frictional force, the ball would slide, not roll.

$$\Sigma \tau_{\text{ext}} = I_{CM}\alpha$$

$$f_s R = \left(\frac{2}{5} mR^2 \right) \left(\frac{a}{R} \right)$$

or

$$f_s = \frac{2}{5} ma$$

When we substitute this relation for f_s in the other equation here also containing f_s, we get

$$mg \sin \theta - \frac{2}{5} ma = ma$$

or

$$a = \frac{5}{7} g \sin \theta$$

so that

$$f_s = \frac{2}{5} ma = \frac{2}{5} m \left(\frac{5}{7} g \sin \theta \right) = \frac{2}{7} mg \sin \theta$$

Note first that all uniform rolling spheres have the same linear acceleration, $\frac{5}{7}g \sin \theta$, down the incline, quite apart from their specific radius or mass.

(b) The maximum static-frictional force is determined, in turn, by the coefficient of static friction μ_s between the ball and the incline. Indeed,

$$f_s \leq \mu_s N \qquad \text{(6-2)}$$

But the normal force is $N = mg \cos \theta$, and the frictional force is $f_s = \frac{2}{7}mg \sin \theta$. Therefore, the inequality above becomes

$$\frac{2}{7} mg \sin \theta \leq \mu_s \, mg \cos \theta$$

or

$$\tan \theta \leq \frac{7}{2} \mu_s$$

The maximum incline angle θ_{max} that will allow the ball to roll without slipping is then given by

$$\tan \theta_{\text{max}} = \frac{7}{2} \mu_s$$

Suppose that the static coefficient is $\mu_s = 0.5$. Then this relation gives $\theta_{\text{max}} = 60°$. This result may be contrasted with that for the maximum incline angle that will keep an object from sliding down the incline. From Example 6-2, we have $\tan \theta_{\text{max}} = \mu_s$; therefore, for $\mu_s = 0.50$, the maximum angle is only $27°$.

(c) When the ball rolls up the incline, the directions of the linear velocity v and angular velocity ω are reversed. But the directions of linear acceleration a and angular acceleration α are not changed. The torque must still be clockwise; the frictional force must still be up the incline, now in the *same* direction as the motion of the ball's center. The lesson: one must be very careful in assigning a direction to a frictional force applied to a rolling object.

Example 13-10. A 15-cm pencil, initially standing on end, falls over [Figure 13-15(a)]. With what speed does the eraser strike the horizontal surface? Assume that the surface is rough enough that the pencil tip does not slip.

The problem is best solved by energy-conservation methods. The pencil is a uniform rod of mass M and length L. The pencil's center of gravity is at its center (originally a vertical distance $L/2$ above the surface). The pencil's weight Mg can be regarded as a single force applied at the center of gravity. When the pencil falls, its center of gravity descends $L/2$, so that the gravitational potential energy decreases by $Mg(L/2)$. The increase in rotational kinetic energy of the pencil about the fixed point equals the loss in potential energy:

$$\frac{1}{2}I\omega^2 = \frac{MgL}{2}$$

where I is the pencil's moment of inertia and ω is its angular speed, both with respect to an axis of rotation passing through the fixed point. If v represents the linear speed of the eraser on striking, then

$$v = \omega L$$

The moment of inertia of a uniform rod about an axis at an end (Section 13-3) is

$$I = \frac{1}{3}ML^2$$

Substituting the relations for I and ω in the energy equation above yields

$$\left(\frac{1}{2}\right)\left(\frac{1}{3}\right)ML^2\left(\frac{v}{L}\right)^2 = \frac{MgL}{2}$$

$$v = \sqrt{3gL} = \sqrt{3(9.8 \text{ m/s}^2)(0.15 \text{ m})} = 2.1 \text{ m/s}$$

Note a curious circumstance. If the eraser had been detached and fallen freely from rest through the same vertical distance, its speed on striking the surface would have been only $\sqrt{2gL} = 1.7$ m/s.

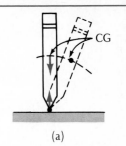

(a)

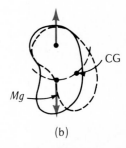

(b)

Figure 13-15. (a) An example of unstable equilibrium. The object's point of support is below the center of gravity (CG). (b) An example of stable equilibrium. An object's point of support is above the center of gravity.

A pencil can, at least in principle, be stood on end. But such an equilibrium is highly unstable; any small angular displacement produces an increasingly large gravitational torque. The unstable equilibrium in Figure 13-15(a) may be associated with lowering the object's center of gravity for a small displacement. Stable equilibrium is illustrated in Figure 13-15(b). Here the point of support is above the center of gravity, so that the center of gravity rises for a small displacement. Note also that the gravitational torque is a restoring torque.

13-6   Pendulums with Rigid Bodies

Physical Pendulum   A simple pendulum consists of a single point mass suspended by massless cord (Section 11-5). A *physical pendulum*, on the other hand, consists of an *extended rigid body* freely pivoted about and undergoing oscillations relative to some point in the rigid body. A simple example is the pendulum in a grandfather's clock.

Figure 13-16 shows a rigid body in which the distance from pivot to center of mass is L_{CM}. The object's weight $m\mathbf{g}$ produces a torque relative to the pivot point with a moment arm $L_{CM}\sin\theta$. Then, applying Newton's second law for rotational motion, we have

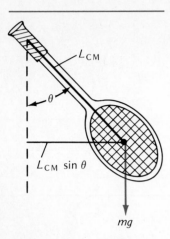

Figure 13-16. *A physical pendulum.*

$$\Sigma\tau_{ext} = I\alpha$$

$$-mg(L_{CM} \sin\theta) = I\frac{d^2\theta}{dt^2}$$

A minus sign appears on the left side because the weight produces a restoring torque. The rigid body's moment of inertia I is measured relative to the rotation axis at the pivot point.

If the angular displacement θ from equilibrium is small, so that $\sin\theta \simeq \theta$, the equation above can be written in the form

$$\frac{d^2\theta}{dt^2} + \omega_0^2\theta = 0 \tag{13-20}$$

where ω_0^2 is a constant defined by

$$\omega_0^2 \equiv \frac{mgL_{CM}}{I} \tag{13-21}$$

Equation (13-20) is (or should be) familiar. It is the basic differential equation for simple harmonic motion, with the variable x in (11-8) replaced by θ.

We can immediately write the solution to (13-20), using (11-9), as

$$\theta = \theta_0 \cos(\omega_0 t + \delta)$$

Angle θ oscillates sinusoidally with time. Here θ_0 is the angular amplitude of the oscillation, δ is the phase constant, and ω_0 the angular frequency. The oscillating period T and frequency f are related in general to ω_0 by

$$T = \frac{1}{f} = \frac{2\pi}{\omega_0}$$

from (11-3). Therefore, for the physical pendulum, we have*

$$T = \frac{1}{f} = 2\pi\sqrt{\frac{I}{mgL_{CM}}} \tag{13-22}$$

All the general properties of an undamped simple harmonic oscillator apply equally to a physical pendulum oscillating with small amplitude:

- Period (or frequency) independent of amplitude.
- Total energy constant, with interchange between kinetic and potential energies.

Torsion Pendulum There are only three basic ways in which any elastic material can be deformed:

- Stretching (or compressing) along a line.
- Twisting, a so-called torsion deformation.
- Squeezing or compressing the material over its entire volume.

---

* It is easy to check that (13-22) reduces, as it must, to the corresponding relation (11-15) for a simple pendulum: $T = 2\pi\sqrt{L/g}$. If all mass is at a distance L_{CM} from the pivot, we have $I = mL_{CM}^2$, so that (13-22) becomes

$$T = 2\pi\sqrt{\frac{mL_{CM}^2}{mgL_{CM}}} = 2\pi\sqrt{\frac{L_{CM}}{g}}$$

Any deformation can be regarded as the superposition of one or more of these basic deformations.

A *torsion pendulum* is one in which elastic material is twisted. Before we consider the basic definitions for a torsional deformation, let us review the properties of an ordinary helical spring because of the close analogies we can then identify.

Suppose that a spring is to be stretched along a line, with forces of magnitude F_a applied to the two ends. See Figure 13-17(a). When the applied force at one end is equal in magnitude to the restoring force F_s of the elastic spring and opposite in direction, the spring will be in translational equilibrium. With a spring stretched an amount x, observation shows that

$$F_s = -kx$$

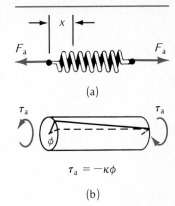

(a)

(b)

Suppose that in analogous fashion torques of magnitude τ_a are applied in opposite directions (or rotation senses) to the two ends of a cylinder, as shown in Figure 13-17(b). The cylinder is twisted. When the applied torque at one end is equal in magnitude to the restoring torque of the elastic material and opposite in direction, the cylinder will be in rotational equilibrium. One end of the cylinder is twisted through an angle ϕ relative to the other end. The rotational form of Hooke's law for a torsional deformation is

$$\tau_s = -\kappa\phi \qquad (13\text{-}23)$$

Figure 13-17. *(a) A spring stretched x by equal and opposite forces, each of magnitude F_a, applied at its ends. (b) A cylinder twisted ϕ by equal and opposite torques of magnitude τ_a applied at its ends.*

Here κ, known as the *torsion constant*, is constant for small deformations; it is a measure of the cylinder's stiffness for twisting.*

Suppose now that a rigid body is attached to the lower end of a vertical torsion fiber or wire, as shown in Figure 13-18; the fiber's upper end is held fixed. Rotation takes place about vertical axis A. When the rigid body undergoes angular oscillations about axis A, the system becomes a *torsion pendulum*.

To find the period of oscillation, we apply the rotational form of Newton's second law to the rigid body at the lower end of the torsion fiber:

$$\Sigma\tau_{ext} = I\alpha$$

$$-\kappa\phi = I\frac{d^2\phi}{dt^2}$$

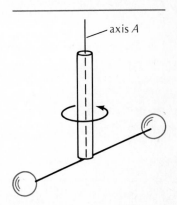

The only torque on the rigid body is that produced by the torsion fiber, $\tau_s = -\kappa\phi$. Moment of inertia I is that of the rigid body with respect to a vertical rotation axis A. Angle ϕ gives the displacement of the rigid body from equilibrium as measured in a horizontal plane.

The differential equation above can be written in the form

$$\frac{d^2\phi}{dt^2} + \omega_0^2\phi = 0$$

Figure 13-18.   A torsion pendulum.

where $\omega_0^2 \equiv \kappa/I$. Again the angular oscillations are simple harmonic, and the period T and frequency f are given by

$$T = \frac{1}{f} = \frac{2\pi}{\omega_0} = 2\pi\sqrt{\frac{I}{\kappa}} \qquad (13\text{-}24)$$

* A cylinder of small radius, or a fiber, is easily twisted. It can be shown that, other things being equal, the constant κ is proportional to the fourth power of the cylinder's radius, $\kappa \propto r^4$. Thus, if a cylinder's radius is reduced by a factor 10, constant κ is reduced by a factor 10^4.

Clearly, if the torsion constant κ is known, (13-24) allows the moment of inertia of any object (even one of peculiar shape for which direct computation might be difficult) to be computed from the measured value of T or f. Conversely, if I for the object at the end of the torsion rod is known, or can easily be computed, then with the measured value of T or f, (13-24) allows the torsion constant κ to be calculated.

Summary

For every rotational quantity there is an analogous translational quantity. Review Table 13-1.

Definitions

Angular acceleration: $\alpha \equiv d\omega/dt$ (13-2)

Moment of inertia: $I \equiv \Sigma m_i r_i^2$ (13-9)

Radius of gyration k: $I = Mk^2$ (13-15)

Torsion constant κ for angular deformation:

$$\tau_s = -\kappa\phi \qquad (13\text{-}23)$$

Important Results

Tangential acceleration component a_t for a point at radius r for angular acceleration α:

$a_t = r\alpha$ (13-8)

Kinetic energy of rotaton: $K = \frac{1}{2}I\omega^2$ (13-10)

Work dW done by torque: $dW = \tau\,d\theta$ (13-11)

Power P transmitted by torque τ: $P = \tau\omega$ (13-13)

Parallel-axis theorem: $I = I_{CM} + Mp^2$ (13-16)

where I_{CM} is the moment of inertia about CM; I, the moment of inertia about the parallel axis distance p away; and M, the total mass.

The total kinetic energy of rigid body in translational and rotational motion is

$$\tfrac{1}{2}MV^2 + \tfrac{1}{2}I\omega^2 \qquad (13\text{-}18)$$

If an object of radius R rolls without slipping,

$$V = R\omega \qquad (13\text{-}17)$$

where V is the velocity of the center and ω is the angular velocity measured relative to the center.

The rotational form of Newton's second law is

$$\Sigma\tau_{ext} = I\alpha \qquad (13\text{-}19)$$

For the physical pendulum with mass m, center-of-mass distance L_{CM} from pivot, moment of inertia I relative to pivot,

$$T = \frac{1}{f} = \frac{2\pi}{\omega_0} = 2\pi\sqrt{\frac{I}{mgL_{CM}}} \qquad (13\text{-}21)$$

For the torsion pendulum:

$$T = \frac{1}{f} = \frac{2\pi}{\omega_0} = 2\pi\sqrt{\frac{I}{\kappa}} \qquad (13\text{-}24)$$

Problems and Questions

Section 13-1   Rotational Kinematics

· **13-1 P** What constant angular acceleration is needed to accelerate an engine from an idling speed of 400 rpm (revolutions per minute) to a final speed of 5200 rpm in 10 seconds?

Figure 13-19.   Problem 13-2.

: 13-2 P The outer diameter of the tires on a bicycle is 70 cm. When the bike is moving 3 m/s, what is the velocity of points A, B, and C in Figure 13-19 with respect to (a) the center of mass of the wheel, and (b) the ground.

· **13-3 P** A sphere of radius 60 cm rotates about a fixed axis through its center. A point on the surface of the sphere 30 cm from the axis moves a distance 12 cm in a time interval of 7.0 s. What is the angular displacement of the sphere during this time interval?

· **13-4 P** A flywheel initially spinning at 1800 rpm is brought to rest in 3 minutes. (a) What is the wheel's average angular acceleration? (b) How many revolutions does the flywheel make in coming to rest?

: 13-5 P A wheel initially rotates *counterclockwise* about a fixed axis, with an initial angular speed of 1.2 radians per second. If it accelerates uniformly to 2.0 radians per second *clockwise* during a 5.0 s interval, what is the magnitude of its displacement during this interval?

: 13-6 P A phonograph turntable rotates at $33\frac{1}{3}$ rpm and has a radius of 0.150 m. It is brought to rest by a friction brake in 15.0 s.

(a) What is the angular acceleration of the turntable? (b) In how many revolutions does it stop? (c) At $\omega = 3.0$ rad/s while slowing down, what are the radial and tangential accelerations of a point on the rim?

: 13-7 P A motorcycle with wheels of diameter 0.60 m accelerates uniformly from rest to 20 m/s in 8 seconds. (a) What is the angular acceleration of the wheels? (b) How many revolutions do the wheels make in this time? (c) What is the angular velocity of the wheels at the end of 8 seconds?

Section 13-2   Rotational Kinetic Energy, Work, and Power

· 13-8 P Schemes have been proposed to store energy in a rotating flywheel. This stored energy could be used to propel a car, for example.

(a) How much energy is stored in a 100 kg disk of radius 30 cm rotating at 20,000 rpm? (b) To what volume of gasoline does the above energy stored in the flywheel correspond? (The energy content of gasoline is approximately 1.8×10^8 J/gal.)

· 13-9 P What is the rotational kinetic energy of the earth about its center?

: 13-10 P As will be proved in detail in Chapter 14, the angular momentum $I\omega$ of an isolated system is constant. Show that, if the mass of a spinning system is somehow redistributed and is shifted farther from the rotation axis, the system's rotational kinetic energy must decrease.

: 13-11 P The engine of a certain automobile produces maximum torque at 5000 rpm. The engine is rated at 100 kW. What is the maximum torque?

Section 13-3   Moment-of-Inertia Computations

· 13-12 Q Through what point must the axis pass if the moment of inertia of a rigid body is to be a minimum?

· 13-13 Q Suppose that a particular rotating rigid body has a given angular velocity, shape, size and mass. If one of these parameters is changed while the others are held fixed, which would result in no change in the moment of inertia of the body?
(A) Angular velocity
(B) Mass
(C) Shape
(D) Size
(E) More than one of the above

: 13-14 Q Two spheres of equal mass are made of materials of different densities. (a) Which sphere will have the greatest moment of inertia about an axis passing through

the center? (b) Which will have the greatest moment of inertia about an axis tangent to the surface?

: 13-15 P Three masses each m are at the corners of a 45°-45°-90° triangle with short sides of length L. What is the moment of inertia for the rotation axis (a) coinciding with a short side of the triangle and (b) coinciding with the long side of the triangle? (c) Where must the rotation axis be located to yield the minimum moment of inertia? (d) For what location, somewhere within or at the boundary of the triangle, is the moment of inertia a maximum?

: 13-16 P Four equal masses each m are placed at the corners of a square of side a. Calculate the moment of inertia of this array about (a) an axis through the center of the square and perpendicular to the plane of the square, (b) an axis through one of the masses and perpendicular to the plane of the square; (c) a line joining two adjacent masses; (d) a line joining two diagonally opposite masses.

: 13-17 P Determine the moment of inertia of a uniform rod of length L about an axis that passes through the midpoint of the rod and makes an angle θ with the rod.

: 13-18 P (a) Consider a planar object lying in the x-y plane. Prove that the moment of inertia I_z about the z-axis is related to the moments of inertia about the x-axis, I_x, and about the y-axis, I_y, by $I_z = I_x + I_y$. This is called the *perpendicular axis theorem*. (b) Use the above result to prove that the moment of inertia of a square sheet of mass M is the same for *any* arbitrary axis lying in the plane of the square and passing through the center. Determine the value for this moment of inertia.

: 13-19 P The methane molecule, CH_4, consists of four hydrogen atoms, each of mass m, placed at the corners of a regular tetrahedron on side a, plus a carbon atom, of mass $12m$, placed at the center of symmetry of the tetrahedron. Calculate the radius of gyration about an axis passing through the carbon and one of the hydrogens.

: 13-20 P Calculate the moment of inertia of a circular disk of mass m and radius R about an axis parallel to the plane of the disk and tangent to the perimeter.

: 13-21 P Determine the moment of inertia of a spherical shell of radius R and mass M about an axis through the center.

: 13-22 Q The five solid objects shown in Figure 13-20 have equal masses and equal maximum dimensions in the x and y directions. With respect to an axis perpendicular to the drawing and passing through the center of mass, rank them according to their moments of inertia.

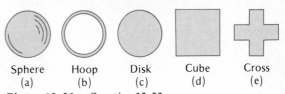

| Sphere | Hoop | Disk | Cube | Cross |
| (a) | (b) | (c) | (d) | (e) |

Figure 13-20.   Question 13-22.

Section 13-4   Energy of a Rolling Object

· **13-23 P** The device that stops a bowling ball is a ramp approximately 40 cm high. What is the maximum speed with which a ball can approach the ramp and just barely make it to the top? A bowling ball has a radius of 23 cm and a mass of 6 kg.

: **13-24 P** The moment of inertia of an irregular object can be measured in the following way (see Figure 13-21). A mass m is attached to a string connected to a turntable. It is released from rest and allowed to drop a distance h to the floor. The time of descent, t_1, is recorded. Next the object whose moment of inertia is to be measured is attached to the turntable. The experiment is repeated and the time of descent t_2 is measured. From this information the moment of inertia can be determined. Calculate the moment of inertia for an object for which the following data are obtained: $m = 50$ gm, $h = 1$ m, $t_1 = 2$ s, $t_2 = 8$ s. Radius = 4 cm.

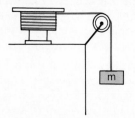

Figure 13-21.   Problem 13-24.

: **13-25 P** A hoop rolls down an inclined plane. What fraction of its total kinetic energy is associated with its rotation about its center of mass?
(A) 1/4
(B) 1/3
(C) 1/2
(D) 2/3
(E) 2/5

: **13-26 Q** Shown in Figure 13-22 are several objects of varying sizes, shapes and masses (but not drawn to scale with respect to size). Each can be attached to an axle of a fixed diameter and be allowed to roll down the parallel

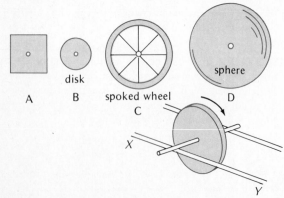

Figure 13-22.   Question 13-26.

inclined rails sketched here. Assume that each object is released from rest at point X. Which will reach point Y most slowly, assuming no slipping?
(A) A
(B) B
(C) C
(D) D
(E) The most massive object
(F) The object of smallest radius
(G) All require the same time

: **13-27 P** A toy car raced in a Cub Scout Pinewood Derby race has a body of mass M to which are attached four wheels, each of mass m and radius R. The car is released from rest and allowed to roll down a slope of height h. (a) Derive an expression for the speed of the car at the bottom of the slope. (b) Compare the performance of two cars with the same total mass, but assume one has essentially "weightless" wheels, using the following parameters: $M = 52$ gm; $R = 1$ cm; $h = 2$ m; $m = 1$ gm. The other car's mass is equally divided between body and the four wheels. Determine by how many seconds the faster car will beat the slower one. (c) How far ahead will the winner be at the finish line?

: **13-28 P** Prove that any hoop (any size, any mass) will always roll down hill slower than any sphere when the two are released under the same conditions.

: **13-29 Q** A hoop and a disk are released from rest at the top of a hill. They roll down without slipping. Which one will reach the bottom first?
(A) The hoop
(B) The disk
(C) The most massive one
(D) The least massive one
(E) The largest one
(F) The smallest one
(G) None of the above answers is correct

: **13-30 Q** If a sphere and a solid cylinder are placed at the top of an inclined plane and released simultaneously, and roll without slipping,
(A) the sphere will arrive at the bottom first.
(B) the cylinder will arrive at the bottom first.
(C) the cylinder and the sphere will arrive at the bottom simultaneously.
(D) it is impossible to tell which will reach the bottom first from this information.

: **13-31 P** A bicycle wheel rolls without slipping down an inclined plane that makes an angle ϕ with the vertical. The wheel's radius is R, and its mass M is concentrated on its outer rim (spokes and hub have negligible mass). The magnitude of its constant acceleration down the plane is a. The coefficients of static and kinetic friction between the wheel and the plane are μ_s and μ_k respectively. The instantaneous speed of the center of the wheel is v.

(a) What is the frictional force on the wheel? (b) What is the ratio of the torque on the wheel to its angular acceleration? (c) What is the magnitude of the angular acceleration of the wheel? (d) What is the direction of the angular acceleration of the wheel? (e) What is the rate of change of the wheel's kinetic energy?

· **13-32 P** A rotating drum on a machine has a mass of 50 kg and a radius of 0.20 m. The drum is initially rotating at 1500 rpm and it is to be brought to rest in 10 seconds. (a) What is the magnitude of the minimum braking force? (b) Where on the drum should this force be applied?

· **13-33 P** A flywheel rotating at 10 revolutions per second is brought to rest by a constant torque in 15 seconds. How many revolutions does it make during this time?

· **13-34 Q** A straight uniform stick is balanced standing upright on a frictionless level surface. When it falls over, where will the bottom end be with respect to where it started?

: **13-35 Q** (a) How can you distinguish between a raw egg and a hard-boiled egg simply by spinning them? (b) Suppose that you momentarily stop a spinning raw egg and quickly release it. The egg will start spinning again. Why?

: **13-36 Q** When crossing a stream on a floating log you may find it helpful to extend your arms to maintain your balance. The reason is that
(A) this increases your moment of inertia about the log.
(B) this raises your center of gravity.
(C) this lowers your center of gravity.
(D) this decreases your moment of inertia for rotation about your feet.
(E) this shifts your center of mass outside your body.

: **13-37 P** The asteroid Icarus has a diameter of 1600 m and an estimated mass of 1.1×10^{13} kg. The F-1 rocket engine develops a thrust of 4.4×10^6 N. How many F-1 engines, firing for one hour, are required to give Icarus a rotational period of 2 hours? Assume that the asteroid is a sphere and that the engines are deployed around it in the most favorable arrangement.

: **13-38 P** A small cylinder of mass m rolls without slipping inside a hollow cylinder of mass M. A belt wrapped around the outer cylinder exerts a constant force tangential to the

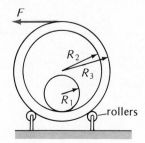

Figure 13-23.   Problem 13-38.

cylinder. See Figure 13-23. If the system is initially at rest, determine the angular velocity of the small cylinder as a function of time.

: **13-39 P** An Atwood's machine consist of two masses, m_1 and m_2, connected by a string passing over a uniform cylindrical pulley of mass m and radius r. The string does not slip on the pulley. Determine the acceleration of the masses and the tensions in the string.

: **13-40 P** A uniform sphere rolls without slipping down a plane inclined at angle θ above horizontal. What is the translational acceleration of the sphere?

: **13-41 P** Objects of masses 3 kg and 4 kg hang from a uniform cylindrical pulley of radius 20 cm and moment of inertia 0.24 kg·m². A person pulls down on the 3 kg object with a constant force of 49 N. Friction on the axle is negligible. See Figure 13-24.
(a) What is the mass of the pulley? (b) What is the linear acceleration of the objects? (c) What is the angular acceleration of the pulley? (d) What is the torque acting on the cylinder?

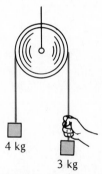

4 kg

3 kg

Figure 13-24.   Problem 13-41.

: **13-42 Q** (a) When a logger fells a tree, he first makes a cut on the side toward the direction he wants the tree to fall. Why does he do this? (b) Why is it dangerous to stand close to the falling tree, even if you are on the side opposite the direction of fall?

: **13-43 P** A vehicle of weight W has wheels of radius R. A torque τ is applied to the rear axle to accelerate the car up a hill with an angle of incline of θ. Determine the normal force exerted on each wheel and the resultant acceleration of the car.

: **13-44 P** A bowling ball of mass M is released with velocity v_0. The ball first slides without rolling down an alley, where the coefficient of kinetic friction is μ_k. How far does the bowling ball advance before it stops sliding and then rolls at constant speed? Graph qualitatively linear velocity versus time and angular velocity versus time.

: **13-45 P** A uniform rod 0.80 m long of mass 0.4 kg is pivoted at one end and allowed to swing as a pendulum. It is pulled 30° to one side and released from rest. (a) What is its maximum angular velocity as it swings back and forth? (b) What is the maximum force exerted by the pivot?

Section 13-6   Pendulums with Rigid Bodies

: 13-46 P In San Francisco's Golden Gate Park a popular swing for children consists of a large metal hoop pivoted at a point on the perimeter. Consider a hoop of 20 kg mass and 2 m diameter. (a) What is the period of oscillation of the hoop alone? (b) What is the period of oscillation with a child of mass 20 kg sitting opposite the pivot, with her center of gravity 1.5 meters from the pivot?

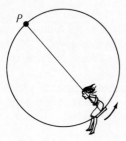

Figure 13-25.   *Problem 13-46.*

: 13-47 P A thin wire of mass m and 20 cm total length is bend at its midpoint through 90°. It is suspended from a frictionless peg and set into oscillation. Determine the frequency of its motion.

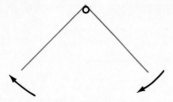

Figure 13-26.   *Problem 13-47.*

: 13-48 P A rigid body of mass M and moment of inertia I_{CM} (about the center of mass) is pivoted about a point

distance L from the CM. The object is pulled to the side through an angle θ_0 and released from rest. What is the maximum velocity of the center of mass as the object oscillates?

: 13-49 P A meter stick is suspended from a pivot at one end. With what frequency will it oscillate as a physical pendulum?

: 13-50 P A circular disk of radius 20 cm is pivoted on a pin placed through a small hole a distance d from the center of the disk. (a) Where should the pin be placed to produce a minimum oscillation period? (b) What is this minimum period?

Supplementary Problems

13-51 P The Society of Automotive Engineers (SAE) gives the following performance characteristics of the engine in a Mazda RX-7: braking horsepower of 100 at 6,000 rpm; torque of 105 lb-ft at 4,000 rpm. Are the two measures of power the same?

13-52 P A world record for domino toppling was set in Japan in 1980 by Erez Klein and John Wickham. They spent five weeks setting up 255,389 dominos in intricate patterns. It took 53 minutes for them all to fall. Is that time reasonable? Do an approximate computation. Estimate the dimensions of a domino, the distance separating adjacent dominos, and still other requisite parameters.

13-53 P An operating windmill in New Jersey (*New York Times,* New Jersey Section, July 11, 1982) produces 40 hp when its vanes are turning at the rate of 8 revolutions per minute. (a) What is the windmill's output torque? (b) Take the diameter of the windmill vanes to be 8 m, use 1.29 kg/m³ as the density of air, make sensible assumptions about efficiency, and compute an order-of-magnitude value for wind speed.

13-54 P A typical skydiver achieves a terminal speed of 120 mph. Assume that a skydiver has a total weight with equipment of 180 lb. What is the rate at which energy is being dissipated as the skydiver descends at the terminal speed?

Angular Momentum

14

Why is the idea of linear momentum $m\mathbf{v}$ introduced into physics? It is because the vector momentum of a single isolated particle is constant or, more importantly, the total momentum of an isolated system of interacting particles is constant. In short, momentum is conserved. Moreover, if a particle's linear momentum changes, we can relate the time rate of change of momentum to the net force on the particle.

Angular momentum is similar. It is constant for an isolated system — there is an angular momentum conservation law. And when angular momentum changes, we may relate the time rate of change of angular momentum to the resultant torque. Therefore, angular momentum is a fundamental physical quantity that, because of its constancy under appropriate conditions, simplifies the description and analysis of many physical situations.

14-1 Angular Momentum of a Particle

In Figure 14-1(a), a particle of mass m has velocity $\mathbf{v}$. Its location relative to the origin O is given at each instant by radius vector $\mathbf{r}$. As the particle travels, it is tracked by vector $\mathbf{r}$.

By definition, the angular momentum $\mathbf{L}$ of the particle relative to the origin O of $\mathbf{r}$ is

$$\mathbf{L} \equiv \mathbf{r} \times m\mathbf{v} \tag{14-1}$$

289

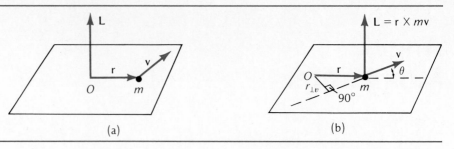

Figure 14-1. *(a) The particle's angular momentum* **L** = **r** × m**v** *is perpendicular to* **v**. *(b) In magnitude* L = r × mvr⊥v.

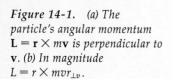

The angular momentum of a particle is the cross product (Section 12-2) of the radius vector **r** with the particle's linear momentum m**v**. (The reason why angular momentum is defined in just this way will become clear in Section 14-3.)

From the general properties of the cross product, we then know that:

• In direction, **L** is perpendicular to the plane of **r** and m**v**. The specific direction of **r** × m**v** is given by the right-hand rule. Accordingly, turn **r** into m**v** through the smaller angle by the right-hand curled fingers; the outstretched right thumb gives the direction of **r** × m**v**.

• In magnitude,

$$L = r_{\perp v}mv = (r \sin \theta)mv \qquad (14\text{-}2)$$

where $r_{\perp v}$ is the component of **r** perpendicular to **v** and θ is the angle between **r** and **v** [Figure 14-1(b)].

Since angular momentum is the product of distance and linear momentum, appropriate units are kilogram meters squared per second since $(m)(kg \cdot m/s) = kg \cdot m^2/s$. An equivalent unit is the joule second, since $J \cdot s = (kg \cdot m^2/s^2)(s) = kg \cdot m^2/s$.

Here are ways to find the rotation sense or direction of the **L** vector:

• Radius vector **r** tracks the particle as it moves. If **r** rotates, the particle has angular momentum. Further, **r** has the same rotation sense as that for the associated angular momentum vector **L**.

• Imagine that an observer at the origin always keeps his eye on the traveling particle. If this observer's eyeball turns, the particle has angular momentum. The sense of rotation of the eyeball is again the rotation sense associated with the cross product and the direction of **L**.

• Imagine the particle to be a bullet that strikes and sticks to an object pivoted at the origin. The rotation sense of the struck object again gives the rotation sense and angular momentum of the particle.

Clearly, then, if the radius vector, or the eyeball, or the pivoted object does not turn, the particle's angular momentum is zero. The particle must then be heading away from or towards the origin O relative to which **L** is measured. In every other instance the particle has angular momentum.

It cannot be emphasized too strongly that the value of a particle's angular momentum depends on what we choose for the origin. The angular momentum vector **L** depends on the point in space at which the tail of **r** is located. Indeed, it is meaningless to give the direction and magnitude of the angular momentum of a particle without at the same time specifying the point, or origin, relative to which it is measured.

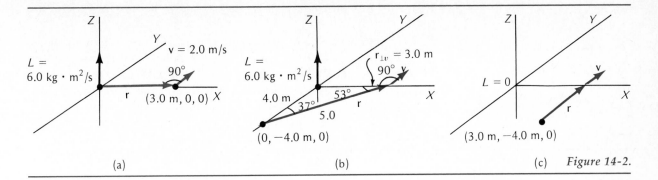

(a)   (b)   (c)   *Figure 14-2.*

Example 14-1.   See Figure 14-2(a). A 1.0-kg particle at the location 3.0 m along the positive X axis, otherwise denoted (3.0 m, 0, 0), has a velocity of 2.0 m/s along the positive Y direction. What is the particle's angular momentum (a) relative to the coordinate origin (0, 0, 0)? (b) relative to point (0, −4.0 m, 0)? (c) relative to point (3.0 m, −4.0 m, 0)?

(a) The origin of the radius vector **r** is the coordinate origin, so that **r** extends along the positive X direction. Since **r** is at right angles to **v**, then $r_{\perp v} = 3.0$ m, and the magnitude of the particle's angular momentum is

$$L = r_{\perp v}mv = (3.0 \text{ m})(1.0 \text{ kg})(2.0 \text{ m/s}) = 6.0 \text{ kg} \cdot \text{m}^2/\text{s}$$

From the right-hand rule for cross products, the direction of **L** is in the positive Z direction.

(b) With the particle's angular momentum now computed relative to the point (0, −4.0 m, 0), the magnitude and direction of the radius vector **r** to the particle are changed. See Figure 14-2(b). The magnitude of **r** is now 5.0 m, and it makes an angle of 53° with the positive X axis.

With $r_{\perp v} = r \sin 37°$, the magnitude of the angular momentum is

$$L = r_{\perp v}mv = (5.0 \text{ m})(0.60)(1 \text{ kg})(2.0 \text{ m/s}) = 6.0 \text{ kg} \cdot \text{m}^2/\text{s}$$

Both the direction and magnitude of **L** are unchanged compared with part (a).

(c) With the origin now at the point (3.0 m, −4.0 m, 0), as shown in Figure 14-2(c), we see that the particle is heading away from the origin for **L**. Vectors **r** and **v** are parallel, and therefore **L** = 0.

14-2   Angular Momentum of a Rigid Body

Here we are concerned with two simple types of rotating rigid objects: (1) a sheet of any shape rotating about an axis perpendicular to the sheet; (2) an extended symmetrical object rotating about a symmetry axis or about an axis that is parallel to a symmetry axis. Under these circumstances, an object's angular-momentum vector **L** is in the *same direction* as its angular-velocity vector ω. On the other hand, **L** may not be parallel to ω if these circumstances are not met.

The total angular momentum of a collection of particles is simply the vector sum of the angular momenta of the individual particles. In a rigid body all particles have the same angular velocity about the rotation axis. Then the total angular momentum of the rigid body takes a particularly simple form. With moment of inertia I and angular velocity ω as the rotational analogs of mass m

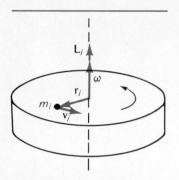

Figure 14-3. *The angular momentum $L_i = m_i v_i r_i$ of particle i of a rigid body rotating at angular velocity ω.*

and linear velocity **v** respectively, we can guess this form to be $\mathbf{L} = I\omega$. Now we prove it.

Consider the *i*th particle of the symmetrical rotating rigid body in Figure 14-3. Its mass is m_i and its velocity $\mathbf{v}_i$, and the radius vector $\mathbf{r}_i$ specifies its location relative to a point on the rotation axis. The particle travels in a circle of radius r_i, and its velocity $\mathbf{v}_i$ is perpendicular to $\mathbf{r}_i$. Therefore, the particle's angular momentum $\mathbf{L}_i = \mathbf{r}_i \times m_i \mathbf{v}_i$ is along the direction of the angular-velocity vector ω, parallel to the rotation axis. In magnitude the particle's angular momentum is

$$L_i = m_i v_i r_i$$

But the particle's speed v_i around the circle of radius r_i is given by $v_i = \omega r_i$, so that the relation above can be written

$$\mathbf{L}_i = m_i r_i^2 \omega$$

Note that this is a vector equation in which we have recognized that $\mathbf{L}_i$ is parallel to ω.

To find the total angular momentum, we sum over all *i* particles. We recognize that they all have the same angular velocity ω:

$$\mathbf{L} = (\Sigma m_i r_i^2)\omega$$

or

$$\mathbf{L} = I\omega \tag{14-3}$$

The rigid body's moment of inertia $I = \Sigma m_i r_i^2$ is measured relative to the rotation axis.

As we have seen, **L** for a single particle depends on the location of the point relative to which **L** is measured; the value of $\mathbf{L} = \mathbf{r} \times m\mathbf{v}$ depends on where the origin for **r** is located. It turns out that the angular momentum of an extended rigid body rotating about a symmetry axis, the object's *spin angular momentum*, has the same value, magnitude and direction, for *any* origin. The detailed proof is given in Example 14-2.

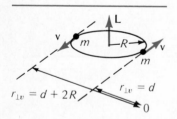

Figure 14-4. *The angular momentum of two particles moving in opposite directions as part of a spinning object.*

Example 14-2. Spin Angular Momentum.   What is the angular momentum of a spinning ring, computed relative to some arbitrary point not necessarily on the rotation axis?

It is sufficient to consider the angular momentum of two particles each of mass *m* on diametrically opposite sides of a ring of radius *R*, as in Figure 14-4. The two particles always move in opposite directions, each with speed *v*, in a circle of radius *R*. Relative to the arbitrary origin *O*, the nearer particle has $r_{\perp v} = d$. This particle's angular momentum has magnitude *mvd*, and its direction is down. For the farther particle, $r_{\perp v} = d + 2R$. This particle's angular momentum has magnitude $mv(d + 2R)$, and its direction is up. The net angular momentum is

$$L = mv(d + 2R) - mvd = R(2m)v$$

in the up direction. Note especially that the distance *d* has canceled out, so that the angular momentum for this pair of particles has the same value for any point in the plane of the ring. The total angular momentum for a ring of mass M is then *MvR*.

Any symmetrical object rotating about its symmetry axis can be considered a collection of symmetrical pairs of particles in the fashion of Figure 14-4. Threrefore, the angular momentum of such a spinning object, or its spin angular momentum, is an *intrinsic property* of the object. Spin angular momentum is independent of the point relative to which the angular momentum is measured.

14-3   Torque and Angular Momentum

The fundamental relation in rotational dynamics is that between a particle's angular momentum $\mathbf{L} = \mathbf{r} \times m\mathbf{v}$ and the torque $\tau = \mathbf{r} \times \mathbf{F}$ on the particle from resultant force $\mathbf{F}$. All other results described in this and the two preceding chapters flow from it.

Once again we have a particle of mass m and instantaneous velocity $\mathbf{v}$ at the location given by radius vector $\mathbf{r}$. The instantaneous resultant force on the particle is $\mathbf{F}$. See Figure 14-5. All three vectors—$\mathbf{r}$, $\mathbf{v}$, and $\mathbf{F}$—may change with time. Moreover, these three vectors need not point in the same direction.

We start with Newton's second law, now written in terms of the time rate of change of linear momentum,

$$\mathbf{F} = \frac{d}{dt}(m\mathbf{v})$$

and take the cross product (Section 12-2) of $\mathbf{r}$ with both sides of this equation:

$$\mathbf{r} \times \mathbf{F} = \mathbf{r} \times \frac{d(m\mathbf{v})}{dt} \tag{14-4}$$

Now consider the following derivative:

$$\frac{d}{dt}(\mathbf{r} \times m\mathbf{v}) = \left(\frac{d\mathbf{r}}{dt} \times m\mathbf{v}\right) + \mathbf{r} \times \frac{d(m\mathbf{v})}{dt} \tag{14-5}$$

(The rule for the derivative of a cross product is exactly analogous to the rule for the derivative of an algebraic product. This may be verified in detail by taking the derivative of the components of a cross product.)

Vector $\mathbf{r}$ locates the particle at every instant; therefore its time derivative is the particle's velocity, $d\mathbf{r}/dt = \mathbf{v}$. The first term on the right of (14-5) then becomes

$$\frac{d\mathbf{r}}{dt} \times m\mathbf{v} = \mathbf{v} \times m\mathbf{v} = 0$$

The cross product of any two parallel vectors, such as $\mathbf{v}$ and $m\mathbf{v}$, is zero.

The first term on the right side of (14-5) is then zero, and (14-4) then reduces to

$$\mathbf{r} \times \mathbf{F} = \frac{d}{dt}(\mathbf{r} \times m\mathbf{v}) \tag{14-6}$$

This is the fundamental relation between torque and angular momentum: torque equals time rate of change of angular momentum. In different symbols, (14-6) is

$$\tau = \frac{d\mathbf{L}}{dt} \tag{14-7}$$

To see the significance of this result, compare the starting point, (14-4),

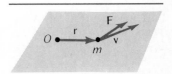

Figure 14-5. *A particle of mass m and velocity $\mathbf{v}$ is subject to a resultant force $\mathbf{F}$ when its displacement relative to point O is $\mathbf{r}$.*

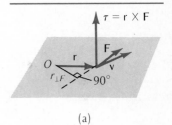

(a)

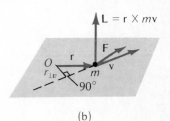

(b)

Figure 14-6. *(a) The torque on the particle $\tau = \mathbf{r} \times \mathbf{F}$. In magnitude, $\tau = Fr_{\perp F}$ where $\mathbf{r}_{\perp F}$ is perpendicular to $\mathbf{F}$ (not $\mathbf{v}$). (b) The particle's angular momentum $\mathbf{L} = \mathbf{r} \times m\mathbf{v}$. In magnitude, $L = mvr_{\perp v}$ where $\mathbf{r}_{\perp v}$ is perpendicular to $\mathbf{v}$ (not $\mathbf{F}$).*

with the ending point, (14-6). The operator d/dt has effectively been shifted to the left. But the meaning of "$\mathbf{r} \times$" on the two sides of (14-6) is not the same. In magnitude, $\mathbf{r} \times \mathbf{F}$ is $r_{\perp F}F$, where $r_{\perp F}$ is the component of $\mathbf{r}$ *perpendicular to* $\mathbf{F}$. On the other hand, the magnitude of $\mathbf{r} \times m\mathbf{v}$ equals $r_{\perp v}mv$, where $r_{\perp v}$ is the component of $\mathbf{r}$ *perpendicular to another vector* $m\mathbf{v}$. Unless $\mathbf{v}$ and $\mathbf{F}$ happen to be pointing in the same direction, $r_{\perp v}$ and $r_{\perp F}$ are not the same. See Figure 14-6.

14-4   Particle with Constant Angular Momentum

The basic relation describing how a particle's angular momentum changes with time is

$$\mathbf{r} \times \mathbf{F} = \frac{d}{dt}(\mathbf{r} \times m\mathbf{v}) \tag{14-6}$$

or written more compactly,

$$\boldsymbol{\tau} = \frac{d}{dt}\mathbf{L} \tag{14-7}$$

In itself this relation says nothing more than what is already contained in Newton's second law (after all, that's how we derived this relation in Section 14-3). What then is its special value? It is this: the preceding relation says that if the *torque* on a particle is *zero*, the *particle's angular momentum does not change.* In symbols, if $\boldsymbol{\tau} = \mathbf{r} \times \mathbf{F} = 0$, then

$$\frac{d}{dt}\mathbf{L} = \frac{d}{dt}(\mathbf{r} \times m\mathbf{v}) = 0$$

Therefore

$$\mathbf{r} \times m\mathbf{v} = \mathbf{L} = \text{constant}$$

Two important situations in which the torque $\boldsymbol{\tau} = \mathbf{r} \times \mathbf{F}$ on a particle *is* zero are these:

• The particle is isolated. The resultant force on the particle is zero, and the particle's angular momentum is then constant. (With $\mathbf{F} = 0$, we have $\mathbf{r} \times \mathbf{F} = 0$ and $\mathbf{r} \times m\mathbf{v} = \text{constant}$.)

• The particle *is* subject to an external force but the force $\mathbf{F}$ happens to be always parallel or antiparallel to radius vector $\mathbf{r}$. Then $\mathbf{r} \times \mathbf{F} = 0$, and the particle's angular momentum $\mathbf{r} \times m\mathbf{v}$ is again constant.

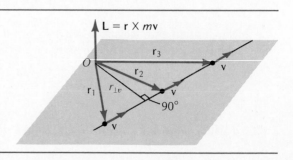

Figure 14-7. A particle with constant velocity has constant angular momentum relative to point O. The radius vector changes from $\mathbf{r}_1$ to $\mathbf{r}_2$ to $\mathbf{r}_3$ but $\mathbf{r}_{\perp v}$ is constant.

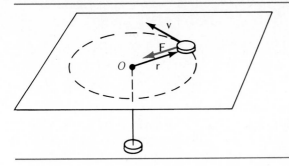

Figure 14-8.

Consider the example of an isolated particle in Figure 14-7. Here we see a particle subject to zero resultant force traveling in a straight line with constant velocity **v**. The radius vector from origin O to the particle changes direction and magnitude from $\mathbf{r}_1$ to $\mathbf{r}_2$ to $\mathbf{r}_3$ . . . , but the component of **r** at right angles to vector **v** remains constant, with a value $r_{\perp v}$. The angular-momentum vector $\mathbf{L} = \mathbf{r} \times m\mathbf{v}$ for the isolated particle measured relative to point O is constant in both magnitude and direction. The particle does not advance along the direction of **L** (it advances along the direction of **v**); nothing is moving in the direction of the angular momentum vector.

Suppose now that $\mathbf{r} \times \mathbf{F}$ is zero because **r** and **F** always point in the same direction or in opposite directions. Figure 14-8 shows an example; here a particle slides on a smooth horizontal surface but is attached to a cord under tension that passes through a small hole. We choose the origin O to be at the *force center*. Then the radius vector **r** is always opposite in direction to the force **F** in the particle. Therefore, $\mathbf{r} \times \mathbf{F} = 0$ and the particle's angular momentum is constant,

$$r_{\perp v} m v = \text{constant}$$

where $r_{\perp v}$ is the component of **r** perpendicular to **v**.

Any force between a pair of particles whose direction is always along the line connecting the particles is known as a *central force*. See Figure 14-9. An example of a central force is the force acting on a particle attached to the end of a stretched rubber band. At a more fundamental level, the electric force between two point electric charges or the gravitational force between two point masses are central forces. The important result then is this: if a particle is subject to a *central force,* the particle's *angular momentum is constant* when the particle's angular momentum is measured *relative to an origin at the fixed force center.*

An alternative way of describing the consequences of a *central force* is to say that the *radius vector* from the force center to the orbiting particle *sweeps through equal areas in equal time intervals.* To prove this we consider Figure 14-10. Over the short time interval dt, the particle with velocity **v** advances through a displacement **v** dt. The area dA swept out by the radius vector in the time interval dt is shown shaded in Figure 14-10; it is half the area of a parallelogram with sides **r** and **v** dt. But the area of the entire parallelogram is $|\mathbf{r} \times \mathbf{v}\, dt|$, so that

$$dA = \tfrac{1}{2}|\mathbf{r} \times \mathbf{v}\, dt|$$

But by definition,

$$\mathbf{L} = \mathbf{r} \times m\mathbf{v} = m(\mathbf{r} \times \mathbf{v})$$

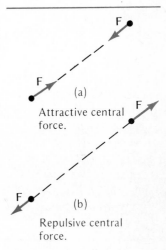

Figure 14-9.   Central forces: (a) Attractive force. (b) Repulsive force.

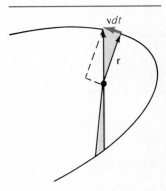

*Figure 14-10.   The area dA (shaded) swept out by the radius vector **r** in time dt is equal to dA = $\tfrac{1}{2}$|**r** × **v**| dt.*

Show Work

$$E = mc^2$$

A. E.

B+
Nice try, Albert.
Next time show
work.

Of course, Albert Einstein (A. E.) never turned in for grading a paper with the best-known equation in physics, $E = mc^2$, on it.* But the point made by the imaginary grader—that you must always "show work"—applies to any equation in physics.

An equation in physics is simply a statement in shorthand, symbolic form of something that can also be expressed in words. There is no special magic in an equation. Unless we know exactly what each of the symbols in an equation means, how the equation is arrived at from more basic principles, and what limitations apply to it, the equation itself is useless.

Physics does not consist solely or even primarily of a set of equations. Learning physics does not consist of memorizing equations. Of course, equations are important in physics: they play a central role because they can say very precisely and economically exactly what we want them to say. Moreover, an equation, even if manipulated mathematically, retains its logical content, and it can give alternative ways of expressing results. In other words, mathematics allows one statement to be transformed into a different equivalent statement.

For example, the statement that the radius vector from a force center to an orbiting particle sweeps out equal areas in equal times can be demonstrated through mathematical analysis to be equivalent to the statement that the force between the particle and the force center lies exactly along the line joining them. (The proof, first done by Isaac Newton, is given in Section 14–4.)

Of course, the equations of physics are important, but what is truly fundamental are the ideas, the principles, and the physics that lie behind the equations.

*This graffito was sighted in 1976 in the Hill Center for Mathematical Sciences, Rutgers University.

Therefore, in magnitude,

$$dA = \frac{1}{2} \frac{L}{m} dt$$

and the rate dA/dt at which the radius vector sweeps out area is

$$\frac{dA}{dt} = \frac{L}{2m} = \text{constant} \tag{14-8}$$

In short, for a particle subject to a central force, not only is the particle's angular momentum **L** constant, but the rate dA/dt at which the radius vector sweeps out the area is also constant.

It works the other way round. Suppose that the radius vector to a particle happens to sweep out equal areas in equal time intervals. It follows that the particle's angular momentum is constant and the force on it is *central.* Astronomical observations show that this is true for a planet moving round the sun; therefore, the gravitational force between the sun and each planet must lie along the line connecting them. (More about this in Chapter 15.)

Example 14-3. A particle slides at constant speed in a circle on a frictionless horizontal surface (see Figure 14-8). The particle is subject to a central force, one that points to the small hole through which the cord to the particle passes. Suppose that the vertical segment of the cord is pulled down slowly until the radius of the particle's circular path is halved. (a) How does the particle's speed in this smaller circle compare with its original speed? (b) Does the particle's kinetic energy change?

(a) The particle is subject to a *central* force, so that its angular momentum $r_{\perp v}\, mv$ is *constant*, where $r_{\perp v}$ is just the radius of the circle. Particle velocity $\mathbf{v}$ is very nearly perpendiuclar to $\mathbf{r}$. When the radius is reduced to half, the particle speed must double to keep the angular momentum constant.

(b) With the particle's speed doubled, its kinetic energy $\frac{1}{2}mv^2$ is *quadrupled*. Clearly, work must have been done in pulling the vertical segment of the cord downward, and the increase in the particle's kinetic energy equals the amount of work done. The force $\mathbf{F}$ on the orbiting particle is always opposite in direction to the radius vector $\mathbf{r}$. But the force vector is not precisely perpendicular to the particle's velocity $\mathbf{v}$, since the particle slowly spirals inward. There is a small component of $\mathbf{F}$ parallel to $\mathbf{v}$, and this force component does work on the particle and increases its kinetic energy.

14-5 Angular Momentum Conservation Law

We have seen that $\tau = d\mathbf{L}/dt$ holds for a single particle. Now we consider any system of particles and prove the most fundamental relation for angular momentum:

$$\Sigma \tau_{\text{ext}} = \frac{d\mathbf{L}}{dt} \qquad (14\text{-}9)$$

where $\Sigma \tau_{\text{ext}}$ is the vector sum of the external torques only on the system and $\mathbf{L}$ is the system's total angular momentum.

These particles are not necessarily parts of a rigid body. The particles interact and they are also subject to external forces. Particle 1 has mass m_1 and velocity $\mathbf{v}_1$, and the resultant force on it is $\mathbf{F}_1$. The radius vector from some arbitrarily chosen origin to particle 1 is $\mathbf{r}_1$. Then the resultant torque on particle 1 is $\tau_1 = \mathbf{r}_1 \times \mathbf{F}_1$, and its angular momentum is $\mathbf{L}_1 = \mathbf{r}_1 \times m_1\mathbf{v}_1$. Similarly, for particles 2, 3, and so on. We have the same *origin* for all radius vectors, angular momenta, and torques. See Figure 14-11.

Applying (14-7) to each of the particles in the system, we have

$$\tau_1 = \frac{d\mathbf{L}_1}{dt}, \quad \tau_2 = \frac{d\mathbf{L}_2}{dt}, \quad \cdots$$

Adding all these equations yields

$$\tau_1 + \tau_2 + \tau_3 + \cdots = \frac{d\mathbf{L}_1}{dt} + \frac{d\mathbf{L}_2}{dt} + \frac{d\mathbf{L}_3}{dt} + \cdots = \frac{d}{dt}(\mathbf{L}_1 + \mathbf{L}_2 + \mathbf{L}_3 + \cdots)$$

The left side of this equation is the resultant of *all* torques, and the term in parentheses on the right side is the system's total vector angular momentum $\mathbf{L}$. Therefore we can write

$$\Sigma \tau_{\text{all torques}} = \frac{d\mathbf{L}}{dt} \qquad (14\text{-}10)$$

Now consider two representative particles 1 and 2 in the system. They interact through the *internal* forces $\mathbf{F}_{2\,\text{on}\,1}$ and $\mathbf{F}_{1\,\text{on}\,2}$, where

$$\mathbf{F}_{2\,\text{on}\,1} = -\mathbf{F}_{1\,\text{on}\,2}$$

by Newton's third law. See Figure 14-12.

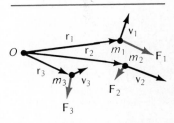

Figure 14-11. *A collection of particles subject to forces. Particle 1 of mass m_1 and velocity v_1 has a displacement $\mathbf{r}_1$ relative to the origin O and is subject to a resultant force $\mathbf{F}_1$. Likewise for particles 2, 3, and so on.*

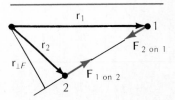

Figure 14-12. *Particles 1 and 2 interacting through a central force. The torques of both forces have the same moment arm $\mathbf{r}_{\perp F}$, and the resultant torque is zero.*

Suppose that particles 1 and 2 interact through a *central force*. Then $\mathbf{F}_{2\,\text{on}\,1}$ and $\mathbf{F}_{1\,\text{on}\,2}$ are not merely in opposite directions but also lie along the line connecting particles 1 and 2. As Figure 14-12 shows, the perpendicular distance $r_{\perp F}$ from any arbitrary origin to the common line of the two internal forces is the same for both forces. The magnitude of the torque $\tau_{2\,\text{on}\,1} = \mathbf{r}_1 \times \mathbf{F}_{2\,\text{on}\,1}$ equals the magnitude of $\tau_{1\,\text{on}\,2} = \mathbf{r}_2 \times \mathbf{F}_{1\,\text{on}\,2}$, since the magnitudes of the two forces are alike. But the directions of the two internal torques are *opposite;* one is clockwise and the other counterclockwise. The internal torques between particles 1 and 2 are equal and opposite, and their sum is zero:

$$\tau_{2\,\text{on}\,1} + \tau_{1\,\text{on}\,2} = 0$$

But if this applies for representative particles 1 and 2 interacting by a central force, it applies for all pairs of particles interacting mutually through central forces. Thus, the sum of the internal torques is zero:*

$$\Sigma \tau_{\text{int torques}} = 0 \qquad (14\text{-}11)$$

Now, the vector sum of all torques on the system of particles is merely the sum of the external torques (those arising from external forces) and the internal torques (from internal forces):

$$\Sigma \tau_{\text{all torques}} = \Sigma \tau_{\text{ext torques}} + \Sigma \tau_{\text{int torques}} \qquad (14\text{-}12)$$

Using (14-11) and (14-12) in (14-10), we have finally

$$\Sigma \tau_{\text{ext}} = \frac{d\mathbf{L}}{dt} \qquad (14\text{-}9)$$

This is the fundamental relation between angular momentum and torque. It says that the vector sum of the external torques equals the time rate of change of the system's total vector angular momentum.

From the general relation, we can easily deduce another important one that applies to a rigid body rotating about a symmetry axis. The system's total angular momentum is then

$$\mathbf{L} = I\omega$$

so that

$$\frac{d\mathbf{L}}{dt} = I\frac{d\omega}{dt} = I\alpha$$

since I is a constant. Substituting this result in (14-9), we get

$$\Sigma \tau_{\text{ext}} = I\alpha \qquad (14\text{-}13)$$

This is the fundamental relation used in Chapter 13. Rigid-body equilibrium is merely that special situation for which $\alpha = 0$ so that $\Sigma \tau_{\text{ext}} = 0$.

Another fundamental result applies to any *isolated* system of particles. No external forces act on such a system, and clearly the resultant external torque on the system is also zero:

$$\Sigma \tau_{\text{ext}} = 0$$

---

* Suppose that the forces between interacting particles were not zero. Then there could be a spontaneous angular momentum change even in the absence of external torques. But this has never been observed, constituting the best in direct proof that all interaction forces are central.

Then from (14-9) we have $d\mathbf{L}/dt = 0$, or

Total angular momentum $= \mathbf{L} = \mathbf{L}_1 + \mathbf{L}_2 + \mathbf{L}_3 + \cdots$
$$= \text{constant} \qquad \text{(14-14)}$$

This is the *law of conservation of angular momentum:* Any system of particles interacting through central forces and subject to *no net external torque* has constant angular momentum.

Everyday examples of angular momentum conservation abound. Suppose, for example, that a symmetrical object is spinning about its symmetry axis and is so mounted (in frictionless gimbals) that no resultant torque can act on it. See Figure 14-13. The spinning object is a *gyroscope* and its angular momentum is constant. If the frame to which the gimbals are attached is moved around or even reoriented in space, the gyroscope continues spinning at the same constant rate, with $\mathbf{L} = I\omega$, and its spin axis continuously pointing in the *same direction*. A freely spinning gyroscope may be used to navigate through space; it is effectively a compass whose direction does not depend on magnetic effects.

Figure 14-13.   A gyroscope, mounted in gimbals.

Consider another example of the constancy of $\mathbf{L} = I\omega$ in the absence of a net external torque. A figure skater executes a spin about a vertical axis (from head to toe). The skater starts spinning with arms outstretched, as shown in Figure 14-14(a). Then, as the skater draws arms in close to the body, her moment of inertia $I = \Sigma m_i r_i^2$ decreases, since mass has been shifted closer to the rotation axis. See Figure 14-14(b). Since the skater's I has decreased, ω must increase correspondingly to maintain $\mathbf{L} = I\omega$ constant. The skater's spin rate increases as she pulls in her arms. Conversely, a spinning skater can put on the brakes, so to speak, or at least slow the spin rate, by extending the arms.

A figure skater is certainly not a rigid body. The change in spin rate arises from the redistribution of mass relative to the rotation axis and the concomitant change in the body's moment of inertia. Another example of a change in spin

(a)

(b)

Figure 14-14.   (a) Large I, low ω. (b) Smaller I, larger ω.

Figure 14-15. Strobe photo of a diver.

rate originating from a change of the human body's mass distribution is seen in a somersault performed by a diver, a trampolinist, or a gymnast.

Figure 14-15 shows the position, at several instants, of a diver executing a somersault about a horizontal rotation axis. We note, first, that the *center of mass* of the diver traces out a *parabolic* path, for the resultant gravitational force on the diver can be thought to act at the diver's center of gravity.

In general, $\Sigma \mathbf{F}_{\text{ext}} = M\mathbf{A}$.

Here we have $M\mathbf{g} = M\mathbf{A}$, so that

$$\mathbf{A} = \mathbf{g}$$

and the center of mass has the same trajectory as a single point mass. The resultant external torque on the diver is zero, and his angular momentum is constant.

Table 14-1 is a summary of corresponding relations in translational and rotational dynamics.

Table 14-1. Analogous translational and rotational quantities

QUANTITY	TRANSLATION	ROTATION
Momentum	$\mathbf{p} = m\mathbf{v}$	$\mathbf{L} = \mathbf{r} \times m\mathbf{v}$ Rigid body about symmetry axis: $\mathbf{L} = I\omega$
Force (torque)	$\mathbf{F}$	$\tau = \mathbf{r} \times \mathbf{F}$
Newton's second law, single particle	$\mathbf{F} = \dfrac{d\mathbf{p}}{dt}$	$\tau = \dfrac{d\mathbf{L}}{dt}$
Newton's second law, any system	$\Sigma \mathbf{F}_{\text{ext}} = M\mathbf{A}$	Rigid body about symmetry axis: $\Sigma \tau_{\text{ext}} = I\alpha$
Momentum conservation	If $\Sigma \mathbf{F}_{\text{ext}} = 0$, then $\mathbf{P} = $ constant	If $\Sigma \tau_{\text{ext}} = 0$, then $\mathbf{L} = $ constant

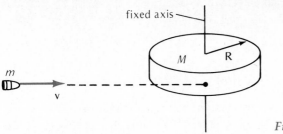

fixed axis

M R

m

v

Figure 14-16.

Example 14-4. A bullet of mass m traveling with a speed v is shot into the rim of a right circular cylinder of radius R and mass M, as shown in Figure 14-16. The cylinder has a fixed vertical rotation axis and is originally at rest. What is the angular speed of the cylinder after the bullet has become imbedded in it?

Our system is the bullet and the cylinder. We choose a point on the cylinder's axis for computing torques and angular momentum. The system is not isolated; a force clearly acts on the cylinder at its rotation axis. But the torque of this force is zero. Consequently, the system of bullet and cylinder is subject to no external torque, and its total angular momentum is constant.

The magnitude of the bullet's angular momentum before the bullet strikes the cylinder is mvR, since R is the perpendicular distance between the point on the axle and the direction of **v**. After the bullet is captured, the total moment of inertia is that for the cylinder alone ($\frac{1}{2}MR^2$) plus that for the bullet (mR^2), or $I = \frac{1}{2}MR^2 + mR^2$.

Equating total angular momentum before the collision with the total after, we have

$$mvR = (\tfrac{1}{2}M + m)R^2\omega$$

where ω is the cylinder's angular speed, or

$$\omega = \frac{mv}{(\tfrac{1}{2}M + m)R}$$

Although angular momentum is conserved in this example, kinetic energy is not, since a nonconservative force acts on the bullet as it is captured.

Example 14-5. A disk with a moment of inertia I relative to a vertical axis through its center is dropped onto an identical disk initially spinning at angular velocity ω_i (like a phonograph record dropped on a spinning turntable). See Figure 14-17. (a) What is the final common angular velocity of the two disks? (b) What fraction of the initial rotational kinetic energy is dissipated?

Our system consists of the two disks. We neglect any *external* friction acting on the disks; there is then no resultant external torque on the disks, and their total angular momentum is constant. Angular momentum conservation requires that total angular momentum before the disks touch equal total angular momentum after both are spinning at the same final rate ω_f, or

$$I\omega_i = (2I)\omega_f$$

$$\omega_f = \tfrac{1}{2}\omega_i$$

With the moment of inertia of the spinning object doubled, the angular velocity is halved.

Internal torques do act within this system. The upper disk produces a frictional torque on the lower disk, slowing its spin; and the lower disk produces a frictional torque on the upper disk, speeding its spin rate. The sum of the two torques is zero at every instant.

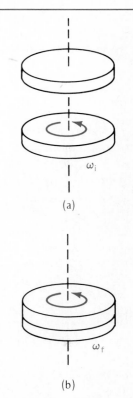

ω_i

(a)

ω_f

(b)

Figure 14-17. A clutch arrangement. (a) Only the lower disk is spinning at the rate ω_i. (b) Both disks are spinning together at rate ω_f.

(b) The kinetic energy initially is

$$K_i = \tfrac{1}{2}I\omega_i^2$$

and the final rotational kinetic energy is

$$K_f = \frac{1}{2}\,(2I)\left(\frac{\omega_i}{2}\right)^2 = \frac{1}{2}\left(\frac{1}{2}\,I\omega_i^2\right) = \frac{1}{2}\,K_i$$

Half the initial kinetic energy is dissipated in what might be called an inelastic rotational collision (in an arrangement more commonly called a clutch). The rotational collision is analogous to an ordinary head-on inelastic collision in which two objects stick together after they collide; that system's linear momentum is also constant, and there too the kinetic energy decreases because of the presence of a nonconservative force.

14-6   Precession of a Top (Optional)

Here we see how the general relation $\Sigma\tau_{ext} = d\mathbf{L}/dt$ can be used to explain the otherwise apparently paradoxical motion of a spinning top. A complete, detailed analysis of top motion is extremely complicated; what follows here is only an approximate explanation of the principal features.

Consider the situation shown in Figure 14-18(a), where a top is free to pivot about the fixed point P. The left end is initially raised to the position shown in Figure 14-18(a). Here the top is not spinning, and its center of mass is initially at rest. On being released, the top simply falls down. How do we describe this in the language of torque and angular momentum? We say that the top is subject to two external forces—its weight $m\mathbf{g}$ acting downward at the top's center of gravity, and the upward force $\mathbf{F}$ at the point of support. As Figure 14-18(a) shows, the resultant torque τ_{ext} of these two forces about the pivot point is horizontal. Since the top is released from rest, its angular momentum is initially zero. The change $\Delta\mathbf{L}$ in angular momentum in the time interval Δt is, from (14-9), given by

$$\Delta\mathbf{L} = \tau_{ext}\,\Delta t$$

Therefore, when the top turns downward about the pivot point, the angular momentum it acquires must be in the same direction as the resultant torque, namely horizontal. More succinctly, the top falls.

Now suppose that the top is initially spinning about its symmetry axis with

Figure 14-18.   A nonspinning top is released from rest. The top falls, the direction of the angular momentum change $\Delta\mathbf{L}$ is along the direction of the resultant torque τ on the top. (b) A spinning top is released: Again, $\Delta\mathbf{L}$ is parallel to τ, but since $\Delta\mathbf{L}$ is perpendicular to $\mathbf{L_s}$, the top undergoes precession at the angular velocity ω_p about a vertical axis.

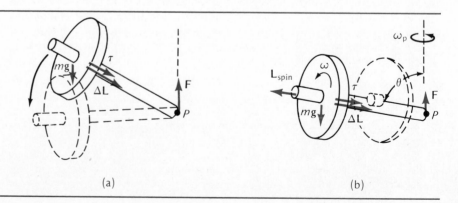

(a)   (b)

angular momentum **L**. See Figure 14-18(b). Again, a resultant torque acts along the horizontal, and the change in the top's angular momentum Δ**L** must again be along the horizontal. Torque τ and angular momentum change Δ**L** are, however, always perpendicular to **L**. This implies that the entire spinning top must turn in a horizontal plane about its point of support, as shown in Figure 14-19(a). For any small angular displacement $\Delta\phi$ in the horizontal plane, both the τ and Δ**L** vectors undergo the same angular displacement. The top continues spinning about its own axis, but it also turns in the horizontal plane. The top does not fall. It undergoes circular motion in this horizontal plane. This is known as *precession.*

What is the angular velocity of precession ω_p, the rate at which the top turns about a vertical axis? Suppose that the top has some arbitrary orientation θ relative to the vertical, as in Figure 14-19(a). The spin angular momentum $\mathbf{L}_{spin}$ points along the symmetry axis. The change in angular momentum over the interval Δt is Δ**L**.

As the top precesses, the radius vector in the horizontal plane undergoes an angular displacement $\Delta\phi$ in time Δt. The geometry of the figure also shows that

$$\Delta L = (L_{spin} \sin \theta) \, \Delta\phi$$

The top's center of mass is a distance d from the point of support, as shown in Figure 14-19(b). Therefore, the resultant torque, computed relative to the fixed end, has the magnitude

$$\tau = mg(d \sin \theta)$$

Substituting the two equations above in the general relation $\Delta L = \tau_{ext} \, \Delta t$, we have

$$L_{spin} \sin \theta \, \Delta\phi = mgd \sin \theta \, \Delta t$$

But by definition, the precession angular velocity ω_p is

$$\omega_p = \frac{\Delta\phi}{\Delta t} = \frac{mgd}{L_{spin}}$$

Note that the precession rate is independent of the top's inclination angle θ. Moreover, since ω_p is inversely proportional to L_{spin}, the higher the spin rate, the lower the precession rate.

It is easy to see that the preceding analysis is only approximately correct. For one thing, if the top undergoes precession about a vertical axis, then the top's angular momentum is not merely that due to spin but must also include the angular momentum of the top as a whole as it precesses about a vertical axis. Moreover, there must be kinetic energy associated with precession, and this implies that the system's potential energy or spin kinetic energy must change. Actually, if a spinning top is released from rest, it initially drops slightly (increasing angle θ) as the precession ensues. More generally, the precession is accompanied by *nutation*, a regular oscillation in angle θ.

Despite these shortcomings, it is clear that the recognition of angular momentum as a *vector* quantity is the key to understanding so complicated a motion as that of a spinning top. Here we have a torque at right angles to (spin) angular momentum, and the angular momentum vector rotates at a constant rate without changing magnitude. This is exactly analogous to the situation in which an external force acts on a particle at right angles to its linear momen-

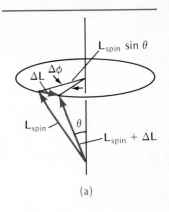

(a)

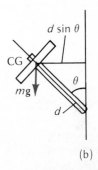

(b)

Figure 14-19. *(a) Changes in the angular momentum $\mathbf{L}_s$ of a spinning and a precessing top. (b) The torque on a top is $mg\,(d \sin \theta)$.*

tum; then the momentum changes direction continuously, but not magnitude. In other words, the particle moves in a circle at constant speed. The question of why a spinning top does not necessarily fall is no more (nor less) mysterious than the question why a particle can move in a circle with constant speed even though it is acted on continuously by a radial force of constant magnitude.

Summary

Definitions

Angular momentum of particle relative to the origin of radius vector **r**:

$$\mathbf{L} \equiv \mathbf{r} \times m\mathbf{v} \tag{14-1}$$

Magnitude: $L = r_{\perp v} mv$
Direction: right-hand rule for cross product.
Central force: a force whose direction lies on the line connecting two interacting point objects.

Units

Angular momentum: $\text{kg} \cdot \text{m}^2/\text{s} = \text{J} \cdot \text{s}$

Fundamental Principles

For a particle, torque is the time rate of change of angular momentum:

$$\mathbf{r} \times \mathbf{F} = \frac{d}{dt}(\mathbf{r} \times m\mathbf{v}) \tag{14-6}$$

$$\tau = \frac{d\mathbf{L}}{dt} \tag{14-7}$$

Angular momentum conservation law: The total (vector) angular momentum of an isolated system is constant,

$$\Sigma\mathbf{L} = \text{total angular momentum}$$
$$= \text{constant for } \Sigma\tau_{\text{ext}} = 0, \tag{14-14}$$

Important Results

The angular momentum of a rigid body spinning about a symmetry axis is

$$\mathbf{L} = I\omega \tag{14-3}$$

A single particle has constant angular momentum when the torque on it remains zero:

$$\mathbf{L} = \mathbf{r} \times m\mathbf{v} = \text{constant} \qquad \text{for } \tau = 0$$

Two important cases for $\tau = \mathbf{r} \times \mathbf{F} = 0$:

- $\mathbf{F} = 0$; an isolated particle.
- **r** and **F** parallel or antiparallel; a central force.

An alternative way of expressing the constancy of a particle's angular momentum when the particle is subject to a central force: The radius vector from the force center to the particle sweeps out equal areas in equal time intervals.
For any system of particles interacting by central forces,

$$\Sigma\tau_{\text{ext}} = \frac{d\mathbf{L}}{dt} \tag{14-9}$$

The time rate of change of system's total angular momentum is determined by the external torques alone on the system.

Problems and Questions

Section 14-1   Angular Momentum of a Particle

· **14-1 Q** A bowler slides a frictionless block at constant speed straight along the alley in the direction north. The block misses the pin for which he aimed by 0.5 m to the east. Which statement is true? (Up means away from the earth's center; down means toward the earth's center.) The angular momentum of the block relative to the pin is
(A) zero.
(B) up and decreases in magnitude to a minimum as the ball passes the pin.
(C) down and decreases in magnitude to a minimum as the ball passes the pin.
(D) up and constant in magnitude.
(E) down and constant in magnitude.

· **14-2 P** A particle of 1 kg mass is traveling left along the line $y = -3$ m at a speed of 2 m/s, as shown in Figure 14-20. The angular momentum of the particle relative to the origin

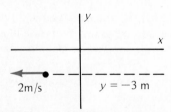

Figure 14-20.   Problem 14-2.

(A) cannot be determined without knowing the particle's x coordinate.
(B) is 6 kg-m²/s along the +x direction.
(C) is 6 kg-m²/s into the paper.
(D) is 6 kg-m²/s out of the paper.
(E) is 2 kg-m²/s along the −x direction.

· **14-3 P** Two identical particles, each of mass m, with equal speed v move in opposite directions along parallel paths separated by distance d. What is the angular momentum of this system? Does it depend on the choice of origin?

· **14-4 P** In magnetic-mirror machines used for plasma containment, the orbital angular momentum of a spiraling electron remains constant. If the circular-orbit diameter is reduced by a factor two, by what factor does the electron's speed change?

: **14-5 P** A particle of mass 2 kg located at $x = 1$ m, $y = 2$ m has velocity components $v_x = 4$ m/s, $v_y = 2$ m/s. What is the particle's angular momentum with respect to the origin?

: **14-6 P** An electron with velocity 4×10^5 m/s is located at (2, 2, 2) in cm. The electron has equal velocity components: $v_x = v_y = v_z$. What is the electron's angular momentum with respect to the origin?

· **14-7 Q** In an angular momentum experiment, a steel ball of mass m traveling with speed v strikes a suspended horizontal disk as shown in Figure 14-21(a). The U-shaped catcher is attached rigidly to the disk. Suppose that we reverse the direction of motion and the point of impact as in Figure 14-21(b), where now the initial speed is $v/2$. Which of the following statements is true?
(A) The angular momentum of the system in (b) has two times the magnitude of (a) and the same direction.

(B) The angular momentum of the system in (b) has two times the magnitude of (a) and the opposite direction.
(C) The angular momentum of the system in (b) has one half the magnitude of (a) and the opposite direction.
(D) The angular momentum of the system in (b) has one half the magnitude of (a) and the same direction.
(E) None of the above.

: **14-8 P** A 3-kg particle at the point $x = 6$ m, $y = 0$, is moving 5 m/s as shown. It is acted on by a force of 7 N directed as shown in Figure 14-22. With respect to the origin, what is the magnitude of (a) the angular momentum of the particle? (b) the torque acting on the particle?

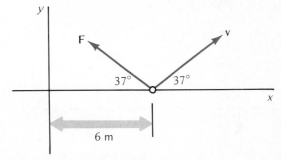

Figure 14-22.   Problem 14-8.

: **14-9 P** Two bullets are shot in succession into a freely pivoted wooden cylinder initially at rest. The first bullet with a mass m and a speed v misses the center of the cylinder by a distance d_1. When the first bullet is captured by the cylinder, it rotates in the counterclockwise sense. The second bullet is to bring the cylinder to rest once more. Its mass is $2m$ and its speed is $2v$. By what distance d_2 (in terms of d_1) must the second bullet miss hitting the center of the cylinder?

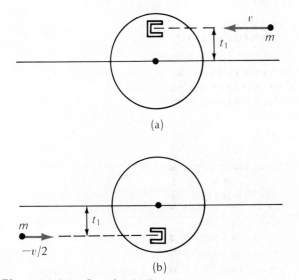

(a)

(b)

Figure 14-21.   Question 14-7.

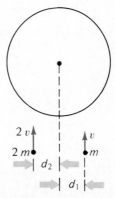

Figure 14-23.   Problem 14-9.

Section 14-2   Angular Momentum of a Rigid Body
: **14-10 P** What is the ratio of the earth's spin angular momentum to its orbital angular momentum, measured with respect to the sun?

: **14-11 P** A planetary gear of mass M and radius R rolls around the inside of a larger stationary gear of inner radius $2R$ with angular frequency ω about its own center. See Figure 14-24. What is the angular momentum of the small gear, for $R = 4$ cm, $\omega = 6\,\text{s}^{-1}$, and $M = 0.2$ kg, (a) about the center of the small gear, (b) about the center of the large gear. The small gear is essentially a uniform disk.

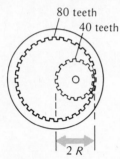

80 teeth
40 teeth
2 R

Figure 14-24. Problem 14-11.

: **14-12 P** A sphere of mass M and radius R, spinning with angular velocity ω_0 about a horizontal axis is dropped onto a rough floor. It takes off, rolling and slipping simultaneously. Eventually it stops slipping and rolls smoothly at constant velocity. If the coefficient of kinetic friction with the floor is μ_k, what is the final velocity of the sphere?

Section 14-3 Torque and Angular Momentum

· **14-13 Q** Why does a helicopter have a small propeller mounted sideways on the tail? What would happen if there were no such propeller?

· **14-14 Q** If the net torque acting on a rigid body is constant in direction and magnitude, then
(A) the force acting on the object is zero.
(B) the force acting on the object cannot be zero.
(C) the linear momentum of the object is constant.
(D) the forces acting on the object are all directed toward a single central point.
(E) the angular momentum of the object varies linearly with time.

: **14-15 Q** Suppose a mass is attached to a metal bar in such a way that it may be secured anywhere along the bar. How will it be easiest to balance the bar on end in your

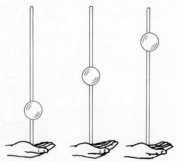

Figure 14-25. Question 14-15.

hand? See Figure 14-25. Should the mass be near your hand, far from your hand, or in the middle of the bar?

: **14-16 Q** If you cross a creek on a floating log and start to lose your balance, you instinctively put your arms out to keep from falling. For the same reason, a tightrope walker often holds a very long pole. Explain.

: **14-17 Q** Is it easier to walk on tall stilts or on short ones? Explain.

· **14-18 Q** A particle of mass m is at the location (x, y, z). The particle's velocity and acceleration components are v_x, v_y, v_z and a_x, a_y, a_z. What is (a) the particle's angular momentum and (b) the resultant torque on the particle, both measured relative to the origin?

: **14-19 P** A rigid rod of length 1.0 m and negligible mass is free to rotate on a horizontal frictionless table about a vertical axle through the center of the rod. See Figure 14-26. The rod is initially at rest. A 3.0-kg disk is attached to the left end of the rod. A 1.0 kg disk is then shot with speed 10 m/s toward the right end of the rod along the perpendicular path shown in the figure. This 1.0-kg disk collides with the right end of the rod and bounces back at 5.0 m/s, in a direction opposite to its initial velocity.

(a) After the collision, what is the angular momentum of the 3.0-kg disk about the vertical axle through the center of the rod? (b) This collision lasts (i.e., the 1.0-kg disk is in contact with the rod) for 0.1 s. What is the magnitude of the average force on the 1.0-kg disk during this time? (c) During the 0.1-s interval of the above collision, what is the magnitude of the average torque (about the vertical axle) on the 3.0-kg disk?

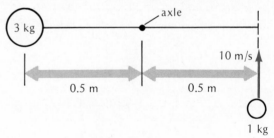

axle
3 kg
10 m/s
0.5 m   0.5 m
1 kg

Figure 14-26. Problem 14-19.

Section 14-4 Particle with Constant Angular Momentum

· **14-20 P** A planet travels in an eccentric elliptical orbit around a star. The planet's maximum distance from the star is 5 times its distance of closest approach. If the minimum speed of the planet is 6 km/s, what is its maximum speed?

: **14-21 P** A planet follows the elliptical orbit shown in Figure 14-27 in its motion about a star. Its period is 10 months. If it is at point A at $t = 0$, estimate graphically its position at one month intervals throughout its motion.

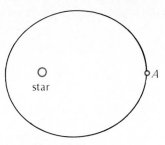

Figure 14-27.   Problem 14-21.

: **14-22 P** A ball of mass 0.5 kg is whirled in a horizontal circle on the end of a string 2 meters long. It is whirled at a frequency of 2 Hz. (*a*) If the string is pulled in until the radius of the circle traversed by the ball is 1 meter, how fast will the ball then be rotating? (*b*) How much work must be done to accomplish this?

Section 14-5   Angular Momentum Conservation Law

· **14-23 P** An ice skater beginning a spin with her arms out makes 2 revolutions per second. She then pulls her arms in and increases her rotation rate to 8 revolutions per second. Assuming the ice exerts no torque on the tip of her skate, what is the ratio of her moment of inertia with arms out to her moment of inertia with arms in?

· **14-24 Q** Suppose that burning fossil fuels created so much carbon dioxide in the air that the "Greenhouse Effect" (the trapping of infrared radiation under the canopy formed by the CO_2) caused the earth's temperature to rise slightly, melting the polar icecaps. If the melted water flowed to the equatorial regions because of the earth's rotation, what effect would this have on the rate at which the earth rotates, i.e., on the length of a day?

· **14-25 Q** Many phenomena, such as a somersaulting gymnast, a hurricane, and a pirouetting ice skater, can be most easily understood by recognizing that
(A) the net external force acting is zero.
(B) the net external torque acting is zero.
(C) the total energy is conserved.
(D) the linear momentum is conserved.
(E) the total mass of the system is constant.
(F) the moment of inertia of the system is constant.

· **14-26 P** Disks A and B are rotating with $\omega_A = 800$ rad/s and $\omega_B = 1200$ rad/s in the same sense as shown in Figure

14-28(a). Their moments of inertia are $I_A = 2$ kg·m² and $I_B = 4$ kg·m². A clutch mechanism brings them together and, after some slipping, they rotate together. (*a*) What is the final angular velocity of the pair? (*b*) How much thermal energy is generated in this process?

· **14-27 Q** A bullet of mass m initially moving at speed v is shot into the rim of a solid cylinder initially at rest. The cylinder's mass is M and its radius is R; the cylinder can spin freely on a frictionless fixed axle. (See Figure 14-29.) After the bullet has been captured by the cylinder, the angular momentum of the cylinder (with the bullet imbedded in it) is
(A) $(m + M)vR$
(B) $mv/(\frac{1}{2}M + m)R$
(C) mvR
(D) $\frac{1}{2}MRv$
(E) $mvR/(M + m)$

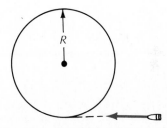

Figure 14-29.   Question 14-27.

: **14-28 Q** One of two identical cylindrical disks is initially spinning about its symmetry axis while the second disk is initially at rest. [See Figure 14-30(a).] Then the spinning disk comes in contact with the second disk until the two disks spin together as a single composite object. The two disks are isolated from the rest of the world. Each disk produces frictional forces on the other disk. [See Figure 14-30(b).] It is then true that
(A) the total kinetic energy and angular momentum are, relative to their initial values, both unchanged.
(B) the total kinetic energy and angular momentum are both reduced to half of their respective initial values.
(C) the total angular momentum is unchanged, but the final kinetic energy is 1/6 of its initial value.
(D) the total angular momentum is reduced to half its initial value, but the total kinetic energy is unchanged.
(E) the total angular momentum is reduced to half and the total energy is reduced to one-quarter of their respective initial values.

(a)   (b)

Figure 14-28.   Problem 14-26.

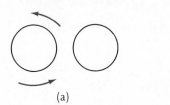

(a)   (b)

Figure 14-30.   Question 14-28.

: **14-29 P** When walking at her normal rate, a girl of mass 40 kg travels at 1 m/s. Suppose that she stands near the outer edge of a merry-go-round of mass 100 kg and radius 5 m. (*a*) If she starts walking around the outer edge of the merry-go-round, which is initially at rest, how long will it take her to return to the point with respect to the earth where she started? (*b*) Through how many revolutions will the merry-go-round have then rotated?

: **14-30 P** Four children, each of mass 20 kg, cling to the outer edge of a merry-go-round that is turning freely at a rate of 1 revolution every 2 seconds. The merry-go-round is a uniform disk of mass 40 kg and 6 m diameter.

(*a*) The children all pull themselves in to the very center of the merry-go-round. What is the resulting frequency of revolution of the merry-go-round? (*b*) What is the kinetic energy of the system before and after the children go to the center? What happens to the energy?

: **14-31 P** A person stands on the outer edge of a rotating merry-go-round, which is a frictionless disk of mass 50 kg. When the person moves into the very center of the merry-go-round the rotation rate doubles. What is the person's mass?

: **14-32 P** A merry-go-round consists of a disk of 5 m radius and 300 kg mass. The merry-go-round is initially at rest, and a child of 30 kg mass stands at a point on the outer edge next to the place where the brass ring is attached to a pole mounted on the ground. The child is able to run at a speed of 3 m/sec. If she runs at this steady speed along the outer edge of the merry-go-round, how long will it take her to return to the brass ring, assuming that friction is negligible?

: **14-33 Q** A cat held upside down and dropped is able to right itself and land on its feet. While falling it is subject to no external torque, and it may seem strange that it is able to rotate its body while in the air. How does a cat do this? How could you rotate yourself while sitting in a swivel chair and holding a weight in your hand?

: **14-34 P** A cylindrical shell of radius R rotates with its axis horizontal, as shown in Figure 14-31. A small ball of radius r rolls without slipping inside the cylinder. The coefficient of static friction between shell and ball is μ_s. (*a*) When the cylinder rotates with constant angular velocity ω, at what

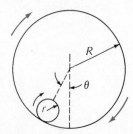

Figure 14-31. Problem 14-34.

angle θ does the center of mass of the ball come to rest? (*b*) When the cylinder rotates with constant angular acceleration α, at what angle θ does the center of mass of the ball come to rest?

: **14-35 Q** A student stands on a turntable and with his right hand holds the axle of a bicycle wheel in a vertical direction so that the rim of the wheel is now directly over his head, as shown in Figure 14-32.

Both the student and the wheel are initially at rest, and all axles are frictionless. The student then reaches directly over his head with his left hand and spins the wheel by pushing tangentially on the rim.

Not only does the wheel start spinning about its axle, but the "student plus turntable" starts rotating about the axle of the turntable. Note that the force of the student's left hand on the wheel, as well as the reaction force of the wheel on the student's left hand, exert no torque about the axle of the turntable.

What torques, if any, act on the "student plus turntable" during this process? Give the line of action and the name of the force causing any such torque.

Figure 14-32.
Question 14-35.

Figure 14-33.
Question 14-36.

: **14-36 Q** A student stands on a turntable and with his right hand holds a bicycle wheel directly above his head so that the vertical axle of the wheel is along the same line as the vertical axle of the turntable. (See Figure 14-33). Both the student and the wheel are initially at rest and all axles are frictionless. The student then reaches out with his left hand and spins the wheel by pushing tangentially on the rim. He gives it an angular momentum $+L_0$ as seen from above.

(*a*) What is the angular momentum L_s of the "student plus turntable" about the vertical axle of the turntable? Express your answer in terms of L_0 and include the proper sign to indicate the sense of rotation as seen from above. (*b*) What torques, if any, act on the "student plus turntable" during this process? Give the line of action and the name of the force causing any such torque.

: 14-37 P Many stars are members of binary systems. A binary system consists of two nearby stars that rotate around each other like spinning dumbbells. One of the stars may not emit enough light to be seen; its presence is inferred from the motion of its partner. A black hole may be a member of a binary system. Such an object is so dense that light cannot escape from it. A good candidate for a black hole occurs in the X-ray source Cygnus X-1. The observable star of this binary system rotates with a period of 5.6 days. It is believed that a black hole may pull off some mass from its partner, and radiation from the resulting accretion layer of matter flowing from one star to the other may indicate the presence of the black hole. Suppose that the black hole in Cygnus X-1 has a mass twice that of the sun and its partner a mass ten times that of the sun. What would the period of rotation become when enough mass had been transferred so that both stars had equal masses?

: 14-38 P Suppose that a star with a mass twice that of the sun and a radius equal to the sun were rotating with a period of 32 days. If such a star were to collapse to form a neutron star of radius 20 km, what would its period of rotation then become?

: 14-39 P When a star burns up much of its nuclear fuel it may undergo a gravitational collapse that changes it into a very dense form called a neutron star. Such stars have a mass about twice that of the sun, and they are extremely dense, with a radius of perhaps 13 km for a mass twice that of the sun. Neutrons are magnetic, and as the star collapses huge magnetic fields build up. As the star rotates it radiates a beam of electromagnetic waves, much like the rotating beacon on a lighthouse. Pulses from such stars have been detected, and these objects are called *pulsars.*

The best known pulsar is one in the Crab nebula. It rotates at 30 revolutions per second. The fastest pulsar discovered as of 1983 is labelled 1937 + 215, buried somewhere in the radio source known as 4C21.53. This star rotates at 642 Hz. Assuming it has a mass comparable to that of the pulsar in the Crab nebula, what would you estimate the ratio of the sizes of these two stars to be?

: 14-40 P A thin iron rod of mass M and length L rests motionless on a frictionless air table. A small magnetic puck of mass M, directed perpendicularly to the rod, strikes one end with velocity v and sticks to it.

(*a*) What is the velocity of the center of mass of the rod-plus-puck system before and after the collision? (*b*) What is the angular momentum of the rod-plus-puck system relative to its center of mass before and after the collision? (*c*) What is the kinetic energy of the system before and after the collision? (*d*) What is the angular velocity of the rod about the center of mass just after the collision?

Supplementary Problems

14-41 P The *New York Times* of July 18, 1982, reported that the 17-year-old aerialist, Miguel Vasquez, of Ringling Brothers and Barnum and Bailey Circus had performed a *quadruple* somersault before 7,000 spectators in Tucson. The news report also stated that Vasquez was "spinning out of the fourth revolution at 75 miles per hour." Is the 75 mph figure reasonable? [The following are approximate moments of inertia for a "typical" human body in several configurations: $I_{straight} = 20$ kg·m^2; $I_{piked} = 6$ kg·m^2; and $I_{tucked} = 1$ kg·m^2. See C. Frohlich, *Am. J. Physics* **47**, 583 (1979). Make reasonable assumptions about the trajectory of the center of mass and other aspects of this remarkable stunt.]

14-42 P The Solar Challenger is a solar-powered aircraft, designed by Paul MacCready who also designed the pedal-powered Gossamer Albatross; the Solar Challenger operates on 16,128 photovoltaic cells on the wings and tail that convert sunlight to electricity and power the 2.7 hp motor turning the propeller. The plane itself weighs 210 lb; its pilot, Janice Brown, weighed 93 lb. The Solar Challenger crossed the English Channel on July 7, 1981, at an average speed of 30 mph. What is the ratio of the total equivalent resistive force on the aircraft to its weight?

14-43 P The essential parts of a baseball-pitching machine consist of two automobile tires driven in opposite rotation senses at the same angular speed while their treads press together over a small area. A baseball, placed near the spot where the two tires touch, is pulled by friction through the region where the tires press together and emerges on the far side with the speed of a top-notch fast ball, about 42 m/s (95 mph). (a) Take each tire to have a diameter of 0.76 m. What is the requisite angular speed of each tire? (b) How might one produce a curve ball?

14-44 P The land speed record was set at Edwards Air Force Base in California on December 17, 1979, by Stan Barrett, who achieved 740.00 mph with the first land vehicle to exceed the speed of sound. He drove a car powered by a 48,000 hp rocket engine and aided by a 12,000 hp Sidewinder Missile. (Chuck Yeager, the first to break the sound barrier with an aircraft, was an observer.) Calculate the combined thrust of the rocket engines and the total resistance force arising from air and road friction.

15 Gravitation

Gravity plays a central role in physics for several reasons.

• Gravity is *all-pervasive*. Every particle in the universe attracts every other particle.

• The gravitational force is one of a very few *fundamental* forces between interacting particles. The others: the electromagnetic force between electrically charged particles, the weak interaction force that governs the slow decay of certain unstable particles (and actually is closely related to the electromagnetic force), and the strong interaction force that dominates the interaction between particles in an atomic nucleus. It is now believed that all may be special examples of a single universal primordial force in a grand unified theory of forces.

• Originating, developing, and applying the law of universal gravitation, one of the most extraordinary intellectual achievements of all time, was largely the work of one physicist, Sir Isaac Newton. He saw that a single universal force of gravity could account equally well for the motion of a dropped apple, of the moon about the earth, and of the planets about the sun. The *laws of physics are universal.* Astronomical objects and objects on earth have their motion governed by the same laws: celestial physics and terrestial physics are the same thing.

• Gravity is still a lively topic in present-day physics and astronomy. The general theory of relativity is basically a comprehensive theory of gravity. Shifts to longer wavelengths of light emitted by massive stars are gravitational red shifts. Black holes are an extreme example of gravitational attraction.

15-1 The Law of Universal Gravitation

Two mass points, m_1 and m_2, are separated by a distance r. Each particle attracts the other one by the gravitational force, as shown in Figure 15-1.

• The magnitudes of the gravitational forces on each of the two interacting particles are the same; their directions are opposite: $\mathbf{F}_{2\,\text{on}\,1} = -\mathbf{F}_{1\,\text{on}\,2}$.
• The gravitational force is a *central force*. It lies on the line connecting the two particles.
• The magnitude of the gravitational force is given by

$$F_g = \frac{Gm_1m_2}{r^2} \tag{15-1}$$

where G is a constant.*

The force magnitude varies inversely with the square of the separation distance $1/r^2$ and is directly proportional to each of the two interacting masses, m_1 and m_2.

• Equation (15-1) applies only to a pair of mass *points*. To compute the gravitational force between objects of finite size, we imagine each object subdivided into a collection of point masses and apply (15-1) to each pair.
• The proportionality constant G, called the *universal gravitational constant*, has the value

$$G = 6.673 \times 10^{-11} \text{ N} \cdot \text{m}^2/\text{kg}^2$$

(How G is measured is discussed in Section 15-3). The SI units assigned to G assure that with m_1 and m_2 in kilograms and r in meters, the force is given in newtons. Unless one of two interacting objects is relatively massive—for example, the earth and an apple—the gravitational force is very small indeed. Take, for example, $m_1 = 1$ kg, $m_2 = 1$ kg, and $r = 1$ m. Then the force on each mass is $F_g = (6.7 \times 10^{-11} \text{ N} \cdot \text{m}^2/\text{kg}^2)(1 \text{ kg})(1 \text{ kg})/(1 \text{ m})^2 = 6.7 \times 10^{-11}$ N, which is somewhat less than the weight near the earth's surface of one one-hundredth of a microgram.

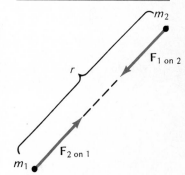

Figure 15-1. *Two point masses attract gravitationally.*

---

* Equation (15-1) is a *scalar* equation; it gives the *magnitude* of the gravitational force on either particle. The *direction* of the gravitational force can be incorporated in a *vector* equation. If, for example, radius vector $\mathbf{r}_{1\,\text{to}\,2}$ has its tail at m_1 and its head at m_2, then the vector gravitational force on m_2 can be written as

$$\mathbf{F}_{1\,\text{on}\,2} = \frac{-Gm_1m_2\,\mathbf{r}_{1\,\text{to}\,2}}{r_{1\,\text{to}\,2}^3}.$$

The minus sign implies that $\mathbf{F}_{1\,\text{on}\,2}$ is *opposite* in direction to $\mathbf{r}_{1\,\text{to}\,2}$. Note that an additional factor $r_{1\,\text{to}\,2}$ appears in the denominator to compensate for the magnitude of $\mathbf{r}_{1\,\text{to}\,2}$ introduced into the numerator.

We can write this vector equation in another way, using the concept of unit vector (Section 2-5): Let $\mathbf{u}_{1\,\text{to}\,2}$ be a *unit vector* parallel to $\mathbf{r}_{1\,\text{to}\,2}$. Then, $\mathbf{F}_{1\,\text{on}\,2} = -Gm_1m_2\,\mathbf{u}_{1\,\text{to}\,2}/r^2$.

• As discussed in detail in Section 15-5, the gravitational force is *conservative;* the energy-conservation principle applies to it.

From Isaac Newton's later account of accomplishments at age 23, we know at least roughly how he first came to believe that the gravitational force varied with $1/r^2$. The arguments went something like this.

Any free object, dropped from rest or thrown near the earth's surface has the *same gravitational acceleration,* 9.8 m/s² toward the center of the earth. Suppose that we could somehow squeeze the moon down to the size of an apple and drop it at the earth's surface. It too would fall at 9.8 m/s². And if this tiny moon were thrown horizontally with the appropriate speed (7.9 km/s), it would again accelerate at the rate of 9.8 m/s² toward the center of the earth, as it now circled the earth as a satellite close to its surface. See Figure 15-2(a).

At its usual location, the moon is acted on by a force of gravity, and it accelerates toward the earth as it orbits. We can readily compute the moon's centripetal acceleration a_M in circular orbit from the values of its orbital radius, $R_M = 3.8 \times 10^8$ m, and its period, $T_M = 2.33 \times 10^6$ s = 27 days, using (4-18):

$$a_M = \frac{4\pi^2 R_M}{T_M^2} \qquad a_M = \frac{4\pi^2 (3.8 \times 10^8 \text{ m})}{(2.3 \times 10^6 \text{ s})^2} = 2.7 \times 10^{-3} \text{ m/s}^2 \qquad \text{(4-18)}$$

If the moon were not in orbit about the earth but somehow "dropped" from rest at its present location, it would have the very same acceleration there of 2.7×10^{-3} m/s² toward the earth's center.

The moon's orbital radius is very close to 60 times the earth's radius, $R_M = 60R$. Furthermore, the accelerations at the two locations are related in a simple way to the square of 60:

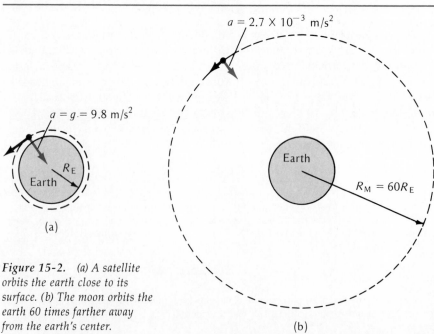

Figure 15-2. (a) A satellite orbits the earth close to its surface. (b) The moon orbits the earth 60 times farther away from the earth's center.

$$2.7 \times 10^{-3} \text{m/s}^2 = \frac{9.8 \text{ m/s}^2}{(60)^2}$$

In words, the gravitational acceleration falls off with $1/r^2$. Since the gravitational acceleration at any location is proportional to the gravitational force there, one concludes that the gravitational force of the earth varies with $1/r^2$. As discussed in Section 15-4, Newton made a far more detailed and stringent test of the gravitational force law by applying it to the motions of the planets of the solar system.

Newton's analysis, as given earlier, is based on an important assumption — that the earth, which certainly cannot be regarded as a particle when it acts on a small object at its surface, is nevertheless equivalent in its gravitational effects to a particle of mass M_E at its center. In other words, when one adds (as vectors) the gravitational forces on an external object from all particles within the earth of mass M and radius R, the resultant gravitational force on the object of mass m:

- Points to the earth's center.
- Has the magnitude GmM/R^2.

See Figure 15-3.

Actually, to carry out the proof, Newton invented the calculus. A detailed proof using calculus is given in the optional section 15-9. The general results for any homogeneous spherical shell of mass M are these:

- When a particle of mass m is *outside* the shell, the force on the particle is toward the center of the shell and of magnitude GmM/r^2, where r is the distance of the particle from the shell's center.
- When the particle is anywhere *inside* the shell, the resultant gravitational force on it is zero. See Figure 15-4.

The shell is equivalent gravitationally to a mass point M at its center when a particle m is outside; when the particle m is inside the shell, the shell is gravita-

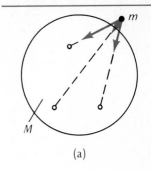

(a)

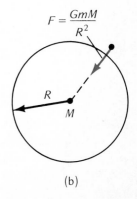

$$F = \frac{GmM}{R^2}$$

(b)

Figure 15-3. (a) Three particles in the earth attract mass m at the surface. (b) The resultant gravitational force is the same as that from a point mass M at the center.

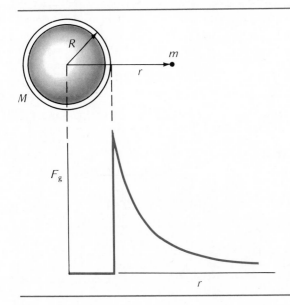

Figure 15-4. Gravitational interaction between a spherical shell and a mass point. The gravitational force is inverse-square for the particle outside the shell and zero for the particle inside.

tionally nonexistent. Any solid sphere can be thought to consist of concentric shells of different radii, like an onion. A solid sphere and a particle of equal mass at its center have equivalent gravitational effects on a particle external to the sphere. Two spheres interact as two particles do, provided that the density of each depends only on the distance from the center. This simple result holds only for spherical shells, not for other shapes; a cubical shell, for example, is not equivalent gravitationally to a mass point at its center.

That F_g varies with separation distance as $1/r^2$ does is the result of *observation*. That the magnitude of F_g is directly proportional to each of the masses m_1 and m_2 of the interacting particles is, however, a matter of *definition*. We can see this through a simple example. Suppose that two objects have weights in the ratio 2 to 1; that is, when each is brought in turn to the same location, the gravitational attraction of earth for one object is twice the amount for the other. On this basis, then, we say that one has twice the mass of the other. Mass ratios are, by definition, the same as the force ratios when each particle in turn interacts gravitationally with the same third object.*

Now consider two interacting particles with masses m_1 and m_2, as in Figure 15-1. The force $\mathbf{F}_{2\text{ on }1}$ on particle 1 is directly proportional to m_1, and $\mathbf{F}_{1\text{ on }2}$ is directly proportional to m_2. But by Newton's third law, $\mathbf{F}_{2\text{ on }1} = -\mathbf{F}_{1\text{ on }2}$; the force magnitudes are the same. The gravitational force on each of the two particles must then be directly proportional to both m_1 and m_2; that is, $F_g \propto m_1 m_2$.

The gravitational force is the weakest of the fundamental forces in physics; yet it is the only fundamental force we experience directly. The attractive electric force (also a central, inverse-square, conservative force) between a proton and an electron is 10^{39} times greater than the attractive gravitational force between them! The reason why the much stronger electric force is not encountered in ordinary experience is simply that all large-scale objects ordinarily are electrically neutral; they contain equal amounts of positive and negative electric charge. The gravitational force cannot be neutralized; there is no such thing as a "gravity insulator" or a "gravity shield." The still stronger nuclear force, which acts between the particles in the nuclei of atoms, is also never evident in ordinary experience, since this force is effective only over a very short range (10^{-15} m).

15-2   Gravitational Acceleration **g**

The acceleration **g** due to the gravitational attraction of the earth does not have a constant magnitude. There are several reasons.

• The gravitational force varies with $1/r^2$. The weight w of an object with mass m interacting with a hypothetical uniform spherical earth of mass M can be written in two ways:

---

* The mass appearing in the gravitational force law, $F_g = Gm_1m_2/r^2$, is really the *gravitational mass*, whereas the mass appearing in Newton's second law, $\mathbf{F} = m\mathbf{a}$, is really the *inertial mass*. The distinction between the two types of mass and the reasons why they can nevertheless be taken to be equivalent are discussed in the optional section 15-7.

$$w = mg \quad \text{or} \quad w = \frac{GmM}{r^2}$$

where r is the distance to the earth's center. Equating the two values above, we have

$$g = \frac{GM}{r^2} \tag{15-2}$$

As the elevation increases, g decreases. At an elevation of 1.6 km (Denver, Colorado, for example), g is smaller than the corresponding value at sea level by 1 part in 2000. At the location of the moon, g is, as we have seen, down by a factor of $60^2 = 3600$.

• The earth is not a sphere; it is squashed somewhat, with the poles closer to the center than points on the equator. For this reason, g varies with latitude. The value for g at the poles (9.83 m/s²) exceeds the value at the equator (9.78 m/s²) by about $\frac{1}{2}$ percent.

• The earth is rotating. A particle at rest with respect to the earth's surface at the equator must, as it moves in a circle relative to the earth's center, have a centripetal acceleration towards the earth's center of

$$a = \left(\frac{2\pi}{T}\right)^2 R_{\mathrm{E}} = \left(\frac{2\pi}{24 \times 60 \times 60 \text{ s}}\right)^2 (6.37 \times 10^7 \text{ m}) = 0.34 \text{ m/s}^2$$

It follows that for a particle at rest on earth at the equator, the inward gravitational force must exceed the net outward force, to yield a resultant force toward the center of the circle. See Figure 15-5. The effect of the earth's rotation is less pronounced at higher latitudes because an object at rest on earth there moves in a smaller circle.

• The earth does not have a uniform density; g at any one location is influenced by how mass is distributed nearby in local mineral deposits. One way, now obsolete, for prospecting for oil was to time the motion of a pendulum at various locations to detect variations in g. (For a simple pendulum of length L, the period is $T = 2\pi\sqrt{L/g}$.)

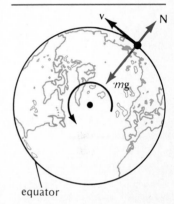

Figure 15-5. *For a particle at rest relative to the earth at the equator, the inward gravitational force m**g** exceeds the outward force **N**.*

15-3   The Cavendish Experiment

Do two objects of ordinary size actually attract one another by gravity? This was first shown directly in a historic experiment of extraordinary delicacy by Henry Cavendish in 1798, more than a hundred years after Newton. The Cavendish experiment is also significant because it allowed the numerical value of the constant G to be computed.

To detect and measure very small forces, Cavendish used a torsion fiber, a thin fiber of quartz that could easily be twisted. Cavendish secured two small spheres, each of mass m_1, to the opposite ends of a light rigid rod and suspended the rod at its midpoint from the fiber, to form a torsion balance as shown in Figure 15-6. Two large lead spheres, each of mass m_2, were then placed near the masses m_1, the distance between the center of m_1 and the nearby m_2 being r. In such an arrangement, each lead sphere attracts the mass m_1 close to it with a force $\mathbf{F}_g$, and a clockwise torque then acts on the rod (as one

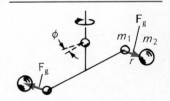

Figure 15-6. *Schematic representation of the Cavendish experiment.*

looks downward). (The attraction between one lead sphere and the more distant mass m_1 is much more feeble.) A clockwise gravitational torque rotates the rod. As the rod rotates, the wire twists. This deformation produces a counterclockwise restoring torque that for small angles of twist is proportional to the twist angle ϕ (Section 13-6). Equilibrium is reached when the gravitational torque equals (in magnitude) the restoring torque. By measuring the restoring torque (which is proportional to ϕ), one finds the attractive force between m_1 and m_2. Knowing F, m_1, m_2, and r then allows the constant G to be computed from (15-1).

The proportionally constant between the restoring torque and the angular displacement ϕ can be determined by measuring the fiber's period of oscillation with the rod and attached spheres suspended from it (Section 13-6).

Example 15-1.   Compute the earth's mass using the measured value of G. From (15-2), we have

$$M = \frac{g r_e^2}{G} = \frac{(9.8 \text{ m/s}^2)(6.4 \times 10^6)^2}{6.7 \times 10^{-11} \text{ N} \cdot \text{m}^2/\text{kg}^2} = 6.0 \times 10^{24} \text{ kg}$$

Cavendish is sometimes said to have "weighed" the earth in his famous experiment. As we see, a knowledge of the constant G permits the earth's mass to be calculated.

Knowing the earth's mass and radius allows its average density to be computed. The result: the earth has a mean density 5.5 times that of water, and exceeding that of material (mostly water, at 1 g/cm³) near the earth's surface. The inevitable conclusion: the earth's interior consists largely of material of high density, most probably metals.

15-4   Kepler's Laws of Planetary Motion

Newton did not, as in popular legend, merely sit thinking in an orchard one day, see an apple drop, associate its attraction toward the earth with that of the moon, make a couple of computations, and — *voila!* — create and verify the theory of universal gravitation. To show that the force of gravity was indeed universal, exactly central, and precisely inverse-square, Newton had to show that the motions of the sun's planets could be accounted for in detail. In addition, Newton analyzed some subtle astronomical effects also arising from gravity: the earth's tides, arising from the gravitational influence of the moon, and the phenomenon known as the precession of equinoxes, whereby the earth's spin axis slowly changes direction.

Newton knew that the right reference frame for describing the planets' motion was not the earth, but an inertial reference frame in which the sun and other stars were effectively at rest. The heliocentric, or Copernican, cosmology had replaced the geocentric, or Ptolemaic, cosmology.

At the time of Newton, the detailed motions of our sun's planets had been summarized in three empirical rules developed over a lifetime of study by Johannes Kepler (1571–1630). Kepler used data that had been carefully compiled over many years of observation (without a telescope; some observations were through a crack in the wall) by his teacher, Tycho Brahe (1546–1601). These data pinpointed where the planets were located in the sky as a function of time.

The three empirical rules of Kepler, usually referred to as Kepler's three laws of planetary motion, are these.

- *Orbital rule.* Each planet moves in an ellipse with the sun at one focus.
- *Area rule.* A line joining the sun to the orbiting planet sweeps out equal areas in equal time intervals.
- *Period rule.* The square of the period divided by the cube of the ellipse's semimajor axis has the same constant value for all planets.

Newton's achievement consisted in showing that these three specialized rules were the necessary consequence of an attractive, central, inverse-square gravitational force between the sun and each planet. The detailed mathematical analysis is too involved to be included here. The second rule has already been proved in detail for a central force (Section 14-5). The first will be shown to be at least plausible, and the third is shown to work in detail for a circular orbit.

Orbital rule   Figure 15-7(a) shows a planet in an elliptic orbit about the sun at one focus. Since the sun's mass M greatly exceeds the planet mass m, the sun can be considered to remain at rest. Strictly, however, both of two gravitationally interacting objects orbit about their center of mass. The long dimension of the ellipse, its major axis, is $2a$; the short dimension of the ellipse is its minor axis $2b$. (A circular orbit corresponds to the special case $a = b$, with the two focal points of the ellipse coalescing to become the center of the circle.)

Figure 15-7(b) shows the gravitational force **F** on the planet at one instant. Component $F_{\parallel}$, which is parallel to the planet's velocity **v**, changes the planet's speed; force component $F_{\perp}$, which is perpendicular to the planet's velocity,

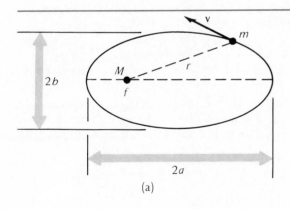

(a)

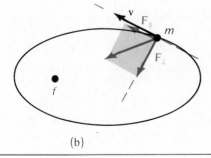

(b)

Figure 15-7.   *(a) Particle orbiting in an ellipse with semi-major axis a and semi-minor axis b about a force center at a focus. (b) The gravitational force shown as the vector sum of force components parallel and perpendicular to velocity* **v.**

But Why Does It Fall?

One standard answer: physics does not pretend to answer the question of purpose—*why* things happen as they do—but instead concentrates on questions of *how*. The law of universal gravitation tells us what happens to a falling apple. As a well established law, it accounts for what has already been observed and we are confident that it will predict successfully what has not yet been observed, the flight path of a Voyager spacecraft to distant planets, for example. The question of *why* is outside the realm of science.

The question cannot be disposed of that easily! Let's put it this way: How does the apple know, so to speak, where the earth is and how big its mass is so that, if the apple is released, it accelerates toward the earth and at the right rate rather than remaining as it is initially, floating weightless?

The answer of contemporary physics is that an apple near the earth is not in truly *empty* space. The earth creates a gravitation *field* in the space surrounding it, and the apple, finding itself immersed in this field, responds by accelerating toward the earth.* The field then becomes the intermediary in the interaction between objects A and B: A creates a gravitational field in the space surrounding it; and B, finding itself in the field from A, responds to the field at its location.

Is the field concept more than a useful fiction? How do we know that the field is really there, that

The gravitational field **g** at the location where mass *m* is acted on by gravitational force **F$_g$** is defined as **g** = **F$_g$**/*m*.

space is not actually empty? The question as to how objects in otherwise empty space can produce forces on one another has had a long history in physics; it is known as the problem of "action at a distance".

The field concept certainly works for another fundamental interaction, the electromagnetic interaction between electrically charged particles, which takes place through the intermediary of an electromagnetic field. Here electric charge A produces an electromagnetic field at the site of charge B, and charge B responds to the electromagnetic field at its location. The proof that the electromagnetic field concept is not only useful but necessary comes from the observation that electromagnetic fields can become detatched from their sources, electric charges, and propagate through space as independent entities. Light and other forms of electromagnetic radiation are just such fields. That the fields are *real*—or as real as anything else in physics—is supported by the fact that such physical attributes as energy, momentum, and angular momentum can be assigned consistently to an electromagnetic field. (How this works in some detail is the subject of Chapters 23 through 35 on electromagnetism.) When a broadcast radio antenna radiates an electromagnetic field, the circuit with the transmitting antenna loses energy; later, when the electromagnetic field is captured by a receiving antenna, the circuit at the receiver gains energy.

Back to the apple. How does it know enough to fall toward the earth? The well understood example of electromagnetic fields strongly suggests that the concept of a gravitational field is valid here too, that

---

deflects the planet. Therefore, the planet's speed is greatest when it is closest to the sun (at the perihelion position) and least at its farthest position from the sun (the aphelion position.)

Area rule As a planet orbits the sun, its speed and distance from the sun change with time. But the rate dA/dt at which the radius vector **r** from the sun to the planet sweeps out area does not change. See Figure 15-8. Here it is also evident that the planet's speed is greatest when the planet is closest to the force center.

The implications of the area rule were discussed in Section 14-4. There it was shown that the area rate dA/dt is equal in magnitude to

$$\frac{dA}{dt} = \frac{L}{2m}$$

(14-9)

it is not merely a pedantic invention or mathematical fiction that helps to visualize the interaction. (The gravitational field concept plays a central role in Einstein's general theory of relativity, which is really a modern theory of gravity. In general relativity a mass actually deforms the space-time surrounding it, and this deformation is the gravitational field.) Just as the existence of electromagnetic waves is the most compelling evidence for the existence of electromagnetic fields, so too the most dramatic proof that the gravitational field concept is valid would be direct observation of gravitational waves—waves somehow detatched from the masses that create them and that, like light, propagate through otherwise empty space at the speed of light.

What would an "antenna" for gravitational waves look like? One simple form is two masses connected by a spring, or even more simply, a big solid cylinder that would be compressed and stretched along its length as a gravitational wave impinges on it.

Detecting a gravitational wave is far, far more difficult than detecting an electromagnetic wave. The basic reason lies in the relative strengths of the electromagnetic and the gravitational forces—a proton and an electron attract one another by gravity with a force that is smaller than their electric attraction by the absolutely enormous factor of 10^{39}. As waves go, gravitational waves are extraordinarily weak. It is estimated that only a cosmic event of absolutely catastrophic proportions—the explosion of a supernova, for example—could create a wave sufficiently energetic to be detected with even the most exquisitely sensitive apparatus on earth (capable of detecting a change in length of about one-millionth of a proton

diameter in a cylinder one meter long). Thus far there has been *no direct* evidence for gravity waves, even though a number of highly sensitive gravity-wave antennas have been constructed and are "on the air". There is, however, fairly strong *indirect* evidence for gravity waves that comes from recent precision measurements in radio astronomy: the binary pulsar officially known as 1913 + 16 is losing energy at a rate consistent with the radiation of gravitational waves.

A pulsar is a small, rapidly spinning, extremely dense star, made up mostly of neutrons, the remnant of a supernova explosion. A pulsar has a mass about that of the sun; its diameter, however, is no more than 30 km. The pulsar emits a radio beam, which because of the spin of the pulsar, sweeps the sky and produces pulses as the beam sweeps past the earth. Pulsar 1913 + 16 makes 16.94 revolutions per second and this is also the pulse rate observed at earth. Actually, the pulsar has a silent co-orbiting companion; the two objects orbit their common center of mass. The presence of the companion is deduced by observing that the pulse rate changes periodically, corresponding to the pulsar's (and the companion's) orbiting one another with a 7.75 year period. More striking, however, is a very subtle effect: the orbital period is decreasing very slowly (about one-tenth of a second over one year). Why? The pulsar and its companion are speeding up in their orbital motion because they are radiating gravitational energy.*

*For a comprehensive account of binary pulsar 1913 + 16 and gravity waves see J. M. Weisberg, J. H. Taylor, and L. A. Fowler, *Scientific American* **245,** 74 (October, 1981)

Here m is the planet's mass and L is its angular momentum measured relative to the force center. Constant dA/dt means constant angular momentum L. But a planet's angular momentum L remains constant only if no torque acts on it; this

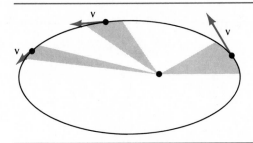

Figure 15-8.   The area rule: a radius vector from the force center sweeps out equal areas in equal time intervals.

in turn implies that the gravitational force on a planet is a *central force*, one that points at each instant from the planet towards the sun.

This result is truly remarkable; the direction of the gravitational force can be deduced, from mathematical analyses and theoretical physics, simply from the observed constancy of dA/dt. Imagine how difficult it would be to measure directly and with high accuracy the *direction* of the force on a planet. Whereas Kepler's first rule implies that $F_g \propto 1/r^2$, the second implies merely that the force is *central*.

Period rule Consider the special case of circular orbit of radius r. The planet's period is T. Then the planet's acceleration toward the center of the circle is $(4\pi^2/T^2)r$, from (4-18), and Newton's second law yields

$$\Sigma\mathbf{F} = \mathbf{ma}$$

$$\frac{GmM}{r^2} = m\left(\frac{4\pi^2}{T^2}\right)r$$

or

$$T^2 = \left(\frac{4\pi^2}{GM}\right)r^3 \qquad (15\text{-}3)$$

The quantity $4\pi^2/GM$ is the *same constant* for all planets, so that T^2/r^3 *is the same constant* for all of them. Newton was the first to show that the equation of (15-3) applied also to an elliptic orbit, with r replaced by the ellipse's

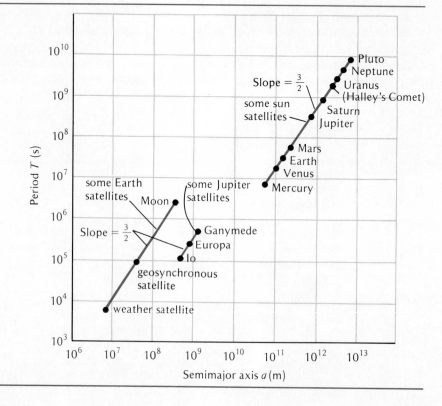

Figure 15-9. *Kepler's third law illustrated for several force centers with orbiting satellites. The logarithm of the period is plotted against the logarithm of the semi-major axis. All three lines have the same slope, 3/2. Therefore, each system obeys the relation log T = 3/2 log a + 1/2 log K, which is equivalent to $T^2 = Ka^3$, with K a different constant for each system.*

semimajor axis a. Surprisingly perhaps, all orbits with the same semimajor axis have the same period, even though they differ in the size of the semiminor axis b. The period rule applies to any force center with two or more satellites; see Figure 15-9.

Example 15-2. A *geosynchronous, or stationary, satellite* has a seemingly fixed position in the sky. Viewed by an observer on earth, it is stationary. Viewed by an observer at rest with respect to the fixed stars, the satellite orbits the earth from west to east at the equator with the period of 24 hours, so that its motion is synchronized with the earth's spin. See Figure 15-10. A communications satellite (operated by the COMSAT Corporation in the United States) is a stationary satellite; it acts as a relay station for telephone, television, and other signals carried by radio waves between two points on earth that are not on a direct line of sight.

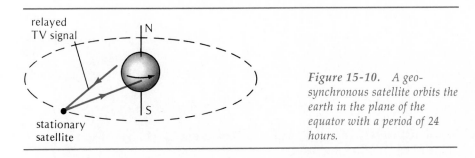

relayed
TV signal

stationary
satellite

Figure 15-10. A geosynchronous satellite orbits the earth in the plane of the equator with a period of 24 hours.

What is the orbital radius for a stationary satellite? (Use the fact that a satellite close to the earth has an orbital radius of $1.025r_e$, where r_e is the earth's mean radius, and a period of 88 minutes.)

Kepler's third law relates the period T to the radius a for satellites about the same force center, and (15-3) can be written as

$$T^2 = Ka^3$$

where K is a constant. Equivalently, we can write for two different orbits

$$\left(\frac{T_1}{T_2}\right)^2 = \left(\frac{a_1}{a_2}\right)^3$$

or

$$a_2 = a_1 \left(\frac{T_2}{T_1}\right)^{2/3} = (1.025r_e)\left(\frac{24\text{ h}}{88\text{ min}/(60\text{ min}/\text{h})}\right)^{2/3} = 6.50\ r_e$$

The orbital radius for the geosynchronous satellite is 6.50 times the earth's mean radius (or about 22,000 miles from the earth's surface).

15-5   Gravitational Potential Energy

Suppose mass m is lifted through a small vertical distance y. Then the gravitational potential energy increases by mgy because the object's weight mg can be taken to be *constant*.

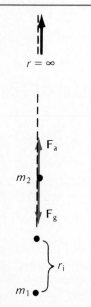

Figure 15-11. *Mass m_2 is lifted away from fixed mass m_1. The gravitational force F_g is balanced at each point by applied force F_a.*

But what if the object is lifted away from the earth's center through any distance? What then is the change in gravitational potential energy? We must take into account that the weight varies inversely with the square of the object's distance from the earth's center.

Consider Figure 15-11, where mass m_1 (think of it as the earth) is fixed in position. Mass m_2 is lifted from an initial separation distance r_1 to an infinite distance from m_1. The particle is moved at constant speed, by an outward applied force $\mathbf{F}_a$ that balances at each location the inward gravitational force $\mathbf{F}_g$ on m_2. The total work done by $\mathbf{F}_a$ will equal the change in the system's gravitational potential energy. This follows from at least two considerations:

• Since m_2 moves at constant speed, the system's kinetic energy does not change, so that any work done by the external force $\mathbf{F}_a$ must equal the change in potential energy.

• The formal mathematical definition for potential energy change from some initial (i) state to a final (f) state is, from (10-4),

$$U_f - U_i = -\int_i^f \mathbf{F} \cdot d\mathbf{r}$$

where $\mathbf{F}$ here is the gravitational force. Since $\mathbf{F}_a = -\mathbf{F}_g$, we have

$$U_f - U_i = \int_i^f \mathbf{F}_a \cdot d\mathbf{r}$$

In words, work done by $\mathbf{F}_a$ equals potential-energy change. The total work done is

$$W_a = \int_{r_i}^\infty F_a \, dr = \int_{r_i}^\infty \frac{Gm_1m_2}{r^2} \, dr = \frac{Gm_1m_2}{r}$$

(We have dropped the subscript i in the equation above, so that r now represents the initial separation distance.)

As we have already seen, W_a is equal to the potential energy change, so that

$$W_a = U_f - U_i = \frac{Gm_1m_2}{r}$$

Now we must choose the arrangement of the particles that corresponds to zero gravitational potential energy. For a force $\mathbf{F}$ that varies with position, it is simplest to choose $U = 0$ for $\mathbf{F} = 0$. (For example, with a spring we have $F = -kx$ and $U = \frac{1}{2}kx^2$, so that when $F = 0$, then also $U = 0$.) The gravitational force is zero for infinite separation distance; therefore, potential energy U_f for m_1 and m_2 infinitely separated is set equal to zero. With $U_f = 0$ in the equation above, we have finally

$$U_g = -\frac{Gm_1m_2}{r} \tag{15-4}$$

for the gravitation potential energy U_g (replacing U_i) of mass points m_1 and m_2 separated by distance r.

The following remarks apply to this fundamental result:

• Potential energy is defined so that we can use the energy-conservation principle. When potential energy given in (15-4) is added to the kinetic energy K, the system's total energy $E = K + U_g$ is constant.

- As required by our choice of zero potential energy for infinite separation, (15-4) yields $U_g = 0$ for $r \to \infty$.

- For any finite separation distance r, the potential energy is *negative*. This is merely a consequence of the choice of zero for U_g; there is nothing peculiar about it. (After all, if ground level is chosen to correspond to zero potential energy, any object below ground level has negative potential energy.)

We can see that U_g must be negative from an example like this: Suppose that two mass points are very far separated and initially at rest. Their kinetic energy is zero; their potential energy is zero; and therefore their *total energy is zero*. When the particles are released, they attract one another, move closer together, and gain kinetic energy. The system's total energy must *remain constantly zero*. Therefore, since the kinetic energy will necessarily be positive, the potential energy *must be negative*.

- As will be proved in Section 15-9, the relation (15-4) also gives the gravitational potential energy between uniform *spherical shells* of masses m_1 and m_2 whose centers are separated by r.

- For a system with three or more particles (or separated spherical shells), the total gravitational potential energy is merely the sum of the potential energies between all pairs of particles in the system. For the particles shown in Figure 15-12, we have

$$U = U_{12} + U_{13} + U_{23}$$

$$U = -\frac{Gm_1m_2}{r_{12}} - \frac{Gm_1m_3}{r_{13}} - \frac{Gm_2m_3}{r_{23}} \tag{15-5}$$

One more item must be examined. We must prove that a potential-energy function can properly be defined for the general gravitational force because this force is conservative. As we know (Section 9-6), any force is conservative if the net work it does on a particle completing a closed loop is zero ($\oint \mathbf{F} \cdot d\mathbf{r} = 0$), or what amounts to the same thing, if the work done by the interaction force is independent of the path followed by the particle between two fixed end points.

Consider Figure 15-13, where m_1 is fixed while m_2 is imagined to move from point 1 to point 3 by two different paths:

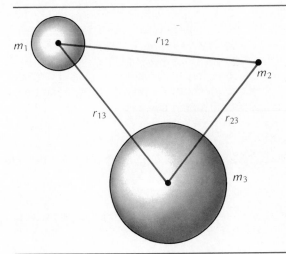

Figure 15-12.   System with three interacting objects.

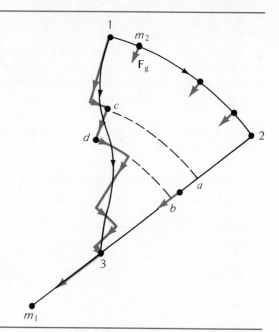

Figure 15-13. Mass m_2
moved from point 1 to point 3
by two different paths: 1 to 2 to
3, and an arbitrary path
approximated by small steps.

• From point 1 along a circular arc of constant radius to point 2; then radially inward to point 3.
• From point 1 to point 3 along some arbitrary general curved path.

It is easy to show that the work done by gravitational force $\mathbf{F_g}$ on m_2 is the same for these two paths (and therefore the same for any two paths). Consider first route $1 \rightarrow 2 \rightarrow 3$. From 1 to 2, no work is done; the particle's displacement along the circular arc is always perpendicular to $\mathbf{F_g}$. From 2 to 3, the gravitational force is parallel to the particle's inward displacement, and positive work is done on m_2 as it moves closer to m_1. The magnitude of $\mathbf{F_g}$ increases as m_2 gets closer to m_1.

The general route $1 \rightarrow 3$ can be approximated by a series of small segments (shown with colored lines in Figure 15-13). Each small segment is either a straight line radially towards m_1 or a circular arc with its center at m_1. Again, no work is done as the particle moves along a small circular arc. Furthermore, each radial displacement, such as cd in Figure 15-13, can be matched against a corresponding displacement ab along the straight-line path $2 \rightarrow 3$, and the work done by the interaction force is the same for all such pairs. In short, the work done is the same for any path between the end points; the gravitational force is conservative; and a gravitational potential-energy function can be defined.

Example 15-3.   A particle is dropped from rest from an elevation equal to the earth's radius. If air resistance is neglected, with what speed does the particle strike the earth's surface?

See Figure 15-14, where the initial state is labelled 1 and the final state 2. Initially, the particle of mass m is at a distance $r_1 = 2r_e$ from the earth's center; its speed is then $v_1 = 0$. When it strikes the earth of mass M, the particle's speed is v_2 and distance from

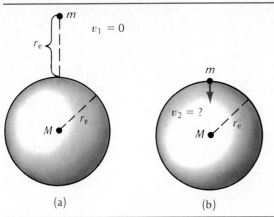

(a)   (b)   *Figure 15-14.*

the earth's center is $r_2 = r_e$. The system's total energy, $E = K + U$, is constant, and we have

$$K_1 + U_1 = K_2 + U_2$$

$$\frac{1}{2} mv_1^2 - \frac{GmM}{r_1} = \frac{1}{2} mv_2^2 - \frac{GmM}{r_2}$$

$$0 - \frac{GmM}{2r_e} = \frac{1}{2} mv_2^2 - \frac{GmM}{r_e}$$

The particle's mass cancels out, and its speed on hitting the earth is

$$v_2 = \sqrt{\frac{GM}{r_e}}$$

If the particle had been projected vertically upward from the earth's surface at this speed, v_e, it would rise, come to rest at an elevation r_e, and then fall back to earth.

Example 15-4.   An object of mass m close to the earth's surface is raised through a vertical height Δy, where Δy is small compared with the earth's radius r_e. Show that the general relation for gravitational potential energy reduces under these circumstances to the simple, familiar relation $\Delta U = mg\,\Delta y$.

See Figure 15-15. In going from r_1 to r_2, the gravitational potential energy changes by ΔU_g, where

$$\Delta U_g = U_2 - U_1 = \left(-\frac{Gmm_e}{r_2}\right) - \left(-\frac{Gmm_2}{r_1}\right) = Gmm_e\left(\frac{1}{r_1} - \frac{1}{r_2}\right)$$

We choose $r_1 = r_e$ and $r_2 = r_e + \Delta y$. See Figure 15-15. If $\Delta y \ll r_e$, we can write $-1/r_2$ as

$$-\frac{1}{r_2} = -\frac{1}{(r_e + \Delta y)} = -\frac{1}{r_e}\left[\frac{1}{(1 + \Delta y/r_e)}\right] = -\frac{1}{r_e}\left[1 + \frac{\Delta y}{r_e}\right]^{-1}$$

Then, by the binomial expansion, $(1 + x)^n \simeq 1 + nx + [n(n-1)/2!]\,x^2 + \cdots$, this becomes

$$-\frac{1}{r_2} = -\frac{1}{r_e}\left[1 - \frac{\Delta y}{r_e} + \left(\frac{\Delta y}{r_e}\right)^2 - \cdots\right] \simeq -\frac{1}{r_e} + \frac{\Delta y}{r_e^2}$$

Using this result in the relation for ΔU_g above, we have

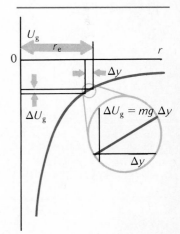

Figure 15-15.   For objects close to the earth, the gravitational potential energy difference is given by $\Delta U_g = mg\Delta y$.

$$\Delta U_g = Gmm_e \left[\frac{1}{r_e} + \left(-\frac{1}{r_e} + \frac{\Delta y}{r_e^2} \right) \right] = m \left(\frac{Gm_e}{r_e^2} \right) \Delta y$$

But by definition, (15-2),

$$g = \frac{Gm_e}{r_e^2},$$

so that

$$\Delta U_g = mg \, \Delta y$$

15-6 Bound Orbits and Escape

Two important situations in which a light particle interacts gravitationally with a second, far more massive object are these:

- The particle is *bound* to a force center and orbits it in a circle.
- The particle is projected with enough energy to *escape* from the force center.

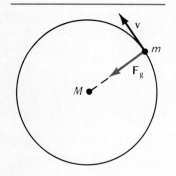

Figure 15-16. *Particle m orbits sphere of mass M.*

Bound Orbit A light planet of mass m travels at constant speed v in a circular orbit of radius r about a sun of much larger mass M. See Figure 15-16. We want to find a relation for the total energy of the planet-sun system in terms of the orbital radius r.

The system's total energy E is just the kinetic energy of the planet, $K = \frac{1}{2}mv^2$ (the kinetic energy of the sun can be neglected) added to the system's gravitational potential energy, $U_g = -Gm_1m_2/r = -GmM/r$:

$$E = K + U_g = \frac{1}{2} mv^2 - \frac{GmM}{r} \tag{15-6}$$

We get another relation between v and r from Newton's second law. Here we recognize that the planet's acceleration toward the center of the circle has a magnitude v^2/r:

$$\Sigma F_r = ma_r$$

$$G\frac{mM}{r^2} = \frac{mv^2}{r}$$

Multiplying both sides of this equation by $r/2$ yields

$$\frac{GmM}{2r} = \frac{mv^2}{2} \tag{15-7}$$

Eliminating $\frac{1}{2}mv^2$ between (15-6) and (15-7), we then get

$$E = \frac{GmM}{2r} - \frac{GmM}{r}$$

$$E = -\frac{GmM}{2r} \tag{15-8}$$

Equation (15-8) gives the total energy of the bound system for any orbital radius r.* We see that for a *bound* system the total energy is *negative;* energy must be added to bring the total *up to zero.* The name *binding energy* is sometimes used to denote the amount of energy, $GmM/2r$, that must be added to the bound system to bring its total energy to zero and "unbind" the system, so to speak.

Equation (15-8) shows further that as the radius r decreases, energy E also decreases (becomes more negative.) An orbiting satellite is more tightly bound to its force center in a small orbit than in a large orbit.

Escape Speed   Is it true that "everything that goes up must come down"? Not if an object is thrown fast enough. At what minimum speed, or *escape speed* v_{es}, must a particle be projected from the earth's surface so that it escapes entirely from the gravitational influence of the earth?

The problem is easily solved with energy conservation. The particle is to be projected from the earth's surface with enough kinetic energy that the system's total energy is *zero.* Then when the particle is infinitely far from the earth (and the potential energy is zero), the particle will just have come to rest (and the kinetic energy is also zero). The minimum kinetic energy is $\frac{1}{2}mv_{es}^2$. If the particle is projected with less kinetic energy, it will turn around at its farthest point and fall back. If projected with more than the minimum kinetic energy, the particle will, when it is an infinite distance from the earth, still have kinetic energy remaining.

With the earth's mass denoted by M and its radius by r_e, we have then

$$E = K + U_g = 0$$

$$= \frac{1}{2}\,mv_{es}^2 + \left(\frac{-GmM}{r_e}\right) = 0$$

or

$$v_{es} = \sqrt{\frac{2GM}{r_e}} \tag{15-9}$$

The relation for the escape speed, (15-9), has several consequences.

• The escape speed does not depend on the particle's mass m.
• The escape speed can be written in another, equivalent form by using the relation (15-2) for the gravitational acceleration at the earth's surface, $g = GM/r_e^2$:

$$v_{es} = \sqrt{\frac{2GM}{r_e}} = \sqrt{2gr_e}$$

• The escape speed is related in a simple way to the speed v_{orb} of a satellite circling the earth near its surface. Since $g = v_{orb}^2/r_e$, we have $v_{orb} = \sqrt{gr_e}$, so that

$$v_{es} = \sqrt{2gr_e} = \sqrt{2}\,v_{orb} = 1.41v_{orb}$$

---

* The most general orbit for a bound system under a central inverse-square, attractive force is an ellipse, with the force center at one focus. It turns out that the total energy for an elliptic orbit is given by $E = -GmM/2a$, a relation just like (15-8), but with radius r replaced by semimajor axis a. For an elliptic orbit, both the kinetic and potential energies change with time, but their sum is constant. All elliptic orbits with the same semimajor axis a have the same energy.

Table 15-1

SPEED	ENERGY	STATE	PATH
$v < v_{es}$	$E < 0$	Bound	Ellipse
$v = v_{es}$	$E = 0$	Escape threshold	Parabola
$v > v_{es}$	$E > 0$	Unbound	Hyperbola

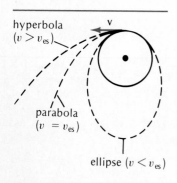

hyperbola $(v > v_{es})$

parabola $(v = v_{es})$

ellipse $(v < v_{es})$

*Figure 15-17. A particle subject to a force center is projected at velocity **v**. The particle's path depends on how the projection speed compares with the escape speed v_{es}.*

The escape speed is only 41 percent greater than the speed of a close satellite. This implies that when a society has developed the technology for placing a satellite in orbit, it is a relatively small step to have a spacecraft leave the planet entirely.

- The minimum speed for escape from the earth's surface is

$$v_{es} = \sqrt{2gr_e} = \sqrt{2(9.80 \text{ m/s}^2)(6.37 \times 10^6 \text{ m})} = 11.2 \text{ km/s} = 6.94 \text{ mi/s}$$

Even though hydrogen is the most abundant element in the universe, the earth has no molecular hydrogen in its atmosphere. Hydrogen escapes rapidly from the earth because at least a small fraction of any molecular hydrogen molecules that reach the top of the atmosphere have speeds exceeding 11 km/s. The moon, with a far smaller mass, has no atmosphere at all. The sun, on the other hand, is far more massive than the earth, and even hydrogen cannot escape from it.

- Equation (15-9) gives the escape speed for particles projected at *any angle* (above the horizontal). After all, our derivation in no way implied projection along the vertical. It can be shown that a particle's path for projection at v_{es} is a parabola. See Figure 15-17. The paths (all conic sections) for particles with three possible ranges of projection speeds are shown in Table 15-1.

Example 15-5. An earth satellite passes through the earth's atmosphere; it is acted on by a resistive force that dissipates energy. Show that as the satellite spirals inward, it does not slow down, but rather *speeds up*.

A dissipative force acts on the satellite, so that the *total* energy of the earth-satellite system must decrease. From (15-8),

$$E = -\frac{GmM}{2r}$$

We see that if E decreases (becomes more negative), r must also decrease. The satellite spirals *inward*.

But from (15-7), we see that the kinetic energy for a circular orbit is given by

$$\frac{1}{2} mv^2 = \frac{GmM}{2r}$$

Therefore, as r decreases, the satellite speed v must *increase*.

We can look at this result from another point of view. The preceding equation can also be written in the form

$$K = -\tfrac{1}{2}U$$

so that any change in kinetic energy ΔK is related to a potential-energy change ΔU by

$$\Delta K = -\tfrac{1}{2}\Delta U$$

Suppose a particle's kinetic energy increases by ΔK. Then the system's potential energy must *decrease by a larger amount*, $\Delta U = -2\Delta K$. The net effect is a *decrease* in total energy.

A satellite subject to a resistive force speeds up and moves in a smaller orbit as its total energy decreases. The resistive force can arise from the atmosphere. But suppose that a satellite is above the atmosphere and it too is to speed up. Then rocket thrusters on the forward side of the rocket—*retro*rockets—must be turned on.

Example 15-6 (Optional).   A *black hole* is a stellar object from which no particles or even light can escape because of the very strong gravitational attraction. Since a black hole absorbs light striking it but lets none escape, a black hole would actually appear totally black. It sucks up light and matter but emits none. Finding an approximate relation for the radius of a black hole of mass M involves the same considerations as in computing an escape speed.

Visible light and other types of electromagnetic radiation consist of particlelike photons. Through the Einstein mass-energy relation, $E = mc^2$, we can associate with a photon of energy E_{ph} a mass given by $m = E_{ph}/c^2$, where c is the speed of light, 3×10^8 m/s.

Suppose that a photon is fired outward from the surface of a black hole of radius R but just fails to escape. The photon's kinetic energy at the surface is $K = E_{ph}$. The system's gravitational potential energy with the photon at the surface is

$$U_g = -\frac{Gm_1m_2}{r} = -\frac{G(E_{ph}/c^2)M}{R}$$

where M is the mass of the black hole. The threshold for escape corresponds to a total energy E equal to or less than zero:

$$E = K + U \le 0$$

$$E_{ph} - \frac{GE_{ph}M}{c^2R} \le 0$$

so that the black hole's radius can be no larger than

$$R \le \frac{GM}{c^2}$$

Note that the photon energy E_{ph} drops out; if light of one wavelength (or energy) cannot escape, all other wavelengths of electromagnetic radiation also fail to escape.

The derivation here is only approximate. A more rigorous derivation, based on the general theory of relativity, gives a more accurate result, which differs from that given earlier by a factor of 2:

$$R_{min} = \frac{GM}{2c^2}$$

How small would our sun ($M = 2 \times 10^{30}$ kg) have to be to become a black hole?

$$R_{min} = \frac{GM}{2c^2} = \frac{(6.67 \times 10^{-11} \text{ N} \cdot \text{m}^2/\text{kg}^2)(2 \times 10^{30} \text{ kg})}{2(3 \times 10^8 \text{ m/s})^2} \approx 0.7 \text{ km}$$

The sun's radius would have to be reduced by a factor of 10^6 to go from its present value of 7×10^5 km to 0.7 km. At the same time the sun's volume would be reduced by a factor of 10^{18} and its density would increase by a factor of 10^{18}.

15-7   Inertial Mass and Gravitational Mass (Optional)

We have used the single term *mass* to denote what are actually two distinct physical properties—an object's inertial mass m_i and its gravitational mass m_g.

The *inertial mass* of a particle appears in Newton's second law, $\Sigma \mathbf{F} = m_i \mathbf{a}$, and in such related mechanical quantities as momentum, $\mathbf{p} = m_i \mathbf{v}$. Any particle's inertial mass is then simply the magnitude of the resultant force on it, of whatever type, divided by its acceleration. This mass is termed *inertial*, since it is a measure of the degree to which the particle shows inertia to changing its velocity under the influence of any external force.

The *gravitational mass* m_g, on the other hand, arises in the gravitational force law, $F_g = G m_{g1} m_{g2}/r^2$ or in such related quantities as an object's weight, $w = m_g g$. The gravitational mass is then that physical property of a particle that is the origin of its interaction through gravity with another particle.

Gravitational mass in the gravitational interaction between two particles has an exactly analogous role to that of electric charge in the electric interaction between two electrically charged particles. The basic source of the electric force one electron exerts on another electron is the electric charge of each. (Nature has, so to speak, directed, exhorted, or *charged* the particles to behave thus.) In similar fashion, the basic source of the gravitational force between a pair of particles is that property of each that we call the gravitational mass.

Any particle's gravitational mass can be defined and measured through a simple operational procedure. Suppose that particles 1 and 2 with gravitational masses m_{g1} and m_{g2} are brought, in turn, to the same location relative to a third object, such as the earth. The gravitational forces on each, F_{g1} and F_{g2}, are measured; in other words, the weights of the two objects are found to be F_{g1} and F_{g2}. Then, by *definition*, the ratio of respective gravitational masses equals the ratio of the corresponding gravitational forces:

$$\frac{F_{g1}}{F_{g2}} \equiv \frac{m_{g1}}{m_{g2}} \tag{15-10}$$

Suppose now that particles 1 and 2 are dropped at the same location and their accelerations a_1 and a_2 are measured (as in the legendary experiment at the Tower of Pisa, or in the actual experiment performed on the moon during the Apollo 15 mission in 1971). Both particles are found to have the same acceleration g. For particles at the earth's surface.

$$a_1 = a_2 = g$$

We apply Newton's second law to the motion of each particle and note that it is the *inertial* mass that must be multiplied by the acceleration to yield the resultant force on it:

$$F_{g1} = m_{i1} a_1 = m_{i1} g$$

and

$$F_{g2} = m_{i2} a_2 = m_{i2} g$$

Now if we take the ratio of the two equations above and use (15-10), we arrive at the simple result

$$\frac{m_{g1}}{m_{g2}} = \frac{m_{i1}}{m_{i2}}$$

By *experiment* the ratio of the gravitational masses of two particles equals the ratio of their inertial masses to nearly 1 part in 10^{12}. This result, although simple, must be regarded as an extraordinary coincidence of nature, not predictable on theoretical grounds alone. After all, two particles with identical electric charges need not have identical inertial masses, and generally they do not; for example, the proton and the positron have precisely equal electric charges, but their inertial masses differ by a factor of nearly 2000.

Since any particle's gravitational mass is directly proportional to its inertial mass, we may use the standard 1-kg object as the basis for measuring not only inertial mass but also gravitational mass. Then, a 1-kg object has an inertial mass of 1 kg and a gravitational mass also of 1 kg. For this reason, we typically drop the adjectives *inertial* and *gravitational* and refer simply to a particle's mass:

$$m_i = m_g \tag{15-11}$$

15-8   The Principle of Equivalence (Optional)

A particle's inertial mass m_i, which is a measure of its inertia to a change in velocity, and its gravitational mass m_g, which is a measure of the strength of its gravitational interaction with a second particle, are distinct attributes. There is no a priori reason why, for a given particle, m_i should equal m_g; yet a variety of highly precise experiments show that they do: $m_i = m_g$.

In classical physics, the equality, or equivalence, of m_i and m_g must be regarded as an extraordinary coincidence. But this equality is taken as a basic assumption in the *principle of equivalence,* a fundamental part of the general theory of relativity, formulated by Albert Einstein. Because the principle of equivalence relates to accelerated reference frames, let us first recall some earlier results.

In Section 6-5, we found that an observer in an accelerated reference frame can actually use Newton's second law to describe a particle's motion. To do so, such an accelerated observer must invoke, in addition to the "real" forces acting on the particle, a fictitious inertial force, $F_i = -m_i a_i$. The inertial force depends on the particle's inertial mass, m_i and on the acceleration $\mathbf{a}_i$ of the observer's reference frame relative to an inertial frame. The noninertial observer writes Newton's second law as

$$\Sigma \mathbf{F} + \mathbf{F}_i = m_i \mathbf{a}$$

with

$$\mathbf{F}_i = -m_i \mathbf{a}_i \tag{6-6}$$

Here **a** is the particle's acceleration as measured in the accelerated frame.

Now, if a particle's gravitational mass m_g is precisely the same as its inertial mass m_i, the inertial force can be written equally well as

$$\mathbf{F}_i = -m_i \mathbf{a}_i = -m_g \mathbf{a}_i$$

Then the "fictitious" inertial force is, like any gravitational force, proportional to the observed particle's gravitational mass. This implies in turn that *an inertial force arising in a noninertial frame is altogether equivalent to, and indistinguishable from, a gravitational force as perceived by an observer at rest in this accelerated frame.* This is the principle of *equivalence.*

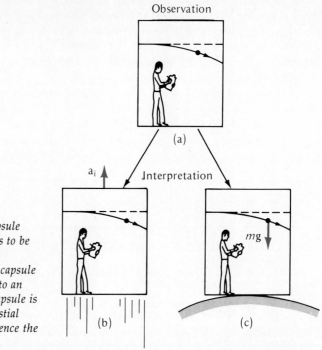

Figure 15-18. *(a) An observer in a closed capsule finds a beam of particles to be deflected. Two possible interpretations: (b) The capsule is accelerating relative to an inertial frame; (c) the capsule is at rest near a large celestial body under whose influence the particles fall by gravity.*

Consider, as an example of the equivalence of an inertial force to a gravitational force, an observer in a closed capsule through which a beam of particles is directed. See Figure 15-18. Suppose the observer finds that the particles depart from straight-line motion and, in fact, travel in a parabolic path. Why? The observer in the capsule may draw either of two conclusions:

• His capsule is accelerating upward relative to an inertial frame in gravity-free space. The particles, free of any real force but acted on by an inertial force, depart from motion at constant velocity.

• His capsule is at rest in an inertial frame near a large celestial body, and the particles fall under the influence of its gravitational field.*

According to the principle of equivalence, there is no way of telling the difference between the two alternative viewpoints by any experiment in physics.

Consider another example. Suppose that Newton is up in an apple tree and that he falls from a limb at the very instant an apple becomes detached. During the fall, Newton sees the apple *at rest.* Again there are two alternative explanations:

---

* What if the beam of particles consists of a beam of light? If the capsule's acceleration is large enough, the path of light will surely be parabolic. But this implies, if we take the principle of equivalence seriously, that a beam of light is deflected by a gravitational field—in short, that light has weight. Einstein's general theory of relativity, of which the principle of equivalence is a basic part, implies just this. The deflection of light by the sun and also other indirect experiments, confirm the prediction in detail.

Happy Birthday, dear Albert!

A gadget like the one pictured here was given to Albert Einstein on March 14, 1955, his seventy-sixth birthday. The inventor of this birthday present and its donor was Einstein's neighbor, Eric M. Rogers, a physics professor at Princeton University.*

What is the surefire way of getting the ball into the cup without touching the ball? (You're not likely to do it by jiggling the pole up and down.) When Einstein got the device, he solved the puzzle immediately, and then confirmed his answer with a little experiment.

Can you figure out the easy way to get the ball in the cup? (*Hint:* Einstein used a fundamental idea he made famous.) For the answer, look at the last page of this chapter.

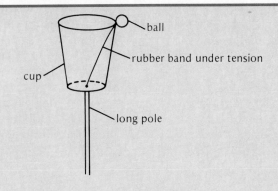

*For Rogers's account of the birthday present, see A. P. French, ed., *Einstein, A Centenary Volume*, (Cambridge: Harvard University Press, 1979), p. 131.

- The apple is in equilibrium under the action of two balancing forces — its weight $m_g\mathbf{g}$ down and an equal inertial force, $m_i\mathbf{a} = -m_i\mathbf{g}$, up.
- The apple floats because it is in gravity-free space.

Or consider still another example, an astronaut in a spacecraft that is circling the earth as a satellite. He sees unattached objects floating freely. If the astronaut is unaware of his motion relative to objects external to the spacecraft, he could equally well conclude that:

- He is in gravity-free space.
- He is in orbit, with objects within the spacecraft in equilibrium under the action of an inward gravitational force and an outward centrifugal force of the same magnitude.

An external observer in an inertial frame would say that the spacecraft and all objects within it, subject only to gravitational forces, have the same acceleration toward the gravitational center and therefore are not in motion relative to one another.

15-9 Gravitational Effect of a Spherical Shell (Optional)

To be proved: a uniform spherical shell attracts a particle on its exterior as if all the shell's mass were located at its center.

To find the resultant gravitational *force* of a spherical shell on a particle outside, we must take into account that the direction as well as the magnitude of the force differs according to which small piece of the shell we take as a particle [Figure 15-3(a)]; that is, we must add up *vector* contributions from all pieces of the shell. An easier task, which we follow here, is to find the gravitational *potential energy* between the shell and a particle. Then we deal with a *scalar* quantity and the summation (integration) is simpler. Once the potential energy relation is found, it is simple enough to find the corresponding force relation.

We want to calculate the gravitational potential energy between a point mass m and a thin, spherical shell of mass M, radius R, and thickness ΔR, as shown in Figure 15-19(a). The various mass elements dM composing the spherical shell are at various distances s from the point mass m and make different contributions to the potential energy. From one mass element dM, there is a contribution to the potential energy of

$$dU = -\frac{Gm\,dM}{s} \tag{15-12}$$

where s is the distance from dM to m. To find the potential energy for the entire shell and the point mass m, we must add the contributions from all mass elements composing the shell.

We first take the point mass m to be outside the spherical shell, as shown in Figure 15-19(b). All points on the thin, circular ring, whose axis is the line joining the center of the shell and the mass m, are the *same* distance s from the mass m. The radius of this ring is $R \sin \theta$, its width is $R\,d\theta$, and its thickness is ΔR; therefore, the ring's volume is

$$dV = (2\pi R \sin \theta)(R\,d\theta)(\Delta R)$$

We take the mass density ρ to be constant throughout the spherical shell. Then the mass dM of the ring is just $\rho\,dV$. The gravitational potential energy between the ring and the point mass m is

$$dU = -\frac{Gm\rho\,dV}{s} = -\frac{Gm\rho(2\pi R^2 dR)\sin\theta\,d\theta}{s} \tag{15-12}$$

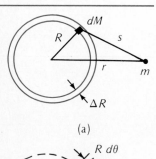

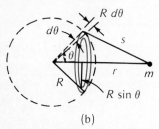

Figure 15-19. (a) A particle of mass m interacting with a spherical shell. (b) A ring of the shell interacting with the point mass.

The volume of the entire shell is $V = 4\pi R^2\,\Delta R$.

To find the total potential energy, we merely sum over all the rings that make up the spherical shell. These rings differ in s and θ. We may add the rings from $\theta = 0$ to $\theta = \pi$, or equivalently, from $s = r - R$ to $s = r + R$, where r is the distance from the center of the shell to the point mass m.

The variables s and θ are related to one another through the law of cosines:

$$s^2 = R^2 + r^2 - 2Rr\cos\theta$$

Both R and r are fixed. When we take the differentials of both sides of the equation above, we get

$$2s\,ds = 2Rr\sin\theta\,d\theta$$

$$\sin\theta\,d\theta = \frac{s\,ds}{Rr}$$

We use this relation in (15-12) and then integrate:

$$U = \int dU = -Gm\rho(2\pi R^2\,\Delta R)\int_{r-R}^{r+R}\frac{s\,ds/Rr}{s}$$

$$= -\frac{Gm\rho 2\pi R\Delta R}{r}\int_{r-R}^{r+R}ds = -\frac{4\pi R^2\,\Delta R\rho Gm}{r} = -\frac{Gm\rho V}{r} \tag{15-13}$$

The density of the shell was taken to be constant. Therefore, we have $\rho V = M_s$, where M_s is the total mass of the shell. We can write the equation above more simply as

$$U = -\frac{GmM_s}{r} \qquad (15\text{-}14)$$

which is just the gravitational potential energy of *point masses* m and M_s separated by a distance r. All the shell's mass M_s can be imagined to be at its center. It follows, then, that the gravitational force between an exterior point mass and a spherical shell is equivalent to the force on each of two point masses a distance r apart.

We now consider point mass m to be inside the spherical shell. To do so we merely need to change the limits of the variable s in (15-13). Now s varies from $R - r$ to $R + r$, and

$$\int_{R-r}^{R+r} ds = 2r$$

Then the results from (15-13) become

$$U = -\frac{GmM_s}{R}$$

Since R is a constant, the potential energy is constant throughout the interior of the shell; in fact, the potential energy inside has the *same* value as when the particle is at the shell, and $r = R$. Since the potential energy does not change with position, the force on the particle anywhere inside the shell, $F = -dU/dr$, (10-9), is *zero*.

In summary, when the particle is outside the shell, the shell acts like a point mass at its center; when the particle is inside the shell, it has no gravitational effect on the particle.

Summary

Fundamental Principles

The law of universal gravitation: A central, inverse-square, conservative, attractive force acts between each pair of interacting masses. For point masses m_1 and m_2 separated by r, the magnitude is

$$F_g = \frac{Gm_1m_2}{r^2} \qquad (15\text{-}1)$$

where G is the constant 6.672×10^{-11} N·m²/kg².

Other Important Results

The gravitational potential energy for a pair of mass points separated by a distance r:

$$U_g = -\frac{Gm_1m_2}{r} \qquad (15\text{-}4)$$

Kepler's laws, observed for planets about the sun and deducible from properties of the gravitational force:

- *Orbital rule.* Path is ellipse, force center at focus.
- *Area rule.* **r** sweeps out equal areas in equal times (equivalent to angular-momentum conservation, applicable for any central force).
- *Period rule.* $T^2/a^3 =$ constant with period T and semi-major axis a.

A uniform spherical shell acting on a mass point external to the shell is equivalent to point mass at its center. The resultant gravitational force on a particle inside a spherical shell is zero.

Problems and Questions

See Appendix C for astronomical data.

Section 15-1   The Law of Universal Gravitation

· **15-1 Q** Which graph in Figure 15-20 best describes how the weight, W (solid line), and mass, M (broken line) of an object change as the object is moved very slowly from the earth to the moon?

(A) A
(B) B
(C) C
(D) D
(E) E
(F) F

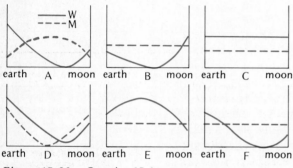

Figure 15-20.   *Question 15-1.*

· **15-2 P** A typical distance between Mars and the earth is 10^8 km. The mass of Mars is 6.5×10^{23} kg. What force does Mars exert on a 50 kg person on earth?

· **15-3 P** *(a)* What is the gravitational force of the moon on the earth? *(b)* Of the sun on the earth? *(c)* What is the ratio of these two forces?

: **15-4 P** At what fraction of the distance from the center of the earth to the center of the moon is the net gravitational force from these two bodies zero?

: **15-5 P** A lead sphere of mass M and radius R contains a spherical void of radius $\frac{1}{2}R$, positioned as shown here. With what force will the sphere attract a small point mass m a

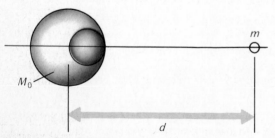

Figure 15-21.   *Problem 15-5.*

distance d away? The centers of the large sphere, the void and the point mass are collinear.

: **15-6 P** Before Cavendish first determined the value of G, the masses of the sun and the planets could not be calculated. One could, however, find the ratio of the masses of any two bodies having orbiting satellites. For example, calculate the ratio of Jupiter's mass to that of the earth (without using the value of G) given the following observations:

	ORBIT RADIUS	PERIOD
Earth-moon	3.80×10^5 km	27 days
Jupiter-moon	6.8×10^5 km	3.5 days

Section 15-2   Gravitational Acceleration g

· **15-7 P** What would be the acceleration due to gravity on the surface of a planet whose radius is twice that of the earth, assuming the planet had the same average density as the earth?

: **15-8 P** The period of a certain pendulum varies by 0.010% between the base and the peak of a mountain. What is the height of the mountain?

: **15-9 P** What is the acceleration due to gravity on the surface of Jupiter? Its mass is 318 times that of earth and its radius is 71,800 km.

: **15-10 P** The earth's polar radius is 6.357×10^6 m; its equatorial radius is 6.378×10^6 m. Taking into account that the earth is not a sphere, as well as the rotation of the earth, what is the ratio of the apparent weight of an object at the equator to the apparent weight at one of the poles?

: **15-11 P** Artificial gravity can be created in a cylindrical spacecraft by rotating it about its axis of symmetry. At what rate must a ship of 10 meters radius rotate in order to create at the outer radius of the ship the same gravitational effect experienced on earth?

Section 15-3   The Cavendish Experiment

· **15-12 P** Suppose that you were to determine the value of G by direct measurement, as Cavendish did. What force would be exerted between two spheres, each of mass 100 kg, separated by 1 meter? This gives some idea of the difficulty in carrying out such a measurement.

Section 15-4   Kepler's Laws of Planetary Motion

: **15-13 P** Ganymede, the largest moon of Jupiter, is observed to move in a nearly circular orbit of period 7.16 days and radius 1.07×10^6 km. From this information deduce the mass of Jupiter, expressed as a multiple of the earth's mass.

: **15-14 P** A satellite is observed to orbit a planet close to the planet's surface in a circular orbit with period T. Show that the average density of the planet is $3\pi/GT^2$.

: **15-15 P** What is the period of a satellite in circular orbit about the earth at a distance of 300 km above the earth's surface?

: **15-16 P** Halley's comet moves in a highly eccentric orbit around the sun. It has been observed in the vicinity of the earth every 75.5 years since 87 B.C. Its perihelion distance (the point of closest approach to the sun) is 8.9×10^{10} m. What is the greatest distance between the comet and the sun?

Section 15-5   Gravitational Potential Energy

· **15-17 P** A particle is dropped from rest at a distance r_E above the earth's surface. The earth's radius is r_E; its mass is M_E. The speed with which the particle hits the earth's surface (neglecting air resistance) is

(A) $\sqrt{2GM/r_E}$
(B) $\sqrt{GM/r_E}$
(C) $\sqrt{GM/2r_E}$
(D) $\sqrt{4GM/3r_E}$
(E) GM/r_E^2

· **15-18 Q** Two planets are on a collision course in outer space a long way from all other bodies. The masses of the two planets are m_1 and m_2; they have the same radius R. At a given instant of time ($t = 0$), their speeds are v_1 and v_2 and at that time their centers of mass are separated by a distance d. At this time the total energy (the total macroscopic kinetic and potential energy) of the system is

(A) $\frac{1}{2}m_1v_1^2 + \frac{1}{2}m_2v_2^2$
(B) $\frac{1}{2}m_1v_1^2 + \frac{1}{2}m_2v_2^2 - Gm_1m_2/R$
(C) $\frac{1}{2}m_1v_1^2 + \frac{1}{2}m_2v_2^2 - 2Gm_1m_2/R$
(D) $\frac{1}{2}m_1v_1^2 + \frac{1}{2}m_2v_2^2 - Gm_1m_2/d$
(E) $\frac{1}{2}m_1v_1^2 + \frac{1}{2}m_2v_2^2 - 2Gm_1m_2/d$

· **15-19 Q** Two identical spheres each of radius R and mass M are initially held at rest with their centers separated by a distance d. Then the spheres are released, they attract one another through the gravitational force, and collide. If the speed of each sphere at the moment of collision is v, then the equation expressing the energy-conservation principle for this situation is:

$$A = B + 2(\tfrac{1}{2}Mv^2)$$

Where

(A) $A = 0, B = -\dfrac{GM^2}{d}$

(B) $A = -\dfrac{GM^2}{d}, B = 0$

(C) $A = \dfrac{GM^2}{d^2}, B = 0$

(D) $A = -\dfrac{GM^2}{2R}, B = -\dfrac{GM^2}{d}$

(E) $A = -\dfrac{GM^2}{d}, B = -\dfrac{GM^2}{2R}$

· **15-20 Q** A planet orbits a star in an elliptical orbit. Which of the following combinations of quantities are each constant throughout the orbit?
(A) Angular momentum and total energy.
(B) Kinetic energy and potential energy.
(C) Angular momentum and kinetic energy.
(D) Kinetic energy and total energy.
(E) Speed and kinetic energy.

: **15-21 Q** An object of mass m is fired vertically upward from the earth's surface with speed v. The object rises a distance $2R$ above the earth's surface, where R is the radius of the earth. The mass of the earth is M. The equation describing this situation in terms of energy conservation is

(A) $\dfrac{1}{2}mv^2 = \dfrac{GmM}{R}$

(B) $\dfrac{1}{2}mv^2 = \dfrac{GmM}{2R}$

(C) $\dfrac{1}{2}mv^2 + \dfrac{GmM}{R} = \dfrac{GmM}{3R}$

(D) $\dfrac{1}{2}mv^2 - \dfrac{GmM}{R} = -\dfrac{GmM}{2R}$

(E) $\dfrac{1}{2}mv^2 - \dfrac{2GmM}{3R} = 0$

: **15-22 P** A 50-kg meteorite falls to the earth from far out in space. (a) If it lost no energy to friction, with what speed would it strike the earth? (b) What is its kinetic energy then?

: **15-23 P** Two planets, with masses m_1 and m_2, whose centers are a distance d apart (at $t = 0$) move toward each other with speeds v_1 and v_2. They are not initially rotating. They collide and shatter. (a) Which of the following quantities are conserved in this process?
(A) The total kinetic energy.
(B) The total kinetic energy plus the gravitational potential energy before the collision.
(C) The linear momentum during the collision.
(D) The total energy (kinetic plus potential) during the collision.
(E) The angular momentum about a fixed star.

(F) The angular momentum about the center of mass.

(G) The angular momentum of m_1 about a fixed star.

(H) The angular momentum of m_1 about its center.

Find the following in terms of the gravitational constant G and the given quantities: (b) The total angular momentum about the center of mass before the collision. (c) The gravitational potential energy at the moment of impact (the planets have radii R_1 and R_2). (d) The gravitational potential energy at $t = 0$ when the centers of the planet are a distance d apart. (e) The total energy at the moment of impact. (f) The velocity of the center of mass at $t = 0$. (g) The total momentum of the debris after the collision.

⋮ **15-24 P** A particle of mass m is positioned at the center of a ring of mass M and radius R. (a) What force acts on m? (b) What is the system's gravitational potential energy?

⋮ **15-25 P** A particle of mass m is positioned along the line of a uniform rod of mass M and length $2a$. The particle is a distance x from the center of the rod. (a) What is the gravitational force acting on m? (b) What is the gravitational potential energy of the system?

⋮ **15-26 P** It has been suggested that manufacturing and mining operations be established on the moon as an intermediary step in building space colonies. Less energy is required to lift a payload from the surface of the moon than from the surface of the earth. (a) To investigate this, calculate the energy needed to lift a 1,000 kg payload from the surface of the earth to a point midway between the earth and the moon. (b) Repeat the calculation for a payload lifted from the moon's surface to the same point.

⋮ **15-27 P** A missile is fired with velocity v_0 at an angle of θ with respect to the vertical from the earth's surface. Determine (a) its maximum distance from the center of the earth and (b) its speed at that point. (Assume for simplicity that the earth is not rotating.)

⋮ **15-28 P** A diametrical tunnel is drilled through the earth. The earth has mass M and radius R. A rapid transit vehicle subject to no friction travels through this evacuated tunnel.

(a) Show that the gravitational force acting on a vehicle of mass m is $(GMm/R^3)r$ when the vehicle is a distance r from the earth's center. (b) With what period would the vehicle oscillate back and forth in the tunnel? (c) Show that a straight frictionless tunnel between *any* two points on the earth's surface, not necessarily passing through the earth's center, will yield the same period as a diametrical tunnel.

⋮ **15-29 P** When the first nuclear weapons were exploded, there was some concern that a huge nuclear chain reaction would be set up that would completely blow the earth to pieces. How much energy would be required to do this? Show that the energy required to disassemble completely the earth of mass M_E and radius R_E is $(9/15)\,GM_E^2/R_E$. (Hint:

imagine layers of the earth are peeled off one by one like layers of onion skin.)

⋮ **15-30 P** Many galaxies have the form of a spiral disk, so the following calculation is of practical interest: Consider a thin disk of radius R and mass M. (a) What is the force on a particle of mass m on the axis a distance z from the disk? (b) What is the gravitational potential energy of such a particle interacting with the disk? (c) Show that the results reduce to the expression for two particles when $z \gg R$.

· **15-31 Q** In order to understand the behavior of an artificial satellite circling the earth, it is useful to recognize that

(A) the satellite is continually accelerating toward the earth.

(B) the satellite is "floating" in the earth's atmosphere which, though very thin, still provides the needed buoyant force to keep the satellite up.

(C) the satellite doesn't fall to earth because it is so far away that it is unaffected by the earth's gravity.

(D) the gravitational force on the satellite due to the earth is counteracted by gravitational forces due to the sun and the moon.

(E) the satellite is in equilibrium and hence it can coast almost indefinitely without falling to earth.

⋮ **15-32 P** What is the escape velocity from the moon?

⋮ **15-33 P** What would be the period of a spacecraft in a very low orbit above the surface of the moon?

⋮ **15-34 P** What is the speed of a satellite in circular orbit 240 km above the surface of the earth?

· **15-35 P** A rocket is fired vertically upward from the north pole with a velocity of 5 km/s. How far above the earth's surface will it rise?

· **15-36 P** The geostationary operational environmental satellite (GEOS) monitors weather conditions in the Atlantic and Eastern Pacific regions. It moves in a geosynchronous orbit that keeps it always above the same point on the earth. How far above the earth's surface is GEOS, and with what speed does it move?

⋮ **15-37 P** Planets A and B are in circular orbits about a star. Planet A is moving half as fast as planet B. What is the ratio of the radii, R_A/R_B, for their orbits?

⋮ **15-38 Q** Is it possible to launch a projectile from the surface of the earth in such a way that it will move periodically around the earth in a stable orbit? The projectile has no engine of its own. In what direction should the projectile be fired, and with what speed, if this is to be accomplished? If this cannot be done, explain why.

⋮ **15-39 Q** A satellite moves in a circular orbit about a planet of mass M. The satellite's mass is unknown but is very small compared with the planet's mass. The satellite's orbital radius is R, its period of revolution is T, its angular velocity is ω, and its speed is v. It is possible to calculate the satellite mass from the following quantities (and the universal gravitational constant G):

(A) T and R
(B) ω, R, and M
(C) v and M
(D) v and T
(E) none of the above

: **15-40 Q** A person in an earth satellite is sometimes said to be weightless. How do you explain this statement?

(A) A person in a satellite is so far away that the effect of the earth's gravitational field on him is negligible.
(B) He is going so fast that gravity effects are negligible.
(C) He is falling freely with acceleration **g**.
(D) He is at a point in space where the gravitational effects of the earth, sun, and moon all cancel.
(E) He is weightless because the downward gravitational force is cancelled by the upward orbital force.

: **15-41 Q** A spaceship in a circular orbit about the earth is to be transferred to a larger circular orbit by firing the engines once at A and once at B, and following an elliptical orbit from A to B. See Figure 15-22. This can best be done by firing the engines so that the ship's speed

(A) increases at A and decreases at B.
(B) increases at A and increases at B.
(C) decreases at A and increases at B.
(D) decreases at A and decreases at B.
(E) None of the above will work, since there is only one circular orbit for a given spaceship.

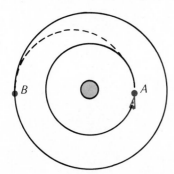

Figure 15-22.   Question 15-41.

: **15-42 Q** When a satellite orbits a planet in a circular orbit,

(A) the total energy of the system of planet and satellite is positive.
(B) a larger orbit means greater satellite speed.
(C) $|P.E.| = 2|K.E.|$
(D) both the kinetic and potential energies are positive.
(E) the total energy is independent of orbit size.

: **15-43 P** Two stars in a binary system orbit each other with a period of 30 days. Each moves with velocity 30 km/s. What is (a) the mass of each, and (b) their separation?

: **15-44 P** (a) What is the speed of the earth in its orbit about the sun? (b) What increase in speed would be needed for the earth to escape completely from the sun's gravitational field?

: **15-45 P** Two binary stars, of masses m and $2m$, rotate about each other in stable circular orbits, with a constant separation d. (a) What is the ratio of their speeds as they rotate about their center of mass? (b) What is the total energy of this system? (c) How much energy is needed to separate the stars completely?

: **15-46 P** In order to escape from the earth's gravitational field, a spacecraft must be given a certain minimum velocity. The rotation of the earth can be used to help meet this requirement. Consider the launch of a spacecraft of mass 100,000 kg. What initial velocity and kinetic energy is required if the launch is made at (a) the north pole? (b) the equator in the most favorable direction?

: **15-47 P** A satellite moves in an elliptical orbit with total energy E, angular momentum L and radial velocity v_r (the component of velocity directed radially outward). Show that

$$E = \tfrac{1}{2}mv_r^2 + L^2/2mr^2 - GMm/r$$

: **15-48 P** The distance from the center of the earth to a satellite in an elliptical orbit around the earth varies from a minimum R_1 (at perigee) to a maximum value of R_2 (at apogee). Show that the energy of a satellite in such a bound orbit is

$$E = -\frac{GM_E M_S}{R_1 + R_2}$$

Note that $R_1 + R_2$ is the major axis of the ellipse, so this is a very useful way of estimating the energy in a given orbit. (Hint: It may be useful to use the result given in Problem 15-47.)

: **15-49 Q** A satellite can orbit the earth in either orbit A or orbit B. See Figure 15-23. The speed of the vehicle at point P is

(A) the same for orbits A and B.
(B) greater in orbit A than in orbit B.
(C) dependent on the mass of the satellite.

Figure 15-23.   Question 15-49.

(D) possibly greater for A than for B, but this cannot be determined without more information.

(E) None of the above is true.

⋮ **15-50 P** Four identical space colonies, each of mass M, are positioned at the corners of a square of side a. At what speed must each move if they all move in the same circular orbit under the influence of the mutual gravitational forces between them? (The four objects maintain their positions on the square relative to each other.)

⋮ **15-51 P** A satellite of mass 180 kg is in circular orbit 600 km above the earth's surface. It experiences an average frictional force of 0.03 N. (a) Determine its speed. (b) Determine its period. (c) How much energy does it lose in 200 revolutions? (d) What are its speed, period, and distance from the earth after 200 revolutions? (Assume the path is essentially circular even while diminishing in radius.)

⋮ **15-52 P** A satellite is moving with speed $v = \sqrt{4GM_e/3r}$ in a direction such that its velocity is directed at $30°$ with respect to the radius vector $\mathbf{r}$ from the earth's center to the satellite. (a) How close to the earth does the satellite come? (b) What is the maximum distance from the earth to the satellite in this orbit? (c) What is the period of the satellite?

Supplementary Problem

15-53 P The Space Studies Institute founded by Princeton University physicist Gerald O'Neill is exploring the prospects of colonizing space with human beings. One plan is to manufacture such required space hardware as satellites, solar converters, and construction materials from minerals mined on the moon. These minerals would be sent directly from the moon to the factories in space. Professor O'Neill believes that manufacturing such items in effectively gravity-free space would be cheaper than sending manufactured hardware into orbit from the earth. The basis, he says, is that it takes 20 times more energy to orbit a given payload from earth than it does from the moon. Where does the factor of 20 come from?

Answer to Puzzle on Page 333

With the pole vertical, simply drop the gadget. By the principle of equivalence, the weight of the falling object (the ball) is effectively zero in a reference frame of the falling object. Then, with the ball subject only to the tension force of the stretched rubber band, the ball is easily pulled inside the cup.

Fluid Mechanics

16

Fluid mechanics is a very large branch of physics that is in general very complicated. In this chapter we consider only two fairly simple and elementary situations: a perfect fluid at rest, and a perfect fluid in steady streamline flow.

16-1   Fluids Defined

It is customary to divide materials into three states of matter: solid, liquid, and gas.* Solids and liquids collectively represent *condensed matter*; their densities are relatively greater than those of gases. Liquids and gases collectively represent fluids.

The states of matter can be distinguished as follows. A liquid or a solid has definite volume, but the volume of a gas is always that of its container. A solid has definite shape, but a liquid or a gas always assumes the shape of its container.

To see more precisely how a fluid is defined and distinguished from a solid, consider the deformable block of material shown in Figure 16-1. Its lower

---

* A *plasma*, consisting of separated positive and negative electrically charged particles, is sometimes referred to as a fourth state of matter.

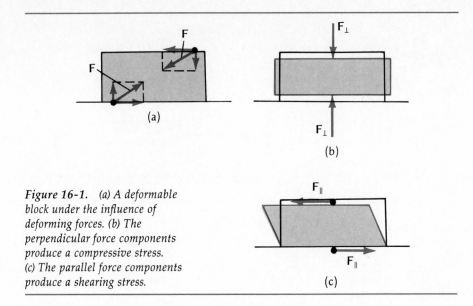

Figure 16-1. (a) A deformable block under the influence of deforming forces. (b) The perpendicular force components produce a compressive stress. (c) The parallel force components produce a shearing stress.

surface is tightly secured to a rigid horizontal surface on a table. An external force **F** is applied to the top surface of the block. The table then exerts a force of equal magnitude and opposite direction to the block's lower surface, so that the block is in equilibrium. It is useful to consider separately the effects of the force components perpendicular [Figure 16-1(b)] and parallel [Figure 16-1(c)] to the surfaces over which they are applied.

• The *perpendicular* force components $F_\perp$ *compress* the material, bringing its upper and lower surfaces closer together (and also increasing slightly the transverse dimensions). The $F_\perp$ components produce a *compressive stress,* or pressure. For a surface area A, this compressive stress is defined as $F_\perp/A$.

• The *parallel* force components $F_\parallel$ tend to shift the top surface relative to the lower one. Indeed, every horizontal layer of the material is shifted parallel to an adjoining layer of material. The $F_\parallel$ components produce a *shear stress;* its magnitude is $F_\parallel/A$.

A deformable solid subjected to a compressive stress or a shearing stress or both will eventually come to equilibrium, but a fluid will not. Indeed a *fluid* is, by definition, a material that *cannot permanently sustain a shear stress.* In a fluid the adjacent layers slide relative to one another. In short, fluids can always flow. The flow may be slow, as in the case of molasses (or even more slowly in uncrystallized glass, which can be classified as a liquid). Although we can generally tell a solid from a liquid or a gas, under some circumstances it may be difficult or even impossible to differentiate between the states of matter.

16-2   Pressure

Since a fluid cannot sustain a shear stress, external forces acting on the exterior surface of a fluid at rest must be *perpendicular* to the surface. See Figure 16-2(a), where the forces on the external surfaces of a liquid in a container are at each point normal to the surface. Now consider forces acting *within* a fluid. More

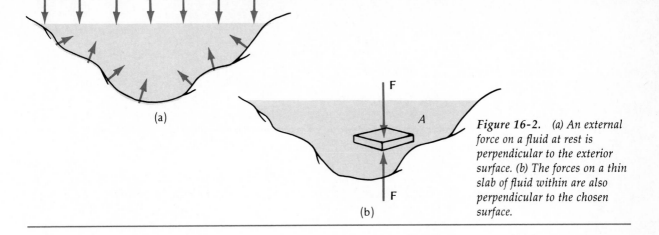

(a)

(b)

Figure 16-2. *(a) An external force on a fluid at rest is perpendicular to the exterior surface. (b) The forces on a thin slab of fluid within are also perpendicular to the chosen surface.*

specifically, consider a small, thin slab of fluid as shown in Figure 16-2(b). On a small area A of the slab, the force of the surrounding fluid is always perpendicular to the surface.

Indeed, wherever we choose a surface — at the external boundary of a fluid at rest or somewhere in the interior — the force on the fluid (or the equal and opposite force of the fluid on its surroundings) is *perpendicular* to the chosen surface. If the chosen surface area A is doubled, so too is the magnitude of the force on it; the force-to-area ratio is unchanged. It is therefore useful to define the quantity *pressure p* as the perpendicular fluid force $F_\perp$ per unit surface area A. In symbols,

$$p \equiv \frac{F_\perp}{A} \qquad\qquad (16\text{-}1)$$

Important aspects of pressure are these:

- Pressure is a *scalar* physical quantity, not a vector quantity. The pressure at any location has magnitude only, not direction. A *force* produced by a pressure in a fluid *has* direction; the force is always perpendicular to the surface of area A that we choose and its magnitude is, from (16-1), $F_\perp = pA$.
- In the SI system, the fundamental unit for pressure is the *pascal* (abbreviated Pa), where by definition

$$1 \text{ Pa} \equiv 1 \text{ N/m}^2$$

- Many other pressure units are in common use. The most important are related to the pascal and to one another as follows:*

---

* A meteorologist typically expresses atmospheric pressure in *millibars*; a weather reporter is more likely to use *inches of mercury* (typically close to 30 inches of mercury). The gauge on the air pump at an automobile service station registers pressure over that of the atmosphere in *pounds per square inch* (1 atm $= 14.7$ lb/in^2). A plumber may give the "head" (pressure) in pipes by the height in *feet of a column of water*. A blood pressure reading of, say, 120/80, as read by a physician on a sphygmomanometer (blood-pressure gauge), gives the maximum (120) and minimum (80) pressure in a blood vessel at the same elevation as the heart in units of mmHg above atmospheric pressure.

$$1 \text{ bar} = 1000 \text{ millibars} \equiv 10^5 \text{ Pa}$$

$$1 \text{ torr}^* \equiv 1 \text{ mmHg}$$

$$1 \text{ standard atmosphere} = 1 \text{ atm} = 1.013 \times 10^5 \text{ Pa}$$
$$= 1.013 \text{ bar} = 760 \text{ mmHg} = 760 \text{ torr}$$

How the height of a column of fluid may be used to measure pressure, as given for example in millimeters of mercury (mmHg), is discussed in Section 16-4.

• It is sometimes convenient to specify a pressure *relative* to atmospheric pressure p_0. This relative pressure is also called the *gauge pressure* p_g because it is what is read directly on a pressure gauge. By definition

$$p_g = p - p_0 \tag{16-2}$$

where p is the total, or absolute, pressure. For example, an ordinary gauge for measuring the pressure in an automobile tire reads gauge pressure. With p_g positive, the tire is inflated; with $p_g = 0$, the tire is flat. A vacuum corresponds to the value zero for absolute pressure p.

16-3   Density

An important property of any material is its *density* ρ, defined as the *mass* (m) of the material *per unit volume* (V):

$$\rho \equiv \frac{m}{V} \tag{16-3}$$

The density of a gas depends markedly on its pressure and temperature (Section 19-4), but the density of solids or liquids typically changes only

Table 16-1

MATERIAL	DENSITY, kg/m³
Gases (at 20°C, 1 atm)	
Air	1.29
Hydrogen	9.0×10^{-2}
Liquids	
Water	1.0×10^3
Mercury	1.36×10^4
Ethyl alcohol	8.1×10^2
Solids	
Ice (at 0°C)	0.92×10^3
Aluminum	2.7×10^3
Copper	8.9×10^3
Iron	7.8×10^3
Lead	11.3×10^3
Platinum	21.4×10^3

* The pressure unit *torr*, defined as the pressure corresponding to a height of 1 mmHg, is derived from the name of the Italian physicist Evangelista Torricelli (1608–1647), inventor of the mercury barometer.

slightly with a change in temperature or pressure. For example, the density of water is 1 gm/cm³ = 10^3 kg/m³ within 4 percent over the whole temperature range from freezing to boiling at atmospheric pressure. The density of water is also relatively insensitive to changes in pressure. A pressure of 10 atmospheres, for example, increases the density of water by only 5 parts in 10,000. Liquids are so nearly incompressible that (in this chapter) we take all liquids to have a constant density, independent of pressure.

The densities of a variety of materials are shown in Table 16-1.

16-4   Variation of Static Pressure with Elevation

Here we derive the relation for pressure as a function of vertical location within the fluid. The fluid is assumed to be of constant density, at rest, and under the influence of a gravity.

Consider a small volume element within the fluid in the shape of a thin disk at an elevation y, having thickness dy and face area A. See Figure 16-3a. The mass within the volume $A\,dy$ is $\rho A\,dy$, where ρ is the constant fluid density. This disk is in static equilibrium, and the resultant force on it must be zero. We take the upward direction to be positive. The specific forces on the disk are:

- Its weight $dw = -(\rho A\,dy)g$
- The forces on all external surfaces arising from the pressure of the surrounding fluid. At the lower face, where the pressure is p, the upward force is $+pA$. At the upper face, where the pressure is $p + dp$ and the elevation is $y + dy$, the downward force is $-(p + dp)A$. The horizontal forces on the thin vertical surface add to zero.

With a zero resultant vertical force on the disk, we then have

$$\Sigma F_y = 0$$

$$pA - (p + dp)\,A - \rho g A\,dy = 0$$

or

$$\frac{dp}{dy} = -\rho g \qquad (16\text{-}4)$$

The minus sign indicates simply that as the elevation increases (y positive *upward*), the pressure decreases. The pressure difference dp arises from the weight of fluid between the lower and upward surfaces. The quantity ρg is the fluid's *weight density*, the magnitude of the fluid's weight per unit volume.

Let p_1 be the pressure at elevation y_1 and p_2 the pressure at y_2. Then the pressure difference can be found, using (16-4), as follows:

$$\int_{p_1}^{p_2} dp = -\int_{y_1}^{y_2} \rho g\, dy$$

$$p_2 - p_1 = -\int_{y_1}^{y_2} \rho g\, dy$$

If we take the fluid's density ρ to be constant, integrating the preceding equation gives

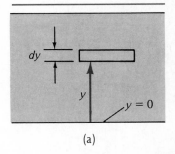

(a)

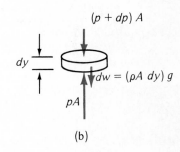

(b)

Figure 16-3. (a) *Volume element at height y and of thickness dy within a fluid. (b) Forces acting on the volume element.*

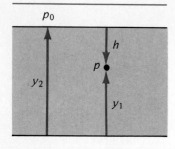

Figure 16-4. *Pressure p at depth h from the upper surface of a fluid, where the pressure is p_0.*

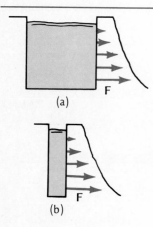

Figure 16-5. *(a) The force on a side wall of a dam increases with depth. (b) The force on the side wall does not depend on the amount of liquid held by the dam.*

$$p_2 - p_1 = -\rho g(y_2 - y_1) \qquad \text{(16-5)}$$

It is often most convenient to express the pressure p at some point in a fluid in terms of its depth h, measured vertically *downward* from surface of the liquid. See Figure 16-4. Let the elevation at the top surface of the liquid be y_2. The pressure there is $p_2 = p_0$ (most often atmospheric pressure). At a depth h, we have $p_1 = p$ and $y_1 = y_2 - h$. Making these substitutions in (16-5), we get

$$p_0 - p = -\rho g h$$

or

$$p = p_0 + \rho g h \qquad \text{(16-6)}$$

From this relation follow several important consequences applicable to any liquid of constant density at rest and under the influence of gravity:

- The static pressure increases *linearly with depth.* Consider Figure 16-5(a), which shows a dam in cross section. Because the pressure increases with depth, the horizontal force on a fixed small area of the dam wall also increases with depth; as a consequence, the dam must be made structurally stronger at its base than at its top. The amount of liquid retained by the dam does not determine these forces. The forces are exactly the same in Figure 16-5(b), where the same dam wall now holds back a far smaller volume of liquid. Static pressure depends on depth only, not on the volume or the horizontal extent of the liquid.

- *All points at the same depth h*, or all points along the same horizontal line, are at the *same pressure*, quite apart from the shape or volume of the container. That is why the top surface of a liquid at rest is a flat horizontal surface. Even within the same liquid, all points at the same level have the same pressure. For example, in Figure 16-6, points A and B are at the same pressure. (To see the result a little differently, imagine that we can go from A to B by several straight horizontal and vertical displacements. There is no pressure change along the horizontal, and compensating pressure increases or decreases arise along the vertical. The overall result: no net change in pressure from A to B.)

- The *height of a column of liquid* of known density can be used as *a measure of pressure.* The simplest example is the common mercury barometer shown in

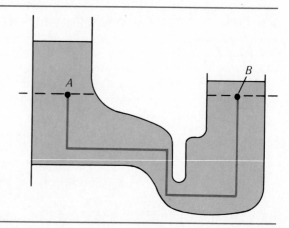

Figure 16-6. *Points A and B at the same horizontal level are at the same pressure.*

Figure 16-7. Above the mercury column is a near vacuum. (Strictly, the pressure at the upper end is equal to the very low vapor pressure of mercury.) The pressure at the surface of the mercury exposed to the atmosphere is the same as the pressure within the tube at the same elevation, which is a distance h vertically below the upper mercury surface within the barometer tube. By definition, a pressure of *one standard atmosphere* corresponds to a mercury height of exactly 76 cm (at a temperature of 27°C), with mercury of density 13.5950×10^3 kg/m³, and with the standard gravitational acceleration, $g = 9.80665$ m/s². Therefore, from (16-6), with $p_0 = 0$ at the top of the mercury column, and $p = p_{atm}$ a vertical distance h below, we have

$$p_{atm} = \rho g h$$

$$1 \text{ atm} = (13.5950 \times 10^3 \text{ kg/m}^3)(9.80665 \text{ m/s}^2)(0.760000 \text{ m})$$

$$1 \text{ atm} = 1.01324 \times 10^5 \text{ N/m}^2 \equiv 1.01324 \times 10^5 \text{ Pa}$$

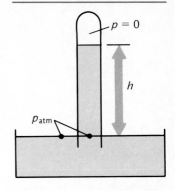

Figure 16-7. *A barometer with a column of liquid of height h. At the top of the column is a vacuum; at the bottom the pressure is atmospheric.*

Within the barometer tube of Figure 16-7, the pressure p_{atm} at a distance h below the top surface is merely the weight per unit area of the column of mercury. Outside the barometer tube and at the surface of mercury, the pressure is again merely the weight per unit area of all material (air) lying above. Atmospheric pressure is the weight per unit area of the entire atmosphere of the earth; the height of the earth's atmosphere is much greater than 76 cm because air is much less dense than mercury.

• Any additional pressure applied at one point to an enclosed fluid of constant density is transmitted undiminished to every portion of the fluid and to the container walls. This result, called *Pascal's principle,* for Blaise Pascal (1623–1662) after whom the pressure unit is named, is not a fundamental independent result, but one that follows directly from (16-6). From this equation, we know that the pressure p at a depth h within a liquid always exceeds the pressure p_0 at the surface by $\rho g h$. Consider, for example, the liquid in Figure 16-8. On the left side, a tightly fitted piston with cross-sectional area a is pressed downward with force f, thereby increasing by f/a the pressure at the liquid surface at the left side over the atmospheric pressure. The pressure at the larger piston on the right must also increase by the same amount, F/A, where A is the cross-sectional area of the larger piston and F is the force this larger piston applies to the liquid (or from Newton's third law, the force the liquid applies to the piston). Thus, $f/a = F/A$. The device in Figure 16-8 is a simple form of *hydraulic press.* An example is the hydraulic brake system of an automobile; here a relatively small force f applied at the brake pedal produces a far larger force $F = (A/a)f$ on the larger brake surface at the wheel.

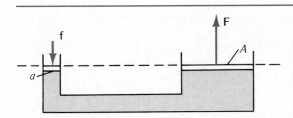

Figure 16-8. *A hydraulic press with a small force f over a small area a and larger force F over larger area A.*

Figure 16-9. *The demonstration of atmospheric pressure by Otto von Guericke in 1654.*

Example 16-1.   Otto von Guericke, inventor of the vacuum pump and mayor of Magdeburg, Germany, put on a spectacular demonstration in 1654 of the effects of atmospheric pressure. Two hollow bronze hemispheres, two feet in diameter, were placed in contact and evacuated with his pump. Then two teams of eight horses each pulled in opposite directions on the two hemispheres. They could not separate them. See Figure 16-9. What minimum force would have separated the Magdeburg spheres? (Take atmospheric pressure to be 15 lb/in².)

The area A involved here in the relation $F_\perp = pA$, for the separation force $F_\perp$ is the *cross-sectional* area πR^2 of a hemisphere; $F_\perp$ is perpendicular to that area. Assume that a near vacuum existed at the interior of the spheres (a bit unrealistically for von Guericke's primitive pump); then we have $F_\perp = pA = p(\pi R^2) = (15 \text{ lb/in}^2)$ $(\pi \, 144 \text{ in}^2) = 6.8 \times 10^3$ lb $\doteq 3$ tons.

Von Guericke used sixteen horses altogether. Would it have been equivalent—although possibly not quite as spectacular—for him to have tied one hemisphere to a strong wall and had only eight horses pull on the other hemisphere?

Example 16-2.   An advertisement for a man's wristwatch claims that the watch is waterproof to a depth of 100 m. What force is applied over the glass crystal of the watch (diameter 30 mm) when it is immersed to this depth?

The watch is assembled under atmospheric pressure, so that under immersion in water to a depth $y = 100$ m, the additional hydrostatic pressure is $p = \rho g y$. The additional force $F_\perp$ on the crystal of diameter $d = 30$ mm is then

$$F_\perp = pA = (\rho g y)\left(\frac{\pi d^2}{4}\right)$$

$$= (1.0 \times 10^3 \text{ kg/m}^3)(9.8 \text{ m/s}^2)(100 \text{ m})\left(\frac{\pi}{4}\right)(30 \times 10^{-3} \text{ m})^2$$

$$\doteq 7 \times 10^2 \text{ N}$$

Example 16-3.   How high is the earth's atmosphere if we assume air to have the same temperature at all elevations?

The density ρ of an ideal gas at constant temperature is proportional to its pressure p (Section 19-4), so that we can write

$$\rho = \left(\frac{\rho_0}{p_0}\right) p$$

where ρ_0 and p_0 represent the density and pressure of the gas at the earth's surface.

Then, from (16-4), we have for the variation of pressure with height

$$\frac{dp}{dy} = -\rho g = -\frac{\rho_0 g}{p_0} p$$

or with terms rearranged,

$$\frac{dp}{p} = -\left(\frac{\rho_0 g}{p_0}\right) dy$$

We integrate the pressure from p_0 at elevation $y = 0$ to the pressure p at elevation y:

$$\int_{p_0}^{p} \frac{dp}{p} = -\left(\frac{\rho_0 g}{p_0}\right) \int_0^y dy$$

$$\ln \frac{p}{p_0} = -\left(\frac{\rho_0 g}{p_0}\right) y$$

$$p = p_0 e^{-(\rho_0 g / p_0) y}$$

The result shows that if the earth's atmosphere were actually at the same temperature for all elevations—which it is not—the atmospheric pressure would decrease exponentially with height. The pressure would then never reach exactly zero; it would fall off exponentially with height as the air thinned out. To get some idea of atmospheric height, let us compute the height at which the pressure drops to $(1/e)$ of its value p_0 at sea level. The height is $y = (p_0/\rho_0 g)$. From the relation above we see that for this value of y, we get $p = p_0 e^{-1} = p_0/e = 0.37\, p_0$. The computed height is

$$y = \frac{p_0}{\rho_0 g} = \frac{(1.01 \times 10^5 \text{ N/m}^2)}{(1.29 \text{ kg/m}^3)(9.8 \text{ m/s}^2)} = 8.0 \times 10^3 \text{ m} = 8.0 \text{ km}$$

How can you show that if the density of air were constantly 1.29 kg/m³ throughout, then 8.0 km would also be the *overall* height of the atmosphere?

16-5 Archimedes' Principle

Nineteen hundred years before Newtonian mechanics was developed, the mathematician and scientist Archimedes discovered the buoyancy principle: any object partially or wholly immersed in a fluid is subject to an upward, or buoyant, force that is equal in magnitude to the weight of the fluid displaced by the object.

One proof of Archimedes' principle is simple and nonmathematical, but altogether rigorous. We focus our attention first on a solid object of volume V but arbitrary shape immersed in fluid, as shown in Figure 16-10(a). Because of the pressure of the surrounding fluid, the object is subject to forces perpendicular to its entire surface and extending over it. The fluid pressure increases with depth, so that the upward force on a lower part of the object is greater than the downward force on an upper part of equal area. The resultant force of the fluid on the object is upward. What is its magnitude?

The Sicilian

The place: the island of Sicily in the Mediterranean. The time: 75 B.C. The scene: Cicero, then a young government official later to become the great Roman statesman and orator, hacks through underbrush until he finally spots what he's been looking for—carved on a slab of stone is a sphere fitted within a cylinder, and the ratio 3/2. It is actually the tombstone, constructed in accord with his instructions, of Archimedes (287–212 B.C.), the greatest mathematician of antiquity.

It was typical of Archimedes to flaunt his accomplishments. (His race from the bath down the street caused some comment.) He was rightly proud of being able to show that the volume of the cylinder divided by the volume of the sphere was the ratio 3/2. Showing this is easy enough when you know the formulas for the volumes of a cylinder and of a sphere. But suppose that you know only the relation for the volume of a cylinder and from it are to find the volume of a sphere? That's exactly what Archimedes did. In fact, he wrote a book simply titled, *On the Sphere and the Cylinder.*

We can see the kind of reasoning Archimedes used in the cylinder-sphere problem by looking at another problem Archimedes solved: how to find the area of a circle from its circumference C and radius r. Imagine a circle to be subdivided into many small wedges, or sectors. Each sector is effectively a triangle with base r and height h, so that its area is $\frac{1}{2}rh$.

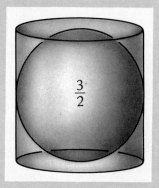

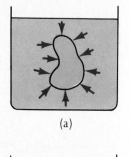

(a)

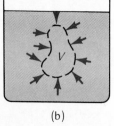

(b)

Figure 16-10. (a) Forces acting on a solid immersed in a fluid. (b) Force acting on fluid of the same shape replacing the solid.

To answer that question, we consider Figure 16-10(b). Here the immersed object is imagined to be replaced by an equal volume V of fluid with the same shape. What had been the boundaries of the solid object previously immersed are now shown as dashed lines. The forces of the surrounding fluid at the boundary are *exactly the same* as they were with the solid object in place. What are the forces on the liquid at rest within the dashed volume? They are its weight downward and, necessarily, an upward force of equal magnitude. These forces add to zero resultant force on the fluid within V. We conclude that the resultant of the forces from the surrounding fluid on the fluid within the dashed boundary is equal in magnitude to the weight of fluid it displaces. So too must be the buoyant force on the solid of the same shape.

Whether a body immersed in a fluid accelerates up, down, or not at all depends on the resultant of all the acting vertical forces, of which the buoyant force is but one. Clearly, if the only external forces acting are its weight and the buoyant force, the body will be accelerated upward if the buoyant force exceeds the weight. This is to say, a body totally immersed in a fluid will be accelerated upward toward the surface if the body is less dense than the fluid. It can float at the surface, partially immersed, when the buoyant force has been reduced in magnitude to match the object's weight. On the contrary, a body with a density greater than that of the fluid will sink.

Example 16-4. A glass tumbler is full to the brim with water. An ice cube floats on the top surface. (a) What fraction of the ice cube is immersed? (b) What happens to the water level after the ice cube is melted?

Adding up contributions from all such wedges around the circle amounts to letting height h become circumference C, so that the area of the circle is $\frac{1}{2}rC$ ($= \frac{1}{2}r(2\pi r) = \pi r^2$). We see that Archimedes' basic strategy—imagining an item to be cut into tiny pieces and then adding up the pieces—bears a strong resemblance to the procedures of integral calculus.

His father was an astronomer. He studied Euclid at the great library of Alexandria. Archimedes devel-

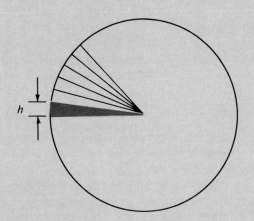

oped the foundations of statics in mechanics: the equilibrium of a rigid body, levers ("Give me a place to stand and I can move the world"), and center of gravity, as well as hydrostatics and the principle associated with his name. Galileo was greatly indebted to Archimedes; he mentions him more than a hundred times in his writings. Not merely a theoretician, Archimedes invented the water snail (a helical tube to raise water for irrigation), a water clock, and a model planetarium. He also acted as scientific consultant to the defense establishment, although it is doubtful that his idea of using mirrors with sunlight to set enemy warships on fire was ever tried.

Archimedes was Sicily's most celebrated citizen. His portrait later appeared on a Sicilian coin.

Archimedes' first love, however, was figuring. He would pursue his geometrical studies on any available surface—sand, dust on the floor, even oil on his body. Indeed, Archimedes was working, according to one account, on a figure in the sand during the time when the Roman general, Marcellus, attacked Sicily. When a Roman soldier stepped on his figure, Archimedes protested and the soldier speared him.

(a) Let the volume of the entire ice cube be V_i and the volume under water V_w. The density of ice is ρ_i, of water, ρ_w. The weight of the ice cube is

$$W_i = \rho_i V_i g$$

The upward buoyant force, equal to the weight of water displaced by ice, is

$$F = \rho_w V_w g$$

Since the two forces above must have the same magnitude with the ice cube in equilibrium, we have

$$\rho_i V_i = \rho_w V_w$$

or

$$V_w/V_i = \rho_i/\rho_w = 0.91$$

Ninety-one percent of the ice cube's volume is under water.

(b) As the ice cube melts, its volume shrinks from V_i to V_w as its density goes from ρ_i to ρ_w. The final water level is just the same as before.

16-6 Streamline Flow

The most general motion of particles in a fluid may be very complicated. It is possible, in principle, to trace out in detail the path of, say, some one particle in a turbulent sea. You must know all forces acting on the particle and then apply Newton's laws. The problem is forbiddingly difficult. Only the simplest kind

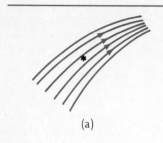

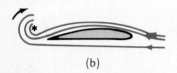

Figure 16-11. *Examples of (a) irrotational and (b) rotational fluid flow.*

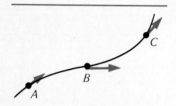

Figure 16-12. *A streamline. The vectors represent the velocities of the fluid particles at A, B, and C.*

of fluid motion can be analyzed easily. This is exemplified by the special case of the *steady flow* of a *nonviscous fluid* of *constant density* in *irrotational streamline motion*.

• Fluid motion is *steady* if the velocity of all particles passing any given fixed point is always the same. An example is the flow of water past the side of a canoe moving through still water at a constant velocity *as viewed by an observer on the canoe*. Rather than concentrate on some one particle as it travels through space, we focus on the velocity of all particles passing through any chosen point. For steady motion, the velocity is always the same at a given location. *Nonsteady* motion is far more common: for example, the rapid, chaotic, and turbulent motion of particles in a pounding surf or even the flow of water past the side of a moving canoe *as viewed by an observer on the bank*.

• We take the fluid to be *nonviscous*. This implies that the fluid flows freely, and no energy is dissipated within the liquid. This also implies that there is no frictional force between the fluid and the constraining tube through which it flows. Actually, when water flows through a pipe, the water in contact with the pipe sticks to the walls and is at rest there while the velocity increases toward the center of the pipe.

• Most liquids (and gases under certain circumstances) can be considered nearly incompressible. Furthermore, if there are no cavities in a fluid, its density may be assumed to have the same value throughout.

• *Irrotational* motion is best illustrated by an experimental test. Suppose that a small paddle wheel, free to turn on its axis, is inserted into a moving fluid. If the wheel does not rotate, the motion at that point is *irrotational*. On the other hand, when a paddle wheel turns when it is placed in a whirlpool or eddy, the fluid's motion is said to be rotational. See Figure 16-11.

Now consider steady streamline motion. Figure 16-12 shows the path traced out by all particles that begin at point *A* and later pass through points *B* and *C*. Every particle starting at *A* follows the same path, or *streamline*. At each point, the direction of the velocity is tangent to the streamline. Streamlines indicate the direction of the fluid's velocity. They can also indicate the magnitude of the velocity as follows: the number of streamlines, passing through any small area oriented perpendicular to the streamlines, is proportional to the magnitude of the velocity, at that point. A collection of streamlines is said to constitute a *velocity field*; the density and direction of the streamlines give,

Figure 16-13. *(a) A collection of streamlines representing a velocity field. (b) The magnitude of the velocity is given by the number of streamlines passing through a unit of area oriented at right angles to the streamlines.*

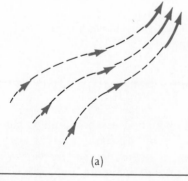

(a)

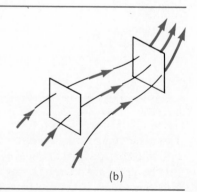

(b)

respectively, the magnitude and direction of the fluid's velocity. See Figure 16-13.

Streamlines cannot cross. If they were to do so, particles reaching an intersection would not have a unique velocity at that point in space.

16-7   The Equation of Continuity

In this section we derive a fundamental relation for fluid flow based on the mass conservation principle.

The *mass flux* of a fluid through a transverse loop at some point is defined as the mass Δm passing per unit time through the area perpendicular to the fluid's velocity at that point. See Figure 16-14(a). Thus,

$$\text{Mass flux} = \Delta m / \Delta t \qquad \text{(16-7)}$$

Similarly, the *volume flux* is the volume ΔV of fluid passing through a loop perpendicular to the fluid's velocity in the time Δt:

$$\text{Volume flux} = \Delta V / \Delta t \qquad \text{(16-8)}$$

Since $\Delta m = \rho \, \Delta V$, where ρ is the fluid's density at the point P at which Δm and ΔV are measured, we can write

$$\text{Mass flux at } P = \rho \times \text{volume flux at } P$$

The mass conservation principle leads to a useful relation, known as the *equation of continuity*, between the velocity and the cross-sectional area for ideal fluid flow. Consider the situation shown in Figure 16-14(b) where a fluid moves in steady streamline flow through a tube of varying cross section. The tube may be an actual pipe constraining the fluid, or it may be simply a collection of adjacent streamlines within a fluid. In any event, no fluid passes through the sides of the tube.

At point 1, the cross-sectional area is A_1, the mass density is ρ_1, and the velocity of all particles crossing A_1 is $\mathbf{v}_1$; at point 2 the corresponding quantities are A_2, ρ_2, and $\mathbf{v}_2$. In steady flow these quantities remain constant in time. Then in the time interval Δt, the mass entering the tube at point 1 is Δm_1, where

$$\Delta m_1 = \rho_1 \, \Delta V_1 = \rho_1(A_1 v_1 \, \Delta t)$$

The mass of fluid Δm_2 emerging from the tube at point 2 during the same time interval Δt is

$$\Delta m_2 = \rho_2 \, \Delta V_2 = \rho_2(A_2 v_2 \, \Delta t)$$

Fluid cannot leak through the walls, and there are no "sources" or "sinks" within. The total mass of fluid in the tube between points 1 and 2 must remain constant. This implies that the mass Δm_1 entering must equal the mass Δm_2 emerging in the same time interval:

$$\Delta m_1 = \Delta m_2$$

and

$$\text{Mass flux} = \rho_1 A_1 v_1 = \rho_2 A_2 v_2 \qquad \text{(16-9)}$$

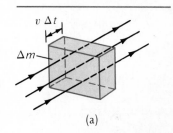

(a)

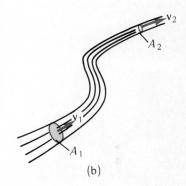

(b)

Figure 16-14. (a) *Mass element Δm passing at speed v through a volume element of thickness $v\Delta t$.* (b) *The fluid speed is large where the cross sectional area is small;* $A_1 v_1 = A_2 v_2$.

The quantity $\rho A v$, which represents the mass per unit time of fluid crossing the area A perpendicular to $\mathbf{v}$, is the *same* at all points along the tube. Equation (16-9) is called the *equation of continuity.**

Density ρ is constant for an incompressible fluid. Then (16-9) reduces to

$$\text{Volume flux} = A_1 v_1 = A_2 v_2 \qquad (16\text{-}10)$$

For constant density, the volume flux $A v$ is conserved; it has the same value at all points along the tube. This implies that the speed of the fluid is great where the cross section is small, and the opposite.

Note that (16-10) is also consistent with the use of streamlines to represent the direction and magnitude of the fluid velocity. Since v is a measure of the number of streamlines passing through a unit cross-sectional area, the quantity $v A$ is the total number of streamlines at any cross section. Equation (16-10) then shows that the total number of streamlines within the tube is conserved; streamlines crowd together at a constriction and diverge where the cross section is large.

Example 16-5. A fluid of constant density flows steadily through a pipe that narrows down to half its original diameter. How do the (a) speed and (b) kinetic energy of a fluid particle at the narrow portion compare with the corresponding quantities at the wider portion?

(a) With the diameter reduced to half, the circular cross-sectional area is reduced by a factor of 4. Since, by the equation of continuity, $A v$ is constant, the speed of the fluid particle increases by a factor of 4.

(b) The speed of the fluid particle is up by a factor of 4 in the narrow portion of the pipe. Therefore, the kinetic energy must have increased by a factor of 16.

16-8   Bernoulli's Theorem

As we have seen, the equation of continuity for fluids is nothing more than an indirect statement of the mass-conservation principle. The energy-conservation principle — or its equivalent, the work-energy theorem — also leads to a useful hydrodynamic relation of great generality. First derived in 1738 by Daniel Bernoulli, it is known as *Bernoulli's theorem,* or the *Bernoulli equation.* This equation relates the pressure, velocity, and height of moving fluid at one location to the corresponding quantities at a second location.

The circumstances are special; an incompressible nonviscous fluid is in streamline flow through a pipe, as in Figure 16-15. The pipe's cross-sectional area at any point is small enough that all particles flowing past a given cross section have the same velocity there. At point 1, where fluid enters the tube, its cross-sectional area is A_1, the fluid's speed is v_1, the pressure is p_1, and the vertical elevation above some arbitrary horizontal level is y_1. At point 2, where fluid leaves the tube, the corresponding quantities are designated A_2, v_2, p_2, and y_2. The fluid's density ρ is the same at all points. We shall concentrate on the fluid between points 1 and 2, shown shaded in Figure 16-15.

Now we apply the work-energy theorem to this fluid. The total work W done on the fluid must equal the change in kinetic energy ΔK and in gravitational potential energy ΔU_g:

---

* Strictly, (16-9) is a special, simplified form of a more general equation of continuity that allows for sources and sinks.

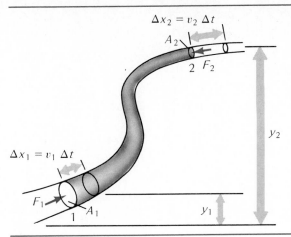

Figure 16-15. Irrotational flow of an incompressible nonviscous fluid through a pipe of varying cross section.

$$W = \Delta K + \Delta U_g \qquad (16\text{-}11)$$

Work is done on the fluid only at points 1 and 2 at the ends. At other locations, the force of the wall on the fluid is at right angles to the fluid velocity, and no work is done there. We are left then with the work done by force $\mathbf{F}_1$ at point 1 and force $\mathbf{F}_2$ at point 2. Force $\mathbf{F}_1$ is in the *same* direction as the fluid's motion and it does *positive* work. Force $\mathbf{F}_2$ is opposite to the fluid motion, and $\mathbf{F}_2$ does *negative* work. In the time interval Δt, the left end of the shaded fluid advances a distance $\Delta x_1 = v_1 \Delta t$ to the right. During the same time interval the right end moves to the right a distance $\Delta x_2 = v_2 \Delta t$. The net work done on the system by external forces is then

$$W = F_1 \Delta x_1 - F_2 \Delta x_2 = (p_1 A_1)(v_1 \Delta t) - (p_2 A_2)(v_2 \Delta t)$$

where we have replaced F_1 by $p_1 A_1$ and F_2 by $p_2 A_2$.

But the equation of continuity, (16-10), requires that $A_1 v_1 = A_2 v_2$, so that the equation above can be written

$$W = A_1 v_1 \Delta t(p_1 - p_2)$$

Now $A_1 v_1 \Delta t$ is the volume of fluid displaced during time Δt at point 1; the same volume is displaced at point 2. The mass Δm displaced at either of the two points is $\Delta m = \rho(A_1 v_1 \Delta t)$, so that the net work done may then be written

$$W = \frac{\Delta m}{\rho} (p_1 - p_2) \qquad (16\text{-}12)$$

Now consider the changes ΔK and ΔU_g in the fluid's kinetic and gravitational potential energies. One circumstance makes this easy to deal with. We need concern ourselves only with fluid entering the system at point 1 in time Δt and the fluid leaving at point 2 in the same time interval. We can ignore the energy of the fluid within the region between points 1 and 2, because the total energy of this fluid remains constant in time. The kinetic energies and potential energies of fluid within any small volume are unchanged, even though matter flows continuously into and out of each volume element. The change in energy of the entire system is just the change in energy of element of mass $\Delta m = \rho A_1 v_1 \Delta t = \rho A_2 v_2 \Delta t$.

The kinetic energy changes by

$$\Delta K = \tfrac{1}{2}\Delta m(v_2^2 - v_1^2) \tag{16-13}$$

and the gravitational potential energy changes by

$$\Delta U_g = \Delta mg(y_2 - y_1) \tag{16-14}$$

Substituting (16-12), (16-13), and (16-14) in (16-11) yields

$$\frac{\Delta m}{\rho}(p_1 - p_2) = \frac{1}{2}\,\Delta m(v_2^2 - v_1^2) + \Delta mg(y_2 - y_1)$$

Multiplying both sides of this equation by $\rho/\Delta m$ and rearranging terms gives the *Bernoulli equation*:

$$p_1 + \rho g y_1 + \tfrac{1}{2}\rho v_1^2 = p_2 + \rho g y_2 + \tfrac{1}{2}\rho v_2^2 \tag{16-15}$$

This relation and the equation of continuity are the foundations of the fluid mechanics of incompressible nonviscous fluids. The continuity equation and Bernoulli's theorem are nothing more than the conservation principles of mass and of energy, now expressed as the conservation of the quantities ρAv and $p + \rho g y + \tfrac{1}{2}\rho v^2$:

- Equation of continuity (mass conservation): $Av = $ constant.
- Bernouilli's theorem (energy conservation): $p + \rho g y + \tfrac{1}{2}\rho v^2 = $ constant.

Suppose the fluid is at rest, with $v_1 = 0$ and $v_2 = 0$. Then (16-15) can be written

$$p_2 - p_1 = -\rho g(y_2 - y_1)$$

This is just (16-5), the relation for variation in static pressure derived earlier.

In applying (16-15), one must note that p represents absolute, not gauge, pressure. Indeed, all three terms in the equation carry the dimensions of pressure, and the units for these terms must be consistent. Thus, if ρ is in kg/m³ and v in m/s, p must be in Pa.

Bernoulli's equation holds only when all the assumptions made in its derivation are satisfied. But even if these conditions are not met exactly — for example, with a somewhat viscous gas of variable density — some qualitative conclusions are valid. Chief of these is the reciprocal relation between particle speed and pressure. From (16-15), we see, with $y_1 = y_2$, that $p_1 > p_2$ if $v_1 < v_2$. In words, the pressure is relatively high at a location where the particle speed is relatively low (and the streamlines are far apart), and the pressure is relatively low at a location where the speed is high (and streamlines are crowded together).

This effect is partly responsible for the *lift force* in an air foil (an airplane

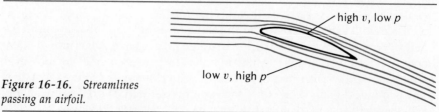

high v, low p

low v, high p

Figure 16-16.   Streamlines passing an airfoil.

wing or the propeller of a helicopter, for example). Consider Figure 16-16, where the wing is tipped relative to the orientation of incident streamlines, so that the lines are crowded on the top side relative to those on the lower side. With a relatively higher particle speed above the wing, the pressure there is reduced below atmospheric pressure. Consequently, the upward force on the lower surface exceeds the downward force on the top, and the net effect is an upward lift. Note in Figure 16-16 that the streamlines leaving the wing foil are turned downward; this corresponds to the fact that with the wing acquiring momentum upward, the particles of air must acquire an equal downward momentum component.

Example 16-6. Water flows through a horizontal pipe that narrows to half its original diameter at a constriction. In the wider section of pipe, the water's speed is 2.0 m/s. By how much does the pressure drop at the constriction?

The narrow part of the pipe with a diameter d_1 is labeled 1; the wide portion with diameter $d_2 = 2d_1$ is labeled 2. See Figure 16-17. Since elevations y_1 and y_2 are the same, Bernoulli's equation reduces to

$$p_1 + \frac{1}{2}\rho v_1^2 = p_2 + \frac{1}{2}\rho v_2^2$$

$$p_2 - p_1 = \frac{1}{2}\rho(v_1^2 - v_2^2) = \frac{1}{2}\rho v_2^2\left(\frac{v_1^2}{v_2^2} - 1\right)$$

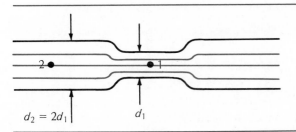

$d_2 = 2d_1$ d_1 *Figure 16-17.*

From the equation of continuity we have $v_1/v_2 = A_2/A_1 = d_2^2/d_1^2$ and the relation above becomes

$$p_2 - p_1 = \frac{1}{2}\rho v_2^2\left(\frac{d_2^4}{d_1^4} - 1\right)$$

$$= \tfrac{1}{2}(1.0 \times 10^3 \text{ kg/m}^3)(2.0 \text{ m/s})^2(16 - 1)$$

$$= 3.0 \times 10^4 \text{ N/m}^2 = 0.3 \text{ atm}$$

Example 16-7. A large tank with liquid of constant density ρ has a small hole a vertical distance y below the liquid surface. With what speed will the liquid emerge from the hole?

The situation is shown in Figure 16-18. We designate a point at the surface as 1 and a point in the emerging stream as 2. Vertical heights are measured from the horizontal plane of the hole. Bernoulli's theorem, (16-15), then becomes

$$p_1 + \rho g y + \tfrac{1}{2}\rho v_1^2 = p_2 + 0 + \tfrac{1}{2}\rho v_2^2$$

Both p_1 and p_2 are essentially at atmospheric pressure; they cancel. The equation of continuity requires that $A_1 v_1 = A_2 v_2$, where A_1 is the cross-sectional area of the tank and

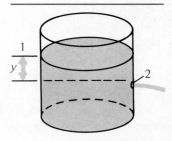

Figure 16-18. Liquid escaping from a hole in the side of a container.

A_2 of the hole. Since $A_1 \gg A_2$, therefore, $v_1 \ll v_2$. The liquid level descends so slowly that it can be taken to be at rest ($v_1 = 0$). The equation above becomes

$$\rho g y = \tfrac{1}{2}\rho v_2^2$$

or

$$v_2 = \sqrt{2gy}$$

This result is known as *Torricelli's theorem*. According to it, liquid emerges with the same speed that droplets would attain after falling freely from rest through a height y.

Summary

Definitions

Fluid: material that cannot permanently sustain a shear stress.

Density ρ: mass m per unit volume V,

$$\rho \equiv m/V \qquad (16\text{-}3)$$

Pressure p: perpendicular force $F_\perp$ over an area A,

$$p \equiv F_\perp/A \qquad (16\text{-}1)$$

Streamlines: graphical representations of velocities of fluid particles in streamline flow. Direction of streamline gives velocity direction; density of streamlines gives velocity magnitude.

Units (of pressure)

Pascal = 1 Pa $\equiv$ 1 N/m²
Torr $\equiv$ 1 mmHg
Bar $\equiv$ 10^5 Pa
Standard atmosphere $\equiv$ 760 mmHg = 1.013×10^5 Pa

Fundamental Principles

Equation of continuity (the mass-conservation law):

$$\text{Mass flux} = \rho A v = \text{constant} \qquad (16\text{-}9)$$

Bernoulli's equation (the energy-conservation law):

$$p + \rho g y + \tfrac{1}{2}\rho v^2 = \text{constant} \qquad (16\text{-}15)$$

Important Results

Pressure p as a function of elevation y in a fluid of constant density ρ under the influence of gravity:

$$p_2 - p_1 = -\rho g(y_2 - y_1) \qquad (16\text{-}5)$$

Pressure p at depth h below a surface of fluid of constant density ρ at which pressure is p_0:

$$p = p_0 + \rho g h \qquad (16\text{-}6)$$

Archimedes' principle: Upward buoyant force on an object immersed in a fluid is equal in magnitude to the weight of the fluid displaced.

Problems and Questions

Section 16-2   Pressure

· **16-1 P** You can estimate the weight of a car like this: Measure the pressure in the tires and the area of each tire in contact with the ground. For example, suppose the gauge pressure is 35 psi and each tire contacts an area 10 cm × 12 cm. What is the mass of the car?

· **16-2 P** The pressure at the center of a violent hurricane may drop as low as 920 millibars. Express this pressure in (a) mmHg, (b) torr, (c) pascals, and (d) atmospheres.

· **16-3 P** To what gauge pressure does an absolute pressure of 780 mmHg correspond?

· **16-4 P** A typical automobile tire is inflated to a gauge pressure of 35 pounds per square inch. What is the absolute pressure in SI units?

: **16-5 P** The cutting edge of a chisel has an effective area

of 0.8 cm². When struck with a hammer, it exerts a momentary force of 80 N on the working surface. What is the pressure under the chisel when this happens?

: **16-6 P** What is the total mass of the earth's atmosphere (atmospheric pressure, 1.0×10^5 Pa; earth's radius, 6.4×10^6 m)? (Neglect the variation in g with elevation above the earth's surface.)

Section 16-3   Density

· **16-7 P** The density of blood may be determined by observing drops of blood suspended in a mixture of two liquids (xylene and bromobenzene) of known density. What is the density of a blood sample that is suspended in a mixture of 24% bromobenzene (relative density 1.50) and 76% xylene (relative density 0.87)?

: **16-8 P** A solid object with a mass of 120 gm is immersed in a container filled with water to the brim. The overflow of water has a mass of 44 gm. Of what material is the solid made?

Section 16-4   Variation of Static Pressure with Elevation

· **16-9 Q** Water is poured to the same level in each of the three vessels shown in Figure 16-19. Each has the same base area. Which statement is correct?

(A) The pressure on the bottom of each is not the same.

(B) Since the total force on the bottom of each is the same, each will weigh the same when placed on a scale.

(C) The total force on the bottom of each vessel is the same, but they do not weigh the same because of the additional forces exerted by the sloping sides in vessels X and Y.

(D) Although the pressure on the bottom of each vessel is the same, they will not weigh the same because the total pressure is different for each.

(E) Vessel X will weigh the most because it has the greatest surface area exposed to the air, and hence a greater downward force acts on it from atmospheric pressure.

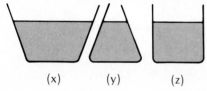

Figure 16-19.   *Question 16-9.*

· **16-10 Q** Consider the gauge pressure at point X in the apparatus shown in Figure 16-20. When liquid is added to level h_1 the gauge pressure at X is P_1. If more liquid is added, raising the level to $h_2 = 2h_1$, the gauge pressure P_2 at X will be

(A) $P_2 = \frac{1}{2}P_1$

(B) $P_2 = P_1$

(C) $P_1 < P_2 < 2P_1$

(D) $P_2 = 2P_1$

(E) $P_2 > 2P_1$

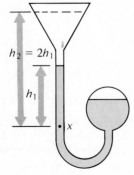

Figure 16-20.   *Question 16-10.*

· **16-11 Q** In a famous experiment, water was poured into a barrel through a small tube as shown in Figure 16-21.

Figure 16-21.   *Question 16-11.*

When the water level in the tube was high enough, the barrel burst. Which of the following is an accurate statement?

(A) When the barrel burst, the upward force on the top of the barrel was equal to the weight of water in the tube.

(B) The height of water needed in the tube is dependent on the diameter of the tube.

(C) The height of the tube is to the height of the barrel as the weight of water in the tube is to the weight of water in the barrel.

(D) When the barrel burst, the pressure on the sides of the barrel was the same as the pressure on the top of the barrel.

(E) None of the above is true.

· **16-12 Q** The convection of water in lakes is vital to marine life, since it results in distribution of nutrients and oxygen. This convection results primarily because of

(A) currents caused by the wind.

(B) the increase in pressure with increasing depth.

(C) buoyancy effects.

(D) the rotation of the earth.

(E) variations in atmospheric pressure.

: **16-13 P** How thick must the earth's atmosphere be in order to create a pressure of 1 atmosphere at the earth's surface, if the density were uniform throughout and had the value it has presently at the earth's surface, about 1.3 kg/m³?

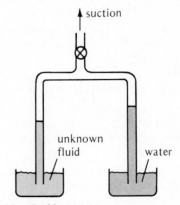

Figure 16-22.   *Problem 16-14.*

: **16-14 P** Hare's apparatus, shown in Figure 16-22, is used to measure the relative density of a fluid. Suction in a common, center tube causes water to rise in the tube on the right and a fluid of unknown density to rise in the tube on the left. When the valve in the center tube is closed, the water has risen to a level of 10.5 cm, and the fluid on the left has risen to a level of 9.2 cm. Calculate the density of the fluid on the left. The density of water is 1000 kg/m³.

· **16-15 P** A research submarine has a circular observation port of diameter 20 cm. It is mounted so that its plane is inclined at 45° to horizontal. What force does seawater exert on it when it is at a depth of 2000 meters?

· **16-16 P** A viewing port used to observe migrating salmon is circular with a diameter of 60 cm. Its center is 4 m below the surface of the river. What is the total force acting on the window?

: **16-17 P** Figure 16-23 shows a side view of a tank filled with water to a depth of 1.0 m. Two sides of the tank (not shown) are vertical, parallel, and separated by 4.0 m. *(a)* What is the force of the water on the bottom of the tank? *(b)* What is the force of the water on the vertical end? *(c)* What is the force of the water on the inclined end?

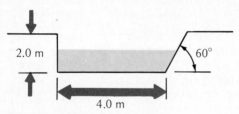

Figure 16-23. Problem 16-17.

: **16-18 P** By sucking hard on a straw, you can reduce the gauge pressure in your lungs to −80 mmHg. What is the greatest height that you can lift water in this fashion?

: **16-19 P** The manufacturer of a rotating crop irrigation device recommends that a minimum water pressure of 2000 mmHg (gauge) be used. How high above the field must a tank be placed to provide this pressure, assuming no pressure drop in the lines?

: **16-20 P** *(a)* Estimate the atmospheric pressure on top of Mt. Everest, elevation 29,028 ft, assuming the atmosphere has the same temperature throughout. *(b)* If the atmosphere were the same temperature throughout, at what elevation would the pressure be half what it is at the surface of the earth?

: **16-21 P** What load can be lifted with the hydraulic jack (Figure 16-8) when a force of 500 N is applied? The pistons are circular with diameters of 2 cm and 20 cm.

: **16-22 P** The tip of a hypodermic needle has a cross-sectional area of 0.1 mm², the area of the barrel is 1 cm². What force must you exert on the plunger in order to make an injection into a vein where the pressure is 14 mmHg?

: **16-23 P** Show that when a liquid in a bucket is rotated about a vertical axis at constant angular speed, the surface of the liquid assumes the shape of a paraboloid (a cross section through the center is a parabola). See Figure 16-24. (*Hint:* View the liquid as a rotating observer at rest with respect to the liquid. In this reference frame each particle of liquid is subject to, in addition to its weight mg, an outward centrifugal force $m\omega^2 x$. The effective weight of a particle on the surface of the liquid must be perpendicular to the surface. To show that $y = Kx^2$, the equation for a parabola, we need merely show that $dy/dx = 2Kx$, where K is a constant.)

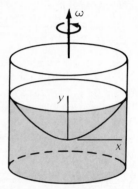

Figure 16-24. Problem 16-23.

Section 16-5 Archimedes' Principle

· **16-24 Q** A rock is thrown into a swimming pool filled with water at uniform temperature. Which statement is correct?
(A) The buoyant force on it is zero while it is sinking.
(B) The buoyant force on it increases as it sinks.
(C) The buoyant force on it decreases as it sinks.
(D) The buoyant force on it is constant as it sinks.
(E) The buoyant force on it is non-zero at first but becomes zero once the terminal velocity is reached.

· **16-25 Q** A ping pong ball floats in water contained in a cylinder fitted with an airtight piston. What will happen when the piston is pressed down?
(A) The ball will continue to float at the same level.
(B) The ball will sink down deeper into the water.
(C) The ball will rise up out of the water slightly.
(D) Whether the ball rises or sinks will depend on its density relative to that of air.

: **16-26 Q** Attached to a balloon is a weight just sufficient to submerge the balloon. What will happen if the balloon is pushed a meter or so below the surface and released?
(A) It will sink deeper and deeper.
(B) It will remain where it was released.
(C) It will rise back to the surface.
(D) It will sink a slight added distance and then come to rest in equilibrium.

(E) It will rise a slight amount and come to rest somewhere between where it was released and the surface.

: **16-27 P** A certain object weighs W_a in air, W_w when immersed in water, and W_u when immersed in a liquid of unknown density. What is (a) the object's density relative to that of water and (b) the unknown liquid's density relative to that of water?

: **16-28 P** A balloon of volume V is filled with gas of low density ρ_g. The surrounding air has density ρ_a. (The mass of the balloon itself, excluding the mass of gas inside, is negligible.) The maximum mass (of small size) that can be lifted by the balloon is

(A) $V(\rho_a - \rho_g)$
(B) $V(\rho_g - \rho_a)$
(C) $V\rho_g$
(D) $V\rho_a$

(E) $V\left(\dfrac{\rho_g}{\rho_a} - 1\right)$

: **16-29 P** During World War II a Canadian project titled "Habbakuk" investigated the possibility of creating floating ice islands for use as antisubmarine bases. How much ice would be required to build a base of area 200 m × 1000 m capable of supporting a mass of 10^8 kg?

: **16-30 P** A helium-filled balloon of negligible mass is to lift a mass of 200 kg, whose volume is small compared to that of the balloon. The density of air is 1.29 kg/m³; the density of helium in the balloon is 0.178 kg/m³. What is the minimum radius of the balloon?

: **16-31 P** A concrete piling of density 2.5 × 10³ kg/m³ has a mass of 2000 kg. It is in the shape of a cylinder 8 meters long. How far out of the water can it be lifted by a force equal to 60 percent the weight of the piling?

: **16-32 P** The Navy recently recovered a jet fighter that sank to the bottom of the Atlantic. If the average density of the plane was four times that of sea water (assume the density to be homogeneous and neglect temperature variations with depth) and the weight of the plane was 10,000 pounds, what force would be required to lift the plane, using a cable attached to a boom on a surface ship?

: **16-33 P** How much work must be done to submerge a submarine of average density 800 kg/m³ to a depth of 100 meters, if the mass of the submarine is 2 × 10⁶ kg?

: **16-34 P** A rectangular ore barge 10 meters long and 6 meters wide has a depth of 4 meters. Its density is 750 kg/m³. (a) What depth of the barge is below water? (b) It is necessary to lift the barge partially out of the water to do some repair work. What vertical force is needed to lift the barge so that only 1 meter of the vessel is below water?

: **16-35 P** A cylindrical object floats in a liquid with its symmetry axis vertical. Show that if the floating object is disturbed slightly, it will undergo simple harmonic motion.

Section 16-7   The Equation of Continuity

· **16-36 Q** Arteriosclerosis is a condition in which the walls of an artery thicken, thereby decreasing the area available for blood flow. This puts an added strain on the heart because

(A) this reduces the total volume of blood in the body.
(B) when the cross-sectional area of a tube is reduced by a factor of 2, the resistance of the tube doubles
(C) when the cross-sectional area of a tube is reduced by a factor of 2, the resistance of the tube increases by a factor of 4.
(D) this increases the viscosity of the blood.
(E) the pulmonary circulation and the systemic circulation are essentially independent.

: **16-37 P** Fluid flows with a velocity of 60 cm/s through a tube of diameter 12 mm. This tube branches into three identical tubes of diameter d, in each of which the fluid velocity is 80 cm/s. What is the value of d?

: **16-38 P** In an average adult, blood flows through the aorta at a speed of 0.33 m/s. What is the flow, in cm³/s, through an aorta of radius 8 mm?

: **16-39 P** In a chemical processing plant a solvent flows through a 10 cm diameter pipe with a flow of 12 m³/minute. What is (a) the flow rate and (b) the fluid velocity when this line branches into three smaller pipes, each of radius 3 cm?

: **16-40 P** What power level is required to pump water through a pipe of 15 cm diameter at a flow rate of 3 m³/minute, if the pressure difference between the two ends of the pipe is 0.5 atm?

: **16-41 P** Water flows out of a faucet with speed v_0 and falls under the influence of gravity. The initial diameter of the stream is D. Determine the diameter of the stream after it has fallen a distance x, assuming droplets do not form.

: **16-42 P** The Pitot tube, shown in Figure 16-25, is a device used to measure the flow speed of a gas. A manometer containing fluid of density ρ_0 is connected as shown. The fluid at point X is stagnant (not flowing) and the pressure here is called the *ram pressure*. The velocity at point Y is the fluid velocity to be measured. Show that the velocity at this

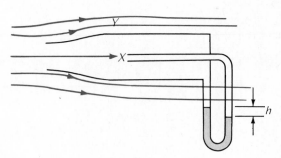

Figure 16-25.   Problem 16-42.

point is $(2gh\rho_0/\rho)^{1/2}$ where ρ is the gas density and h is the difference in height of the fluid in the two arms of the manometer.

Section 16-8   Bernoulli's Theorem
· **16-43 Q** It can be dangerous to stand near the edge of a subway platform when a train whizzes by, even if no one is standing near you. What happens in such a situation is similar to
(A) what happens when you balance a ping pong ball on a stream of air rushing out of a reversed vacuum cleaner hose.
(B) what makes a baseball curve.
(C) what happens when you blow between two sheets of paper.
(D) the operation of a perfume atomizer.
(E) all of the above.
(F) none of the above.

· **16-44 Q** The ability of a pitcher to curve a baseball is most closely related to which of the following?
(A) An object moving through a viscous fluid experiences a drag force dependent on its velocity.
(B) Flow is proportional to pressure difference in a fluid.
(C) The pressure in a fluid is low in regions where the velocity is high, and high where the velocity is low.
(D) The gain in KE of a falling object is equal to the loss in PE.
(E) For a fluid, (Velocity) × (Area) = (Flow rate).

· **16-45 Q** Which of the following statements best describes what happens when the roof blows off a house in a wind storm?
(A) The house explodes.
(B) The force of the wind on the roof pushes it off.
(C) The wind creates a reduced pressure inside the house.
(D) The wind creates a high pressure point at the peak of the roof.
(E) The wind drives the roof into vibrational resonance.

: **16-46 P** Two standpipes shown in Figure 16-26 indicate a difference in height of 2 mm when water flows in the constricted line shown here. The diameter of the large section is 4 cm and that of the small section 2 cm. What is the velocity of flow in the large section?

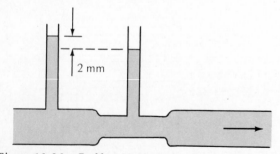

Figure 16-26.   Problem 16-46.

: **16-47 P** In a horizontal pipeline of constant cross-section the pressure drops by 2.5×10^4 Pa in a distance of 300 meters. What is the energy loss per unit volume of oil per meter along the pipeline?

: **16-48 P** The outlet of a large water tank of diameter 6 meters and water depth 15 meters is directed vertically upward, as shown in Figure 16-27. The diameter of the outlet spout is 4 cm. Assuming negligible friction, how high will the stream of water go?

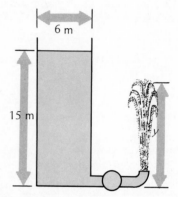

Figure 16-27.   Problem 16-48.

: **16-49 P** At an elevation where the air density is 1.2 kg/m³, the air flow beneath an airplane wing is 150 m/s. What air velocity over the upper surface is needed to give a lift pressure of 10^5 Pa?

: **16-50 P** A rectangular gate of width w and height h is opened at the top of a dam at the mouth of a large reservoir. (a) Calculate the volume flow through the opening. Since the reservoir is very large, the water velocity is essentially zero once one is a short distance away from the gate. Also, the level in the reservoir does not drop appreciably. Note the fluid gauge pressure is zero once it flows through the gate, since there is then only air surrounding it. (b) Show that the volume flow is $Q = (2w/3)\sqrt{2g}\,h^{3/2}$.

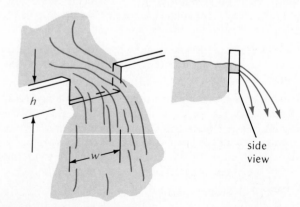

Figure 16-28.   Problem 16-50.

Mechanical Waves I

17

Sometimes physics is said to be the study of mass and energy. That's an oversimplification. Physics is also described as the study of particles and waves, and their interactions. That's a better definition. Various types of waves play an absolutely central role and show up in all areas of physics. Sound and light are wave phenomena. The atomic and subatomic world is governed by the quantum theory, also known as wave mechanics.

17-1   Basic Wave Properties

Energy can be transported from point A to point Z by sending a particle directly from A to Z. Another way of transporting energy is through *wave motion*. Suppose a disturbance from equilibrium occurs first at A. It influences B. Then B influences C, and so on. Finally, what happened first at A happens later at Z. This is the essential feature of wave motion; some disturbance from equilibrium travels from one location to more distant locations because adjoining regions in the medium through which the wave is propagated are coupled to one another.

Consider a uniform, perfectly flexible string under tension. Its right end is fixed in position somewhere far to the right. Suppose the left end is suddenly displaced laterally and then returned to its initial equilibrium position. A wave

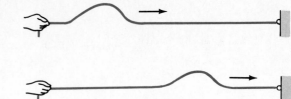

Figure 17-1. *A wave pulse traveling along a stretched string.*

pulse—a disturbance over a limited region of space—travels to the right at constant speed, as shown in Figure 17-1. Waves on a string are a prototype of wave motion because the shape of the wave is exactly the shape of the string. For a wave traveling along a string under tension, the coupling between one particle of the string and its immediate neighbors arises from elastic forces between particles in the string.

The wave disturbance—an unchanged wave shape—travels along the string at constant speed. Energy is also propagated along the string; it consists of the kinetic energy of a particle in motion on the string, and the potential energy of the string particles displaced from equilibrium. No particle travels to the right or left, however. As the wave pulse passes some fixed point, each particle is displaced transversely from its equilibrium position and then returns to it. Each particle moves at right angles, or transversely, to the direction of wave propagation, and such a wave is said to be a *transverse wave.* The speed c at which the wave shape is propagated is different at any instant from the speed of a particle of the string (and usually much larger).

Several other simple types of elastic waves are shown in Figure 17-2:

• Transverse wave on a stretched spring (a long helical spring, or "Slinky"), generated by displacing one part laterally.

• Longitudinal wave on a stretched spring. Here some part of the spring is displaced longitudinally, so that particles are displaced along the line in which the wave pulse is propagated.

• A compressional wave in some elastic medium—a compressible gas, liquid, or solid. With the particles displaced longitudinally from their equilibrium positions, this type of wave is also longitudinal; the wave disturbance may also be characterized by a local change in the density of the medium.

• A torsional wave. Here one part of a deformable medium is twisted relative to neighboring regions. For example, a torsional wave can be propagated with equally spaced transverse rods attached to a deformable strip lying along the line of wave propagation. The disturbance is an angular displacement, and any particle on a rod moves transversely in a circular arc.

The simplest type of periodic traveling wave is one produced by a simple harmonic oscillator. Figure 17-3 shows a wave generated by an oscillator with amplitude A, frequency f, and period $T = 1/f$ that is oscillating transversely at the left end of a string. The wave shape is sinusoidal—that of a sine or a cosine. Any particle along the string oscillates transversely in the same simple harmonic motion as the generator but not necessarily with the same phase. The distance between adjacent points on the wave in the *same phase*—the distance between adjacent crests, for example—is called the *wavelength λ.* The oscilla-

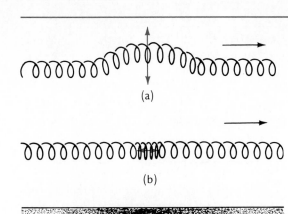

(a)

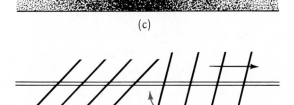

(b)

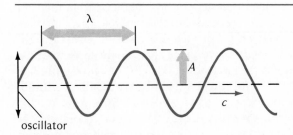

(c)

(d)

Figure 17-2.   Several types of elastic waves: (a) a transverse wave in a Slinky; (b) a longitudinal wave in a Slinky; (c) a compressionable wave in a deformable medium; and (d) a torsional wave.

tor completes one oscillation in a time interval T. During this time, the waveform advances a distance λ. Therefore, the speed c of wave propagation is

$$c = \frac{\lambda}{T} = f\lambda \qquad (17\text{-}1)$$

For a constant wave speed c, high-frequency waves have short wavelengths, and the opposite is also true.

For all elastic waves we shall be considering, we assume the wave disturbance to be *small*, small enough that the restoring force on any particle displaced from equilibrium is proportional to the particle's displacement (and Hooke's law, $F_s = -kx$, is satisfied). This means that the wave shape must be relatively "flat" and smooth; for sinusoidal waves, the amplitude A must be small compared with the wavelength λ. Then, and only then, will the following equivalent basic wave properties be obeyed exactly.

λ

A

c

oscillator

Figure 17-3.   A sine wave of amplitude A and wavelength λ.

- All wave shapes propagate at the same *wave speed*.
- As the wave propagates, its shape is *unchanged* (although damping may reduce the amplitude). A snapshot of a wave at some one instant is just like a snapshot at any other instant but with the shape displaced along the direction of wave propagation.
- What happens at one location — how the displacement varies with time — is repeated later, without change, at any other location along the line of wave propagation.

In the detailed derivations of wave speeds to be given later, we shall see how these conditions are met.

When the disturbance from equilibrium is too large, these properties no longer apply. One consequence is that individual waves differing in wavelength do not have the same wave speed; therefore, the resultant wave shape is spread out, or dispersed, as the disturbance travels through space. Although the most familiar types of mechanical waves are those on the surface of water, these waves are actually complicated; they exhibit *dispersion* (wave speed depends on wavelength), and individual particles are displaced both longitudinally and transversely as they execute motion in a loop. Because of these complications, we do not treat water waves.

Example 17-1.   The lowest and highest frequencies to which the human ear is sensitive are approximately 20 Hz and 20 kHz. At room temperature, the speed of sound waves through air is 344 m/s. What are the wavelengths in air of sounds at the extremes of the audible range?

From (17-1), we have for $f = 20$ Hz,

$$\lambda = \frac{c}{f} = \frac{344 \text{ m/s}}{20 \text{ s}^{-1}} = 17.2 \text{ m}$$

and for $f = 20$ kHz,

$$\lambda = \frac{c}{f} = \frac{344 \text{ m/s}}{20 \times 10^3 \text{ s}^{-1}} = 1.7 \text{ cm}$$

It is no accident that a tuba and a small whistle have sizes roughly comparable to these values.

17-2   Speed of a Wave on a String

Here we derive the relation that shows, in terms of the string's physical characteristics, how fast a wave propagates on a taut string. We assume the string to be perfectly flexible. It has a constant mass per unit length, or linear density represented by ρ, and the tension in the undeformed string is T.

Consider Figure 17-4, in which a wave pulse is shown traveling to the right at speed c. A vertical force component T_y was first applied to the left end of the string at time $t = 0$. At a time t later, the leading edge of the wave pulse has progressed a distance ct to the right. All parts of the pulse are moving vertically upward at speed v, so that at time t the left end is a vertical distance vt from its undisplaced position. The string is stretched slightly because of the presence of the pulse, and the tension in the string is slightly greater than its undeformed value of T. As Figure 17-4 shows, the resultant force on the end of the perfectly

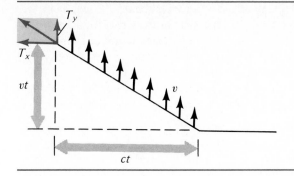

Figure 17-4. *A wave pulse traveling at speed c. The transverse speed of the pulse is v. The components of the string tension are T_x and T_y.*

flexible string is parallel to the string; its horizontal component is T_x, and its much smaller vertical component is T_y.

According to the impulse-momentum theorem, any impulse Ft on an object is equal to the change $\Delta(mv)$ in its momentum (Section 7-3). Here the impulse produced by vertical force component F_y over a time t equals the momentum gained by the segment of the string that is moving vertically upward with velocity **v**. The wave disturbance is assumed to be small—very nearly flat—so that $v \ll c$ and $T_y \ll T$. The length of string in motion at speed v is essentially the same as ct; the mass of the string is ρct and its momentum $(\rho ct)v$.

The impulse momentum theorem then gives

$$Ft = \Delta(mv)$$

$$T_y t = (\rho ct)v$$

From the similar triangles for force components and velocities in Figure 17-4, we have

$$\frac{T_y}{T_x} \simeq \frac{T_y}{T} = \frac{vt}{ct} = \frac{v}{c}$$

or

$$T_y = \frac{v}{c} T$$

Using this result in the relation for $T_y t$ above gives

$$\frac{v}{c} T = \rho cv$$

$$c = \sqrt{\frac{T}{\rho}} \qquad (17\text{-}2)$$

Note the cancellation of v. It implies that every small wave disturbance will be propagated at the speed c given by (17-2).

The relation for the wave speed along a taut string illustrates a feature applicable to all types of waves in elastic media. The wave speed is proportional to the square root of the ratio of some characteristic measure of *elasticity* (here tension T) divided by a characteristic measure of mass or *inertia* (here linear density ρ). We are reminded of properties of simple harmonic oscillators; as we saw in Chapter 11, the frequency of any oscillator is also proportional to the square root of an elastic parameter divided by an inertial parameter. [For example, for a mass m attached to a spring of stiffness k, we have frequency $\propto (k/m)^{1/2}$, (11-13).] Indeed, each particle of the string interacts elastically with its immediately neighboring particles and constitutes an oscillator, so that we can think of a wave traveling along the string to be, in effect, an excitation in turn of many oscillators located along the string.

> **Example 17-2.**   A uniform flexible rope is 2.0 m long and has a mass of 49 gm. Its upper end is fixed, and a mass of 5.0 kg is hung from its lower end. If the lower end is displaced suddenly along the horizontal, how long does it take for a wave pulse to travel to the upper end?
>
> We denote the rope length L. Then the time for the pulse to travel all the way up is $t = L/c$. From (17-2), we have
>
> $$c = \sqrt{\frac{T}{\rho}} = \sqrt{\frac{Mg}{m/L}}$$
>
> where m is the mass of the rope and M the mass of the suspended weight. Then
>
> $$t = \frac{L}{c} = \frac{L}{\sqrt{Mg/m/L}} = \sqrt{\frac{mL}{Mg}} = \sqrt{\frac{(49 \times 10^{-3}\ \text{kg})(2.0\ \text{m})}{(5.0\ \text{kg})(9.8\ \text{m/s}^2)}} = 0.045\ \text{s}$$
>
> (We have ignored the 1.0 percent difference in tension between the top and the bottom; strictly, the wave speed is greater by 0.5 percent at the top than at the bottom.)

17-3   The Superposition Principle and Interference

What happens when two small wave disturbances traveling in opposite directions along the same string "collide"? Figure 17-5 shows the results of observation. As the two pulses merge, the resultant displacement at each point along the string and at any instant is the simplest possible, the algebraic sum of the separate wave disturbances. This is the *principle of superposition*. A consequence is that any wave will pass through another wave completely unscathed, with *unchanged shape.* Said differently, there is no wave-wave interaction; each wave carries away from the "collision" precisely the energy and transverse linear momentum it carried into the collision.

Superposition applies only for small deformations (here, small string displacements). If the string were displaced great distances, or were deformed with sharp corners, or if the tension were not constant, the approximations made in deriving the relation for wave speed would no longer apply, and the principle of superposition would be inapplicable.

Superposing separate wave displacements to arrive at the resultant displacement is known as *interference.* (The term is an unhappy one, because the superposition principle says in effect that the separate waveforms actually do not interfere with, or change, one another.) When two superposed waves add, so that the resultant displacement is greater than the amount for each wave separately, we have what is called *constructive interference.* On the other hand, when two waves of opposite displacements are superposed and the magnitude of the resultant displacement is then less than the amount for each wave separately, the waves are said to show *destructive interference.*

One special case of wave interference is shown in Figure 17-5(c). Here two wave pulses, one to the left, the other to the right, have identical shapes but are inverted both up-down and left-right. We see that when the two waves inter-

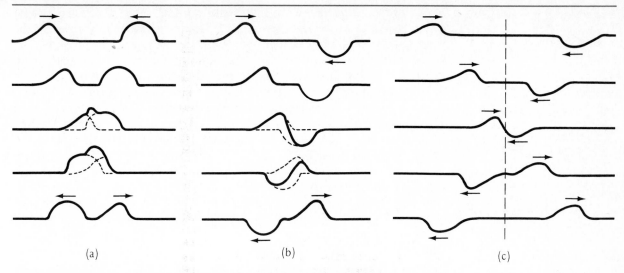

(a)   (b)   (c)

Figure 17-5.   *The "collision" of wave pulses traveling in opposite directions. (a)*
Constructive interference. (b) Destructive interference. (c) As the pulses interfere, a single
point on the string remains undisplaced at all times. Each group shows snapshots taken
at equal time intervals, time increasing downward.

fere, there is a single point along the string at which the resultant displacement
always remains zero.

17-4   Reflection of Waves

What happens when a wave collides with a boundary, a point at which the
medium propagating the wave (here, the string) changes? First imagine a string
attached firmly to a rigid wall, as in Figure 17-6. Since the wall cannot be
displaced, we may describe this situation formally by saying that the string
displacement y must always be zero at the point at which the string joins the
wall. Now, if a wave pulse is propagated to the right, we find it reflected to the
left from the boundary. After reflection, the shape is reversed *left-right;* the
initial leading edge of the pulse is still the leading edge after reflection. In
addition, the sign of the wave is reversed; that is, the wave is inverted *up-down,*
a positive transverse displacement becoming negative on reflection.

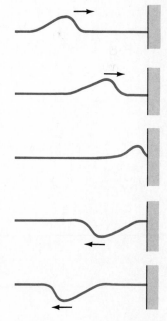

We can see the physical basis for this behavior. When the leading edge of
the wave disturbance arrives at the boundary, the tension of the string produces an upward force on the infinitely massive, and therefore immovable,
wall. By Newton's third law, the wall applies an equal downward force on the
string. This force is greater than the force applied by an adjoining segment of
string in the absence of the wall, because the string undergoes a larger change
in curvature at the boundary. The force of wall on string is so great that it does
not merely return it to $y = 0$; the wall pulls so hard on the string that the string
is brought below the line $y = 0$. Thus, an inverted wave to the left is generated.

Now imagine that the end of the string, rather than being tied down, is
perfectly free to move in the transverse direction. For definiteness, suppose
that the string is terminated with a small massless ring that can slide freely
along a smooth vertical post. The results of a reflection are shown in Figure

Figure 17-6.   *Reflection of a*
wave pulse at an infinitely
massive boundary.

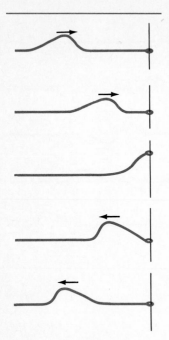

Figure 17-7. *Reflection of a wave pulse from a free end.*

17-7. Here again we have a left-right shape reversal, as the direction of propagation is changed from right to left. But there is *no* change in the sign of the wave shape. An upright incident wave is reflected as an upright wave. The physical basis for reflection from a free boundary is this—since there is no string to the right of the free end to provide a downward force component through the tension, the string overshoots as the disturbance reaches the end.

Thus far we have considered the two extreme types of reflection: a string attached to an infinitely rigid or massive second medium and a string attached to a soft or massless second medium. A more general case is one in which one string with linear density ρ_1 is connected to a second string of linear density ρ_2. The tension T is the same in both strings. We see from (17-2) that if $\rho_2 > \rho_1$, the wave speed c_1 in the first string exceeds the wave speed c_2 in the second string. Figure 17-8(a) shows what happens when a wave pulse incident from the left with speed c_1 encounters the boundary. The wave is partially transmitted into the second medium and partially reflected into the first medium. Here, the boundary moves laterally as the waves reach it. The transmitted wave undergoes no change in sign, but the reflected wave is reversed. Moreover, the transmitted wave shape is, because of the decreased speed c_2, shrunk longitudinally. That reversal in sign of the reflected wave follows from the behavior

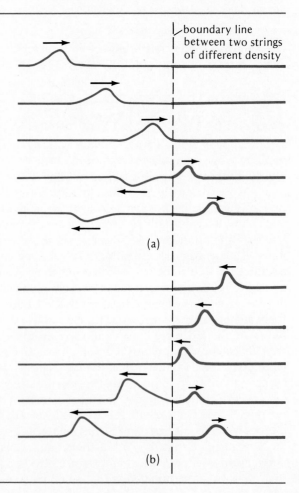

Figure 17-8. *Reflection and transmission of a wave pulse incident upon (a) a string of greater linear density and (b) a string of lesser linear density.*

found earlier for reflection from an infinitely massive second medium. Here the second medium is not infinitely massive but has a greater inertia than the first. Figure 17-8(b) shows an incident wave traveling from the more massive string into the less massive string. Here the polarity is not reversed for either the transmitted or the reflected wave.

17-5   Sinusoidal Traveling Waves

A sinusoidal wave—one with the shape of a sine or cosine—is the most important type of traveling wave. The reasons:

• A sinusoidal wave is easily produced; simply oscillate one part of the elastic medium in simple harmonic motion.
• Any persistent wave shape can be regarded as the superposition of strictly sinusoidal waves.*

Suppose that the left end ($x = 0$) of a very long string oscillates along y in simple harmonic motion with amplitude A and frequency f (and period T, where $T = 1/f$).† Then (Section 11-1) the transverse displacement at this location is given by

$$y(x, t) = y(0, t) = A \sin \omega t$$

where $\omega = 2\pi f = 2\pi/T$. From the arguments of Section 17-2, we know that the displacement of any other point on the string to the right will show later exactly the same variation with time. At x there will be a delay, or phase *lag*, δ:

$$y(x, t) = A \sin (\omega t - \delta)$$

The wave disturbance travels along the $+x$ direction at constant speed; therefore, the phase lag δ is proportional to x, and we can write

$$\delta = kx \qquad (17\text{-}3)$$

where k is known as the *wave number* (for reasons soon to be seen). The transverse displacement $y(x, t)$ can then be written

Wave in $+x$ direction: $y(x, t) = A \sin (\omega t - kx)$ $\qquad$ **(17-4a)**

This is one form of the equation for a traveling sinusoidal wave. It gives the transverse displacement y as a function of both the coordinate x along the direction of propagation and the time t. If the wave were to travel in the $-x$ direction, the displacement at point x would lead, rather than lag, the displacement at $x = 0$. Then the equation for the traveling wave would be written

Wave in $-x$ direction: $y(x, t) = A \sin (\omega t + kx)$ $\qquad$ **(17-4b)**

We see that the displacement y oscillates in simple harmonic motion at every location along the string. The equations also show, for any fixed time t,

---

* Recall that any persistent periodic motion can be regarded as the summation of strictly simple harmonic oscillations (Section 11-6).

† The conventional symbol T is used for period. From the context, it will be clearly distinguishable from T used earlier to represent string tension.

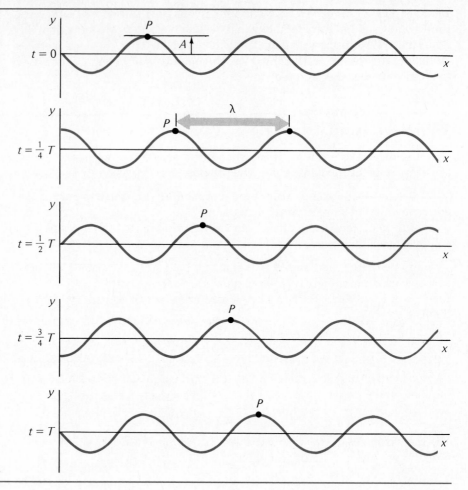

Figure 17-9. *A sinusoidal wave for several different times.*

that y varies sinusoidally with x; that is, a snapshot of a wave generated by a simple harmonic oscillator is a sine or a cosine, as shown in Figure 17-9.*

Points with both the same displacement and the same velocity are in phase, for example, the dots P of Figure 17-9. The distance between any two such adjacent points at the same phase is the *wavelength*, λ. Thus, if x changes by λ, the phase must change by 2π, or from (17-3),

$$2\pi = k\lambda$$

$$k = \frac{2\pi}{\lambda} \qquad (17\text{-}5)$$

The wave number k gives the number of wavelengths per unit length multiplied by 2π; also, k is the rate of change in phase with distance.

Using the definition $\omega = 2\pi/T$, we may write (17-4a) in another useful form:

* For clarity, the amplitude of a sinusoidal wave in Figure 17-9 is shown as comparable to the wavelength. But waves propagate with unchanged shape, and superposition principle is followed exactly only for small—that is, relatively flat—wave disturbances. Consequently, the amplitude must be small compared with the wavelength to satisfy these conditions.

$$y(x, t) = A \sin 2\pi \left(\frac{t}{T} - \frac{x}{\lambda} \right) \qquad \text{(17-6)}$$

Equation (17-6) applies for a wave traveling along $+x$; as before, for a wave along $-x$, replace the minus sign by a plus sign.

As we have seen, the frequency f, period T, and wavelength λ for a sinusoidal wave are related to the wave speed c as follows:

$$c = \frac{\lambda}{T} = f\lambda \qquad \text{(17-1)}$$

An alternative form of this relation is $c = (2\pi f)(\lambda/2\pi) = \omega/k$. Since the *phase* of oscillation advances along the propagation direction at the same speed as the wave shape, the wave speed is also called the *phase speed*.

When we write the wave speed as $c = \omega/k$, still another form in which (17-4) can be written is

$$y = A \sin(\omega t - kx) = A \sin k(ct - x)$$
$$y(x, t) = -A \sin k(x - ct) \qquad \text{(17-7)}$$

again for a wave propagated along $+x$.

The three forms of the equation for a traveling sinusoidal wave given in (17-4), (17-6), and (17-7) are equivalent; they all give y as a function of x and t. The equations differ merely by which of the parameters k, λ, ω, T, f, and c appear. We have, for simplicity, taken the displacement y to be zero when $x = 0$ and $t = 0$. To allow for any initial displacement at $x = 0$ and $t = 0$, we merely incorporate a phase constant ϕ, so that (17-7) can be written

$$y(x, t) = A \sin[k(x - ct) - \phi] \qquad \text{(17-8)}$$

Suppose that two sinusoidal waves with the same wavelength travel in the same direction at the same speed but are out of phase by angle ϕ. What is the resultant wave? It is easy to see that it is another sinusoidal wave with the same wavelength and speed.

Consider the waves

$$y_1 = A \sin k(x - ct) \qquad \text{and} \qquad y_2 = A \sin[k(x - ct) - \phi]$$

The individual waves have the same amplitude A, but differ in phase by angle ϕ.

Using the trigonometric identity

$$\sin a + \sin b = 2 \cos\left(\frac{a - b}{2}\right) \sin\left(\frac{a + b}{2}\right)$$

we get

$$y = y_1 + y_2 = 2A \cos\frac{\phi}{2} \sin\left[k(x - ct) - \frac{\phi}{2} \right]$$

The composite wave has an amplitude of $2A \cos(\phi/2)$; its phase $\phi/2$ is midway between the two separate waves. See Figure 17-10(a), where $\phi = 60°$. Two cases are of special interest.

• No phase shift, or $\phi = 0$. The resultant wave has *twice* the amplitude of the individual waves. The waves interfere *constructively*. See Figure 17-10(b).

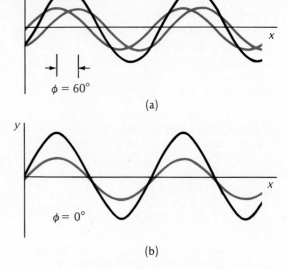

$\phi = 60°$

(a)

$\phi = 0°$

(b)

Figure 17-10. Superposition of two sinusoidal waves of equal amplitude and wavelength for relative phase differences: (a) 60°; (b) 0°; (c) 180°. The heavy black line represents the resultant wave.

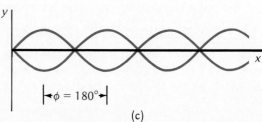

$\phi = 180°$

(c)

• A phase shift of $\phi = 180°$. Here the individual waves are out of phase by *one half-wavelength.* The resultant amplitude is zero. The waves cancel out in complete *destructive interference.* See Figure 17-10(c).

Example 17-3.   A rope with linear density of 70 gm/m is under a tension of 10 N. A sinusoidal wave is generated by a simple harmonic oscillator at the point $x = 0$. The oscillator executes 4.0 oscillations per second with an amplitude of 2.0 cm. (a) What is the wave speed? (b) What is the wavelength? (c) Assume that the oscillator is at the upper amplitude position at time $t = 0$. Write the equation for the traveling sinusoidal wave as a function of x and t. (d) What is the magnitude of the maximum transverse linear momentum of a small segment of string 1.0 mm long? (e) What is the maximum resultant force on such a segment?

(a) From (17-2),

$$c = \sqrt{\frac{T}{\rho}} = \sqrt{\frac{10 \text{ N}}{0.070 \text{ kg/m}}} = 12 \text{ m/s}$$

(b) From (17-1),

$$\lambda = \frac{c}{f} = \frac{12 \text{ m/s}}{4.0 \text{ Hz}} = 3.0 \text{ m}$$

(c) We use the general equation for a traveling wave, (17-8):

$$y(x, t) = A \sin [k(x - ct) - \phi]$$

Since $y = A$ at $t = 0$ and $x = 0$,

$$A = A \sin(-\phi) = -A \sin\phi$$

or

$$\phi = -\frac{\pi}{2}$$

Then

$$y(x, t) = A \sin\left[k(x - ct) + \frac{\pi}{2} \right]$$

or

$$y(x, t) = A \cos k(x - ct)$$

Using (17-5), $k = 2\pi/\lambda$, and substituting the numerical values of λ, A, and c in the last equation, we have

$$y(x, t) = (2.0 \times 10^{-2}) \cos\frac{2\pi}{3}(x - 12t)$$

where y and x are in meters and t is in seconds.

(d) To find the maximum transverse linear momentum of a small rope segment, we must find the maximum transverse speed $(\partial y/\partial t)_{max}$ and multiply it by the mass of the 1.0-mm segment, $0.070 \text{ kg/m} \times 1.0 \times 10^{-3} \text{ m} = 7.0 \times 10^{-5} \text{ kg}$. The length of the segment is so small compared with the wavelength (1 in 3000), that we can properly regard all parts of the small segment as having the same speed. Taking the time derivative of y (with x held constant), we have, from the equation in part (c),

$$\frac{\partial y}{\partial t} = kcA \sin k(x - ct)$$

Therefore,

$$\left(\frac{\partial y}{\partial t}\right)_{max} = kcA = \left(\frac{2\pi}{3} \text{ m}^{-1}\right)(12 \text{ m/s})(2.0 \times 10^{-2} \text{ m}) = 0.50 \text{ m/s}$$

The magnitude of the maximum transverse linear momentum of the 7.0×10^{-5} kg segment is

$$p_{max} = m\left(\frac{\partial y}{\partial t}\right)_{max} = (7.0 \times 10^{-5} \text{ kg})(0.50 \text{ m/s}) = 3.5 \times 10^{-5} \text{ kg} \cdot \text{m/s}$$

(e) To find the magnitude of the maximum resultant force on the small rope segment, we first compute the maximum transverse acceleration $\partial^2 y/\partial t^2$:

$$\frac{\partial^2 y}{\partial t^2} = -(kc)^2 A \cos k(x - ct)$$

$$= (2\pi \times 4.0 \text{ Hz})^2(2.0 \times 10^{-2} \text{ m}) = 13 \text{ m/s}^2$$

The magnitude of the maximum resultant force is

$$(F_y)_{max} = m\left(\frac{\partial^2 y}{\partial t^2}\right)_{max} = (7.0 \times 10^{-5} \text{ kg})(13 \text{ m/s}^2) = 9.1 \times 10^{-4} \text{ N}$$

Note that the maximum transverse force is much less than the rope's tension of 10 N.

17-6   Standing Waves and Resonance

What happens when two sinusoidal waves of the same wavelength and amplitude travel at the same speed but in opposite directions? A graphic represen-

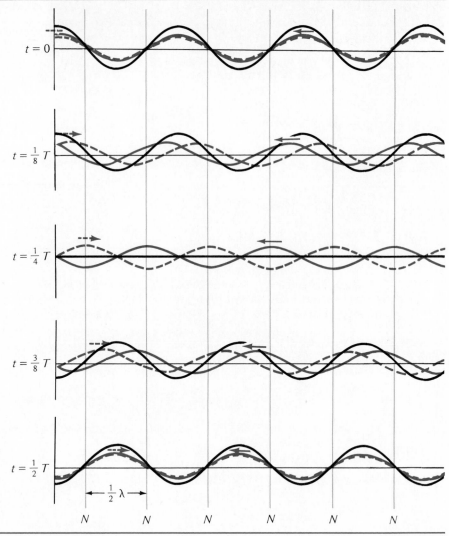

$t = 0$

$t = \frac{1}{8} T$

$t = \frac{1}{4} T$

$t = \frac{3}{8} T$

$t = \frac{1}{2} T$

$\leftarrow \frac{1}{2} \lambda \rightarrow$

N   N   N   N   N   N

Figure 17-11. Standing waves. The resultant waveform (heavy black line) for two waves traveling in opposite directions for a succession of times. Adjacent nodes (N) and adjacent antinodes (or loops, L) are separated by one half-wavelength.

tation of the results, in Figure 17-11, separately shows one wave traveling to the right and the other to the left, and their resultant, for a succession of times.

At certain times, the two waves fall exactly on top of one another, namely, at $t = 0$ or $t = \frac{1}{2}T$, where T is the period. At these instants, the resultant wave has twice the amplitude of each wave separately. The waves annul completely at $t = \frac{1}{4}T, \frac{3}{4}T, \ldots$. At the instant of complete cancellation, the potential energy of the string is zero, so that the total energy at these instants consists solely of kinetic energy. Note especially from Figure 17-11 that at certain locations along the string the two traveling waves *always interfere destructively.* These points, where the string never undergoes a displacement, are known as nodal points, or *nodes.* The string has its largest amplitude of oscillation at other locations, known as *antinodes* (or sometimes, loops), midway between adjoining nodes. As Figure 17-11 shows, *adjacent nodes are separated by one half-wavelength ($\frac{1}{2}\lambda$); likewise adjacent antinodes are separated by $\frac{1}{2}\lambda$.* The resultant oscillating disturbance is called a *standing wave.* This term is appropriate, since no resultant waveform travels left or right, and no energy is transferred left or right. The oscillating pattern stands in place; the nodes and antinodes are fixed.

It is easy to derive these results analytically. The resultant displacement can be written as

$$y = A \sin (\omega t - kx) + A \sin (\omega t + kx)$$

We use the trigonometric identity

$$\sin a + \sin b = 2 \cos \left(\frac{a - b}{2}\right) \sin \left(\frac{a + b}{2}\right)$$

in the equation above, so that it becomes

$$y = 2A \cos kx \sin \omega t = 2A \cos \frac{2\pi x}{\lambda} \sin \omega t \qquad (17\text{-}9)$$

This is the equation for a standing wave.

Equation (17-9) shows that every particle on the string undergoes simple harmonic motion, $\sin \omega t$, at the angular frequency $\omega = 2\pi f = 2\pi/T$. The oscillation amplitude [$2A \cos (2\pi x/\lambda)$] depends on the position x. The displacement y is always zero for those positions for which $\cos (2\pi x/\lambda)$ is zero. This implies that $2\pi x/\lambda$ is an odd multiple of $\frac{1}{2}\pi$ radians. Therefore, nodes come at the positions

$$\frac{2\pi x}{\lambda} = n \frac{\pi}{2} \qquad \text{where } n = 1, 3, 5, \ldots$$

$$\text{Nodes:} \quad x = \frac{\lambda}{4}, \frac{3\lambda}{4}, \frac{5\lambda}{4}, \ldots$$

We see again that adjacent nodes are separated by $\lambda/2$.

Antinodes exist at those positions for which the amplitude $2A \cos (2\pi x/\lambda)$ has its maximum value, $2A$. This occurs for $\cos (2\pi x/\lambda) = \pm 1$, which in turn implies that $2\pi x/\lambda$ is an integral multiple of π. Antinodes come at the positions

$$\frac{2\pi x}{\lambda} = m\pi \qquad \text{where } m = 0, 1, 2, \ldots$$

$$\text{Antinodes:} \quad x = 0, \frac{\lambda}{2}, \frac{2\lambda}{2}, \ldots$$

Antinodes lie midway between nodes; adjacent antinodes are also separated by $\lambda/2$.

How do you produce identical sinusoidal waves traveling in opposite directions and thereby create standing waves? One way is by placing transverse simple harmonic oscillators, or wave generators, at opposite ends of a string. Even more simply, reflect an incident sinusoidal wave at a hard boundary and thereby produce also a reflected wave traveling in the opposite direction. A standing-wave pattern can, for example, be generated on a string by oscillating one end laterally with a mechanical oscillator, keeping the other end fixed. See Figure 17-12. The oscillations may be so rapid that you see only the envelope of the standing-wave pattern.

Standing waves can also be produced with both ends of the string fixed. Then the standing-wave pattern must fit between the two ends of the string.

Figure 17-12. *An arrangement for demonstrating standing waves on a string. One string end is attached to a vibrating tuning fork; the other end is attached to a weight and hung over a pulley.*

This in turn means that the string length L must be an integral multiple of half-wavelengths. Only then will the *boundary conditions* at the string ends, no displacement at $x = 0$ and at $x = L$, be satisfied. With the two string ends fixed, a node must be located at each end. The allowed wavelengths for standing waves on a string fixed at both ends are given by

$$n\frac{\lambda}{2} = L \qquad \text{where } n = 1, 2, 3, \ldots$$

The allowed standing-wave patterns are those for which 1, 2, 3, or any integral multiple of $\lambda/2$ fits along the string length L.

If the wavelengths of standing waves are restricted by the conditions given above, so too are the frequencies. Using the relation for allowed wavelengths above, we have that the allowed, or characteristic, frequencies are

$$f = \frac{c}{\lambda} = \frac{c}{2L/n}$$

$$f = n\left(\frac{c}{2L}\right) = nf_1 = f_1, 2f_1, 3f_1, \ldots \qquad (17\text{-}10)$$

where $n = 1, 2, 3, \ldots$.

The lowest frequency, also called the *fundamental frequency,* is $f_1 = c/2L$. The higher frequencies are *integral multiples* of the fundamental frequency and are therefore said to be *harmonics* of the fundamental. More specifically, $2f_1$ is the second harmonic, $3f_1$ is the third harmonic, and so on. See Figure 17-13.

Suppose that a standing-wave pattern is established for a string fixed at both ends. We then have two equivalent ways of describing what is happening to particles of the strings:

• Oppositely directed waves are traveling along the string. If an integral multiple of half-wavelengths fits between the string ends, the resultant wave pattern stands in place.

• All particles of the string (except at the locations of the nodes) are undergoing simple harmonic motion at the same allowed frequency. The oscillation amplitude differs according to the particle's location along the string, with the amplitude a maximum at antinodes and zero at nodes. Each particle is coupled through the tension in the flexible string to its immediate neighbors.

According to the superposition principle, several waves can exist simultaneously on a string without one wave's disturbing another. Waves pass through one another without interaction. Thus, two or more standing waves of different frequencies and wavelengths can exist simultaneously on a single string with fixed ends. See Figure 17-14. It shows standing waves for the first

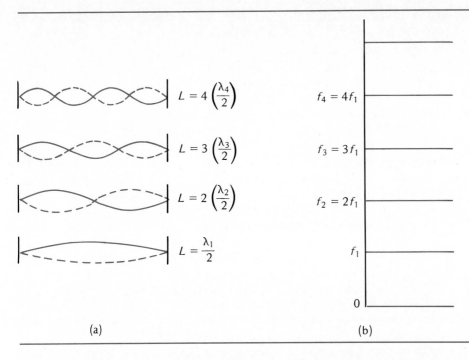

(a)

(b)

Figure 17-13. *(a) Allowed oscillation modes and (b) allowed frequencies for a string attached at both ends.*

and second harmonics, and also their resultant, for several instants. The resultant waveform is more complicated than with a simple standing-wave pattern, but the waveform repeats periodically at the fundamental frequency f_1.

Any periodic disturbance on a string—any oscillations of a string that persist in time—must consist of one or more of the allowed standing waves. On the other hand, any wave disturbance that is not at one of the allowed frequencies—any oscillation that does not correspond to one of the allowed modes of oscillation—must die out, its energy being dissipated at the boundaries. Thus, when a string is struck or bowed, as in the piano or violin, it typically

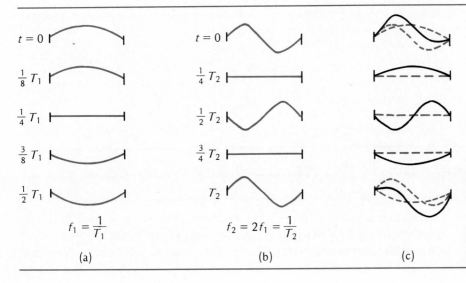

(a)

(b)

(c)

Figure 17-14. *Wave pattern for a succession of times corresponding to the simultaneous excitation of the first and second harmonics. (a) First harmonic alone; (b) second harmonic alone; (c) resultant waveform.*

oscillates simultaneously at several of the allowed frequencies or harmonics. Typically, the amplitude of the fundamental exceeds that of higher harmonics. The higher harmonics are sometimes referred to as "overtones"; they have a frequency larger than ("over") that of the fundamental. For a string with fixed ends, the first overtone is the second harmonic; the second overtone is the third harmonic; and so on.

What does it take to set a single oscillating particle at resonance? As shown in Section 11-8, the oscillator must be driven at its natural oscillation frequency. Only when the external driver and the oscillator have the *same* frequency will the oscillator's amplitude of oscillation be increased appreciably.

In similar fashion, resonance can be produced for the standing waves on a string fixed at its ends. The string has many constituent particles and many natural frequencies of oscillation, not just one; there are, in fact, an infinite number of characteristic frequencies and characteristic modes of oscillation. Every resonant frequency corresponds to the one particular type of standing wave. The oscillations of an actual string are damped, either by friction with the surrounding air or by internal friction arising from the stretching of the string. Suppose that a string is excited by some external driving force, for example, by variations in the air pressure from a sound wave. The string exhibits resonance and the oscillation amplitude may be large if the exciting frequency is at one of the string's natural frequencies.

The discrete frequencies and characteristic modes of oscillation for standing waves have an important analogy in the quantum theory (also known as wave mechanics). Indeed, the stability and structure of atoms, molecules, and nuclei are understood through the wave properties of electrons, protons, and neutrons.

Example 17-4. A violin with a string 31.6 cm long, of linear density 0.65 gm/m, is placed near a loudspeaker fed by an audio-oscillator of variable frequency. The frequency of the sound waves is varied continuously over the range 500–1500 Hz. The string oscillates only at 880 and 1320 Hz. What is the tension in the string?

The violin string, fixed at both ends, oscillates in resonance at its characteristic frequencies, which are in the ratio of the integers 1, 2, 3, ... The ratio of the two resonance frequencies here is $1320/880 = 3/2$. Therefore, these two resonances correspond to the string oscillating in its second and third harmonics. (If the two harmonics were, instead, the fourth and sixth, their ratio would again be $3/2$, but then there would also be a resonance at the fifth harmonic, 1100 Hz.) Consequently, the fundamental, or first harmonic, is 440 Hz (A above middle C on the concert scale). The length of the string for this frequency is one half-wavelength:

$$L = \frac{\lambda}{2} = \frac{c}{2f} = \frac{\sqrt{T/\rho}}{2f}$$

$$T = 4f^2 L^2 \rho = 4(440 \text{ Hz})^2 (0.316 \text{ m})^2 (6.5 \times 10^{-4} \text{ kg/m}) = 50 \text{ N}$$

17-7 Power of a Wave (Optional)

Wave motion is a mode of energy transfer. Particles in transverse motion along the string have kinetic energy; the deformed string has potential energy. Here we find the rate at which energy is transferred along the string for a sinusoidal wave.

Each string point executes simple harmonic motion at the same angular frequency $\omega = 2\pi f$ and with the same amplitude A. We consider the string to be made up of small adjoining particles, each of mass m. The total energy of an oscillating particle, the sum of its kinetic and potential energies, is

$$E = \tfrac{1}{2}k'A^2 \qquad\qquad (11\text{-}14)$$

where k' represents the equivalent force constant:

$$k' = \omega^2 m \qquad\qquad (11\text{-}7)$$

Therefore, the energy for each particle may be written

$$E = \tfrac{1}{2}\omega^2 m A^2$$

Now if the wave amplitude A is small compared with the wavelength, the total mass in a segment of string one wavelength long is $\rho\lambda$. The total energy E_λ of all oscillating particles in the one-wavelength segment is, with $m = \rho\lambda$,

$$E_\lambda = \tfrac{1}{2}\omega^2 \rho\lambda A^2$$

Energy is transferred by the wave motion, but the total energy of any segment is constant because energy enters the segment at the same rate as that at which energy leaves. More specifically, all the energy originally contained in a one-wavelength segment will have left that segment in the time T it takes for the sinusoidal wave to advance a distance λ, where T is the period, both of the wave and of any simple harmonic oscillator. The power P of the wave is therefore

$$P = \frac{E_\lambda}{T} = \frac{\tfrac{1}{2}\omega^2 \rho\lambda A^2}{T}$$

$$P = \tfrac{1}{2}\rho c(\omega A)^2 = 2\pi^2 \rho c f^2 A^2 \qquad\qquad (17\text{-}11)$$

where $\omega = 2\pi f$. As one would expect, the rate of energy flow is proportional to the wave speed c. Equation (17-11) also shows that the power is proportional to the *square* of both the *frequency f* and the *amplitude A*. A high-frequency, or short-wavelength, sinusoidal wave may carry appreciable power even though its amplitude is relatively small.

Summary

Definitions

Wave: disturbance propagated through a system of coupled particles.

Wavelength λ: for a sinusoidal wave, distance between two adjacent points in the same phase of oscillation,

Wave number k: $2\pi/\lambda$

Transverse wave: particle displacement at right angles to line along which wave is propagated.

Longitudinal wave: particle displacement parallel to line along which wave is propagated.

Constructive interference: condition in which resultant wave disturbance exceeds in magnitude the separate wave disturbances.

Destructive interference: condition in which resultant wave disturbance is less in magnitude than the separate wave disturbances.

Standing waves: the oscillating wave pattern with fixed nodes and antinodes produced by sinusoidal waves of the same wavelength traveling in opposite directions.

Node: location in a standing-wave pattern where the wave disturbance is always zero; adjacent nodes are separated by $\tfrac{1}{2}\lambda$.

Antinode: location in a standing-wave pattern where the wave disturbance has a maximum oscillation amplitude; adjacent antinodes are separated by $\tfrac{1}{2}\lambda$.

Fundamental frequency: lowest natural oscillation frequency for system of particles.

Harmonic: an oscillation mode for which the frequency is an *integral* multiple of the fundamental frequency.

Overtone: an oscillation mode of greater frequency than the fundamental, but not necessarily an integral multiple thereof.

Important Results

Superposition principle: The resultant of two or more separate waves is the algebraic sum of the separate disturbances.

Speed of wave along a uniform string of linear density ρ and under tension T:

$$c = \sqrt{\frac{T}{\rho}} \qquad (17\text{-}2)$$

Reflection of a wave at a boundary, or change in medium for wave propagation: from less massive ("soft") sec-ond medium, no phase change; from more massive ("hard") second medium, 180° phase change.

A sinusoidal wave, generated by a simple harmonic oscillator of amplitude A, angular frequency ω, and period T, and traveling along the positive x axis, may be represented in any of the following, equivalent forms:

$$y = A \sin (\omega t - kx) \qquad (17\text{-}4)$$

$$y = A \sin 2\pi \left(\frac{t}{T} - \frac{x}{\lambda} \right) \qquad (17\text{-}6)$$

$$y = -A \sin k(x - ct) \qquad (17\text{-}7)$$

The product of the frequency f and wavelength λ is the wave speed c:

$$c = f\lambda \qquad (17\text{-}1)$$

Problems and Questions

Section 17-1 Basic Wave Behavior

· **17-1 P** What is the velocity of a wave of wavelength 2 m and frequency 170 Hz?

· **17-2 P** Sound travels at 344 m/s in air at room temperature. What is the wavelength of a 440 Hz sound wave in air?

· **17-3 Q** The speed, or phase velocity, of a wave is
(A) the speed at which a point of constant phase moves.
(B) the speed with which vibrating particles move.
(C) the speed with which energy oscillates back and forth from kinetic energy to potential energy.
(D) the first time derivative of the displacement of the wave.
(E) the time rate of change of the phase.

: **17-4 P** A transverse wave of frequency 20 Hz travels down a string. Two points 5 m apart are 120° out of phase. What is the wave velocity?

: **17-5 P** A wave of frequency 800 Hz has a wavelength 0.5 m. (a) How far apart are two points that differ in phase by 45°? (b) What is the phase difference between two displacements at a given point at times 0.25 ms apart? (c) What is the wave velocity?

Section 17-2 Speed of a Wave on a String

· **17-6 Q** Show that the units of $\left(\dfrac{\text{Tension}}{\text{linear density}} \right)^{1/2}$ are m/s in the SI system.

: **17-7 P** A garden hose 20 meters long has a mass of 3 kg. If you pull on it with a tension of 80 N while one end is attached to a faucet, how long will it take a pulse generated by you to travel down the hose to the faucet and back to your hand?

· **17-8 Q** Why are the strings used for low notes on a stringed instrument thicker than the ones used for high notes?

: **17-9 P** Prove that when a transverse wave travels along a string, the slope of the string at any point is numerically equal to the ratio of the particle speed to the wave speed at that point.

Section 17-3 The Superposition Principle and Interference

· **17-10 Q** When we say that two waves "interfere" we mean that
(A) they interact.
(B) they necessarily cancel each other out.
(C) one is reflected or scattered by the other.
(D) their frequencies are multiples of each other.
(E) they are traveling in opposite directions.
(F) none of the above.

· **17-11 Q** When two waves moving in opposite directions interfere, what effect does one have on the other?
(A) They cancel each other out.
(B) They create a standing wave.
(C) The one with larger amplitude slows the other one down.
(D) Both waves slow each other down.
(E) Neither wave has an effect on the other.

Section 17-4   Reflection of Waves

· **17-12 Q** When transverse waves are propagated on a string tied at one end,

(A) standing waves are always set up.
(B) the tied end of the string is always a node of displacement.
(C) the tied end of the string is always an antinode of displacement.
(D) the reflected wave is not inverted at the tied end.
(E) there is no reflected wave unless the other end is also tied.

: **17-13 P** Two strings of lengths L_1 and L_2 and linear densities ρ_1 and ρ_2 are joined end to end and fixed at their ends, as shown in Figure 17-15. A standing wave with five segments is set up. If $\rho_1/\rho_2 = \frac{3}{4}$, what is the ratio L_1/L_2?

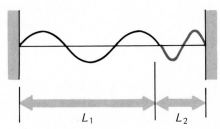

Figure 17-15. *Problem 17-13.*

Section 17-5 Sinusoidal Traveling Waves

· **17-14 Q** If the displacement of a transverse wave is $y = 0.01 \sin(200\pi t - 3x)$, the phase of the wave is
(A) 200
(B) 3
(C) $3x$
(D) $200\pi t - 3x$
(E) impossible to specify unless t and x are given numerical values.

· **17-15 P** A transverse wave on a string is described by

$$y = 0.02 \sin(1800t - 6x)$$

(a) In what direction is the wave traveling? (b) What is the frequency of the wave? (c) What is the wavelength of the wave? (d) What is the velocity of the wave?

: **17-16 P** A traveling wave is described by $y = 0.02 \sin 800(t - 3x)$. Determine (a) the amplitude, (b) the frequency, (c) the wavelength, (d) the wave velocity, (e) the maximum particle velocity, (f) the maximum particle acceleration.

· **17-17 P** Two traveling waves are described by

$$y_1 = 0.02 \sin\left(4\pi t - \frac{\pi}{6}x\right)$$

$$y_2 = 0.04 \sin\left(8\pi t - \frac{\pi}{3}x\right)$$

Plot the resultant of these two waves at $t = 0.5$ s for $0 \leq x \leq 12$ m.

: **17-18 P** Four component waves have amplitudes 1 cm, 2 cm, 3 cm, and 4 cm and, at $t = 0$ and $x = 0$, phases 0°,

45°, 120°, and 180° respectively. Plot the resultant wave as a function of x at $t = 0$. All have the same wavelength λ.

: **17-19 P** Two waves traveling on a string have displacements given by

$$y_1(t) = 0.02 \sin(400\pi t - 2\pi x)$$

and

$$y_2(t) = 0.03 \sin(400\pi t - 2\pi x + \tfrac{1}{4}\pi)$$

(a) Write an expression for the resultant wave. (b) How do the frequencies, wavelengths, and velocities of the two component waves and the resultant wave compare?

· **17-20 P** Write the equation of a traveling transverse wave traveling on a string in the negative x-direction with amplitude of 0.02 m, frequency 400 Hz and velocity 343 m/s.

: **17-21 P** Two strings of linear density $\rho_1 = 1$ gm/m and $\rho_2 = 4$ gm/m are joined together end to end and placed under tension. (a) If a wave has a wavelength of 4 cm in the first string, what is the wavelength in the second string? (b) If the frequency is 50 Hz in string #1, what is the frequency in string #2? (c) If the amplitude of the wave incident on the boundary from string #1 is 3 cm, and the amplitude of the transmitted wave in string #2 is 2 cm, what is the amplitude of the reflected wave?

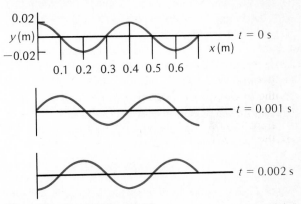

Figure 17-16. *Problem 17-22.*

: **17-22 P** Shown in Figure 17-16 are snapshots of a sinusoidal traveling wave taken at three different times. What are (a) the wave velocity? (b) the amplitude? (c) the frequency? (d) the wavelength? (e) the direction of travel? (f) Write an equation for this traveling wave.

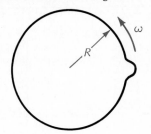

Figure 17-17. *Problem 17-23.*

: 17-23 P A loop of rope is whirled with high angular velocity ω so that it forms a circular loop of radius R. A kink forms in the rope, as shown in Figure 17-17. *(a)* Show that the tension in the rope is $T = \rho\omega^2 R^2$, where ρ is the linear density of the rope. *(b)* What is the velocity of the kink with respect to the ground?

Section 17-6 Standing Waves and Resonance

· 17-24 Q In order for two traveling waves to set up standing waves, it is necessary that
(A) they have the same frequency.
(B) they have the same wavelength.
(C) they have the same velocity.
(D) they travel in opposite directions.
(E) all of the above be true.

· 17-25 Q Resonance occurs when an oscillating driving force is applied to a system such that
(A) the driving force has the same wavelength as does a natural wave in the system.
(B) the driving force is applied at a natural frequency of the system.
(C) energy flows into the system on one half cycle and out on the next.
(D) all boundary conditions are satisfied.

· 17-26 P The tension in a string 2 meters long is adjusted until the fundamental frequency is 200 Hz. What is the wave velocity on the string?

· 17-27 P A string vibrating in five segments has a frequency of 120 Hz. What is the fundamental frequency of the string?

: 17-28 P Two 40-cm waves of equal amplitude travel in opposite directions in a string, with wave velocity 250 m/s. If the string is 2 m long, how many nodes and antinodes are there, not counting the fixed ends?

: 17-29 Q *(a)* One can play different notes on a guitar by pressing the strings against frets placed along the stem. What parameters are being changed when this is done? Fundamental frequency? Fundamental wavelength? Wave velocity? *(b)* If you press a fret on a guitar and shorten the effective string length by 20% (so the new length is $0.8L$), by what factor does the fundamental frequency change? *(c)* One tunes a guitar by adjusting the tension of the thumb screw to which a string is attached. Explain what is being varied. Fundamental wavelength? Fundamental frequency? Wave velocity? Linear density? Tension?

: 17-30 P The fundamental frequency of a violin string of length 42 cm is 400 Hz. Where should a musician place her finger to increase the fundamental frequency to 440 Hz?

: 17-31 P A violin string 50 cm long has a fundamental frequency of 440 Hz. If it is bowed exactly at its midpoint, what frequency standing waves can be excited?

: 17-32 P A suspension bridge is essentially a stretched loaded wire, analogous to a violin string. If transverse waves travel 120 m/s along the bridge and the central span is 1200 m, what is the natural frequency of oscillation?

: 17-33 P A string of 0.5 gm mass, 1.2 meters long, is tied at one end. The other end passes over a small pulley and has a mass of 400 gm attached to the end. What is the fundamental frequency and the frequency of the first harmonic of this string?

: 17-34 P A steel piano wire is 50 cm long and has a mass of 2 gm. The tension in the wire is 640 N. *(a)* What is the fundamental wavelength of the wire? *(b)* What is the fundamental frequency? *(c)* If the maximum frequency a person can hear is 15 kHz, how many harmonics can this person hear from this string?

: 17-35 P A string clamped at both ends vibrates in four sections with a frequency of 640 Hz. The tension in the string is 600 N and its linear density is 0.5 gm/m. *(a)* What is the length of the string? *(b)* If the amplitude of vibration is 0.02 m, write an expression for the displacement as a function of time.

: 17-36 P A steel piano wire 1.2 meters long is to have a fundamental frequency of 280 Hz when under a tension of 600 N. What diameter wire should be used?

: 17-37 P A standing wave is

$$y = 0.4 \sin \frac{\pi x}{4} \cos 80\,\pi t$$

(a) Write down two standing waves which when superimposed will result in this standing wave. *(b)* What is the distance between nodes of the standing wave? *(c)* What is the frequency of vibration? *(d)* What are the velocity and acceleration of a particle at the position $x = 5$ meters?

: 17-38 P For a series of strings of fixed length and tension, prove that the fundamental frequencies are inversely proportional to the wire diameters.

: 17-39 P A string is tied at one end and at the other end attached to a ring, which can slide up and down without friction on a vertical rod. What is the condition for standing waves to be set up in the string for a given tension, linear density, and length?

: 17-40 P A string of length L is fixed at both ends and stretched with tension T. Its midpoint is displaced a distance $d \ll L$ from equilibrium and released. What frequencies of standing waves can give rise to this displacement?

Mechanical Waves II

18

Compressional elastic waves through a solid, a liquid, or a gas are another type of wave. All the wave effects discussed for transverse waves on a string — superposition, reflection, standing waves, resonance — apply also to compressional waves. The only important difference is that for compressional waves, the wave disturbance is not a transverse displacement but a longitudinal displacement, a density variation, or a pressure variation.

18-1   Longitudinal Waves

Figure 18-1 shows a simple structure for propagating longitudinal elastic waves; several identical masses coupled by springs form a chain. Each spring follows Hooke's law for small elongations or compressions. This arrangement is the prototype of a compressional wave through any deformable elastic medium.

When any one mass is displaced longitudinally, the springs attached to it are stretched or compressed; these deformations produce a force not only on the displaced mass but also on the neighboring masses. Thus, neighboring masses are set in motion, and a compressional disturbance is propagated along the chain. Each mass undergoes, in turn, a *longitudinal* displacement and then returns to its equilibrium position, as a *longitudinal*, or a *compressional*, wave

Figure 18-1. *A simple model for the medium for longitudinal waves.*

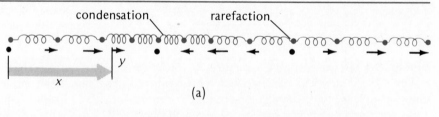

travels along the chain. The essential condition for the existence of longitudinal waves is that the *medium,* in this case the chain of masses and springs, possess *inertia* and be *elastically deformable.* Energy is transported longitudinally — along the line of the mass's motion — as the kinetic energy of the masses and the elastic potential energy of the deformed springs. Furthermore, a traveling longitudinal wave carries linear momentum *along* the direction of wave propagation. (A transverse wave on a stretched string has transverse linear momentum; the momentum is at right angles to the direction of energy propagation.)

Suppose that one end of the chain in Figure 18-1 is moved longitudinally in simple harmonic motion. Then all the masses will eventually be set in simple harmonic motion, with the same amplitude and frequency as those of the source. Each oscillating mass lags in phase relative to the source; the phase lag at each location is proportional to its distance from the source. How can we represent such a longitudinal wave graphically and describe it analytically?

As before, coordinate x gives the equilibrium location of each mass. We now use y to represent the *longitudinal* displacement of a particle from its equilibrium position along the x axis, the direction of wave propagation. Figure 18-2(a) shows the y displacements of the particles along the chain, and Figure 18-2(b) is a plot of y against x, both for a sinusoidal wave. Just as before, the shape of the function $y(x)$ for any given time t is a sinusoidal wave. This wave shape travels to the right at the wave speed c, and we can represent it analytically by any of the relations, (17-4), (17-6), or (17-7), that describe a traveling sinusoidal wave. For example, we can write

$$y = A \sin (\omega t - kx) \tag{18-1}$$

where A is the longitudinal oscillation amplitude, $\omega = 2\pi f$, and $k = 2\pi/\lambda$. As before, the wavelength λ is the distance between adjoining oscillators having the same phase, and the wave speed is $c = f\lambda$.

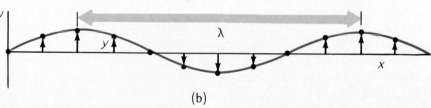

condensation   rarefaction

(a)

Figure 18-2. *(a) A sinusoidal longitudinal wave. (b) The corresponding wave shape, with displacement from equilibrium plotted as a function of distance.*

λ

(b)

Figure 18-2(b), which graphs the wave shape $y(x)$, is not a snapshot of the longitudinal wave. We may, however, in describing a longitudinal wave use exactly the same mathematical expressions as for a transverse wave if we recognize that the particle displacement y becomes a longitudinal displacement. The regions in Figure 18-2(a) (a snapshot) in which the particles are crowded together, the regions of maximum density, are known as *condensations.* Regions in which the particles have their greatest relative separation, or regions of minimum density, are called *rarefactions*. Rarefactions and condensations travel in the direction of wave propagation with the wave speed c. In fact, a longitudinal wave may be described as a disturbance for which the propagated property, or wave function, is the density variation of the medium.

Our prototypical model for a longitudinal wave consists of discrete masses coupled by identical springs. A still simpler physical arrangement for longitudinal waves is a single long stretched helical spring. Here the mass is distributed continuously throughout the length of the spring, and displacement y is the shift from its equilibrium position of any turn of the spring.

18-2   Superposition, Reflection, and Standing Waves

For elastic waves, superposition, reflection, and standing waves are very much like the corresponding effects for transverse waves on a stretched string. The difference is the wave property, which for elastic waves is a longitudinal displacement, or a pressure or density change, rather than a transverse displacement. Two compressional waves interfere constructively when the density or pressure is enhanced by superposition of the separate changes; and the opposite is true for destructive interference.

Consider the boundary conditions for the reflection of a compressional wave. Suppose that the deformable medium is terminated by an infinitely rigid and massive second medium. When a compressional wave reaches the boundary and is reflected, particles at the boundary cannot be displaced longitudinally. The longitudinal *displacement y* must be *zero* at the boundary; a *displacement node* occurs at the boundary between a deformable medium and a nondeformable medium. Now if y is zero at the boundary, then the pressure difference Δp must be a maximum (or a minimum) there: a pressure *antinode* exists at the boundary. This follows since the particles will tend to pile up at the boundary (or move away from it), thereby producing a high density and pressure (or low density and pressure) at this point. On the other hand, if the deformable medium is terminated with a less dense medium, the displacement y will be a maximum, or antinode, and the pressure Δp will be zero, or a node, at this boundary. As with transverse waves, a compressional wave undergoes a 180° change in phase when reflected from a boundary leading to a second medium in which the wave speed is less, whereas no phase change occurs for waves traveling in the reverse direction, from the second to the first medium.

Standing waves are produced in a medium whenever waves of the same wavelength travel in opposite directions. The resultant disturbance has alternate nodes and antinodes, with adjacent nodes (and adjacent antinodes) separated by half-wavelengths.

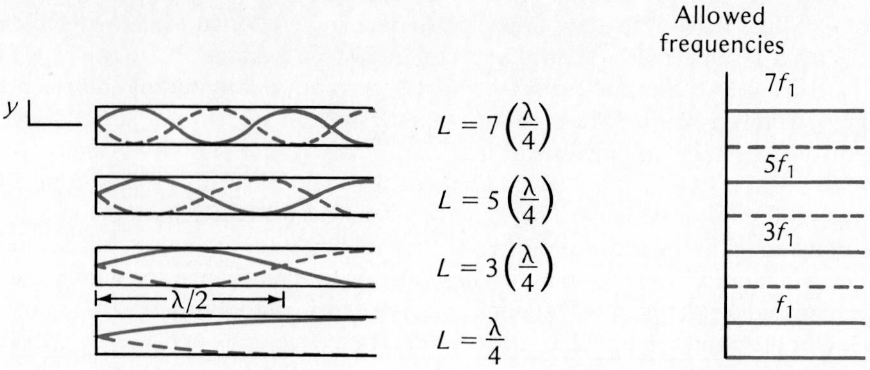

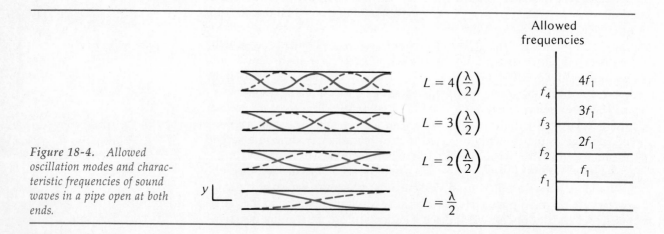

Figure 18-3. *Allowed oscillation modes and characteristic frequencies for sound waves in a pipe closed at one end and open at the other. Note that the standing wave patterns show the longitudinal displacements, rather than the pressure variations.*

$$L = 7\left(\frac{\lambda}{4}\right)$$

$$L = 5\left(\frac{\lambda}{4}\right)$$

$$L = 3\left(\frac{\lambda}{4}\right)$$

$$L = \frac{\lambda}{4}$$

Allowed frequencies

$7f_1$

$5f_1$

$3f_1$

f_1

Example 18-1. A sound wave is propagated through a pipe closed at one end. What are the allowed oscillation modes and characteristic frequencies?

The boundary conditions are these. At the closed end there is a displacement node or pressure antinode. The allowed oscillation modes, or standing-wave patterns, are as shown in Figure 18-3 (the actual displacement antinode lies somewhat beyond the open end of the pipe). Note particularly the quantity plotted here; it is the longitudinal *particle displacement y.* If one were to plot Δp, the antinodes and nodes would be reversed.

The length L of the tube is an odd multiple of $\frac{1}{4}\lambda$; in symbols, $L = n(\lambda/4)$, where $n = 1,\ 3,\ 5,\ \ldots$. The allowed frequencies are $f = c/\lambda = n(c/4L) = nf_1 = f_1,\ 3f_1,$ $5f_1,\ \ldots$. The only overtones are odd harmonics of the fundamental frequency f_1. As with transverse waves, two or more characteristic oscillation modes may exist simultaneously, and any disturbance persisting in time must consist of a superposition of allowed nodes.

If the length of the closed pipe is 0.50 m, roughly the length of a clarinet, the fundamental frequency is

$$f_1 = \frac{c}{4L} = \frac{344 \text{ m/s}}{4(0.50 \text{ m})} = 172 \text{ Hz}$$

Allowed frequencies

$$L = 4\left(\frac{\lambda}{2}\right)$$

$$L = 3\left(\frac{\lambda}{2}\right)$$

$$L = 2\left(\frac{\lambda}{2}\right)$$

$$L = \frac{\lambda}{2}$$

f_4 $4f_1$

f_3 $3f_1$

f_2 $2f_1$

f_1 f_1

Figure 18-4. *Allowed oscillation modes and characteristic frequencies of sound waves in a pipe open at both ends.*

Example 18-2.   A sound wave is propagated through a pipe open at both ends. What are its allowed oscillation modes and characteristic frequencies?

The boundary conditions are now these: a displacement antinode (or a pressure node) at each of the two open ends. Then the allowed oscillation modes are as shown in Figure 18-4. The tube length is always an integral multiple of $\frac{1}{2}\lambda$; the allowed frequencies consist of all harmonics of the fundamental.

18-3   Sound and Acoustics

Mechanical vibrations over the frequency range from 20 to 20,000 Hz (a factor of 10^3), with an intensity (power per unit transverse area) lying between 10^{-16} and 10^{-4} W/cm^2, are perceived by a typical human ear as sound. (The eye detects electromagnetic vibrations only over a frequency range of a factor of about 2, or one octave.)

Roughly speaking, nonperiodic variations in the air pressure correspond to noise, whereas oscillations at frequencies in the ratios of simple integers are recognized as musical tones. The first 16 harmonics are displayed in musical notation in Figure 18-5, starting with C below the bass clef as fundamental.* One tone is an octave above another if its fundamental frequency is up by a factor of 2; a fifth and its tonic have a frequency ratio of 3 : 2; a major triad (do, mi, sol) involves three tones with frequency ratios 4 : 5 : 6. In the well-tempered musical scale, all 12 half-tones within one octave are equally spaced, so that there is a frequency ratio between adjoining half-tones of $(2)^{1/12}$. Most musical instruments are based on the ability of strings or air columns (for woodwind and brass instruments) to excite various harmonics.

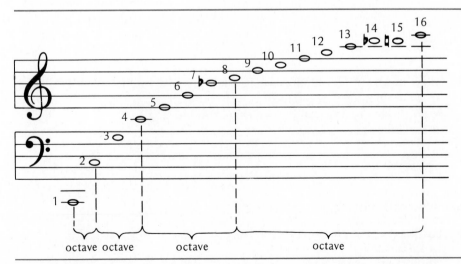

Figure 18-5.   The first 16 harmonics in musical notation.

* Strictly, harmonics 7 and 14 are somewhat flat for a B♭; harmonic 11 is roughly midway between F♮ and F♯.

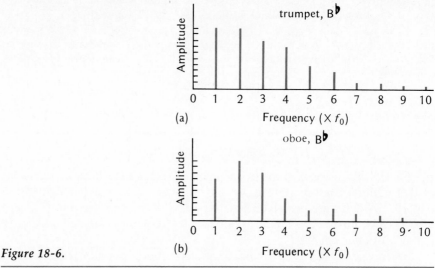

Figure 18-6.

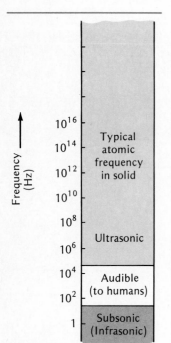

Figure 18-7. *The acoustic spectrum.*

By the *pitch* of a musical tone is meant the lowest, or fundamental, frequency. The relation of perceived pitch to fundamental frequency is not precise, however. The tonal *quality*, by which the ear can distinguish, for example, between an oboe and a trumpet playing the same note, is a measure of the relative amplitudes of the overtones, most often harmonics, excited simultaneously with the fundamental oscillation. See Figure 18-6.

The acoustic spectrum is shown in Figure 18-7. Frequencies alone are given; the wavelengths depend on the medium (the wave speed) through which the elastic vibrations travel.

The relation between the intensity I of a sound wave, the power of the sound wave per unit transverse area, and its *sound level* β is defined to be

$$\beta \text{ (in decibels)} \equiv 10 \log \frac{I}{I_0} \tag{18-2}$$

where the logarithm is to the base 10. Here I_0 has the value 10^{-12} W/m^2; this particular reference for intensity is chosen because it is close to the faintest sound audible to a typical human ear. Sound level β in (18-2) is specified in

Table 18-1

SOURCE	SOUND LEVEL β (db)	INTENSITY I (W/m^2)
Saturn rocket (at 50 m)	200	10^8
Threshold of pain	120	1
Rock concert (close to the speakers)	100	10^{-2}
City street with traffic	70	10^{-5}
Ordinary conversation	60	10^{-6}
Whisper	20	10^{-10}
Rustling leaves	10	10^{-11}
Threshold of hearing	0	10^{-12}

units of *decibels* (abbreviated db and named for the inventor of the telephone, Alexander Graham Bell). The relation between sound level β and intensity I is taken to be logarithmic because this corresponds to the actual behavior to the ear; the sensation of loudness transmitted from the ear to the brain is proportional to the logarithm of the sound intensity.

The faintest sound corresponds to $I = I_0 = 10^{-12}\,\text{W/m}^2$. Then, from (18-2), we see that $\beta = 0$ db. At the other extreme, a sound with $I = 1\,\text{W/m}^2$ is beginning to be painfully loud, and (18-2) gives $\beta = 120$ db. The typical sound levels of other sources is given in Table 18-1.

Example 18-3.   The intensity of a certain sound is doubled. By how many decibels does the sound level increase?

Let the two intensity levels be I_1 and I_2. Then we can write

$$\log\left(\frac{I_2}{I_1}\right) = \log 2$$

Therefore,

$$\beta_2 - \beta_1 = 10 \log 2 = 3\ \text{db}$$

With $I_2 = 2I_1$, the sound level difference is 3 db. We also see at once that increasing the intensity by a factor of 10 increases the sound level by 10 db.

18-4   Wave Fronts, Rays, and Huygens's Principle

The simplest configuration for wave propagation in two or three dimensions is shown in Figure 18-8. Here the wave source is spread over a plane, like an oscillating flat diaphragm, and the waves progress to the right in a single direction. The direction in which waves radiate from the source is indicated by straight lines with arrows to show propagation direction, or *rays*. A *wave front* is defined as a surface on which all points have the same phase of oscillation. The wave fronts of a plane wave source consist of planes at right angles to the rays. The perpendicular distance between adjoining wave fronts in the same phase is the wavelength. The wave fronts are surfaces of constant phase that advance at the phase speed, or wave speed, *c*. A line source generates cylindrical wave fronts centered about it, as shown in Figure 18-9. Rays again intercept wave fronts at right angles. Figure 18-10 shows spherical wave fronts in three dimensions emanating from a point source. In general, rays are always perpendicular to the associated wave fronts.

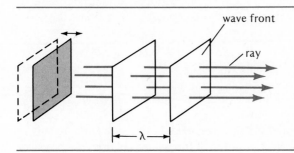

Figure 18-8.   *Plane wave source and plane wave fronts.*

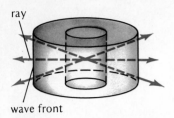

ray

wave front

Figure 18-9. *Line source and cylindrical wave fronts.*

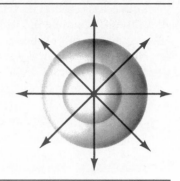

Figure 18-10. *Point source and spherical wave fronts. (Rays are shown only for one slice.)*

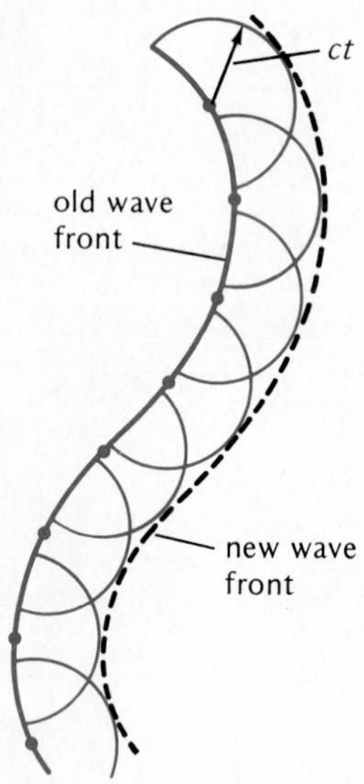

ct

old wave front

new wave front

Figure 18-11. *Huygens's construction for advancing wave fronts.*

Any small portion of a cylindrical or spherical surface may be regarded as a plane; over a very small solid angle, a cylindrical or spherical wave front closely approximates a plane wave front.

Given one wave front for a progressing wave, how do you find a future wave front? A remarkably simple geometrical procedure, *Huygens's principle,* can be used to chart the progress of a wave front through a medium. Devised by C. Huygens (1629–1695) in 1678, this principle asserts that each point on an advancing wave front may be regarded as a new point source generating spherical *Huygens wavelets* in the forward direction of wave propagation. To find the wave front at a time t later, simply draw circular arcs of radius ct centered at points along the wave front; the new wave front at time t is merely the envelope of these wavelets. See Figure 18-11. Thus, a plane wave front generates another plane wave front in a uniform medium; a spherical wave front generates another spherical wave front, of larger radius, about the same point source.

Huygens's construction is a geometrical procedure, not a physical method. Clearly, if one is to find a wave front at some *future* time, one must draw the envelope of the wavelets along the *leading,* rather than the trailing, side of the wave front. Huygens's method gives only the possible wave fronts at some future time; it does not give the distribution of energy over the wave fronts.

18-5   Intensity Variation with Distance from Source

The *intensity* of a wave is defined as the energy passing per unit time through a unit cross-sectional area. Intensity I is then the power P per unit transverse surface area S:

$$I = \frac{P}{S} \tag{18-3}$$

For a sound wave, intensity is related to loudness [see (18-2)]; for a light wave, intensity is a measure of brightness.

In every type of wave generated by a sinusoidal oscillator, the intensity is proportional to the square of the amplitude of the wave function, the physical property (Δp for an elastic wave) characterizing the wave disturbance. This result, proved in Section 17-7, corresponds to the fact that the total energy $\frac{1}{2}kA^2$ of a simple harmonic oscillator is proportional to A^2, the square of the amplitude.

How does the intensity of a wave vary with distance? The results for the three simplest geometrical configurations shown in Figures 18-8, 18-9, and 18-10 are these:

- *Plane source* produces *plane waves* with

$$I = \text{constant} \qquad \textbf{(18-4a)}$$

- *Line source* produces *cylindrical waves* with

$$I \propto \frac{1}{r} \qquad \textbf{(18-4b)}$$

- *Point source* produces *spherical waves* with

$$I \propto \frac{1}{r^2} \qquad \textbf{(18-4c)}$$

The distance r is measured from the source to the location at which I is measured. For cylindrical waves, r is the radius of a cylindrical wave front; for spherical waves, r is the radius of a spherical wave front.

The variation of I with distance from the source comes directly from energy conservation and the way in which the energy of a wave is spread over a transverse area as the wave propagates. An oscillating plane source produces plane wave fronts (Figure 18-8). The same energy passes through every unit transverse area so that intensity remains unchanged as we move away from the source. For the cylindrical wave fronts shown in Figure 18-12(a), surface area S is proportional to the cylinder's circumference and therefore proportional to r. Since $S \propto r$, we have for constant power output P that $I = P/S \propto 1/r$.

The energy is diluted more severely in the expanding spherical wave fronts from a point source. See Figure 18-12(b). The surface area of a sphere is $S = 4\pi r^2 \propto r^2$. Therefore, with constant power P, we have that $I = P/S \propto 1/r^2$, an inverse-square decrease with distance r.

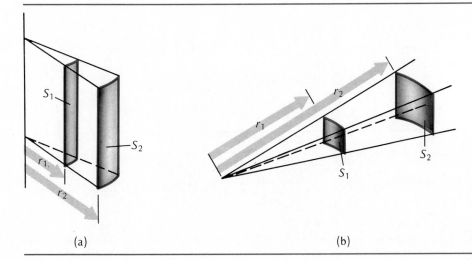

(a)   (b)

Figure 18-12. (a) Angular segment for cylindrical wave fronts. (b) Angular segment for spherical wave fronts.

Example 18-4. A point source radiates isotropically (same in all directions) at the rate of 10 kW. What is the intensity of the radiation 10 m from the source?

The intensity I at a radius r from a point source is equal to the power P through a spherical area $S = 4\pi r^2$:

$$I = \frac{P}{S} = \frac{(1.0 \times 10^4 \text{ W})}{4\pi(10 \text{ m})^2} = 8.0 \text{ W/m}^2$$

18-6 Beats

When two tones of nearly the same pitch are sounded simultaneously, the ear senses a pulsation in loudness known as *beats*. For example, two sinusoidal oscillations have frequencies f_1 and f_2, and for simplicity, the same amplitude A. Then the resultant displacement at some location is

$$y = A \sin 2\pi f_1 t + A \sin 2\pi f_2 t$$

Using the trigonometric identity

$$\sin a + \sin b = 2 \cos \left(\frac{a - b}{2}\right) \sin \left(\frac{a + b}{2}\right)$$

we can rewrite the relation above as

$$y = \left[2A \cos \frac{2\pi t(f_1 - f_2)}{2} \right] \sin 2\pi t \left(\frac{f_1 + f_2}{2}\right) \tag{18-5}$$

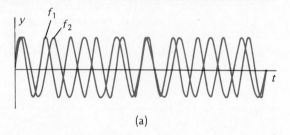

(a)

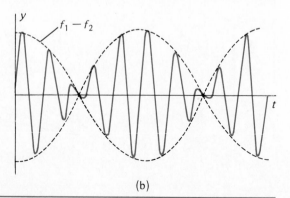

Figure 18-13.   (b)

The two factors in (18-5) have the following meanings:

- sin $2\pi t(f_1 + f_2)/2$ represents sinusoidal oscillation at the *average* frequency, $(f_1 + f_2)/2$.
- The amplitude, or envelope, of the resultant oscillation is equal to $2A \cos 2\pi t(f_1 - f_2)/2$. It oscillates at the frequency $(f_1 - f_2)/2$, and alternates between zero and the maximum amplitude of $2A$ at *twice* that rate. The frequency at which the envelope oscillates, or *beat frequency*, is then $|f_1 - f_2|$.

The separate oscillations and their resultant are shown as a function of time in Figure 18-13.

18-7   Doppler Effect

Everyone has heard the change in pitch of an automobile horn that occurs when the listener, or the horn, or both are in motion relative to the medium (air) through which the sound is propagated. This wave phenomenon, first explained by C. J. Doppler (1803–1853), is known as the *Doppler effect.*

We consider only the simple case in which the observer, traveling at speed v_0 relative to a uniform, isotropic medium, and the source, traveling at v_s relative to air, move along the same straight line. Figure 18-14 shows the circular wave fronts emitted by the source between time $t = 0$ and later time t. Because the source moves to the right, the centers of the wave fronts do not coincide. The wave fronts are crowded to the right of the source and stretched out on the left. Once launched from the source, every circular wave front expands outward through the medium at the same speed c; the waves do not "remember" that they were emitted from a moving source.

The source at frequency f_s was at the center of the largest circle at time $t = 0$. After a time interval t, the source has advanced to the right a distance $v_s t$, and it has generated $f_s t$ waves. During this same time, a wave front has covered a distance ct. Therefore, the total distance between the source and the observer

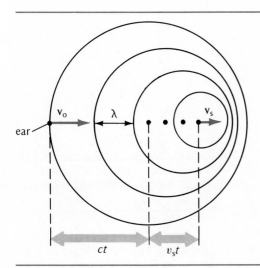

Figure 18-14.   *A source moving at velocity v_s generates wave fronts. An observer has velocity v_o.*

at time t is $ct + v_st$, as shown in Figure 18-14. The wavelength λ in the region between the observer and the source is simply the total distance $ct + v_st$ between them divided by the total number of waves f_st lying in the distance:

$$\lambda = \frac{ct + v_st}{f_st} = \frac{c + v_s}{f_s}$$

What is the frequency f_o perceived by the moving observer? In the region between observer and source, the waves move left relative to the medium at speed c; the observer moves right, also relative to the medium, at speed v_o. The speed of the waves relative to the moving observer is then $c + v_o$, by the simple rule for adding relative velocities (Section 4-6). The frequency f_o perceived by the observer is simply the wave speed, relative to the observer of $c + v_o$ divided by the wavelength λ as given in the relation above:

$$f_o = \frac{c + v_o}{\lambda} = \frac{c + v_o}{c + v_s} f_s$$

or in a form that is easier to remember,

$$\frac{f_o}{c + v_o} = \frac{f_s}{c + v_s} \tag{18-6}$$

This equation implies a convention for the signs of velocities: $\mathbf{v_o}$ and $\mathbf{v_s}$ are in the same direction, and the observer's ear is headed toward the source. For other directions of motion, one simply reverses the sign of $\mathbf{v_o}$ or $\mathbf{v_s}$ or both in (18-6). Note the two situations for which the perceived and true frequencies are alike, with $f_o = f_s$:

- Both source and observer are at rest; $v_s = 0$ and $v_o = 0$.
- Source and observer are moving with the same velocity; $v_s = v_o$.

The general Doppler relation, (18-6), can be written in a useful, simpler form when $v_o \ll c$ and $v_s \ll c$. We represent the change in frequency, or Doppler shift, by $\Delta f = f_o - f_s$. Then (18-6) becomes

$$\frac{f_s + \Delta f}{f_s} = \frac{c + v_o}{c + v_s} = \frac{1 + (v_o/c)}{1 + (v_s/c)}$$

$$1 + \frac{\Delta f}{f_s} = \left(1 + \frac{v_o}{c}\right)\left(1 - \frac{v_s}{c}\right)$$

Or more succinctly,

$$\frac{\Delta f}{f} = \pm \frac{v}{c} \qquad \text{for } v \ll c \tag{18-7}$$

where, it must now be understood, v can stand for either v_o or v_s. The plus sign applies when source and observer approach one another and the frequency is increased; the minus sign applies for relative recession and a drop in frequency. In every case, the fractional change in frequency $\Delta f/f$ equals the speed of observer or source divided by the speed of waves through the medium.

The Doppler effect occurs also for light or other forms of electromagnetic waves; indeed, Christian Doppler himself predicted a change in the color of an object in motion. Special considerations arise, however, in the propagation of electromagnetic waves; in particular, all observers, whatever their state of motion, measure the same speed for light through a vacuum, quite apart from the state of motion of the source. A brief account of the Doppler effect for light is given in Chapter 35. The approximate relation (18-7) applies for electromagnetic waves only when the source or the observer or both have a speed much less than that of light.

Example 18-5. A police Doppler radar unit directs a continuous microwave radio beam at an automobile presumed to be speeding down a highway. The beam reflected to the police unit is shifted in frequency by the Doppler effect. The echo beam is mixed with a portion of the outgoing beam and beats are produced. The radar unit indicates that the echo signal has a lower frequency than that of the transmitted signal by 1.90 kHz. The radar unit operates at 9.375 GHz (G $\equiv$ *giga* $\equiv 10^9$). What is the speed of the car?

We must first recognize that when a continuous wave is reflected from a moving target, the Doppler effect is apparent *twice*—first when the incident wave strikes the reflector and then when the reflected wave leaves the reflector:

- The number of wave crests per unit time that strike a reflector moving at speed v is simply the frequency f_r that would be measured by an observer traveling with the reflector. For a reflector retreating from the transmitter, f_r is less than the transmitted frequency f_t by $\Delta f_1 = f_t - f_r$, where from (18-7),

$$\frac{\Delta f_1}{f_t} = \frac{v}{c}$$

- The frequency of the beam emitted from the moving reflector is f_r. But frequency f_o observed back at the transmitting unit is, for a retreating reflector, less than f_r by $\Delta f_2 = f_r - f_o$, where again from (18-7),

$$\frac{\Delta f_2}{f_r} = \frac{v}{c}$$

The automobile's speed v is far less than the speed of the radar beam, $c = 3.00 \times 10^8$ m/s. Consequently, each frequency shift is relatively small, so that $f_r \simeq f_t$ and $\Delta f_1 \simeq \Delta f_2$. The overall frequency shift is $2\Delta f$, where

$$2\frac{\Delta f}{f} = 2\frac{v}{c}$$

so that the speed of the retreating car is*

$$v = \left(\frac{2\Delta f}{f}\right)\left(\frac{c}{2}\right)$$

Using the numbers given above, we have

$$v = \frac{(1.90 \times 10^3 \text{ Hz})}{(9.375 \times 10^9 \text{ Hz})}\left(\frac{3.00 \times 10^8 \text{ m/s}}{2}\right) = 30.4 \text{ m/s} \simeq 68 \text{ mi/h}$$

* This relation applies only when the incident and reflected radar beams are along the line in which the car is heading.

Summary

Definitions

Longitudinal, or compressional, wave: particle displacement parallel to the direction of wave propagation; propagated wave property is density or pressure change.

Condensation: region of maximum density.

Rarefaction: region of minimum density.

Intensity I of a wave: the power P per unit transverse area S,

$$I \equiv P/S \qquad (18\text{-}3)$$

Loudness β in *decibels* for a sound of intensity I:

$$\beta \text{ (in decibels)} = 10 \log \frac{I}{I_0} \quad (18\text{-}2)$$

where $I_0 = 10^{-12} \text{ W/m}^2$.

Wave front: surface on which all points of a traveling wave have the same phase.

Ray: directed line segment giving the local direction of wave propagation.

Beat: periodic pulsation in loudness from slightly different frequencies f_1 and f_2; beat frequency $= |f_1 - f_2|$.

Important Results

Huygens's principle, a procedure for finding future wave front: Take envelope on the forward side of the Huygens spherical wavelets at all points of the wave front.

Intensity I variation with distance r from source:

$$\left.\begin{array}{l}\text{Plane source, plane wave fronts, } I = \text{constant} \\ \text{Line source, cylindrical wave fronts, } I \propto 1/r \\ \text{Point source, spherical wave fronts, } I \propto 1/r^2\end{array}\right\} \quad (18\text{-}4)$$

Doppler Effect:

$f_s =$ source frequency
$v_s =$ source velocity (relative to the medium)
$v_o =$ observer velocity
$c =$ sound velocity

$$\frac{f_o}{c + v_o} = \frac{f_s}{c + v_s} \quad (18\text{-}6)$$

(v_o and v_s are in the *same* direction with the observer headed toward the source.)

For $v_o \ll c$ or $v_s \ll c$, the fractional Doppler shift in frequency $\Delta f / f$ is

$$\frac{\Delta f}{f} = \pm \frac{v}{c} \quad (18\text{-}7)$$

Problems and Questions

Section 18-1   Longitudinal Waves

· **18-1 Q** When a sound wave goes from air into water, which of the following is unchanged?
(A) Frequency
(B) Wavelength
(C) Velocity
(D) Amplitude
(E) None of the above, i.e. they all change.

· **18-2 P** A depth finder used on ships sends out pulses of sound that are reflected back from the ocean floor. What is the depth at a point where the pulse comes back after 0.6 seconds, assuming the velocity of sound in sea water to be 1420 m/s?

· **18-3 Q** Which of the following is the best example of a pressure wave?
(A) A lightning bolt.
(B) A vibrating string.
(C) An ocean wave.
(D) Green light.
(E) Drifting sand.
(F) A sound wave.
(G) None of the above are pressure waves.

· **18-4 P** Whales can communicate with sound over long distances. The speed of sound in water is 1450 m/s. What is the time delay between the emission of a sound by one whale and its reception by another whale 10 km away?

· **18-5 P** Under conditions where the velocity of sound in air is 340 m/s, a bat hears an echo 200 ms after emitting a sound pulse. How far away was the object that reflected the sound?

: **18-6 P** The *Mach number N* gives the speed of an object, such as an aircraft, relative to the speed of sound in the medium through which the object travels. Show that the angle θ of a shock wave produced by an object with a Mach number N is given by $\sin^{-1}(1/N)$.

Section 18-2   Superposition, Reflection, and Standing Waves

: **18-7 Q** What is the purpose of the valves in a trumpet or the slide in a trombone or the holes in a flute, oboe, clarinet, or bassoon?

· **18-8 P** What frequency standing waves can be set up in an open organ pipe at one end, if the pipe's length is 3 meters? ($v_{so} = 340$ m/s.)

: **18-9 P** The human ear canal will resonate as a pipe with one closed end and one open end at a frequency of about 2000 Hz. (*a*) Estimate the length of the ear canal. (*b*) What effect would there be on the resonant frequency if your ear were filled with water, as might be the case when swimming. (*c*) What effect might this have on your hearing?

· **18-10 P** A tuning fork of frequency 440 Hz is held above a vertical tube partly filled with water. For what lengths of the air column in the tube will resonances occur if the speed of sound in air is 344 m/s?

: **18-11 P** A pipe open at both ends resonates at 840 Hz. When one end is closed it oscillates at 210 Hz. What is the minimum pipe length satisfying these conditions? (Note that you do not know which harmonics these are.)

: **18-12 P** The fundamental note of an open organ pipe is middle D, 293.7 Hz. The third harmonic of a different closed pipe has the same frequency. What are the lengths of the two pipes if the velocity of sound in air is 340 m/s?

Section 18-3 Sound and Acoustics

· **18-13 Q** When sound traveling in air strikes a sheet of glass,
(A) no sound will enter the glass.
(B) the wavelength of the sound in the air will be the same as the wavelength of the sound in the glass.
(C) the wavelength of the sound in the air will be greater than the wavelength of the sound in the glass.
(D) the wavelength of the sound in the air will be less than the wavelength of the sound in the glass.
(E) the wavelength of the sound in the glass will depend on the thickness of the sheet of glass.

· **18-14 Q** Why does a given note sound different when played on different instruments?

· **18-15 P** If the amplitude of oscillation of a loudspeaker diaphragm is doubled, what effect will this have on the sound power radiated?

: **18-16 P** A 12-in $33\frac{1}{3}$-rpm phonograph record can reproduce sound with frequencies of up to 16 kHz. (a) What is the maximum separation distance in the record groove (near the beginning of the record) between the oscillations corresponding to the highest frequency? (b) What is the corresponding separation distance on a magnetic tape played at the speed of $1\frac{7}{8}$ in/s?

: **18-17 P** A bolt of lightning can act as a line source of sound. If the sound intensity at a distance of 500 meters from a clap of thunder is 10^{-10} W/m^2, what would the intensity be at a distance of 2000 meters from the source, in a direction to the line of the lightning bolt?

· **18-18 P** What is the change in sound level when the sound intensity changes (a) by a factor of 4? (b) By a factor 8?

: **18-19 P** In normal conversation, the sound intensity level is 70 db at a distance of 1.5 meters. Estimate the total sound power emitted by the speaker.

: **18-20 P** The sound level 3 meters from a jackhammer is 110 db. Suppose that the sound is radiated uniformly over a hemisphere. How far away must one move to reduce the sound level to 60 db?

: **18-21 P** Two hundred watts of sound power is emitted uniformly in all forward directions from the speakers at a rock concert in the park. What is the intensity of the sound at a distance of 50 meters from the source, assuming minimal absorption in the air?

: **18-22 P** What would be the difference in sound intensity level in db between a 2 W speaker and a 40 W speaker when you are 4 meters from the speakers?

: **18-23 P** It is not uncommon for a rock group to use speakers that radiate 300 watts of sound power into a hemispherical region facing the audience. How far from such speakers must you sit in order that the sound intensity level not exceed the pain threshold of 120 db?

Section 18-5 Intensity Variation with Distance from Source

· **18-24 Q** One may readily deduce how sound intensity varies with distance from a uniformly radiating sound by making use of
(A) the relation $v = f\lambda$.
(B) Newton's laws of motion.
(C) the law of conservation of energy.
(D) the law of conservation of momentum.
(E) the fact that sound is a longitudinal wave.

: **18-25 P** The sound intensity 10 meters from a point source of sound is 10^{-4} W/m^2. How far from the source is the intensity 10^{-6} W/m^2?

: **18-26 P** A shell explodes at high elevation, and releases 2×10^7 J of energy in 0.1 s. If all of this energy appears as sound, and if the sound is absorbed at a rate of 7 db/km, what would be the sound intensity at a distance of 6 km from the explosion?

: **18-27 P** The intensity at the earth of electromagnetic radiation from the sun is 1.4 kW/m^2. What is the electromagnetic power output of the sun? (earth-sun distance $= 1.5 \times 10^{11}$ m)

: **18-28 P** How does the amplitude of a sound wave fall off with distance for (a) a point source and (b) a uniform line source?

: **18-29 P** What is the minimum power output of an isotropic point source of sound that will produce an intensity of 10^{-16} W/cm^2 (the minimum threshold of hearing) at the ear of a listener 1.0 km distant from the source?

: **18-30 P** A 100-gm billiard ball rolling 2 m/s strikes another ball and makes a 0.1 s click, heard with an intensity level of about 30 db at a distance of 2 meters. Estimate the fraction of the ball's energy lost as sound.

Section 18-6 Beats

· **18-31 P** Two orchestral instruments are slightly out of tune. One plays a fundamental frequency of 384.0 Hz; the other plays 384.5 Hz. What frequencies will be heard by a listener?

Section 18-7 Doppler Effect

· **18-32 Q** You are standing on a street corner and hear a police car coming toward you with its siren screaming. Which of the curves in Figure 18-15 best represents the frequency of the sound you hear as the car approaches you, whizzes past, and then vanishes down the street?

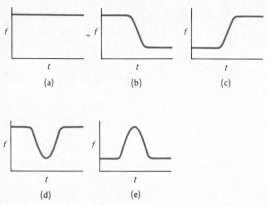

Figure 18-15. *Question 18-32.*

: **18-33 P** A person hears the siren of a police car as it approaches, passes, and moves away. The frequency heard varies from 272 Hz to 340 Hz. What is the speed of the car? What is the frequency heard by someone in the car? $v_{so} = 340$ m/s.

: **18-34 P** A police radar unit uses a frequency of 10.00 MHz. What frequency will be received when the signal is reflected from a car approaching at 100 km/h?

: **18-35 P** A 10 GHz radar signal is shifted in frequency by 60 kHz when reflected from a receding missile. What is the velocity of the missile?

: **18-36 P** An earth satellite carries a radio transmitter operating at 100 MHz. What is the maximum observed frequency change by an observer on earth when the satellite is orbiting close to the earth's surface?

: **18-37 P** Two cars are traveling in the same direction at 25 m/s. One of the cars sounds a horn at 120 Hz. What frequency does a person in the other car hear if the velocity of sound is 340 m/s? Consider two cases: *(a)* the horn is in the front car; *(b)* the horn is in the back car.

: **18-38 Q** Will a Doppler effect be observed if the observer and the source of sound are moving at 90° to each other? Explain qualitatively.

: **18-39 Q** Will a Doppler effect be observed if the source and the observer are at rest with respect to each other, but the medium through which the sound is transmitted is moving? Explain.

Supplementary Problems

18-40 P The automatic rangefinder on the Polaroid SX-70 Sonar Land camera measures the time it takes for an ultrasonic pulse to travel out from the camera to the subject and back again to the camera. When the button is pressed, a 420 kHz quartz crystal clock generates a chirp of 1.0 ms duration and broadcasts it outward. The receiver starts to listen for the echo pulse only after 0.6 ms have elapsed from the sending of the pulse. What is the time interval between transmission and reception of the pulse for a subject 3.0 m away?

18-41 P The horseshoe bat has a highly specialized echolocation system for hunting insects. Most bats use sonar and measure the time interval between the transmission and reception of a pulse. The horseshoe bat is unusual because it also uses the Doppler effect. The bat's ear is highly sensitive only to sound pulses with a frequency centered close to 83 kHz. If the bat is in motion (up to its maximum speed of 12 m/s) toward the target, it reduces the frequency of the transmitted pulse, so that the reflected pulse received at the ear has a frequency close to 83 kHz. (The feedback system is somewhat sluggish; it takes 0.34 ms for the bat to correct for a 1 kHz shift. Because of this effect, the bat is sensitive both to frequency and amplitude modulation of the echo arising from wing motion by the prey.) *(a)* What frequency shift is required for the bat approaching stationary prey at 10 m/s? *(b)* What roughly is the minimum size of the prey that can be detected by the bat? [See G. Neuweiler, *Physics Today* **33**, 34 (August, 1980)]

18-42 P In the Labor Day hurricane of 1935, the "most powerful" on record in the U. S. A., the barometric pressure in the eye of the hurricane was only 892 millibars. Suppose that such a hurricane passes over the top of a building with a roof area of 100 m². What is the upward force on the roof?

18-43 P The performer Komar can lie on a bed of nails while a small truck weighing 4.8150 kN (0.541 tons) rolls onto a plywood sheet resting on Komar's chest and abdomen. Adjacent nails in the bed are separated from one another by about 1/4 inch. Make reasonable assumptions about the area of Komar's body that rests on the nails and show that, although it may not be altogether pleasant, the feat does not necessarily involve the penetration of skin by nail. (Komar holds the Guinness record for lying between two beds of nails with a load of 7.306 kN (1642.5 lb) on top of him. After this feat in 1977, the category was retired.)

Thermal Properties of an Ideal Gas, Macroscopic View   19

19-1   Temperature and the Zeroth Law of Thermodynamics

The most rudimentary conception of temperature arises from our sense of touch. We can tell by touch, although only qualitatively and roughly, whether two bodies have the same temperature or whether one is relatively hot or cold. But we must have a more precise and quantitative means for asserting that two bodies have the *same* temperature and a quantitative measure of *difference* in temperature. That is, we need a thermometer.

Equality of temperature is established as follows. Suppose that we bring two bodies together, place them in contact, isolate them from external influences (that is, insulate them)—and wait. After enough time has elapsed, the bodies are said to be in *thermal equilibrium*, and the property the two bodies then have in common—whatever their differences in size, mass, material, and past history—is *temperature*. Two bodies in thermal equilibrium have the same temperature, by definition.

Now suppose that we have three bodies: vessel A, vessel B, and some thermometer. The thermometer is first placed in vessel A. After we have waited and these two objects have come to thermal equilibrium, we read their common temperature on the thermometer. (A thermometer reads, first of all, its own temperature; but the thermometer also gives the temperature of a body with which it is in equilibrium.) Suppose that the thermometer is then placed in vessel B and we read the very same final temperature as for vessel A. We know that the thermometer has the same temperature as that of both vessels A and B because it has achieved thermal equilibrium with each in turn. Do A and B have the same temperature? Are A and B in thermal equilibrium? *The zeroth law of thermodynamics* asserts that they are.* That is, *if bodies A and B are separately in thermal equilibrium with a third body, C, then A and B are in thermal equilibrium with each other.*

19-2   Thermometry

A physicist may define length as the quantity that one measures with a meterstick, and time as the quantity that one reads on a clock. Similarly, we must first say that temperature is simply what one reads on a thermometer. But what is a thermometer? A thermometer is any device having some measurable physical property that changes with the degree of relative hotness or coldness of the body. The change in the volume of a liquid with temperature, for example, is the basis of the common liquid-in-glass thermometer. We shall consider this type of thermometer in some detail, to see how the temperature scales are defined and the thermometers calibrated, and also to see some important problems in thermometry that apply to all thermometers.

Suppose that our liquid-in-glass thermometer consists of a spherical bulb of glass filled to the brim with mercury. If the temperature is then raised, the mercury expands more rapidly than the glass container. Some of the liquid overflows. In principle, one can read the temperature by measuring the volume of mercury overflow. It is simpler, however, to attach a cylindrical stem of uniform cross section to the bulb. Then, when mercury overflows from the bulb into the stem, one can measure this overflow volume indirectly by noting the level of mercury along the stem.

Now we must calibrate the thermometer. The thermometer is placed in a mixture of ice and water under standard atmospheric pressure. We mark the position of the meniscus and label it "0." By definition, the *ice point* at standard pressure has a temperature of 0° on the *Celsius* temperature scale. The thermometer is then brought into thermal equilibrium with a mixture of water vapor and boiling water, again under standard atmospheric pressure. We make a second mark at the new position of the meniscus and label it 100°C. By definition, the steam point of water at standard pressure is 100 degrees Celsius. The calibration is completed when we add *uniformly* spaced graduations between 0 and 100°C, so that there are 100 graduations (centigrade) between the two chosen *fixed points*. See Figure 19-1. We can extend the scale by adding graduations of the same size below the 0° and above the 100° marks.

In the familiar *Fahrenheit* scale of temperature, the ice point is defined as 32°F and the steam point as 212°F. Thus, 180 graduations separate the two fixed points on the Fahrenheit scale, as against 100 graduations on the Celsius scale. The temperature readings are different at the fixed points on the two scales; furthermore, the sizes of the degrees differ. As Figure 19-1 shows, any

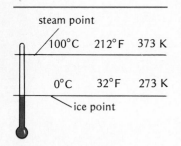

steam point
100°C   212°F   373 K

0°C   32°F   273 K
ice point

Figure 19-1. Definition of the fixed points for the Celsius, Fahrenheit, and Kelvin temperature scales.

---

* The zeroth law of thermodynamics, which is fundamental to all thermodynamics, is so named because it has priority over the first and second laws of thermodynamics and because it was recognized as a law only after the first and second laws had been assigned their numbers.

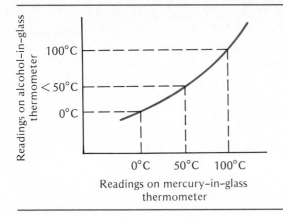

Figure 19-2. *Readings of a uniformly graduated alcohol-in-glass thermometer against the corresponding readings for a mercury-in-glass thermometer. (The departure from linearity is greatly exaggerated.)*

temperature t_C on the Celsius scale is related to the same temperature t_F registered on the Fahrenheit scale, by

$$\frac{t_C}{100} = \frac{t_F - 32}{180} \tag{19-1}$$

Although there is no upper limit on temperature, the lowest possible temperature is $-273.15\,°C$.*

How do we know that the mercury in our thermometer expands *uniformly* between 0 and 100°C? We don't know—in fact, we can't know—since we can test for uniform thermal expansion only with a thermometer. Suppose then that we construct two thermometers using different liquids, one with mercury and the second with alcohol (colored with a pink dye). We calibrate this second thermometer in the same fashion as the first, again using the two fixed points of water and marking uniformly spaced graduations.

How do the readings on the two thermometers compare when we bring both thermometers into thermal equilibrium with a third body? (From the zeroth law of thermodynamics, we know that they must then have the same temperature.) At the two fixed points, the thermometers agree perfectly. They must, of course. But if we compare the readings at other temperatures, we find that the two readings, although they may be nearly the same, do not agree precisely. See Figure 19-2. Thus, the behavior of the thermometer depends on the *thermometric substance* (here, mercury or alcohol). We cannot, at this stage, say which of the two thermometers is more nearly correct. We know only that mercury differs from alcohol in thermal expansion.

A thermometer may be based on a physical effect other than thermal expansion. Some physical properties that change with temperature and may therefore be used as the basis for a thermometer are:

• The electrical resistance (resistance thermometer).
• The electromotive force developed by a pair of unlike wires with their junctions at different temperatures (thermocouple).

---

* To distinguish temperature differences from temperatures, it is customary to represent temperature differences as Celsius degrees (C°) instead of degrees Celsius (°C). Thus 60°C differs from 40°C by 20 C°. Furthermore, 100 C° = 180 F°, although 100°C is not the same as 180°F. In general, the relative sizes of degrees on the two scales is given by 1 C° = $\frac{9}{5}$ F°.

Table 19-1

THERMOMETER AND PHYSICAL PROPERTY OR EFFECT	TEMPERATURE RANGE C°
Magnetic properties of paramagnetic salts	−273 to −272
Pressure of helium vapor in equilibrium with liquid helium	−272 to −269
Resistance thermometer (electrical resistance of substances)	−261 to 600
Thermocouple (electromotive force of two wires of dissimilar materials, the junctions maintained at different temperatures)	−250 to 1000
Liquid-in-glass thermometers (difference in expansion properties)	−196 to 500
Gas thermometers (constant pressure or constant volume, ideal gas of very low density)	>≈300
Optical pyrometer (visible electromagnetic radiation from body to be measured—the color of an incandescent filament is compared with that of the body)	>600

- The color of light emitted from a solid, or from a gas (pyrometer).
- The pressure of a gas (gas thermometer).

Table 19-1 lists several different types of thermometers, the physical property or effect on which each is based, and the temperature ranges over which each is most commonly used.

Constant-Volume Gas Thermometer Is there a thermometer whose readings do not depend on the thermometric substance? Yes, a *constant-volume gas thermometer* using a gas of very low density is one. A simple type, consisting merely of a tank of constant volume with a pressure gauge, is shown in Figure 19-3(a). The *pressure* of the gas is the physical quantity we use to register the temperature of the gas. The temperature of the gas must be high enough, and its pressure low enough, that the gas has a very low density and approaches infinite dilution. In other words, the gas must not be close to conditions for which it would condense into a liquid.

Our procedure for calibrating the gas thermometer is the same as what is used for the liquid-in-glass thermometer. With gas volume maintained constant, "0°C" is marked on the gauge for the ice point and "100°C" for the steam point. Equally spaced graduations are added, and these markings may again be extended beyond the 0°–100°C range.

We have not said what kind of low-density gas was in the constant-volume gas thermometer. Experiment shows that it does not matter! All very low density gases have, to a first approximation, exactly the same behavior. We come to this conclusion by comparing the temperature readings on two gas thermometers that differ in kind of gas within, and also possibly in volume of the container. If we plot the temperature readings for one thermometer against temperature readings on a second thermometer (actually, both are basically pressure readings), we get a *straight* line, as in Figure 19-4. The various constant-volume low-density gas thermometers agree at all temperatures, not merely at the fixed points. In short, a low-density gas thermometer is a good thermometer; its temperature readings do not depend on the thermometric substance.

In the gas thermometer, readings on the pressure gauge are effectively reinterpreted as temperature readings. If pressure is plotted against tempera-

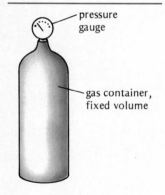

pressure gauge

gas container, fixed volume

Figure 19-3. *A constant-volume gas thermometer: a tank of low-density gas of constant volume with a pressure gauge.*

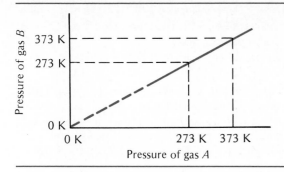

Figure 19-4. *Pressure (that is, absolute temperature) for one constant-volume gas thermometer against the corresponding pressure (temperature) for a second constant-volume gas thermometer. Both gas densities are very low. Compare with Figure 19-2.*

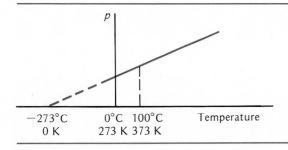

Figure 19-5. *A pressure-temperature graph for a constant-volume gas thermometer.*

ture, the results for any low-density gas are as shown in Figure 19-5, a linear relation. Extrapolating the straight line to low temperatures, we find that the pressure is zero when the Celsius temperature is $-273.15°$. Expressed a little differently, if the behavior of the gas at elevated temperatures persisted unchanged to the lowest temperatures (which it does not, because gases liquefy), the gas pressure would be zero at $-273.15°C$. The temperature $-273.15°C$ then corresponds to a *possible* lower limit of temperature. It is therefore convenient to introduce a new temperature scale with its zero at this point. The *Kelvin temperature scale* is so chosen that its zero is at $-273.15°C$. The size of the Kelvin degree is the same as the Celsius degree. Kelvin temperatures are represented by T. Then, because of the linear pressure-temperature relation, the Kelvin temperature is given in the limit of very low density gases by*

$$T = \left(\frac{p_T}{p_{273}}\right) 273.15 \text{ K} \qquad (19\text{-}2)$$

with

$$T = t_C + 273.15°$$

where p_{273} is the pressure of the gas at the ice point and p_T the gas pressure at temperature T.

The ice and steam points of water under standard atmospheric pressure are the bases for *approximate* temperature-scale determinations. The two fixed points used officially in defining the Kelvin temperature scale are (1) zero degrees Kelvin and (2) the triple point of water. (The procedures for using these fixed points are discussed in Chapter 22.) The triple point of water, which

---

* Kelvin temperatures are written without a degree mark.

occurs at a pressure of 610 Pa and a temperature of 273.16 K, corresponds to that unique condition under which water can exist simultaneously in thermal equilibrium in the liquid, solid, and vapor states.

Zero on the Kelvin scale, the so-called *absolute zero,* is indeed the lowest possible temperature. This must be proved through thermodynamic arguments (Chapter 22); it is suggested by Figure 19-5, but not deduced therefrom. We shall see that the Kelvin, or absolute, scale of temperature has a simple interpretation in the mechanics of molecular motion. Suffice it to say here that the independence of the thermometric substance found for low-density gases is not entirely unexpected. In any low-density gases, the molecules are distant from one another and interact with one another and the container walls only during infinitesimally small periods.

19-3   Thermal Expansion of Solids and Liquids

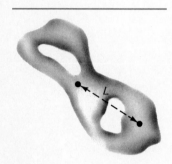

Figure 19-6.   Thermal expansion of a body of arbitrary shape. The distance between an arbitrary pair of points is L.

Most solids and liquids, that are free (not constrained by an external container) expand as their temperature increases. We first consider the parameters that are used to describe thermal expansion quantitatively. Then we consider briefly how thermal expansion is related to atomic properties.

Suppose that we have an object of any shape (including cavities and holes) that is free to expand when its temperature changes. See Figure 19-6. The length L is the distance between any two points in the body. When the temperature changes, so does L, by an amount ΔL. Experiment shows that for a given material, $\Delta L/L$ is the *same for all pairs of points*. Thermal expansion is like photographic enlargement; the size changes, but not the shape. The length increase ΔL is proportional to the original length L.

In addition, over a small enough temperature range, the fractional change in length $\Delta L/L$ varies *linearly* with the temperature change Δt, as shown in Figure 19-7. The fractional increase in length follows the rule

$$\Delta L/L \propto \Delta t$$

This proportionality can be expressed as an equation when we introduce α, the *linear coefficient of thermal expansion.* Over a small enough temperature range, the expansion coefficient can be taken to be constant. The value for α depends only on the particular material:

$$\Delta L/L = \alpha \Delta t \qquad\qquad (19\text{-}3)$$

Figure 19-7.   Length as a function of temperature for a typical thermal expansion. Over a small range, the change in length ΔL is directly proportional to the temperature difference Δt.

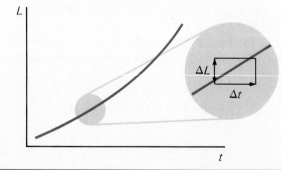

Table 19-2

MATERIAL	TEMPERATURE RANGE, $C°$	COEFFICIENT OF LINEAR THERMAL EXPANSION, $(C°)^{-1}$	COEFFICIENT OF VOLUME THERMAL EXPANSION, $(C°)^{-1}$
Aluminum	20–600	28×10^{-6}	
Copper	−25.3–10	16×10^{-6}	
Copper	0–100	17×10^{-6}	
Glass (Pyrex)	20–100	3.3×10^{-6}	
Ice	−250	-6.1×10^{-6}	
Ice	0	53×10^{-6}	
Invar			
(64% Fe, 36% Ni)	0–100	$\approx 1 \times 10^{-6}$	
Platinum	0–100	9.1×10^{-6}	
Silver	0–100	19.4×10^{-6}	
Steel (average carbon)	20–100	12×10^{-6}	
Tungsten (wolfram)	0–100	4.6×10^{-6}	
Carbon (diamond)	27		3.2×10^{-6}
Alcohol (ethyl, 30%)	18–39		293×10^{-6}
Mercury	0–100		182×10^{-6}
Water	0		-68×10^{-6}
	4		0.0×10^{-6}
	20		207×10^{-6}

We see from the definition of α that its units must be those of reciprocal temperature difference. For example, steel at room temperature has an α of 12×10^{-6} $(C°)^{-1}$; this means that the distance between any two points of a steel object increases by 12 parts in a million for every one Celsius degree rise in its temperature.

Table 19-2 lists experimentally determined values of α for several solids. Note that the expansion coefficient may change with temperature. A negative α implies that the material shrinks with a temperature rise.

Equation (19-3) may be written differently. Let L be the length of an object at some original temperature and L_t its length at a temperature higher by Δt. Then

$$L_t = L + \Delta L = L + \alpha L \, \Delta t$$

$$L_t = L(1 + \alpha \, \Delta t) \qquad \text{(19-4)}$$

It is easy to define area and volume coefficients of thermal expansion in terms of the linear expansion coefficient α. Consider a square whose edge length goes from L to L_t and whose area goes from $A = L^2$ to $A_t = L_t^2$ when its temperature changes by Δt. Using (19-4), we have

$$L_t^2 = L^2(1 + \alpha \, \Delta t)^2 = L^2(1 + 2\alpha \, \Delta t + \alpha^2 \, \Delta t^2)$$

We can neglect the very small term in α^2, and the relation reduces to

$$A_t = A(1 + 2\alpha \, \Delta t)$$

The coefficient of area expansion is 2α. This result is altogether general since any flat surface can be regarded as a collection of small squares.

Now suppose that a cube expands from volume $V = L^3$ to $V_t = L_t^3$. We have

$$L_t^3 = L^3(1 + \alpha\,\Delta t)^3 = L^3(1 + 3\alpha\,\Delta t + \cdots)$$

$$V_t = V(1 + 3\alpha\,\Delta t)$$

We see that the volume coefficient of thermal expansion is 3α. Again, the result is general because any solid can be regarded as a collection of small cubes.

For a liquid, one can define a thermal coefficient β of volume expansion, but not linear or area coefficients:

$$\Delta V/V = \beta\,\Delta t$$

The thermal expansion of water is anomalous. Indeed, water contracts with a temperature rise over a small range of temperatures. Figure 19-8 is a plot of the volume of a 1-gm mass of water as a function of its temperature. The line is curved, not straight. One cannot define a *single* coefficient of volume expansion for water over the range $0-100°$C. Note especially that from the ice point $0°$C to $4°$C, the volume of water *decreases* as the temperature increases; in this range, β is *negative*. Another way of describing the expansion and contraction behavior of water is to say that the density of water increases from $0°$C to a maximum at $4°$C and then decreases for higher temperatures.

The anomalous expansion property of water causes lakes to freeze first at the upper surface. As the temperature of water in a lake is reduced by a colder atmosphere, the most dense water goes to the bottom. As the temperature drops from $4°$C to zero, the most dense (and warmest) water remains at the lake bottom. Therefore, the water that first reaches the ice point is at the top, and the top of the water freezes first.

Expansion and Atomic Properties   A change in the external dimensions of a solid implies that corresponding changes occur in the average separation distances between the atoms or molecules composing the body. Thermal ex-

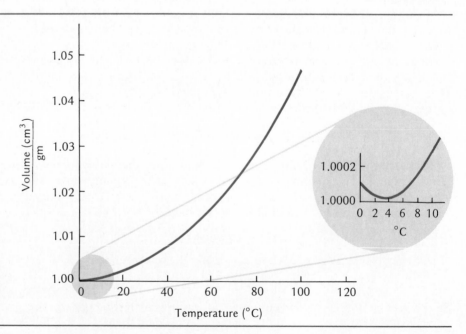

Figure 19-8.   Thermal expansion of water: the volume of a 1-gm mass as a function of temperature.

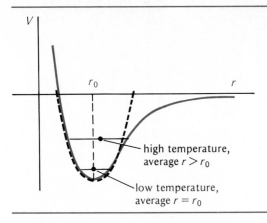

Figure 19-9. *A typical interatomic potential-energy curve. The potential curve is nearly parabolic (dashed line) in the vicinity of the equilibrium position r_0.*

pansion is intimately related to the interatomic and intermolecular forces that determine, in turn, the separation distances between the atoms.

Figure 19-9 shows a typical interatomic potential energy between two neighboring atoms as a function of their separation distance r. Near the equilibrium position r_0, the potential curve is parabolic and the restoring force is proportional to the atomic displacement $(r - r_0)$ from equilibrium. In this region, the atomic vibrations are simple harmonic oscillations about r_0 (Section 11-4). At low temperatures the atoms vibrate, with relatively low energies, near the minimum in the potential-energy curve. Any one atom then spends as much time at positions for which $r > r_0$ as at positions for which $r < r_0$. On the average, the separation distance between adjacent atoms is just r_0.

When the temperature of a solid or a liquid rises, so does the energy of the atomic vibrations. At higher temperatures, the atoms make excursions into the nonparabolic portions of the potential curve. As Figure 19-9 shows, this curve bends downward for $r > r_0$ and upward for $r < r_0$, both relative to the parabola (dashed line) that describes the potential-energy curve near its minimum. The intermolecular force is proportional to the slope of the potential-energy-displacement curve [$F_r = -dU/dr$, from (10-9)]. The interatomic restoring force is less than the force arising from a parabolic potential for $r > r_0$, but greater for $r < r_0$. The atom oscillates periodically, but not in simple harmonic motion; it spends more time at large separation distances than at small ones. On a time average, the separation distance is then larger than r_0. The distances between adjacent atoms increase with temperature, and the object as a whole expands. (Objects that contract with a temperature rise have potential-energy curves that depart from a parabolic shape in the opposite sense.) If the interatomic potential energy remained strictly parabolic at all energies, then a temperature rise would cause the atoms to oscillate with greater amplitude about their equilibrium positions, but the object would neither expand nor contract.

19-4   Thermal Properties of an Ideal Gas

We are concerned here with an *ideal*, or *perfect*, gas. Any gas behaves like an ideal gas if its density is very low. This requires, in effect, that the gas's temperature be relatively high and its pressure relatively low, so that the gas is not close

The End of the Caloric Theory

Benjamin Thompson, who was born in Massachusetts in 1753, served as a colonel in the Revolutionary War, on the side of the British redcoats. Because of his strong Tory views, he was forced to leave the colonies, and he persuaded George III to make him a knight in recognition of his services to the Crown.

Thompson was a turncoat, an opportunist, a professional soldier of fortune, a generous benefactor, and a remarkably talented amateur scientist. His observations showed that the caloric theory of heat could not be right.

Actually, the caloric theory had a lot going for it. Why was one object hot and another, otherwise like the first, cold? The hot one contained more of the massless, invisible fluid called caloric; the cold one had less caloric. Bring a hot object in contact with a cold one and caloric goes (we still say "the heat flows") from the hot to the cold object until thermal equilibrium is reached at some intermediary temperature.

Presumably you could also get caloric out of a solid simply by cutting open its "pores" so that the fluid could easily flow out. Clearly, then, a sharp tool would work better than a blunt one in opening pores. Further, once all the caloric originally contained in the solid had been released, that was it. No further cutting or rubbing could generate more heat.

Thompson showed that things could not work this way. He had worked his way up to heading the Bavarian Army and had finagled for himself an appointment by the Duke of Bavaria as Count Rumford, count of the Holy Roman Empire. One day he was supervising the boring of cannon barrels. The cutting tool was blunt. The cannon, Count Rumford noted, got just as hot with a blunt tool as with a sharp one.

Moreover, heat was produced without limit so long as the tool rubbed against the cannon. These observations contradicted the caloric theory, and Thompson was an early champion of the "vibratory" theory of heat.

Always a man on the make, his career was off to an advantageous start when he married a wealthy widow; he was 19, she 33. Later he married the widow of Antoine Lavoisier, the French chemist who discovered oxygen and lost his head in the French Revolution. (Some wags said that Antoine was actually better off than his wife.) Thompson donated a portion of his considerable fortune to establish the Royal Institution of Great Britain, a museum of technology for the public. Later, Thomas Young and Michael Faraday had their careers considerbly boosted when they were employed by the Royal Institution. In 1814, Count Rumford left his estate to Harvard University to establish the Rumford professorship.

Count Rumford was responsible for many other technological and scientific advances. He invented a reliable photometer for measuring light intensity; he showed that mat surfaces radiate more effectively than polished ones; he spurred the development of the potato as a staple crop in Central Europe when he invented the double boiler for cooking it; he used the ballistic-pendulum method (Example 10–10) for measuring bullet speed; he studied the relative insulation properties of materials. Also, he introduced the damper into fireplace design and designed a steam heating system; he designed a drip-coffee machine; and he laid out the park, now known as the English Gardens in Munich, Germany.

to the conditions under which it condenses into a liquid. Most gases at room temperature and atmospheric pressure can be regarded approximately as ideal gases.

The general-gas law gives the relationship among the macroscopic measurable properties of an ideal gas: the absolute temperature T, the pressure p, the volume V, and the mass m. It is customary to specify the mass m indirectly through the amount of substance n expressed in moles. Here M is the *atomic mass* measured relative to that of carbon-12 as exactly 12. One mole (abbreviated mol) is the amount of any pure substance that contains as many elementary entities as there are atoms in exactly 0.012 kg of carbon-12. In general,

$$n = \frac{m}{M} \tag{19-5}$$

For example, water (H_2O) has a molecular mass of approximately 18; therefore, 1 mol of water consists of 18 gm.

The standard form of the general-gas law, based on experiment with low-density gases, is

$$pV = nRT \tag{19-6}$$

where R is the *universal gas constant.*

It is an experimental fact that 1 mol of an ideal gas at standard temperature (273.15 K or 0°C) and standard atmospheric pressure (1.0132×10^5 Pa) occupies a volume of 22.415×10^3 cm$^3 \approx 22.4 \times 10^3$ liters.* Therefore, from (19-6),

$$R = \frac{pV}{nT} = \frac{(1.0132 \times 10^5 \text{ N/m}^2)(22.415 \times 10^{-3} \text{ m}^3)}{(1 \text{ mol})(273.15 \text{ K})}$$

$$= 8.314 \text{ J/mol} \cdot \text{K}$$

The numerical value of R depends on the units used for p and V. When p is given in atmospheres and V in liters, then

$$R = 0.08206 \text{ liter} \cdot \text{atm/mol} \cdot \text{K}$$

An alternative form of the general-gas law, (19-6), is

$$\frac{p_1 V_1}{m_1 T_1} = \frac{p_2 V_2}{m_2 T_2} \tag{19-7}$$

where subscripts 1 and 2 denote two different states of the gas.

If the volume and mass of gas are both constant, (19-7) becomes

$$\text{For constant } V \text{ and } m: \quad \frac{p_1}{T_1} = \frac{p_2}{T_2}$$

The pressure is directly proportional to the absolute temperature; this is the basis of the constant-volume gas thermometer (Section 19-2).

If the pressure and mass are both constant, (19-7) reduces to

$$\text{For constant } p \text{ and } m: \quad \frac{V_1}{T_1} = \frac{V_2}{T_2}$$

The proportionality of volume and temperature can be used as the basis of a constant-pressure gas thermometer; the relation is sometimes referred to as the law of J. L. Gay-Lussac and J. A. C. Charles, its discoverers.

Boyle's law describes an ideal gas with a constant temperature and mass:

$$\text{For constant } T \text{ and } m: \quad p_1 V_1 = p_2 V_2$$

An alternative form, using the mass density $\rho = m/V$, is

$$\text{For constant } T \text{ and } m: \quad \frac{p_1}{\rho_1} = \frac{p_2}{\rho_2}$$

---

* Conditions of standard pressure and standard temperature are often abbreviated STP.

Example 19-1.   What is the density of helium under STP?

We compute ρ directly by recalling that for a pressure of 1 atm and a temperature of 0°C, the volume of 1 mol of any gas is 22.415×10^{-3} m³. Since helium is monatomic and has an atomic mass of 4.003,

$$\rho = \frac{4.003 \text{ gm}}{22.415 \times 10^{-3} \text{ m}^3} = 0.1786 \text{ kg/m}^3$$

Example 19-2.   A tank with a volume of 30.0 liters contains nitrogen (N_2) gas at 20.0°C at a gauge pressure of 3.00 atm. The tank's valve is opened momentarily, and some nitrogen escapes. After the valve is closed and the gas has returned to room temperature, the tank's pressure gauge reads only 2.40 atm. How much nitrogen leaked out?

We designate amount of substance (number of moles) and the pressure before and after the leak by subscripts 1 and 2, respectively. Then, from (19-6),

$$n_1 - n_2 = \frac{p_1 V_1}{RT_1} - \frac{p_1 V_2}{RT_1} = \frac{V}{RT}(p_1 - p_2)$$

$$= \frac{(30 \text{ liters})(0.60 \text{ atm})}{(0.08206 \text{ liter} \cdot \text{atm/mol K})(293 \text{ K})} = 0.749 \text{ mol}$$

Since nitrogen has an atomic mass of 14, one mole of N_2 has a mass of 28 gm. The mass escaping is, then,

$$m = nM = (0.749 \text{ mol})(28 \text{ gm/mol}) = 21 \text{ gm}$$

19-5   Changes in State

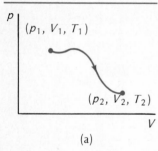

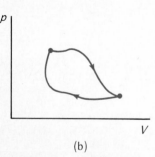

Figure 19-10.   (a) A process involving the change in state of a gas from (p_1, V_1, T_1) to (p_2, V_2, T_2). (b) A cycle.

To characterize the macroscopic properties of a gas, we usually specify the three quantities p, V, and T (there are still other possible macroscopic state variables). We may ignore variations in the mass of gas by taking it always to be 1 mol; alternatively, we may use V to indicate the volume per unit mass, or *specific volume*, of gas.

The *state* (at equilibrium) of a gas is completely specified by knowing p, V, and T (and m). The state of a gas does not depend, for example, on its past history. If a gas is ideal, so that $pV = nRT$, we need specify only two of the three variables, because the general-gas law permits the third variable to be computed. The general-gas law is the *equation of state* of an ideal gas. If a certain gas is not ideal and the general-gas law does not describe it, a more complicated relation between p, V, and T, again known as the "equation of state," gives the gas's behavior.

The state of a gas of constant mass m can be portrayed as a point on a diagram in which the pressure is plotted as a function of the volume, a so-called pV diagram. Two different points on a pV diagram correspond to two states of the gas — for example, (p_1, V_1, T_1) and (p_2, V_2, T_2), which differ in at least *two* of the variables p, V, and T.

A continuous change in the state of a gas is called a *process*. It is shown on the pV diagram as a line, or a "path," going from the initial to the final state. See Figure 19-10(a). Clearly, many possible paths or processes can connect a given pair of initial and final states. A process in which a gas is returned to its initial state is called a *cycle*; on a pV diagram [Figure 19-10(b)], it is a closed loop.

Work Done By A Gas   The area under the curve on a pV diagram has a simple meaning—it is work done by the gas. To see this, suppose that a gas undergoes an expansion. To be more specific, suppose that an expanding gas displaces a piston fitted into a cylindrical container. The gas's volume increases by dV. See Figure 19-11. The expanding gas does work dW on the piston as it applies a force F over an infinitesimal displacement dx. The force of the gas on the piston of area A is $F = pA$. From the definition of work,

$$dW = F\,dx = (pA)\,dx = p(A\,dx)$$

But $A\,dx = dV$, the volume swept out by the displaced piston, and also the change in gas volume. We then have

$$dW = p\,dV$$

The work dW done by an expanding gas is $p\,dV$. We write the work in differential form to allow for the possibility that the pressure may change as the gas expands. The total amount of work $W_{1\to2}$ done as the volume goes from V_1 to V_2 is, then,

$$W_{1\to2} = \int_{V_1}^{V_2} p\,dV \qquad (19\text{-}8)$$

The work may be computed, and the integral in (19-8) evaluated, by integrating when p is known as a function of V.

The work done by an expanding gas on the piston is positive. (The work done by the piston on the gas is negative.) A gas being compressed does negative work (positive work is done on the gas by the piston).

Figure 19-12 shows an expansion process. The work done by the gas as it expands is equal in magnitude to the area under the curve on the pV diagram. This area depends on the path between the end points. Suppose that a gas is compressed, so that its volume is decreased. Then the path on the pV diagram is to the left, and the area under the curve is the work done *on the gas*. Therefore, when a gas is taken through an entire cycle, the *net* amount of work done *by the gas* over the cycle is represented by the area enclosed by the loop, as shown in Figure 19-13. (Strictly, the area of the loop gives the net work done by the gas when the loop is traversed in the clockwise sense.)

Consider a gas taken through a cycle; it returns to its initial state of pressure, volume, and temperature. It is reasonable to suppose (and we shall later prove in detail) that the total energy of the gas cannot have changed. Yet the gas has done net work over the cycle. This implies that energy cannot be conserved

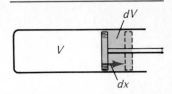

Figure 19-11.   Work done by a gas in expanding through a volume change. ΔV as the piston is displaced dx.

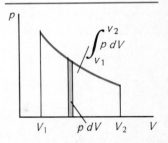

Figure 19-12.   Work done by an expanding gas, as represented by the area under the pressure-volume graph.

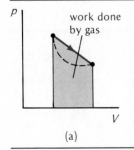

(a)

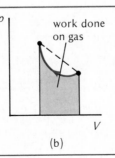

(b)

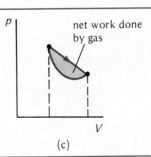

(c)

Figure 19-13.   Graphic representation on a pressure-volume diagram of the work done by and on a gas. (a) Work done by the gas in expanding. (b) Work done on the gas as it is compressed. (c) Net work done by the gas over the cycle.

unless there exists an additional form of energy transfer besides the work done by the gas. This form of energy transfer is heat. (Much will be said about heat in subsequent chapters.)

Changes in State   Several types of change in the state of a gas are particularly simple, because one variable in the equation of state remains constant. Each process is briefly described below for the simple case of an ideal gas.

● *Isothermal process:* the *temperature* is constant. Lines of constant T, corresponding to pV = constant, are called *isotherms*. (On a pV diagram, an isotherm for an ideal gas is an equilateral hyperbola.) The pressure decreases in an isothermal expansion and increases in an isothermal compression. See Figure 19-14(a).

● *Isobaric process:* the *pressure* is constant. We see from Figure 19-14(b) that when a gas expands isobarically, its temperature rises, since it goes from a low-T isotherm to a higher-T isotherm.

● *Isovolumetric process:* the *volume* is constant. Figure 19-14(c) shows that the temperature of a gas in a container of fixed size drops as its pressure falls.

Another important process is an *adiabatic process.* (We shall study this type of process in more detail in Section 20-5.) Suffice it to say here that a gas will undergo an adiabatic process if there is *no heat transfer* into or out of the gas. This will happen if the container is a heat insulator, which isolates the gas thermally from its surroundings. As we shall prove later, for an ideal gas, an adiabatic process is described by the relation

$$pV^\gamma = \text{constant} \tag{19-9}$$

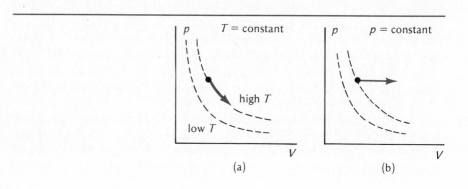

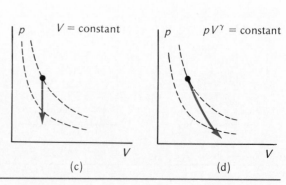

Figure 19-14.   Several important changes (represented by colored lines) in the state of a gas. The dashed lines represent two isotherms. (a) Isothermal, (b) isobaric, (c) isovolumetric, and (d) adiabatic processes.

where γ is a constant. The numerical value of γ depends on the nature of the gas, but it is always greater than 1. Therefore, adiabatic lines on a pV diagram are more steeply inclined than isotherms. Figure 19-14(d) shows that the temperature falls in an adiabatic expansion.

Example 19-3. (a) An ideal gas expands isothermally. How much work is done by the expanding gas? (b) Suppose that instead the gas were to expand isobarically. How much work would be done by the gas?

(a) From (19-9), the work done by the gas in going from state 1 to 2 is

$$W_{1 \to 2} = \int_{V_1}^{V_2} p \, dV$$

Since the gas is ideal and $p = nRT/V$,

$$W_{1 \to 2} = n \, RT \int_{V_1}^{V_2} \frac{dV}{V} = nRT \ln \frac{V_2}{V_1} \qquad (19\text{-}10)$$

Note that T could be brought to the left of the integral sign only because the temperature is constant in an isothermal process.

(b) The work done by the gas when expanding isobarically is

$$W_{1 \to 2} = \int_{V_1}^{V_2} p \, dV = p \int_{V_1}^{V_2} dV = p(V_2 - V_1)$$

Here p can be brought to the left of the integral sign because it remains constant in the isobaric process.

Summary

Definitions

Thermal equilibrium: the state achieved by any collection of objects isolated from external influence after a sufficiently long time has elapsed; objects in thermal equilibrium have in common the property of temperature.

Linear thermal expansion coefficient:

$$\alpha = \frac{\Delta L/L}{\Delta t} \qquad (19\text{-}3)$$

Ideal gas: any low-density gas whose equation of state is the general-gas law

$$pV = nRT \qquad (19\text{-}6)$$

Fundamental Principles

Zeroth law of thermodynamics: if A and C are in thermal equilibrium and B and C are in thermal equilibrium, then A and B are in thermal equilibrium.

Important Results

The work done by an expanding gas on its surroundings is $\int p \, dV$, the area under the curve describing the process on a pV diagram.

Problems and Questions

Section 19-1   Temperature and the Zeroth Law of Thermodynamics

· **19-1 Q** If two bodies are in *thermal equilibrium*, they

(A) have zero thermal acceleration.

(B) have zero net thermal force acting between them.

(C) have equal amounts of thermal energy.

(D) are at the same temperature.

(E) are in a thermal steady state condition so that their

temperatures, although possibly different, are not changing.

Section 19-2   Thermometry

· **19-2 Q** How does one use a constant volume gas thermometer?

(A) You measure the pressure of the gas.

(B) You measure small changes in the volume of the gas.

(C) You measure the temperature of the gas with a second-ary thermometer, which you then calibrate against the gas thermometer.

(D) You measure the density of the gas.

(E) You measure the velocity distribution of the individual gas molecules and then use the ideal gas law to determine the temperature.

· **19-3 Q** Which of the following best explains why a constant-volume gas thermometer is so useful for temperature measurement?

(A) It is easier to use than most other types of thermometers.

(B) It is very portable.

(C) It is very inexpensive compared to other devices.

(D) It is easy to use for remote sensing of temperature.

(E) Its readings do not depend on the thermometric properties of the specific substance used.

· **19-4 P** Express the following temperatures in °F, °C, and K. *(a)* 20°C, *(b)* 98°F, *(c)* 10°F, *(d)* 4.2 K

· **19-5 P** Express the following temperatures in °C, K, and °F. *(a)* 23°C, *(b)* −196°C, *(c)* −40°F, *(d)* 72°F, *(e)* 300 K, *(f)* 77 K

: **19-6 P** At what temperature is the reading of a thermometer in degrees Fahrenheit twice the reading of a Celsius thermometer?

: **19-7 P** At what temperature do the Celsius and Fahrenheit scales give the same numerical reading?

Section 19-3   Thermal Expansion of Solids and Liquids

· **19-8 Q** If on a very hot day a surveyor uses a metal tape to measure a building site, his readings will probably be

(A) smaller than the true dimensions.

(B) larger than the true dimensions.

(C) accurate, assuming the earth is at the same temperature as his tape.

(D) too large for small dimensions and too small for large dimensions.

· **19-9 Q** In order to ignite the air-gasoline mixture in the cylinder of your car's engine, the spark plug gap must be adjusted to the correct spacing. This is done while the spark plug is removed from the car, of course. What happens when the spark plug is in operation in the car and it becomes very hot?

(A) The gap will increase when the plug heats up.

(B) The gap will decrease when the plug heats up.

(C) The gap will not change when the plug heats up.

(D) Depending on the amount of temperature rise, the gap may either increase or decrease, but this will not affect its operation.

(E) None of the above.

· **19-10 Q** Shown in Figure 19-15 are several possible graphs of potential energy as a function of interatomic spacing for a particular solid. Which would you expect to result in the largest coefficient of linear thermal expansion?

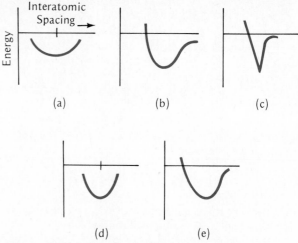

Figure 19-15.   Question 19-10.

· **19-11 Q** Illustrated in Figure 19-16 is a bimetallic strip, a widely used temperature measuring device. What is the principle upon which it operates?

(A) A stressed material will distort, when heated, to relieve the stress.

(B) The coefficient of thermal expansion of different metals is different.

(C) The two sides of the active element are at different temperatures.

(D) If two dissimilar metals are joined and their two junctions are at different temperatures, a voltage will be generated.

(E) The thermal conductivity of different metals is different.

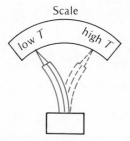

Figure 19-16.   Question 19-11.

· **19-12 P** When laying steel railroad track it is necessary to allow expansion space in the joints to prevent buckling on hot days. If a section of track is 20 m long at 0°C, how long will it be when the temperature rises to 35°C?

· **19-13 P** A steel tire is mounted on a wooden wagon wheel by heating it to 700°C before putting it on the wheel. The wooden rim has a diameter of 120 cms, and the steel tire just barely slips on when hot. What is the diameter of the steel tire when it has cooled to 20°C? (It compresses the wooden wheel in cooling.)

: **19-14 P** A mercury barometer indicates atmospheric pressure to be 761.1 torr when the temperature is 35°C. What would the reading have been for the same atmospheric pressure if the temperature dropped to −10°C?

: **19-15 P** If Figure 19-9 were drawn to scale, and the two energy levels shown corresponded to temperatures of 200°C and 400°C, estimate graphically the coefficient of linear thermal expansion for the material.

: **19-16 P** A mercury thermometer has a quartz bulb of volume 0.35 cm³ attached to a tube of cross-sectional area 1.0×10^{-8} m². If the expansion of the quartz is negligible, how far in the stem will the mercury rise when the temperature changes from 24°C to 26°C?

: **19-17 P** The pendulum on a grandfather's clock ticks seconds (two ticks for one oscillation). The pendulum is made of brass ($\alpha = 19 \times 10^{-6}$°C^{-1}). How many seconds would it gain or lose in one month if the temperature were 30°C instead of 20°C?

: **19-18 P** A surveyor measures a plot of land in the New Mexico desert with a steel tape on a day when the temperature of the tape is 110°F. The tape was calibrated to give an accurate reading at 20°C. If the reading obtained with the tape is 264.000 meters, what is the true dimension of the plot?

: **19-19 P** Suppose you wish to design a mercury-in-glass thermometer, to consist of a glass bulb of volume 0.4 cm³ filled with mercury, to which is connected a thin capillary tube in which the mercury will rise. You want the mercury to rise 1 mm per °C.
    (a) If you neglect any thermal expansion of the glass, what cross-sectional area should the capillary tube have? (b) If the coefficient of linear expansion of the glass is 9×10^{-6} per °C, what error will this cause in your readings?

: **19-20 P** A long uniform bar of length L_0 at 0°C has one end maintained at 0°C and the other at a fixed temperature T_0. The temperature along the bar varies linearly with the distance from the cold end. The coefficient of linear thermal expansion is α. Show that when the bar is heated from 0°C to temperature T_0 at one end in this fashion, the change in length is $\Delta L = \alpha L_0 T_0 / (2 - \alpha T_0)$. (Note: The approximation $\ln(1 + x) = x - \frac{1}{2}x^2 + \ldots$ is useful.)

: **19-21 P** A bimetallic strip (Figure 19-16) is used as the active element in many thermostats. Such a device can be used to turn a furnace on and off to regulate the temperature. Two different metals with coefficients of linear thermal expansion α_1 and α_2 are riveted or welded together. At some temperature T_0, both have the same length L_0 and thickness t. When the strip is heated, one part expands more than the other and the strip bends to one side. The motion of the end of the strip can operate an electrical switch to turn a furnace on or off. Calculate the radius R of the bend as a function of temperature T, measured to the interface between the two strips. (Hint: Calculate the expanded length down the center of each element and then deduce the radius of bend.)

Section 19-4   Thermal Properties of an Ideal Gas

· **19-22 P** How many molecules are there in 6 grams of CO_2 gas?

· **19-23 P** What is the mass of 8.60×10^{22} molecules of N_2 gas?

· **19-24 P** How many molecules of CH_4 are there in 2.4 moles?

: **19-25 P** Under normal diving conditions a scuba diver inhales the same volume of air and takes the same number of breaths per minute, independent of the depth at which he is swimming. In view of this, if a tank of air will last 2 hours at the surface, how long will it last at a depth of 30 meters?

: **19-26 P** A motorist fills her tires to a gauge pressure of 33 psi when the temperature is 5°C. After driving for several hours, she again checks the pressure and finds it has risen to 37 psi. Assuming the volume of the tire did not change, what was the temperature of the air in the tire at 37 psi?

: **19-27 P** Air is approximately 80% N_2 and 20% O_2. What is the average density of air at 1 atm and 23°C?

: **19-28 P** A tire is inflated to a gauge pressure of 30 lb/in² (psi) when the temperature is 35°F. The temperature in the tire rises to 104°F after the car has been driven for a while. Assuming the volume of the tire doesn't change, what will be the gauge pressure at this temperature?

: **19-29 P** One torr is considered a good high vacuum for many laboratory purposes. At this pressure at 27°C for N_2 gas, calculate (a) the number of molecules per cm³, and (b) the density of the gas, in kg/m³.

: **19-30 P** The temperature of interstellar space is 2.7 K. This temperature characterizes the electromagnetic radiation and the sparse population of hydrogen atoms found there. In the interstellar regions of our galaxy, the Milky Way, it is estimated from radiation detected that there is about one neutral hydrogen atom per cm³. What pressure would this give rise to? Note that 10^{-6} torr is considered a high vacuum in the laboratory, and 10^{-9} torr is a very high vacuum.

: **19-31 P** The volume of a nitrogen molecule is approximately 8×10^{-30} m³. What fraction of the volume occupied by nitrogen gas at STP is occupied by the molecules themselves?

: **19-32 P** An experimenter building a toy rocket motor places 4 gm of solid dry ice (CO_2) in a metal container of volume 6 cm³. All of the CO_2 sublimes and becomes a gas at 23°C. What pressure does it reach?

: **19-33 P** An air bubble of volume 2 mm³ is formed at the bottom of a lake where the temperature is 4°C and the depth 120 m. What is the volume of the bubble just as it reaches the surface of the lake, where the temperature is 15°C?

Section 19-5 Changes in State

: **19-34 P** A gas is expanded from $p_1 = 5$ atm, $V_1 = 1$ liter, to $p_2 = 1$ atm, $V_2 = 2$ liters, $T_2 = 60°C$, in the process sketched here. How much work is done by the gas in the process? Is the work positive or negative?

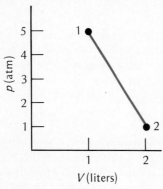

Figure 19-17.   **Problem 19-34.**

: **19-35 P** A gas initially at STP is carried through the following cycle:

- Expanded at constant pressure from 2 liters to 4 liters.
- Cooled at constant volume to 100°C.
- Compressed at constant pressure to a volume of 2 liters.
- Heated at constant volume to a pressure of 1 atmosphere.

(a) Draw the process on a *pV* diagram. *(b)* Calculate the work done by the gas in one cycle.

: **19-36 P** A hot air balloon has a volume of 960 m³ and a total mass of 180 kg, including the pilot. Assume the temperature of the gas in the balloon is uniform and that the pressures inside and outside the balloon are equal. *(a)* What gas temperature is needed when the air is at 1 atm and 27°C? *(b)* At an altitude of 3500 meters the temperature drops to 10°C and the atmospheric pressure to 530 torr. What gas temperature is needed in the balloon under these conditions?

: **19-37 P** One mole of an ideal gas starting at point #1 is carried around the cycle drawn here on a *pV* diagram. In the process the gas does 3×10^5 J of work. What are the highest and lowest temperatures reached by the gas?

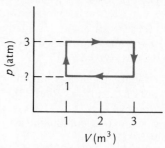

Figure 19-18.   **Problem 19-37.**

: **19-38 P** One mole of an ideal gas at 600 K is expanded isothermally until its volume doubles. It is then compressed isobarically to its original volume, then heated isovolumetrically to its original state. *(a)* How much work does the gas do in the process? *(b)* Draw the process on a *pV* diagram.

: **19-39 P** A membrane separates two compartments of volumes V_1 and V_2 containing CH_4 and H_2 respectively at pressures p_1 and p_2. Both compartments are at the fixed temperature of the surroundings, 27°C. The membrane ruptures and the gases mix together. What is the equilibrium pressure reached if *(a)* $V_1 = V_2$, $p_1 = 2p_2 = 1$ atm? *(b)* $V_1 = 2V_2$, $p_1 = p_2 = 1$ atm? *(c)* $V_1 = 2V_2$, $p_1 = 4p_2 = 280$ torr?

· **19-40 P** A gas expands isothermally from p_1, V_1 to p_2, V_2. What is the work done by the gas if $p_1 = 8$ atm, $V_1 = 0.1$ m³, $p_2 = 1$ atm, $V_2 = 0.8$ m³?

Thermal Properties of an Ideal Gas, Microscopic View

20

When we looked at the macroscopic properties of a very simple system — an ideal gas — we saw, among other things, how it is possible to construct reliable thermometers. But apart from the obvious statement that temperature is what you read on a thermometer, what is it really? Does temperature have a less abstract, more mechanical meaning?

The answer is yes, and it comes from looking at an ideal gas from a microscopic point of view, one in which we examine the motions of individual gas molecules. Here we use the kinetic theory of gases to elucidate in molecular terms the meaning of gas pressure (molecular bombardment of container walls) and temperature (average molecular kinetic energy).

In this chapter we first encounter the term *heat* and the first law of thermodynamics, which is really just a generalization of the energy-conservation principle. We consider qualitatively the second law of thermodynamics, also as applied to an ideal gas, and especially its interpretation in terms of relative order and disorder.

20-1   Molecular Properties

Here, for the record, are some well-known facts from chemistry.

Avogadro's Number   The basic number in molecular physics, the number for counting molecules, is *Avogadro's number*, N_A. One mole (abbreviated mol) of *any* pure substance contains N_A atoms or molecules. One mole is the amount of material equal to the atomic mass or molecular mass expressed in grams.[*]

The numerical value of Avogadro's number is

$$N_A = 6.022045 \times 10^{23}/\text{mol}$$

Measuring N_A amounts effectively to counting atoms or molecules, a very subtle undertaking.

Molecular Size   Knowing N_A, it is simple to compute the approximate size of a molecule. Consider ordinary water. Its density is close to $1\text{gm/cm}^3 = 10^3 \text{ kg/m}^3$. The molecular mass of H_2O is close to 18, so that 1 mol of H_2O amounts to 18 gm. With density defined as $\rho = m/V$, we have for the volume occupied by 1 mol of water

$$V = \frac{m}{\rho} = \frac{18 \times 10^{-3} \text{ kg/mol}}{10^3 \text{ kg/m}^3}$$

The volume occupied by one molecule of water is then

$$\frac{V}{N_A} = \frac{1.8 \times 10^{-5} \text{ m}^3/\text{mol}}{6.02 \times 10^{23} \text{ molecules/mol}} = 30 \times 10^{-30} \text{ m}^3/\text{molecule}$$

For simplicity, imagine the molecules to be packed cubes with an edge length L. Each molecule has a volume L^3, and we get

$$L = (30 \times 10^{-30} \text{ m}^3)^{1/3} \simeq 3 \times 10^{-10} \text{ m} = 0.3 \text{ nm}$$

This result is typical of all molecules that comprise no more than a few atoms.

In our approximate computation of molecular size, we assumed effectively that the molecules were "touching." This assumption is supported by the fact that liquids and solids are relatively hard to compress. To reduce the volume of water by just 1 part in 10^3, for example, requires a pressure of 20 atm.

Actually, we computed a molecular diameter by precisely the same procedure one might use to find the diameter of a large sphere such as a billiard ball. The diameter of any one of identical billiard balls is the distance between the centers of a pair of such balls in contact with one another; that is, the diameter is that separation distance below which the force between the balls becomes strongly repulsive. In like fashion, we took the molecular diameter to be the separation distance between adjacent molecules that marked the onset of a strongly repulsive intermolecular force.

No one has ever seen a molecule directly, not with visible light. But individual molecules, even individual atoms, can be observed indirectly. See Figures 20-1 and 20-2.

---

[*] In the SI unit system, the mole is defined in terms of grams, not kilograms.

Figure 20-1. *Electron micrograph with geometrical array.*

Any gas that can be regarded as approximately ideal has a density that is lower than that of a liquid or a solid by a factor of at least 1000. So if the molecules are thought of as touching in the condensed states, then in the gaseous state the molecules must be separated from one another, on the average, by at least $1000^{1/3}$, or 10, times a molecular diameter.

Figure 20-2. *Picture, taken with a field ion microscope, of individual atoms at the surface of a finely pointed needle acting as cathode with a fluorescent screen as anode. The method was invented by the late Erwin W. Müller of the Pennsylvania State University.*

Molecules as Hard Spheres A molecule is certainly not a tiny, hard billiard ball. Yet for most of our considerations in this chapter, we shall take a molecule to be just that, an infinitely hard sphere. We shall ignore altogether that a molecule is composed of atoms that may be in relative motion, that an atom is composed of electrons in motion about a nucleus, and that a nucleus comprises protons and neutrons also in motion. It turns out that for temperatures no greater than a few thousand degrees Kelvin, the atomic structure and the nuclear structure within a typical molecule do not undergo any change, even though the molecules very frequently collide with one another and with the walls of a container. The reasons lie in the quantum theory of atomic structure (Chapter 41).*

This result — that a molecule in a low-density gas at moderate temperature may be treated like a hard sphere, even though it has internal structure and internal parts in motion — is extraordinary. It is as if we had a collection of ordinary mechanical clocks, each with a complicated internal structure, colliding continuously with one another, without any effect at all on the works of any clock from these collisions. A typical molecule in nitrogen at STP makes about 10 billion collisions per second.

20-2 Kinetic Theory and Gas Pressure

In deriving the fundamental relation between the pressure of a gas and the microscopic properties of gas molecules, we shall be led to a *mechanical* interpretation of the concept of temperature.

The following basic assumptions of the kinetic theory of gases are satisfied for a low-density ideal gas:

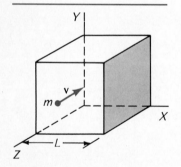

Figure 20-3. Molecule of mass m and speed v striking the right-hand wall of a cubical container of edge length L.

- The molecules of a gas are, on the average, separated by distances that are large compared with the molecular diameters. For an ideal gas, any *molecule* has a *volume* that is *negligible* compared with the volume of the container. A gas effectively consists of empty space with the molecules as point masses.

- The molecules are in constant *random motion.* The number of molecules is enormous; their motion is utterly chaotic. Molecules move in all directions with equal probability and with a variety of speeds. The center of mass of the gas as a whole then remains at rest and the total momentum of all molecules is zero.

- The collisions of the molecules with one another and with the walls are *perfectly elastic.* The total kinetic energy of all molecules is constant.

- Between collisions the molecules are free of forces and move with a constant velocity.† A collision takes place when one molecule is within the

---

* Briefly, atomic energies are quantized, or restricted to certain discrete values. For temperatures less than ∼ 10^4 K, molecules cannot gain enough energy from a collision to allow for an upward quantum jump in atomic energy level. The atoms and nuclei must therefore remain in their lowest energy states.

† Strictly, the molecular speeds are constant between collisions only if the molecules move in gravity-free space. All molecules in a gas at the earth's surface are subject to a constant acceleration g downward. But, we shall see, the change in velocity arising from gravity is small compared with the molecular velocities and can therefore be ignored. Said differently, the change in the kinetic energy of a molecule arising from its rise or fall in a gravitational field is negligible compared with the kinetic energy of a typical molecule.

short-range intermolecular force of a second molecule. The *duration of any collision is negligibly short compared with the time between collisions.* A pair of molecules will momentarily lose kinetic energy and gain corresponding potential energy during a collision, but the *potential energy can be ignored* because a molecule spends a negligible fraction of its time in collisions.

• *Newtonian mechanics applies* to molecular collisions, and computing the gas pressure is merely an exercise in mechanics.

We wish to compute the pressure, a macroscopic property, of an ideal gas in terms of the microscopic properties of the molecules: N, the number of molecules in the container; m, the mass of each molecule; and v, the speed of a molecule. The pressure of the gas comes from the bombardment of the container walls by molecules. There are N molecules in a cubical box of edge length L.

Consider the collision of a molecule with the right-hand wall shown in Figure 20-3. The molecule approaches with velocity components v_x, v_y, v_z and collides elastically with the wall. It rebounds with the same speed. The molecule's v_y and v_z components are not changed, but the v_x velocity component becomes $-v_x$ after the collision. See Figure 20-4. Therefore, in this collision, the component of the molecule's momentum changes from $+mv_x$ to $-mv_x$; the overall momentum change is $(-mv_x) - (+mv_x) = -2mv_x$. At the same time, the container wall has *its* momentum changed by $+2mv_x$. The molecule travels a total distance $2L$ along the x axis over a round trip at the speed v_x, so that the time interval Δt between collisions with the right-hand wall is $\Delta t = 2L/v_x$. (You can easily convince yourself that this time interval is not altered by collisions with top, bottom, and side walls.)

The average force $\overline{F}$ on the wall from one molecule over the time between successive collisions is, from Newton's second law,

$$\overline{F} = \frac{\Delta(mv)}{\Delta t} = \frac{2mv_x}{2L/v_x} = \frac{mv_x^2}{L} \tag{20-1}$$

The force on the molecule is to the left; the force of equal magnitude on the wall is to the right. Each time a molecule strikes the container wall, it imparts an impulsive force to it; the total force on the wall is the resultant of numerous discrete and abrupt molecular collisions. See Figure 20-5. The molecular speeds may vary, and the molecules may strike the wall from different directions. Therefore, the individual impulses *differ* in size. But because the number of molecules is large, the total force ΣF on the wall is essentially constant. We divide the total force on the wall by its area L^2 to get the pressure p on the wall:

$$p = \frac{\Sigma F}{L^2} = \frac{F_1 + F_2 + F_3 + \cdots}{L^2}$$

From (20-1), this becomes

$$p = \frac{m(v_{x1}^2 + v_{x2}^2 + v_{x3}^2 + \cdots)}{L^3} \tag{20-2}$$

Here we have used subscripts 1, 2, 3, and so on to refer to the various individual molecules.

The speeds of the molecules are distributed over a continuous range of values. And so too are the x components of the velocities. Let $\overline{v_x^2}$ represent the average of the squares of the x components of the speeds above:

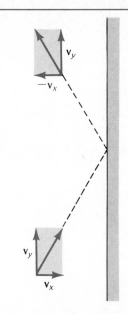

Figure 20-4. *A molecule collides obliquely with a container wall. Only the velocity component at right angles to the surface changes.*

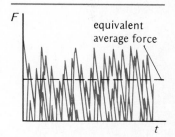

Figure 20-5. *The force on a container wall, arising from discrete molecular collisions, as a function of time. The equivalent average force is constant.*

$$\overline{v_x^2} = \frac{v_{x1}^2 + v_{x2}^2 + v_{x3}^2 + \cdots}{N} \tag{20-3}$$

Then (20-2) can be written

$$p = \frac{Nm\overline{v_x^2}}{L^3} \tag{20-4}$$

Averages for the y and z velocity components are given by relations analogous to (20-3).

Now the average of the squares of the molecules' speeds is, by definition,

$$\overline{v^2} = \frac{v_1^2 + v_2^2 + v_3^2 + \cdots}{N}$$

where for any one particle,

$$v_1^2 = v_{x1}^2 + v_{y1}^2 + v_{z1}^2$$

It then follows that

$$\overline{v^2} = \overline{v_x^2} + \overline{v_y^2} + \overline{v_z^2} \tag{20-5}$$

The molecules move in truly random directions; one direction is like any other. Therefore, the average in one direction is the same as in another direction, and

$$\overline{v_x^2} = \overline{v_y^2} = \overline{v_z^2}$$

Equation (20-5) can then be written

$$\overline{v^2} = 3\overline{v_x^2} \qquad \text{or} \qquad \overline{v_x^2} = \tfrac{1}{3}\overline{v^2}$$

so that (20-4) becomes

$$pV = \tfrac{1}{3}Nm\overline{v^2} \tag{20-6}$$

where the container volume is $V = L^3$.

It is useful to introduce the *root-mean-square speed*, v_{rms}. By definition, v_{rms} is the square *root* of the *mean* of the *squares* of the molecular speeds; in symbols,

$$v_{\text{rms}} \equiv \sqrt{\overline{v^2}}$$

Then (20-6) can be written as

$$pV = \tfrac{1}{3}Nmv_{\text{rms}}^2 \tag{20-7}$$

We have related the macroscopic properties of a gas (pressure p and volume V) to the microscopic molecular properties (N, m, and v_{rms}). Equation (20-7) can be written more compactly by introducing the mass density ρ of the gas, where ρ is the total mass Nm of molecules divided by the volume V they occupy, $\rho = Nm/V$. Then we have

$$p = \tfrac{1}{3}\rho v_{\text{rms}}^2 \tag{20-8}$$

This equation permits us to compute the rms speed of the molecules by knowing merely the pressure and density of the gas.

Example 20-1.   What is the rms speed of molecular hydrogen (H_2) at STP? The density of H_2 is

$$\rho = \frac{2.016 \text{ gm}}{22.4 \text{ liters}} = \frac{2.016 \text{ kg}}{22.4 \text{ m}^3} = 9.00 \times 10^{-2} \text{ kg/m}^3$$

The pressure is $p = 1.013 \times 10^5$ Pa. From (20-8),

$$v_{\text{rms}} = \sqrt{\frac{3p}{\rho}} = \sqrt{\frac{3(1.013 \times 10^5 \text{ N/m}^2)}{9.00 \times 10^{-2} \text{ kg/m}^3}} = 1.84 \text{ km/s}$$

A typical hydrogen molecule at STP has a speed of nearly 2 km/s ($\sim$ 4000 mi/h).

20-3 Kinetic-Theory Interpretation of Temperature

We now have two relations for the product pV of the pressure and volume of an ideal gas:

• A macroscopic relation, the general gas law,

$$pV = nRT \tag{19-6}$$

• A microscopic relation from the kinetic-theory,

$$pV = \tfrac{1}{3}Nmv_{\text{rms}}^2 \tag{20-7}$$

By equating the two equations we arrive at a relation giving a mechanical interpretation of absolute temperature T. From the two equations,

$$\tfrac{1}{3}Nmv_{\text{rms}}^2 = nRT$$

The total number of molecules N can be written as the product of the number of moles n and the number of molecules per mole N_A; that is, $N = nN_A$. Using this result in the equation above and solving for $\tfrac{1}{2}mv_{\text{rms}}^2$, we get

$$\tfrac{1}{2}mv_{\text{rms}}^2 = \frac{3}{2}\frac{R}{N_A}T \tag{20-9}$$

We can write this relation in simpler form by introducing the *Boltzmann constant k,* which is defined as

$$\text{Boltzmann constant} = k \equiv \frac{R}{N_A} = 1.380662 \times 10^{-23} \text{ J/K} \tag{20-10}$$

The Boltzmann constant is the universal gas constant per molecule.
   Equation (20-9) can then be written

$$\tfrac{1}{2}mv_{\text{rms}}^2 = \tfrac{3}{2}kT$$

Now, by definition, the rms molecular speed is $v_{\text{rms}} = \sqrt{\overline{v^2}}$, where v^2 represents the average of the squares of the individual molecular speeds. It follows that $\tfrac{1}{2}mv_{\text{rms}}^2 = \tfrac{1}{2}m\overline{v^2}$. Therefore, the relation above becomes simply

$$\text{Average translational kinetic energy per molecule} = \tfrac{1}{2}m\overline{v^2}$$
$$= \tfrac{3}{2}kT \tag{20-11}$$

The average kinetic energy of a molecule in random chaotic motion in a low-density gas with many molecules is directly proportional to the absolute temperature of the gas in thermal equilibrium. This fundamental relation tells us what temperature means for an ideal gas; temperature is proportional to, and therefore a measure of, the translational kinetic energy per molecule.

Before considering implications of (20-11), let us emphasize that this simple relation is special. It applies only for gases that are, in macroscopic terms, of very low density or, in equivalent microscopic terms, satisfy all the assumptions made for the kinetic theory of gases (Section 20-2). Equation (20-11) does not apply to solids or liquids; in these states of matter the atoms are close together and interact continuously. Nor does this relation apply to a dense gas, any gas that departs from the general-gas law. For such a nonideal gas, the molecules spend an appreciable time interacting, and their total energy cannot therefore be taken to consist only of translational kinetic energy. For any of these more complicated systems, there is a relation between the system's energy and the absolute temperature of the material, but this relation is more complex than (20-11). In short, the absolute temperature measures average kinetic energy per molecule *only for a low-density ideal gas.*

Note, furthermore, that even for an ideal gas it is the average *translational* kinetic energy of the molecule as a whole that is proportional to *T*. For a gas of monatomic molecules, that is the only kinetic energy we need be concerned with. But for polyatomic gas molecules there may also be energy associated with the motions of the atoms relative to the center of mass of the molecule. In the diatomic molecule H_2, for example, the two hydrogen atoms may vibrate in simple harmonic motion along the interatomic axis under the influence of a restoring interatomic force; we then have both kinetic and potential energy associated with the vibrational simple harmonic motion. In addition, an H_2 molecule may also rotate about an axis perpendicular to the interatomic axis in the fashion of a dumbbell, and there is then kinetic energy associated with the molecular rotation. Quite apart from whether there is molecular vibration or rotation, however, for ideal gas it is the average translational kinetic energy of the molecule that measures the gas's absolute temperature.

The implications of (20-11) are these:

• For any ideal gas in thermal equilibrium, the average translational kinetic energy per molecule is constant. Any temperature change is reflected in a corresponding change in average molecular translational kinetic energy. If the gas could continue to have the properties of an ideal gas all the way down to the lowest temperatures—a strictly hypothetical situation, since every gas becomes a liquid at a sufficiently low temperature—then the *absolute zero* of temperature would correspond to that state in which all molecules are *at rest.*

• When two different kinds of gas are in thermal equilibrium and therefore have a common temperature, *all* the molecules, quite apart from differences in their masses, have the *same* average translational kinetic energy. Therefore, molecules of low mass have relatively high speeds and molecules of high mass have relatively lower speeds. For example, in a mixture of molecular hydrogen H_2 (mass $\simeq 2$) and molecular oxygen O_2 (mass $\simeq 32$) in thermal equilibrium, the mass ratio of oxygen to hydrogen is 16. Both types of gas have the same value for $\frac{1}{2}m\overline{v^2}$, and for hydrogen the rms speed must be 4 times that of oxygen. Suppose that a mixture with both types of gas molecules had to fight its way through a long maze. Then the fast-moving, low-mass molecules would get through more quickly than the slower-moving, high-mass molecules. This process of separating gases of different mass is known as *thermal diffusion.* To cite another example, there is essentially no hydrogen in the earth's atmosphere, since molecules of hydrogen at the top of the atmosphere have speeds exceeding the escape speed from earth (Section 15-6).

Example 20-2.   Hydrogen gas becomes ionized when the average kinetic energy of a hydrogen molecule is of the order of 1×10^{-19} J. At what approximate temperature does hydrogen ionize?

The average kinetic energy K per molecule is, from (20-11),

$$K = \tfrac{3}{2}kT$$

The ionization temperature is then

$$T = \frac{K}{\tfrac{3}{2}k} = \frac{1 \times 10^{-19}\,\text{J}}{(1.5)(1.4 \times 10^{-23}\,\text{J/K})} = 5 \times 10^{3}\,\text{K}$$

The temperature at the surface of the sun is considerably higher than 5000 K, and the hydrogen there is completely ionized.

20-4   Internal Energy

In thermal physics the term *internal energy,* represented by U, is used to designate the *total energy content* of a system of particles. Internal energy is, however, measured in a reference frame in which the system's *center of mass is at rest.* The total momentum of the system, in other words, must be zero.

For an ideal gas consisting of monatomic hard-sphere molecules, the system's internal energy is particularly simple. It is just the total translational kinetic energy of the molecules, which is equal to the average translational kinetic energy per molecule $\tfrac{3}{2}kT$, multiplied by their total number N:

$$\text{Hard-sphere ideal gas:}\quad U = N(\tfrac{3}{2}kT) = \tfrac{3}{2}nRT \qquad (20\text{-}12)$$

As (20-12) shows, the internal energy of an ideal gas is a function of its *absolute temperature only.* In symbols,

$$U = U(T \text{ only}) \qquad \text{for } pV = nRT$$

For a system more complicated than an ideal gas with hard-sphere molecules, a system in which the particles interact much of the time — a nonideal gas, a solid, or a liquid, for example — the internal-energy relation is more complicated.

A central feature of thermal physics is this: a system with many constituent particles may have internal energy that does not show up macroscopically. If, for example, we look at gas in a bottle, we don't "see" internal energy, not directly anyway. For an ideal gas, the internal energy is the kinetic energy of molecules in random, disordered motion. The molecules have kinetic energy but no net momentum. This comes, basically, from the circumstance that the net momentum $\Sigma \mathbf{p}$ (a vector quantity) of a pair of molecules that move in opposite directions may be zero, whereas the net kinetic energy ΣK (a scalar quantity) of such a pair of molecules cannot be zero.

We can say the same thing as follows. If a system of particles has net momentum, this system certainly also has kinetic energy. An example would be the particles moving together in *ordered* fashion, all particles with same velocity, in Figure 20-6(a). But the contrary is not true. If a system of particles as a whole has no net momentum — for example, the particles in random motion in Figure 20-6(b) — the kinetic energy need not be zero.

As we shall later see in some detail (Section 21-1), the same thing can be said about potential energy. A stretched spring obviously has potential energy.

Figure 20-6. (a) Molecules with the same velocity display ordered motion. (b) Molecules with randomly distributed velocities display disordered motion.

But a macroscopically *undeformed* object may nevertheless have potential energy on a microscopic scale, not manifest macroscopically. A central concern of thermal physics is, in fact, with what might be called "invisible" energy—kinetic and potential energy, to be sure, but on a microscopic scale.

The internal energy U of any system is said to be a *state variable*, or a *state function*, of the system; it is a characteristic property of the system. (Other examples of state variables are temperature, volume, and pressure.) Internal energy is not merely an alternative name for what in everyday language is called simply *heat*.

20-5 Heat

The term *work* as used in physics has a precise meaning, which differs from its usage in ordinary speech. The same is true of *heat*. As used in physics, and especially as used in the laws of thermodynamics, *heat* has a special and restricted meaning that differs from its informal meaning in everyday language.

First, heat concerns energy transfer, a process in which energy is transferred because of a difference in temperature. Water in a pan on the stove is being heated; because of a temperature difference between the pan and the water in contact with it, energy is transferred from the pan to the water. *Heat Q* is defined as the *amount of energy transferred* from one object or system to another object or system *by virtue of a temperature difference* between them.

Since heat measures an amount of energy transferred, it has energy units. In the SI system, with the joule as the official energy unit, heat is also most properly measured in joules.* The term *heat* is defined formally as the process of energy transfer arising from a temperature difference, and that strictly, this is the only proper use of heat in thermodynamics. The word *heat* is, however,

---

* Traditionally, heat has been measured in calories. The calorie (abbreviated cal) is now officially defined by the joule, as follows:

$$1 \text{ cal} \equiv 4.1840 \text{ J}$$

The calorie has a simple meaning in the thermal properties of water; one calorie added to one gram of water raises its temperature by 1 C°. Strictly, the energy required to raise the temperature of 1 gm of water by precisely one Celsius degree differs according to the initial temperature of the water. It is a minimum in the temperature interval from 14.5°C to 15.5°C, but greater for lower or higher temperatures.

The term *calorie* when used in discussions of human diets actually means *kilocalorie* (kcal). A person on a "1000-calorie diet" consumes food that releases a total of 10^3 kcal when oxidized in digestion.

The thermal-energy unit for the U.S. Customary or English system of units is the *British thermal unit*, abbreviated Btu. It too has a simple interpretation in terms of the thermal properties of water: 1 Btu added to 1 lb of water increases its temperature by 1 F°. 1 Btu = 1054 J. Comprehensive conversion factors for various energy units appear in Appendix D.

often used (really misused) in a broader context. The trouble is that heat is commonly used to mean two different things:

- The energy-transfer process itself, its formal definition.
- Loosely, as energy content, as in "Friction produces heat," or "Heat is a form of energy."

What's wrong with this? As we shall see in our discussions of the first law of thermodynamics, heat cannot be a property of a system itself. The temperature and the internal energy of a system are state variables or state functions, but heat is not. We can heat a system from some initial state 1 to some final state 2 in different ways, just as there are various ways of doing work on a system in going from state 1 to state 2 (Section 19-5), and the total heat depends on the specific process used to get from state 1 to state 2. Roughly speaking, heat tells us something about how we go from state 1 to state 2, not where we are.

Both *work* and *heat* denote energy transfer. You cannot speak about the heat content of a system any more than you can speak of the work content. The continuing misuses of *heat* are rooted in the idea, now thoroughly rejected, that heat (or caloric, as it was first called) is somehow a substance. Alas, the misuses of *heat* are thoroughly entrenched in much of the parlance of thermal physics.

If it is improper to use *heat* to mean the energy content of a system, what term then can we use? We prefer *thermal energy* to represent the disordered microscopic energy content of any system, and we shall use that term consistently in what follows.

20-6 First Law of Thermodynamics, Microscopic Interpretation

The first law of thermodynamics is a simple and fundamental statement about the internal energy U of a system, especially about changes in a system's internal energy. It is the energy-conservation principle generalized to include two modes of energy transfer:

- Work W, the *macroscopic* energy transfer in which a force component acts along the direction of the displacement of a large-scale object.
- Heat Q, the transfer of *microscopic* thermal energy associated with a temperature difference.

The formal statement of the first law of thermodynamics is given in symbols that have the following meanings:

- dU, a differential *increase* in the system's internal energy U.
- dQ, a small amount of energy transferred *to* the system as heat.
- dW, a small amount of work done *by* the system ($+dW$ is work done by the system; $-dW$ represents work done on the system).*

---

* Although it might seem more sensible to concentrate on energy that is entering the system because of work done on it, rather than on energy that is leaving the system as work done by the system, the convention in physics is to have *positive dW* represent energy *leaving* the system and negative dW as energy entering the system. The reason for this apparently odd choice is that the first law was applied mainly to heat engines after it was first formulated. The principal function of an engine is to have the expanding gas in a cylinder do work on the piston as it pushes it out, so the convention was established that $+dW$ means work by the system.

"It's Only a Theory"

It's a common put-down, to say that some idea is *only* a theory. The implication, of course, is that *theory* necessarily means an idea that is conjectural, un-proved, untested, and therefore quite possibly wrong.

But that is not what theory means in science, and especially in physics. Theory is the conceptual frame-work, the basic suppositions, the guts of the idea, the way we try to make sense out of the behavior of the natural world. A theory may be expressed in math-ematical terms—that's often the easiest way to sum-marize relations in physics—but theory is not merely mathematics. In a very real sense, physics is just a collection of theories—the right ones, the ones that have been found by experimental test to best describe physical phenomena.

Theory and experiment play complementary roles in physics: experiment discloses the facts of nature; theory tries to make sense out of them. Are theory and fact then opposites? Absolutely not! The opposite of fact is ignorance, lack of knowledge, fiction, even fantasy (but *not* theory); the opposite of theory is no idea, no guiding principle (but *not* fact).

The way in which the labels *theory, law,* and *principle* are applied (and stick) is often a matter of historical accident. One way to think of a law of physics is this: it is a broad organizing principle that has worked so well in the past that we are willing to bet very heavy odds that it will work one more time. The Newtonian laws of motion immediately spring to mind. But we could just as well, perhaps even pref-erably, call them the Newtonian theory of mechanics. These "laws" actually have only limited applicability; they fail when applied to the domain of the very fast (relativity physics is required here) and to the domain of the very small (quantum physics is required here). The "theory of special relativity," on the other hand, has been known by that label ever since Einstein first propounded it eighty years ago. But special relativity has been checked by so many stringent experimental tests that it could more accurately be called "the law of special relativity."

What are the hallmarks of a good theory?

• First of all, it must *work*. It must agree with experiment. But the limitations of experimental test and the range of applicability must always be rec-ognized. (An elementary example: the change in grav-itational potential energy is *mgy*, but only if *g* can be taken to be constant and *y* is small compared to the earth's radius.)

Heat going into the system is $+dQ$. Work going into the system is $-dW$. The corresponding increase in internal energy is $+dU$. Therefore, with work and heat as the only modes of energy transfer, we have

$$dQ + (-dW) = dU$$

or as the first law of thermodynamics is usually written,

$$dQ = dU + dW \qquad (20\text{-}13)$$

Figure 20-7 is a schematic representation of the first law.

The first law of thermodynamics applies to any thermal system. But in this

---

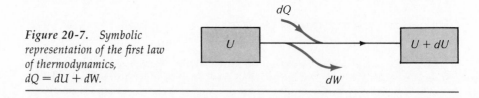

Figure 20-7. Symbolic representation of the first law of thermodynamics, $dQ = dU + dW$.

• It has an *extensive area of applicability*. (Example: Newton's theory of gravity is universal; it applies to apples, the moon, the planets.)

• It relates a *wide variety of disparate phenomena* (Example: the kinetic theory of gases not only accounts for the mechanical origin of the pressure of a gas; it can also explain at once and in remarkably simple fashion why, when a capsule of gas is broken in an evacuated chamber, the gas fills the chamber almost instantaneously.)

• It is *simple, and therefore beautiful*. The grand theories of physics have been extraordinarily simple, and it is both an esthetic preference as well as an act of faith for physicists to believe that when a theory is right it is also simple and elegant. (Example: Observations in 1919 showed that light was bent by the sun in accordance with predictions of the general theory of relativity, an especially elegant theory. Einstein was asked, "But professor, what if the results had *not* agreed with your theory?" "That," said Einstein, "would have been too bad for the Dear Lord." Einstein's confidence in the ultimate simplicity of science and in the way it sees the universe is expressed in what has become perhaps Einstein's most celebrated saying: "Subtle is the Lord, but nasty He is not."*)

*In the original German, "Raffiniert is der Herrgott, aber boshaft is er nicht." The statement is engraved on the mantle of a fireplace in Fine Hall, Princeton University. For an extensive list of quotations

When a theory is found to be defective, it can sometimes be fixed by minor patchwork. But at crucial turning points in the development of physics, radical surgery is often required. Indeed, a new theory may differ drastically from the one it supplants. (Example: the older view that a gas continuously fills space was replaced, in the kinetic theory of gases, by the idea of a low-density gas consisting of empty space populated with an occasional molecule.) The changes in basic conceptions produced by relativity theory and quantum theory at the beginning of the 20th century were so dramatic that even today, more than eight decades after these ideas were first introduced, people still refer to these fields of physics as "modern" physics.

Physicists can easily become hung up on theories they have become accustomed to, and it is sometimes hard to be receptive to new ideas. The great Danish physicist, Niels Bohr, who was the first to apply quantum theory to atomic structure, once heard a young physicist present his rather wild new ideas at a meeting of physicists. After the talk had ended and comments were invited, Bohr got up and said, "Young man, you sure have a crazy idea there. The trouble is, it's not crazy enough."

by Einstein on physics and science see *Nature* **278,** 215 (March 15, 1979).

chapter we consider only the simplest thermal system — an ideal gas — from both a microscopic and a macroscopic point of view. We wish especially to see first what dQ, dU, and dW mean in terms of the translational motion of gas molecules of a monatomic ideal gas.

Suppose that a low-density gas is in a cylindrical container with a tightly fitted, leakproof and frictionless piston. The gas may expand, pushing the piston outward and doing positive work on the piston. Or the gas may be compressed by moving the piston inward; then the gas does negative work. We suppose that the temperature of the container may be controlled to be the same as the temperature of the gas inside, or higher or lower. Here the system itself is the gas, or in microscopic terms, the molecules in random motion inside the container.

Consider first the simplest situation; the piston is locked in position, and the entire system is in thermal equilibrium. The gas is then at the same fixed temperature as the container. With no temperature difference between the gas and the container, there can be no heat in or out of the gas, and $dQ = 0$. Furthermore, with the piston fixed in position, no work is done by the gas, and $dW = 0$. From the first law of thermodynamics, (20-13), it follows then that

$$dQ = dU + dW$$

$$0 = dU + 0$$

Since $dU = 0$, the internal energy does not change. A constant U implies, from (20-12),

$$U = \tfrac{3}{2}nRT$$

that the absolute temperature is constant. And a constant T implies, from (20-11),

$$\tfrac{1}{2}m\overline{v^2} = \tfrac{3}{2}kT$$

that the average translational kinetic energy per molecule is also constant.

What is going on at the molecular level? The molecules collide frequently with one another. Some molecules gain kinetic energy in a collision; others lose kinetic energy. But on the average there is no change in molecular kinetic energy. Indeed, from the molecular point of view, this state of thermal equilibrium can be described as follows: on the average, a molecule neither gains nor loses kinetic energy in collision.

Gas molecules also collide with the container walls. Now any container wall, viewed microscopically, is not perfectly hard and smooth. The wall consists of atoms in vibratory motion. So a gas molecule hitting the container wall typically strikes an atom in motion. The gas molecule can gain kinetic energy or lose it in such a collision. But if the gas and container are at the same temperature, the gas molecule hitting a wall cannot, on the average, either gain or lose kinetic energy. From a molecular point of view, equality of gas and container temperatures means simply that here again, a typical gas molecule rebounds from a collision with the wall with the same kinetic energy. See Figure 20-8.

How do we interpret heat in molecular terms? Now suppose that the container walls are of a higher temperature than the gas, so that $dQ > 0$. The piston is still locked in position, so that $dW = 0$. From the first law,

$$dQ = dU + dW$$

We then have with $dW = 0$, $dQ > 0$, that

$$dU > 0$$

What's happening to the gas molecules in this heating process? The gas's internal energy increases; so does its temperature and the average molecular kinetic energy. How do the molecules gain the additional kinetic energy? It

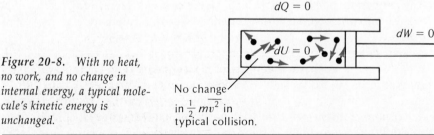

Figure 20-8. With no heat, no work, and no change in internal energy, a typical molecule's kinetic energy is unchanged.

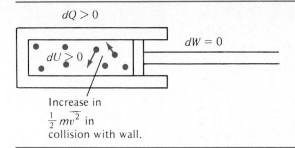

Figure 20-9. *The gas is heated, no work is done, and the internal energy increases. A typical molecule gains kinetic energy on colliding with a wall.*

must be in collisions with atoms in the container walls. With the wall hotter than the gas, a typical molecule must, in colliding with an oscillating atom in the wall, rebound with greater kinetic energy. From a microscopic point of view, heat means a change, on the average, in the kinetic energy of a gas molecule hitting the wall. See Figure 20-9. Suppose that the gas temperature were higher than the temperature of the container, so that the "direction" of heat would be reversed. Then the drop in internal energy of the gas would correspond to the circumstance that a typical molecule *loses* kinetic energy when it hits a wall.

Finally, what does work mean in molecular motion? Now we suppose that the gas expands relatively slowly and does positive work on the piston, so that $dW > 0$. To keep things simple, we suppose further that there is no energy transfer by heat, so that $dQ = 0$. This means that the temperature of the container walls is somehow always maintained at the same temperature as the gas within. From the first law

$$dQ = dU + dW$$

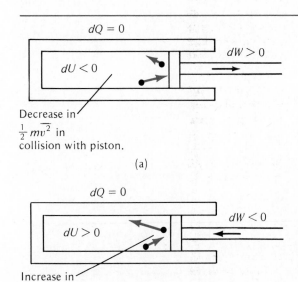

(a)

(b)

Figure 20-10. *(a) A typical molecule loses kinetic energy when it strikes an outwardly moving piston. (b) The piston is moved inward and work is done on the gas. The inwardly moving piston increases the average kinetic energy of a molecule striking it.*

With $dW > 0$ and $dQ = 0$, we have

$$dU < 0$$

The internal energy of the expanding gas decreases; so do the gas temperature and the average molecular kinetic energy.

What slows the molecules? It's a matter of simple mechanics. Those molecules colliding with the piston are hitting an object *moving outward*, so that these molecules rebound, on the average, at a lower speed than that with which they approached the piston. See Figure 20-10(a). The energy lost by molecules colliding directly with the piston is quickly shared among other molecules through intermolecular collisions.

If the piston were to move inward, rather than outward, compressing the gas, the situation would be exactly reversed. See Figure 20-10(b). Then the inwardly moving piston would speed up molecules that hit it; consequently, the internal energy and gas temperature would rise. The rise in gas temperature and internal energy would simply mean that a very massive moving object (the piston) was hitting many light objects (the molecules)!

Macroscopically, work and heat may appear to be different energy-transfer processes. At the microscopic level, however, they cannot be distinguished. Indeed, the only thing happening at the molecular level is that one molecule (or atom) collides with another one and they interact.

20-7   First Law of Thermodynamics, Macroscopic Interpretation

A system's *internal energy* is a characteristic property of the system, or in more formal language, internal energy is a *state function*, or a *state variable*. It is easy to see that both heat and work are not state functions, but rather describe processes. We consider just two of the many possible ways in which an ideal gas can go gradually from some initial state i to some final state f, as shown in Figure 20-11. In Figure 20-11(a), a gas expands isothermally. Since T is constant, the internal energy does not change.

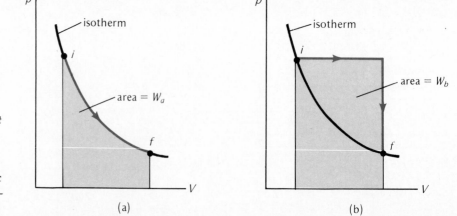

Figure 20-11. *Two different processes in which there is no net change in the internal energy of an ideal gas: (a) isothermal expansion; (b) isobaric expansion followed by isovolumetric compression.*

The expanding gas does positive work W_a, whose magnitude equals the area under the line representing the process on a pV diagram (Section 19-5). The first law of thermodynamics, (20-13), written in the form appropriate for a finite change ΔU in internal energy, is

$$Q = \Delta U + W \qquad\qquad (20\text{-}14)$$

where Q is total heat in and W is total work out of (or work done by it) the system. For the isothermal process, we have $\Delta U_a = 0$, so that

$$Q_a = W_a$$

Energy *leaves* the system as work, and an equal amount *enters* as heat.

Now consider Figure 20-11(b). Here the gas first expands slowly at constant pressure and then experiences a gradual drop in pressure at constant volume until it reaches the *same* final state f as in Figure 20-11(a). This process is not an isothermal one, but the final temperature is again the same as the initial one. Here again $\Delta U_b = 0$, so that $Q_b = W_b$. The areas under the pV curves clearly are not the same in Figures 20-11(b) and 20-11(a), so that $W_a \neq W_b$. The heats Q_a and Q_b must also be different for the two different routes on the pV diagram between the same initial and final states. The conclusion: the *change in internal energy does not depend on how we get from an initial to a final state, but the heat and work do depend on the path.* This result is confirmed by experiment for all thermal systems. Sometimes the essential content of the first law of thermodynamics is said to be merely that a system's internal energy is a state function.*

Example 20-3.   An ideal gas, initially with pressure p_1 and volume V_1, expands slowly at constant pressure until its volume doubles. Then, with volume held constant, the pressure of the gas is slowly reduced to half. See Figure 20-12. (a) What is the overall change in the gas's internal energy? (b) What is the overall work done by the gas? (c) By how much is the gas heated overall?

(a) The changes in the state of the gas shown in Figure 20-12 can be represented as follows:

$$(p_1, V_1) \rightarrow (p_1, 2V_1) \rightarrow (\tfrac{1}{2}p_1, 2V_1)$$

In the final state the product of pressure and volume is $(\tfrac{1}{2}p_1)(2V_1) = p_1V_1$, exactly the same as initially. With pV the same at the beginning and end, we know that the gas's final temperature is the same as its initial temperature. As the gas expands its temperature rises; then, as its pressure decreases, the temperature falls and returns to the initial value. The results here are the same as those shown in Figure 20-11(b).

The internal energy depends only on the gas temperature. There is, therefore, *no change* overall in internal energy.

(b) The work done by the expanding gas is equal in magnitude to the area under the curve on a pV diagram. From Figure 20-12, we see that $W_{1\rightarrow2} = p_1(2V_1 - V_1) = p_1V_1$.

(c) From the first law of thermodynamics, (20-14),

$$Q_{1\rightarrow2} = (U_2 - U_1) + W_{1\rightarrow2}$$

Here $U_2 - U_1 = \Delta U = 0$ and $W_{1\rightarrow2} = p_1V_1$. Therefore, $Q_{1\rightarrow2} = p_1V_1$.

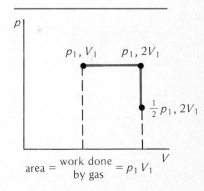

Figure 20-12.

* The first law of thermodynamics is often written in symbolic form as $dU = dQ - dW$, to show that whereas dU is a true differential, dQ and dW are not.

20-8 Specific Heats of an Ideal Gas

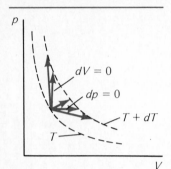

Figure 20-13. Several changes in the state of an ideal gas, leading to an increase in its temperature from T to T + dT. The vertical displacement corresponds to isovolumetric heating; the horizontal displacement, to isobaric heating.

The *specific heat* of any substance is a measure of how hard or easy it is to change the material's temperature by adding thermal energy to or removing it from the material. More precisely, the specific heat of any material is defined as the heat required to raise the temperature of one unit mass of that material by one degree.* The *molar specific heat* is the heat required to raise the temperature of one mole by one degree. In this section we are concerned with the specific heats of an ideal gas.

We use the plural, specific *heats*, because there are actually an indefinite number of processes that can raise the temperature of a gas by dT through the addition of heat dQ. The various specific heats differ according to how other properties of the gas are affected as its temperature changes. In all the processes shown on a pV diagram in Figure 20-13, the temperature of a gas is raised. The two specific heats of the greatest interest are the molar specific heat C_v at *constant volume* and the molar specific heat C_p at *constant pressure*.

It is easy to see that C_p must always be larger than C_v; that is, more heat is required to raise the temperature of an ideal gas when its pressure remains constant than to raise the temperature of the gas by the same amount with its volume unchanged. Consider Figure 20-14, in which a gas in a chamber is confined by a leakless piston. The gas is heated (a) with the piston fixed in position and the volume constant, and (b) with the gas expanding, the piston moving upward so that the pressure of the gas remains constant.

For constant volume (a), *all* the heat added to the gas goes into raising its temperature; no energy leaves the gas, since the gas is not expanding and therefore doing work on the piston. On the other hand, if the pressure (b) is to be maintained constant, the gas must expand. But in expanding, the gas does work on the piston. More heat must then enter the gas for a given temperature change to compensate for the energy leaving the gas as work it does on the container. In short, C_p must exceed C_v.

The particular values for C_p and C_v differ according to the type of gas. For an ideal gas, however, the two molar specific heats always differ by the same value R, the constant of the general-gas law:

$$C_p - C_v = R \qquad (20\text{-}15)$$

To prove this, we first consider heating at constant volume. From the first law of thermodynamics,

$$dQ_v = dU_v + dW_v \qquad (20\text{-}16)$$

$$dQ_v = dU_v + 0$$

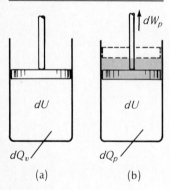

Figure 20-14. (a) Isovolumetric heating. All heat dQ_v entering the gas raises the internal energy dU of the gas. (b) Isobaric heating. Some of the heat dQ_p entering the system raises the internal energy dU of the gas, while the remainder corresponds to work dW_p done by the gas.

* "Specific heat" is shorthand for what is known more completely as a material's specific heat capacity. The *heat capacity* of any object is defined as the amount of thermal energy required to increase the object's temperature by $1\,C°$; heat capacity is a property of a particular object. *Specific heat capacity* is the heat capacity per unit mass; it is a property of the material of which the object is composed. The word *specific* is generally prefixed to indicate that the physical quantity is measured relative to some standard unit; in the case of specific heat capacity, this is the heat capacity per unit mass. The term *specific* may also be used to designate a physical property relative to some standard material; for example, the "specific gravity" of 13.6 for mercury means that mercury has a density 13.6 times that of water.

where we recognize that with volume held constant (subscript v), $dV = 0$ and $dW_v = p\, dV = 0$.

By definition, the molar specific heat at constant volume is

$$C_v = \frac{dQ_v}{n\, dT}$$

or

$$dQ_v = nC_v\, dT$$

Therefore, (20-16) can be written

$$\text{Isovolumetric heating:}\quad dU_v = nC_v\, dT \qquad \text{(20-17)}$$

Now the same ideal gas is heated through the same temperature difference dT, but at constant pressure. The gas must expand, and the expanding gas does work in the amount

$$dW_p = p\, dV_p$$

The molar specific heat at constant pressure is defined as

$$C_p = \frac{dQ_p}{n\, dT}$$

or

$$dQ_p = nC_p\, dT$$

so that first law of thermodynamics is now written as

$$\text{Isobaric heating:}\quad nC_p\, dT = dU_p + p\, dV_p \qquad \text{(20-18)}$$

The internal energy of an ideal gas depends only on its temperature T, and the temperature difference dT is the same for isovolumetric and isobaric heating. Therefore, the internal energy changes are also the same:

$$dU_p = dU_v$$

If we then eliminate dU between (20-17) and (20-18), we have

$$nC_p\, dT = nC_v\, dT + p\, dV_p$$

This result is general; it applies to any gas, whether the gas is ideal or not.

If the gas is ideal, the general-gas law, $pV = nRT$, applies. In an isobaric process, with p constant, we then have

$$p\, dV_p = nR\, dT$$

With this result substituted in the relation above, we have finally

$$n(C_p - C_v)\, dT = nR\, dT$$

$$C_p - C_v = R \qquad \text{(20-15)}$$

We have shown that for an ideal gas, at constant pressure the molar specific heat always exceeds that at constant volume by the amount $R = 8.31\,\text{J/mol} \cdot \text{K}$, a result that agrees with experiment for all low-density gases.

Although C_p and C_v always differ by an amount R for all ideal gases, the specific heats for particular gases depend on the molecular structure of the gas

Table 20-1

STATE OF MOTION	ENERGY PER MOLECULE (FOR AN IDEAL GAS)	U	C_v	$C_p = C_v + R$	$\gamma = \dfrac{C_p}{C_v}$
Monatomic molecule, or a diatomic molecule undergoing *translation* only	$\frac{3}{2}kT$	$\frac{3}{2}nRT$	$\frac{3}{2}R$	$\frac{5}{2}R$	$\frac{5}{3}$
Diatomic molecule in *translational* and *rotational* motion	$\frac{3}{2}kT + kT$	$\frac{5}{2}nRT$	$\frac{5}{2}R$	$\frac{7}{2}R$	$\frac{7}{5}$
Diatomic molecule in *translational, rotational, and vibrational* motion	$\frac{3}{2}kT + kT + kT$	$\frac{7}{2}nRT$	$\frac{7}{2}R$	$\frac{9}{2}R$	$\frac{9}{7}$

molecules. If the gas is monatomic, the internal energy consists *solely* of the *translational* kinetic energy of the molecules, and (20-12) gives

$$U = \tfrac{3}{2}nRT$$

so that

$$C_v = \frac{1}{n}\frac{dU}{dt} = \frac{3}{2}R$$

For such a gas, we then have

$$C_p = C_v + R = \tfrac{3}{2}R + R = \tfrac{5}{2}R$$

A diatomic molecule may have additional kinetic energy associated with the rotation of the molecule as a rigid body about a rotation axis perpendicular to the line joining the two atoms. It can be shown that because of molecular rotation, the internal energy is increased by R (and the kinetic energy per molecule by kT), so that $C_v = \tfrac{3}{2}R + R = \tfrac{5}{2}R$. Furthermore, the vibration of atoms in a diatomic molecule along the interatomic axis increases the internal energy by R (and the kinetic energy per molecule by kT), so that for molecules with vibrational motion, as well as rotational and translational motion, we have $C_v = \tfrac{3}{2}R + R + R = \tfrac{7}{2}R$. Table 20-1 summarizes the specific heats for diatomic molecules. At $T = 40$ K, for example, molecular hydrogen has a measured value of C_p equal to 2.5R (translation only); at 400 K, $C_p = 3.5R$ (translation and rotation only); and at 4000 K, $C_p = 4.5R$ (translation, rotation, and vibration).

Example 20-4. An ideal gas undergoes an adiabatic process, one in which no thermal energy is transferred into or out of the system by heat. Show that an *adiabatic process* is described by $pV^\gamma =$ constant, where $\gamma \equiv C_p/C_v$.

For an adiabatic process, with $dQ = 0$, the first law of thermodynamics gives

$$dQ = dU + dW$$

$$0 = dU + dW$$

An expanding gas does work $dW = p\,dV$. The internal energy of the gas changes by $dU = nC_v\,dT$. Note that the specific heat at constant *volume* is used for dU inasmuch as we take into account separately work done dW, and no work is done by a gas heated at constant volume. Then the equation above can be written

$$0 = n\,C_v\,dT + p\,dV$$

Note that an adiabatic expansion $(dV > 0)$ is accompanied by a drop in temperature $(dT < 0)$.

To integrate the equation above, we must eliminate one of the three variables p, V, and T that appear in it. We use the equation of state, $pV = nRT$, to eliminate pressure p, and arrive at

$$nC_v\,dT = -\frac{nRT}{V}\,dV$$

$$\int \frac{dT}{T} = -\frac{R}{C_v} \int \frac{dV}{V}$$

$$\ln T = -\frac{R}{C_v} \ln V$$

$$\ln T + \ln V^{R/C_v} = 0$$

$$TV^{R/C_v} = K_1 \tag{20-19}$$

where K_1 is a constant. This equation describes an adiabatic process by the variables T and V.

To get the equation of an adiabatic process in terms of the variables p and V, we use the fact that $pV \propto T$ for an ideal gas. Eliminating T from (20-19), we then have

$$(pV)V^{R/C_v} = K_2$$

$$pV^{R/C_v + 1} = K_2$$

where K_2 is another constant.

Now since $C_p - C_v = R$, according to (20-15), the equation above becomes

$$pV^{(C_p - C_v)/C_v + 1} = pV^{C_p/C_v} = K_2$$

$$pV^\gamma = K_2 \tag{20-20}$$

where $\gamma \equiv C_p/C_v$ is the specific heat ratio.

Since C_p always exceeds C_v, γ always exceeds 1. An adiabatic process, with $pV^\gamma = $ constant, as portrayed on a pV diagram, differs from an isothermal process, with $pV = $ constant. As Figure 19-14(d) shows, an adiabatic process falls off more sharply with an increase in pressure than an isothermal process does, starting from the same point. Moreover, the adiabatic expansion always involves a temperature drop. Since, by definition, an adiabatic process is one in which no heat enters or leaves, a gas can change its state adiabatically only when it is effectively in an insulated container.

20-9   Disorder and the Second Law of Thermodynamics

A fundamental principle of physics — the second law of thermodynamics — is based on chance. We here consider this proposition semiquantitatively and only for the relatively simple system of the molecules in an ideal gas. (A more complete and formal treatment of the second law is given in Chapter 22.) For our present purposes, the second law of thermodynamics can be stated as follows:

An isolated system, free of external influence, will if it is initially in a state of relative order, always pass to states of relative disorder until it eventually reaches the state of maximum disorder, thermal equilibrium.

Consider, first, how the molecules of an ideal gas are distributed spatially within a container. We know by intuition that the distribution is essentially uniform; the number of molecules per unit volume in any small volume is the same at any location throughout the volume of the container.

To see why this must be so, we make just one basic assumption: *one location is like any other,* so that the location of any one molecule is determined strictly by the laws of chance. This means that if we have some container divided — in our imagination — into two parts of equal volume, the probability that some one molecule is in the left half of the container is the same as the probability that it will be in the right half. A molecule shows no preference for left or right; therefore, the probability that one molecule will be in the left side is $\frac{1}{2}$ and the probability that it will be in the right side is also $\frac{1}{2}$. (We could equally well decide where a molecule will go by flipping a coin, heads meaning left side, tails meaning right side.)

Suppose that we have a gas of very low density — one with only four molecules — and we are to find the various ways in which the four molecules can be located in our container. Each molecule can be in either the left- or the right-hand sides; both sides have the same volume. It is easy to find the probability that all four molecules will be found in the left side, with none on the right. The probability that the first molecule is in the left is $\frac{1}{2}$. The probability that a second molecule is in the left is also $\frac{1}{2}$, so that the probability that both are on the left is $(\frac{1}{2})(\frac{1}{2}) = \frac{1}{4}$. Clearly, then, the probability that all four molecules will be found simultaneously in the left half is $(\frac{1}{2})(\frac{1}{2})(\frac{1}{2})(\frac{1}{2}) = (\frac{1}{2})^4 = \frac{1}{16}$, just one chance in 16. By the same token, the probability that all four molecules will be found in the right side is also $\frac{1}{16}$. It is improbable that we shall find one side empty of molecules and the other side full of them.

It is also easy to find the relative probabilities for all the possible distributions of the four molecules. We simply write, as shown in detail in Table 20-2, all the possible ways in which the four molecules, now identified for convenience as molecules 1, 2, 3, and 4, can be located in the two equal parts of the container.

It is clear what is improbable and what is probable. The most ordered state, the one with all molecules in one half, has the lowest probability; it is only one part in 16 (confirming what we also found earlier). On the other hand, the most probable arrangement — with a probability of 6 out of 16 — is the most disordered state, in which there are equal numbers of molecules in the two halves. In short, the *disordered state is probable* and the *ordered state is improbable* simply because *there are more disordered states than ordered states.*

How about the relative probabilities of ordered and disordered states relating to the motions of the molecules?

We need consider only the *sign* of the horizontal component of any molecule's velocity. Moving to the right is just as probable as moving to the left, so that the two possible directions of horizontal motion correspond to what was earlier the two possible halves of the container. "Being in the left half" is now replaced by "moving to the left." The various possibilities are just like those listed in Table 20-2. With four molecules in random motion, the chance that all molecules will be moving left simultaneously (in a fairly ordered state of motion) is 1 part in 16, whereas the probability will be greatest (6 parts in 16) for the more disordered state, with equal numbers of molecules moving left and right.

What are the chances of having all four molecules in the left half, and at the

Table 20-2.   *Possible distributions of four molecules in left- or right-hand sides of container.*

DISTRIBUTION	NUMBER OF MOLECULES IN LEFT HALF	NUMBER OF EXAMPLES
(1)(2)(3)(4) |	4	1
(2)(3)(4) | (1) ⎫		
(1)(4)(3) | (2) ⎪		
(1)(4)(2) | (3) ⎬	3	4
(1)(3)(2) | (4) ⎭		
(1)(2) | (3)(4) ⎫		
(1)(3) | (2)(4) ⎪		
(2)(3) | (1)(4) ⎪		
(1)(4) | (3)(2) ⎬	2	6
(2)(4) | (3)(1) ⎪		
(3)(4) | (2)(1) ⎭		
(1) | (2)(4)(3) ⎫		
(2) | (1)(4)(3) ⎪		
(3) | (1)(4)(2) ⎬	1	4
(4) | (1)(3)(2) ⎭		
| (1)(2)(3)(4) ⎫	0	1
		‾‾
	Total =	16

same time, all four also moving left? The probability is $(\frac{1}{16})(\frac{1}{16})$, or 1 part in 256. Clearly, if all molecules were to be located within a very small region of space within the container and all molecules were also to move together, with exactly the same velocity, the probability for such a highly ordered state would be extraordinarily small. In other words, the chance that all four gas molecules will coalesce to form a little lump of solid that moves through the container is vanishingly small. On the other hand, the reverse process, in which a lump of solid evaporates into four molecules that become dispersed throughout the container, is highly probable.

We have been discussing a gas with only four molecules. For a more realistic example, consider the air molecules in an empty 12-ounce beer can. There are about 10^{22} of them; you would have to count molecules at a rate of 16,000 molecules per second for a total time of 20 billion years (the age of the universe) to get them all. For such a sample of gas, the probability that all the 10^{22} molecules would at one time be in just one half of the container would be $\frac{1}{2}$ raised to the power 10^{22}, or 7×10^{-322}, a very small number indeed.

From these simple examples, we conclude that if all microscopic states are equally probable, then:

- The *ordered* states are *few*.
- The *disordered* states are *many*.

If all states are equally probable, then any system with a large number of particles will almost inexorably move from the less numerous, and therefore less probable, ordered states to the more numerous and therefore more probable, disordered states until it reaches finally the most probable and therefore most disordered state—thermal equilibrium. For any system with a reasonably large number of particles, there are so many more microscopic states representing disorder than representing order that the state of the system will eventually become, almost certainly, the state of maximum disorder. The second "law" of thermodynamics is based, in terms of microscopic behavior, on overwhelming probabilities rather than on absolute certainty.

Individual molecular collisions are reversible in time. We cannot tell whether a moving-picture film portraying an intermolecular collision is being run forward or backward. But this is not the case when one deals with large numbers of particles. In the macroscopic world, processes are essentially irreversible. Every system with many constituent particles free of external influence moves inexorably from states of relative order to disorder. From our experience, we *can* tell at once when a moving picture of some ordinary large-scale phenomenon, such as an exploding bomb, is run backward. The second law of thermodynamics implies a directionality of time. At the macroscopic level, time's arrow points to the future. Order turns to disorder; ordered energy is degraded into disordered or thermal energy. So too the direction of thermal-energy flow—the "direction" of heat, as it were—is from the higher- to the lower-temperature body, it being thereby ensured that two isolated bodies, initially at different temperatures, achieve a common final temperature.

20-10   Molecular Speed Distribution (Optional)

The kinetic theory relates the root-mean-square speed v_{rms} of the molecules of an ideal gas in thermal equilibrium to the absolute temperature of the gas:

$$\tfrac{1}{2}mv_{rms}^2 = \tfrac{3}{2}kT \tag{20-9}$$

The speeds of the various molecules differ from one another, and for a gas in thermal equilibrium, are distributed in the fashion shown in Figure 20-15. Some molecules have very high speeds and others lower speeds, and a very few may actually be at rest.

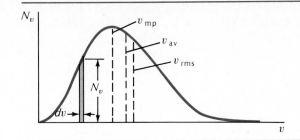

Figure 20-15. *Molecular-velocity distribution. The most probable speed is v_{mp}, the root-mean-square speed is v_{rms}, and the average speed is v_{av}.*

It is important to be clear on what exactly is being plotted in a speed-distribution curve like Figure 20-15, with N_v, as ordinate, plotted as a function of speed v, as abscissa. Suppose we choose a small range of speeds dv extending from v to $v + dv$. Then the area of the thin rectangle $N_v \, dv$ gives the number of molecules with speeds between v and $v + dv$. The total area under the curve equals the total number of molecules N. Any ideal gas in thermal equilibrium has a molecular speed distribution like that of Figure 20-15. If, for a given gas, the temperature is increased, the distribution curve for the new state of thermal equilibrium shifts to the right in accordance with Equation 20-9.

The mathematical form for the speed distribution function N_v was first derived by J. C. Maxwell, and is known as the *Maxwellian speed distribution*. A triumph of nineteenth-century theoretical physics, this relation has been confirmed in detail by experiment. The relation is

$$N_v \, dv = Av^2 e^{-mv^2/2kT} \, dv \qquad (20\text{-}18)$$

$$= Av^2 e^{-K/kT} \, dv$$

where m is the mass of a molecule, K is its kinetic energy $\tfrac{1}{2}mv^2$, and T is the absolute temperature. The quantity A is, for a fixed temperature, a constant, independent of speed v, and has the value $A = (4N/\sqrt{\pi})(m/2kT)^{3/2}$; this specific value can be confirmed by computing the integral $\int_0^\infty N_v \, dv = N$, where N is the total number of molecules.

We see from Figure 20-15 that the speed distribution is not symmetrical with respect to its peak; it falls off relatively slowly to the right of the peak. Three characteristic molecular speeds, all shown in Figure 20-15, can be given for any equilibrium distribution:

- The *most probable* molecular speed v_{mp}. By definition, it corresponds to the peak in the curve, or the maximum value of N_v in (20-21). If one were to sample many molecular speeds, more molecules would have this particular speed than any other.
- The *root-mean-square speed* v_{rms}. It is defined by $v_{rms} = \sqrt{\overline{v^2}}$ and related to temperature T by (20-9). The rms speed v_{rms} exceeds the most probable speed v_{mp} because high speeds carry greater weight than low speeds when one computes an average depending on the *square* of the speeds. It can be shown that $v_{rms} = 1.22 v_{mp}$.
- The *average* speed v_{av}. The simple average is defined by $v_{av} = (v_1 + v_2 + \cdots + v_n)/N$. The average speed lies between v_{mp} and v_{rms}; more specifically, $v_{av} = 1.12 v_{mp}$.

If molecular speeds are distributed for an ideal gas in thermal equilibrium, so too are the molecular kinetic energies. We know that the *average* molecular kinetic energy is directly proportional to the absolute temperature through the simple relation $\bar{K} = \frac{3}{2}kT$. We assert without proof that the most probable molecular kinetic energy K_{mp} is also directly proportional to the absolute temperature and is given in detail by $K_{mp} = \frac{1}{2}kT$.

Summary

Definitions

rms (root-mean-square) molecular speed:

$$v_{rms} \equiv \sqrt{\overline{v^2}}$$

Boltzmann constant k:

$$k \equiv \frac{R}{N_A} \tag{20-10}$$

Internal energy U: total energy of any system (measured relative to a reference frame in which the system's center of mass is at rest).

Heat Q: The process in which energy is transferred by virtue of temperature difference and the amount of energy transferred.

Specific heat: a property of a material giving the heat required to raise the temperature of a unit mass of material one degree. (For a gas especially, the conditions for heating — constant pressure, constant volume, for example — must be specified.)

Units

1 cal ≡ 4.1840 J

Fundamental principles

First law of thermodynamics:

$$dQ = dU + dW \tag{20-13}$$

where dU = increase in system's internal energy, dQ = heat to system, and dW = work done by system. The system's internal energy U is a state function, whereas heat and work depend on the particular process. The first law is the energy-conservation generalized to include heat as a mode of energy transfer.

Second law of thermodynamics (qualitative formulation for microscopic description of a system of many particles): An isolated system will pass from an initial state of relative order to a state of relative disorder until it reaches the state of maximum disorder, thermal equilibrium.

Important results

From the kinetic theory of gases, the pressure of an ideal gas of mass density ρ is

$$p = \frac{1}{3}\rho v_{rms}^2 \tag{20-8}$$

The average translational kinetic energy per molecule in an ideal gas at absolute temperature T is

$$\frac{1}{2}m\overline{v^2} = \frac{3}{2}kT \tag{20-11}$$

The internal energy of ideal gas consisting of hard-sphere molecules is

$$U = \frac{3}{2}nRT \tag{20-12}$$

The difference between the molar specific heats at constant pressure and at constant volume for any ideal gas is

$$C_p - C_v = R \tag{20-15}$$

where

$$C_p \equiv \frac{dQ_p}{n\,dT} \quad \text{and} \quad C_v \equiv \frac{dQ_v}{n\,dT}$$

An adiabatic process for an ideal gas is described by the alternative relations

$$pV^\gamma = K_2 \tag{20-20}$$

or

$$TV^{R/C_v} = K_1 \tag{20-19}$$

where $\gamma \equiv C_p/C_v$.

Problems and Questions

Section 20-1   Molecular Properties

· **20-1 P** What is the volume of an ideal gas at STP that contains the same number of molecules as the population of the United States, about 240 million?

: **20-2 P** Estimate how many marbles of diameter 12 mm can be placed in a cylindrical beaker of diameter 9 cm and height 12 cm. What fraction of the volume is occupied by marbles?

: 20-3 P What is the average separation between molecules in an ideal gas at STP?

Section 21-2   Kinetic Theory and Gas Pressure

: 20-4 P An atomic beam with N particles per unit volume, each with velocity v and mass m, collides perpendicularly with a wall and rebounds elastically. Calculate the average pressure on the wall.

: 20-5 P Compute the number of molecules per second of N_2 gas at STP that strike an area of $1 \, m^2$.

: 20-6 P What is the ratio of the loss in gravitational potential energy for a single molecule undergoing a vertical fall of 1 meter to the average translational kinetic energy per molecule of hydrogen in a gas at 300 K?

: 20-7 P Find the average pressure exerted by a gas consisting of identical non-interacting molecules, each of mass 6×10^{-27} kg, with speed 1300 m/s. There are 3×10^{22} molecules per liter in the gas.

· 20-8 P Suppose that 10 particles had the following speeds, given in m/s. $v = 1, 2, 5, 6, 6, 7, 9, 10, 12, 15$. What is (a) the average speed, (b) the rms speed, and (c) the most probable speed?

: 20-9 P Show that the rms speed of a gas of molecular mass m at the absolute temperature T is given by $v_{rms} = \sqrt{3RT/m}$.

: 20-10 P Show that the following are consequences of the kinetic theory of gases. (a) Avogadro's law: Under the same conditions of temperature and pressure, equal volumes of gas contain equal numbers of molecules. (b) Dalton's law of partial pressures: When two or more gases which do not interact chemically are present together in the same container, the total pressure is the sum of partial pressures contributed independently by each of the several gases.

: 20-11 P The weight of a vessel containing gas exceeds the weight of the same vessel evacuated; a balance indicates a weight for a filled vessel equal to the sum of the weight of the evacuated vessel and weight of the gas alone. Explain why this is so, even though any one molecule of the gas is, according to the kinetic theory, moving freely in space except during collisions of negligible duration. Use the following considerations: A vessel of height h contains one molecule of mass m moving vertically up and down and making perfectly elastic collisions with the top and bottom walls. Its speed is v as it moves upward immediately after striking the lower wall, and it is constantly accelerated downward at the rate g. Show that the average force of the molecule on the bottom wall exceeds that on the top wall by mg; that is, the average force of the molecule on the container is just its weight.

: 20-12 P At STP the molecules of nitrogen make approximately 7×10^9 intermolecular collisions per second. As-

sume that, contrary to the assumptions of the kinetic theory, the intermolecular collisions are *not* perfectly elastic and that one billionth of the molecular kinetic translational energy is lost in each collision. Approximately how long would it take for the temperature of the gas to fall by 1.0 K?

Section 20-3   Kinetic-Theory Interpretation of Temperature

· 20-13 Q We can gain a good intuitive understanding of the meaning of absolute temperature if for a dilute ideal gas we associate T with
(A) the average velocity of a molecule.
(B) the average translational kinetic energy of a molecule.
(C) the average potential energy of a molecule.
(D) the average linear momentum of a molecule.
(E) the total number of degrees of freedom of the system.

· 20-14 Q A flask contains one gram of H_2 and one gram of O_2 gas in thermal equilibrium. In this case,
(A) each gas exerts an equal pressure.
(B) the average velocity of an H_2 molecule is the same as the average velocity of an O_2 molecule.
(C) the average energy of an H_2 molecule is the same as the average energy of an O_2 molecule.
(D) if the temperature of the mixture is T, the average temperature of one species of molecules is greater than T and the average temperature of the other species is less than T.

: 20-15 P What is the average energy, in joules and in electron volts, of a gas molecule at (a) room temperature, $27°C$; (b) liquid nitrogen temperature, 77 K; (c) liquid helium temperature, 4.2 K?

· 20-16 P Helium gas can be ionized when the average kinetic energy per atom is 24.6 eV. To what temperature must helium gas be raised for this to occur?

· 20-17 P The fragments from uranium nuclei that have undergone fission have kinetic energies of about 1.1×10^{-11} J. What would be the temperature of a gas of such particles?

· 20-18 P A thermal neutron is one whose average kinetic energy is equal to that of a molecule of gas at 300 K. What is the rms speed of a thermal neutron (mass, 1.67×10^{-27} kg)?

· 20-19 P A gas consists of a mixture of CH_4 and H_2 at 300 K. What is the ratio of the rms speeds for the two molecules?

· 20-20 P At what temperature will the molecules in an ideal gas have an rms speed 10 times as great as their rms speed at $20°C$?

· 20-21 P The speed required for a molecule to escape the earth's gravitational speed is about 11.2 km/s. (a) At what temperature would a hydrogen molecule have an rms speed as large as the above escape velocity? (b) At what temperature would a nitrogen molecule have an rms speed this large?

: **20-22 Q** Is the root-mean-square speed for the molecules in a gas always greater than their average speed, no matter how the molecular speeds are distributed? Explain.

: **20-23 P** Show that if the earth's atmosphere is assumed to be at a constant temperature, the variation in atmospheric pressure as a function of distance y above the earth's surface is given by $P = P_0 \, e^{-(Mg/RT)y} = P_0 \, e^{-y/h_0}$ where M = molecular weight of air and P_0 = atmospheric pressure at $y = 0$. Experimentally $h_0 \cong 8.6$ km.

Section 20-5   Heat

· **20-24 P** What is the internal energy of one mole of an ideal monatomic gas at STP?

· **20-25 Q** Internal energy, heat, and work are all measured in energy units, but they are quite different kinds of quantities conceptually. One of these quantities is a state function, the other two are not. A key distinction here is that
(A) the state function is influenced by temperature, whereas the other two are not.
(B) when the system is carried from p_1, V_1, T_1 to p_2, V_2, T_2 the change in the state function does not depend on the path followed on a pV diagram.
(C) the state function is measured in calories (a thermal unit) but the other two are measured in joules (a mechanical unit).
(D) a state function is an intensive property, whereas the other two are extensive properties.

: **20-26 Q** Consider an ideal gas in a well insulated container. If the gas expands and does work dW against a piston, $dW = -dU$. We can understand this relation on a microscopic basis by observing that
(A) in an elastic collision, the kinetic energy of a particle does not change.
(B) in an elastic collision, the momentum of the system is conserved.
(C) when a molecule rebounds from a collision with a receding wall, its speed is reduced.
(D) whenever a molecule collides with a fixed wall it exerts a force.
(E) temperature is proportional to internal energy U for an ideal gas.

· **20-27 Q** *Thermal energy* is
(A) the energy that flows into a cold object from a hot object.
(B) the same as internal kinetic energy.
(C) essentially the temperature of an object.
(D) the energy of a system, both kinetic and potential, associated with the disordered, random motions of the constituents of the system.
(E) heat.

· **20-28 Q** Heat is essentially the same as
(A) temperature.
(B) the "hotness" of an object.
(C) internal energy.
(D) thermal energy.
(E) the energy transferred from one object to another when they are at different temperatures.

· **20-29 P** When 20 cal of heat are added to a gas it does 12 calories of work. What is the change in internal energy of the gas?

· **20-30 P** When 500 J of heat are removed from a gas, and 100 J of work are done on the gas, what is the change in internal energy of the gas?

· **20-31 P** Two moles of N_2 at 1 atmosphere and 300 K is expanded adiabatically until its volume has increased by a factor of 12. What is the final temperature of the gas?

: **20-32 P** One mole of monatomic ideal gas goes from (p_1, V_1) to $(p_1, 3V_1)$ to $(\frac{1}{4}p_1, 3V_1)$. (a) Draw the process on a pV diagram. (b) What is the overall change in the gas's internal energy? (c) What is the overall work done by the gas? (d) How much net heat is added in the overall process?

: **20-33 P** A gas is carried around the cycle shown on the pV diagram of Figure 20-16. What is the work done by the gas in this process?

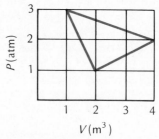

Figure 20-16.   Problem 20-33.

: **20-34 Q** Which of the following are the same kind of process if examined on the microscopic level?
(A) Work and isothermal expansion.
(B) Temperature and total internal energy.
(C) Heat flow and mass transfer.
(D) Work and heat.
(E) Pressure and thermal energy.

Section 20-8   Specific Heats of an Ideal Gas

· **20-35 P** What are C_v and C_p (in kcal/kg°C) for argon gas?

: **20-36 P** The molar specific heat at constant pressure for a certain ideal gas is 20.8 J/mol·°C. How much heat is required to increase the temperature of 1 mole of this gas by 80 C° at constant pressure?

: **20-37 P** How much heat must be added to 2 moles of a monatomic ideal gas to increase its volume from 0.1 m³ at 300 K to 0.4 m³ at constant pressure?

: **20-38 Q** Generally speaking, we expect that as the number of atoms in a gas molecule increases, the following will happen to the molar specific heats, C_p and C_v.
(A) the difference between C_v and C_p will increase.
(B) the difference between C_v and C_p will decrease.
(C) both C_v and C_p will increase, but their difference will not change.
(D) both C_v and C_p will decrease, but their difference will not change.
(E) C_p will increase and C_v will decrease.

: **20-39 Q** Which of the following is an accurate statement concerning an ideal gas?
(A) C_p and C_v are equal if the gas is monatomic.
(B) C_p is greater than C_v because when heat is added to a gas at constant pressure some work is done by the gas, whereas this is not the case at constant volume.
(C) C_p should be greater than C_v because there will be a smaller temperature rise in a gas at constant pressure than in one at constant volume according to the ideal gas law.
(D) C_v is greater than C_p because added heat must be added to oppose the increase in pressure that results when the volume is held constant.
(E) C_p is greater than C_v because the temperature of an ideal gas will always rise when heat is added to it.

: **20-40 P** How much work must be done to compress 20 gm of CO at STP to one-sixth of its original volume? (Hint: Note that the work done is equal to the increase in internal energy, $mC_v\Delta T$ in an adiabatic process, where C_v is the specific heat per unit mass).

: **20-41 P** Assuming oxygen and nitrogen behave as ideal gases, calculate C_v for air (78% N_2, 22% O_2) and compare your result with the experimental value of 0.171 kcal/kg·°C.

· **20-42 Q** Is it true that the temperature of a gas *always* drops during an adiabatic expansion? Explain.

· **20-43 P** A container of O_2 gas at STP is compressed adiabatically to one tenth its initial volume. What is the temperature of the gas after compression?

: **20-44 P** Two liters of helium at 1 atmosphere is suddenly compressed to one liter so that the compression is adiabatic. How much work is done in the process?

: **20-45 P** Show that the work done by an ideal gas expanding adiabatically from temperature T_1 to T_2 is $nR(T_1 - T_2)/(\gamma - 1)$.

: **20-46 P** One mole of ideal gas is initially at STP. It expands adiabatically to twice its initial volume. What is (a) the work done by the gas, (b) the final temperature, (c) the change in internal energy of the gas?

: **20-47 Q** Which of the following is a true statement concerning a given ideal gas?

(A) Through a given point on a pV diagram there can pass more than one adiabatic curve.
(B) Through a given point on a pV diagram there can pass more than one isothermal curve.
(C) All physically allowed processes on a pV diagram have negative slope.
(D) There is no heat flow in or out of the gas in an isothermal process.
(E) The adiabatic curve through a given point on a pV diagram always has steeper slope than an isothermal curve through the same point.

Section 20-9   Disorder and the Second Law of Thermodynamics

· **20-48 P** A container with a volume of 1,000 cm³ holds 10^8 molecules. What is the probability of (a) finding one particular molecule located within a certain 1-cm³ region of the container, (b) finding all 10^8 molecules within this region, (c) finding a particular molecule to be moving to the right, (d) finding all molecules moving to the right, and (e) finding all molecules located within a 1-cm³ region and moving to the right?

· **20-49 Q** The second law of thermodynamics is
(A) based to some extent on Newton's laws of motion.
(B) essentially a statement of probability theory.
(C) a necessary consequence of the first law of thermodynamics.
(D) applicable to thermal, but not mechanical, phenomena.
(E) just as useful for a system consisting of a single particle as for a gas of 10^{23} particles.

: **20-50 P** Suppose that you throw six coins into a box. What is the most probable distribution of head and tails, and what is its probability?

Supplementary Problems

20-51 P The lowest weather temperature ever recorded, -90 °F, was observed at Verkhoyansk in Siberia in 1887. By what factor does the rms speed of a molecule of air change when the temperature drops from normal room temperature (68 °F) to $-90°$ F?

20-52 P In April, 1982, scientists at Bell Laboratories announced that they had produced a pulse of laser light with a duration of 30 fs (30 femto s = 30×10^{-15} s). With such a short laser pulse, researchers can investigate physical and chemical changes that occur in very short time intervals. (a) How far does the laser pulse travel in a time interval equal to its pulse duration? (b) Suppose that you were going to use the laser light to take a flash picture of a typical hydrogen atom in a gas at room temperature. Would the picture be blurred? (The size of a hydrogen atom in its ground state is about 10^{-10} m.)

Thermal Properties of Solids and Liquids

The basic thermodynamic concepts can be extended from the simple system of an ideal gas to the more complicated systems of solids and liquids. Relevant in this connection are the fundamental experiments of Joule, which established the first law of thermodynamics. Also important are the specific heats and heats of transformation. Two modes of thermal-energy transfer we shall examine are conduction and radiation.

21-1 Solids and Liquids as Thermal Systems

Ideal gases are simple thermal systems. At low densities, all gases follow the same equation of state, the general-gas law; the absolute temperature of any ideal gas is merely a measure of the average molecular translational kinetic energy; and one may relate the microscopic molecular behavior to the macroscopic properties of a gas directly through a mechanical model, the kinetic theory.

Why do gases show such relatively simple behavior? The reason is that apart from perfectly elastic intermolecular collisions of negligible duration, a molecule in a low-density gas is always in force-free motion; it travels along a straight line at constant speed. The intermolecular collisions merely maintain molecular chaos and thermal equilibrium; otherwise, one need not (for strictly ideal gases) be concerned with interactions between the molecules.

Solids and liquids are different. In these condensed states of matter, the molecules are not separated, on the average, by distances that are large compared with the range of the intermolecular force. Neighboring molecules interact continuously. As a consequence, the total energy of the particles is not merely the sum of their kinetic energies. One must also include intermolecular *potential* energies. What, then, does internal energy or thermal energy mean for the solid and liquid states?

Recall first that for a monatomic gas in thermal equilibrium, the only contribution to internal energy is the disordered *translational* kinetic energy of the molecules. The total linear momentum of all molecules is zero. For every molecule with a momentum component in one direction, there is another molecule with momentum of the same magnitude in the opposite direction. For diatomic or polyatomic molecules, there may be additional contributions to the system's internal energy. If a molecule rotates, the kinetic energy of rotation is a part of the gas's thermal energy. But this *rotational* kinetic energy is disordered because the molecules rotate about axes oriented in all directions; the total angular momentum of all rotating molecules is zero. Moreover, if the atoms of a molecule also undergo vibrational motion, the energy associated with the molecular oscillations contributes to the gas's internal energy. The *vibrational* energy — a combination of both kinetic energy and potential energy for any simple harmonic motion — is disordered because the vibrational axes are randomly oriented.

Now consider the contributions to the disordered, internal, or thermal, energy for a solid. A solid may be thought of as a collection of atoms bound together by interatomic forces, each atom oscillating about an equilibrium position at any finite temperature.*

We must distinguish between ordered and disordered energy. Suppose that all the atoms of a solid had the same velocity. Then the solid as a whole would be in motion in the direction of all the velocity vectors, and the kinetic energy of the body would be ordered. See Figure 21-1(a). Likewise, if the body were at rest and a pair of external forces were applied in opposite directions to the solid to stretch it, the atoms would increase their separation distances along the direction of the applied forces and the body as a whole would have what could be called ordered elastic potential energy. See Figure 21-1(b). The ordered energy would be manifest macroscopically — you could see it, the kinetic energy as an object in motion and the potential energy as a deformed solid.

The *thermal* energy of a solid is something else. It consists partly of the kinetic energy of atomic vibrations; this kinetic energy can be called disordered because the velocities of atoms vary, in both magnitude and direction, as shown in Figure 21-1(c). The resultant linear momentum of the atoms is zero. Furthermore, there is thermal energy associated with the intermolecular potential energy. This potential energy can be said to be disordered. The atoms

---

* In addition, some essentially free atomic electrons may wander throughout the solid. Except for extremely low temperatures, the translational kinetic energy of these electrons is not important thermodynamically, for reasons explained by the quantum theory. Therefore, for a solid, the thermal energy consists (almost entirely) of the kinetic and potential energy of atoms vibrating about equilibrium positions.

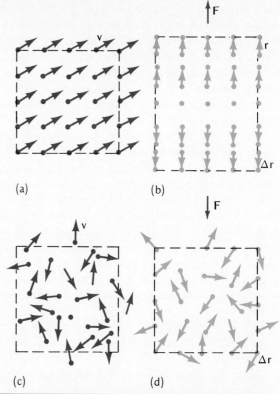

Figure 21-1. *Ordered vs disordered energy for the particles of a solid. (a) The body as a whole is in motion and has ordered kinetic energy, since all atoms have the same velocity. (b) The body as a whole is stretched and has ordered potential energy, since the atoms are displaced along the common direction of the applied forces. (c) Disordered kinetic energy: The vectors show the atomic velocities, which are distributed at random in magnitude and direction. (d) Disordered potential energy: The vectors show the atomic displacements, which are distributed at random in magnitude and direction.*

are displaced at random from their equilibrium positions; these displacements differ in magnitude and in direction, and the solid as a whole shows no net deformation (apart from thermal expansion). In short, the random oscillations of atoms in a solid are the origin of the solid's thermal energy. Note that in our discussion we have intentionally avoided using the term *heat*. Recall (Section 20-5) that in physics heat is energy transfer, a process that necessarily involves a temperature difference; heat does not designate the energy content of any system.

In like fashion, the thermal energy of a liquid is associated with disordered molecular kinetic and potential energy. In this state, however, the molecules are not bound to an equilibrium position, and the disordered translational kinetic energy of the molecules also contributes to the thermal energy.

21-2   Specific Heats

The specific heat for a solid or a liquid is defined as for a gas (Section 20-8). Suppose that some material of mass m has its temperature increased by dT because heat in the amount dQ is transferred to it. Then, by definition, the *specific heat* c of the material is

$$c \equiv \frac{dQ}{m\,dT} \tag{21-1}$$

or

$$dQ = cm \, dT \qquad \text{(21-2)}$$

The numerical value of c tells how easy or hard it is to change the temperature of any particular material. Since c is, by its definition, energy per unit mass per unit temperature change, the fundamental units for specific heat in the SI system are joules/kilogram·Celsius degree $= \text{J/kg} \cdot \text{C}°$.*

For some purposes it is useful to know the energy *per mole* per unit temperature change, or *molar specific heat C*. Then, by definition,

$$C \equiv \frac{dQ}{n \, dT} \qquad \text{(21-3)}$$

where n is the number of moles of the material. For any material of mass m and atomic mass M, the number $n = m/M$. Therefore, the molar specific heat C (21-3) is related to the ordinary specific heat c (21-1) by

$$C = M c \qquad \text{(21-4)}$$

The constant R of the general-gas law (19-6) has the same units as molar specific heats (energy per mole per unit temperature change), and molar specific heats are frequently expressed in units of R. For example, the molar specific heat of water (molecular mass $= 18$) can be given, using (21-4), as

$$C_{\text{water}} = M c = (18 \text{ gm/mol})(4.18 \text{ J/gm} \cdot \text{C}°) \left(\frac{R}{8.314 \text{ J/mol} \cdot \text{K}} \right) = 9.05 \, R$$

We found in Section 20-7 that the molar specific heat C_p of a gas heated at constant pressure differed markedly from its molar specific heat C_v at constant volume. For liquids and solids, however, the specific heats at constant pressure and constant volume are so nearly alike that it is ordinarily not necessary to distinguish between them. The reason basically is that liquids and solids expand only slightly when heated.

Table 21-1 lists specific heats for a number of materials. Note these features:

• The specific heat depends in general on temperature. For example, at $20°C$ copper has a specific heat of $3.89 \times 10^2 \text{ J/kg} \cdot \text{C}°$, but its specific heat at $-263°C$ is only $3.6 \text{ J/kg} \cdot \text{C}°$. Indeed, the specific heats of all substances approach zero at the absolute zero of temperature.

• As specific heats go, the value for water is relatively high. It is hard to heat water, and hard to cool water, once heated. This circumstance, in addition to ready availability and chemical stability, is responsible for the common use of water in heating systems.

---

* Other common units for specific heat involve the calorie (cal) or kilocalorie (kcal), where by definition

$$1 \text{ cal} \equiv 10^{-3} \text{ kcal} \equiv 4.18400 \text{ J}$$

The specific heat of water is close to $1 \text{ cal/gm} \cdot \text{C}° = 1 \text{ kcal/kg} \cdot \text{C}°$. Specific heat may also be measured in units of British thermal units per pound per Fahrenheit degree, or Btu/lb·F°, where the specific heat of water has the numerical value 1 Btu/lb·F°. Clearly, then, $1 \text{ kcal/kg} \cdot \text{C}° = 1 \text{ Btu/lb} \cdot \text{F}°$.

Table 21-1

SUBSTANCE	SPECIFIC HEAT $(kJ/kg \cdot C°)$	TEMPERATURE $(°C)$
Gases (at constant pressure)		
Air	1.005	27
Helium	20.6	27
Hydrogen	20.6	27
Liquids		
Mercury	0.138	0–100
Water	4.2177	0
	4.1840	17
	4.2160	100
Solids		
Aluminum	0.900	25
Copper	0.0036	−260
Copper	0.38	25
Copper	4.60	1000
Iron	0.47	20–100
Lead	0.130	20–100
Ice	2.04	−10

Example 21-1. One kilogram of lead just solidified from the liquid state is dropped into an insulated container holding one liter of water initially at 10.2°C. The mixture comes to a final temperature of 19.7°C. What is the approximate melting point of lead?

The lead heats the water, and the water cools the lead. The lead loses thermal energy, and the water gains an equal amount of thermal energy. The heat from the lead, initially at temperature t_i is,

$$Q_{lead} = mc \, \Delta T_{lead} = (1 \text{ kg})(0.13 \text{ kJ/kg} \cdot C°)(t_i - 19.7°C)$$

The heat to the one liter (or one kilogram) of water is

$$Q_{water} = mc \, \Delta T_{water} = (1 \text{ kg})(4.2 \text{ kJ/kg} \cdot C°)(19.7° - 10.2°C)$$

Equating Q_{water} and Q_{lead}, we then have, in solving for t_i,

$$t_i = 3.3 \times 10^2 \text{ C}°$$

Note that in this example the 1 kg of lead changed temperature by $(330° - 20°)C = 310 \text{ C}°$, whereas an equal mass of water changed temperature by only $(20° - 10°) \text{ C} = 10 \text{ C}°$. The difference merely reflects that water has a far larger specific heat than lead.

The procedure of this example is one basis for measuring specific heats. The general experimental procedure is known as *calorimetry*.

21-3   Heats of Transformation

Heating a substance may change its temperature; the process is characterized by the specific heat of the substance. But heating a substance may also produce a change in its state with no change in temperature. Such transformations in state are characterized by *heats of transformation*.

Suppose that we have 1 kg of ice at atmospheric pressure and initially at a temperature of −40°C. The ice is heated at a constant rate, and we observe its

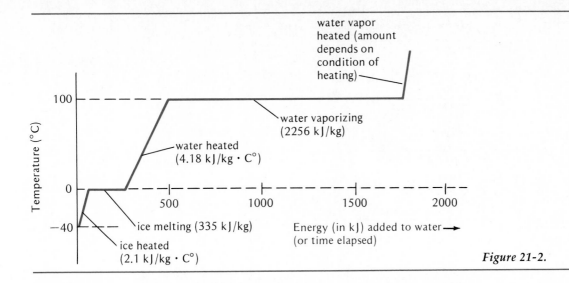

water vapor
heated (amount
depends on
condition of
heating)

water vaporizing
(2256 kJ/kg)

water heated
(4.18 kJ/kg · C°)

ice melting (335 kJ/kg)

ice heated
(2.1 kJ/kg · C°)

Energy (in kJ) added to water →
(or time elapsed)

Figure 21-2.

temperature as a function of time. If the heater has a constant power output and all its energy goes into heating the sample, the temperature varies with time, as shown in Figure 21-2. (Since the power P is constant, the abscissa is also a measure of the total energy $E = Pt$ delivered after time t as heat).

The plot has the following features (reading from left to right):

• The temperature of ice increases until it reaches 0°C, its rise being determined by the specific heat of ice, approximately 2.1 kJ/kg·C° in this temperature range.

• Ice at 0°C is transformed by melting into water at 0°C. First we have all ice, then a mixture with the fraction of water increasing and that of ice decreasing, and finally all water still at 0°C. The *specific heat of transformation,* called the specific heat of fusion, for this change in state of water from solid to liquid is 335 kJ/kg.

• The temperature of water increases uniformly from 0°C to 100°C at a rate that is controlled by the essentially constant specific heat of water, 4.18 kJ/kg·C°.

• Liquid water at 100°C is transformed to gaseous water vapor also at 100°C. Here the specific heat of transformation, called the specific heat of vaporization, is 2256 kJ/kg, an amount of heat that exceeds substantially the total heat required to bring the 1 kg of ice at −40°C to water at 100°C. If bubbles of water vapor form within the liquid, the process is, of course, boiling.

• The temperature of the water vapor rises above 100°C. The particular heating process (constant pressure or constant volume, for example) determines the rate of temperature increase.

In general, the *specific heat of transformation L for any change in state,* always taking place at a *constant temperature,* is defined as the *heat per unit mass of material.* The heat H required to change the state of a mass m is then*

$$H = mL \qquad (21\text{-}5)$$

---

* A somewhat old-fashioned name still in use for what is here termed the specific heat of transformation is *specific latent heat.*

Table 21-2. *Specific Heats of Transformation*

SUBSTANCE	MELTING POINT (°C)	HEAT OF FUSION (kJ/kg)	BOILING POINT (°C)	HEAT OF VAPORIZATION (MJ/kg)
Water	0	335	100	2.26
Water			40	2.40
Water			180	2.00
Water	−10	280		
Water	−22	230		
Oxygen	−218	13.8	−183	0.21
Mercury	−39	11.8	357	0.27
Copper	1083	176	2300	7.32
Lead	327	24.5	1620	0.73

Primary units for L are J/kg. Values of L for a few materials are shown in Table 21-2.

It is easy to see why energy must be added to a solid to melt it or to a liquid to vaporize it. Consider what happens to molecules in these transformations. A molecule in a solid is bound to its equilibrium position. If the molecule is to become free enough to wander about in the liquid state, it must have additional energy delivered to it. In the liquid state, a molecule is still influenced strongly by neighboring molecules, whereas in the gaseous state a molecule is essentially free of other molecules, at least between collisions. Energy must then be added to molecules in a liquid to break the coupling with neighboring molecules.

The specific heats of transformation depend on the temperature at which the change in state occurs. For example, if water boils at 40°C (with a pressure much lower then atmospheric pressure), the specific heat of transformation is 6 percent greater than that for 100°C.

Heat must be *added* for the process solid → liquid → vapor. In the reverse processes, heat must be removed, of course. Thus evaporation cools a liquid, whereas condensation from vapor to liquid releases energy. Terms used to denote particular changes in state are shown in Figure 21-3.

Example 21-2. One kilogram of water at atmospheric pressure and 100°C is transformed into water vapor with a density of 0.60 kg/m³. What is the change in the internal energy of the matter?

The internal energy change here is not merely the specific heat of vaporization L for water. From the first law of thermodynamics, we have

$$Q = \Delta U + W \tag{20-13}$$

or for a change in state,

$$mL = \Delta U + W$$

where Q is heat to the system, ΔU is the change in the system's internal energy, and W is the work done by the system on its surroundings.

Water *expands* as it is transformed into a vapor with a lower density. The work done by the expanding system is $W = p\,\Delta V$. The volume change ΔV can be computed from the densities of water (1.0×10^3 kg/m³) and of water vapor (0.60 kg/m³):

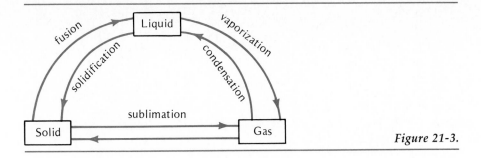

Figure 21-3.

$$\Delta V = \frac{1\ kg}{0.60\ kg/m^3} - \frac{1\ kg}{1.00 \times 10^3\ kg/m^3} = 1.67\ m^3$$

Then the internal energy change is found as follows:

$$mL = \Delta U + p\,\Delta V$$

$$\begin{aligned} \Delta U &= mL - p\,\Delta V \\ &= (1\ kg)(2256\ kJ/kg) - (1.01 \times 10^5\ Pa)(1.67\ m^3) \\ &= (2256 - 169)\ kJ \\ &= 2087\ kJ \end{aligned}$$

The internal-energy change per kilogram differs significantly from the specific heat of transformation.

21-4 The Joule Experiment

How does the first law of thermodynamics apply to liquids and solids? For an infinitesimal process,

$$dQ = dU + dW \tag{20-13}$$

Here again dU is the change in the internal energy of the system. Heat dQ is the disordered energy entering the system by virtue of a temperature difference between the system and its surroundings; work dW is ordered energy transferred from the system to its surroundings. The essence of the first law is this: the change in a system's internal energy depends on the *net amount* of energy entering the system, *not on whether* the internal-energy change arises *by work* or *by heat*.

The classic experiments of J. Joule (1818–1889) carried out over more than thirty years, first established the validity of the first law of thermodynamics. Consider the situation shown in Figure 21-4(a), where a liquid has its internal energy changed by heating. Here an electric generator (with whose details we need not be concerned) is run by a descending weight. The generator sends an electric current through the coil of wire, and the hot wire heats the liquid. The generator's internal energy does not change, so that the work $\Delta W = mg\,\Delta y$ done by weight mg descending a distance Δy is equal in magnitude to the heat ΔQ introduced into the liquid. The value ΔQ also represents the change in the liquid's internal energy ΔU_{liq}, since the liquid does no work on its surroundings:

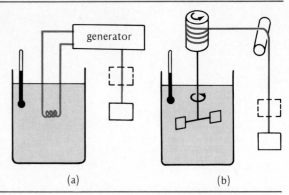

Figure 21-4. *The Joule experiment. (a) A descending weight runs an electric generator, which operates an electric heater. This in turn raises the temperature of a liquid. (b) A descending weight runs a stirrer, which stirs the liquid and raises its temperature.*

(a)   (b)

$$\Delta W = \Delta Q = \Delta U_{liq}$$

Now consider the situation shown in Figure 21-4(b). Here energy enters the liquid as work, not heat. The descending weight now turns a mechanical stirrer immersed in the liquid. The liquid soon comes to rest because of internal friction; its ordered rotational kinetic energy is converted into disordered thermal energy. The work ΔW done by the descending weight equals the work done by the stirrer on the liquid, which is, in turn, equal to the increase ΔU in the liquid's internal energy:

$$\Delta W = \Delta U_{liq}$$

The change in the liquid's internal energy shows up as a change ΔT in its temperature. Joule demonstrated by experiments as shown in Figures 21-4(a) and (b), that the size of this temperature change was the same whether energy was added by heat or by work — so long as the work ΔW done by the descending weight was the same.

> **Example 21-3.**   Two identical lead bullets, each with an initial speed of 500 m/s, happen to collide head-on and stick together. By how much is the temperature of the bullets raised in this completely inelastic collision?
>
> The energy transformation is this: the macroscopic, ordered kinetic energy of the two lead bullets becomes, after the collision, the microscopic, disordered thermal energy of lead atoms. The loss in kinetic energy for the two bullets, each of mass m, is
>
> $$\Delta K = 2(\tfrac{1}{2}mv^2) = m(500 \text{ m/s})^2$$
>
> The gain in thermal energy is
>
> $$\Delta Q = 2(mc\,\Delta T) = 2m(0.13 \text{ kJ/kg} \cdot \text{C}°)\,\Delta T$$
>
> We see that when the two relations are equated, the bullet mass drops out. The final result is a temperature rise of $\Delta T = 96 \text{ C}°$. If the speed of the bullets were doubled, the initial kinetic energy then would be up by a factor 4; then the energy dissipated would be enough to melt some of the lead.

21-5   Thermal Conduction

The three modes of thermal-energy transfer are:

- *Thermal conduction* (the topic of this section).
- *Radiation* (the topic of Section 21-7).

• *Convection*, the bulk transport of heated fluids. (This does not lend itself to simple quantitative analysis and therefore is not treated in this book.)

When a hot object is in contact with a cold object, thermal energy flows from the hot to the cold one. Thermal energy can flow within a single object when different locations are at different temperatures; this thermal-energy transfer is *thermal conduction*.

The thermal energy of a solid typically consists of the energy of atoms vibrating about their equilibrium positions. Adjoining atoms in a solid interact; as one atom oscillates, it influences the motion of a neighboring atom. Suppose that one region of a solid is at a higher temperature than an adjoining region. Then the amplitudes (and energies) of the atomic oscillations are greater at the hot region than at the nearby colder regions. Thermal energy is then transferred from the hot to the cold regions by the coupling between neighboring oscillators.

When a thermal conductor is a metal, the solid has, in addition to vibrating atoms, free electrons, and these free electrons also contribute to the thermal conduction. That free electrons are significant in thermal conductivity is shown by the fact that good thermal conductors, such as metals, are usually also good electrical conductors. Free electrons are the origin of electric currents in electrical conductors; these same free electrons also transfer thermal energy in thermal conduction.

For the quantitative aspects of thermal conduction, consider the situation shown in Figure 21-5. Here we have a rod of uniform cross-sectional area A surrounded by an insulating material; no heat leaks into or out of the rod through its sides. The hot left end is maintained at a constant high temperature T_h while the cold right end is maintained at a constant lower temperature T_c. The heat rate into the rod from the hot reservoir is dQ/dt. At a steady-state condition, with the temperature constant at all points within the conductor, dQ/dt is also the heat per unit time leaving the rod. In fact, dQ/dt represents the thermal energy crossing area A per unit time at *any* point along the axis of the rod, and sometimes dQ/dt is called the *heat current*. The net heat entering the rod, or entering any small volume of the rod, is zero; the rod's internal energy remains constant. A thermal conductor acts merely as a "heat pipe" between the hot and cold reservoirs; it degrades thermal energy by sending it to a lower temperature. This behavior can continue only so long as the two ends are maintained at different temperatures. When $T_h = T_c$, then $dQ/dt = 0$.

We concentrate now on the temperature drop dT occurring across an infini-

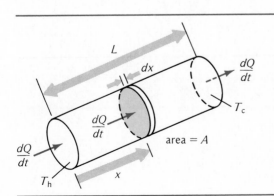

Figure 21-5. Thermal conduction through a rod of length L and cross section A, having its left and right ends at the constant hot and cold temperatures T_h and T_c respectively. The rate of heat through any cross section is dQ/dt.

tesimally thin section of thickness dx. The quantity dT/dx, called the *temperature gradient*, measures the temperature change per unit displacement along the direction of heat flow. If we take x to increase along the direction of dQ/dt, as in Figure 21-5, the temperature gradient dT/dx is *negative*. That is, the temperature drops as x increases.

Experiment shows that the heat rate dQ/dt is directly proportional to both the cross-sectional area A and the temperature gradient dT/dx. That is,

$$\frac{dQ}{dt} = -\lambda A \frac{dT}{dx} \tag{21-6}$$

Quantity λ, a positive constant called the *thermal conductivity*, is characteristic of the material of the thermal conductor.

Now we integrate (21-6) over the entire rod of length L; it extends from $x = 0$ to $x = L$. The corresponding temperatures are T_h and T_c, respectively.

$$\frac{dQ}{dt} = -\lambda A \frac{dT}{dx}$$

$$\frac{dQ}{dt} \int_0^L dx = -\lambda A \int_{T_h}^{T_c} dT$$

Note that dQ/dt, λ, and A are all constants. Then,

$$\frac{dQ}{dt} = \frac{\lambda A (T_h - T_c)}{L} \tag{21-7}$$

According to (21-7), the temperature drops uniformly along the rod. The differential form of the thermal-conduction equation in (21-6) is more general than (21-7). Equation (21-6) can be applied to *all* shapes of conductors, not merely to uniform rods. (One can imagine the conductor to consist of a collection of infinitesimally thin sheets for each of which the differential form holds exactly.)

Measured values of the thermal conductivity for various materials are given in Table 21-3. A good thermal conductor has a high λ value; a low λ characterizes a poor thermal conductor, or a good thermal insulator.

Table 21-3

SUBSTANCE	TEMPERATURE ($^\circ$C)	THERMAL CONDUCTIVITY λ (W/m·C$^\circ$)
Conductors		
Aluminum	25	237
Aluminum	-173	300
Copper	25	398
Iron	25	80
Lead	25	33
Silver	25	427
Water	25	6
Insulators		
Air (dry, still)	0	0.024
Hydrogen	0	0.14
Ice	0	0.92
Rock wool	0	0.041

21-6   Thermal Resistance

The effectiveness of an insulator in impeding the flow of thermal energy is typically given in the building industry by the insulator's thermal resistance, or R value. The R value is so defined that with two or more insulating layers, the R value for the combination is simply the sum of the R values of the individual slabs. An insulator rated R10, for example, placed on top of an insulator rated R5 is equivalent to a single R15 insulator.

To see how thermal resistance is defined and related to a material's thermal conductivity, consider Figure 21-6. Here we have two layers of different insulating materials placed between a constant high-temperature reservoir on the left and a constant lower-temperature reservoir on the right. The temperature difference ΔT across both insulators is simply the sum of the temperature differences across each of the two separately:

$$\Delta T = \Delta T_1 + \Delta T_2$$

In the steady equilibrium state, we have from (21-7) that

$$\Delta T_1 = \left(\frac{L_1}{\lambda_1 A}\right)\frac{dQ}{dt}$$

and

$$\Delta T_2 = \left(\frac{L_2}{\lambda_2 A}\right)\frac{dQ}{dt}$$

Here L_1 and L_2 are the thicknesses of the two slabs, and λ_1 and λ_2 their respective thermal conductivities. The heat rate dQ/dt is the same through any cross-sectional area A, and this area is the same for both insulators.

Combining the relations above then gives

$$\Delta T = \Delta T_1 + \Delta T_2 = \left(\frac{L_1}{\lambda_1} + \frac{L_2}{\lambda_2}\right)\frac{1}{A}\frac{dQ}{dt}$$

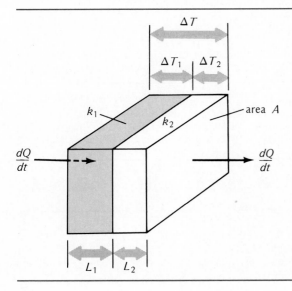

Figure 21-6.

Clearly the quantity in parentheses becomes, for any number of insulating slabs, simply $\Sigma(L/\lambda)$.

To assure the additivity property, the thermal resistance R of a single insulator is then defined as

$$R \equiv \frac{L}{\lambda} \tag{21-8}$$

The basic thermal-conduction relation, (21-7), can then be written in the form

$$\frac{dQ}{dt} = \frac{A}{R} \Delta T \tag{21-9}$$

with $R = R_1 + R_2 + R_3 + \cdots$. Here $R_1, R_2, \ldots$ are the thermal resistances of individual insulators of the same area A stacked in series, and R is the single equivalent thermal resistance of the combination.

From its definition in (21-8), we see that the R value for a particular insulator is inversely proportional to its thermal conductivity λ; an effective insulating slab, one with a low thermal conductivity λ, will have a high value for its thermal resistance R. Note that R depends also on the insulator thickness L. Thermal conductivity λ is a property of the insulating material, but R is a property of a particular insulator.

Engineers and others concerned with the R values of insulators in building materials almost invariably use the following bizarre units:

- Heat rate dQ/dt in Btu/h.
- Temperature difference ΔT in Fahrenheit degrees.
- Area A in square feet.
- Insulator thickness in inches (in.).
- R-value units are then $\text{ft}^2 \cdot \text{h} \cdot \text{F}°/\text{Btu}$.

(Most often the units for R are not even shown on commercial products!)

Table 12-4 lists R values for a 1-in. thickness of various insulating materials. We see from the table that still dry air has a particularly high R value. This is why fiberglass, rock wool, polyurethane foam, fur, and thermal underwear, all of which depend on trapped air, are good insulators.

Example 21-4.   A moose hunter builds a small cabin in the north woods. It is basically a cube 10 ft on an edge. He uses 10 in. of fiberglass insulation on all six sides of the cabin. What is the minimum power rating of a heater kept inside the cabin if the temperature there is not to drop below 60°F when the temperature outside is $-20°$F? (Assume, unrealistically, that there are no cracks, no openings, no air leaks.)

From Table 21-4, we see that 10 in. of fiberglass produces an R30 insulator. The total surface area for six sides is $6 \times 10\ \text{ft} \times 10\ \text{ft} = 600\ \text{ft}^2$. The temperature difference is $60 - (-20)°\text{F} = 80\ \text{F}°$. From Eq. (21-9), we have

$$\frac{dQ}{dt} = \frac{A}{R} \Delta T = \frac{600\ \text{ft}^2}{30\ \text{ft}^2 \cdot \text{F}° \cdot \text{h}/\text{Btu}} (80\ \text{F}°) = 1600\ \text{Btu/h} \approx 0.5\ \text{kW}$$

The heater requirement, only 0.5 kW, is surprisingly modest. A person of normal size consuming about 2500 kcal in food per day would generate energy at a rate of about 120 W, even if sedentary (that would be the person's basal metabolism). It is not surprising, then, that a small family of Eskimos can manage quite comfortably in an igloo, built with walls of snow, even though the only heater inside is body heat.

Table 21-4

MATERIAL (at 20°C, 1 atm)	R value for a 1-in. slab (in ft²·F°·h/Btu)
Vacuum	∞
Gases	
Air (dry, still)	5.6
Helium	1.0
Hydrogen	0.8
Building Materials	
Foam (urea-formaldehyde)	4.8
Rock wool	3.6
Cork board	3.7
Balsam wood	3.7
Fiberglass	3.0
White pine (cross grain)	1.3
Building brick	0.22
Concrete	0.13
Glass	0.18
Silver	3.4×10^{-3}
Liquid helium (superfluid)	0

21-7 Thermal-Energy Transfer through Radiation

Some basic facts about radiation are these. Any material at a finite temperature is made up of electrically charged particles in motion. Accelerated charged particles lose energy by emitting electromagnetic radiation. Electromagnetic radiation consists of oscillating electric and magnetic fields that propagate through empty space at the speed of 3.00×10^8 m/s, the speed of light.

Absorption is the reverse of emission. In absorption, the energy of electric and magnetic fields is transferred to electrically charged particles in the absorbing material. All materials can radiate and absorb electromagnetic energy. (These and still more specific properties of electromagnetic waves will be discussed in some detail in Chapter 35.) Our concern here is with the relation between electromagnetic radiation and the properties, particularly the temperature, of materials emitting and absorbing it.

A result of both experiment and theory is that the total electromagnetic radiation from an emitting object in thermal equilibrium at absolute temperature T is proportional to T^4. The relation, known as the Stefan-Boltzmann law, may be written as

$$H = \epsilon \sigma T^4 \qquad (21\text{-}10)$$

The symbols here have the following meanings:

- H is the electromagnetic energy emitted per unit time by a unit area of the surface of the emitter; that is, H is emitted electromagnetic power per unit area.
- σ is the fundamental constant of the Stefan-Boltzmann law. Its value, derived from theoretical analysis and verified by experiment, is 5.670×10^{-8} W/m²·K⁴.
- ϵ is the *emissivity* of the emitting material. Possible values for ϵ depend on

the emitting surface and temperature; they range from 0 (no emission) to 1 (a perfect emitter).

Note the strong temperature dependence—the *fourth power* of absolute temperature T—of the radiated electromagnetic power per unit area. Suppose, for example, that an object made of lead is brought from room temperature to its relatively low melting point of 330°C, thereby approximately doubling its absolute temperature. The rate at which the lead object radiates energy increases by a factor of $2^4 = 16$. Even a slight change in the temperature of the sun would have profound consequences on the weather on earth because of the change in radiation reaching the earth.

Whereas emission strongly depends on the emitter's temperature, experiment shows that *absorption of electromagnetic radiation is independent of the absorber's temperature.* Any object will simultaneously absorb and emit radiation. Suppose an object is in a vacuum. Then the absorption of radiation shining on it and the emission of radiation from it are the only processes that change the object's energy content. If the isolated object's temperature falls, radiation is emitted more rapidly than radiation is absorbed. If the object's temperature rises, absorption outweighs emission. And if the isolated object's temperature remains constant, the emission and absorption rates must be the same.

We typically describe an object's absorption properties for visible light by the terms *black* and *white.* An object appears black when it absorbs all visible light striking it, reflecting none. And an object is white when it reflects all visible light, absorbing none. In thermodynamics, the term *blackbody* is used to denote an object that absorbs *all* electromagnetic radiation, visible and invisible, that impinges on it. A blackbody is a perfect absorber of radiation. One simple, nearly perfect blackbody consists merely of a hole leading to the interior of an enclosure made of any material; electromagnetic waves entering the enclosure through the hole undergo multiple reflections inside the enclosure until all the radiation is absorbed. (The opening to a high-temperature furnace of any material would be a good approximation of a blackbody.)

Suppose that we have an isolated object at some finite temperature in empty space. It radiates. We know that if the object maintains a constant temperature, it must also be absorbing energy at the same rate to keep its total energy and temperature unchanged. A blackbody, defined to be a perfect absorber, must then also be a perfect emitter of radiation. Its emissivity in (21-10) must be $\epsilon = 1$. A blackbody therefore need not appear black; as a perfect radiator, it may appear bright. Our sun and other stars are close approximations of blackbodies.

Example 21-5.   An iron sphere of 3.0-cm radius is placed in a furnace until its temperature reaches 800°C. At this temperature, the ball is glowing red-hot. Its emissivity is 0.85. (a) At what rate is the red-hot ball emitting radiation? (b) At what rate is the ball absorbing radiation?

The power P radiated by a ball of radius r is, from (21-10),

$$P = H(4\pi r^2) = \epsilon\sigma T^4(4\pi r^2)$$

Here we have multiplied the power radiated per unit area H by the surface area $4\pi r^2$ of a sphere to yield P. Using the parameters given above, we get

$$P = (0.85)(5.67 \times 10^{-8}\ \text{W/m}^2 \cdot \text{K}^4)[(800 + 273)\text{K}]^4(4\pi)(3 \times 10^{-2}\ \text{m})^2$$

Radiated power $= 0.72$ kW

(b) The rate at which the ball absorbs radiation can be determined only if we know the temperature of its surroundings. If the red-hot ball were to remain inside the furnace, itself at a temperature greater than $800°$C, then the ball would absorb radiation at a rate greater than the rate 0.72 kW at which it is emitting radiation. The ball's temperature would increase.

If on the other hand, the red-hot ball were out in the room, it would radiate energy at a greater rate than that at which it absorbs radiation from its lower-temperature surroundings. The ball's temperature would fall.

Summary

Definitions

Specific heat c: the heat dQ per unit mass m per unit temperature change dT,

$$c \equiv \frac{dQ}{m\ dT} \qquad (21\text{-}1)$$

Specific heat of transformation L: the heat H required to change the state of a material of mass m with no change in temperature,

$$H = m\ L \qquad (21\text{-}5)$$

Thermal conductivity: the parameter λ appearing in the relation

$$\frac{dQ}{dt} = -\lambda A\ \frac{dT}{dx} \qquad (21\text{-}6)$$

where dQ/dt is the time rate of heat through any cross-sectional area in a thin slab of thickness dx across which there exists a temperature difference dT.

Thermal resistance R: for an insulating slab of thickness L and thermal conductivity λ,

$$R \equiv \frac{L}{\lambda} \qquad (21\text{-}8)$$

Blackbody: a perfect absorber (and radiator) of electromagnetic radiation.

Important Results

The first law of thermodynamics:

$$dQ = dU + dW \qquad (20\text{-}13)$$

where dQ is the heat to a system, dW the work done by the system, and dU the change in the system's internal energy, applies to all thermal systems, as first shown in the Joule experiment.

The power of electromagnetic waves radiated per unit surface area H from an object at absolute temperature T is given by the Stefan-Boltzmann law:

$$H = \epsilon \sigma T^4 \qquad (21\text{-}10)$$

where ϵ is the emissivity of the radiating surface $(0 < \epsilon < 1)$ and σ is a constant.

Problems and Questions

Section 21-2   Specific Heats

· **21-1 Q** You may have noticed if you try to eat a hot apple pie right out of the oven before it has cooled, the filling burns your tongue but the crust does not. This is partly because the crust, being on the outside, cools first, but an even more important reason is that
(A) the filling gets hotter than the crust in the oven.
(B) the filling, being mostly water, has a higher specific heat capacity than does the crust.
(C) your mind tricks you into thinking the filling is hotter.
(D) the filling is sweeter.
(E) the crust is solid.

: **21-2 P** When 100 gm of copper at $100°$C is added to 100 gm of water at $20°$C, the resulting temperature of the mixture will be
(A) between $20°$C and $60°$C.
(B) $60°$C.
(C) between $60°$C and $100°$C.
(D) dependent on the heat of vaporization of water.
(E) None of the above.

: **21-3 P** One hundred gm of water is contained in a 50-gm copper calorimeter container at $20°$C. What will be the final temperature of the mixture when 150 gm of iron pellets at $250°$C are added to the water?

: **21-4 P** When burned, coal yields 2.8×10^7 J/kg. How many kilograms of coal must be burned in a boiler in order to raise the temperature of 500 kg of water by 80 C° if the boiler has an efficiency of 30%?

: **21-5 P** The metabolic rate (rate of using energy) of a person of mass 55 kg increases to 400 W while running. If she were able to dissipate only 350 W of this, how much would the temperature of her body rise in 1 hour? (The average specific heat of the body is 3.5 kJ/kg·C°.)

: **21-6 P** A domestic hot water heater with an efficiency of 50% burns 3.0 m³ of natural gas per hour. The heat of combustion of natural gas is 5.6×10^7 J/m³. How many kg of water could be heated from 10°C to 95°C per hour by such a heater?

: **21-7 P** In doing push-ups, an 80 kg person lifts his center of mass 20 cm. If he could not get rid of the heat generated, by how much would his body temperature rise if his muscle efficiency is 20%? The specific heat of his body is 3.5 kJ/kg·C°, and its thermal conductivity is 2.0 J/m·s·C°.

: **21-8 P** The law of Dulong and Petit says that the molar specific heat capacities of solids at high temperatures are all the same, namely 3R per mole. During the nineteenth century, this law was used to determine the atomic mass of an element by measuring its heat capacity. Check the Dulong-Petit law for aluminum (atomic mass 27), iron (56), lead (204), and copper (64). For each, calculate the ratio of the experimental to the theoretical value.

: **21-9 P** Three liquids, X, Y, and Z are at initial temperatures T_X, T_Y, and T_Z. When equal masses of X and Y are mixed, their final temperature is T_1. When equal masses of Y and Z are mixed, their final temperature is T_2. What temperature, in terms of T_1 and T_2 and specific heats C_x, C_y, and C_z, would result if equal masses of X and Z were mixed?

: **21-10 P** The specific heat of a substance is given by $C = 0.8$ kJ/kg·C° $+ 0.002$ kJ/kg·(C°)³T^2 for 0°C $< T <$ 50°C. (a) How much thermal energy must be added to raise the temperature of 5 gms of the substance from 10°C to 20°C? (b) What fractional error would be made if the thermal energy in (a) were calculated using the constant value for C equal to the value it has at 15°C?

Section 21-3 **Heats of Transformation**
· **21-11 Q** If a substance is heated,
(A) it will always expand.
(B) its temperature will always rise.
(C) the substance will always eventually melt.
(D) the substance will always change phase.
(E) none of the above.

: **21-12 P** An amount of 120 grams of ice at −6°C is mixed with 30 grams of water at 40°C and 5 grams of steam at 100°C and 1 atm. What is the final temperature of the mixture when it is in equilibrium?

· **21-13 Q** While riding a ski lift, you may have noticed that your face feels much colder when the wind blows. Which of the following is the most accurate explanation?
(A) This effect is psychological; there is no actual reduction in the temperature of your skin.
(B) The moving air (the wind) comes from higher altitudes where the air is colder.
(C) The wind brings dry air in contact with your skin, and removes the quiescent layer of air that acts as an insulator.
(D) Even though the wind is at the same temperature as still air, in a wind more cold molecules hit your face every second.
(E) The fact that the wind is blowing has nothing to do with the effect. If you move into a region of cold air, you will feel colder. Also, if the temperature of the air changes, this will cause a wind to blow. Thus one associates the cold with the wind, when in fact the wind is caused by the cold, and not the other way around.

· **21-14 P** The cooling condenser at a coal-fired electric generating plant changes steam at 100°C to water at 50°C. The heat released is carried away by cooling water, which enters the plant at 10°C from a river and is discharged at 40°C. How much water is needed for each kilogram of steam condensed?

: **21-15 P** A human can generate as much as 1.5 kg of perspiration per hour for moderately short periods of time. What is the maximum rate of evaporative heat loss for such a person?

: **21-16 P** A skier of mass 80 kg descends a 30° slope at a constant speed of 16 m/s. What is the maximum amount of snow per second that could be melted by friction with his skis if all of the energy dissipated goes into melting snow?

: **21-17 P** A hailstone of radius 1 cm reaches a terminal velocity of about 20 m/s. What fraction of the hailstone would be melted if all of the kinetic energy lost on impact with the ground were to melt ice?

: **21-18 P** Estimate the binding energy (in eV) per molecule between water molecules, knowing that the heat of transformation for the vaporization of water is 330 kJ/kg.

: **21-19 P** Heat is added at a steady rate of 2 watts to a

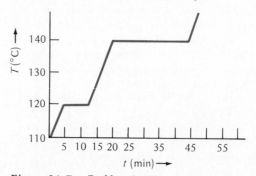

Figure 21-7. Problem 21-19.

100-gm sample and the temperature of the sample as a function of time is measured, yielding the curve shown in Figure 21-7. Determine (a) the specific heats of the solid and liquid phases. (b) the heat of fusion. (c) the heat of vaporization. (d) the boiling temperature.

⁝ **21-20 P** A great crater in Arizona is believed to have been created by a giant meteor whose mass has been estimated to have been 5×10^8 kg. (a) If the meteorite had fallen from very far away, what would have been its kinetic energy when it hit the earth if air friction were negligible? (b) Large amounts of fused silica have been found in the area of the crater. Could this have been created by the meteorite? The specific heat of silica is about 1.3 kJ/kg·C°, and silica fuses at 1650°C. Suppose that all of the kinetic energy of the meteorite went into fusing silica. What mass could have been fused? (c) Significant amounts of ground water could have been boiled by the heat generated by the meteorite. The clouds created might have had important consequences. Suppose that half the meteorite's energy went into boiling ground water. What mass of steam would have been created?

Section 21-4   The Joule Experiment

· **21-21 Q** Which of the following aspects of water is most important in the regulation of body temperature by perspiration?
(A) Thermal conductivity.
(B) Emissivity.
(C) Specific heat.
(D) Heat of fusion.
(E) Heat of vaporization.
(F) Heat of sublimation.
(G) Boiling temperature.

· **21-22 Q** Ice is placed in a picnic cooler to keep the contents cold. To facilitate cooling, one should
(A) open the cooler frequently to allow the moisture-saturated air to escape.
(B) crush the ice to enable it to melt more effectively.
(C) drain the ice water from the cooler occasionally.
(D) wrap the ice in newspaper, because the newspaper is an insulator that will keep the ice from melting too quickly.
(E) keep the ice and food apart, or at least have the food wrapped with newspaper for insulation.
(F) shake the contents vigorously frequently in order to keep the ice and ice water at the same equilibrium temperature.

· **21-23 P** (a) What is the kinetic energy of a 1200-kg car traveling 90 km/hr? (b) What temperature increase could this energy give rise to if added to 1200 kg of water?

· **21-24 P** An immersion heater is able to heat water with minimal heat loss in a well-insulated container. What is its power output if it can raise the temperature of 500 gm of water 10 C° in 2 minutes?

⁝ **21-25 P** In his investigations of the nature of heat and work, James Joule tried to measure the temperature rise in water that had fallen down a waterfall. If there were no heat loss, what temperature rise would you expect for water that had fallen a distance of 40 m?

⁝ **21-26 P** Yosemite Falls is 436 m high. If all of the kinetic energy of the falling water were to go into heating the water, by how much would the temperature be raised?

⁝ **21-27 P** (a) How much thermal energy is generated by a friction brake used to stop a 20-kg flywheel with a radius of gyration of 30 cm and an initial angular velocity of 200 rad/sec? (b) If the brake, of mass 12 kg and specific heat 0.45 kJ/kg·C°, absorbs all of the heat generated, by how much will its temperature rise?

⁝ **21-28 P** A 5-gm steel spring of force constant 800 N/m is compressed 5 cm and placed in 50 gm of water at 20°C in a well insulated container. Once the spring comes to thermal equilibrium with the water at 20°C, it is released and oscillates briefly until frictional forces damp its motion. By how much does the temperature of the water rise?

⁝ **21-29 P** A 10-gm lead bullet with a speed of 400 m/s is shot horizontally into a 100-gm block of lead resting on a horizontal frictionless surface. The bullet sticks in the block. What is the temperature rise of the block and bullet as a result of the collision, assuming each is in thermal equilibrium at a uniform temperature?

Section 21-5   Thermal Conduction

· **21-30 Q** An elevator's metal door felt much colder to touch than did the plaster wall next to it. Since both were in the same room, they should be at the same temperature. Which of the following is the best explanation of this phenomenon?
(A) The door and wall were not at the same temperature even though they were in thermal equilibrium.
(B) The wall was black underneath the paint and the metal was silver underneath the paint.
(C) The metal had a higher thermal conductivity than did the wall and hence conducted heat away from my hand more rapidly.
(D) The wall had a higher specific heat than did the metal door and hence held the heat of my hand better.
(E) In one case heat was transferred by conduction and in the other by radiation or convection.

· **21-31 Q** If you hold a nail against a piece of ice, the end in your hand will soon become cold. Why is this?
(A) Thermal energy flows from hand through nail to ice.
(B) Cold flows from hand through nail to ice.
(C) Answers A and B are equivalent; both are correct.
(D) Thermal energy flows from the nail into your hand, leaving the nail cold.
(E) None of the above is correct.

· **21-32 P** A concrete wall is 20 cm thick and has dimensions 2.5 m × 4 m. What is the heat loss through such a wall when the inside temperature is 23°C and the outside temperature is 0°C? The thermal conductivity of concrete is 2×10^{-4} kcal/m·s·C°.

: **21-33 P** Coal with heat of combustion of 7000 kcal/kg is burned in a furnace with 40% efficiency. How much coal must be burned each day to provide for the heat lost through a glass window pane 0.9 m × 1.2 m if the glass is $\frac{1}{4}$ in thick and the temperature difference across the window is 20°C? The thermal conductivity of glass is 2×10^{-4} kcal/m·s·C°.

: **21-34 P** Geophysical exploration has revealed a temperature increase of approximately 1 C° for each 30 meters into the earth's crust from the surface. The thermal conductivity of the crust is of the order of 0.8 J/m·s·C°. What is the rate of heat flow out through 1 m² of the earth's surface? Compare this with the incidence of sunlight, which is of the order of 1 kW/m².

: **21-35 P** An iron rod 20 cm long with cross-sectional area of 2 cm² is welded to a circular ring of copper of diameter 50 cm and cross-sectional area 1 cm². One end of the iron bar is at 20°C, and point P on the copper ring is held at 150°C. At what rate does heat flow along the iron rod? Assume no heat loss through the walls of the bars.

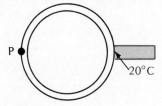

Figure 21-8. Problem 21-35.

: **21-36 P** Two aluminum plates are separated by an insulator whose thermal conductivity is 0.04 W/m·C°. The thickness of the insulator is 1 mm. Each metal plate is 4-mm thick. The outer surface of one plate is held at 120°C while the other is fixed at 20°C. Determine (a) the heat flow per cm² and (b) the temperature at the two surfaces of the insulator.

: **21-37 P** A long cylindrical hot water pipe has radius R_1 and is at temperature T_1. It is wrapped with a thickness a of insulator of thermal conductivity λ. The outer surface of the insulator is at temperature T_2. Derive an expression for the rate of heat loss per unit length of insulated pipe.

: **21-38 P** The air above the surface of a lake is at a temperature $-T$ (below freezing). The lake is covered by a layer of ice of thickness x at $t = 0$. (a) Show that this layer increases in thickness at the rate $dx/dt = \lambda T/L\rho x$, where $\rho =$ density of ice $= 0.92 \times 10^3$ kg/m³, $\lambda =$ thermal conductivity of ice $= 0.92$ J/m·s·C°, and $L =$ heat of fusion of water. (b)

How long would it take for a layer of ice to build up to a thickness of 15 cm starting from no ice?

: **21-39 P** A double pane storm window consists of two panes of glass separated by an air space. See Figure 21-9. If the thermal conductivity of the glass is λ_1 and that of the air is λ_2, and the pane thickness is t_1 and the air space thickness t_2, (a) show that when the outside temperature is T_1 and the inside temperature is T_2 the heat loss per unit area per second is

$$\frac{dQ}{dt} = \frac{\lambda(T_2 - T_1)}{2t_1 + t_2}, \text{ where } \lambda = \frac{2t_1 + t_2}{2\dfrac{t_1}{\lambda_1} + \dfrac{t_2}{\lambda_2}}.$$

(b) Determine the temperature at the inner surface of each pane of glass.

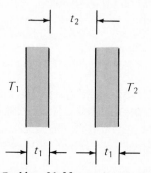

Figure 21-9. Problem 21-39.

: **21-40 P** Heat is generated at a rate P at the center of a nuclear fuel containment vessel of spherical shape. The fuel element has radius R_1 and the surrounding vessel, whose thermal conductivity is λ, has outer radius R_2. The outer surface is fixed at temperature T_2. What will be the temperature at the surface of the fuel element under these conditions?

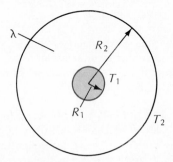

Figure 21-10. Problem 21-40.

: **21-41 P** An insulating material is made of alternating layers of equal thicknesses of sheets with high strength and high thermal conductivity λ_1 and sheets of low strength and low thermal conductivity λ_2, with $\lambda_1 = 7\lambda_2$. Such a laminate consists of 10 layers total, with the strong material on the outside. What is the temperature at the first interface from the outside if the outside surface is at 700°C and the inner surface is at 100°C?

(A) 685°C
(B) 667°C
(C) 640°C
(D) 595°C
(E) 580°C

Section 21-6   Thermal Resistance

: 21-42 P Commercial insulation is rated in terms of its R value, expressed in units of F°·ft²·h/Btu. (a) What is the R value of a standard 5 in. layer of rock wool? (b) In heating zone 5 in the U.S. (northern Minnesota, Wisconsin and Michigan) R-38 is recommended for ceilings and R-19 for walls. What thicknesses of rock wool are needed to provide these degrees of insulation?

: 21-43 P The total surface area of a house is 260 m². The average R value for the structure is 20. The house is heated with natural gas, which costs 6¢/kW·h. If the temperature difference between the inside and the outside of the house is 20 C°, how much will it cost to heat the house for a month?

: 21-44 P In the southernmost parts of the U.S., an R value of 19 is recommended for home wall insulation. What thickness of wood is required to provide this insulation? The thermal conductivity of wood is typically about 3×10^{-4} cal/cm·s·C°.

Section 21-7   Thermal-Energy Transfer through Radiation

· 21-45 Q If the absolute temperature of the filament of a light bulb were doubled, the energy radiated per second by the filament would

(A) remain the same.
(B) double.
(C) increase by a factor of 4.
(D) increase by a factor of 8.
(E) increase by a factor of 16.

· 21-46 P What surface area is needed in a tungsten lamp filament radiating power at 100 watts, if the filament temperature is 3000 K and the emissivity is 0.28?

· 21-47 Q Any human body

(A) continuously radiates electromagnetic radiation, principally in the ultraviolet region of the spectrum.
(B) continuously radiates electromagnetic radiation, principally in the visible region of the spectrum.
(C) continuously radiates electromagnetic radiation, principally in the far infrared region of the spectrum.
(D) continuously radiates electromagnetic radiation, principally in the microwave region of the spectrum.
(E) does not emit electromagnetic radiation.

· 21-48 P A radiation pyrometer used to measure the temperature of a furnace receives 2 W from a furnace at 700°C. What will be the power incident on the pyrometer when the temperature of the furnace is raised to 1400°C?

: 21-49 P A furnace is effectively radiating like a blackbody (emissivity = 1). The energy radiated through a port of area 1.8 cm² in the furnace is measured to be 100 watts. What is the temperature of the furnace?

: 21-50 P Sunlight incident on the top of the earth's atmosphere has an intensity of about 1400 W/m². Suppose that the earth were a blackbody and that its surface were all at one temperature. What would its surface temperature then be if the earth were to radiate away as much heat as it absorbs?

: 21-51 P The sun radiates like a blackbody with emissivity = 1 and temperature 6000 K. The diameter of the sun is approximately 1.39×10^9 m. (a) What is the total power radiated from the sun? (b) How much of this power is intercepted by the earth? (c) Worldwide power use is about 2×10^{10} W. What is the ratio of the total solar power falling on the earth to this value?

: 21-52 Q Which of the following is an accurate statement?

(A) An object is not considered to be radiating unless it is emitting visible light.
(B) Most of the radiation from a tungsten light bulb is in the visible region of the spectrum.
(C) The human body emits radiation that is peaked in the infrared region of the spectrum.
(D) Infrared rays transfer energy, whereas ultraviolet rays do not.
(E) Energy cannot be transferred by radiation through a vacuum.

: 21-53 P An aluminum sphere of radius 4 cm is heated to 500°C and suspended from a fine thread in an enclosure maintained at 23°C. The emissivity of the sphere is 0.8. (a) At what rate is the sphere radiating energy? (b) Estimate the time for the sphere's temperature to drop to 490°C, assuming it is always at a uniform temperature throughout.

: 21-54 Q In years of heavy snowpack in the mountains it is sometimes desirable to induce early melting of the snow, rather than to wait until the snow all melts suddenly and causes floods. It has been suggested that a way to accomplish this might be to have planes fly over the snow fields and sprinkle them with black soot. Would this work?

(A) Yes, it would probably work because the black surface would be a better absorber of sunlight than would the white snow.
(B) Yes, it would work because the soot would raise the melting point of the snow.
(C) Yes, it would work because the soot would decrease the specific heat capacity of the snow.
(D) No, it would not work, because it is infrared radiation not visible radiation that melts snow.
(E) No, it would not work, because sunlight has very little effect on how fast the snow melts.

21-55 P The temperature of the earth increases with depth at about 3 C° per 100 m. Since the deepest explorations are only to depths of a few thousand meters or so, what happens deep within the earth is uncertain. The fact that the surface of the earth is cooler than the layers just below indicates that the earth is loosing thermal energy in the form of infrared radiation. Long ago when the earth was hotter inside the rate of loss would have been greater than it is today. Geologists estimate that at one time all of the earth was at a temperature of about 2000°C.

(a) Assuming that the average temperature of the earth is now of the order of 0°C, how much thermal energy has the earth lost in the cooling process, assuming the earth to have an average specific heat of 1000 J/kg C°? (b) Approximately how long would the earth have been cooling in order to lose this much thermal energy? Take the thermal conductivity of rock to be about 2 W/m C°, and assume that the present thermal gradient of 3 C° per 100 m applies for earlier periods.

Supplementary Problems

21-56 P (a) The R values of insulators used in the building industry are expressed in a curious mixture of units. Show that the numerical value of R, expressed in these units, is equal to the time in hours required for 1 BTU of thermal energy to leak through one square foot of insulation 1 inch thick when the temperature difference between the two sides of the insulator is maintained at 1 F°. (b) What is the R value of concrete 1.0 ft thick whose thermal conductivity is equal to 8.0 mW/cm · K?

21-57 P A well streamlined automobile may lose its "grip" on the roadway at high speeds. Why? What is the function of a spoiler?

21-58 P Skilled discus throwers claim that a properly thrown discus will travel several meters farther if it is thrown *against* the wind than if it is thrown along the direction of the wind. Can you explain the effect in qualitative terms? For a more detailed explanation, see C. Frohlich, *Am. J. Physics* 49, 1125 (1981).

The Second Law of Thermodynamics and Heat Engines

22

As we have seen, the kinetic theory is basically a microscopic mechanical approach to thermal behavior. Formal thermodynamics, on the other hand, concentrates on the macroscopic behavior of systems. It takes no account of the details of the system's microscopic behavior. And therein lies its power. We can deduce remarkably sweeping results using only a few general thermodynamic principles.

We consider the most general properties of heat engines and heat pumps, not so much because of their practical importance, but rather because we are thereby led to an operational formulation of the second law of thermodynamics. We shall see that it is possible, using only the properties of heat engines, to define a temperature scale completely independent of the nature of the thermometric substance. Finally, we discuss the thermodynamic concept of entropy, which gives a quantitative measure of a system's disorder. The entropy concept then leads to still another formulation of the second law.

22-1 Energy Convertibility

In this chapter "the system" may consist of any well-defined collection of objects. The system could be a single spring, or a magnet, or some complicated

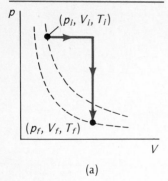

(a)

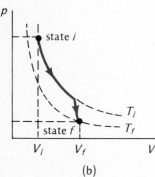

(b)

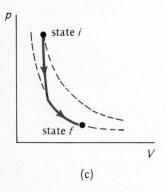

(c)

Figure 22-1. Three reversible paths leading from state i to state f at a lower temperature. (a) Isobaric expansion followed by isovolumetric cooling. (b) Isothermal expansion followed by adiabatic expansion. (c) Adiabatic expansion followed by isothermal expansion.

gadget with many internal moving parts. Or the system can simply be an ideal gas in a container. Actually, we shall use an ideal gas to illustrate thermodynamic processes because the behavior of such an ideal gas is well known; its equation of state is the general-gas law.

A system is said to be in thermodynamic equilibrium when:

- It is in *mechanical* equilibrium with the resultant force, and the resultant torque on it is zero.
- It is in *chemical* equilibrium, and no chemical reactions take place.
- It is in *thermal* equilibrium, and all its parts as well as its environment are at the same temperature.

We shall concentrate here on the last requirement.

A process is a change in the state of a system (Section 19-5). Two types of processes must be distinguished, reversible processes and irreversible processes.

- A reversible process can be made to run backward merely by reversing the direction in which the thermodynamic variables change.
- An irreversible process cannot.

For example, suppose that the pressure of a gas in a container of fixed volume is increased by an infinitesimally small amount dp. The temperature then rises by dT. Then, if one wishes, the gas can be returned to its original state simply by lowering the pressure by dp. As a consequence, the temperature falls by the same amount dT. Such a process is reversible. Suppose that on the other hand, a gas is allowed to expand freely and suddenly to fill an empty container of larger size. No work is done on the gas, and heat neither enters nor leaves the gas, But this free expansion process is irreversible. The gas will not return to its original state by spontaneously contracting. A reversible process must in general consist of a succession of infinitesimal changes taking place slowly, so that at each stage the system is in thermodynamic equilibrium. Then, and only then, can a temperature be defined for all intermediate stages.

Any isothermal process is reversible. The temperature remains constant throughout the system, and the system is always in thermal equilibrium as its state changes. A nonisothermal process, on the other hand, may or may not be reversible. Consider an adiabatic process. If a process is to be both adiabatic and reversible, the process must take place fast enough that no thermal energy enters or leaves the system, but slowly enough that the system is at all times in thermal equilibrium. Such conditions can be achieved when the system is inside a good thermal insulator. There are many examples of irreversible processes: a bursting balloon, a dropped egg, an overstretched spring, a magnetic material suddenly demagnetized by the turning off of an external magnetic field. A perfectly reversible process is virtually unattainable; in the real macroscopic world, all transformations are somewhat irreversible.

Consider an ideal gas undergoing a reversible process from some initial state (p_i, V_i, T_i) to some final state (p_f, V_f, T_f). According to the first law of thermodynamics, the heat Q_{if} supplied to the system, the work W_{if} done by the system, and the change in the system's internal energy ΔU_{if} in going from i to f are related by

$$Q_{if} = \Delta U_{if} + W_{if} \tag{20-1}$$

Both Q_{if} and W_{if} depend on the process leading from i to f, but $\Delta U_{if} = U_f - U_i$ does not. Recall that the internal energy of an ideal gas is a function of the state of the system, not of its history.

Figure 22-1 shows three possible reversible paths on a pV diagram between the same pair of initial and final states:

- The gas expands at first isobarically and then isovolumetrically.
- The gas expands at first isothermally and then adiabatically.
- The gas expands at first adiabatically and then isothermally.

Since i and f are the same for all three processes, the internal-energy change is the *same* for all three. The temperature falls, and therefore ΔU_{if} is negative. The work done by the gas in going from i to f, represented by the area under the pV curve, differs for (a) to (c). So does the total heat going into the system for the three processes. Since ΔU_{if} is negative, the first law of thermodynamics requires that the work done in all three processes exceed the net heat added.

The three processes shown in Figure 22-1 are reversible. Indeed, a process can be represented by a continuous line on a pV diagram only if it is reversible. Only then does the system progress through a succession of states of thermal equilibrium with a well-defined temperature at each stage. For an irreversible process, one can show the two end points on a pV diagram, but *no continuous line* can be drawn connecting them.

Now suppose that the three processes of Figure 22-1 are exactly reversed. The system now undergoes compression, rather than expansion. From state f to i, the internal energy change is now positive, heat is removed from the system, and work is done on it. Whenever a system undergoes a reversible process in which heat is converted to work, one may rerun the process to return the system to its initial state, with work then being converted to heat. The overall change in the internal energy over this special cycle is then zero. Moreover, since the overall process is represented by a single closed line on a pV diagram, rather than a loop, the heat entering the system over the entire cycle is zero, and the work done by the system over the entire cycle is also zero.

22-2   Heat Engines, Heat Pumps, and the Second Law of Thermodynamics

To use the energy stored in chemical or nuclear fuels, such as coal, oil, natural gas, or uranium, one typically first converts the fuel's potential energy to thermal energy and then converts some of the thermal energy to work. Chemical or nuclear potential energy can be converted to thermal energy with 100 percent efficiency (for example, by oxidation or nuclear fission). We shall see, however, that a process in which heat is converted to work through the use of a heat engine operating over a cycle can never have 100 percent efficiency. Before we consider how to achieve the maximum efficiency for a heat engine, we first note some general properties of any heat engine and heat pump.

A *heat engine* is defined as any device that in *operating through a cycle, converts (some) heat to work and discards the remainder into a cold reservoir*. A heat engine is always returned to its initial state in a cycle. Most ordinary heat

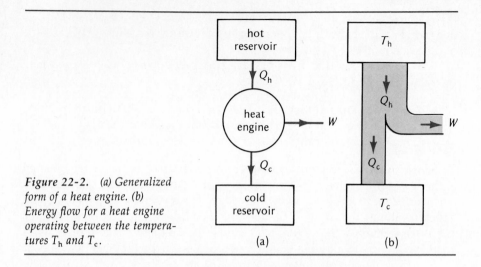

Figure 22-2.   *(a) Generalized form of a heat engine. (b) Energy flow for a heat engine operating between the temperatures T_h and T_c.*

(a)   (b)

engines contain a gas as the *working substance.** The heat engines we consider are idealized; they have no friction. But even if somehow no energy at all is dissipated in friction, an engine can never be perfectly efficient. The reasons are far more fundamental.

A *heat pump* is just a heat engine run backward. More specifically, a heat pump is a device that in *operating through a cycle, converts work into heat,* at the same time *transferring heat* from a low- to a high-temperature reservoir. A familiar example of a heat pump is an ordinary refrigerator.

The most general type of heat engine is shown schematically in Figure 22-2. We skip any mechanical details of construction. This general engine, represented by a simple circle, is a system into and out of which heat can flow, and which, when taken through a complete cycle, does net work on its surroundings. It converts (some!) heat to work. The engine may be connected to a heat reservoir at some high temperature T_h. We suppose that this reservoir contains so large an amount of thermal energy that even as it loses or gains heat from the engine, its temperature T_h remains unchanged. A second heat reservoir, also of large thermal-energy capacity, remains at the low temperature T_c.

Here are the steps that take place as an engine is run through one complete cycle. The engine is brought in thermal contact with the hot reservoir, and heat in the amount Q_h enters the engine. Some of the heat Q_h is converted to mechanical energy, or work W, that leaves the system; the remainder becomes thermal energy Q_c passing from the engine to the low-temperature reservoir. After the engine is returned to its initial state, the engine has completed the cycle. The heat Q_c discarded to the low-temperature reservoir, or exhaust, is here represented as a positive quantity. For convenience, we use Q_h, Q_c, and W to present merely the *magnitudes* of heat and work; whether energy enters or leaves the system will be indicated in diagrams by the directions of arrows.

---

* An engine need not use gas, even though this is typical. Although the process might seem bizarre and awkward, an ordinary elastic spring can actually act as a heat engine. You can heat or cool a spring; the spring will do work, or have work done on it, as it expands and contracts.

The ratio of useful work done by any heat engine over a complete cycle to heat supplied to it is called the engine's *thermal efficiency* e_{th}:

$$e_{th} = \frac{\text{work out}}{\text{heat in}} = \frac{W}{Q_{in}} \qquad (22\text{-}1)$$

Here W is the work out per cycle and Q_{in} is the heat in per cycle. This definition makes sense in that it compares what we get out of the engine in useful work with what we pay for in net heat in. Since we suppose that all the heat enters at the same high temperature T_h, we can write $Q_{in} = Q_h$. When we apply the first law of thermodynamics to one cycle, we get

$$Q = \Delta U + W$$
$$Q_h - Q_c = 0 + W$$

This says simply that work out equals *net* heat in. In completing one cycle, the system returns to its initial state. Therefore, $\Delta U = 0$. Using this result in (22-1), we then have

$$e_{th} = \frac{Q_h - Q_c}{Q_h} = 1 - \frac{Q_c}{Q_h} \qquad (22\text{-}2)$$

This relation shows that an engine can have 100 percent thermal efficiency only if $Q_c = 0$. An engine can be perfectly efficient only if *no* thermal energy is exhausted to the cold reservoir. A perfectly efficient heat engine would convert all the thermal energy entering it to work output when operated over a complete cycle, discarding none. This is impossible.

The reason? *The second law of thermodynamics.* Like any other fundamental law in physics, it is confirmed by the circumstances that no exception to it has ever been found. In this chapter we shall encounter the second law of thermodynamics in several different but equivalent formulations. We have already encountered the second law as it relates to the behavior of a system of numerous particles, which always proceeds to states of greater disorder (Section 20-9). Our first statement of the second law of thermodynamics is as follows: *No heat engine, reversible or irreversible, operating in a cycle, can take in thermal energy from its surroundings and convert all this thermal energy to work.* For any cyclic engine, $Q_c > 0$ and $e_{th} < 100$ percent. Lord Kelvin first propounded this statement of the second law in 1851.

Consider a heat engine run in reverse as a heat pump. See Figure 22-3. During each cycle, work W is done on the system, heat in the amount Q_c is extracted from the low-temperature reservoir, and heat in the amount Q_h is exhausted to the high-temperature reservoir. The net effect is that heat is pumped from the low- to the high-temperature reservoir. Note that the thermal energy Q_h delivered to the hot reservoir is greater than the thermal energy Q_c extracted from the cold reservoir. This follows from the first law of thermodynamics. With $\Delta U = 0$,

$$Q_h = Q_c + W$$

The heat pump is effectively a refrigerator; it removes thermal energy from the cold reservoir. If this reservoir were to have a noninfinite heat capacity, its temperature would fall.

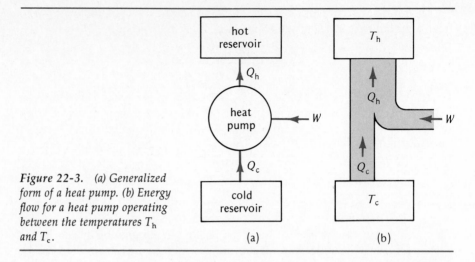

Figure 22-3. (a) Generalized
form of a heat pump. (b) Energy
flow for a heat pump operating
between the temperatures T_h
and T_c.

An equivalent statement of the second law of thermodynamics can then be given in terms of the general properties of a heat pump. It was first given by R. Clausius in 1850, as follows. *No heat pump, reversible or irreversible, operating over a cycle, can transfer thermal energy from a low-temperature reservoir to a higher-temperature reservoir without having work done on it.* For any cyclic heat pump, $W_{in} > 0$. This statement of the second law tells us that if a hot body and a cold body are placed in thermal contact and isolated, it is impossible for the hot body to get hotter while the cold body gets colder, even though this would not violate energy conservation, or the first law of thermodynamics. The observed fact that when a hot object and a cold object are brought together, they reach a final temperature *between* the initial temperatures is an illustration of the second law.

Example 22-1.   (a) Over each cycle, a heat engine removes 1000 J of heat from a hot reservoir and exhausts 400 J to a cold reservoir. What is the thermal efficiency of this engine? (b) When the same engine is run in reverse, with the same reservoirs, it is found that to pump 1000 J of heat per cycle to the hot reservoir requires that 2000 J of work be done on the pump. Is this engine a reversible or an irreversible engine?

(a) We have $Q_h = 1000$ J and $Q_c = 400$ J. Therefore, $W = Q_h - Q_c = 600$ J. From (22-2),

$$e_{th} = 1 - \frac{Q_c}{Q_h} = 1 - \frac{400}{1000} = 60\%$$

(b) If the engine were reversible, then when it was operated as a heat pump, we should again have $Q_h = 1000$ J, $Q_c = 400$ J, and $W = 600$ J. But we are given that the work W is actually 2000 J. The engine is clearly irreversible.

22-3   Reversible and Irreversible Heat Engines

Even if there were no frictional losses, no heat engine can be perfectly efficient. What, then, is the least inefficient cycle for the conversion of heat to work? In 1824 (actually before the validity of the first law of thermodynamics had been fully accepted!), the engineer S. Carnot (1796–1832) analyzed this question from a theoretical viewpoint and arrived at the following results, all consistent with the first and second laws of thermodynamics.

Of all possible heat engines operating cyclically between any two extremes of temperature T_h and T_c:

- A reversible engine taking in heat Q_h at a fixed temperature T_h and rejecting heat Q_c at a fixed temperature T_c is always more efficient than any irreversible engine operating between the same temperatures.
- Of all reversible engines operating between the two temperature extremes, the most efficient is the Carnot cycle (Section 22-4), one that takes in all its heat at a common high temperature and discharges all its heat at a common low temperature.
- The efficiency of a reversible Carnot engine (Section 22-4) is independent of the working substance of the engine.

The first and third statements can be proved here directly from the first and second laws of thermodynamics. The second statement will be proved in the next section.

Proof of Statement 1. Recall first that a reversible engine can be run forward or backward with no change in the magnitudes of Q_h, Q_c, or W. One need merely change their "directions." Suppose, for the sake of argument, that an irreversible engine can actually have a greater efficiency than a reversible engine. We shall find that, if true, this would violate the second law and therefore cannot happen.

Consider the two engines R (for reversible) and I (for irreversible) in Figure 22-4 operating between the same two reservoirs. In Figure 22-4(a), we see the engine R operating with $Q_h = 10$, $Q_c = 8$, and $W = 2$. (We choose specific values for Q_h, Q_c, and W to avoid an abstract argument.) The engine I in Figure 22-4(c) is assumed to have the same $Q_h = 10$, but $Q_c = 6$ and $W = 4$. Thus, the

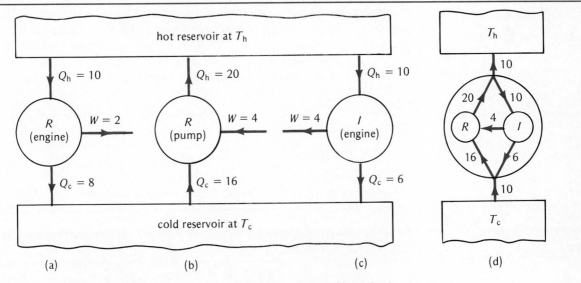

Figure 22-4. *(a,b) A reversible engine R (and pump) and (c) an irreversible engine I operating between the same hot and cold reservoirs. (d) A hypothetical combined reversible pump and irreversible engine operating together between T_h and T_c.*

irreversible engine I would have a thermal efficiency $e_{th} = 1 - (Q_c/Q_h) = 40$ percent; this *exceeds* the efficiency of the reversible engine R, which is only 20 percent. Figure 22-4(b) shows the reversible engine R now run backward as a heat pump; for convenience, here Q_h, Q_c, and W are all doubled in magnitude over the values in Figure 22-4(a). Next we imagine the heat pump R of Figure 22-4(b) to be coupled to the engine I of Figure 22-4(c). Then the work output of I is the work input of R. The composite is shown in Figure 22-4(d). We see that the work into the composite engine-pump system, made up of I and R together, is *zero*. The net result is the pumping of thermal energy from the low-temperature reservoir to the high-temperature *without* work on the system. This would not violate the first law of thermodynamics, but would obviously violate the second law, as formulated by Clausius. The conclusion: an irreversible engine cannot have an efficiency greater than that of a reversible engine operating between the same hot and cold temperatures.

The hypothetical device shown in Figure 22-4(d) is sometimes called a perpetual-motion machine of the *second kind*.* Such a device would pump heat to higher temperatures without having work done on it. If such a device could be made to work, it would be possible to convert the thermal energy in the universe to ordered energy, or work. One could run an engine by using the thermal energy in the ocean, for example. The "energy crisis" would be solved simply by dropping the temperature of the ocean a couple of degrees. An alternative statement of the second law of thermodynamics is then that *it is impossible to construct a perpetual-motion machine of the second kind.*

Proof of Statement 3. The efficiency of a reversible engine is independent of the engine's working substance (the particular gas used in the engine).

Consider two engines operating in the same reversible cycle, but having different working substances (one a steam engine and the other a diesel engine, for example). We assume that one engine is more efficient than the other. Clearly, we can run the lower-efficiency engine in reverse as a heat pump, couple it to the higher-efficiency engine, and then have a net result like that of Figure 22-4(d) — a transfer of heat from a low- to a high-temperature reservoir without any work being done. By the second law of thermodynamics, this is impossible. The conclusion: *The efficiency of a reversible engine is completely independent of the nature of the working substance.*

22-4   The Carnot Cycle and the Thermodynamic Scale of Temperature

Carnot recognized that of all possible heat engines operating between two temperature extremes, the most efficient was a reversible one that would operate as follows:

• Receive thermal energy isothermally from some hot reservoir maintained at a constant temperature T_h.

---

* A perpetual-motion machine of the first kind is one in which the mechanical energy remains constant. The device runs indefinitely because it is totally free of frictional or other dissipative losses. Such a device, which would not involve heat reservoirs, would be basically a toy rather than a useful machine.

- Reject thermal energy isothermally to a cold reservoir maintained at a constant temperature T_c.
- Change temperature in reversible adiabatic processes.

Such a cycle, which consists of two isothermal processes bounded by two adiabatic processes, is known as a *Carnot cycle*. See Figure 22-5.

As shown in the last section, the thermal efficiency of any reversible cycle, including the Carnot cycle, is independent of the working substance. From (22-2), we have for the Carnot cycle

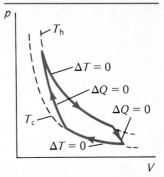

Figure 22-5.   *A Carnot cycle, consisting of two reversible adiabatic and two isothermal processes, operating between T_h and T_c.*

$$e_{th} = 1 - \frac{Q_c}{Q_h} \qquad (22\text{-}2)$$

The ratio Q_c/Q_h does not depend on the working substance. If the engine operates in a Carnot cycle, the ratio Q_c/Q_h can depend only on the temperatures T_h and T_c at which the heat enters and leaves the system.

It is possible, therefore, to use the Carnot cycle to define an *absolute thermodynamic temperature scale*, one that is independent of the thermometric substance. The scale is defined merely in terms of the heats, Q_h and Q_c, entering and leaving during a Carnot cycle. This was first done by Lord Kelvin in 1848, and the thermodynamic temperature scale is sometimes called the Kelvin temperature scale. Because the thermodynamic temperature does not depend on the nature of the thermometric substance, it is also the *absolute* temperature. By *definition*, for a reversible Carnot cycle,

$$\frac{T_h}{T_c} = \frac{Q_h}{Q_c} \qquad (22\text{-}3)$$

The size of the unit of the temperature scale is still arbitrary. For consistency with the size of the Celsius unit (defined in Chapter 19), the triple point of water is assigned the temperature 273.16 K.

In principle, the temperature of any substance (a solid or a liquid, as well as a gas) can be found as follows.

- Cycle the substance through a Carnot cycle.
- Measure the heat Q entering (or leaving) the system at the unknown temperature T.
- Measure the heat Q_{tp} leaving (or entering) the system when the system is in thermal equilibrium with water at its triple point.
- Use the relation

$$T = (273.16 \text{ K})\frac{Q}{Q_{tp}}$$

By combining (22-2) and (22-3), we can write the thermal efficiency of a Carnot cycle in terms of temperatures as

$$\text{Carnot cycle:} \quad e_{th} = 1 - \frac{T_c}{T_h} \qquad (22\text{-}4)$$

Equation (22-4) gives the maximum thermal efficiency attainable for *any* engine operating between the temperatures T_h and T_c. We see that it is 100 percent only if the engine exhausts heat to a cold reservoir at, and remaining at, the absolute zero of temperature—clearly an impossibility. Heat engines typi-

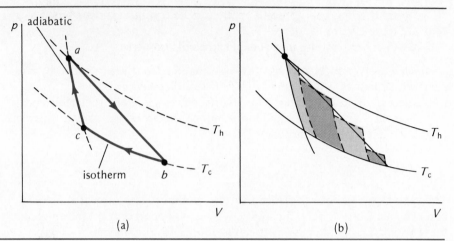

Figure 22-6. (a) *A non-Carnot cycle operating between* T_h *and* T_c. (b) *The reversible expansion can be approximated closely by a series of adiabatic and isothermal expansions.*

cally have very low efficiency. For example, if an engine takes in heat at the high temperature 200°C and exhausts heat at a room temperature of 30°C ($T_h = 473$ K, $T_c = 303$ K), its maximum efficiency is $e_{th} = 1 - \frac{303}{473} = 36\%$. In any real engine, friction is present, the processes are not perfectly reversible, and the operating cycle is not a Carnot cycle. Consequently, the actual efficiency is even less.

Now we show that the Carnot cycle is the most efficient of all reversible cycles operating between two fixed temperature extremes. Consider the cycle shown in Figure 22-6(a). From point a, the system expands reversibly along the line ab (neither an adiabatic nor an isothermal path), as the temperature decreases from T_h to T_c. This is followed by an isothermal compression to point c, and then adiabatic compression, which returns the system to starting point a. How does this reversible cycle compare in efficiency with a Carnot cycle between the same temperature extremes? Figure 22-6(b) shows how the reversible expansion can be approximated as closely as we wish by a series of small isothermal and adiabatic steps. We then replace the reversible cycle of Figure 22-6(a) by the small, adjacent Carnot cycles shown in Figure 22-6(b). The efficiency $e_{\text{non-C}}$ of any one of these small Carnot cycles depends on its upper and lower temperatures, T'_h and T'_c, according to $e_{\text{non-C}} = 1 - T'_c/T'_h$. But in Figure 22-6(b), the upper temperature T'_h of any small Carnot cycle is generally not as high as T_h (similarly, the lower temperature T'_c need not be as low as T_c). With $T'_h \leq T_h$ and $T'_c \geq T_c$, the overall efficiency of the whole reversible cycle must be less than the efficiency of a Carnot cycle between T_h and T_c. Thus, we can write

$$\text{Any non-Carnot reversible engine:} \quad e_{th} < 1 - \frac{T_c}{T_h}$$

where T_c and T_h are the temperature extremes of the working substance in the engine.

Example 22-2. Show that the temperature scale defined by the general-gas law for an ideal gas is equivalent to the Kelvin, or thermodynamic, temperature scale (which holds for any thermometric substance, ideal gas or otherwise).

We take an ideal gas around the closed Carnot cycle, shown in Figure 22-7. We wish to show that $Q_h/Q_c = T_h/T_c$, where T_c and T_h are now the temperatures defined by the general-gas law.

For the isothermal processes,

$$pV = \text{constant}$$

and the work done in expansion is

$$W_{i \rightarrow f} = nRT \ln \frac{V_f}{V_i} \qquad \textbf{(19-10), (22-5)}$$

For the adiabatic processes,

$$pV^\gamma = K \qquad \textbf{(19-9)}$$

or equivalently,

$$TV^{\gamma-1} = K \qquad \textbf{(22-6)}$$

Note that the T appearing in (22-5) and (22-6) is the temperature defined by the general-gas law.

The internal energy of an ideal gas does not change in an isothermal process. Therefore, during the isothermal expansion, the heat Q_h entering the system at the high temperature must equal the work done by the gas in expanding from a to b:

$$Q_h = W_{a \rightarrow b} = nRT_h \ln \frac{V_b}{V_a} \qquad \textbf{(22-7)}$$

where we have used the results of (22-5). Likewise, the heat Q_c leaving the system is

$$Q_c = W_{d \rightarrow c} = nRT_c \ln \frac{V_c}{V_d} \qquad \textbf{(22-8)}$$

We wish to show that $V_b/V_a = V_c/V_d$. Now for any adiabatic expansion of an ideal gas between the same two isotherms, the ratio of the final volume to the initial volume depends only on the two fixed temperatures T_h and T_c. This follows from (22-6). Thus, the ratio of final to initial volume is constant:

$$\frac{V_d}{V_a} = \frac{V_c}{V_b}$$

$$\frac{V_b}{V_a} = \frac{V_c}{V_d} \qquad \textbf{(22-9)}$$

Using (22-9) in (22-7) and (22-8) to find the ratio of Q_h to Q_c, we have finally

$$\frac{Q_h}{Q_c} = \frac{T_h}{T_c}$$

This is identical with (22-3), which defines the absolute temperature scale. An ideal-gas thermometer does indeed register the absolute thermodynamic temperature.

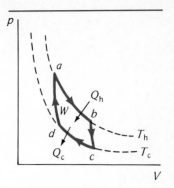

Figure 22-7. *Energy transfers in a Carnot cycle between temperatures T_h and T_c.*

22-5 Entropy

Here we introduce the important thermodynamic variable, the entropy of a system. As we shall see later (Section 22-7), the entropy is a quantitative measure of the disorder of the many particles that compose any thermodynamic system.

First, a matter of nomenclature. The absolute thermodynamic temperature was defined for a reversible Carnot cycle in such a way that the ratio of heat to temperature was the same for both the isothermal expansion and the isothermal compression. That is, from (22-3), $Q_h/T_h = Q_c/T_c$ for a *Carnot* cycle, where Q_h and Q_c represent, respectively, the *magnitudes* of the heat in and the heat out. In the analysis that follows, it will be important to adhere to the sign convention used originally in developing the first law of thermodynamics: Heat *entering* the system is *positive;* heat *leaving* is *negative.* Using this convention, we then have for the Carnot cycle

$$\frac{Q_h}{T_h} = -\frac{Q_c}{T_c}$$

or

$$\frac{Q_h}{T_h} + \frac{Q_c}{T_c} = 0 \qquad (22\text{-}10)$$

Thus, for a Carnot cycle, the sum of the quantities Q/T around a closed cycle is zero. This rule is actually more general. It holds for any reversible cycle, as we shall now show.

Consider the reversible cycle shown in Figure 22-8. Any reversible cycle can be approximated as closely as we wish by a series of isothermal and adiabatic processes; that is, a reversible cycle is equivalent to a series of junior Carnot cycles. We can, for example, roughly approximate the cycle in Figure 22-8 by several adjacent Carnot cycles. Equation (22-10) holds for each of these. Adding the equations for the individual small Carnot cycles that approximate the original reversible cycle, we have

$$\left(\frac{Q_1}{T_1} + \frac{Q_1'}{T_3}\right) + \left(\frac{Q_2}{T_2} + \frac{Q_2'}{T_4}\right) + \left(\frac{Q_3}{T_3} + \frac{Q_3'}{T_4}\right) + \cdots = 0$$

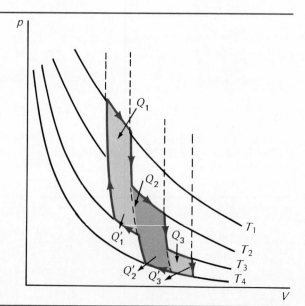

Figure 22-8.   A reversible cycle approximated by Carnot cycles.

A little thought shows that no net heat enters or leaves the system apart from the processes at the periphery (dashed lines in Figure 22-8). Therefore, we can write the last equation more generally as

$$\sum \frac{Q}{T} = 0$$

The summation is taken around the periphery of the original cycle. In the limit, we can then write

$$\oint \frac{dQ}{T} = 0 \qquad \text{for any } \textit{reversible} \text{ cycle} \qquad \text{(22-11)}$$

The circle on the integral sign indicates that the integration is to be taken around a closed path.

In words, (22-11) says that for any reversible cycle, the sum of the quantities giving the ratio dQ/T of the heat dQ entering the system to the temperature T at which the heat enters is zero around the cycle. This is equivalent to saying that the integral of dQ/T between *any* initial state i and *any* final state f is the same for all reversible paths from i to f. For example, in Figure 22-9, $\int_i^f (dQ/T)$ along the path P_1 equals $\int_i^f (dQ/T)$ along the path P_2 between the same end points i and f.

Now we exploit a mathematical property found earlier in the relation between a conservative force **F** and the associated potential energy U of the system. The potential energy difference between two end points i and f is related to the conservative force by

$$U_f - U_i = - \int_i^f \mathbf{F} \cdot d\mathbf{r} \qquad \text{(10-4)}$$

This relation can be written, however, only if the force is conservative and

$$\oint \mathbf{F} \cdot d\mathbf{r} = 0 \qquad \text{(10-2)}$$

with the net work done by the conservative force equal to zero over a closed loop.

In like fashion we may define a thermodynamic quantity, called the *entropy* S, whose difference depends only on the end points. By definition,

$$S_f - S_i \equiv \int_i^f \frac{dQ}{T} \qquad \text{(22-12)}$$

Note that the integration may be carried out along *any* reversible path leading from i to f. Equation (22-12) reduces to (22-11) when $i = f$ around a closed loop.

Now suppose that some system proceeds irreversibly from state i to f. We cannot represent any irreversible process by a path on a pV diagram. Nevertheless, we can determine the entropy difference between the states i and f. We simply imagine the system to pass from i to f along a reversible path connecting the two end points and compute the change in entropy, using (22-12). This is allowed because the entropy difference depends only on the end points, not on the path. (Analogously, in mechanics we can evaluate the potential-energy difference between two points, even when a nonconservative dissipative force also acts and the system is not able to pass reversibly between the end points; we do this by computing the work done by the conservative force alone.)

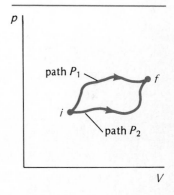

Figure 22-9. Two reversible paths, P_1 and P_2, leading from state i to state f.

In general, when a system is taken round a complete reversible cycle and returned to its initial state i, the following changes occur. The net change in internal energy is zero ($\Delta U = 0$). The net change in entropy is zero ($\Delta S = 0$). The work W done by the system is equal to the area enclosed by the loop on the pV diagram. By the first law of thermodynamics, the net heat Q entering the system is then $Q = W$. Now suppose that a system is taken through an irreversible cycle and returned to its initial state i. The change in internal energy again is zero ($\Delta U = 0$). The change in entropy is also zero ($\Delta S = 0$). The system has done work in the amount W, but it is *not* representable by any area on a pV diagram. Once again, net heat entering the system is $Q = W$.

Example 22-3.   One mole of an ideal gas is initially in the state (p_1, V_1, T_1). Find the change in entropy of the system when the gas expands to twice its initial volume (a) along a reversible adiabatic path, (b) along a reversible isothermal path, and (c) along a reversible isobaric path.

(a) The three processes are shown in Figure 22-10(a). Using (22-12), we have, for the adiabatic path from 1 to 4,

$$S_4 - S_1 = \int_1^4 \frac{dQ}{T} = 0$$

since dQ is zero for an adiabatic process. Therefore, $S_4 = S_1$. The entropy is the same at all points along a reversible adiabatic path; therefore, a reversible adiabatic process is sometimes referred to as an *isentropic* process.

(b) For the isothermal path from 1 to 3,

$$S_3 - S_1 = \int_1^3 \frac{dQ}{T} = \frac{Q_{1 \to 3}}{T_1}$$

Since the temperature is constant along an isothermal path, T could be taken outside the integral sign. We recall that the change in the internal energy of the gas is zero along an isothermal path. Therefore, the first law of thermodynamics gives

$$Q_{1 \to 3} = W_{1 \to 3} = RT_1 \ln \frac{V_3}{V_1} \qquad \text{(19-10)}$$

Combining the last two equations and using the fact that $V_3 = 2V_1$, we have

$$S_3 - S_1 = R \ln 2$$

(c) The entropy change $S_2 - S_1$ for the isobaric process is independent of the reversible path we choose between points 1 and 2. Therefore, to simplify the computation, let us first proceed along the isothermal path to point 3' and then along the adiabatic path from 3' to 2, as shown in Figure 22-10(b).

From part (a), there is no entropy change along the adiabatic compression from 3' to 2. We have, then, using the results of part (b),

$$S_2 - S_1 = S_{3'} - S_1 = R \ln \frac{V_{3'}}{V_1} \qquad \text{(22-13)}$$

For the adiabatic path between 3' and 2, we have, from (20-20),

$$pV^\gamma = (pV)V^{\gamma-1} \propto TV^{\gamma-1} = \text{constant}$$

Therefore,

$$T_1 V_{3'}^{\gamma-1} = T_2 V_2^{\gamma-1} \qquad \text{(22-14)}$$

The ideal-gas law relates points 1 and 2:

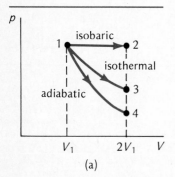

(a)

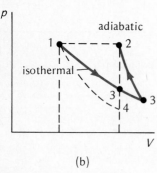

(b)

Figure 22-10.   *(a) Three reversible expansions. (b) The entropy change in the expansion from 1 to 2 is most easily computed by considering an isothermal expansion from 1 to 3 to 3' followed by an adiabatic compression to 2.*

$$\frac{p_1 V_1}{T_1} = \frac{p_2 V_2}{T_2}$$

Since $p_2 = p_1$ and $V_2 = 2V_1$, we have $T_2 = 2T_1$. Using these relations in (22-14) yields

$$\frac{V_{3'}}{V_1} = \frac{2V_{3'}}{V_2} = 2 \times 2^{1/(\gamma-1)} = 2^{\gamma/(\gamma-1)}$$

and (22-13) becomes

$$S_2 - S_1 = R \ln 2^{\gamma/(\gamma-1)} = R \left(\frac{\gamma}{\gamma - 1} \right) \ln 2$$

The results show that the entropy is greater at point 2 than at point 3, and in turn, greater at point 3 than at point 4.

22-6 Entropy and the Second Law of Thermodynamics

We state the second law of thermodynamics once more, now in terms of entropy:

For an isolated system the total entropy remains constant in time if all processes occurring within the system are reversible; on the other hand, the total entropy of an isolated system increases with time if any process within the system is irreversible. Since all actual macroscopic systems undergo irreversible processes, the *total entropy of any real system always increases with time.*

It is easy to show the equivalence between this statement of the second law and the one given in Section 22-2. Consider a system composed of a hot reservoir at temperature T_h, a cold reservoir at temperature T_c, and a heat engine operating between the two heat reservoirs, as shown in Figure 22-11. The engine may be either reversible or irreversible.

For each complete cycle of the heat engine, the total change in the entropy of the entire system — the heat engine and its surroundings — is accounted for as follows.

- The hot reservoir loses entropy because heat Q_h leaves the reservoir: ΔS (hot reservoir) $= -Q_h/T_h$.
- The cold reservoir gains entropy because heat Q_c enters the reservoir: ΔS (cold reservoir) $= Q_c/T_c$.
- The heat engine alone undergoes no change in entropy since it is returned to its initial state with the same entropy after completing a cycle, whether the engine is reversible or not.
- Work is done by the engine, and energy leaves the system. But no entropy leaves or enters the system since ordered energy associated with work has no entropy content.

Adding all contributions, we find for the total entropy change ΔS of the system:

$$\Delta S = \frac{Q_c}{T_c} - \frac{Q_h}{T_h}$$

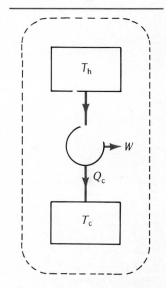

Figure 22-11. The system, consisting of the heat engine together with the hot and cold reservoirs, chosen in applying the second law of thermodynamics to heat engines and in computing entropy changes.

Now recall the definition of the thermal efficiency of a heat engine:

$$\text{\textit{Any} engine:}\quad e_{\text{th}} = 1 - \frac{Q_c}{Q_h}$$

We proved that no engine operating between the two fixed temperatures T_h and T_c could be more efficient than a reversible Carnot engine, whose efficiency is

$$\text{\textit{Carnot} engine:}\quad e_{\text{th}} = 1 - \frac{T_c}{T_h} \tag{22-4}$$

Therefore,

$$\text{\textit{Any} engine:}\quad e_{\text{th}} = 1 - \frac{Q_c}{Q_h} \le 1 - \frac{T_c}{T_h}$$

or rewritten,

$$\frac{Q_c}{T_c} \ge \frac{Q_h}{T_h}$$

The first equation above then gives

$$\text{Entire system:}\quad \Delta S \ge 0$$

which is what we set out to prove. The equality sign applies for reversible processes; the inequality sign, for irreversible processes.

Example 22-4.   What is the entropy change arising in the free expansion of an ideal gas from volume V_i to V_f?

No heat enters or leaves the system in this irreversible process ($Q = 0$), and the gas does no work when it expands freely ($W = 0$). Therefore, from the first law, $\Delta U = Q - W = 0$. Since the temperature is directly proportional to the internal energy for an ideal gas, the free expansion represents an irreversible adiabatic expansion in which the temperature is the same in the final equilibrium state as in the initial state. (This free expansion can not be described, however, as an isothermal expansion, even though the initial and final temperatures are the same. The expansion proceeds irreversibly; during this expansion, thermal equilibrium does not exist and a temperature cannot be defined for the system.) The initial and final states are shown in Figure 22-12.

We can find the entropy change of the gas between the states i and f by recalling that ΔS depends only on the end points, not on the path. For the purpose of computing ΔS, it is simplest to imagine the gas as proceeding from i to f via an isothermal path:

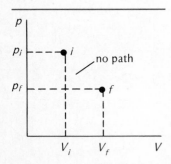

Figure 22-12.   *The initial and final states for a free (irreversible) expansion. Note that no path can be drawn from i to f.*

$$S_{i \to f} = \int_i^f \frac{dQ}{T} = \frac{1}{T} \int_i^f dW$$

It was shown in part (b) of Example 22-3 that $W_{i \to f} = nRT \ln (V_f/V_i)$ in an isothermal expansion. This being so,

$$\Delta S(\text{gas}) = nR \ln \frac{V_f}{V_i}$$

This is the entropy change whenever an ideal gas expands from (V_i, T_i) to (V_f, T_i), whether reversibly or irreversibly. The gas does not interact with its surroundings in the irreversible free expansion. Therefore, the total entropy change of the system (gas and surroundings) is that of the gas alone. Thus,

$$\Delta S(\text{system}) = nR \ln \frac{V_f}{V_i} \qquad (22\text{-}15)$$

The entropy change is, as always, greater than zero and independent of the temperature.

Example 22-5.   One mole of an ideal monatomic gas is held in each of two halves of an insulated container. The gas in one half has an initial temperature T; the other gas is initially at the temperature $T/2$. No heat leaves or enters through the walls. Now imagine that the partition separating the two gases has become a thermal conductor, so that thermal energy can flow from the hot gas to the cold until both gases reach thermal equilibrium at the common final temperature $\frac{3}{4}T$. What is the total entropy change for the system?

The initial and final states are shown in Figure 22-13. The change in entropy is,

for the hot gas, $\displaystyle\int_{T}^{(3/4)T} \frac{dQ}{T}$ for the cold gas, $\displaystyle\int_{(1/2)T}^{(3/4)T} \frac{dQ}{T}$

The total entropy change for the system is then

$$\Delta S = \int_{T}^{(3/4)T} \frac{dQ}{T} + \int_{(1/2)T}^{(3/4)T} \frac{dQ}{T}$$

Neither gas does work: $W = 0$. From the first law,

$$Q = \Delta U = \tfrac{3}{2} R\, \Delta T$$

Therefore,

$$\Delta S = \frac{3}{2} R \int_{T}^{(3/4)T} \frac{dT}{T} + \frac{3}{2} R \int_{(1/2)T}^{(3/4)T} \frac{dT}{T}$$

$$= \tfrac{3}{2} R (\ln \tfrac{3}{4} + \ln \tfrac{3}{2}) = \tfrac{3}{2} R \ln \tfrac{9}{8}$$

Once again, the entropy of the system *increases* in this irreversible process.

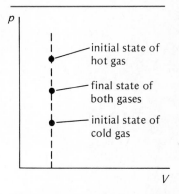

Figure 22-13. *Changes in the states of a hot and a cold gas brought into thermal equilibrium.*

22-7   Entropy and Disorder (Optional)

One macroscopic formulation of the second law of thermodynamics is that the entropy of any real system increases with time. In the microscopic considerations (Chapter 21), we saw that the disorder of any real system increases with time. The second law can then be expressed in terms of an increase either in the entropy or in the disorder of a system. How is entropy related quantitatively to disorder?

We assert, and will prove for one special case, that the entropy S of a system can be written

$$S = k \ln P \qquad (22\text{-}16)$$

where P is the probability that the system will be the state it is actually in compared with all other possible states, and k is the Boltzmann constant. Probability P thus is a measure of the relative disorder of the system, since the system's disorder is high when the probability of the state it occupies is also high.

To show that (22-16) is the appropriate relation between entropy and disorder, we consider one simple case in which the entropy of a system and its disorder both increase — the free expansion of an ideal gas.

As shown in Example 22-4, when an ideal gas expands freely from an initial volume V_i to a final volume V_f in an irreversible process, the entropy difference is given by

$$S_f - S_i = nR \ln \frac{V_f}{V_i} \qquad (22\text{-}15)$$

where n is the number of moles of gas. By definition, the Boltzmann constant is $k = R/N_A$, where N_A is Avogadro's number. The total number of molecules of the gas N is nN_A. Therefore,

$$nR = n(kN_A) = (nN_A)k = Nk$$

Equation (22-15) then can also be written in the form

$$S_f - S_i = Nk \ln \frac{V_f}{V_i} \qquad (22\text{-}17)$$

Now consider the free expansion of an ideal gas from a microscopic point of view, especially the change in the relative disorder of the system. We can be sure that the disorder of this system does increase on the following grounds; we know that before expansion, any one molecule was necessarily somewhere within the initial volume V_i, whereas we can say only for some molecule after expansion that it is within a larger volume V_f. Since we are less certain about the location of some one typical molecule, the disorder of the gas must increase. (An ideal gas undergoing a free expansion has the same final as initial temperature, and so there is no change in disorder as measured by a change in molecular kinetic energy.)

Let p_1 represent the probability of finding a single molecule within a volume V_1, where V_1 is *less* than the volume of the container for the gas. Now the probability p_1 is directly proportional to volume V_1, that is,

$$p_1 = cV_1$$

where c is a constant. Suppose that V_1 is doubled; then the probability of finding a single molecule somewhere within twice the initially chosen volume is also doubled. The probability of finding the molecule becomes 100 percent — we are certain to find the molecule — only when V_1 is made equal to the entire volume of the container; then we know that the molecule is somewhere inside the container.

Now consider the probability p_2 of finding two molecules both within the same volume V_1. The probability for two independent but simultaneous events is generally the product of their separate probabilities; here we have

$$p_2 = p_1^2 = (cV)^2$$

For example, if V_1 is chosen as half the volume of the container, the probability of finding a first molecule within V_1 is $\frac{1}{2}$. The probability of finding a second molecule within the same V_1 is also $\frac{1}{2}$, and the probability of finding both molecules simultaneously within V_1 is then $(\frac{1}{2})(\frac{1}{2}) = (\frac{1}{2})^2 = \frac{1}{4}$. It follows then that the probability P of finding all N molecules of the ideal gas in a volume V is

$$P = p_N = (p_1)^N = (cV)^N$$

If we take the natural logarithm of both sides of the relation above, we get

$$\ln P = N \ln (cV) = N(\ln c + \ln V)$$

Using the relation (22-16) connecting entropy with probability, we can write the relation above as

$$S = k \ln P = Nk (\ln c + \ln V)$$

Then the entropy difference between the initial and final state is

$$S_f - S_i = Nk (\ln V_f - \ln V_i)$$

$$S_f - S_i = Nk \ln \frac{V_f}{V_i}$$

which is just (22-17). The connection between entropy and probability is just that assumed at the beginning. Although proved here for a special case, the relation between entropy and probability given by (22-16) can be shown to be altogether general. The microscopic and macroscopic formulations of the second law of thermodynamics are in fact equivalent:

- Microscopic formulation—the system's *disorder increases.*
- Macroscopic formulation—the system's *entropy increases.*

Summary

Definitions

Reversible process: one that can be made to run backward by an infinitesimal reversal of thermodynamic variables (a *line* on a pV diagram).

Irreversible process: one that cannot be made to run backward (only the *end points* on a pV diagram).

Heat engine: any device that, over a cycle, converts (some) heat into work.

Heat pump: any device that, over a cycle, converts work into heat.

Thermal efficiency of an engine over a cycle:

$$e_{\text{th}} = \frac{\text{work out}}{\text{work in}} = 1 - \frac{Q_c}{Q_h} \qquad (22\text{-}1)$$

Carnot cycle: reversible cycle with two isothermals bounded by two adiabatics.

Absolute thermodynamic temperature: for a Carnot cycle with Q_h heat in at temperature T_h and Q_c heat out at temperature T_c,

$$\frac{T_h}{T_c} \equiv \frac{Q_h}{Q_c} \qquad (22\text{-}3)$$

Entropy difference:

$$S_f - S_i \equiv \int_i^f \frac{dQ}{T} \qquad (22\text{-}12)$$

between initial state i and final state f.

Fundamental Principles

Alternative and equivalent statements of the second law of thermodynamics:

- No heat engine operating in a cycle can convert to work all thermal energy taken in (Kelvin).
- No heat pump operating in a cycle can transfer thermal energy from a low- to a high-temperature reservoir without work; it is impossible to construct a perpetual-motion machine of this kind (Clausius).
- The total entropy of any isolated system cannot decrease; the entropy of any real system increases with time.

Other Important Results

A reversible engine is always more efficient than an irreversible engine operating between the same temperatures.

The efficiency of a Carnot engine is independent of the working substance.

A Carnot engine is more efficient than any other engine operating between the same temperatures.

For any closed reversible cycle the entropy change is zero:

$$\oint \frac{dQ}{T} = 0 \qquad (22\text{-}11)$$

Problems and Questions

Section 22-2   Heat Engines, Heat Pumps, and the Second Law of Thermodynamics

· **22-1 Q** An example of a reversible process is
(A) the free expansion of a gas.
(B) an isothermal expansion of a gas.
(C) the bursting of a balloon.
(D) any process that forms a closed cycle on a pV diagram.
(E) one carried out at constant pressure.

· **22-2 Q** Suppose you have available 50 J of energy with which to do work. Is there any way you could get more than 50 J of heat from a device if you have only this much energy available for work?

· **22-3 Q** In any process, the amount of heat that can be converted to work is
(A) 100%.
(B) dependent on the temperatures of the hot and cold reservoirs.
(C) dependent on the working substance used.
(D) dependent on the amount of friction present.

: **22-4 P** A heat engine absorbs 200 J and rejects 130 J on each cycle. (a) What is the engine's power output if the engine completes 6 cycles per second? (b) What is the efficiency of the engine?

· **22-5 P** A car uses 8 kg of gasoline per hour; its efficiency is 20%. What horsepower does it develop if the gasoline's heat of combustion is 4.8×10^4 kJ/kg?

: **22-6 Q** Consider a process in which an ideal gas is compressed from volume V to volume $\frac{1}{2}V$. Is more work done, when the compression is isothermal, or when it is adiabatic?

: **22-7 Q** A heat engine operates by extracting heat at temperature T_1 and
(A) converting some into work and rejecting the rest at temperature T_1.
(B) converting some into work and rejecting the remainder at temperature $T_2 < T_1$.
(C) converting some into work and rejecting the rest at a temperature $T_2 > T_1$.
(D) converting it all into work.

: **22-8 Q** A certain gas does work as it expands its container. The temperature of the gas remains constant during the expansion. It necessarily follows then that
(A) The gas is not an ideal gas.
(B) The average molecular speed decreases.
(C) The container is at a higher temperature than the gas within.
(D) The container cannot be a perfect insulator.
(E) The process described would violate the second law of thermodynamics.

: **22-9 P** A heat engine performs 20 cycles per second and does 400 J of work each cycle with 25% efficiency. How much heat is (a) absorbed and (b) rejected each cycle?

· **22-10 P** A refrigerator consists of a reversible heat pump. It takes heat from a low temperature reservoir at the rate of 1.6 kW when work is done at the rate of 2.4 kW. At what rate is heat rejected at the high temperature reservoir?

: **22-11 P** (a) A cyclic heat engine removes 1200 J of heat from a hot reservoir and exhausts 500 J to a cold reservoir. What is the thermal efficiency of the engine? (b) When the same engine is run in reverse with the same reservoirs, it requires 1700 J of work to pump 1200 J per cycle to the hot reservoir. Is this engine reversible or irreversible?

: **22-12 P** Starting at p_0, V_0, T_0, an ideal gas undergoes n successive adiabatic expansions, with the volume doubled each time. Show that the final temperature and pressure are

$$T = \frac{T_0}{2^{n(\gamma-1)}} \quad \text{and} \quad P = \frac{P_0}{2^{n\gamma}}$$

: **22-13 P** The internal combustion gasoline engine used in a car may be approximated by the *Otto cycle,* shown in Figure 22-14. (a) Show that the efficiency of the Otto cycle is $e = 1 - (T_b - T_c)/(T_a - T_d)$. (b) Show that the efficiency can be expressed in terms of the compression ratio $V_b/V_a = V_c/V_d = r$.

$$e = 1 - (r)^{1-\gamma}$$

(c) What is the efficiency for a compression ratio of $r = 7$ with $\gamma = 1.4$ (air)? A higher compression ratio cannot be used because the high temperature reached at point a will cause pre-ignition.

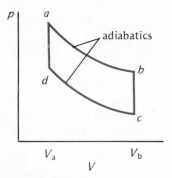

Figure 22-14.   Problem 22-13.

: **22-14 P** An ideal gas (with $\gamma = 1.4$) completes the cycle shown in Figure 22-15. Its maximum temperature is 600 K. Determine the temperature at each corner of the cycle and the efficiency of the cycle.

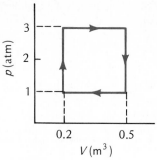

Figure 22-15. Problem 22-14.

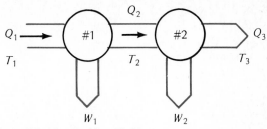

Figure 22-16. Problem 22-21.

Section 22-4 The Carnot Cycle and the Thermodynamic Scale of Temperature

· **22-15 Q** Is a heat engine less than 100% efficient because friction is always present, or are other considerations involved?

· **22-16 Q** In an adiabatic process,
(A) the temperature of the gas cannot change.
(B) if the gas is compressed, the work done on it is equal to the increase in internal energy.
(C) heat is conserved.
(D) the internal energy of the gas does not change.
(e) none of the above is true.

: **22-17 Q** The Carnot cycle is very important because
(A) most real heat engines utilize this cycle.
(B) it clarifies the first law of thermodynamics.
(C) it is the only irreversible cycle that can be completely analyzed theoretically.
(D) it provides a way of determining the maximum possible efficiency achievable with reservoirs of given temperatures.
(E) it is the only cycle that is independent of the properties of the specific working substance used.

· **22-18 P** A Carnot engine operates between temperatures of 500 K and 200 K. In one cycle it rejects 600 J of energy. How much heat does it absorb?

· **22-19 P** What is the maximum possible theoretical efficiency of a heat engine operating between two reservoirs at 400°C and 200°C?

· **22-20 P** Ocean thermal gradients are being used in attempts to generate electricity. Such generators use ammonia gas as a working substance in a heat engine which runs an electric generator. The hot reservoir is surface sea water at temperatures as high as 35°C and the cold reservoir is deep water at 4°C. What is the maximum possible theoretical efficiency for such an engine?

: **22-21 P** Two Carnot engines are run in series. See Figure 22-16. The heat exhausted from engine #1 is used as the input heat for engine #2. What is the overall efficiency, the output work divided by the input heat Q_1?

: **22-22 P** An engine uses one mole of an ideal gas with $\gamma = 1.4$. At point 1 the gas is at STP. It is heated at constant volume to 400 K at point 2. Next, it is expanded isothermally until the pressure drops back to 1 atm at point 3. Finally, it is compressed at constant pressure to its starting point. See Figure 22-17.

(a) Determine the temperatures at points 2 and 3. (b) Determine the heat flow into or out of the system for each part of the cycle. (c) Determine the work done in the cycle. (d) Determine the efficiency. (e) Determine the efficiency of a Carnot engine operating between the same temperature extremes as this cycle.

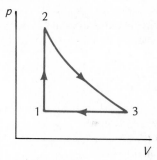

Figure 22-17. Problem 22-22.

Section 22-6 Entropy and the Second Law of Thermodynamics

· **22-23 Q** Is entropy something like energy, which depends on the amount in the system (that is, an extensive property) or is entropy like temperature, which does not (an intensive property).

· **22-24 Q** A container is divided into two equal parts. In one side is N_2 gas and in the other O_2 gas. The partition is broken and the gas particles mix. Has the entropy of the system changed?

· **22-25 Q** When you make ice cubes, the entropy of the water
(A) does not change.
(B) increases.
(C) decreases.
(D) may either increase or decrease, depending on the specific process used.

: **22-26 P** What is the increase of entropy in 1 kg of water when it is heated from 20°C to 80°C?

: **22-27 P** Two moles of an ideal gas expand isothermally, starting at STP, until the volume has doubled. What is the change in entropy of the gas?

: **22-28 P** A mass M of a substance with specific heat C per unit mass is heated from T_1 to T_2. Show that the change of entropy is $MC \ln(T_2/T_1)$.

: **22-29 P** What is the change in entropy of 2 moles of an ideal gas that experiences a reversible isothermal compression from 6 liters to 3 liters?

: **22-30 P** An isolated system that consists of 50 grams of water in equilibrium with 10 grams of ice has 600 J added to it. What is the change in the system's entropy?

: **22-31 P** Consider an isolated system consisting initially of 20 gm of ice at 0°C and 200 gm of water at 10°C. The system comes to thermal equilibrium. Determine the entropy change of the system, and show that it is positive for this irreversible process.

: **22-32 P** A Carnot engine operates between reservoirs at 600 K and 300 K and does 2400 J of work each cycle. (a) Draw the cycle on a pV diagram. (b) Determine the efficiency of the process. (c) What is the entropy change of the gas for the cycle?

: **22-33 Q** When a cup of hot coffee is allowed to sit in the room and cool, its entropy decreases. Explain why this situation does not violate the second law of thermodynamics.

Supplementary Problem

22-34 P A *solar salt pond* is a solar energy source whose operation depends on the fact that a temperature difference can be maintained between salt water at various elevations in a pond illuminated with sunlight. The density of salt water with 300 gm of NaCl added to 1 kg of water is about 15 percent greater than salt water with only 40 gm of NaCl to a kg of water. The more dense salt water stays at the bottom of the pond. When sunlight strikes the pond, it goes through the less dense, more transparent upper layer and is absorbed in the more dense salt water at the bottom. As a result, the salt water at the bottom is hotter than that at the top. In the simplest application, the solar salt pond can simply act as a heater and produce hot water for domestic use (now used in one-quarter of the homes in Israel). In more sophisticated applications, the temperature difference can run an engine.

Find the minimum surface area of a solar salt pond that would produce 1,000 kW of electric power by means of a turbine that takes in energy of 130 °C and exhausts it at 50 °C with solar radiation incident on the pond surface with the intensity of 0.120 W/cm².

Point Electric Charges

All the known forces in physics arise from four fundamental interaction forces: the strong nuclear force, the electromagnetic and the closely related weak interaction force, and the gravitational force. The nuclear and the weak interaction forces are important only within the atomic nucleus, in certain collisions between nuclei, and in the decay of unstable elementary particles. The familiar gravitational force is important only when one of two interacting objects has a mass comparable to that of a planet. This leaves the *electromagnetic force.* Except for the force due to gravity, all the forces of ordinary experience — the restoring force of a stretched string, the tension in a cord, the force between colliding automobiles — indeed, all the forces acting between the atomic nucleus and its surrounding electrons, or between atoms in molecules, are ultimately electromagnetic in origin. Electromagnetic forces dominate the interactions of all systems from the size of atoms to the size of planets.

The term *electromagnetism* emphasizes that electric and magnetic phenomena are not separate or unrelated. Both electrical and magnetic effects are a consequence of the property of matter called *electric charge.*

23-1 Some Qualitative Features of the Electric Force

The distinctively new concept introduced in electromagnetism is *electric charge.* First, how do ordinary objects become charged? Whenever any two

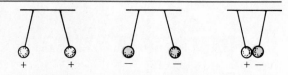

Figure 23-1. *Charged objects suspended from insulated strings. Like charges repel, unlike charges attract.*

dissimilar nonmetallic objects are brought into intimate contact — for example, glass rubbed with silk — and then separated, they show a mutual attraction, which far exceeds their gravitational attraction. Such objects are said to be electrically charged, and the presence of electric charge on each object is responsible for their interaction through an electric force. We assume that you are familiar with this effect and the qualitative experiments that establish the following fundamental facts about these electric interactions:

- *Like charges repel* each other.
- *Unlike charges attract* each other.

In this chapter we deal with electrostatics, the study of electric charges at rest. Electric charges in motion, besides interacting by the electric force, also interact through the so-called magnetic force. In the next several chapters we deal only with those situations in which the speeds of any charges are so much smaller than the speed of light (3.0×10^8 m/s) that the magnetic force is negligible compared with the electric force.

All electrical phenomena exist because fundamental particles in physics may have the property of electric charge. Thus, the electron has a negative charge, the proton a positive charge, and the neutron a zero charge. The use of the algebraic signs $+$ and $-$ to denote the two kinds of charge is appropriate, since combining equal amounts (to be defined precisely below) of positive and negative charges produces a zero electric force on an object. As we know, the nucleus of an atom consists of protons and neutrons bound together by the nuclear force within a volume whose length dimensions are never much greater than 10^{-14} m; the nucleus is surrounded by electrons that are bound to it by the electric force. An atom is electrically neutral as a whole when the number of electrons surrounding the nucleus equals the number of protons in the nucleus.

In solids, atoms are closely packed, their nuclei being separated from one another by distances of the order of 10^{-10} m. In *conductors,* of which many metals are common examples, most of the electrons are bound to their parent nuclei and remain with them. But in a conductor, typically one electron per atom is a *free electron.* A free (or conduction) electron, although bound to the conductor as a whole, may wander throughout the interior of the material and can easily be displaced by external electric forces. In *insulating,* or *dielectric,* materials on the other hand, all atomic electrons are *bound,* to a greater or lesser degree, to their parent nuclei. Electrons are removed from an insulating material or added to it only with the expenditure of significant energy. Examples of common conductors are metals, liquids having dissociated ions (electrolytes), the earth, moist air, and the human body. Good insulators are very often transparent materials: plastics, glass, and a vacuum, which is a perfect insulator. The best electrical conductors are better than the worst conductors (or best insulators) by enormous factors, up to 10^{20} (this conductivity ratio is given precise quantitative meaning in Chapter 27). Lying between these extremes are

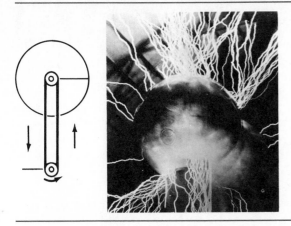

Figure 23-2. *(a) Schematic diagram for a Van de Graaff generator, a device for separating charge. The moving belt continuously carries charge from sharp points near it at the lower roller to points on the interior of a spherical conductor. (b) Photograph of a Van de Graaff accelerator.*

the so-called semiconductors; their conductivity is intermediate between conductors and insulators. Silicon is an example of a semiconductor. In a semiconductor, only a very small fraction of the electrons are free.

Qualitative electrostatic effects can be understood on the basis of the atomic model and of the properties of conductors and dielectrics. Suppose two unlike dielectrics are rubbed together. Electrons at the interface will leave the material to which they are less tightly bound for the other material, to which they become more tightly bound. On separation, one object carries excess electrons and is negatively charged, and the other has a deficiency of electrons and is positively charged. More generally, whenever a large-scale object has its electrical neutrality disturbed by losing electrons, the object is positively charged. Similarly, a negatively charged object has excess electrons. "Charging" an object consists simply in adding or subtracting electrons. Of course, when one type of charge is produced on an ordinary object, the other type must appear in equal amounts on a second object. The charging of any large-scale body results from the separation of charged particles (see Figure 23-2).

A charged body has acquired or lost electrons, and it is natural sometimes to speak of the net charge *on* a body. Of course the body acquires (or loses) not only the charge of electrons added to (or removed from) it but also the mass of the added (or removed) electrons. The additional mass is usually so trivial as to be negligible.

A simple device for detecting electric charge, often used in classroom demonstrations of electrostatic effects, is the *electroscope.* See Figure 23-3. The angular deflection from alignment with the vertical of a light pivoted conducting rod and the vertical conducting rod to which it is connected measures the charge on the electroscope. When charge is added to the conducting plate at the top of the vertical rod by touching it, the charge spreads, because of mutual repulsion, over the entire conductor. The pivoted rod is repelled by like charge on the fixed conducting rod. An electroscope is not a precise charge-measuring device; and we shall later learn of other devices through which charge magnitude is measured indirectly but far more precisely.

When a charged object touches an initially neutral conductor, the conductor acquires some of the object's net charge. It is possible, however, to charge a conductor without ever touching it directly with a charged object. The procedure is known as *charging by induction.* The steps are shown in Figure 23-4.

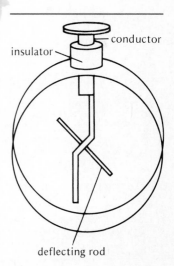

Figure 23-3. *An electroscope that shows its state of net electric charge by the deflection of the pivoted conducting rod.*

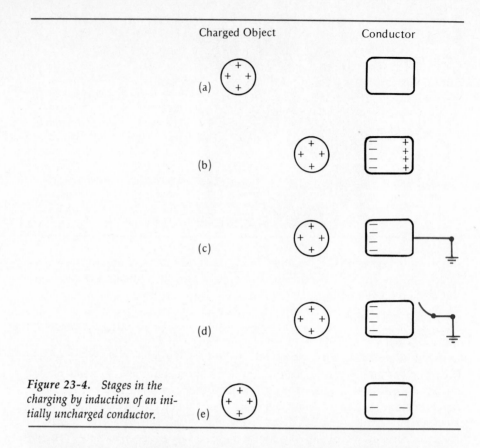

Figure 23-4. Stages in the charging by induction of an initially uncharged conductor.

(a) The charged object (here chosen as positive) is far away, and the conductor is electrically neutral, both as a whole and on all its surfaces.

(b) The charged object comes close to the conductor's left side. Electrons within the conductor are attracted to and move readily to the left side. This leaves an equal amount of positive charge on the conductor's right side. Bringing the charged object close to the conductor has merely separated positive from negative charge on the conductor; strictly, electrons have simply shifted left. If the charged object were removed at this stage, the situation would return to that shown in part (a).

(c) The conductor is *grounded.* It is connected by a conducting wire to the largest and most readily accessible good conductor nearby—the entire earth. (The electrical symbol for ground is ⏚.) Now the conductor has effectively become a part of a much larger conductor, the earth, and the positive charge that had been located on the conductor's right side is now distributed over the entire earth. The earth, because it is such a large conductor and acquires a relatively modest amount of positive charge from the conductor, is still effectively neutral. The negative charge does not leave the left side, however. It is held there by the attractive force from the positive charge on the nearby object.

(d) The conducting wire is disconnected and things remain as in (c).

(e) The charged object is taken away. Because of the mutual repulsion of electrons, the negative charge that had been concentrated on the left side is now distributed over the entire conductor. Note that the conductor has a charge that is *opposite* in sign to that on the charging object.

23-2   Coulomb's Law

The basic relation for the electric force between a pair of electrically charged particles is *Coulomb's law*. Consider Figure 23-5, where two point charges q_1 and q_2 (assumed to be of the same sign) are separated by distance r. Each charge repels the other by the electric interaction. The principal features, based on experiment, are these:

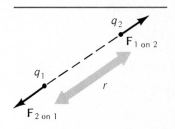

- The *magnitudes* of the Coulomb force on each charge are the *same*; their *directions* are *opposite*:

$$\mathbf{F}_{2 \text{ on } 1} = -\mathbf{F}_{1 \text{ on } 2}$$

- The electric force is a *central* force. It lies on the line connecting the two point charges.
- The magnitude of the electric force is given by

$$F_e = \frac{k q_1 q_2}{r^2} \tag{23-1}$$

Figure 23-5.   Two point electric charges interacting through the Coulomb force.

where k is constant.*

- Equation (23-1) applies only to *point charges*. This means that the physical size of charge q_1 or q_2 must be small compared with their separation distance r. In this chapter we consider point charges only. We consider continuous distributions of charge in the next chapter; the procedure for dealing with a charge distribution is simply to imagine it subdivided into a collection of effective point charges, and to apply (23-1) to each pair of charges, summing the contributions by the techniques of integral calculus.

- That the electric force is inverse-square with distance was first established by C. A. de Coulomb (1736–1806). He used a torsion balance like that Cavendish used later in studying the gravitational force (Section 15-3). It is now known through indirect experiments (Section 24-4) that the exponent of r is 2 to within a few parts in 10^{16}. Indeed, Coulomb's law is perhaps the most

---

* Equation (23-1) is a *scalar* equation; it gives the *magnitude* of the electric force on either charged particle. If the signs of q_1 and q_2 are included in (23-1), the sign we compute for F_e tells whether the force is repulsive or attractive: positive F_e for like charges and repulsion, negative F_e for unlike charge and attraction.

The *direction* of the electric force can be incorporated in a *vector* equation. Let radius vector $\mathbf{r}_{1 \text{ to } 2}$ have its tail at q_1 and its head at q_2. Then the vector force on q_2 can be written as

$$\mathbf{F}_{1 \text{ on } 2} = \frac{k q_1 q_2 \mathbf{r}_{1 \text{ to } 2}}{r_{1 \text{ to } 2}^3}$$

When q_1 and q_2 have the same sign, $\mathbf{F}_{1 \text{ on } 2}$ is in the same direction as $\mathbf{r}_{1 \text{ to } 2}$, corresponding to a repulsive force; the direction of $\mathbf{F}_{1 \text{ on } 2}$ is opposite to $\mathbf{r}_{1 \text{ to } 2}$ when q_1 and q_2 have opposite signs. Note that an additional factor $r_{1 \text{ to } 2}$ appears in the denominator to compensate for the magnitude of $\mathbf{r}_{1 \text{ to } 2}$ introduced into the numerator.

We can write the vector relation a little differently by using a unit vector. Let $\mathbf{u}_{1 \text{ to } 2}$ be a unit vector parallel to $\mathbf{r}_{1 \text{ to } 2}$. Then,

$$\mathbf{F}_{1 \text{ on } 2} = \frac{k q_1 q_2 \mathbf{u}_{1 \text{ to } 2}}{r_{1 \text{ to } 2}^2}$$

Force $\mathbf{F}_{2 \text{ on } 1}$ is, of course, just the negative to $\mathbf{F}_{1 \text{ on } 2}$. As a practical matter, it is usually best to compute just the magnitude of the electric force from Coulomb's law and indicate the direction of the force by a vector in a diagram.

thoroughly tested basic relation in physics. It works down to separation distances as small as 3×10^{-18} m, about one-thousandth the size of a proton.

● In the SI system of units, the unit for the scalar quantity called electric charge is the *coulomb* (abbreviated C). The definition of the coulomb is a bit complicated. First, by definition, a net charge of 1 C passes through the cross section of an electric conductor when an electric current of one ampere (1 A) exists in the conductor for one second: $1 \text{ C} \equiv (1 \text{ A})(1 \text{ s})$. The ampere is defined in turn by the magnetic force between two current-carrying conductors (Section 30-4).

With the coulomb unit for charge defined as given above, the constant in the Coulomb-law relation has the value

$$k = 8.987\ 55 \times 10^9 \text{ N} \cdot \text{m}^2/\text{C}^2$$

For most computations, it is satisfactory to use the rounded value, $k \approx 9.0 \times 10^9 \text{ N} \cdot \text{m}^2/\text{C}^2$.

The units assigned to k assure that with q_1 and q_2 both given in coulombs and r in meters, the force is in newtons. Thus, with two point charges each of 1 C separated by 1 m, the electric force on each is

$$F_e = (9.0 \times 10^9 \text{ N} \cdot \text{m}^2/\text{C}^2)(1 \text{ C})(1 \text{ C})/(1 \text{ m})^2 = 9.0 \times 10^9 \text{ N}$$

As static charges go, one coulomb is enormous. A laboratory device of ordinary size would usually not have a charge larger than about $10^{-6} \text{ C} = 1\ \mu\text{C}$ ($\mu = micro \equiv 10^{-6}$); this might be the charge on the sphere of a classroom Van de Graaff generator. A small conductor—say, a dime or a paper clip—touched to the terminal of a 9-V battery picks up a charge no larger than about $10^{-11} \text{ C} = 10 \text{ pC}$ (p = $pica \equiv 10^{-12}$).

● The constant k of Coulomb's law* is also written as

$$k \equiv \frac{1}{4\pi\epsilon_0} \tag{23-2}$$

where ϵ_0 is called the *electric permittivity of the vacuum*. The zero subscript indicates that as written in (23-1) and (23-2), Coulomb's law applies to electric charges in a *vacuum*. As we shall see in later chapters, it is sometimes preferable to write equations related to Coulomb's law using the constant ϵ_0 instead of k.

● Two fundamental forces in physics—the electric and the gravitational—are similar in some ways. Both are *central, conservative, inverse-square forces*. Because of this, many of the concepts and relations developed for gravity carry over unchanged into electricity. But there are also important differences. For one thing, there are *two* kinds of electric charge, but only one kind of gravitational charge (more often called gravitational mass). Electric charges may attract or repel; gravitational charges attract only. Further, there is no such thing as a gravitational conductor or shield. Another important difference is the relative magnitudes of these forces. The electric force is immensely larger than the gravitational force; for example, between an electron and

---

* Constant k relates to the *electric* interaction between charged particles. We shall encounter later (Chapter 32) a second constant k for the *magnetic* interaction between charged particles. To keep straight which is which, we shall later designate the constant associated with Coulomb's law as k_e and the constant associated with the magnetic interaction as k_m.

proton the electric attraction is 10^{39} times greater than the gravitational attraction (Example 23-3).

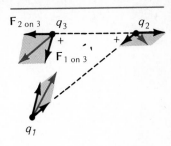

At the atomic level the gravitational force is so very small, compared with the electric force, that it can be ignored entirely. How is it then — if the electric force is so much greater than the force of gravity — that the only force a person is aware of in ordinary experience is gravity? It is that there are equal amounts of opposite charge in ordinary materials. The strong attraction between the opposite charges makes ordinary materials electrically neutral. With the electric force balanced, only the far weaker gravitational force remains.

Figure 23-6.   Three interacting point charges. The principle of superposition applies for Coulomb forces.

• The *superposition principle* for forces holds for the coulomb interaction. Suppose that a charge q_3 is in the presence of two other charges q_1 and q_2, as shown in Figure 23-6. Experiment shows that the force on q_3 is just the *vector sum* of the separate forces on it from q_1 and q_2. Said differently, the force between any two charges is independent of the presence of other charges; to find the resultant force, we merely add the individual forces as vectors. The superposition principle, although simple, is not self-evident; it is a result of observation for electric interactions. As we shall see, its consequences are many and important.

Example 23-1.   What is the electric force exerted on each of two identical small conductors each with a charge of $0.10 \ \mu C$ and separated by 1.0 cm?

$$F_e = \frac{kq_1q_2}{r^2} = \frac{(9.0 \times 10^9 \text{N} \cdot \text{m}^2/\text{C}^2)(1.0 \times 10^{-7} \text{ C})^2}{(1.0 \times 10^{-2} \text{ m})^2} = 0.90 \text{ N}$$

We expect then, that electric forces between charged objects will be obvious when charges are of the order of $1 \mu C$ and the charged objects are separated by the order of 1 cm.

Example 23-2.   Two small identical conducting spheres first attract one another. Then the conductors are touched together and brought back to the same separation distance as initially. The conductors now repel one another with the same force magnitude as initially. What was the ratio of the original charge magnitudes?

Let the initial charge magnitudes be q_A and q_B. Since the spheres attract initially, the charges must be of opposite sign. The net charge magnitude initially of the two spheres together is $q_A - q_B$. After the spheres have touched, the net charge is shared equally between the identical conducting spheres. The magnitude of the charge of each sphere is now $\frac{1}{2}(q_A - q_B)$.

The magnitude of the initial attractive force is

$$F = \frac{kq_Aq_B}{r^2}$$

The magnitude of the final repulsive force at the same separation distance is

$$F = \frac{k[(q_A - q_B)/2]^2}{r^2}$$

Equating the two forces yields

$$q_Aq_B = \left(\frac{q_A - q_B}{2}\right)^2$$

which can be written as a quadratic equation in q_A/q_B:

$$\left(\frac{q_A}{q_B}\right)^2 - 6\left(\frac{q_A}{q_B}\right) + 1 = 0$$

whose two roots are

$$\frac{q_A}{q_B} = 3 + \sqrt{8}, \, 3 - \sqrt{8}$$

There should be just one answer to the question, but there are two different roots to the quadratic equation. Actually, the two values are equivalent, since one is the reciprocal of the other: $3 + \sqrt{8} = 1/(3 - \sqrt{8})$.

23-3   Further Characteristics of Electric Charge

How is Charge Magnitude Measured?   Suppose that you observe two charges, q_A and q_B, not of the same magnitude, attracting one another. Which is the larger charge? With only two charges, you cannot tell, since the force magnitudes on the two charges are the same. In fact, you can't tell which of the two is positive and which is negative.

It takes a *third* charge. Suppose that this third charge is positive, and we first bring q_A close to it and measure the force F_A on it, as shown in Figure 23-7. With a repulsive force on q_A, we know that q_A is also *positive*. Now suppose that q_A is taken away, and q_B is brought to the same location to interact with our positive third charge. The force F_B on q_B is attractive, as shown in Figure 23-7(b). We conclude that q_B is negative. Moreover, the ratio of the force magnitudes gives us by *definition* the ratio of the charge magnitudes:*

$$\frac{q_A}{q_B} \equiv \frac{F_A}{F_B} \tag{23-3}$$

For the situation shown in Figure 23-7, we have $|F_A| = 2\,|F_B|$; therefore, $|q_A| = 2|q_B|$.†

Electron Charge   The electron's charge is designated by e; this is also the magnitude of the charge of any other ordinary charged "elementary" particle. The best current value for e is

$$e = 1.602\ 189\ 2 \times 10^{-19}\ C$$

The direct measurement of the electronic charge e, as first made in the Millikan experiment of 1909, is discussed in Problem 26-7.

An electron's charge is extremely small. One microcoulomb, a fairly typical charge by laboratory standards, corresponds to an excess or deficiency of 6×10^{12} electrons. Thus, we can ordinarily assume electric charge to be infinitely divisible and continuous and ignore its essential "graininess." We can, for example, imagine a negatively charged surface to have charge spread

---

* The procedure used here to define electric-charge magnitude is exactly like the one used earlier to define the ratio of the magnitudes of gravitational mass (Section 15-7).

† Does the magnitude of the charge of a particle depend on its speed? The evidence that a particle's charge always has the same magnitude, independent of speed, comes from a simple observation: All atoms that are expected to be electrically neutral because they have equal numbers of electrons and protons actually have no net electric charge, despite the fact that the electrons differ drastically in their speeds in the various elements.

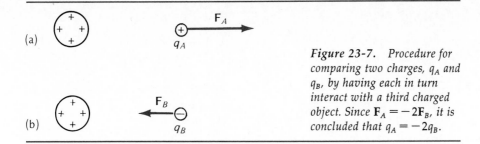

Figure 23-7. *Procedure for comparing two charges, q_A and q_B, by having each in turn interact with a third charged object. Since $\mathbf{F}_A = -2\mathbf{F}_B$, it is concluded that $q_A = -2q_B$.*

continuously over it, rather than concern ourselves with the actual finite number of discrete electrons, acting as point charges, that reside on it.

When electrons are transferred to or from a laboratory object being charged, the difference in mass is trivial. For example, when an object acquires a charge of 1 μC (or 6×10^{12} electrons), its mass changes by only $(9.1 \times 10^{-31}$ kg/electron) $(6 \times 10^{12}$ electrons) $= 5 \times 10^{-18}$ kg.

Charge Conservation   Electric charge is conserved. According to the fundamental law of conservation of charge, the net charge, or *algebraic sum of the charges, in any isolated system is constant.* This simple but important experimental result is illustrated very simply by the processes in which *two* neutral objects become charged. Electrons are transferred from one object to the other, and the result is one object with positive charge and the second with an equal amount of negative charge. No violation of charge conservation has ever been observed.

Electric-charge conservation does not imply, however, that charged particles can be neither created nor destroyed; it implies only that the creation of one of positive charge must be accompanied by the creation of another with an equal amount of negative charge. An important example is the phenomenon of *pair production.* When a sufficiently energetic photon, an electrically neutral particle of electromagnetic radiation, enters a closed container, as in Figure 23-8, it may be annihilated when close to an atom, and in its stead two particles appear, an electron with charge $-e$ and a positron with charge $+e$. The *net* charge within the container has not changed.

The positron, identical with an electron in all respects except for the sign of its charge (and the consequences of the difference in sign), is called the *antiparticle* of the electron. The electron and positron are but one example of a particle-antiparticle pair. Other examples are the proton and antiproton (charges $+e$ and $-e$, respectively) and more exotic elementary particles such as the π^+ meson and the π^- meson. Just as particle-antiparticle pairs can be created, a particle and its antiparticle can *annihilate* each other, producing two or more photons, or pairs of other particles. No matter what processes take place within a system—whether charge transfer between bodies in contact, nuclear transformations, creation of matter, or annihilation of particles—the total charge is always conserved. With the conservation of electric charge, the *classical* list of conservation laws is complete. They refer to the conservation of linear momentum (Section 7-2), angular momentum (Section 14-6), mass (Section 5-2), energy (Section 10-5), and now, electric charge.

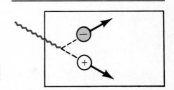

Figure 23-8.   *A photon enters a closed chamber and produces an electron-positron pair.*

Famous Physicist, Founding Father

When Benjamin Franklin (1706–1790) arrived in Europe in 1757, he was already world-renowned, not as a representative of the American colonies, but as a physicist. For his fundamental contributions to physics, Franklin had received honorary degrees from Harvard, Yale, and the College of William and Mary, and his book on electricity was then in its third edition.

He was an extraordinarily accomplished experimentalist, who constructed special apparatus to perform a variety of experiments that clarified many aspects of electrostatics. His "single fluid" theory eliminated the necessity of speaking of two distinct types of electric charge. Thus, an object with an excess of the basic charge, or fluid, was "positive," and another object with a deficiency was "negative."

The idea of electric-charge conservation originated with Franklin. He clarified the behavior of capacitors and the role of the dielectric. He showed that unbalanced charge resides on the outside of conductors and is especially concentrated at points. He explained what happens in the process we now call charging by induction. Anticipating electrons as the mobile carriers of electric charge, he believed that electric charge consisted of "extremely subtle particles." And Franklin showed in the famous kite and related experiments that electricity is not merely a laboratory curiosity but also the basis for such a large-scale natural phenomenon as lightning; he showed further that the lightning stroke typically goes from earth to the cloud, instead of the reverse. Franklin gave the sensible advice, "Never take shelter under a tree in a thunder gust [a thunderstorm]," and he invented the lightning rod. Other inventions: bifocals, the rocking chair, the Franklin stove, Daylight Saving Time.

Franklin's scientific interests ranged widely. He devised a thermometer to measure temperatures to depths in water of 100 ft. He studied cloud formation. He advanced arguments against the idea that light consisted of particles. He performed experiments to measure thermal conductivity. He advocated the caloric theory of heat. He wrote on such subjects as lead poisoning, gout, the physiology of sleep, deafness.

And not the least, Franklin had quite a hand in founding a nation.

Charge Quantization   The net electric charge of any object is just the algebraic sum of the charges of the particles that make up the object. All elementary particles, although they may differ greatly in mass and other properties, are found to have just one of *three possible charge values:*

$$+e, \qquad 0, \qquad \text{or} \qquad -e$$

where e is the magnitude of the charge of an electron.* For example, the charges of the electron and positron are $-e$ and $+e$, respectively; and the charges of the three kinds of mesons, the π^+, π^0, and π^-, are $+e$, 0, and $-e$, respectively. The neutron and its antiparticle, the antineutron, have charges of exactly zero.

Since every charged object is nothing more than a collection of particles, the only possible values of the total charge Q of any object are integral multiples of e:†

---

* One exception: The very short-lived exotic particle Δ^{++} has the charge $+2e$.

† There is indirect but nevertheless compelling evidence that some of the so-called elementary particles actually consist of still smaller parts called *quarks.* A quark has a fractional electric charge: $+\frac{2}{3}e, -\frac{2}{3}e, +\frac{1}{3}e$, or $-\frac{1}{3}e$. A proton, for example, is believed to consist of two "up" quarks, each with a charge of $+\frac{2}{3}e$, together with a "down" quark with charge $-\frac{1}{3}e$. A neutron consists of one up quark with charge $+\frac{2}{3}e$ and two down quarks, each with a charge $-\frac{1}{3}e$. A positive pion π^+ consists of an up quark (charge $+\frac{2}{3}e$) plus an "antidown" quark (charge $+\frac{1}{3}e$).

Quarks come in six types: up, down, charm, strange, top, and bottom. Furthermore, each quark comes in one of three "colors": red, blue, green. Finally, for each quark there exists an anti-quark. Adding up all possibilities gives a total of 36 distinct varieties.

$$Q = \pm Ne \qquad \text{where } N = 0, 1, 2, \ldots \tag{23-4}$$

Electric charge is said to be *quantized*. It appears only as *integral* positive and negative *multiples* of the charge of the electron, and no others; see Figure 23-9.

The discreteness, or granularity, of electric charge is evident only through subtle experiments, simply because most charged objects have a charge very much larger than e; that is, the integer N in (23-4) is typically very much larger than 1. Charge quantization shows up in chemistry in the chemical idea of atomic number; this integer is merely the total number of electrons (or protons), and hence, the total negative (or positive) charge in a neutral atom. Charge quantization is also implied in valence, which can have only an *integral* value.

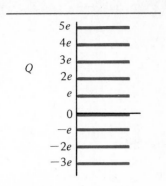

Figure 23-9.   Charge quantization. The only possible values of any charge Q are integral multiples of the electron charge e.

Example 23-3.   What is the ratio of the electric to the gravitational force between a proton and an electron?

The gravitational force (15-1) is $F_g = Gm_1 m_2 / r^2$, and the electric force is $F_e = ke^2/r^2$, where e is the charge magnitude for both electron and proton. The force ratio for any separation distance r is

$$\frac{F_e}{F_g} = \frac{ke^2}{Gm_1 m_2}$$

$$= \frac{(9.0 \times 10^9 \text{ N·m}^2/\text{C}^2)(1.6 \times 10^{-19} \text{ C})^2}{(6.7 \times 10^{-11} \text{ N·m}^2/\text{kg}^2)(1.7 \times 10^{-27} \text{ kg})(9.1 \times 10^{-31} \text{ kg})} \approx 10^{39}$$

(The electron and proton masses come from Appendix C.)

23-4   Electric Field Defined

The concept of field—here the electric field, later the magnetic field—is absolutely central to electromagnetism. Indeed, electromagnetism can be said to be the study of electric and magnetic fields and their relation to electric charge. At the first level, the field concept makes it easier to visualize, to graph, and to compute electromagnetic interactions; at a more fundamental level, the fields themselves are endowed with such physical properties as energy, momentum, and angular momentum.

To see why it is useful to define an electric field, consider first Figure 23-10(a). Here we have some point charges—we shall call them the *source charges*—fixed in position. Still one more charge, a *test charge* q_t, is brought to point P. The resultant electric force $\Sigma\mathbf{F_e}$ acting on q_t is found by applying Coulomb's law to its interaction with each of the source charges and adding the individual forces as vectors.

To make the circumstances more specific, suppose that the test charge happens to be $q_t = +1\ \mu\text{C}$ and the resultant force on it at P is found to be $\mathbf{F} = 1 \times 10^{-2}$ N in the direction east, as shown in Figure 23-10(b). Now suppose we take away the $+1$-μC charge and replace it by a charge of $-1\ \mu$C, again at point P, as shown in Figure 23-10(c). What is the force on the -1-μC charge? There is no need to repeat detailed computations. We can immediately say that with the charge magnitude unchanged but the sign reversed, the resultant force is still 1×10^{-2} N, but in the direction west. Similarly, if the charge magnitude is doubled so that we have $q_t = +2 \times 10^{-6}$ C, as in Figure 23-10(d), the force is also doubled to 2×10^{-2} N in the direction east. In short, if we know the resultant electric force on some *one* charge at P, we can immedi-

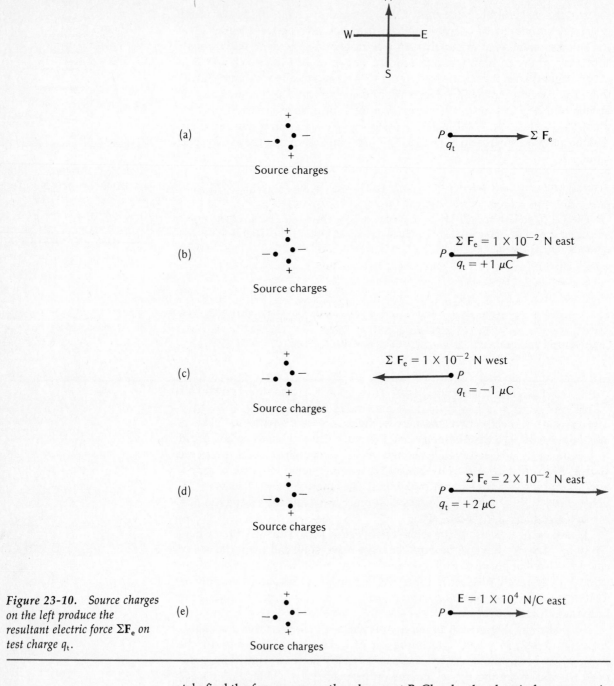

Figure 23-10. Source charges on the left produce the resultant electric force $\Sigma \mathbf{F}_e$ on test charge q_t.

ately find the force on *any other* charge at P. Clearly, the electric force per unit positive charge at point P is

$$\frac{\Sigma \mathbf{F}_e}{q_t} = \frac{1 \times 10^{-2} \text{ N east}}{+1 \times 10^{-6} \text{ C}} = 1 \times 10^4 \text{ N/C east}$$

This is the electric field **E** at P.

More formally, the *electric field* **E** at point *P* arising from source charge(s) *fixed* in position is given by

$$\mathbf{E} \equiv \frac{\Sigma \mathbf{F}_e}{q_t} \qquad (23\text{-}5)$$

where $\Sigma\mathbf{F}_e$ is the resultant electric force from the source charges on test charge q_t at point *P*. More succinctly, the electric field is the electric force per unit positive charge.

Said a little differently, the resultant electric force acting on charge q_t at a location where the electric field is **E** is given by

$$\Sigma\mathbf{F}_e = q_t\mathbf{E}$$

The electric field is a vector whose magnitude and direction may change from one point in space to another. As we shall see, we should think of the electric field as existing at point *P* even though there may be no actual charge q_t at this location to feel an electric force.

Why must the source charges be fixed in position? If the charges creating the electric field were not nailed down, then bringing an additional test charge q_t to *P* would produce a force on the source charges that might shift their positions and therefore *change* the electric field they create. This consideration is especially important for a conductor. Suppose that we have an uncharged conductor. Clearly, it produces zero electric field at all external locations. But if a test charge is brought near the conductor, this additional charge will redistribute the free electrons in the conductor, as shown in Figure 23-4, and the conductor may then produce electric fields different from zero at external locations. Rather than insist that the source charges remain fixed, we could instead imagine test charge q_t to be infinitesimally small. Then the forces that q_t would produce on the source charges would be so small that the locations of the source charges would not be changed.

Sometimes the quantity **E** is referred to as the intensity, or strength, of the electric field. Here we shall call it simply the electric field. From its definition, we see that the units for **E** are newtons per coulomb (N/C); we shall see later (Section 25-6) that equivalent (and more common) units are volts per meter.

What is the electric field a distance *r* from a single point charge q_s? With a test charge q_t a distance *r* from q_s, we have, from Coulomb's law,

$$F = \frac{kq_sq_t}{r^2}$$

But $\mathbf{E} = \mathbf{F}/q_t$. Therefore

$$E = \frac{kq_s}{r^2} \qquad (23\text{-}6)$$

The magnitude of the electric field from a single point charge falls inversely with the square of the distance *r* from it. The direction of the electric field is along the line connecting q_s and q_t. For *positive* q_s, then, **E** is radially *outward* from q_s; for a negative q_s, **E** is radially *inward*.* See Figure 23-11. The electric field for a single point charge is especially important, because any more com-

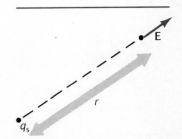

*Figure 23-11. Source charge q_s produces electric field **E** at a distance r.*

---

* A vector equation for **E** is $\mathbf{E} = kq_s\mathbf{r}/r^3$.

plicated charge distribution can always be regarded as a collection of point charges with (23-6) applying to each one.

The electric-field concept gives a new way of looking at the interaction between two charges q_a and q_b. We can say that charge q_a acts as a source of, or creates, an electric field $\mathbf{E}_a$ that surrounds it. Charge q_b, immersed in this field, is then subject to an electric force $\mathbf{F}_{a\,on\,b}$. That is,

$$q_a \text{ creates } \mathbf{E}_a; \qquad \mathbf{F}_{a\,on\,b} = q_b\mathbf{E}_a$$

(Alternatively, we can interpret the electric force of q_b on q_a by saying that q_b generates electric field $\mathbf{E}_b$ and that q_a immersed in the field $\mathbf{E}_b$ then experiences the force $\mathbf{F}_{b\,on\,a} = q_a\mathbf{E}_b$.)

This two-stage process—the production of the field by one charge and the response to the field by the second charge—may seem at first sight to be pedantry. But there is physical justification for the field concept, that is, for visualizing electric interactions as taking place via the electric field. For one thing, when q_a is moved, so that its separation distance from q_b is changed, q_b is not subject to a different force *instantaneously*. Rather, q_b continues to be influenced by the original force (therefore, electric field) for the time required for light to travel from a to b. That is, disturbances in the electric field arising from accelerated charges are propagated at the *finite* speed of light. This is no mere accident. As we shall see, light consists of electric (and magnetic) fields traveling through space.

Example 23-4.   Point charges $+q$ and $-q$ are separated by a distance d. What is the electric field of the two charges at point P in Figure 23-12? (Point P is on a line perpendicular to the line separating the two charges and at a distance d above charge $+q$.)

The electric field from charge $+q$ has the magnitude [from (23-6)] of

$$E_+ = \frac{kq}{d^2}$$

with $\mathbf{E}_+$ along the $+y$ direction.

Charge $-q$ is a distance $\sqrt{2}d$ from point P. The magnitude of its electric field at P is

$$E_- = \frac{kq}{(\sqrt{2}d)^2} = \frac{kq}{2d^2}$$

Field E_- is at angle of $45°$ below the x axis in the third quadrant, as shown in Figure 23-12(a).

To find the vector sum of $\mathbf{E}_+$ and $\mathbf{E}_-$, we use their rectangular components. The resultant electric field $\mathbf{E}$ has the components

$$E_x = -E_-\cos 45° = -\frac{kq}{2d^2}\left(\frac{1}{\sqrt{2}}\right)$$

$$E_y = E_+ - E_-\sin 45° = \frac{kq}{d^2} - \frac{kq}{2d^2}\left(\frac{1}{\sqrt{2}}\right)$$

so that

$$E = \sqrt{E_x^2 + E_y^2} = \frac{kq}{d^2}\left(\frac{5}{4} - \frac{1}{\sqrt{2}}\right)^{1/2} = 0.74\frac{kq}{d^2}$$

The direction of $\mathbf{E}$ relative to the negative x axis [see Figure 23-12(c)] is given by

$$\tan\phi = \frac{E_y}{E_x} = \frac{1 - 1/(2\sqrt{2})}{1/(2\sqrt{2})} = \tan 61°$$

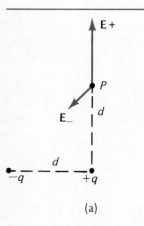

(a)

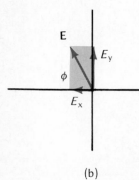

(b)

Figure 23-12. (a) Point charges $+q$ and $-q$ produce respective electric fields $\mathbf{E}_+$ and $\mathbf{E}_-$ at point P. (b) The resultant field $\mathbf{E}$ as the vector sum of its x and y components.

Example 23-5.   Charge $+q$ is at the origin. Charge $+3q$ is on the x axis at $x = d$. See Figure 23-13. Where is the electric field of the two charges zero?

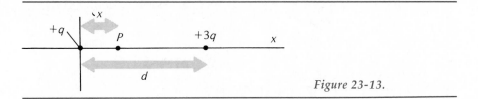

Figure 23-13.

Apart from being an infinite distance from the two charges, the point P at which $\mathbf{E} = 0$ must be on the x axis. Only along this line can the forces of $+q$ and $+3q$ on a third charge cancel. The coordinate of P is x.

Equating the magnitudes of the electric fields produced at P by the two charges separately then gives

$$kq/x^2 = 3kq/(d-x)^2$$

where we have used the fact that charge $+3q$ is at distance $d - x$ from point P.

The equation above can be recast in the form

$$2(x/d)^2 + 2(x/d) - 1 = 0$$

which is a quadratic equation whose roots are

$$\frac{x}{d} = -\tfrac{1}{2} + \tfrac{1}{2}\sqrt{3}, \ -\tfrac{1}{2} - \tfrac{1}{2}\sqrt{3}$$

or

$$\frac{x}{d} = 0.366, \ -1.37$$

The answer we want is $x/d = 0.366$. This corresponds to a location *between* the two charges, where the two *positive* charges contribute electric fields in *opposite* directions with equal magnitudes. The second root, $x/d = -1.37$ corresponds to a location to the *left* of both charges; there the two charges contribute electric fields of the same magnitude but in the same direction (to the left).

23-5   Electric-Field Lines

How can you represent the electric field of a point charge? Since $\mathbf{E}$ is a vector, one way is to use several vectors, as shown in Figure 23-14(a), with the tail of each vector at the location where $\mathbf{E}$ is specified. For a positive charge, the vectors are all radially outward. The magnitude of an $\mathbf{E}$ vector is inversely proportional to the square of the distance r from the charge.

An equivalent way of mapping the electric field is shown in Figure 23-14(b). Here we have uniformly spaced, outwardly directed, straight *electric-field lines* radiating from the positive point charge. (Strictly, the lines go outward in three dimensions.) Clearly, at any point the direction of the field line is also the direction of $\mathbf{E}$.

What about the magnitude of $\mathbf{E}$, which falls off with distance as $1/r^2$ does? Imagine the point charge to be surrounded by a sphere of radius r. All the same field lines will penetrate the sphere's surface, whatever its radius. The sphere's surface area ($4\pi r^2$) *is proportional to* r^2, so that the number of electric-field lines per unit area falls off with $1/r^2$ in exactly the same way as the magnitude of $\mathbf{E}$.

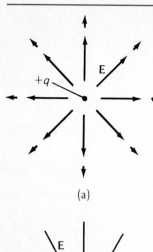

(a)

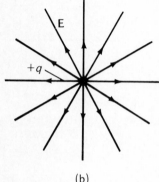

(b)

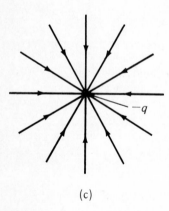

(c)

Figure 23-14. *(a) Electric-field vectors for a positive point charge q. (b) Electric-field lines for a positive point charge and (c) electric-field lines for a negative point charge.*

We can use the number of electric-field lines per unit area to indicate the magnitude of **E** at any point. It turns out (the general proof is given in Section 24-7) that the density of electric-field lines can always be used to represent the magnitude of the electric field. What clearly works for a single point charge, works for any charge distribution.

The electric-field lines surrounding a negative point charge are shown in Figure 23-14(c). Since the direction of **E** always gives the direction of the electric force on a *positive* charge placed in the field, the electric-field lines here are radially *inward* and converge on a negative charge.

Sketches of electric-field lines are especially useful because we can tell at a glance what will happen to still one more charge introduced into the field. A positive charge will be accelerated in the direction of **E**; a negative charge will be accelerated in the opposite direction. Important properties of electric-field lines representing the electric field **E** are these:

- Field lines **E** originate from positive charges and terminate on negative charges. Furthermore, as will be proved in Chapter 24, **E** lines are always continuous.
- At any point, the direction of an **E** line gives the direction of the electric field at that point.
- The density of **E** lines at *any* point—the number of lines per unit transverse area—is proportional to the magnitude of the electric field at that point. Therefore, the electric field becomes weaker where the **E** lines diverge and stronger where they converge.
- Electric-field lines give the direction of the *force* that will act on a unit positive charge introduced into the field. The field lines do not portray the paths or velocities of charged particles, although particle velocities, displacements, and paths can be computed from a knowledge of **E**. Only a particle's *acceleration* is along (or opposite to) the direction of **E**.
- The lines representing **E** are a useful fiction; the electric field itself is real, or as real as any other measurable physical property.

Figure 23-15 shows electric-field configurations produced by two point charges. In Figure 23-15(a) we have two separated point charges of the same magnitude but opposite sign—an *electric dipole* (see also Section 24-6). All field lines originating from the positive charge terminate on the negative charge. Point *P* corresponds to the location for which **E** was computed in Example 23-4. Indeed, the electric field at any location can be computed in exactly the fashion of that example. Note that close to either point charge the field lines are essentially those of a single point charge, since the second charge is then relatively far away and uninfluential. It is important to appreciate that Figure 23-15 shows the electric force acting on a *third* positive charge introduced into the presence of the two charges creating the electric field.

In Figure 23-15(b), with two separated positive charges of the same magnitude, the location midway between the two charges is one at which **E** = 0, and there are no field lines there. All field lines diverge outward. Viewed from afar, the field lines approximate those of a *single* positive charge of +2q. If the two charges were both negative and of the same magnitude, then the configuration of the field lines would not change, but the arrows would be reversed, with the field lines then converging on the negative charges.

In Figure 23-15(c), we have charges −q and +3q. Three times as many field lines originate from the +3q charge as go to the −q charge. In Figure 23-15(d)

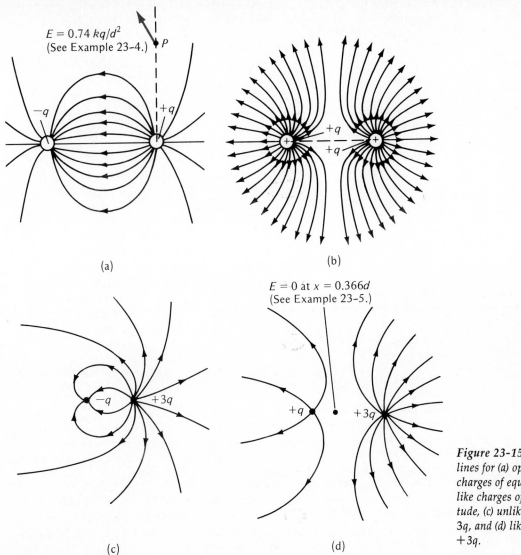

$E = 0.74\,kq/d^2$
(See Example 23-4.)

$-q$ $+q$

(a)

$+q$
$+q$

(b)

$-q$ $+3q$

(c)

$E = 0$ at $x = 0.366d$
(See Example 23-5.)

$+q$ $+3q$

(d)

Figure 23-15.   Electric-field lines for (a) opposite point charges of equal magnitude, (b) like charges of equal magnitude, (c) unlike charges $-q$ and $3q$, and (d) like charges $+q$ and $+3q$.

the spot where $\mathbf{E} = 0$ (computed in detail in Example 23-5) is closer to the $+q$ charge than the $+3q$.

Summary

Definitions

Electric field $\mathbf{E}$: the resultant electric force $\Sigma\,\mathbf{F}_e$ per test charge q_t,

$$\mathbf{E} \equiv \frac{\Sigma\mathbf{F}_e}{q_t} \qquad (23\text{-}5)$$

Electric-field lines may be used to present the vector field $\mathbf{E}$.

Units

Electric charge: Coulomb (C)
Electric field: Newtons per coulomb (N/C)

Fundamental Principles

Coulomb's law: for the electric force F_e between point

electric charges q_1 and q_2 separated by distance r

$$F_e = \frac{kq_1q_2}{r^2} \qquad (23\text{-}1)$$

where

$$k = \frac{1}{4\pi\epsilon_0} \approx 9.0 \times 10^9 \text{ N}\cdot\text{m}^2/\text{C}^2$$

Electric charge conservation: the net electric charge of an isolated system is constant.

Important Results

Electric charge is quantized; the observed charge magnitude is always an integral multiple of the electronic charge, $e = 1.60 \times 10^{-19}$ C.

Problems and Questions

Section 23-1   Some Qualitative Features of the Electric Force

· 23-1 Q When a positively charged rod is brought near a negatively charged electroscope, the pivoted conductor will
(A) deflect more
(B) deflect less
(C) become positively charged.
(D) stay the same.

· 23-2 Q A balloon that has been rubbed on your sweater and then pressed to a wall will often stick there. Which of the following is the best explanation?
(A) Rubbing removes a surface layer of grease, allowing the balloon to come close enough to the wall that air pressure holds it there.
(B) Rubbing the balloon charges it electrostatically, and this charge on the balloon induces an opposite charge on the wall. The attraction between the induced charge and the charge on the balloon holds the balloon to the wall.
(C) A wall typically has a net electric charge on it, and rubbing the balloon charges it electrostatically. If the wall happens to have a charge opposite to that on the balloon, the balloon will stick.
(D) Rubbing the balloon causes moisture to condense on it, and surface tension causes the balloon to stick to the wall.
(E) Rubber molecules form weak polymers with molecules in the wall.
(F) Rubbing the balloon surface causes it to become slightly conducting. When the balloon is touched to the wall, electrons flow from the balloon to the wall. This sets up an electric field that bonds the balloon weakly to the wall.

· 23-3 Q A charged particle is fired at right angles toward an electrically neutral flat conducting surface. Does the particle speed up or slow down as it approaches the surface? *Speed up*

: 23-4 Q Figure 23-4 shows how a single, initially uncharged conductor is given a charge by the process of "charging by induction." The positively charged object could actually be used over and over again to give an indefinitely large number of initially uncharged conductors neg-ative charges. Does this violate energy conservation? Where does the energy come from?

: 23-5 Q Two objects can exert an electric force on each other
(A) only if they are both conductors.
(B) only if they are both insulators.
(C) only if each carries a net nonzero charge.
(D) only if each contains some electrons.
(E) even if only one of the objects has a net charge.

: 23-6 Q An uncharged metal sphere hangs from an insulating thread. If a positively charged glass rod is brought near the sphere (but not touching), what will happen?
(A) The sphere will be attracted to the rod.
(B) The sphere will be repelled by the rod.
(C) The sphere will experience no net force since it is electrically neutral.
(D) The sphere will acquire a net charge.

: 23-7 Q Two identical negative charges are placed along the x axis. Midway between them is placed a small positive test charge. What can be said about the stability of the test charge?
(A) It is stable for motion along any axis.
(B) It is stable for motion only along the x axis.
(C) It is stable only for motion along the y or z axis.
(D) It is stable for motion along any direction perpendicular to the z axis.
(E) It is not stable for motion in any direction.

Section 23-2   Coulomb's Law

· 23-8 Q Two charged conducting spheres of the same size and mass are suspended by threads. Initially they attract each other. Then they touch and are electrically repelled. One can then conclude that before the spheres touched
(A) both were positively charged.
(B) both were negatively charged.
(C) the spheres had charges of equal magnitude and opposite sign.
(D) the spheres had charges of unequal magnitude and opposite sign.
(E) None of the above conclusions can be drawn.

· 23-9 Q Two identical spheres hold charges of different

magnitudes. After these spheres are allowed to touch they will always
(A) repel each other. *if conducting spheres*
(B) attract each other.
(C) sometimes attract and sometimes repel depending on the circumstances.
(D) be electrically neutral.
(E) attract at small distances and repel at large distances.

· **23-10 P** Two point charges experience an attractive force of 4 N when they are separated by 1 m. What force do they experience when their separation is (a) 0.5 m? (b) 2 m? (c) 10 m?

: **23-11 P** Charges q_1, q_2, q_3, and q_4 are placed at the corners of a square of side 20 cm (Figure 23-16). Determine the magnitude and direction of the force on q_1, given $q_1 = q_2 = q_3 = 5 \ \mu C$ and $q_4 = -5 \ \mu C$.

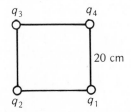

Figure 23-16.   Problem 23-11.

: **23-12 P** Three charges are placed as shown in Figure 23-17. What is the force on each charge?

Figure 23-17.   Problem 23-12.

: **23-13 P** A charge of $+2 \ \mu C$ is placed at the origin and a charge of $-4 \ \mu C$ is placed at $x = 6$ cm. Where can a third charge of $+1 \ \mu C$ be placed (not at infinity) so that the electric force on it will be zero?

: **23-14 P** Charges $+2Q$, $+2Q$, and $-Q$ are placed at the corners of an equilateral triangle of side a. Determine the magnitude and direction of the force on each charge.

: **23-15 P** Three charges are placed as follows: $+Q$ at $x = a$, $+Q$ at $x = -a$, and $-Q$ at $y = +a$. Determine the force on the charge $-Q$.

: **23-16 P** Five charges $+Q$ are placed in a line with equal spacing a between adjacent charges. Determine the force on each.

: **23-17 P** Two small spherical conductors are separated by a distance d, which is large compared with their radii. How should a fixed amount of charge Q be distributed between them to maximize the force of repulsion between them?

: **23-18 P** Two small spherical conductors, each of mass 10 gm, are suspended from two threads attached to a common point. The threads are each 60 cm long. A charge Q is placed on one conductor. This conductor is then touched to the second conductor, which was initially uncharged. The two then repel each other. When they come to rest, each string makes an angle of 37° with the vertical. What is the initial charge Q?

: **23-19 P** Charges $+Q$ and $-Q$ are placed at the corners of a cube of side a so that nearest-neighbor charges all have opposite signs. Determine the magnitude and direction of the force on a charge $+Q$ and on a charge $-Q$.

: **23-20 P** Charge Q_1 is placed at $z = 1 \times 10^{-10}$ m and charge Q_2 at $z = -1 \times 10^{-10}$ m. Determine the force on a test charge q placed at $x = 1 \times 10^{-8}$ m and at $x = 2 \times 10^{-8}$ m. (a) First take $Q_1 = Q_2 = 1.6 \times 10^{-19}$ C. (b) Then take $Q_1 = -Q_2 = 1.6 \times 10^{-19}$ C. This combination of charges is called an *electric dipole*. Note that the force due to it decreases with increasing distance much more rapidly than the force due to two like charges (which are effectively a "monopole"). (c) Calculate the ratio of the force at 10^{-8} m and at 2×10^{-8} m in each case.

: **23-21 P** Two equal positive charges $+Q$ are placed on the x axis at $x = +a$ and $x = -a$. Where should a charge q be placed on the z axis to experience the maximum force?

Section 23-3   Further Characteristics of Electric Charge

· **23-22 P** A rubber rod that has been rubbed with cat's fur acquires a charge of -4×10^{-10} C. How many electrons were transferred from the fur to the rod?

· **23-23 P** Two electrons are initially at rest with a separation of 2 cm. What acceleration will they experience when released?

· **23-24 P** The electron and the proton in a hydrogen atom are separated by about 5.3×10^{-11} m. (a) What is the magnitude of the electric force between them? (b) What is the ratio of the electric force to the gravitational force of attraction?

· **23-25 P** In an electroplating process, the ions Ag^+ deposit a total charge of 10 C. To what mass of silver does this correspond? (Atomic mass of silver, 108.)

· **23-26 P** A cell membrane has a typical thickness of 10^{-8} m. What is the force of attraction between ions Na^+ and Cl^- separated by this distance?

· **23-27 Q** A negative electric charge
(A) interacts only with positive charges.
(B) interacts only with negative charges.
(C) interacts with both positive and negative charges.
(D) may interact with either positive or negative charges, depending on the circumstances.
(E) can always be subdivided into two equal negative electric charges.

· **23-28 P** The 92 protons in the uranium nucleus have an

average separation of 2.5×10^{-15} m. What is the electric force of repulsion between two adjacent protons? (Another force, the so-called strong force, or nuclear force, pulls them together.)

· **23-29 P** The fission of $^{236}_{92}$U can produce the fragments $^{146}_{56}$Ba and $^{90}_{36}$Kr. These two nuclei have charges $+56e$ and $+36e$ respectively. Determine the Coulomb force acting on each just after their formation when their centers are separated by 1.6×10^{-14} m.

: **23-30 P** In the Bohr model of the hydrogen atom, an electron is a point charge orbiting the proton, which is the atom's nucleus. For an electron moving in a circular orbit of radius 5.3×10^{-11} m, what is the frequency of revolution if its Coulomb attraction to the proton provides the required centripetal force?

: **23-31 P** One theory of the expanding universe imagines that a hydrogen atom, instead of being electrically neutral, carries a slight positive charge (that is, the charge on the nucleus is slightly greater than the charge on the orbiting electron). This causes an electric force of repulsion between hydrogen atoms that counteracts the gravitational attraction. If these two forces are in balance, what value of $\Delta q / q$ is required, where $-q$ = charge on the electron and $q + \Delta q$ = charge on the hydrogen nucleus (the proton)? In other words, Δq is the net charge on the atom.

: **23-32 P** A conducting sphere with a diameter of 1.0 cm has a net charge of only -1.0 pC. Imagine for simplicity that the excess electrons are uniformly spread in a single layer over the sphere's outer surface. What is the approximate distance between adjacent electrons?

: **23-33 P** What fraction of the electrons in the earth would have to be removed from it and placed on the moon so that the electrostatic attraction between the earth and moon would have the same magnitude as their gravitational attraction? (To simplify the computation make the approximation that the nucleus of any element has equal numbers of protons and neutrons, each with a mass of 1.7×10^{-27} kg.)

Section 23-4   Electric Field Defined

· **23-34 P** A spherical drop of latex of mass 2×10^{-4} gm is suspended in a vertically downward electric field of 600 N/C. What is the net charge on the droplet?

· **23-35 P** A dust particle of mass 2×10^{-6} kg has a charge of $3 \mu C$. What vertical electric field is required to suspend the particle against the downward force of gravity?

: **23-36 P** The surface of the earth carries a negative charge, and this charge gives rise to an electric field of about 30 N/C downward at an elevation of 1.5 km. (*a*) What net charge would a very small aircraft of only 500 kg mass have to carry to be supported against the force of gravity? (*b*) To how many extra electrons would this correspond?

: **23-37 Q** In a perfect conductor, charge carriers can move freely. Because of this property one is led to conclude that in electrostatic equilibrium
(A) the net charge on the conductor is distributed uniformly throughout its volume.
(B) there is a constant nonzero electric field throughout the volume of the conductor.
(C) the electric field is zero everywhere throughout the volume of the conductor.   *See page 526*
(D) a conductor cannot carry a net charge.

: **23-38 P** Charge $+Q$ and $-2Q$ are separated by a distance d. Where is the electric field due to them zero?

: **23-39 P** A charge of $+4 \mu C$ is 2 cm from a charge of $+6 \mu C$. Where is the electric field of the two charges zero?

: **23-40 Q** Charges $+Q$ and $-2Q$ are placed as shown in Figure 23-18. Near which of the points indicated is the electric field due to these charges most likely to be zero?

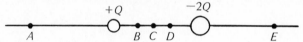

Figure 23-18. Problem 23-40.

: **23-41 Q** Charges Q, Q, and $-2Q$ are placed at the corners of an equilateral triangle, as shown in Figure 23-19.

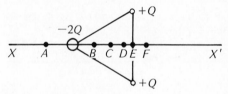

Figure 23-19. Problem 23-41.

The point along the line XX' where the electric field is most likely to be zero is
(A) A
(B) B
(C) C
(D) D
(E) E
(F) F
(G) None of the above, since the field vanishes only at infinity.

: **23-42 P** The electric field between the electrodes of the gas discharge tube used in a certain neon sign has a magnitude of 5×10^4 N/C. What acceleration will a neon ion experience? A neon ion's mass is 3.3×10^{-26} kg; its charge is $+e$?

: **23-43 Q** The electric field at a point in space near electric charges is the
(A) force per unit charge acting on a small test charge placed at the point.
(B) work done per unit test charge against the force vectors in carrying a charge from infinity to the point.

(C) electric force at the point.

(D) work done against electric forces in carrying a test charge from infinity to the point.

: **23-44 Q** You have a collection of small spheres. You can adjust the sign and magnitude of the charge on each, and the mass of each, to whatever values you wish. Can you then, using only electrostatic forces, place the spheres so that they would be in stable equilibrium?

: **23-45 Q** The electric field lines arising from two charges Q_1 and Q_2 are shown in Figure 23-20. From this drawing we can see that

(A) the electric field could be zero at P_1.

(B) the electric field could be zero at P_2.

(C) both Q_1 and Q_2 have the same sign.

(D) $|Q_1| > |Q_2|$.

(E) none of the above is true.

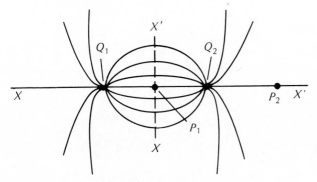

Figure 23-20.   Problem 23-45.

: **23-46 P** A charge $+2Q$ is a distance d from a charge $-Q$. (*a*) Where is the electric field from the two charges zero? (*b*) Sketch the electric-field lines from the charges.

: **23-47 P** Charges $+Q$ are placed at two adjacent corners of a square of side a, and charges $-Q$ are placed at the other two corners. (*a*) Determine the magnitude and direction of the electric field at the center of the square. (*b*) Sketch the electric-field lines.

: **23-48 P** Which of the following is the most accurate statement?

(A) *Electric field* and *electric force* are two terms with the same meaning.

(B) Electric-field lines never cross.

(C) Electric-field lines indicate the direction but not the magnitude of the electric field.

(D) An electric field line at a given point indicates the direction an electron placed at that point would start to move if released from rest.

(E) An electric field is strictly a theoretical construct without out physical reality.

: **23-49 Q** Charges $+q$, $+Q$, and $-Q$ are placed at the corners of an equilateral triangle of side a. (*a*) Determine approximately where the electric field is zero. (*b*) Sketch some of the electric field lines.

Supplementary Problem

23-50 An *electrostatic precipitator* is a device for removing dust particles from a smoke stack. It works by giving an electric charge to dust particles in a smoke stack, deflecting them in an electric field, and then collecting the charged particles on an electrode with the opposite charge. Suppose a dust particle rises through the precipitator with a constant velocity of 5 m/s, and it is given a specific charge of 10^{-5} C/kg. What horizontal electric field would deflect the dust particle 0.5 m horizontally as it ascends 20 m through the stack?

24

Continuous Distributions of Electric Charge

In continuing with electric fields, we see first how a continuous distribution of electric charge can be subdivided into a collection of effective point charges, and their individual contributions added, to yield a resultant electric field. Through the concepts of electric flux and Gauss's law, we consider an alternative procedure for finding the electric field from charge distributions. We examine special properties for the electric fields produced by charged conductors. The characteristics of electric dipoles in an electric field and the rigorous proof of Gauss's law are given in optional sections.

24-1   Electric Field for Three Simple Geometries

Here we consider the electric field produced by three particularly simple geometrical arrangements of electric charge.*

- A single point charge.
- A uniform, infinite straight line of continuous charge.
- A uniform, infinite flat sheet of continuous charge.

* The results of Section 24-1 are also derived in Section 24-4, using Gauss's law.

More complicated charge distributions can frequently be thought of as superpositions of simple distributions.

Point Charge   We have already found the electric-field magnitude to be

$$\text{Point charge:} \qquad E = \frac{kq}{r^2} \propto \frac{1}{r^2} \qquad\qquad (23\text{-}6)$$

The field **E** is radial (outward for positive q, inward for negative q) and its value falls off inversely with the square of the distance r from the charge.

Line of Charge   Electric charge is distributed uniformly along the length of an infinitely long straight line. The constant charge per unit length, or *linear charge density,* is represented by λ (the Greek letter L for "linear"); SI units for λ are coulombs per meter.

The line of charge lies along the x axis. See Figure 24-1(a). We want to find the direction and magnitude of **E** at point P (shown on the y axis), which is a perpendicular distance r from the charged wire. It is easy to show, on the basis of *symmetry alone,* that **E** must have the following characteristics:

- **E** is perpendicular to the charged wire (outward for positive charge).
- **E** has the same magnitude at all points the same distance from the wire.
- **E** is uniformly distributed in angular position in a plane transverse to the wire [in Figure 24-1(a), the yz plane].

The kind of proof to be used here will appear several times in this chapter and subsequent ones. We assume the contrary proposition and then show that it would lead to a contradiction and therefore cannot be true.* Consider the first assertion, that **E** is perpendicular to the wire. Suppose, for the sake of argument, that **E** at some point were tipped toward the right. This would mean that the side to the right of P differed from the left side. But this cannot be true because the wire is *uniformly* charged throughout and extends to *infinity* in both directions. The two other properties of **E** follow from symmetry considerations likewise.

To deal with a *continuous* charge distribution, we simply imagine it subdivided into effectively infinitesimal *point* charges dq; then we add, in vector fashion, the contribution $d\mathbf{E}$ to the resultant electric field **E** from all the point charges in the continuous distribution.†

We concentrate on the charge element dq lying within the infinitesimal length element dx [Figure 24-1(b)], where $dq = \lambda dx$ at the coordinate x. This effective point charge is a distance R from point P, and along line R it produces an electric field $d\mathbf{E}$, whose magnitude is

$$dE = \frac{k\,dq}{R^2} = \frac{k\lambda\,dx}{R^2}$$

---

* This type of argument, originated by the ancient Greeks, is officially known as *reductio ad absurdum,* "reduction to an absurdity."

† By "point" here we mean, strictly, a very small volume. The volume must be very small compared with overall dimensions of the charged wire, yet large enough to contain many atoms. If the volume were truly submicroscopic, the electric field would fluctuate violently as we approached individual electrons or nuclear particles and receded from them.

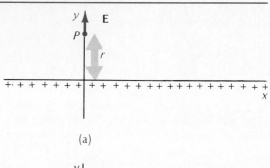

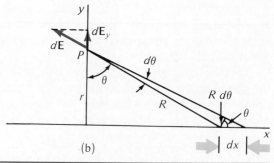

Figure 24-1. *(a) The electric field* **E** *at point P a distance r from a uniformly charged infinite straight line. (b) The electric field d***E** *at point P contributed by the element dx of an infinite, uniformly charged wire.*

We know that only the transverse component dE_y will contribute to the resultant field (the longitudinal component dE_x is cancelled by an equal but opposite component from a charge element of the same distance from the origin on the *negative x* axis). Therefore,

$$dE_y = dE \cos \theta = \frac{k\lambda \, dx \cos \theta}{R^2}$$

We have three variables — x, R, and θ. The integration we must carry out is simplest with θ as the variable. To eliminate R and x, consider the geometry of Figure 24-1(b), where

$$R = \frac{r}{\cos \theta} \quad \text{and} \quad dx = \frac{R \, d\theta}{\cos \theta}$$

Using these relations to eliminate x and R in the equation above, we get

$$dE_y = \frac{k\lambda \cos \theta \, d\theta}{r}$$

To find the resultant field E from an infinitely long wire, we integrate θ from $-\pi/2$ to $+\pi/2$:

$$E = \int dE_y = \frac{k\lambda}{r} \int_{-\pi/2}^{\pi/2} \cos \theta \, d\theta$$

Line of charge: $$E = \frac{2k\lambda}{r} \propto \frac{1}{r} \qquad (24\text{-}1)$$

The magnitude of **E** falls off inversely as the *first* power of the distance r from the uniformly charged wire.

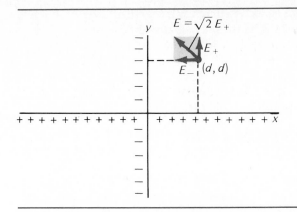

Figure 24-2.

Example 24-1.   There is a uniform *positive* linear charge density λ along the x axis and a uniform *negative* charge density of the same magnitude along the y axis. What is $\mathbf{E}$ at the coordinates (d, d)?

See Figure 24-2, where the electric fields, $\mathbf{E}_+$ and $\mathbf{E}_-$, contributed separately by the positive and negative lines of charge, are shown. We exploit the superposition principle for electric forces, and now for electric fields; the resultant field is simply the vector sum of the contributions from all the parts. In the present problem, we simply find the electric fields separately from the two infinite lines of charge and then add them as vectors to find the resultant $\mathbf{E}$. From (24-1), with $r = d$, we have

$$E_+ = \frac{2k\lambda}{d}$$

Similarly,

$$E_- = \frac{2k\lambda}{d}$$

The resultant field $\mathbf{E}$ has the magnitude

$$E = \sqrt{2}\,E_+ = \frac{2\sqrt{2}\,k\lambda}{d}$$

and makes an angle of $45°$ with the negative x axis, as shown in Figure 24-2.

The lines of charge in this example could certainly *not* be on two long and initially uniformly charged *conducting* wires, even if somehow they could be kept from touching at the origin. Charges on one wire would influence the charge distribution on the other, and neither could remain uniform.

Infinite Surface of Charge   The infinite plane surface is uniformly charged, with constant *surface charge density* σ (Greek *s* for surface); in SI units, σ would be given in coulombs per square meter.

Again, the direction of $\mathbf{E}$ is determined by the geometry of a uniform, charged infinite plane; $\mathbf{E}$ must be perpendicular to the surface (outward for positive charge). See Figure 24-3(a). Suppose it were not, so that $\mathbf{E}$ at some point was tipped, say, to the right. Then, the region to the right would have to be somehow different from the left. But this would violate the assumption of *uniform* charge over an *infinite* plane. Furthermore, the magnitude of $\mathbf{E}$ can depend, at most, on the perpendicular distance to the plane, not on location along the plane.

To sum all contributions to $\mathbf{E}$, we could imagine the plane subdivided into small patches $dx\,dy$, each effectively a point charge. Still more simply, however,

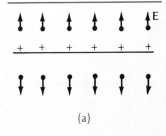

(a)

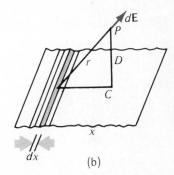

(b)

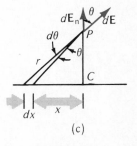

(c)

Figure 24-3.   *Electric field from a uniformly charged infinite flat sheet. (a) On the basis of symmetry, $\mathbf{E}$ must be perpendicular to the sheet. (b) and (c) An infinitely long charged strip of width dx produces field $d\mathbf{E}$ at point P.*

we can think of the plane as a collection of infinitely long, uniformly charged, and parallel wires. In Figure 24-3(b), the field $d\mathbf{E}$ at point P a distance D from the surface is contributed by one wire of width dx and a distance x from point C on the plane directly below P. Over a length L along one wire, there is a charge λL; we can also think of this charge as spread over an area $L\,dx$ and having the magnitude $\sigma L\,dx$. Therefore, $\lambda L = \sigma L\,dx$, or $\lambda = \sigma dx$. One strip contributes an electric field $d\mathbf{E}$, whose magnitude is, from (24-1),

$$dE = \frac{2k\lambda}{r} = \frac{2k\sigma\,dx}{r}$$

Only the normal component dE_n of $d\mathbf{E}$ will not be cancelled, and we have

$$dE_n = dE\cos\theta = \frac{2k\sigma\,dx\cos\theta}{r}$$

Again, the integration is most easily carried out with θ as the variable. From the geometry of Figure 24-3(c), we have

$$dx = \frac{r\,d\theta}{\cos\theta}$$

Using this relation in the one above it, we then get

$$dE_n = 2k\sigma\,d\theta$$

We integrate θ from $-\pi/2$ to $+\pi/2$ to include all strips covering the infinite plane:

$$E = \int dE_n = 2k\sigma \int_{-\pi/2}^{+\pi/2} d\theta = 2\pi k\sigma$$

Plane of charge: $E = 2\pi k\sigma$ (24-2)

The electric field is *constant* in magnitude; for a truly infinite plane of uniform charge, **E** does not depend on the distance from the plane.*

24-2   Uniform Electric Field

The simplest electric-field configuration is this: **E** is *uniform, constant,* or *homogeneous.* The direction and magnitude of **E** is the same at all points in a uniform field, and electric-field lines are straight, uniformly spaced, and parallel.

As we have seen in Section 24-1, an infinite, uniformly charged sheet produces a uniform field. But even a uniformly charged sheet of finite size produces an effectively constant **E** at points that are close to the sheet and relatively far from the edge of the sheet. It is simply that if we are very close to a flat sheet of finite size, the sheet appears to be effectively infinite.

Now let us find the electric field produced by two uniformly charged sheets with the same surface charge density but of opposite sign. We merely superpose the fields produced by each sheet separately. First, with a single positively

---

* We shall later find it advantageous to replace the constant k by $1/4\pi\epsilon_0$. Then (24-2) can be written as $E = \sigma/2\epsilon_0$.

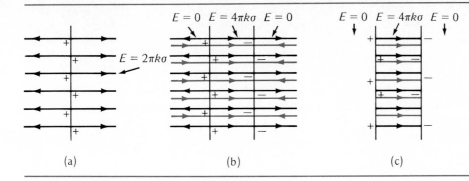

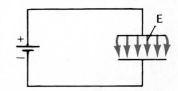

Figure 24-4. (a) Electric field of a single uniformly charged sheet. (b) Electric fields of two parallel uniformly charged sheets of opposite sign. (c) Resultant electric field of part (b).

Figure 24-5. A uniform electric field **E** between parallel charged conducting sheets connected to the two terminals of a constant-voltage source such as a battery.

charged sheet, a uniform outward electric field of magnitude $E = 2\pi k\sigma$ is produced on *both sides* of the sheet, as shown in Figure 24-4(a). Of course, for a negatively charged sheet, field **E** is toward the surface on both sides of the charged sheet. Now, consider the net field from two parallel charged sheets with *opposite* surface charges, both of the same magnitude σ, as in Figure 24-4(b). Between the sheets, the resultant field has equal contributions from the two surfaces. The resultant field between the sheets is $E = 4\pi k\sigma$. Outside this region, the fields from the two oppositely charged surfaces cancel and $E = 0$ [see Figure 24-4(c)].

The most common way to produce a uniform **E** in the laboratory is with two parallel conducting sheets or plates connected to the oppositely charged terminals of a battery (or some other source maintaining a constant "voltage"). See Figure 24-5. If the separation distance between plates is small compared with the width of either plate, the electric field is very nearly uniform at interior points, far from the edges of the plates. A fringing of the electric field, or a bending of **E** lines, takes place at the plate edges.

What is the motion of a particle with charge q and mass m in a uniform field **E**? The electric force $\mathbf{F}_e$ on the particle is

$$\mathbf{F}_e = q\mathbf{E}$$

The force is constant in both magnitude and direction. The particle's acceleration **a**, also constant, is given by

$$\mathbf{a} = \frac{\mathbf{F}_e}{m} = \frac{q}{m}\,\mathbf{E}$$

We already know that a particle or projectile in a uniform gravitational field has a constant acceleration **g** downward and traces out, in general, a parabolic path (Chapter 4). So too, a charged particle in a uniform **E** field traces out a parabolic path. All the relations derived for a projectile apply also to the electrically charged particle. We merely replace **g** by the acceleration $(q\mathbf{E}/m)$. Of course, a *positively* charged particle has a constant acceleration *in* the direction of **E**; a *negatively* charged particle is accelerated in the directon *opposite* to **E**.

Example 24-2. An electron with charge e, mass m, and initial horizontal velocity $\mathbf{v}_0$ is fired horizontally between two parallel, oppositely charged conductors, producing

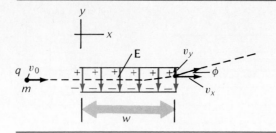

Figure 24-6. *A charged particle is fired into the uniform electric field between two oppositely charged parallel plates.*

a constant, vertically downward electric field **E**. See Figure 24-6. The electron travels a horizontal distance w in the uniform electric field; the gravitational force on the electron is negligible compared with the electric force on it. What are the electron's velocity components (v_x and v_y) as it emerges from the plates?

While the electron is in the uniform field, it has the following acceleration components:

$$a_x = 0$$

$$a_y = +\frac{qE}{m}$$

Note that the acceleration is *upward;* we have a *negative* charge in a *downward* **E**.

The velocity components of the electron as it leaves the uniform field are

$$v_x = v_0 = \frac{w}{t}$$

$$v_y = a_y t = \left(\frac{qE}{m}\right)t$$

where t is the time the electron spends between the plates. Along the horizontal, the electron has a constant velocity component v_0 and travels a horizontal distance w. Eliminating t between the two equations above gives

$$v_y = \left(\frac{qE}{m}\right)\left(\frac{w}{v_0}\right)$$

The angle ϕ at which the electron emerges is then specified by

$$\tan\phi = \frac{v_y}{v_x} = \frac{qEw/mv_0}{v_0} = \frac{qEw}{mv_0^2}$$

After emerging from the plates, the electron coasts in a straight line.

Precisely the arrangement shown in Figure 24-6 can be used to deflect an electron beam in a CRT (cathode ray tube, or oscilloscope), using electrostatic deflection. The vertical position of electrons striking the fluorescent screen is controlled by horizontal charged plates; the horizontal position, by vertical charged plates. Note that for small ϕ (with $\tan\phi \approx \phi$), the deflection angle $\phi \approx qEw/mv_0^2$ is directly proportional to the electric field E between the plates and inversely proportional to the electron's kinetic energy. The displacement of electrons on the screen is a direct measure of the electric field applied to the deflecting plates.

Example 24-3. An electron with an initial speed of 3.0×10^6 m/s ($\frac{1}{100}$ the speed of light, a relatively low speed for an electron) is fired in the same direction as a uniform electric field with a magnitude of 100 N/C. How far does the electron travel before being brought to rest momentarily and turned back?

The constant force on the electron is opposite to its initial velocity $\mathbf{v}_0$. The acceleration magnitude is $a = F_e/m = -eE/m$. If the electron travels a distance x before coming to rest, we have

$$x - x_0 = \frac{v^2 - v_0^2}{2a} \qquad (3\text{-}10)$$

$$x = -\frac{v_0^2}{2a} = \frac{v_0^2\, m}{2eE}$$

$$= \frac{(3.00 \times 10^6 \ m/s)^2 \ (9.11 \times 10^{-31} \ kg)}{2(1.60 \times 10^{-19} \ C)\, (100 \ N/C)} = 0.26 \ m$$

The last equation can also be written as

$$\tfrac{1}{2}mv_0^2 = (eE)x = F_e\, x$$

and we can describe the situation above in terms of work and kinetic energy as follows. An electric force $F_e = eE$ does negative work on the electron in slowing it to rest as the electron's kinetic energy drops from $\tfrac{1}{2}mv_0^2$ to zero.

24-3 Electric Flux

Gauss's law, our principal concern in Section 24-4, is a statement about electric flux, so we first define this quantity.

Imagine a closed surface of arbitrary shape — a so-called Gaussian surface; we can think of it as an imaginary balloon that can be pushed into any shape we wish. We focus on some very small patch of a Gaussian surface, as shown in Figure 24-7(a). An infinitesimally small patch of surface is effectively flat, and we can represent both its orientation and the size of its area by a vector $d\mathbf{S}$. The element of surface vector $d\mathbf{S}$ is *perpendicular* to the patch and points *outward* from the closed surface. The magnitude of $d\mathbf{S}$ represents the area of the patch. Clearly, the orientation of $d\mathbf{S}$ will differ from one part of the Gaussian surface to another, as shown in Figure 24-7(b).

Now consider the electric field $\mathbf{E}$ penetrating some surface element $d\mathbf{S}$, as shown in Figure 24-8. (What is responsible for $\mathbf{E}$ does not concern us at the moment.) Electric field $\mathbf{E}$ makes an angle θ with $d\mathbf{S}$; equivalently, $\mathbf{E}$ makes angle θ with the outward normal of the surface patch. Since $d\mathbf{S}$ is vanishingly small, $\mathbf{E}$ can be taken to be constant over $d\mathbf{S}$. By definition, the *electric flux* through the surface element is

$$d\phi_E \equiv E \cos\theta \, dS$$

In words, the electric flux is the component $E\cos\theta$ of the electric field parallel to $d\mathbf{S}$ (or perpendicular to the surface) multiplied by area dS. We can express the same thing more compactly by using the dot product (Section 9-4) of vectors $\mathbf{E}$ and $d\mathbf{S}$:

$$d\phi_E \equiv \mathbf{E} \cdot d\mathbf{S} \qquad (24\text{-}3)$$

Figure 24-9 shows three particularly simple situations for finding $d\phi_E$:

(a) $\theta = 0$; $\mathbf{E}$ is parallel to $d\mathbf{S}$; and $d\phi_E = +E\, dS$.
(b) $\theta = 90°$; $\mathbf{E}$ is perpendicular to $d\mathbf{S}$ (and lies in the surface); and $d\phi_E = 0$.
(c) $\theta = 180°$; $\mathbf{E}$ is opposite to $d\mathbf{S}$ (and points into the Gaussian surface); and $d\phi_E = -E\, dS$.

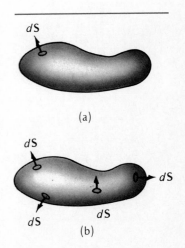

Figure 24-7. (a) A closed Gaussian surface. Vector $d\mathbf{S}$ represents a small patch of area on the surface. (b) Every differential surface element $d\mathbf{S}$ is outwardly directed from the Gaussian surface.

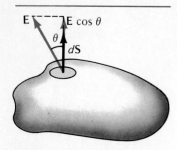

Figure 24-8. *Electric field* **E** *at the location of surface element dS. The component of* **E** *perpendicular to the surface (and parallel to dS) is E cos θ.*

Electric-field lines coming *out* of the surface correspond to *positive* electric flux; lines going *in* correspond to *negative* flux. When the field lines lie in the surface, there is no flux through that surface.

The magnitude of **E** can be imagined as represented by the number of electric-field lines per transverse cross section (Section 23-5); therefore, electric flux $d\phi_E$ is a measure of the number of electric-field lines coming through dS.*

A Gaussian surface can be chosen, we shall see, to have any shape. To make the calculation of electric flux easy, we shall always try to choose the shape so that:

- **E** is parallel to dS, and the flux is $d\phi_E = +E\,dS$.
- **E** is perpendicular to dS, and the flux is $d\phi_E = 0$.

To find the total electric flux ϕ_E over an entire closed Gaussian surface, we merely sum up the contributions from all surface elements, taking into account, of course, that both **E** and dS may vary in magnitude and direction from one point on the surface to another. The total electric flux is

$$\phi_E = \oint d\phi_E = \oint \mathbf{E} \cdot d\mathbf{S} \qquad (24\text{-}4)$$

The total flux $\oint \mathbf{E} \cdot d\mathbf{S}$ is sometimes referred to as the surface integral of vector field **E** over the closed Gaussian surface. The little circle on the integral sign is there simply to remind us that the integration is taken over the entire Gaussian surface. Although evaluating such an integral might, at first sight, seem formidable, we shall see that it is remarkably easy in situations with a high degree of geometrical symmetry.

> **Example 24-4.** What shape of Gaussian surface will yield *zero* total electric flux in a *uniform* electric field?
>
> Two orientations of a Gaussian surface are easily dealt with: **E** perpendicular to dS, and **E** parallel to dS. The Gaussian surface shown in Figure 24-10—a right cylinder

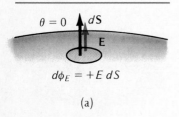

$$d\phi_E = +E\,dS$$

(a)

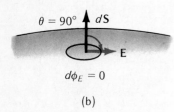

$$d\phi_E = 0$$

(b)

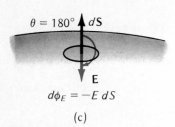

$$d\phi_E = -E\,dS$$

(c)

Figure 24-9. *Electric flux* $d\phi_E$ *through surface element dS for three simple orientations of the electric field* **E**: *(a)* **E** *parallel to dS and* $d\phi_E$ *positive; (b)* **E** *perpendicular to dS and* $d\phi_E$ *zero; and (c)* **E** *anti-parallel to dS and* $d\phi_E$ *negative.*

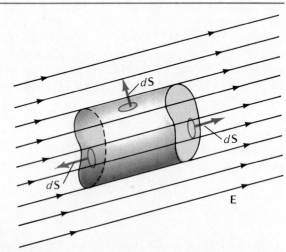

Figure 24-10. *Electric flux through a cylinder aligned with uniform field* **E**.

* Why call $\mathbf{E} \cdot d\mathbf{S}$ "flux," which implies something flowing? For the electric flux of the vector electric field, nothing flows. But consider the corresponding quantity for the streamlines of a fluid's velocity field (Sections 16-6 and 16-7). The flux $d\phi_v = \mathbf{v} \cdot d\mathbf{S}$ is actually the volume of fluid per unit time, or volume flux, flowing "out" through dS.

with ends of area S perpendicular to $\mathbf{E}$—satisfies these conditions. The flux through the end from which electric-field lines emerge is $+ES$. The flux at the other end is $-ES$. There is no flux through the side wall of the cylinder. The total electric flux is zero.

It turns out, as we shall see, that the electric flux is zero for a Gaussian surface of any shape in a uniform field. Indeed, the result is still more general. The total electric flux through a closed Gaussian surface of any shape is zero for *any* configuration of electric field—uniform or nonuniform—so long as there is no net electric charge inside the surface.

24-4   Gauss's Law

Elegant, powerful, neat. That's how Gauss's law is often described. With it, one can for simple geometries perform easily—sometimes with just one line of mathematics—calculations that would otherwise be long and complicated (like the one for the electric field from an infinitely long charged wire given in Section 24-1). The derivation of Gauss's law here is simple and special; a general proof is given in Section 24-7.

Let us find the total electric flux from a single point charge q (assumed to be positive for definiteness) under the simplest possible circumstances. We imagine q to be surrounded by a spherical Gaussian surface of radius r with q at its center. See Figure 24-11. The electric field $\mathbf{E}$ produced by the single point charge is radially outward and perpendicular to the spherical surface at each point. Finding the total electric flux is then particularly simple.

The magnitude of $\mathbf{E}$ at a distance r from q is

$$E = \frac{kq}{r^2} \qquad (23\text{-}6)$$

For reasons soon to be obvious, it is a good idea to replace the constant k in Coulomb's law by its equivalent.

$$k \equiv \frac{1}{4\pi\epsilon_0} \qquad (23\text{-}2)$$

Written in terms of ϵ_0, the permittivity of free space, the equation above for E becomes

$$E = \left(\frac{1}{4\pi\epsilon_0}\right)\frac{q}{r^2}$$

The total flux through the Gaussian sphere, with a surface area of $4\pi r^2$, is

$$\phi_E = \oint \mathbf{E}\cdot d\mathbf{S} = \frac{1}{4\pi\epsilon_0}\frac{q}{r^2}(4\pi r^2)$$

$$\phi_E = \frac{q}{\epsilon_0} \qquad (24\text{-}5)$$

That's the end of the derivation. The crucial point: distance cancels out, because the way in which the Coulomb force falls off with distance ($\propto 1/r^2$) is matched exactly by the way a spherical surface area grows with increased radius ($\propto r^2$).

What (24-5) says in words is this: The total electric flux ϕ_E through the

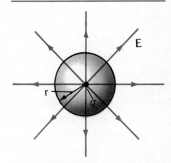

Figure 24-11. *Point charge q at the center of a Gaussian surface of radius r.*

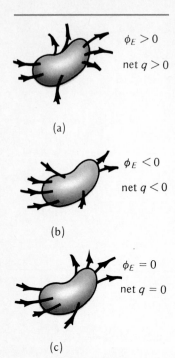

$\phi_E > 0$

net $q > 0$

(a)

$\phi_E < 0$

net $q < 0$

(b)

$\phi_E = 0$

net $q = 0$

(c)

Figure 24-12. *To find the net charge within a Gaussian surface, one can simply count the net number of* **E** *lines out of the surface: (a) positive electric flux, positive charge within; (b) negative electric flux, negative charge within; and (c) zero net flux, zero net charge within.*

closed surface with charge q inside is equal to the value of that charge (including sign) divided by the permittivity constant ϵ_0. *Gauss's law* is just the statement that this simple result holds *in general*—the total electric flux through a closed surface of *any* shape is simply the *net* charge q enclosed *anywhere* within that surface divided by ϵ_0 (and does not depend on any charge that may be outside the Gaussian surface). Proving Gauss's law in general (Section 24-7) depends simply on invoking geometry and two fundamental characteristics of the Coulomb force: (a) it is exactly inverse-square; and (b) the superposition principle for forces applies.

Gauss's law says, in effect, that if we know what is happening to electric-field lines over a closed surface, we can deduce the net charge within. Recall that the flux out of a surface is a direct measure of the number of electric-field lines emerging through the surface. Then Gauss's law in graphical terms is illustrated in Figure 24-12 as follows:

• Net positive flux; more E lines out of the surface than in; net charge within positive.
    • Net negative flux; more E lines in than out; net charge within negative.
    • Zero net flux; same number E lines in and out; zero net charge within.

Since ϕ_E is proportional to the net charge q, and ϕ_E is also proportional to the number of **E** lines, it is proper to draw the number of electric-field lines from a charge that is proportional to q.

Gauss's law may be thought of as a fundamental statement about the properties of electric fields and the lines through which they are portrayed. It incorporates the inverse-square character of Coulomb's law, but it also has this to say about **E** lines:*

• **E** lines originate from positive charges.
• **E** lines terminate on negative charges.
• **E** lines are continuous.

Applied to charge distributions in general, Gauss's law involves a surface integral whose evaluation may be difficult. But for uniform charge distributions of high geometrical symmetry—a uniform, infinite line of charge, for example—it is a simple, almost trivial matter to compute the integral. The key is this; since the Gaussian surface can have any shape, we choose the shape to match the symmetry of the charge distribution.

Example 24-5 Electric Field of a Uniformly Charged Infinite Wire. We have already worked this problem (Section 24-1) by summing the contributions to the resultant electric field from infinitesimal charge elements. The result:

$$E = \frac{2k\lambda}{r} = \frac{\lambda}{2\pi\epsilon_0 r} \tag{24-1}$$

where λ is the constant linear charge density.

To apply Gauss's law, especially to choose the shape of the Gaussian surface to make applying Gauss's law easy, we exploit what we already know about the electric

---

* As treated here, Gauss's law applies to electric fields produced by electric charges at rest. But as we shall see (Chapter 31), electric fields can also be created by a changing magnetic flux. Such an electric field consists solely of loops (each **E** line ends on its own tail). Loops of **E** contribute nothing to the flux, and Gauss's law is valid for all situations.

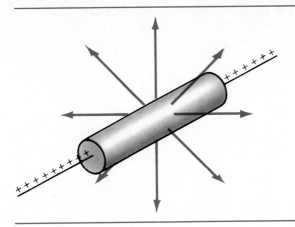

Figure 24-13. A uniform infinite line of charge with a coaxial cylindrical Gaussian surface.

field on the basis of symmetry alone. Arguments given in Section 24-1 (not repeated here) show that all **E** lines are perpendicular to the charged wire, uniformly spaced and radially outward from the wire in a plane perpendicular to the wire, and dependent only on the distance r to the wire. The shape of Gaussian surface that will make it easy to find flux ϕ_E is a right circular cylinder of radius r and length L whose axis coincides with the wire, as shown in Figure 24-13. The electric field at the ends of the cylinder lies in the planes of the ends; consequently the flux through the two ends is zero. This leaves only the cylindrical surface with a total area of $2\pi rL$; the **E** lines are perpendicular everywhere to this curved surface, and the flux through it is $E(2\pi rL)$. Therefore, the total flux through the Gaussian surface is

$$\phi_E = \oint \mathbf{E} \cdot d\mathbf{S} = E(2\pi rL)$$

The charge inside the cylinder of length L is λL. Therefore, Gauss's law gives

$$\phi_E = \frac{q}{\epsilon_0}$$

$$E(2\pi rL) = \frac{\lambda L}{\epsilon_0}$$

or

$$E = \frac{\lambda}{2\pi\epsilon_0 r}$$

precisely the result found before, but with almost no computation.

Example 24-6 Electric Field from a Uniformly Charged Infinite Sheet. We also worked this problem (Section 24-1) by summing contributions to E from elements of the surface. The result:

$$E = 2\pi k\sigma = \frac{\sigma}{2\epsilon_0} \qquad (24\text{-}2)$$

where σ is the constant surface charge density.

From the symmetry arguments given in Section 24-1, we know that **E** is everywhere perpendicular to the infinitesimally thin sheet of charge and could depend, at most, on the distance from the sheet. Again, the Gaussian surface is chosen for easy flux computation. It is a cylinder with its axis perpendicular to the sheet and its two ends, each of area A, at the same distance from the sheet, as shown in Figure 24-14. The **E** lines now lie in the curved cylindrical surface, and there is then no flux through this surface. We

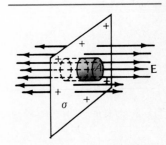

Figure 24-14. The cylindrical Gaussian surface used for computing the electric field from a uniformly charged sheet.

are left with contributions to ϕ_E from the two ends, where **E** is perpendicular to the surfaces and outward. We count a contribution to ϕ_E of EA from each of the two ends. The total charge within the cylindrical Gaussian surface is σA. From Gauss's law, we have

$$\phi_E = \frac{q}{\epsilon_0}$$

$$2EA = \frac{\sigma A}{\epsilon_0}$$

or

$$E = \frac{\sigma}{2\epsilon_0}$$

again in agreement with our earlier result.

Example 24-7 Spherical Shell of Charge.   Here we have a total charge Q (assumed positive for definiteness) spread uniformly over a thin spherical shell of radius R. We shall find the electric field, using Gauss's law, at a distance r from the center of the shell with (a) $r > R$ and (b) $r < R$.

(a) By symmetry, we know that any electric field from the spherical shell of charge must be in the radial direction and depend, at most, on the distance r. Imagine the shell surrounded by a spherical Gaussian surface of radius r, as shown in Figure 24-15(a), so that **E** is at any point on the Gaussian surface perpendicular to the surface. Over the entire spherical surface, the flux ϕ_E is then $\phi_E = E(4\pi r^2)$, and Gauss's law yields

$$\phi_E = \frac{q}{\epsilon_0}$$

$$E(4\pi r^2) = \frac{Q}{\epsilon_0}$$

$$E = \left(\frac{1}{4\pi\epsilon_0}\right)\frac{Q}{r^2} = \frac{kQ}{r^2}$$

Anywhere *outside* the shell, the electric field is just like that from a single point charge Q at the center of the shell; **E** does not depend on the shell's radius R.

(b) To find **E** inside the spherical shell, we suppose that the Gaussian surface is a concentric spherical surface *inside* the shell of charge, as in Figure 24-15(b). Since *no charge* is enclosed by this surface, the electric flux through it must be *zero*. But if the shell of charge is truly spherically symmetric, any **E** must be radial and with the same magnitude at all points on the Gaussian surface. The net charge q inside is zero, so that Gauss's law gives

$$\phi_E = \frac{q}{\epsilon_0}$$

$$E(4\pi r^2) = 0$$

or

$$E = 0$$

for all points inside the shell. There is no electric field in the interior of a uniform sphere of charge, and therefore no electric force on any point electric charge placed inside a uniformly charged shell.

The results here are just like those found earlier (Section 15-9) for the gravitational effects of a uniform shell of mass: (a) equivalent on the outside to a point mass at the center of the shell; and (b) no gravitational force at all on the inside. The reason is that

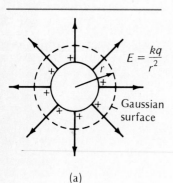

$$E = \frac{kq}{r^2}$$

Gaussian surface

(a)

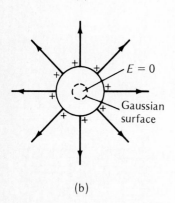

$$E = 0$$

Gaussian surface

(b)

Figure 24-15.   (a) Spherical shell of charge with a concentric Gaussian surface. (b) The concentric spherical Gaussian surface is now inside the shell of charge.

both Coulomb and gravitational forces are inverse-square. The special advantage of Gauss's law is seen if one compares the simple way (one line of mathematics) we arrived at the results for the spherical shell compared with the involved calculus derivation (Section 15-9).*

* Gauss's law can be formulated for the universal gravitational force as follows: $\int \mathbf{g} \cdot d\mathbf{S} = -4\pi Gm$. Here the gravitational field $\mathbf{g}$ is the gravitational force per unit mass, G is the universal gravitational constant, and m is the mass inside any Gaussian surface. A minus sign appears on the right side of the equation because any two masses attract one another whereas electric charges of the same sign repel one another.

24-5   Electric Field and Charged Conductors

Consider an electric conductor of any shape. It may carry a net charge, and there may be other charged objects nearby. If electrostatic equilibrium has been achieved, so that any net charges remain fixed in position, the following properties, summarized in Figure 24-16(a), apply to the charged conductor:

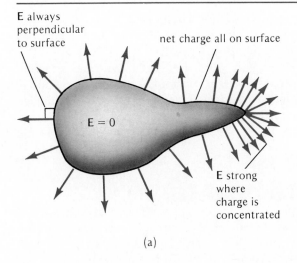

E always perpendicular to surface

net charge all on surface

E = 0

E strong where charge is concentrated

(a)

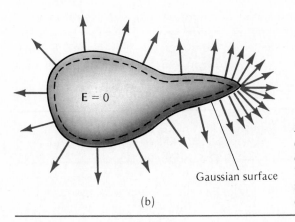

E = 0

Gaussian surface

(b)

Figure 24-16.   (a) Charged conductor of arbitrary shape in electrostatic equilibrium. (b) The Gaussian surface lies just inside the outer boundary of the charged conductor.

- **E** is *zero* at any point *inside* the conducting material.
- At the exterior surface, **E** is *perpendicular* to the surface at each point.
- Any *net charge* on the conductor is entirely at the *surface*.
- **E** is the strongest at those points on the surface at which the net charge is most heavily concentrated.

We prove each of these assertions in turn.

E = 0 on Interior of Conducting Material Suppose that **E** were not zero at some point within the conducting material. Then a free electron there would be subject to a force. The free electron would be accelerated, in contradiction to the assumption that the charged conductor is in electrostatic equilibrium. Therefore, **E** must be zero.

E Perpendicular to Surface Suppose that **E** were not perpendicular to an external surface. There would then have to be a component of **E** parallel to the surface. A charged particle at the surface would be accelerated, again in contradiction to electrostatic equilibrium. Therefore, **E** must be perpendicular to the surface.

Net Charge at Surface Here we exploit Gauss's law. We choose the Gaussian surface to lie just within the outer surface of a charged conductor of any shape, as shown in Figure 24-16(b). Since **E** = 0 at all points on the Gaussian surface, the net electric flux ϕ_E is zero. But from Gauss's law, if $\phi_E = 0$, then $q = 0$, and there is *no* charge (net) inside the charged conductor. If there is no charge inside the Gaussian surface, any net charge on the conductor must be *outside* the Gaussian surface, or on its surface. Within the conducting material, atoms are neutral; any excess or deficiency of electrons exists only in a thin layer of atoms at the conductor's surface. We saw earlier that the electric field within a *spherically symmetrical* shell of charge is zero. Our present result for charged conductors in equilibrium is more general; the interior field is zero for *any* shape, and for *any* charges on the exterior surface.

Our result — that quite apart from the electric fields or charges that may exist at the external surface of a conductor, there is *no net charge inside* — has profound consequences. It is, first of all, a result derived from Gauss's law, which was in turn derived from Coulomb's law. Testing by experiment to see whether there is actually no net charge inside a conductor amounts to testing indirectly whether the electric interaction between point charges is inverse-square. Indeed, a sensitive experimental test for net charge inside a conducting shell is by far the most sensitive test of the exponent of r in Coulomb's law. Many experimenters have done the experiment — Benjamin Franklin, Michael Faraday, Henry Cavendish. The most recent and precise tests, by E. R. Williams, J. E. Faller, and H. A. Hill in 1971, show that in the relation $F_e = kq_1q_2/r^n$, the exponent n differs from 2 by fewer than three parts in 10^{16}.

The traditional form of the experiment is the ice-pail experiment first performed by Faraday (who used an actual bucket for ice) in the mid–nineteenth century, and illustrated in Figure 24-17. The conductor is an ice pail, and the charge on its outer surface, indicated by the charge on the attached electroscope, is initially zero [Figure 24-17(a)]. A positively charged conducting object, suspended from an insulating thread, is introduced into the conductor. Negative charges appear on the inner surface and positive charges necessarily

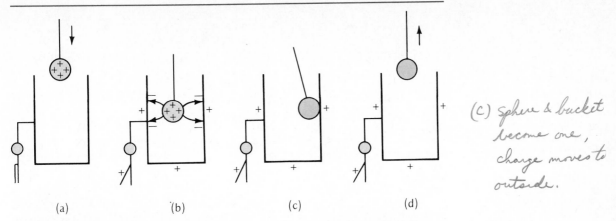

(a)   (b)   (c)   (d)

(C) sphere & bucket become one, charge moves to outside.

Figure 24-17. *Stages in the transfer of charge from a charged object to the outside of a hollow conductor, whose state of charge is indicated by the attached electroscope. (a) The charged object is outside. (b) The charged object is within the conductor, and equal and opposite charges are found on the inside and outside of the conductor. (c) When the object is touched to the conductor's inside, it annuls the charge on the conductor's inner surface and the exterior charge is unchanged. (d) The object, now electrically neutral, is removed.*

of the same magnitude on the outer surface of the pail [Figure 24-17(b)]. It is customary to say that the charged object "induces" charges on the inner and outer surfaces of the conductor. Then the introduced conducting object is touched to the inside of the bucket [Figure 24-17(c)] and—this is the crucial point—the electroscope shows no change. The charged object introduced annuls *all* the charges on the inside of the conductor, so that at the end [Figure 24-17(d)], *all* the net charge is on the conductor's outer surface, *none* is on the inside.

Inside a conducting shell, there is *electric* shielding. No matter how strong the electric fields may be at or outside the surface of a conducting shell, inside the shell of whatever shape, a charge feels nothing. The interior is shielded completely from what goes on outside.

We can see, by considering Figure 24-18, why electric shielding works on a basis less formal than Gauss's law. Here we have a conducting shell first [Figure 24-18(a)] with no external electric field. On the interior of the conducting shell, $\mathbf{E} = 0$. Now an external $\mathbf{E}_{ext}$ field (from a point charge) is applied, as in Figure 24-18(b). Positive charge, and equal amounts of negative, appear on the surface of the conductor, and the electric field within is still zero. We can look at this in a different way, as shown in Figure 24-18(c); the external field $\mathbf{E}_{ext}$ still exists within the conducting shell but it is exactly cancelled by the electric field $\mathbf{E}_{sep}$ in the opposite direction arising from the separated charges on the surface of the conducting shell.

Dependence of E on Local Surface-Charge Density For a *spherical* conducting surface, the charge per unit area σ has the *same* value at all points on the surface. The electric field also has the same magnitude at all points on a charged spherical conductor.

Not so for any other shape of charged conductor. The surface-charge density σ and electric field $\mathbf{E}$ both may change in magnitude (and in sign) from

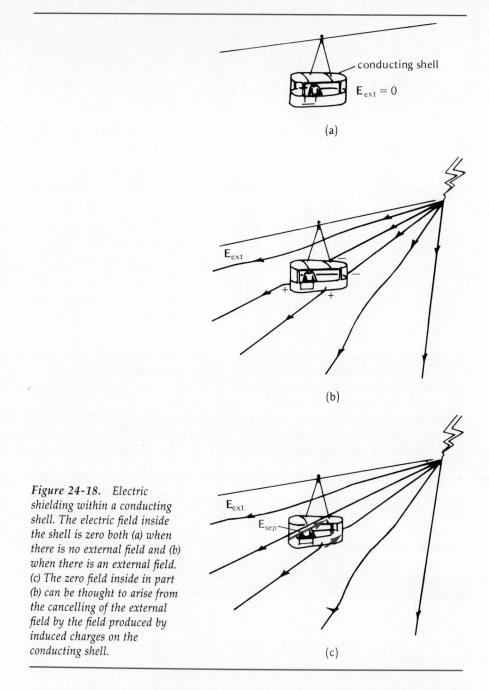

Figure 24-18. *Electric shielding within a conducting shell. The electric field inside the shell is zero both (a) when there is no external field and (b) when there is an external field. (c) The zero field inside in part (b) can be thought to arise from the cancelling of the external field by the field produced by induced charges on the conducting shell.*

point to point. (We shall show in detail in Section 25-7 that σ and $\mathbf{E}$ have maximum values at points, or at regions of small radius of curvature, and minimum values at flat portions of a conducting surface.) Here we derive the general relation between σ and $\mathbf{E}$.

Consider the conducting surface of arbitrary shape shown in Figure 24-19. We concentrate on a portion of the surface so small that it can be regarded as flat. We know already that $\mathbf{E}$ on the outer surface is perpendicular to the surface and that $\mathbf{E} = 0$ within the conductor. We choose as a Gaussian surface a small flat cylinder whose axis is parallel to the normal to the surface and whose ends lie, respectively, within the conductor and just outside its surface. We can regard both σ and $\mathbf{E}$ to be constant over the cylinder's surface area A.

The total charge enclosed by the Gaussian surface is $q = \sigma A$. The only contribution to the electric flux is through the outer surface of the cylinder, for which $\phi_E = EA$. The flux is zero over the inner surface ($\mathbf{E}$ is zero there) and over the curved cylinder surface ($\mathbf{E}$ is parallel to the surface there). Gauss's law then gives

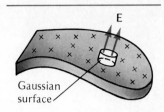

Figure 24-19. *Surface of a charged conductor with a Gaussian surface used to evaluate the electric field* **E**.

$$\phi_E = \frac{q}{\epsilon_0}$$

$$EA = \frac{\sigma A}{\epsilon_0}$$

$$E = \frac{\sigma}{\epsilon_0} \tag{24-6}$$

The magnitude of the electric field at any point of the surface of a charged conductor is directly proportional to the surface charge density at that point.

Example 24-8. A hollow conductor initially has a net charge of $+3Q$; all of this charge is, of course, on its outer surface. See Figure 24-20(a). Then a charged object of $+4Q$ is introduced into the interior of the conductor without touching the inner surface. What is the net charge on the inner surface and on the outer surface of the conductor?

When the $+4Q$ charge is introduced into the interior of the hollow conductor, it induces an equal but opposite charge of $-4Q$ on the *inner* surface of the conductor. (To satisfy yourself on this point, starting with fundamental principles, first choose a Gaussian surface lying entirely within the conductor material, as shown in Figure 24-20(b). The electric field $\mathbf{E}$ is zero all over this surface; the electric flux is zero from Gauss's law; and therefore the net charge within the Gaussian surface is zero.) Since the hollow conductor still has a net charge of $+3Q$ but now has $-4Q$ on its inner surface, the outer surface must have a net charge of $+7Q$.

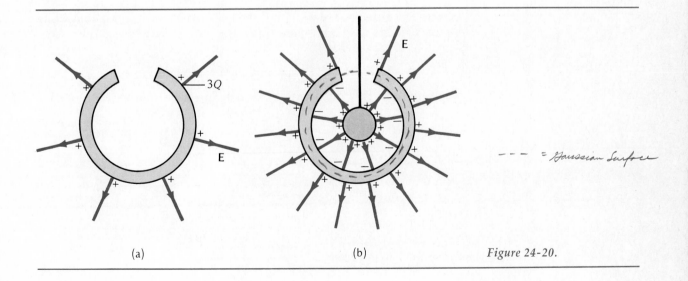

(a)

(b)

--- = Gaussian Surface

Figure 24-20.

Note again the effect of electric-field shielding. With the $+4Q$ charge on the inside and the $-4Q$ charge on the conductor's inner surface there *is* an electric field on the interior of the hollow conductor. But this field is produced solely by charges *within* the conductor, not at all by electric effects on the exterior of the conductor. Even if still another charged object were to be brought close to the exterior of the charged conductor, its outer surface would continue to have a net charge of $+7Q$ (even though this charge might be redistributed over the outer surface) but nothing inside the conductor would change.

24-6   An Electric Dipole in an Electric Field (Optional)

An electric dipole consists of two point charges of equal magnitude but opposite sign separated by a distance d. As a whole, the dipole is electrically neutral. It is useful to define the *electric dipole moment* **p** as

$$\mathbf{p} \equiv q\mathbf{d} \tag{24-7}$$

where q is the magnitude of either of the two charges and **d**, a vector, is the displacement of the positive point charge relative to the negative point charge (see Figure 24-21). The magnitude of the electric dipole moment is simply qd.

What happens when an electric dipole is placed in a uniform electric field **E**? In Figure 24-22 we have an electric dipole **p** making an angle θ with the electric-field lines. Each of the two point charges is acted on by an electric force of magnitude qE. The two forces are in opposite directions. Consequently, the *resultant* force on the dipole is zero, and the dipole is not accelerated translationally in a *uniform* electric field.

There is, however, a resultant torque on the dipole. We use the general definition for the torque τ,

$$\boldsymbol{\tau} = \mathbf{r} \times \mathbf{F} \tag{12-5}$$

or in magnitude,

$$\tau = r_\perp F \tag{12-4}$$

It is convenient to choose the tail of the vector **d** as the axis for computing torques (actually, we get the same torque for any choice of axis). Then, from Figure 24-22, we see that

$$\tau = r_\perp F = (d \sin \theta)(qE) = pE \sin \theta \tag{24-8}$$

This relation can be written more compactly in vector form as

$$\boldsymbol{\tau} = \mathbf{p} \times \mathbf{E} \tag{24-9}$$

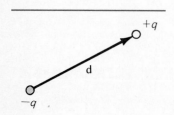

Figure 24-21.   An electric dipole.

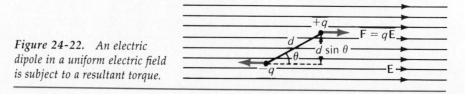

Figure 24-22.   An electric dipole in a uniform electric field is subject to a resultant torque.

The torque on the dipole is zero when it is aligned with the uniform field and the maximum when it is at right angles ($\theta = 90°$) to the field.

What is the potential energy U of an electric dipole $\mathbf{p}$ in electric field $\mathbf{E}$? The potential energy is equal to the work done, W, in reorienting the dipole. In general, the work done by the torque is

$$U = W = \int \tau \, d\theta \tag{13-11}$$

Using (24-8), we have

$$U = \int_{\pi/2}^{\theta} pE \sin\theta d\theta = -pE \cos\theta \tag{24-10}$$

The zero for measuring potential energy is chosen arbitrarily, and the lower limit of the integral was here chosen as $\frac{1}{2}\pi$ to give a simple final result. The zero for the potential energy then corresponds to the angle $\theta = 90°$. We can write (24-10), using the dot product, as

$$U = -\mathbf{p} \cdot \mathbf{E} \tag{24-11}$$

The potential energy has its minimum ($U = -pE$) when the dipole is aligned with the field; and the maximum ($U = +pE$) corresponds to $\theta = 180°$.

The cross product of $\mathbf{p}$ with $\mathbf{E}$ gives the torque of an electric dipole in an electric field, and the dot product gives the potential energy. We can regard (24-9) and (24-11) as *defining* an electric dipole moment. We can attribute an electric dipole moment to *any* object that is subject to a torque and has potential energy of orientation when in a uniform electric field. For example, a neutral molecule may have an electric dipole moment that is reoriented in an electric field. Here there are *not* two separated *point* charges of equal magnitude and *opposite* sign; a neutral molecule has an electric dipole moment because its "centers" of positive and negative charge do not coincide. The water molecule, H_2O, has a permanent electric dipole moment, and is said to be a *polar molecule,* because the "end" of the molecule with the hydrogen atoms has a net positive charge while the "end" with the oxygen has a net negative charge. A *nonpolar molecule* has no permanent electric-dipole moment but acquires a dipole moment when immersed in an electric field. It is not necessary to specify the charge separation distance d for a polar molecule or any other electric dipole; indeed, it is not possible to give q and d separately, since it is their product qd that enters into the torque and potential-energy relations.

When an electric dipole is in a *nonuniform*, or inhomogeneous, electric field, the forces acting on the two opposite charges are *not* of equal magnitude; the dipole is then subject to a resultant force.

24-7 General Proof of Gauss's Law (Optional)

In Section 24-4, we found that the relation $\oint \mathbf{E} \cdot d\mathbf{S} = q/\epsilon_0$ holds for a *single* point charge q at the *center* of a *spherical* Gaussian surface. To prove Gauss's law in general, we shall show that this relation holds, first, for a single charge inside or outside a Gaussian surface of *any shape,* and then, for *any number* of point charges.

Consider Figure 24-23, where point charge q is at the center of a spherical Gaussian surface and also inside a closed surface of arbitrary shape. We con-

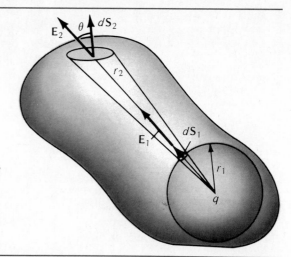

Figure 24-23. *A point charge q surrounded by a volume of arbitrary shape and by a concentric sphere. The electric flux through the two shaded areas is exactly the same.*

centrate on the small solid angle, or cone, that intercepts the surface element dS_1, at a distance r_1 from q and also surface element dS_2 at a distance r_2. We want to show that the electric flux is the same through these two matched surface elements. The electric fields at $\mathbf{r}_1$ and $\mathbf{r}_2$ are $\mathbf{E}_1$ and $\mathbf{E}_2$.

Note that $\mathbf{E}_2$ and dS_2 are not parallel in general; the angle between them is θ. The flux $d\phi_2$ through dS_2 is

$$d\phi_2 = \mathbf{E}_2 \cdot d\mathbf{S}_2 = E_2 \cos \theta \, dS_2$$

Area dS_2 is larger than dS_1, for two reasons: dS_2 is farther from q, and dS_2 is inclined relative to the direction of $\mathbf{E}_2$. From the geometry of Figure 24-22, we have that

$$dS_2 \cos \theta = \left(\frac{r_2}{r_1}\right)^2 dS_1$$

where the factor $(r_2/r_1)^2$ comes from the difference in distances and the factor $\cos \theta$ from the inclination of dS_2.

The electric field from a point charge falls off inversely with the square of distance, so that

$$E_2 = \left(\frac{r_1}{r_2}\right)^2 E_1$$

Substituting the two equations immediately above in the relation for $d\phi_2$ then gives

$$d\phi_2 = E_2(\cos \theta \, dS_2) = E_1 dS_1 = d\phi_1$$

The flux is the *same* for this pair of matched surfaces. Indeed, the flux is the same for all pairs of matched surface elements. The total flux $\oint \mathbf{E} \cdot d\mathbf{S}$ has the same value (q/ϵ_0) for *any* shape of surface enclosing q.

Suppose now that a point charge is *outside* some arbitrary Gaussian surface, as in Figure 24-24. Nothing changes. For every straight electric-field line into the surface there is a corresponding line out. A charge outside the surface does not contribute to the net flux.

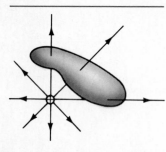

Figure 24-24. *The total electric flux through a closed surface not enclosing a net charge is zero.* $\oint \mathbf{E} \cdot d\mathbf{S} = 0$ *for q outside.*

Thus far we have proved Gauss's law for a *single* point charge. Now suppose that we have a collection of point charges—q_1, q_2, q_3, . . . —at various locations. We apply Gauss's law to each charge in turn with, however, the *same* Gaussian surface for all. We then can write

$$\oint \mathbf{E}_1 \cdot d\mathbf{S}_1 = \frac{q_1}{\epsilon_0}, \quad \oint \mathbf{E}_2 \cdot d\mathbf{S}_2 = \frac{q_2}{\epsilon_0}, \quad \cdots$$

where $\mathbf{E}_1$ is the field produced by q_1 alone, $\mathbf{E}_2$ by q_2 alone, and so forth.

Now we add all the equations above, to get

$$\oint (\mathbf{E}_1 + \mathbf{E}_2 + \mathbf{E}_3 + \cdots) \cdot d\mathbf{S} = \frac{q_1 + q_2 + q_3 + \cdots}{\epsilon_0}$$

We have added the several **E**'s—*under* the integral sign—and it's all right because the same surface element applies to all of them. Furthermore, because of the superposition principle for electric forces, we *can* add as vectors the fields produced by separate charges to find the resultant electric field, $\mathbf{E} \equiv \mathbf{E}_1 + \mathbf{E}_2 + \mathbf{E}_3 + \cdots$, produced by the net algebraic charge $q = q_1 + q_2 + q_3 + \cdots$, *inside* the Gaussian surface.

The relation above is Gauss's law in its general form, which we usually write more compactly as

$$\phi_E = \frac{q}{\epsilon_0} \qquad (24\text{-}5)$$

Summary

Definitions

Uniform electric field: **E** = constant, magnitude and direction.

Electric flux $d\phi_E$ through an outwardly directed surface element $d\mathbf{S}$, where the electric field is **E**:

$$d\phi_E \equiv \mathbf{E} \cdot d\mathbf{S} \qquad (24\text{-}3)$$

Gaussian surface: a closed surface of arbitrary shape.

Fundamental Principles

Gauss's law, a fundamental principle relating electric fields to electric charges and equivalent to Coulomb's law, states: For any closed Gaussian surface, the net electric flux $\phi_E = \oint \mathbf{E} \cdot d\mathbf{S}$ is equal to q/ϵ_0, where q is the algebraic net charge enclosed by the surface.

$$\phi_E = \frac{q}{\epsilon_0} \qquad (24\text{-}5)$$

Important Results

E for various charge distributions

• Point charge:

$$E = \frac{kq}{r^2} \propto \frac{1}{r^2} \qquad (23\text{-}6)$$

• Infinite uniform line of charge (linear charge density λ):

$$E = \frac{2k\lambda}{r} = \frac{\lambda}{2\pi\epsilon_0 r} \propto \frac{1}{r} \qquad (24\text{-}1)$$

• Infinite uniformly charged sheet of charge (surface-charge density σ):

$$E = 2\pi k\sigma = \frac{\sigma}{2\epsilon_0} = \text{uniform } E \qquad (24\text{-}2)$$

• Spherical shell of charge Q: Outside of shell—E same as that from point charge Q at center of shell. Inside of shell—$E = 0$.

E and Charged Conductors:

• Within conducting material, **E** is zero.
• At exterior surface, **E** is perpendicular to surface.
• Net charge is on conductor's surface.
• Magnitude of **E** at surface is proportional to the local charge density:

$$E = \frac{\sigma}{\epsilon_0} \qquad (24\text{-}6)$$

Problems and Questions

Section 24-1 Electric Field for Three Simple Geometries

· **24-1 P** What is the electric field at a distance of 4 cm from a long straight wire carrying a linear charge density of $2 \, \mu C/m$?

· **24-2 P** What is the electric field at a distance of 1 mm from a 1 m $\times$ 1 m sheet of aluminum that carries a charge of 2×10^{-6} C?

· **24-3 P** Consider a sphere having a radius of 4 cm and carrying a charge of $8 \, \mu C$. What would the surface charge density be if the charge were distributed (a) uniformly over the surface? (b) Uniformly throughout the volume?

24-4 P Charge Q is distributed uniformly along the z axis from $z = -a$ to $z = +a$. Determine the electric field at a point on the x axis.

: **24-5 P** An infinite insulating plane sheet of charge has a charge density of $+8 \times 10^{-9}$ C/m². At a distance of 10 cm from the plane is a point charge of $+0.01$ C. What is the magnitude and direction of the electric field at a point midway between the charge and the plane?

: **24-6 P** The earth carries a negative charge, which is created by lightning strikes. On a fair day, a typical value of the electric field at the earth's surface is 130 N/C. Assuming that the earth is a conductor, determine the surface charge density needed to create this field.

: **24-7 P** A small foam sphere of mass 1×10^{-4} kg carries a charge 2×10^{-9} C. It hangs from a thread 60 cm long, attached to a vertical conducting wall carrying a uniform surface charge of 2×10^{-6} C/m². What angle does the thread make with the vertical?

: **24-8 Q** Charge Q is distributed uniformly along the x axis from $-L$ to L (Figure 24-25). What is the electric field at the point $(0, L)$?

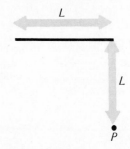

Figure 24-25. Question 24-8.

(A) $\dfrac{Q}{8\pi\epsilon_0} \displaystyle\int_{-L}^{L} \dfrac{dx}{(x^2 + L^2)^{3/2}}$

(B) $\dfrac{Q}{4\pi\epsilon_0} \displaystyle\int \dfrac{dx}{x^2}$

(C) $\dfrac{Q}{8\pi\epsilon_0} \displaystyle\int \dfrac{dx}{x^2}$

(D) $\dfrac{Q}{8\pi\epsilon_0 L} \displaystyle\int \dfrac{dx}{x^2 + L^2}$

(E) $\dfrac{Q}{4\pi\epsilon_0 L^2}$

(F) $\dfrac{Q}{\epsilon_0} \displaystyle\oint dS$

: **24-9 P** Charge Q is distributed uniformly along the z axis from $z = -a$ to $z = +a$. Determine the electric field at all points on the z axis for $z > +a$ and $z < -a$.

: **24-10 P** Charge Q is distributed uniformly along a circular loop of wire of radius R. (a) What is the electric field on the axis of the loop a distance z from its center? (b) For what value of z is the field a maximum?

: **24-11 P** Charge Q is distributed uniformly throughout a thin disk of radius R. Determine the electric field at a point on the axis of the disk a distance z from the center. (*Hint:* Consider concentric rings of charge and find the field due to each. Integrate to find the total field.)

: **24-12 P** A rod of length L carries charge Q uniformly distributed along its length. Determine the electric field at point P indicated in Figure 24-26.

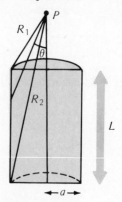

Figure 24-26. Problem 24-12.

: **24-13 P** Charge is distributed with uniform surface-charge density over the strip of the yz plane extending from $y = -a$ to $y = +a$. Determine the electric field at a point on the x axis.

: **24-14 P** A cylindrical shell of radius a and length L has a uniform surface-charge density σ (Figure 24-27). Show that the electric field at point P on the axis is

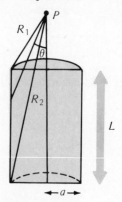

Figure 24-27. Problem 24-14.

$$E = \frac{\sigma}{4\pi\epsilon_0} \frac{1}{L} \left(\frac{1}{R_1} - \frac{1}{R_2} \right)$$

(*Hint:* To carry out the necessary integration, change variables from z to θ.)

Section 24-2   Uniform Electric Field

· **24-15 P** What charge must a water droplet of radius 1 mm carry if it is not to fall under the force of gravity when placed in a vertical electric field of 5000 N/C?

24-16 P Three parallel insulating infinite sheets of charge have surface-charge densities of $\sigma_1 = +1$ nC/m², $\sigma_2 = +2$ nC/m², and $\sigma_3 = -3$ nC/m². Determine the magnitude and direction of the electric field in regions a, b, c, and d of Figure 24-28.

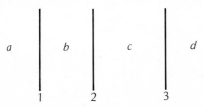

Figure 24-28.   Problem 24-16.

: **24-17 P** In a CRT (cathode ray tube), a narrow beam of electrons is projected horizontally between two oppositely charged horizontal parallel plates. The electric field between the plates is 600 N/C. The beam consists of a line of electrons each separated by 1×10^{-5} m from its immediate neighbors. Calculate the force on an electron from (*a*) the electric field of the plates; (*b*) the electric field of a neighboring electron; (*c*) its weight.

: **24-18 P** Two large parallel metal plates, each of area A and separated by a small distance d, carry charges $+Q$ and $-2Q$ (distributed over both sides of each plate). Determine the electric field between the plates and outside the plates on each side.

24-19 P An electron with an initial velocity of 2×10^6 m/s is slowed by a uniform electric field of 200 N/C. (*a*) How long does it take for the electron to come momentarily to rest? (*b*) How far does the electron travel before it comes to rest?

: **24-20 P** A uniform electric field exists between two parallel conducting plates. The magnitude of the surface-charge density on each plate is σ. A third conducting plate is now placed between the two charged plates, oriented parallel to them and one-fourth of the way from one plate to the other. (*a*) Determine the surface-charge density induced on each side of the third plate. (*b*) Does the result in (*a*) depend on how far the third plate is from either of the other two?

: **24-21 Q** A metal sphere is placed in the region between

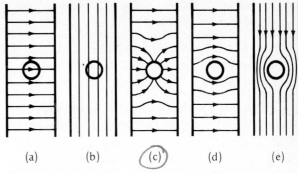

(a)   (b)   (c)   (d)   (e)

Figure 24-29.   Question 24-21.

two oppositely charged parallel plates. Which of the sketches in Figure 24-29 best represents the electric field in this case?

: **24-22 P** Suppose you want to bend a beam of particles, each of mass m and charge q, through 90°. The particles are to exit at the same speed they enter the field. In practice this is often done using magnetic fields, but it can also be done with static electric fields. What is the electric field required in each of these two schemes: (*a*) The beam passes between two curved electrodes, which set up an electric field of constant magnitude directed in toward the center of curvature. See Figure 24-30. (*b*) An electric field constant in both magnitude and direction is used.

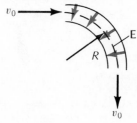

Figure 24-30.   Problem 24-22.

Section 24-3   Electric Flux

24-23 P What is the electric flux through a plane surface of area 2 m² whose normal is at an angle of 30° to a uniform electric field of 300 N/C?

: **24-24 P** Consider a cube of edge length L. What can you say about the outward electric flux through each face when a positive point charge Q is positioned (*a*) at the center of the cube? (*b*) inside the cube very near the center of one face? (*c*) outside the cube very near the center of one face?

24-25 P A hemispherical surface of radius 4 cm is placed in a uniform electric field of 200 N/C. What is the maximum electric flux that can pass through the surface?

· **24-26 Q** Which of the following has the meaning closest to "electric flux through an open surface"?
(A) The flow of electric potential through the surface.

(B) The flow of electric charge through the surface.

(C) The total charge on a conducting body.

(D) The number of electric field lines passing through the surface.

(E) The net electric force acting on a charged object.

Section 24-4 Gauss's Law

24-27 P A point charge Q is placed at the center of a cube of side a. What is the electric flux through one face of the cube? $Q/6\epsilon_0$

· 24-28 Q In using Gauss's law to find the electric field due to a charge distribution, which of the following is the most accurate statement?

(A) We should like a Gaussian surface over which $\mathbf{E}$ is constant and perpendicular to the surface.

(B) We should like to find a surface over all of which $\mathbf{E}$ is parallel to the surface.

(C) We first find the electric potential, then differentiate to find the electric field.

(D) We find it is easier to solve problems involving an array of point charges, rather than a continuous distribution of charge, such as a sphere.

(E) We try to find a Gaussian surface that will enclose zero net charge.

· 24-29 Q In using the equation $\oint_s \mathbf{E} \cdot d\mathbf{S} = q/\epsilon_0$ to find the electric field, it is important to keep in mind which of the following?

(A) All charges have to be included in q.

(B) The surface S may be open or closed, depending on the particular problem.

(C) The surface S is the surface of the charge distribution of interest.

(D) The surface S should be chosen so that $\mathbf{E}$ and $d\mathbf{S}$ are either parallel or perpendicular over the surface.

(E) This law does not apply to conductors, where $\mathbf{E} = 0$.

· 24-30 Q A Gaussian surface encloses zero net charge. Therefore,

(A) Gauss's law requires that the electric field be zero everywhere on the surface.

(B) the vector sum of all electric fields passing through the surface is zero.

(C) external charges cannot change the flux through the surface.

(D) the electric field is zero everywhere inside the surface.

(E) none of the above.

· 24-31 Q Which of the following is an accurate statement about Gauss's law?

(A) It requires that if a closed surface contains no net charge, the electric field must be zero everywhere over the Gaussian surface.

(B) It requires that if the electric field is everywhere zero over a closed surface, the net charge enclosed by the surface must always be zero.

(C) It applies only to symmetric charge distributions.

(D) It applies only to continuous distributions of charges and not to point charges.

: 24-32 Q (a) Suppose that charge is uniformly distributed over the surface of a spherical balloon. What can you say about the electric field inside and outside the balloon? (b) Suppose the balloon is then distorted into some sausagelike shape, still with constant surface-charge density. How does this change the field from case (a)?

: 24-33 Q A point charge is placed at the center of a spherical Gaussian surface and the electric flux ϕ_E through this sphere is calculated. Which, if any, of the following would cause a change in the flux ϕ_E?

(A) The sphere is replaced by a cube of the same volume.

(B) The sphere is replaced by a cube of the same surface area.

(C) The point charge is moved off center of the original sphere but remains inside the sphere.

(D) The charge is moved just outside the original sphere.

(E) A second charge is placed near, and outside, the original sphere.

: 23-34 Q In Gauss's law, $\epsilon_0 \oint_s \mathbf{E} \cdot d\mathbf{S} = q$

(A) The integral is always carried out over the surface of a conductor.

(B) The $\mathbf{E}$ in the integral is due only to the charge q.

(C) q is the charge inside the surface.

(D) $d\mathbf{S}$ is an element of displacement (measured in meters).

(E) q is the charge lying on surface S.

: 24-35 P Charge is distributed uniformly with density ρ throughout the volume of an insulating sphere of radius R. Determine the electric field at all points, inside and outside the sphere's surface.

: 24-36 P A spherical shell of inner radius R_1 and outer radius R_2 has a uniform volume-charge density ρ. Determine the electric field everywhere.

: 24-37 P Use Gauss's law for the gravitational field to find the gravitational field inside and outside the surface of a sphere of radius R and constant mass density. Sketch the gravitational field as a function of r.

: 24-38 P Charge is distributed uniformly with density ρ throughout the volume of an infinite slab of insulator defined by the planes $z = +a$ and $z = -a$. Use Gauss's law to find the electric field at all points.

: 24-39 P A long insulating cylinder of radius R has a uniform volume-charge density ρ. Determine the electric field inside and outside the cylinder.

: 24-40 P A long cylinder of radius R_1 has a uniform volume charge density ρ (Figure 24-31). A long cylindrical void of radius R_2 runs parallel to the axis of the cylindrical charge; the center of the void is a distance a from the center of the cylinder of charge; and $R_1 > R_2$. Show that the elec-

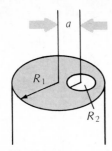

Figure 24-31.   Problem 24-40.

tric field in the void is constant and determine its value. (*Hint:* Superimpose a positive and a negative charge distribution to arrive at the given geometry, and calculate separately the field due to each.)

⋮ **24-41 P** An insulating slab of thickness t has a charge density given by

$$\rho(x) = \rho_0 \sin \pi x/t$$

where x is the distance from one face of the slab. Find E at points inside and outside the slab.

⋮ **24-42 P** A spherically symmetric charge distribution of radius R gives rise to an electric field

$$E = \frac{q_0}{4\pi\epsilon_0}\left[\frac{(1-e^{-ar})}{r^2}\right] \quad \text{for } r \le R$$

(*a*) What is the charge density of the distribution? (*b*) What is the electric field for $r > R$? (*c*) Sketch $E(r)$ and $\rho(r)$ vs r.

⋮ **24-43 P** Charge Q is distributed uniformly over an insulating spherical shell of radius R (Figure 24-32). A very small piece of the sphere is cut out and removed (along with the charge that was there). Show that the electric field on the outside of the sphere is $Q/4\pi\epsilon_0 R^2$ everywhere over the outside of the sphere except at the hole, where the field has the value $Q/8\pi\epsilon_0 R^2$. (*Hint:* Recall the idea of superposition of charge distributions.)

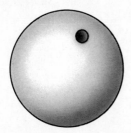

Figure 24-32.   Problem 24-43.

⋮ **24-44 P** In a simple model of the hydrogen atom, a point nucleus of charge $+e$ is assumed to be surrounded by a spherically symmetric cloud of negative charge $-e$ whose density is $\rho(r) = \rho_0 e^{-2r/a}$. Determine the electric field at a distance r from the nucleus for this model.

Section 24-5 Electric Field and Charged Conductors

· **24-45 P** Air will undergo electrical breakdown if subjected to an electric field in excess of about 3×10^6 N/C (the exact value depends on the pressure and humidity). What surface-charge density on a conductor will give rise to this field?

· **24-46 P** The electric field at a particular point on the surface of an irregularly shaped conductor is 5000 N/C. What is the surface-charge density at the point? If this cannot be determined, explain.

· **24-47 Q** Charge is placed on the lump of aluminum sketched in Figure 24-33. How will the charge be distributed on the object?

Figure 24-33.   Problem 24-47.

(A) Uniformly throughout the volume.
(B) Uniformly over the surface.
(C) With greatest density near point C on the surface.
(D) With greatest density near point D in the interior.
(E) With greatest density near point E on the flat surface.

· **24-48 P** A sufficiently large electric field will cause an electric breakdown of air. For this reason, a given amount of charge on a spherical conductor of small radius is more likely to result in a "spark" than a larger sphere is. To see this, calculate the field at the surface of a conducting sphere when a charge of 12 μC is placed on a sphere of radius (*a*) 2 cm, (*b*) 20 cm, (*c*) 2 m.

· **24-49 Q** A spherical conducting shell carries a charge Q. The electric field in the space inside the shell is
(A) directed radially inward.
(B) zero
(C) of the same magnitude as at a point just outside the shell.
(D) dependent on the position inside the shell.

⋮ **24-50 Q** Identify the incorrect statement.
(A) The electric field inside a charged conductor can never be zero.
(B) The electric field is a measure of the force per unit charge.
(C) Gauss's law is an alternative formulation of Coulomb's law.
(D) The electric field due to two charges can only be zero somewhere on the line joining them.
(E) All charges are made up of multiples of either the electron or the proton charge.

⋮ **24-51 P** The surface of the earth carries a negative

charge that, it is believed, is maintained by negatively charged lightning striking the ground and by positive corona discharge. In clear weather, the electric field near the earth's surface is approximately 100 V/m. To what surface charge does this correspond?

: 24-52 P Three concentric conducting spheres carry charges $+Q$ (on the inner sphere), $-2Q$ (on the middle sphere), and $+3Q$ (on the outer sphere). Determine the charge on each surface (inner and outer) of each sphere.

: 24-53 Q Three concentric conducting spherical shells carry charges as follows: $+4Q$ on the inner shell, $-2Q$ on the middle shell, and $-5Q$ on the outer shell. The charge on the inner surface of the outer shell is

(A) zero.
(B) $4Q$.
(C) $-Q$.
(D) $-2Q$.
(E) $-3Q$.
(F) impossible to determine unless the radii of the shells are known.

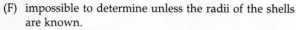

: 24-54 P A spherical conducting shell of inner radius 2 cm and outer radius 4 cm carries a charge of $+4\,\mu C$. Determine the electric field at the following distances from the center: (a) 1 cm, (b) 3 cm, (c) 5 cm, (d) 8 cm.

Supplementary Problems

24-55 P An electrostatic precipitator is a device for removing dust particles from a smoke stack. The rising dust particles are given an electric charge, then they are deflected in an electric field, and finally they are collected on an electrode with the opposite charge before they can leave the stack. A typical precipitator produces an electric field between a cylindrical conductor and a thin coaxial conductor inside it. Choose reasonable order-of-magnitude values for the various parameters that would enter into the design of a precipitator and show that it can work effectively—that is, that charged dust particles can be precipitated out of the emerging smoke in a stack of reasonable dimensions.

24-56 P When a charge $+q$ is placed a distance d from a plane metal surface, a negative surface charge will be induced on the metal. We have seen that the resulting electric field lines must be perpendicular to the metal surface, and that the magnitude of the electric field at the metal surface is proportional to the surface charge induced there. Electric-field lines would be so distributed, entering the metal normal to the surface, if the field were due just to the charge $+q$ and a second, fictitious "image" charge, placed

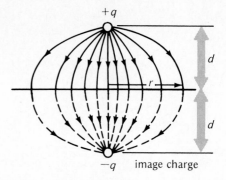

Figure 24-34. Problem 24-56.

as shown in Figure 24-34. (The charge $+q$ appears thus to be "reflected" in the mirror of the metal surface.) Using this method of images, determine (a) the electric field at the metal surface as a function of r; (b) the surface-charge density as a function of r; and (c) the total induced surface charge on the metal. Determine the total charge by integrating the surface-charge density over the entire infinite surface. You should find the total induced charge to be $-q$.

24-57 Q What accounts for the purple glow (the corona discharge) one sometimes sees around high-voltage apparatus, such as transformers?

(A) When an electric field is sufficiently intense, the field becomes visible. The purple color exists because it is nearest the UV, or high-energy, end of the spectrum.

(B) Intense electric fields accelerate stray electrons or ions to high speeds. These particles collide with air molecules, ionizing them. These fragments are in turn accelerated, forming an avalanche. Some of the ions and electrons recombine and in so doing emit a characteristic light.

(C) Electrons, which are normally invisible, can be seen when present in high concentrations. Generation of high voltage requires high charge concentration, and electrons are the carriers of electricity in metals.

(D) The intense electric fields cause nuclei to undergo fission, much as in a nuclear reactor, and as a result radiation is given off. This radiation is also known as Cerenkov radiation.

(E) Intense electric fields are known to focus light beams. The purple light seen is the result of focusing stray light that is usually unnoticed. The effect is dependent on light wavelength and is most pronounced for short wavelengths.

Electric Potential

We have already found in our study of mechanics how much easier it is to analyze physical situations and solve problems with energy-conservation ideas than with forces. Energy is a scalar quantity; force, and such force-related concepts as the electric field, are vectors. We now concentrate on electric potential, a central concept relating to the energy of interacting electric charges.

25-1   The Coulomb Force as a Conservative Force

Is energy conserved for the Coulomb force? Is the Coulomb force a conservative force? We know that the electric force between point charges must be conservative because it is of the same form as the universal gravitational force between point masses; both are central and inverse-square forces. Since the gravitational force is conservative (Section 15-5), the electric force must also be conservative.

It is easy to check separately, however, that the net work done by the electric force between a pair of point charges is zero when one charge is moved around any closed path while the other charge remains at rest. Then the

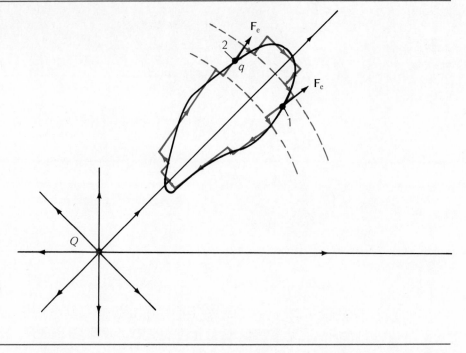

Figure 25-1. *Charge q is moved in a closed path in the vicinity of fixed charge Q. Charge q is subject to an outward electric force* **F**$_e$ *over the inward displacement at 1 and to the same electric force over the outward displacement at segment 2.*

requirement for a conservative force,

$$\oint \mathbf{F} \cdot d\mathbf{r} = 0 \qquad (9\text{-}14)$$

is met, and we can be sure that energy conservation applies to the Coulomb interaction.

Consider Figure 25-1. Charge Q is fixed in position and produces the electric field shown; charge q is moved around some arbitrary closed path. Actually, we imagine the path to consist of two types of small line segments:

- Circular arcs centered on Q.
- Radial lines pointing towards or away from charge Q.

We are to find the net work done on q by the electric force $\mathbf{F}_e$ on it from Q. (For definiteness we imagine both q and Q to be positive.) The results are simple:

- Along any circular arc, the force $\mathbf{F}_e$ is perpendicular to the displacement $d\mathbf{r}$, and the work done is zero.
- Along a radial line, every inward displacement, such as that at 1 in Figure 25-1, can be matched with a corresponding outward displacement along a radial line, as at 2 in Figure 25-1. The force magnitude is the same at the two locations; both 1 and 2 are the same distance from Q. The displacement magnitudes are also the same. But $\mathbf{F}_e$ and $d\mathbf{r}$ are antiparallel for the inward displacement and parallel for the outward. Positive work is done by $\mathbf{F}_e$ on q for an outward displacement; an equal amount of negative work is done for the matched inward displacement. The net work is zero, for this and every other matched pair of radial displacements.

The net work $\oint \mathbf{F}_e \cdot d\mathbf{r}$ done around any closed path is zero. Another way of saying the same thing is this: the net amount of work done on a charge by the Coulomb force as it moves from some initial location A to a final location B does not depend on the path followed between A and B, but *only on the end points.*

25-2   Electric Potential Defined

To see how electric potential is defined, consider Figure 25-2. Here charges Q_1, Q_2, Q_3, . . . are fixed while test charge q_t is moved at constant speed from A to B by some external agent. The agent applies a force $\mathbf{F_a}$ that balances in magnitude and direction at each location the electric force $\mathbf{F_e}$ of the fixed charges on q_t. As q_t is moved from A to B, the work done on it by the applied force is designated $W_{A\rightarrow B}$. We know that it does not matter how q_t gets from A to B; the work $W_{A\rightarrow B}$ is the same for all paths between the same end points.

Then, by definition, the *electric potential difference* between A and B produced by charges Q_1, Q_2 . . . is

$$V_B - V_A \equiv \frac{W_{A\rightarrow B}}{q_t} \qquad (25\text{-}1)$$

In words, the electric potential difference at the final location (B) relative to the initial location (A) is the work done per unit positive test charge by an external agent that takes the charge at constant speed from the initial location to the final.

Now, if we know the work required to transport a *unit* positive charge from one location to another, we can immediately specify the work required to move *any* charge q at constant speed from A to B. From (25-1), it is

$$W_{A\rightarrow B} = q(V_B - V_A) \qquad (25\text{-}2)$$

If q is a positive charge and the work done is also positive, then the electric potential V_B at point B is higher (more positive) than V_A at A.

In Figure 25-2, we had an external agent doing work on charge q_t without, however, changing its kinetic energy. Therefore, the potential energy of the entire system of charged particles must have increased. In fact, the change in electric potential energy between the two configurations, $U_B - U_A$, is just equal to the work done, $W_{A\rightarrow B}$. Therefore, we can relate the change in electric *potential*, $V_B - V_A$, between locations A and B to the change in the system's electric *potential energy*, $U_B - U_A$. From the relation above, it follows that

$$U_B - U_A = q(V_B - V_A) \qquad (25\text{-}3)$$

In applying the energy-conservation principle, we are always concerned with energy *differences;* the choice of the zero of energy is arbitrary. Similarly, we are

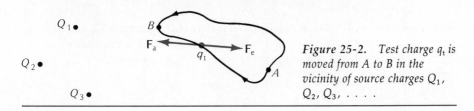

Figure 25-2.   Test charge q_t is moved from A to B in the vicinity of source charges Q_1, Q_2, Q_3, ...

free to choose arbitrarily the zero for electric potential. The most common choices are these:

- The electric potential *zero* at an *infinite* distance from one or more charged particles.
- The electric potential of the *earth* (ground) *zero*, particularly for electric circuits.

With starting point A in Figure 25-2 chosen to have zero potential ($V_A = 0$), the electric potential at any point V_p can then be written from (25-1) as

$$V_p = \frac{W_p}{q_t} \tag{25-4}$$

where W_p is now the work required to move q_t at constant speed from the location where the electric potential is zero to p.

The SI unit for electric potential is the *volt* (abbreviated V), named for Alessandro Volta (1745–1827), the inventor of the voltaic pile (a primitive form of energy cell, or battery). By definition,

$$1 \text{ volt} \equiv 1 \text{ joule/coulomb}$$

$$1 \text{ V} \equiv 1 \text{ J/C}$$

Common derived units are the microvolt ($\mu V = 10^{-6}$ V), the millivolt (mV = 10^{-3} V), and the megavolt (MV = 10^6 V).

Electric potential V and electric potential energy U have nearly the same names. Although closely related, the two terms have different meanings and different units, and they must be carefully distinguished:

- *Electric potential V*, measured in volts, is *a property at a location in space of the source charges.* This algebraic number tells us how much work must be done per unit positive charge to move the charge at constant speed to the location where V is measured from the place where the electric potential is chosen to be zero. Equivalently, electric potential change is electric potential energy change per unit positive charge.
- *Electric potential* energy U, measured in joules, is *a property of a system of two or more charged particles.* This algebraic number tells us how much net work must be done to assemble the system.

Example 25-1. Points A, B, and C have the following electric potentials:

$$V_A = 0 \text{ V} \qquad V_B = -9 \text{ V} \qquad V_C = +3 \text{ V}$$

The way in which a 9-V and a 12-V battery can be connected, using conducting wire, to

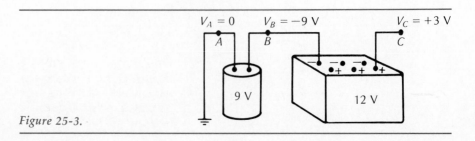

Figure 25-3.

Table 25-1

CHARGE MOVED, IN μC	FROM	TO	ROUTE	WORK DONE	COMMENT
+0.1	A	A	$A \rightarrow B \rightarrow C \rightarrow A$	0	$\oint \mathbf{F}_e \cdot d\mathbf{r} = 0$
+0.1	A	B	$A \rightarrow C \rightarrow B$	$-0.9\ \mu J$	$W_{A \rightarrow B} = q(V_B - V_A)$ $= (0.1\ \mu C)(-9\ V - 0)$
+0.1	B	A	$B \rightarrow A$	$+0.9\ \mu J$	$W_{B \rightarrow A} = -W_{A \rightarrow B}$
+0.2	B	C	$B \rightarrow A \rightarrow C$	$+2.4\ \mu J$	$W_{B \rightarrow C} = q(V_C - V_B)$ $= (0.2\ \mu C)[3\ V - (-9V)]$
−0.2	B	C	$B \rightarrow A \rightarrow C$	$-2.4\ \mu J$	Sign reversed for W, compared with entry above, because sign of q is reversed.
−0.2	A	B	$A \rightarrow B$	$+1.8\ \mu C$	$W_{A \rightarrow B} = (-0.2\ \mu C)(-9V - 0)$
−0.2	A	C	$A \rightarrow B \rightarrow C$	$-0.6\ \mu C$	$W_{A \rightarrow C} = W_{A \rightarrow B} + W_{B \rightarrow C}$ $= (+1.8 - 2.4)\ \mu J$

produce these electric potentials is shown in Figure 25-3. (A battery maintains a constant potential difference between its terminals. All points on a conducting wire are at the same potential.) How much work must be done to move the charge specified in the first column of Table 25-1 from the initial location to the final one, following the route specified? The answers, together with pertinent comments, are given in the last column of Table 25-1.

In all entries, we recognize that the net work done, which is independent of the route between the end points, p to r, is given by

$$W_{p \rightarrow r} = q(V_r - V_p)$$

25-3   Electric Potential for Point Charges

Here we derive the relation for the electric potential V at a distance r from a single point charge q. We choose $V = 0$ for $r \rightarrow \infty$.

Consider Figure 25-4, where q is fixed in position while test charge q_t is brought in along a radial line from infinity to its final location, a distance r from q. We compute the work done by the force $\mathbf{F}_a$ applied by an external agent who moves charge q_t at constant speed. The inward force $\mathbf{F}_a$ is matched in magnitude at each location by the oppositely directed electric force $\mathbf{F}_e$. The origin of the radius vector $\mathbf{r}$ is at the location of q. Therefore, $+dr$ represents an *outward* incremental radial displacement; an inward infinitesimal displacement is given by $-dr$.

The incremental work dW done by $\mathbf{F}_a$ as q_t moves inward over an infinitesimal displacement $d\mathbf{r}$ is

$$dW = \mathbf{F}_a \cdot d\mathbf{r} = (F_e)\,(-dr) = \frac{kqq_t}{r^2}\,(-dr)$$

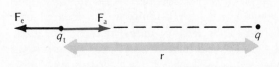

Figure 25-4.   *Test charge q_t is moved at constant speed relative to fixed charge q under the action of electric force $\mathbf{F}_e$ and applied force $\mathbf{F}_a$.*

The total work done is

$$\int_\infty^r dW = -\int_\infty^r \frac{kqq_t}{r^2}\,dr = \frac{kqq_t}{r}$$

$$W = \frac{kqq_t}{r} \tag{25-5}$$

The electric potential V is, by definition, the work per unit charge, so that

$$V = \frac{W}{q_t} = \frac{kqq_t}{q_t r}$$

$$V = \frac{kq}{r} \tag{25-6}$$

Electric potential is an algebraic quantity and can have positive and negative values. The potential from a single positive charge is positive; its magnitude increases as one approaches the charge. Negative charges produce negative potentials. See Figure 25-5, where the electric potential of a single (a) positive and (b) negative point charge is plotted as a function of distance r. For example, at a distance of 1.0 cm from a point charge of 0.1 μC, the electric potential is

$$V = \frac{kq}{r}$$

$$= \frac{(9.0 \times 10^9 \ \text{N} \cdot \text{m}^2/\text{C}^2)(1.0 \times 10^{-7}\text{C})}{1 \times 10^{-2} \ \text{m}}$$

$$= 9.0 \times 10^4 \text{V} = 90 \ \text{kV}$$

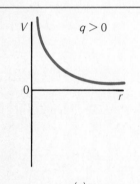

(a)

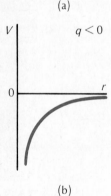

(b)

Figure 25-5. Electric potential V as a function of distance r from a point charge q that is (a) positive and (b) negative.

where we have used the fact that newton meters per coulomb equal joules per coulomb equal volts.

What is the electric potential produced by two or more point charges? See Figure 25-6, where charge q_A is a distance r_A from point P, and similarly for charges q_B, q_C, ... The potential V at P is simply the algebraic sum of the potentials produced separately by individual point charges:

$$V = \frac{kq_A}{r_A} + \frac{kq_B}{r_B} + \frac{kq_C}{r_C} + \cdots$$

Writing this result more compactly, we get

$$V = \Sigma \frac{kq_i}{r_i} \tag{25-7}$$

Each charge produces its own potential at P. This result follows because the work done by q_A, q_B, q_C, . . . on still another charge brought to P equals the sum of the work done by the electric force of each charge separately, a result that in turn follows from the superposition principle for the electric force. The electric potential for any collection of charge can be found simply, by applying (25-7). To find the *electric field* from a collection of charges may, on the other hand, be far more complicated — you would have to add *vectors*. Computing electric potential is easier because you add algebraic, *scalar* quantities.

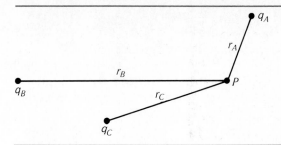

Figure 25-6. *Point P has respective distances r_A, r_B, r_C, . . . from point charges q_A, q_B, q_C, ...*

Example 25-2.   Two opposite point charges, each of magnitude q, are separated by a distance $2d$. For the point midway between the charges, what is (a) the electric potential? (b) the electric field? See Figure 25-7.

(a) The electric potential at the midpoint is

$$V = \frac{k(+q)}{d} + \frac{k(-q)}{d} = 0$$

(b) The *magnitude* of the electric field produced at the midpoint by either point charge alone is

$$E_{\text{one charge}} = \frac{kq}{r^2} = \frac{kq}{d^2}$$

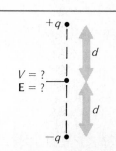

Figure 25-7.

The direction of **E** produced by both $+q$ and $-q$ is from $+q$ to $-q$ (downward in Figure 25-7). Therefore, the net electric field at the midpoint is twice the value from either charge separately, or

$$\mathbf{E} = \frac{2kq}{d^2}$$

in the direction from $+q$ to $-q$.

The potential V is zero at the midpoint, but **E** is not. A zero value for V does not imply that **E** must also be zero at that location. Zero V at the midpoint means that no work need be done to bring a third charge from infinity to that location. Nonzero **E** at the midpoint means that an electric force would act on a third charge placed at that location.

25-4   Electric Potential Energy

Point charges q_1 and q_2 are separated by a distance r_{12}. What is their electric potential energy U_e?

We found in Section 25-3 that the work required to bring charge q_t at constant speed from infinity to a distance r from q is kqq_t/r [(25-5)]. This is then also the potential energy of the system. With $q = q_1$, $q_t = q_2$, and $r = r_{12}$ we have

$$U_e = \frac{kq_1q_2}{r_{12}} \qquad (25\text{-}8)$$

The significance of the potential-energy function is that it appears in the general energy-conservation principle. A system of two or more isolated but mutually interacting charged particles has a *constant* total energy E, where

$$E = K + U_e = \text{constant}$$

and K is the total kinetic energy of the particles.

The sign of the potential energy depends on the sign of q_1 and q_2. If the two charges have *like* signs and repel one another, U_e is *positive*. On the other hand, if the two charges have *unlike* signs and attract one another, U_e is negative.* There is nothing wrong with a negative potential energy. Suppose, for example, that a positive and negative charge are far separated and initially at rest. Then $K = 0$, $U_e = 0$, and $E = K + U_e = 0$. The total energy is initially zero and must remain so. But these charges attract one another and they will later have gained kinetic energy as they approach one another. Consequently, with $E = 0$ and $K > 0$, we must have $U_e < 0$.

The matter of signs arises often in this chapter. You can always deal properly with this matter by scrupulously assigning the right signs to q, V, and U. Still better, physical arguments will always confirm that the signs have been correctly assigned.

What is the total electric potential energy for a system with three or more charges? Let r_{12} be the distance separating q_1 and q_2, and r_{23} the distance between q_2 and q_3, and so on. Then the potential energy of the system is

$$\text{Total } U = U_{12} + U_{23} + U_{13} + \cdots$$
$$= \frac{kq_1q_2}{r_{12}} + \frac{kq_2q_3}{r_{23}} + \frac{kq_1q_3}{r_{13}} + \cdots \tag{25-9}$$

The system's total potential energy is just the sum of terms for all possible pairs of particles. This simple result, like that for gravity, is a direct consequence of the superposition principle for the Coulomb force.

Because the electric force is of the same form as the gravitational force, detailed results derived for gravity (Chapter 15) carry over to the corresponding situations involving charged particles. For example, a charged particle attracted electrically to an oppositely charged particle of far greater mass will orbit it in an elliptical path, and the Kepler rules for planetary motion will have exact counterparts.

Electric potential energy in the SI system is expressed in joules. For atomic particles or atomic systems, a more convenient energy unit is the *electron volt* (abbreviated eV). Consider an electron or any other particle with charge magnitude e that is moved across a potential difference ΔV of 1 V. From (25-3), the potential energy changes by ΔU, where

$$\Delta U = q\,\Delta V$$

Here, with $q = e$ and $\Delta V = 1$ V, we can write

$$\Delta U = (e)(1 \text{ V}) = 1 \text{ eV} = 1 \text{ electron volt}$$

Alternatively, we can express the energy change in joules as

$$\Delta U = (1.602\ 189\ 2 + 10^{-19}\text{C})(1 \text{ V}) = 1.602\ 189\ 2 \times 10^{-19} \text{ J}$$

---

* The form of (25-8) for the electric potential energy of a pair of point electric charges is very much like the relation $U_g = -Gm_1m_2/r_{12}$, for the gravitational potential energy of a pair of point masses [(15-4)]. We simply have $-Gm_1m_2$ replaced by kq_1q_2. The difference in sign arises as follows. Two interacting masses (with both m_1 and m_2 positive) *attract* one another and the potential energy is *negative*; two interacting unlike charges (with one of q_1 and q_2 positive and the other negative) also attract one another, and the potential energy is again negative.

Therefore, the conversion relation for the electron volt can be written as

$$1 \text{ eV} \equiv 1.602\ 189\ 2 \times 10^{-19} \text{ J}$$

The electron volt can therefore be defined as the work done on a particle with charge e as it goes across a potential difference of 1 V. Equivalently, a particle of charge e will have its kinetic energy change by 1 eV when it is accelerated across a potential difference of 1 V.

Example 25-3. An electron (charge, $-e$) is released from rest at a distance of 0.2 nm $= 2.0 \times 10^{-10}$ m from a proton (charge, $+e$) also initially at rest. What is the electron's kinetic energy when it is .1 nm from the proton?

We can regard the proton as remaining essentially at rest because it has a far greater mass than the electron. Only the electron gains kinetic energy. With subscript 1 denoting the initial state and subscript 2 the final state, we have from energy conservation

$$E = K_1 + U_1 = K_2 + U_2$$

$$= 0 + \frac{k(+e)(-e)}{0.2 \text{ nm}} = K_2 + \frac{k(+e)(-e)}{0.1 \text{ nm}}$$

$$K_2 = \frac{ke^2}{0.2 \text{ nm}}$$

$$= \frac{(9.0 \times 10^9 \text{ N} \cdot \text{m}^2/\text{C}^2)(1.60 \times 10^{-19} \text{ C})^2(1 \text{ eV})}{(2.0 \times 10^{-10} \text{ m})(1.60 \times 10^{-19} \text{ J})} = 7.2 \text{ eV}$$

The result here is typical of subatomic particles separated by atomic dimensions; kinetic and potential energies are of the order of a few electron volts.

Example 25-4. Three charges $+Q$, $+Q$, and $-Q$ are brought to the corners of an equilateral triangle of side L. See Figure 25-8(a). What is the total energy required to assemble these charges?

The work W done in assembling the system equals the system's total potential energy. Therefore, from (25-9),

$$W = U = +\frac{kQ^2}{L} - \frac{kQ^2}{L} - \frac{kQ^2}{L} = -\frac{kQ^2}{L}$$

The total energy is *negative*. This means that the three charges form a *bound* system. The total U turns out to be equal to that of just a single pair of unlike charges. We can see that this must be the result from another angle. Imagine that first the charges $+Q$ and $-Q$ are separated by L and therefore have a potential energy $-kQ^2/L$. Then the third charge is brought to its final position along a line perpendicular to the line joining the first two charges; see Figure 25-8(b). This third charge is acted on by the electric field of the first two charges [see Figure 23-15(a)]. But that field is always perpendicular to the displacement of the third particle, and no work is required to move the third charge to its final location along this route (or any other)! From another point of view, the potential of the first two charges is zero at all points on the dashed line in Figure 25-8(b).

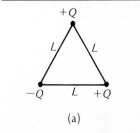

(a)

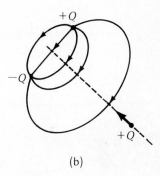

(b)

Figure 25-8. *(a) A system consisting of three point charges at the corners of an equilateral triangle. (b) No work is required to bring the third charge midway between the other two charges (constituting an electric dipole) along the dashed path, or along any other path.*

25-5 Equipotential Surfaces

Consider a single point charge q, taken to be positive for definiteness. The electric potential V at any distance r from the point charge is $V = kq/r$. All points at a fixed distance r from q are on a sphere of radius r, and all these points are at the *same* potential V. The concentric spherical surface about a point charge constitutes an *equipotential surface*. Indeed, we can imagine the point

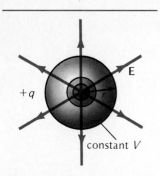

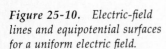

Figure 25-9. Electric-field lines and equipotential surfaces for a point charge.

charge to be surrounded by a number of concentric spherical surfaces; every point on a given surface has the same electric potential.

Now consider the electric field produced by a single positive point charge and its relation to equipotential surfaces. The electric field is radially outward and therefore perpendicular at every point to an equipotential surface. Furthermore, as one moves along the direction of **E**, radially outward for a positive charge, one moves to surfaces at progressively lower potentials. In a two-dimensional figure, the equipotential surfaces are represented by equipotential lines perpendicular to electric-field lines. See Figure 25-9.

More generally, an equipotential surface for any charge distribution is the locus of points all at the same electric potential. By definition, a charge can be moved along an equipotential surface with no work done on it by the electric field. Therefore, **E** must at every location be perpendicular to the constant *V* surface; otherwise, a component of **E** would lie on the *V* surface and do work on and change the energy of a charge lying on that surface. Moreover, the direction of **E** is always toward surfaces of *lower* potential.

The equipotential surfaces for a uniform electric field consist of flat planes perpendicular to **E**. See Figure 25-10.

Field lines and the associated equipotential lines for two equal but opposite electric charges (an electric dipole) are shown in Figure 25-11. The constant *V* loops are reminiscent of contour lines drawn on an ordinary map to indicate points at the same elevation. Just as one can walk in a loop around a mountain peak at constant elevation without doing work against the gravitational force, a

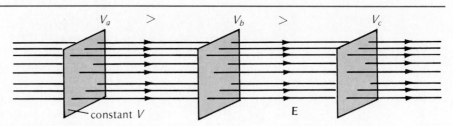

Figure 25-10. Electric-field lines and equipotential surfaces for a uniform electric field.

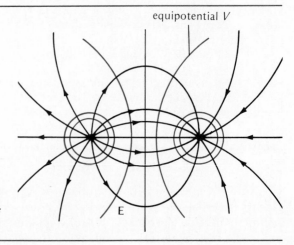

*Figure 25-11. Electric-field lines **E** and the associated equipotential lines in the plane containing two point charges of equal and opposite charges.*

charge can traverse an equipotential line with no work done on it. In the potential-hill analogy, where elevation represents electric potential, a positive point charge produces an infinitely high "spike" ($V = kq/r$), and a negative point charge an infinitely deep "hole in the ground." A uniform electric field is represented by an inclined plane. As will be proved in the next section, the electric field in the potential-hill model corresponds to the steepest slope between points at different elevation (or potential).

25-6   Relations between V and $\mathbf{E}$

The electric properties of a region of space can be mapped in two different ways:

- By the *vector electric field* $\mathbf{E}$, the *force* per unit positive charge, with $\mathbf{E}$ mapped by *electric field lines*.
- By the *scalar electric potential V*, the *work* per unit positive charge, with V mapped by *equipotential surfaces.*

What are the general relations between $\mathbf{E}$ and V? If we know $\mathbf{E}$ at some location, how does V vary as we go to nearby points? Or the converse question, if we know V over a small region of space, how do we find $\mathbf{E}$ there?

Consider first an important special situation: a *uniform electric field*. Electric-field lines and mutually perpendicular associated equipotential lines are shown in Figure 25-12. Suppose that a positive charge q is moved a distance d along the direction of $\mathbf{E}$ from an initial electric potential V_i to a final potential V_f. Charge q moves in the direction of $\mathbf{E}$; therefore, V_i is higher than V_f. The change in potential from i to f is $\Delta V = V_f - V_i$; with $V_i > V_f$, the change ΔV is a *negative.*

The electric force on q has the magnitude $F = qE$ and the work done by this constant force over distance d is

$$W = Fd = qEd$$

The change in electric potential ΔV is also related to the work done by the electric force through (25-2):

$$W = -q \, \Delta V$$

(The minus sign here simply reflects the circumstance that, since ΔV is the potential *increase*, $-\Delta V$ is the potential *drop.*)

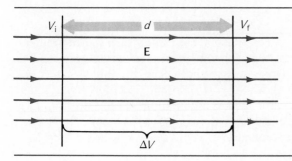

Figure 25-12. *Drop in electric potential ΔV over a distance d along a uniform electric field $\mathbf{E}$.*

Equating the two relations above then yields

$$\Delta V = -Ed$$

or

$$E = -\Delta V / d \qquad (25\text{-}10)$$

The minus sign tells us that if we travel in the direction of **E** over a distance d, the electric potential *decreases;* the change ΔV is negative because the potential drops.

Equation (25-10) shows that appropriate units for the electric field are *volts per meter,* V/m. For example, a uniform electric field with a magnitude of, say, 100 V/m is one in which the electric potential drops by 100 V for each meter traversed in the direction of the electric-field lines (or a 1-V drop for each centimeter). It is easy to see that the two units for electric field, originally newtons per coulomb and now also volts per meter, are equivalent: V/m = (J/C)/m = (N·m/C)/m = N/C.

To get the general relations between V and **E**, we simply consider an infinitesimal displacement *d***s** in any direction in any electric field **E**. The work done on q can be expressed either as

$$dW = \mathbf{F}_e \cdot d\mathbf{s} = q\mathbf{E} \cdot d\mathbf{s}$$

or

$$dW = -q\,dV$$

Again equating the two relations gives

$$dV = -\mathbf{E} \cdot d\mathbf{s} \qquad (25\text{-}11)$$

If we now integrate this relation from starting point 1 to final point 2, we have for the difference in electric potential between the two points:

$$\int_1^2 dV = -\int_1^2 \mathbf{E} \cdot d\mathbf{s}$$

$$V_2 - V_1 = -\int_1^2 \mathbf{E} \cdot d\mathbf{s} \qquad (25\text{-}12)$$

The quantity $\int \mathbf{E} \cdot d\mathbf{s}$ is described in mathematical language as the *line integral* of the vector field **E**. We see from (25-12) that if the electric field **E** is known as a function of position, the potential V can be computed by integration.

Displacement *d***s** was considered above to be in an arbitrary direction. In what direction must *d***s** point to produce the maximum change dV in electric potential? Obviously, *d***s** must then be parallel to **E**. Under these circumstances we can write (25-11) as

$$E = -\frac{dV}{ds} \qquad (25\text{-}13)$$

where the derivative dV/ds is computed for an infinitesimal displacement *along the direction* in space for which V changes *most rapidly* with distance; **E** is in the direction for which V drops most rapidly. The quantity dV/ds is described in mathematical language as the *gradient* of the scalar function V; thus, (25-13) says that the static electric field at any point in space is the negative gradient of the electric potential in the neighborhood of that point. The electric potential can be expressed as a function of coordinates x, y, and z. Then the rectangular components of the electric field are

$$E_x = -\frac{\partial V}{\partial x}, \qquad E_y = -\frac{\partial V}{\partial y}, \qquad E_z = -\frac{\partial V}{\partial z} \qquad (25\text{-}14)$$

where the partial derivative $\partial V/\partial x$ of V with respect to x implies that variables y and z are held constant, and likewise for $\partial V/\partial y$ and $\partial V/\partial z$.

Example 25-5. As shown in Section 24-1, an infinite plane sheet of charge with a constant surface-charge density σ produces a uniform electric field. On both sides of the sheet, $\mathbf{E}$ is perpendicular to the sheet and has a magnitude

$$E = 2\pi k \sigma \qquad (24\text{-}2)$$

How does the electric potential vary with distance from the sheet?

For definiteness, assume the sheet to be positively charged and to lie in the yz plane. Then $\mathbf{E}$ points along the x axis, as shown in Figure 25-13.

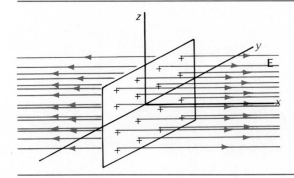

Figure 25-13. *Electric field $\mathbf{E}$ from a uniformly charged infinite plane sheet.*

The electric potential V is given in general in terms of $\mathbf{E}$ by (25-12):

$$V_2 - V_1 = -\int_1^2 \mathbf{E} \cdot d\mathbf{s}$$

We take displacement $d\mathbf{s}$ to be along the x axis, the only direction in which $\mathbf{E}$ has a nonzero component. The two relations above yield

$$V_x - V_0 = -\int_0^x 2\pi k \sigma \, dx$$

where we integrate variable x from 0 to x, and designate the arbitrarily chosen zero of potential (at $x = 0$) by V_0. Then,

$$V_x - V_0 = -2\pi k \sigma x$$

We see that the electric potential drops uniformly with distance x from the charged sheet.

Example 25-6. Figure 25-14 shows a uniformly charged ring of radius R and total charge Q. What is the electric field at P, a distance x from the ring's center measured outward along the symmetry axis?

One way to find $\mathbf{E}$ is to add *as vectors* the contributions to $\mathbf{E}$ from small charge elements dq distributed around the ring. An easier way is first to find the electric potential V at P by adding *as scalar quantities* the contributions from all charge elements and then applying (25-14).

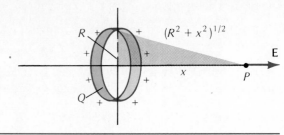

Figure 25-14. *Electric field* **E** *from a uniformly charged ring of radius R.*

Every charge element on the ring is the same distance $(R^2 + x^2)^{1/2}$ from P. The total electric potential at P is then

$$V = \Sigma \, \frac{kq_i}{r_i} = \frac{kQ}{(R^2 + x^2)^{1/2}}$$

Since V depends only on the variable x, the electric field has an x component only; **E** is along the symmetry axis. From (25-14), we have

$$E_x = -\frac{dV}{dx} = \frac{kQx}{(R^2 + x^2)^{3/2}}$$

For positive Q, field **E** points away from the ring.

We can check that the relation above is correct for two extreme values of x:

• $x = 0$, at the center of the ring; then $E_x = 0$. A point charge brought to the center of the uniformly charged ring would be surrounded equally in all directions by identical charges; at this location there is no force on a charge, and necessarily, **E** = 0.

• $x \rightarrow \infty$, far from the ring; then $E_x \simeq kQ/x^2$ with $R^2 \ll x^2$. Viewed from afar, the ring is effectively a point charge, and the field a distance x away is that of point charge Q.

25-7   Electric Potential and Conductors

Consider an insulated conductor with a net charge. We have already seen (Section 24-5) that after electrostatic equilibrium has been achieved:

• The net charge appears only on the conductor's outer surface.
• Inside the conducting material, **E** = 0.
• Just outside the conductor's surface, **E** at every location is perpendicular to the surface.

We can now add one more general property:

• *All points* on the charged conductor's surface and interior are at the *same electric potential*; the entire volume of the conductor is an equipotential volume. See Figure 25-15.

The proof is straightforward. First, just outside the conductor's surface, **E** is perpendicular at all points to the surface, so that this surface must be an equipotential surface. Said a little differently, a charge can be carried along the surface with no work done on the charge because there is no component of **E** along the displacement of the charge. Indeed, the charge can be carried into the conducting material and around anywhere *within* it, and again no work is done

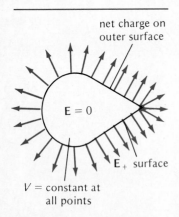

net charge on outer surface

E = 0

E $_+$ surface

V = constant at all points

Figure 25-15. *A charged conductor in electrostatic equilibrium.*

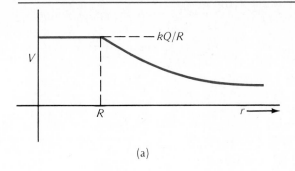

(a)

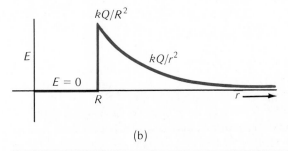

(b)

*Figure 25-16.   For a spherical conducting shell of radius R with charge Q, as a function of distance r from the center, (a) electric potential V and (b) electric field **E**.*

on the charge, because inside $\mathbf{E} = 0$. The entire volume of an insulated charged conductor is equipotential.

Consider a spherical conducting shell of radius R with a net charge Q. How do the electric potential V and the electric field $\mathbf{E}$ each vary in magnitude with distance r from the center of sphere? We know that from the outside, a spherical shell of charge is equivalent to a single point charge at its center. Therefore,

$$\text{For } r \geq R: \quad V = \frac{kQ}{r}$$

$$E = \frac{kQ}{r^2}$$

At any point within the charged shell, the field $\mathbf{E}$ is zero, and therefore, the potential is the same as on the shell itself.

$$\text{For } r < R: \quad V = \frac{kQ}{R}$$

$$E = 0$$

Figure 25-16 shows V and E as functions of r.

Suppose that some well-designed Van de Graaff lecture-demonstration electrostatic generator has a spherical conducting shell of 15-cm radius at an electric potential of 250 kV. Then the charge on the sphere is, from the relation above,

$$Q = \frac{VR}{k} = \frac{(250 \times 10^3 \text{ V})(15 \times 10^{-2} \text{ m})}{(9.0 \times 10^9 \text{ N} \cdot \text{m}^2/\text{C}^2)} = 4.2 \ \mu\text{C}$$

and the electric field at the shell's surface is

$$E = \frac{kQ}{R^2} = \frac{V}{R} = \frac{250 \times 10^3 \text{ V}}{15 \times 10^{-2} \text{ m}} = 1.7 \times 10^6 \text{ V/m}$$

This value for E on the surface is close to the critical value, called the *dielectric strength* (equal to 3×10^6 V/m for dry air at atmospheric pressure), at this critical value, the very strong electric field starts to rip electrons from air molecules, ionizing the air. Any initially charged conductor will lose its charge when E at any point on the surface exceeds the dielectric strength of the medium in which the conductor is immersed.

The relation above, $E = V/R$, gives the magnitude of E at the surface of any spherical conductor of radius R with electric potential V. Suppose that several charged spherical conductors all have the same potential. With $E \propto 1/R$, then E is greatest for small R. A large charged sphere is less likely to discharge than a small one at the same potential. This inverse relation, $E \propto 1/R$, applies to any single charged conductor of arbitrary shape, all points of which are necessarily at the same potential. We can now interpret R to mean the local radius of curvature over any convex small region of the conductor's surface. The electric field is a minimum at a flat region (large R) and a maximum at a sharp point (small R). We saw earlier (Section 24-5) that the electric-field magnitude on the surface of a charged conductor is proportional to the local surface-charge density, $E \propto \sigma$. It follows then that surface charge is also concentrated at sharp points. See Figure 25-15. This result has important practical consequences: to avoid discharge, make a charged conductor's surface as flat and smooth as possible (a sphere has the flattest surface for a given volume); to encourage discharge, use sharp points (for example, the lightning rod).

Summary

Definitions

Electric potential (V): the electric potential difference between points A and B is the work done $W_{A \to B}$ by an applied force to move charge q_t at instant speed from A to B:

$$V_B - V_A = \frac{W_{A \to B}}{q_t} \qquad (25\text{-}1)$$

Equipotential surface: locus of points at the same electric potential.

Units

Electric potential: Volt (V)

$$1 \text{ V} \equiv 1 \text{ J/C}$$

Energy: electron volt (eV)
1 eV = energy to move charge e across a potential difference of 1 V = 1.60×10^{-19} J
Electric field: Volt/meter (V/m)

$$1 \text{ V/m} = 1 \text{ N/C}$$

Important Results

The Coulomb force is a conservative force: $\oint \mathbf{F}_e \cdot d\mathbf{r} = 0$.

The electric potential V arising from point charge(s) q_i at distance(s) r_i:

$$V = \Sigma \frac{kq_i}{r_i} \qquad (25\text{-}7)$$

The electric potential energy U_e of point charges q_1 and q_2 separated by distance r_{12}:

$$U_e = \frac{kq_1 q_2}{r_{12}} \qquad (25\text{-}8)$$

Electric-field lines are perpendicular at every location to the associated equipotential surfaces and point to regions of lower electric potential.

Computing V from $\mathbf{E}$ gives

$$V_2 - V_1 = -\int_1^2 \mathbf{E} \cdot d\mathbf{s} \qquad (25\text{-}12)$$

Computing $\mathbf{E}$ from V gives

$$E = -\frac{dV}{ds} \qquad (25\text{-}13)$$

where the infinitesimal displacement $d\mathbf{s}$ is along the direction for which V decreases most rapidly with distance, also the direction of **E**.

For a uniform **E**:

$$E = -\frac{\Delta V}{d} \qquad (25\text{-}10)$$

for a displacement d along the direction of **E**.

All points on the surface and interior of a charged conductor in equilibrium are at the same potential.

At the surface of a charged conductor, **E** is inversely proportional to the radius of curvature of a convex region. Surface charge and electric field are large at sharp points of a charged conductor.

Problems and Questions

Section 25-2   Electric Potential Defined
· **25-1 Q** In comparing electric and gravitational forces, we find that electric potential is analogous to
(A) mass m.
(B) mgh.
(C) g.
(D) mg.
(E) gh.
(F) h.

$V = \frac{U_E}{q}$

$gh = \frac{U_g}{m} = \frac{mgh}{m}$

· **25-2 P** An electron with initial velocity 4×10^6 m/s is accelerated through a potential difference of 100 V. What is its final velocity?

: **25-3 P** In a typical lightning flash, the potential difference between discharge points is about 10^9 V, and about 30 C of charge is transferred. How much ice could a lightning bolt melt if all the energy were used for that purpose?

: **25-4 Q** Which of the following is an accurate statement?
(A) A person raised to a potential of a few thousand volts will almost certainly suffer serious injury.
(B) It is possible for clouds to reach potentials of the order of half a million volts.
(C) By walking across a nylon rug, you cannot raise your body to a potential of more than a few millivolts.
(D) Electrostatic demonstration experiments are unaffected by humidity.
(E) It is now believed that our galaxy carries a net electric charge.

Section 25-3   Electric Potential for Point Charges
: **25-5 P** A charge of $+2\ \mu C$ is placed at the origin and a second charge of $-4\ \mu C$ at the point (2 m, 0). Determine the electric potential at the following points: (a) (1 m, 0) (b) (1 m, 1 m) (c) (−1 m, 0) (d) (0, 1 m).

: **25-6 P** Charge Q is distributed uniformly along the x axis from $x = 0$ to $x = L$. Determine the potential at a point $x > L$ on the x axis.

: **25-7 P** A beam of charged particles forms a cylindrical column of radius R. If the charge density is uniform throughout the beam, what is the potential (a) inside and (b) outside the beam?

: **25-8 P** Electric charge is distributed with constant den-

sity throughout an insulating sphere of radius R. Determine the electric potential inside and outside the sphere as a function of distance from the center.

Section 25-4   Electric Potential Energy
· **25-9 Q** An electron volt is a unit used to measure
(A) energy.
(B) potential difference.
(C) electric field.
(D) force.
(E) electric charge.

· **25-10 P** An electron passes a proton at a distance of 2×10^{-10} m. What is the minimum kinetic energy of the electron that will prevent its being bound to the proton?

: **25-11 P** Three charges q_1, q_2, and q_3 are placed at positions x_1, x_2, and x_3 on the x axis. What is the total potential energy of the charges if $q_1 = +1\ \mu C$, $x_1 = 0$; $q_2 = -2\ \mu C$, $x_2 = 2$ cm; $q_3 = +3\ \mu C$, $x_3 = 3$ cm?

: **25-12 P** Two $1\text{-}\mu C$ point charges are separated by 2 cm. How much work is needed to bring an electron from a point far away to the point midway between the charges?

: **25-13 P** An electron travels in a circular orbit of radius r with constant speed v around a stationary proton. The charge on the electron is $-e$ and on the proton is $+e$. The electron mass is m. The system's total energy is
(A) $\frac{1}{2} mv^2$
(B) $\dfrac{-k\, e^2}{r}$
(C) $\dfrac{mv^2}{2} + \dfrac{ke^2}{r}$
(D) $\dfrac{-ke^2}{2r}$
(E) $\dfrac{ke^2}{r}$

: **25-14 P** Four charges $+Q$ and four charges $-Q$ are placed at the corners of a cube of side a in such a way that nearest-neighbor charges have opposite signs. How much energy is needed to assemble this charge array?

· **25-15 Q** Point A is at an electric potential of -4.0 V and point B at an electric potential of $+6.0$ V. An electron passes point A with a kinetic energy of 12.0 eV. When it reaches point B, its kinetic energy will be

(A) 2 eV.
(B) 6 eV.
(C) 18 eV.
(D) 22 eV.
(E) indeterminable unless we know the direction the electron was moving at point A.

handwritten: $U_A + K_A = U_B + K_B$
handwritten: EQN (25-3) $-4 + 12 = -6 + K_B$
handwritten: $U_A - U_B = q(V_A - V_B)$ $\Rightarrow K_B = 22$
handwritten: $q = -1$

: 25-16 P Two fixed point charges, each $+Q$, are separated by a distance d. (a) Graph the potential energy of a third charge $+q$ as a function of position along the line joining the two fixed charges. (b) Show that for small displacements x from the midpoint, the potential energy E_p is parabolic, that is, $E_p \propto x^2$. (c) When the charge q is at the midpoint, is it in stable or unstable equilibrium with small displacements along the line joining the fixed charges? along a direction perpendicular to this line?

: 25-17 P A proton initially very far away from a nucleus with charge $+Ze$ is projected toward the nucleus with velocity v_0. When the proton is a distance R from the nucleus, its velocity has decreased to $\frac{1}{2}v_0$. (a) When its velocity has dropped to $\frac{1}{4}v_0$, how far will it be from the nucleus? (b) What is the distance of closest approach to the nucleus?

: 25-18 P An alpha particle (a helium nucleus with charge $+2e$) with an initial kinetic energy 10.0 MeV approaches a gold nucleus (charge $+79e$) along a path that would bring it within 2.0×10^{-14} m of the gold nucleus if it were not for the Coulomb repulsion. Assume that the heavy gold nucleus remains at rest. (a) What is the distance of closest approach of the two particles? (b) What is the kinetic energy of the alpha particle at the point of closest approach?

: 25-19 P Assuming that the density of the earth is constant (and equal to M_e/V_e), determine how much energy would be needed to blow the earth apart, that is, to send each piece of it off to infinity. (*Hint:* Gravitational potential energy may be handled much like electric potential energy.)

: 25-20 P Electrons are emitted from a cathode with negligible kinetic energy and accelerated through a potential difference of 500 V to an anode 20 cm from the cathode. Take the beam to have circular cross section of diameter 1 mm, and assume that a constant electric field acts on the electrons. What is the charge density in the beam as a function of the distance from the cathode?

Section 25-5   Equipotential Surfaces

· 25-21 Q Equipotential lines, in a two-dimensional plot, are most nearly analogous to

(A) flow lines in a fluid.
(B) the tracks left by a charged particle moving in a cloud chamber.
(C) the path followed by a particle undergoing Brownian motion.

(D) the direction a stream would follow on a map.
(E) contour lines on a topographic map.
(F) a map of regions in a room in which there is no sound.

: 25-22 P Charges of $+1$ nC and -1 nC are separated by 2 cm. Using approximate numerical methods, carefully sketch the equipotential lines that pass through points at distances of 8 mm, 9 mm, 10 mm, 11 mm, and 12 mm from the positive charge along the line joining the two charges.

Section 25-6   Relations between V and E

· 25-23 P An electron is accelerated by an electric field of 2000 V/m over a distance of 2 cm. By how much, in electron volts and in joules, does its energy increase?

· 25-24 P A 3-keV electron is shot through a hole in one of two parallel plate electrodes. A potential difference of 10,000 V is applied between the electrodes, as shown in Figure 25-17. How far past the first electrode will the electron go before stopping?

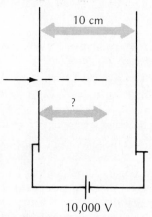

10 cm

?

10,000 V

Figure 25-17.   Problem 24-25.

· 25-25 P In a given region, the electric field in the x direction has the constant value -100 V/m. If the potential at $x = 4$ cm is 6 V, what is the potential at $x = 10$ cm?

· 25-26 P If the electric potential varies from 100 V to 150 V in a distance of 5 cm, what is the magnitude of the average electric field in this region?

: 25-27 P An infinite conducting sheet carries a uniform charge density of 10^{-7} C/m². How far apart are equipotentials that differ by 5 V?

: 25-28 P In a tandem Van de Graaff accelerator, a hydrogen ion is accelerated from rest through a potential difference of 10 MV. Then the two electrons are stripped from the ion and the resulting bare proton is accelerated back to ground potential through a potential difference of -10 MV. What is (a) the final kinetic energy and (b) the speed of the proton?

· 25-29 P Sketched in Figure 25-18 is the potential arising from a hypothetical charge distribution. At which indicated point or points on the x axis does the electric field E_x (a) have its largest positive value? (b) have its largest negative value? (c) have the value zero?

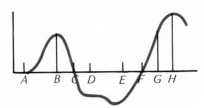

Figure 25-18.   *Problem 25-29.*

: **25-30 Q** Which of the following is the most accurate statement?

(A) The electric potential difference between two points equals the work that must be done in moving an electron from one point to the other.

(B) The presence of a positive point charge always tends to make the potential in its vicinity more positive.

(C) We associate electric potential with the slope of the electric field.

(E) The potential due to two point charges q and $-q$ will be zero at only one point in a finite region of space.

(F) More than one of the above is true.

· **25-31 P** An electric field varies from 200 V/m at point A to 300 V/m at point B, and A and B are 0.5 m apart. Can you calculate the difference in potential between points A and B with only this information? If so, what is the value of the potential difference?

: **25-32 P** The potential due to a point charge is $V = q/4\pi\epsilon_0 r$ where $r = \sqrt{x^2 + y^2 + z^2}$. Calculate electric-field components E_x, E_y, and E_z from this expression, and show that the expected result $\mathbf{E} = (q/4\pi\epsilon_0)(\mathbf{r}/r^3)$ is obtained.

: **25-33 Q** Consider the electric-field lines and equipotentials arising from an array of positive and negative point charges. Which statement is true?

(A) The electric field does not vanish (except possibly at infinity).

(B) The electric field lines never cross.

(C) The equipotential surfaces never intersect.

(D) No work is required to move a test charge parallel to an electric field line.

(E) Both B and C above are true.

: **25-34 Q** Which of the following statements are true?

(A) The electric potential arising from an array of point charges is the sum of the potentials arising from each individual charge.

(B) The electric field is zero where the potential is zero.

(C) The potential arising from a single point charge may vary from positive to negative in different regions.

(D) The force on a charged particle located on an equipotential surface is zero.

(E) All of the above are true.

· **25-35 P** Charges $+Q, +Q$, and $-2Q$ are at the corners of an isosceles triangle. Make a qualitative sketch of some of the equipotentials and electric-field lines for this configuration.

: **25-36 Q** Which of the following is a true statement?

(A) If the electric field is zero at a point, the electric potential will be zero there also.

(B) If the electric potential is zero at a point, the electric field must be zero there also.

(C) If the electric field is zero throughout some region of space, the electric potential must also be zero throughout that region.

(D) If the electric potential is zero throughout some region, the electric field must also be zero throughout the region. $E = grad\ V$

(E) The electric field is large where the electric potential is large and small where the electric potential is small.

: **25-37 Q** In Figure 25-19 are drawn some equipotential planes or electric-field lines. At which of the points indicated is the electric-field strength greatest?

(A) A.

(B) B.

(C) C.

(D) D.

(E) It is equally large at B and D.

(F) It is equally large at A and C.

(G) The question cannot be answered unless we know whether the lines are equipotentials or electric-field lines.

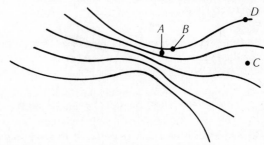

Figure 25-19.   *Question 25-37.*

: **25-38 P** Charge is uniformly distributed over a circular disk of radius a. (a) Show that the potential at a point on the symmetry axis a distance z from the center is

$$V = \frac{\sigma}{2\epsilon_0}(\sqrt{a^2 + z^2} - z)$$

(b) Show that this result gives the electric field on the axis as

$$E_z = \frac{\sigma}{2\epsilon_0}\left(1 - \frac{z}{\sqrt{a^2 + z^2}}\right)$$

(c) Consider what the limiting values for V and E are when $z \gg a$ and when $z = 0$. Do these results agree with what you would expect?

Section 25-7   Electric Potential and Conductors

· **25-39 Q** In learning about electrostatics, it is helpful to use the analogy between the description of electrical phenomena and a contour map. The surface of a piece of copper, then, would be analogous to
(A) the direction in which a stream flows on the map.
(B) a contour line.
(C) a point of high elevation on the map, such as a mountain peak.
(D) a point of low elevation on the map, such as the bottom of a well.
(E) a uniform slope.
(F) a road drawn on the map.

· **25-40 Q** Can there be a potential difference between two adjacent insulated conductors, each of which carries the same charge?
(A) No, since if two conductors carry the same charge, they must be at the same potential.
(B) No. All conductors are equipotential surfaces; hence they are all "grounded," that is, they are at zero potential.
(C) Yes. An example would be two spheres of different radii.
(D) A conductor is by definition electrically neutral; hence this question makes no sense, since one cannot have a "conductor carrying a charge."

· **25-41 Q** Two conductors are connected by a long copper wire. Thus
(A) each conductor must have the same charge.
(B) each conductor must be at the same potential.
(C) the electric field at the surface of each conductor is the same.
(D) one conductor always feels a force due to the other conductor.
(E) each conductor has zero net charge.

· **25-42 P** Trailing from the wings of a jetliner are pointed needles. When static charge accumulates on the airplane, a corona discharge occurs at these needles, thereby preventing dangerous sparks near the fuel tanks. If air breaks down in an electric field of 3×10^6 V/m, about what is the maximum potential to which the plane could be charged electrostatically? (The needles have a tip radius of 0.01 mm.)

⋮ **25-43 Q** A long metal bar of square cross section is given a charge. Using qualitative reasoning, sketch a series of equipotential lines and electric-field lines. No calculation is necessary, but your drawing should be accurate for the areas close to the conductor and very far away from it.

· **25-44 Q** To understand new concepts, it sometimes helps to draw analogies with familiar ideas. In this sense one can think of a two-dimensional plot of equipotential surfaces as a contour map. In such a representation, a charged hollow copper sphere would be most like
(A) a volcano.
(B) a mountain lake.

(C) a very steep cliff.
(D) a sloping hillside of constant slope.
(E) a very narrow well, dug straight down far into the earth.

⋮ **25-45 P** Four spherical raindrops, each of radius r, carry such a charge that the potential of each is V_0. They join to form a single large droplet. (a) What is the potential of the large droplet? (b) What is the change in electrostatic energy of the system when the droplets combine? How do you account for the change?

⋮ **25-46 P** Calculate the work necessary to charge a conducting sphere of radius R with a total charge Q.

⋮ **25-47 P** What charge must be placed on a conducting sphere to raise it to a potential of 2 MV (2×10^6 V) if the radius of the sphere is (a) 1 m? (b) 10 cm?

(c) Since less charge is needed to raise the smaller sphere to the required voltage, why are large spheres used in a Van de Graaff generator? Substantiate your answer by calculating the electric field at the surface of the sphere in each of the above cases. Note that a large electric field is more likely to cause electrical breakdown and also cause a more rapid leakage of charge because of migrating ions in the surrounding atmosphere.

⋮ **25-48 Q** Why would one want to use a large spherical conductor at the high-voltage end of a Van de Graaff generator?
(A) Because the electric field is smaller at the surface of a large sphere than at the surface of a small sphere at the same potential.
(B) Because the larger sphere can store more charge.
(C) Because a given amount of charge placed on a large sphere will raise it to a higher potential than what would happen if the same charge were placed on a small sphere.
(D) Because with a large sphere it is easier to generate the corona discharge needed for acceleration of particles.
(E) The main reason for a large sphere is that this in turn allows one to use a wide belt to carry charge up to the sphere; hence charging can be done more rapidly and efficiently.

⋮ **25-49 Q** Three hollow, conducting aluminum spheres of radius 1 m, 2 m, and 3 m are placed at the corners of an equilateral triangle of side 50 m. A charge of 400 μC is placed on the large sphere and then the spheres are connected by wires. What is the final charge on each sphere?

⋮ **25-50 P** In a lecture demonstration, three conducting spheres, of radius 2 cm, 4 cm, and 6 cm, are used. The largest sphere is charged to a potential of 500 V. It is first touched to the smallest sphere and next it is touched to the intermediate sphere. What will be the potential of each sphere and the charge on each after this is done, each sphere then being separate from the others?

⋮ **25-51 P** A Geiger counter uses two concentric metal cylinders of radius $R_1 = 0.1$ mm and $R_2 = 1.2$ cm; between the cylinders a voltage of 1000 V is applied. Incident radiation particles ionize the air between these electrodes, giving rise to a pulse of electric current (flow of charge) between them. (a) What is the electric field at the surface of the inner conductor? (b) What is the electric field at the surface of the outer conductor? (c) What is the linear charge density along each conductor?

Supplementary Problems

25-52 P The electrostatic charge that can build on the surface of a liquid drop (such as a raindrop or a droplet in an aerosol spray) can cause it to break up. This has many important consequences. Consider a spherical drop of radius R that can break into two identical smaller drops. The energy of the conducting drop consists of electrostatic energy and surface-tension energy, $E_s = 4\pi R^2\sigma$, where σ = surface tension = 0.07 J/m² for water.

(a) Write an expression for the energy of the drop before it breaks up and after. When it has broken up, it consists of two small touching drops. The initial charge Q is shared equally between these two smaller drops. (b) The initial drop will break up when the energy of the two small touching drops is less than the total energy of the large drop. Show that the condition for instability is $Q^2/4\pi\epsilon_0 > 2.5\sigma R^3$. (c) Determine the maximum electric field that can exist on the surface of the drop before it breaks up.

(If you have never noticed that water droplets acquire a charge, try deflecting a fine spray from a faucet by using a comb you have charged by rubbing it on a sweater. This phenomenon is important in meteorology, spray painting, aerosol sprays, and ink jet printers.)

25-53 P In an ink jet printer, a thin stream of conducting ink is projected from a nozzle down the center of a cylindrical electrode. A high voltage is applied between the jet and the surrounding cylinder. The droplets acquire a net charge opposite to that of the cylinder. Electrostatic and surface-tension forces cause the jet to break into charged droplets. These droplets are then deflected by charged parallel metal plates, just as an electron beam is deflected in an oscilloscope.

(a) If the jet stream has radius R_j and the charging cylinder has radius R_c, what is the charge per unit length on the jet in terms of the voltage V between the jet and the cylinder? (b) If the jet breaks into droplets of radius $R_d = 2R_j$ how many droplets per second will be generated when $R_j = 2 \times 10^{-5}$ m and the jet velocity is 25 m/s? (c) What

will be the charge per droplet if $R_c = 5$ mm and the charging voltage is $V = 500$ V? (d) Through what angle θ will the droplets be deflected when they pass through the deflecting plates shown in Figure 25-20? A deflecting voltage of 1000 V is used.

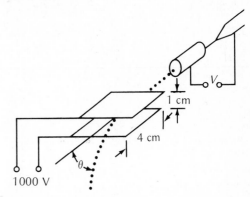

Figure 25-20.   Problem 25-53.

25-54 P In an electrostatic dust precipitator, a long straight wire passes down the center of a cylindrical metal duct. A large potential difference V is maintained between the wire and the duct, with the duct positive. This causes a corona discharge near the wire, and the resulting free electrons tend to attach to dust particles that are then swept out to the outer duct wall. Periodically the voltage is shut off and the dust particles fall to the bottom of the duct, where they are collected. During some industrial processes (smelting, for instance) useful materials such as sulfur are recovered in this way while also reducing air pollution. If it is desired to produce a field of 3×10^6 V/m at the surface of the wire, what voltage must be applied if the radius of the wire is 1 mm and of the duct is 20 cm?

25-55 P The high-voltage electrode of a Van de Graaff generator is a metal sphere of radius 0.8 m. Electrical breakdown will occur in the pressurized gas in the machine at a field of 1.2×10^8 V/m. Charge leaks off the high-voltage electrode at a rate of 2×10^{-4} C/s from a variety of effects. This charge is replenished by a moving rubber belt 60 cm wide. (a) What is the maximum voltage to which the machine can be charged? (b) At what minimum speed must the belt be driven to replace the charge lost by leakage? (c) What then is the surface-charge density on the belt? (Hint: If the charge density on the belt is too great, breakdown will occur.) (d) What power is needed to drive the belt if frictional losses are neglected?

26

Capacitance and Dielectrics

The capacitor is important for several reasons:

- It stores separated electric charge.
- It stores electric potential energy.
- It plays a crucial role in electric and electronic circuits, particularly for varying electric currents. (Anyone who has turned the dial to tune a radio has changed the capacitance of a variable capacitor.)

26-1   Capacitance Defined

A general type of capacitor is shown in Figure 26-1. It consists of two separated conductors—we call them *plates* whatever their shape—carrying opposite charges of equal magnitude Q with a potential difference V between them.* Since all points on the surface and interior of any one conductor in electrostatic equilibrium are at a single potential, V represents the potential difference between *any* point on one plate and *any* point on the oppositely charged plate.

---

* A potential difference is often represented by such a symbol as ΔV. For a capacitor, however, it is customary to write the potential difference simply as V.

The net charge of the entire capacitor is zero, with charges $+Q$ and $-Q$ on the two plates. To charge a capacitor clearly requires work. If it were possible to take just one electron from one uncharged plate and place it on the other plate, this electron would repel all other electrons brought to the negatively charged plate.

The charge magnitude Q on either plate is directly proportional to the potential difference V between the plates, so that we can write

$$Q = CV$$

where C is the proportionality constant for any particular configuration of capacitor plates.* Indeed, C is defined to be the *capacitance* of the capacitor:

$$C \equiv \frac{Q}{V} \qquad (26\text{-}1)$$

Capacitance is a quantitative measure of a capacitor's capacity for storing separated charges and therefore for storing electric potential energy. A relatively large C means that even a small V can produce a large charge magnitude Q. A capacitor is symbolized in electric circuit diagrams by ⊣⊢ or by ⊣∈ ; the symbol ⌁ denotes a variable capacitance.

From its definition in (26-1), we see that capacitance has the units coulombs per volt. A special name, the farad (abbreviated F) is given to this ratio:

$$1\text{F} \equiv 1\text{C}/\text{V}$$

(The capacitance unit is named after the famed nineteenth-century experimentalist in electromagnetism, Michael Faraday, who first investigated the effects of dielectric materials in capacitors.) As we shall see, a capacitance of one farad is enormous (simply because a charge of one coulomb is enormous). More common capacitor units are the microfarad ($1\ \mu\text{F} = 10^{-6}\ \text{F}$) and the picofarad ($1\ \text{pF} = 10^{-12}\ \text{F}$).

Any two oppositely charged conductors make up a capacitor. A practical capacitor, however, is small, has its oppositely charged plates easily insulated from one another, and is so constructed that external fields will not disturb the distribution of charge on the plates. These requirements are met by capacitors with parallel plates or with coaxial cylindrical conductors, and the symmetry of these geometries allows the capacitance to be computed easily.

Example 26-1 Parallel-Plate Capacitor.   A capacitor with two parallel plates, each of area A and separated by a distance d, is shown in Figure 26-2. The plate separation d is small compared with the size of the plates, so that the electric field **E** is uniform and confined almost entirely to the region between the plates. Although fringing of the electric force lines at the boundaries is always present, we take it to be negligible. The two plates must be held apart by some insulating material, not only to prevent shorting of the capacitor but also to hold apart the oppositely charged plates

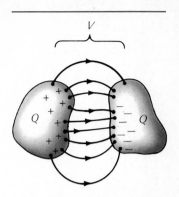

Figure 26-1. Most general form of capacitor; two conductors carrying equal but opposite charges of magnitude Q and having an electric potential difference V between them.

---

* That C is indeed a constant for any capacitor follows directly from the relation ($V = kq/r$) for the electric potential V of a single point charge q. At any particular location (any r), $V \propto q$. But any continuous distribution of charges over the capacitor plates consists basically of a collection of point charges with the proportionality relation $V \propto q$ holding for each point charge. The potential at any point is then proportional to *all* charges. Further, when the net charge on any conductor is changed, the relative distribution of charge on the conductor is not changed but depends only on the geometry of the conductor. Therefore, $Q = CV$, where C is a constant.

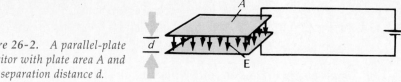

*Figure 26-2.   A parallel-plate
capacitor with plate area A and
plate separation distance d.*

under the action of the attractive force between them. The insulator may be a dielectric material sandwiched between the plates, in which case the capacitance is enhanced over its value when a vacuum exists between the plates. Here we assume the plates to be in a vacuum; the effect of dielectric materials will be treated in Section 26-3.

The most direct way to charge a capacitor is simply to connect its plates with conducting wire to the terminals of a battery. The constant potential difference between the battery terminals then also becomes the same potential difference V across the capacitor.

To find the capacitance C of a parallel-plate capacitor we must know the charge Q on either plate in terms of the potential difference V. The link is the electric field **E**. Since the field is uniform over the plate separation distance d, we have

$$E = \frac{V}{d} \tag{25-10}$$

It was shown earlier (Section 24-5) that the electric field near the surface of any conductor is related to the local surface charge density σ by

$$E = \frac{\sigma}{\epsilon_0} = \frac{Q/A}{\epsilon_0} \tag{24-6}$$

where $\sigma = Q/A$ for a parallel-plate capacitor, with σ constant over the area A of each plate carrying charge Q. Using the relations above in the definition of C, we get

$$C = \frac{Q}{V} = \frac{Q}{Ed} = \frac{Q}{(Q/\epsilon_0 A)d} = \frac{\epsilon_0 A}{d} \tag{26-2}$$

where

$$\epsilon_0 = \frac{1}{4\pi k} = 8.854\ 187\ 8 \times 10^{-12}\ \text{C}^2/\text{N} \cdot \text{m}^2$$

As (26-2) shows, the capacitance is (1) proportional to the plate area and (2) inversely proportional to the plate separation. Both make sense on the basis of simple physical arguments: (1) the larger the plate area (for a given V and d), the larger the amount of charge that can be stored on the plates; (2) with Q and A fixed (and E thereby constant), bringing the plates closer together reduces the distance over which this field exists and hence reduces V.

Notice further from (26-2) that C equals the constant ϵ_0 multiplied by the ratio A/d, a quantity with the dimensions of the length. This result applies for all types of capacitors, and each capacitor's capacitance depends only on the geometry of the conductor. We can see that one farad of capacitance is enormous directly from (26-2). Suppose that each of the two plates has an area of 1.0 m² and that they are separated by 1.0 mm = 10^{-3} m. Then,

$$C = \frac{\epsilon_0 A}{d} = \frac{(8.85 \times 10^{-12}\ \text{C}^2/\text{N} \cdot \text{m}^2)(1.0\ \text{m}^2)}{10^{-3}\ \text{m}} = 8.85 \times 10^{-9}\ \text{F}$$

or somewhat less than $\frac{1}{100}\ \mu$F. With this separation distance, one would need square plates 10 km along a side to have a capacitance of 1 F.

Example 26-2 Coaxial Capacitor.   A coaxial capacitor consists of two concentric cylindrical conductors of radius r_1 and r_2, as shown in Figure 26-3. A cylindrical capacitor of extended length, often constructed with a dielectric material filling the space between the conductors, is known as coaxial cable (colloquially, "coax" or sometimes merely "cable," as in "cable TV").

Each of the two cylindrical surfaces is an equipotential surface; between these exists a potential difference $V_1 - V_2$. For definiteness, the inner conductor is taken to be positive, so that **E** is radially outward and potential V_1 is higher than V_2. To compute the capacitance, we shall use the relation giving the electric field E in terms of the charge per unit length λ on either conductor and then find the potential difference from (25-12):

$$V_2 - V_1 = -\int_1^2 \mathbf{E} \cdot d\mathbf{r}$$

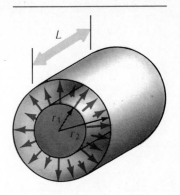

Figure 26-3.   *A coaxial cylindrical capacitor.*

It is easy to see that the electric field between the two cylindrical shells is exactly the same as that from a single infinite, straight, uniformly charged wire at the axis of the cylinders with the same linear charge density λ as either cylindrical shell. The infinite charged wire produces a radial field **E** whose magnitude is given by $E = 2k\lambda/r$ (Section 24-1); equipotential surfaces, necessarily at right angles to the radial field **E**, are also right circular cylinders. (That the field between the conductors of the coaxial capacitors is $E = 2k\lambda/r$ can also be deduced directly from Gauss's law; one chooses a concentric cylindrical Gaussian surface lying between the two conductors.)

When $E = 2k\lambda/r$ is substituted into the relation above, we get

$$V_2 - V_1 = -2k\lambda \int_{r_1}^{r_2} \frac{dr}{r} = -2k\lambda \ln \frac{r_2}{r_1}$$

With $\lambda = Q/L$, where Q is the charge on one conductor of length L, we find the capacitance per unit length C/L to be

$$\frac{C}{L} = \frac{Q}{(V_1 - V_2)L} = \frac{1}{2k \ln(r_2/r_1)} = \frac{2\pi\epsilon_0}{\ln (r_2/r_1)} \tag{26-3}$$

For example, any air-filled coaxial cable with $r_2 = 2r_1$ has a capacitance per unit length of $1/(2)(9.0 \times 10^9 \text{ N} \cdot \text{m}^2/\text{C}^2) \ln 2 = 80 \text{ pF/m}$.

Figure 26-4 shows several types of commercial capacitors.

Figure 26-4.   *Various types of capacitors. The capacitor at back left is a variable capacitor of the type used to tune a radio receiver; in the middle back is an old-fashioned Leyden jar, the original capacitor, named for the town in the Netherlands where it first originated.*

26-2   Capacitor Circuits

Electric circuits composed entirely of capacitors are not very interesting. They are worthy of attention, however, because they illustrate fundamental concepts common to all circuits.

First, these fundamental principles:

- *Electric charge conservation.* For our immediate purposes this implies that the *net* charge of an isolated conductor does not change, although the charge may be redistributed.
- *Energy conservation.* For an equilibrium distribution of electric charges, the net work done by electric forces on a point charge carried around a closed loop (a circuit) is zero, $\oint \mathbf{F}_e \cdot d\mathbf{r} = 0$. Another way of saying the same thing is that the *net change* in *electric potential* around any circuit is *zero*, or in still other words, the sum of all of electric potential drops ΣV around any circuit is zero:

$$\sum_{a \to a} V = 0 \qquad (26\text{-}4)$$

A drop in potential over one portion of the circuit must be compensated for by an equal potential rise somewhere else in the circuit.

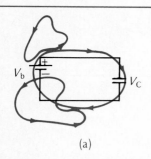

(a)

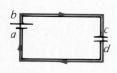

(b)

Figure 26-5.   Circuit consisting of a battery and a capacitor. (a) For electrostatic equilibrium, the net change in electric potential around any closed loop (shown with color lines) is zero. (b) A loop abcda around the circuit passing along the conducting wires.

Consider the circuit of Figure 26-5(a) with a battery (symbolized by 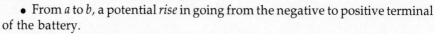) connected across a single capacitor. If we travel along any closed loop, not necessarily coinciding with the conducting wires connecting the battery to the capacitor, the net change in potential is zero. This of course holds also for the special route shown in Figure 26-5(b), one that coincides with the conducting wire:

- From *a* to *b*, a potential *rise* in going from the negative to positive terminal of the battery.
- From *b* to *c*, along a conductor in electrostatic equilibrium, no change in potential.
- From *c* to *d*, from the positive to the negative plate of the charged capacitor, a potential drop.
- From *d* to *a*, again along a conductor in equilibrium, no change in potential.

Here we have

$$\Sigma V = 0$$
$$V_{ab} + V_{bc} + V_{cd} + V_{da} = 0$$
$$V_{ab} + 0 + V_{cd} + 0 = 0$$
$$V_{cd} = -V_{ab}$$

The potential *drop* across the capacitor, V_{cd}, is equal to the negative potential *rise*, $-V_{ab}$, across the battery. The potential drops across the two are the same.

We now consider capacitors in two simple circuit arrangements:

- *In series* [Figure 26-6(a)]: in going from *A* to *B*, we encounter in turn each of the circuit elements *a*, *b*, *c*.
- *In parallel* [Figure 26-6(b)]: there are alternative ("parallel") routes from *A* to *B*, each one containing one circuit element.

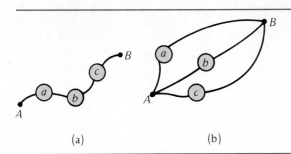

Figure 26-6. *Circuit elements a, b, and c connected (a) in series and (b) in parallel.*

(a)   (b)

Capacitors in Series   Capacitors C_1, C_2, C_3 are connected in series and attached across battery terminals as shown in Figure 26-7(a). We wish to find the value of C of the single capacitor that is equivalent to this group of capacitors in the sense that it has the same charge as all series capacitors taken together when connected to the same battery.

From Figure 26-7(a), we see that the *net* charge on the single conductor shown within the color loop must remain zero; this section is isolated electrically from everything else. The battery separates the charges on this conductor; equal amounts of positive and negative charge appear on the plates of adjoining capacitors. Thus, all capacitors in series have the *same charge:*

$$Q_1 = Q_2 = Q_3$$

The potential across the battery V_b is the same as the total drop across the capacitors in series,

$$V_b = V_1 + V_2 + V_3 = \frac{Q_1}{C_1} + \frac{Q_2}{C_2} + \frac{Q_3}{C_3}$$

where we have used $V_1 = Q_1/C_1$, and similarly for C_2 and C_3.

The relation may be written

$$V_b = Q\left(\frac{1}{C_1} + \frac{1}{C_2} + \frac{1}{C_3}\right)$$

where we have used Q to represent any one of $Q_1 = Q_2 = Q_3$. With a single equivalent capacitor C across the battery terminals, the potential difference

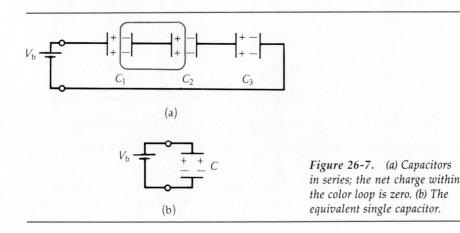

(a)

(b)

Figure 26-7. *(a) Capacitors in series; the net charge within the color loop is zero. (b) The equivalent single capacitor.*

Knowing the Connections

What is this experimenter doing in the physics lab?

As the figure shows, two oppositely charged plates are attracting one another; the electric force on the upper charged plate is balanced by a weight placed on the pan on the other side of the beam balance. In this fairly simple arrangement, we have, basically, two electrically charged objects attracting one another. The formula for the electric force on a plate ($F = Q^2/2\,\epsilon_0 A$, where A is plate area) is a little different from Coulomb's law for a couple of point charges, but here also the force depends on essentially the same quantities: the magnitude of charge, the dimensions, and the constant ϵ_0. So what is being measured?

There a few possible interpretations:

• The experimenter might be merely checking that the relation for the force is right, a pretty dull experiment.

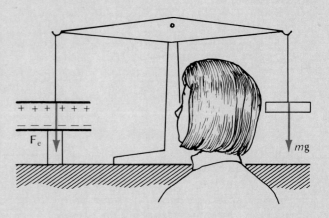

• She might be comparing an electric force with a gravitational force. Now that could be interesting. We can see from the setup above that the electric interaction is enormously larger than the gravitational interaction. After all, it takes the entire earth pulling

across its plates, $V = Q/C$, would equal V_b. Therefore,

$$\frac{Q}{C} = Q\left(\frac{1}{C_1} + \frac{1}{C_2} + \frac{1}{C_3}\right)$$

$$\frac{1}{C} = \frac{1}{C_1} + \frac{1}{C_2} + \frac{1}{C_3} = \Sigma \frac{1}{C_i} \qquad (26\text{-}5)$$

For capacitors in series, the reciprocal of the single equivalent capacitance is the sum of the reciprocals of the individual capacitances. As a consequence, the equivalent capacitance is always *less* than the smallest of the series capacitors. For example, with four capacitors, each of 20 μF and connected in series, the equivalent capacitance is 5 μF.

Capacitors in Parallel For capacitors connected in parallel, see Figure 26-8(a). What single capacitor, as in Figure 26-8(b), is equivalent for this group of capacitors? The charges and potential differences for C_1, C_2, and C_3 are again designated, respectively, Q_1, Q_2, Q_3 and V_1, V_2, V_3. By taking potential differences along the closed loops shown by color lines in the figure, we find that the

Figure 26-8. *(a) Capacitors in parallel; the net potential difference around any color loop is zero. (b) The equivalent single capacitor.*

on the object on the right-hand pan to balance the electric force between two objects with very modest amounts of charge.

• The experimenter might be—as surprising as this might seem—measuring the speed of light! Not directly, of course. But, as we shall later see (Section 34–3), light is an electromagnetic phenomenon, and its speed through empty space is related directly to ϵ_0. Know one and you can compute the other.

That's the kind of remarkable thing you can do when you know the connections that physics reveals. When you know basic physics—that things hang together and how they hang together—you have gained not only extraordinary insight, but also power.

We've already seen examples of how quantities that might at first seem absolutely inaccessible can be computed indirectly from basic relations, and we shall see many more.

• What's the mass of the earth? Certainly you can't measure it directly. But if you know the law of uni-versal gravitation and the value of the gravitational constant G, then computing the earth's mass is almost ridiculously easy (Section 15–3 and Example 15–1).

• At the other extreme, what's the mass of an electron? Again, any direct measurement is ruled out. But if we know how electric and magnetic fields influence a moving charged particle, we can set up what amounts to an obstacle course for electrons (Section 29–4). If an electron makes it all the way through the course, we can compute its mass directly.

The list could go on. The conclusion is simple but important: when you know how things fit together in the physical universe—when you know basic physics—then you can do astonishing things with it.

And that, of course, is why, if you're a student in engineering, physics is almost certainly a required course. To do the challenging and difficult job of applying basic scientific ideas to practical use—to be an effective engineer—you must first have straight the fundamental insights that physics provides.

potential difference is the same across each capacitor as across the battery terminals. Parallel capacitors have the *same potential difference;* that is,

$$V_b = V_1 = V_2 = V_3$$

The circuit consists, effectively, of only two conductors attached to the battery terminals (all the upper plates with their connecting wires, and all the lower plates with theirs). Therefore, the total charge Q held on a single capacitor equivalent to those in parallel is given by

$$Q = Q_1 + Q_2 + Q_3$$

or

$$Q = C_1 V_1 + C_2 V_2 + C_3 V_3 = V_b(C_1 + C_2 + C_3)$$

But $C = Q/V_b$. Therefore,

$$C = C_1 + C_2 + C_3 = \Sigma\, C_i \qquad (26\text{-}6)$$

For parallel capacitors, the single equivalent capacitance equals the sum of the individual capacitances. The equivalent capacitance always exceeds the largest capacitance in parallel.

26-3   Dielectric Constant

First we describe how the introduction of a dielectric material changes the capacitance of a capacitor. Then (Section 26-6) we discuss in qualitative terms why a dielectric changes a capacitor's capacitance.

The dielectric constant κ of an insulating, or dielectric, material can be defined in terms of the capacitance C of a capacitor whose plates are immersed in a vacuum and the capacitance C_d of the same capacitor but with dielectric material filling the region between the plates:

$$\kappa \equiv \frac{C_d}{C} \qquad (26\text{-}7)$$

The capacitance always *increases* with the introduction of dielectric. $C_d > C$, and κ is always *greater than one*.

Table 26-1 lists dielectric constants and dielectric strengths (explained below) at room temperature for several materials. Note that κ for air is very close to 1.00, which is, by definition, the dielectric constant for a vacuum.

Since a capacitor's capacitance changes with the introduction of a dielectric, other associated properties of the capacitor may also change. Here we distinguish two simple circumstances:

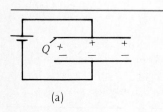

(a)

• Capacitor maintained at a *constant potential difference V* (by being attached to a battery). See Figures 26-9(a) and (b). Then from the definitions,

$$\text{Same } V: \quad \kappa = \frac{C_d}{C} = \frac{Q_d/V}{Q/V} = \frac{Q_d}{Q}$$

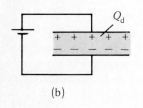

(b)

The *charge* Q_d on a plate *increases* with the introduction of a dielectric; the *electric field* $E = V/d$ between the plates is *unchanged*. The additional charge is supplied by the battery.

• Capacitor maintained at *constant charge Q* (by being detached from other circuit elements). See Figures 26-9(c) and (d). Again from the definitions,

$$\text{Same } Q: \quad \kappa = \frac{C_d}{C} = \frac{Q/V_d}{Q/V} = \frac{V}{V_d} = \frac{E}{E_d}$$

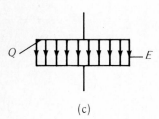

(c)

Both the potential difference V_d and electric field E_d are reduced from their respective values for a vacuum.

The *electric permittivity* ϵ of a dielectric material is defined by

$$\epsilon = \kappa \epsilon_0 \qquad (26\text{-}8)$$

With $\kappa = 1$, we have $\epsilon = \epsilon_0$, the permittivity of the vacuum.* To generalize any relation we have derived so as to include filling by a dielectric material, we simply replace ϵ_0 by ϵ. Thus, from (26-2) the capacitance of a parallel-plate capacitor filled with dielectric is $C_d = \kappa C = \kappa \epsilon_0 A/d = \epsilon A/d$.

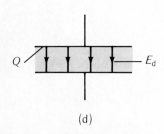

(d)

Figure 26-9. Charged capacitor maintained at constant potential difference (a) before and (b) after dielectric is introduced. Insulated charged capacitor (c) before and (d) after dielectric is introduced.

Dielectric materials serve essential functions in actual capacitors. They insulate the plates from one another, and they enhance the capacitance by a factor κ. An ordinary "paper capacitor" is formed by sandwiching thin paper between two metallic foils and rolling an essentially parallel-plate capacitor into cylindrical shape.

A capacitor can maintain its insulating properties only up to some maximum electric field $E_{\max}$, called the *dielectric strength* of the insulating material. Above this electric field, electric discharge, or arcing, occurs through the di-

---

* The term *permittivity* is used for ϵ and ϵ_0 because it gives a measure of the degree to which a dielectric "permits" an external electric field to pass through it. More on this in Section 26-6.

Table 26-1

MATERIAL	DIELECTRIC CONSTANT κ	DIELECTRIC STRENGTH E_{max} AT ROOM TEMP. (V/m)
Air (1 atm)	1.00059	30×10^6
Air (100 atm)	1.0548	——
Germanium	16	——
Mylar	3.1	——
Plexiglas	3.4	40×10^6
Pyrex	5.6	15×10^6
Quartz	3.8	8×10^6
Paraffined paper	2	40×10^6
Mica	5	200×10^6
Barium titanate	1200	300×10^6

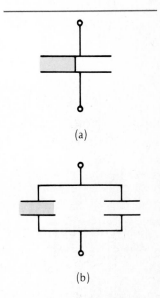

(a)

(b)

Figure 26-10. (a) Parallel-plate capacitor half-filled with dielectric; (b) the equivalent circuit of part (a).

electric. Therefore, a dielectric material is specified by the maximum electric field E_{max} (for example, 200 kV/mm for mica and 40 kV/mm for paraffined paper) as well as by its dielectric constant (see Table 26-1). The dielectric strength of the insulating material determines in turn the maximum potential difference that can be applied to the capacitor terminals without breaking down the dielectric material. Commercial capacitors are characterized by two numbers: capacitance in farads, and maximum allowable potential difference to avoid breakdown.

Example 26-3. A parallel-plate capacitor is half-filled with a dielectric, as shown in Figure 26-10(a). What is its capacitance?

The half-filled capacitor is equivalent to two capacitors in parallel, since they have a common potential difference. One capacitor is completely filled with dielectric and has a capacitance C_d; the other is empty and has a capacitance C_e, shown in Figure 26-10(b). The two capacitances are, from (26-2) and (26-8),

$$C_d = \frac{\kappa \epsilon_0 A}{2d} \quad \text{and} \quad C_e = \frac{\epsilon_0 A}{2d}$$

where A represents the plate area of the whole capacitor. Then, from (26-6),

$$C = C_d + C_e = \frac{\epsilon_0 A}{2d}(1 + \kappa)$$

26-4   Energy of a Charged Capacitor

To charge a capacitor, work is required. After the first small charge is placed on one plate, this charge repels other charges of like sign subsequently added. We want to find the total work U_e needed to charge a capacitor to a final potential difference V_f with a final charge Q on each plate.

The potential difference between the plates, always proportional to charge Q, grows from zero to V_f as the capacitor is being charged. On the average, the work per unit charge, or potential difference between the plates, is $\frac{1}{2}V_f$, so that the work done in putting a total charge Q on each plate is just

$$U_e = Q(\tfrac{1}{2}V_f) = \tfrac{1}{2}QV_f$$

The same result follows from summing the contributions to the work $dW = V\,dq$ done on charge dq as it goes across a potential difference V. We have

$$U_e = \int_0^Q V\,dq = \int_0^Q \frac{q}{C}\,dq = \frac{Q^2}{2C}$$

where U_e is the electric potential energy of the separated charges on the capacitor plates. From the definition, $C = Q/V_f$, the total energy can be written in three equivalent forms as

$$U_e = \frac{Q^2}{2C} = \frac{1}{2}\,CV_f^2 = \frac{1}{2}\,QV_f \tag{26-9}$$

Example 26-4.   The spherical conducting shell on a Van de Graaff electrostatic generator has a radius of 15 cm, and it is charged to a potential (relative to ground as zero) of 250 kV. (a) What is the capacitance of the spherical shell? (b) What is the energy of the charged conducting shell?

(a) We can regard an isolated spherical conductor as one plate of a capacitor in which the other plate is the earth, effectively an infinite distance away and at zero potential.

The potential of a spherical conductor of radius R with charge Q is

$$V = \frac{kQ}{R} \qquad \textit{See page 553}$$

so that its capacitance is

$$C = \frac{Q}{V} = \frac{R}{k}$$

The capacitance is proportional to the conductor's radius. For a 15-cm conducting shell, we have

$$C = \frac{R}{k} = \frac{15 \times 10^{-2}\ \text{m}}{9.0 \times 10^9\ \text{N}\cdot\text{m}^2/\text{C}^2} = 17 \times 10^{-12}\ \text{F} = 17\ \text{pF}$$

(b) The energy of the charged sphere is

$$U_e = \tfrac{1}{2}CV^2 = \tfrac{1}{2}(17 \times 10^{-12}\ \text{F})(250 \times 10^3\ \text{V})^2 = 0.53\ \text{J}$$

roughly equal to the kinetic energy of a 0.1-kg object after it falls from rest over a distance of 0.5 m.

Example 26-5.   A capacitor C_1 is charged initially with a potential difference V_i, as shown in Figure 26-11(a). Then this capacitor is connected to an initially uncharged capacitor C_2, as shown in Figure 26-11(b). What is (a) the final potential difference across each capacitor? (b) the ratio of the final to the initial potential energy of the capacitors?

(a) The charge Q_i initially on C_1 is

$$Q_i = C_1 V_i$$

After the switch has been closed, this charge is shared between the two capacitors

$$Q_i = Q_1 + Q_2$$

where Q_1 and Q_2 are respectively the final charges on C_1 and C_2. But $Q_1 = C_1 V_f$ and $Q_2 = C_2 V_f$, so that the relation can be written

$$C_1 V_i = C_1 V_f + C_2 V_f$$

and the final potential difference is

$$V_f = V_i\left(\frac{C_1}{C_1 + C_2}\right)$$

We see that V_f is *always less* than V_i.

(b) The energy U_i initially of C_1 is, from (26-9),

$$U_i = \tfrac{1}{2}C_1 V_i^2$$

and the energy U_f finally on the two capacitors is

$$U_f = \tfrac{1}{2}C_1 V_f^2 + \tfrac{1}{2}C_2 V_f^2$$

The ratio of final to initial energy is then

$$\frac{U_f}{U_i} = \frac{\tfrac{1}{2}(C_1 + C_2)V_f^2}{\tfrac{1}{2}C_1 V_i^2} = \frac{(C_1 + C_2)V_f^2}{C_1 V_i^2}$$

Using the result for part (a) to eliminate V_f/V_i, we can write the energy ratio as

$$\frac{U_f}{U_i} = \frac{C_1}{C_1 + C_2}$$

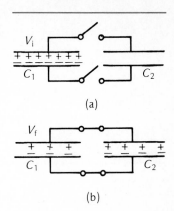

Figure 26-11.

For any capacitors, the final potential energy of the charged capacitors is less than the initial energy. Where does the energy go? It is dissipated mostly as thermal energy, and to a lesser degree as radiation, as the charges redistribute themselves over the conductors when the switch is closed, as will be discussed in later chapters.

26-5   Energy Density of the Electric Field

A charged capacitor has electric potential energy because work is required to assemble the charges at the plates. We can say the same thing differently: A charged capacitor has energy because an electric field has been created in the region between its plates. In this view we ascribe energy to the electric field itself. (To avoid double counting, we include either the electric potential energy of separated charges, or now its equivalent, energy in the electric field, but not both.)

Using (26-9) and (26-2) for a parallel-plate capacitor, we have for the energy

$$U_e = \frac{1}{2}CV_f^2 = \frac{1}{2}\left(\frac{\epsilon_0 A}{d}\right)(Ed)^2 = \frac{1}{2}\epsilon_0 E^2(Ad)$$

We neglect fringing and suppose that the uniform field **E** is confined entirely to the volume Ad between the plates. Then the electric energy per unit volume, the *electric energy density* u_e, is

$$u_e = \frac{U_e}{Ad} = \frac{1}{2}\epsilon_0 E^2 \tag{26-10}$$

for an electric field of magnitude E in a vacuum. Note that the energy density is proportional to the *square* of the electric field. Although derived here for the specific case of a parallel-plate capacitor (with negligible fringing), the result above gives the electric energy density for any electric field.

We now have two ways of describing the energy associated with electric charges:

- The potential energy associated with the relative positions of the charges.
- The electric field created by the charged particles.

Actually, the more fundamental view is to identify energy with the field itself. The reason is that we *can* have a *field without an electric charge* even though we cannot have a charge without an associated field. As we shall see, an electric field (and also a magnetic field) may become detached from the charge or charges that created it and propagate through empty space as an electromagnetic wave. Such unattached fields traveling at the speed of light have such physical attributes as energy, linear momentum, and angular momentum.

26-6   Electric Polarization and Microscopic Properties (Optional)

In this section we discuss qualitatively how the microscopic properties of dielectric materials are responsible for the increase in a capacitor's capacitance when a dielectric material is inserted between the plates. More broadly, we are concerned with the behavior of dielectric materials in an external electric field.

Recall first what happens when insulating material with dielectric constant κ is inserted into an isolated and initially charged capacitor (Section 26-4). (For simplicity, we suppose the capacitor to have parallel plates.) The capacitance is increased by the factor κ. Since charge Q is unchanged, the *potential difference* between the plates, $V_d = Q/C$, must *decrease* (by the factor κ). The uniform *electric field, $E_d = V_d/d$,* between plates separated by distance d must also *decrease* with the introduction of dielectric. Since the electric field between the charged plates is reduced, even though the free charge, $+Q$ and $-Q$, on each of the two plates is unchanged, we must conclude that the dielectric material between the charged plates itself produces an electric field that cancels at least partially the electric field between the plates.

How this occurs is shown in Figure 26-12. In Figure 26-12(a), there is no

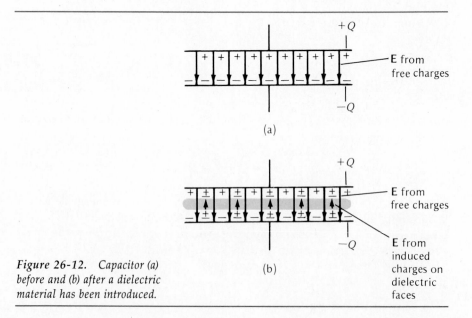

Figure 26-12.   Capacitor (a) before and (b) after a dielectric material has been introduced.

dielectric within the capacitor and the electric field, shown downward, arises solely from the free charges $+Q$ and $-Q$ on the capacitor plates. In Figure 26-12(b), a dielectric slab has been inserted between the plates of the capacitor, still with free charges $+Q$ and $-Q$. (For clarity, a gap is shown between each of the capacitor plates and the nearby face of the dielectric.)

Under the influence of the electric field between the capacitor plates, the dielectric has become *polarized*. That is to say, negative charges appear on its upper face and positive charges on its lower face. The charges *on the capacitor plates* are called *free charge*; they are able to move freely within a conductor. The charges *on the dielectric faces* are called *induced charges,* or *polarization charges;* they are induced by the external electric field and appear on the dielectric *surfaces* only. The magnitude of the induced charge on either dielectric face is less than Q. We see from Figure 26-12(a) that the electric field from the free charges is downward whereas the electric field (within the dielectric) from the induced charges is upward. The net effect: reduction of the electric-field magnitude between the capacitor plates.

The question now is how the microscopic properties of dielectric materials, under the influence of an external electric field, produce polarization charges on the dielectric faces.

First, we must distinguish between two general types of molecules, polar molecules and nonpolar molecules. A *polar* molecule has a *permanent electric dipole* moment (Section 24-6); although electrically neutral as a whole, its "centers" of positive and negative charge do not coincide. For example, in the polar molecule NaCl, the end with the sodium ion is positive while the end with the chlorine ion is negative. On the other hand, a *nonpolar* molecule has *no electric dipole* moment *in the absence of an external field.*

What happens when a nonpolar dielectric material is placed in an external electric field? The field polarizes the molecules, that is, it displaces the electrons (on the average) in the direction opposite to that of **E**, as shown in Figure 26-13. The external electric field induces electric dipole moments in the molecules, shifting the center of negative charge of electrons relative to the positive charge at the nucleus. This polarization is manifest, however, only at the surface of the

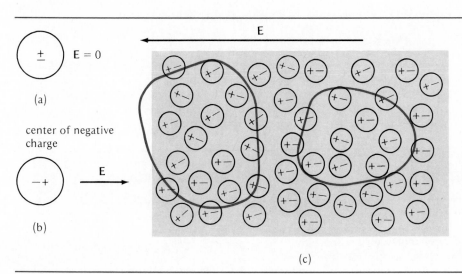

Figure 26-13. (a) In the absence of an external electric field, the centers of positive and negative charge in a molecule coincide. (b) An external electric field separates the centers of positive and negative charge. (c) A dielectric material polarized by an external electric field. A net charge appears at the surface of the dielectric material, but not within the material.

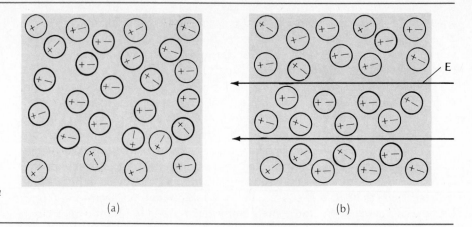

Figure 26-14. *(a) Polar molecules in the absence of an external field. (b) Polar molecules in the presence of an external field.*

(a)   (b)

dielectric. As the figure shows, for any small volume lying entirely *within* the dielectric material the positive and negative charges, although not coincident, appear in equal magnitudes. This does not happen however, when we choose a small volume that encloses one external surface of a dielectric. A net charge appears at the surface. The degree of polarization, measured by the amount of induced charge at the dielectric surface, is typically proportional to the magnitude of the field.

In the absence of an external electric field, a polar dielectric, one with molecules having a permanent electric dipole moment $\mathbf{p}$, shows no net polarization. Because of the thermal agitation within the material, the molecular dipoles are randomly oriented, as shown in Figure 26-14(a). When an external field $\mathbf{E}$ is applied, each dipole is subject to a torque $\tau = \mathbf{p} \times \mathbf{E}$ [(24-9)] that tends to align the dipole moments with the external field (Figure 26-14(b)). The polarizing influence of the external field competes with the depolarizing influence of thermal motion. As the temperature is lowered, the polarization of the dielectric increases for a given external field. In addition to aligning the permanent dipoles, the external field induces electric dipole moments in the molecules, just as for nonpolar dielectrics. Again, the polarization of the dielectric produces net charges only at the surface of the material.

Summary

Definitions

Capacitance:

$$C \equiv Q/V \qquad (26\text{-}1)$$

where Q is the charge magnitude on each of the two plates between which a potential difference V exists.

Dielectric constant:

$$\kappa \equiv C_d/C \qquad (26\text{-}7)$$

where C_d is the capacitance of a capacitor filled with dielectric material with dielectric constant κ, and C is the capacitance of the same capacitor in a vacuum.

Dielectric strength: the maximum electric field that can be applied to a dielectric material without electrically breaking down the material.

Units

Farad, the unit for capacitance:

$$1\,\text{F} \equiv 1\,\text{C/V}$$

Fundamental Principles

For any circuit in electrostatic equilibrium,

● From energy conservation: the net change in electric potential around any circuit is zero.

$$\Sigma V = 0 \qquad (26\text{-}4)$$

- From charge conservation: the net charge on any isolated circuit element is constant.

Important Results

Parallel-plate capacitor — plate area A, plate separation d:

$$C = \frac{\epsilon_0 A}{d} \qquad (26\text{-}2)$$

Coaxial cylindrical capacitor — inner radius r_1, outer radius r_2, length L:

$$\frac{C}{L} = \frac{2\pi\epsilon_0}{\ln(r_2/r_1)} \qquad (26\text{-}3)$$

Equivalent capacitance C for capacitors (C_i) in *series* (same Q):

$$\frac{1}{C} = \Sigma \frac{1}{C_i} \qquad (26\text{-}5)$$

Equivalent capacitance C for capacitors (C_i) in *parallel* (same V):

$$C = \Sigma C_i \qquad (26\text{-}6)$$

Energy U_e of a capacitor charged with final charge Q and final potential difference V_f:

$$U_e = \frac{Q^2}{2C} = \frac{1}{2} CV_f^2 = \frac{1}{2} QV_f \qquad (26\text{-}9)$$

Energy density (energy per unit volume) u_e of electric field E:

$$u = \tfrac{1}{2}\epsilon_0 E^2 \qquad (26\text{-}10)$$

Problems and Questions

Section 26-1   Capacitance Defined

· **26-1 Q** One farad is the same as one
(A) ohm/meter.
(B) coulomb/volt.
(C) volt/coulomb.
(D) joule/coulomb.
(E) ampere·second.

· **26-2 P** (*a*) What is the capacitance of an air-filled parallel-plate capacitor of plate area 30 cm × 30 cm (about 1 ft^2) and plate separation 1 mm? (*b*) Air breaks down at about 3×10^6 V/m. What is the maximum voltage that could be applied to such a capacitor before breakdown will occur?

· **26-3 Q** Show that ϵ_0 has units of farads per meter.

: **26-4 Q** The capacitance between two conductors depends on
(A) the charge on them.
(B) the potential difference between them.
(C) the electric field between them.
(D) the energy stored between them.
(E) none of the above.

: **26-5 Q** Two large parallel copper plates carry charges $+Q$ and $-Q$ respectively. They are separated by a small distance of 10 mm. They differ in potential by 12 V. If the plates are moved together slowly to a spacing of 6 mm, the potential difference between them will
(A) increase.
(B) decrease.
(C) remain unchanged.
(D) increase or decrease according to what is chosen as the "zero" of potential.

: **26-6 P** What is the capacitance of an isolated conducting sphere of radius R? (*Hint:* If some charge were on the sphere, electric-field lines would be directed out radially. They would end on the other "plate" of this capacitor. Where is the other plate?)

: **26-7 P** The first precision measurement of the charge of an electron was made by Robert A. Millikan in 1909 in his famous *oil-drop experiment.* In this experiment an oil drop having an excess or deficiency of no more than a few electrons is in the uniform vertical electric field of a parallel-plate capacitor. An oil drop of charge q, radius r, and velocity v is subject to three forces in the vertical direction: (1) its weight, (2) the electric force qV/d, where V is the potential difference between plates separated by d, and (3) a resistive force, $F_r = -6\pi\eta rv$, where η is the viscosity of the medium (for air, $\eta = 1.82 \times 10^{-5}$ N·s/m^2).

In its simplest form the experiment can be done as follows: (a) the charged drop is brought to rest by balancing its weight against an upward electric force; (b) the drop is allowed to fall at a constant measured velocity with the electric field turned off, so that the weight is now balanced against an upward resistance force. Part (b) yields the drop's radius and therefore its weight; part (a) then allows q to be computed.

Suppose that some particular oil drop falls at constant speed over a vertical distance of 1.00 mm (as viewed through a microscope) in a time interval of 50 s. (*a*) What is the radius of the drop? (*b*) Suppose that this drop had a net charge of $-2e$. What electric potential difference would have to be applied across the capacitor plates, separated by 1.75 mm, in order to bring the drop to rest? (Oil density, 850 kg/m^3)

: **26-8 P** Show that the expression for the capacitance of a coaxial capacitor approaches that for a parallel-plate capacitor as $r_1 \to r_2$. Use the approximation $\ln(1 + x) \approx x$ if $x \ll 1$, and write $r_2 = r_1 + \delta$, where $\delta \ll r_1$.

⋮ 26-9 Q Is it possible for a single conductor to have capacitance? Which of the following statements best answers and explains this question?

(A) No. A single conductor can have no capacitance, because although charge can be placed on it, there is no second conductor against which to measure a potential difference.

(B) No. A single conductor can have no capacitance because the unit of capacitance, the farad, is specifically defined for the case of two capacitors.

(C) No. A single conductor can have no capacitance, because a conductor is electrically neutral and can carry no net charge.

(D) Yes. A single conductor can have capacitance, because one part of the conductor can be considered one plate and another part the second plate.

(E) Yes. A single conductor can have capacitance, since capacitance is only a measure of the ability to store charge and is unrelated to other factors such as potential difference or the proximity of other conductors.

(F) Yes. A single conductor can have capacitance. The "second plate" of the capacitor may be thought of as the equipotential surface at infinity.

⋮ 26-10 P Consider two isolated conductors of arbitrary shape, one carrying charge $+Q$ and the other $-Q$. In general, the electric field between them might be complicated, as suggested in Figure 26-15. Suppose that along one of the field lines, say the one labeled XX', the electric field intensity varies according to

$$E = E_0\left(1 - \frac{x}{2L}\right)Q$$

This is the electric field at a point P a distance x measured along the line XX', and L is the length of this line. The quantity E_0 is a constant, which depends on the shape and separation of the two conductors. What is the capacitance of these two conductors?

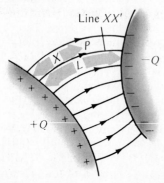

Figure 26-15. Problem 26-10.

Section 26-2 Capacitor Circuits

· 26-11 P Three capacitors (1-pF, 2-pF, and 4-pF) are to be connected to a 120-V power supply. Determine the potential difference across each and the charge on each when (a) all three are connected in series; (b) all three are connected in parallel.

· 26-12 P Determine the equivalent capacitance of the combinations of capacitors shown in Figure 26-16.

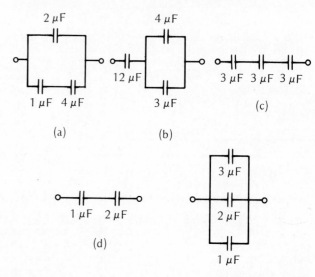

Figure 26-16. Problem 26-12.

⋮ 26-13 P A capacitor of known capacitance C_1 is charged to a potential difference V_1. It is then connected across an uncharged capacitor of unknown capacitance C_2. The final potential difference across the two capacitors is V_2. Show that the capacitance of the unknown capacitor is

$$C_2 = \left(\frac{V_1 - V_2}{V_2}\right)C_1$$

⋮ 26-14 Q A sheet of aluminum foil is placed between the plates of a parallel-plate capacitor as shown in Figure 26-17. How will this affect the capacitance?

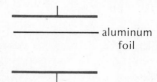

Figure 26-17. Question 26-14.

(A) C will increase.
(B) C will decrease.
(C) C will not change.

Suppose that the aluminum foil is then connected to one of the plates with a wire. How will this affect the capacitance?

(D) C will increase.
(E) C will decrease.
(F) C will not change.

: **26-15 P** An air-filled variable capacitor shown in Figure 26-18, of the type used to tune a radio, is made by varying the overlapping area between two sets of plates through rotation of one set of plates (attached to the knob you turn). Such a capacitor consists of n adjacent plates, half of which are attached to one pole and half to the other, with a gap d between adjacent plates. Show that the capacitance is $C = (n - 1)\epsilon_0 A/d$, where A is the overlapping area of a plate.

Figure 26-18.   Problem 26-15.

: **26-16 P** A metal slab of thickness t and area A is inserted into a parallel-plate capacitor of area A and separation d, as shown in Figure 26-19. (a) What is the new capacitance between the plates? (b) Does the result in (a) depend on where the slab is placed? Explain.

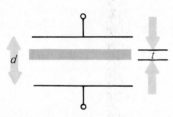

Figure 26-19.   Problem 26-16.

: **26-17 Q** Suppose that a capacitor is connected to a battery. Is it true that each plate will receive a charge of the same magnitude? If so, why? If not, why not?
(A) Each plate will receive equal amounts of charge only if they are of the same area.
(B) A capacitor is designed so that all the electric-field lines emanating from one plate terminate on the other plate. From Gauss's law, we then know that there is an equal amount of charge on each plate.
(C) No, each plate will not receive an equal amount of charge, whether or not they have equal areas. The charge on a plate depends on the shape of the plate as well as on its area.
(D) Yes, each plate receives the same charge because each is at the same potential.

: **26-18 P** (a) Determine the capacitance of two concentric spherical conductors of radius R_1 and R_2. (b) Show that if $R_2 = R_1 + \delta$, the result from (a) reduces to the formula for the capacitance of a parallel-plate capacitor, where $\delta \ll R_1$.

: **26-19 P** Consider the circuit shown in Figure 26-20 (a) What is the charge on each capacitor and the potential difference across each? (b) Suppose that the 4-μF capacitor breaks down electrically and becomes a conductor. What then would be the potential differences across each capacitor and the charge on each?

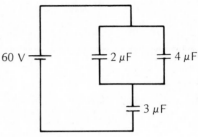

Figure 26-20.   Problem 26-19.

: **26-20 P** Two capacitors (2-μF and 4-μF) are each charged to a potential difference of 100 V and then joined in parallel, with plates of opposite polarity connected. What is the final potential difference across each capacitor, and how much charge is on each?

: **26-21 P** The capacitors shown in Figure 26-21 are initially uncharged. The switch is first thrown to position 1, then to position 2. What is the final charge on each of the three capacitors?

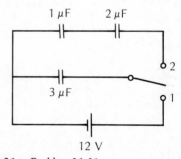

Figure 26-21.   Problem 26-21.

: **26-22 P** What is the equivalent capacitance of the five identical capacitors shown in Figure 26-22? (*Hint:* From symmetry, what is the charge on the middle capacitor?)

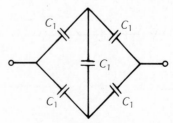

Figure 26-22.   Problem 26-22.

: **26-23 P** How many different values of capacitance can

you get by connecting three identical capacitors in various ways? What are the values in terms of C_0, the value of an individual capacitance?

: **26-24 P** Consider two spheres of radius R_1 and R_2 whose centers are separated by a large distance d, where $d \gg R_1, R_2$. What is the capacitance of this system? (*Hint:* Imagine charge $+Q$ on one and $-Q$ on the other. Write down the potential of each due to both charges. The difference in potential will be found to be proportional to Q; and from the constant of proportionality, one can deduce C.)

: **26-25 P** Three capacitors $C_1, C_2,$ and C_3 are given charges $Q_1, Q_2,$ and Q_3. They are then connected in series, with plates of opposite polarity being joined. What is the final charge on each capacitor?

Section 26-3   Dielectric Constant

· **26-26 P** What is the capacitance per unit length of a coaxial cable filled with polyethylene (dielectric constant 2.3) if the inner-conductor radius is 2 mm and the outer-conductor radius is 7 mm?

: **26-27 P** One-fourth of the area of a parallel-plate capacitor is filled with material with dielectric constant κ that extends from one plate to the other. What is the capacitance of this capacitor if its capacitance with no dielectric is C_0?

: **26-28 Q** Consider the three arrangements of capacitors shown in Figure 26-23. When a dielectric slab is inserted into one capacitor, the capacitance between the terminals X and Y will increase

(A) only in case A.

(B) only in case B.

(C) only in case C.

(D) in all three cases, independent of whether or not X and Y are connected to a battery.

(E) in all three cases, provided X and Y are connected to a battery.

(F) in all three cases, provided X and Y are not connected to a battery.

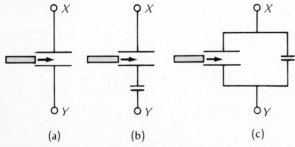

(a)   (b)   (c)

Figure 26-23.   Question 26-28.

: **26-29 Q** An air-filled parallel-plate capacitor is charged by a 12-V battery. The capacitor is then disconnected from

the battery. What will happen if a dielectric slab is then slid between the plates of the capacitor?

(A) The amount of free charge on the metal plates of the capacitor will change.

(B) The potential difference between the plates will decrease.

(C) The electric field in the region between the plates will increase.

(D) The capacitance of the capacitor will be unchanged.

(E) The potential difference between the plates and the electric field in the region between the plates will both be unchanged.

: **26-30 Q** Suppose that while a capacitor is connected to a battery a dielectric slab is placed between the plates.

(A) The capacitance will not be affected, since the battery holds the potential difference fixed.

(B) More charge will flow on to the plates from the battery.

(C) Charge will flow from the capacitor plates back into the battery.

(D) The effective capacitance will be reduced.

(E) The potential difference between the plates will increase because of induced charges on the surface of the dielectric.

: **26-31 Q** An air-filled parallel-plate capacitor is charged by a battery and then disconnected from the battery. A dielectric slab is then placed between the plates. (*a*) Will the charge on the plates increase, decrease, or remain unchanged? (*b*) Will the potential difference between the plates increase, decrease, or remain unchanged? (*c*) Will the capacitance increase, decrease, or remain unchanged? (*d*), (*e*), (*f*) Answer the above questions for the case in which the capacitor remains connected to the battery.

: **26-32 P** (*a*) The earliest capacitors were called Leyden jars. They were glass jars coated inside and outside with metal. To estimate the capacitance of the jars used, calculate the capacitance of a cylinder that is 30 cm tall and has a diameter of 20 cm; its ends are flat, and the whole is made of glass that is 3 mm thick. The dielectric constant of glass is about 5. (*b*) If the dielectric strength of glass is 15×10^6 V/m, what is the maximum potential difference that can be maintained across the glass? (*c*) What is the maximum charge that can be stored in this capacitor?

: **26-33 Q** You are designing a capacitor to be used in an experiment for which a particular value C_0 of capacitance is required. You decide to use a pair of parallel metal plates immersed in an oil bath. Unfortunately, you find that on your first test run, the plates are oversized and will not fit in the container with the oil. What should you do to redesign the capacitor?

(A) Don't immerse the plates in oil; just use them air-filled with the same spacing.

(B) Increase the area of the plates and move them closer together.

(C) Increase the area of the plates and move them farther apart.

(D) Decrease the area of the plates and move them farther apart.

(E) Decrease the area of the plates and move them closer together.

⋮ **26-34 P** Suppose that you want to design a 1-μF oil-filled capacitor to be used at 5000 V. Oil with a dielectric constant of 5.0 and dielectric strength of 15×10^6 V/m is to be used. As a safety factor, the electric field is to be limited to 80 percent of the breakdown value. What is the minimum amount of oil required, expressed in liters?

⋮ **26-35 P** A parallel-plate capacitor is charged by a battery and then disconnected. A slab of dielectric is then inserted between the plates, part way in, as shown in Figure 26-24. The plates are horizontal, and friction is negligible. What will happen when the slab is released?

(A) It will shoot straight through the plates and fly out on the other side.

(B) It will be pushed out of the capacitor.

(C) It will be drawn into the capacitor and oscillate back and forth horizontally.

(D) It will be drawn into the capacitor and come to rest with its midpoint at the midpoint of the plates.

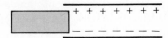

Figure 26-24.   Problem 26-35.

⋮ **26-36 P** A parallel-plate capacitor is filled with equal volumes of materials with dielectric constants κ_1 and κ_2. What is the capacitance if the plate area is A and the plate separation is d, for the two arrangements shown in Figure 26-25?

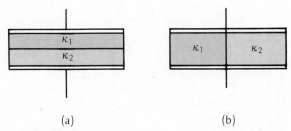

(a)   (b)

Figure 26-25.   Problem 26-36.

⋮ **26-37 P** An air-filled coaxial capacitor has an inner radius of 1 cm and an outer radius of 2 cm. What is the minimum volume (per unit length of the capacitor) of material of dielectric constant 3 that can be added to the space between the plates to double the capacitance? Sketch where the dielectric material should be placed. (*Hint:* You can do this if you think about how a dielectric works. It is most effective if placed in the strongest electric field.)

⋮ **26-38 Q** Which of the following is the best explanation of how (or why) a dielectric increases the capacitance of a capacitor?

(A) The electric field acting on the dielectric sets up elastic stresses. This means that more energy is stored in the capacitor as elastic potential energy, and hence the capacitance is increased (since $U = \frac{1}{2}CV^2$).

(B) Free charge can flow from the dielectric on to the plates of the capacitor. This means that the battery must provide less charge; hence it does less work, and the capacitance is increased.

(C) Charge displacement in the dielectric tends to reinforce the electric field due to the charge on the plates, thereby increasing the potential difference across the capacitor and increasing the capacitance.

(D) Induced charge on the surface of the dielectric sets up electric fields that tend to cancel the field due to charge on the plates. This means that more charge must be placed on the plates to cause a given potential difference, and more coulombs per volt means more capacitance.

(E) None of the above is correct. A dielectric *decreases* capacitance.

Section 26-4   Energy of a Charged Capacitor

· **26-39 P** Suppose that one square centimeter of a muscle-cell membrane has a capacitance of 8×10^{-7} F and that a potential difference of 70 mV exists across it (these are typical values). How much energy is stored in one square centimeter of membrane?

26-40 P What is the ratio of the stored energy in capacitors C_1 and C_2 when they are connected (*a*) in parallel? (*b*) in series? See Figure 26-26.

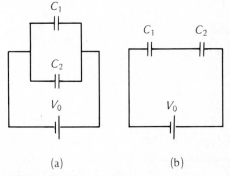

(a)   (b)

Figure 26-26.   Problem 26-40.

⋮ **26-41 P** While the capacitors shown in Figure 26-27 are connected to a 12-V battery, a slab of dielectric constant 3 is inserted into the 4-μF capacitor. Determine for each capacitor the (*a*) potential difference, (*b*) charge, and (*c*) stored energy before and after the slab is inserted. The slab completely fills the capacitor.

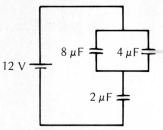

Figure 26-27.   *Problem 26-41.*

: 26-42 P (*a*) How should three capacitors C_1, C_2, and C_3 be connected to provide the maximum energy storage when connected to a single power supply? The capacitor C_1 is 1-μF rated at 500 V, C_2 is 4-μF rated at 100 V, and C_3 is 2-μF rated at 300 V. (*b*) What is the maximum stored energy?

: 26-43 P The potential drop across the plates of a capacitor has decreased to one-half of its initial value. By what factor has the energy stored in the capacitor been reduced?

: 26-44 P Show that a plate of a parallel-plate capacitor experiences a force $F = q^2/(2\epsilon_0 A)$ due to the electric field between the plates. (*Hint:* Consider the work done and the gain in energy when the separation of the plates is increased by a small amount.)

: 26-45 P Integrate the energy density $\frac{1}{2}\epsilon_0 E^2$ over the volume of a cylindrical capacitor, to show that the energy stored in the capacitor is $Q^2/2C$.

: 26-46 Q Two identical capacitors are connected in series to a battery. While they are connected to the battery, a dielectric slab is inserted into the lower capacitor. What happens—increases, decreases, or remains the same—to each of the following when this is done?

   (*a*) Charge on upper capacitor. (*b*) Charge on lower capacitor. (*c*) Potential difference across upper capacitor. (*d*) Potential difference across lower capacitor. (*e*) Total energy stored in the two capacitors. (*f*) Total capacitance.

: 26-47 P An air-filled parallel plate capacitor in Figure 26-28 has capacitance C_0. It is charged to a potential difference V_0. A slab of dielectric constant κ is inserted a distance x into the capacitor. As a function of x, what is (*a*) the voltage across the plates? (*b*) the energy stored in the capacitor?

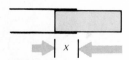

Figure 26-28.   *Problem 26-47.*

: 26-48 P An air-filled parallel-plate capacitor has capacitance C_0. It is charged to a potential difference V_0 and disconnected from the battery. A dielectric slab (which just fills the capacitor) is released from rest at the edge of the capacitor and allowed to be drawn into the capacitor. What is the maximum velocity the slab can achieve if its mass is m?

: 26-49 P Einstein showed that energy and mass are equivalent and related by the expression $E = mc^2$. It was once speculated that the rest mass of an electron might be just the energy associated with the electric field created by the electron. What is the radius of a conducting sphere with charge e whose total electric field energy is just mc^2, where m is the mass of the electron? (As far as we know the electron acts essentially like a point object, and its radius, if it has one, is less than 10^{-18} m.)

Section 26-5   Energy Density of the Electric Field

· 26-50 P The energy content of 1 liter of gasoline is about 4.4×10^7 J. What would be the volume of a mica-filled capacitor that would store this much energy?

: 26-51 P Calculate the energy stored in the electric field of charge Q when the charge is distributed (*a*) uniformly throughout the volume of a sphere of radius R; (*b*) uniformly over the surface of a sphere of radius R. (*c*) Compute the ratio of the result of (a) to the result of (b) and explain qualitatively the reason for such a result.

: 26-52 P A spherical conductor of radius R carries a charge Q. (*a*) What is the radius of an imaginary concentric sphere that would enclose half the stored energy in the electric field surrounding the charged sphere? (*b*) At what radius does the potential due to the sphere drop to half its value at the surface of the sphere. Compare this result with that found in (*a*) and comment on the relation.

: 26-53 P A parallel-plate capacitor of area A and separation d contains a dielectric slab of dielectric constant κ, area A, and thickness αd, where $\alpha < 1$. A potential difference V_0 is established between the plates of the capacitor. Determine (*a*) the total energy stored in the capacitor and the fraction of the energy stored in the air gap and in the dielectric; (*b*) the potential difference across the air gap and across the dielectric. (*c*) Evaluate each of the above for the case $\alpha = \frac{1}{2}$, $\kappa = 2$, $V_0 = 100$ V, $A\epsilon_0/d = 2 \times 10^{-6}$ F.

: 26-54 P Consider a spherical conductor of surface charge density σ. Since the charges try to repel each other, they give rise to an outward "electric pressure." (*a*) Show that the electric pressure is $P = \frac{1}{2}\epsilon_0 E^2 =$ electric field energy density. Consider first the work that would have to be done to compress the radius of the sphere by a small amount δr. Then equate this to the increase in the energy stored in the electric field (because then there would be slightly more space around the smaller sphere than initially). (*b*) Would it be possible to keep a conducting balloon spherical with no air inside just by charging it? Remember that the surrounding air will break down if E exceeds 3×10^6 V/m. What pressure in atmospheres could one create in this way?

Supplementary Problems

26-55 P A useful model of a nerve fiber (an axon) treats it as a cylindrical coaxial capacitor filled with myelin, which has a dielectric constant of 7. For a frog axon, typical values of the inner and outer radii are 5×10^{-6} m and 7×10^{-6} m. A potential difference of approximately 70 mV exists across the dielectric layer in the resting state. (*a*) What is the capacitance per meter of such an axon? (*b*) What is the surface-charge density on the outer surface? (*c*) Assuming that a molecule occupies about 25×10^{-20} m², determine how many electrons per molecule the surface-charge density found in (*b*) corresponds to.

26-56 P Energy storage is very important. The "energy density" measured in joules per kilogram is an important parameter, for example, in the design of an electric car. How to store the electrical energy needed by such a car is the main problem in the development. For example, whereas gasoline provides about 4.4×10^7 J/kg, a lead-acid battery stores only about 1.8×10^5 J/kg. Could one do better with a capacitor than with a battery? Calculate the energy density (in joules per kilogram) for a mica-filled capacitor charged almost to breakdown. The density of mica is about 2900 mg/m³. Assume the plates contribute negligible mass and volume to the capacitor.

26-57 P Capacitors play an essential role in microelectronics technology. An individual capacitor can act as a storage cell for one bit of information (charged = yes = 1; uncharged = no = 0). In a silicon semiconductor memory chip, for example, a pair of conducting plates (combined with a transistor) is a capacitor storage cell. One large-scale integrated memory chip with RAM (random access memory) has 16,384 bits spread in a single layer over an area less than 0.5 cm on an edge. (*a*) Make reasonable assumptions to show that the size of one capacitor plate is of the order of 20 microns (1 micron = 1 μm = 10^{-6} m). (Recently developed storage cells are as small as about 1 micron). (*b*) Using 12 as the dielectric constant for the silicon dioxide insulator between the plates, show that the capacitance of the storage cell is of the order of 50 fF (where f = femto = 10^{-15}). (*c*) If the storage cell is fully charged with a potential difference of 10 V, what is the net charge on one plate in multiples of the charge of an electron? (*d*) Assuming that all electrons on the negatively charged plate are spread uniformly over the plate, what is the average distance between adjacent electrons? (*e*) What is the energy required to charge all 16,384 storage cells in the chip? [For more on "Microelectronics Memories" see the article with that title by David A. Hodges in *Scientific American* **237**, 130 (September, 1977).]

27

Electric Current and Resistance

Thus far we have studied electrostatics — electric charges at rest. Now we turn to electric currents — electric charges in motion. Electric resistance and resistivity are defined. A circuit consisting of a capacitor and resistor is discussed. Finally, we give a brief and qualitative account of the microscopic properties influencing a material's resistivity.

27-1   Electric Current

Electric charges in motion constitute an *electric current*. By definition, the current i through a chosen surface is the total *net* charge dQ passing through that surface divided by the elapsed time dt:

$$i = \frac{dQ}{dt} \tag{27-1}$$

Electric current is always given in terms of charges passing through a surface bounded by a loop; for an ordinary conducting wire the loop would ordinarily be the wire's circumference.

The direction of the so-called *conventional current* is taken to be that in which positive charges move. Thus, protons (charge, $+e$) traveling to the right constitute a current to the right, while electrons (charge, $-e$) traveling to the right constitute a current to the left. To obtain the total current i through a loop, we must always count the *net* charge passing through per unit time. For example, if positive ions move to the right through a surface and at the same time negative ions move to the left through the same surface, as might happen for currents in liquids, *both* types of ions contribute to an electric current to the *right*; see Figure 27-1.

The SI unit of current is the *ampere* (abbreviated A). The unit is named after André M. Ampère (1775–1836), who made significant contributions to understanding the relation between an electric current and its associated magnetic field. One ampere corresponds to a net charge of 1 C passing through a chosen surface in the time of 1 s:*

$$1A \equiv 1 \ C/s$$

Whereas the coulomb is an uncommonly large amount of charge by laboratory standards, laboratory currents of a few amperes are typical. Related units are the milliampere (1 mA $= 10^{-3}$ A) and the microampere (1 μA $= 10^{-6}$A).

A direct current (dc) is one in which net positive charges move in one direction only (but *not* necessarily at a constant rate). For an alternating current (ac), on the other hand, the current changes direction periodically.

How is the current i related to the properties of the charge carriers? Suppose that a current consists of n identical charge carriers per unit volume drifting at a speed v_d along a single direction. See Figure 27-2. Each carrier has a positive charge q. In the time dt a charge carrier advances a distance $v_d \ dt$, and all the charge carriers within a cylindrical volume of length $v_d \ dt$ and cross-sectional area A will have passed through one end of the cylinder. By the same token, an equal number of charge carriers will have entered the cylindrical column through the other end. The total charge dQ passing through the area A in time dt is the charge per unit volume nq multiplied by the cylinder's volume $Av_d \ dt$:

$$dQ = nqAv_d \ dt$$

Consequently, the current i contributed by these charge carriers is

$$i = \frac{dQ}{dt} = nqAv_d \qquad (27\text{-}2)$$

The charge carriers need not move at a *constant* velocity. They may have a very complicated motion. The free electrons in an ordinary conductor are in random thermal motion at speeds of the order of 10^6 m/s. The drift speed v_d, on the other hand, is the average displacement per unit time along the direction of the current. As we shall see the drift speed is typically *very* small compared with the thermal speed.

The most common device used for measuring current is an *ammeter*; its operation depends on the torque produced by a magnet on a current-carrying coil of wire (Section 29-6).

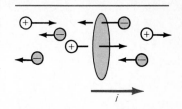

Figure 27-1. *Both the positive charges to the right and the negative charges to the left through the loop contribute to a conventional current i to the right.*

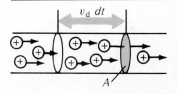

Figure 27-2. *Charges drifting at speed v_d through a cylinder of cross-sectional area A in the time dt.*

---

* Strictly, the coulomb is defined as one ampere-second and the ampere itself, one of the base units in the SI system, is defined in terms of the magnetic interaction between a pair of current-carrying conductors. See Appendix A and Section 30-2.

Example 27-1.   A current of 1.0 A exists in a conducting copper wire whose cross-sectional area is 1.0 mm². What is the average drift speed of the conduction electrons under these conditions?

In copper, as in other typical metallic conductors, there is approximately one free (conduction) electron per atom. We can find the density n of charge carriers by computing the number of atoms per unit volume from Avogadro's number N_A (the number of atoms per mole), the atomic mass m (the number of grams per mole), and the mass density ρ_m. Clearly,

$$n = \frac{\text{charge carriers}}{\text{volume}} = \left(\frac{1 \text{ charge carrier}}{\text{atom}}\right)\left(\frac{\text{atoms}}{\text{volume}}\right)$$

$$= \left(\frac{1 \text{ charge carrier}}{\text{atom}}\right)\left(\frac{\text{atoms}}{\text{mole}}\right)\left(\frac{\text{moles}}{\text{kilogram}}\right)\left(\frac{\text{kilograms}}{\text{volume}}\right)$$

or

$$n = (1)\,(N_A)\left(\frac{1}{m}\right)(\rho_m)$$

$$= \left(\frac{1 \text{ ch car}}{\text{atom}}\right)(6.0 \times 10^{23} \text{ atoms/mol})\left(\frac{1}{64 \times 10^{-3} \text{ kg/mol}}\right)(9.0 \times 10^3 \text{ kg/m}^3)$$

$$= 8.4 \times 10^{28} \text{ charge carriers/m}^3$$

Then from (27-2), we have

$$v_d = \frac{i}{nqA}$$

$$= \frac{1.0 \text{ A}}{(8.4 \times 10^{28} \text{ m}^{-3})(1.6 \times 10^{-19} \text{ C})(1.0 \times 10^{-6} \text{ m}^2)}$$

$$= 7.4 \times 10^{-5} \text{ m/s} = 0.074 \text{ mm/s}$$

The free electrons drift through the conductor with an average speed that is actually less than 0.1 mm/s, a remarkably low speed. It should *not* be inferred, however, that when such a current is established at one end of a conducting copper wire it takes almost 10 s for the signal to travel a mere 1 mm. The speed at which the electric field driving the free electrons is established is close to the speed of light. One must distinguish here between the speed with which the charged particles drift and the speed at which the signal is propagated, just as one must distinguish between the speed (possibly very low) at which a liquid drifts through a pipe and the much higher speed at which a change in pressure is propagated along the pipe. The *drift* speed of conduction electrons is much less than the *random* thermal speeds of electrons at any finite temperature. (We shall explore this point further in Section 27-5).

Equation (27-2) can be written in an alternative form by introducing the *current-density vector* **j**, whose magnitude is

$$j = \frac{i}{A} = nqv_d \tag{27-3}$$

and whose direction is determined by the direction of the velocity vector $\mathbf{v}_d$ of the charge carriers:

$$\mathbf{j} = nq\mathbf{v}_d \tag{27-4}$$

with **j** parallel to $\mathbf{v}_d$.

To see the utility of the current-density vector, consider Figure 27-3. Here vector *d***S** is a small element of a surface through which charges flow. We find

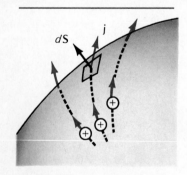

Figure 27-3.   Current-density vector **j** *and an element of surface d***S**.

the electric-charge flux, or electric current di through $d\mathbf{S}$, by multiplying the component of $\mathbf{j}$ along the normal to the surface by the area $d\mathbf{S}$; that is,

$$di = \mathbf{j} \cdot d\mathbf{S}$$

The total current i through the surface S is

$$i = \int_S \mathbf{j} \cdot d\mathbf{S} \qquad (27\text{-}5)$$

When $\mathbf{j}$ and $d\mathbf{S}$ are parallel and the surface is flat, (27-5) reduces to (27-3).

27-2   Current and Energy Conservation

Here we set down the fundamental relation for the rate at which electric energy is transferred to any device through which charges flow. Our arrangement is altogether general; some energy source—for example, a battery, a solar cell, or a gasoline generator—maintains a potential difference $V_{ab} = V_a - V_b$ across the terminals of what we call simply *the load* and through which current i passes. See Figure 27-4. The load might be an electric motor, a lamp, or a heater. All that matters is that current i goes through the load while potential difference V_{ab} is maintained across its terminals. The input potential V_a is higher than the output potential V_b, so that conventional current is *in* at a and *out* at b. (Electrons would go "uphill" in potential, from b to a.)

When positive charge dQ goes through the load, the work dW done on it by electric forces associated with potential difference V_{ab} is

$$dW = (dQ)V_{ab}$$

Since $i = dQ/dt$,

$$dW = iV_{ab}\, dt$$

and the rate of doing work on the load dW/dt, or delivering power P to it, is

$$P = \frac{dW}{dt} = iV_{ab} \qquad (27\text{-}6)$$

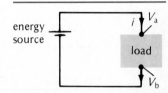

Figure 27-4.   An energy source delivering energy to a load.

The instantaneous power is just the product of i and V_{ab}. Note that power P has units of watts: $(A)(V) = (C/s)(J/C) = J/s = W$.

If i and V_{ab} are constant, so that P is also constant with time, the total work W done on the charges transported through the load over a time interval t is

$$W = iV_{ab}t$$

If i and V vary with time, the work is

$$W = \int P\, dt = \int iV\, dt$$

27-3   Resistance and Ohm's Law

Figure 27-5 shows voltage-current plots for three dissipative current elements: (a) a gas in a closed container with two electrodes, (b) a semiconducting device, and (c) a homogeneous solid conductor. Only for the conductor is the V-i

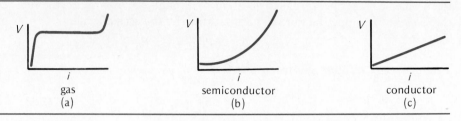

Figure 27-5. *Voltage-current characteristics for (a) a gas, (b) a semiconducting device, and (c) a conductor obeying Ohm's law.*

relation simple: a straight line. The ratio of V to i for any purely dissipative circuit element is *defined* to be the device's *electrical resistance R*

$$R \equiv V/i \qquad (27\text{-}7)$$

The experimental finding that R, so defined, is a *constant* for a given pair of connections to a conductor (maintained at constant temperature), so that R is independent of the magnitude of either i or V, is known as *Ohm's law:*

$$\text{Ohm's law:} \quad R = \text{constant for metallic conductors} \qquad (27\text{-}8)$$

Georg S. Ohm (1787–1854) found that R was a constant for a large variety of materials and for an enormous range of currents and potential differences (a factor of 10^{10} for some materials).

The SI unit for resistance is the ohm (abbreviated by Greek capital omega, Ω), and it is defined by

$$1\Omega \equiv 1 \text{ V/A}$$

Related units are the megohm (1 MΩ = 10^6 Ω) and the microhm (1 $\mu\Omega$ = 10^{-6} Ω). The reciprocal of resistance is called electrical *conductance*. The official SI unit for conductance is the *siemens* (abbreviated S):

$$1 \text{ S} \equiv 1 \text{ A/V} = 1/\Omega = 1 \text{ mho}$$

The siemens unit is seldom used in practice; the common but unofficial unit for conductance is the reciprocal ohm, or mho.

In circuit diagrams, a resistor is depicted by the symbol —ww—; and a variable resistor, or rheostat, is represented by —ww— or —ww— . A dissipative circuit element whose resistance is not independent of V (and therefore also not independent of i) is known as a nonohmic or nonlinear resistor (its V-i plot is not a straight line).

The power delivered to any circuit element is iV. When that circuit element is a resistor the power dissipated can also be written

$$P = iV = i^2 R = \frac{V^2}{R} \qquad (27\text{-}9)$$

That the thermal energy dissipated per unit-time in a resistor (the so-called i^2R loss) is, in fact, exactly equal to the electric energy delivered to it was first established in the nineteenth century by the historic experiments of James Joule. The effect is sometimes referred to as Joule heating.

Example 27-2. A transistor radio operates at 10 mW with a 9-V dry cell. (a) What is the current into and out of the radio? (b) What is the radio's electric resistance?

$$(a) \quad i = \frac{P}{V} = \frac{10 \times 10^{-3} \text{ W}}{9.0 \text{ V}} = 1.1 \text{ mA}$$

(b) $R = \dfrac{V}{i} = \dfrac{9.0\ \text{V}}{1.1 \times 10^{-3}\ \text{A}} = 8.2\ \text{k}\Omega$

27-4 Resistivity

Ohm's law holds for a conductor of any shape and with the leads attached to any two points. Consider, however, the simple arrangement in which a potential difference is established between the ends of a cylindrical conductor of length L and cross-sectional area A, as in Figure 27-6. For a given material, experiment shows that the resistance for this simple geometrical configuration is directly proportional to the length and inversely proportional to the cross-sectional area. We may write the resistance as

$$R = \rho\,\frac{L}{A} \qquad (27\text{-}10)$$

where the constant ρ, called the resistivity, is a property of the material of which the conductor is made but does not depend on the conductor's physical shape. Resistivity carries the units ohm meters ($\Omega \cdot \text{m}$).

Table 27-1 lists resistivities for several materials, both good and poor conductors. Note the great range in resistivities; the best conductors, metals, are better than the worst conductor by a factor of about 10^{24}. (The best *thermal* conductors are also metals; the low thermal and electrical resistivity arises from the free electrons in metallic conductors.) From (27-10), we see that the resistance of a cube 1 m along an edge, with a current between opposite faces maintained at different potentials, is equal numerically to the material's resistivity in ohm meters.

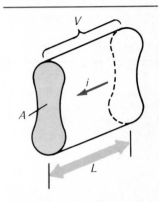

Figure 27-6. A potential difference V is maintained across the ends of a cylindrical conductor of length L and cross-sectional area A through which a current i passes.

Table 27-1.

MATERIAL	RESISTIVITY (MICROHM · CM)	TEMPERATURE COEFFICIENT OF RESISTIVITY AT 20°C ($C^{\circ -1}$)
Metal		
Aluminum	2.824	0.0039
Brass	7	0.002
Constantan	49	0.00001
Copper (annealed)	1.724	0.0039
Gas Carbon	5,000	−0.0005
Manganin	44	0.00001
Nichrome (trade mark)	100	0.0004
Silver	1.59	0.0038
Steel (manganese)	70	0.001
Tungsten (drawn)	5.6	0.0045
Insulator, semiconductor		
Germanium	4.3×10^{7}	
Silicon	2.6×10^{11}	
Wood (maple)	4×10^{19}	
Mica	9×10^{21}	
Quartz	5×10^{24}	

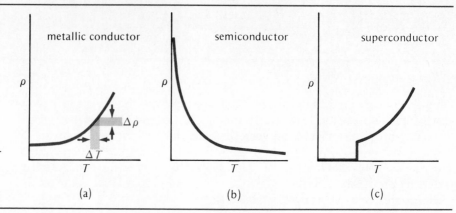

Figure 27-7. *Variation in electrical resistivity ρ with temperature T for (a) a metallic conductor, (b) a semiconductor, and (c) a superconductor.*

The reciprocal of the resistivity ρ is called the *conductivity* σ:

$$\sigma \equiv \frac{1}{\rho} \tag{27-11}$$

Figure 27-7 shows how the resistivity of three types of materials varies with temperature: (a) an ordinary metallic conductor, (b) a semiconductor, and (c) a superconductor.

The resistance, and therefore also the resistivity, of metallic conductors increases with temperature. Over a temperature range ΔT small enough for the curve in Figure 27-7(a) to be approximated by a straight line, the change in resistivity $\Delta \rho$ is proportional to ΔT. The fractional change in resistivity $\Delta \rho / \rho$ can be written as

$$\frac{\Delta \rho}{\rho} = \alpha \, \Delta T \tag{27-12}$$

where α is a constant (over the small temperature range) called the *temperature coefficient of resistance.* Values of α for some selected conductors are shown in Table 27-1. Copper at room temperature has $\alpha = 0.003\ 93\ (\text{C}°)^{-1}$; this means that the resistivity of copper increases by 0.39 percent for an increase in temperature of one Celsius degree. The change in resistance with temperature can be used as the basis of a resistance thermometer. Note that the alloy constantan (60 percent copper, 40 percent nickel), with an extremely small value of α, has a resistivity that is nearly constant over a range of temperatures.

The resistivity of semiconductors [Figure 27-7(b)] decreases with a temperature rise, primarily because the number of charge carriers increases as the temperature is raised. Temperature coefficients of resistivity are negative for these materials, of which common examples are carbon, silicon, and germanium.

The behavior of superconductors [Figure 27-7(c)] is extraordinary. Below a certain critical temperature (below 7.175 K for lead), the resistivity of a superconductor drops to *zero*—not merely a small resistivity, but exactly zero. A superconductor offers no resistance to the flow of electric charge and dissipates no energy. Once a superconducting current is established in a superconducting loop, it persists indefinitely. Currents of several hundred amperes induced in a superconducting lead ring have been observed to persist with no measurable diminution for several years! Although bizarre, the phenomenon of supercon-ductivity is now well understood on the basis of the quantum theory.*

Example 27-3. A current of 1.0 A is maintained in a copper wire 10 m long and having a cross-sectional area of 0.25 mm^2. (a) What is the potential difference between its ends? (b) What is the magnitude of the electric field within the wire driving electrons? (c) At what distance from a proton is the electric field from it of the same magnitude as the driving electric field?

(a) From Ohm's law and the definition of resistivity, we have

$$V = iR = i\rho \frac{L}{A}$$

$$= (1.0 \text{ A})(1.72 \times 10^{-8}\Omega \cdot \text{m}) \left(\frac{10 \text{ m}}{0.25 \times 10^{-6} \text{ m}^2} \right)$$

$$= 0.69 \text{ V}$$

Ordinarily the potential difference along short lengths of connecting wire in a circuit is small enough to be taken as zero.

(b) The electric field produced within the wire by the potential difference across its ends is constant in magnitude along the length of the wire, so that

$$E = \frac{V}{L} = \frac{0.69 \text{ V}}{10 \text{ m}} = 0.069 \text{ V/m}$$

(c) The electric field at a distance r from a point charge is $E = kq/r^2$ [(23-6)]. There-fore, the distance from a proton, with $q = 1.6 \times 10^{-19}$ C, at which $E = 0.069$ V/m is

$$r = \sqrt{\frac{kq}{E}} = \sqrt{\frac{(9.0 \times 10^9 \text{N} \cdot \text{m}^2/\text{C}^2)(1.6 \times 10^{-19}\text{C})}{0.069 \text{ V/m}}} = 0.14 \text{ mm}$$

Compared with the electric forces acting on a conduction electron from other charged particles nearby (atoms in a solid might be separated by $\sim 10^{-7}$ mm), the electric force producing the steady current is relatively feeble.

27-5 RC Circuits

Consider the circuit shown in Figure 27-8. It consists of an initially charged capacitor C connected across a resistor R. The capacitor has an initial charge magnitude Q_0 on each plate, and the potential difference between the plates is initially $V_0 = Q_0/C$.

It is easy to describe qualitatively what happens immediately after the switch is closed. Free electrons move counterclockwise around the circuit and through the resistor until finally the capacitor is electrically neutral. While charges are moving, a current i exists in the circuit; the current is zero when charge Q and potential difference V of the capacitor reach zero. At the same time, the electric-potential energy associated with the opposite charges on the capacitor plates is dissipated into thermal energy as charges move through the resistor.

---

* Actually, we are not surprised—but perhaps should be—to learn that electrons within an atom can remain in motion indefinitely without dissipating energy. The reason for this microscopic behavior, as well as for the macroscopic phenomenon of superconductivity, lies with quantum theory.

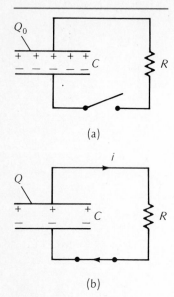

(a)

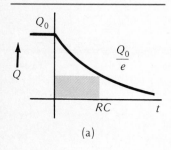

(b)

Figure 27-8. *An RC circuit. (a) Capacitor initially charged; (b) capacitor discharging through the resistor.*

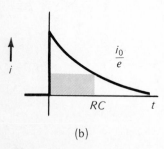

(a)

(b)

Figure 27-9. *Variation with time of (a) the capacitor charge and (b) the current for an RC circuit.*

We wish to find how charge Q and current i vary with time. The net change in potential around the circuit is zero at each instant of time. Going clockwise around the loop, we have a potential drop iR across the resistor and a potential drop $-Q/C$ across the capacitor. (We go from negative to positive plate through the capacitor—a potential *rise*). Therefore,

$$iR - \frac{Q}{C} = 0$$

The current at each instant is

$$i = -\frac{dQ}{dt}$$

The minus sign indicates that positive current corresponds to a *decrease* in charge Q on the capacitor plates. The first equation can then be written

$$-R\frac{dQ}{dt} - \frac{Q}{C} = 0$$

This differential equation holds at any instant. Rearranging terms and recognizing that R and C are independent of time, we have

$$\int_{Q_0}^{Q} \frac{dQ}{Q} = -\frac{1}{RC}\int_0^t dt$$

where we are integrating charge from its initial value Q_0 to any later value Q at time t and integrating time from zero to time t:

$$\ln \frac{Q}{Q_0} = -\frac{t}{RC}$$

$$Q = Q_0 e^{-t/RC} \qquad (27\text{-}13)$$

The charge Q on each capacitor plate decreases exponentially with time, as shown in Figure 27-9(a). The rate of decay is controlled by the quantity RC, the *time constant* τ of the RC circuit:

$$\tau \equiv RC \qquad (27\text{-}14)$$

As (27-13) shows, when $t = \tau = RC$, then $Q = Q_0/e$. The constant RC is the time elapsing until the capacitor's charge (and also potential difference) is $(1/e)$, or 37 percent, of its initial value. For example, in a circuit with $R = 1.0$ MΩ and $C = 1.0$ μF, the time constant is $\tau = (1.0 \times 10^6$ V/A) $(1.0 \times 10^{-6}$ C/V) $= 1.0$ s.

How does current vary with time? Using (27-13), we have

$$i = -\frac{dQ}{dt} = \left(\frac{Q_0}{RC}\right)e^{-t/RC}$$

But the initial current is $i_0 = V_0/R = Q_0/RC$. The equation above can then be written

$$i = i_0 e^{-t/RC} \qquad (27\text{-}15)$$

The current, zero until the switch is closed, rises abruptly to i_0 and then decays exponentially with time, again with characteristic time constant $\tau = RC$, as shown in Figure 27-9(b). The rate of energy dissipation in the resistor is i^2R,

Figure 27-10.   A sawtooth wave, with the sloped straight-line portion derived from an RC circuit.

proportional to the *square* of i. Therefore, the rate at which energy is dissipated in the resistor (also the rate at which the charged capacitor loses energy) is more rapid than the decay in Q or i. The time constant for energy decay is $\frac{1}{2}RC$.

An RC circuit provides a means of producing a *sawtooth wave*, a variation in voltage with time shown in Figure 27-10. The change in potential across the capacitor or resistor with time is, as we have seen, exponential. But a small piece of an exponential line is effectively a *straight* line. The linear variation in voltage with time is obtained by allowing a capacitor to discharge (or charge) in an RC circuit for a time short compared with τ and then be abruptly recharged back to its initial state. The potential difference across V and R can thereby be made to vary as shown in Figure 27-10. Sawtooth waves are used frequently in electric circuits. For example, the steady and repeated progression of an electron beam across the face of an oscilloscope (or TV picture tube) can be produced by a sawtooth wave applied to horizontal deflecting plates.

27-6   Electric Resistance from a Microscopic Point of View

When there is no external electric field, the free, or conduction, electrons of a conductor are in random thermal motion. They move at relatively high speed ($\sim 10^6$ m/s), collide with one another and with the atoms of the conductor. In such a collision, a free electron is just as likely to gain as to lose kinetic energy. On the average, its kinetic energy is unchanged, and it undergoes no net displacement within the conductor.

Now consider what happens to free electrons in a conductor across which an electric potential difference is maintained. The simple cylindrical conductor in Figure 27-11 of length L and cross-sectional area A has a constant potential difference V between its ends. A constant electric field **E**, pointing along the length of the conductor from high to low potential, then exists within the conductor, with $E = -V/L$. Free electrons still make frequent collisions, and are driven by this field. Electron drift is responsible for the current in the conductor. When a conductor is in *electrostatic* equilibrium, $\mathbf{E} = 0$ everywhere within a conductor, but when charges flow, and a current exists, in the conductor the electric field can no longer be zero.

The approximate behavior of free electrons in a conductor under the influence of a driving electric field is illustrated by the mechanical model shown in Figure 27-12. Here balls (the free electrons) accelerate down a series of inclines driven by gravitational force (the driving electric field). Each ball gains kinetic energy as it goes down a ramp, but essentially all this kinetic energy is lost when the ball collides inelastically with a bumper (an atom in the conductor). After each collision, the ball starts from rest again and makes further inelastic collisions. At the bottom of the zigzag ramps (the resistor), each ball has the

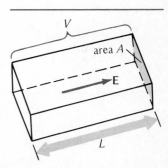

*Figure 27-11.   Constant electric field **E** established on the interior of a conductor by electric potential difference ΔV.*

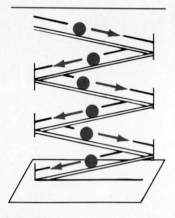

Figure 27-12. *A mechanical model for a resistor: marbles rolling down a succession of inclined planes.*

same kinetic energy it started with at the top—zero. All the energy gained in accelerating down the ramps becomes thermal energy in collisions. For any ball to go down the ramp again, it must be lifted back to the top (a battery, or other energy source) and its gravitational potential energy (electric potential) must be increased.

As the parenthetic entries above imply, the crucial aspects of conduction are these:

- *Inelastic* collisions with atoms of the conductor lattice.
- After each collision, each electron starts over again, essentially with zero *drift velocity*.
- Between collisions, electrons are driven at *constant acceleration* by an external electric field.

Superimposed on this drift is random thermal motion. Indeed, the average drift velocity acquired by free electrons is so very small compared to their average thermal speed that the time between collisions is effectively independent of the applied electric field.

Written as $V = iR$, Ohm's law is a macroscopic relation. We wish to express Ohm's law in microscopic form, that is, in terms of the driving electric field $\mathbf{E}$ and the current density $\mathbf{j}$ at any point within a conductor. For the uniform cylindrical conductor of Figure 27-11, we have $V = EL$ from (25-10) and $R = \rho L/A$ from (27-9̶). Therefore

(27-10)

$$V = iR$$

$$EL = i\,\rho\,\frac{L}{A}$$

Since the current density has the magnitude $j = i/A$, the relation above can be written

$$\mathbf{E} = \rho\mathbf{j} \qquad (27\text{-}16)$$

Equation (27-16) is written in vector form, since the charge carriers have drift velocities in the direction of the electric field.

Written in terms of conductivity $\sigma = 1/\rho$, Ohm's law is

$$\mathbf{j} = \sigma\mathbf{E} \qquad (27\text{-}17)$$

Although derived here for the special case of a uniform field, (27-16) and (27-17) are general relations that hold at any point within any conducting ohmic material.

Summary

Definitions

Electric current i:

$$i \equiv \frac{dQ}{dt} \qquad (27\text{-}1)$$

where net charge dQ passes through a chosen surface in time dt.

Current density j with current i passing transversely through area A:

$$j = \frac{i}{A}$$

Resistivity ρ of the material in a cylindrical conductor of resistance R, length L, and cross-sectional area A:

$$\rho \equiv \frac{RA}{L} \qquad (27\text{-}10)$$

Electric conductivity σ of a material with electric resistivity ρ:

$$\sigma = \frac{1}{\rho} \qquad \text{(27-11)}$$

Thermal coefficient of resistivity α for a material whose resistivity changes by the fractional amount $(\Delta\rho/\rho)$ for a temperature change ΔT:

$$\alpha \equiv \frac{\Delta\rho/\rho}{\Delta T} \qquad \text{(27-12)}$$

Units

Resistance: 1 ohm = $1\Omega \equiv 1\text{ V/A}$
Conductance: 1 siemens = 1 S = 1 A/V = 1 mho

Important Results

The power P delivered to a load across which the electric potential difference is V_{ab} and carrying current i:

$$P = iV_{ab} \qquad \text{(27-6)}$$

Ohm's law states that resistance R of a conductor is independent of current i and potential difference V:

$$R \equiv \frac{V}{i} = \text{constant} \qquad \text{(27-7)}$$

For microscopic application, Ohm's law is written

$$\mathbf{E} = \rho\mathbf{j} \qquad \text{(27-16)}$$

or

$$\mathbf{j} = \sigma\mathbf{E} \qquad \text{(27-17)}$$

where $\mathbf{E}$ is the driving electric field and $\mathbf{j}$ the current density at some point in a conductor with resistivity ρ and conductivity σ.

When a charged capacitor C is discharged through a resistor R (an RC circuit), the charge on either capacitor plate decays exponentially with time according to

$$Q = Q_0 e^{-t/\tau} \qquad \text{(27-13)}$$

where the time constant is

$$\tau = RC \qquad \text{(27-14)}$$

Problems and Questions

Section 27-1 Electric Current

· **27-1 P** How much charge passes in 2 min through a junction through which a steady current of 5 A exists?

· **27-2 P** In a typical lightning bolt, about 1 C of negative charge (a huge amount) is delivered to the earth in about 4×10^{-5} s. What current does this represent?

· **27-3 P** What is the current density in a copper wire of diameter 1.2 mm in which a current of 14 A flows?

· **27-4 P** A plasma is a neutral mixture of negatively charged electrons and positive ions, such as a very hot gas in a star. Attempts to create thermonuclear reactions like those in the sun are important in energy-source research and development. For the needed high temperatures, the plasma is often heated by passing a current through it. In one such device (a toroidal stellarator), the current is carried primarily by electrons whose density is $10^{19}/\text{m}^3$. What is the drift velocity required to produce a current density of 10^5 A/m^2?

· **27-5 P** A typical current used in a household circuit is 10 A. Estimate the drift velocity of the carriers in this case, assuming that No. 14 copper wire (diameter 0.163 mm) is used.

· **27-6 P** In aluminum there are 1.8×10^{29} conduction electrons per cubic meter. What is the drift velocity in an aluminum wire of diameter 1.2 mm in which a current of 2 A flows?

: **27-7 P** A neutral beam has equal proton and electron densities, each $6.0 \times 10^6/\text{cm}^3$. Both electrons and protons have energies of 12 eV, with protons moving to the right and electrons moving to the left. If the cross-sectional area of the beam is 2 mm², what are the net (a) current and (b) current density in the beam?

: **27-8 P** Silver, with a density of 10.5 gm/cm³, has about one conduction electron per atom. In a silver wire of diameter 0.1 mm, what is (a) the density of conduction electrons (in electrons/m³)? (b) the drift velocity of the electrons when a current of 1 A exists in the wire?

: **27-9 P** What current is required in an electroplating process if 3 mg of copper (atomic mass 64) is to be deposited in 5 min? A copper sulfate solution containing the ions Cu^{2+} is used as the electrolyte.

: **27-10 P** What mass of gold (atomic mass 197) will be deposited in an electroplating process in which the ions Au^+ carry a current of 10 A for 5 minutes?

: **27-11 P** An experimenter wants to deposit a layer of silver of thickness 0.02 mm over a microwave component of area 4 cm². With a solution of Ag^+ and a current of 2 A, how long will it take to deposit the desired amount of silver? (Silver density, 10.5 gm/cm³; atomic mass, 108).

: **27-12 P** In a simple model of the hydrogen atom, a single electron circles a fixed proton in an orbit of radius 0.5×10^{-10} m under the action of the Coulomb force from the proton. (a) What is the frequency of rotation of the electron? (b) What current is associated with the electronic motion?

Section 27-2 Current and Energy Conservation

· **27-13 Q** In buying a battery you may find that one of the specifications provided is the "ampere-hour" rating. This is a measure of

(A) the maximum current the battery can provide.
(B) the maximum potential difference the battery can provide.
(C) the maximum electric charge the battery can deliver.
(D) the maximum power the battery can deliver.
(E) the maximum electrical energy the battery can deliver.

· **27-14 Q** The kilowatt is a unit used to measure
(A) electric current.
(B) potential difference.
(C) resistivity.
(D) electrical energy.
(E) the rate at which work is done or energy is transferred.
(F) the rate of creation of electrical charge.

· **27-15 Q** For you to be electrocuted,
(A) your body must acquire a large net electric charge.
(B) your body must be raised to a high electrical potential.
(C) a current must pass through your body.
(D) a potential difference of at least 1000 V must be applied across your body.
(E) you must be in contact with a dc source of voltage.
(F) at least part of your body must be wet.

· **27-16 Q** A certain new, freshly charged car battery is rated at "12 volts, 83 ampere hours." This means
(A) each terminal has an initial charge of 83 C.
(B) the battery's internal resistance is 0.14 Ω.
(C) the battery will be dead after 83 h.
(D) the maximum current the battery can deliver is 83 A.
(E) the energy stored in the battery is 3.6×10^6 J.

: **27-17 P** A 4.0-MeV proton beam has a density of $2.5 \times 10^{11}/m^3$ and an area of 2.0 mm². (a) What is the current in the beam? (b) What is the power of the beam (that is, the energy delivered by the beam per second)? (c) How many years would it take such a beam to deposit 1 gm of protons?

: **27-18 P** Electrons are emitted from a hot cathode at the rate of 5×10^{18} per second. They have negligible initial kinetic energy, but are accelerated by a potential difference V to an anode, where they are collected. Calculate (a) the current when $V = 100$ V; (b) the power dissipated at the anode when $V = 100$ V; (c) the current when $V = 200$ V; (d) the power when $V = 200$ V.

Section 27-3   Resistance and Ohm's Law

· **27-19 P** A 100-Ω resistor is rated at 2.0 W. (a) What is the maximum potential difference that can be applied across it? (b) What maximum current can it carry?

· **27-20 P** What current does a hair drier draw if it is rated at 800 W and is intended for use with a 120-V supply?

· **27-21 Q** If the length and diameter of a wire of circular cross-section are both doubled, the resistance
(A) is unchanged.
(B) is doubled.

(C) is increased fourfold.
(D) is halved.
(E) None of the above.

· **27-22 Q** Which of the following is most likely to cause you to suffer electrical injury?
(A) A large electric charge accumulates on your body.
(B) Your body is raised to a very high electric potential.
(C) A large electric current passes through your body.
(D) The resistance of your body becomes very large.
(E) The cells in your heart become polarized.

· **27-23 Q** Suppose that a bird lands on a bare 14,000-V power line. What is likely to happen?
(A) It will be cooked.
(B) It will not be hurt because its feet are good insulators.
(C) It will not be hurt because it has a high body resistance.
(D) It will not be hurt because there is no potential difference across its body.
(E) It will not be hurt because the power line cannot deliver enough charge to cause injury.
(F) None of the above is true.

· **27-24 P** A typical 20-A circuit in a home uses No. 12 copper wire with a resistance of 5.2×10^{-3} Ω/m. For such a 2-wire conductor of length 20 m, calculate (a) the potential drop when 20 A flows; (b) the power dissipated in the wire.

: **27-25 P** An immersion electric heater used in the lab will change the temperature of 400 ml of water by 40C° in 5 min when operated with a 120-V supply. (a) What power does the heater deliver? (b) What current does it draw?

: **27-26 P** A typical cost of electricity in the United States is $0.08 per kW·h. At this rate, assuming 100 percent efficiency, how much does it cost to (a) cook a piece of toast using a 1-kW toaster for 3 min? (b) heat a bathtub full of water (about 80 L) enough to increase the temperature 30°C?

: **27-27 P** To minimize the danger of fire due to overheating, building codes limit the current in No. 10 (diameter 2.59 mm) rubber-coated copper wire to 25 A. Observing this condition, determine for a 100-m length of such wire (a) the wire resistance; (b) the current density; (c) the potential drop; (d) the electric field; (e) the rate at which heat is dissipated in the wire.

: **27-28 P** The circuit breaker in a typical household circuit will open (and thus disconnect the circuit) if the current exceeds 15 A. Suppose that a person tries to plug the following appliances into a 120-V outlet, starting first with the lamp and then the following items in order. Which appliance will cause the circuit breaker to open?
(A) 150-W lamp (plugged in first).
(B) 800-W hair drier (plugged in second).
(C)   40-W radio.
(D) 700-W television set.

(E) 150-W typewriter.
(F) 50-W calculator.
(G) 100-W video game.

: 27-29 P When a lamp is connected to a 12-V battery, it dissipates energy at the rate of 10 W. Assuming that the lamp resistance does not change, determine the rate at which the lamp will dissipate energy when connected to a 6-V battery? Assume that the batteries have negligible resistance.

: 27-30 P A fully charged battery is rated at 12 V and 10 A·h. It is connected to a 5-Ω resistor. Assuming that the battery has negligible internal resistance, determine the total energy dissipated in the resistor by the time the battery has been completely discharged, or is "dead."

: 27-31 P A radiant heater that is used to speed the drying of sheet rock mortar in cold weather is rated at 5000 W when the voltage supply is 120 V. (*a*) What is the resistance of the heating filaments? (*b*) What current does the heater draw? (*c*) How many kilocalories are generated in one hour by the heater? (*d*) If the change in resistance of the heating element with temperature is neglected, what would be the power output if the voltage were to drop by 10 percent to 108 V? In view of the actual variation of resistance with temperature for a metal filament, would the actual power be greater than the value calculated here, or less?

: 27-32 P A fuse, such as that used in a car or a stereo set, is a conductor with a low melting point; it will thus get hot and melt if too much current exists in it. Such devices are used to limit the current in a circuit, and thereby prevent overheating from too much current.

When current exists in the fuse, its temperature will rise until the thermal losses equal the power P dissipated in the fuse. The temperature rise will generally be proportional to the power dissipation P, $\Delta T = aP$, where a is a constant that can be determined experimentally for a given geometry. As the fuse gets hot, its resistance will increase according to $R = R_{20}(1 + \alpha \Delta T)$. Show that if α is large enough, the fuse will "blow" (ΔT will become very large) when the current reaches the value $I = 1/\sqrt{\alpha a R_{20}}$.

Section 27-4 Resistivity

· 27-33 Q Which of the following is the most accurate statement?
(A) *Resistance* and *resistivity* have the same meaning.
(B) The resistivity of a given conductor depends on its size and shape and on the material of which it is composed.
(C) A piece of copper has lower resistivity than a piece of iron.
(D) A thick copper wire has greater resistance than a thin copper wire of the same length.
(E) Resistance in a metal conductor is due to the "back" internal electric fields that the conduction electrons exert on each other.

· 27-34 P What length of No. 18 copper wire (diameter = 0.040 in.) will have a resistance of 1 Ω?

· 27-35 P What is the resistance of a copper bar of cross-sectional area 2 mm² and length 3 m at 20°C?

: 27-36 P Consider an aluminum wire and a copper wire, each of the same length and resistance. What is the ratio of the weight of the copper wire to the weight of the aluminum wire? (Aluminum density, 2.7 gm/cm³; copper density, 8.9 gm/cm³.)

· 27-37 P Nichrome is an alloy used as a heater wire in constructing electric heaters and furnaces. What resistance should such a heater have at 20°C if it is to dissipate 1000 W at 800°C when operated from a 120-V supply? (Assume that the temperature coefficient of resistivity of nichrome remains equal to 0.0004 C°⁻¹ over the entire temperature range.)

: 27-38 P At what temperature will the resistance of a copper conductor be twice its value at 20° C, assuming its temperature coefficient of resistivity to remain constant? If you neglect the thermal expansion of the wire, will this affect seriously the accuracy of your result?

: 27-39 P Carbon resistors find many applications in electronics. Consider such a resistor whose resistance at 20°C is 5.00 Ω. The mass of the resistor is 50 gm, its specific heat capacity is 0.10 kcal/kg°C, and its temperature coefficient of resistivity is $-5.0 \times 10^4/C°$ (note that resistance *decreases* when temperature is increased). A fixed potential difference of 12 V is applied to the resistor at $t = 0$. (*a*) If all the electric power dissipated goes into heating the resistor, how long will it take for the resistor to reach 40°C? (*b*) What will be the current through the resistor at 40°C?

: 27-40 P A power station delivers 1 MW of power at voltage V to a transmission line made of aluminum cables of diameter 6 cm (Figure 27-13). The power is to be delivered to a town 80 km distant. The resistivity of aluminum is 2.8×10^{-6} Ω·cm.

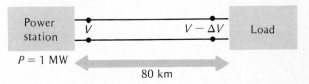

Figure 27-13. Problem 27-40.

(*a*) What is the resistance of the line? (Note that two wires are used.) (*b*) What is the voltage drop along the line in terms of the voltage V and the output power P at the power station? (*c*) If the voltage drop along the line is to be no greater than 4 percent of V, what minimum value of V is required? (*d*) What power is dissipated in the line? (*e*) Explain qualitatively why, for a given power P, line losses can be reduced by using high line voltage.

: 27-41 P An axon membrane has a typical resistance conductance of 5 mho/m^2 and a thickness of 7.5×10^{-9} m. (*a*) What is the resistivity of the membrane? (*b*) Suppose that the membrane conductivity is due to pores of diameter 8×10^{-10} m filled with conducting fluid of resistivity 0.16 $\Omega \cdot$m, with the bulk of the membrane material of very high resistivity. Approximately how many pores per meter squared would be required to give the observed resistance? (*c*) Estimate the spacing of the pores.

: 27-42 P What is the resistance of a conductor having resistivity ρ and the shape of a slightly tapered truncated cylindrical cone (see Figure 27-14)? Assume that the current density is uniform across any cross section.

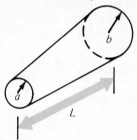

Figure 27-14. Problem 27-42.

Section 27-5 RC Circuits

· 27-43 P A charged 2-μF capacitor is discharged through a 2-M Ω resistor. (*a*) What is the time constant for the decay? (*b*) After how many time constants will the energy stored in the capacitor decrease to half its initial value?

· 27-44 Q The capacitor shown in Figure 27-15 is initially uncharged. The leads XX' are connected to the vertical axis of an oscilloscope, so a plot of the potential difference V_R across the resistance is obtained as a function of the time after the switch is closed. Which graph best describes what one would observe?

: 27-45 Q In working with electronic instrumentation, it is often necessary to use fast switching circuits. In such applications it is important to take care that the instrumentation and circuitry does not decrease the speed of response of the instruments. Which of the following is most likely to help attain this aim?

(A) Keep the capacitance in the circuits small.
(B) Keep the resistance in the circuits large.
(C) Keep the voltages used large.
(D) Keep the Joule heating small.
(E) Work in a room with very low humidity.

: 27-46 Q In a lecture demonstration intended to show the time needed for charging and discharging a capacitor, the circuit shown in Figure 27-16 was used. The switch was thrown to connect the capacitor to the dc power supply for charging, and it was then thrown to discharge the capacitor. The time to discharge was observed to be less than the time for charging. How do you best explain this result?

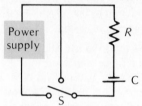

Figure 27-16. Question 27-46.

(A) Circuit analysis shows that it takes longer to charge a capacitor through a given resistor R than to discharge the capacitor through the same resistance.
(B) The potential difference across the capacitor aided the discharging, whereas it opposed the charging.
(C) The power supply had appreciable internal resistance.
(D) In discharge, the potential difference across R added to the potential difference across the capacitor, whereas this was not the case during charging.
(E) The capacitor was not fully charged when the switch was thrown; hence one would not expect the charging and discharging times to be equal.

: 27-47 P A parallel-plate capacitor is filled with mica and has the following characteristics:

Plate separation $d = 2 \times 10^{-2}$ cm
Plate area = 200 cm^2
Dielectric constant = 6
Resistivity of mica = $1.7 \times 10^{12} \ \Omega \cdot$m

(*a*) What is the capacitance of the capacitor? (*b*) What is the leakage resistor of the capacitor? (*c*) How long will it

Figure 27-15. Question 27-44.

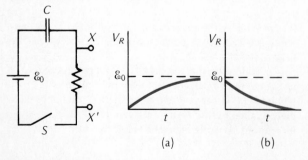
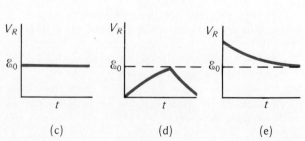

(a) (b) (c) (d) (e)

take for the voltage on the capacitor to drop to half its initial value?

: 27-48 P A $4\text{-}\mu F$ capacitor is charged to 100 V. After 24 h, it is found that the voltage has decreased to 98 V. (*a*) What is the leakage resistance of the capacitor? (*b*) What will the voltage be 72 h after it has first been charged?

: 27-49 P The region between the plates of a parallel-plate capacitor is filled completely with a material having resistivity ρ and dielectric constant κ. Show that the time constant for this capacitor is given by $(\rho\epsilon) = (\rho\kappa\epsilon_0)$, which is independent of the dimensions of the capacitor plates. The result is, in fact, a general one: all capacitors filled between their plates have a time constant that depends only on the filling material's resistivity and dielectric constant.

: 27-50 P Using energy conservation, show that the differential equation describing the time variation in the energy U_C of a charged capacitor in an RC circuit is given by $dU_C/dt = i^2R$.

: 27-51 P Show that when a charged capacitor is discharged through a resistor, the total energy initially stored in the capacitor, $Q^2/2C$, equals the energy dissipated in the resistor, $\int_0^\infty i^2R\,dt$.

: 27-52 P A neon lamp has the following interesting property, which makes it useful in electronic circuits. The lamp consists of a small glass envelope filled with neon gas. There are two electrodes in the envelope. If the voltage applied to the electrodes exceeds 70 V, the neon becomes ionized and the resistance between the electrodes drops essentially to zero. The gas discharge is evident from the orange light given off. If the voltage drops below 70 V, the discharge stops in a few milliseconds and the tube is again an insulator. (Assume that the few milliseconds after the voltage drops below 70 V is enough time for the capacitor to be discharged completely, inasmuch as the circuit's resistance is then effectively zero.)

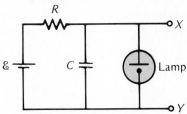

Figure 27-17. *Problem 27-52.*

(*a*) Determine the periodic frequency of the voltage between terminals X and Y for the *relaxation oscillator* shown in Figure 27-17. $V = 100$ V; $R = 100$ kΩ; $C = 2$ μF. (*b*) Sketch the voltage between X and Y as a function of time.

Supplementary Problems

27-53 P A common type of memory unit in a computer chip is a capacitor, whose state of charge corresponds to

one bit of information. See supplementary problem 26-57. Even though the silicon dioxide between the plates of a memory capacitor is an excellent insulator, the charge on the plates does leak away, and it is therefore necessary to regenerate, or "refresh," the stored charge periodically. Indeed, the recharging must be done every 2 or 3 milliseconds in a typical microcomputer chip if the information stored in memory is not to be lost. Given that the dielectric constant of silicon dioxide is 12, compute the order of magnitude of the resistivity of silicon dioxide. (The result given in Problem 27-49 may be useful.)

27-54 P A diode is a device that allows current flow in only one direction. The circuit shown in Figure 27-18 is a *staircase generator*. A negative square pulse is applied to the input. This allows a current to flow through diode D_1, charging C_1. When the pulse ends, the positive charge that was on the right plate of C_1 can pass through D_2 and accumulate on C_2. The process is then repeated. Again C_1 is charged, and when the voltage drops again to zero, once more charge is passed on from C_1 to C_2.

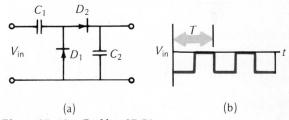

Figure 27-18. *Problem 27-54.*

(*a*) Plot output voltage vs time for $C_1 = 0.1$ μF, $C_2 = 1$ μF, $V_{in} = -2$ mV, $T = 1$ ms. From your result you will see why this is called a staircase generator. (*b*) If an ammeter is connected across the output terminals, a current I will flow in it, where $I = Q_2 f$. Here Q_2 is the charge that accumulates on C_2 every time a pulse of voltage V_{in} is applied to the input terminals, and f is the frequency of the input pulses. In this form, the circuit is a rudimentary frequency meter. Show that the frequency of the input pulses is $f = I/C_1 V_{in}$.

27-55 P An incandescent lamp uses a tungsten filament heated to a temperature of about $3000°$C. A lamp operated with voltage V dissipates power P as radiated light and infrared radiation, where P is related to the Kelvin temperature T of the filament by $P = \sigma T^4$ (σ is a known constant). The resistance varies with temperature as $R = R_{29}(1 + \alpha\,\Delta T)$. Note that $\Delta T \approx T$ since the operating temperature of about 3000 K is so high. For tungsten $\alpha = 4.5 \times 10^{-7}$ per Celsius degree.

(*a*) What is the ratio of the filament resistance at 3000 K to the resistance at room temperature? (*b*) Show that the temperature T varies with the voltage V as $T \propto V^{2/5}$. (*c*) Show that the power varies with V as $P \propto V^{8/5}$ approximately. (*d*) How much does the power output increase if the voltage is increased by 10 percent?

DC Circuits

28-1   EMF

Consider the simple dc circuit of Figure 28-1: resistor R connected across battery terminals. The positive battery terminal is a and the negative terminal b. Potential V_a is higher than V_b. The battery maintains the constant potential difference $V_a - V_b = V_{ab}$ across its terminals.

Positive charges enter the resistor at a, go downhill in potential, and leave at b. (Strictly, electrons enter the resistor at b, go uphill in potential, and leave at a.) The connecting wires have so small a resistance that the potential drop across them is negligible compared with that across R. Every time charge leaves any one point in the circuit, it is immediately replaced by other charge of the same magnitude. Consequently, there is no net change in charge at any point in the circuit, and each point in the circuit maintains a constant electrostatic potential.

Suppose that we follow a positive charge from a, through the resistor to b, and then through the battery back to a. From a to b the charge is acted on by an electric field within the conductor. It thereby gains energy, but because of frequent inelastic collisions with lattice atoms, it disposes of this electric energy as thermal energy in the conductor and emerges at point b with exactly the same energy it had at a. The charge now enters the negative battery terminal

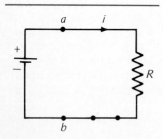

Figure 28-1.   *A simple dc circuit.*

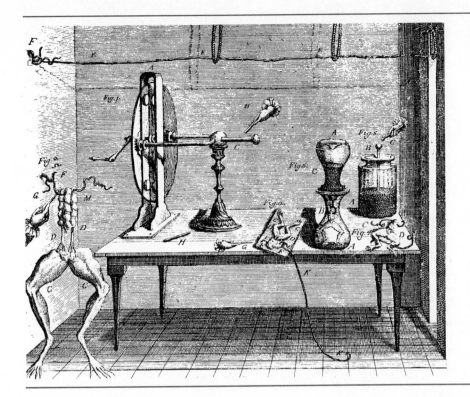

Figure 28-2. Luigi Galvani (1737–1798) first observed the existence of current through the nerve of a frog's leg. This engraving shows the frog legs hanging by the nerves (at left). When the electrostatic machine on the table revolved, or the Leyden jar at the right-hand end of the table was discharged, Galvani observed that the legs jerked when a scalpel touched the nerve. Current electricity, also once called "animal" electricity, was first thought to be a phenomenon entirely separate from static electricity.

and is transported through the battery back to the positive battery terminal; that is, the positive charge must move from a point at the potential V_b to the higher potential V_a. Clearly, this is not possible if the only force acting on the charge, as it passes through the interior of the battery, is the electric force derived from the potential difference (the force associated with the positive and negative charges residing at the battery terminals). There must exist, in addition to this electric force, a force not derived from the electrostatic charge distribution, which somehow drives the positive charge *uphill* in potential from V_b to V_a. In an electrochemical cell, or battery, the force that drives the charges against the opposing electrostatic force is sometimes called the *chemical force*.

When any two dissimilar metals are immersed in a conducting medium, a potential difference is found to exist between them; this is the most rudimentary form of an electrochemical cell.* The chemical reactions that are the origin of this potential difference and the details of what goes on within a battery lie in the area of electrochemistry and will not be dealt with here. Suffice it to say, however, that the potential difference has its origin in the differences in the binding energy of electrons to atoms of different types. In this sense, the nonelectrical, or chemical, forces are ultimately electrical in origin, while still being non-electrostatic. See Figure 28-2.

* The term *battery* means, strictly, a battery of electrochemical cells; for example, a 12-V automobile battery consists of six 2-V cells.

We may characterize any energy source that is capable of driving charges around a circuit against opposing potential differences as an *emf* (pronounced "ee-em-eff"). This is the abbreviation for *electromotive force*, a term so misleading (emf is *not* a force) that we shall hereafter refer to it simply as the emf, or symbolically, as $\mathscr{E}$.* By the $\mathscr{E}$ of a battery is meant the energy per unit positive charge gained (or the work per unit charge done by the chemical forces within a battery) in the transfer of a charge within the battery from the negative to the positive terminal:

$$\mathscr{E} = \text{work per unit positive charge done by the } \textit{chemical forces}$$

Since energy per unit charge, or joules per coulomb, is equivalent to volts, an emf, like a potential difference, is expressed in volts.

Some energy is always dissipated within any actual battery. One accounts for this energy loss by ascribing an *internal resistance r* to the battery, where the rate at which energy is dissipated within the battery is i^2r. A battery with emf $\mathscr{E}$ and internal resistance r, connected across a load resistor R, is represented by the circuit diagram shown in Figure 28-3. Since the emf is by definition the energy per unit charge gained from the battery, the time rate at which nonelectric chemical potential energy is transformed into electric energy, or the power associated with the emf, is $\mathscr{E}i$. Some of the electric energy is dissipated to thermal energy within the battery at the rate i^2r; the remaining energy is delivered to the load resistance at the rate $iV_{ab} = i^2R$. Therefore, energy conservation gives

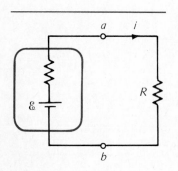

$$\mathscr{E}i = i^2r + i^2R = i(ir + iR)$$

$$\mathscr{E} = i(r + R)$$

Since iR is the potential difference V_{ab} across the load resistor and also across the battery terminals, the equation above can be written

$$V_{ab} = \mathscr{E} - ir \qquad (28\text{-}1)$$

Figure 28-3. *The color line shows what is within the battery: an emf and an internal resistance.*

Thus, the potential difference across the battery terminals is the battery's emf less the potential drop across its internal resistance. The potential difference V_{ab} across the battery terminals will equal the emf $\mathscr{E}$ only if the current i is zero, but as we see from (28-2), this will happen only when the load resistor R is infinite. Said differently, the potential difference appearing across the battery terminals on *open circuit* ($R = \infty$) is the battery's emf. A *short circuit* corresponds to $R = 0$.

The emf of a particular type of cell depends only on the chemical identity of its parts. As a battery ages or loses its "charge" (strictly, a battery loses only its internal chemical potential energy), the internal resistance increases but the emf remains unchanged. When the battery is "discharged," V_{ab} goes to zero as r increases, even for a relatively small load resistance R.

An electrochemical cell, or battery, is but one example of a device characterized by an emf. Other arrangements that can convert nonelectric energy to electric energy are these:

• A *thermocouple* consisting of two dissimilar conducting wires connected together in a circuit, with the two junctions maintained at different tempera-

---

* The term *electromotance* is sometimes used for this quantity.

tures. An electric current then exists in such a circuit, driven by a *thermoelectric* emf.

- A *photovoltaic cell*, exemplified by a photographic exposure meter, where visible light strikes a sensitive material and generates an electric current.
- An electric generator, with mechanical energy transformed into electric energy.
- Electromagnetic induction effects, where an emf is produced by a changing magnetic flux (to be treated in Chapter 31).

Example 28-1. A transistor radio runs on a battery rated 9 V and 500 mA·h, and costing $2. (a) What is the battery's energy content? (b) What does this energy cost in cents per kilowatt-hour?

(a) The quantity 500 mA·h is electric *charge*, the product of current and time. The total electric energy delivered by the electrochemical cell is

$$\mathscr{E}Q = (9 \text{ V})(500 \text{ mA·h}) = 4.5 \text{ W·h} = 4.5 \times 10^{-3} \text{ kW·h}$$

Not all this energy is delivered to the load, however. As the battery ages and its internal resistance increases, an increasingly larger fraction of the total energy is dissipated within the battery. (A fresh battery might have an internal resistance much less than 1 Ω. After a typical zinc-carbon battery has been connected to a 1000-Ω load for 4 h per day for two weeks, its internal resistance has grown to about 1000 Ω, and the potential drop across its terminals has dropped to half the initial value.)

(b) The total energy cost, only a portion of which goes to the load, is

$$\frac{\$2}{4.5 \times 10^{-3} \text{ kW·h}} = 4.4 \times 10^4 \text{ ¢/kW·h}$$

Compare this with a typical cost of 10¢/kW·h for household electricity. One pays dearly for the convenience of portable batteries.

28-2 Single-Loop Circuits

We wish to find the general relation between the emf's in a circuit and the potential differences. Consider, as a specific example, the circuit of Figure 28-4, which contains two batteries with emf's $\mathscr{E}_1$ and $\mathscr{E}_2$ and three resistors r_1, r_2, and R connected in series (r_1 is the internal resistance of $\mathscr{E}_1$ and r_2 of $\mathscr{E}_2$). The arrows associated with the two batteries, pointing from the negative to the positive terminal in both cases, indicate the directions of the electric field associated with the nonelectrostatic forces; they give, so to speak, the "directions" of the emf. The direction, or sense, of the current i is precisely the same at each point in the circuit loop because of electric-charge conservation, which implies that electric charge does not pile up or become depleted at any one point in the circuit.

Consider how energy conservation applies to the circuit. Proceeding in a clockwise direction around the loop, in the direction chosen for current i, we simply match the total energy delivered *by* the energy sources with the total energy delivered *to* the circuit elements (here, resistors). The power delivered *by* the batteries is

$$\mathscr{E}_1 i + \mathscr{E}_2 i = i \Sigma \mathscr{E}$$

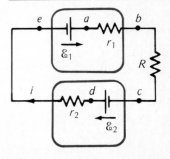

Figure 28-4.

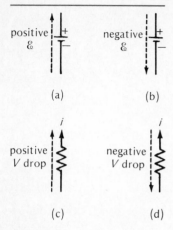

Figure 2-5. *Sign conventions for emf's and potential drops. The dashed arrows indicate the direction in which the circuit element is traversed. (a) The emf is taken as positive if the battery is traversed from the negative to the positive terminal. (b) The emf is negative if traversed from the positive to the negative terminal. (c) The potential drop is positive if the resistor is traversed in the same direction as the current. (d) The potential drop is negative if the resistor is traversed in the direction opposite to that of the current.*

We take both terms on the left side to be positive, since the directions of $\mathcal{E}$ and i are the same for both batteries. The total power delivered *to* the circuit elements is

$$iV_{ab} + iV_{bc} + iV_{de} = i\Sigma V$$

Note especially that ΣV includes only the potential differences across circuit elements, not the potential rises across the batteries.

Equating power delivered by the energy sources to the power delivered to circuit elements, we have

$$\Sigma\mathcal{E} = \Sigma V \tag{28-2}$$

This is the fundamental relation for solving single-loop dc circuits. One must be careful about signs in applying this relation. Note that $\Sigma\mathcal{E}$ is the *algebraic* sum of the emf's in the circuit. An emf is taken as positive if the battery is transversed from the negative to the positive terminal; but the emf must be taken as negative if the battery is traversed in the other direction. See Figures 28-5(a) and (b). Similarly, ΣV is the *algebraic* sum of potential *drops* across circuit elements. Any V is taken as positive if we traverse the circuit element from one potential to a lower potential in the direction of the conventional current; V is negative if we traverse the circuit element in the direction opposite that of the current. See Figures 28-5(c) and (d).

Example 28-2. Two batteries are connected in opposition, as shown in Figure 28-6; the emf's and internal resistances are 18.0 V and 2.0 Ω, 6.0 V and 1.0 Ω, respectively. (a) What is the current in the circuit? (b) What is the potential difference across the battery terminals? (c) At what rate does the discharging battery charge the charging battery? (d) At what rate is energy dissipated as heat in the 6.0-V battery?

(a) We decide to traverse the circuit of Figure 28-6 in the *clockwise* sense. Moreover, we take this clockwise sense as the direction of the current i, knowing that the current will be clockwise, since its direction will be controlled by the emf of the larger, 18.0-V battery. But it is not necessary to do so. We may choose the current direction arbitrarily, and then if it is not correct, the current will appear *negative* in the solution; that is, the current will have been shown actually to exist in the sense opposite to that chosen initially.

With the circuit traversal and current directions taken as shown in Figure 28-6, we have

$$\Sigma\mathcal{E} = \Sigma V$$

$$18V - 6V = 2i + 1i$$

Note that the emf of the 6-V battery must be assigned a minus sign, since we pass through this battery from the positive to the negative terminal. The potential drops across the two resistors, $2i$ and $1i$, are both positive, since we traverse each resistor in the same direction as the current. Solving for the current i in the relation above gives

$$i = \frac{12.0 \text{ V}}{3.0 \text{ }\Omega} = 4.0 \text{ A}$$

(b) Within the 18.0-V battery there is a potential drop across the 2-Ω internal resistance of

$$V = iR = (4.0 \text{ A})(2.0 \text{ }\Omega) = 8.0 \text{ V}$$

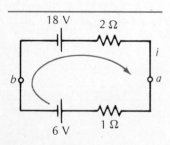

Figure 28-6.

Therefore, across the 18.0-V battery terminals the potential difference is 18.0 V − 8.0 V = 10.0 V.

If a 10.0-V potential difference exists across the 18-V battery terminals, this same potential difference must also appear across the 6.0-V battery terminals. Along the lower path from a to b in Figure 28-6, we have a potential drop $(4.0 \text{ A})(1.0 \ \Omega) = 4.0 \text{ V}$ across the internal resistance and a *rise* of 6.0 V from the emf, giving a total potential drop of $(4.0 + 6.0)\text{V} = 10.0 \text{ V}$.

(c) The rate at which any emf delivers energy is $\mathcal{E}i$. If $\mathcal{E}i$ is positive — that is, if the emf and current are both in the same direction — chemical energy from the battery is delivered *to* other circuit elements. On the other hand, if $\mathcal{E}$ and i are of opposite sign, as for the 6-V battery here, energy is delivered by other sources *to* it. Thus, the 18-V battery delivers energy (that is, loses chemical potential energy) at the rate

$$\mathcal{E}i = (18.0 \text{ V})(4.0 \text{ A}) = 72 \text{ W}$$

The 6-V battery has energy delivered to it (that is, it gains chemical potential energy) at the rate $\mathcal{E}i = (6.0 \text{ V})(4.0 \text{ A}) = 24 \text{ W}$. The battery with the larger emf is being discharged at the rate of 72 W, while the smaller battery is being charged at the rate of 24 W. What happens to the difference, $72 \text{ W} - 24 \text{ W} = 48 \text{ W}$? It must be dissipated in the two resistors.

(d) The rate at which energy is dissipated within the 6-V battery is

$$P = i^2R = (4.0 \text{ A})^2(1.0 \ \Omega) = 16 \text{ W}$$

and the rate at which energy is dissipated within the 18-V battery is $P = i^2R = (4.0 \text{ A})^2(2.0 \ \Omega) = 32 \text{ W}$. The total power dissipated in the resistors, $16 \text{ W} + 32 \text{ W} = 48 \text{ W}$, is just the difference between the power delivered *by* the discharging battery (72 W) and the power delivered *to* the charging battery (24 W).

28-3   Resistors in Series and in Parallel

Series and parallel arrangements of circuit elements were defined in Figure 26-6.

Series Resistors   What is the equivalent resistance of resistors in series? By the equivalent resistance is meant the resistance of that single resistor which, when it replaces the separate resistors, does *not* change the current drawn from the energy source.

Clearly, in Figure 28-7, the same current i passes through each resistor, so that

$$V = V_1 + V_2 + V_3$$

$$= iR_1 + iR_2 + iR_3 = i(R_1 + R_2 + R_3)$$

where V_1, V_2, and V_3 are the potential drops across resistors R_1, R_2, and R_3, and V is the potential difference across the battery terminals. The same current i will exist in the circuit when the single equivalent R replaces the group in series:

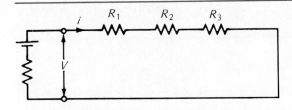

Figure 28-7.   *Resistors in series.*

$$V = iR$$

Comparing the two equations above gives

Resistors in series: $R = R_1 + R_2 + R_3 = \Sigma R_i$ (28-3)

The equivalent resistance is the sum of the separate resistances in series. The *same current* exists in each series resistor; the potential drop is the same across the entire group as across the single equivalent resistance.

Parallel Resistors   Recall first that in a parallel arrangement of circuit elements [Figure 26-6(b)] a single conductor splits into two or more conductors, so that there are alternative ("parallel") routes between two points in a circuit. All the arrangements of resistors shown in Figure 28-8 are effectively the *same*.

We can see that the qualitative effect of adding resistors in parallel is to reduce the effective total resistance, since every additional parallel resistor provides an additional path for charges to flow from the input to the output location.

What is the equivalent resistance of resistors shown in Figure 28-8? We designate the potential drops across the resistors R_1, R_2, and R_3 as V_1, V_2, and V_3. The input current is i, and the currents through the resistors are i_1, i_2, and i_3. Current i is divided into three currents through the resistors:

$$i = i_1 + i_2 + i_3$$

$$= \frac{V_1}{R_1} + \frac{V_2}{R_2} + \frac{V_3}{R_3}$$

where $i_1 = V_1/R_1$, and so on.

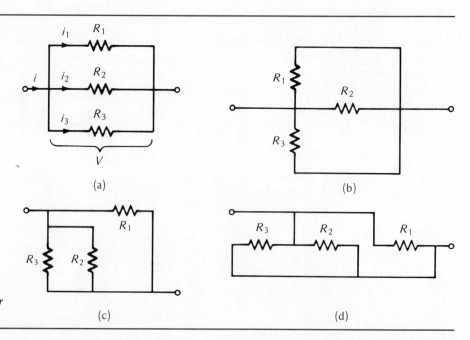

Figure 28-8.   *All four resistor circuits are the same circuit.*

In traversing any loop in Figure 28-8(a), we see that the potential differences all have the *same* value V:

$$V = V_1 = V_2 = V_3$$

The equation above can then be written as

$$i = V\left[\frac{1}{R_1} + \frac{1}{R_2} + \frac{1}{R_3}\right]$$

A single equivalent resistor R replacing the parallel resistors must satisfy the relation

$$i = \frac{V}{R}$$

Comparing the last two equations gives, finally,

$$\text{Resistors in parallel:} \quad \frac{1}{R} = \frac{1}{R_1} + \frac{1}{R_2} + \frac{1}{R_3} = \Sigma\frac{1}{R_i} \qquad \text{(28-4)}$$

The reciprocal of the equivalent resistance equals the sum of the reciprocals of the separate parallel resistances. The equivalent resistance is always less than the smallest of the parallel resistances. The essential feature of parallel connections is that all elements have the *same potential difference*. It follows that for any two resistors in parallel, the ratio of the currents is inverse to the ratio of the respective resistances. For example, with 1.0- and 3.0-Ω resistors in parallel, the 1.0-Ω resistor always has three times the current of the 3.0-Ω resistor, which ensures that the potential difference across both is always the same.

Example 28-3.   Consider the circuit of Figure 28-9. Find (a) the current in the battery, (b) the current in the 3.0-Ω resistor, (c) the potential difference across the 6.0-Ω

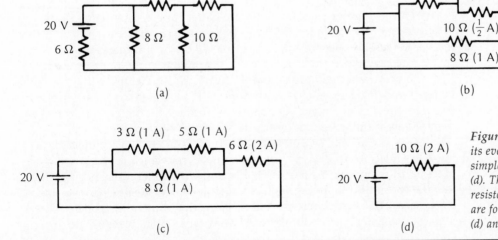

(a)

(b)

(c)

(d)

Figure 28-9.   A circuit (a) and its evolution into progressively simpler equivalent forms (b) to (d). The currents in the various resistors (shown in parentheses) are found by starting with part (d) and working backward to (a).

resistor, and (d) the rate at which thermal energy is dissipated in the 8.0-Ω resistor. The currents shown in parentheses are not given, but calculated.

We first recognize that any of the questions that can be asked about a circuit like that in Figure 28-9(a) (potential differences, power dissipation, and the like) require that we first find the current through each resistor. Other quantities then can be easily computed. To find the current through each resistor, we must first reduce the complex of resistors, through the rules for combining resistors in series and in parallel, until we are left with a single equivalent resistor connected across the battery.

Figures 28-9(b) to (d) show the evolution of the circuit, Figure 28-9(a), into progressively simpler forms. (The numerical values of the resistances have been so chosen here that the computations can, without difficulty, be carried out in one's head.) We see that the current through the single equivalent resistor is 2 A. Now we work backward, through Figures 28-9(c), (b), and (a), in turn, to find the current in each resistor, Here we use the facts that the current through all series resistors is the same, and that the potential difference across all parallel resistors is the same. Of course, as soon as we find the current through a given resistor, we can immediately compute the potential drop iR across it and the power $i^2R = iV$ dissipated in it. In this way we have, finally:

(a) Current through battery = 2 A.
(b) Current through 3-Ω resistor = 1 A.
(c) Potential difference across 6-Ω resistor = 12 V.
(d) Power dissipated in the 8-Ω resistor = 8 W.

28-4   DC Circuit Instruments

Here we consider the essential features of four instruments common in dc circuits:

- Ammeter, for measuring electric current.
- Voltmeter, for measuring electric potential difference.
- Wheatstone bridge, for comparing resistances.
- Potentiometer, for comparing emf's.

Ammeter and Voltmeter   Let us be clear on how the ammeter and voltmeter are connected in a circuit to measure current and potential difference. Figure 28-10(a) shows a simple circuit of a battery and resistor with a current i. When an ammeter (symbolized by —Ⓐ—) is connected in series with the resistor to measure the current, as in Figure 28-10(b), it changes the circuit so that the measured current i' is less than i, the original current. If the resistance R_a of the ammeter is *much less* than R, then $i' \simeq i$. In Figure 28-10(c), a voltmeter (symbolized by —Ⓥ—) is connected in parallel across the resistor. Again the current is changed. The current i'' through the battery is now greater than i. Only if the voltmeter resistance is *large* compared with R will $i'' \simeq i$.

Ideally, no current or potential difference is altered by connecting ammeters or voltmeters in a circuit; an ammeter resistance is so low that the potential difference across it is negligible, and a voltmeter resistance is so high that the current through it is negligible. Actually, any measuring instrument, electrical or otherwise, interferes somewhat with the quantity to be measured. In designing an ammeter or voltmeter, we want to minimize these perturbing effects, or at least to be aware of them.

Ordinary ammeters and voltmeters both use a pivoted needle whose angular position registers the current or potential difference. The needle is attached to a coil through which current passes, and this coil is immersed in the magnetic field of a magnet. Such a device is known as a *galvanometer* (an instrument we shall discuss in more detail in Section 29-7). A galvanometer (symbolized by —Ⓖ—) registers small currents. For example, a galvanometer showing full-scale deflection for a current of 1.0×10^{-6} A $= 1.0$ μA and having a resistance of 1,000 Ω might be typical (very much more sensitive galvanometers give full-scale deflection for currents as small as 10^{-12} A). Such a galvanometer, when placed into a circuit, will certainly register a full-scale deflection when 1.0 μA passes through it; at the same time, there will exist a potential difference across its terminals

$$V = iR = (1.0 \times 10^{-6} \text{ A})(1.0 \times 10^3 \text{ }\Omega) = 1.0 \times 10^{-3} \text{ V} = 1.0 \text{ mV}$$

If this 1.0-mV potential difference is small compared with potential differences existing across other circuit elements, the galvanometer's influence is negligible.

By itself, such a galvanometer could be used as a microammeter to measure currents up to 1 μA, or as a millivoltmeter to measure potential differences up to 1 mV.

Suppose, however, that we are to construct an ammeter that registers 1.0 A full-scale, using this galvanometer. To do this, we connect a very small resistance R_p in parallel with the galvanometer, as shown in Figure 28-11; said differently, one "shunts" the galvanometer with a small resistance R_p. What is R_p? Nearly all the current, 1.0 A, through the ammeter will go through the shunt resistance R_p since only 1.0 μA is permitted through the galvanometer itself. Moreover, the potential differences across the galvanometer and its shunt must both be 1.0 mV at full-scale deflection. It follows that

$$R_p = \frac{V}{i} = \frac{1.0 \times 10^{-3} \text{ V}}{1.0 \text{ A}} = 1.0 \times 10^{-3} \text{ }\Omega$$

The total resistance of the ammeter is also close to 10^{-3} Ω.

Suppose, now, that we use this same galvanometer to construct a voltmeter registering 10 V full-scale. To construct a voltmeter from a galvanometer, we place a high resistance R_s in series with it, as shown in Figure 28-12. Since the galvanometer alone can have a potential difference of only 1.0×10^{-3} V across its terminals, the potential difference V across the resistor R_s must be essentially 10 V, when the current i through it is 1.0×10^{-6} A. Therefore,

$$R_s = \frac{V}{i} = \frac{10 \text{ V}}{1.0 \times 10^{-6} \text{A}} = 1.0 \times 10^7 \text{ }\Omega = 10 \text{ M}\Omega$$

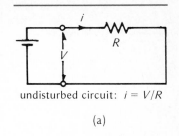

undisturbed circuit: $i = V/R$

(a)

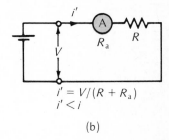

$i' = V/(R + R_a)$
$i' < i$

(b)

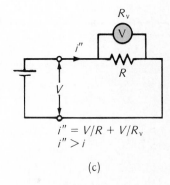

$i'' = V/R + V/R_v$
$i'' > i$

(c)

Figure 28-10.   A circuit (a) undisturbed, (b) with an ammeter in series with the resistor, and (c) with a voltmeter in parallel with the resistor.

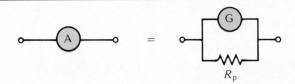

Figure 28-11.   An ammeter is a galvanometer in parallel with a small resistance.

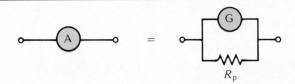

Figure 28-12.   A voltmeter is a galvanometer in series with a large resistance.

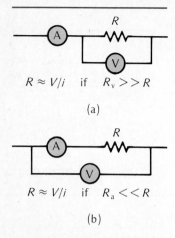

$R \approx V/i$ if $R_v >> R$

(a)

$R \approx V/i$ if $R_a << R$

(b)

Figure 28-13. Two arrangements for measuring the resistance R with an ammeter and a voltmeter: (a) voltmeter across R; (b) ammeter in series with R.

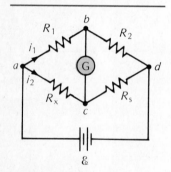

Figure 28-14. A Wheatstone bridge circuit.

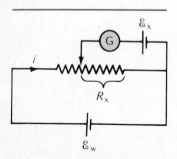

Figure 28-15. A potentiometer circuit for comparing emf's.

The voltmeter has so large an internal resistance that it produces little perturbation of any circuit with elements whose resistance is small compared with 10 MΩ.

A simple way of measuring resistance is shown in Figure 28-13(a); one measures the current i through the device with an ammeter and the potential difference V across it with a voltmeter, and then applies Ohm's law, $R = V/i$. Strictly, however, the current measurement is too high, since the ammeter of Figure 28-13(a) registers the current through both the resistor and the voltmeter. You could correct this by placing the ammeter inside the connections to the voltmeter, as shown in Figure 28-13(b). The ammeter then reads the current through R alone, but the voltmeter reads the potential difference, not across R alone, but across the resistance and ammeter.

Wheatstone Bridge   The difficulties mentioned above are eliminated when resistance is measured, or more properly compared, with a known standard resistance R_s, using the bridge circuit devised by C. Wheatstone (1802–1875). See Figure 28-14. In its simplest form, the circuit consists of four resistors, a battery, and a sensitive galvanometer. The values of R_1, R_2, and R_s are all known; R_x is the unknown resistance. Like the ordinary beam balance, which indicates equal masses on its two pans when the needle shows no deflection from the vertical, the Wheatstone bridge is a *null instrument*. With a given unknown resistance R_x, the resistors R_1, R_2, and R_s are so adjusted that the galvanometer registers no current. Then points b and c are at the same potential. The current i_1 through resistor R_1 is the same as the current through R_2; likewise, R_x and R_s carry the current i_2. The potential differences V_{ab} and V_{ac} are equal, as are V_{bd} and V_{cd}. It follows that

$$V_{ab} = i_1 R_1 = V_{ac} = i_2 R_x$$
$$V_{bd} = i_1 R_2 = V_{cd} = i_2 R_s$$

Eliminating i_1 and i_2 from these relations yields

$$\frac{R_x}{R_s} = \frac{R_1}{R_2} \tag{28-5}$$

With the bridge balanced, the unknown resistance R_x is computed in terms of the standard resistance R_s and the ratio R_1/R_2, using (28-5).

Potentiometer   Suppose that the emf of a battery is to be measured. A voltmeter placed across the battery terminals reads the *approximate* emf. The voltmeter reading V is always less than the true emf, since the potential difference appearing across a battery's terminals is the emf $\mathcal{E}$ less the potential drop ir across the battery's internal resistance r:

$$V = \mathcal{E} - ir \tag{28-1}$$

The emf $\mathcal{E}$ and potential difference V can be the same only if no current passes through the battery.

The potentiometer circuit permits the emf of a battery to be measured under the condition that the battery current is actually zero. Strictly, the potentiometer permits an unknown emf $\mathcal{E}_x$ to be compared with the precisely known emf $\mathcal{E}_s$ of a *standard cell* by a *null* method. The circuit is shown in Figure 28-15. A so-called working battery with emf $\mathcal{E}_w$, which need not be known but must be

Physics is fundamental and powerful and immensely practical and beautiful . . .

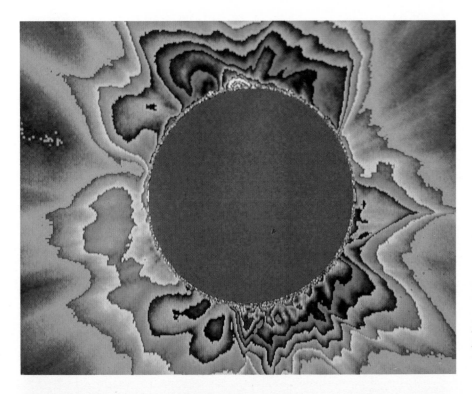

This picture of the sun was reconstructed from the image recorded by the Solar Max satellite. Near 12 o'clock is a solar flare; charged particles in the flare are deflected into an arc by a strong magnetic field. (Dan McCoy/Rainbow)

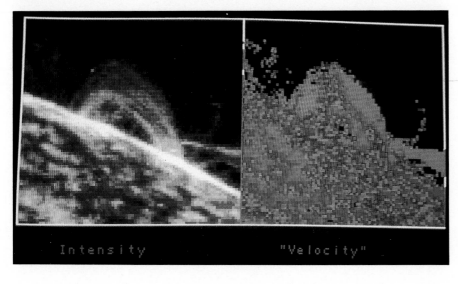

Intensity   "Velocity"

Other images recorded by the Solar Max satellite of the explosion of particles in a solar flare on the sun's surface: on the left, intensity; on the right, particle velocity. (Dan McCoy/Rainbow)

Since Wilhelm Röntgen won the first Nobel prize in physics in 1901 for the discovery of X-rays, advances in physics have often had medical applications. The 1952 Nobel prize in physics went to Felix Bloch and Edward M. Purcell for contributing to the discovery of nuclear magnetic resonance. NMR is now used to produce images like the cross-section of the human head below. Magnetic resonance imaging *measures the rate at which the magnetic moments of inverted nuclear spins recover to their equilibrium configurations; this rate depends strongly on the spins' immediate surroundings and differs from one kind of tissue to another. (Dan McCoy/ Rainbow)*

The lead-coated ball is floating above a coil of niobium wire. Cooled by liquid helium to a temperature of 4.2 K, these materials lose all electrical resistance and become superconductors (Section 27-4). Once induced in the coil, the current persists indefinitely (witness the disconnected leads to the power supply). A mere laboratory curiosity, this example of magnetic levitation? Not at all! A prototype maglev passenger train, floating off the ground through the use of superconductors, has run in Japan at 480 km/h. Superconducting magnets can maintain very large magnetic fields at virtually no cost in electric energy and are commonly used in industry; they may well become household items within a couple of decades. (Photograph © Michael Freeman)

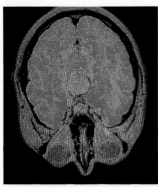

Fermilab, located about 50 km west of Chicago, is an international center for research in high-energy physics and the properties of elementary particles. In the final stages of acceleration, protons race through an evacuated underground tube at nearly the speed of light around the circle 6.3 km in circumference; the particles are deflected by magnetic fields and energized by electric fields. The protons finally fly off the circle at a tangent to enter target areas on the left. (The Fermilab herd of buffalo grazes nearby.) (Courtesy Fermi National Accelerator Laboratory)

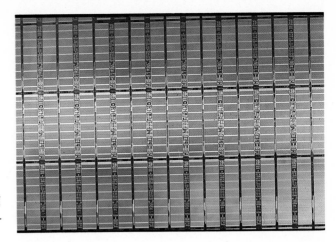

A close-up of a section of a microelectronic circuit, the IBM 256k random-access-memory chip (Problems 26-57, 27-53). (Dan McCoy/Rainbow)

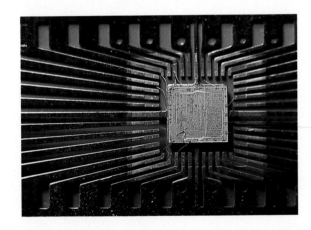

A Xilog computer chip with its attached leads. (Dan McCoy/Rainbow)

Are the lead wires really connected firmly to the circuit within the computer chip? One way to find out is to take a color acoustic micrograph, a very recently developed technique. First, an image is formed when a beam of inaudible ultrasonic radiation having a wavelength less than 0.1 mm travels through the chip and through a layer of liquid and finally reaches a thin gold film on the far side. The acoustic picture consists of tiny dimples and protuberances on the gold film. This image is read by a fine laser beam that scans the gold surface in the same fashion as an electron beam drawing a picture on a television picture tube. Any displacement on the gold surface causes the laser beam to be deflected, and this deflection is finally translated into color corresponding to that one point in the final picture. The picture here shows that there are indeed some loose connections. (© T. E. Adams)

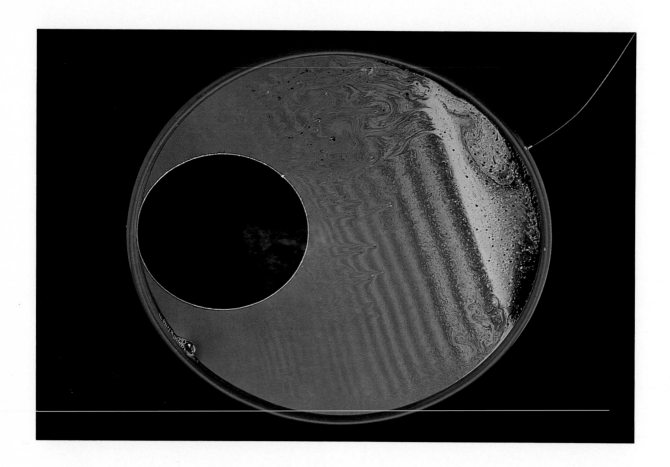

(Above) A soap film in a wire ring; the black region inside the loop of thread is open. The parallel bands (interference fringes) are produced by light reflected from the back and front surfaces of the thin film (Section 38-8). The film is thinner at top than at bottom. Where is the bottom? The soap film minimizes its surface area, so that the hole in the loop of thread is perfectly circular (it looks slightly elliptical because the wire ring was tipped in this picture). (Gordon Gahan/Prism)

The laser: it reads Uniform Product Code numbers at the supermarket check-out counter and the audio signal on a compact disc; it performs surgery in the operating room and transmits telephone calls along optical fibers. The laser (Section 42-6) is also commonly used for delicate measurements in the place where it first originated — the physics laboratory. Here a laser is used to monitor rotation in a latter-day Cavendish experiment (Section 15-3) to test whether gravity is slowly weakening ("constant" G decreasing by as little as 1 part in 10^{10} per year). (Photograph © Michael Freeman)

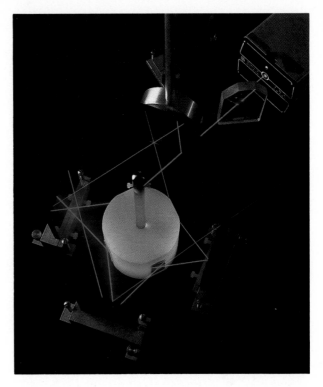

An array of special retro-reflecting prisms of the sort left on the moon by astronauts. The device looks something like a bicycle reflector, and it works in the same way. Behind each curved surface is a square corner of glass that, through total internal reflection, acts as a corner mirror (Sections 36-4 and 36-7). Any light ray reaching the corner is reflected back in a direction precisely opposite to the incident ray. Timing a laser pulse from earth to moon and back allows small changes in the earth-moon distance — even a few centimeters — to be detected (panel on p. 779). (Photograph © Michael Freeman)

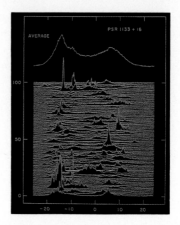

(Above) Signals from a pulsar (panel on pages 318-319) picked up on the Arecibo telescope. We see pulses in the individual traces and in their summation. (Dan McCoy/Rainbow)

(Left) The world's largest single-dish radio telescope is at the Arecibo Observatory in Puerto Rico. Fitted into a natural mountain basin, the dish has a diameter of 0.33 km. Microwave signals from astronomical sources are reflected from the parabolic mirror and focused into a moveable antenna supported 160 m above the dish from three towers. Why the large size? A stronger signal, of course, but also to give a sharper image by improving the resolution (Section 39-4). (Dan McCoy/Rainbow)

One way to achieve high resolution with a radio telescope is to build a single large dish. Another is to use many individual smaller antennas spread over a large distance and combine their signals through interference. It's rather like having a diffraction grating (Sections 38-5 and 39-7) — with its well resolved, sharp lines — run backward. The world's biggest and most sensitive radio telescope, at Socorro, New Mexico, part of the National Radio Astronomy Observatory, is known simply as the Very Large Array. See also Figure 39-11. (Dan McCoy/Rainbow)

Innerland, a dream world in crystal, was created for Steuben Glass by the Scottish sculptor, Eric Hilton, to express his conception of "the unity of life, the oneness of time and space." This dazzling work consists of twenty-five crystal cubes, each 9.8 cm on an edge. A master cutter and four engravers worked four years to complete it. Note the interplay of refraction, dispersion, and total internal reflection (Sections 36-5, 36-6, 36-7). (Courtesy of Steuben Glass, New York)

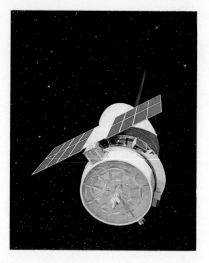

When Galileo Galilei first looked at Saturn in 1610 with the telescope he invented, he saw that the image was not symmetrical; Saturn had "ears." In 1656 Christian Huygens identified the rings. In 1856 James Clerk Maxwell (panel on pages 742-743) began a long study of the stability of the ring system; Maxwell determined that the rings must consist of independent particles. This close-up picture of Saturn was taken on October 5, 1980, by the Voyager spacecraft when it was 5.1×10^{10} m from the planet (and 1.5×10^{12} m from earth). (NASA photo.)

Visible light gets through the earth's atmosphere, and so do radio waves at some frequencies. Other electro-magnetic radiation from stellar sources is mostly absorbed before it can reach the earth's surface. This satellite probe launched by NASA in 1981 carries an infrared telescope. The IRAS (infrared astronomical satellite) is an international venture involving the United States, the Netherlands, and the United Kingdom. (NASA illustration.)

A conception by the artist Robert McCall of future solar sails. Each sail, perhaps several kilometers across, is propelled slowly but with high efficiency through the vacuum of interstellar space by the radiation pressure of sunlight (Section 35-6). Solar sails would be effective cargo vehicles for transporting — serenely but very cheaply — minerals from, say, a mine in the asteroid belt to a space station. (© Smithsonian Institution. Courtesy of the National Air and Space Museum.)

greater than $\mathscr{E}_x$ (or $\mathscr{E}_s$), maintains a constant current i through the resistor. The adjustable tap (see Figure 28-15 where the left end of the galvanometer wire connects to the resistor) is set so that the current through the sensitive galvanometer is zero. The potential drop across the galvanometer must be zero, and the potential drop across the internal resistance of the unknown battery is then also zero. Thus, the total potential drop across the branch in the circuit containing the unknown battery equals the battery's emf $\mathscr{E}_x$. But this is also the potential drop iR_x across the resistor from the tap to the right end. Therefore, $\mathscr{E}_x = iR_x$.

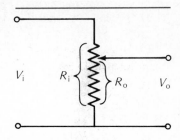

Figure 28-16. A potentiometer used as a voltage divider.

Now if the unknown battery is replaced by the standard cell and the resistor tap is again adjusted for balance (zero galvanometer current), we have $\mathscr{E}_s = iR_s$. Here R_s is the corresponding resistance from the tap to the end of the resistor and, because there is no current through the galvanometer, the current is still the same i. Eliminating i from these relations yields

$$\frac{\mathscr{E}_x}{\mathscr{E}_s} = \frac{R_x}{R_s} \qquad (28\text{-}6)$$

From this equation we see that comparing emf's with a potentiometer circuit consists in comparing resistances (or if the adjustable resistor is a wire of uniform cross section, in comparing lengths).

The term *potentiometer* is used in another sense in electric circuits — that of a *voltage divider;* see Figure 28-16. An input voltage V_i is applied across a variable resistor of total resistance R_i with a center tap. The output voltage is V_o, and R_o represents the resistance between the tap and the lower end of the resistor. Clearly, for negligible current in the output circuit, $V_o/V_i = R_o/R_i$.

28-5 Multiloop Circuits

Some circuits with more than one current loop, such as the one shown in Figure 28-9, can be solved simply by applying the rules for combining resistances in series and in parallel. Generally, however, this is not possible. Consider, for example, the relatively simple circuit shown in Figure 28-17. It is a circuit with *three* loops (a left inside loop, a right inside loop, and an outside loop going all the way around the circuit). There is no way to reduce this multiloop circuit into one involving a single battery and resistor.

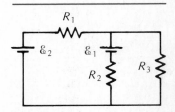

Figure 28-17. A simple multiloop circuit.

One general procedure for solving multiloop circuit problems is expressed by *Kirchhoff's rules.* These two rules are simply statements, in the language of electric circuits, of the fundamental *conservation laws* of (1) *electric charge,* and (2) *energy.*

Junction Rule Consider a *junction,* a point in the circuit where three or more conducting wires are joined, as shown in Figure 28-18. Currents in the several wires are labeled $i_1, i_2, i_3,$ and i_4; current i_1 is into the junction and the other currents are out of the junction. Net charge cannot accumulate or be depleted at a junction. Therefore, from electric-charge conservation we know that the net charge per unit time into any junction must equal the net charge per unit time out of the junction. The net current into the junction equals the net current out of the junction. For the situation shown in Figure 28-18, this implies that

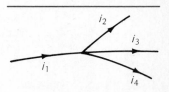

Figure 28-18. A circuit junction. The total current into the junction is zero.

$$i_1 = i_2 + i_3 + i_4 \qquad \text{or} \qquad i_1 - i_2 - i_3 - i_4 = 0$$

The negative terms appearing in the right-hand relation above may be interpreted as representing *negative* currents *into* the junction, which are equivalent to positive currents *out* of the junction. A general formulation of the junction theorem for currents is, then,

$$\Sigma i = 0 \tag{28-7}$$

where it is understood that currents into a junction are identified as positive and currents out of a junction as negative. The current rule is also known as *Kirchhoff's first rule.*

Loop Rule   Recall the general relation

$$\Sigma \mathscr{E} = \Sigma V \tag{28-2}$$

which says that the algebraic sum of the emf's $\Sigma \mathscr{E}$ equals the algebraic sum of the potential drops ΣV around the loop. Equation (28-2) is merely an expression of the energy conservation law. The left-hand side represents the energy per unit charge supplied by energy sources in the circuit loop, and the right-hand side represents the energy per unit charge delivered to circuit elements around the loop. Equation (28-2) is known as the *loop equation,* or *Kirchhoff's second rule.*

Remember the conventions to be observed in applying (28-7) and (28-2):

• Having decided arbitrarily on the sense in which a particular loop will be traversed (clockwise or counterclockwise), we never reverse this sense through any circuit element around the loop.

• We choose the direction for any unknown current arbitrarily, but observe the sign convention denoting currents into or out of a junction in using (28-7).

• We take the emf of a battery to be *positive* when it is traversed from the *negative to the positive* terminal, and we take a potential drop across a resistor to be *positive* when the resistor is traversed in the *same* direction as that of *current* flow.

Example 28-4.   Find the currents for the circuit of Figure 28-17, shown again in Figure 28-19. Currents in the *three branches* i_1, i_2, and i_3 are assigned directions arbitrarily. The three currents are the unknowns to be solved for. From (28-7), we have

$$i_1 - i_2 - i_3 = 0 \tag{A}$$

This is one equation in the three unknown currents. Applying (28-2) to each of the

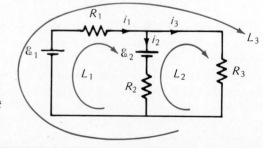

Figure 28-19.   Three circuit loops (L_1, L_2, and L_3) for applying Kirchhoff's rules.

loops in Figure 28-19 gives three more equations; it is readily apparent, however, that one of these equations is not independent of the other two. A procedure for removing redundancies of this sort is to choose only loops in which one always goes in the same direction as the current. In Figure 28-19, for example, we can choose loops L_1 and L_3. Applied to loop L_1, Equation (28-2) yields

$$\mathcal{E}_1 - \mathcal{E}_2 = R_1 i_1 + R_2 i_2 \qquad \textbf{(B)}$$

Applied to loop L_3, it gives

$$\mathcal{E}_1 = R_1 i_1 + R_3 i_3 \qquad \textbf{(C)}$$

We now have three simultaneous linear equations (A, B, C) in the three unknowns (i_1, i_2, and i_3). Solving for the currents is straightforward. The results are

$$i_1 = \frac{R_3(\mathcal{E}_1 - \mathcal{E}_2) + R_2\mathcal{E}_1}{R_1 R_2 + R_1 R_3 + R_2 R_3}$$

$$i_2 = \frac{R_3(\mathcal{E}_1 - \mathcal{E}_2) - R_1\mathcal{E}_2}{R_1 R_2 + R_1 R_3 + R_2 R_3}$$

$$i_3 = \frac{R_2\mathcal{E}_1 + R_1\mathcal{E}_2}{R_1 R_2 + R_1 R_3 + R_2 R_3}$$

The equation above shows that current i_3 is always positive; it has the direction shown in Figure 28-19. On the other hand, i_1 and i_2 can be either negative or positive, depending on the relative sizes of the emf's and resistances. With negative current, the actual direction is opposite that chosen in Figure 28-19.

Solving any multiloop circuit, however complicated, means applying Kirchhoff's rules and solving some simultaneous linear equations. One can work this in reverse, using electric circuits to solve linear equations. The emf's and resistances in a circuit are made to correspond to the parameters of the linear equations; solving for the unknowns then consists merely in measuring the currents in the various branches of the circuit. This is one simple example of an *analog computer* in which one studies the physical behavior of a system obeying a well-known mathematical relationship, to solve for mathematical unknowns.

Summary

Definitions

Emf: the work done per unit positive charge by an energy source.

Important Results

Around any loop

$$\Sigma V = \Sigma \mathcal{E} \qquad (28\text{-}2)$$

The emf $\mathcal{E}$ is positive if the source is traversed from negative to positive terminal. The potential drop V across any circuit element is positive if the element is traversed in the same direction as the conventional current through it.

Resistors in *series*, all with the *same current*:

Equivalent single resistor $R = R_1 + R_2 + R_3 + \cdots$

$$(28\text{-}3)$$

Resistors in *parallel*, all with the *same potential difference*:

$$\frac{1}{R} = \frac{1}{R_1} + \frac{1}{R_2} + \frac{1}{R_3} + \cdots \qquad (28\text{-}4)$$

Ammeter: registers current, connected in series with circuit element, low resistance.

Voltmeter: registers potential difference, in parallel with circuit element, high resistance.

Wheatstone bridge: null instrument for comparing resistances.

Potentiometer: null instrument for comparing emf's.

Multiloop circuits (Kirchhoff's rules):

● Junction rule:

$$\Sigma i = 0 \qquad (28\text{-}7)$$

where i is positive into a junction.

● Loop rule:

$$\Sigma \mathcal{E} = \Sigma V \qquad (28\text{-}2)$$

with the conventions given above.

Problems and Questions

Section 28-1 EMF

· **28-1 P** The manufacturer's specifications for a 12-volt storage battery rate the maximum current the battery can deliver as 80 A. What is the internal resistance of the battery?

· **28-2 P** The internal resistance of a 1.5-V zinc-carbon battery is found to be 1.5 Ω. What is the maximum current it can deliver?

· **28-3 P** A 6-V car battery provides a current of 4 A to a load for 8 min. By how much is the chemical energy of the battery decreased during this time?

· **28-4 P** The open-circuit voltage of a particular solar cell is 2.2 V, and the short-circuit resistance (that is, a load resistor with $R_L = 0$) is 1.1 A. (*a*) What is the internal resistance of the battery? (*b*) What is the emf of the battery?

: **28-5 P** Suppose that a load resistor R_L is connected across the terminals of a real battery. The power dissipated in the load resistor
(A) will be independent of the value of R_L.
(B) will increase when R_L is increased.
(C) will decrease when R_L is increased.
(D) may increase or decrease when R_L is increased, depending on the initial value of R_L.
(E) may decrease or stay the same, depending on the initial value of R_L.
(F) None of the above is true.

: **28-6 Q** In some of our considerations, we have neglected any internal resistance of a source of emf, such as a battery. In fact, batteries have appreciable internal resistance. This internal resistance can be taken into account by including a resistor R_i in series with the battery, as shown in Figure 28-20. In this case, one could deduce that the potential difference between the battery terminals would
(A) be independent of the load resistance R_L.
(B) increase as the load resistance R_L is increased.
(C) decrease as the load resistance R_L is increased.
(D) either increase or decrease as the load resistance is varied, depending on whether the load resistance is larger or smaller than the internal resistance.
(E) drop to zero if the load resistance were to exceed the internal resistance.

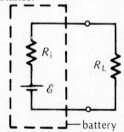

Figure 28-20. Question 28-6.

: **28-7 P** When a resistor is connected to a battery, a current of 2.0 A exists. When an additional 15 Ω is added to the circuit in series, the current drops to 0.4 A. What is the emf of the battery?

: **28-8 Q** A load resistance R_L is connected to a 6-V battery whose internal resistance is 0.1 Ω. The battery is rated at 120 A·h. Which of the following is the most accurate statement?
(A) The maximum current the battery can deliver is 30 A.
(B) The potential difference across the load will be a maximum when $R_L = 0.1$ Ω.
(C) The potential difference between the battery terminals will be a maximum when $R_L = 0$.
(D) The maximum power the battery can deliver to the load is 90 W.
(E) The maximum time for which the battery can deliver current without being recharged is 2 h.

Section 28-2 Single-Loop Circuits

· **28-9 P** A battery with an emf of 12 V and an internal resistance of 1.0 Ω is connected to a load resistor of 2 Ω. How much power is delivered to the load?

· **28-10 P** A 12-V car battery is rated at 200 A·h. If we assume that the potential difference stays constant and the internal resistance does not rise within the time for delivering the rated charge, for how many hours can the battery light a 50-W light bulb?

· **28-11 P** Two batteries are connected as shown in Figure 28-21. (*a*) What current flows in each battery? Indicate the direction with an arrow. (*b*) At what rate is heat generated in the 6-Ω resistor?

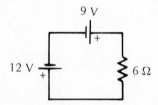

Figure 28-21. Problem 28-11.

· **28-12 P** A current of 2 A flows in the 4-Ω resistor in

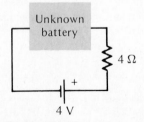

Figure 28-22. Problem 28-12.

Figure 28-22. What are the possible emf's of the unknown battery?

: 28-13 P Two batteries are connected in series to a load resistor. A current of 4.0 A exists. The polarity of one battery is then reversed by reconnecting it, and the current drops to 1.0 A. One battery has an emf of 12 V. What is the emf of the other?

: 28-14 P (*a*) Show that a battery of emf $\mathcal{E}$ and internal resistance *r* will deliver maximum power to a load resistor R_L when $R_L = r$. (*b*) Plot the power delivered to the load as a function of R_L/r.

: 28-15 P Show that the resistance of the infinite network shown in Figure 28-23 is $(1 + \sqrt{3})\, R$.

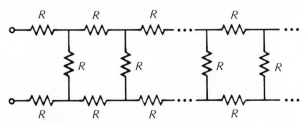

Figure 28-23.   Problem 28-15.

Section 28-3   Resistors in Series and Parallel

· 28-16 P Find the equivalent resistance of each combination of resistors shown in Figure 28-24.

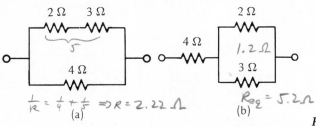

$\frac{1}{R} = \frac{1}{4} + \frac{1}{5} \Rightarrow R = 2.22\,\Omega$
(a)

$R_{eq} = 5.2\,\Omega$
(b)

$2\,\Omega \quad 3\,\Omega \quad 4\,\Omega$

$R_{eq} = 9\,\Omega$

(c)

$\frac{1}{R} = \frac{1}{2} + \frac{1}{3} + \frac{1}{4} = 1.3.$
(d)

$R_{eq} = .923\,\Omega$

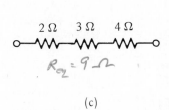

Figure 28-24.   Problem 28-16.

· 28-17 Q When several resistors are connected in parallel,
(A) the power dissipated in each is the same.
(B) the current through each is the same.
(C) the potential difference across each is the same.
(D) the net electric charge on each is the same.

· 28-18 P What is the resistance between terminals *X* and *Y* for the network in Figure 28-25?

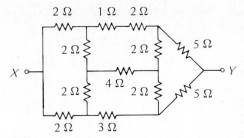

Figure 28-25.   Problem 28-18.

· 28-19 P What is the equivalent resistance of the network shown in Figure 28-26?

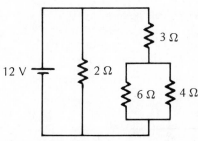

Figure 28-26.   Problem 28-19.

· 28-20 P What is the potential drop across the 4-Ω resistor in the circuit of Figure 28-27 for 1 A through the 6-Ω resistor?

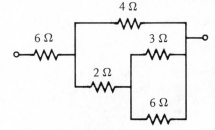

Figure 28-27.   Problem 28-20.

· 28-21 P When the switch in the circuit of Figure 28-28 is closed, what change will occur in the current in the 1-Ω resistor?

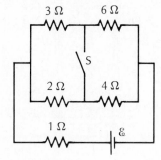

Figure 28-28.   Problem 28-21.

(A) There will be no change.
(B) The current will increase.

(C) The current will decrease.

(D) Whether or not the current will increase depends on the value of the battery emf.

: 28-22 P A car battery that is about to fail has an emf of 11.6 V and an internal resistance of 0.02 Ω. It is providing current to a 2-Ω load. A second battery, of emf 12.6 V and internal resistance 0.01 Ω, is connected in parallel to it to help it out. Determine (a) the circuit that describes the situation; (b) the current in each battery and in the load; (c) the total power delivered by the 12.6-V battery and the power delivered to the load; (d) the power delivered by the 11.6-V battery. (*Hint:* Pay attention to the sign of the power and the physical meaning of it.)

: 28-23 P Four heating coils, each rated at 150 W at 120 V, are to be operated in combination from a 120-V supply. What different power outputs can be obtained by connecting them in various arrangements?

: 28-24 P What is the electric potential at point *P* in the circuit of Figure 28-29?

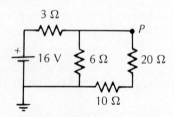

Figure 28-29. *Problem 28-24.*

: 28-25 Q If the circuit were broken at point *P* in Figure 28-30,

(A) the current in R_1 would not change.

(B) the potential difference between point *X* and ground would increase.

(C) the current provided by the battery would increase.

(D) the emf provided by the battery (assumed to have no internal resistance) would change.

(E) the current in R_4 would increase.

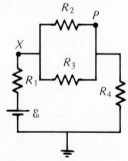

Figure 28-30. *Problem 28-25.*

: 28-26 Q If the circuit shown in Figure 28-31 were broken at point *F*,

(A) the current in R_1 would not change.

(B) the current in R_2 would increase.

(C) the current in R_6 would decrease.

(D) the potential difference between *G* and *P* would decrease.

(E) the potential difference between *G* and *Q* would change.

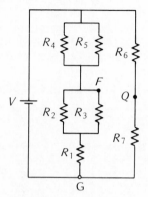

Figure 28-31. *Question 28-26.*

: 28-27 P (a) Find the equivalent resistance of the network shown in Figure 28-32. (b) What potential difference applied to the terminals will result in a current of 1 A in the 6-Ω resistor?

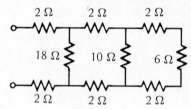

Figure 28-32. *Problem 28-27.*

: 28-28 P A three-way light bulb has two filaments connected as shown in Figure 28-33. It is possible to apply 120 V across connections *ab, ac,* or *bc.* (a) What resistances are required for a 150 W-75 W-50 W lamp? (b) What power outputs would be obtained if $R_1 = 48$ Ω and $R_2 = 144$ Ω?

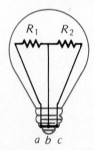

Figure 28-33. *Problem 28-28.*

: 28-29 P Twelve identical resistors, each of resistance *R*, are connected to form a cube, with each resistor on an edge

(Figure 28-34). Determine the equivalent resistance between corners A and D.

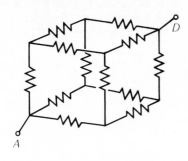

Figure 28-34.   Problem 28-29.

: 28-30 Q Sketched in Figure 28-35 is a circuit that has been designed so that the potential difference between points X and Y is 10 V. In trying to find the source of a difficulty that has arisen, you measure the potential between X and Y and find a difference of 15 V. You suspect that this is due either to a short circuit (from two bare wires touching) or to an open circuit (perhaps from a faulty solder joint). In the sketch, possible short circuits are indicated as dashed lines. Which of the following possibilities, if any, could account for your finding of $V_{XY} = 15$ V?
(A) Short circuit between T and Q.
(B) Short circuit between P and Z.
(C) Break in circuit at Q.
(D) Break in circuit at Z.
(E) Break in circuit at P.
(F) More than one of the above (indicate which).
(G) None of the above.

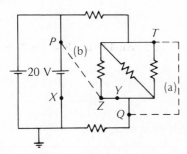

Figure 28-35.   Question 28-30.

: 28-31 Q Ordinary household light bulbs are meant to be connected in parallel to a 120-V supply. Suppose, however, that a 50-W lamp and a 100-W lamp are connected in series to a 120-V source. Which of the following statements best describes what will then happen?
(A) Both will be brighter than normal, with the 100-W lamp brighter than the 50-W lamp.
(B) Both will be dimmer than normal, with the 100-W lamp brighter than the 50-W lamp.
(C) The 100-W lamp will be brighter than normal, and the 50-W lamp will be dimmer than normal.

$$R = \frac{V^2}{P} \qquad P_{50} = 22.2\ W$$
$$P_{100} = 11.1\ W$$

(D) The 100-W lamp will not be so bright as the 50-W lamp.
(E) Both lamps will still have their normal brightness.

: 28-32 P What is the resistance of the network shown in Figure 28-36? (*Hint:* Use symmetry.)

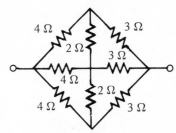

Figure 28-36.   Problem 28-32.

: 28-33 P Thermal energy is to be supplied to an experimental apparatus under changing conditions that may cause a change in the heater resistance R. It is desired to have the power dissipated in R independent (as much as possible) of the value of R. This can be achieved with the circuit shown in Figure 28-37. Determine the relation between R_1, R_2, and R if the power dissipated in R is independent, to first order in R, of R. The applied voltage V_0 is constant.

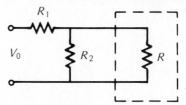

Figure 28-37.   Problem 28-33.

Section 28-4   DC Circuit Instruments

· 28-34 P The Wheatstone bridge shown in Figure 28-38 is balanced (there is no current in the galvanometer) when the variable resistance is 240 Ω. What is the value of the unknown resistance?

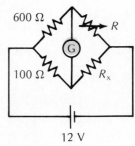

Figure 28-38.   Problem 28-34.

· 28-35 Q To say that a circuit is "grounded" at point P means that

(A) point P in the circuit is connected to the negative terminal of a battery.

(B) no current will flow through point P.

(C) point P is connected to a very large conductor, whose potential we arbitrarily choose to be zero volts.

(D) point P is electrically insulated from all other points, just as the earth is so insulated.

(E) point P is the point at which charge drains out of the circuit.

· **28-36 P** A galvanometer that has a resistance of 1000 Ω gives full-scale deflection for a current of 1.0 mA. (*a*) What shunt resistance is needed to construct an ammeter that deflects full-scale for a current of 100 A? (*b*) What length of No. 24 copper wire (diameter 0.0201 in., resistance 28.4 Ω per 1000 ft) would be needed to make the shunt resistor?

: **28-37 Q** In the circuit in Figure 28-39, the black box contains only resistors. If I make some changes inside the box and observe that the voltmeter reading decreases, then it is possible that I

(A) added some resistors.

(B) removed some resistors.

(C) changed resistors from parallel to series connection.

(D) decreased the current from the battery.

(E) All of the above are correct.

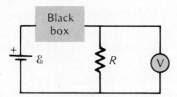

Figure 28-39. *Question 28-37.*

: **28-38 P** Shown in Figure 28-40 are two ways of measuring an unknown resistance R. If the resistance of the ammeter is 0.12 Ω and the resistance of the voltmeter 20 000 Ω, for what values of R will the resistance be given as V/i to within 1 percent under either arrangement?

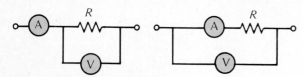

Figure 28-40. *Problem 28-38.*

: **28-39 P** A galvanometer has a resistance of 60 Ω and deflects full-scale for a current of 400 μA. Show how to use it to construct (*a*) a voltmeter that deflects full-scale for a voltage of 100 V; (*b*) an ammeter that deflects full-scale for a current of 5 A.

: **28-40 P** A galvanometer with a resistance of 800 Ω shows full-scale deflection when a current of 5 mA is passed through it. How would you use this galvanometer to construct (*a*) an ammeter registering 10-A full-scale deflection? (*b*) a voltmeter registering 500-V full-scale deflection?

: **28-41 P** The output of the voltage divider circuit shown in Figure 28-41 is to be measured with voltmeters of varying resistance. What will be the reading when the meter resistance is (*a*) 100 Ω? (*b*) 1 kΩ? (*c*) 50 kΩ? (*d*) 1 MΩ?

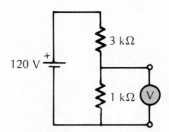

Figure 28-41. *Problem 28-41.*

: **28-42 P** A multirange ammeter is constructed using a galvanometer that gives a full-scale deflection when 1.0 mA flows in it (Figure 28-42). The galvanometer resistance is 1000 Ω. What are the values of the resistors (*a*) R_a? (*b*) R_b? (*c*) R_c?

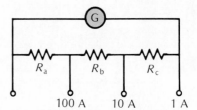

Figure 28-42. *Problem 28-42.*

: **28-43 P** A multirange voltmeter is constructed using a galvanometer that gives full-scale deflection for a current of 0.8 mA (Figure 28-43). The resistance of the galvanometer is 1200 Ω. What are the values of the resistors (*a*) R_a? (*b*) R_b? (*c*) R_c?

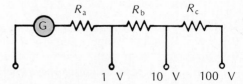

Figure 28-43. *Problem 28-43.*

: **28-44 P** Determine the ratio V/V_0 for the potential divider circuit shown in Figure 28-44.

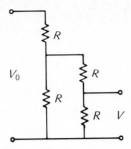

Figure 28-44. Problem 28-44.

: **28-45 Q** If the circuit shown in Figure 28-45 is broken at point P,
(A) the power delivered by the battery will increase.
(B) the current in the ammeter, I, will increase.
(C) the current in the ammeter will not change.
(D) the potential difference read by the voltmeter, V, will increase.
(E) the potential difference read by the voltmeter will decrease.
(F) the potential difference read by the voltmeter will not change.
(G) The answer cannot be determined without knowing R_x.
(H) More than one of the above is true.

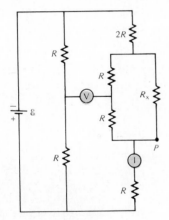

Figure 28-45.   Question 28-45.

Section 28-5   Multiloop Circuit
: **28-46 Q** For any network with current, it is always true that
(A) the sum of the currents around a closed loop is zero.
(B) the sum of the currents entering a junction equals the sum of the currents leaving the junction.
(C) the sum of the emf's around any closed loop is zero.
(D) the sum of the internal-resistance drops around any closed loop is zero.
(E) the total current, including all branches, is zero.

· **28-47 P** What is the value of the emf $\mathcal{E}_1$ (Figure 28-46)?

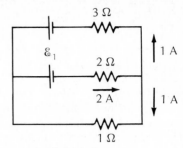

Figure 28-46.   Problem 28-47.

: **28-48 P** What is the potential difference between points X and Y in the circuit of Figure 28-47?

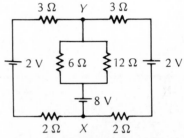

Figure 28-47.   Problem 28-48.

: **28-49 P** Suppose you have three 1.5-V flashlight batteries, each with internal resistance 1 Ω. What is the maximum power you can deliver to a load resistor, and how should the batteries be connected to do this when the load resistance has the value (a) 6 Ω? (b) 2 Ω? (c) 0.5 Ω?

Supplementary Problems
28-50 P A resistor network consists of an infinite square array of resistors, as shown in Figure 28-48. The resistance between any two adjacent junctions is R. What is the equivalent total resistance between two adjacent terminals, such as A and B in Figure 28-48? (*Hint:* the only easy way to solve this problem is to use two powerful ideas—symmetry and superposition. Superposition implies that the actual arrangement in which charge is injected into A and extracted

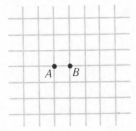

Figure 28-48.   Problem 28-50.

trom B may be considered as equivalent to taking the superposition of two situations: (a) injection of charge into A and collection at infinity and (b) extraction of charge from B with injection from infinity.)

28-51 P One of a pair of telephone wires is shorted out to ground in a storm. The pair of wires is 20 km long and runs through rough and inaccessible country. See Figure 28-49. The resistance of a single wire is 39 Ω/km. To determine the location of the short, the line crew must join the ends of the wires at one end of the line and establish a bridge circuit at the other. The bridge is balanced when $R_1 = 62 \ \Omega$ and $R_2 = 86 \ \Omega$. How far from point P has the short occurred?

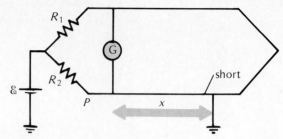

Figure 28-49.   Problem 28-51.

The Magnetic Force

29

The *fundamental* magnetic effect is this: An electric charge in motion produces a force on a second moving charge in addition to the electric (Coulomb) force. This velocity-dependent force between electrically charged particles is the *magnetic force.*

As we have seen, the electric force between charged particles may be thought to act through the intermediary of the electric field as follows:

- Charge q_a creates an electric field **E** at the site of charge q_b.
- Charge q_b, finding itself in field **E**, is acted on by an electric force.

In similar fashion, the magnetic interaction between charged particles is most readily described as taking place via a magnetic field:

- *Moving* charge q_a creates a *magnetic field* **B** at the site of moving charge q_b.
- *Moving* charge q_b, finding itself in field **B**, is acted on by a *magnetic force.*

The magnetic force is more complicated than the electric force since the magnetic force depends on the velocities of the two interacting charges as well as the charge magnitudes and signs and the charge separation distance. For this reason, it is useful to separate the magnetic interaction into two parts:

• One moving charge creates a magnetic field (the subject of Chapter 30).
• The magnetic field acts on a second moving charge (the subject of this chapter).

29-1 Magnetic Field Defined

The basic relation for the magnetic force $\mathbf{F_m}$ on a particle with charge q moving at velocity $\mathbf{v}$ in a magnetic field $\mathbf{B}$ — the relation on which all else in this chapter is based — is this:

$$\mathbf{F_m} = q\mathbf{v} \times \mathbf{B} \qquad (29\text{-}1)$$

This equation involves the cross product (Section 12-2) of vectors $\mathbf{v}$ and $\mathbf{B}$; therefore, magnetic force $\mathbf{F_m}$ has the following properties:

• *Direction of the magnetic force.* The force $\mathbf{F_m}$ is perpendicular to the plane containing $\mathbf{v}$ and $\mathbf{B}$. More specifically, the direction of $\mathbf{F_m}$ is given by the right-hand rule for cross products: imagine $\mathbf{v}$ to be turned through the smaller angle θ by the right-hand curled fingers to align with $\mathbf{B}$; then the right thumb points in the direction of $\mathbf{F_m}$. See Figure 29-1, which shows the force direction for a *positive* charge; with a negatively charged particle, the direction of $\mathbf{F_m}$ is reversed.

• *Magnitude of the magnetic force.* From the definition of the cross product, we have

$$F_m = qv\,B \sin\theta = qv_\perp B \qquad (29\text{-}2)$$

where θ is the smaller angle between $\mathbf{v}$ and $\mathbf{B}$. The component of $\mathbf{v}$ perpendicular to $\mathbf{B}$ is $v_\perp = v \sin\theta$.

Equation (29-1) and Figure 29-1 say it all. The properties of the magnetic force and the definition of the magnetic field come from experiments in which charged particles are observed as they move through a magnetic field. (How the magnetic field is produced is not our concern here; it is treated in Chapter 30.) For example, one might study the deflection of a beam of electrons in an oscilloscope under the influence of a permanent magnet.

Characteristics of $\mathbf{F_m}$ and $\mathbf{B}$, all implicit in (29-1), are as follows:

• $F_m \propto q$. When a magnetic force acts, it is *proportional to charge magnitude.* Replacing a positive charge by a negative will, with other things unchanged, reverse the direction of $\mathbf{F_m}$.

• $F_m \propto v_\perp$. Only the *velocity component perpendicular* to $\mathbf{B}$ influences the magnitude of $\mathbf{F_m}$.

• $\mathbf{F_m} \perp \mathbf{v}$. When a magnetic force acts, it is always *perpendicular to the particle's velocity* and deflects the moving charged particle.

• $\mathbf{F_m} = 0$ for $\mathbf{v} \parallel \mathbf{B}$. A particle of either sign fired along the direction of $\mathbf{B}$ (or opposite to $\mathbf{B}$) is subject to no magnetic force. Indeed, the line of the magnetic field $\mathbf{B}$ is defined as that line in space along which a charged particle can be fired without being subject to a magnetic force.

• $\mathbf{F_m}$ has its maximum magnitude for $\mathbf{v} \perp \mathbf{B}$. With $\theta = 90°$ in (29-2), we have

$$F_{m,\text{max}} = qv_\perp B$$

or with $\mathbf{v} \perp \mathbf{B}$,

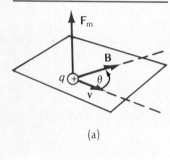

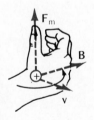

Figure 29-1. Magnetic force $\mathbf{F_m}$ on a particle with electric charge q and velocity $\mathbf{v}$ in magnetic field $\mathbf{B}$. (a) Relative orientation of the vectors; (b) the right-hand rule for $\mathbf{F_m} = q\mathbf{v} \times \mathbf{B}$.

$$B \equiv \frac{F_{m,max}}{qv} \qquad (29\text{-}3)$$

This relation *defines* the *magnitude* of **B** as the maximum magnetic force per unit charge divided by the particle speed.

● The direction of **B** comes from the right-hand rule in $\mathbf{F}_m = q\mathbf{v} \times \mathbf{B}$. With the *line* of **B** — the *undirected* line — defined as that for zero force, there are *two* possible opposite directions that might be chosen for **B**. The one universally used is shown in Figure 29-1 and summarized by the right-hand rule for cross products.*

Magnetic force $\mathbf{F}_m$ and magnetic field **B** *differ greatly* from their electric counterparts:

● An *electric force* $\mathbf{F}_e$ acts on charge q, whatever the state of its motion. The magnitude of $\mathbf{F}_e$ in field **E** depends only on the charge magnitude, $\mathbf{F}_e = q\mathbf{E}$. The direction of $\mathbf{F}_e$ is parallel (or antiparallel) to **E**.

● A *magnetic force* acts only on a moving charge q that has a velocity component perpendicular to **B**. The magnitude of $\mathbf{F}_m$ depends on the perpendicular velocity component $v_\perp$, as well as the charge q. The direction of $\mathbf{F}_m$ is *never* along **B**.

Equation (29-1) implies that magnetic field **B** is a vector. We have already discussed how to assign magnitude and direction to **B**. But a directed quantity is a true vector only if it obeys the rules of vector addition found to hold for displacement vectors (Section 2-1). Experiment shows that **B** is indeed a vector in the following sense. Two or more magnetic fields from separate sources acting simultaneously on a charged particle produce a magnetic force given by $\mathbf{F}_m = q\mathbf{v} \times \mathbf{B}$ if by **B** is meant the *vector sum* of the separate fields. Said a little differently, magnetic fields follow the superposition principle.

From (29-2) we see that appropriate units for the magnetic field are newtons per ampere·meter (N/A·m) as follows:

$$\frac{N}{C \cdot m/s} = \frac{N}{(C/s) \cdot m} = \frac{N}{A \cdot m}$$

In the SI system, this combination of units for the magnetic field is called the *tesla* (abbreviated T), named for N. Tesla (1856–1943). By definition,

$$1 \text{ T} \equiv \text{N/A} \cdot \text{m}$$

By laboratory standards, a magnetic field of one tesla is large. A very large electromagnet with an iron core can produce a magnetic field of only about 2 T. Magnetic fields are often given in the smaller units of the *gauss* (abbreviated G), where

$$1 \text{ T} \equiv 10^4 \text{ G}$$

The earth's magnetic field at the surface, at a latitude of 45°, has a magnitude of about 0.6 G. A typical small permanent magnet might produce a field of a few hundred gauss near its poles.

The quantity that we have been calling the magnetic field goes by other

---

* This direction for **B** agrees with the familiar behavior of a magnet in an external magnetic field. If freely pivoted, a magnet becomes aligned with **B**. For a compass in the earth's magnetic field, the so-called north pole of the compass points towards geographic north.

names — *magnetic induction field* and *magnetic induction.* For reasons to be seen in Section 29-2, **B** is also termed the *magnetic flux density.* Mostly we shall stick with the simple name *magnetic field.**

Example 29-1. Figure 29-2(a) is a bubble-chamber photograph showing the creation of electron-positron pairs by the annihilation of highly energetic photons. The antiparticle of an electron, the positron, is like an electron but has positive electric charge of magnitude e. Charged particles, such as electrons or positrons, leave a wake of bubbles along their paths as they travel in a bubble chamber through liquid hydrogen close to its boiling point. A photon, a "particle" of electromagnetic radiation, is electrically neutral and leaves no track. Each electron-positron pair appears as an inverted V with curled ends. An external magnetic field, into the plane of the photograph in Figure 29-2(a), deflects the negative electrons and positive positrons in opposite directions. Which is which?

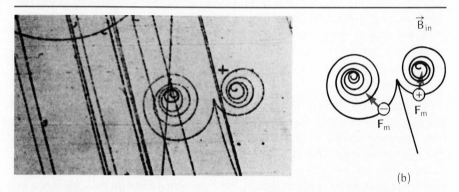

(b)

Figure 29-2. (a) *Bubble-chamber photograph showing tracks of bubbles left by electron-position pairs as the particles travel in a magnetic field through liquid hydrogen. Each electron-positron pair appears as an inverted V with curled ends. The electron and positron are oppositely charged and therefore deflected in opposite directions. As each electron and positron loses kinetic energy, its radius of curvature decreases.* (b) *Applying the relation* $\mathbf{F}_m = q\mathbf{v} \times \mathbf{B}$ *to the paths of the two particles moving in an inwardly directed magnetic field shows that the negatively charged electron is on the left and the positively charged positron is on the right.*

Figure 29-2(b) shows the directions of the magnetic forces on the two oppositely charged particles of one pair. The electron is deflected to the left, the positron to the right.

29-2   Magnetic-Field Lines and Magnetic Flux

Electric-field lines are used to show the direction and magnitude of **E**. Similarly, magnetic-field lines can show the direction and magnitude of **B**.† A magnetic field is strong where the magnetic lines are crowded and weak where

---

* It is not proper to call **B** the magnetic field *intensity*. That term is reserved for a related magnetic vector quantity, usually symbolized by **H**, with a physical significance and units distinct from **B**.
† We say magnetic-*field* lines, never magnetic "lines of force." After all, the magnetic force on a moving charged particle is never parallel to the lines representing **B**.

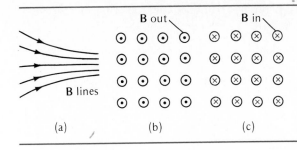

B out **B in**

B lines

(a)　　(b)　　(c)

Figure 29-3. (a) Magnetic field lines. Field lines (b) out of the paper and (c) into the paper.

they are far apart. A constant, or uniform, magnetic field is represented by uniformly spaced, parallel, straight lines.

When magnetic-field lines are perpendicular to the plane of the paper, we use the following representation:

- Symbol $\odot$ to show a magnetic field out of the paper.
- Symbol $\otimes$ to show a magnetic field into the paper.

These symbols are chosen because they remind us, respectively, of an arrow point emerging from the paper and the feathers on the tail of an arrow going into the paper. See Figure 29-3.

The electric flux $d\phi_E$ over an infinitesimal surface element dS is defined as

$$d\phi_E \equiv \mathbf{E} \cdot d\mathbf{S} \qquad (24\text{-}4)$$

Similarly, the magnetic flux $d\phi_B$ through an infinitesimal surface element dS is related to the magnetic field $\mathbf{B}$ through that small patch of area by

$$d\phi_B \equiv \mathbf{B} \cdot d\mathbf{S} = B \cos \theta \, dS \qquad (29\text{-}5)$$

where dS is a vector perpendicular to the surface and θ is the angle between $\mathbf{B}$ and $d\mathbf{S}$. To find the magnetic flux through surface area dS, we simply multiply dS by the component $B \cos \theta$ of the magnetic field perpendicular to the surface element. See Figure 29-4. When the $\mathbf{B}$ lines lie in the surface and $\theta = 90°$, the magnetic flux through dS is zero.

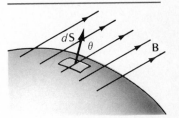

Figure 29-4. Magnetic flux through a surface element dS.

The total magnetic flux ϕ_B through a finite surface area is just the algebraic sum of the infinitesimal contributions

$$\phi_B = \int \mathbf{B} \cdot d\mathbf{S} \qquad (29\text{-}6)$$

where the integration is carried out over the surface area through which the magnetic flux is to be computed.

Suppose that magnetic field $\mathbf{B}$ is uniform over a flat area A oriented at right angles to $\mathbf{B}$. Then the flux through A is simply

$$\phi_B = BA$$

and the magnetic field B can be expressed as

$$B = \frac{\phi_B}{A} \qquad (29\text{-}7)$$

From (29-7), magnetic field B is the *magnetic flux density*, the magnetic flux divided by the area through which the magnetic-field lines penetrate at right angles.

The relations among magnetic field $\mathbf{B}$, magnetic flux ϕ_B, and the magnetic-field lines are exactly like the corresponding relations among electric field, electric flux, and electric-field lines:

- The number of magnetic-field *lines per unit transverse area* is a measure of the *magnitude of the magnetic field* **B**.
- The total *number of magnetic-field lines* through the chosen transverse area is a measure of magnetic *flux* $\phi_B = BA$.

In the SI system, the unit for *magnetic flux*—a measure of the *total number* of magnetic lines crossing a transverse area—is the *weber* (abbreviated Wb), named for W. Weber (1804–1890). A corresponding unit for the magnetic flux density B is then Wb/m². By definition, one weber per square meter is one tesla:

$$1 \text{ Wb/m}^2 \equiv 1 \text{ T}$$

Example 29-2.   Cosmic radiation consists of highly energetic, positively charged particles (mostly protons) that rain upon the earth in all directions from outer space. How does the magnetic field of the earth affect protons approaching it (*a*) toward the North Pole or the South Pole and (*b*) at the equator?

The earth's magnetic field lines are shown in Figure 29-5. Protons approaching the earth at the poles travel along magnetic field lines and are consequently undeflected [Figure 29-5(a)]. On the other hand, protons approaching the earth at the equator cross magnetic field lines at right angles. These magnetic field lines go from geographic south to geographic north. Using (29-1), we see that the particles are acted on by a magnetic force that deflects them toward the east [Figure 29-5(b)].

The incoming cosmic-ray particles have a large range of energies. Some are highly energetic (energies up to 10^{18} MeV), and the earth's magnetic field is feeble (of the order of 0.6 G near the surface). Yet, this feeble magnetic field is able to influence the motion of these particles because it extends far out into space. Low-energy particles arriving at the equator are deflected so strongly that they miss hitting the earth's atmosphere. As a consequence, the intensity of the cosmic radiation is greater at the North and South poles than near the equator (the so-called latitude effect), proving that the incoming particles are electrically *charged*. Moreover, those particles arriving at the equator come preferentially from the west (the so-called east-west effect), proving that the primary particles are *positively* charged.

Example 29-3.   A uniform magnetic field of 0.10 T makes an angle of 30° with the plane of a square that is 20 cm on a side. See Figure 29-6. What is the magnetic flux through the square?

The angle between **B** and the *normal* to the plane is $\theta = 60°$, so that the flux through

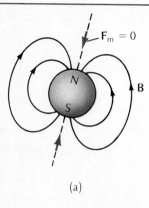

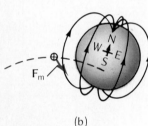

(a)

(b)

Figure 29-5.   *(a) Magnetic field of the earth; charged particles approaching along the magnetic axis are undeflected. (b) Positively charged particles approaching the earth at the equator are deflected toward the east.*

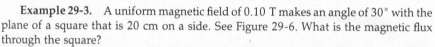

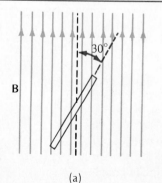

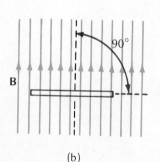

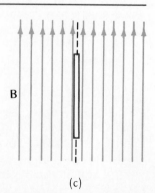

Figure 29-6.   (a)   (b)   (c)

the square loop is, from (29-5),

$$\phi_B = BS \cos \theta = (0.10 \text{ T})(0.20 \text{ m})^2 \cos 60° = 2.0 \times 10^{-3} \text{ Wb/m}^2$$

Suppose that the square is turned so that **B** becomes perpendicular to the plane. Then the magnetic flux and the number of magnetic field lines passing through the square are doubled [Figure 29-6(b)]. On the other hand, turning the square so that **B** lies in the plane of the square gives $\phi_B = 0$; no field lines then penetrate the square [Figure 29-6(c)].

29-3 Charged Particle in a Uniform Magnetic Field

The magnetic force is different from such other fundamental forces as the gravitational force or the electric force; F_m is *velocity-dependent*. It is therefore not possible to associate a scalar potential energy with the magnetic interaction. Furthermore, the magnetic force can do no work; the speed and kinetic energy of a charged particle are constant as it travels in a constant **B**. This follows directly from the direction of F_m relative to **v**; the force F_m is always perpendicular to **v**. Since no component of F_m acts along the direction of the particle's motion, F_m does no work.

What then is the effect of F_m on a moving charged particle? Since F_m is always at right angles to **v**, the particle is *deflected* by this sideways force. The direction of **v** changes — but not the speed.

What is the general path of a charged particle in a uniform magnetic field? We first consider two simple special cases and then the general case:

- **v∥B**. As (29-1) shows — and the definition of **B** requires — a particle moving initially *parallel* (or antiparallel) to **B** coasts at *constant velocity*. See Figure 29-7(a).
- **v ⊥ B**. The magnitude of F_m is constantly equal to qvB; the direction of F_m remains perpendicular to **v** as the velocity changes direction. These are just the conditions for *uniform circular motion*. The force F_m points always to the center of the circle, and the particle's speed is constant. The charged particle encircles magnetic-field lines in a plane transverse to **B**. See Figure 29-7(b).
- **v** neither ∥**B** nor ⊥**B**. The velocity component $v_\parallel = v \cos \theta$ parallel to **B** remains constant in magnitude and direction. The perpendicular velocity com-

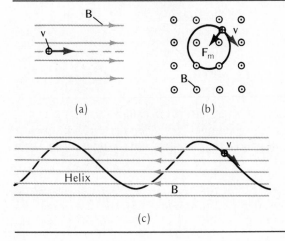

(a) (b) (c)

Figure 29-7. *A charged particle moving in a uniform field: (a) with **v** parallel to **B**, the path is a straight line; (b) with **v** perpendicular to **B**, the path is a circle; (c) for other angles between **v** and **B**, the path is a helix.*

ponent $v_\perp = v \sin \theta$ also remains constant in magnitude but changes direction continuously. The resulting motion? Simply the superposition of drift at constant speed $v_\parallel$ along **B** and circling at constant speed $v_\perp$ around **B**. The path is a *helix* whose symmetry axis coincides with the magnetic field. See Figure 29-7(c).

Electrically charged particles follow paths that are wrapped around **B** lines even when the magnetic field varies (slowly) with position. See Figure 29-8.

Consider a particle of mass m, charge q, and speed v circling at radius r in a plane perpendicular to a uniform magnetic field **B**. From Newton's second law, we have

$$\Sigma \mathbf{F} = m\mathbf{a}$$

$$qv_\perp B = \frac{mv_\perp^2}{r}$$

or

$$mv_\perp = qrB \tag{29-8}$$

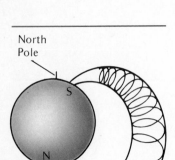

Figure 29-8. *Charged particles trapped in the earth's Van Allen belt. Note that the south magnetic pole is near the geographic North Pole.*

North Pole

S

N

B

We see that the magnitude of the particle's linear momentum is directly proportional to the magnitude of B and to r, the radius of curvature of the circular arc. This gives a basis for measuring a charged particle's momentum: measure **B**, measure r, know q, and apply (29-8). The momentum magnitudes of the electron and positron in the bubble-chamber photograph of Figure 29-2 are readily found in this fashion.

Any charged particle projected into a uniform magnetic field travels at constant speed in a helix wrapped around the magnetic field lines. Suppose that numerous charged particles of one type, such as electrons, which differ in initial velocity and energy, are injected into the same constant magnetic field. Each electron goes in some helical path (or in special cases, a circle or a straight line). It is a remarkable fact that *all* such electrons, despite differences in their energies, speeds, and paths, will complete one loop in precisely the *same time*! All cycling particles will have the *same frequency*. Let us prove it.

The perpendicular velocity component $v_\perp$ can be written as $v_\perp = \omega r$, where ω is the circling particle's angular speed and r the radius of the helix. Using this relation in (29-8) gives

$$\omega = \frac{q}{m} B \tag{29-9}$$

The ordinary frequency $f = \omega/2\pi$ of the circling particle and its period T are then given by

$$f = \frac{1}{T} = \frac{q}{2\pi m} B \tag{29-10}$$

The characteristic frequency given in (29-10) is called the *cyclotron frequency* (for reasons to be evident in Example 29-5). The cyclotron frequency depends on the charge-to-mass ratio q/m but does not involve the particle speed $v_\perp$ or radius r.

Example 29-4. A proton (mass, 1.67×10^{-27} kg) is observed in a circular arc of 30-cm radius when it travels transverse to a magnetic field of 1.5 T. What is the proton's (a) momentum? (b) cyclotron frequency?

(a) From (29-8), we have

$$p = mv = qrB$$
$$= (1.60 \times 10^{-19} \text{ C})(0.30 \text{ m})(1.5 \text{ T})$$
$$= 7.2 \times 10^{-20} \text{ kg·m/s}$$

(b) From (29-10), we have

$$f = \frac{qB}{2\pi m}$$
$$= \frac{(1.60 \times 10^{-19} \text{ C})(1.5 \text{T})}{2\pi(1.67 \times 10^{-27} \text{ kg})}$$
$$= 2.3 \times 10^{7} \text{ s}^{-1} = 23 \text{ MHz}$$

Example 29-5   The Cyclotron. The cyclotron, a machine for accelerating charged particles to relatively high kinetic energies, is based on the fact that the cyclotron frequency of a charged particle circling magnetic-field lines does not depend on the radius of the orbit or the particle's speed. The cyclotron was invented in 1932 by E. O. Lawrence (1901–1958) and M. S. Livingston (b. 1905), and it is the simplest of a whole class of particle accelerators based on the cyclotron principle.

The essential parts of a cyclotron are shown in Figure 29-9. Positive particles such as protons, deuterons (nuclei of heavy hydrogen atoms, or deuterium), or still more massive ions are injected into the evacuated central region at central point C between two flat D-shaped hollow metal conductors. An alternating high-frequency voltage at the particle's cyclotron frequency is applied to the dees and a uniform magnetic field is applied transverse to the dees. The charged particle is then subject to two forces:

- A continuous deflecting magnetic force lying in the plane of the dees, always pointing toward central point C.
- When the charged particle is in the gap between the dees, an accelerating electric force from the instantaneous potential difference between the dees. When the particle is within either dee, it is electrically shielded inside a hollow conductor and feels no electric force.

The particle is accelerated across the gap between the dees. Actually, the particle is injected so that it arrives at the gap when the sinusoidally varying potential difference reaches its maximum value. After the particle has traveled a half-circle and is at the gap again, one half-cycle has elapsed and the particle is again accelerated across the gap. The process continues, with the particle accelerated at each gap crossing. The circulating charged particles remain synchronized with the alternating voltage because the time for completing a half-circle does not depend on the particle's speed or orbital radius.

The particles are deflected by ejector plate E at the outer edge of the dee and strike target T.

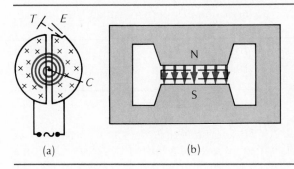

(a)   (b)

Figure 29-9. (a) Top view of cyclotron dees; a uniform magnetic field is directed into the paper. (b) Side view of cyclotron dees in magnetic field.

Suppose that a certain cyclotron with dees of 30-cm radius accelerates protons in a magnetic field of 1.5 T. What is the maximum kinetic energy of accelerated protons? We found in Example 29-4 that the cyclotron frequency of such protons is 23 MHz and their momentum is 7.2×10^{-20} kg·m/s. Therefore, the corresponding final kinetic energy is

$$K = \frac{p^2}{2m} = \frac{(7.2 \times 10^{-20} \text{ kg·m/s})^2}{2(1.67 \times 10^{-27} \text{ kg})} = 1.55 \times 10^{-12} \text{ J}$$

Electron volts are more appropriate as units for specifying the energy of such particles and we have that

$$K = 1.55 \times 10^{-12} \text{ J} \left(\frac{1 \text{ MeV}}{1.60 \times 10^{-13} \text{ J}} \right) = 9.7 \text{ MeV}$$

29-4 Charged Particles in Uniform **B** and **E** Fields

If the gravitational force is neglected, the resultant force on any charged particle is merely the vector sum of the electric force $\mathbf{F}_e = q\mathbf{E}$ and the magnetic force $\mathbf{F}_m = q\mathbf{v} \times \mathbf{B}$:

$$\mathbf{F} = \mathbf{F}_e + \mathbf{F}_m = q\mathbf{E} + q\mathbf{v} \times \mathbf{B} = q(\mathbf{E} + \mathbf{v} \times \mathbf{B}) \qquad (29\text{-}11)$$

Equation (29-11) is usually referred to as the *Lorentz force relation*, after H. E. Lorentz (1853–1928), who made important contributions to the theory of electromagnetism. The Lorentz relation can be regarded as giving the basic definitions of **E** and **B**—the magnetic force is that part of the electromagnetic interaction that depends on the particle's velocity; the electric force is that part of the interaction that does not. If we know a particle's charge and initial velocity, then from knowledge of **E** and **B** at each point in space, we can predict in detail the particle's future motion.

Here we consider only the special situations in which both **E** and **B** are uniform and constant with time. We know already that with just a single field:

- *Uniform E only*, the particle has a *constant acceleration* and moves in a *parabola* whose symmetry axis is along the **E** direction (Section 24-2). See Figure 29-10(a).
- *Uniform B only*, the particle has a *constant speed* and moves in a *helix* whose symmetry axis is along the direction of **B**. See Figure 29-10(b).

How should uniform **E** and uniform **B** be arranged so that a charged particle goes through both fields undeflected? The only effect of a magnetic force is to deflect a charged particle, so that the sideways magnetic deflection must be cancelled by an oppositely directed electric deflection. The required arrangement is shown in Figure 29-11, where **v** is perpendicular to both **E** and **B** and the directions of **E** and **B** are also mutually perpendicular. The forces $\mathbf{F}_e$ and $\mathbf{F}_m$ are of equal magnitude when

$$F_e = F_m$$
$$qE = qvB$$
$$v = \frac{E}{B} \qquad (29\text{-}12)$$

Notice the remarkable result. If the directions of **E** amd **B** are as shown in Figure 29-11 and their relative magnitudes are adjusted to satisfy (29-12), a particle of any charge magnitude traveling at speed E/B passes through the

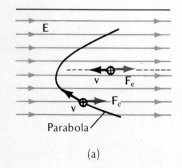

(a)

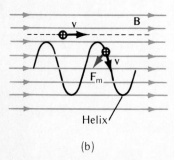

(b)

Figure 29-10. (a) In a uniform electric field, the path of a charged particle is a parabola. (b) In a uniform magnetic field, the path of a charged particle is a helix.

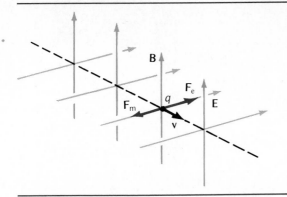

Figure 29-11. *A particle with charge q and velocity* **v** *moving undeflected through uniform crossed* **E** *and* **B** *fields. The electric and magnetic forces,* **F**ₑ *and* **F**ₘ*, cancel.*

region of crossed fields undeflected. (Even the sign of q does not enter; with a negatively charged particle, the directions of *both* $\mathbf{F}_e$ and $\mathbf{F}_m$ in Figure 29-11 are reversed.) A device with crossed electric and magnetic fields arranged as in Figure 29-11 is appropriately known as a *velocity selector*. From a beam of polyenergetic particles, all particles save those of the design speed $v = E/B$ will be deviated from a straight path.

The relation for a velocity selector shows directly that compared with electric force, the magnetic force is usually quite feeble. Suppose, for example, that $E = 10^4$ V/m (a relatively weak electric field with a drop in potential of 100 V over 1 cm) and that $B = 1$ T (a relatively strong magnetic field). We then have from (29-12) that $v = (10^4 \text{ V/m})/1 \text{ T} = 10^4$ m/s. Only with a relatively high-speed charged particle, one moving at 10 km/s, will the magnetic force match the electric force in size.

We already have considered a *momentum selector*. It consists basically of a beam of particles directed at right angles to a uniform **B**. The particles travel in circular arcs with a linear momentum magnitude $p = mv = qrB$.

A *kinetic-energy selector* of the simplest type consists of charged particles moving along or opposite to a uniform electric field **E**. The electric potential drops by V over a distance d measured in the direction of **E**, so that

$$V = Ed \qquad\qquad (25\text{-}10)$$

The kinetic energy of a charged particle traveling with or against electric-field lines changes by ΔK, where

$$\Delta K = qV = qEd$$

For example, an electron accelerated from rest across a potential difference of 1.0 kV acquires a kinetic energy of 1.0 keV. Similarly, a particle with charge e passing through a retarding potential difference of 50 V has its kinetic energy reduced by 50 eV.

In summary,

- With **B** alone, we can measure momentum mv.
- With **E** alone, we can measure kinetic energy $\frac{1}{2}mv^2$ (or more generally, a change in kinetic energy).
- With crossed **B** and **E**, we can measure velocity v.

Clearly, then, there are several ways of measuring the mass of a charged particle. We simply combine the separate measurements given above:

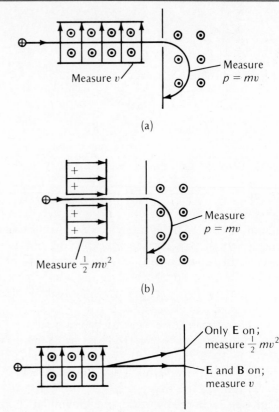

Figure 29-12. *Arrangements for measuring the mass of a charged particle: (a) a velocity selector followed by a momentum selector, (b) an accelerating electric field followed by a momentum selector, and (c) a velocity selector followed by a deflecting electric field (the Thomson arrangement for measuring the electron q/m ratio).*

- Measure v and mv, as shown in Figure 29-12(a).
- Measure $K = p^2/2m$, and $p = mv$, as shown in Figure 29-12(b).
- Measure v and $K = \frac{1}{2}mv^2$, as shown in Figure 29-12(c).

An arrangement like that shown in Figure 29-12(c) was used by J. J. Thomson in 1897 in the first measurement of the mass of the electron; indeed, Thomson is said to have discovered the electron through this experiment. Strictly, any of these procedures yields the charge-to-mass *ratio q/m*, not the mass alone. With charge measured independently, the mass can then be computed. Mass spectrometers using these procedures have been so refined that measuring atomic masses to within a few parts in 10^7 has become a routine laboratory procedure.

29-5   Magnetic Force on a Current-Carrying Conductor

To find the magnetic force on a current-carrying conductor, we first concentrate on a single charge-carrier with charge dq moving at drift velocity **v**. The magnetic force $d\mathbf{F}_m$ on dq in field **B** is

$$d\mathbf{F}_m = dq\, \mathbf{v} \times \mathbf{B}$$

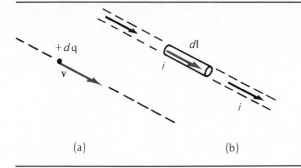

(a)   (b)

Figure 29-13. *Charge +dq with velocity* **v** *constitutes conventional current i along displacement d***l**.

We can write **v** as $d\mathbf{l}/dt$, where $d\mathbf{l}$ is the particle's displacement over time interval dt. We consider **v** the drift velocity of a positive charge-carrier within a thin conductor. Then $d\mathbf{l}$ is also a length vector, pointing along the conductor in the direction of conventional current. See Figure 29-13. The quantity $dq\,\mathbf{v}$ can be written as

$$dq\,\mathbf{v} = dq\left(\frac{d\mathbf{l}}{dt}\right) = \left(\frac{dq}{dt}\right)d\mathbf{l} = i\,d\mathbf{l} \tag{29-13}$$

where $i = dq/dt$ is the conventional current associated with charge dq.

Using (29-13) in the magnetic force relation, we then have

$$d\mathbf{F}_\mathrm{m} = i\,d\mathbf{l} \times \mathbf{B} \tag{29-14}$$

Current i can now represent the *total* current from *all* charge carriers in a segment of conductor of length $d\mathbf{l}$, and $d\mathbf{F}_\mathrm{m}$ is the *total* magnetic force on this infinitesimal segment. See Figure 29-14.

The resultant magnetic force on a finite conductor of any shape is found by applying (29-14) to each infinitesimal length segment and adding the infinitesimal forces as vectors. If the conductor is straight, then the force on length **L** is

$$\mathbf{F}_\mathrm{m} = i\,\mathbf{L} \times \mathbf{B} \tag{29-15}$$

where **L** points in the direction of the current.

We can write the relation above in algebraic form as

$$F_\mathrm{m} = iLB\sin\theta$$

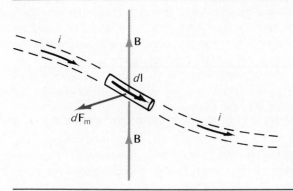

Figure 29-14. *Magnetic force* $d\mathbf{F}_\mathrm{m}$ *on a current element d***l** *with current i in magnetic field* **B**: $d\mathbf{F}_\mathrm{m} = i\,d\mathbf{l} \times \mathbf{B}$.

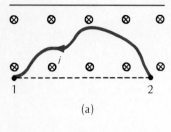

(a)

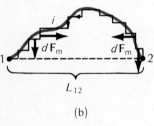

(b)

Figure 29-15. *(a) A current-carrying conductor of arbitrary shape in an external magnetic field. (b) Equivalent conductor comprising segments parallel and perpendicular to the line joining the end points.*

where θ is the angle between **B** and **L**. Note that $\mathbf{F_m} = 0$ when the conductor is aligned with the external field lines.

Example 29-6.   A conductor of arbitrary shape carrying current i runs from point 1 to point 2 in two dimensions, as shown in Figure 29-15(a). A constant magnetic field **B** acts into the plane of the paper. What is the resultant force on this segment of conductor?

We can approximate the continuously varying conductor as follows. We use instead a series of short, straight-line segments lying alternately parallel and perpendicular to the straight line L_{12} from point 1 to 2. The magnetic forces on these segments are, from (29-14), in the directions shown in Figure 29-15(b). We see that the total force component parallel to the line L_{12} is zero; each force to the right is matched by an equal force to the left. This leaves only the magnetic forces perpendicular to L_{12}. The resultant force perpendicular to line L_{12} is exactly the same as the magnetic force on a straight conductor running from 1 to 2. Therefore, the net magnetic force has a magnitude $F_m = iL_{12}B$; its direction is perpendicular to the line joining the end points. The arbitrarily shaped conductor is equivalent to a straight wire between its end points carrying the same current.

Now suppose that we form a current loop by imagining points 1 and 2 brought together. From the arguments above, the resultant magnetic force on the loop must be *zero*. A little thought will show that this conclusion applies whatever the shape of the loop — and whether the loop lies entirely in a plane or not — and that it is also independent of the loop's orientation relative to the magnetic-field lines. *The magnetic field must, however, be uniform.* Although a closed current loop is subject to no resultant magnetic force in a uniform magnetic field, it may be subject to a resultant torque, as we shall see in Section 29-7.

29-6   The Hall Effect

When a magnetic field is applied at right angles to a current-carrying conductor, there is not only a magnetic force on the conductor, but also a *transverse electric potential difference* across the conductor. This phenomenon, discovered in 1879 by E. H. Hall, is known as the *Hall effect*.

Consider Figure 29-16, where magnetic field **B** is applied at right angles to a conducting slab carrying current i. We suppose that the charge carriers composing the current are electrons; they drift with velocity $\mathbf{v_d}$ in a direction opposite to the conventional current. Then the deflecting magnetic force $\mathbf{F_m}$ on a negatively charged particle has the direction shown in Figure 29-16(a). Electrons are deflected to one edge of the slab until the outward magnetic force is balanced by an equal and inward electric force arising from excess electrons already at the edge. Excess electrons accumulate at one conductor edge, so that it acquires a net negative charge. The opposite edge, with a corresponding deficiency of electrons, acquires a net positive charge. The transverse electric potential difference between the opposite edges of the conductor is known as the *Hall potential difference* V_H.

Which side of the conductor is at the higher electric potential depends on the sign of the charge carriers. If the electrons in Figure 29-16(a) were replaced by positive charge carriers, with the direction of **B** and i kept unchanged, then the charge carriers would drift in the same direction as the current, the direction of the deflecting magnetic force $\mathbf{F_m}$ would be reversed, charges of reversed signs would accumulate on the two edges of the conductor, and the polarity of

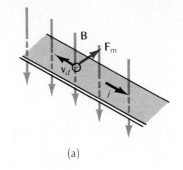

(a)

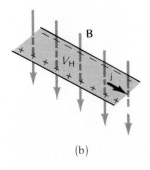

(b)

Figure 29-16.   *The Hall effect. (a) An electron with drift velocity* **v**_d *(opposite to the conventional current i) is deflected by magnetic force* **F**_m *when the conductor is subject to transverse magnetic field* **B**. *(b) Opposite charges appear at the conductor edges and produce a Hall potential difference* V_H.

the Hall potential difference in Figure 29-16(b) would be *reversed*. The sign of the Hall potential difference indicates the sign of the charge carriers; it can be shown that the magnitude of V_H is a measure of the concentration of charge carriers.

The sign of the observed Hall effect for ordinary metallic conductors confirms that the charge carriers are indeed negatively charged particles, electrons. Semiconducting materials, however, may exhibit either polarity. Silicon doped with arsenic is an *n-type semiconductor*; here the charge carriers, electrons, are negatively charged. Silicon doped with gallium is a *p-type semiconductor*; in such a material the effective charged carriers, known as *holes*, are positively charged.

A hole in the semiconducting material is a site at which an electron is missing because impurity atoms have been introduced. A semiconductor with electron holes remains electrically neutral. When an electron adjoining a hole shifts position to fill the hole, it creates a new hole at its former location. The electron moves in one direction; the hole shifts in the opposite direction, thereby effecting a current in the direction in which the holes move. Electron holes in a sea of electrons behave like bubbles in liquid. Bubbles are holes in a sea of liquid; as a bubble rises, liquid descends to fill the space occupied by the hole, and the situation can be described by assigning a negative mass to the bubble.

29-7   Magnetic Torque on a Current Loop

To find the resultant magnetic torque that acts on a loop of current, first we concentrate on a rigid rectangular loop with sides W and L carrying current i.

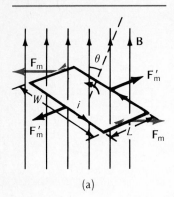

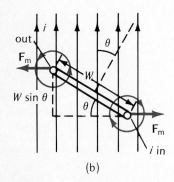

Figure 29-17. (a) Magnetic forces on a rectangular current-carrying conducting loop in a magnetic field. (b) A side view of part (a).

Consider Figure 29-17(a), where external field **B** makes an angle θ with the normal to the rectangular loop. We first find the directions of the magnetic forces on each of the four sides. As Figure 29-17(a) shows, we have two pairs of oppositely directed magnetic forces of the same magnitude. The resultant *force* on the loop is *zero*. The forces on the sides of length W are collinear; they produce no torque. The equal and opposite forces on the sides of length L are not collinear; these forces produce a torque on the loop that aligns its normal with the direction of **B**. Let us find the torque's magnitude.

From (29-15), the magnitude of the magnetic force on a side of length L is

$$F_m = iLB$$

For convenience, we compute torques relative to the lower of the two sides of length L. Then, using the definition of torque magnitude (12-4), we have

$$\tau = r_\perp F = (W \sin \theta)(iLB)$$

We can write this relation in simpler form by recognizing LW as the area A of the loop:

$$\tau = iAB \sin \theta \tag{29-16}$$

For the situation shown in Figure 29-17(b), the torque is counterclockwise. From the vector properties of torque [(12-5), $\tau = \mathbf{r} \times \mathbf{F}$], we see that the direction of torque in Figure 29-17(b) is out of the paper. We can incorporate the vector character of torque by writing (29-16) in vector form as

$$\tau = i\mathbf{S} \times \mathbf{B} \tag{29-17}$$

Here **S** is a vector whose magnitude represents the loop's area A. The direction of **S** is perpendicular to the plane of the loop and related to the sense of current around the loop by a right-hand rule. See Figure 29-18, which shows that the direction of τ is such as to align **S** with external field **B**.

Equation (29-17) is actually a general relation; it gives the magnetic torque on a plane loop of current of any shape. We can simply imagine the conducting wire to be approximated by small straight segments at right angles. Of course, if a loop has N identical turns (all with current in the same sense) rather than just a single turn, the torque in (29-16) and (29-17) is increased by the factor N.

The magnetic torque on a current-carrying conductor immersed in a magnetic field is the basis of many important devices. One is the *galvanometer*, a sensitive current-measuring device used in some ammeters and voltmeters. A galvanometer has a multiturn coil typically in the magnetic field of a perma-

Figure 29-18. A current loop subject to a magnetic torque τ in magnetic field **B**. (a) Vector **S** represents the normal to the plane of the loop with current i. (b) The direction of **S** is related to the sense of current i through a right-hand rule.

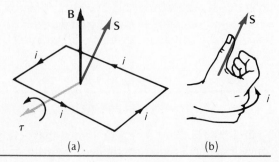

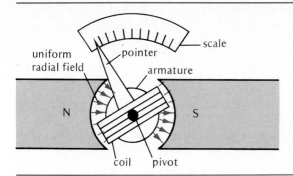

Figure 29-19. *The coil of a galvanometer is subject to a magnetic torque proportional to the current in the coil.*

nent magnet, as shown in Figure 29-19. The magnet is so shaped that the magnetic field at the location of the coil is not uniform but always at right angles to the plane of the coil; in effect, angle θ of (29-16) and Figure 29-19 is always 90°. The galvanometer coil is also subject to an elastic linear restoring torque from a spring ($\tau = -\kappa\phi$), which balances out the magnetic torque. Therefore, in equilibrium, the angular deflection ϕ of the coil and a needle attached to it is directly proportional to the current i through the galvanometer coil.

The magnetic torque on a current-carrying loop is also the fundamental principle behind the operation of an electric motor.

29-8   Magnetic Dipole Moment

Equation (29-17), for the magnetic torque on a current loop in a magnetic field, is like the relation for the electric torque on electric dipole moment **p** in electric field **E**:

$$\tau = \mathbf{p} \times \mathbf{E} \qquad (24\text{-}9)$$

This suggests that we attribute a *magnetic dipole moment*, symbolized by vector μ, to a current loop, where by definition,

$$\mu \equiv i\mathbf{S} \qquad (29\text{-}18)$$

See Figure 29-20. The direction of μ is related to the sense of conventional

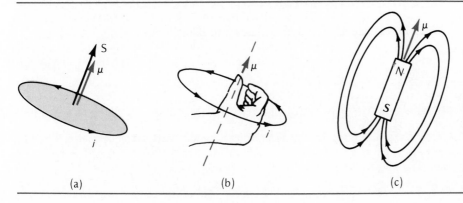

(a)   (b)   (c)

Figure 29-20. *Magnetic dipoles. (a) Magnetic dipole moment μ is aligned with vector **S** representing the enclosed area. (b) The direction of μ is related to the sense of current i through a right-hand rule. (c) Magnetic dipole for a permanent magnet.*

current in the loop by a right-hand rule. With the right-hand fingers curled in the sense of the current, the right thumb points in the direction of μ. Using (29-18) in (29-17), we get

$$\tau = \mu \times \mathbf{B} \tag{29-19}$$

The torque is a maximum for μ at right angles to $\mathbf{B}$ and zero for a dipole aligned with the external field. (Although the torque is zero for μ opposite to $\mathbf{B}$, this orientation is highly unstable.)

Just as an electric dipole in a *uniform* field is subject to no resultant force, a magnetic dipole in a uniform magnetic field is subject to no resultant magnetic force. Both electric and magnetic dipoles are, however, subject to resultant forces in nonuniform fields. Dipoles are attracted to the region in which the field is strongest.

The electric potential energy of an electric dipole in an external electric field is given by:

$$U_E = -\mathbf{p} \cdot \mathbf{E} \tag{24-11}$$

The corresponding relation for the magnetic potential energy U_B of a magnetic dipole μ in field $\mathbf{B}$ is

$$U_B = -\mu \cdot \mathbf{B} \tag{29-20}$$

Note that zero potential energy corresponds to μ perpendicular to $\mathbf{B}$. Potential energy U_B is minimum $(-\mu B)$ when the dipole is aligned with the field and the angle between μ and $\mathbf{B}$ is zero; U_B is maximum $(+\mu B)$ when the dipole points opposite to $\mathbf{B}$.

The behavior of a rigid current-carrying loop in a magnetic field reminds us of the behavior of a magnet in a magnetic field. A magnet (such as a compass needle) not aligned with a magnetic field is subject to a magnetic torque; it takes work to turn a magnet away from alignment with the field. Indeed, one may attribute a magnetic dipole moment to any object — a current-carrying loop or a magnet — that follows (29-19) and (29-20) for the torque and potential energy in an external magnetic field. See Figure 29-20(c). Although the term *dipole* may seem especially appropriate for a magnet with its magnetic effects concentrated at its two ends, we must emphasize that single magnetic poles, or magnetic monopoles, do not exist in nature (Section 30-4).

Example 29-7.   A freely pivoted coil is initially aligned with an external magnetic field of 1.5 T. The coil has 100 turns and a 10.0-cm radius, and carries 1.0 A. (a) How much work must be done on the coil to turn it 180°? (b) What minimum force applied directly to the coil will keep $\mathbf{B}$ lines lying in the plane of the loop?

(a) From (29-20), we see that the coil's potential energy goes from $-\mu B$ to $+\mu B$ as it is turned from $\theta = 0$ to 180°. The total work done is $2\mu B$. The magnetic moment of the coil is [(29-18)] $\mu = Ni\pi r^2$. Therefore,

$$\text{Work done} = 2\mu B = 2(Ni\pi r^2)B$$
$$= 2(100)(1.0 \text{ A})\pi(0.10 \text{ m})^2(1.5 \text{ T}) = 9.4 \text{ J}$$

(b) The torque is maximum and the force minimum with a force applied at the outer edge of the loop and perpendicular to $\mathbf{B}$ so that the moment arm $r_\perp$ equals the loop's radius. From (29-19), we then have, with an angle of 90° between μ and $\mathbf{B}$,

$$F_{min} = \frac{\mu B}{r} = \frac{\frac{1}{2}(\text{work from part } a)}{r} = \frac{\frac{1}{2}(9.4 \text{ J})}{0.10 \text{ m}} = 47 \text{ N}$$

Example 29-8.   A fixed length of conducting wire is to be used to construct a coil. What form of coil—a single turn of large radius or many turns of small radius—will produce a maximum torque when the coil is placed in the earth's magnetic field?

For maximum magnetic torque, $\tau = \mu \times \mathbf{B}$, the coil's magnetic moment $\mu = NiS$ must be a maximum. For conducting wire of length L, the number of turns is $N = L/2\pi r$, where r is the coil radius. The coil area is $S = \pi r^2$. Therefore,

$$\mu = NiS = \left(\frac{L}{2\pi r}\right) i(L\pi r^2) = \left(\frac{L^2 r}{2\pi}\right) i$$

The coil current i is determined only by the potential difference applied across the conductor ends. With $\mu \propto r$, we have maximum magnetic moment and therefore maximum torque for a *single* turn.

Summary

Definitions

The observed *magnetic force* $\mathbf{F}_m$ on charge q with velocity $\mathbf{v}$ defines the magnetic field $\mathbf{B}$:

$$\mathbf{F}_m = q\mathbf{v} \times \mathbf{B} \qquad (29\text{-}1)$$

For a charged particle moving at right angles to the direction for which $\mathbf{F}_m = 0$, the magnetic force has its maximum magnitude $F_{m(max)}$, and the magnitude of $\mathbf{B}$ is defined as

$$B \equiv \frac{F_{m,(max)}}{qv_\perp}$$

The *Lorentz force* on charge q and velocity $\mathbf{v}$ in electric field $\mathbf{E}$ and magnetic field $\mathbf{B}$ is

$$\mathbf{F} = q\mathbf{E} + q\mathbf{v} \times \mathbf{B} \qquad (29\text{-}11)$$

Magnetic flux $d\phi_B$ of magnetic field $\mathbf{B}$ over surface element $d\mathbf{S}$:

$$d\phi_B \equiv \mathbf{B} \cdot d\mathbf{S} \qquad (29\text{-}5)$$

Magnetic flux through a surface is a measure of the net number of magnetic-field lines through that surface. Magnetic field, or magnetic-flux density, is a measure of the net number of magnetic-field lines per unit transverse area.

The magnetic dipole moment μ of a loop of current i around the periphery of surface $\mathbf{S}$ is

$$\mu \equiv i\mathbf{S} \qquad (29\text{-}18)$$

The direction of μ is related to the sense in which current goes around the edge of $\mathbf{S}$ by a right-hand rule.

Units

Magnetic field (also magnetic flux density, magnetic induction) $\mathbf{B}$

$$1 \text{ tesla} \equiv 1 \text{ newton/ampere} \cdot \text{meter}$$
$$1 \text{ T} \equiv 1 \text{ NA}^{-1}\text{m}^{-1}$$
$$1 \text{ T} = 10^4 \text{ gauss} = 10^4 \text{ G}$$

Magnetic flux

$$\text{Weber} \equiv \text{Tesla} \cdot \text{meter}^2$$
$$1 \text{ Wb} \equiv 1 \text{ T} \cdot \text{m}^2$$

Important Results

Properties of a particle of charge q and velocity $\mathbf{v}$ in a uniform magnetic field $\mathbf{B}$:

- Magnetic force $\mathbf{F}_m$ does *no* work on q.
- Path for $\mathbf{v} \parallel \mathbf{B}$: straight line.
- Path for $\mathbf{v} \perp \mathbf{B}$: circle.
- Path in general: helix.
- The charged particle circles magnetic-field lines at the *cyclotron frequency f*, where

$$f = (q/2\pi m)B \qquad (29\text{-}10)$$

- The magnitude of the particle's momentum in a plane perpendicular to $\mathbf{B}$ is

$$mv_\perp = qrB \qquad (29\text{-}8)$$

Velocity selector: the speed of any charged particle that can travel undeviated through crossed $\mathbf{E}$ and $\mathbf{B}$ fields:

$$v = \frac{E}{B} \qquad (29\text{-}12)$$

The magnetic force $d\mathbf{F}_m$ on an element dl of a conductor carrying current i in magnetic field $\mathbf{B}$ is

$$d\mathbf{F}_m = i\, dl \times \mathbf{B} \qquad (29\text{-}14)$$

where dl is a length element of the current-carrying conductor in the direction of conventional current.

Hall effect: the appearance of a transverse electric potential difference, whose sign indicates the sign of charge carriers, when a magnetic field is applied transversely to a current-carrying conductor.

The magnetic torque τ and potential energy U_B of a magnetic dipole moment μ in magnetic field $\mathbf{B}$:

$$\tau = \mu \times \mathbf{B} \qquad (29\text{-}19)$$

$$U_B = -\mu \cdot \mathbf{B} \qquad (29\text{-}20)$$

Problems and Questions

Section 29-1 Magnetic Field Defined

· **29-1 Q** In the equation $\mathbf{F} = q\mathbf{v} \times \mathbf{B}$,
(A) force and velocity vectors can have any angle between them.
(B) force and magnetic-field vectors can have any angle between them.
(C) velocity and magnetic-field vectors can have any angle between them.

· **29-2 Q** Charged particles are influenced by a uniform magnetic field
(A) under all circumstances.
(B) only if they have a component of velocity parallel to the magnetic field.
(C) only if they have a component of velocity perpendicular to the field.
(D) only if they are moving in a current loop.
(E) only if they are negatively charged.

Section 29-2 Magnetic Field Lines and Magnetic Flux

· **29-3 P** A circular plane loop of radius 2 cm is oriented so that its plane makes an angle of 30° with a uniform magnetic field of 0.30 T. What is the magnetic flux through the loop?

· **29-4 Q** A magnetic field of 40 G acts parallel to the x axis in a certain region. A 20 cm × 20 cm square of cardboard is oriented with its plane at an angle of 30° to the x axis (Figure 29-21). What is the magnetic flux through the cardboard?

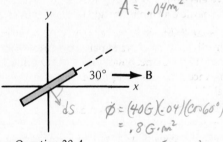

$$A = .04 m^2$$

$$\phi = (40G)(.04)(\cos 60°)$$
$$= .8 \, G \cdot m^2$$
$$= 8 \times 10^{-5} \, T \cdot m^2$$

Figure 29-21. Question 29-4.

(A) Zero, since cardboard is not a conductor.
(B) Zero, but not because cardboard is not a conductor.
(C) 1.6×10^{-4} T·m²
(D) 1.4×10^{-4} T·m²
(E) 8.0×10^{-5} T·m²

: **29-5 P** At a point on the earth's surface at a latitude of 50° N, the earth's magnetic field has a magnitude of 0.4 G and is inclined downward below the horizontal at an angle of 70°. A compass at this site points due north geographi-

cally. What is the magnetic flux at this location through a loop of area 20 cm² when the loop is oriented (a) horizontally? (b) vertically in an east-west plane? (c) vertically in a north-south plane?

: **29-6 P** A magnetic field points in the $+x$ direction. It does not vary with z, but does vary in the y direction according to $B_x(y) = K/y$. Determine the magnetic flux through a loop of wire in the shape of a square of side 4 cm oriented as shown in Figure 29-22, given $K = 0.001$ T cm.

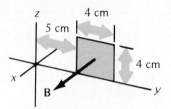

Figure 29-22. Problem 29-6.

: **29-7 P** A hemispherical shell of radius R is oriented so that the maximum flux due to a uniform magnetic field **B** passes through it (Figure 29-23). Show by direct integration over the spherical surface that the flux passing through that surface is indeed $\pi R^2 B$.

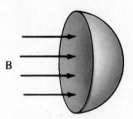

Figure 29-23. Problem 29-7.

: **29-8 P** A loop of area 10 cm² is oriented so that the components of a unit vector perpendicular to the loop are $(\frac{1}{3}, \frac{1}{3}, \frac{1}{3})$. It is placed in a magnetic field with x, y, and z components (0.010 T, 0.010 T, 0.005 T). What is the magnetic flux through the loop? (*Hint:* try vector methods using the scalar product.)

Section 29-3 Charged Particle in a Uniform Magnetic Field

· **29-9 P** Find the direction of the magnetic force acting on the charged particle moving as shown in each of the situations in Figure 29-24.

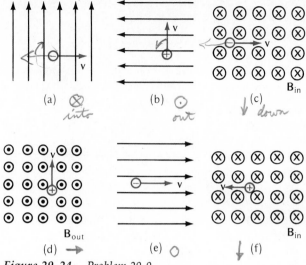

(a) $\otimes$ *into*   (b) $\odot$ *out*   (c) $\downarrow$ *down*

(d) $\rightarrow$ (e) $\bigcirc$ (f) $\downarrow$

Figure 29-24.   *Problem 29-9.*

29-10 Q An electron moving in the uniform magnetic field indicated in Figure 29-25 will experience a force in the direction

(A) toward *A*.
(B) toward *B*.
(C) toward *C*.
(D) toward *D*.
(E) out of the paper.
(F) into the paper.

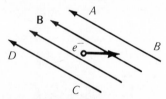

Figure 29-25.   *Question 29-10.*

· **29-11 Q** In a mass spectrograph, a positively charged ion traveling in a horizontal path is injected into a region where a uniform vertical magnetic field is present. In such a situation, the magnetic field will

(A) increase the kinetic energy of the particle.
(B) decrease the kinetic energy of the particle.
(C) not change the kinetic energy of the particle.
(D) exert a force parallel to the magnetic field lines.
(E) exert a force parallel to the velocity of the particle.

· **29-12 P** A positively charged deuterium nucleus of momentum 2×10^{-19} kg·m/s enters a region of uniform magnetic field. The magnetic-field lines are perpendicular to the deuteron's momentum vector. For what magnitude of magnetic field will the particle move in a circle of 1.8-m radius?

· **29-13 Q** Which of the following is the most accurate statement concerning the operation of a cyclotron?
(A) The speed of the particles remains constant.

(B) The energy of the particles is increased by the magnetic field.
(C) The energy of the particles is increased by the electric field.
(D) The energy of the particles is increased by both the electric and magnetic fields.
(E) The particles travel in circular orbits of fixed radius.

: **29-14 P** A beam of 10-keV electrons moving along the positive x axis is deflected by a magentic field of 0.002 T directed along the positive z axis. (*a*) Toward what axis are the electrons deflected? (*b*) What is the radius of curvature of the path?

: **29-15 P** (*a*) What is the orbit radius of 100-eV proton that moves perpendicular to a magnetic field of 1 G? (*b*) What is its frequency of revolution?

: **29-16 Q** An electron moves in a circular orbit of radius r in a uniform magnetic field **B**. The acceleration of the electron is thus
(A) zero.
(B) eBr.
(C) e^2B^2r/m^2.
(D) eBr/m.
(E) impossible to determine without more information.

: **29-17 P** Consider a 100-eV electron moving perpendicular to the earth's magnetic field of 0.00010 T. (*a*) What is the frequency of revolution? (*b*) What is the radius of the orbit?

: **29-18 P** A particle detected in a bubble chamber is found to have the same charge-to-mass ratio as an electron. It bends to its right in a circle of radius 2 mm while traveling horizontally in a vertical magnetic field of 0.01 T directed upward. (*a*) Is the particle an electron or a positron (a positive electron)? (*b*) What is the particle's speed? (*c*) What is the particle's kinetic energy, expressed in electron volts?

: **29-19 P** At what angle with a uniform magnetic field direction must a charged particle be projected into the field so that the distance it moves parallel to field lines equals the distance it moves along a circular arc perpendicular to the field?

: **29-20 Q** In a synchrotron accelerator, charged particles are accelerated at a fixed radius. For a given accelerator magnetic field and for a given type of particle, how will the final energy depend on R, the radius of the orbit?
(A) E is proportional to R.
(B) E is proportional to $R^{3/2}$.
(C) E is proportional to R^2.
(D) E is proportional to R^{-1}.
(E) E is independent of R.

: **29-21 P** In a mass spectrometer, a beam of chloride ions, Cl^-, each with a kinetic energy of 1 keV, enter perpendicularly a magnetic field of 0.30 T. After the beam bends through 180°, the ions are detected when they strike a photographic plate. The beam consists of two isotopes of chlorine, ^{35}Cl (mass 35 u) and ^{37}Cl (mass 37 u, where $1\ u = 1.67 \times 10^{-27}$ kg). What is the separation distance between these two components of the beam when they strike the detector plate?

⋮ **29-22 P** A proton with velocity components $v_x = 2 \times 10^6$ m/s, $v_y = 0$, $v_z = 2 \times 10^6$ m/s passes through the origin at $t = 0$ into a region where a constant magnetic field $B_z = 0.010$ T. (a) Sketch the projection of the path of the particle in the xy plane. (b) What is the maximum value of x and of y for the particle's motion? (c) Where does the particle first pass the z axis again after leaving the origin?

⋮ **29-23 P** A particle velocity selector can be made in the following way. A beam of particles, all with the same charge and mass but varying velocities, passes through a small hole in a metal plate. A uniform magnetic field **B** is directed perpendicular to the plate. The particle velocities all make angle θ with the field. See Figure 29-26. A second plate is placed a distance d from the first. It also contains small hole, in line with the hole in the first plate. Show that only particles with velocity $v = qBd/2\pi m\ \cos\ \theta$ will pass through the second hole.

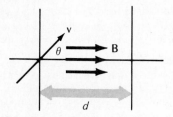

Figure 29-26. Problem 23-23.

⋮ **29-24 P** Several particles emitted in a cosmic-ray shower are detected in a bubble chamber, yielding the tracks sketched in Figure 29-27. The tracks are shown actual size, and they lie in the plane of the paper. A magnetic field was directed perpendicularly into the paper when the tracks were photographed. Two of the tracks were due to electrons. (a) Which tracks were made by electrons? (b) What is

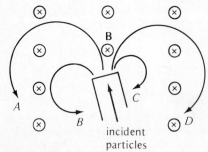

Figure 29-27. Problem 29-24.

the value of E_f/E_s, where E_f and E_s are the energies of the fast and slow electron respectively?

Section 29-4 Charged Particles in Uniform B and E Fields

· **29-25 Q** An electron is shot into a region of constant non-zero magnetic field and zero electric field. Its path will most likely be
(A) a straight line.
(B) a circle.
(C) a parabola.
(D) a hyperbola.
(E) an ellipse.
(F) a helix.
(G) a sine curve.

· **29-26 Q** An electron is not deflected in moving through a certain region of space. We can be certain that
(A) no magnetic field is there.
(B) no electric field is there.
(C) neither a magnetic nor an elecric field is there.
(D) the velocity of the electron remains constant.
(E) none of the above is true.

· **29-27 Q** In a mass spectrometer, the purpose of the magnetic field is
(A) to cause the particles to travel in a circular path.
(B) to increase the speed of the particles.
(C) to compensate for the electrical force acting on the particles.
(D) to decelerate the particles at the detector.
(E) to ionize the particles so that they can then be accelerated.

⋮ **29-28 P** A charged particle has velocity components $v_x = 1 \times 10^5$ m/s and $v_z = 2 \times 10^5$ m/s, and is subject to a uniform mangetic field $B_z = 0.020$ T. Determine the magnitude and direction of an electric field that will produce no deflection of the particle.

⋮ **29-29 P** A beam of 20-keV electrons passes between two plates separated by 2 cm. A uniform magnetic field of 200 G acts into the plane of the paper. What polarity and magnitude of voltage must be applied to the plates in Figure 29-28 so that the beam is undeflected?

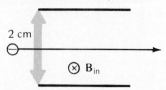

Figure 29-28. Problem 29-29.

⋮ **29-30 P** A particle of mass m and charge q is released from rest at one plate of a parallel-plate capacitor across which an accelerating potential V is applied. A uniform magnetic field B is into the paper. (a) Show that there will be no current between the plates if $V < qB^2d^2/m$. (b) With the electron just grazing the upper plate, how far from its starting point will the electron hit the lower plate? (See Figure 29-30).

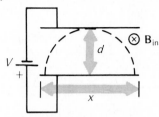

Figure 29-30. Problem 29-34.

Section 29-5 Magnetic Force on a Current-Carrying Conductor

· **29-31 P** A long, straight wire carrying a current of 3.0 A makes an angle of 60° with a uniform magnetic field of 0.040 T. What force, magnitude and direction, is exerted on a 3-m-long segment of the wire?

: **29-32 Q** A straight wire, of length L, carries a current. (The vector **L** is directed in the direction of the conventional current.) In calculating the force **F** such a wire will experience when placed in a magnetic field **B**, we recognize that for given vectors **B** and **L**,
(A) the direction of **F** will depend on the sign of the charge carriers in the wire.
(B) the angle between **F** and **B** can have any value.
(C) the angle between **F** and **L** can have any vaue.
(D) the total net force on the segment of wire of length L will be zero.
(E) the angle between **F** and **L** will always be 90°.

: **29-33 P** A long, straight transmission line carries a current of 100 A. It makes an angle of 50° with the earth's magnetic field, with magnitude 0.80×10^{-4} T. What is the magnetic force per unit length on the wire?

: **29-34 P** A conducting rod of mass m slides on two parallel fixed conducting bars separated by d and inclined at angle θ above horizontal. See Figure 29-30. A constant magnetic field B acts vertically upward. What magnitude of current must exist in the conductor if once started, the rod is to slide up the bars with constant speed?

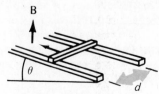

Figure 29-29. Problem 29-30.

: **29-35 P** (a) What current density would be needed to cause an aluminum conductor (density, 2.7 gm/cm³) to "float" in the earth's magnetic field at the equator, where the field strength is about 10^{-4} T? (b) How should the conductor be oriented for best results, and in what direction should the current flow? (c) Estimate whether this would be a practical undertaking. At what rate would thermal energy be generated in the aluminum? Is it likely to melt?

: **29-36 P** The horizontal arm of the beam balance shown in Figure 29-31 carries current I. A uniform magnetic field **B** acts perpendicularly out of the paper. (a) Determine the magnetic field B needed to keep the scale in balance. Express your answer in terms of m_1, m_2, g, x_1, x_2, and I. (b) What is the force on the pivot when B has the value found above?

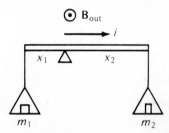

Figure 29-31. Problem 29-36.

: **29-37 P** A circular loop of wire of radius R carries current I. A uniform magnetic field **B** is directed perpendicular to the plane of the loop. What is the tension in the wire?

: **29-38 P** A wire 8 m long carries a current of 2 A. It lies on a horizontal table, 3m × 4m, as shown in Figure 29-32. A uniform magnetic field of 0.050 T acts parallel to the long edge of the table. What is the net magnetic force on the wire?

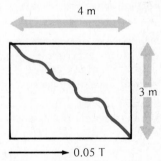

Figure 29-32. Problem 29-38.

Section 29-7 Magnetic Torque on a Current Loop

· **29-39 P** A rectangular loop of wire of 10 turns has its long edge parallel to the z axis, and the plane of the loop makes an angle of 30° with the y axis. See Figure 29-33. A uniform field of 1.5 G is directed along the y axis. A cur-

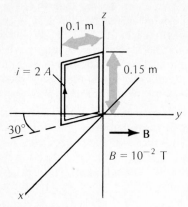

Figure 29-33.   Problem 29-39.

rent of 2 A exists in the coil. (*a*) What is the net force on the coil? (*b*) What is the net torque on the coil?

· **29-40 P** A circular coil of diameter 2.4 m and 150 turns is wrapped around a satellite to provide attitude control. This is done by passing a current through the coil and allowing the resulting torque due to the earth's magnetic field to rotate the spacecraft. What torque could be produced by a current of 1 A at an elevation of 550 km, where the earth's field is 8.0×10^{-5} T? Although small, the torque can produce appreciable rotation if allowed to act long enough, inasmuch as no friction acts on the satellite.

· **29-41 P** A square 2 cm × 2 cm coil of 100 turns carries 2 A. The plane makes an angle of 60° with a uniform magnetic field of 0.050 T. What is the torque on the coil?

: **29-42 P** A circular loop of wire with current I experiences a torque τ when placed in a uniform magnetic field **B**. Suppose that this same wire is wrapped into a coil of N small loops, oriented in the same way with the magnetic field. What torque now acts on the coil?

: **29-43 P** A circular coil of diameter 6 cm and 500 turns carrying a current of 10 mA experiences a maximum torque of 0.028 m·N when rotated in a uniform magnetic field. What is the minimum value of the magnetic field?

: **29-44 P** A thin disk carries a uniform surface-charge density σ on one side. It is rotated with angular velocity ω about a diameter. A uniform magnetic field **B** is present and makes an angle θ with the axis of rotation. What is the magnitude of the torque on the disk?

: **29-45 P** A thin rod of length L has charge Q uniformly distributed along its length. It is rotated with angular velocity ω about an axis passing through its center and perpen-

dicular to the rod. A uniform magnetic field **B** acts at an angle θ to the axis of rotation. What torque acts on the rod?

Section 29-8   Magnetic Dipole Moment

· **29-46 Q** When placed in a uniform magnetic field, a current-carrying coil of wire will
(A) experience a net force perpendicular to the magnetic field.
(B) experience a net force in the direction of the magnetic field.
(C) tend to align itself so that it encompasses the maximum number of magnetic-field lines.
(D) experience a torque $\mu \cdot \mathbf{B}$.
(E) experience no torque and no net force.

: **29-47 P** A charged particle moving in a circle constitutes a loop of electric current, and a magnetic dipole moment can be associated with it. For example, in a simple atomic model, the electron in a hydrogen atom circles the nucleus, and the atom can have a magnetic dipole moment from the orbiting electron. Suppose an electron with mass m and charge magnitude e is in a circular orbit and has orbital angular momentum of magnitude L. Show that the magnetic dipole moment of the orbiting electron is given by $-(e/2m)\mathbf{L}$. The quantity $e/2m$ is known as the *magnetogyric ratio*. It turns out that the ratio of the orbital magnetic dipole moment to orbital angular momentum has the same value for any orbital path.

Supplementary Problems

29-48 P Clouds can accumulate significant amounts of electrostatic charge (this is what gives rise to lightning). As the cloud drifts through the earth's magnetic field, forces can act, affecting the shape of the cloud. Using the following crude model, estimate the magnitude of the forces that would arise. Suppose the cloud has a charge +1 C on its upper half and −1 C on its lower half (the order of magnitude of the charge released in a lightning bolt). Suppose the cloud is drifting westward at the equator with a speed of 10 m/s at a point at which the earth's field is horizontal and directed north-south and has magnitude 1 G. What force acts to separate the two halves of the cloud?

29-49 P What is the magnetic force on an arc 40 mm long that carries a current of 500 A if it is acted on by a perpendicular magnetic field of 0.10 T? Such forces on arcs find many applications, such as in a rail gun or in a high-current circuit-breaker, where a magnetic field is used to "blow away" the arc if the current becomes too great.

Sources of the Magnetic Field

30

In Chapter 29 we dealt with one part of the magnetic interaction: a moving charged particle acted on by a magnetic force when it is in a magnetic field. The other part of the magnetic interaction involves a moving charged particle creating a magnetic field.

30-1   The Oersted Effect

An important scientific discovery is often the result of long and careful experiment. Sometimes it comes as a lucky accident. But how about a fundamental finding made with a demonstration experiment during a physics lecture?

It actually happened at the University of Copenhagen in 1820. Until then, electricity and magnetism were considered entirely separate phenomena; after all, a charged object does not attract or repel a magnet. But H. C. Oersted (1777–1851) believed that electricity and magnetism were related. Oersted's experiment (he said later that he did not have time to check it before the lecture) was simple enough; he sent an electric current through a conducting wire and looked to see whether a compass placed near the conductor was affected. The compass did indeed respond (strictly, it twitched a bit because of

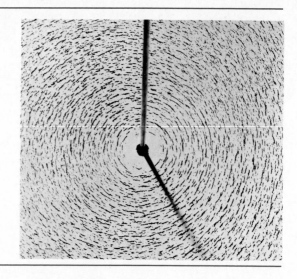

Figure 30-1. *The large current in a long, straight conductor causes iron filings to arrange themselves in circular rings.*

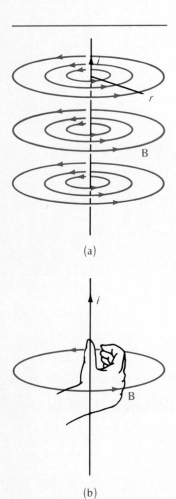

(a)

(b)

Figure 30-2. *(a) A long, straight conductor with current i produces circular magnetic-field loops. (b) The right-hand rule for relating the direction of conventional current i and the sense of the associated **B** loops.*

the low-current sources available then, and the people at Oersted's lecture were unimpressed). But repetitions of the experiment soon confirmed what has since that time been known as *the Oersted effect—a magnetic field is created by an electric current, by electric charges in motion.*

Figure 30-1 shows the effect on iron filings of the magnetic field from a relatively large current in a straight conductor. The magnetic field **B** produced by current i in a long straight conductor is shown in more detail in Figure 30-2(a). The **B** lines consist of circles of radius r concentric with the conductor and lying in planes transverse to the line of the conductor. The sense of the **B** loops is related to the direction of conventional current i by the right-hand rule shown in Figure 30-2(b): with the outstretched right thumb pointing in the direction of i, the curled right-hand fingers give the sense of the **B** loops. Reversing the current direction reverses the direction of **B** at each location.

How does the magnitude of **B** vary with the perpendicular distance r from the long straight conductor, where r is also the radius of the magnetic-field loop? As J. B. Biot (1774–1862) and F. Savart (1791–1841) first showed shortly after Oersted's discovery, $B \propto 1/r$; the magnetic field at any point is inversely proportional to the distance r from the conductor. Furthermore, the magnitude of **B** is directly proportional to current i, so that $B \propto i/r$. For reasons to become evident later in this chapter, it is customary to write the magnitude of **B** from a long, straight conductor as

$$B = k_m \frac{2i}{r} \qquad (30\text{-}1)$$

The quantity $k_m \equiv \mu_0/4\pi$ is a magnetic interaction constant whose role in magnetism is analogous to what the constant $k_e = 1/4\pi\epsilon_0$ does in electricity. Suffice it to say at this point that k_m is *assigned* the numerical value of *exactly*

$$k_m \equiv \frac{\mu_0}{4\pi} \equiv 10^{-7} \frac{\text{T} \cdot \text{m}}{\text{A}} \qquad (30\text{-}2)$$

Constant μ_0 itself is officially called the *permeability of free space.* The units assigned to $k_m = \mu_0/4\pi$, tesla meters per ampere, assure that B will have the units teslas. [Check that with (30-1).] Why the factor 4π? It is introduced here, so that it will not appear later in the mathematical formulation of Ampère's law (Section 30-5), just as $1/4\pi\epsilon_0$ appears in Coulomb's law but only ϵ_0 in Gauss's law for electricity.

We can see from (30-1) that it takes a fairly large current to produce a magnetic field of even moderate size. What is the magnetic field 2 cm from a straight conductor carrying 10 A? From (30-1), we find that $B = 10^{-4}$ T $= 1$ G, roughly the magnitude of the earth's magnetic field near the surface.

An *infinitely* long, *perfectly* straight conductor can never be constructed, if only because we must always have a complete loop, a return conductor, to maintain a steady current. Nevertheless, (30-1) can be applied for a straight conductor of finite length. We must simply be sure that the point at which B is computed is close to the straight conductor and relatively far from the locations where the conductor deviates from a straight line.

Example 30-1.   A straight conductor lies along the x axis and carries a conventional current in the positive x direction. What is the direction of the magnetic force on an electron fired as shown in Figure 30-3: (a) in the first quadrant and in the $+x$ direction; and (b) in the first quadrant and in the $-y$ direction?

From the right-hand rule for the magnetic field from a straight conductor, we find that **B** is out of the paper (in the $+z$ direction) in the first quadrant. We apply the magnetic-force relation, $\mathbf{F_m} = q\mathbf{v} \times \mathbf{B}$, recognizing that an electron has negative charge, to find the results shown in Figure 30-3: (a) $\mathbf{F_m}$ along $+y$; and (b) $\mathbf{F_m}$ along $+x$.

Example 30-2.   See Figure 30-4(a), where each of two parallel straight conductors separated by a horizontal distance d carries a current i into the paper. What is the magnetic field at point P, a distance d from each of the two conductors?

The two conductors and point P are at the corners of an equilateral triangle of side d. As shown in Figure 30-4(b), the magnetic field $\mathbf{B_1}$ produced at P by conductor 1 is perpendicular to the line from 1 to P and therefore at an angle of 30° below the horizontal. In similar fashion, $\mathbf{B_2}$ at P from conductor 2 is, as shown in Figure 30-4(c),

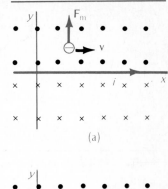

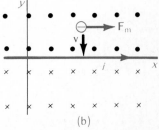

Figure 30-3. *Electron with velocity* **v** *subject to magnetic force* **F$_m$** *when traveling through the outwardly directed magnetic field* **B**.

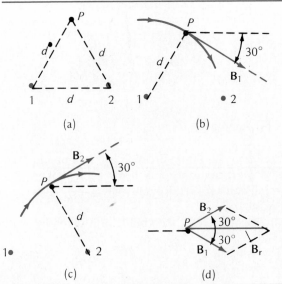

Figure 30-4. (a) Magnetic field at point P from conductors 1 and 2. (b) Conductor 1 alone produces magnetic field **B$_1$**. (c) Conductor 2 alone produces magnetic field **B$_2$**. (d) Resultant field **B$_r$** is the vector sum of **B$_1$** and **B$_2$**.

30° above the horizontal. Since P is equally distant from 1 and 2, then $B_2 = B_1 = k_m$ $(2i/d)$, from (30-1). The resultant magnetic field $\mathbf{B_r}$ at P is shown in Figure 30-4(d). The vertical components of $\mathbf{B_1}$ and $\mathbf{B_2}$ cancel, and their vector sum along the horizontal has the magnitude

$$B_r = 2B_1 \cos 30° = 2k_m\left(\frac{2i}{d}\right)\left(\frac{\sqrt{3}}{2}\right)$$

$$= 2\sqrt{3}\, k_m \frac{i}{d}$$

30-2   The Magnetic Force between Current-Carrying Conductors

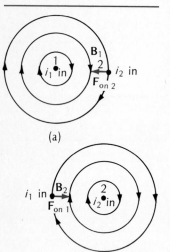

(a)

(b)

Figure 30-5. *(a) Straight conductor 1 with current i into the paper produces magnetic field **B** at the location of conductor 2, with current i_2 also into the paper. The magnetic force on conductor 2 is $\mathbf{F}_{on\,2}$. (b) The converse of part (a) with current $\mathbf{i_2}$ producing magnetic field $\mathbf{B_2}$ and force $\mathbf{F}_{on\,1}$.*

Consider two long, straight, parallel, current-carrying conductors. They interact magnetically because each conductor produces a magnetic field at the site of the other. This is sometimes referred to as *the fundamental* magnetic interaction, since both conductors are electrically neutral and there is no electric force between them. We wish to find magnetic force per unit length on either conductor in terms of the currents i_1 and i_2 in the two conductors and their separation distance d.

Conductor 1 with current i_1 produces magnetic field $\mathbf{B_1}$. At the location of conductor 2, a distance d away, we have from (30-1) that

$$B_1 = k_m \frac{2i_1}{d}$$

The field lines for $\mathbf{B_1}$ are shown in Figure 30-5(a). To find the direction of the magnetic force $\mathbf{F}_{on\,2}$ on conductor 2, we apply the rule for the magnetic force on a current element [(29-14)], $d\mathbf{F_m} = i\, d\mathbf{l} \times \mathbf{B}$. In Figure 30-5(a), with current i_2 in the same direction as i_1, we find that $\mathbf{F}_{on\,2}$ is to the left. The force magnitude on a length L of conductor 2 is

$$F_{on\,2} = B_1 i_2 L = k_m \frac{2i_1 i_2 L}{d}$$

where we have used the result for B_1 given above.

In exactly analogous fashion, we find that the magnetic force $\mathbf{F}_{on\,1}$ on conductor 1 is to the right, as shown in Figure 30-5(b), and that it has the magnitude

$$F_{on\,1} = B_2 i_1 L = k_m \frac{2i_2 i_1 L}{d} = F_{on\,2}$$

Two conductors with currents in the same direction attract one another magnetically. On the other hand, for currents in opposite directions, it is easy to see that the two conductors repel one another. Roughly speaking—like currents attract, unlike currents repel.

The force magnitude per unit length on either conductor is given by

$$\frac{F_{on\,1}}{L} = \frac{F_{on\,2}}{L} = k_m \frac{2i_1 i_2}{d} \qquad (30\text{-}3)$$

It is directly proportional to each of the currents, i_1 and i_2, and is inversely proportional to their separation distance d.

Suppose that each of two straight conductors carries a current of exactly 1 A and that they are separated by exactly 1 m. The force per unit length is, then, from (30-3),

$$\frac{F}{L} = k_m \frac{2i_1 i_2}{d} = \left(10^{-7} \frac{T \cdot m}{A}\right) \frac{(2)(1A)^2}{(1\ m)} = 2 \times 10^{-7}\ N/m$$

where we have used $1\ T = 1\ N/A \cdot m$. The force per meter of conductor is exactly $2 \times 10^{-7}\ N$.

Now recall that the magnetic constant $k_m \equiv \mu_0 / 4\pi$ is assigned the value of exactly $10^{-7}\ T \cdot m/A$. Actually, the relation for the magnetic force between two parallel straight conductors defines the ampere as the basic electric unit for current in the SI system of units. At last we have the definition of the ampere (and therefore all other electromagnetic units that depend on the ampere):

The equal currents in two parallel straight conductors separated by 1 m in a vacuum are each, *by definition, exactly one ampere* when the magnetic force per meter on either conductor is precisely $2 \times 10^{-7}\ N$.

30-3   The Magnetic Field from a Current Element; the Biot-Savart Relation

Although its magnetic field is relatively simple, an *infinitely long, straight,* current-carrying conductor is a special configuration. The basic current element is one of infinitesimal length $d\mathbf{l}$ carrying current i, where vector $d\mathbf{l}$ points in the direction of the conventional current. (The infinitesimal current element does for currents and their associated magnetic fields what point electric charges do in electrostatics and their associated electric fields.)

Figure 30-6(a) shows the magnetic field originating from an infinitesimal current element. The magnetic field consists of concentric circular loops lying in planes transverse to $d\mathbf{l}$; the magnitude of $\mathbf{B}$ falls off with distance from the element in all directions. The direction of $\mathbf{B}$ is related to the direction of the current by the right-hand rule.

The magnitude of the magnetic field $d\mathbf{B}$ at any point P in Figure 30-6(b) is given by

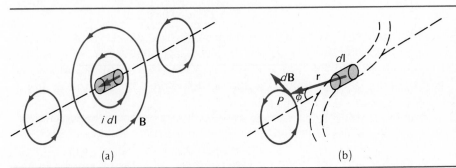

(a)

(b)

Figure 30-6. (a) Magnetic field loops from an infinitesimal current element. (b) The magnetic field $d\mathbf{B}$ at point P from current i through an infinitesimal element of conductor $d\mathbf{l}$. Radius vector $\mathbf{r}$ gives the location of P relative to $d\mathbf{l}$, and ϕ is the angle between $\mathbf{r}$ and $d\mathbf{l}$.

And the Beat Goes On

His routine never varied. During his lecture tour in America, the English novelist Charles Dickens would, upon arriving at an inn for the night's stay, whip out the compass he always carried, find north, and have his bed set along that line. The reason? The salubrious effect of having the human body aligned with the earth's magnetic field. After all, the 19th century argument went, if a compass is under least stress when it is aligned with a magnetic field, surely the human body must also enjoy minimum stress when it is aligned with a magnetic field?

Not only has magnetism confused novelists, the magnetic interaction has also been perplexing to scientists. Well, the magnetic force *is* different, even peculiar: it involves particle velocities, a couple of fairly complicated cross products, and energy conservation does not apply to it. The gravitational and the electric forces were never like this. Would it not

be far better—if physics is truly to be simple at its most basic level—if there simply were no magnetic force?

Even that can be accomplished. We can, so to speak, "turn off" the magnetic force on a charged particle by a simple strategem: we merely ride with the particle as an observer in the reference frame in which the particle is always at rest. Then, since the charged particle is at rest, the magnetic force on it must be zero. It's quite extraordinary. We've made the magnetic field disappear simply by changing reference frame. But the particle's motion cannot be affected by whether we see it in motion (with a magnetic force acting on it) or travel along with it (and have no magnetic force). The resultant force on the particle must still be the same even though the magnetic force may have been turned off. So we are compelled to reach this conclusion: what had been a

$$dB = k_\mathrm{m} \frac{i\, dl \sin \phi}{r^2} \tag{30-4}$$

where **r** is a radius vector from $d\mathbf{l}$ to P, and ϕ is the angle between $d\mathbf{l}$ and **r**. Note that the magnetic field falls off inversely with the square of the distance r from the current element; the *magnetic force* between a pair of infinitesimal current elements is *inverse-square*. (It is no surprise, then, that B from a long conductor varies with $1/r$, in the same fashion as E from a long, uniformly charged wire.) Note the effect of the $\sin \phi$ factor in (30-4); other things being equal, the magnetic field dB is a maximum in the transverse plane containing the current element ($\phi = 90°$) but falls to zero ahead of ($\phi = 0$) or behind ($\phi = 180°$) the current element.

The direction of $d\mathbf{B}$ shown in Figure 30-6(b) can be built into a vector relation, using the cross product (Section 12-2) as follows:

$$d\mathbf{B} = k_\mathrm{m} \frac{i\, d\mathbf{l} \times \mathbf{r}}{r^3} \tag{30-5}$$

Satisfy yourself that (30-5) gives the direction of $d\mathbf{B}$ shown in Figure 30-6(b) in terms of the directions of $d\mathbf{l}$ and **r**. The additional factor of r in the denominator—r^3 in (30-5) as against r^2 in (30-4)—is needed to compensate for the additional magnitude of r introduced into the numerator of (30-5) by vector **r**.

It is easy to verify that the magnetic field consists of circular loops centered on the current element. Imagine that all quantities on the right side of (30-5) are fixed except for the direction **r**, which we imagine as rotating about $d\mathbf{l}$ at a constant angle ϕ. Vector $d\mathbf{B}$, unchanged in magnitude, then turns through a circle in a transverse plane.

magnetic field (with the particle in motion) has been transformed into an electric field (with the particle at rest). It turns out that an electric field can similarly be transformed into a magnetic field.

To pursue in more detail these fascinating aspects of electromagnetism would take us far beyond the intent of this text. But we should know that there is more to the story. The magnetic field and the magnetic force *do* look simple when electromagnetic effects are re-examined from the point of view of relativity theory.* To identify just one rather amazing result, recall what we found for electrically neutral conductors: they interacted *only* by the magnetic force (Section 30–2). Relativity gives a different way of looking at this. Suppose that you ride with a conduction electron in one conductor; a second conductor, with current in the same direction, is nearby. Conduction electrons are at rest in the second conductor, positive ions are in motion. Because of the relativistic space-contraction effect, the positive ions are somewhat

bunched together, and the conductor as a whole carries a *net positive* charge. *That* is why conductors with like current attract. In this sense, the magnetic force is not a distinctive type of force but really just another aspect of the electric force.

So relativity gives the final insight into basic electromagnetism? Not so. Quantum theory (Chapter 41) shows electromagnetic radiation is quantized into particle-like photons. So that's the last word? Not so. Recent developments in quantum field theory show that the electromagnetic force is very closely related to the so-called "weak interaction" that shows up in the radioactive decay of certain unstable nuclei. Indeed, the two forces are separate manifestations of an even more primordial electro-weak force.

*For an introductory, intermediate-level treatment of classical electromagnetism, that is relativistic right from the start, see *Electricity and Magnetism*, Second Edition (New York: McGraw-Hill, 1985) by Nobel laureate E. M. Purcell.

Equation (30-5) for the magnetic field of an infinitesimal current element is usually referred to as the *Biot-Savart law.* How do we know it is right? There is, after all, no such thing as a single tiny piece of current unconnected to the rest of the world; a current element must always be part of a complete loop of current. Equation (30-5) is correct because it always works; that is, when all the $d\mathbf{B}$ contributions, each given by (30-5), from the infinitesimal elements that make up an actual current loop are added (as vectors), the computed resultant magnetic field at each location always agrees with experiment.

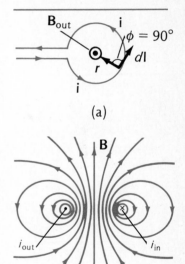

(a)

(b)

Example 30-3   B at Center of Circular Loop.   The loop has radius r and carries current i, as shown in Figure 30-7. We suppose that the two lead wires that carry current to and from the loop are placed side by side. Because we then have nearly coincident equal currents in opposite directions, the magnetic effects of the lead conductors cancel, and we are left with the magnetic field of the ring of current alone. Element $d\mathbf{l}$ is on the circumference of the loop and radius vector $\mathbf{r}$ points to the center of the loop. For the counterclockwise current of Figure 30-7, the magnetic field $d\mathbf{B}$ produced by every element $d\mathbf{l}$ along the circumference is out of the plane of the paper and perpendicular to it. Therefore, we can sum the $d\mathbf{B}$ contributions algebraically. With angle ϕ between $d\mathbf{l}$ and $\mathbf{r}$ equal to $90°$, we have, from (30-4), for the magnitude of $\mathbf{B}$ at the center of the loop,

$$B = \int dB = k_m \frac{i}{r^2} \int dl$$

But $\int dl = 2\pi r$, the distance around the circumference, so that the final result is

$$B = k_m \frac{2\pi i}{r} \qquad (30\text{-}6)$$

For example, with a current of $i = 1$ A in a loop with $r = 1$ cm $= 10^{-2}$ m, Equation

Figure 30-7. (a) The magnetic field $\mathbf{B}$ at the center of a circular current loop of radius r with current i. (b) Magnetic field lines in a plane transverse to a circular conducting loop and containing its symmetry axis.

(30-6) gives for the field at the center, $B = 0.6 \times 10^{-4}$ T $= 0.6$ G. For a coil of N identical circular loops, B at the center (and all other locations) is increased by a factor N.

Computing **B** from a circular loop at locations other than its center is far more complicated. The results for a plane transverse to the plane of the loop and containing the symmetry axis is shown in Figure 30-7(b). Close to the conducting wire, the field lines are nearly circular, since more distant current elements of the loop contribute little to the field there. The field lines are symmetrical with respect to the loop's symmetry axis. The direction of **B** at the center can be found by applying the right-hand rule to any current element. There is another way, however, to relate the direction of **B** at the center to the sense of current in the loop. Let the *curled fingers* of the right hand give the sense of the *current;* then the right *thumb* points in the direction of the *field* at the loop's center.

Example 30-4   B from a Long, Straight Conductor.   Here we confirm that (30-1) for the magnetic field from an infinitely long, straight, current-carrying conductor follows from the Biot-Savart relation, (30-5), for an infinitesimal current element.

See Figure 30-8, where current i flows along the positive y axis. At point P, a distance R from the conductor, the field $d\mathbf{B}$ produced by the element dy is into the paper. The angle between current element $d\mathbf{l}$ and radius vector $\mathbf{r}$ to point P is ϕ. From (30-4), we have

$$dB = k_m \frac{i \, dy \sin \phi}{r^2} \tag{30-7}$$

Every current element of the straight conductor produces a field into the paper at P, so that we merely integrate the equation above algebraically over the entire length of the conductor to find the total field. Quantities y, r, and ϕ are not independent, however; we must first write (30-7) in terms of a single variable, this variable chosen as the one that will be the simplest to integrate. That variable is angle α, the complement of ϕ.

From the geometry of Figure 30-8, we see that

$$y = R \tan \alpha$$

so that

$$dy = R \sec^2 \alpha \, d\alpha$$

We also see that

$$r = R \sec \alpha$$

and

$$\sin \phi = \cos \alpha$$

Using these relations in (30-7), we get

$$dB = k_m \frac{i(R \sec^2 \alpha \, d\alpha)(\cos \alpha)}{(R \sec \alpha)^2} = k_m \frac{i \cos \alpha \, d\alpha}{R}$$

For a conductor of infinite length, we integrate α from $-90°$ to $90°$. (To find the field of a straight conductor of finite length, we should have to choose different limits.) Therefore,

$$B = \int dB = \frac{k_m i}{R} \int_{-\pi/2}^{\pi/2} \cos \alpha \, d\alpha = k_m \frac{2i}{R}$$

To conform to our earlier usage, we now replace R by r for the distance from the conductor, and have finally

$$B = k_m \frac{2i}{r} \tag{30-1}$$

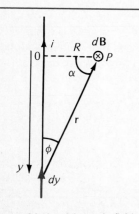

Figure 30-8.   *Magnetic field from a long straight current-carrying conductor.*

30-4   Gauss's Law for Magnetism

First recall Gauss's law for electricity (Section 24-4):

$$\phi_E = \oint \mathbf{E} \cdot d\mathbf{S} = \frac{q}{\epsilon_0} \qquad (24\text{-}5)$$

The total electric flux ϕ_E out of any closed surface is proportional to q, the net charge enclosed by the surface. Gauss's law says essentially that for electric charges at rest, *electric-field lines* **E**:

- Originate from positive charges.
- Terminate on negative charges.
- Are continuous between charges.

Therefore if we know how the electric field lines are arranged over any arbitrary closed Gaussian surface, we can immediately deduce the net charge inside.

Gauss's law for magnetism is similar. It deals with the total magnetic flux $\phi_B = \oint \mathbf{B} \cdot d\mathbf{S}$ through an imaginary closed Gaussian surface of any shape by adding the contributions $\mathbf{B} \cdot d\mathbf{S}$ of magnetic field **B** over a small, outwardly directed surface element $d\mathbf{S}$. See Figure 30-9. Positive magnetic flux corresponds to **B** lines out of the surface, negative magnetic flux to inward **B** lines. If the total magnetic flux over a closed surface is zero, equal numbers of **B** lines enter and leave the surface. This is precisely what is found in every case.* Experimental findings are consistent with the statement:

$$\phi_B = \oint \mathbf{B} \cdot d\mathbf{S} = 0 \qquad (30\text{-}8)$$

This is *Gauss's law for magnetism*. It says, in effect:

- **B** lines always form closed loops.
- **B** lines do not originate from and terminate on magnetic "charges" (or what are known more formally as *magnetic monopoles*). The reason is simply that isolated magnetic poles do not exist in nature.† Every **B** line ends on its own tail.

Gauss's law for the magnetic fields produced by current-carrying conductors follows directly from the Biot-Savart relation (Section 30-3). Every infinitesimal current element produces circular magnetic-field lines; the magnetic flux with circular loops is clearly zero over a closed surface of any shape. Then any finite collection of current elements must also produce continuous magnetic field loops, so that $\oint \mathbf{B} \cdot d\mathbf{S} = 0$ in all situations.

What about magnets (magnetic dipoles), where **B** lines appear to emanate from the north pole and terminate at the south pole, the two ends of the magnet? See Figure 30-10. Here the magnetic field lines pass through the interior of the magnet and again form closed loops.

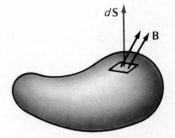

Figure 30-9.   *Magnetic field* **B** *through patch of surface represented by vector* $d\mathbf{S}$.

---

* Well, almost every case. See the next footnote.

† In very subtle experiments, B. Cabrera of Stanford University in 1982 found evidence for a single magnetic monopole, most likely one surviving from the primordial fireball at the creation of the universe.

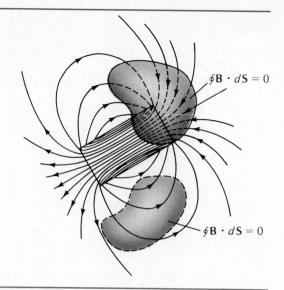

Figure 30-10. *Gauss's law for magnetism: The net magnetic flux through any closed surface is zero.*

30-5   Ampère's Law

Recall the two ways of relating electric field to electric charge:

• Coulomb's law. Each point charge dq creates an electric field $d\mathbf{E}$ at the location given by radius vector $\mathbf{r}$, where $d\mathbf{E} = k\, dq\, \mathbf{r}/r^3$ (Section 23-4).

• Gauss's law of electricity. The electric flux $\oint \mathbf{E}\cdot d\mathbf{S}$ over any closed surface is proportional to the net charge q within, where $\oint \mathbf{E}\cdot d\mathbf{S} = q/\epsilon_0$ (Section 24-4).

Coulomb's law is in *differential* form, Gauss's law in *integral* form. There are also differential and integral forms for relating magnetic field to electric current (see Table 30-1):

• The Biot-Savart law. Each current element $i\, d\mathbf{l}$ creates a magnetic field $d\mathbf{B}$ at the location given by radius vector $\mathbf{r}$, where $d\mathbf{B} = k_m i\, d\mathbf{l} \times \mathbf{r}/r^3$.

• *Ampère's law*, which relates the magnetic field around the edge of any imaginary sheet to the net current passing through the sheet.

Table 30-1.   *Fields and their Sources*

	ELECTRICITY	MAGNETISM
Differential form	Electric field, point charge from Coulomb's law $$d\mathbf{E} = k\,\frac{dq\,\mathbf{r}}{r^3}$$	Magnetic field, infinitesimal current element from Biot-Savart law $$d\mathbf{B} = k_m\,\frac{i\,d\mathbf{l} \times \mathbf{r}}{r^3}$$
Integral form	Gauss's law for electricity $$\oint \mathbf{E}\cdot d\mathbf{S} = \frac{q}{\epsilon_0}$$	Ampère's law $$\oint \mathbf{B}\cdot d\mathbf{l} = \mu_0 i$$

More specifically, Ampère's law can be expressed by the relation

$$\oint \mathbf{B} \cdot d\mathbf{l} = \mu_0 i \qquad (30\text{-}9)$$

Here $d\mathbf{l}$ is an infinitesimal displacement vector lying along the outer edge of a sheet of arbitrary shape, and $\mathbf{B}$ the local magnetic field at that location. See Figure 30-11. Contributions of the component of $\mathbf{B}$ along $d\mathbf{l}$ are added, going all the way around the loop along the outer edge of sheet, and the complete line integral of the magnetic field around the loop $\oint \mathbf{B} \cdot d\mathbf{l}$ is thereby computed. It turns out, as we shall see, that this line integral is always equal to $\mu_0 i$, where i is the net current through the sheet. Figure 30-11 implies a convention, corresponding to the right-hand rule, relating directions; when a loop is traversed in the *counterclockwise* sense (with the interior of the loop on the left) and the *line integral* is *positive,* current coming *out* of the sheet is *positive.*

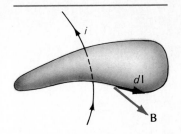

Figure 30-11. Magnetic field $\mathbf{B}$ *at infinitesimal displacement vector* $d\mathbf{l}$ *along the edge of a sheet through which current* i *passes.*

Like Gauss's law, Ampère's law is a very general principle relating a field (here $\mathbf{B}$) to its source (here i). And like Gauss's law, Ampère's law is easy to apply in concrete situations only if there is a high degree of geometrical symmetry.

Let us see how Ampère's law works in one familiar simple situation — a long, straight current-carrying conductor. Consider Figure 30-12, where as we know, the magnitude of $\mathbf{B}$ at a distance r from the conductor is

$$B = k_{\mathrm{m}} \left(\frac{2i}{r} \right) = \left(\frac{\mu_0}{4\pi} \right) \left(\frac{2i}{r} \right) \qquad (30\text{-}1)$$

(Here we have replaced k_{m} by its equivalent, $\mu_0/4\pi$.) The relation above can also be written as

$$B(2\pi r) = \mu_0 i$$

Here we have chosen the arbitrarily shaped sheet in Ampère's law to consist of a flat circular plate of radius r, concentric with and perpendicular to the conductor, as shown in Figure 30-13. Then, at each point along the sheet's outer circular edge, $\mathbf{B}$ is constant in magnitude and parallel to $d\mathbf{l}$. Therefore, the line integral around the chosen closed loop, $\oint \mathbf{B} \cdot d\mathbf{l}$, is here just equal to B multiplied by the circumference $2\pi r$. Clearly, the relation above is just what Ampère's law gives.

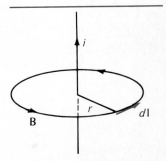

Figure 30-12. Magnetic field $\mathbf{B}$ *at a distance* r *from a long, straight conductor with current* i.

Now we shall see that we get the very same result for the long straight conductor for any loop enclosing a sheet of *any* shape. Consider the loop of Figure 30-13(a), where the path consists of two circular arcs coinciding with field lines and two radial lines at right angles to $\mathbf{B}$. For radius r_1, the field has magnitude B_1; at r_2 it is B_2. But (30-1) shows that the field falls off inversely with distance r from the conductor, so that $B_1 r_1 = B_2 r_2$. Along the radial lines, $d\mathbf{l}$ is perpendicular to $\mathbf{B}_1$, so that there is no contribution to the line integral. The decrease from B_1 to B_2 in going from r_1 to a larger r_2 is exactly matched by the increase in arc length. For the loop of Figure 30-13(a), with current i enclosed, we again have

$$\oint \mathbf{B} \cdot d\mathbf{l} = \mu_0 i$$

$$B(2\pi r) = \mu_0 i$$

Suppose the loop does not enclose the conductor; this is shown in Figure 30-13(b). Again there is no contribution to the line integral along the radial

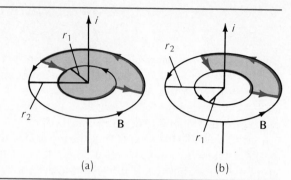

Figure 30-13. *Magnetic field* **B** *from a long straight conductor for a loop (a) enclosing and (b) not enclosing current i.*

(a)   (b)

lines. Once again *in magnitude*, $B_1 r_1 = B_2 r_2$; contributions along the two circular arcs have the same magnitude. But their signs differ. At r_2 the path is traversed in the *same* direction as **B**; at r_1, the directions of $d\mathbf{l}$ and **B** are *opposite*. Consequently, the total line integral is now zero,

$$\oint \mathbf{B} \cdot d\mathbf{l} = 0$$

which corresponds to no current passing through the loop.

To deal with more complicated loops, we simply replace the arbitrary path by small circular arcs and radial segments. See Figure 30-14(a). Circular sections subtending the same angle give the same contribution; radial segments all make zero contribution. In every case, the line integral around a closed loop is equal to the current threading the loop multiplied by the constant μ_0. Note further that if the direction of the current is reversed, and therefore also the direction of **B** at each location, a factor of -1 is thereby introduced. Even a sheet of arbitrary shape, one that does not lie in a plane perpendicular to the straight conductor, introduces no complications: **B** is entirely in planes transverse to the conductor, so that a path element parallel to the conductor does not contribute to the line integral.

Thus far we have proved that Ampère's law applies for a single long, straight conductor. Suppose that we have two or more infinitely long, straight conductors. This introduces nothing new; magnetic fields add as vectors, so that if Ampère's law holds for one straight conductor, it also applies for a collection of straight conductors. We must, however, take the current *i* to be the net current through the sheet whose edge constitutes the loop for the line integral, using different signs for current out of and into the sheet. One way to

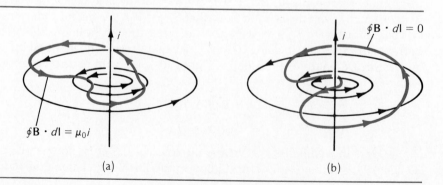

Figure 30-14. *Ampère's law applied to a general loop shape (a) enclosing the conductor and (b) not enclosing the conductor.*

(a)   (b)

prove that Ampère's law holds for conductors of any shape, not merely long, straight ones, is to show that any current distribution and its associated magnetic field can be produced by equivalent straight conductors.*

Although Ampère's law is altogether general, it can be used to calculate the magnetic field only when the geometry of **B** can be deduced in advance on the basis of symmetry. Then we know what sort of path to choose. (This is reminiscent of Gauss's law for electricity: If the geometry of **E** is known in advance on the basis of symmetry, we know how best to choose the Gaussian surface.)

One final remark on Ampère's law. As written in (30-9), it is not complete. In Section 34-2 we shall see that a changing electric flux, as well as a current, can contribute to the line integral of the magnetic field.

Example 30-5   Cylindrical Coaxial Conductors.   A coaxial conductor consists of a solid inner circular cylinder of radius r_1, carrying current I, and a concentric outer conductor of inner and outer radii r_2 and r_3 and carrying current I in the opposite direction. See Figure 30-15. The current in both conductors is uniformly distributed over their cross sections. What is the magnetic field (a) outside the outer conductor? (b) between the two conductors? (c) within the inner conductor?

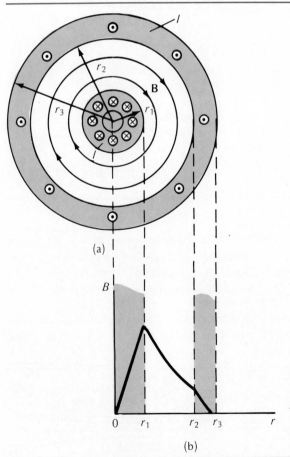

(a)

(b)

Figure 30-15.   (a) Magnetic field lines for a coaxial conductor; (b) the magnitude of the field as a function of radial distance r.

---

* See F. Reines, *American Journal of Physics* 39:838 (1971).

First, we know on the basis of symmetry alone that the magnetic field must, at any place where it is not zero, consist of circular loops concentric with the conductors. This means that in applying Ampère's law we shall always choose a concentric circular loop.

(a) Outside both conductors the net current, $I - I$, within the loop is *zero*. Consequently, the field there must, from Ampère's law, also be zero.

For $r > r_3$, we get $B = 0$.

(b) For a circular path in the region between the two conductors, only the current I *within the loop* contributes to the field. (The current of the outer conductor has no effect whatsoever on the region inside; it merely cancels the field of the inner conductor for $r > r_3$.) From Ampère's law we have, then,

$$\oint \mathbf{B} \cdot d\mathbf{l} = \mu_0 i$$

$$B(2\pi r) = \mu_0 I$$

For $r_1 < r < r_2$,

$$B = \frac{\mu_0 I}{2\pi r}$$

(c) Our circular loop of radius r now lies within the inner conductor. Only a fraction of the total current I is *within* this loop. With current distributed uniformly over the cross section, the fraction of I inside of r is the ratio of the circular areas $\pi r^2 / \pi r_1^2$, so that the current within r is $(r/r_1)^2 I$. Ampère's law then gives

$$B(2\pi r) = \mu_0 \left(\frac{r}{r_1}\right)^2 I$$

For $0 < r < r_1$,

$$B = \frac{\mu_0 r\, I}{2\pi r_1^2}$$

The magnetic field falls off linearly with r. The field within the outer conductor can be computed in similar fashion.

The variation of B with r is shown in Figure 30-15(b).

30-6 . The Solenoid

A solenoid consists of a tightly wound helix of conducting wire. It is, in effect, a series of adjacent circular conducting loops arranged into a cylindrical shell. Near the center, the magnetic field within the solenoid is essentially uniform, as shown in Figure 30-16. The external magnetic field has a configuration like that of a long permanent cylindrical magnet.

We can find the magnitude of **B** at the center of an infinitely long solenoid —or at least one whose length is much greater than its diameter—from

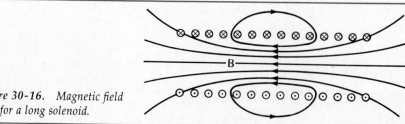

Figure 30-16. Magnetic field lines for a long solenoid.

Ampère's law. Each loop carries current i_1, and the number of turns per unit length is n. Consider, then, the rectangular closed path shown dashed in Figure 30-17. Just outside the solenoid, **B** must be essentially zero; the field lines passing through the interior are very thinly spread on the exterior. There is no contribution to the line integral of **B** along portion A. The field inside, along line B, must by symmetry be parallel to the solenoid axis; if it were not, this would imply a noninfinite solenoid length. For this reason, **B** is perpendicular to the path along the transverse portions C and D, and these parts of the closed path also do not contribute to the line integral. The only contribution comes from the inside segment of length L. The number of turns within length L is nL, and the total current enclosed is nLi_1. Therefore, from Ampère's law we have

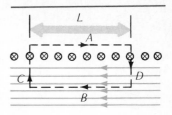

Figure 30-17. *Ampère's law applied to a solenoid.*

$$\oint \mathbf{B} \cdot d\mathbf{l} = \mu_0 i$$

$$BL = \mu_0 nLi_1$$

$$B = \mu_0 ni_1 \tag{30-10}$$

The magnetic field at any interior point near the center of a long solenoid is uniform, independent of the transverse distance from the axis; the magnitude of **B** depends only on the number of turns per unit length and i_1, the current through each.

The solenoid is a device for combining the magnetic effects of many loops to produce a strong, uniform magnetic field. As the Biot-Savart relation or Ampère's law shows, B is always $\propto i$; large currents create big fields. Since the resistance of a superconductor is zero, the very large currents that can be

Figure 30-18. *The tunnel of the main accelerator at Fermilab. The upper ring of magnets (in a square casing) is the 400 GeV accelerator. The lower ring is superconducting magnets for the Tevatron. In the Tevatron, protons accelerated to a kinetic energy of 1 TeV (10^{12} eV) collide head-on with antiprotons of the same energy. The Tevatron employs the world's largest collection of superconducting magnets, more than a thousand of them. Each magnet, kept at a temperature below 4.6 K by liquid helium, produces a magnetic field of 4.5 T.*

achieved in superconducting materials are now used increasingly to produce large magnetic fields. See Figure 30-18.

30-7 Magnetic Materials

Figure 30-19(a) shows the magnetic field of a solenoid, Figure 30-19(b) the magnetic field of a magnet with same dimensions. The two fields are the same. We may, in fact, think of the magnet's field as coming from currents. Such currents would encircle the magnet's external cylindrical surface.* Actually,

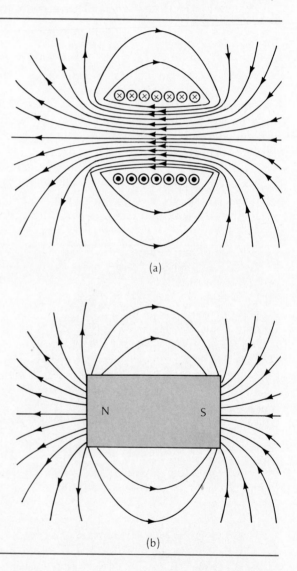

(a)

Figure 30-19. Magnetic field produced by (a) a solenoid and (b) a permanent magnet with the same dimensions.

(b)

---

* This insight—that the external magnetic field of a magnet can be attributed to circulating currents—was first given by André M. Ampère (1775–1836). He developed essentially all the fundamental relations between currents and magnetic fields within several weeks of learning of Oersted's discovery that an electric current creates a magnetic field.

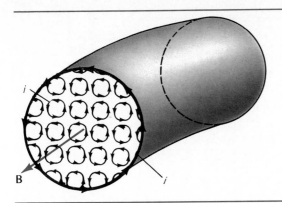

Figure 30-20. *Cross section of core material within toroid, showing current loops.*

separate loops of current distributed over the entire cross section of a magnet are equivalent to a single current loop around the periphery; at any interior location, the opposite currents from adjoining loops cancel, so that we are left with only the current (sometimes called the *Ampèrian current*) around the outside. See Figure 30-20.

What exactly are the atomic currents that produce magnetism in iron? Because atomic properties are understood adequately only on the basis of the quantum theory, what follows can be only qualitative and approximate. Certainly atoms contain electrically charged particles and microscopic currents; we have electrons moving in orbits about nuclei. But in ordinary materials, electron orbits are so paired that their magnetic effects cancel; for each electron going clockwise, there is another going counterclockwise in a similar orbit. The origin of the magnetism of magnetic materials is *electron spin.* There is no complete classical analog to electron spin, a strictly quantum-mechanical effect, but we may visualize it as follows. An electron regarded as a sphere of negative charge spins perpetually at a constant rotation rate about an internal rotation axis. See Figure 30-21. The circulating charge produces an electron-spin magnetic moment μ as shown (because an electron has negative charge, its spin magnetic moment is opposite in direction to its spin angular momentum). The special—and from the point of classical electromagnetism, peculiar—property of electron spin is this: the electron-spin magnetic moment has the same magnitude (1.61×10^{-23} J/T) for all electrons, and this magnetic moment (and the associated spin rate) is unaffected by an external magnetic field. In this sense, a spinning electron is the only example of a perfect permanent magnet.

In nonmagnetic materials, not only are the electron orbital magnetic moments paired off to yield no net magnetic effects. So too are electron spins, with pairs of electron spins in antialignment to yield zero electron-spin magnetization. In feebly magnetic, or *paramagnetic*, materials (for example, copper, manganese, chromium) there are unpaired electron spins, however. In zero external field, the magnetic moments of these unpaired spins produce no net magnetization, because the spin magnetic moments of the magnetic ions are oriented at random through the material. But when an external magnetic field is turned on, more electron-spin magnetic moments align themselves with the field than against the field, so that there is net magnetization along the direction of the external field. The alignment of electron-spin magnetic moments by

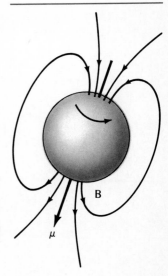

Figure 30-21. *Magnetic field surrounding a spinning charged sphere, a model for electron spin.*

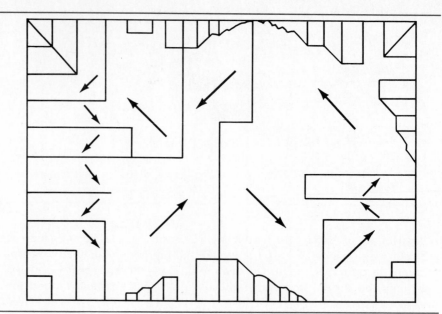

Figure 30-22. *Ferromagnetic domains in a single crystal of nickel. The direction of the magnetic field in each domain is shown with an arrow.*

the external field competes with the disorganizing effect of thermal motion within the paramagnetic material; the atomic vibrations tend to produce random spin orientations and no net magnetization. Therefore as the temperature is lowered and the thermal vibrations diminish, the magnetization of a paramagnetic material typically increases.

Ferromagnetic materials—such as iron, nickel, and cobalt—are strongly magnetic in the sense that the electron spins from many adjoining ferromagnetic ions are strongly coupled together by a quantum-mechanical force for which there is no classical analog, known as an *exchange force.* As a consequence large groups of atoms, or *magnetic domains,* are formed in which many electron spins are aligned together to form a fairly large magnetic moment for the domain. A typical domain has a size of about 10^{-7} m (about 10^3 atomic diameters). In an unmagnetized ferromagnetic material, the domains are oriented at random so as to give no net magnetization for the material as a whole. See Figure 30-22.

Now suppose that an external magnetic field is turned on. The boundaries of the domains change, so that those domains whose magnetic moment is pointed along the direction of the external field grow in size at the expense of other domains. In addition, the orientation of the magnetic moment within any one domain many change direction toward alignment with the external field. In fact, a very large fraction of the electron-spin magnetic moments may become aligned with the external field, and the material acquires a net magnetic moment. When the external magnetic field is turned off, the magnetic material may retain a smaller magnetic moment, and a "permanent" magnet is produced. Ferromagnetism is complicated by the fact that the properties of ferromagnetic materials depend not only on external magnetic fields but also on the past magnetic (and thermal) history of the material.

Summary

Definitions

The magnetic interaction constant k_m and permeability of the vacuum μ_0 have values that are defined to be exactly

$$k_m \equiv \frac{\mu_0}{4\pi} \equiv 10^{-7} \text{ T·m/A}$$

The *ampere* as the SI unit for current and the basis for all other electromagnetic SI units: that current in each of two infinitely long, parallel, straight conductors that produces a magnetic force on one meter of conductor of exactly 2×10^{-7} N [(30-3)].

Fundamental Principles

The Oersted effect: electric charges in motion, or currents, create magnetic-field loops. The directions are related by the right-hand rule.

Biot-Savart relation for the magnetic field $d\mathbf{B}$ at radius vector $\mathbf{r}$ from an infinitesimal element of current i in a displacement $d\mathbf{l}$:

$$d\mathbf{B} = k_m \frac{i \, d\mathbf{l} \times \mathbf{r}}{r^3} \tag{30-5}$$

Gauss's law for magnetism:

$$\phi_B = \oint \mathbf{B} \cdot d\mathbf{S} = 0 \tag{30-8}$$

The net magnetic flux ϕ_B out of any closed surface is zero.

Equivalently, **B** lines form closed loops; magnetic monopoles do not exist.

Ampere's law:

$$\oint \mathbf{B} \cdot d\mathbf{l} = \mu_0 i \tag{30-9}$$

where $d\mathbf{l}$ is a displacement along the edge of a sheet of arbitrary shape and i is the net current through the loop formed by the edge of the sheet.

Important Results

Magnetic field B at distance r from an infinitely long, straight conductor with current i:

$$B = k_m \left(\frac{2i}{r} \right) \tag{30-1}$$

Magnetic force per unit length between two parallel conductors with current i_1, and i_2 and separated by distance d:

$$\frac{F}{L} = k_m \left(\frac{2i_1 i_2}{d} \right) \tag{30-3}$$

"Like currents" (same direction) attract; "unlike currents" (opposite directions) repel.

Magnetic properties of magnetic materials have their origin in electron spin.

Problems and Questions

Section 30-1 The Oersted Effect

· **30-1 Q** What is the character of the magnetic-field lines produced by an infinitely long, straight wire that carries a current?

(A) They spiral in toward the wire in planes perpendicular to the wire.
(B) They spiral outward from the wire in planes perpendicular to the wire.
(C) They are radial lines perpendicular to the wire.
(D) They are concentric circles in planes perpendicular to the wire.

· **30-2 Q** Current is established along the x axis toward positive x in a long, straight wire. An electron passes through the point $y = 0$, $z = a$ moving toward positive y parallel to the y axis. The force acting on the electron will be

(A) directed toward the $+z$ direction.
(B) directed toward the $-z$ direction.
(C) directed toward the $+x$ direction.
(D) parallel to the y axis.
(E) zero.

: **30-3 Q** A long, straight wire is oriented perpendicular to the plane of the paper. In the plane of the paper is a uni-

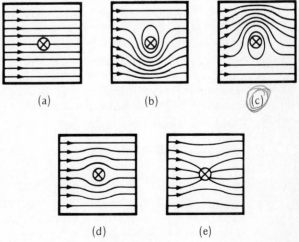

Figure 30-23. *Question 30-3.*

form magnetic field directed from left to right. Which of the diagrams in Figure 30-23 best describes the magnetic field lines resulting from this arrangement when current in the wire is directed into the paper?

· **30-4 Q** Current exists in the irregular loop shown in Figure 30-24. The loop lies in the plane of the paper. Thus the magnetic field at point P is
(A) directed toward A.
(B) directed toward B.
(C) directed toward C.
(D) directed toward D.
(E) directed out of the paper.
(F) directed into the paper.
(G) zero.
(H) oriented in a direction which can't be determined without more information.

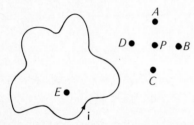

Figure 30-24.   Questions 30-4, 30-5.

· **30-5 Q** An irregular loop of wire carrying a current lies in the plane of the paper (Figure 30-24). Suppose that the loop is then distorted into some other shape while still lying in the plane of the paper. Point E is still within the loop. Which of the following is a true statement concerning this situation?
(A) The magnetic field at E will not change in magnitude.
(B) The magnetic field at E will not change in direction.
(C) It is possible that the magnetic field at E may be zero.
(D) The magnetic field at point E will always lie in the plane of the paper.
(E) None of the above is true.

: **30-6 P** A current of 2 A exists along the z axis in the positive z direction, and a current of 3 A exists along the line $y = 0.02$ m in the xy plane in the positive x direction. What is the magnetic field at the point (0, 0.01 m, 0)?

: **30-7 Q** Consider a square loop of current in which a

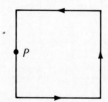

Figure 30-25.   Question 30-7.

sinusoidally alternating current exists, as shown in Figure 30-25. Which of the following is an accurate statement?
(A) The loop will experience a torque that will tend to make it rotate.
(B) A magnetic force to the left will act on point P.
(C) A magnetic force to the right will act on point P.
(D) A magnetic force out of the paper will act on point P.
(E) A magnetic force oscillating in direction will act on point P.
(F) The magnetic force on point P will be zero.

: **30-8 P** In a simple model of the hydrogen atom, the electron orbits the proton as nucleus in a circle of radius 5.3×10^{-11} m at a speed of 2.2×10^{6} m/s. What is the magnitude of the magnetic field produced at the site of the proton by the orbiting electron?

Section 30-2   The Magnetic Force between Current-Carrying Conductors

· **30-9 Q** Two long parallel straight wires carry currents of 2 A and 4 A as indicated in Figure 30-26. Thus the right-hand wire will experience a force that is
(A) directed to the right.
(B) directed to the left.
(C) zero.
(D) directed out of the paper.
(E) directed into the paper.
(F) different in magnitude from the force exerted on the left-hand wire.

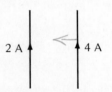

Figure 30-26.   Question 30-9.

· **30-10 P** Each of two long, straight, parallel wires carries a current of 2 A. The wires are separated by 4 cm. What force is experienced by a 3-cm segment of one of the wires?

: **30-11 P** The ampere is defined so that it is the current that exists in each of two parallel wires separated by 1 m if one meter of wire is to experience a magnetic force of 2×10^{-7} N/m drawing them together. What linear electric-charge density would each of two such wires have to carry to produce the same attractive force?

: **30-12 Q** Two concentric circular loops of flexible wire lie on a table. There is a sudden pulse of current in the outer wire, and then a second pulse of current in the opposite direction. How will the inner loop of wire be affected?

(A) It will first contract slightly and then expand slightly.
(B) It will first expand slightly and then contract slightly.
(C) It will contract with each surge of current.
(D) It will expand with each surge of current.
(E) It will not be affected by the outside loop.

: **30-13 Q** Sketch the resultant magnetic field for two parallel, straight current-carrying conductors with currents (a) in the same direction and (b) in opposite directions. The result seen here, one that is altogether general, is that the direction of the magnetic force on a current-carrying conductor is in the direction from strong-to-weak resultant magnetic field.

: **30-14 P** Four long straight wires each carry a current of I. They are arranged at the corners of a square of side d (see Figure 30-27). What is the magnitude and direction of the force per unit length acting on one of the wires? The current is in the same direction in all wires.

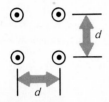

Figure 30-27.   *Problem 30-14.*

: **30-15 P** Three long, straight wires run parallel to each other and carry currents of I, two with currents in the same direction and the third with its current in the opposite direction. Each wire passes through a corner of an equilateral triangle of side d. Determine the magnitude and direction of the magnetic force per unit length on the wires.

: **30-16 P** Two protons move parallel to each other with velocity $\mathbf{v}$. Show that the ratio of the magnetic force $\mathbf{F}_m$ to the electric force $\mathbf{F}_e$ on either proton is $F_m/F_e = v^2\epsilon_0\mu_0 = (v/c)^2$.

Section 30-3 The Magnetic Field from a Current Element; the Biot-Savart Relation

· **30-17 P** What is the magnetic field at the center of a loop of wire, of radius 10 cm, carrying a current of 1 A?

: **30-18 P** Two semicircles of wire are joined to form a loop carrying 4 A (Figure 30-28). What is the magnetic field at point C, the center of the concentric semicircles? The inner semicircle has radius 2 cm; and the outer radius, 4 cm.

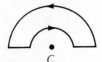

Figure 30-28.   *Problem 30-18.*

: **30-19 P** Determine **B** at the center C of the semicircular loop carrying current I shown in Figure 30-29.

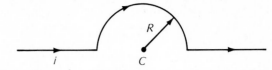

Figure 30-29.   *Problem 30-19.*

: **30-20 P** Current I exists in the wire drawn in Figure 30-30. Determine the magnetic field at the center C of the arc.

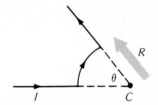

Figure 30-30.   *Problem 30-20.*

: **30-21 P** Derive an expression for the magnetic field at a point P on the axis of a circular current loop of radius R carrying current i, in the case for which the distance z from the center of the loop to P is much greater than R.

: **30-22 Q** Two beams of protons are fired in the same direction along parallel paths. What happens to the motion of the two beams when the initial velocity of one of the beams is increased?
(A) The motion will be unaffected by this change.
(B) The beams will tend to draw together more.
(C) The beams will repel each other other more.
(D) A spiral motion of the beams will be induced.
(E) The speed of particles in the second beam will also increase.

: **30-23 P** Using the following methods, determine the magnetic field at the center of a square loop of side $2a$ with current I. Use the law of Biot-Savart to (a) calculate **B** due to one side. Multiply by 4 to get the desired result. (b) Fit a circular loop just outside the square and find **B** at its center. Then fit a circular loop just inside and again find **B**. Average these values to get an approximate value for the field due to the square. (c) What is the ratio of the values found in (a) and (b)? Does the approximation improve for larger loops?

: **30-24 P** Charge Q is uniformly distributed over the surface of a circular disk of radius R. The disk is rotated with constant angular velocity ω about its symmetry axis. (a) Determine the magnetic field at the center of the disk. (b) Where is the magnetic field greatest? (Give a qualitative argument.)

Section 30-4 Gauss's Law for Magnetism

: 30-25 Q Which of the following are accurate statements?

(A) Magnetic-field lines would "sprout" out of magnetic monopoles in the same way that static electric-field lines emanate from positive electric charges.

(B) The magnitude of a magnetic field is proportional to the spacing between its magnetic-field lines; that is, the larger the separation between lines, the stronger the magnetic field.

(C) Magnetic-field lines are analogous to whirlpool flow lines in a fluid, where the "paddlewheels" stirring up the magnetic fields are electric currents.

(D) A proton moving in a magnetic field experiences a magnetic force directed tangent to a magnetic-field line.

(E) Magnetic fields exert forces only on magnetic objects, not on electrically charged particles.

Section 30-5 Ampère's Law

· 30-26 Q It is important, in using the equation $\oint \mathbf{B} \cdot d\mathbf{l} = \mu_0 i$ to find the field due to a long, straight wire, to recognize that

(A) the integral is evaluated on a path along the wire.

(B) the integral is evaluated on a circular path in a plane perpendicular to the wire.

(C) $d\mathbf{l}$ is a small vector pointing from the wire to the point at which the field is to be evaluated.

(D) $d\mathbf{l}$ is a small vector directed along the wire.

(E) $\mathbf{B}$ will not have constant magnitude along the path C.

· 30-27 Q The equation $\oint \mathbf{B} \cdot d\mathbf{l} = \mu_0 i$ is

(A) based on experimental observation.

(B) one in which $d\mathbf{l}$ is a small vector pointing along a wire carrying current i.

(C) a result of Newton's second law of motion.

(D) a consequence of Gauss's law, $\int \mathbf{E} \cdot d\mathbf{S} = q/\epsilon_0$

(E) a consequence of $\mathcal{E} = -d\phi_B/dt$

: 30-28 P Current exists with uniform density across the cross-sectional area of a long, straight wire of radius R. The magnetic field at the surface of the wire is B_0. What is the value of the magnetic field inside the wire at a point a distance $\frac{1}{2}R$ from the axis?

: 30-29 P A uniform current I exists in a cylindrical pipe with inner radius R_1 and outer radius R_2. The current density is uniform over the cross section of the pipe. Determine the magnetic field everywhere.

: 30-30 P The current density in a long straight cylindrical conductor of radius R is given by $j = j_0 e^{-r/r_0}$ at a distance r from the cylinder axis. Find the magnetic field as a function of r.

: 30-31 P What is the magnetic field just outside an infinite plane sheet that carries a uniform surface current density $\mathbf{j}$?

: 30-32 P Uniform, constant surface current densities are established parallel to each other in two thin sheets separated by a distance d. Determine the magnetic field at an arbitrary point between the two sheets and outside the two sheets when (a) the current densities are in the same direction; (b) the current densities are in opposite directions (antiparallel).

Section 30-6 The Solenoid

· 30-33 Q Consider the magnetic field set up in space by a coil of N turns carrying current i. No other sources of magnetic field are present. If the number of turns is doubled, the magnetic field will

(A) be unchanged.

(B) increase in some regions and decrease in others, so that the total magnetic flux will be doubled.

(C) double everywhere.

(D) increase everywhere, but not by the same factor at every point.

(E) None of the above.

· 30-34 Q The easiest way to calculate the magnetic field within a long solenoid of N turns is to

(A) note that inductance depends only on geometric factors.

(B) use $B = \mu_0 i / 2\pi r$.

(C) use $\mathcal{E} = -d\phi/dt$.

(D) use $\oint \mathbf{B} \cdot d\mathbf{l} = \mu_0 i$.

(E) use $d\mathbf{B} = (\mu_0/4\pi)i d\mathbf{l} \times \mathbf{r}/r^3$.

(F) use $d\mathbf{F} = i\, d\mathbf{l} \times \mathbf{B}$.

· 30-35 P A solenoid that is 80 cm long and 2 cm in diameter is wound with 5000 turns of wire. What is the field near the center of the solenoid when it carries a current of 2 A?

· 30-36 Q An electron is fired along the axis of a current-carrying solenoid. It is aimed at point A on the fluorescent screen shown in Figure 30-31. It strikes the screen at one of the indicated points. Which point does it strike?

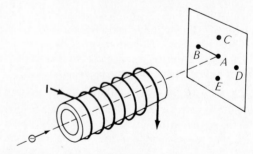

Figure 30-31. *Question 30-36.*

: 30-37 Q The pattern of magnetic field lines shown in Figure 30-32 is set up by a little black box. Such a magnetic field might result if the black box contained

(A) a toroid (a solenoid in the shape of a doughnut) carrying a current.

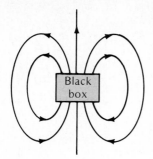

Figure 30-32.   *Question 30-37.*

(B) a single small loop of wire carrying a current.
(C) a current-carrying wire wound into a sphere like a ball of yarn.
(D) a small bar magnet.
(E) none of the above.
(F) More than one of the above could be correct.

: **30-38 P** A toroid (a solenoid bent into a circle to form a doughnut) has inner radius r_1 and outer radius r_2. Show that the magnetic field inside the toroid depends only on r, the distance from the center, and not on the shape of the cross section of the toroid (for example, the cross section may be circular, elliptical, or rectangular).

30-39 P A toroid of circular cross section has an inner radius of 4 cm and an outer radius of 6 cm (Figure 30-33). It is wound with 400 turns of wire. What is the magnetic field at the center of the toroid (that is, at a radius of 5 cm) when a current of 2 A exists in it?

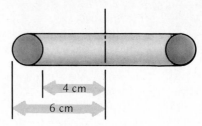

Figure 30-33.   *Problem 30-41.*

: **30-40 P** Show that the magnetic field at the end of a very long solenoid is just half the value at the center. (*Hint:* Think of an infinitely long solenoid as consisting of two long solenoids placed end to end.)

Supplementary Problem

30-41 P The very high currents that can exist with charged particles in motion in a plasma can give rise to the effect in magnetohydrodynamics known as the *pinch effect.* Parallel currents in the same direction attract one another. In similar fashion, current over the cross section of a beam of particles can produce inwardly directed magnetic forces over the outer surface of the beam. Such a magnetic pressure can pinch the beam of charged particles. (*a*) Show that the magnetic pressure varies as the square of the current in the beam. (*b*) Show that the pressure can reach 1 atmosphere when the current in a beam of 1 cm radius reaches about 25,000 A. (For 10^6 A, the pressure is 1600 atm.)

31

Electromagnetic Induction

Basic features of the interactions between electrically charged particles are these:

- Part of the force does not depend on the particle velocities. This is the *electric*, or Coulomb, force.
- A velocity-dependent force also acts between charged particles. This is the *magnetic* force.

We might be inclined to think that so far as basic electromagnetism is concerned, this is the whole story. It is not! Electromagnetism is concerned with electrically charged particles and their interactions, of course, but equally important are electric and magnetic fields and their interrelation. Electric and magnetic fields are created by electric charges at rest and in motion. But in addition, we shall see that:

- A changing magnetic field can create an electric field. (That is the fundamental phenomenon of electromagnetic induction to be treated in this chapter.)
- A changing electric field can create a magnetic field. (That topic is treated in chapter 34.)

Beyond its role in classical electromagnetic theory, electromagnetic induction also has enormous practical importance. Such common devices as inductors, transformers, alternating-current generators — and indeed, a large fraction of all applied electric power — are based directly in the induction effect.

31-1   Induced Currents and EMF's

The phenomenon of *electromagnetic induction* was discovered in England in 1831 by Michael Faraday (1791 – 1867), regarded as the greatest experimental physicist of all time. The same effects were discovered independently in the United States about the same time by Joseph Henry (1797 – 1878), the principal scientific advisor to Alexander Graham Bell in the development of the telephone.

Figure 31-1 shows a variety of experimental situations in which electromagnetic induction is observed. Each involves a loop of conductor with a galvanometer that can register the existence of an *induced current* in the loop; we call this the detector loop. Some *changing* condition induces the current:

• Closing the switch in a nearby primary loop containing a battery, Figure 31-1(a).
• Moving an entire current-carrying primary loop toward a detector loop, Figure 31-1(b).
• Moving a permanent magnet toward a detector loop, Figure 31-1(c).
• Reorienting a detector loop when it is close to the current-carrying primary loop, Figure 31-1(d).
• Deforming a detector loop when it is close to the current-carrying primary loop, Figure 31-1(e).

The general qualitative experimental observations are these (the current direction is discussed in Section 31-3):

• An induced current exists only while the change shown in Figure 31-1 is occurring (while the switch is being closed, while the magnet is moving, and so on).
• Only relative motion matters. If, for example, in Figure 31-1(c) the magnet remains at rest with the detecting loop moved left, instead of the magnet's being moved right with the loop at rest, a current is again induced. On the other hand, with both magnet and loop at rest or with both moving at the same velocity, there is no induced current.
• The more rapid the change or motion, the greater the magnitude of induced current.
• If the change is reversed (switch opened, rather than closed; motion to left rather than right; and so on) the direction of the current is also reversed.

The effects are peculiar, at least in terms of what we have learned thus far of electromagnetism. The detector loop has no battery in it. The loop is unconnected to any other circuit and sits in apparently empty space. Yet an emf drives charges around the conductor to produce an induced current.

What common feature did Faraday identify in these varied circumstances? What was the basic underlying change associated with every induced current? Faraday was an early champion of the field concept. For him electric-field lines and magnetic-field lines (he always called them lines of force) were a palpable

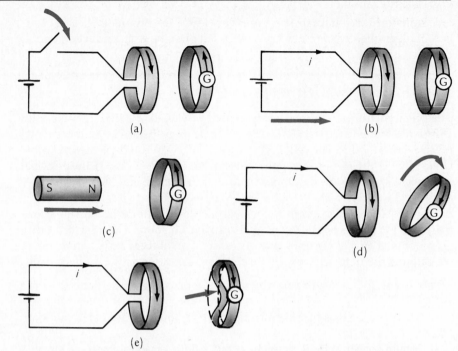

Figure 31-1. *Experimental situations for observing electromagnetic induction. (a) Closing a switch in a primary loop with a battery. (b) Moving the entire current-carrying primary loop. (c) Moving a permanent magnet. (d) Reorienting the detector loop. (e) Deforming the detector loop.*

presence. In Faraday's mind's eye, every charged object bristled with electric-field lines, every current-carrying conductor was wreathed with magnetic-field loops. Faraday could perceive in the situations portrayed in Figure 31-1 what cannot be shown in any single static picture:

• In graphical terms, the *number of magnetic-field lines* threading through the detector loop *changes with time*.

• Or in equivalent mathematical terms, the *magnetic flux* through the detector loop *changes with time*.

In Figure 31-1(a), for example, closing the switch and establishing a current in the primary loop produces a growing magnetic field, not only in the region of the primary loop, but also through the detector loop. Similarly, the motion of a current-carrying primary loop [Figure 31-1(b)] or the motion of a magnet [Figure 31-1(c)] changes the number of magnetic-field lines through the detector loop.

Recall that the magnetic flux (Section 29-2) $d\phi_B$ through some small patch of surface element dS at a location where the magnetic field is $\mathbf{B}$ is defined as

$$d\phi_B = \mathbf{B} \cdot d\mathbf{S} = BS \cos \theta = B_\perp \, dS \qquad (31\text{-}1)$$

where θ is the angle between $\mathbf{B}$ and $d\mathbf{S}$, and $B_\perp$ is the component of $\mathbf{B}$ perpendicular to the surface. To find the net magnetic flux over some finite loop, one simply sums the contributions from infinitesimal surface elements. We can think of the magnitude of $\mathbf{B}$ as equal to the number of magnetic-field lines per unit transverse area; the net magnetic flux ϕ_B through a loop is then equal to the net number of magnetic-field lines threading the chosen loop.

A magnetic flux change through an element of surface d**S** can then arise from any of the following circumstances, individually or in combination:

- A change in the *magnitude* of the *magnetic field* **B** [Figure 31-1(a), (b), (c)].
- A change in the *magnitude* of *area* d**S** of the loop [Figure 31-1(e)].
- A change in the angle θ between **B** and d**S**, that is, in their *relative orientation* [Figure 31-1(d)].

Although induced current is observed directly in a conducting loop, the more fundamental induced quantity is the induced emf $\mathcal{E}$ that drives charged particles around the conducting loop to create the induced current. Experiment shows that all electromagnetic induction effects can be summarized in what is also called *Faraday's law*. This fundamental principle relates the emf $\mathcal{E}$ induced around any loop to the time rate of change of the magnetic flux through that loop as follows:

$$\mathcal{E} = -\frac{d\phi_B}{dt} \tag{31-2}$$

First, let us check that the units in this relation are all right. With $d\phi_B/dt$ given in the units webers per second, does $\mathcal{E}$ have the units volts? The unit conversions go as follows:

$$\text{Wb/s} = \text{T} \cdot \text{m}^2/\text{s} = (\text{N}/\text{A} \cdot \text{m})\text{m}^2/\text{s} = \text{N} \cdot \text{m}/\text{A} \cdot \text{s} = \text{J}/\text{C} = \text{V}$$

Implications of Faraday's law of electromagnetic induction are these:

- Whether an emf is induced around some loop depends only on the change with time in the net magnetic *flux* through that loop. The magnetic field is not the crucial quantity. The magnetic field at all locations on the periphery of a loop may be constant, even constantly zero, but if the net flux within the loop changes, an emf is created (see Example 31-2).
- An induced emf, however, cannot be localized, like a battery; an emf is associated with the entire loop and is a property thereof.
- Suppose that the conducting loop has N turns. An emf is induced in each of the N coincident loops, and the net emf is larger than that from a single loop by a factor N. In that situation, Faraday's law may be written as

$$\text{Loop of } N \text{ identical turns:} \quad \mathcal{E} = -\frac{d(N\phi_B)}{dt} \tag{31-3}$$

where ϕ_B is the flux ~~change~~ through a single loop and $\mathcal{E}$ is the net emf. The product $N\phi_B$ is called the *flux linkage*.

The minus sign in (31-2) relates to the direction or sense of the emf. This is treated in Section 31-3.

Example 31-1. A uniform magnetic field of 0.20 T is applied transverse to a flexible circular loop of a conductor 10 cm in radius. See Figure 31-2(a). The opposite points along a diameter of the loop are pulled outward so that 0.20 s later, the loop lies along a straight line, as shown in Figure 31-2(b). What is the average emf induced in the loop over the 0.20-s interval?

In this example the magnetic flux changes by ϕ_B because the area of loop is reduced from πr^2 to zero in a time interval Δt. The time average of the emf is then

$$\bar{\mathcal{E}} = -\frac{\Delta\phi_B}{\Delta t} = -\frac{B(\pi r^2)}{\Delta t} = -\frac{(0.20 \text{ T})\pi(0.10 \text{ m})^2}{0.20 \text{ s}} = -31 \text{ mV}$$

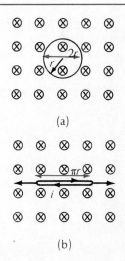

Figure 31-2. (a) A circular conducting loop in a magnetic field is pulled along opposite points along a diameter. (b) The loop lies along a straight line.

Example 31-2.   A very long current-carrying solenoid passes through a square conducting loop. See Figure 31-3. The solenoid has a 1.0-cm radius, 20 turns/cm, and the current through it changes at the rate of 100 A/s. What is the magnitude of the emf induced in this square loop?

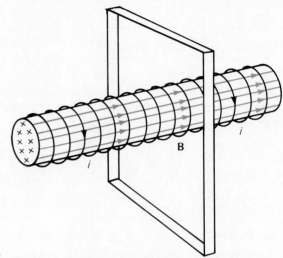

Figure 31-3.   A very long current-carrying solenoid within a square conducting loop.

Inside the solenoid (Section 30-6) the magnetic field is uniform and lies along the solenoid axis; in magnitude [(30-10)], $B = \mu_0 ni$, where n is the number of turns per unit length and i the current in each turn. We assume that the solenoid's length is so great compared with the size of the square loop that the only magnetic field through the square loop is that confined to the interior of the solenoid. The magnetic field at all points of the square loop is *zero* and remains so.

The magnetic flux through a cross section of the solenoid is

$$\phi_B = B_\perp A = (\mu_0 ni)(\pi r^2)$$

where r is the solenoid radius. This is also the total magnetic flux through the entire square loop, since the only magnetic-field lines penetrating the square are the lines inside the solenoid. The emf induced in the square loop is then

$$\mathcal{E} = -\frac{d\phi_B}{dt} = -\frac{d}{dt}(\mu_0 ni\pi r^2) = -\mu_0 \pi n r^2 \frac{di}{dt}$$

$$= (4\pi \times 10^{-7} \text{ Wb/A·m})\pi(20 \times 10^2/\text{m})(10^{-2}\text{ m})^2(100 \text{ A/s})$$

$$= -7.9 \times 10^{-5} \text{ V} = -79\mu\text{V}$$

(The minus sign relates to the sense of the emf.)

Note especially what the induced emf does not depend on: the dimensions of the loop, or even its shape; where the solenoid goes through the square. Even the angle between the plane of the loop and the axis of the solenoid does not enter. If the solenoid axis is not perpendicular to the rectangle, the area intercepted by the solenoid on the plane of the square is increased by a factor $(1/\cos \theta)$, but the magnetic flux and the number of magnetic-field lines is unchanged because of the appearance also of factor $\cos \theta$ in the relation for magnetic flux. All that matters in determining the emf is how rapidly the flux penetrating the loop changes with time.

Example 31-3 Electric Generator. A coil of N turns, each of area A, is rotated at constant angular speed ω in a uniform magnetic field of magnitude B. What is the emf induced in the coil?

The arrangement here is that of a simple electric generator. The magnetic flux through each turn of the coil is given by

$$\phi_B = AB \cos \theta$$

where θ is the angle between the normal to the plane of the coil and the direction of **B**; see Figure 31-4(a). With the coil rotated at constant angular speed ω, we can write $\theta = \omega t$, and the instantaneous flux is

$$\phi_B = AB \cos \omega t$$

We have N identical conductor turns connected in series, so that the total emf is

$$\mathcal{E} = \frac{-d(N\phi_B)}{dt}$$

$$= \omega NAB \sin \omega t = \mathcal{E}_0 \sin \omega t$$

The output of the coil is a sinusoidally varying alternating emf (ac), as shown in Figure 31-4(b).

It is interesting to note how the coil is oriented when the instantaneous emf is a maximum and when it is zero:

● Maximum $\mathcal{E}$ corresponds to $\theta = \omega t = 90°$. At this instant, the magnetic flux throughout the coil is zero; $\phi_B = AB \cos \omega t = 0$. No magnetic field lines pass through the coil. But a small change in θ from $90°$ allows the lines to penetrate the coil, and the time rate of change of ϕ_B is a maximum.

● Zero $\mathcal{E}$ corresponds to $\theta = \omega t = 0°$. At this instant the magnetic flux has its maximum value; $\theta_B = AB \cos \omega t = AB$. Field lines pass perpendicularly through the coil but their number is unchanged for a small change in θ from $0°$.

In the electric generator described here, mechanical power supplied by an external source turning the coils is transformed into electric power. Conversely, one can operate the arrangement in Figure 31-4 as an electric motor by supplying a sinusoidal electric current to the coil; then a magnetic torque acts on the coil, and electric power is converted to mechanical power. An electric generator operated in reverse is a motor.

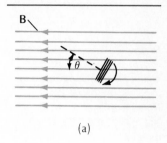

(a)

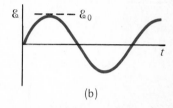

(b)

Figure 31-4. (a) A coil rotating at constant angular speed ω in a uniform magnetic field. (b) The sinusoidally varying emf induced in the coil.

31-2 EMF in a Moving Conductor

Here we investigate one example in which a current is induced in a moving conductor, from two points of view:

● Faraday's law of electromagnetic induction.
● The detailed forces acting on conduction electrons.

Consider Figure 31-5, where a rectangular conducting loop is moving into a uniform transverse magnetic field **B** directed into the paper. For simplicity, we assume that the magnetic field drops abruptly to zero at its left boundary. The loop's height is l; its velocity to the right is v. After time t, the right end of the loop has advanced a distance vt into the field, so that an area lvt is within the field. The instantaneous magnetic flux through the loop is

$$\phi_B = Blvt$$

so that the emf induced in the loop is

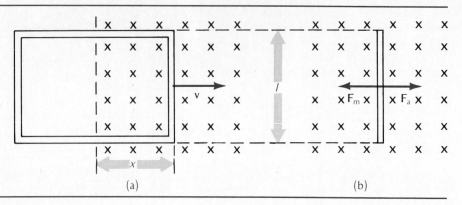

Figure 31-5. *(a) Rectangular conducting loop of height l moving at velocity* **v** *into a uniform magnetic field* **B**. *(b) Forces acting on the right-hand end of the conducting loop: retarding magnetic force* **F**$_m$ *and applied force* **F**$_a$.

(a)   (b)

$$\mathcal{E} = -\frac{d\phi_B}{dt} = -Blv \qquad (31\text{-}4)$$

The minus sign relates to the sense of the induced emf, which is, as we shall see, counterclockwise. The induced current exists so long as the loop is in motion and the flux through it is changing. After the left end of the loop has entered the magnetic field and the loop is then entirely within the region of **B**, flux ϕ_B is constant and therefore $\mathcal{E} = 0$. As the loop's right end leaves the region of magnetic field, there is again an induced current, but in the reversed sense.

Now we examine the forces on conduction electrons. First we consider, not the entire loop in Figure 31-5, but only its right end, a rod of length l, shown in Figure 31-6. With the rod moved to the right at velocity v, so too is every conduction electron within it. The magnetic force on such a moving free electron has the direction, from $\mathbf{F}_m = q\mathbf{v} \times \mathbf{B}$, vertically downward.* As a consequence, free electrons are forced to the lower end of the moving rod; it acquires a net negative charge, so that the upper end of the rod is positively charged. A magnetic force also acts on the positive charges within the rod, but these charges remain locked in place. The charge separation continues until the downward magnetic force on a free electron is balanced by an equal upward electric force from electrons that are already at the rod's lower end. An electric field is created within the rod by opposite charges at the two ends, and an electric potential difference is established between the rod ends. This potential difference exists only so long as the rod is in motion and a magnetic force drives free electrons as they are pulled through the magnetic field. If the rod is brought to rest, the potential difference drops to zero; if the rod changes direction of motion, the potential difference reverses polarity.

Back to Figure 31-5. The entire rectangular conducting loop has only its right end moving through **B**. Although charges are separated on the end within **B**, there is no charge separation on the left end, which is outside of **B**. Excess electrons no longer accumulate on the lower end of the right end. The electrons are driven around the loop in the clockwise sense.

With current in the right end of the loop as it is moved through the magnetic field, a magnetic force acts on this current-carrying conductor. From $d\mathbf{F}_m = i\, d\mathbf{l} \times \mathbf{B}$, we see that the magnetic force on the right end is in the direction

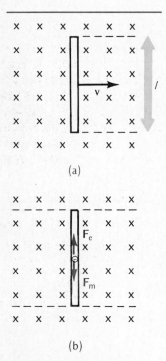

(a)

(b)

Figure 31-6. *(a) Rod of length l moving at velocity* **v** *through a magnetic field* **B**. *(b) The forces on a conduction electron moving to the right: a downward magnetic force* **F**$_m$ *and an upward electric force* **F**$_e$.

---

* Strictly, the electron moves *obliquely* downward in the rod, not just to the right. To simplify the analysis, we neglect the downward velocity component.

How Does Physics Advance?

A parable:

The room is dark. Objects don't fit, don't make sense. At first everything seems chaotic. But then the investigator finds two pieces that fit together, then another pair. Finally, after great effort, everything is in place. What had seemed to be disorderly parts now make up useful furniture; light illuminates the room brilliantly. As the investigator makes one more tour around the room, congratulating himself on his skill in bringing order out of chaos, he notices an imperfection. On the floor he sees a dark line — no, it is actually a rectangle that on further inspection turns out to be a trapdoor. Descending the stairs, the investigator finds that the room is dark. Objects don't fit, don't make sense.

Well, that's one way. Physics can progress by gradually, painstakingly fitting things together to form a coherent picture and then finding a defect that leads to a whole new and possibly quite different view of nature. That sort of thing happened when quantum theory was first introduced in 1900 by Max Planck to repair a defect in the theory of radiation from a black body; quantum physics gave the first sensible picture of atomic structure and led to an entirely new and different way of doing physics.

There are other ways also. Serendipity, the lucky accident in which an investigator stumbles upon a new effect he or she had not been looking for, usually happens to those thorough investigators who deserve it. For example, X-rays were discovered in 1895 by Wilhelm C. Röntgen, the first recipient of a Nobel prize in physics, when he noticed some materials glowing in the dark and also a mysterious darkening of photographic plates. It did not take long for this discovery to be put to practical use; X-ray pictures were being taken of bones within weeks.

Other major discoveries have come from someone's asking, "What if?" Albert Einstein, working alone in relative obscurity in 1905 kept asking himself, as he had ever since he first studied electromagnetism as a teen-ager, "What would an electromagnetic wave look like if you could ride along with the electric and magnetic fields?" The answer came in the special theory of relativity.

---

opposite the motion of the loop. Now if the loop is to be moved at constant speed v, the retarding force $F_m = Bil$ must be balanced by an external agent applying force $\mathbf{F_a}$ to the right with the magnitude $F_a = F_m = Bil$. The external agent does work dW as the loop advances a distance dx, where

$$dW = F_a \, dx = Bil \, dx$$

We can write current as $i = dq/dt$, so that

$$dW = B \frac{dq}{dt} l \, dx = Bl \frac{dx}{dt} \, dq$$

where $v = dx/dt$. The work per unit charge done by the applied force must equal the emf in the circuit, so that we get

$$\mathcal{E} = \frac{dW}{dq} = Bvl \tag{31-4}$$

which is the same result as that found earlier by applying Faraday's law for electromagnetic induction.

The agent does work continuously on the conducting loop; its kinetic energy does not change, since it moves at constant speed. Where does the energy go? The power P_a delivered to the loop by the agent is

$$P_a = F_a v$$

$$P_a = Bilv \tag{31-5}$$

Let R represent the electric resistance of the entire conducting loop. Then we can write

$$i = \frac{\mathcal{E}}{R} = \frac{Bvl}{R}$$

This result, when substituted in (31-5), gives finally

$$P_a = i^2 R$$

We see that the rate at which the agent does work is just equal to the rate at which thermal energy is dissipated in the loop.

Note a curious circumstance; we got the relation $\mathcal{E} = Blv$ simply by considering the magnetic force on conduction electrons. We did not need to use the electromagnetic induction principle at all. Does this mean the induction effects are not really new and fundamental? Not at all. To see that electromagnetic induction is a distinctive phenomenon, we need merely to take another look at the conductor loop moving through the magnetic field from the point of view of another observer, one riding with the conductor.

In this reference frame, the conduction electrons are at rest. They are subject to no magnetic force driving them through the conductor. Yet the observer traveling with loop does find an induced current and therefore an induced emf produced by changing magnetic flux.*

Example 31-4. An airplane flies horizontally at 270 m/s over the location in northern Canada where the earth's magnetic field is vertically downward and has a magnitude of 0.58 G. (This location is above the earth's magnetic pole at latitude 76.1°N and longitude 100°W.) (a) What, if any, is the electric potential difference between the wing tips, which are separated by 40 m? (b) How might the potential difference be observed?

(a) The situation is like that in Figure 31-6, with a conductor moving transversely through a magnetic field. The charge separation on the airplane's wings produces a potential difference V between the tips given, from (31-4), by

$$V = Blv = (0.58 \times 10^{-4} \text{ T})(40 \text{ m})(270 \text{ m/s}) = 0.63 \text{ V}$$

The left wing tip is positive, and the right wing tip is negative.

(b) Suppose that one tries to measure the electric potential difference with an ordinary voltmeter that rides with the airplane and has two conducting lead wires connected to the wing tips. No current goes through the voltmeter, and it reads zero. We can see this from two angles:

- The voltmeter with its lead wires also constitutes a conductor extending between the wing tips, and therefore the same electric potential difference exists across it as across the airplane wings.
- The voltmeter with leads and the airplane wings constitute a conducting loop. The magnetic flux through this loop does not change, and there can be no induced current.

Suppose that somehow the voltmeter remained at rest on the ground with its leads still connected to the airplane wing tips. Then the area of the conducting loop would change with time, as would the magnetic flux through it, and an emf would be induced in the loop. The voltmeter would register this emf.

* The traveling observer finds, in effect, the magnetic field turned off and an electric field turned on; or more fundamentally, in the new reference frame a magnetic field has been transformed into an electric field.

This example illustrates a special consideration that must always be kept in mind in dealing with moving conductors and induced emfs; an emf or an electric potential difference is specified meaningfully only if, at the same time, we specify the reference frame relative to which it is measured.

Example 31-5. As shown in Figure 31-7, a conducting rod of mass m slides to the right at initial speed v_0 over parallel conducting tracks separated by distance l and connected at their left ends across a resistance R. A uniform magnetic field **B** is applied perpendicular to the plane of the tracks. (a) Find the instantaneous velocity of the rod as a function of time. (b) What is the total thermal energy dissipated in the resistor?

(a) The induced emf $\mathcal{E} = Blv$ from (31-4) produces an induced current $i = \mathcal{E}/R = Blv/R$. The retarding magnetic force on the rod is then

$$F = Bil = B\left(\frac{Blv}{R}\right)l = \frac{B^2l^2v}{R}$$

From Newton's second law, we have

$$\Sigma\mathbf{F} = m\mathbf{a}$$

$$-\left(\frac{B^2l^2}{R}\right)v = m\frac{dv}{dt}$$

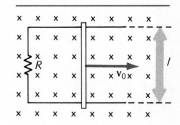

Figure 31-7.

The minus sign indicates that the magnetic force retards the motion. The equation above is simplified when we put

$$k \equiv \frac{B^2l^2}{mR}$$

so that it can be written as

$$-\int_0^t k\,dt = \int_{v_0}^v \frac{dv}{v}$$

Evaluating the definite integral yields

$$v = v_0 e^{-kt}$$

The speed decreases exponentially with time.

(b) Energy is dissipated in resistor R at the rate

$$i^2R = \left(\frac{\mathcal{E}}{R}\right)^2 R = \frac{(Blv)^2}{R} = \frac{B^2l^2v_0^2}{R}e^{-2kt}$$

and the total energy dissipated is

$$\int_0^\infty i^2R\,dt = \frac{B^2l^2v_0^2}{R}\int_0^\infty e^{-2kt}dt = \frac{B^2l^2v_0^2}{2kR}$$

Substituting the value for k given above yields finally

$$\int_0^\infty i^2R\,dt = \tfrac{1}{2}mv_0^2$$

The total energy dissipated is just equal to the rod's initial kinetic energy.

31-3 Lenz's Law

For any loop of conducting material, there are always two possible directions, or senses, for the electric current. But when a current is induced by a changing magnetic flux, the current goes in one direction only. Which is it? The basis for finding the direction of the induced current, or the induced emf, is Lenz's law, named for H. F. E. Lenz (1804–1865).

Consider again the situation of Figure 31-5, where a conducting loop is pulled into a magnetic field. As we have seen, an external agent had to apply a force to balance the retarding magnetic force, and the work done by the agent appeared as thermal energy dissipated in the loop's resistance. The current in the loop of Figure 31-5 was found to be counterclockwise.

Suppose, for the sake of argument, that it went the other way, clockwise current in the loop of Figure 31-5. Then the magnetic force would be in the direction of the loop's motion, and an agent would not be required to drag the loop through the magnetic field. Once given a little push to get it started, the conductor would be accelerated by the magnetic force. Consequently, the current and the magnetic force would grow; the conductor would go still faster; and all the while thermal energy would be dissipated at an ever increasing rate. Clearly, this is impossible. It violates the principle of energy conservation. In this, as in all other examples of induced currents, the *direction of the current is always such as to preclude a violation of energy conservation.* That is one way of stating Lenz's law.

Here is another. *The direction of the induced current is always such as to oppose the change (in magnetic flux) that produces it.* For the situation in Figure 31-5, with a conducting loop pulled into a magnetic field, the opposition to magnetic flux change is manifest as follows. The magnetic field arising from the induced current (*out* of the paper for counterclockwise induced current) opposes the external magnetic field (*into* the paper) and thereby "tries" to maintain the magnetic flux through the loop unchanged. Opposition is also manifest in the retarding force.

In every example of induced emf, Lenz's law implies that:

• When a force acts on a conductor, it is a *retarding* force, one that opposes the relative motion.
• The magnetic field produced by an induced current is always in such a direction as to *oppose* the change in magnetic flux.

Another situation illustrating Lenz's law is shown in Figure 31-8(a). Here one current-carrying loop is being moved toward a second, fixed loop, in which a current is then induced. The magnetic flux through the fixed loop on the right increases with time to the *right.* Consequently, the situation in the fixed loop must be as follows. The magnetic field produced by the induced current must be in such a direction that it annuls (or tends to annul) the magnetic field producing the increasing flux to the right through the fixed loop.

Figure 31-8. *An example of Lenz's law. (a) The left-hand current-carrying conductor moves to the right, inducing a current in the fixed right-hand coil in the opposite sense. (b) Magnets equivalent to the coils of part (a).*

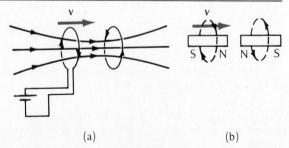

(a)   (b)

That is, the induced current must itself produce a magnetic field to the *left*. Clearly then, from the right-hand rule relating the current to the magnetic field, the current in the fixed loop must be in the sense shown in Figure 31-8(a).

Here is another way of looking at it. Imagine a current-carrying loop to be equivalent to a magnet (see Figure 30-19). Then, we see from Figure 31-8(b) that the equivalent magnets associated with the two loops repel each other. Actually, it is the magnetic force between the two current-carrying conducting loops that is responsible for the repulsion between them. We have already found that the two currents are in opposite senses. This is in accord with the general result that two parallel current-carrying conductors repel one another when the currents are in opposite directions (Section 30-2).

For practice in applying Lenz's law, check that the directions of induced current shown in Figure 31-1 are correct.

The minus sign that appears in the mathematical statement of Faraday's law, $\mathcal{E} = -d\phi_B/dt$, is a symbolic representation of Lenz's law. We discuss it in Section 31-4.

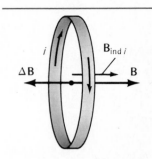

Figure 31-9.   Magnetic field **B** *decreases and changes by* Δ**B**. *The magnetic field* **B**$_{ind\ i}$ *produced by induced current i is opposite to* Δ**B**.

Example 31-6.   Magnetic field **B** is applied along the normal of the conducting loop, shown in Figure 31-9; **B** is decreasing with time. What is the sense of the current induced in the loop?

Because **B** is decreasing in magnitude, the change in this field Δ**B** is opposite in direction to **B**. From Lenz's law, the field **B**$_{ind\ i}$ produced by the induced current must be opposite to Δ**B**. (Said a little differently — **B**$_{ind\ i}$ points in the direction of **B** to make up for its decreased magnitude.) The right-hand rule applied to field **B**$_{ind\ i}$ and the induced current *i* that produces it then show the current to be in the direction displayed in Figure 31-9.

31-4   Faraday's Law and the Induced Electric Field

According to the electromagnetic induction principle, a current is induced in a loop of conductor whenever the loop is exposed to a changing magnetic flux. But a current driven by an emf means, basically, that charged particles are driven by an electric field. Does a changing magnetic flux create an electric field in space, even when no electric charges are present? The answer: yes. Indeed, the fundamental electromagnetic induction effect is this: *A changing magnetic flux creates an electric field.*

First let us relate the induced emf $\mathcal{E}$ to the associated electric field **E**. The work dW done on charge q by electric field **E** as the charge is carried around a closed loop is

$$W = \oint \mathbf{F_e} \cdot d\mathbf{l} = \oint q\mathbf{E} \cdot d\mathbf{l}$$

But emf $\mathcal{E}$ is defined as work done per unit charge in taking it around the loop, so that

$$\mathcal{E} = \frac{W}{q} = \oint \mathbf{E} \cdot d\mathbf{l} \qquad (31\text{-}5)$$

Using (31-5) in (31-2), we get the fundamental form of Faraday's law:

$$\oint \mathbf{E} \cdot d\mathbf{l} = -\frac{d\phi_B}{dt} \qquad (31\text{-}6)$$

The loop chosen for computing the line integral $\oint \mathbf{E} \cdot d\mathbf{l}$ of the electric field is arbitrary, but the chosen loop must also be that through which magnetic flux change $d\phi_B/dt$ is computed.

We must distinguish carefully two general types of electric field: $\mathbf{E}$ originating directly from electric changes (Coulomb's law), and $\mathbf{E}$ created by a changing magnetic flux (Faraday's law):

- $\mathbf{E}$ *from static charges* is a *conservative* electric field. The line integral $\oint \mathbf{E} \cdot d\mathbf{l}$ is always zero. That is to say, if one transports an electric charge around a closed loop in a conservative electric field, no net work is done on the charge. Consequently, we can associate an electric potential with a conservative $\mathbf{E}$. The electric field lines originate from positive charges, terminate on negative charges, and are continuous in between.

- $\mathbf{E}$ *from changing ϕ_B* is a *nonconservative* electric field. Electric field is still defined as force per unit charge, but the line integral $\oint \mathbf{E} \cdot d\mathbf{l}$ around a closed loop is not zero. Instead, it is equal to the rate of magnetic flux change through the loop. A scalar electric potential cannot be associated with a nonconservative electric field; the work done per unit charge around a closed loop is not zero but equal to the emf around that loop. Moreover, since the electric-field lines do not originate from and terminate upon electric charges but are continuous nevertheless (Gauss's law for electricity), the *electric-field lines* from a changing magnetic flux form *closed loops*.

Example 31-7.   The magnetic field produced by a long solenoid is uniform and confined to the circular region inside the windings. A cross section is shown in Figure 31-10(a), where $\mathbf{B}$ is directed into the paper. Suppose that the magnitude of $\mathbf{B}$ increases with time. Find the electric field induced by the changing magnetic field.*

By symmetry we know that the closed $\mathbf{E}$ loops consist of circles concentrically surrounding the magnetic field and lying in a plane perpendicular to the solenoid symmetry axis. To find the sense of $\mathbf{E}$ loops, we can imagine a circular conducting wire coinciding with an $\mathbf{E}$ loop and determine the direction of the induced current (hence, the direction of $\mathbf{E}$) from Lenz's law. Since ϕ_B is increasing into the paper, the field produced by the induced $\mathbf{E}$ must be out of the paper. From the right-hand rule connecting current and magnetic-field directions, this means that the $\mathbf{E}$ loops must be counterclockwise, as shown.

To find how the magnitude of $\mathbf{E}$ depends on the distance r from the center of the magnetic field, we consider separately two regions: first outside the boundary of $\mathbf{B}$ ($r > R$) and then inside ($r < R$).

We evaluate the line integral in Faraday's law around a circular loop of radius r:

$$\oint \mathbf{E} \cdot d\mathbf{l} = -\frac{d\phi_B}{dt}$$

$$E(2\pi r) = -\left(\frac{d\phi_B}{dt}\right)_R$$

Here $(d\phi_B/dt)_R$ represents the total time rate of flux change within R, the region to which the field is confined. We have, then,

$$\text{For } r > R, \quad E = -\left(\frac{1}{2\pi r}\right)\left(\frac{d\phi_B}{dt}\right)_R$$

The significance of the minus sign is given (finally!) at the end of this section.

---

* This example is very similar in mathematical form to Example 30-5 for finding the magnetic field from a current-carrying cylindrical conductor by applying Ampère's law.

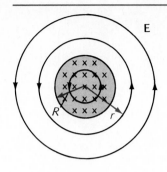

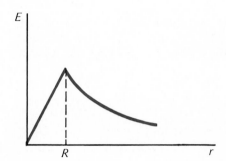

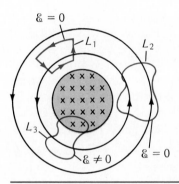

Figure 31-10. (a) A uniform magnetic field is into the paper over a region of radius R, and the field increases with time. The induced electric-field lines consist of counterclockwise circular loops. (b) Magnitude of **E** as a function of r. (c) Some loops for computing the induced electric field.

For any point outside the region of changing magnetic field, the electric field falls off inversely with distance r. Note a curious circumstance. We have a nonzero **E** in a region where **B** remains zero. What matters is not whether there is a changing magnetic field *at* the location of the loop, but whether the total magnetic *flux* changes *within* the loop. Put an electric charge in such an electric field and it is accelerated.

Now we go inside, with $r < R$. Only a fraction of the total flux penetrates a circle of radius r. Indeed, the fraction of flux within r equals the area ratio, $\pi r^2/\pi R^2$, since **B** is uniform within R. The rate of flux change within r is then $(r/R)^2(d\phi_B/dt)_R$, and Faraday's law gives

$$\oint \mathbf{E} \cdot d\mathbf{l} = -\frac{d\phi_B}{dt}$$

$$E(2\pi r) = -\left(\frac{r}{R}\right)^2\left(\frac{d\phi_B}{dt}\right)_R$$

$$\text{For } r < R, \quad E = -\frac{r}{2\pi R^2}\left(\frac{d\phi_B}{dt}\right)_R$$

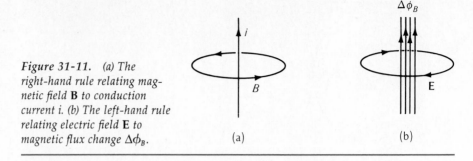

Figure 31-11. *(a) The right-hand rule relating magnetic field* **B** *to conduction current i. (b) The left-hand rule relating electric field* **E** *to magnetic flux change* $\Delta\phi_B$.

Within R, the electric field is directly proportional to r. The magnitude of the circumferential induced electric field is shown in Figure 31-10(b) as a function of r.

Now consider still other closed loops, different from circles, as shown in Figure 31-10(c). Loop L_1 has two circular arcs and two radial segments. The electric field **E** falls off inversely with r, and the line integral around L_1 is zero, because the contribution to $\oint \mathbf{E} \cdot d\mathbf{l}$ is positive for one circular arc and negative but of the same magnitude for the other circular arc. Similarly, loop L_2, through which there is no change in magnetic flux, has no emf around it. Loop L_3, with ϕ_B changing through at least some of it, has a nonzero emf.

The reason for the minus sign in Faraday's law is shown in Figure 31-11. Recall first the right-hand rule for relating the directions of an electric current to the magnetic-field loops that surround it; see Figure 31-11(a). The relation between **B** and i is given, in mathematical terms, by Ampère's law, $\oint \mathbf{B} \cdot d\mathbf{l} = \mu_0 i$, (30-9). No minus sign appears in this equation. The relative "directions" of the induced electric field **E** and magnetic-flux change $\Delta\phi_B$ are shown in Figure 31-11(b) [also shown in Figure 31-10(a)]. The right-hand rule does not apply here; if the thumb gives the direction of $\Delta\phi_B$, the curled right-hand fingers are opposite to the direction of the induced electric-field lines. This is indicated formally by the minus sign in Faraday's law.

31-5   Eddy Currents

Consider Figure 31-12(a), where a permanent magnet is moved toward a circular conducting loop. A current is induced in the loop, and its sense is, from Lenz's law, counterclockwise. Now consider a similar situation in which the north pole of a permanent magnet is moved toward a conducting sheet. As Figure 31-12(b) shows, there are now many concentric induced current loops, in each of which the current is counterclockwise.

These currents, induced by a changing magnetic flux in an extended conductor, are called *eddy currents* because they resemble the whirling eddies in a liquid.

Different forms of eddy currents are produced when the magnet moves parallel to the conducting sheet, as shown in Figure 31-13. We now have two sets of loops, one clockwise and the other counterclockwise. Their associated equivalent magnets oppose the motion of the magnet creating them.

Eddy currents are created not merely by moving a magnet near a conductor, but whenever the magnetic flux changes over a conducting sheet or bulk conductor. For every eddy current:

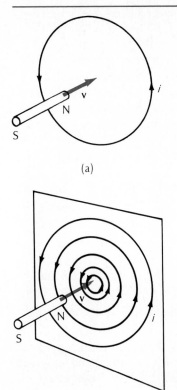

(a)

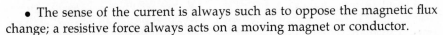

(b)

Figure 31-12. (a) Magnet moving toward a conducting loop induces a current. (b) Magnet moving toward a conducting sheet induces eddy currents.

- The sense of the current is always such as to oppose the magnetic flux change; a resistive force always acts on a moving magnet or conductor.
- Energy is dissipated.

The resistive force arising from an eddy current may be used, for example, in the magnetic dumping of an analytical balance. The energy losses from eddy currents may be highly undesirable in electrical devices, for example, in the conducting magnetic materials used as cores of transformers. The eddy-current losses can be reduced by cutting slots in conducting materials, or equivalently, by using laminations with high-resistance outer surfaces.

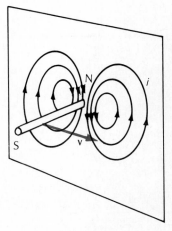

Figure 31-13. Magnet moving transverse to a conducting sheet induces two sets of oppositely directed eddy currents.

31-6   Diamagnetism

We have already seen (Section 30-7) that when a paramagnetic or a ferromagnetic material is immersed in an external magnetic field, such materials enhance the magnetic field because electron spins are reoriented by the applied field. A material is *diamagnetic* however, if its presence reduces the magnitude of an applied magnetic field.

Lenz's law tells us that any material must show some degree of diamagnetism. There are, after all, mobile electric charges in any type of material. When the external magnetic flux over a diamagnetic material goes from zero to some final value, the motion of the charged particles must be such as to *oppose* this flux change and *reduce* the applied magnetic field. As a consequence, the net magnetic field is less than the value it would have reached in the absence of material with mobile charged particles.

On a simple classical atomic model, the electrons in a material can be thought of as orbiting their parent nuclei; each such orbiting electron constitutes a current loop, and it produces a magnetic moment (Section 29-8). In the absence of an external magnetic field, the electron orbits in any material are so paired off that the magnetic moments cancel. For every electron orbiting clockwise in one plane, there is another orbiting counterclockwise in the same plane. The magnetic moments from orbital motion are, in the absence of an external field, oppositely directed and of the same magnitude. They produce no net magnetic effect.

Now suppose that an external field **B** is turned on. For simplicity, we suppose that **B** is perpendicular to the planes of electron orbits. Changing the magnetic field from zero to **B** has two effects:

- The changing magnetic flux through the current loop (the circular electron orbit) induces an emf; the induced electric field is tangent to the orbit, so that the electron's linear speed v and angular speed ω are changed.
- A magnetic force acts on the orbiting electron, in addition to the electrostatic force from the force center; this magnetic force is radially inward or outward.

The effects are shown in Figure 31-14. We see that the electron speed for the left orbit (with its magnetic moment μ aligned with **B**) is reduced, as is also its angular speed from ω_0 to $\omega_0 - \Delta\omega$. The effect for the orbit on the right (with μ opposite B) is just the reverse; here the electron speed is increased, and so is the angular velocity to $\omega_0 + \Delta\omega_0$. A curious result that we simply state without detailed proof: although the linear speeds, angular speeds, and net radial forces change as **B** changes, the orbital radii are essentially unchanged. As Figure 31-14(b) shows, with **B** turned on, the orbital magnetic moments no longer cancel. Both magnetic moments change, and the net induced magnetization opposes the external field. This is diamagnetism.

All materials exhibit diamagnetism, although the effect may be masked by stronger para- and ferromagnetic effects.

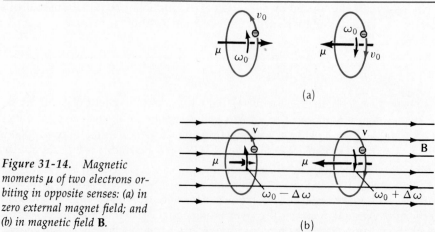

*Figure 31-14. Magnetic moments μ of two electrons orbiting in opposite senses: (a) in zero external magnet field; and (b) in magnetic field **B**.*

Summary

Definitions

Flux linkage $\equiv N\phi_B$ for a coil of N identical turns with magnetic flux ϕ_B through each.

Eddy current: loops of current induced in an extended conducting material by a changing magnetic flux.

Diamagnetism: the phenomenon whereby introducing a material into an external magnetic field reduces the magnitude of the net field.

Fundamental Principles

Law of electromagnetic induction (also known as Faraday's law):

$$\mathcal{E} = -\frac{d\phi_B}{dt} \qquad (31\text{-}2)$$

where $\mathcal{E}$ is the emf induced around a loop by net magnetic flux changing over that loop at the rate $d\phi_B/dt$.

In more fundamental form, the electromagnetic induction law is written as

$$\oint \mathbf{E} \cdot d\mathbf{l} = -\frac{d\phi_B}{dt} \qquad (31\text{-}6)$$

Here $\mathbf{E}$ is the nonconservative induced electric field. The line integral $\oint \mathbf{E} \cdot d\mathbf{l}$ is taken around the same loop as that through which the rate of change of magnetic flux $d\phi_B/dt$ is computed.

Lenz's law: the direction of the induced current (or induced emf, or induced electric field) is always such as to oppose the change in magnetic flux that produces it. The minus signs in the relations above symbolize Lenz's law; in effect, $\mathbf{E}$ and $\Delta\phi_B$ are related by a *left*-hand rule.

Problems and Questions

Section 31-1   Induced Currents and EMF's

· **31-1 Q** A coil of wire is placed in the field of an electromagnet so that magnetic flux passes through the coil. In what ways could you induce an emf in the coil?
(A) Rotate the coil.
(B) Pull the coil out of the electromagnet.
(C) Change the shape of the coil.
(D) Change the current in the electromagnet.
(E) None of the above will work.

· **31-2 P** A small coil of 4 cm² area and 10 turns is jerked out of a magnetic field of 0.040 T in 20 ms. What is the induced emf?

· **31-3 P** A coil has 80 turns and an area of 2800 cm². It rotates in a magnetic field of 0.75 T. At what frequency should it be rotated to produce a peak emf of 150 V?

· **31-4 Q** The number of magnetic field lines passing through a surface is a measure of the
(A) magnetic moment of the current flowing around the boundary of the surface.
(B) magnetic flux through the surface.
(C) induced emf along a path bounding the surface.
(D) magnetic field at the surface.

· **31-5 P** An anemometer used to measure wind velocity is constructed of four light cups mounted on a shaft to which is attached a coil. See Figure 31-15. When the shaft rotates, the coil also rotates in the field of a small permanent magnet that provides a magnetic field **B** perpendicular to the axis of rotation. Experimentally it is found that the cups rotate with a velocity that is 80 percent of the wind velocity. With this device, what emf would you expect to generate in

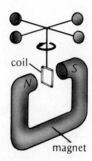

Figure 31-15.   Problem 31-5.

a 60-km/h wind for the following design parameters: area of coil = 4 cm²; 500 turns; cup arm radius = 15 cm; magnetic field = 300 G.

· **31-6 Q** A beam of electrons is shot through a gap in a toroidal coil, as shown in Figure 31-16. When there is no

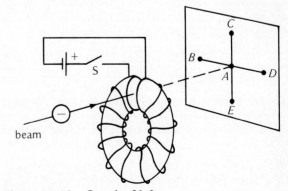

Figure 31-16.   Question 31-6.

current in the coil, the beam strikes the screen at point *A*. When the current is turned on, the beam strikes the screen at one of the indicated points. Which is it? (*A* is a possible answer also.)

: 31-7 Q A bar magnet is suspended as a pendulum, as shown in Figure 31-17. A coil of *N* turns lies on the table below the point of suspension. The pendulum is pulled to one side and released. It swings back and forth with frequency *f*. Which of the following is the most accurate statement?
(A) An emf oscillating at frequency *f* will be induced in the coil.
(B) An emf oscillating at frequency 2*f* will be induced in the coil.
(C) Whether or not an emf will be induced in the coil depends on whether or not the switch is closed.
(D) The magnitude of the induced emf will depend on the value of the resistance *R*.
(E) The induced emf will be independent of the number of turns in the coil.

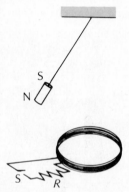

Figure 31-17.   Question 31-7.

: 31-8 P A coil of wire is hung as a pendulum with the plane of the coil vertical in an east-west plane. The pendulum is pulled to the south and released from rest at $t = 0$. It subsequently swings back and forth in a north-south direction. Carefully sketch the current induced in the coil as a function of time after it is released. Let *T* be the period of the pendulum. Take the current to be positive if it flows counterclockwise when viewed from the north.

: 31-9 P What is the amplitude of the emf produced by a generator coil of area 0.1 m² and 500 turns that rotates at 60 Hz in a magnetic field of 0.1 T?

: 31-10 P A circular coil of 6-cm diameter has 200 turns. The coil resistance is 100 Ω. At what rate must a magnetic field normal to the plane of the coil change to produce Joule heating in the coil at the rate of 2.0 mW?

: 31-11 P A large coil and a small coil lie in the same plane, with the small coil at the center of the large one. The large coil has 4000 turns and a radius of 40 cm, and the small coil

has 2000 turns and a radius of 2 cm. What is the induced emf in the small coil when the current in the large coil is changed at the rate of 8000 A/s? Assume that the field due to the large coil is uniform over the area of the small coil.

: 31-12 P An air-core solenoid has 2000 turns, a length of 60 cm, and a diameter of 1.8 cm. What emf is induced in it if the current in the windings increases from 2 A to 8 A in 3 s at a steady rate?

: 31-13 P A flip coil consists of a circular coil of 2.0-cm diameter and 100 turns. It is situated in a magnetic field of 0.050 T with the field at an angle of 60° with the plane of the coil. What emf will be induced in the coil if it is pulled out of the field in 10 ms?

: 31-14 P The perpendicular magnetic field through a single loop of area 2.0×10^{-4} m² changes with time as shown in Figure 31-18. The loop has resistance 0.0030 Ω. Make graphs as a function of time of (*a*) the emf induced in the loop; (*b*) the current in the loop; (*c*) the rate at which energy is dissipated in the loop.

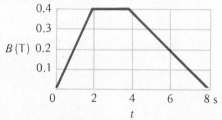

Figure 31-18.   Problem 31-14.

: 31-15 P A coil of 2000 turns and resistance 2.0 Ω is wrapped around a solenoid with 10,000 turns per meter and of radius 0.20 m. What current is produced in the coil when the current in the solenoid is changed at the rate of 2.0 A/s?

: 31-16 P A rectangular loop of wire of resistance *R* is placed next to a long straight wire carrying current, as shown in Figure 31-19. The current in the long wire changes at a rate *di/dt*. What is the current in the loop?

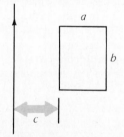

Figure 31-19.   Problem 31-16.

: 31-17 P Current *i* exists along the *z* axis. A circular loop of radius *R* has its center at a point *x* along the *x* axis. The plane of the loop lies in the *xz* plane. What emf is induced in the loop when the current changes at a rate *di/dt*?

: 31-18 P Current i exists in a long straight wire placed along the z axis. A circular loop of radius a has its plane parallel to the xz plane. The loop moves away from the wire along the x axis with constant velocity v. At $t = 0$, the center of the loop is at position x_0. Determine the emf induced in the loop as a function of time.

: 31-19 P A small search coil of 250 turns and radius 1 cm has a resistance of 40 Ω. In order to measure the earth's magnetic field, the coil is suddenly rotated through 180°. The coil is connected to a ballistic galvanometer, and a total charge of 3.2×10^{-7} C is found to flow through the coil because of the induced emf. Determine the magnitude of the earth's magnetic field. Assume that the coil was initially oriented for maximum flux through it.

: 31-20 P A coil of N turns and area A is suspended as a pendulum of length L. It is displaced through an angle θ_0 at $t = 0$, as shown in Figure 31-20. A uniform horizontal magnetic field **B** acts in the plane in which the pendulum swings. Determine the induced emf in the coil as a function of time.

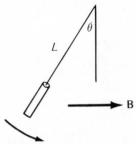

Figure 31-20. *Problem 31-20.*

: 31-21 P Shown in Figure 31-21 are three identical light bulbs lighted by a battery. The shaded circle represents the end view of a solenoid in which the magnetic flux can be increased at a constant rate to produce an induced emf. What rate of change of magnetic flux, expressed in terms of $\mathcal{E}_0$, is required to shut off lamp No. 1? What is the sense of the change in flux required?

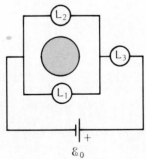

Figure 31-21. *Problem 31-21.*

: 31-22 P The planar circuits shown in Figure 31-22 are divided into two areas, A_1 and A_2, as indicated. A magnetic

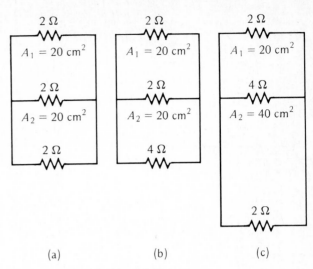

Figure 31-22. *Problem 31-22.*

field directed out of the paper is increasing at a constant rate of 0.10 T/s. Determine, for each case, the magnitude and direction of the induced current in each resistor.

: 31-23 P A coil of 400 turns and area 4.0 cm^2 is connected to a so-called "ballistic" galvanometer of resistance 24 Ω. The maximum angular displacement of the galvanometer coil is proportional to the charge that flows through it. The flux through the coil changes suddenly, and a charge of 0.8 C flows through the galvanometer. What is the change in flux through the coil?

: 31-24 P Three long straight wires pass through the corners of an equilateral triangle of side a. The plane of the triangle is perpendicular to the wires. A sinusoidal current of amplitude I_0 and angular frequency ω passes through each wire, but there is a phase difference of 120° between the current in each wire; that is,

$$I_1 = I_0 \sin \omega t \qquad I_2 = I_0 \sin \left(\omega t + \frac{2\pi}{3} \right)$$

$$I_3 = I_0 \sin \left(\omega t + \frac{4\pi}{3} \right)$$

(*a*) Show that the magnetic field at the center of the triangle can be represented by a vector of constant magnitude rotating with angular frequency ω. (*b*) Determine the emf induced in a small coil of N turns and area A placed at the center of the triangle, with its plane parallel to the wires. (*c*) What emf will be induced if the coil is rotated clockwise with frequency ω_1? (*d*) What emf will be induced if the coil is rotated counterclockwise with frequency ω_1?

Section 31-2   EMF in a Moving Conductor

· 31-25 Q A uniform magnetic field directed into the plane of the paper acts in the shaded region of Figure 31-23(a). A circular loop of wire is moved with constant

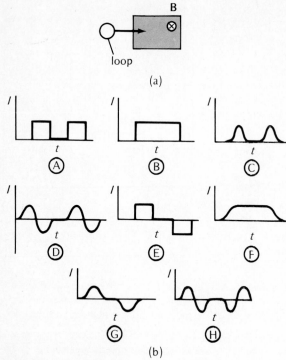

(a)

(b)

Figure 31-23. *Question 31-25.*

velocity through the magnetic-field region and out the other side, starting at $t = 0$ at the position where it is drawn. Which of the graphs A–H in Figure 31-22(b) depict the induced current in the loop as a function of time? A counterclockwise current is taken as positive.

· **31-26 P** An electromagnetic flowmeter works like this. A conducting fluid (such as blood) moves with velocity v through a magnetic field **B** oriented perpendicular to the direction of flow. Suppose that the fluid flows through a rectangular tube of cross section ab, with b parallel to the magnetic field. Write an expression for the fluid velocity as a function of the voltage between electrodes attached to opposite sides of the tube.

· **31-27 P** A square wire loop of side 2.0 cm is moved at a constant velocity of 6.0 cm/s along the x axis. In the region from $x = 8.0$ cm to $x = 14.0$ cm, a uniform magnetic field of 0.010 T acts perpendicular to the plane of the loop. Graph the induced current in the loop as a function of time, starting and ending when the loop is outside the magnetic field. The loop lies in the xz plane with one side parallel to the x axis. It moves in the $+x$ direction, and the magnetic field acts in the $+y$ direction.

: **31-28 P** The conducting blades of a helicopter rotor are 5.0 m long and rotate at 640 rpm in normal flight. The earth's magnetic field has a horizontal component of 0.60 G and a vertical component of 0.30 G. What potential difference is induced between the tip of a blade and the shaft on which it rotates?

: **31-29 P** A rectangular loop of width a and height b is released from rest with its plane vertical. See Figure 31-24. As it falls, its lower end enters a uniform magnetic field **B** oriented normal to the plane of the loop. Dimension b is great enough that the loop can reach a constant terminal falling velocity (with the top end of the loop remaining out of the magnetic field). What is this terminal velocity? The mass of the loop is m and its resistance is R.

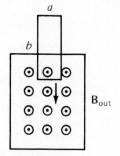

Figure 31-24. *Problem 31-29.*

: **31-30 Q** A bar magnet is placed inside a coil, as shown in Figure 31-25. In the following, "work" refers to any work done because the bar is magnetic. If the bar is suddenly pulled out of the coil,
(A) no work will be done, independent of whether or not the switch is closed.
(B) more work will be done if the switch is open.
(C) more work will be done if the switch is closed.
(D) equal (nonzero) amounts of work will be done whether or not the switch is open.
(E) whether the work done is positive or negative depends on whether the magnet is pulled out of the right end of the coil or the left end.

Figure 31-25. *Question 31-30.*

: **31-31 P** A prototype model of a "mass driver" to be used to launch objects into orbit around the moon is sketched in Figure 31-26. A battery and a resistor are connected to two copper rails and a bar, which have negligible electric resistance. The bar and projectile have a mass of 5.0 gm and slide without friction on the rails. A magnetic field of 0.20 T is directed perpendicularly into the paper. (a) What is the initial acceleration of the bar and projectile when the switch is closed? (b) What is the maximum velocity reached by the projectile, if the length of the rails is not a limitation?

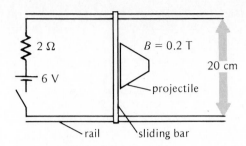

Figure 31-26.   Problem 31-31.

: 31-32 P A uniform magnetic field of 0.050 T acts vertically upward. A conducting rod 60 cm long is rotated clockwise (as viewed from above) at 100 Hz about one end in a horizontal plane. (*a*) Which end of the rod is positive as a result of the induced emf? (*b*) What is the potential difference between the ends of the rod?

: 31-33 P A rectangular frame is rotated at angular frequency ω about the axis XX^1 shown in Figure 31-27. The two sides of the frame have resistances R_1 and R_2. A uniform magnetic field **B** acts normal to the paper. Find the ratio of the power dissipated in the two arms.

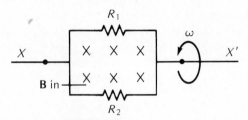

Figure 31-27.   Problem 31-33.

Section 31-3   Lenz's Law
31-34 Q A magnet is placed above a horizontal conducting ring, as sketched in Figure 31-28. If the magnet is moved upward at constant speed,
(A) no current will exist in the ring.
(B) a clockwise current, as viewed from above, will exist in the ring.
(C) a counterclockwise current, as viewed from above, will exist in the ring.

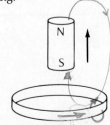

Figure 31.28.   Question 31-34.

· 31-35 Q Verify the directions of the induced currents in Figure 31-1.

· 31-36 Q All the effects associated with electromagnetic induction (induced currents and emf's and magnetic forces) always act in the sense of opposing the change that caused them. This is because
(A) of the law of conservation of energy.
(B) of the law of conservation of momentum.
(C) magnetism and electricity are two different aspects of the same phenomenon.
(D) opposites attract and likes repel.
(E) magnetic fields decrease as the distance to the source increases.

· 31-37 Q Two wire loops are oriented with their axes parallel, as shown in Figure 31-29. If a current is suddenly established in one of the loops in the direction shown, what force acts on the second loop?
(A) No net force acts on the loop.
(B) The loop experiences a force to the right.
(C) The loop experiences a force to the left.

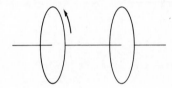

Figure 31.29.   Question 31-37.

Section 31-4   Faraday's Law and the Induced Electric Field
· 31-38 Q A copper ring and a wooden ring of the same dimensions are placed so that there is the same changing magnetic flux through each. How do the induced electric fields in each compare?
(A) The induced electric fields are the same in both.
(B) The induced fields are greatest in the copper.
(C) The induced fields are greatest in the wood.

· 31-39 Q Two coils are wound on a wooden dowel as shown in Figure 31-30. Immediately after the switch is closed, the current in the resistor will be
(A) from right to left.
(B) from left to right.

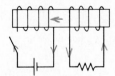

Figure 31.30.   Question 31-39.

(C) zero, since wood is nonmagnetic.

(D) zero, but not because wood is nonmagnetic.

· 31-40 P The magnetic field in a solenoid is uniform and increases at a constant rate of 50 G/s. The radius of the solenoid is 0.20 m. What is the electric field in the plane normal to the magnetic field at a distance from the axis of the solenoid of (a) 0.10 m? (b) 0.20 m? (c) 0.40 m?

: 31-41 Q When a bar magnet is dropped lengthwise down a long, vertical copper pipe in the absence of air friction, it will

(A) fall with constant acceleration less than g.

(B) fall with acceleration g.

(C) alternately speed up and slow down.

(D) eventually reach a constant terminal velocity.

: 31-42 Q Use the answers (A) left, (B) right, or (C) no current for the following questions: (a) Immediately after the switch S in Figure 31-31 is closed, in what direction is the current in resistor R? (b) Quite a long time after the switch is closed, what is the direction of the current in the resistor? (c) After the switch has been closed for a long time, it is then opened. In what direction is the current is resistor R?

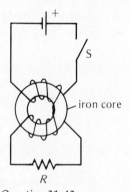

Figure 31.31. *Question 31-42.*

Section 31-5 Eddy Currents

: 31-43 P A plane conducting object falls through a transverse magnetic field. Show that the power dissipated in eddy currents is proportional to the square of the speed of the falling object.

: 31-44 P A metal disk of radius a, thickness d, and resistivity ρ experiences a magnetic field $B_0 \sin \omega t$ parallel to its axis. Determine (a) the total current in the disk; and (b) the average power dissipated in the disk.

: 31-45 P A metal disk of radius a, thickness t, and resistivity ρ is oriented normal to a uniform magnetic field **B**. The disk is rotated about its axis with constant angular velocity

ω. What is (a) the current in the disk? (b) the potential difference between the center of the disk and a point on the perimeter?

: 31-46 P An electromagnetic brake consists of a conducting disk, of thickness d and resistivity ρ, rotating with angular velocity ω about its axis. A magnetic field **B** is applied normal to the disk only over a small area a^2 at a distance x from the axis. The induced current experiences both a force and a magnetic torque that slows the disk. Show that this torque has the approximate magnitude of $(\omega/\rho)B^2x^2a^2d$.

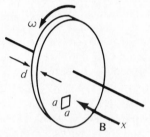

Figure 31-32. *Problem 31-46.*

Supplementary Problem

31-47 P Very large magnetic fields can be obtained with a technique called flux compression. Consider Figure 31-33, with a thin metal tube positioned inside a solenoid where the field is B_0. Between the tube end and the solenoid is a layer of high explosive (HE), and outside the solenoid is thick armor plate to strengthen the solenoid and keep it from blowing apart when the explosive is detonated. When the explosive is set off, the tube collapses. The flux through it is drastically reduced in a short time, thereby inducing huge currents in the tube wall. This current creates a very large **B** field.

(a) Show that if the tube shrinks very rapidly, the magnetic field reaches a value $B \cong B_0(R_0/R)^2$ when the tube radius is reduced from R_0 to R. (b) What surface current is needed in the tube (approximately) to produce a field of the order of 10^9 T? (c) What B can be obtained if $B_0 = 10$ T, $R_0 = 6$ cm, and $R = 3$ mm?

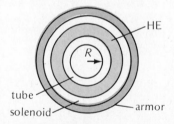

Figure 31-33. *Problem 31-47.*

Inductance and Electric Oscillations

32

Inductance is the property of a circuit element that derives from electromagnetic induction. We shall treat an inductor in combination with one other circuit element, first with a resistor in an *LR* circuit and then with a capacitor in an *LC* oscillator circuit. Close analogies can be identified between electric-circuit elements and their mechanical counterparts.

32-1   Self-Inductance Defined

Consider Figure 32-1(a), where current i goes through a coil with several turns. The current is driven by some external energy source, such as a battery with emf $\mathcal{E}_b$. Suppose that the current is changing—more specifically, that it is increasing. The time rate of change of current, di/dt, is positive. This would happen just after a switch connecting a battery to the coil is closed. As the current i through the conducting wires changes, so too does the magnetic flux through turns of the coil, so that an emf is induced in the coil. The *induced emf* here opposes the emf of the battery and is sometimes called a *back emf*.

An emf is also induced if the current decreases, and $di/dt < 0$, See Figure 32-1(b). We know that the induced emf is always in the direction opposing the *change* that produces it, so that here the induced emf is in the same direction as

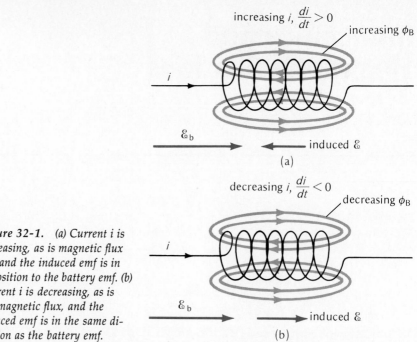

Figure 32-1. (a) Current i is increasing, as is magnetic flux ϕ_B, and the induced emf is in opposition to the battery emf. (b) Current i is decreasing, as is the magnetic flux, and the induced emf is in the same direction as the battery emf.

the applied emf $\mathcal{E}_b$. The coil "tries," so to speak, to maintain the current unchanged. A constant current of any magnitude through the coil produces no induced emf. Self-induction depends on a changing magnetic flux and therefore, for conductors fixed in position, on changing a current.

Self-induction occurs in every electric circuit, not merely in a multiple-turn coil. Every circuit with a continuous current must have at least one loop, so that when the current changes, the magnetic flux through this single loop will change and induce an emf in the circuit. For example, merely closing the switch on a simple circuit with a battery and a resistor will change the magnetic flux through the loop and thereby induce an emf, possibly very feeble. It is easy to see that using N turns enhances the self-induction effect by a factor N^2. First, the magnetic flux through each turn is increased by factor N; additionally, an emf is induced in each of N turns.

Any device showing the self-induction effect, such as a coil of conducting wire, is called an *inductor.* It is represented in a circuit diagram by the symbol -⦚⦚⦚- .

It is useful to compare general characteristics of an inductor with the corresponding properties of a capacitor:

A *capacitor:*

• Stores separated charge of opposite sign.
• Stores potential energy associated with separated electric charges; alternatively, we can say that a capacitor stores energy in the electric field in the region between oppositely charged plates.

As we shall see, an *inductor:*

• Acts to maintain a constant electric current.

• Stores energy associated with electric charges in motion; alternatively, we can say that an inductor stores energy in the magnetic field in the region surrounding the current-carrying conductor.

An inductor is characterized by its *self-inductance L*; unless there is need to avoid confusion with mutual inductance (Section 32-7), L is usually called simply the inductance.

In defining inductance, we first recognize that the magnitude of the magnetic field B produced by any current element at any point in space is directly proportional to the current magnitude i (Ampère's law). Further, the magnetic flux ϕ_B is proportional to B, which is in turn proportional to i. It follows that $\phi_B \propto i$. For an inductor of N identical turns, we write

$$N\phi_B \equiv Li \qquad (32\text{-}1)$$

where ϕ_B is the flux enclosed by one loop of the inductor and L is a proportionality constant that depends on the size and shape of the conductor. The product $N\phi$ is referred to as the *flux linkage*.

From Faraday's law, (31-3), we have the induced emf given by

$$\mathcal{E} = -\frac{d}{dt}(N\phi) \qquad (32\text{-}2)$$

Substituting (32-1) in (32-2), we get

$$\mathcal{E} = -L\frac{di}{dt} \qquad (32\text{-}3)$$

This relation is another equivalent definition of L, where

$$L \equiv -\frac{\mathcal{E}}{di/dt} \qquad (32\text{-}4)$$

Inductance may be defined in two ways:

• In terms of the magnetic flux and its relation to current, $L \equiv N\phi_B/i$, (32-1). This form is most useful in computing the self-inductance of a particular conductor arrangement.

• In terms of the emf produced by a change in current, $L \equiv -\mathcal{E}/(di/dt)$, (32-4). This form is most useful in describing the behavior of an inductor in an electric circuit or measuring inductance experimentally.

From (32-4) we see that inductance has the SI units volts per ampere per second. A special name, the *henry* (abbreviated H) is given to this combination of units:*

$$\text{Henry} \equiv \frac{\text{volt}}{\text{ampere per second}}$$

$$1\ \text{H} \equiv \frac{\text{V} \cdot \text{s}}{\text{A}}$$

---

* The unit for inductance honors the American physicist Joseph Henry (1797–1878), who discovered electromagnetic induction independently of Faraday. Henry was Professor of Physics at Princeton University, and later he was appointed the first director of the Smithsonian Institution in Washington, D.C.

Thus, an inductor producing an emf of 1 V when the current through it is changed at the rate of 1 A/s has an inductance of 1 H. Related units are the millihenry (1 mH $= 10^{-3}$ H) and the microhenry (1 μH $= 10^{-6}$ H). Air-filled laboratory inductors of moderate size have inductances typically of the order of several millihenries. With cores of magnetic material (Section 30-7), the inductance may rise to several henries. As examples below show, any inductor's inductance is the product of μ_0, some characteristic dimension, and a dimensionless constant relating to the geometry of the magnetic field.

Example 32-1.   Every coil has both inductance and resistance. When the current through a certain coil is 0.50 A and increasing at the rate of 100 A/s, the potential difference across the coil terminals is 6.0 V. When the current through the coil is again 0.50 A and in the same direction, but decreasing at the rate of 100 A/s, the potential difference across the coil terminals is only 4.0 V. What are the inductance and the resistance of the coil?

The two situations may be described by the relations

$$V_1 = iR + L\frac{di}{dt}$$

$$V_2 = iR - L\frac{di}{dt}$$

where V_1 is the larger and V_2 the smaller potential difference.

We have then that

$$V_1 + V_2 = 2iR$$

$$R = \frac{V_1 + V_2}{2i} = \frac{(6.0 + 4.0)\ \text{V}}{2(0.50\ \text{A})} = 10\ \Omega$$

$$V_1 - V_2 = 2L\frac{di}{dt}$$

$$L = \frac{V_1 - V_2}{2di/dt} = \frac{(6.0 - 4.0)\ \text{V}}{2(100\ \text{A}\cdot\text{s}^{-1})} = 10\ \text{mH}$$

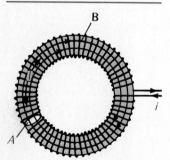

Figure 32-2.   A toroidal solenoid with current i, magnetic field B, and cross-sectional area A.

Example 32-2.   A toroidal inductor consists basically of a long solenoid bent into a circle so that its ends meet and its external shape is that of a toroid (a doughnut). See Figure 32-2. What is the inductance for a toroidal inductor of n turns per unit length with central circumference l and transverse area A? Assume that the radius of any turn is small compared with the radius of the toroid.

The toroid is one of a very few inductor configurations in which the geometry of the magnetic field is simple enough for the magnetic flux, and therefore the inductance, to be readily computed. Here all the magnetic field is confined within the turns of the toroid. Moreover, since the turn radius is small compared with the toroid radius, the magnetic field can be taken to be uniform through any turn.

The magnetic field at any point near the center of a long solenoid, and therefore any interior point for the toroid, is

$$B = \mu_0 ni \qquad\qquad\qquad \text{(30-10), (32-5)}$$

The turns per unit length n can be written as $n = N/l$, where N is the total number of turns and l is the circumference of the toroid.

From the definition of inductance, (32-1), we have

$$L = \frac{N\phi_B}{i} = \frac{NBA}{i} = \frac{nlBA}{i}$$

and using the relation for B above, we get

$$L = \mu_0 n^2 (Al)$$

or

$$L = \mu_0 n^2 V \qquad\qquad \textbf{(32-6)}$$

where $V \equiv Al$ is the total volume within the windings. Note that as we anticipated earlier, the inductance varies with the *square* of the number of turns.

Suppose, for example, that a toroidal inductor has 2000 turns, a turn radius $r = 2.0$ cm, and a toroid mean radius $R = 20$ cm. Then $A = \pi r^2$, $l = 2\pi R$, and $n = N/2\pi R$; and the relation above yields

$$L = \mu_0 n^2 Al = \mu_0 \left(\frac{N}{2\pi R}\right)^2 (\pi r^2)(2\pi R)$$

$$= \frac{\mu_0 N^2 r^2}{2R} = \frac{(4\pi \times 10^{-7}\ \text{Wb/A}\cdot\text{m})(2000)^2(2.0 \times 10^{-2}\ \text{m})^2}{2(20 \times 10^{-2}\ \text{m})}$$

$$= 5.0\ \text{mH}$$

The inductance would be increased substantially by winding the turns on a doughnut of strongly magnetic material, such as iron.

Example 32-3. An air-filled coaxial cable consists of concentric cylindrical conductors of radius r_1 and r_2 with equal currents, each of magnitude i, in opposite directions in the inner and outer conductors. What is the inductance per unit length of the coax cable?

As shown in Example 30-5, a magnetic field exists only in the region between the inner and outer conductors. The magnetic-field loops are circular and lie in planes transverse to the symmetry axis of the cylindrical conductors. Between r_1 and r_2, the magnitude of the magnetic field is

$$B = \frac{\mu_0 i}{2\pi r}$$

(The magnetic field is the same as that from a single, infinitely long conducting wire at the symmetry axis carrying current i.) See Figure 32-3.

In applying the definition for inductance,

$$L = \frac{N\phi_B}{i}$$

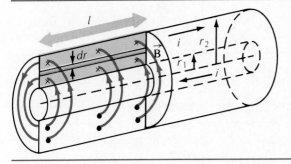

Figure 32-3.   A coaxial conductor with inner radius r_1 and outer radius r_2. The magnetic flux is computed over the shaded area.

we choose the surface for computing magnetic flux ϕ_B to be a rectangle of length l along the symmetry axis, and of width $r_2 - r_1$. All **B** lines in a length l along the axis are perpendicular to this rectangle and pass through it. For an infinitesimal radial displacement dr, the magnetic flux is

$$d\phi_B = Bldr = l\,\frac{\mu_0 i}{2\pi r}\,dr$$

Therefore

$$\phi_B = \int_{r_1}^{r_2} d\phi_B = \frac{\mu_0 il}{2\pi}\int_{r_1}^{r_2}\frac{dr}{r} = \frac{\mu_0 il}{2\pi}\ln\frac{r_2}{r_1}$$

Here $N = 1$, so that the definition above gives, for the inductance per unit length.

$$\frac{L}{l} = \frac{\phi_B}{il} = \frac{\mu_0}{2\pi}\ln\frac{r_2}{r_1}$$

For example, any air-filled coaxial cable with $r_2 = 2r_1$ has an inductance per unit length of

$$\frac{L}{l} = \frac{(4\pi \times 10^{-7}\ \text{Wb/A·m})\ln 2}{2\pi} = 13.9\ \mu\text{H/m}$$

As shown in Example 26-2, the capacitance per unit length of coax cable was of a similar form: $C/l = 2\pi\epsilon_0/\ln(r_2/r_1)$.

32-2   The *LR* Circuit

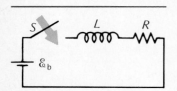

Figure 32-4. *An LR circuit, with inductor L, resistor R, battery emf $\mathcal{E}_b$, and switch S in series.*

When a charged capacitor C is connected across a resistor R, the charge on either plate does not fall to zero instantaneously. Rather, the charge Q decays exponentially with a characteristic time constant RC, the time required for the charge to fall to $1/e$ of its initial value (Section 27-5).

The same sort of behavior is found when a current-carrying inductor is connected with a resistor. The inductor exhibits inertia as a consequence of Lenz's law: When the current through an inductor changes, the inductor opposes the change, and the induced emf is always such as to tend to offset the change in current.

Consider the circuit in Figure 32-4, where an inductor L and a resistor R are in series with a switch and a battery of emf $\mathcal{E}_b$ (we neglect its internal resistance). The current i in the circuit does not reach its steady-state value instantaneously, because as the current grows from an initial zero value, the back emf $-L\,di/dt$ from the inductor opposes the emf of the battery $\mathcal{E}_b$, and therefore, the buildup of current.

Applying the loop theorem (Kirchhoff's second circuit rule) we have

$$\Sigma\mathcal{E} = \Sigma V$$

$$\mathcal{E}_b - L\frac{di}{dt} = iR$$

(Note that the term $-L\,di/dt$ is put on the left side of the equation above; it is an emf, not a potential drop. This is equivalent to placing a term $+L\,di/dt$ on the right side of the equation and then counting it as an equivalent potential drop.)

The equation above may be written as

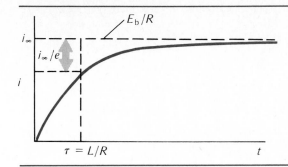

Figure 32-5. *Current i in the LR circuit of Figure 32-4 plotted as a function of the time t after the switch is closed. The time constant $\tau = L/R$ represents the time it takes for the current to come within $1/e$ of its steady-state final value $i_\infty = \mathcal{E}_b/R$.*

$$\mathcal{E}_b = L\frac{di}{dt} + iR \qquad (32\text{-}7)$$

It is not difficult to solve the differential equation for the current i as a function of time t, and we shall merely state that the solution is

$$i = \frac{\mathcal{E}_b}{R}\left[1 - \exp\left(-\frac{t}{L/R}\right)\right] \qquad (32\text{-}8)$$

That (32-8) is indeed the solution is easily verified by substituting it in (32-7).

Figure 32-5 is a plot of (32-8). The current grows steadily with time, and approaches its steady value $i_\infty = \mathcal{E}_b/R$. The quantity L/R is the characteristic time constant τ of the LR circuit where

$$\tau \equiv \frac{L}{R} \qquad (32\text{-}9)$$

We see from (32-8) that the current differs from its final value i_∞ by $1/e$, or 36.8 percent, after a time interval $\tau = L/R$ elapses following the initial closing of the switch. For example, with $L = 1$ mH, $R = 1$ MΩ, we have

$$\tau = \frac{L}{R} = \frac{10^{-3}\text{ H}}{10^{6}\text{ }\Omega} = \frac{10^{-3}\text{ V}\cdot\text{s/A}}{10^{6}\text{ V/A}} = 1 \times 10^{-9}\text{ s} = 1\text{ ns}$$

Example 32-4. Both the inductance and the internal resistance of a very large induction coil are unknown and are to be determined indirectly. An observer finds that when the inductor alone is connected to a source maintained a constant voltage, the current through the inductor rises to half its final value in 0.60 s. When a 4.0-Ω resistor is placed in series with inductor, it takes only 0.30 s for the current to reach half its new final value. What are the inductor's inductance L and resistance R?

We see from (32-8) that the current reaches half its final value at time $t_{1/2}$, where

$$i = \frac{1}{2}i_\infty = i_\infty\left[1 - \exp\left(-\frac{t_{1/2}}{L/R}\right)\right]$$

This equation simplifies to

$$\exp\left(\frac{t_{1/2}}{L/R}\right) = 2$$

or

$$t_{1/2} = \frac{L}{R}\ln 2 = 0.693\frac{L}{R}$$

The time it takes to reach one-half the final current is 69.3 percent of the characteristic time constant, L/R.

With the inductor alone, we have in the equation above

$$0.60 \text{ s} = 0.693\frac{L}{R}$$

and with the additional 4.0-Ω resistor in series,

$$0.30 \text{ s} = 0.693\frac{L}{R + 4.0 \text{ } \Omega}$$

Solving for L and R in the two simultaneous equations above yields

$$L = 3.5 \text{ H}$$

$$R = 4.0 \text{ } \Omega$$

32-3 Energy of an Inductor

To find the energy associated with an inductor through which a current is passing, we consider again the circuit of Figure 32-4, in which a battery is connected to a circuit containing an inductor and a resistor. The loop theorem yielded (32-7),

$$\mathcal{E} = L\frac{di}{dt} + iR$$

With this equation multiplied by the instantaneous current i, we get

$$\mathcal{E}i = Li\frac{di}{dt} + i^2R$$

The terms in this equation are interpreted as follows. The left-hand term $\mathcal{E}i$ represents the rate at which the energy source of emf $\mathcal{E}$ delivers energy to the circuit elements; it is the input power. The term i^2R is the rate at which electric energy is being dissipated into thermal energy in the circuit's resistance; it is, so to speak, the output power. This leaves $Li \, (di/dt)$. This term is positive, since both the current i and its time rate of change di/dt are positive. The term $Li \, (di/dt)$ must therefore represent electric power delivered to the circuit but not yet dissipated. It is the rate at which energy is supplied to and stored in the inductor. Labeling the instantaneous power into the inductor P_L and the energy associated with the inductor U_L, we can then write

$$P_L = Li\frac{di}{dt}$$

and

$$U_L = \int_0^t P_L \, dt = \int_0^t Li\frac{di}{dt} \, dt = \int_0^i Li \, di$$

$$U_L = \tfrac{1}{2}Li^2 \tag{32-10}$$

The energy stored in an inductor with current i is proportional to the inductance and to the square of the current. Note the similarity to the relation for the electric energy U_C stored in a charged capacitor of capacitance C with charge of magnitude Q on each plate:

$$U_C = \frac{1}{2}\frac{Q^2}{C} = \frac{1}{2}CV^2 \qquad (26\text{-}9)$$

32-4   Energy of the Magnetic Field

A charged capacitor establishes an electric field between its plates, and we may speak of the capacitor's energy as residing in the electric field between the plates (Section 26-5). Similarly, the energy of a current-carrying inductor may be considered to reside in its magnetic field. We wish to compute the energy density u_B or magnetic energy per unit volume, of magnetic field B.

The energy U stored in the magnetic field of an inductor carrying current i is

$$U = \tfrac{1}{2}Li^2 \qquad (32\text{-}10)$$

We take the conductor to be a toroid; the magnetic field is then confined entirely to the region within the windings. The inductance of a toroid is given by (32-6), and the current i by (32-5). Then the equation above becomes

$$U = \frac{1}{2}(\mu_0 n^2 V)\left(\frac{B}{\mu_0 n}\right)^2 = \frac{1}{2}\frac{B^2}{\mu_0}V$$

where V is the volume of the inductor. The magnetic-energy density $u_B = U/V$ is then

$$u_B = \frac{B^2}{2\mu_0} \qquad (32\text{-}11)$$

The energy density of the magnetic field is proportional to the *square* of the magnetic field.

Although derived for the special case of a toroidal inductor, (32-11) holds in general. The relation for the magnetic energy density is like that for the energy density u_E of an electric field:

$$u_E = \tfrac{1}{2}\epsilon_0 E^2 \qquad (26\text{-}10)$$

32-5   Electrical Free Oscillations

Here we treat the oscillations of a simple circuit consisting of an initially charged capacitor connected across an inductor. The more general case of an electric oscillator with resistance also driven by a sinusoidally varying emf of arbitrary frequency is treated in Chapter 33, on ac circuits.

Consider the circuit of Figure 32-6(a), where a capacitor C, initially carrying a charge of magnitude Q_m on each plate, is connected across an inductor L by closing a switch. For simplicity we assume that the circuit contains no resistance.

We can see on a qualitative basis that the circuit will oscillate. Immediately after the switch is closed, charges start to leave the capacitor plates, thereby creating a current; this current changes with time, and an emf is also set up in

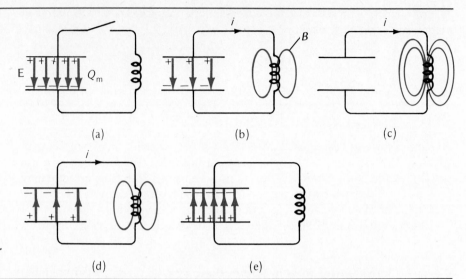

Figure 32-6. *Stages in the oscillation of an LC circuit over one-half cycle.*

the inductor [Figure 32-6(b)]. When the charge on each capacitor plate has reached zero [Figure 32-5(c)], the current continues because of the inertia of the inductor. The inductor's emf now opposes a decrease in current, and charges accumulate on the capacitor plates in the reverse sense (Figure 32-5(d)]; the current now has decreased in magnitude as the first charges to arrive on the capacitor plates repel other charges arriving later. Still later, the current falls to zero and the capacitor is again fully charged, but with opposite polarity and again with charges of magnitude Q_m on each plate [Figure 32-5(e)]. At this point, the electric oscillator has completed exactly *one-half* of a cycle. The process is then repeated but in the opposite sense (for the sign of charge and the direction of current), until the capacitor reaches its initial charge state [Figure 32-5(a)].

Oscillations in the electric charge and electric current continue. No energy is dissipated; the circuit has no resistance. The oscillations can also be characterized by the continuous alternation of energy stored in the electric field of the charged capacitor and energy stored in the magnetic field of the current-carrying inductor. The two circuit elements play different roles:

• The capacitor C stores energy in its electric field when charged, but tends to lose its charge and be restored to its equilibrium state of electric neutrality.

• The inductor L stores magnetic energy when carrying a current, and it displays electrical inertia in that its self-induced emf tends to maintain the charges in motion.

Now we analyze the electric oscillator with mathematics. Applying Kirchhoff's second rule (energy conservation) to the circuit loop of Figure 32-6, we have

$$\Sigma \mathcal{E} = \Sigma V$$

$$-L \frac{di}{dt} = \frac{q}{C}$$

The inductor's emf is $-L \, di/dt$; the electric potential difference across the capacitor is q/C with q as the charge on one capacitor plate at any instant. By definition, $i = dq/dt$; substituting this in the equation above, we have

$$-L\frac{d^2q}{dt^2} = \frac{q}{C}$$

$$\frac{d^2q}{dt^2} = -\frac{1}{LC}q$$

We make the substitution

$$\omega_0^2 \equiv \frac{1}{LC} \tag{32-12}$$

Then the equation above can be written as

$$\frac{d^2q}{dt^2} + \omega_0^2 q = 0 \tag{32-13}$$

This is a particularly important linear second-order differential equation, that for the simple harmonic oscillator [(11-8)]. Its solution is

$$q = Q_m \cos \omega_0 t \tag{32-14}$$

as can easily be verified by substituting (32-14) in (32-13). The charge Q_m is the maximum on either capacitor plate; it is also, in this problem, the initial charge

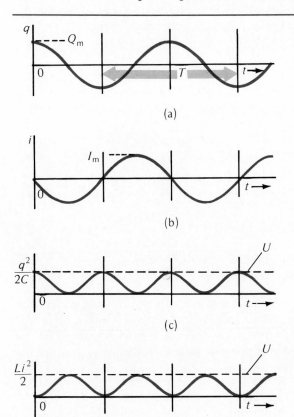

(a)

(b)

(c)

(d)

Figure 32-7. *Time variation for an electric oscillator of (a) charge on a capacitor plate, (b) current, (c) capacitor energy, and (d) inductor energy.*

($q = Q_m$ at $t = 0$). The zero subscript on ω_0 is a reminder that the circuit contains no resistance.

The charge varies sinusoidally with time, as shown in Figure 32-7(a). The angular frequency ω_0 of the free oscillations is given from (32-12) by

$$\omega_0 = \frac{1}{\sqrt{LC}}$$

The (ordinary) frequency $f = \omega_0/2\pi$ and its reciprocal, the period of oscillation T, are given by

$$f = \frac{1}{T} = \frac{1}{2\pi\sqrt{LC}} \tag{32-15}$$

Instantaneous current $i = dq/dt$ also oscillates sinusoidally. We see this by taking the time derivative of (32-14):

$$i = -\omega_0 Q_m \sin \omega_0 t = -I_m \sin \omega_0 t \tag{32-16}$$

where $I_m = \omega_0 Q_m$ is the maximum value of i.

Comparing (32-16) and (32-14) [and Figures 32-6(a) and (b)], we see that the charge (here varying directly with the cosine) and the current (here varying directly with the sine) are 90° out of phase. That is, when the capacitor is fully charged and the energy resides entirely in the capacitor's electric field, the current through the inductor and the magnetic field associated with it are zero. Conversely, when the capacitor is discharged, the inductor has all the energy.

The circuit's total energy U consists of the capacitor's energy $U_C = q^2/2C$ and the inductor's energy $U_L = Li^2/2$:

$$U = U_C + U_L = \frac{q^2}{2C} + \frac{Li^2}{2} = \frac{Q_m^2 \cos^2 \omega_0 t}{2C} + \frac{LI_m^2 \sin^2 \omega_0 t}{2} \tag{32-17}$$

To see that this total energy remains constant with time, we use $I_m = \omega_0 Q_m$ from (32-16) and $\omega_0^2 = 1/LC$ from (32-12) in LI_m^2. We have then

$$LI_m^2 = L\omega_0^2 Q_m^2 = \frac{Q_m^2}{C}$$

so that (32-17) becomes

$$U = \frac{Q_m^2}{2C} (\sin^2 \omega_0 t + \cos^2 \omega_0 t) = \frac{Q_m^2}{2C} \tag{32-18}$$

The energies of the capacitor and inductor vary sinusoidally with time (at *twice* the frequency of q and i), as shown in Figures 32-7(c) and (d); their sum is constant.

Any actual circuit has some resistance, even if only the resistance of the inductor's windings. With resistance included, we have a damped harmonic electrical oscillator. The energy then decreases with time exponentially (in the fashion of Figure 11-13 for a damped mechanical oscillator) because of the ever-present i^2R loss in the resistor.

Example 32-5.   An electric oscillator is a part of every radio receiver. One tunes to a station typically by adjusting a variable capacitance with a tuning knob so that the radio oscillator's frequency is brought to the frequency of the incoming radio wave.

The oscillator of a certain radio receiver has an inductor of 1.0 mH and is tuned to an AM station at 710 kHz. (a) What is the capacitance in this oscillator? (b) In what sense, clockwise or counterclockwise, must the knob attached to the variable capacitor in Figure 32-8 be turned to tune to a station at 880 kHz?

(a) From (32-13), we have

$$f = \frac{1}{2\pi\sqrt{LC}}$$

or

$$C = \frac{1}{4\pi^2 f^2 L} = \frac{1}{4\pi^2(710 \times 10^3 \text{ s}^{-1})^2(1.0 \times 10^{-3} \text{ H})} = 50 \text{ pF}$$

(b) Tuning to a higher frequency implies, from the equation above, that the capacitance must decrease. For a parallel-plate capacitor of adjustable plate area [$C = \epsilon_0 A/d$, (26-2)], the area of the overlapping adjacent capacitor plates must be reduced to decrease C. The knob must be tuned clockwise.

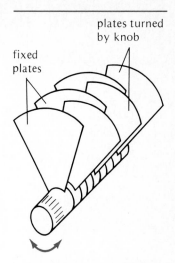

Figure 32-8. A variable parallel-plate capacitor. Turning the knob clockwise reduces the area that overlaps between adjacent capacitor plates.

As (32-15) shows, the frequency of an electric oscillator increases as the magnitudes of L and C decrease. It follows that if one is to construct an electric oscillator of very high frequency, the capacitor and inductor must be small, not merely in the magnitudes of C and L, but in the actual dimensions of these circuit elements.

Figure 32-9 shows the evolution of an ordinary LC circuit, with obvious, lumped capacitance and inductance elements, into two varieties of high-frequency oscillators, in which these circuit elements are less easily recognized. In Figure 32-9(a), first the inductance is reduced drastically by replacing the coil with a single conducting wire; then the capacitance also is reduced drastically by shrinking the area of the capacitor plates. What remains is a single conducting loop broken by a gap at one point. This is indeed an electric oscillator, and if the loop's size is of the order of 1 m with a gap of perhaps 1 cm, it oscillates at a frequency of tens of megahertz, a frequency lying in the radiofrequency region of the electromagnetic spectrum.

An oscillator of just this type was used in the historic experiments of Heinrich Hertz (1857–1894), who first demonstrated the existence of electromagnetic radiation in 1887. Hertz used two such oscillators, both resonant at the same frequency. A spark at the gap served to identify the oscillations. Electromagnetic radiation was detected by observing that when the second oscillator was moved relatively far from the first oscillator, the sparks across its

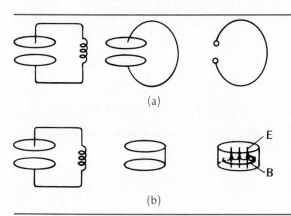

(a)

(b)

Figure 32-9. Evolution of an ordinary electric oscillator into (a) a high-frequency oscillator consisting of a single conducting loop with a gap and (b) a microwave cavity oscillator consisting of a hollow right circular cylindrical conductor.

gap persisted, although this effect could not be attributed to the direct action of the electric and magnetic fields of the first oscillator on the second.

Figure 32-9(b) shows the evolution of a simple *LC* circuit into a different type of high-frequency oscillator. Here the inductance is reduced first by connecting the capacitor plates with a single straight conducting wire. Then the inductance is reduced still further by connecting additional wires in parallel with the first. (Inductors in parallel follow the same rule as resistors.) Indeed, one constructs an entire cylindrical surface between the two capacitor plates, and thereby forms a hollow, closed right circular cylinder of conducting material. Superficially at least this certainly does not look like an *LC* oscillator of the ordinary variety. It is, in fact, one simple type of a *microwave* oscillator. For dimensions of the order of a few centimeters, the free oscillations occur at microwave frequencies of the order of a few gigahertz (or electromagnetic waves having wavelengths of a few centimeters). The oscillating electric field, as well as the oscillating magnetic field, is confined entirely within the closed cylinder. Here it becomes more useful to describe the electric oscillations in terms, not of the current through the circuit or of the potential difference across various pairs of points, but of the electric and magnetic fields within the closed resonating cylinder.

32-6 Electrical-Mechanical Analogs

An electric oscillator reminds us in many respects — the sinusoidal variation in time, the constant total energy, to name just two — of an ordinary mechanical oscillator. There is indeed an exact analogy between mechanical quantities and properties and their counterparts in electromagnetism and electric circuit elements.

Consider, for example, a mechanical oscillator with mass m attached to spring of force constant k. We get the equation of motion from Newton's second law:

$$\Sigma \mathbf{F} = m\mathbf{a}$$

$$-kx = m\,\frac{d^2x}{dt^2}$$

or

$$\frac{d^2x}{dt^2} = -\frac{k}{m}\,x$$

This equation has *exactly* the same form as the equation for an electric oscillator:

$$\frac{d^2q}{dt^2} = -\frac{1}{LC}\,q$$

One equation can be transformed into the other simply by making the substitutions listed in Table 32-1.

The correspondence is more than formal. Like mass m, which is a measure of the inertial tendency of a particle to maintain constant velocity v, inductance L is a measure of an inductor's tendency to maintain a constant current i.

Table 32-1

MECHANICAL QUANTITY		ELECTRIC (OR MAGNETIC) QUANTITY	
Displacement	x	q	Electric charge
Velocity	$v = \dfrac{dx}{dt}$	$\dfrac{dq}{dt} = i$	Electric current
Acceleration	$a = \dfrac{dv}{dt} = \dfrac{d^2x}{dt^2}$	$\dfrac{d^2q}{dt^2} = \dfrac{di}{dt}$	Rate of current change
Mass	m	L	Inductance
Force	$F = ma$	$L\dfrac{di}{dt} = -\mathcal{E}$	EMF, potential difference
Stiffness	k	$\dfrac{1}{C}$	Reciprocal of capacitance
Force of spring	$F = -kx$	$\dfrac{q}{C} = V$	Potential difference, capacitor
Potential energy of spring	$\dfrac{1}{2}kx^2$	$\dfrac{1}{2}\left(\dfrac{1}{C}\right)q^2$	Potential energy of capacitor
Kinetic energy	$\tfrac{1}{2}mv^2$	$\tfrac{1}{2}Li^2$	Energy of inductor
Mechanical (linear) resistive force	$F_r = -rv$	$iR = V$	Potential difference across electrical resistor
Constant weight	$W = mg$	$\mathcal{E}_b$	Constant emf battery
Rate of energy dissipation	$P = rv^2$	$i^2R = P$	Rate of energy dissipation

Similarly, a spring's restoring constant k corresponds to what might be called the capacitor's restoring constant $1/C = V/q$. (Be careful about the symbols, however. Both an inductor and a helical spring are typically represented in a diagram by the same sort of symbol. But a capacitor is analogous to a spring, and an inductor to mass.)

Note, further, that the electric potential difference q/C corresponds to the elastic restoring force $-kx$, and the induced emf $L(di/dt)$ to the force $m(dv/dt)$. The parallel to the mechanical dissipative force $F = -rv$, proportional to the particle's velocity, is the electrical resistance R, with potential difference $V = iR$. Similarly, the potential energy $\tfrac{1}{2}kx^2$ of the stretched spring corresponds to the electric energy $q^2/2C$ of the charged capacitor, and the kinetic energy $\tfrac{1}{2}mv^2$ of the moving particle to the magnetic energy $Li^2/2$ of the inductor.

The analogs are complete, in both mathematical and physical behavior, between mechanical and electrical "circuit" elements. It may be useful to analyze a complicated mechanical system by constructing its electrical analog with ordinary electric circuit elements. Then the measured charges, currents, and potential differences give the corresponding particle displacements, velocities, and forces.

Example 32-6. An object falls from rest under the influence of gravity and it is also subject to linear resistive force (proportional to the object's velocity). How does the velocity of the falling object change with time?

This problem was dealt with in Section 6-4 as an example of Newton's laws. Now we can write down the solution immediately, simply by recognizing its electric analog: an *LR* circuit, with an inductor and resistor in series connected to a battery of constant

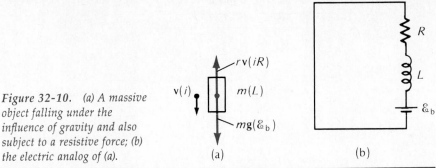

Figure 32-10. *(a) A massive object falling under the influence of gravity and also subject to a resistive force; (b) the electric analog of (a).*

emf $\mathcal{E}_b$. See Figure 32-10. Using the translation dictionary in Table 32-1 and (32-8) for the current as a function of time in an *LR* circuit, we have at once

$$v = \frac{mg}{r}\left[1 - \exp\left(-\frac{t}{m/r}\right)\right]$$

The time constant m/r now represents the time elapsing until the falling object comes within $(1/e)$th of its terminal velocity mg/r.

32-7   Mutual Inductance (Optional)

Self-inductance L relates the emf induced in an inductance coil to the rate at which current changes in that same coil. *Mutual* inductance, on the other hand, has to do with the emf induced in one inductor by a current change in another inductor, not a part of the same circuit. The term *transformer* (Section 33-7) is also used, especially in connection with ac circuits, to denote a mutual inductor.

   Consider Figure 32-11, where we have two inductors at rest, each a part of separate circuits. The coils are, however, linked by magnetic flux: an electric

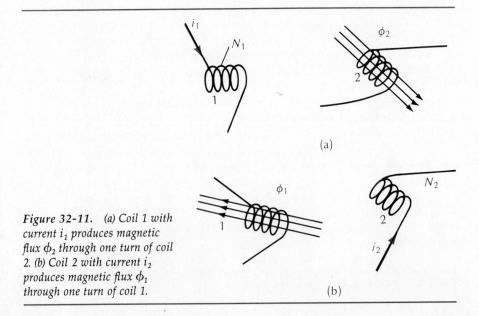

Figure 32-11. *(a) Coil 1 with current i_1 produces magnetic flux ϕ_2 through one turn of coil 2. (b) Coil 2 with current i_2 produces magnetic flux ϕ_1 through one turn of coil 1.*

current in either inductor produces a magnetic flux through the other one. Current i_1, in coil 1 with N_1 identical turns, produces a magnetic flux ϕ_2 through one turn at coil 2. Similarly, current i_2 in coil 2 of N_2 identical turns produces a flux ϕ_1 through one turn of coil 1. The emf $\mathcal{E}_2$ induced in coil 2 when the flux ϕ_2 changes with time is, from Faraday's law,

$$\mathcal{E}_2 = -N_2 \frac{d\phi_2}{dt} \qquad (32\text{-}19)$$

Magnetic flux ϕ_2 is proportional to the current i_1 that creates it; and the *mutual inductance* M_{21} is defined, in analogy with self-inductance [(32-1)], by

$$N_2\phi_2 \equiv M_{21}i_1 \qquad (32\text{-}20)$$

We see that M_{21} is a quantitative measure of how the flux linkage $N_2\phi_2$ in coil 2 is influenced by current i_1 in coil 1. When we take the time derivative of (32-20) and substitute the result in (32-19), we get

$$\mathcal{E}_2 = -M_{21} \frac{di_1}{dt} \qquad (32\text{-}21)$$

a relation analogous to (32-3) for self-inductance. For any pair of inductors fixed in position, the mutual inductance for the pair is a constant whose value depends on the geometry of the arrangement. Like self-inductance, mutual inductance has units of henries.

Mutual inductance can be considered defined either by (32-20) or (32-21). The first equation is most useful for computing the value for M_{21} from a knowledge of the flux linking the two inductors. The second equation is most useful for describing the behavior of a mutual inductor in coupled circuits.

In arriving at (32-20) and (32-21), we began with changing current i_1 producing a magnetic flux ϕ_2 and emf $\mathcal{E}_2$. The reciprocal arrangement is that in which changing current i_2 produces a magnetic flux ϕ_1 and emf $\mathcal{E}_1$. The equations corresponding to (32-20) and (32-21) are, with subscripts 1 and 2 interchanged,

$$N_1\phi_1 \equiv M_{12}i_2 \qquad (32\text{-}22)$$

and

$$\mathcal{E}_1 = -M_{12} \frac{di_2}{dt} \qquad (32\text{-}23)$$

We assert without proof a remarkable result. The two mutual inductances, M_{21} and M_{12}, are exactly equal for every mutual inductor. We can then use a single symbol M for either mutual inductance:

$$M \equiv M_{12} = M_{21} \qquad (32\text{-}24)$$

Summary

Definitions

Self inductance L for a coil of N identical turns with magnetic flux ϕ_B through each turn produced by current i:

$$N\phi_B \equiv Li \qquad (32\text{-}1)$$

When the current through an inductor changes at the rate di/dt, a back emf $\mathcal{E}$ is produced:

$$\mathcal{E} = -L \frac{di}{dt} \qquad (32\text{-}3)$$

Units

Henry, the unit for inductance:

$$1\,H \equiv 1\,V\cdot s/A$$

Important Results

Time constant τ for an LR circuit:

$$\tau = \frac{L}{R} \qquad (32\text{-}9)$$

Energy U_L of an inductor of inductance L carrying current i:

$$U_L = \tfrac{1}{2}Li^2 \qquad (32\text{-}10)$$

Energy density (energy for unit volume) u_B of magnetic field **B**:

$$u_B = \frac{B^2}{2\mu_0} \qquad (32\text{-}11)$$

LC electric oscillator: the charge (on a capacitor plate) and current oscillate sinusoidally with time at frequency f and period T, where

$$f = \frac{1}{T} = \frac{1}{2\pi\sqrt{LC}} \qquad (32\text{-}15)$$

Electric-mechanical analogs (see Table 12-1): electric-circuit elements are exactly parallel, in mathematical form and physical properties, to mechanical quantities.

Problems and Questions

Section 32-1   Self-Inductance Defined

· **32-1 P** An emf of 4.0 V is induced in a certain coil when the current in the coil changes at a rate of 4000 A/s. What is the self-inductance of the coil?

· **32-2 P** A current of 100 A has been established in a large electromagnet with an inductance of 20 H. Suppose that the circuit is broken by accident and the current falls to zero in 1 ms. What emf would be induced across the magnet?

· **32-3 P** A 5-H inductor carries a current of 2 A. How can an emf of 200 V be made to appear across the inductor?

· **32-4 Q** Show that μ_0 has units of henries per meter.

· **32-5 Q** The inductance of a coil depends on
(A) the current in it.   *See sentence below*
(B) the flux in it.   *eqn (32-1).*
(C) the emf applied to it.
(D) geometrical factors such as size, shape, and number of turns.
(E) none of the above.

· **32-6 P** What is the inductance of a solenoid of length 50 cm and diameter 2.0 cm wound with 10,000 turns of wire?

· **32-7 P** An inductor consists of a closely wound, long solenoid. Its inductance is 200 mH. What is the total magnetic flux through the middle cross section of the coil when a current of 5 A exists in the solenoid?

: **32-8 Q** Sometimes when you first turn on a large electric motor in your garage, you find that the house lights dim momentarily. Why is this?
(A) A large current is used briefly to charge the capacitor attached to the motor.
(B) The large back emf generated by the motor reduces the voltage available to the house.
(C) Before the motor armature reaches full speed, it does not generate much back emf and as a result a large current is drawn in the wires coming into the house.
(D) This is a transient effect associated with the fact that the current in the motor is ac.
(E) A large torque is needed to cause the armature to start rotating; but once it reaches full speed, essentially very little torque needs to be applied.

: **32-9 P** Two concentric circular loops lie in a plane. They have radii R_1 and R_2, where $R \gg R_2$. What emf is induced in the small loop when the current in the large loop changes at the rate di/dt?

: **32-10 P** Determine the inductance of a toroid of rectangular cross section $a \times b$ of N turns, with inner radius R and outer radius $R + a$. See Figure 32-12.

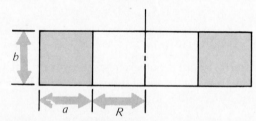

Figure 32-12. Problem 32-10.

: **32-11 P** Two long, straight, parallel wires of radius r are separated by a distance d. Show that if they each carry current I in opposite directions, their self-inductance per meter length is, with the flux within the wires themselves neglected,

$$L = \frac{\mu_0}{\pi}\ln\left(\frac{d - r}{r}\right)$$

Section 32-2   The LR Circuit

· **32-12 P** A coil with a inductance of 2.5 H is suddenly connected to a 24-V battery. The inductor has a resistance of 12 Ω. Graph current as a function of time.

32-13 P How long would it take for the voltage across the resistor in an *LR* circuit to drop to 1 percent of its initial value if $L = 1$ mH and $R = 1000$ Ω?

· **32-14 P** An inductor is connected to a 6-V battery with an internal resistance of 0.5 Ω. The inductor has 4.2-Ω resistance. The current is found to grow to 95 percent of its final value in 15 ms. What is the inductance of the coil?

· **32-15 P** Show that equation (32-8) *is* a solution to (32-7).

: **32-16 P** After how many time constants will the current in an *LR* circuit be within 1 percent of its final value?

: **32-17 Q** You have momentarily forgotten the relation for the time constant in an *RL* circuit. (Is it L/R, or LR, or R/L?) Show what the correct relation is by using dimensional analysis.

: **32-18 Q** An inductance L and a resistance R are connected in a circuit as shown in the text in Figure 32-4. When the switch is closed,

(A) the current in the resistor will initially be greater than the current in the inductance.

(B) the current in the inductance will initially be greater than the current in the resistance.

(C) the current in the battery will decay exponentially.

(D) the current wll rise slowly at first, and then more rapidly.

(E) the time required to reach the final current in the resistor is proportional to R/L.

(F) None of the above is true.

: **32-19 P** A coil with an inductance L and a resistance R is connected to a battery with emf $\mathcal{E}$. How long after the switch is closed will (*a*) the current reach 99.9 percent of its steady-state value? (*b*) the energy stored in the inductor reach 99.9 percent of its steady-state value?

: **32-20 P** A short time after the switch is closed in an *LR* circuit, the current is 2 mA. Ten ms later the current is 6 mA, and finally it is 10 mA. What is the time constant of the circuit?

: **32-21 P** A solenoid of inductance L is connected to a power supply that applies a potential difference V_0. (*a*) Suppose the switch to the supply is opened in a short time t. Show that the voltage that appears across the solenoid is $V = (\tau/t)V_0$, where $\tau = L/R =$ the time constant for the charging circuit. (*b*) Suppose that when the switch is opened, an arc 2 cm long jumps across the two poles of the switch. The breakdown strength of air is 30×10^6 V/m. Estimate the sparking time t.

: **32-22 P** For the circuit shown in Figure 32-4, (*a*) show that at $t = 0$, $di/dt = i_f/\tau$, where $\tau = L/R$ and i_f is the final value of the current. (*b*) Find what the ratio is of voltage drop across inductor to voltage drop across resistor when $i = \frac{1}{2}i_f$.

: **32-23 Q** A large inductance (such as a big electromagnet) is connected to a battery as shown in Figure 32-13. Which of the following is the most accurate statement? Assume that the circuit has reached equilibrium before the switch is opened or closed.

(A) Just before the switch is closed the potential difference between its contacts is $\frac{1}{2}\mathcal{E}$.

(B) Just after the switch is opened the potential difference across it is $\frac{1}{2}\mathcal{E}$.

(C) Just after the switch is opened, the potential difference across it can never be greater than $\mathcal{E}$.

(D) Just after the switch is opened, the potential difference across it can be much greater than $\mathcal{E}$.

(E) Corrosion of the switch contacts due to arcing is more likely to occur in closing than in opening the switch.

(F) None of the above is true.

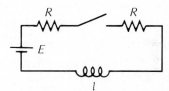

Figure 32-13.   Question 32-23.

: **32-24 Q** Suppose that an electromagnet with a very large inductance and a fairly low resistance are connected in series to a 100-V dc power supply, as shown in Figure 32-4. If the switch were suddenly opened, the potential difference between the contacts of the swtich

(A) would be zero initially.

(B) would be less than 100 V.

(C) would be equal to 100 V.

(D) could be much greater than 100 V.

(E) None of the above is true.

: **32-25 P** For the circuit shown in Figure 32-14, find what the values of i_1 and i_2 are (*a*) immediately after the switch is closed; (*b*) long after the switch is closed; (*c*) just after the switch is then opened; (*d*) long after the switch has been opened. ($L = 3.0$ H, $R_1 = 5$ Ω, $R_2 = 12$ Ω, $R_3 = 20$ Ω, $\mathcal{E} = 60$ V)

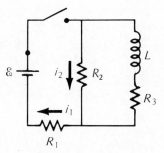

Figure 32-14.   Problem 32-25.

Section 32-3   Energy of an Inductor

32-36 P The current in a circuit with inductance L and resistance R grows to $1 - 1/e$ (63.2%) of its final value in a time interval L/R. What is the time interval for the energy of the inductor to grow to 63.2% of its final value?

: **32-27 P** For the circuit shown in Figure 32-4, $\mathcal{E}_b = 12$ V, $R = 3 \, \Omega$ and $L = 6$ mH. Six milliseconds after the switch is closed, what is (a) the power being dissipated in the resistor? (b) the power being supplied by the battery? (c) the rate at which energy is being stored in the magnetic field of the inductor?

· **32-28 P** A current of 1 A has been established in an inductor of 1.0 H. What capacitance would store the same amount of energy when charged to a voltage of 1 V?

· **32-29 P** A solenoid of 4-cm radius is 80 cm long and has 5000 turns. (a) What is the self-inductance of the coil? (b) What magnetic energy is stored in the coil for a current of 1 A?

: **32-30 P** Show that the equivalent self-inductance for two inductors in series is $L_s = L_1 + L_2$, and that the equivalent self-inductance for two inductors in parallel is L_p, where $1/L_p = 1/L_1 + 1/L_2$. Assume that there is negligible magnetic coupling between the two inductors (zero mutual inductance).

: **32-31 P** (a) Two solenoids of inductance 2 mH and 6 mH are connected in series. They are well separated. What is the total magnetic energy stored in them for a current in each of 1 A? (b) The two coils are connected in parallel and a total current of 1 A enters the junction with the two inductors. The inductors are still well separated. What is the total energy stored in the inductors?

: **32-32 P** Two coils of inductance 2.0 H and 4.0 H are connected in parallel. They are well separated. What current flows in each when a total magnetic energy of 6 J is stored in them?

Section 32-4   Energy of the Magnetic Field

· **32-33 Q** The current through a conducting loop is doubled. The total energy of the magnetic field associated with the circuit is thus changed by a factor of
(A) $\frac{1}{4}$.
(B) $\frac{1}{2}$.
(C) 2.
(D) 4.
(E) The change cannot be determined without more information about the shape and size of the loop.

· **32-34 P** A magnetic field of 0.10 T is set up in a certain region of space. What electric field would have the same energy density?

· **32-35 P** A long, straight wire carries current i. What is the magnetic-field-energy density at a distance r from the wire?

· **32-36 P** Very large magnetic fields can be obtained by discharging a capacitor through a low-inductance coil. The capacitor leads must have very low inductance themselves. Such "fast" capacitors cost about $3 per joule of stored energy capacity. (a) Estimate the cost of a capacitor that could store as much energy as contained in a 200-T field in a volume of one liter. (b) Assuming electricity costs $0.10 per kW·h, what does the electricity for charging the capacitor cost?

: **32-37 P** A flat circular coil of N turns and radius R carries current i. What is the magnetic-field energy density at the center of the coil?

: **32-38 P** The switch in the circuit of Figure 32-4 is closed at $t = 0$. Plot as a function of time (a) the current in the circuit; (b) the voltage across the inductance; (c) the energy stored in the magnetic field of the inductance.

: **32-39 P** Energy density and pressure have the same dimensions, and it can be shown that a magnetic pressure equal to the magnetic energy density $B^2/2\mu_0$ can be associated with magnetic field B. More specifically, a surface with magnetic field B on one side and zero field on the other is subject to magnetic pressure $B^2/2\mu_0$. For what magnetic field is the magnetic pressure equal to one atmosphere?

: **32-40 P** A long, straight wire carries current i uniformly distributed over its circular cross section. Find the magnetic energy stored per unit length within the conducting wire and show that it is independent of the wire radius.

32-5   Electrical Free Oscillations

· **32-41 P** An inductor with an inductance of 10 mH is available for use in an electric oscillator that is to resonate over the range of the AM radio dial from 530 kHz to 1600 kHz. A suitable variable capacitor is to be chosen. What is its (a) minimum and (b) maximum capacitance?

· **32-42 P** An electric oscillator resonates at frequency f_0. Then the original air-filled capacitor is filled with a dielectric having a dielectric constant κ. What is the new resonance frequency?

: **32-43 P** An inductor of fixed inductance and two identical variable capacitors are available to construct an electric oscillator. Each capacitor's capacitance can be changed continuously by a factor 2. The capacitors can be connected in series, or in parallel, or used singly. Over what range of frequencies can the oscillator be used?

: **32-44 Q** Suppose that you have temporarily forgotten how the resonance frequency of an LC circuit depends on L and C. Find the functional dependence of frequency on L and C by using dimensional analysis.

: **32-45 P** A certain electric oscillator consists of a parallel-plate capacitor and long cylindrical solenoid. (a) What happens to the resonance frequency if *every* dimension is reduced by a factor 2? (*b*) What then happens to the total resistance of the solenoid?

Section 32-6   Electrical-Mechanical Analogs

: 32-46 Q The analogy between an *LC* circuit and a harmonic oscillator is helpful in gaining insight into the system. In this analogy, the energy stored in the inductor is like
(A) the mass of the harmonic oscillator.
(B) the spring constant of the oscillator.
(C) the potential energy of the oscillator.
(D) the velocity of the oscillating mass.
(E) none of the above.

: 32-47 Q As Example 32-6 shows, the mechanical analog of an *LR* circuit is a massive object moving through a resistive medium. What is the mechanical analog of an *RC* circuit? How might one achieve such a mechanical device in practice?

Supplementary Problem

32-48 P A magnetic rail gun is represented schematically in Figure 32-15. A power supply charges a large capacitor bank, which is then discharged by closing the switch S. Current in the right-hand loop creates magnetic forces, which push the movable armature to the right. In guns being studied at the Los Alamos and Livermore Laboratories, a plasma arc is the moving armature, and it is contained in the bore of the gun in such a way that it is able to push a projectile along in front of it. The velocity obtainable by a conventional gun is limited by the rate at which a hot gas can expand, but there is no such limitation here. It is possible to get velocities well in excess of 10 km/s. These devices are being investigated for many applications, including high-pressure research, industrial manufacturing processes, impact thermonuclear fusion, and space propul-

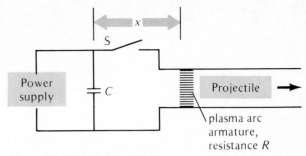

Figure 32-15.   *Problem 32-48.*

sion. These guns could be used to propel material into space for use in manufacturing. Such a device, called a mass driver, is planned for use on the moon. It would hurl material out to a manufacturing space station.

To see what is involved in the design of such a system, consider a rail gun with the following characteristics:

$$L_1 = \text{inductance per meter of}$$
$$\text{rail length} = 4 \times 10^{-7} \text{ H/m}$$

$$I = 2.5 \times 10^6 \text{ A}$$

$$m = \text{projectile mass} = 4 \text{ kg}$$

$$\text{Energy efficiency of device} = 11\% = \frac{E_{k,\text{projectile}}}{E_{\text{stored,capacitor}}}$$

(*a*) Calculate the force on the projectile in terms of L_1 and I. (*Hint:* Consider what happens to the energy provided by the capacitor. It goes into magnetic energy, kinetic energy, and heat.) (*b*) What length of rails would be needed to accelerate the projectile to 9.5 km/s, a typical velocity needed for an orbit of the earth? (*c*) How much energy must the power supply provide for each launch? (*d*) What is the average power delivered by the capacitor bank?

33

AC
Circuits

Some electric devices run on direct current (dc) — on batteries, or other sources of constant emf. Nearly all other electric devices operate on alternating current (ac). A particularly important ac circuit is one involving a resistor, an inductor, and a capacitor, all in series and driven by a sinusoidally varying emf of variable frequency.

33-1   Some Preliminaries

A constant danger in physics is that the basic ideas, which are usually straightforward, can become obscured in mathematical analysis — sometimes messy, and always potentially distracting. This is particularly true of ac circuits. Here we deal with sinusoidally varying currents and a circuit with all three basic circuit elements — resistance R, capacitance C, and inductance L. It is easy to understand these elements taken singly; their combined behavior is far more complicated.

After we have made some basic definitions, we shall consider separately the qualitative behavior of resistor, capacitor, and inductor for a current that changes with time.

We have already discussed a simple form of ac generator (Example 31-3). It consists of a coil of wire rotated at a constant angular speed ω in a uniform magnetic field perpendicular to the rotation axis of the coil. The generator produces a sinusoidally varying emf $\mathcal{E}$, which has the same frequency as the rotating coil:

$$\mathcal{E} = \mathcal{E}_m \sin (\omega t + \phi)$$

where the phase constant ϕ determines the value of the emf at the arbitarily chosen zero of time:

$$\mathcal{E} \text{ (at } t = 0) = \mathcal{E}_m \sin \phi$$

Figure 33-1 illustrates the variation in time of a constant dc source of emf and a sinusoidal ac source. It is customary to represent the dc source by the symbol ⊣⊢ and the ac source by ∿. The special importance of the *sinusoidal* variation in time of the emf is that first, most ordinary ac sources, such as household current, are of this type, but more fundamentally, *any* variation in emf with time can be regarded as the superposition of strictly sinusoidal variations (Optional Section 11-6).

We now consider in turn the qualitative behavior of R, C, and L for a constant emf and also a time-varying emf.

The potential drop V across a resistance R carrying current is $V = iR$. This relation holds for dc certainly. It also applies at each instant for ac, provided only that the frequency of the current is not too high. To see this, we recall first that the origin of resistance in any conductor is basically the collision of conduction electrons with the remaining atomic lattice and with other conduction electrons. The average time between collisions in a typical conductor at room temperature is of the order of 10^{-14} s. Therefore, a change in current with time, and hence a change within a conductor in the electric field driving conduction electrons, will appear to a conduction electron to occur so slowly that the current is effectively constant, so long as the frequency of the changing current is small compared with 10^{14} Hz (roughly the frequency of infrared light). Consequently, for all ordinary ac frequencies, the behavior of a resistor is the simplest possible; at each instant, $V = iR$. As the current changes with time, the potential drop across R keeps pace. In more formal language, i and V are exactly in phase for a resistor.

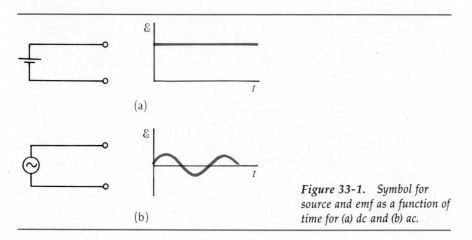

Figure 33-1. *Symbol for source and emf as a function of time for (a) dc and (b) ac.*

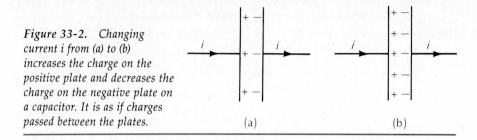

Figure 33-2. Changing current i from (a) to (b) increases the charge on the positive plate and decreases the charge on the negative plate on a capacitor. It is as if charges passed between the plates.

(a)   (b)

Now consider an inductor. If the current through an inductor is constant so that $di/dt = 0$, the inductor produces no induced emf, and $\mathcal{E} = -L\, di/dt = 0$; an inductor does not oppose a constant direct current. An inductor opposes a changing current, however; the higher the frequency of the varying current, the greater the rate of change of current and therefore the larger the opposing emf. In short, an inductor passes a *constant* current without opposition, but impedes the flow of charge through it increasingly as the frequency of the current increases.

How does a capacitor behave in a circuit with varying current? Strictly, there can never be any real current through a capacitor since no charged particles actually move from one charged plate to the other, oppositely charged plate through the space between them. With a capacitor maintained at a constant potential difference, the charge magnitude on each plate is also fixed. Then, not only is there no current between the plates; the current is also zero in the lead wires connecting the capacitor to other parts of an electric circuit. At constant voltage, a capacitor is an open circuit.

When the potential difference across a capacitor changes with time, however, there is actually an *equivalent* current through C. We can see this in Figure 33-2, where the potential difference across the capacitor plates increases from (a) to (b). With voltage V increased, the charge magnitude q on each plate must also increase by the same factor ($q = CV$). But if the positive plate becomes more positive, a current in the conductor connected to this plate must have brought additional positive charges to this plate; at the same time, an equal current removes positive charges from the negatively charged plate and makes it more negative. In short, when the potential drop across a capacitor is changing, the same instantaneous current exists in both lead wires connected to the capacitor—it is as if charges passed between the plates—and we can regard this current as effectively "passing through" the capacitor. Whereas a capacitor impedes completely constant direct current, it passes an alternating current.

Table 33-1

	DC (OR LOW FREQUENCIES)	AC (OR HIGH FREQUENCIES)
R	Passes	Passes
L	Passes	Impedes
C	Impedes	Passes

The qualitative features of R, L, and C circuit elements for dc and ac are summarized in Table 33-1. Note especially the reciprocal behavior of an inductor and a capacitor:

- L passes low frequencies but impedes high frequencies.
- C passes high frequencies but impedes low frequencies.

Example 33-1.   A sinusoidal current superimposed on constant current is applied to the input terminals of the two circuit arrangements shown in Figure 33-3: (a) a low-pass filter and (b) a high-pass filter. Show on the basis of the qualitative behavior of an inductor and a capacitor that (a) the low-pass filter "passes" the low-frequency, or dc, signal to the output terminals but impedes the ac signal, whereas (b) the high-pass filter passes the ac signal but impedes the dc signal.

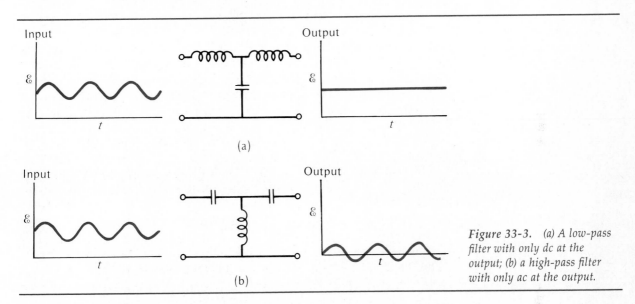

Figure 33-3.   *(a) A low-pass filter with only dc at the output; (b) a high-pass filter with only ac at the output.*

(a) The inductors act as "choke coils"; the ac signal is impeded by them from reaching the output terminals. The ac signal passes readily through the "by-pass" capacitor, however, and is returned to the input terminals. The signal at the output terminal is more nearly a constant emf.

(b) In the high-pass filter, the circuit elements are reversed. Now dc is blocked from reaching the output terminals by the capacitors, but dc passes readily through the inductor. The output signal is more nearly a purely sinusoidal emf.

33-2   Series *RLC* Circuit

The situation: An ac emf of variable frequency is applied to a circuit of R, L, and C all in series. The question: What is the magnitude of the current and its phase relative to the sinusoidal applied emf?

Before beginning the formal analysis, let us consider qualitatively what the general result must be. We know that a simple LC circuit (Section 32-5) oscillates at its characteristic angular frequency $\omega_0 = 1/\sqrt{LC}$, (32-10), with an undamped current amplitude. A more realistic electric oscillator, one with at least some small resistance, would oscillate at essentially the same frequency. Because of energy lost in the resistor, the oscillations would die out with time in the same fashion as a damped mechanical oscillator (Optional Section 11-7). Finally, an *RLC* circuit driven by an external emf would respond appreciably —that is, oscillate at resonance with large current amplitude—only when the driving frequency matches the characteristic oscillation frequency of the damped electric oscillator (Optional Section 11-8).

The series *RLC* circuit is shown in Figure 33-4. Applying Kirchhoff's loop rule (energy conservation) to this circuit, we have

$$\Sigma \mathcal{E} = \Sigma V$$

$$\mathcal{E} - L\frac{di}{dt} = iR + \frac{q}{C}$$

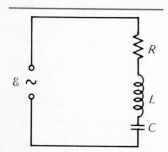

Figure 33-4. A series RLC circuit driven by a sinusoidal emf of variable frequency.

where $\mathcal{E}$ is the instantaneous applied emf; $-L\,di/dt$, the emf induced in the inductor; iR, the potential drop across the resistor; and q/C, the potential drop across the capacitor.

We shift the term $-L\,di/dt$ to the right side of the equation above and it becomes

$$\mathcal{E} = L\frac{di}{dt} + iR + \frac{q}{C} \tag{33-1}$$

The instantaneous current i is the *same* at each instant through all circuit elements (Kirchhoff's junction rule, charge conservation). We take the current to be of the form

$$i = I_m \sin \omega t \tag{33-2}$$

where I_m is the maximum instantaneous value of current and ω is the angular frequency of both the current and the applied sinusoidal emf.

The rate of change of current is then

$$\frac{di}{dt} = \omega I_m \cos \omega t \tag{33-3}$$

The charge q on the positively charged capacitor plate is related to the current in the circuit by

$i = \frac{dq}{dt}$

$$q = \int i\,dt = \int I_m \sin \omega t\,dt$$

$$q = -\frac{I_m}{\omega}\cos \omega t \tag{33-4}$$

So much for the physics; what follows is strictly mathematics. We substitute (33-2), (33-3), and (33-4) in (33-1), to get

$$\mathcal{E} = \omega L I_m \cos \omega t + R I_m \sin \omega t - \frac{1}{\omega C} I_m \cos \omega t$$

$$\mathcal{E} = I_m\left[R \sin \omega t + \left(\omega L - \frac{1}{\omega C}\right)\cos \omega t\right] \tag{33-5}$$

The following general definitions allow this equation to be written more compactly:

$$\text{Reactance} = X \equiv X_L - X_C \tag{33-6a}$$

where

$$\text{Inductive reactance} = X_L \equiv \omega L \tag{33-6b}$$

and

$$\text{Capacitive reactance} = X_C \equiv \frac{1}{\omega C} \tag{33-6c}$$

Equation (33-5) then can be written as

$$\mathcal{E} = I_m(R \sin \omega t + X \cos \omega t) \tag{33-7}$$

We see that the term with R varies directly with the sine of ωt, whereas the term with X varies directly with the cosine of ωt. We wish to manipulate the equation so that it contains a single sinusoidal function.

Consider the trigonometric identity:

$$\sin (A + B) = \sin A \cos B + \cos A \sin B$$

We set $A = \omega t$ and $B = \phi$, so that this equation becomes

$$\sin (\omega t + \phi) = \sin \omega t \cos \phi + \cos \omega t \sin \phi \tag{33-8}$$

For reasons soon to be apparent, it is a good idea to define quantities Z and ϕ as follows:

$$\text{Impedance} = Z \equiv \sqrt{R^2 + X^2} = \sqrt{R^2 + \left(\omega L - \frac{1}{\omega C}\right)^2} \tag{33-9}$$

$$\text{Tangent of } phase\ angle = \tan \phi \equiv \frac{X}{R} = \frac{\omega L - 1/\omega C}{R} \tag{33-10}$$

where we have used (33-6a), (b), and (c).

These definitions correspond to the simple geometrical construction shown in Figure 33-5, which is a helpful mnemonic for keeping straight the relations among Z, ϕ, X, and R. (This geometrical construction acquires additional significance in the discussion of Section 33-4).

From Figure 33-5 we see that

$$R = Z \cos \phi \quad \text{and} \quad X = Z \sin \phi$$

By these relations, (33-7) can then be written as

$$\mathcal{E} = I_m Z(\sin \omega t \cos \phi + \cos \omega t \sin \phi)$$

which simplifies, with the use of (33-8), to

$$\mathcal{E} = I_m Z \sin (\omega t + \phi)$$

We write this instantaneous applied emf as

$$\mathcal{E} = \mathcal{E}_m \sin (\omega t + \phi) \tag{33-11}$$

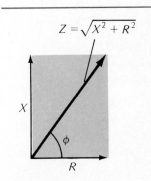

Figure 33-5.   Geometrical construction serving as a mnemonic for the relations among Z, φ, X, and R.

where $\mathcal{E}_m$ is the maximum value of $\mathcal{E}$, and ϕ is the phase angle by which the sinusoidal emf leads the sinusoidal current. Comparing the two equations immediately above then gives for the *magnitude of the maximum value of instantaneous current*

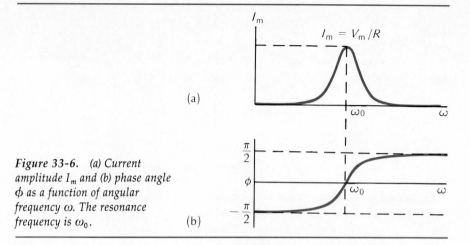

Figure 33-6. (a) Current amplitude I_m and (b) phase angle ϕ as a function of angular frequency ω. The resonance frequency is ω_0.

$$I_m = \frac{\mathcal{E}_m}{Z} = \frac{\mathcal{E}}{\sqrt{R^2 + (\omega L - 1/\omega C)^2}} \qquad (33\text{-}12)$$

The *phase ϕ of current i relative to the applied emf $\mathcal{E}$ is*, as we write (33-10) once more,

$$\tan \phi = \frac{\omega L - 1/\omega C}{R} \qquad (33\text{-}10)$$

That ends the mathematical analysis.* Now let us see what the results mean.

• Impedance Z plays a role analogous to that of resistance in a dc circuit. Impedance Z and reactance X have the units ohms. From (33-12), we see that the *maximum* value $\mathcal{E}_m$ of the applied emf is related to the *maximum* value I_m of the current by $\mathcal{E}_m = ZI_m$ (reminiscent of $\mathcal{E} = RI$). Note especially that we cannot write $\mathcal{E} = Zi$, since $\mathcal{E} = \mathcal{E}_m \cos(\omega t + \phi)$ and $i = I_m \sin \omega t$ do not reach their maximum values simultaneously — there is the phase difference ϕ between $\mathcal{E}$ and i. The impedance is frequency-dependent. From (33-9), we see that Z grows increasingly large at high frequencies, where the inductive reactance [(33-6b)] ωL is large. Impedance Z also becomes large at low frequencies, where the capacitive reactance [(33-6c)], $1/\omega C$, inversely proportional to angular frequency, dominates. Resistance is the only frequency-independent part of impedance.

• From (33-12), the amplitude I_m of the current varies with ω, as shown in Figure 33-6(a). The peak current $I_m = \mathcal{E}_m/R$ occurs at that angular frequency ω_0, for which the impedance has its minimum value. From (33-9), we see that minimum Z corresponds to the condition

$$\omega_0 L - \frac{1}{\omega_0 C} = 0$$

---

* Another way to get the same final results: use the electrical-mechanical analogs (Table 32-1) in the solution for a driven, damped mechanical oscillator, (11-21). The solution given above is the *steady-state* solution; in addition, there is a transient solution, applicable, however, only for a short period after the emf is first applied.

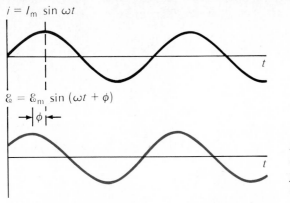

$i = I_m \sin \omega t$

$\mathcal{E} = \mathcal{E}_m \sin (\omega t + \phi)$

Figure 33-7. *Instantaneous current i and emf $\mathcal{E}$ as a function of time showing that $\mathcal{E}$ leads i by phase angle ϕ.*

or

$$\omega_0 = \frac{1}{\sqrt{LC}} \qquad (33\text{-}13)$$

The resonant frequency ω_0 of the driven *RLC* circuit is the same as for a simple *LC* circuit [(32-10)]; resonance corresponds to the circuit's being driven at its natural resonance frequency, $\omega = \omega_0$.

   • At the resonant angular frequency ω_0, the inductive and capacitive reactances have the same magnitude, and cancel, so that the impedance consists of resistance only: Z (for $\omega = \omega_0$) = R.

   • The resonance is also manifest in the variation of the phase angle ϕ with ω. See Figure 33-6(b), which is a plot of (33-10).

   First, let us be clear on the meaning of ϕ, by considering Figure 33-7, where both *i* and $\mathcal{E}$ are plotted as a function of time.* We see that the applied emf $\mathcal{E}$ *leads* the current *i* by phase angle ϕ because $\mathcal{E}$ reaches a peak, for example, *before i* does. (For a more specific example, take $t = 0$. Then *i* is zero, whereas $\mathcal{E}$ has already reached a positive value.) Of course, if $\mathcal{E}$ leads *i* by ϕ, then *i* lags $\mathcal{E}$ by ϕ.†

   As Figure 33-6(b) shows, resonance, $\omega = \omega_0$, corresponds to $\phi = 0$. At this one frequency, the current and applied emf are exactly *in phase*; both *i* and $\mathcal{E}$ reach peak values at the same instant. For higher frequencies, $\omega > \omega_0$, at which the inductive reactance dominates, (33-10) shows that ϕ is positive. Angle ϕ reaches its maximum value, $\pi/2$, at infinitely high frequency; the emf then *leads* the current by $\pi/2$. At frequencies below resonance, $\omega < \omega_0$, the capacitive reactance dominates, and (33-10) shows that ϕ is negative. Angle ϕ reaches its minimum value, $-\pi/2$ rad, at the lowest frequencies; the emf then *lags* the current by $\pi/2$.

   Still further aspects of an *RLC* series resonant circuit remain to be treated:

---

* What is shown plotted in Figure 33-7 can be seen directly on the screen of a dual-trace oscilloscope. The applied emf controls one trace; the potential drop across the resistor *iR*, which is proportional to *i*, controls the second trace.

† Be very careful about the phase angle, especially what is leading or lagging what. In some treatments of ac circuits, the phase angle is so defined that it represents the lag of $\mathcal{E}$ behind *i* rather than, as here, the lead of $\mathcal{E}$ ahead of *i*.

the relative phases of voltage drops across circuit elements; rms values for current and voltage; and power considerations (Sections 33-4 through 33-6).

Example 33-2.   An overhead electric transmission line has a series inductive reactance 0.5 Ω/km at the frequency of 60 Hz. What is the inductance per unit length of the transmission line?

From (33-6b),

$$X_L = \omega L = 2\pi f L$$

$$L = \frac{X_L}{2\pi f} = \frac{0.5 \ \Omega/\text{km}}{2\pi(60 \text{ s}^{-1})} = 1.3 \text{ mH/km}$$

Example 33-3.   What is the capacitive reactance of a 500-pF capacitor in a high-fidelity audio amplifier at the extremes of the audible range, (a) 20 Hz and (b) 20 kHz?

(a) From (33-6c),

$$X_C = \frac{1}{\omega C} = \frac{1}{2\pi f C} = \frac{1}{2\pi(20 \text{ s}^{-1})(500 \times 10^{-12} \text{ F})} = 16 \text{ M}\Omega$$

(b) With the frequency increased by a factor 10^3, the capacitive reactance is reduced by a factor 10^3 to 16 kΩ.

Example 33-4.   An *RLC* series circuit is connected across the terminals of a sinusoidal emf with an amplitude of 170 V and a frequency of 60 Hz. The circuit elements are a resistor of 50 Ω, a large capacitor of 27 μF, and a large inductor of 133 mH. (a) What is the total impedance of the circuit? (b) What is the phase angle by which the instantaneous applied emf leads the instantaneous current? (c) What is the peak value of the current? (d) Is the *RLC* circuit being driven above its resonance frequency, or below?

(a) We first compute the angular frequency

$$\omega = 2\pi f = 2\pi(60 \text{ s}^{-1}) = 377 \text{ s}^{-1}$$

The inductive reactance is

$$X_L = \omega L = (377 \text{ s}^{-1})(133 \times 10^{-3} \text{ H}) = 50 \ \Omega$$

The capacitive reactance is

$$X_C = \frac{1}{\omega C} = \frac{1}{(377 \text{ s}^{-1})(27 \times 10^{-6} \text{ F})} = 100 \ \Omega$$

With $R = 50 \ \Omega$, we then have for the impedance, from (33-9),

$$Z = \sqrt{R^2 + (X_L - X_C)^2} = \sqrt{(50 \ \Omega)^2 + (100 \ \Omega - 50 \ \Omega)^2} = 71 \ \Omega$$

(b) From (33-10),

$$\phi = \tan^{-1}\left(\frac{X_L - X_C}{R}\right) = \tan^{-1}\left(\frac{50 \ \Omega}{50 \ \Omega}\right) = \frac{\pi}{4} \text{ rad}$$

(c) The maximum value of current is, from (33-12),

$$I_m = \frac{\mathcal{E}_m}{Z} = \frac{170 \text{ V}}{71 \ \Omega} = 2.4 \text{ A}$$

(d) From (33-13),

$$f_0 = \frac{\omega_0}{2\pi} = \frac{1}{2\pi\sqrt{LC}} = \frac{1}{2\pi\sqrt{(0.133 \text{ H})(27 \times 10^{-6} \text{ F})}} = 84 \text{ Hz}$$

The circuit is driven at 60 Hz, which is *below* its resonance frequency. At the resonance frequency with a voltage amplitude still equal to 170 V, we get $Z = R$, and the peak current rises to $I_m = \mathcal{E}_m/R = 170\text{ V}/50\ \Omega = 3.4$ A.

33-3 Phasors

Phase relations between voltage and current in an ac circuit are made especially easy to visualize and compute through the use of phasors. Phasors are, in fact, useful in any physical situation involving two or more sinusoidal oscillations at the same frequency but with a difference in phase. (Later we shall use phasors in optics.)

Recall how a rotating vector—a *phasor*—can be used to represent a sinusoidal oscillation (Section 11-1 and Figure 11-2). Consider, for example, the alternating voltage

$$\mathcal{E} = \mathcal{E}_m \sin{(\omega t + \phi)}$$

It corresponds in Figure 33-8(a) to a vector of length $\mathcal{E}_m$ rotating counterclockwise at the constant angular speed ω. At $t = 0$, the vector makes an angle ϕ with the positive x axis. Figure 33-8(b), a plot of $\mathcal{E}$ as a function of time t, is derived from Figure 33-8(a) simply by taking the component of the vector along the y axis.

The special advantage of phasors is apparent when we have two or more vectors such as **A** and **B** in Figure 33-9(a), rotating together at the *same* angular frequency ω in the same sense. The angle between **A** and **B**, the relative phase difference between the two oscillations, remains fixed with time. Suppose that vector **B** lags vector **A** in phase by the angle ϕ. Vector **A** lies along the x axis at time $t = 0$, so that the equation describing the y component of vector **A** is

$$y_A = A \sin{\omega t}$$

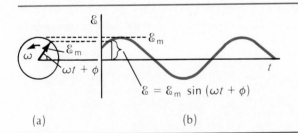

(a) (b)

Figure 33-8. *(a) Rotating phasor $\mathcal{E}_m$ and (b) sinusoidal $\mathcal{E}$ as a function of time.*

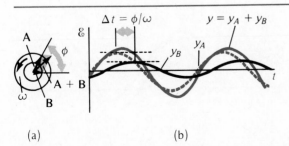

(a) (b)

Figure 33-9. *(a) Rotating vectors **A**, **B**, and **A + B**; (b) the corresponding displacements as a function of time.*

The corresponding equation for the y component of **B** is then

$$y_B = B \sin (\omega t - \phi)$$

Figure 33-9(b) shows both y_A and y_B plotted as a function of time. Here again y_B lags behind y_A by the angle ϕ; y_B reaches a peak later than y_A does. The phase lag is shown by the circumstance that the peak of the curve for y_B is shifted, relative to y_A, to the *right* (to *later* times) by Δt. The phase shift measured in terms of the time interval Δt between the two curves can be related to the phase difference ϕ and the common period T and angular frequency ω of the oscillation by

$$\frac{\Delta t}{T} = \frac{\phi}{2\pi}$$

$$\Delta t = \frac{\phi}{2\pi} T = \frac{\phi}{\omega} \tag{33-14}$$

Suppose that we are to find the sum of two sinusoidal functions with the same frequency ω but differing in phase by ϕ. The resultant y would be written as

$$y = y_A + y_B = A \sin \omega t + B \sin (\omega t - \phi)$$

Computing an actual value for y or finding it graphically from Figure 33-9(b) can be tedious, since the two oscillations are not in phase.

There is an easy way to find the sum. We simply recognize that the projection of the vector sum of two vectors is equal to the sum of the projections of the two vectors. Thus, we can immediately find the resultant oscillations of **A** and **B** by adding these as vectors and then taking the component along the y axis of their vector sum. The resultant oscillation is at the same frequency as the component oscillations; it differs from them both in amplitude and in relative phase, as shown in Figure 33-8. The three vectors **A**, **B**, and **A** + **B** remain locked together as they all rotate at the same rate ω. The amplitude of each oscillation is merely the length of the respective vectors, **A**, **B**, or **A** + **B**. In general, the amplitude of the resultant oscillation is not the sum of the amplitudes of the component oscillations.

Example 33-5. Use phasors to find the sum of the sinusoidal oscillations $y_A = A \cos \omega t$ and $y_B = B \sin \omega t$.

Phasors for y_A and y_B are shown in Figure 33-10(a). Vector **A** leads vector **B** by 90°;

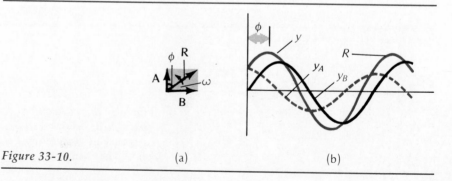

Figure 33-10.   (a)   (b)

A reaches a maximum one-quarter-cycle before vector **B**. To find the resultant oscillation, we merely add vectors **A** and **B** to obtain **R**, and then take the y component of this phasor to obtain $y = y_A + y_B$. From Figure 33-9(a), we see that **R** lags vector **A** by angle ϕ, where

$$\tan \phi = \frac{B}{A}$$

and the amplitude of **R** is

$$R = \sqrt{A^2 + B^2}$$

Therefore, the equation for the resultant oscillation is

$$y = R \cos (\omega t - \phi)$$

Figure 33-9(b) shows y_A, y_B, and y as functions of time.

33-4   Series *RLC* Circuit with Phasors

The phase relationships make ac circuits somewhat complicated. Such oscillating quantities as emf and current do not reach their maximum values simultaneously. But phasors, as we have seen, simplify the representation of oscillating quantities, and they are especially helpful in displaying the phase relationships in ac circuits. We resume our consideration of the *RLC* series resonance circuit driven by an external sinusoidal emf, but now with the aid of phasors and with special attention to the potential drops, or voltages, across the several circuit elements.

We again write the applied emf $\mathcal{E}$ with a maximum value $\mathcal{E}_m$ as

$$\mathcal{E} = \mathcal{E}_m \sin (\omega t + \phi) \tag{33-15a}$$

From the general relation $\Sigma \mathcal{E} = \Sigma V$, we see that V is the instantaneous potential difference across all three circuit elements (R, L, and C); V_m is its maximum value, and we can write, using (33-15a),

$$V = V_m \sin (\omega t + \phi) \tag{33-15b}$$

The current through all three elements is still written as

$$i = I_m \sin \omega t \tag{33-2}$$

Plots of i and of V as functions of time are shown respectively in Figures

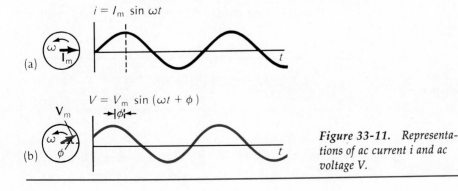

Figure 33-11. *Representations of ac current i and ac voltage V.*

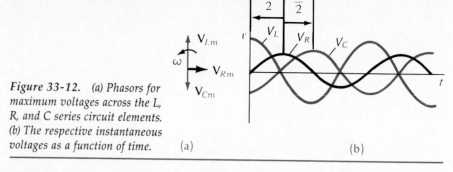

Figure 33-12. *(a) Phasors for maximum voltages across the L, R, and C series circuit elements. (b) The respective instantaneous voltages as a function of time.*   (a)   (b)

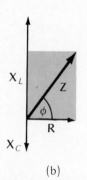

(a)

(b)

Figure 33-13. *(a) Relative orientations for voltage phasors of the three circuit elements, together with their "vector sum"* $\mathbf{V}_m$, *the instantaneous voltage across all three circuit elements. (b) Impedance vector diagram, derived from part (a) by dividing each vector there by the maximum current amplitude* I_m.

33-11(a) and 33-11(b); the phasors also show the same information. Again, V leads i by phase angle ϕ.

The instantaneous potential drops across the three circuit elements are, by (33-2),

$$V_R = iR = RI_m \sin \omega t \tag{33-16a}$$

$$V_L = L\frac{di}{dt} = \omega L I_m \cos \omega t = X_L I_m \sin\left(\omega t + \frac{\pi}{2}\right) \tag{33-16b}$$

$$V_C = \frac{q}{C} = -\frac{1}{\omega C} I_m \cos \omega t = X_C I_m \sin\left(\omega t - \frac{\pi}{2}\right) \tag{33-16c}$$

We used (33-6) in the above.

The three voltages are plotted as a function of time in Figure 33-12(b). The corresponding phasors in Figure 33-12(a) display the same information, especially the phase relationships, in far simpler fashion. Here the magnitudes of the three phasor vectors are V_{Rm}, V_{Lm}, and V_{Cm}, where from (33-16) we have

$$V_{Rm} = RI_m \tag{33-17a}$$

$$V_{Lm} = X_L I_m \tag{33-17b}$$

$$V_{Cm} = X_C I_m \tag{33-17c}$$

All three voltages V_R, V_L, and V_C, oscillate at the same angular frequency, so that all three phasors in Figure 33-12(a) rotate at the same rate. The three phasors are locked together, with V_L always leading V_R by $\pi/2$ and V_C always lagging V_R by $\pi/2$. For our purposes, all that really matters is the relative orientation of the three phasors, not their rotation as such.

The phasors of Figure 33-12(a) are shown again in Figure 33-13(a), where an additional vector $\mathbf{V}_m$, equal to their "vector sum,"

$$\mathbf{V}_m = \mathbf{V}_{Rm} + \mathbf{V}_{Lm} + \mathbf{V}_{Cm} \tag{33-18}$$

is also shown. The vector $\mathbf{V}_m$ is the phasor for the voltage across all three circuit elements [(33-15b)]. In *magnitude only* this maximum voltage is

$$V_m = ZI_m \tag{33-19}$$

where Z is the impedance of the series *RLC* circuit.

Phase angle ϕ in Figure 33-13(a) has exactly the same meaning as given earlier: ϕ is the angle by which the total voltage $\mathbf{V}_m$ leads the voltage across the

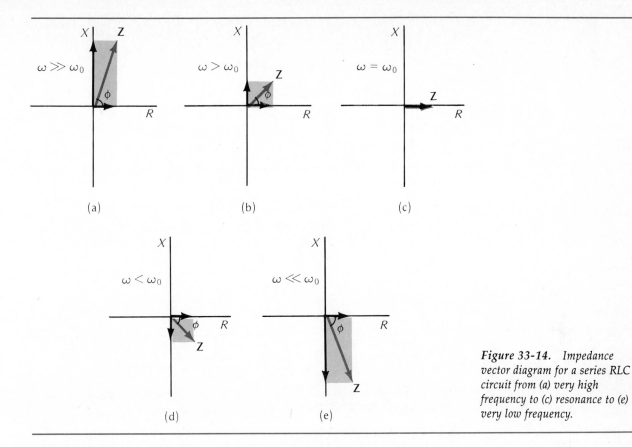

Figure 33-14. *Impedance vector diagram for a series RLC circuit from (a) very high frequency to (c) resonance to (e) very low frequency.*

resistor $\mathbf{V}_{Rm}$; this is equivalent to saying that ϕ is the angle by which the applied emf $\mathcal{E} = V$ leads the current i (proportional to iR).

All four of the phasor vectors in Figure 33-13(a) have a magnitude proportional to I_m. See (33-17) and (33-19). It is therefore still simpler to remove this common factor and redraw the vectors as shown in Figure 33-13(b). Here we have an *impedance vector diagram*, identical to Figure 33-5. Although the displayed quantities are resistance, inductive and capacitive reactance, and total impedance, it must be recognized that what the diagram really portrays are the *phase relationships among the voltages across the various circuit elements*. (In no way does Figure 33-13(b) say, for example, that inductance is "perpendicular" to resistance.)

An impedance diagram summarizes very neatly a lot of information on the properties of an ac circuit. Consider Figure 33-14, which shows how the magnitude and direction (really, the phase) of the impedance vector changes as the frequency of the applied $\mathcal{E}$ is reduced from an initial high value and the circuit goes through resonance. In Figure 33-14(a) at high ω, the inductance reactance $X_L = \omega L$ is large, and ϕ is close to $\pi/2$. In Figure 33-14(b) at a lower frequency, X_L is smaller and so is ϕ. Figure 33-14(c) corresponds to the resonance peak with $\omega = \omega_0$; here impedance has its minimum value, $Z = R$, and $\phi = 0$. At still lower frequencies, Figure 33-14(d) and (e), the capacitance reactance has grown larger, the impedance is also larger, and phase angle ϕ is negative and approaches $-\pi/2$.

Example 33-6. An *RLC* series circuit is driven by a sinusoidal emf with maximum value 170 V at 60 Hz. The circuit parameters are $R = 50\ \Omega$, $C = 27\ \mu F$, and $L = 133$ mH. (These values are the same as those given in Example 33-4). Find (a) the peak voltage drop across each element separately and (b) the peak voltage across the inductor and resistor together in series.

(a) In Example 33-4, we found that

$$X_L = 50\ \Omega$$

$$X_C = 100\ \Omega$$

$$I_m = 2.4\ A$$

From (33-17), we have for the maximum voltage drops

$$V_{Rm} = RI_m = (50\ \Omega)(2.4\ A) = 120\ V$$

$$V_{Lm} = X_L I_m = (50\ \Omega)(2.4\ A) = 120\ V$$

$$V_{Cm} = X_C I_m = (100\ \Omega)(2.4\ A) = 240\ V$$

Note that these values are quite different from the peak value 170 V of the applied emf. The voltages combine as *vectors* (phasors), as shown in Figure 33-12a.*

(b) The peak voltage across L and R in series with L is given by

$$Z_{R,L}\, I_m = \sqrt{R^2 + X_L^2}\, I_m = \sqrt{(50\ \Omega)^2 + (50\ \Omega)^2}\,(2.4\ A) = 170\ V$$

Note the curious result; the peak voltage across all three circuit elements is the same as that across L and R by themselves.

---

* The fact that in this example $V_{Rm} = V_{Lm}$ and $V_{Cm} = 2V_{Rm}$ is strictly accidental; it is a consequence of the particular parameters chosen.

33-5 RMS Values for AC Current and Voltage

An ordinary ac voltmeter is connected to a household outlet. It reads a steady 120 V. The alternating voltage has a frequency 60 Hz, so that the reading cannot be the rapidly changing instantaneous voltage. After all, an alternating voltage makes equal excursions to positive and negative values, so that the time average value of the sinusoidally oscillating voltage is zero. What kind of average does 120 V represent?

We show below why ordinary ac voltmeters and ammeters are calibrated to read effective, or rms (root-mean-square) values, where

$$I_{rms} \equiv \frac{I_m}{\sqrt{2}} \tag{33-20a}$$

$$V_{rms} \equiv \frac{V_m}{\sqrt{2}} \tag{33-20b}$$

Here I_m and V_m are, as before, the maximum values of the instantaneous current and voltage, respectively. For 120 V, 60 Hz, household voltage, a voltmeter reads $V_{rms} = 120$ V, so that $V_m = \sqrt{2}\, V_{rms} = \sqrt{2}\,(120\ V) = 170$ V. (The peak-to-peak value from positive to negative extreme is 340 V.)

Effective values of current and voltage are so defined that familiar relations for dc circuits work also for ac circuits. Consider power dissipated in a resistor. It is given by i^2R for constant current i through a resistor R. A relation of the same form, $I_{rms}^2\, R$, is to hold for ac.

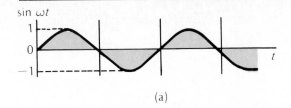

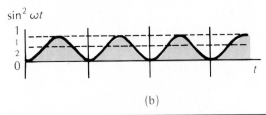

Figure 33-15.   The time average value (a) of sin ωt is zero and (b) of sin² ωt is ½.

The *instantaneous* power p dissipated in a resistor through which ac current i passes is

$$p = i^2 R$$

$$= (I_m \sin \omega t)^2 R$$

$$= I_m^2 R \sin^2 \omega t$$

To find the time average $\bar{p}$ of this power, we must find the time average of sin² ωt. The result follows immediately from Figure 33-15, where (a) sin ωt and (b) sin² ωt are plotted as a function of time t. Clearly, the time average of sin ωt is zero in Figure 33-15(a), because over one cycle, there are equal areas above and below the horizontal zero line. In Figure 33-15(b), the areas are all positive, and the time average of sin² ωt is equal to ½, since every area above the ½ line can be fitted exactly into a corresonding empty space below the ½ line. We have then that

$$\bar{p} = \tfrac{1}{2} I_m^2 R$$

$$= \left(\frac{I_m}{\sqrt{2}}\right)^2 R$$

Using (33-20a), we have finally what we set out to prove:

$$\bar{p} = I_{rms}^2 R$$

The term "root-mean-square" is appropriate because we have computed the square *root* of the time *average* value of the *square* of the oscillating current. The proof that voltage follows the same rule, $V_{rms} = V_m / \sqrt{2}$, is given in Section 33-6.

33-6   Power in AC Circuits

The power p delivered to any circuit is given by

$$p = iV \qquad\qquad (27\text{-}6)$$

where V is the voltage across and i is the current through the load. Here p is the *instantaneous* power, i the *instantaneous* current [(33-2)], and V the *instantaneous* voltage [(33-15b)] across all three elements in a series RLC circuit. The equation above can be written in more detail as

$$p = iV = (I_m \sin \omega t)[V_m \sin (\omega t + \phi)]$$

From the trigonometric identity in (33-8), this equation can also be written

$$p = I_m V_m \cos \phi \sin^2 \omega t + I_m V_m \sin \phi \sin \omega t \cos \omega t$$

We are interested in the time average value $\bar{p}$ of the power delivered to the circuit. The time average of the second term in the equation above is zero because $\sin \omega t \cos \omega t = \frac{1}{2} \sin 2\omega t$; the time average of $\sin^2 \omega t$ in the first term is $\frac{1}{2}$ (both results proved in Section 33-5). Therefore, the time-average power is

$$\bar{p} = \frac{1}{2} I_m V_m \cos \phi$$

$$\bar{p} = \left(\frac{I_m}{\sqrt{2}}\right)\left(\frac{V_m}{\sqrt{2}}\right) \cos \phi$$

$$\bar{p} = I_{rms} V_{rms} \cos \phi \tag{33-21}$$

where we have used (33-20). Equation (33-21) is just like the relation for power in a dc circuit except for the additional factor, $\cos \phi$, called the *power factor*, that enters in an ac circuit.

Equation (33-21) can be written in other useful forms by introducing the circuit's impedance Z. From (33-12), we have

$$V_m = ZI_m$$

which can also be written

$$V_{rms} = ZI_{rms} \tag{33-22}$$

Using (33-22) in (33-21), we get

$$\bar{p} = I_{rms}^2 Z \cos \phi \tag{33-23}$$

From Figure 33-5, we have that

$$R = Z \cos \phi$$

so that (33-23) can also be written as

$$\bar{p} = I_{rms}^2 R \tag{33-24}$$

Equations (33-21), (33-23), and (33-24) are alternative forms for the same basic relation giving the time-average power delivered to a circuit with R, L, and C in series. The implications of these relations are as follows:

• On a time average, *no power is delivered to a purely reactive load,* that is, to a circuit with inductance or capacitance, or both, but no resistance. Suppose that $R = 0$. Suppose further that $C = 0$, so that the load then consists of an inductor L only. Then $\phi = \pi/2$ [(33-10)] and $Z = \omega L$ [(33-9)]. Voltage leads current by $\pi/2$, and the power factor $\cos \phi$ is zero. From any of the equations (33-21), (33-23), or (33-24), we have that $\bar{p} = 0$. We can see in more detail that the time average power is zero from Figure 33-16. As Figure 33-16(d) shows, power alternately goes into and comes out of the inductor as the inductor

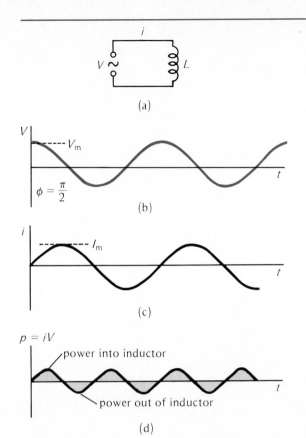

Figure 33-16.   *(a) Circuit with pure inductive reactance. As a function of time, (b) voltage, (c) current, and (d) power to the inductive load.*

temporarily gains and loses energy. The inductor merely reacts (hence, *reactance*) to the applied oscillating voltage; it does not extract energy permanently. The same sort of behavior is exhibited with a single capacitor as load. In that situation, $\phi = -\pi/2$. Once again the power factor and average power to the load are zero.

• On a time average, *power is delivered only to the resistance in the circuit.* Equation (33-24), $\bar{p} = I_{rms}^2 R$, shows this directly. A power factor different from zero implies that phase angle ϕ is not as large as $\pi/2$ nor as small as $-\pi/2$. Said differently, the impedance vector **Z** has some nonzero "component" along the resistance axis [Figure 33-13(b)]. On the average, more power goes into the load than comes out; see Figure 33-17.

• *Resonance shows up in the power delivered to an RLC circuit* as a function of the driving frequency. Using (33-12) in (33-24), we have, with $\mathscr{E}_m = V_m$,

$$\bar{p} = I_{rms}^2 R = \frac{V_{rms}^2 R}{Z^2}$$

$$\bar{p} = \frac{V_{rms}^2 R}{R^2 + (\omega L - 1/\omega C)^2} \tag{33-25}$$

The *resonance* frequency, $\omega = \omega_0$, with $\omega_0 L = 1/\omega_0 C$, corresponds to *maximum power* to the load, $\bar{p}_{max} = V_{rms}^2/R$. The power dissipated in the resistor is less at lower and higher frequencies, as shown in Figure 33-18.

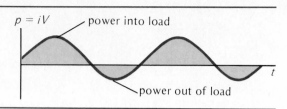

Figure 33-17. *Instantaneous power as a function of time to a load with some resistance.*

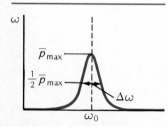

Figure 33-18. *Average power dissipated in the resistor of a series RLC circuit as a function of frequency. The full width in angular frequency at half of the maximum power is $\Delta\omega$.*

• *Both the height and width of the resonance peak in Figure 33-18 are controlled by the magnitude of the resistance.* At the resonance peak, $\bar{p}_{\max} = V_{\text{rms}}^2/R$, so that the power absorbed at the resonance frequency decreases as R is increased, with all other parameters unchanged. At the same time, an increase in R broadens the width of the resonance peak. As (33-25) shows, the average power has half its peak value when $R = \omega L - 1/\omega C$. If the half-power points are designated by $\omega = \omega_0 \pm \frac{1}{2}\Delta\omega$, so that the full width in angular frequency at half maximum power is $\Delta\omega$, then it follows from the relation $R = (\omega_0 \pm \frac{1}{2}\Delta\omega)L - [1/(\omega_0 \pm \frac{1}{2}\Delta\omega)C]$ after a little algebra that

$$\frac{\Delta\omega}{\omega_0} = \frac{R}{\omega_0 L} \tag{33-26}$$

As (33-26) shows, the smaller the resistance, the narrower the peak.

The ratio of the inductive reactance $\omega_0 L$ at the resonance peak to the resistance R is defined as the Q (for "quality") of the circuit:

$$Q \equiv \frac{\omega_0 L}{R} \tag{33-27}$$

Comparing (33-26) and (33-27), we see that the dimensionless Q factor gives a direct measure of the sharpness of the resonance peak, and the capacity of the resonance circuit for differentiating the resonance frequency ω_0 from other nearby frequencies. Since the higher the Q, the narrower the peak.

Example 33-7.   Design an *RLC* series resonance circuit with the following properties: resonance at 200 kHz, a Q of 100, and power dissipation at the resonance peak of 2.0 μW, with an ac source of 10 mV rms.

The average power dissipated at the resonance peak is, from (33-25),

$$\bar{p}_{\max} = \frac{V_{\text{rms}}^2}{R}$$

so that

$$R = \frac{V_{\text{rms}}^2}{\bar{p}_{\max}} = \frac{(10 \times 10^{-3}\text{ V})}{2.0 \times 10^{-6}\text{ W}} = 50\ \Omega$$

The inductance can then be computed from (33-27) as

$$L = \frac{QR}{\omega_0} = \frac{(100)(50\ \Omega)}{2\pi\,(200 \times 10^3\text{ s}^{-1})} = 4.0\text{ mH}$$

Finally, the capacitance is, from (33-13),

$$C = \frac{1}{L\omega_0^2} = \frac{1}{(4.0 \times 10^{-3}\text{ H})[2\pi(200 \times 10^3\text{ s}^{-1})]^2} = 158\text{ pF}$$

33-7 The Transformer

As electric energy is transmitted from a generating station to such a simple household electric device as a doorbell, the magnitude of the alternating voltage changes several times. The turbo generator output, typically several kilovolts, is transformed into very high voltage (up to 765 kV rms) for long-distance transmission over the countryside along lines suspended from tall towers. The ac coming down the street along lines on utility poles may be at several to tens of kilovolts. Ordinary household current is 120 V and the doorbell operates on 6 V. The several voltage transformations are readily accomplished for ac by the circuit element known as a transformer and symbolized in circuit diagrams by ⊒⫴⊏ .* Indeed, the principal advantage of alternating current over direct is the relative ease of changing voltage amplitude with transformers.

A transformer allows the ac voltage amplitude to be changed with little energy dissipation, so that the voltage amplitude can be made most appropriate for the specific application. To minimize i^2R losses in long-distance transmission lines, the current i must be very low; with fixed power $p = iV$, the voltage V must therefore be made as large as possible. On the other hand, safety considerations dictate low voltages for household applications.

A transformer, a circuit device operating on the basis of electromagnetic induction, consists of two multiturn coils typically wound around a laminated core of soft iron. See Figure 33-19. The input coil, known as the primary, has N_1 turns; the output coil, the secondary, has N_2 turns. When a current exists in, say, the primary coil, a magnetic field is created in the core. Ideally, the field lines lie entirely within the soft iron core. Therefore, the same number of field lines pass through every turn of primary and secondary coil, which is to say, the magnetic flux is the same through any one turn of either coil.

The basic transformer effect is that a changing emf $\mathcal{E}_1$ applied at the primary coil changes the magnetic flux through the primary and secondary coils and thereby creates a varying emf $\mathcal{E}_2$ at the secondary coil. More specifically, from Faraday's law of induction,

$$\mathcal{E}_1 = -N_1 \frac{d\phi_1}{dt}$$

and

$$\mathcal{E}_2 = -N_2 \frac{d\phi_2}{dt}$$

But the rate of magnetic flux change per turn is the same for primary and secondary, so that $d\phi_1/dt = d\phi_2/dt$. The two equations above then yield

$$\frac{\mathcal{E}_2}{\mathcal{E}_1} = \frac{N_2}{N_1}$$

The input and output voltages are, it can be shown for an ideal, lossless

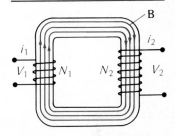

Figure 33-19. *Transformer with equal magnetic flux through any turn of input or output coils. At the input, N_1 turns, current i_1, and voltage V_1 with corresponding terms for the output with subscript 2.*

---

* The transformer is one important type of mutual inductor (Optional Section 32-7).

transformer, in the same ratio as the corresponding emf's, so that the relation above can be written as

$$\frac{V_2}{V_1} = \frac{N_2}{N_1} \tag{33-28}$$

where V_1 and V_2 can represent either the peak or rms values at the primary and secondary.

If there are more secondary than primary turns, with $N_2 > N_1$, then the output voltage V_2 exceeds the input voltage V_1, and we have a *step-up trans-former*. With $N_2 < N_1$, and therefore $V_2 < V_1$, the device is a *step-down trans-former*.

Transformers can be made so highly efficient that the output power differs from the input power by less than 1 percent. Eddy currents within the core are drastically reduced by using laminated sheets; the iron oxide that forms on the surfaces of the laminations has a very high resistance. A more detailed analysis than we give here shows that the power factor cos ϕ (Section 33-6) can be close to 1. Then, if no losses take place in the transformer, the input power $i_1 V_1$ is related to the output power $i_2 V_2$ for a lossless transformer by

$$i_1 V_1 = i_2 V_2$$

where i_1 and i_2 are the currents in the primary and secondary circuits. Using this result in (33-26), we then have

$$\frac{V_2}{V_1} = \frac{i_1}{i_2} = \frac{N_2}{N_1} \tag{33-29}$$

The current ratio equals the inverse turns ratio. To the degree that the output voltage is raised, the output current is reduced correspondingly; the converse is also true.

Example 33-8.   A doorbell operates on 6 V rms at 0.5 A rms. What current would be required in a high-"tension" transmission line operating at 765 kV rms to run the doorbell?

Apart from small resistive losses, the power supplied by the transmission line is nearly the same as power supplied to the doorbell, so that the current is

$$\frac{0.5 \text{ A}}{765 \times 10^3 \text{ V}/6 \text{ V}} = 3.9 \ \mu\text{A rms.}$$

Summary

All results apply to an ac circuit with R, L, and C in series. The angular frequency of current and voltage is $\omega = 2\pi f$.

A capacitor passes ac, impedes dc.

$$\text{Impedance} = Z \equiv \sqrt{R^2 + (X_L - X_C)^2} \tag{33-9}$$

Impedance is the ac analog of resistance.

Definitions

Phase angle by which applied emf leads current:

Inductive reactance $= X_L \equiv \omega L$ (33-6b)

An inductor passes dc, impedes ac.

$$\phi \equiv \tan^{-1}\left(\frac{X_L - X_C}{R}\right) \tag{33-10}$$

Capacitive reactance $= X_C \equiv \dfrac{1}{\omega C}$ (33-6c)

where $\mathcal{E} = \mathcal{E}_m \sin(\omega t + \phi)$ and $i = I_m \sin \omega t$.

Phasor: rotating vector whose projection along a diameter is used to represent a sinusoidal variation with time.

$$Q \text{ value} \equiv \frac{\omega_0 L}{R} \tag{33-27}$$

Units

Reactance X and impedance Z are in ohms.

Important Results

Nomenclature:

Subscript m denotes maximum, or peak, value.

Subscript rms denotes root-mean-square, or effective, value.

Current-voltage-impedance relation:

$$I_m = \frac{\mathscr{E}_m}{Z} \tag{33-12}$$

$$\text{or} \quad V_m = ZI_m \tag{33-19}$$

$$\text{or} \quad V_{rms} = ZI_{rms}$$

RMS versus peak values:

$$I_{rms} = \frac{I_m}{\sqrt{2}} \tag{33-20a}$$

$$V_{rms} = \frac{V_m}{\sqrt{2}} \tag{33-20b}$$

Time-average power $\bar{p}$ dissipated in the load:

$$\bar{p} = I_{rms} V_{rms} \cos \phi \tag{33-21}$$

$$= I_{rms}^2 Z \cos \phi \tag{33-23}$$

$$= I_{rms}^2 R \tag{32-24}$$

where $\cos \phi$ is called the *power factor*.

Phase relations among the voltage drops across the several circuit elements are summarized in an impedance diagram, shown in Figure 33-12(b). The *resonance peak* in a series *RLC* circuit is characterized, for an applied emf of constant voltage amplitude but variable frequency ω, by:

- $\omega = \omega_0 = \dfrac{1}{\sqrt{LC}}$ (33-13)

- Maximum current
- Maximum power
- Purely resistive load, $Z = R$
- Instantaneous current in phase with instantaneous applied emf; $\phi = 0$.

Full width $\Delta \omega$ in angular frequency for half maximum power:

$$\frac{\Delta \omega}{\omega_0} = \frac{R}{\omega_0 L} = \frac{1}{Q} \tag{33-26}$$

Transformer with primary coil of N_1 turns, current i_1, voltage V_1, and secondary coil of N_2 turns, current i_2, and voltage V_2:

$$\frac{V_2}{V_1} = \frac{i_1}{i_2} = \frac{N_2}{N_1} \tag{33-29}$$

Problems and Questions

Section 33-1   Some Preliminaries

· **33-1 P** Commercial power in the United States is at a frequency of 60 Hz. To what angular frequency does this correspond?

· **33-2 P** A sinusoidal voltage of amplitude 10 V is applied with a frequency of 400 Hz. (*a*) What is the period of the voltage? (*b*) What is the maximum instantaneous rate of change of the voltage?

· **33-3 Q** For which of the following would ac or dc voltages be equally acceptable? (None or more than one may be correct.)
(A) Incandescent light bulb.
(B) Electric toaster.
(C) Electric stove.
(D) Neon sign transformer.
(E) Battery charger.
(F) Chrome plating a metal object.

: **33-4 P** A 20-μF capacitor is charged to 100 V and connected to a resistanceless inductance of 40 mH. What is the maximum current in the circuit after the switch is closed?

: **33-5 P** An 8.0 μF capacitor charged to 40 V is connected across a 2.0 mH inductor. (*a*) What is the maximum value of the current in the inductor? (*b*) Sketch the voltage across the capacitor, V_C, the voltage across the inductor, V_L, and their sum $V_C + V_L$ as a function of time. (*c*) What is the total energy of the oscillator at any time?

· **33-6 P** A sinusoidal potential difference is applied to an *RLC* circuit. An experimenter observes on an oscilloscope the potential difference across the resistor in the circuit and finds that successive positive peaks occur at intervals of 1.0 ms. What is the angular frequency of the applied emf?

Section 33-2   Series *RLC* Circuit

· **33-7 Q** When the switch of the circuit in Figure 33-20 is closed, the potential difference across the resistor R is found to vary as shown. From this we should deduce that the black box probably contains
(A) a capacitor.
(B) an inductor.
(C) a neon lamp.

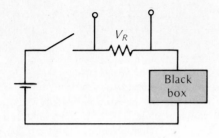

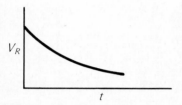

Figure 33-20. Question 33-7.

(D) a battery.
(E) a resistor.
(F) a resonant circuit.

· **33-8 Q** The circuit sketched in Figure 33-21 is used in a radio receiver. Which of the following best describes it?

(A) Since the current in the circuit is not in phase with the voltage across either the inductor or the capacitor, no energy will be dissipated in them and thus the detector will have maximum possible efficiency.

(B) This circuit is a demodulator, which enables us to separate the desired audio signal from the high-frequency carrier signal.

(C) This is a so-called flip-flop circuit, which inverts the polarity of the detected signal on alternate half-cycles.

(D) One adjusts the capacitance until the natural resonant frequency of the circuit matches the modulation frequency of the signal to be detected, thereby causing very large currents to flow in the resistance R.

(E) One adjusts the capacitance until the natural resonant frequency of the circuit matches the carrier frequency of the signal to be detected, thereby causing very large currents to flow in the resistance R.

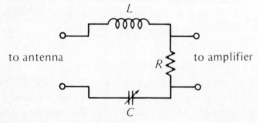

Figure 33-21. Question 33-8.

· **33-9 P** At a microwave frequency of 10 GHz, (a) what capacitance has a reactance of 100 Ω? (b) what inductance has a reactance of 100 Ω?

· **33-10 P** At what frequency will the phase angle be 30° for the circuit in Figure 33-22?

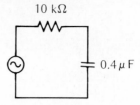

Figure 33-22. Problem 33-10.

· **33-11 P** What capacitance must be used with an inductance of 200 μH if the circuit is to resonate at 2.0 MHz?

: **33-12 P** The circuit shown in Figure 33-23 can be used to attenuate a narrow band of frequencies. Here R is the resistance of the inductor. Make a schematic sketch of V_{out} versus frequency and determine the ratio of the minimum value to the value far from resonance.

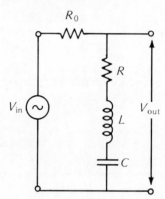

Figure 33-23. Problem 33-12.

: **33-13 P** Shown in Figure 33-24 is a high-pass filter, which attenuates low-frequency signals. (a) What is the voltage gain, V_{out}/V_{in}, for the filter? (b) Make a qualitative sketch of $\log (V_{out}/V_{in})$ versus angular frequency. (c) Show that the voltage gain is $\frac{1}{2}$ at the break-point frequency, $\omega_B = 1/RC$.

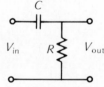

Figure 33-24. Problem 33-13.

: **33-14 P** In Figure 33-25 is diagrammed a low-pass filter, which attenuates high-frequency signals. (a) What is the voltage gain, V_{out}/V_{in}, of the circuit? (b) Make a qualitative sketch of $\log (V_{out}/V_{in})$ versus $\log \omega$. (c) Show that the voltage gain is $1/\sqrt{2}$ when $\omega = \omega_B = 1/RC$. This is called the break point of the curve found in (b).

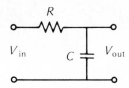

Figure 33-25.   Problem 33-14.

: **33-15 Q** The reason that inductive reactance varies proportionately with frequency is that
(A) as the frequency increases, the current decreases because electron motion is not able to respond to the driving force.
(B) more rapid changes in current induce larger back emf's.
(C) the potential difference across an inductance is proportional to the current through it.
(D) the more rapidly the current through an inductance changes, the more rapidly the voltage across it changes.
(E) capacitive reactance varies inversely with frequency, and the product $X_L X_C = 1/LC$ is independent of frequency.

: **33-16 P** A series RLC circuit has $R = 10\ \Omega$, $L = 1.0$ H, and $C = 20\ \mu F$. (a) For what angular frequency of applied emf will the current be a maximum? (b) At what angular frequencies will the current be half maximum?

: **33-17 P** An initially charged capacitor is connected across an inductor at time $t = 0$. The magnetic energy stored in the circuit first reaches a maximum at $t = 1\ \mu s$. At what time does the capacitor first return to its initial state of charge?

: **33-18 P** An emf of 20 V at a frequency of 2 kHz is applied to a series RLC circuit with $R = 10\ \Omega$, $L = 6.0$ mH, and $C = 2.4\ \mu F$. (a) For what applied frequency will a maximum current exist in the circuit? (b) Should capacitance be added in series or in parallel to the 2.4 μF to maximize the current? (c) What is this capacitance?

Section 33-4   Series *RLC* Circuit with Phasors

: **33-19 Q** A series RLC circuit is driven by an alternating applied potential difference. If V_R, V_C, and V_L represent the voltages across the resistor, capacitance, and inductor, then
(A) V_R, V_C, and V_L will all be in phase.
(B) V_C and V_L will always be 180° out of phase.
(C) the current will always be in phase with the applied emf.
(D) the current in the resistor will not be in phase with the current in the capacitor.
(E) equal amounts of power will be dissipated in the resistor, the capacitor, and the inductor.

: **33-20 Q** Consider a series RLC circuit driven by an alternating supplied emf. Quantities V_R, V_L, and V_C represent

the potential differences across the resistance, inductance, and capacitance and i is the current in the circuit. Then
(A) V_R, V_L, and V_C will all be in phase.
(B) i will always be in phase with the applied emf.
(C) i will always be in phase with V_R.
(D) i will always be in phase with V_L.
(E) i will always be in phase with V_C.

: **33-21 P** An RLC circuit has $R = 20\ \Omega$, $C = 4$ pF, $L = 1.0$ H. Calculate (a) the impedance at resonance; and (b) the impedance at 60 kHz. (c) At what other frequency besides 60 kHz will the impedance have the same value it has at 60 kHz?

: **33-22 P** A voltage $V_0 \sin \omega t$ is applied to the input terminals of the phase shifter in Figure 33-26. (a) What is the phase of the output voltage with respect to the input voltage? (b) What is the ratio of the output voltage amplitude to the input voltage amplitude?

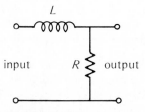

Figure 33-26.   Problem 33-22.

: **33-23 P** Draw the phasor diagram at resonance for a series RLC circuit, with $R = 10\ \Omega$, $C = 1\ \mu F$, $L = 20$ mH, to which a peak voltage of 12 V is applied.

: **33-24 P** At an instant when the applied voltage has the value zero in the circuit of Figure 33-27, what is the instantaneous voltage across each of the elements R, L, and C?

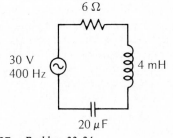

Figure 33-27.   Problem 33-24.

: **33-25 P** The impedance of a certain coil is 30 Ω at 100 Hz and 60 Ω at 500 Hz. (a) What is the inductance of the coil? (b) What is the resistance of the coil? (c) Find the phase angle between the current and the instantaneous voltage across the inductor for each of these frequencies.

: **33-26 P** The applied emf and potential differences for a series RLC circuit are shown in Figure 33-28. The resistance is 10 Ω. What is the capacitive reactance of the circuit?

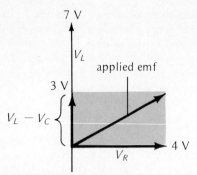

Figure 33-28.   *Problem 33-26.*

: 33-27 P A rudimentary tone control for a radio consists of a variable low-pass *RC* filter placed between the audio frequency source of ac and the load. In the circuit of Figure 33-29, the load is 600 Ω, and the filter resistance can be varied from 100 Ω to 2000 Ω. (*a*) At what frequency will there be equal currents in the load and through the capacitor when the filter resistance is 1000 Ω? (*b*) For a frequency of 400 Hz, sketch the ratio I_L/I_C as a function of filter resistance that is varied over its full range; I_L = load current and I_C = current in capacitor.

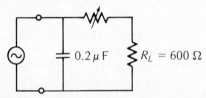

Figure 33-29.   *Problem 33-27.*

: 33-28 P A tunnel diode has the current-voltage characteristic shown in Figure 33-30a. It is used in the tunnel-diode oscillator circuit shown in Figure 33-30b. For the circuit to oscillate, what is the minimum value of $\mathcal{E}_0$ (in terms of i_a, V_a, and R)?

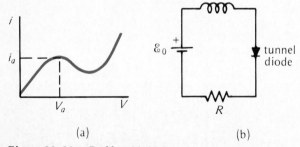

(a)   (b)

Figure 33-30.   *Problem 33-28.*

: 33-29 P A phasor diagram for a series *RLC* circuit is shown in Figure 33-31. From this picture we can see that the applied emf is
(A) 4 V.
(B) 6 V.
(C) 9 V.

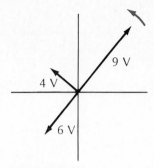

Figure 33-31.   *Problem 33-29.*

(D) more than 9 V.
(E) oscillating at a frequency below resonance.
(F) in phase with the current in the circuit.
(G) None of the above is true.

Section 33-5   RMS Values for AC Current and Voltage

· 33-30 P An emf with an rms value of 115 V is applied at a frequency of 50 Hz to a circuit. The voltage is a maximum at $t = 5$ ms. Write an expression giving the instantaneous emf as a function of time.

· 33-31 P A circuit has current $i = 2 \sin(288\pi t + 3\pi/8)$ ampere. (*a*) What is the peak current? (*b*) What is the rms current? (*c*) What is the frequency? (*d*) What is the phase of the current at $t = 1.0$ s?

· 33-32 P The current through a 2-mH inductance is $20 \sin 500t$ ampere. What is the rms voltage across the inductor?

: 33-33 Q Suppose that you apply an alternating potential difference to an inductor and measure the rms value of the current in the inductor. You find that the current decreases as you increase the frequency. The reason for this is that
(A) capacitive reactance varies inversely with frequency, and the product $X_L X_C = L/C$ is independent of frequency.
(B) the more rapidly the current in an inductor changes, the more rapidly the potential difference across it changes.
(C) energy is conserved.
(D) the induced emf in the inductor is proportional to the rate of change of the magnetic field in the inductor.
(E) in any tuned circuit, the current decreases as you move off resonance.

: 33-34 P An rms voltage of 12 V is applied at a frequency of 3200 Hz to a series *RLC* circuit with $R = 50$ Ω, $L = 2.0$ mH, and $C = 1.0$ μF. Determine the phase difference between the applied voltage and the current in the circuit. What is the rms current?

: 33-35 P Determine the rms voltages for each of the waveforms (*a*) square pulse, (*b*) rectified sine wave, and (*c*) ramp (saw-tooth) voltage shown in Figure 33-32.

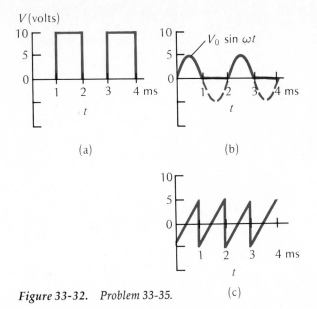

Figure 33-32.   Problem 33-35.

(a)   (b)   (c)

Section 33-6   Power in *AC* Circuits

· **33-36 P** What is the average power dissipated in a circuit where $i = 2 \cos \omega t$ ampere and $V = 12 \cos (\omega t + 60°)$ volt?

· **33-37 P** The current in a circuit is $i = 4 \sin (600\pi t + \frac{1}{4}\pi)$ ampere and the applied emf is $V = 60 \sin (600\pi t)$ volt. What is the average power delivered to the circuit?

· **33-38 P** A pure inductor of inductance 2.4 H is connected to a 60-Hz supply, with $V_{rms} = 115$ V. Find (*a*) the inductive reactance; (*b*) the rms current; (*c*) the maximum power delivered to the inductor; (*d*) the maximum energy stored in the inductor.

: **33-39 P** A series *RLC* circuit has $R = 8\ \Omega$, $L = 2.0$ mH, and $C = 30\ \mu F$. A 40-V ac potential difference is applied across the three elements in the circuit. What are the current and the phase angle when the applied frequency is (*a*) equal to the resonant frequency? (*b*) half the resonant frequency? (*c*) three-halves the resonant frequency?

: **33-40 Q** A series *RLC* tuned circuit is driven by the signal from an antenna in a radio receiver. A particular station is usually "tuned in" by varying the capacitance in the circuit. The purpose of doing this is to
(A) make $1/\omega C \gg \omega L$
(B) make $1/\omega C \ll \omega L$.
(C) maximize the current in the resistor.
(D) minimize the power dissipated in the resistor.
(E) decrease the time constant of the circuit and thereby increase the response speed.
(F) increase the phase difference between the current and the applied emf.

: **33-41 P** In a series *RLC* circuit, what happens to I_{max}, ω_0, and $\Delta\omega$, the width of the resonance, when (*a*) *L* is increased? (*b*) *C* is increased? (*c*) *R* is increased?

: **33-42 P** A dual-trace oscilloscope shows traces for the current and the applied emf in a circuit as shown in Figure 33-33. What is the average power delivered to the circuit?

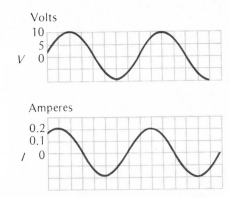

Figure 33-33.   Problem 33-42.

: **33-43 P** A 60-Hz rms voltage of 100 V rms is applied to a series *RLC* circuit. A current of 2.5 A rms exists in the circuit, and the rms voltages across the elements are 60 V across *R*, 40 V across *L*, and 120 V across *C*. (*a*) Find *R*, *L*, and *C*. (*b*) What is the power dissipated in *R*, in *L*, and in *C*?

: **33-44 P** Design an *RLC* circuit that has the following properties for $V_{rms} = 10$ V: $Q = 500$, $p_{max} = 2$ W (average power at resonance peak), and $\omega_0 = 10^4$ rad/s.

: **33-45 P** A generator applies a peak voltage of 100 V at a frequency of 400 Hz to an RLC circuit with $R = 10\ \Omega$, $L = 60$ mH, and $C = 20\ \mu F$. As the frequency is increased, the power dissipated in the resistor will
(A) not change.
(B) continuously decrease.
(C) decrease at first and then later increase.
(D) continuously increase.
(E) increase at first and then later decrease.
(F) None of the above is correct.

: **33-46 P** For a series *RLC* circuit near the resonance peak, express each of the following quantities in terms of the *Q* factor for the circuit: (*a*) The impedance Z. (*b*) The phase angle. (*c*) Magnitude of *i*. (*d*) Power dissipated. (*e*) The reactance *X*.

: **33-47 P** Show that the curve of current amplitude versus frequency is broader at half-maximum than is the curve of time average power dissipated versus frequency.

Section 33-7   The Transformer
: **33-48 Q** A certain transformer has thin primary wires

but thick secondary wires. Which of the following is the most probable statement concerning the transformer's characteristics?

(A) The transformer may operate on ac or dc.

(B) The output power of the transformer exceeds the input power.

(C) The primary and secondary voltages of the transformer are equal.

(D) Any device connected to the secondary operates on less than 110 V.

(E) The transformer probably steps up the voltage delivered to the apparatus.

Maxwell's Equations

34

The fundamentals of electromagnetism tell how electric and magnetic fields are related to one another and to electric charge and current. These fundamentals are summarized by Gauss's laws for electricity and magnetism, Ampère's law, and Faraday's law.* The four mathematical relations expressing these laws are known as Maxwell's equations because of the epochal contributions around 1865 of the Scottish physicist, James Clerk Maxwell (1831–1879). More specifically,

• Maxwell showed that Ampère's law as it had been formulated up to that time was incomplete, and he suggested how to remedy the shortcoming.

• He formulated for the first time in precise and comprehensive mathematical language what are now known as Maxwell's equations.

• Maxwell predicted that electric and magnetic fields, unattached to electric charge, propagate at the speed of light through otherwise empty space as electromagnetic waves.

34-1 The General Form of Ampère's Law

According to Ampère's law, a magnetic field **B** is created by an electric current i. More specifically, the line integral $\oint \mathbf{B} \cdot d\mathbf{l}$ around any closed loop equals the

* Sections 24-4, 30-4, 30-5, and 31-4.

net current i (multiplied by constant μ_0) crossing *any surface* bounded by the loop:

$$\oint \mathbf{B} \cdot d\mathbf{l} = \mu_0 i \tag{30-9}$$

As given above, Ampère's law recognizes only one source of a magnetic field: moving electric charges. Here we shall see that Ampère's law must be generalized to include another source of magnetic field: a changing electric field (or more precisely, a changing electric flux).

Consider the simple circuit shown in Figure 34-1. Here a capacitor is being charged by connecting it across a battery with the closing of a switch. Immediately after the switch is closed we have this transient effect: the current rises from zero to some maximum value and then falls to zero again when the capacitor has become fully charged. We are interested in that interval during which the current in the circuit is changing. Actually, to say it a little more carefully, we are interested in the time during which the current *through the conducting wire* is changing, since a real current does not exist at any time in the region between the capacitor plates. After all, no charged particles ever pass through this region.

Suppose that we apply the equation to the circular loop shown in Figure 34-2(a); the loop is centered on the conducting wire and is perpendicular to it. The simplest surface bounded by the circular loop is the flat circular plate shown in Figure 34-2(a). Elecric current penetrates this surface, and according to Ampère's law, there is then a magnetic field circling in the loop around the conductor. But recall that Ampère's law relates the line integral of the magnetic field around a chosen loop to the net current through a closed surface of any shape bounded by that loop (Section 30-4). Suppose then that we keep the

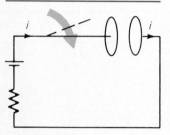

Figure 34-1. *A parallel-plate capacitor being charged by switching it across a battery. Current i changes with time.*

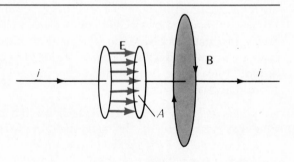

(a)

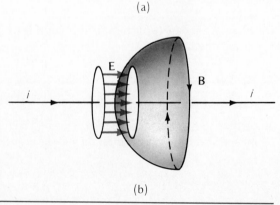

(b)

Figure 34-2. *Ampère's law applied to a battery being charged. (a) The loop is circular and the surface flat. A magnetic field **B** circles the loop, and a real current passes through the surface. (b) The same circular loop as in part (a) but with a surface that passes between the capacitor plates. A magnetic field **B** still circles the loop but no real current passes through the surface.*

circular loop at the location shown in Figure 34-2(a) but choose the surface to be a hemisphere that passes through the region between the capacitor plates. The real current through the hemispherical surface is now zero. But there must still be a magnetic field at the circular loop bounding this surface, just as before. In short, Ampère's law as given thus far does not work when it is applied to the situation shown in Figure 34-2(b). What must be done to fix it?

Enter Maxwell. He conjectured that when there is no real current, a changing electric flux can act as an effective current, or *displacement current*, in creating magnetic field loops.* Why a changing electric flux? Maxwell's argument was basically esthetic, one based on symmetry: if a changing magnetic flux creates electric-field loops (Faraday's law), should we not expect that a changing electric flux might create magnetic-field loops?

To allow for displacement current i_d, as well as real current i, we write Ampère's law as

$$\oint \mathbf{B} \cdot d\mathbf{l} = \mu_0(i + i_d) \tag{34-1}$$

For the situation shown in Figure 34-2 we must have, because of the continuity of current, that

$$i \text{ (in conductor)} = i_d \text{ (between capacitor plates)}$$

Electric flux ϕ_E is defined by

$$\phi_E \equiv \oint \mathbf{E} \cdot d\mathbf{S} \tag{34-2}$$

where $\mathbf{E}$ is the electric field through surface element $d\mathbf{S}$.

Instantaneous current $i = dq/dt$ gives the rate at which charges pass through any cross section of the conductor; dq/dt is also the rate at which charges accumulate on a capacitor plate. The uniform electric field $\mathbf{E}$ between capacitor plates with area A is related to the charge q on one plate by

$$E = \frac{q}{\epsilon_0 A} \tag{24-6}$$

so that $q = \epsilon_0 EA = \epsilon_0 \phi_E$ where we have identified EA as the electric flux ϕ_E through a transverse area between the capacitor plates.

The displacement current is then given by

$$i_d = i = \frac{dq}{dt} = \frac{d}{dt}(\epsilon_0 \phi_E) = \epsilon_0 \frac{d\phi_E}{dt}$$

Using this result in (34-1), we get for the generalized form of Ampère's law

$$\oint \mathbf{B} \cdot d\mathbf{l} = \mu_0 \left(i + \epsilon_0 \frac{d\phi_E}{dt} \right) = \mu_0 i + \epsilon_0 \mu_0 \frac{d\phi_E}{dt} \tag{34-3}$$

This relation implies that magnetic field loops can be produced by a changing electric flux even in the absence of electric charges. Magnetic field loops can of course be created by either a steady current or a changing real current, but only a *changing* electric flux creates $\mathbf{B}$ loops. Although "derived" from the special

* The term *displacement current* originated in the nineteenth-century concept of an invisible medium, or ether, pervading all space. The ether concept has been thoroughly discredited and has been discarded, but the term *displacement current* lingers on.

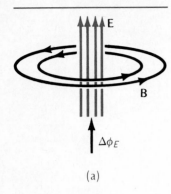

(a)

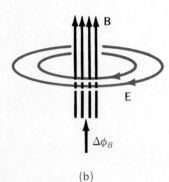

(b)

Figure 34-3. *(a)* **B** *loops surrounding a region in which electric flux* ϕ_E *is changing (right-hand rule); (b)* **E** *loops surrounding a region in which magnetic flux* ϕ_B *is changing (left-hand rule).*

case of a parallel-plate capacitor, the generalized form of Ampère's law, with the effects of changing electric flux included, holds in all situations.

Suppose some region of space is entirely free of electric charges and currents. Then a magnetic field can be created there only by a changing electric flux, and (34-3) reduces to

$$\oint \mathbf{B} \cdot d\mathbf{l} = \epsilon_0 \mu_0 \frac{d\phi_E}{dt} \qquad (34\text{-}4)$$

This relation is similar in form to Faraday's law, in which an electric field is created by a changing magnetic flux:

$$\oint \mathbf{E} \cdot d\mathbf{l} = -\frac{d\phi_B}{dt} \qquad (31\text{-}6), (34\text{-}5)$$

The two situations are illustrated in Figure 34-3. In Figure 34-3(a), **B** loops circle a region in which ϕ_E is changing; in Figure 34-3(b), **E** loops circle a region in which ϕ_B is changing. The sense of the **B** loops is related to the direction in which ϕ_E increases by a *right*-hand rule, just as the sense of **B** loops is related to the direction of real current by a right-hand rule. On the other hand, the **E** loops in Figure 34-3(b) surrounding a region in which ϕ_B is changing is related to the direction of ϕ_B change by a *left*-hand rule (a consequence of Lenz's law). This difference in the sense of the field loops is reflected in (34-4) and (34-5); one has a plus sign, the other a minus sign.

Because of the close parallel between (34-4) and (34-5), the conclusions we drew earlier from Faraday's law concerning the **E** field accompanying a change in ϕ_B (Section 31-4) hold equally well for the **B** field accompanying a change in ϕ_E. For example, the line integral $\oint \mathbf{B} \cdot d\mathbf{l}$ of the magnetic field depends only on the time rate of change of the total electric flux through a surface enclosed by the path about which we evaluate the line integral, not on whether there is flux at every interior point within the loop. Thus, if we choose concentric circular paths of various radii outside a region in which the electric flux is changing, the integral $\oint \mathbf{B} \cdot d\mathbf{l}$ is the same for all such closed paths. This implies that the magnitude of **B** falls off inversely with r, where r is the radius of the path and also the distance from the center of the region of changing electric flux. This result corresponds exactly to the fact that the magnetic field from a long straight wire falls off inversely with the distance from the wire.

Example 34-1. A parallel-plate capacitor with circular plates of radius R is being charged at a constant rate. Assume that the electric field is confined entirely to the region between the capacitor plates; see Figure 34-4. What is the magnitude of the magnetic field **B** at any distance r from the center of the capacitor plates?

We worked an exactly analogous problem with Faraday's law (as Example 31-7), so that we need not go through it again in detail. Comparing (34-4) with (34-5), we see that we need merely interchange **E** and **B** and introduce the additional factor $\epsilon_0 \mu_0$. The results are shown in Figure 34-4(c). The magnitude of **B** increases linearly with r for $r < R$ and decreases inversely with r for $r > R$. At $r = R$, $B = \frac{1}{2}\epsilon_0 \mu_0 R(dE/dt)$.

To get an idea of the magnitude of the magnetic field induced by the changing electric flux, suppose that the capacitor-plate radius R is 10 cm and that the electric field changes at the rate $dE/dt = 10^{10}$ V·m/s (corresponding to a time rate of change in electric potential difference across the capacitor of 10 V/μs for capacitor plates separated by 1 mm). Then the maximum magnetic field (at $r = B$) is

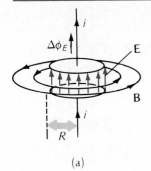

(a)

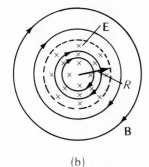

(b)

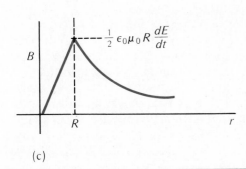

(c)

Figure 34-4. (a) Parallel-plate
capacitor of radius R with a
changing electric field **E**
producing magnetic field **B.** (b)
Magnetic field loops corresponding to (a). (c) Magnetic
field magnitude B as a function
of distance r from the center.

$$B = \frac{1}{2} \epsilon_0 \mu_0 R \frac{dE}{dt}$$

$$= \tfrac{1}{2}(8.9 \times 10^{-12} \text{ C}^2/\text{N}\cdot\text{m}^2)(4\pi \times 10^{-7} \text{ T}\cdot\text{m/A})(0.10 \text{ m})(10^{10} \text{ V/m}\cdot\text{s})$$

$$= 5.6 \times 10^{-9} \text{ T} = 0.056 \text{ mG}$$

Even for the relatively large rate of electric-flux change in this example, the induced
magnetic field is very small indeed. This contrasts with induced electric fields produced
by a changing magnetic flux, where emf's of the order of volts are relatively easily
obtained.

34-2 Maxwell's Equations

Faraday invented the field concept as a useful and picturesque means of
visualizing electric and magnetic effects. Maxwell took the electric and mag-

The Odd Couple

He was something of a fitness freak. While he was in college at Cambridge University he used, at least for a while, to jog an hour every day. He would repeatedly run along the dormitory corridor, down a flight of steps, back along a lower corridor, and up stairs to finish the loop—starting at 2:00 AM.

James Clerk Maxwell (1831–1879) is remembered especially for fixing Ampère's law, for the equations in classical electromagnetism that bear his name, and for his prediction of electromagnetic waves at the speed of light. But he had many other accomplishments. Born to a family of very comfortable means with a country estate in Scotland, Maxwell showed extraordinary talent at an early age. He was only 15 when he submitted his first paper to the Edinburgh Royal Society. Its title: "On the Description of Oval Curves and Those Having a Plurality of Foci." His college tutor remarked that, "It appears impossible for him to think incorrectly on physical subjects."

Maxwell was only 26 when he won the prestigious Adams prize for the best paper on a scientific subject; Maxwell's was on the rings of Saturn, a topic that would concern him for many years. He was, first of all, a mathematical physicist; he deduced, strictly on theoretical grounds, the distribution of molecular speeds (henceforth to be known as the Maxwell dis-

tribution). But he also did experiments: measuring both the viscosity of gases and the standard ohm. He became director of the Cavendish Laboratory at Cambridge University at its founding, which was to become a world center for advances in atomic and nuclear physics.

One of Maxwell's delightful inventions has become known as the *Maxwell demon*. Visualize gas in thermal equilibrium in a container with a partition between the two halves. A small hole in the partition allows a molecule headed for the hole to pass from one side to the other. The demon sits near the hole with a paddle, and he keeps his eye on approaching molecules. If a fast molecule is heading toward the left half, he lets it pass through the hole, but if he sees that a slow molecule would go through the hole to the left he blocks it with his paddle. In similar fashion, he keeps fast molecules from going from the left to the right side, but allows slow ones to pass through. The upshot, of course, is that eventually the left side has mostly fast molecules and is at a higher temperature than the right side. It costs almost no energy for the demon to put the tiny paddle over the hole or

netic fields seriously, and he developed the general mathematical expressions for their properties and interrelations. These four fundamental relations, which say everything there is to say about classical electromagnetism, are known as *Maxwell's equations*.*

---

* *Classical* electromagnetism excludes quantum effects, especially the basic quantum phenomenon that electromagnetic radiation consists not of continuous fields, but of discrete particlelike photons. Photons are discussed in Chapter 41.

remove it, so that Maxwell's demon has foiled the second law of thermodynamics!

Well, not quite. For the demon to perform his feat, he needs to know whether a molecule approaching the hole is a fast one or a slow one. He needs information. Actually, information can be contrasted with ignorance in the same way that we contrast order, or low entropy, with disorder, noise, or high entropy. Indeed, information theory, a fundamental discipline that relates to the transmission of intelligence, is altogether analogous to the statistical thermodynamics that Maxwell knew so well.

Maxwell also acknowledged his profound indebtedness to Faraday for clarifying a wide variety of electromagnetic effects through his experiments and especially for Faraday's championing what has ever after been a central concept in physics—the field.

Michael Faraday (1791–1867) had humble beginnings; his father was a blacksmith. Faraday was first apprenticed to a bookseller, and within a couple of years he had two boys working for him. But his real interest was chemistry, on which he had read considerably. Faraday's big break came when he was 22. He wanted to become an assistant to Sir Humphrey

Davy, chemist of the Royal Institution, so he wrote up and bound very neatly notes he had kept while attending public lectures by Davy.

In Faraday's words, "My desire to escape from trade, which I thought vicious and selfish, and to enter into the service of Science, which I imagined made its pursuers amiable and liberal, induced me at last to take the bold and simple step of writing to Sir H. Davy, expressing my wishes . . . at the same time I sent the notes I had taken of his lectures."

Faraday landed the job. At first his duties were fairly onerous—washing out test tubes, and that sort of thing. But, by looking over Davy's shoulder and diligent self-study, Faraday soon became an expert chemist himself. Indeed, he is acknowledged to have been perhaps the most talented experimentalist in chemistry and physics of all time.

Everybody knows about Faraday's discovery of and interpretation of electromagnetic induction. It is easy to miss what a remarkable accomplishment it really was. After all, Faraday had to do everything from scratch—make the batteries, the meters, and even the conducting wire. Faraday never really studied mathematics, and he was a bit suspicious about it. He was not really sure that he understood what Maxwell had done. Faraday concentrated on electric and magnetic fields; for him the fields were the most real part of electromagnetism. But he did other things of great significance. He clarified electrolysis and the notions of atomic mass and valence. He showed by experiment that a magnetic field can change the direction of polarization of a beam of light, the very first direct evidence that light and electromagnetism are closely related.

Faraday did other things besides physics and chemistry. Among his special enthusiasms were poetry, the beauties of nature, various games, acrobatics, Punch and Judy shows, children. He never lost his sense of joy and wonder, not only in the laboratory, but in everything around him.

Let us first be clear on the meanings of electric field **E** and magnetic field **B**. They are given by the relation for the total force on an electric charge q from electric and magnetic fields:

$$\mathbf{F} = q(\mathbf{E} + \mathbf{v} \times \mathbf{B}) \tag{29-11}$$

This equation defines **E** and **B**. That part of the force $q\mathbf{v} \times \mathbf{B}$ that depends on the charge's velocity **v** is the magnetic force; the remaining part $q\mathbf{E}$ is the electric force.

Table 34-1

NAME	EQUATION	EXPERIMENTAL EVIDENCE
Gauss's law for electricity (34-6a)	$\epsilon_0 \oint \mathbf{E} \cdot d\mathbf{S} = q$	Electric (or Coulomb) force is inverse-square; no net charge on interior of hollow charged conductor under steady-state conditions. **E** lines either originate from and terminate on charges, or form closed loops.
Gauss's law for magnetism (34-7b)	$\oint \mathbf{B} \cdot d\mathbf{S} = 0$	No isolated magnetic poles. **B** lines form closed loops.
Faraday's law **(34-6c)**	$\oint \mathbf{E} \cdot d\mathbf{l} = -\dfrac{d\phi_B}{dt}$	Electromagnetic induction effects.
Ampère's law **(34-6d)**	$\oint \mathbf{B} \cdot d\mathbf{l} = \mu_0 i + \epsilon_0 \mu_0 \dfrac{d\phi_E}{dt}$	Magnetic force between current-carrying conductors; electromagnetic waves.

Table 34-1 lists the four Maxwell equations for charged particles and currents in a vacuum and gives the common name and the primary experimental evidence for each.

Gauss's Law for Electricity Gauss's law for electricity says that the net electric flux through any closed surface depends only on the net charge within:

$$\epsilon_0 \oint \mathbf{E} \cdot d\mathbf{S} = q \tag{34-6a}$$

Gauss's law (Section 24-4) can be derived from Coulomb's law; indeed Gauss's law for electricity is merely an alternative way of saying that the force between point charges varies inversely with the square of the distance between them. Gauss's law is confirmed by experiments showing that the Coulomb force is inverse-square. A more precise verification of Gauss's law comes from the observation that for static conditions there is never any net charge within a conductor.

Gauss's Law for Magnetism Gauss's law for magnetism says that the net magnetic flux through any closed surface is always zero:

$$\oint \mathbf{B} \cdot d\mathbf{S} = 0 \tag{34-6b}$$

If it were not zero, one would have "magnetic charges," or single magnetic monopoles, on which magnetic field lines would originate and terminate. Gauss's law for magnetism is based on the observation that isolated magnetic poles do not exist in nature.

Faraday's Law Faraday's law of electromagnetic induction says that a changing magnetic flux generates electric-field loops:

$$\oint \mathbf{E} \cdot d\mathbf{l} = -\frac{d\phi_B}{dt} \tag{34-6c}$$

Faraday's law is confirmed by electromagnetic induction effects, for example, a current can be induced in a conducting loop by a nearby and separate conducting loop in which the current is changing.

Ampère's Law According to Ampère's law, a magnetic field has two origins: electric current i, from electric charges in motion, and changing electric flux (displacement current):

$$\oint \mathbf{B} \cdot d\mathbf{l} = \mu_0 i + \epsilon_0 \mu_0 \frac{d\phi_E}{dt} \qquad \text{(34-6d)}$$

The magnetic force between straight, parallel current-carrying conductors, varying inversely with the separation distance, is a direct experimental test of the magnetic field originating from moving charges. The most convincing evidence that a magnetic field is generated also by a changing electric flux comes from the observed properties of electromagnetic waves.

The equations for Ampère's law and Faraday's law are very nearly symmetrical, but not completely. We do not have in Faraday's law a term corresponding to the current i. This merely reflects the fact that since isolated magnetic poles do not exist, there can be no "magnetic current" arising from magnetic poles in motion. Since all electromagnetic phenomena can be accounted for without magnetic monopoles, we see that in this instance Nature chose economy over symmetry.

Other relations in electromagnetism are not fundamental. Ohm's law, for example, merely describes the properties of certain conducting materials.

34-3 Electromagnetic Waves from Maxwell's Equations (Optional)

One of the most notable predictions of theoretical physics of all time was that made by James Clerk Maxwell in 1865.* Maxwell predicted from the four equations of classical electromagnetism that electric and magnetic fields may exist in space far removed from electric charges and currents and that these fields propagate at the speed of light as electromagnetic waves. For empty space, with both q and i equal to zero, Maxwell's equations (34-6) reduce to the following:

$$\oint \mathbf{E} \cdot d\mathbf{S} = 0 \qquad \oint \mathbf{E} \cdot d\mathbf{l} = -\frac{d\phi_B}{dt}$$

$$\oint \mathbf{B} \cdot d\mathbf{S} = 0 \qquad \oint \mathbf{B} \cdot d\mathbf{l} = \epsilon_0 \mu_0 \frac{d\phi_E}{dt}$$

The first two equations say, in effect, that both electric and magnetic fields must form closed loops. The third equation (Faraday's law) says that a changing magnetic field (strictly, a changing magnetic flux) generates an electric field, and the fourth equation (Ampère's law) says just the reverse, that a changing electric flux generates a magnetic field.

The derivation that follows shows that electric and magnetic fields unat-

* The principal results of this section are summarized in Section 35-1.

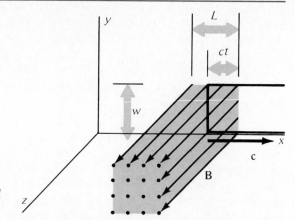

*Figure 34-5. Constant magnetic field **B** along +z, of thickness L, advancing at speed c along +x.*

tached to electric charges and currents do propagate through empty space at the speed of light. The strategy will be this:

- We first suppose that a pulse of **B** field can travel through space and we find that an **E** field must accompany it.
- Then we just do the reverse: imagine an **E** pulse to travel through space and find that **B** must accompany it.

Suppose then that a constant pulse of magnetic field **B** directed along z moves through otherwise empty space at speed c along the x axis. See Figure 34-5. How such a magnetic field can be separated from currents and launched into space does not concern us for the moment.

The thickness of the **B** pulse from its leading to its trailing edge is L; along the y and z axes we take **B** to be effectively infinite in extent.* Outside this region, $\mathbf{B} = 0$.

Now consider an imaginary rectangular loop lying in the xy plane. The loop has width w along the y direction and is indefinitely long along x. We suppose that the leading edge of the magnetic pulse reaches the left end of the loop at time $t = 0$. Then, after a time interval t has elapsed, the pulse will have progressed into the loop a distance ct. The magnetic flux through the loop is changing so that there must be an induced electric field whose magnitude we get from Faraday's law.

The magnetic flux penetrating the loop at time t is

$$\phi_B = BA = B(wct)$$

The rate at which the flux changes is, then,

$$\frac{d\phi_B}{dt} = Bwc$$

After a time $t = L/c$, the trailing edge of the magnetic pulse will have entered the loop, and thereafter ϕ_B will be constant. The induced electric field is in the y

---

* The **B** field cannot be strictly uniform in magnitude and truly infinite in extent because the **B** lines must form closed loops. We are here considering a limited region of space in which the **B** lines are effectively straight.

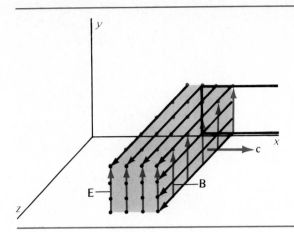

Figure 34-6. *The magnetic field* **B** *of Figure 34-5 is accompanied by a transverse electric field* **E** *along* $+y$.

direction. The only contribution to the line integral $\oint \mathbf{E} \cdot d\mathbf{l}$ comes from the left end of the loop:

$$\oint \mathbf{E} \cdot d\mathbf{l} = -Ew$$

Equating $\oint \mathbf{E} \cdot d\mathbf{l}$ and $-d\phi_B/dt$, as required by Faraday's law, then yields

$$Ew = Bwc$$

$$B = \frac{E}{c} \tag{34-7}$$

The magnetic field is smaller than the electric field by the factor c.

What is the direction of **E**? We get that from Lenz's law. Imagine the loop to be conducting; then an induced current must, from Lenz's law, circulate in the clockwise sense. This means, in turn, that at the loop's left end, **E** points in the $+y$ direction. Now **E** exists only so long as the flux ϕ_B is changing. We see that the original magnetic pulse must be accompanied by an electric pulse extending over the same region of space, as shown in Figure 34-6. The **E** and **B** fields are mutually perpendicular, and both fields are perpendicular to the direction in which what we must call the electromagnetic pulse is traveling.

Let us review what we have thus far. We began with a moving **B** and found, from Faraday's law, that **E** must accompany it. Now we do just the reverse: start with an **E** pulse and find, from Ampère's law, that **B** must accompany it.

Figure 34-7(a) shows an electric pulse **E** pointed along the $+y$ direction traveling at speed c along the $+x$ direction. We consider an imaginary indefinitely long rectangular loop of width l lying in the xz plane. The distance from the leading to trailing edge of the pulse is again L. We use Ampère's law to find the magnetic field produced by the changing electric flux ϕ_E through the loop. The electric flux penetrating the loop at time t is

$$\phi_E = EA = E(lct)$$

so that the electric flux changes at the rate

$$\frac{d\phi_E}{dt} = Elc$$

This flux changes only so long as the trailing edge is outside the loop; once the

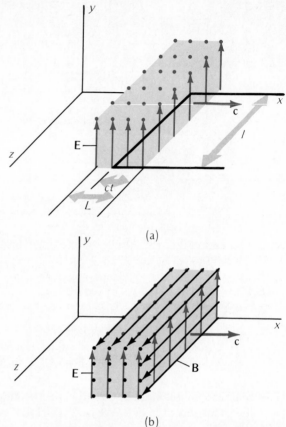

Figure 34-7. (a) Constant electric field **E** along $+y$, of thickness L, advancing at speed c along $+x$. (b) The electric field is accompanied by a transverse magnetic field along $+z$. Compare with Figure 34-6.

pulse is entirely within the loop and ϕ_E is constant, we no longer have a magnetic field induced.

The only contribution from the line integral $\oint \mathbf{B} \cdot d\mathbf{l}$ comes again from the left end of the loop, so that we have

$$\oint \mathbf{B} \cdot d\mathbf{l} = Bl$$

Substituting these results in Ampère's law then gives

$$\oint \mathbf{B} \cdot d\mathbf{l} = Bl = \epsilon_0 \mu_0 \frac{d\phi_E}{dt}$$

$$Bl = \epsilon_0 \mu_0 Elc$$

$$B = \epsilon_0 \mu_0 Ec \tag{34-8}$$

The direction of **B** at the loop's left end is along $+z$. This result follows from taking an electric flux to be equivalent to a current and then applying the right-hand rule. Again the induced **B** field is confined entirely to the region of the traveling **E** pulse. We see, moreover, by comparing Figures 34-6 and 34-7(b) that the relative directions of **E**, of **B**, and of pulse propagation are the *same*. The magnitudes must also be consistent. We use (34-7) in (34-8) and find that

$$\frac{E}{c} = \epsilon_0 \mu_0 Ec$$

$$c = \frac{1}{\sqrt{\epsilon_0 \mu_0}} \qquad\qquad (34\text{-}9)$$

The speed of the electromagnetic pulse depends solely on the two fundamental constants of electromagnetism:

$$\epsilon_0 = 8.854\ 187\ 82 \times 10^{-12}\ \mathrm{C}^2/\mathrm{m}^2 \cdot \mathrm{N}$$

$$\mu_0 \equiv 4\pi \times 10^{-7}\ \mathrm{N} \cdot \mathrm{s}^2/\mathrm{C}^2$$

so that (34-9) yields*

$$c \equiv 2.997\ 924\ 58 \times 10^8\ \mathrm{m/s} \approx 3.00 \times 10^8\ \mathrm{m/s}$$

The speed of electromagnetic waves is the same as the speed of light. More fundamentally, light is one type of electromagnetic wave.

In the preceding derivation, it was assumed that the electric and magnetic fields change with time in a very simple way; the fields are "turned on and off" abruptly at the leading and trailing edges of a constant pulse. Far more typical is an electromagnetic wave in which the electric and magnetic fields vary continuously with time, especially monochromatic sinusoidal waves. But any more general waveform involves no basically new considerations, however, since any continuously varying field can be approximated by a succession of short constant pulses, or step functions.

* The speed of light is now *assigned* its numerical value because the meter is *defined* in terms of c. The value for ϵ_0 is *computed*, using (34-9), from c and the assigned value for μ_0.

Summary

Definitions

Electric flux: the flux $d\phi_E$ of the electric field $\mathbf{E}$ through a surface element $d\mathbf{S}$ is

$$d\phi_E \equiv \mathbf{E} \cdot d\mathbf{S} \qquad (34\text{-}2)$$

Fundamental Principles

Ampère's law in complete form is

$$\oint \mathbf{B} \cdot d\mathbf{l} = \mu_0 i + \epsilon_0 \mu_0 \frac{d\phi_E}{dt} \qquad (34\text{-}3)$$

The second term on the right implies that a changing electric flux, as well as an electric current i, creates magnetic field loops. The quantity $\epsilon_0 d\phi_E/dt$ is sometimes referred to as the displacement current.

Important Results

Maxwell's equations for classical electromagnetism: see Table 34-1.

Properties of electromagnetic waves are summarized in Section 35-1.

Problems and Questions

Section 34-1   The General Form of Ampère's Law

· **34-1 Q** Which of the following is a true statement concerning displacement current?

(A) It always has the same magnitude as a nearby conduction current.

(B) It is a magnetic current, whereas the conduction current is an electric current.

(C) Its importance is relatively greater at lower frequencies.

(D) Its importance is relatively greater at higher frequencies.

(E) It is a measure of the rate at which charge is transported across the gap in a parallel-plate capacitor.

· **34-2 Q** Which of the following has a meaning closest to that of "displacement current"?

(A) The current associated with the small displacements of bound charges in a dielectric medium.

(B) The rate of change of electric flux.

(C) The rate of change of magnetic flux.

(D) A current that is changing with time, as opposed to a "dc" current.

(E) The current associated with the motion of charged particles through a vacuum.

· **34-3 Q** Maxwell, in his equations, introduced the concept of the displacement current and the resulting term $\epsilon_0(d\phi_E/dt)$, using
(A) an intuitive guess.
(B) experimental evidence of its existence.
(C) the mathematical necessity of its inclusion to make the generalized equation for Ampère's law consistent.
(D) dimensional analysis.
(E) an attempt to explain the absence of magnetic monopoles in nature.

· **34-4 P** Show that $\epsilon_0(d\phi/dt)$ has the dimensions of amperes.

· **34-5 P** Show that the displacement current in a parallel-plate capacitor is given by the expression $i_d = C(dV/dt)$, where C is the capacitance and dV/dt the rate at which the potential difference between the plates changes with time.

· **34-6 P** What is the magnitude of the maximum displacement current in a capacitor consisting of two parallel plates, each of area 0.25 m², separated by an air gap of 1.0 mm, when a sinusoidal potential difference with amplitude $V_m = 2.5$ kV and at frequency 1.0 kHz is applied across the capacitor plates?

⋮ **34-7 P** An alternating voltage of 10 V rms is applied at frequency f to an RC circuit with $R = 10$ Ω, $C = 20$ μF. Determine the real rms current in R and the displacement rms current in C for (a) $f = 60$ Hz and (b) $f = 2$ MHz.

⋮ **34-8 P** An AC voltage of 100 V (rms) is applied at angular frequency ω to a resistor $R = 2$ Ω in series with capacitance $C = 10$ μF. The parallel-plate capacitor is air filled and has circular plates of radius 12 cm separated by 1.0 mm. Determine the magnetic and electric fields (rms values) at a distance of 10 cm from the center of the capacitor for (a) $\omega = 400$ s⁻¹ and (b) $\omega = 6.0 \times 10^6$ s⁻¹.

Section 34-2   Maxwell's Equations

· **34-9 Q** Fields **E** and **B** are defined by
(A) the energy they store.
(B) potential differences and currents.
(C) the forces they exert on electric charges.
(D) vector fields.
(E) two of the fundamental constants of nature, ϵ_0 and μ_0.

· **34-10 Q** The equation $\oint \mathbf{B} \cdot d\mathbf{S} = 0$ leads us to the conclusion that
(A) the magnetic field **B** is zero.
(B) magnetic field lines have no beginnings or ends; that is, they end on themselves.

(C) magnetic field lines have constant separation; that is, they are parallel lines.
(D) the universe is comprises equal numbers of positive and negative magnetic charges.

⋮ **34-11 Q** Are Maxwell's equations fully symmetric for **E** and **B**?
(A) Yes. They are symmetric in all respects.
(B) No, because **B** exerts a force perpendicular to **B** lines, whereas **E** exerts a force along **E** lines.
(C) No, because there is no source of **E** analogous to the displacement current as a source of **B**.
(D) No, because no magnetic monopoles have been discovered.
(E) No, because they are measured in different units.

⋮ **34-12 Q** In the comparison with fluid flow, we can think of magnetic-field lines as analogous to
(A) the flow lines when sources and sinks are present.
(B) the flow lines when energy is not conserved in a viscous fluid.
(C) lines of constant pressure in a fluid.
(D) the flow lines describing flow in a whirlpool.

⋮ **34-13 Q** An important difference between electric and magnetic fields is that
(A) magnetic fields do not act on electrically charged particles, whereas electric fields do.
(B) a magnetic field is created by a changing electric flux, but an electric field is not created by a changing magnetic flux.
(C) the laws describing magnetic phenomena are deduced from the experimental observation of electric phenomena, but the reverse is not true.
(D) magnetic fields due to a current of electric charge have been observed in nature, but electric fields due to a current of magnetic charge have not been found.
(E) one field stores energy, whereas the other does not.

⋮ **34-14 Q** The law $q = \epsilon_0 \int \mathbf{E} \cdot d\mathbf{S}$ is equivalent to
(A) Ampère's law.
(B) the fact that a time-varying magnetic flux creates an electric field.
(C) the fact that the force between two charges varies inversely with the square of the distance between them.
(D) a statement that all the flux lines into a closed volume must also flow out.
(E) a statement of conservation of charge.

⋮ **34-15 P** Suppose that magnetic monopoles were discovered and their existence well confirmed experimentally. Let q_m be the "magnetic charge" analogous to the electric charge q. (a) What units would q_m have? (b) Write new Maxwell equations incorporating the magnetic monopoles.

Electromagnetic Waves

35

35-1   Basic Properties of Electromagnetic Waves

For those who skipped the derivation of the basic properties of electromagnetic waves from Maxwell's equations (Optional Section 34-3), here are some important results:

- Electromagnetic (abbreviated EM) waves can exist in otherwise empty space because a changing electric flux creates a magnetic field (Ampère's law) and a changing magnetic flux creates an electric field (Faraday's law).
- EM waves are *transverse.* At each instant and in each location in space, the instantaneous **E** and **B** fields are mutually perpendicular vectors that lie in a plane transverse to the direction of wave propagation. See Figure 35-1.
- The speed of an EM wave through a vacuum is

$$c = \frac{1}{\sqrt{\epsilon_0 \mu_0}} \equiv 2.997\ 924\ 58 \times 10^8 \text{ m/s} \simeq 3.00 \times 10^8 \text{ m/s} \quad (35\text{-}1)$$

- The relative magnitudes of **E** and **B** at each instant and location in space are

$$B = \frac{E}{c} \quad (35\text{-}2)$$

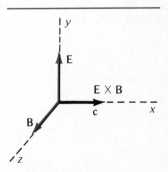

Figure 35-1. *Relative directions for an electromagnetic wave of electric field* **E**, *magnetic field* **B**, *and direction of wave propagation* **c**.

751

35-2   Sinusoidal Electromagnetic Waves and the Electromagnetic Spectrum

The simplest EM wave is a monochromatic, sinusoidal, plane wave (Section 17-5).

For a sinusoidal wave, **E** varies sinusoidally with time at each location. Furthermore, at each instant, the magnitude of **E** varies sinusoidally with position along the direction of wave propagation. Magnetic field **B** varies in similar fashion. A wave traveling along the $+x$ axis can be represented (Section 17-5) by

$$E_y = E_0 \sin k(x - ct)$$
$$B_z = B_0 \sin k(x - ct) \tag{35-3}$$

with the other rectangular components of **E** and **B** equal to zero.*

Wave number k is

$$k = \frac{2\pi}{\lambda} \tag{17-5}$$

and the wavelength λ and frequency v are related by

$$c = v\lambda \tag{17-1}$$

Figure 35-2 shows the EM wave of (35-3) at one instant in two different representations:

- By **E** and **B** *vectors.* The envelopes of the tips of the vectors are sine waves.
- By **E** and **B** *field lines* whose density is a measure of the field magnitude.

All sinusoidal plane EM waves are identical in their basic properties. The complete spectrum of EM radiation is shown in Figure 35-3, with frequency and wavelength plotted on logarithmic scales. The various regions differ in:

- How the radiation is produced and detected.
- How the radiation affects materials.

The waves of lowest frequency (and longest wavelength) are radio waves; they are generated by oscillating electric currents. Short-wavelength radio waves, or microwaves, have wavelengths comparable to those of audible sound through air; they are generated by specialized electron vacuum tubes. Infrared radiation is produced by heated solids or the molecular vibrations in gases and liquids. Visible light is produced by rearrangement of the outer electrons in atoms. The very narrow range of wavelengths, between 400 nm and 700 nm (from violet to red light), to which the human eye is sensitive corresponds, in musical terminology, to slightly less than one octave (a factor of 2 in frequency) and is to be contrasted with the enormous frequency range (20 to 20,000 Hz) to which the human ear is sensitive. Ultraviolet radiation immediately adjoins the visible spectrum; it is easily absorbed by ordinary glass and many other materials transparent to visible light. X-rays have wave-

---

* The EM wave given by (35-3) is *linearly polarized* along the y axis; at each point in space, **E** is always along y.

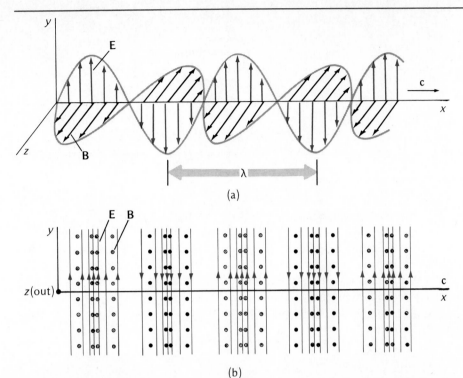

(a)

(b)

Figure 35-2. Two represen-tations of the **E** and **B** fields of a linearly polarized, plane, monochromatic electromagnetic wave of wavelength λ: (a) **E** and **B** vectors. (b) **E** and **B** field lines.

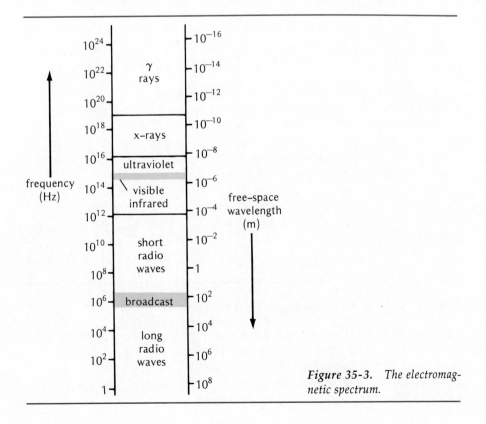

Figure 35-3. The electromag-netic spectrum.

Table 35-1. *The Visible Spectrum*

COLOR	WAVELENGTH IN VACUUM (nm)
Ultra Violet (UV)	<400
Violet	400–424
Blue	424–491
Green	491–575
Yellow	575–585
Orange	585–647
Red	647–700
Infra-red (IR)	>700

lengths of the approximate size of atoms, and they originate in the rearrangement of innermost electrons of atoms. Gamma rays are the electromagnetic waves of the highest frequency and shortest wavelength; they originate in the rearrangement among the particles within the atomic nucleus.

Table 35-1 shows the visible spectrum delinated by color and wavelength.

The boundaries between the adjoining regions are not sharply defined. For example, one cannot distinguish between a short-wavelength x-ray and a long-wavelength γ ray.

35-3   Energy Density, Intensity, and Poynting Vector

The heating effect of sunlight is obvious to anyone standing outside on a bright clear day. Any EM wave represents the transport of energy through space, and in this section we derive some important results relating to the energy of an EM wave.

First, where does the energy come from? An EM wave is created by an accelerated electric charge and the energy in the wave simply equals the total work done by the agent accelerating the charge less the energy acquired by the charged particle itself. By the same token, the energy extracted from an EM wave by an absorber is the work done, primarily by the electric field **E**, on a charged particle within the absorbing material.

Energy Density   Consider now the energy density, or energy per unit volume, of an EM wave. The energy density u_E of the electric field **E** is, from (26-10) in Section 26-5,

$$u_E = \tfrac{1}{2}\epsilon_0 E^2 \tag{35-4}$$

The energy density u_B of magnetic field B is, from (32-11) in Section 32-4,

$$u_B = \frac{1}{2\mu_0} B^2 \tag{35-5}$$

If **E** and **B** vary with time, so do the energy densities u_E and u_B. The relations above give the *instantaneous* energy densities at any point in space. Using (35-2) in (35-5) allows us to write u_B as

$$u_B = \frac{1}{2\mu_0}\left(\frac{E}{c}\right)^2 = \frac{\epsilon_0\mu_0}{2\mu_0}E^2 = \frac{1}{2}\epsilon_0 E^2$$

Comparing this relation with (35-4), we see that

$$u_B = u_E \tag{35-6}$$

The energy densities of the electric and the magnetic fields of an EM wave are equal; the energy transported is shared equally between the electric and magnetic fields.

The total instantaneous energy density u of an EM wave is

$$u = u_E + u_B = 2u_E = \epsilon_0 E^2 \tag{35-7}$$

Equivalently, we can write $u = 2u_B = B^2/\mu_0$. The energy density is proportional to the *square* of either the electric field or the magnetic field.

For a sinusoidal EM wave, electric field **E** at any point in space varies with time as $E = E_0 \cos \omega t$, where E_0 is the electric-field amplitude and ω is the angular frequency. Then the time average of E^2 is given by

$$\overline{E^2} = E_0^2 \,\overline{\cos^2 \omega t} = \tfrac{1}{2}E_0^2$$

In the last step, we have used the fact that the time average of the square of any sinusoidal function is $\tfrac{1}{2}$ (proved in detail in Section 33-5). The time average of the total energy density of an EM wave can then be written as

$$\bar{u} = \tfrac{1}{2}\epsilon_0 E_0^2 \tag{35-8}$$

Intensity   The intensity I of any wave is defined as the energy flow per unit time, or power P, passing through a unit area oriented at right angles to the direction of wave propagation. In symbols,

$$I = \frac{P}{A} \tag{35-9}$$

We wish to relate the intensity I of any wave to its energy density u and propagation speed c. Suppose that a wave is propagated along the axis of a cylinder. The cylinder's cross-sectional area is A and its length is L, as shown in Figure 35-4. If the cylinder's thickness L is small, the wave's energy density is constant throughout the cylinder. We can write the thickness L as $L = ct$, where t is the time for the wave to travel from the front to the back cylinder face. Over the interval t, all energy originally contained in the cylinder's volume $AL = Act$ will have passed through the area A. We can write

$$\text{Intensity} = \frac{\text{energy}}{\text{area} \times \text{time}} = \frac{(\text{energy/volume}) \times \text{volume}}{\text{area} \times \text{time}}$$

$$I = \frac{u \times Act}{At}$$

$$I = uc \tag{35-10}$$

Equation (35-10) relates the instantaneous energy density u and propagation speed c of *any* wave to its instantaneous intensity I. For an EM wave we have, using (35-7),

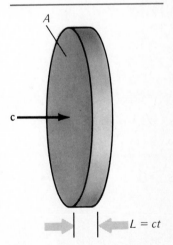

Figure 35-4.   Wave propagating at speed c through a cylinder of thickness L and cross-sectional area A.

$$I = \epsilon_0 c E^2 \tag{35-11}$$

The instantaneous intensity is proportional to the *square* of the electric field. Note that neither the energy density nor the intensity of an EM wave depends on its frequency.

Poynting Vector   Equation (35-11) for the intensity can be written in an interesting alternative form by using (35-1) and (35-2):

$$I = \epsilon_0 E^2 c = \epsilon_0 E(Bc)c = \frac{\epsilon_0 EB}{\epsilon_0 \mu_0} = \frac{EB}{\mu_0}$$

This relation can be written as a vector equation,

$$\mathbf{I} = \mathbf{E} \times \frac{\mathbf{B}}{\mu_0} \tag{35-12}$$

where the intensity vector **I** points in the direction of wave propagation and of energy transport. See Figure 35-1. Vector **I** is referred to as the *Poynting vector*,* named for J. H. Poynting (1852–1914). To find the energy flow per unit time, or power P, from EM fields through a surface with a differential element $d\mathbf{S}$, we merely integrate $\int \mathbf{I} \cdot d\mathbf{S}$ over the surface

$$P = \int \mathbf{I} \cdot d\mathbf{S} = \int \left(\mathbf{E} \times \frac{\mathbf{B}}{\mu_0} \right) \cdot d\mathbf{S} \tag{35-13}$$

Example 35-1.   A radio antenna has a very modest power output of 1.0 W and broadcasts uniformly in all directions. At a location 1000 km from the source, what is (a) the intensity? (b) the energy density of the EM wave? (c) the electric field? (d) the magnetic field?

(a) The power P of the transmitting antenna is spread uniformly over a spherical area of $4\pi r^2$, so that at a distance r from the source, the intensity is

$$I = \frac{P}{A} = \frac{P}{4\pi r^2} = \frac{1.0 \text{ W}}{4\pi(1.0 \times 10^6 \text{ m})^2} = 7.9 \times 10^{-14} \text{ W/m}^2$$

(b) The energy density u is, from (35-10),

$$u = \frac{I}{c} = \frac{7.9 \times 10^{-14} \text{ W/m}^2}{3.0 \times 10^8 \text{ m/s}} = 2.7 \times 10^{-22} \text{ J/m}^3 = 2.7 \times 10^{-10} \text{ }\mu\text{J/cm}^3$$

(c) Electric field E is, from (35-7),

$$E = \sqrt{u/\epsilon_0} = \sqrt{(2.7 \times 10^{-22} \text{ J/m}^3)(8.85 \times 10^{-12} \text{ C}^2/\text{N} \cdot \text{m}^2)} = 5.5 \text{ }\mu\text{V/m}$$

It is relatively easy to detect the electric field with a modern radio receiver. Suppose that the electric field of the incoming EM wave is parallel to a receiving antenna 1 m long; then a signal of 5.5 μV is produced along it. Note that the computed value for E is the instantaneous value; if the antenna transmits a sinusoidal wave with an *average* power output of 1.0 W, then the computed value above is the rms value of the oscillating electric field.

(d) The magnetic field at the receiver location is, from (35-2),

---

* The Poynting vector can be written in still simpler form by introducing the vector **H** for *magnetic field intensity*, defined as $\mathbf{H} \equiv \mathbf{B}/\mu_0$. Then $\mathbf{I} = \mathbf{E} \times \mathbf{H}$.

$$B = \frac{E}{c} = \frac{5.5 \times 10^{-6} \text{ V/m}}{3.0 \times 10^8 \text{ m/s}} = 1.8 \times 10^{-14} \text{ T} = 1.8 \times 10^{-10} \text{ G}$$

The magnetic field of the received EM wave is indeed small. It is smaller than the earth's magnetic field by a factor of more than 10^{10}.

35-4   Electric-Dipole Oscillator

Every electric charge has electric fields attached to it, and every charged particle in motion has magnetic-field loops surrounding the line of its velocity. But an EM wave in otherwise empty space consists of **E** and **B** lines unattached to charges. How is an EM wave launched? How do the **E** and **B** lines become detached from charges?

The essential requirement is *acceleration;* a charged particle must be accelerated to generate an EM wave. A positively charged particle at rest or in motion at constant velocity has in effect electric-field lines rigidly attached to it and extending straight outward. But if such a particle initially at rest is accelerated briefly and then brought to rest again, the acceleration shows up as a kink, with a transverse component, in the **E** lines, traveling outward at speed c. This is shown in Figure 35-5. The transverse **E** component is part of an outwardly expanding EM wave. Detailed analysis (not to be given here) shows not only that acceleration produces *transverse* **E** and **B** field components but also that the magnitude of these transverse components falls off with distance less rapidly than the radial electric field characteristic of Coulomb's law.

The simplest generator of sinusoidal EM radiation is an *electric-dipole* oscillator. Recall that an electric dipole consists of two separated point charges of opposite sign and equal magnitude (Section 24-6). One simple form of an electric-dipole oscillator is an electric dipole in which the two charges oscillate at the same frequency but 180° out of phase, so that the electric-dipole moment oscillates sinusoidally with time. Another equivalent electric-dipole oscillator consists of two straight-line conductors to which a sinusoidally alternating voltage is applied. When the tip of one conductor is positive, the other conductor tip is negative with the same charge. See Figure 35-6.

The electric- and magnetic-field patterns from an electric-dipole oscillator,

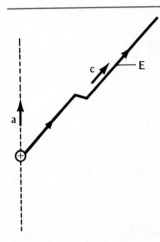

Figure 35-5. If a charged particle is given an acceleration **a**, *a kink, or transverse component, is produced in the electric field* **E** *that travels outward at speed c.*

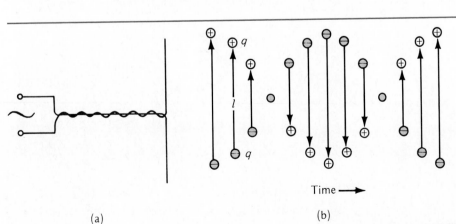

Time ⟶

(a)   (b)

Figure 35-6. (a) Electric-dipole oscillator. (b) Equivalent oscillating electric charges.

Everything Was Big but the Particles

Its nickname, the Z⁰. Its full name, the Neutral Intermediate Vector Boson. Forty years ago this fundamental exotic particle was first predicted to exist; contemporary elementary-particle theory said that a Z⁰ would have, among others, these specific properties: a mass about 120 times that of a proton, and a lifetime so short (10^{-10} s) that it can not be observed directly but only by finding the oppositely charged particles into which the highly unstable Z⁰ can decay (either an electron-positron pair or a muon-antimuon pair).

Is there really such a creature? A research report in 1983 said yes [*Physics Letters* **126B** 398 (7 July 1983)]. After an enormous effort, a total of five Z⁰'s were positively identified. This work, together with research on finding the closely related W⁺ and W⁻ particles (see Figure 10–20), was the basis of the 1984 Nobel prize in physics.* The recipients were

*A large fraction of the November, 1983, issue of the *CERN Courier* **23** is devoted to this story. This international journal of high energy physics is available in most physics libraries.

Carlo Rubbia of Harvard University, who masterminded the two projects, and Simon van der Meer, of CERN (the European Organization for Nuclear Research, near Geneva, Switzerland), who devised the ingenious devices for handling — steering, deflecting, accelerating, bunching — vast numbers of antiprotons traveling at nearly the speed of light for many hours at a time. The antiprotons were then directed to slam head-on into protons of the same energy. The hoped-for result of such a collision: one of the quarks within a proton would combine with an antiquark within an antiproton to create a Z⁰.

It was a remarkable feat. Everything about the project was big:

- *The team of scientists.* A total of 128 physicists from a dozen research centers in Europe and the U.S., together with many more hundreds of technicians and assistants in supporting roles, participated in the project.
- *The accelerating machine.* It was the SPS Collider, a Super Proton Synchrotron accelerating ma-

for one instant, are shown in Figure 35-7. We can see several features in the figure:

- Close to the dipole, electric-field lines go from positive to negative charge; this is the Coulomb field. [See Figure 23-15(a).] At larger distances, the electric-field lines have become detached from the oscillating dipole and form closed **E** loops of the radiated EM wave. Similarly, **B** loops become detached from the oscillating charges.
- The **E** and **B** fields are mutually perpendicular at each location and also perpendicular to the radially outward direction of the radiated EM wave.

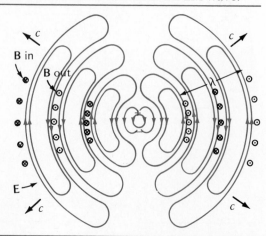

Figure 35-7. Electric and magnetic fields produced by a sinusoidally varying electric dipole.

chine modified to allow protons and antiprotons to circulate and be accelerated while traveling in opposite directions inside an evacuated ($\sim 10^{-13}$ atm) underground tube of 7 km circumference. The 6×10^{11} antiprotons in a pulse, each particle with a kinetic energy of 270 GeV (2.7×10^{11} eV), were made to collide head-on a few thousand times a second with a roughly equal number of protons of the same kinetic energy.

• *The detecting apparatus.* Known as the UA 1 (Underground Area #1), the detecting device was really a collection of many thousands of individual particle detectors. The largest ever constructed for an accelerating machine, the detector was located in a vast cavern 20 m underground and weighed 2000 tons, but was nevertheless moveable. It alone cost many hundreds of man-years of effort in design and construction, and $20 million.

• *The data handling system.* Each time a sufficiently energetic proton hits an equally energetic antiproton head-on, the two particles mutually annihilate one another and many newly created particles splatter off in all directions. The data from just one such event would fill a large telephone directory. All told about

10^9 such collisions were analyzed in detail, while the experiment was running, to find just the five Z^0 events. Closely articulated microprocessors and computers were essential parts of the apparatus.

• *The stakes.* A lot was riding on this experiment; the Nobel committee had handed out a few Nobel prizes in physics a few years earlier to people who said that intermediate vector bosons surely existed and would be observed in the laboratory. It would have been awkward, to say the least, if the Z's and W's had not been found. More importantly, the grand unified field theories, according to which the fundamental forces (the gravitational, the weak, the electromagnetic, and the strong) are united in one primordial interaction, would have been put in very serious jeopardy if these peculiar and highly transient particles were missing. Elementary-particle physicists now have enough confidence in the essential correctness of their ideas of what the truly fundamental particles are and how they interact with one another that they can tell in remarkable detail what happened very shortly—within even 10^{-43} s—of the initial Big Bang that marked the instant when the Universe came into being.

• At great distances from the dipole, the wave fronts, which are surfaces of constant phase, are spheres centered on the dipole. The dipole as an effective point source radiates a spherical wave.

• The frequency of the EM wave is the same as the frequency of the dipole oscillator. It takes a full period for **E** at any location to return to its initial value, and, therefore, a half period of oscillation to produce just one **E** loop.

• The wave is *linearly polarized;* that is, the transverse electric field at any location in space oscillates along a fixed line. This direction is called the *direction of polarization* of the EM wave.

• The radiation pattern is symmetric relative to the vertical electric-dipole axis (think of it here as a north-south axis). The radiated intensity is a maximum at the equator, and the intensity reaches zero in the directions outward from the north and south poles. No EM radiation is emitted along the direction of a charged particle's acceleration. See Figure 35-8. It can be shown that the intensity I of EM waves from an electric-dipole oscillator varies with distance r from the source and angle θ from the dipole axis according to $I \propto \sin^2 \theta / r^2$.

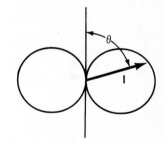

*Figure 35-8. Radiation pattern from an electric-dipole oscillator. Intensity **I** in a polar plot as a function of angle θ from the line of the dipole's acceleration direction.*

35-5   The Speed of Light

The fundamental constant of electromagnetism is c. It relates through (35-1), $c^2 = 1/\epsilon_0 \mu_0$, the basic constants for the electric (ϵ_0) and the magnetic (μ_0) interactions. Atomic structure is governed by the EM interaction between electrically charged particles, and c is also a fundamental constant of atomic theory. Furthermore, relativity physics is intimately connected with c; the most famous equation in physics, $E = mc^2$, contains it.

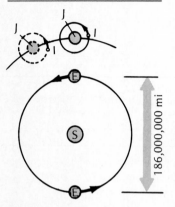

Figure 35-9. *Roemer's measurement of the speed of light. The sun (S), earth (E), Jupiter (J), and one of Jupiter's moons, Io (I).*

The Standard Meter and c   Since the action in Paris on October 20, 1983, by the International Committee of Weights and Measures, the speed of light is, in effect, *defined* as having the magnitude 299 792 458 m/s in the SI system of units. Strictly speaking, the standard meter is now defined as the distance traveled in a vacuum by an electromagnetic wave in a time interval equal to 1/299 792 458 s.

Measuring c   Timing light signals is an important means of measuring distance indirectly. For example, laser pulses directed to mirrors placed on the moon during the Apollo flights make it possible to measure very small changes (to within a few centimeters) in the distance from the earth's surface to the moon's. Still greater planetary distances have been measured very precisely by timing signals from Voyager spacecraft that have reached distant planets.

Recognizing that light might not be transmitted instantaneously, with an infinite speed, was an important advance in early science, but measuring such a high speed as c requires great sublety. Since c is now defined to have its present numerical value, the ways in which it has been measured over the last several hundred years are mostly of historical interest. Here, briefly, are some of the principal methods of measuring c:

• Use astronomical distances so that the travel time is long enough to be measured readily. This was done by Ole Roemer (1644–1710), first in 1666. He measured the time (about 1000 s) for light to traverse the diameter (186,000,000 miles) of the earth's orbit around the sun. The signals came from an eclipsing moon orbiting the slower-moving and more distant planet Jupiter. As the earth moved farther and farther from Jupiter, the disappearance of a moon behind Jupiter was found to be progressively later because of the additional distance the light signals had to cover to reach the observer on earth. See Figure 35-9.

• *Stellar aberration* was the method used in 1725 by the astronomer James Bradley (1692–1762). This method depends on combining one very high velocity, that of light, with another high velocity, that of the earth as it orbits the sun at 3×10^4 m/s. To see a star directly overhead, one does not point a telescope straight up when the earth is moving at right angles to the line joining the earth and the distant star. The telescope must be tilted a little (about 20 seconds of arc). See Figure 35-10. Six months later, with the telescope moving in the opposite direction, the telescope must be tilted the other way. The angle of tilt, together with the known speed of the earth around the sun, yields c.

• To measure the time for light to travel several kilometers to a distant mirror (by terrestial standards) and back again was first done with rotating mechanical devices. The first determinations were made by A. H. L. Fizeau (1819–1896) in 1849; the American physicist, A. H. Michelson (1852–1931) carried the method to its greatest refinement through several decades. Basically, the rotating device chops a continuous light beam into a series of pulses and its speed of rotation serves also to measure the time interval for a round trip. See Figure 35-11.

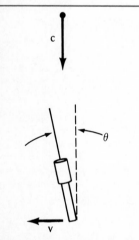

Figure 35-10. *Stellar aberration: when a telescope is moving at right angles to the direction of light propagation, the telescope must be tipped at angle θ in the direction of its velocity **v**.*

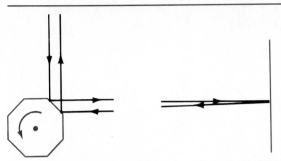

Figure 35-11. *A rotating octagonal mirror used by A. A. Michelson in measuring the speed of light. If the mirror makes an eighth turn in the time it takes for a light pulse to travel to a distant mirror and back, the light beam is reflected in the same direction as if there were no rotation of the mirror.*

• Until the recent definition of c, the most recent measurements of c depended on measuring the wavelength of nearly monochromatic radiation against the standard meter, defined in terms of wavelength of light, and also measuring the frequency of the same radiation against the standard second as defined in terms of a standard atomic clock. Then $c = \nu\lambda$. This method was limited, finally, by the uncertainty in the measurement of wavelength (about 4 parts in 10^9); the uncertainty in the measurement of time was far less (about 1 part in 10^{10}, which is equivalent to an uncertainty of 3 s in a millennium). With the meter defined in terms of c, uncertainties are limited to those in time intervals. Comparing the frequencies of highly monochromatic light sources over the visible and adjoining regions of the EM spectrum depends on two basic measurement effects: *nonlinear materials*, which produce harmonics, oscillations at integral multiples of the incident radiation (see Section 11-6); and *beats* (Section 18-6), in which a measurement of the beat frequency between two oscillations yields their frequency difference.

• Speed c has such a large numerical value and its direct measurement is so difficult that you can easily overlook simple indirect means for determining it. Suppose, for example, that you measure the attractive electric force between the two oppositely charged plates of a parallel-plate capacitor. Then, if you also know the charges on the plates, and the dimensions of the capacitor, you can compute the value for ϵ_0. This means you can compute the value of c from $\epsilon_0 = 1/\mu_0 c^2$ using (35-1).

Doppler Effect for Light When we say that the speed of sound through air is, say, 344 m/s, we mean that that is how fast a pressure disturbance travels relative to the medium—air—in which sound waves are propagated. An observer in motion relative to air measures a different speed. Not so for an EM wave, however. The speed of light through a vacuum is the *same* for *all* observers, quite apart from their state of motion or the motion of the source emitting the light. A basic postulate for the special theory of relativity (Chapter 40), confirmed in detail by experiment, is that c has the very same value for all observers.

One important consequence of the constancy of c is that the simple relations for the shift in frequency, the Doppler effect, as derived in Section 18-7 for mechanical waves, do not hold for light. For mechanical waves, we distinguish between the wave source in motion and the observer of waves in motion, both relative to the medium propagating the waves. But for EM waves propagated through empty space, the speed is always c, so that we cannot distinguish between source in motion and observer in motion. We can speak only of the speed of the observer relative to the source; and therefore a single relation gives the relativistic Doppler shift for EM waves. It is, as derived from the special theory of relativity,

$$f_o = f_s \frac{1 + v/c}{\sqrt{1 - (v/c)^2}} \qquad (35\text{-}14)$$

where v is the relative speed of approach between source and observer, f_s is the source frequency, and f_o is the observed frequency. (The observed wavelength is always $\lambda_0 = c/f_o$.) For low speeds, with $v/c \ll 1$, the radical in the denominator of (35-14) is essentially equal to 1, and the relation for the relativistic Doppler effect reduces to the simple classical result

$$\frac{\Delta f}{f} = \pm \frac{v}{c} \qquad (18\text{-}7)$$

where Δf is the frequency shift.

Equation (35-14) is confirmed directly in detail for the light emitted along the direction of their motion by high-speed atoms. This equation is also used by astrophysicists to measure the speed of stellar objects from the frequency shifts in their emitted light — a red shift to longer wavelengths for receding objects and a blue shift to shorter wavelengths for approaching objects. Indeed, Doppler-shift observations indicate that some very distant stellar objects are in motion away from us at speeds approaching c.

Example 35-2.   At what speed would a star have to recede from the earth so that light emitted from the star at the violet limit of the visible spectrum would, because of the Doppler shift, be observed on earth as light at the red limit of the visible spectrum?

Wavelengths at the extremes of the visible spectrum differ by a factor of 2, so that the Doppler-shifted wavelength would be twice the emitted wavelength, or the observed frequency would be half the source frequency.

With $f_o = f_s/2$ and $v/c \equiv x$, the equation (35-14) becomes

$$\frac{1}{2} = \frac{1 + x}{\sqrt{1 - x^2}}$$

After a little algebra and the solving of a quadratic equation, we get

$$x = \frac{v}{c} = 0.6$$

For stars receding from us at speeds greater than 60 percent c, the "visible" spectrum from the high-speed emitter is shifted to the infrared region of the EM spectrum.

35-6   Radiation Force and Pressure, and the Linear Momentum of an Electromagnetic Wave

An EM wave carries energy. It is easy to show that an EM wave also has *linear momentum* in the direction of propagation and that such a wave can exert a force, or *radiation pressure*, on a material upon which it impinges.*

For simplicity, suppose that an EM wave is incident on a material that absorbs all the energy striking it, and reflects and transmits none. This implies

---

* P. N. Lebedev (1866–1912) first measured the pressure of light in 1901.

that when the electric field **E** does work on a charged particle within the material, the energy removed per unit time from the EM wave is exactly the power absorbed by the material. Note that we say the work done by the *electric* field. The magnetic force, since it always acts at right angles to a charged particle's velocity, does *no work*.

Consider the situation in Figure 35-12. Here an EM wave travels along the positive x axis; at one instant, the electric field is **E** along the positive y axis and the magnetic field **B** along the positive z axis. We are interested in the forces produced by **E** and **B** on an electron within the material. The electric force on the electron has a magnitude $F_e = eE$. This force acts in the direction opposite to that of **E** and accelerates the electron in a direction transverse to the wave propagation.

What is the effect of the magnetic field? We take the electron to be moving with speed v along the negative y axis. In general, the direction and magnitude of the magnetic force is given by $\mathbf{F_m} = q\mathbf{v} \times \mathbf{B}$. Here the magnitude of $\mathbf{F_m}$ is evB, and its direction is along the positive x axis. The EM wave produces a *force* on the electron (and therefore also on the material to which the electron is bound) *along the direction of wave motion*. In short, when an EM wave impinges on an electric charge, the **E** field accelerates the charge in the transverse direction and does work on it, while the **B** field, acting on the moving charge, produces a longitudinal force.

We wish to find relations for the radiation force and pressure, and the linear momentum of the EM field, in terms of such quantities as the intensity I and power P of the wave. We found above that the radiation force F_r is given by

$$F_r = evB$$

where all quantities are instantaneous values. Since the magnitudes of **E** and **B** are related by $B = E/c$, this equation can be rewritten as

$$F_r = \frac{v}{c} eE \qquad (35\text{-}15)$$

But eE is just the magnitude of the electric force $\mathbf{F_e}$, the force that does work. In general, the rate of doing work, or the power P, is given by

$$P = Fv$$

where F is the force doing work and v is the speed of the particle acted on. Then (35-15) can be written

$$F_r = \frac{v}{c} F_e$$

Total absorption: $\qquad F_r = \frac{P}{c} \qquad (35\text{-}16)$

The radiation force of an EM wave on a material that *absorbs it completely* is simply the power of the wave divided by the speed of light.

A longitudinal force P/c is exerted by an EM wave on a material that absorbs it completely. It follows that when a material *emits* radiation of power P in one direction, this emitter must recoil under the action of a recoil radiation force of magnitude P/c. We can see this most easily by noting that emission is, so to speak, absorption run backward in time. In emission, charges within the material *lose* energy and create an outgoing EM wave. With time reversal, the

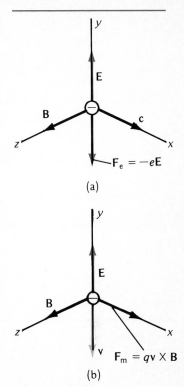

Figure 35-12. *An electromagnetic wave incident upon an electron in an absorbing material. (a) An electric force $\mathbf{F_e} = -e\mathbf{E}$ acts on the electron. (b) A magnetic force $\mathbf{F_m}$, in the direction of wave propagation, acts on electron in motion with velocity $\mathbf{v}$ in magnetic field $\mathbf{B}$.*

directions of the electric field **E**, of the electric force $\mathbf{F_e}$, and of the radiation force $\mathbf{F_r}$ all remain *the same;* but the direction of the *velocity* **v** is reversed, and so is the direction of the *magnetic field* **B**. Under time reversal — that is, with emission rather than absorption — the direction of the Poynting vector is reversed, and energy then flows away from the material rather than toward it.

What is the radiation force on a material that reflects all the radiation striking it, absorbing none? Think of reflection as taking place in two stages: absorption of the incident radiation followed by reemission in the reverse direction. Since a radiation force of magnitude P/c acts on the material both in absorption and in emission, the radiation force for complete reflection is*

$$\text{Complete reflection:} \quad F_r = \frac{2P}{c} \tag{35-17}$$

The radiation force given in (35-16) and (35-17) applies for radiation that is incident in a direction *perpendicular* to the absorbing or reflecting surface. For oblique incidence, with an angle θ between the direction of wave propagation and the normal to the plane of the absorber or reflector, the radiation force normal to the surface is obtained by multiplying F_r by the factor $\cos\theta$. For total reflection, when $\theta \neq 0$, there is no tangential force.

It is also useful to have relations for the *radiation pressure* p_r, the radiation force F_r per unit transverse area A. Since pressure p_r is, by definition, F/A, we have for complete absorption,

$$p_r = \frac{F_r}{A} = \frac{P}{cA}$$

The intensity I is given by $I = P/A$. Therefore,

$$\text{Complete absorption:} \quad p_r = \frac{I}{c}$$

$$\tag{35-18}$$

$$\text{Complete reflection:} \quad p_r = 2\,\frac{I}{c}$$

Example 35-3. A 3-W beam of EM radiation shines on a black object and is completely absorbed by it. (a) What is the radiation force on the absorber? (b) What is the recoil force on the source emitting the beam?

(a) For complete absorption,

$$F_r = \frac{P}{c} = \frac{3\ \text{W}}{3.0 \times 10^8\ \text{m/s}} = 1 \times 10^{-8}\ \text{N}$$

which is a very small force indeed.

(b) The source emitting the 3-W beam, whether it is a source of light or a radio transmitter, will, as long as the emitted waves travel outward in a single direction, recoil

---

* That the radiation force for complete reflection is twice the force for complete absorption has an exact analog in mechanics. When a particle with initial momentum $+mv$ strikes and sticks to an object, the linear momentum transferred to the struck object is $+mv$; but when a particle with initial momentum $+mv$ is "reflected" from the struck object, rebounding with the same speed, the particle's final momentum is $-mv$, its momentum having been changed by $\Delta(mv) = -mv - (+mv) = -2mv$. Thus, for reflection, the struck object acquires a momentum $+2mv$, just *twice* the momentum acquired in absorption. Equivalently, the force (average) on a struck object is twice as great for reflection as for absorption.

under the action of a force of 10^{-8} N. Thus, a flashlight emitting light constitutes a very elementary form of a rocket.

For any sources of moderate intensity or power, the radiation force is negligibly small. In stellar phenomena, where very high intensities are encountered, the radiation force may equal or exceed the gravitational force, as evidenced by an exploding star, or a supernova.

If EM radiation can exert a force and transfer linear momentum to an object upon which it impinges, linear momentum must be associated with the EM field itself. It is easy to derive the expression for the momentum M of an EM wave. (We use the symbol M for linear momentum, rather than the conventional symbol p, to avoid confusion with the pressure p and the power P.) By definition, the force F is related to the momentum M by

$$F = \frac{dM}{dt}$$

Similarly, the power P is related to the energy by

$$P = \frac{d(\text{energy})}{dt}$$

From (35-16),

$$F_r = \frac{P}{c}$$

so that

$$\frac{dM}{dt} = \frac{1}{c} \cdot \frac{d(\text{energy})}{dt}$$

We use the relation for the radiation force in *absorption* since we want to count the energy transfer only *once*. Integrating yields

$$\text{EM momentum} = \frac{\text{EM energy}}{c} \qquad (35\text{-}19)$$

The magnitude of the linear momentum of an EM wave is the energy of the wave divided by c; the direction of the momentum is along the direction of energy propagation.

We have seen that we can attribute energy and linear momentum to an EM wave. We can also attribute *angular* momentum (actually, *spin* angular momentum) to circularly polarized electromagnetic waves. The magnitude of the EM angular momentum of a circularly polarized wave of frequency f is the energy of the wave divided by $\omega = 2\pi f$, and the direction of the angular momentum vector is parallel or antiparallel to the direction of wave propagation.

35-7 Polarization

Types of Polarization Imagine a plane transverse to the direction of propagation of an EM wave. We concentrate on the path traced out on this plane by

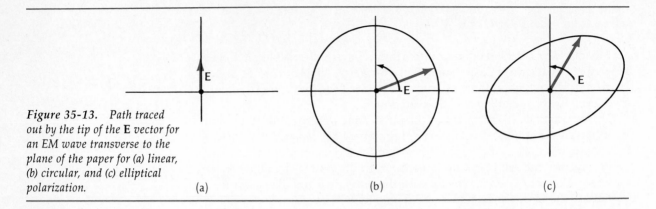

Figure 35-13. *Path traced out by the tip of the* **E** *vector for an EM wave transverse to the plane of the paper for (a) linear, (b) circular, and (c) elliptical polarization.*

(a)   (b)   (c)

the tip of the vector **E**, representing the instantaneous electric field, as shown in Figure 35-13.

- *Linearly polarized* wave. The path is a single straight line. This line gives the *direction* of polarization; the *plane* of polarization contains this line and the line representing direction of wave propagation.
- *Circularly polarized* wave. The path is a circle. It is easy to confirm that two simple harmonic oscillations of the same frequency and amplitude, but out of phase by 90°, yield uniform circular motion. Circular polarization can then be thought of as the superposition of two linearly polarized waves of the same amplitude and frequency but out of phase by 90° traveling in the same direction.
- *Elliptically polarized* wave. The path is an ellipse. Mutually perpendicular simple harmonic oscillations of the same frequency but differing in amplitude or phase by an amount different from 90° produce elliptical polarization.

Unpolarized Light   Visible light from ordinary light sources (electric excitation, heating) comes from the random radiation of many individual atoms, each atom "turned on" about 10^{-8} s. The emitted light is incoherent; the phases of successive wave trains are not the same (Section 38-6). Furthermore, the polarization of the light from the various radiating atoms does not take one direction but is distributed randomly in all possible directions. When the polarization direction changes randomly and rapidly, so that one cannot follow it in time, the light is said to be *unpolarized.*

Figure 35-14. *Representation of states of polarization. Unpolarized light transverse to the plane of the paper. (a) random orientations of the* **E** *vectors, and (b) equivalent polarizations along mutually perpendicular directions. (c) Unpolarized light propagated in the plane of the paper to the right.*

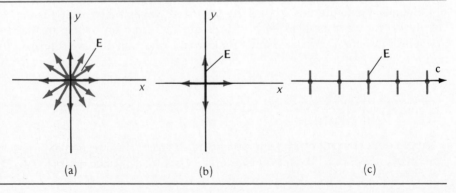

(a)   (b)   (c)

Figure 33-14 shows ways of representing unpolarized light. In (a), we have linear oscillations in all transverse directions. Since each oblique oscillation can be replaced by its rectangular components, the two mutually perpendicular oscillations, which are neither continuous nor coherent, of part (b) also represent unpolarized light. In part (c), the propagation direction is in the plane of the paper rather than perpendicular to the plane.

The Law of Malus Suppose a microwave beam is linearly polarized and encounters a set of parallel conducting wires oriented along the direction of polarization of the wave. Then the electric field of the EM wave will induce currents in the conducting wire, thermal energy will be dissipated in the wires, and the polarized wave will be absorbed.

The commercial material Polaroid behaves in the same way for visible light. The material has needlelike molecules (herapathite) aligned mostly along one direction (by stretching a flexible transparent sheet as it solidifies). Suppose that the electric field of a polarized light wave is parallel to the long axis of molecules; then the polarized wave is absorbed. On the other hand, with the polarization direction perpendicular to the long axis, the wave is transmitted (mostly). The easy transmission direction is called the polarization direction.

Suppose that unpolarized light passes in turn through two Polaroid sheets, with the second one rotatable relative to the first. Light emerging from the first sheet is linearly polarized. If the second Polaroid sheet has its polarization direction at angle θ relative to the first sheet, only the parallel component of the electric field, $E \cos \theta$, will be transmitted though the second sheet; the perpendicular component will be absorbed. See Figure 35-15.

What is the intensity of the light emerging from the second sheet? Since intensity I varies with the square of the electric field amplitude, we have that

$$I = I_0 \cos^2 \theta \qquad (35\text{-}20)$$

where I_0 is the maximum intensity, corresponding to $\theta = 0$.

When $\theta = 90°$ and the polarization directions of the two sheets are at right angles, the trasmitted intensity is zero. Said differently, when the *polarizer* and the *analyzer* are at right angles, we have *extinction*. Equation (35-20) thus governs the intensity of light through *any* two devices, each of which transmits only one polarization direction; it is known as *Malus's law,* after its discoverer, E. L. Malus (1775–1812).

Polarization by Scattering The EM radiation from an electric-dipole oscillator is linearly polarized. It follows that radiation scattered from a material may also be polarized, since scattering means essentially that the electric field of an EM wave is driving a charged particle in a material into ocillation and

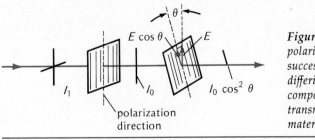

Figure 35-15.   With the polarization directions of two successive polarizing materials differing by angle θ, the component of the electric field transmitted through the second material is E cos θ.

causing it to become effectively an electric-dipole oscillator. For example, sunlight scattered by particles and molecules in the earth's atmosphere is at least partially polarized.

Example 35-4. An unpolarized beam of light passes through a single sheet of Polaroid. Compare the transmitted beam and incident beam in intensity.

If the incident light is truly unpolarized, then the two equivalent perpendicular oscillations into which the beam may be resolved, shown in Figure 35-14(b), must be of equal magnitude. We may choose the orientation of one oscillation to be parallel to the polarization direction of the Polaroid sheet. This component is completely transmitted; the other is completely absorbed. Therefore, the transmitted beam has an intensity half that of the incident beam.

If there were no absorption whatsoever for oscillations parallel to the polarization direction of a Polaroid sheet, adding a second Polaroid sheet with its polarization direction to that of the first would produce no further attenuation in the light intensity. This ideal behavior is not found, however, because partial absorption occurs even in the preferred orientations.

Example 35-5. An unpolarized light beam falls on two Polaroid sheets so oriented that no light is transmitted through the second sheet. A third Polaroid sheet is then introduced between the first two sheets. How does the intensity of transmitted light vary with the orientation of the third sheet? (The intensity is not zero for all orientations!)

The polarization directions of sheets 1 and 2 are mutually perpendicular. We take the angle between the polarization directions of sheets 1 and 3 to be θ. Then, as Figure 35-16(a) shows, if E is the magnitude of the electric field transmitted through the first sheet, the component emerging through the third (the one placed between the first and second sheets) is $E \cos \theta$. From Figure 35-16(b), we see that the component emerging through the last sheet has the magnitude $E \cos \theta \cos (90° - \theta) = E \cos \theta \sin \theta$. Thus, the intensity varies according to

$$I = I_0 \cos^2 \theta \sin^2 \theta = \tfrac{1}{4} I_0 \sin^2 2\theta$$

where I_0 is the intensity transmitted through the first sheet. Note that there are four positions at which the intensity is maximum; $45°$, $135°$, $225°$, and $315°$. Note also that interchanging sheets 1 and 3 does not change the dependence of I and θ.

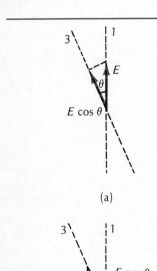

(a)

(b)

Figure 35-16.

Summary

Definitions

Poynting vector **I** of an EM wave:

$$\mathbf{I} = \mathbf{E} \times \frac{\mathbf{B}}{\mu_0} \qquad (35\text{-}12)$$

Vector **I** gives the direction and magnitude of the intensity, the power per unit transverse area.

Polarization state of an EM wave is given by the path traced out by the **E** vector in a plane transverse to wave propagation: linear, circular, or elliptical.

Fundamental Principles

The propagation speed c of an EM wave through a vacuum is

$$c = \frac{1}{\sqrt{\epsilon_0 \mu_0}} \approx 3.00 \times 10^8 \text{ m/s} \qquad (35\text{-}1)$$

Electromagnetic radiation is produced by accelerated electric charges.

Important Results

Relative magnitudes of the **E** and **B** fields of an EM wave:

$$B = \frac{E}{c} \qquad (35\text{-}2)$$

Instantaneous energy density u of an EM wave:

$$u = u_E + u_B = \epsilon_0 E^2 = \frac{B^2}{\mu_0} \qquad (35\text{-}7)$$

Intensity for any wave of energy density u and propagation speed c:

$$I = uc \qquad (35\text{-}10)$$

Radiation force F_r and radiation pressure p_r for total absorption at normal incidence of an EM wave:

$$F_r = \frac{P}{c} \qquad (35\text{-}16)$$

$$p_r = \frac{I}{c} \qquad (35\text{-}18)$$

where P is the power and I the intensity. For total reflection of the EM wave at normal incidence, the force and pressure are doubled.

$$\text{EM momentum} = \frac{\text{EM energy}}{c} \qquad (35\text{-}19)$$

Law of Malus for intensity I of radiation transmitted through two polarizing materials with angle θ between their polarization directions:

$$I = I_0 \cos^2 \theta \qquad (35\text{-}20)$$

Problems and Questions

Section 35-1   Basic Properties of Electromagnetic Waves

· **35-1 Q** When electromagnetic waves travel through a vacuum, they all have the same
(A) frequency.
(B) wavelength.
(C) velocity.
(D) More than one of the above.
(E) None of the above.

· **35-2 Q** Electromagnetic waves can travel
(A) only through a vacuum.
(B) only through gravitational fields.
(C) only through static electric and magnetic fields.
(D) only through gases.
(E) through all of the above.

· **35-3 Q** In a beam of sunlight that is incident on the earth, the **E** and **B** fields are
(A) 180° out of phase.
(B) 90° out of phase, with **E** leading **B**.
(C) 90° out of phase, with **B** leading **E**.
(D) in phase.
(E) varying independently with no fixed phase relation between them.

· **35-4 P** An electromagnetic wave is directed vertically upward from a point on the earth's equator. At an instant when the electric field is directed toward the north, the magnetic field will be directed.
(A) south.
(B) east.
(C) west.
(D) north.
(E) vertically upward.
(F) vertically downward.

· **35-5 Q** Which of the following statements is not true for an electromagnetic wave in free space?
(A) **E** and **B** are perpendicular to each other and to the direction of propagation.
(B) The magnitudes of **E** and **B** are in the same ratio at every instant.
(C) **E** and **B** are out of phase with each other by 90°.
(D) The velocity of the wave is always the same, independent of variations in frequency.
(E) Such waves always travel at 3×10^8 m/s.

: **35-6 Q** Which of the following statements is false?
(A) Changing electric fields create magnetic fields, and changing magnetic fields create electric fields.
(B) A Maxwell displacement current exists only when there is a magnetic field that changes with time.
(C) A Maxwell displacement current gives rise to a magnetic field, just as an ordinary current does.
(D) Without a displacement current, there can be no electromagnetic waves.

Section 35-2   Sinusoidal Electromagnetic Waves and the Electromagnetic Spectrum

· **35-7 Q** In which of the following pairs are the members in the same relation as microwaves and ultraviolet radiation?
(A) A baritone and a tenor.
(B) A whisper and a shout.
(C) An electron and a photon.
(D) A baby whale and an adult whale.
(E) Waves and particles.

· **35-8 Q** Which of the following forms of electromagnetic radiation has the longest wavelength?
(A) Gamma rays.
(B) Visible light.
(C) Microwaves.
(D) Ultraviolet rays.
(E) Infrared radiation.

· **35-9 Q** A certain light wave has a frequency of 6.38×10^{14} Hz. (*a*) What is its wavelength in a vacuum? (*b*) What color is it?

· **35-10 P** Electromagnetic waves with a frequency as low as 100 Hz can be used to communicate with a submerged submarine. Compare the wavelength of this radiation in free space with the earth's diameter.

· **35-11 P** What is the frequency of yellow light with a wavelength of 570 nm?

· **35-12 P** What is the wavelength of a radar wave of frequency 3.0 GHz?

Section 35-3   Energy Density, Intensity, and Poynting Vector

· **35-13 P** If the distance from a point source of light to a detector is doubled, the intensity of light at the detector will
(A) be unchanged.
(B) double.
(C) increase by a factor of 4.
(D) be half as great.
(E) be one-quarter as great.

· **35-14 P** The intensity of sunlight at the top of the earth's atmosphere is 1.4 kW/m². Determine the magnitude of the electric and magnetic fields of this radiation.

: **35-15 P** What is the energy density in the beam from a 2-mW laser if the area of the beam is 2×10^{-4} m²?

: **35-16 P** A 10-W point source of radiation radiates uniformly in all directions. For a distance 4 m from the source, determine (a) the electric field; (b) the magnetic field; (c) the intensity.

: **35-17 P** The intensity of the sun's radiation at the top of the earth's atmosphere is approximately 1400 W/m². If the sun is considered a sphere of radius 7.0×10^{8} m, what is (a) the intensity of sunlight at the surface of the sun? (b) the value of the electric field in the sunlight at the surface of the sun? (c) the value of the magnetic field in the sunlight at the surface of the sun?

Section 35-6   Radiation Force and Pressure, and the Linear Momentum of an Electromagnetic Wave

· **35-18 P** What is the momentum of a laser pulse of 10 MW and with a pulse duration of 1.0 μs?

· **35-19 P** A 10 kW light source is turned on for 10 s and it produces a beam with a plane wavefront. What are (a) the energy and (b) the momentum of the beam? (c) The recoil force on the light source while it is turned on? (d) The distance between the leading and trailing edges of the beam?

· **35-20 P** A continuously operating high-intensity laser can have a power output of 1 kW. Suppose that such a laser is used as a photon rocket on a device with a total mass of only 2 kg. If the rocket starts from rest in interstellar space and operates continuously for one year, what is then the rocket's speed?

· **35-21 P** According to the quantum theory, electromagnetic radiation consists of particle-like photons, where the energy of a photon of frequency v is given by hv, where h is a constant. Show that the momentum of a photon is h/λ, where λ is the photon wavelength.

: **35-22 P** The intensity of electromagnetic radiation from the sun just outside the earth's atmosphere is 1.4 kW/m². (a) What is the sun's power output? (b) What is the intensity of the sun's radiation at the surface of the sun? (c) What is the radiation pressure (in atm) at the surface of the sun?

: **35-23 P** Dye lasers can generate ultra-short pulses. A pulse of 0.65 fs corresponds to just 8 full cycles of oscillation. (a) What is the wavelength of the radiation? (b) Over what distance along the propagation direction does the pulse extend? (c) What is the total energy per pulse if the power of the pulse is 1.0 GW?

: **35-24 P** A beam of electromagnetic radiation of power P is incident normally on a surface. One-third of the incident radiation is absorbed, and the rest is reflected. What is the radiation force on the surface?

: **35-25 P** A beam of electromagnetic radiation of intensity I is incident at a angle of 60° relative to the normal upon a surface that aborbs 25 per cent of the radiation and reflects the rest. What is the radiation pressure on the surface?

: **35-26 P** A 10 kW beam of electromagnetic radiation shines normally on an absorbing surface of a 1.0-kg object. Ten percent of the incident radiation is reflected. (a) What is the total energy absorbed by the object over a period of 1.0 hour? (b) Assuming the object to be initially at rest and subject only to the radiation force, what is the final kinetic energy of the object?

: **35-27 P** (a) What is the momentum of a typical nitrogen molecule (N_2) in air at 300 K? (b) What is the energy of a pulse of light with the same momentum? (c) If the pulse duration is 1.0 μs, what is the average power of the pulse?

: **35-28 P** a horizontal electromagnetic beam with an intensity of 10 MW/m² shines on a perfectly absorbing rectangular sheet of 1.0 m² area that hangs freely and rotates about a horizontal axis at the top edge. The sheet's mass is 0.050 kg. At what angle with respect to the vertical will the sheet remain in equilibrium?

Supplementary Problems

35-29 P A *solar sail vehicle* can be propelled by radiation pressure. (a) Suppose that a solar sail vehicle is in equilibrium because the only gravitational force on it, that from the sun, is exactly balanced by the sun's radiation force on the sail. Show that if the orientation of the sail is not changed, it will be in equilibrium at *any* distance from the sun. (b) Suppose that a sail can be constructed whose total area is 1 km². What is the maximum allowable mass of the solar-sail vehicle? (c) What is the total volume of solid, assuming it be aluminum (2.7×10^{3} kg/m³). (d) If one-fourth of the total mass were used for the capsule and the rest for the sail, what would be the sail's average thickness? (At the earth, the sun has $I = 1.4$ kW/m².)

35-30 P A simple way to tell whether a laser beam can effectively eat away at a material it strikes is to see, first of all, whether the electric field in the electromagnetic field is comparable in magnitude to the interatomic field in an ordinary material. (a) A relatively strong interatomic field would be that at a distance of 10 nm from a proton. What is its magnitude? (*b*) What is the average energy density of an electromagnetic wave with an electric field of this magnitude? (*c*) What is the intensity of the beam? (*d*) Assume that this intensity is produced in a laser pulse that is focused to a cross-sectional area of 0.10 mm². What is the power of the pulse? (*e*) Assume the pulse duration to be 1.0 μs. What is the energy per pulse? (*f*) Assume that each atom in the material upon which the laser pulse shines is bound to its neighbors with an energy of 2.0 eV and that adjacent atoms are separated by 0.2 nm. Assume further that only about one-fifth of the energy of the pulse goes into eating away material. How deep a hole is produced by one pulse? (*g*) How many pulses are required to make the hole 0.1 mm deep?

35-31 P A sufficiently intense laser beam shining on a material may produce, because of *nonlinear* effects, coherent radiation at twice the frequency of the illuminating radiation. In the simplest model of atomic structure, an electron is bound to its parent atom by a linear restoring force. Then, when the electron is driven into simple harmonic motion by a monochromatic electromagnetic field, the electron acts as a kind of subatomic antenna, radiating waves of the same frequency. If the driving electric field is sufficiently strong, however, the equivalent force on the electron is nonlinear, the oscillations are not simple harmonic, and the electron generates, not merely radiation of the same frequency as the driving radiation, but also harmonics at integral multiples of the fundamental frequency (Section 11-6).

To see what characteristics are required of laser radiation capable of generating harmonics, we suppose that the electron oscillations are surely nonlinear when the magnitude of the electric field in a laser beam is comparable in magnitude to an interatomic field, say, the electric field at a distance of 1.0 nm from a proton. (*a*) What is that electric field magnitude? (*b*) What are the energy density and intensity of a laser pulse with the requisite electric field magnitude? (*c*) Suppose that the laser beam can be focused to a cross sectional area of 0.1 mm². What is the required laser power output? (A continuously operating carbon-dioxide laser can have a power output of the order of a kW.)

36 Ray Optics

In this chapter we deal with waves, primarily electromagnetic waves of light, traveling in two and three dimensions and encountering the boundaries between media under the special condition in which the wavelength is small compared with the size of obstacles or apertures. Then the only phenomena occurring at the interfaces—reflection and refraction—can be understood, and the progress of a wave can be charted, by a simple geometrical procedure, ray tracing. In our examination of ray optics, or geometrical optics, we exclude such distinctive wave effects as interference and diffraction.

36-1 Ray Optics and Wave Optics

An opaque object with a sharp boundary casts a sharp shadow when illuminated with visible light from a small source. Light travels strictly along a straight line; and it is not obvious, certainly not to the casual observer, that light is actually a wave phenomenon. (Early physicists, including even Isaac Newton, thought that light consisted of particles, or corpuscles.) Consider first the circumstances under which we can assume light, or any other wave disturbance, to follow the "paths" given by the rays associated with wave fronts (Section 18-4) and ignore distinctive wave effects.

Figure 36-1(a) shows waves of various wavelengths λ impinging on an opaque object with an aperture of width d. When $\lambda > d$, the waves spread outward from the aperture in all directions; the wave fronts are circular. For shorter wavelengths, the spreading, or diffraction, of the waves beyond the limits of the geometrical shadow is less pronounced. Finally, when $\lambda \ll d$, the wave fronts remain straight lines and the wave disturbance lies strictly within the limits of the "shadow" of the opening, so that any ray entering the opening continues through without a change in direction.

A similar behavior is seen for waves encountering an isolated opaque object, as shown in Figure 36-1(b). With the wavelength relatively large compared with d, the rays are bent and the wave fronts are curved as a result of diffraction by the object. For $\lambda \ll d$, however, the object casts a sharp shadow. Diffraction effects are then negligibly small, and we can draw any ray as undeviated. In the remainder of this chapter, we assume that the condition $\lambda \ll d$ is always met. (Diffraction effects, which arise when this condition is not satisfied, are dealt with in Chapter 39.)

Under what conditions, then, may we treat visible light by the procedures of ray (or geometrical) optics rather than wave (or physical) optics? A typical wavelength for visible light is somewhat less than 10^{-3} mm. Therefore, ray optics works all right so long as we deal with objects or openings of ordinary size, that is, much larger than 10^{-3} mm ($= 1\ \mu\text{m} = 10^3$ nm). The requirement for ray optics applies equally well to other wave types. For example, an audible sound wave through air with a frequency of 1000 Hz and a wavelength of 34 cm, or an electromagnetic microwave with a wavelength of several centimeters, can be traced by geometrical "optics" only when such waves strike objects whose characteristic dimensions are much larger than a few centimeters.

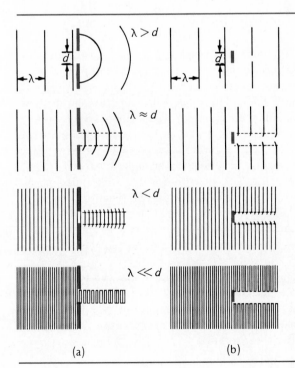

(a)   (b)

Figure 36-1. Waves of decreasing wavelength encountering (a) an aperture and (b) an obstacle of size d. Diffraction is significant for $\lambda \gtrsim d$; ray optics applies for $\lambda \ll d$.

Although we shall deal with wavelengths much shorter than the width of apertures and obstacles, we also assume that the wavelength is large compared with the microscopic objects responsible for the reflection and refraction effects. It is individual atoms and their associated electrons in a solid or a liquid that cause visible light to be reflected and refracted; but the spacing between adjacent atoms, typically of the order of a nanometer, is much less than the wavelength of visible light. By the same token, a sheet of ordinary chicken wire, porous to visible light, acts as an opaque reflector for radio waves several meters in wavelength.

36-2   The Reciprocity Principle

Suppose that we view a motion picture of a wave phenomenon, perhaps a wave traveling along the surface of a liquid. We assume no energy dissipation. Now, if the motion picture is run backward — if we view the wave motion with time reversed — the wave fronts move in the opposite directions and the directions of rays associated with these wave fronts are also reversed. Both wave motions, the first one with time running forward and the second with time reversed, are possible motions. Both are consistent with the laws governing wave motion. Thus, if we know the "path" of a wave from one point to a second point by knowing the configuration of the ray (or rays) connecting the two points, then the reverse path, from the second to the first point, is found simply by reversing the arrows on the rays. In short, if a ray goes from *A* to *B*, a ray also will go from *B* to *A* by the same route. This is the *reciprocity principle*. It asserts that any two points connected by a ray are reciprocal in the sense that the directions of wave propagation can be interchanged with no alteration in the pattern of the wave fronts. Also termed the principle of *optical reversibility*, this principle applies to all nondissipative wave motion, not merely visible light.

Consider Figure 36-2. Here we see a portion of a wave front and the corresponding rays radiating from the point source and reflecting off a parabolic reflector. (The remaining waves emitted by the point source continue to expand outward as spherical waves.) The reflected beam consists of plane waves. A parabolic reflector changes diverging rays into parallel rays; it changes a point source into a plane wave source, as it were. Reversing the ray directions (imagining time as running backward), we see that a beam of plane wave fronts incident upon the parabolic reflector is brought to a focus, and all the rays intersect at a single point. This says that a parabolic reflector can equally well be a transmitter or a receiver of a parallel beam. Even more generally, if we know the radiation pattern of a transmitting antenna — it tells how the intensity radiated varies as a function of direction — we also know the behavior of the same antenna used as a receiver of radiation incident upon it.

Another example of reciprocity is shown in Figure 36-3(a). Here light from a point source passes through a lens to form an image. On ray (or time) reversal, the image becomes the source, and the source becomes the image. Strictly, the arrows on the rays are not necessary, except to indicate for convenience which of the two possibilities is under consideration.

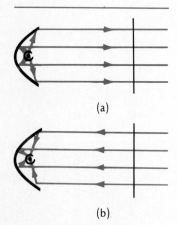

(a)

(b)

Figure 36-2. Example of "optical" reversibility: a parabolic reflector as (a) a transmitter and (b) a receiver.

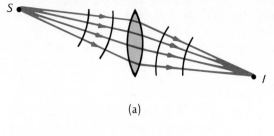

(a)

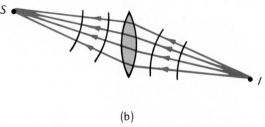

(b)

Figure 36-3.   Example of the reversibility of rays. (a) Source S forms image I when rays pass through a lens. (b) Reversing the ray directions interchanges source and image.

36-3   Rules of Reflection and Refraction

Suppose that a wave encounters a boundary between two media. The incident wave is partially transmitted into the second medium and partially reflected into the first. The transmitted wave is usually bent, or *refracted,* at the interface; it is called the *refracted* ray.

First we give the rules for reflection and refraction. Later we shall see how they follow directly from fundamental principles (Sections 36-4 and 36-6). Figure 36-4 defines terms. It shows a ray incident at the interface between two media; for example, a narrow pencil of light is entering water from air. By convention, the directions of the incident ray θ_1, of the reflected ray θ_1', and of the refracted ray θ_2 are measured relative to the normal to the interface.

The fundamental facts about reflection and refraction of rays are these:

- All four of the following lines lie in a *single plane:* the incident ray, the normal, the reflected ray, and the refracted ray
- Angle of incidence equals angle of reflection:

$$\theta_1 = \theta_1' \tag{36-1}$$

- Angles of incidence θ_1, and of refraction θ_2 are related by

$$n_1 \sin \theta_1 = n_2 \sin \theta_2 \tag{36-2}$$

where n_1 and n_2 are constants characteristic of media 1 and 2, respectively. This relation is known as Snell's law, after its discoverer W. Snell (1591–1626). The constants, called *indices of refraction,* are related to the wave speeds and wavelengths in the respective media, as will be shown in Section 36-5. The principle of reciprocity says that we may reverse the ray directions, so that designating one angle as that of incidence and the other as that of refraction is actually arbitrary.

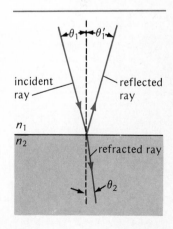

Figure 36-4.   Reflection and refraction of an incident ray at an interface.

This is the long and short of ray optics. If you know the value of n for various materials, you can, at least in principle, trace the rays in detail as they go through a whole succession of surfaces. *Specular* reflection means reflection from a polished surface, for example, a flat or curved mirror with only small surface irregularities. *Diffuse* reflection occurs when the surface is not smooth; then the law of reflection holds exactly at any portion of the surface small enough to be considered flat and smooth. The program of ray optics is simple in principle. It involves nothing more than simple geometrical constructions following the rules given above. But the actual design of optical systems, usually with multiple lenses, is so extraordinarily tedious and difficult that nowadays it is done, trial-and-error fashion, by computers simply because one must trace an extremely large number of rays. (The simpler elements of lens design are discussed in Chapter 37.)

One aspect of reflection and refraction cannot be treated by the methods of ray optics: the relative intensities of the reflected and transmitted beams. The reflection-transmission ratio can, in fact, be computed from the n_1/n_2 ratio by applying Maxwell's equations to electromagnetic waves.

Example 36-1.   A fish pond is in the shape of a right circular cylinder; its diameter is 2.00 m and its depth is also 2.00 m. The pond is first empty of water, and a person standing at a distance 170 cm from the near edge of the pond can just barely see the boundary on the opposite side between the side wall and the bottom. See Figure 36-5(a). The eye for this person is 170 cm from the bottom of his feet, so that the light ray from the boundary to his eye makes an angle of 45° with the vertical.

Now the pool is filled to the brim with water, and the observer can move farther from the edge of the pool and still see the boundary between side and bottom. See

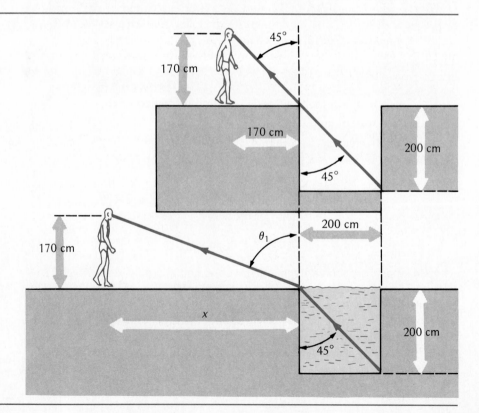

Figure 36-5. *Observer just able to see the boundary between side wall and bottom of a fish pond (a) with pool empty and (b) with pool filled with water.*

Figure 36-5(b). What is his maximum distance from the edge? For water, $n = 1.33$; for air, $n \simeq 1.00$.

Applying Snell's law to the refraction shown in Figure 36-5(b), we find that

$$n_1 \sin \theta_1 = n_2 \sin \theta_2$$

$$(1.00) \sin \theta_1 = (1.33) \sin 45°$$

$$\theta_1 = 70°$$

so that the maximum distance x from the pool edge is given by

$$\tan 70° = \frac{x}{170 \text{ cm}}$$

$$x = 467 \text{ cm}$$

36-4   Reflection

Let us first see that the law of reflection ($\theta_1 = \theta_1'$) can be deduced from general properties of wave propagation. Figure 36-6 shows a succession of plane wave fronts incident upon a plane surface; the angle of each wave front *relative to the surface* is also the angle of incidence θ_1. To find the reflected wave fronts, we use Huygen's principle (Section 18-4): to find a future wave front, we take the envelope of the Huygens wavelets generated along the wave front. At the interface in Figure 36-6, Huygens wavelets are generated in both the forward and the backward directions. The physical basis for Huygens's construction applied to electromagnetic waves is simply that as an electromagnetic wave travels through a medium, its electric field sets electrons in forced oscillation, and these electric oscillators generate electromagnetic waves that propagate both forward and backward. Within the refracting medium the net backward radiation is zero. At a boundary, however, the symmetry required for cancellation of the backward wave no longer obtains, so that both reflected and refracted waves are generated.

Figure 36-6 shows that the left end A of the incident wave front touches the surface when the right end B is a distance BC away from the surface. After time $t = BC/c$ has elapsed (with c as the wavespeed in the incident medium), the right end of the wave front reaches point C. At this same instant, the left end is at D, where $AD = ct$. Therefore $BC = AD$, and $\theta_1 = \theta_1'$. The reflection law has been proved.

Now consider Figure 36-7. Spherical wave fronts from a point source S are

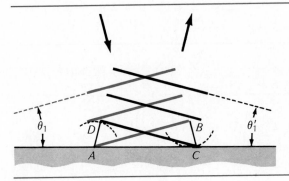

Figure 36-6. *Plane wave fronts incident upon a plane reflecting surface. Incident wave front AB later becomes reflected wave front CD.*

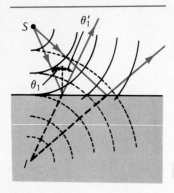

Figure 36-7. Spherical wave fronts from point source S striking a plane reflecting surface. The reflected spherical wave fronts appear to originate from virtual image I.

incident on a plane surface. The reflected rays and the reflected wave fronts are found by applying the rule $\theta_1 = \theta_1'$ to each ray. As the figure shows, the reflected rays appear to diverge from a single point I. To an observer viewing the reflected rays only, these rays and their associated spherical wave fronts seem to originate from the point *image I* rather than from their true origin, the point source S. The human eye is naïve; it interprets all rays reaching it as having always traveled in unbroken straight lines. To the eye (or a camera), the source appears (literally!) to be located at the position of the image. From the geometry of Figure 36-7, it is clear that the image is symmetrically located with respect to the source, with the reflecting boundary midway between them. The image here is said to be *virtual*. The rays appear to originate from location I; they actually come from S.

Example 36-2.   An object is placed near two plane mirrors that are at right angles to one another. What images can be seen in the mirrors?

See Figure 36-8(a), where the object is represented by O and a viewer's eye by E. Applying the rule $\theta_1 = \theta_1'$ at each reflection, we find that there are three virtual images:

- Image I_1, formed by reflection from mirror 1.
- Image I_2, formed by reflection from mirror 2.
- Image I_{12}, formed by reflections from both mirrors. Image I_{12} can be described as the image in mirror 2 of the image I_1 (or equivalently, the image in mirror 1 of image I_2).

The object and its three images are located symmetrically with respect to the lines representing the mirrors. Fold the paper at the mirror lines and the three images and object fall exactly on top of one another.

Note that the ray reaching the eye from image I_{12} is parallel to the same ray leaving the object O. This means that any ray undergoing two reflections at a corner mirror always emerges antiparallel to the ray leaving the object. (The behavior is like that of a billiard ball that makes two "reflections" at a corner of a billiard table.) Three mutually perpendicular mirrors form a corner reflector. *Any* ray undergoing a reflection from each of the three mirrors emerges *parallel* to the incident ray. Any ray reflected in a corner mirror comes straight back, undeviated in direction (but displaced laterally). Corner reflectors, usually made of red glass, are used on the rear fenders of bicycles; any light shining on them is returned toward the light source. Large-scale corner reflectors are in common use as "targets" for radar signals. Corner reflectors left on the moon's

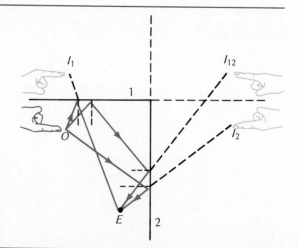

Figure 36-8. Images formed by reflections from two plane mirrors at right angles.

Using a Point Source

If you live near the San Andreas fault in California, your interest in what geologists call plate tectonics is more than casual. You'd like very much to know, if possible, when an earthquake is coming. Even small shifts in the relative positions of earth masses can provide a clue.

Actually, very small shifts in the earth's crust—even as small as 1 cm—*can* be measured with modern techniques. One way is shown in the figure. Here two radio antennas separated by the baseline distance *d* are pointed toward some distant point source. A microwave pulse reaches antenna A_1 first and then a time Δt later it arrives at antenna A_2, because antenna A_2 is farther from the point source than antenna A_1 by the distance $c\Delta t$. Measuring Δt to within one cycle of the microwave oscillation allows (with angle θ known) the distance *d* to be measured with an error of no more than about one microwave wavelength. (In more technical terms, radio-interferometry is used for long base-line geodesy, the measurement of distances between points on earth.)

What is a point source? How do you measure Δt? The point source can be a quasar, a *quasi-stellar astronomical source,* in a distant galaxy that emits pulses of radiation at microwave frequencies. Each pulse has its own characteristic shape, or signature, so that you can tell when the *same* burst of radiation reaches antennas A_1 and A_2. The signal arriving at each antenna is mixed with the signal from a continuously operating microwave oscillator of nearly constant frequency called a *maser* (where "m" for microwave replaces "l" for light in the acronym "laser"). The technique has been developed by a group from MIT,

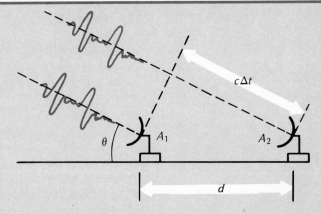

Goddard Space Flight Center, and the Haystack Observatory; the technique goes by the code name ARIES.*

Another similar technique for measuring small shifts in the position between the two locations on earth involves an orbiting satellite. The satellite (code named LAGEOS), launched in 1976 and flying at an altitude of 6000 km, is covered with retroreflectors (corner mirrors); a light signal from earth is sent back in the very same direction. Light pulses (in nanoseconds) from lasers go from earth to the satellite; their return to earth, and especially the delay in their arrival at the more distant receiving site, exploits a technique that can be called "optical radar." This technique has been developed by JPL, the Jet Propulsion Laboratory at CalTech.

*See the cover story for *Physics Today* **34**, 20 (April, 1981).

surface return light pulses sent from earth and allow the distance from earth to moon to be monitored to within a few centimeters.

36-5   Index of Refraction

We first recognize what does not change when a wave travels from medium 1, in which its wave speed is v_1, to a second medium 2, in which the wave speed is v_2. The *frequency f* of the wave is the *same* in all media. With wavelengths λ_1 and λ_2 in the two media we then write

$$v_1 = f\lambda_1 \quad \text{and} \quad v_2 = f\lambda_2$$

so that

$$\frac{v_1}{v_2} = \frac{\lambda_1}{\lambda_2} \tag{36-3}$$

The wavelength is greater in the medium with the higher wave speed.

The speed of an electromagnetic wave *in vacuo* is represented by c. By definition, the index for refraction n for a medium is the ratio of c to the wave speed in that medium. Therefore,

$$v_1 \equiv \frac{c}{n_1} \quad \text{and} \quad v_2 \equiv \frac{c}{n_2} \tag{36-4}$$

where n_1 and n_2 are called the indices of refraction for media 1 and 2. By definition, the index of refraction of a vacuum, or empty space, is exactly 1. Light travels through any medium at a *lower* speed than c: the refraction index always exceeds 1. For example, the speed of light through water is $2.25 \times 10^8 \, m/s$, so that water's index of refraction for visible light is $n = (3.00 \times 10^8 \, m/s)(2.25 \times 10^8 \, m/s) = 1.33$. For air near the earth's surface, n is 1.000 29, nearly the same as in a vacuum.

Visualize light going through a transparent material from a subatomic point of view. Then the electromagnetic wave actually travels *through a vacuum* at speed c, and it occasionally encounters a charged particle that scatters some of the incident beam. The combined effect of the scattering from the charged particles is to yield macroscopic propagation at a speed less than c.

Table 36-1 lists indices of refraction for some common transparent materials.

The *relative* index of refraction of medium 2 to medium 1, represented by n_{21}, is given by

$$n_{21} \equiv \frac{n_2}{n_1} = \frac{v_1}{v_2} \tag{36-5}$$

It follows that

$$n_{21} = \frac{1}{n_{12}}$$

In Section 36-6 we prove that a refraction index defined as the ratio of wave

Table 36-1

MATERIAL (at 20°C)	REFRACTIVE INDEX for yellow sodium light ($\lambda = 589.3$ nm)
Diamond	2.42
Ethyl alcohol	1.36
Glass (crown)	1.52
Glass (light flint)	1.58
Glass (heaviest flint)	1.89
Ice	1.31
Acetone	1.36
Sodium chloride	1.54
Stibnite (Sb_2S_3)	4.46
Water	1.33

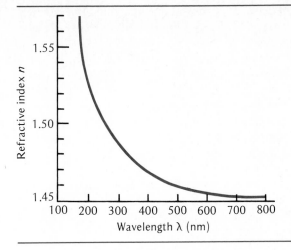

Figure 36-9. *Refractive index n of fused quartz as a function of free-space wavelength λ over the ultraviolet and visible regions of the electromagnetic spectrum.*

speeds in two media is the same as the refractive index appearing in Snell's law, (36-2). One can measure the value of the relative refractive index directly by observing the refraction of waves at an interface. The wavelengths in two media can be related to the respective indices of refraction by means of (36-3) and (36-4):

$$\frac{\lambda_1}{\lambda_2} = \frac{n_2}{n_1} \qquad (36-6)$$

If the index is large, the wavelength is small. Suppose, for example, that blue light of wavelength 400 nm (in free space) enters glass having an index of refraction of 1.50. The wavelength in the glass of this light, still blue, is *less*, namely (400 nm)/1.50 = 267 nm. It is customary to characterize a particular color of visible light by its *wavelength in free space*, rather than by its frequency, simply because the wavelength can be measured directly, whereas the frequency is computed from a knowledge of the wave speed.

The relative index of refraction for a given transparent material usually depends on the frequency. The index of refraction usually decreases with wavelength, as shown in Figure 36-9, or increases with frequency. Thus violet light travels through glass at a lower speed (approximately 1 percent) than red light, which has a longer wavelength. Whereas all the component frequencies of white light travel through a vacuum at the same speed c, the speeds of the various wavelengths differ in a refracting medium. The phenomenon is called *dispersion.*

The surface waves on a liquid of varying depth show a similar behavior. The wave speed decreases as the depth of the water is reduced. Thus, when waves in the ripple enter a region in which the depth is reduced, the waves are compressed because of the decrease in wavelength. See Figure 36-10.

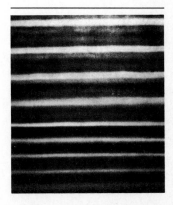

Figure 36-10. *Change in wavelength arising from a change in wave speed for water waves in a ripple tank.*

36-6 Refraction

Here we see that Snell's law for refraction follows directly from a change in wave speed.

Figure 36-11 shows wave fronts from medium 1, in which the wavespeed is

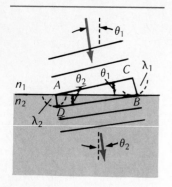

Figure 36-11.   Refraction of plane wave fronts at an interface. Incident wave front AC later becomes refracted wave front DB.

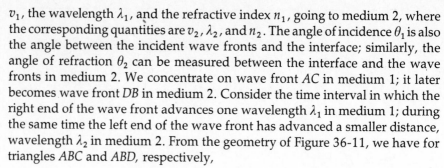

v_1, the wavelength λ_1, and the refractive index n_1, going to medium 2, where the corresponding quantities are v_2, λ_2, and n_2. The angle of incidence θ_1 is also the angle between the incident wave fronts and the interface; similarly, the angle of refraction θ_2 can be measured between the interface and the wave fronts in medium 2. We concentrate on wave front AC in medium 1; it later becomes wave front DB in medium 2. Consider the time interval in which the right end of the wave front advances one wavelength λ_1 in medium 1; during the same time the left end of the wave front has advanced a smaller distance, wavelength λ_2 in medium 2. From the geometry of Figure 36-11, we have for triangles ABC and ABD, respectively,

$$\sin \theta_1 = \frac{CB}{AB} = \frac{\lambda_1}{AB} \quad \text{and} \quad \sin \theta_2 = \frac{AD}{AB} = \frac{\lambda_2}{AB}$$

Eliminating AB gives

$$\frac{\sin \theta_1}{\sin \theta_2} = \frac{\lambda_1}{\lambda_2}$$

Using (36-6), we can write this as

$$n_1 \sin \theta_1 = n_2 \sin \theta_2 \qquad \text{(36-2), (36-7)}$$

which is Snell's law.

Another form of Snell's law, written in terms of the relative refractive index [(36-5)], is

$$\frac{\sin \theta_1}{\sin \theta_2} = n_{21} = \frac{1}{n_{12}} \qquad \text{(36-8)}$$

Refraction of a ray at an interface arises from a change in the wave speed, so that we can compute the relative wave speeds in two media simply by applying Snell's law to find the relative refractive index. In 1862, J. B. L. Foucault performed an experiment highly significant in the history of the theory of light when he showed that the speed of visible light through water is less than through air.*

Consider Figure 36-12(a). It shows a ray incident on a slab of refracting material with *parallel* faces. Snell's law governs the refraction at both interfaces, and both interior angles have the same value θ_2; this means that the emerging ray is exactly parallel to the incident ray. The emerging ray is, however, displaced laterally by an amount that depends on the thickness of the slab and its refraction index, but the ray is not deviated from its initial direction. This means that objects viewed through an ordinary sheet of window glass (one with parallel surfaces) are not distorted but merely displaced laterally by the refraction of light rays.

A ray is deviated from its original direction when it passes through a slab of refracting material with *nonparallel* faces. See Figure 36-13(b). Such a device is

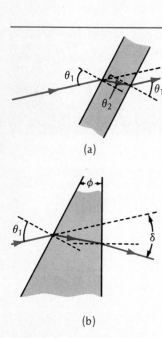

Figure 36-12.   (a) Refractions of a ray through a plate with parallel faces. The emerging ray is not deviated in direction. (b) Refractions of a ray through a plate with nonparallel faces (a prism). The emerging ray is deviated through angle δ.

---

* Foucault's experiment refuted the particle theory of light, in which the refraction of a light ray is attributed to a change in the particle's velocity at the interface. Snell's law can be derived from the particle model, but this model predicts a higher speed in a refracting medium than in a vacuum. See Problem 36-40.

known as a *prism*. The angle of deviation δ between the incident and emergent rays depends, for a given incident angle, on the refractive index of the prism and its apex angle ϕ. Because the refractive index depends on wavelength, the various frequency components of white light are deviated by *different* angles, or dispersed into the visible spectrum. Index n is greater for high frequencies (violet) than for lower frequencies (red); consequently, violet light is deviated most and red least, with all intermediate colors of the visible spectrum lying between.

Isaac Newton observed that in the dispersion of the visible spectrum by a prism, one single color, such as green, cannot be further resolved into component colors. He also observed that two prisms can be used first to disperse and then to reunite the various components into white light. See Figure 36-13.

A prism is commonly used in an instrument known as a prism spectrometer, for analyzing visible light into its component wavelengths. See Figure 36-14. Light from the source goes through a narrow slit and then through the prism. If the emitted light consists of certain discrete frequencies, the eye sees, or a photographic film records, a succession of "lines," where each line is an image of the slit at a particular frequency. A prism spectrometer can be used only for relative wavelength measurements; it must be calibrated against a device, such a diffraction grating (Section 38-4) that can make absolute wavelength measurements.

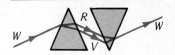

Figure 36-13. Dispersion of white light (W) by a prism into red (R) and violet (V) and its approximate recombination by a second prism.

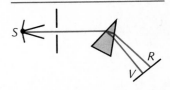

Figure 36-14. Simple elements of a prism spectrometer.

Example 36-3. Viewed from above, an object under water appears to be closer to the surface than its actual depth. Indeed, the *apparent depth* of any object immersed in a medium with relative refractive index n is given by $d' = d/n$, where d is the object's actual distance from the interface. Prove this result.

Figure 36-15 shows a point source O at depth d below the surface in a medium with refractive index n. We concentrate on a ray that emerges from O close to the vertical at angle θ_2 in the medium; this ray makes angle θ_1 relative to the normal in air. A ray directed vertically upward is undeviated. Extending these two rays backwards—as the two eyes do, in effect, in binocular vision—shows that they intersect at point I, which the eyes interpret to be the effective source, or image. The apparent depth d' is the distance of image I to the surface.

We apply Snell's law to the refracted ray, setting $n_1 = 1$ and $n_2 = n$. The angles are small enough that $\sin \theta_1 \simeq \theta_1$ and $\sin \theta_2 \simeq \theta_2$. We then have

$$n_1 \sin \theta_1 = n_2 \sin \theta_2$$
$$\theta_1 = n\theta_2$$

The geometry of Figure 36-15 shows that distances d and d' are related to angles θ_1 and θ_2 by

$$\tan \theta_1 \simeq \theta_1 = x/d' \quad \text{and} \quad \tan \theta_2 \simeq \theta_2 = x/d$$

so that eliminating x gives

$$\theta_1 = (d/d')\theta_2$$

Eliminating θ_1/θ_2 between this equation and the one above gives finally

$$d' = d/n$$

Suppose an object is immersed in water, for which $n \simeq 4/3$. Then, when viewed from above, the object will appear to be $\frac{3}{4} = 75$ percent of its true distance from the surface.

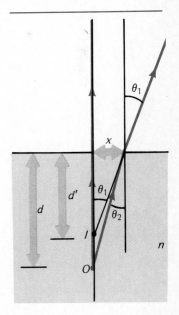

Figure 36-15.

36-7   Total Internal Reflection

When a beam is incident upon the interface between refracting media, some of its energy is transmitted and the rest is reflected. In one situation, however, the incident beam cannot be transmitted; consequently the incident wave is totally reflected.

Consider Figure 36-16. It shows a series of rays traveling from a medium of low optical density into a more dense medium (into a medium with larger n). (For simplicity, reflected rays are not shown in Figure 31-16, although there is at least some reflection.) The largest possible incident angle θ_1 is 90°. From Snell's law, the corresponding angle for θ_2 is

$$\frac{\sin \theta_1}{\sin \theta_2} = \frac{1}{\sin \theta_2} = \frac{n_2}{n_1}$$

Under these conditions, θ_2 is the *critical angle* θ_c, with

$$\sin \theta_c = \frac{n_1}{n_2} \qquad \text{where } \frac{n_1}{n_2} < 1 \tag{36-9}$$

The angle θ_c is the largest possible angle of refraction in medium 2.

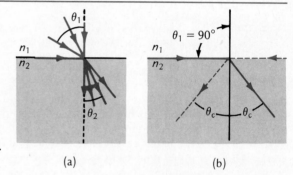

Figure 33-16.   *(a) The range of possible refraction angles θ_2.* *(b) Critical angle θ_c.*

(a)   (b)

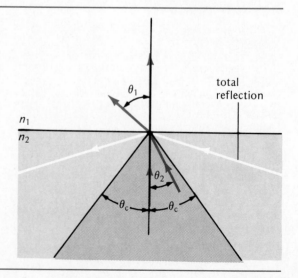

Figure 36-17.   *Total internal reflection occurs for $\theta_2 > \theta_c$.*

Imagine now that all ray directions in Figure 36-16 are reversed, as shown in Figure 36-17. A ray incident upon the interface from medium 2 at angle θ_c travels in medium 1 along the interface. All rays incident at lesser angles go through to medium 1. But for θ_2 greater than θ_c, it is impossible for a ray to be refracted into medium 1. All rays incident at an angle greater than the critical angle will be *totally reflected* into the optically more dense medium. For example, at an air-water interface, $\theta_c = \sin^{-1}(1/1.33) = 49°$.

Suppose you are underwater in a pool with a perfectly flat water-air surface. What do you see when you look up? It would be a transparent circular hole (refraction of all rays above the surface) surrounded by a mirror (total reflection). Total internal reflection has important applications. Suppose light is incident upon a glass prism, as in Figure 36-18 at an angle greater than θ_c. The ray is reflected from the *interior* face as from a perfectly reflecting mirror. Likewise, the particular brilliance of a gem, such as a diamond, arises from its very large refractive index (2.42), its high dispersion, and especially the multiply reflected rays. An *optical fiber* is one form of a *light pipe*, a device for confining light within the refracting material by total internal reflection. See Figure 36-19.

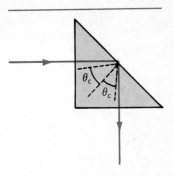

Figure 36-18. *Total internal reflection in a 45°-90°-45° prism.*

Example 36-4. Will total internal reflection take place in a 45°-90°-45° prism (Figure 36-18) made of the heaviest flint glass ($n = 1.89$) and immersed in (a) air? (b) in water?

(a) The critical angle for glass (1.89) in air (1.00) is, from (36-9),

$$\theta_c = \sin^{-1}\left(\frac{n_1}{n_2}\right) = \sin^{-1}\left(\frac{1}{1.89}\right) = 31.9°$$

Since θ_c is less than 45°, total reflection does take place.

(b) The critical angle for glass (1.89) in water (1.33) is

$$\theta_c = \sin^{-1}\left(\frac{1.33}{1.89}\right) = 44.7°$$

The critical angle is just slightly less than 45°, so that with very careful alignment, total internal reflection can still take place.

Figure 36-19. *Optical fiber photograph.*

Summary

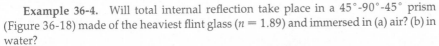

Definitions

Ray optics: optical effects accounted for entirely by ray tracing, with distinctive wave effects ignored; appropriate only when wavelength is far less than dimensions of apertures or obstacles.

Index of refraction n of material in which speed of light is v:

$$n \equiv \frac{c}{v} \tag{36-4}$$

where c is the speed of light in a vacuum.

Relative refractive index of medium 2 with respect to medium 1:

$$n_{21} \equiv \frac{n_2}{n_1} \tag{36-5}$$

Dispersion: variation of refractive index with wavelength (typically n decreases with an increase in λ).

Total internal reflection: reflection within the slow medium of all light incident upon an interface between two media. It arises when there can be no refraction in the faster medium (condition given below).

Fundamental Principles

Reciprocity principle: If light goes from A to B by some route, light goes from B to A along the same route; more simply, arrowheads on rays may be reversed.

Laws of reflection and refraction, both of which can be deduced from Huygens's principle:

$$\text{Reflection} \quad \theta_1 = \theta_1' \tag{36-1}$$

Angles of incidence and reflection are the same.

Refraction (Snell's law) $n_1 \sin \theta_1 = n_2 \sin \theta_2$ (36-2)

where all angles are measured relative to the normal to the interface between media and the n's are the respective refractive indices.

Important Results

When a wave goes from a medium to another, the frequency is unchanged whereas the wavelength is changed by the same factor as the wave propagation speed:

$$\frac{\lambda_1}{\lambda_2} = \frac{v_1}{v_2} \tag{36-3}$$

Critical angle θ_c (within the medium of larger refractive index) for total internal reflection:

$$\sin \theta_c = \frac{n_1}{n_2} \qquad \text{where } \frac{n_1}{n_2} < 1 \tag{36-9}$$

Problems and Questions

Section 36-4   Reflection

· **36-1 P** A person views an object through a periscope as sketched in Figure 36-20. How far away from the observer is the image he sees?

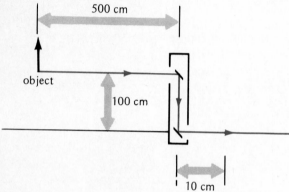

Figure 36-20.   Problem 36-1.

· **36-2 P** A bird flies directly toward its image in a mirror at 10 m/s. What is the speed of the bird relative to its image?

· **36-3 P** A person 180 cm tall stands in front of a vertical wall mirror. What is the minimum height of mirror, measured from top to bottom of the mirror, that will allow him to see all of himself in the mirror?

· **36-4 Q** Suppose that when standing in front of a vertical wall mirror, you can see only half of your body. Which of the following would enable you to see more of your body in the mirror?
(A) Move closer to the mirror.
(B) Move away from the mirror.
(C) Stand on chair.
(D) None of the above.

: **36-5 Q** The image you see in an ordinary mirror is reversed left-right but not up-down. Why?

· **36-6 P** Many sensitive instruments rely on the measurement of a very small angular displacement. Frequently the angle is measured by mounting a small mirror on the rotating component and reflecting a light beam from the mirror.

Show that, if the mirror rotates through angle θ, the light beam is deflected through angle 2θ.

: **36-7 P** (a) Show that the power of the signal received at the focus of a parabolic reflector is proportional to the square of the radius R of the antenna. (See Figure 36-21.) (b) The radius of the parabolic mirror in a reflecting telescope determines the intensity of the image registered on a photographic plate. If it is assumed, for simplicity, that distant stars have the same power output, how does doubling the radius of the mirror affect the range of stars that can be "seen"?

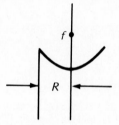

Figure 36-21.   Problem 36-7.

: **36-8 Q** An ellipse can be defined as the locus of points the sum of whose distances from two fixed points is a constant. Show that any wave disturbance originating at one focus is focused at the other (Figure 36-22).

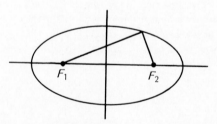

Figure 36-22.   Question 36-8.

: **36-9 P** Two plane mirrors make an angle θ (Figure 36-23). Locate all the images for
(a) $\theta = 90°$, (b) $\theta = 60°$, (c) $\theta = 30°$.

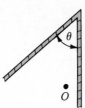

Figure 36-23.   Problem 36-9.

: **36-10 P** Two adjacent walls of a room and the floor are covered with mirrors. What is the maximum number of images that can be seen with this arrangement?

: **36-11 P** A rangefinder can be constructed of two mirrors, as shown in Figure 36-24. Mirror M_1 is fixed and half-silvered; light can pass through it along path PM_1O. Light can also be reflected along path PM_2M_1O. An observer at O sees two images corresponding to light rays that have traveled the two paths. The two images can be made to coincide if mirror M_2 is rotated about a vertical axis; this will occur when the angle between the two mirrors is θ. Show that the distance d to P is then given by $d = a/(\tan 2\theta)$. The separation a of the mirrors and θ can be accurately measured, thereby allowing d to be determined accurately also.

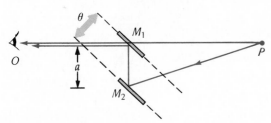

Figure 36-24.   Problem 36-11.

: **36-12 P** A woman whose eyes are 150 cm above the floor stands next to a child 80 cm tall. They stand 2 m from a vertical wall mirror. What is the minimum size of mirror (from top to bottom) that will enable the woman to see all of the child?

: **36-13 P** A ray of light leaves a source at S, is reflected at a mirror surface and then travels to point O (Figure 36-25). Show that (a) the time for the light to travel this path is $t = (1/c)(y_1 \sec \theta_1 + y_2 \sec \theta_2)$. (b) Show that this time is a minimum when $\theta_1 = \theta_2$. (This illustrates a general principle called Fermat's law of least time, according to which the path followed by a light ray in going from one point to another is one for which the time is least. See also Problem 36-41.)

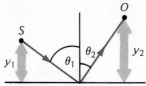

Figure 36-25.   Problem 36-13.

: **36-14 P** A parallel beam of light is incident on the apex of a prism of angle ϕ (Figure 36-26). Show that the angle between the two reflected beams is 2ϕ.

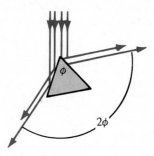

Figure 36-26.   Problem 36-14.

Section 36-5   Index of Refraction

· **36-15 Q** The speed of light in water is about $\frac{3}{4}c$. When a beam of light passes from air into water, its frequency

(A) remains the same but its wavelength is reduced to three-fourths that in air.

(B) is three-fourths that in air but its wavelength remains unchanged.

(C) and wavelength are unchanged.

(D) and wavelength are reduced by a factor of $\frac{3}{4}$.

(E) and wavelength are increased by a factor of $\frac{4}{3}$.

· **36-16 P** Light has a wavelength of 460 nm in crown glass ($n = 1.52$). What are the wavelength and frequency in (a) a vacuum? (b) lithium fluoride ($n = 1.37$)? (c) magnesium oxide ($n = 1.75$)?

· **36-17 P** Light travels through carbon bisulfide at a speed of 1.84×10^8 m/s. What is the index of refraction of this material?

· **36-18 P** The index of refraction of magnesium oxide is 1.75. What is the speed of light in this material?

· **36-19 P** The index of refraction of the eye is 1.33. What is the speed of light of wavelength 540 nm when passing through the eye?

· **36-20 P** Light of wavelength 500 nm in air passes through glass of index of refraction 1.4. What is the frequency of the light when it is in the glass?

: **36-21 Q** When sunlight is refracted by a piece of broken glass, you can often see sparkling colors in the transmitted light. Why is this?

(A) Different colors are absorbed in different amounts by the glass.

(B) Various light rays travel paths of different lengths in the glass.

(C) The speed of light in glass varies slightly with the frequency of the light.

(D) The sunlight stimulates the glass to emit fluorescent light of various wavelengths.

(E) Light of different wavelengths is scattered by different amounts depending on the angle of scattering (that is, on the angle at which you view the light).

: 36-22 Q At a long, straight ocean beach, you see that waves come in almost parallel to the shoreline, no matter what the wind direction on the open sea where the waves are formed. Which of the following best explains this effect?

(A) Water waves move faster in deep water than in shallow water.

(B) Water waves move faster in shallow water than in deep water.

(C) It is only the component of the wind perpendicular to the shore line that causes waves.

(D) Water wave velocity varies with wave frequency.

(E) The amplitude of a wave decreases as it moves away from its source.

: 36-23 Q Figure 36-27 is a sketch of what appears in a satellite photograph of a harbor. The wave speed of water waves in shallow water varies with $h^{1/2}$, where h is the depth of water. What can you deduce about the configuration underwater from the wavefronts shown in this picture?

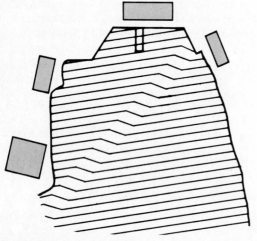

Figure 36-27. *Question 36-23.*

: 36-24 P Any curve can be approximated over a small region by a circular arc whose radius R, the radius of curvature of the curve, is $R = [1 + (dy/dx)^2]^{3/2}/(d^2y/dx^2)$. The derivatives are evaluated at the point at which the curve is to be fitted. The equation of a parabola can be written $x^2 = 4Fy$, where F is the distance between the focus and the vertex of the parabola. Show that the focus of a parabola lies at $\frac{1}{2}R$, where R is the radius of curvature of the parabola at its vertex.

Section 36-6   Refraction

: 36-25 P A penny rests at the bottom of a container that holds a 4-cm depth of water ($n = 1.33$) on top of which

floats a 2-cm layer of oil ($n = 1.60$). What is the apparent depth of the penny when viewed from directly above?

· 36-26 P A ray of light strikes a glass plate at an angle of 60° with the surface. The index of refraction of the glass is 1.45. What is the angle between the refracted ray and the surface?

· 36-27 P A light ray strikes the surface of water in a beaker at an angle of 30° above the horizontal. At what angle with the vertical does the ray travel in the water?

: 36-28 P Two 45°-45°-90° glass prisms are joined as shown in Figure 36-28. Their refractive indices are 1.45 and 1.60. By what angle will a light ray normally incident be deviated?

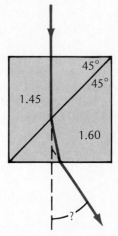

Figure 36-28. *Problem 36-28.*

: 36-29 P Light is incident on a glass of index of refraction n at an angle of incidence θ. What should be the angle of incidence in terms of n for the angle of refraction to be $\frac{1}{2}\theta$?

: 36-30 P A fish rests at the bottom of a pool of water 3 m deep and at a distance of 4 m from the edge of the pool. How far back from the edge of the pool must a fisherman 180 cm tall stand if he is not to be seen by the fish?

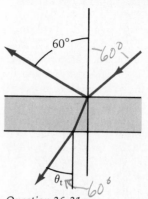

Figure 39-29. *Question 36-31.*

: **36-31** **Q** A laser beam is incident on a plate of glass with parallel sides and index of refraction 1.4. Some light is reflected at an angle of 60° to the normal, and some transmitted through the glass. The situation is sketched in Figure 36-29, although the angles are not necessarily drawn correctly. What is the value of the angle θ_t made by the transmitted beam?

(A) 60°.
(B) 30°.
(C) 45°.
(D) 38°.
(E) θ_t cannot be determined, since the angle of incidence is unknown.

: **36-32** **P** At what angle relative to the normal must a ray strike the interface between a medium with refractive index n_1 and a second medium with refractive index n_2 so that the reflected and transmitted rays will be at right angles?

: **36-33** **P** A nautical buoy 1 m tall floats with 40 cm of its length projecting above a lake at a point where the water is 4 m deep. If the sun is 30° above the horizon, how long is the shadow of the buoy on the bottom of the lake? The index of refraction of water is 1.33.

: **36-34** **P** Light is incident on a parallel plate of glass of thickness t and index of refraction n at an angle of incidence θ_1. Show the transmitted ray undergoes a lateral displacement $t \sin (\theta_1 - \theta_2)/\cos \theta_2$ where $\sin \theta_2 = (1/n) \sin \theta_1$.

: **36-35** **P** A light ray passes through several parallel slabs of material with different refractive indices. Total internal reflection does not occur at any interface. Show that the final transmitted ray is parallel to the incident ray quite apart from the order the various slabs are arranged in; for example, 4, 3, 1, 2 gives the same effect as 1, 2, 3, 4.

: **36-36** **P** A light ray is incident at 45° on a 45°-45°-90° prism as shown in Figure 36-30. The refractive index of the glass is 1.5. What is angle ϕ?

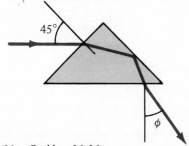

Figure 36-30. *Problem 36-36.*

: **36-37** **P** A ray passing through a prism is deviated by an angle that depends on the ray's angle of incidence with the first surface, on the angle of the prism, and on the index of refraction. Show that the deviation is a minimum when the ray passes symmetrically through the prism, that is, when $\theta_1 = \theta_2$ in Figure 36-31. (*Hint:* Consider reciprocity.)

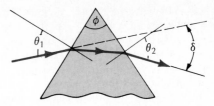

Figure 36-31. *Problems 36-37, 36-38, 36-39.*

: **36-38** **P** When a light ray passes symmetrically through a prism, it experiences minimum deviation (Question 36-37). Show that if δ is the angle of deviation and ϕ is the angle of the prism,

$$n = \frac{\sin \frac{1}{2}(\delta + \phi)}{\sin \frac{1}{2}\phi}$$

This provides a direct way of measuring the refractive index of the material of the prism.

: **36-39** **P** Show that for a thin prism (ϕ small) with the incident light close to the normal (θ small), the deviation angle δ is independent of the angle of incidence and is given by $\delta = (n - 1) \phi$. See Figure 36-31.

: **36-40** **P** Snell's law ($\sin \theta_1 / \sin \theta_2 = $ a constant) may be derived from a particle model of light in which it is assumed that the particles of light travel at constant velocity within any uniform medium and that the component of a particle's velocity parallel to an interface between two refracting media is unchanged. See Figure 36-32. Note that $v_{t1} = v_{t2}$. (*a*) Derive Snell's law from the particle model and show that $\sin \theta_1 / \sin \theta_2 = v_2/v_1$, where v_1 and v_2 are the respective particle speeds in media 1 and 2. (*b*) Show that, according to this particle model, the speed of light propagation through a refracting medium exceeds that through vacuum, in contradiction to the experimental findings.

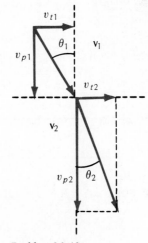

Figure 36-32. *Problem 36-40.*

: **36-41** **P** Ray optics—the laws of reflection and of refraction—are special examples of a remarkable general principle called the *principle of least time,* or *Fermat's principle,* and first propounded by P. Fermat (1608–1665) in 1650. In its simplest form Fermat's principle says that a ray going from point *A* to point *B* will take the path that corresponds to minimum travel time between *A* and *B*. The rectilinear propagation of light through a uniform medium is clearly consistent with this principle since a straight line between two points is the shortest distance between them and also the path of least time. To show that Snell's law is also a consequence of Fermat's principle consider Figure 36-33. Here a ray goes from point *A* in medium 1, to the interface at point *B* and finally to point *C* in medium 2. Think of the distance from fixed point *D* to point *B* to be the variable distance *x*. Write a relation for the total travel time from *A* to *C*, find the value for *x* corresponding to minimum travel time, and show that Snell's law is the consequence.

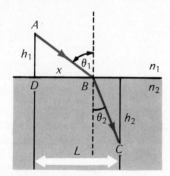

Figure 36-33.   Problem 36-41.

Section 36-7   Total Internal Reflection

· **36-42 P** If you look up at the undisturbed surface of water from the bottom of a swimming pool, you will see a circular transparent hole looking out on the sky surrounded by a shiny mirror. What is the radius of the hole when your eye is 3 m below the surface?

· **36-43 P** What is the maximum angle ϕ of incidence at which light can enter the end of a glass fiber in air shown in Figure 36-34 if the light is not to escape through the sides of the fiber? The index of the fiber is 1.40, and the fiber is straight.

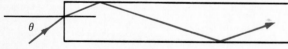

Figure 36-34.   Problem 36-43.

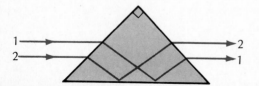

Figure 36-35.   Problem 36-44.

· **36-44 Q** Figure 36-35 shows two parallel rays incident on a 45°-45°-90° prism of index of refraction 1.5. Trace the rays to show that they emerge inverted. Of what practical use is a prism of this type?

: **36-45 P** A ray of light is moving in a plane parallel to one face of a cube of glass of index *n*. The ray strikes another interior wall at an angle of incidence θ, as shown in Figure 36-36. To what range of values is θ restricted if the ray is to remain trapped within the cube?

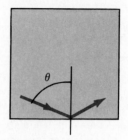

Figure 36-36.   Problem 36-45.

: **36-46 Q** We can deduce that when light in material 1 is incident on the interface with material 2, total internal reflection will occur
(A) whenever $n_1 > n_2$, for all rays.
(B) whenever $n_1 < n_2$, for all rays.
(C) by examining the law of reflection, $\theta_1 = \theta_1'$.
(D) by examining the law of refraction, $n_1 \sin \theta_1 = n_2 \sin \theta_2$.
(E) independent of the values of n_1 and n_2 so long as the angle of incidence is greater than 49°, the critical angle.

: **36-47 Q** Sketched in Figure 36-37 is a light ray passing through a pair of prism binoculars. In such an instrument,
(A) the two perpendicular faces of the prism are silvered to make them reflecting.
(B) the index of refraction of the glass is not an important design consideration, so long as the prism angles are correct.
(C) the index of refraction of the prism should be greater than 1.414.

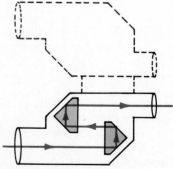

Figure 36-37.   Question 36-47.

(D) the index of refraction of the prism should be greater than 1.522.

(E) the index of refraction of the prism should be less than 1.500.

(F) the prism increases the overall magnification.

: 36-48 P Three light rays are directed into a prism with refractive index n, as in Figure 36-38. (a) What is the minimum value of n if ray No. 2 is to be totally internally reflected? Call this value n_c. (b) If ray No. 1 is to be totally reflected, is the required minimum index larger or smaller than n_c? (c) If ray No. 3 is to be totally reflected, is the required minimum n larger or smaller than n_c?

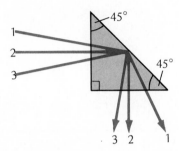

Figure 36-38.   Problem 36-48.

Supplementary Problem

36-49 P On a hot summer day you see mirages ahead of you on the surface of a highway. Typically, you see what

appears to be a pool of water, but you may even see what appears to be a reflection of an approaching car. The effect is a manifestation of total internal reflection. A layer of air just above the hot, black roadway is significantly hotter than the air a meter or so above the pavement. The index of refraction of air is related to its density ρ by a relation of the form $n^2 - 1 = (\text{constant})\rho$.

Thus light angling down from above emerges from cool air (large n) and is reflected from hot air (smaller n). If the angle of incidence is large enough, the light will be totally "internally" reflected. This puts a limit on how close the near edge of the apparent "pool of water" can be to you. As you approach it, it will keep moving away. For simplicity, suppose there is a sharp temperature difference of 12 C° within a few centimeters of the pavement. Assume that the cooler air is at 300 K and has $n = 1.003$.

(a) Show that in Figure 36-39 $\phi \approx \sqrt{2(n-1)(\Delta T/T)}$. (b) If your eyes are 1.5 m above the roadway as you are driving, how far away will the near edge of the mirage appear?

Figure 36-39.   Problem 36-49.

37

Thin
Lenses

Tracing rays through a whole succession of interfaces between transparent materials that differ in refractive index—the basic problem to be solved in designing a high-quality multiple-element thick lens—is easy in principle (just Snell's law) but tediously difficult in practice. We won't tackle it. Instead we consider only the far simpler situation with a single, very thin lens. This special case illuminates most of the fundamental ideas of image formation and allows the basic design principles of optical instruments to be understood.

37-1   Focal Length of a Converging Lens

Suppose a ray passes through a plate of glass with *parallel* surfaces. The ray is refracted at each interface but the emerging ray is *parallel* to the incident ray [Figure 37-1(a)]. The emerging ray is displaced laterally but does not deviate in direction. On the other hand, when a ray encounters a slab of glass with nonparallel faces (a prism), the emerging ray is deviated relative to the incident ray [Figure 37-1(b)].

Now consider the structure shown in Figure 37-2; it consists of two prisms

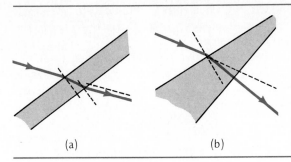

(a)   (b)

Figure 37-1. (a) Refraction through a slab with parallel faces produces a lateral displacement but no deviation. (b) A ray is deviated by a prism.

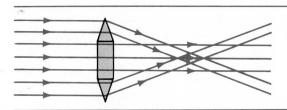

Figure 37-2. Focusing by a crude lens.

and a plate with parallel faces. A beam of horizontal rays incident from the left passes through this rather crude lens. The emerging rays intersect in a relatively small region to the right. Rays through the center are undeviated. Rays through the top prism are deviated down, and rays through the bottom prism are deviated up, and they emerge at a single deviation angle. We wish to devise a structure in which all the parallel incident rays intersect at a single point. It is clear what is required: the rays' deviation must increase gradually upward (or downward) from the center of the lens, not abruptly as in Figure 37-2. In practice, lenses are made with *spherical* surfaces; this shape is easy to grind. Here we consider only a *thin lens*. This means that its thickness is small compared with its radius, as shown in Figure 37-3.

Figure 37-3. A lens with spherical surfaces.

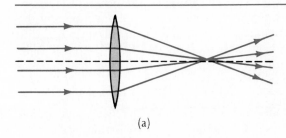

(a)

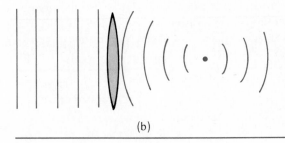

(b)

Figure 37-4. (a) Parallel incident rays and (b) plane wave, focused by a lens.

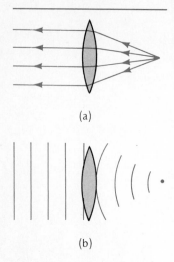

(a)

(b)

Figure 37-5. Ray and wave-propagation reversal of Figure 37-4. (a) Rays from the focal point emerge after refraction as parallel rays. (b) Diverging spherical wave fronts emerge after refraction as plane wave fronts.

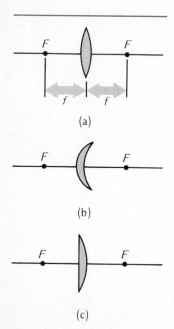

(a)

(b)

(c)

Figure 37-6. Examples of converging lenses: (a) double convex; (b) concave-convex; and (c) plano-convex.

Figure 37-4(a) shows a thin lens with two convex spherical surfaces. Incident rays initially parallel to the horizontal axis will, after passing through the lens, intersect at a single point, or at least in a very much smaller region than shown in Figure 37-2. (The proof from Snell's law is given in Section 37-6.) This intersection point is known as a *principal focal point,* or *principal focus,* of the lens. Suppose that a source is placed at an infinite (or at least a very large) distance from the lens. Then the rays incident upon the lens are effectively parallel over a small solid angle, and these parallel rays will, after traversing the lens, intersect at the principal focus. The focusing originates from the circumstance that as we go up (or down) from the center of the lens, each ray is deviated slightly more than the one below (or above) it. The behavior of wave fronts associated with the rays is shown in Figure 37-4(b). The wavefronts undergo a change in curvature at each of the two interfaces. The incident wave fronts are plane, whereas the emerging wave fronts are converging spherical wave fronts that collapse into a point at the principal focus (and thereafter expand outward as diverging spherical wave fronts). The wave fronts change shape because the relatively thicker central portion of the lens slows a wave front more than the thinner portions near its outer edge do.

Any focal point can be defined in several equivalent ways:

- The point at which *all rays intersect.*
- The point at which *wave fronts collapse to a point* (changing from converging into diverging wave fronts).
- The point at which the beam has *maximum intensity.*

Recall the principle of reciprocity, or time reversal (Section 36-2); it says that we can reverse the directions of rays (or imagine time to run backward) without changing the light paths. Applied to Figure 37-4, this means that if a point source is placed at the principal focus, rays diverge from this point and emerge from the lens as a beam of parallel rays [Figure 37-5(a)]. Equivalently, spherical wave fronts diverging from the principal focus and passing through the lens emerge as plane wave fronts [Figure 37-5(b)].

Suppose that we flip over the thin lens so that its left and right faces are interchanged. Rays then pass through the lens in opposite direction. But the incident parallel rays again converge to a focus at the same distance from the lens's center. Every lens has two principal focal points, one on each side. Both principal foci, denoted by F, are at the same distance f, the *focal length,* from the center of a thin lens. See Figure 37-6(a).

This result is general; it holds not only for the thin lens with two convex surfaces that we have concentrated on thus far, but also for a lens, such as that shown in Figure 37-6(b), with a concave and a convex surface, and for a plano-convex lens (the plane surface with an infinite radius of curvature) as in Figure 37-6(c). All these are converging lenses. The essential requirement is that the lens be thicker at its center than at its edges. (Clearly, the focal length depends on the radius of curvature of the two surfaces and on the refractive index of the lens material. The lens maker's formula, which gives f in terms of these parameters, is derived in Section 37-6.)

Imagine now that the lens is tilted a little, as in Figure 37-7(a), so that its symmetry axis (shown dashed) no longer lies along the direction of incident parallel rays. For a small-tilt angle, nothing changes; the rays focus at essentially the same point as before, a distance f from the center of the lens. Now look at the very same situation but with the lens axis, and all else, turned until it

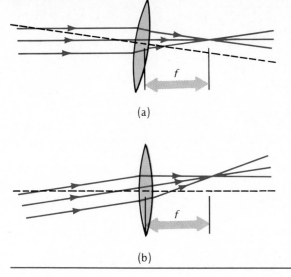

(a)

(b)

Figure 37-7. *(a) The location of the focal point does not change when the lens is tilted slightly. (b) Part (a) drawn with the lens axis horizontal. It shows that oblique rays also come to a focus at distance f from the lens.*

is horizontal, as in Figure 37-7(b). For what are now obliquely incident parallel rays, the focal point is displaced transversely from the lens axis. Clearly, the ray through the lens's center is undeviated; it passes, in effect, through two parallel surfaces. Thus, rays from an infinitely distant source are brought to focus in a *plane*, the *focal plane*, a distance f from the lens. This result holds, however, only if the angle between the oblique rays and the lens axis is small. Such rays, nearly parallel to the lens axis, are called *paraxial rays*. We shall assume hereafter that the lens is very thin and the rays are paraxial.

The type of lens discussed thus far is known as a *converging* lens. (Strictly, a lens thicker at its center than at its edges is a converging lens only if its material is optically more dense than its surroundings; an example is a glass lens for visible light when the lens is immersed in air.) Any converging lens increases the degree of convergence of rays and wave fronts passing through it, or decreases the degree of their divergence.

37-2   Ray Tracing to Locate a Real Image

We are now prepared to deal with a general problem. What happens to the rays from a point source located at any location along the lens axis? The object, some

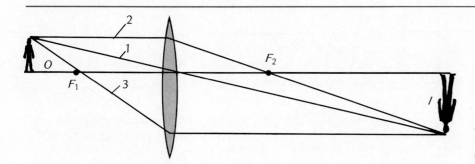

Figure 37-8. *Formation of image I by rays 1, 2, and 3 from object O.*

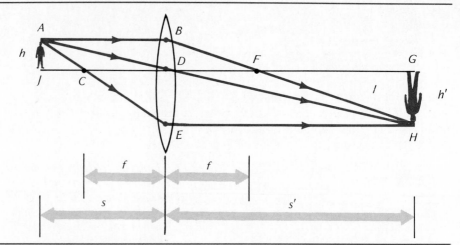

Figure 37-9. Geometrical relations among object distance s, image distance s', and focal length f.

luminous source, is represented by O. It is located near a thin converging lens, but farther than the focal length, as shown in Figure 37-8. We want to find the place where the light from the top of the person representing object O comes to a focus after passing through the lens. Three rays are easy to trace:

- Ray 1 passes through the *center* of the thin lens *unchanged.* This ray is not deviated — the lens faces are parallel. Also, this ray is not displaced laterally by an appreciable distance — the lens is very thin.
- Ray 2 *incident parallel* to the lens axis is deviated by the lens, so that the emerging ray goes *through principal focus F_2* on the far side.
- Ray 3, aimed to pass *through* the *near principal focus F_1*, is deviated and *emerges parallel* to the lens axis.

These three rays intersect, or focus, at a single point.* Indeed, *all* other paraxial rays from the upper tip of O intersect to form a point image of O. We have traced rays from the top of the head representing object O. When we choose some other point on an extended object lying in the same transverse plane, we can again find the corresponding image point. For paraxial rays through a thin lens, we can obtain the entire image I of object O.

Finding the image of an object is simple. All we need is a prior knowledge of the lens's focal length f; then we can locate the image by drawing the three rays, as in Figure 37-8. Ray tracing, which is strictly a geometrical procedure, is not always practicable, however. To deal with general situations, we wish to find the mathematical relation between three important quantities: the focal length f, object distance s of the object from the lens, and image distance s' of the image from the lens.

Consider Figure 37-9. It is merely Figure 37-8 redrawn, with identifying letters added. The object distance s is AB, the image distance s' is $EH = DG$, and the focal length f is $CD = DF$. The object height h is $AJ = BD$; the image height h' is $GH = DE$. Triangles AJD and HDG are similar. Therefore,

---

* The rays in Figure 37-8 are drawn as if the lens were infinitesimally thin, with a single straight ray to the plane of the lens and another single straight ray away from the plane of the lens. In actuality, every ray undergoes *two* refractions at the lens; there is a refraction at each of the two surfaces.

$$\frac{AJ}{JD} = \frac{HG}{DG}$$

$$\frac{h}{s} = \frac{h'}{s'} \tag{37-1}$$

The ratio of image-object *distances*, s'/s, is equal to the ratio of image-object *sizes*, h'/h. The ratio of the linear dimensions of image-to-object, h'/h, is known as the *lateral magnification*.

Triangles *BDF* and *HGF* are also similar. Therefore,

$$\frac{DF}{BD} = \frac{GF}{GH}$$

$$\frac{f}{h} = \frac{s' - f}{h'}$$

Using (37-1) in this relation gives, after a little algebra,

$$\frac{f}{s} = \frac{s' - f}{s'} = 1 - \frac{f}{s'}$$

and dividing by f and rearranging yield

$$\frac{1}{f} = \frac{1}{s} + \frac{1}{s'} \tag{37-2}$$

This is the formula for computing image locations formed by a thin lens. Everything we can do with it can be done equally well by ray construction [that is how we derived (37-2)]. In working any problem in which an image is formed by a lens, it is always advisable to sketch, at least roughly, the rays forming the image, as well as to do the numerical calculation.

Example 37-1. An object is 18 cm from a converging lens with a focal length of 12 cm. Where is the image? How does the object's size compare with the size of the image?

With $f = 12$ cm and $s = 18$ cm, we get from (37-2) that $s' = 36$ cm. The ray diagram is that shown in Figure 37-8 and 37-9.

Since the image distance is twice the object distance, the lateral dimensions of the image are twice the corresponding dimensions of the object. By the same token, the transverse area of the image is four times the area of the object.

Suppose that we move the object far from the lens, so that $s = \infty$ in (37-2). Then we find $s' = f$; that is, the image of a very distant object is at a principal focus, in accord with its definition. Conversely, if $s = f$, then $s' = \infty$.

The image shown in Figures 37-8 and 37-9 is said to be a *real* image; light rays actually pass through this location. If a sheet of paper were placed a distance s' from the lens, we should see an actual focused image on it. This image would be inverted in the transverse focal plane; that is, up and down would be interchanged relative to the object, and so too would be left and right. Equation (37-2) shows that as s increases, s' decreases, and the opposite, so that when the object is shifted along the lens axis, its image is displaced in the same direction. Thus, the image of a three-dimensional object, although inverted in the transverse plane, is not inverted along the lens axis.

Suppose that we apply the reciprocity principle to Figures 37-8 and 37-9. Then the rays are reversed and our original image becomes an object, and vice versa. These two locations, where object and image switch locations, are said to be *conjugate* points. Note also that when $s = 2f$, then $s' = 2f$. For this special case, the lens merely inverts the object without changing its size.

Many familiar optical devices involve a single converging lens forming a real image. Most familiar is the eye; it forms a real image on the retina. The image distance s' from the eye lens to the retina is fixed; objects at various distances from the eye are brought into focus on the central plane of the retina when the eye muscles change the focal length f of the eye lens by changing the radius of curvature of the lens surfaces. A simple camera also forms a real image; it is in the focal plane of the photographic film. On the other hand, a projector forms a much enlarged and inverted image of transparent film on the screen.

Example 37-2.   A normal human eye can focus on objects from 25 cm to infinity by changing the curvature, and therefore the focal length, of the eye lens. The process is called *accommodation*. A far-sighted person can see far objects in focus, but not near ones. Suppose that the eye of a certain far-sighted person can see no objects in focus closer than 200 cm (that person's *near point*). What focal length of contact lens will bring objects as close as 25 cm into focus?

The fixed distance from the eye lens to the surface of the retina is designated s'. Then, with an object 200 cm from the eye and without a corrective lens we have, using lens equation (37-2)

$$1/f_e = 1/200 \text{ cm} + 1/s'$$

where f_e is the focal length of the unaided eye lens.

In Section 37-6 it is shown that the equivalent focal length f of two thin lenses close together is

$$1/f = 1/f_1 + 1/f_2 \tag{37-5}$$

When a contact lens with focal length f_{cl} is used with an object 25 cm away, applying the lens equation again gives

$$1/f_{cl} + 1/f_e = 1/25 \text{ cm} + 1/s'$$

Eliminating f_e and s' from the two equations above gives

$$f_{cl} = 28.6 \text{ cm} = 0.286 \text{ m}$$

Optometrists use the reciprocal of the focal length in meters to express the *lens power* in units of *diopters*, so that the contact lens above would have a lens power of $1/0.286 \text{ m} = +3.5$ diopters.

Converging contact lenses (with positive lens power) correct the vision of a far-sighted person. Without a correcting lens the image is formed behind the retina surface; with a converging lens added, the focussed image is brought forward, closer to the eye lens, to the surface of the retina. A near-sighted person requires a diverging contact lens (with a negative lens power); here the image is moved away from the eye lens and toward the retina.

37-3   Ray Tracing to Locate Virtual Image

A new phenomenon occurs when an object is placed closer to a converging lens than the principal focal point. See Figure 37-10. We can locate the image geometrically by drawing exactly the same three rays as before:

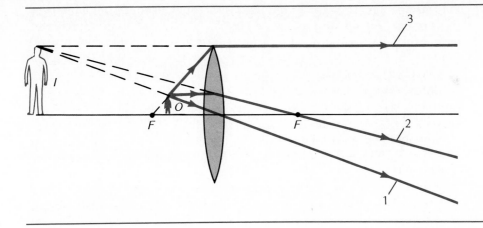

Figure 37-10. Formation of virtual image I by rays 1, 2, and 3 from object O.

- Ray 1 through the lens center, undeviated.
- Ray 2, initially parallel to the lens axis, through the far focal point.
- Ray 3, so drawn that its direction after leaving the object is the same as that of a ray starting at the near focal point and passing through the top of the object. This ray emerges from the lens parallel to the axis.

These three rays do not intersect. What sort of image is now formed? What does an eye see when looking from the right toward the lens?

The eye (or a camera) is naive in the following sense: it interprets any ray reaching it as always having traveled strictly along a straight line. Said differently, the eye recognizes only the final directions of rays entering it. Therefore, the three rays appear to have originated from that point on the left of the lens where their backward extensions intersect. They appear to come from what is called a *virtual image I.* A sheet of paper placed at the location of *I* shows no image. But a person viewing the object through the lens would see an erect, enlarged, and virtual image. Used in this fashion, with an object closer to the lens than the focal length, a converging lens is a *magnifying glass,* or *simple magnifier.*

The lens equation, (37-2), can also be used to compute the image distance s' for a virtual image. Here s' is taken to be *negative;* this can be proved in detail by analyzing the geometry of Figure 37-10 in the fashion of Figure 37-9.

Example 37-3. An object is placed 9 cm from a converging lens of 12 cm focal length. Where is the image and what is its character?

With $f = 12$ cm and $s = 9$ cm, we get by substituting in (37-2) that $s' = -36$ cm. The image is virtual, erect, and with 4 times the lateral dimensions of the object, as shown in Figure 37-10.

37-4   Diverging Lens

Thus far we have considered converging lenses; such lenses are thicker in the center than at the edges (when their material has a higher refractive index than the surrounding medium). Now consider just the reverse: a lens with spherical surfaces thinner at its center than at its edges, as shown in Figure 37-11. Such a lens is a *diverging* lens. It increases the degree of divergence of rays and wave fronts passing through it. Incident parallel rays diverge after passing through

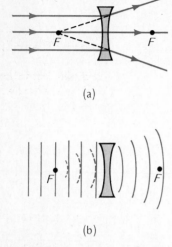

(a)

(b)

Figure 37-11. The divergence of (a) the rays and (b) the wave fronts, by a diverging lens.

the lens and appear to have originated from a point source at the principal focus on the left side of the lens (Figure 37-11(a)]. Said a little differently, incident plane waves emerge from a diverging lens as diverging spherical wave fronts (Figure 37-11(b)]. We can again use the basic lens equation to relate image distance, object distance, and focal length, provided that the focal length of the diverging lens is taken to be *negative*. The object distance s is, as before, taken as positive. Again there are two principal foci, one on each side of the lens, and the focusing of paraxial rays is independent of the tilt angle of the lens or of the face of the lens exposed to the object.

We locate an image produced by a diverging lens with the same ray-construction procedure as before, as shown in Figure 37-12.

- Ray 1 passes undeviated through the lens's center.
- Ray 2, parallel to the lens axis, emerges as if originating from the near principal focus F.
- Ray 3 is "aimed" to go through the far principal focus; therefore this ray is deviated, to emerge parallel to the axis.

The rays emerging from the lens appear to diverge from image I. This image is reduced, erect, and virtual. Indeed, a diverging lens always forms a virtual image of a real object. That I is virtual is again indicated by the fact that the image distance s', as computed from (37-2), is negative.

Example 37-4. An object is placed 15 cm from a diverging lens with a focal length of 12 cm. Where is the image and what is its character?

With $f = -12$ cm and $s = 15$ cm, we find in applying (37-2) that $s' = -6.7$ cm. As shown in Figure 37-12, the image is virtual, erect, and reduced in size relative to the object.

We have concentrated on lenses of the usual variety, which are made of optically dense materials, such as glass, and intended for use with visible light. Lens devices can also be made of optically light materials. For example, Figure 37-13 shows converging and diverging lenses formed by cavities within an optically dense medium. The converging lens here has *concave* surfaces, and

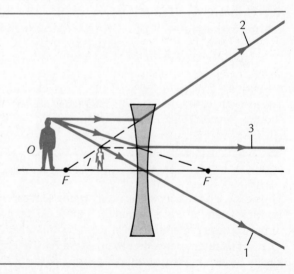

Figure 37-12. Image I formed by rays 1, 2, and 3 from object O for a diverging lens.

the diverging lens *convex* surfaces. As before, a converging lens causes plane wave fronts to collapse to a point, whereas a diverging lens causes plane wave fronts to expand outward.

The term *lens* does not need to be restricted to devices that focus visible light. There are lenses for sound waves and electromagnetic microwaves. Magnetic- and electric-field arrangements that focus a beam of charged particles are referred to respectively as magnetic and electrostatic lenses.

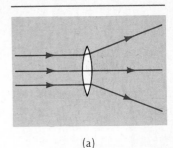

(a)

Example 37-5.   The eye of a certain near-sighted person can see in focus no objects farther away than 500 cm (that person's *far point*). The image for far objects is formed in front of the retinal surface; the rays from far objects converge too much. What focal length of contact lens will bring very far objects into focus?

The fixed distance from eye lens to the surface of the retina is designated s', because s' is the image distance. Then, with an object 500 cm from the eye and without a corrective lens, we have, using the lens equation (37-2)

$$1/f_e = 1/500 \text{ cm} + 1/s'$$

where f_e is the focal length of the unaided eye lens.

When a contact lens with focal length f_{cl} is added and the object is infinitely far away, we have, in applying the lens equation again

$$1/f_{cl} + 1/f_e = 1/\infty + 1/s'$$

where we have also used (37-5) for the equivalent focal length of two thin lenses in contact.

Eliminating f_e and s' from the two equations above gives

$$f_{cl} = -500 \text{ cm}$$

The contact lens must be diverging with a focal length of 5.0 m; the required lens power is $-1/5.0 \text{ m} = -0.2$ diopter.

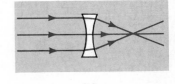

(b)

Figure 37-13.   Lenses formed of a material that is optically less dense than the material in which they are immersed. (a) A double convex but diverging lens. (b) A double concave but converging lens.

37-5   Lens Combinations

To trace rays through two or more lenses in sequence, proceed as follows. Treat the image for the first lens as the object for the second lens, then treat the image for the second as the object for the third lens, and so forth.

Astronomical Telescope   Figure 37-14 shows a combination of two converging lenses with focal lengths f_1 and f_2. Lens 1 forms a real inverted image I_1 of the object O_1. Image I_1 becomes the object O_2 for lens 2. In the particular arrangement shown here, the object O_2 falls just inside the focal point of the second lens, so that lens 2 acts effectively as a magnifying glass. Consequently, the final image I_2 formed by lens 2 is virtual; it is to the left of lens 2. The rays chosen to find the image I_1 are not continued through the second lens; instead, new rays are chosen, whose deviations through lens 2 are found in the fashion shown in Figure 37-10. Note also that for the purposes of ray construction, the lenses are taken to be of infinite transverse size.

An *astronomical telescope* has the lens arrangement in Figure 37-14. Since object O_1 is far from lens 1, its real image I_1 is examined by lens 2 acting as a magnifying glass. The final image is formed at a great distance from the lens. Lens 1 (closer to the object) is known as the *objective* lens; lens 2 (closer to the eye) is known as the *eyepiece*, or *ocular*. With object O_1 at a great distance from a

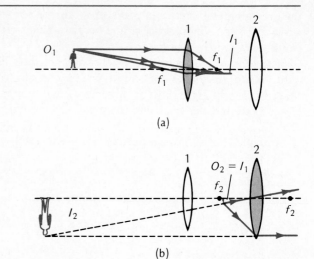

(a)

Figure 37-14. Image formation (a) by lens 1, and then (b) by lens 2, in an astronomical telescope.

(b)

telescope and a final image I_2 also at a great distance, it is clear that the total distance between the objective lens and the eyepiece is close to $f_1 + f_2$. This is the minimum length of the telescope. (One side of a binocular is basically an astronomical telescope with the rays reflected twice from prisms, as in Figure 36-47, one to reverse up-down and the other to reverse left-right.)

What matters in any optical instrument used to magnify an object is not the size of the image itself but rather the size of the image *formed on the retina of the eye.* The retinal image size is determined, in turn, by the angular spread of rays entering the eye, as shown in Figure 37-15. A proper measure of the magnification by an optical instrument is the *angular magnification,* or *magnifying power,* defined as

$$\text{Angular magnification} = \frac{\theta_I}{\theta_O} \tag{37-3}$$

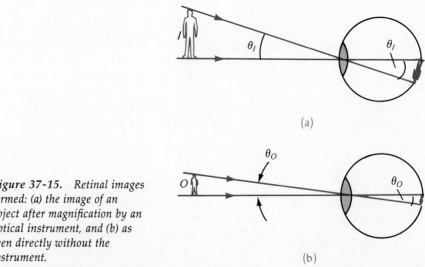

(a)

Figure 37-15. Retinal images formed: (a) the image of an object after magnification by an optical instrument, and (b) as seen directly without the instrument.

(b)

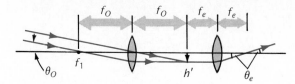

Figure 37-16. *Angular magnification by an astronomical telescope.*

Here angle θ_I is subtended at the eye by the rays coming from the final image, and angle θ_O is subtended at the eye by the object when viewed directly, without the aid of the instrument. Equivalently, the magnifying power is the ratio of the size of the retinal image formed with an optical instrument to the retinal image size without the instrument. Angular magnification is a more useful criterion than lateral magnification, given by (37-1), which is just the ratio of image size to object size. We can see this as follows. There is *no* angular magnification when the image size is doubled while at the same time the image is located at twice the distance of the object from the viewer.

Now we compute the angular magnification of an astronomical telescope. Figure 37-16 shows the rays incident upon a telescope from a very distant object. The angle θ_O subtended by the object at the objective lens is essentially the same as the angle subtended by the object at the unaided eye. The height of the real image of the objective lens is denoted h'. This image is formed at a distance f_O from the objective lens and a distance f_e from the eyepiece, where f_O and f_e are the focal lengths of the objective lens and eyepiece, respectively. From the geometry of Figure 37-16, we have

$$\tan \theta_O = \frac{h'}{f_O} \qquad \text{and} \qquad \tan \theta_e = \frac{h'}{f_e}$$

where θ_e is the angle subtended at the eye by the final image. Since the angles θ_O and θ_e are very small,

$$\frac{h'}{f_O} = \tan \theta_O \simeq \theta_O \qquad \text{and} \qquad \frac{h'}{f_e} = \tan \theta_e \simeq \theta_e$$

But by the definition in (37-3), we have

$$\text{Angular magnification} = \frac{\theta_e}{\theta_O} = \frac{f_O}{f_e} \tag{37-4}$$

A telescope's magnifying power is simply the ratio of the focal lengths of the objective and the eyepiece. High magnification means using a long-focal-length objective lens with a short-focal-length eyepiece.

Microscope The lens combination shown in Figure 37-17 is that for a compound microscope. Here the object O_1 is placed close to but outside of the

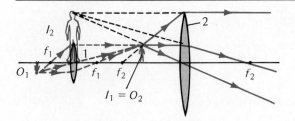

Figure 37-17. *A compound microscope.*

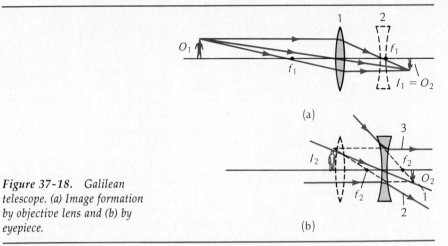

*Figure 37-18.   Galilean
telescope. (a) Image formation
by objective lens and (b) by
eyepiece.*

principal focus f_1 of the objective lens; its image I_1 is enlarged and real. The eyepiece is used as magnifying glass to form a still further enlarged but virtual image I_2. How do we examine an object to see it most clearly? We must make the size of the image *on the retina* as large as possible, so that we bring the object as close to the eye as the focusing properties of the eye will permit, typically 25 cm from the eye. Therefore, a microscope is also most effective if the final image I_2 is located at 25 cm from the eye. The angular magnification of a microscope is θ_I/θ_O, where θ_O is the angle subtended by the object held 25 cm from the unaided eye.

Galilean Telescope   A special situation arises when the image formed by the first lens does not lie between two lenses in combination. Consider Figure 37-18(a). If lens 2 were not present, the image I_1 would be formed at the location shown. To locate the final image after rays traverse the second lens, we must regard this image I_1 as a *virtual object* for lens 2. We find the final image I_2 of the virtual object $O_2 = I_1$ in Figure 37-18(b) as follows:

• Ray 1 is so "aimed" that it passes through the center of the diverging lens without deviation.
• Ray 2, initially parallel to the lens axis and aimed at I_1, is deviated, so as to be directed from the near focal point f_2.
• Ray 3 is aimed to pass through the far focal point and therefore emerges from lens 2 parallel to the axis.

The final image I_2 is greatly enlarged, erect, and virtual. The optical device shown in Figure 37-18 comprising a converging objective lens and a diverging eyepiece, is a *Galilean telescope,* or opera glass.

37-6   The Lens Maker's Formula

Here we derive the lens maker's formula. It gives the focal length of a thin lens in terms of the radius of curvature of its two spherical surfaces and the relative index of refraction of its material.

The underlying assumptions are these:

- The lens is very thin.
- All rays are *paraxial;* that is, the angle θ of any ray relative to the lens axis is so small that we can assume $\sin \theta \simeq \theta$ and $\tan \theta \simeq \theta$.

To simplify the derivation, we imagine a doubly convex lens to be split into two plano-convex lenses, as shown in Figure 37-19. We concentrate first on the right half with a curved surface of radius R_1. See Figure 37-20, where the incident ray strikes the plane surface along the normal at a distance h from the lens axis. This ray is therefore undeviated as it continues into the lens material. Relative to the normal of the curved surface, this ray makes an angle of incidence θ_1. The emerging ray makes angle θ_2 relative to the normal to the exterior curved surface, and it intercepts the lens axis at a distance f_1 from the lens. (Since the lens is assumed to be very thin, we need not specify more precisely exactly from what point f_1 is measured.)

Assume that the lens has refractive index n and is immersed in a vacuum. Then θ_1 and θ_2 are related by Snell's law:

$$n \sin \theta_1 = \sin \theta_2$$

All angles are assumed small, so that $\sin \theta_1 \simeq \theta_1$, and $\sin \theta_2 \simeq \theta_2$. Snell's law reduces to

$$n\theta_1 = \theta_2$$

We see in Figure 37-20 that the emerging ray makes angle $\theta_2 = n\theta_1$ with the

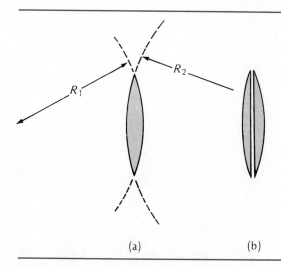

(a)   (b)

Figure 37-19.   *(a) A doubly convex lens with radii R_1 and R_2. (b) This lens imagined as a composite of two plano-convex lenses.*

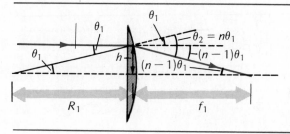

Figure 37-20.   *Refraction of a ray by a plano-convex lens.*

normal. The angle of this ray relative to the horizontal lens axis is $\theta_2 - \theta_1 = n\theta_1 - \theta_1 = (n-1)\theta_1$.

From the geometry of Figure 37-20, we have for small angles

$$\theta_1 = \frac{h}{R_1} \quad \text{and} \quad (n-1)\theta_1 = \frac{h}{f_1}$$

Eliminating θ_1 and h from these equations then yields

$$\frac{1}{f_1} = \frac{n-1}{R_1}$$

The other plano-convex lens of Figure 37-19 has a focal length f_2, given by

$$\frac{1}{f_2} = \frac{n-1}{R_2}$$

Now imagine that the two plano-convex lenses are reassembled to form the original doubly convex lens. We apply the general lens equation to each of the two parts as follows:

$$\frac{1}{s_1} + \frac{1}{s_1'} = \frac{1}{f_1} = \frac{n-1}{R_1}$$

and

$$\frac{1}{s_2} + \frac{1}{s_2'} = \frac{1}{f_2} = \frac{n-1}{R_2}$$

But the image of the first lens is the object of the second one, so that $s_1' = -s_2$ (the minus sign appears because the real image of the first lens becomes a *virtual* object for the second lens). Adding the two equations above and dropping the subscripts give

$$\frac{1}{s} + \frac{1}{s'} = (n-1)\left(\frac{1}{R_1} + \frac{1}{R_2}\right)$$

$$\frac{1}{s} + \frac{1}{s'} = \frac{1}{f_1} + \frac{1}{f_2}$$

The general thin-lens equation is

$$\frac{1}{s} + \frac{1}{s'} = \frac{1}{f} \tag{37-2}$$

so that the equivalent focal length f for two thin lenses with focal lengths f_1 and f_2 in contact is

$$\frac{1}{f} = \frac{1}{f_1} + \frac{1}{f_2} \tag{37-5}$$

and the focal length of the doubly convex thin lens is given by

$$\frac{1}{f} = (n-1)\left(\frac{1}{R_1} + \frac{1}{R_2}\right) \tag{37-6}$$

This relation applies for a lens of refractive index n immersed in a vacuum.

More generally, if a lens with refractive index n_2 is immersed in a medium with index n_1, we replace n in (37-6) by the relative refractive index $n_{21} = n_2/n_1$.

Radii of curvature R_1 and R_2 are taken to be positive quantities for convex surfaces. The radius of curvature for a concave surface is then negative in (37-6). Clearly, then, a doubly concave lens (both R_1 and R_2 negative) has a negative f and is diverging. More generally, the sign of the focal length is controlled by the relative magnitudes and signs for R_1 and R_2 and by whether $n > 1$ or $n < 1$. A plane surface corresponds to an infinite radius of curvature.

37-7   Lens Aberrations

An ideal lens forms an image that is, apart from being inverted or magnified, an exact replica of the object. All parts of the image are in exact focus, with every point in the object rendered as a point in the image. All colors are faithfully rendered. The shape is not distorted. No such ideal lens exists. A variety of aberrations attributable to the lens cause the image to differ from the object not only in size, but in clarity, in color, and in shape.

The most fundamental limitation comes from the circumstance that a lens focuses *waves*. As a consequence of *diffraction*, a point in the object is rendered not as a point, but as a smeared image, the size of the smear depending on the wavelength. As we shall see in Section 39-6, diffraction effects are determined by the ratio of the wavelength to the outside radius of the lens aperture. The shorter the wavelength, the smaller the diffraction. Using short-wavelength violet light, rather than long-wavelength red light, reduces diffraction. The limitation arising from diffraction is obvious when you recognize that you cannot "see" any smaller dimension in the object than one wavelength of the waves used to "look" at the object.

Diffraction is an inherent lens aberration. The following lens aberrations can, however, be corrected to some degree by using two or more lenses in combination.

• *Chromatic aberration* has its origin in dispersion. For transparent materials in the visible region of the spectrum, the refractive index depends on the freespace wavelength, so that the refraction angle at an interface depends on the color of light. As Figure 36-9 shows, the refractive index is higher for violet light than for red. Consequently, if polychromatic light from an object refracts through a converging lens, the violet component will be imaged closer to the lens than the red component, as shown in Figure 37-21.

• *Spherical aberration* arises because a spherical lens surface, although easy to grind, is not the ideal shape for focusing all rays originating at the lens axis. As the angle between the rays and the lens axis increases, so that rays are not all paraxial, the outer edges of the lens focus at a different point from that formed by rays through the central portion of the lens, as shown in Figure 37-21. Because of spherical aberration, a point in the object is rendered as a diffuse circular disc in the image plane.

• Still other aberrations occur when the object is not on the lens axis: *Coma*, an extension of spherical aberration, makes the images of objects off the lens axis appear comet-shaped. *Astigmatism* arises because object points in a single transverse plane are imaged on a spherical surface. *Distortion* comes from the

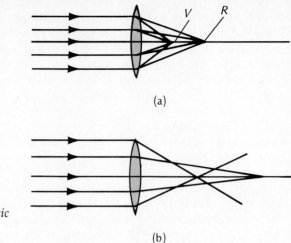

Figure 37-21. *(a) Chromatic aberration. (b) Spherical aberration.*

circumstance that the magnification depends to some degree on the object's distance from the lens axis.

37-8   Spherical Mirrors (Optional)

A mirror consisting of a small portion of a spherical surface has focusing properties similar to those of a thin lens. First consider the beam of parallel rays incident on the concave spherical mirror in Figure 37-22a. The rays intersect at principal focus F at a distance $\frac{1}{2}R$ from the mirror surface, where R is the radius of curvature of the mirror. This result is obvious from the geometry of Figure 37-22b, where a ray is incident parallel to the mirror's symmetry axis, strikes at point P, and makes an angle of incidence i with the normal to the surface, which is also a radius R of the sphere, whose center is at C. The reflection angle

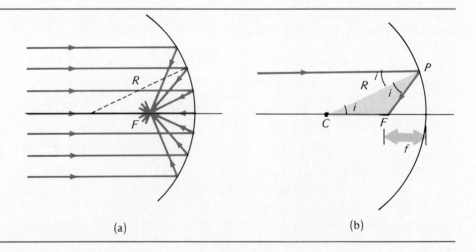

Figure 37-22. *(a) Spherical concave mirror with radius of curvature R brings paraxial rays to focus at F; (b) ray incident on point P is reflected through principal focus F at distance f from the mirror surface.*

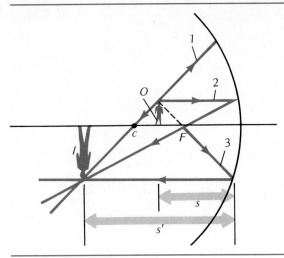

Figure 37-23. *Object O produces real image I.*

is also i and the ray crosses the symmetry axis at focal point F. The two small angles in triangle CPF are equal to i, so that sides CF and FP are equal. For very small i (paraxial rays), each small triangle side is approximately $\frac{1}{2}R$. The focal length f from principal focal point F to the mirror surface is $f = \frac{1}{2}R$.*

A concave spherical mirror is very much like a thin converging lens in focusing an incident beam of parallel rays. The significant difference is this: rays go through a lens, but rays are bent back to the same side as the incident rays after reflection in the mirror.

Just as for image formation by a thin lens, it is easy to draw three principal rays to locate the image I of object O, as illustrated in Figure 37-23.

- Ray 1 passes through (or is directed away from) the *center of curvature C.* This ray necessarily strikes the mirror surface along the normal, and it is reflected back along the *same straight line.*
- Ray 2 is *incident parallel* to the mirror axis so that it is reflected to pass *through the principal focus F.*
- Ray 3 is aimed to pass *through* (or to originate from) the *principal focus,* so that this ray is reflected to go *parallel* to the lens axis.

Not only is the ray-construction procedure the same as for a thin converging lens; the mathematical relation for focal length f, object distance s, and image distance s' is also the same:

$$1/f = 1/s + 1/s' \tag{37-2}$$

The proof is implicit in the geometry of Figure 37-23. A concave spherical mirror is assigned a *positive* focal distance; image distance s' is also positive for a real image.

A virtual image is formed when the object is closer to the surface of a concave mirror than the principal focus. See Figure 37-24, where the same principal rays are used to locate the image. A virtual image — also erect and

---

* Only paraxial rays focus at point F; rays far from the mirror axis undergo spherical aberration. A parabola (or, strictly, a parabola turned around its symmetry axis to form a paraboloid) focuses all rays incident along the symmetry axis at the parabola's focal point.

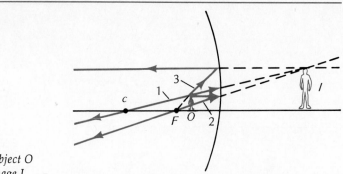

Figure 37-24. *Object O produces virtual image I.*

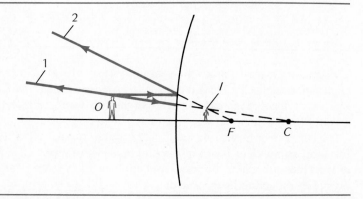

Figure 37-25. *Convex spherical mirror with center of curvature C and principal focus F produces a virtual image I of object O.*

enlarged—corresponds to a negative value for image distance s' in the relation above.

Just as the concave spherical mirror is analogous to the converging thin lens, a convex spherical mirror is the analog of the diverging thin lens. Rays incident on a convex spherical mirror along the lens axis diverge after reflection and appear to originate from a focal point on the far side of the mirror, a distance $\frac{1}{2}R$ from the surface. A negative focal length is assigned to the convex mirror. As Figure 37-25 shows, a convex mirror always forms a virtual image of an object at any distance s from the mirror surface.

Summary

Definitions

Principal focal point (for converging lens): location at which rays from infinitely distant point source intersect, or wave fronts collapse to a point, or intensity has maximum value.

Focal length: distance of principal focal point from thin lens.

Important Results

Thin lens equation:

$$\frac{1}{s} + \frac{1}{s'} = \frac{1}{f} \qquad (37\text{-}2)$$

where object and image distances, s and s', and focal length f, are measured from the lens.

Sign conventions:

$f > 0,$	Converging lens	$f < 0,$	Diverging lens
$s' > 0,$	Real image	$s' < 0,$	Virtual image
$s > 0,$	Real object	$s < 0,$	Virtual object

Lens-maker's formula for the focal length f of a lens of relative refractive index n:

$$\frac{1}{f} = (n - 1)\left(\frac{1}{R_1} + \frac{1}{R_2}\right) \qquad (37\text{-}6)$$

where R_1 and R_2 are the radii of curvature (positive for a convex surface when $n > 1$).

The equivalent focal length f of two thin lenses in contact with focal lengths f_1 and f_2 is given by

$$\frac{1}{f} = \frac{1}{f_1} + \frac{1}{f_2} \qquad (37\text{-}5)$$

Problems and Questions

Section 37-2   Ray Tracing to Locate a Real Image

· **37-1 Q** Which of the following statements best describes the underlying reason that a lens is able to form an image of an object?

(A) The velocity of light is independent of the motion of the source and of the observer.

(B) The angle of incidence equals the angle of reflection.

(C) Two light rays interfere constructively or destructively, depending on the phase difference between them.

(D) Light has both a particle nature and a wave nature.

(E) Light travels at different speeds in different media.

· **37-2 P** A simple box camera has a focal length of 50 mm. When the camera is focused to take a picture of an object 100 m away, what is the distance from the film to the center of the lens?

· **37-3 P** A camera has a focal length of 50 mm. What is the size of the image on the film of a building that is 4 m tall and 40 m from the camera?

· **37-4 P** A projector is used to form an image of a slide on a screen. The slide has dimensions 3×3 cm and is 20 cm from a lens. How far from the lens must the screen be if the image is to be 300×300 cm?

: **37-5 P** A point isotropic radiator of light is located at $x = 0$ along the axis of a converging lens of focal length 20 cm. The lens is located at $x = 40$ cm. Sketch the intensity of light as a function of x from $x = 1$ cm to $x = 80$ cm.

: **37-6 Q** A real object is placed on the axis of a converging lens at a point far away from the lens. As the object is brought closer to the lens, the image

(A) is initially at the focal point but recedes from the lens as the object approaches.

(B) is initially at the focal point but approaches the lens as the object approaches.

(C) remains at the focal point but changes in size.

(D) moves away from the lens in such a way that the distance from the object to the image stays constant.

: **37-7 P** If an object is a distance $2f$ in front of a converging lens of focal length f, the image is thus

(A) twice the size of the object and inverted.

(B) twice the size of the object and erect.

(C) of the same size as the object and inverted.

(D) of the same size as the object and erect.

(E) four times the size of the object.

: **37-8 Q** The radius of curvature R is the same for each of the two lenses in Figure 37-26. The lenses are made of the same kind of glass. One is a thin lens and the other a thick lens. Can you reason qualitatively something about their focal lengths?

(A) They should have equal focal lengths.

(B) The thin lens should have greater focal length.

(C) The thick lens should have greater focal length.

(D) Whether the thick lens has a greater or a smaller focal length than the other lens depends on whether the ratio t/R is greater than 1 or less than 1.

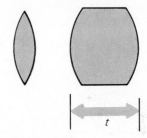

Figure 37-26.   Problem 37-8.

: **37-9 P** Two thin lenses of focal lengths f_1 and f_2 are placed very close to each other. Show that they are equivalent to a single thin lens whose focal length is approximately $f = f_1 f_2 / (f_1 + f_2)$.

: **37-10 P** A screen is placed 10 m from a converging lens that is to be used to project an image on a screen with magnification $\times 4$. (a) Where should the object be placed relative to the lens, and (b) what focal length must the lens have?

: **37-11 P** An object is located 4.0 m from a thin converging lens of focal length 1.0 m. (a) Where is the image formed? Draw the principal rays. (b) Where should the object be placed to form a real image 150 cm from the lens? (c) Where should the object be placed to form a virtual image 25 cm from the lens?

: **37-12 P** A camera lens of focal length 55 mm is positioned so that a distant object is in focus. (a) How far must the lens be moved to focus on an object that is 3 m distant? (b) Must the lens be moved toward the film or away from the film to bring the close object into focus?

: **37-13 P** Show that the lens formula

$$\frac{1}{f} = \frac{1}{s} + \frac{1}{s'}$$

can be written $f^2 = xx'$, where x = distance from the first focal point to the object, and x' = distance from the second focal point to the image.

: **37-14 P** A simple thin lens has a focal length of 30 cm. Find the image and determine whether it is real or virtual, and erect or inverted, for each of the following object distances: (a) 15 cm; (b) 30 cm; (c) 45 cm; (d) 60 cm.

: **37-15 P** The "f-stop" number of a lens is the ratio of the lens focal length to the diameter of the aperture used. On a typical camera, the f stops are numbered 1.4, 2.0, 2.8, 4.0, 5.6, 8.0, 11, 16, 22. (a) By what factor does the amount of light striking the film change when the f stop is changed from f5.6 to f8? (b) Why are the particular f number settings listed above used on cameras?

: **37-16 Q** Shown in Figure 37-27 are three light rays emanating from a source S and being imaged on a screen. If all three left the source at the same time, which would arrive at the screen first?

(A) A.
(B) B.
(C) C.
(D) All would arrive at the same time.
(E) This question cannot be answered without more information.

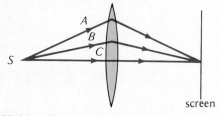

Figure 37-27. Question 37-16.

Section 37-3   Ray Tracing to Locate a Virtual Image

· **37-17 Q** Is it possible to tell just by looking at an image with your eye if the image is real or virtual?

(A) Yes, because real images are always turned upside down.
(B) Yes, because one can see right through a virtual image (since it is not actually there), whereas a real image looks like a solid object.
(C) Yes, because a real image can be seen from the side, whereas a virtual image can only be seen head-on.
(D) No. They both look the same to the eye.

· **37-18 Q** The bright filament of a lamp is placed 30 cm from a converging lens of focal length 20 cm. A piece of white cardboard is held at the proper position on the side of the lens opposite the lamp, so that an image of the filament is seen on the cardboard. What happens to this image if the cardboard is then removed?

(A) Nothing; that is, the image is still at the same place as when the cardboard was present.
(B) The image no longer exists.
(C) The image is present, but now it is at the surface of the lens.
(D) The image is still present, but now it is a virtual image instead of a real image.
(E) The image is still present, but now it is a real image instead of a virtual image.

: **37-19 P** An object 1 cm tall is placed 30 cm from a converging lens, as shown in Figure 37-28. The lens of diameter 2 cm has a focal length of 40 cm. Sketch rays emanating from either end of the object and thereby determine the region of space to which an eye is limited if it is to see the entire virtual image formed by the lens.

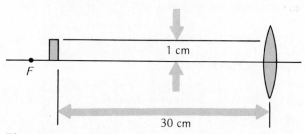

Figure 37-28. Problem 37-19.

Section 37-4   Diverging Lens

· **37-20 Q** A virtual image of a real object formed by a thin lens

(A) always occurs between the lens and a focal point.
(B) can be formed by a diverging lens but not by a converging lens.
(C) cannot be used to expose film and thereby make a photograph.
(D) may be erect or inverted, depending on the position of the object.
(E) cannot be seen with the eye since it does not actually exist.

· **37-21 Q** A diverging lens will always give an image that is

(A) virtual and reduced.
(B) virtual and enlarged.
(C) real and reduced.
(D) real and enlarged.
(E) None of the above is true consistently.

: **37-22 P** For a normal eye, the far point is at infinity and the near point is at 25 cm. What focal-length lens should be used to correct the vision of a person who can read a book clearly when it is 80 cm away but not when it is closer?

· **37-23 P** A girl can see most clearly objects placed 25 cm from her eyes when she wears glasses of power $+1.2$ diopters. Where does she see objects most clearly when not wearing her glasses?

: **37-24 P** A point source of light is placed at $x = 0$ on the axis of a diverging lens of focal length 10 cm. The lens is at $x = 20$ cm. Sketch along the x axis from $x = 1$ cm to $x = 40$ cm, the intensity of light as a function of x.

: **37-25 Q** A certain person can see clearly an object 25 cm from his eyes only when he wears eyeglasses with converging lenses. If this person takes off his glasses, the image formed in his eye is
(A) 25 cm farther behind the eye.
(B) farther from the lens of his eye than the retina.
(C) closer to the lens of his eye than to the retina.
(D) virtual.
(E) erect, rather than inverted.

: **37-26 Q** Which of the following is more likely to be able to see clearly underwater?
(A) A person with normal vision.
(B) A farsighted person.
(C) A nearsighted person.
(D) A person with stigmatism.
(E) All of the above could see equally well.

: **37-27 Q** You are sitting in a theatre behind a person wearing eyeglasses. You notice that when you view the scene on stage through one of the eyeglasses of the person ahead of you, the image is smaller than when you view the scene directly. Is the person ahead of you near-sighted or far-sighted?

Section 37-5 Lens Combinations

· **37-28 Q** Suppose that the top half of the objective lens of a telescope is covered. What effect, if any, will this have on the image you see when looking through the telescope?
(A) The top half of the image will be blacked out.
(B) The bottom half of the image will be blacked out.
(C) The image will look the same as before except that it will be dimmer.
(D) The magnification will be half as large as before.
(E) There will be no noticeable effect.

: **37-29 Q** See Figure 37-29, where a convex-convex lens of focal length 8 cm is placed 2 cm to the left of a concave-concave lens of focal length 6 cm. A 1-mm beam of parallel light rays is incident on the convex lens from the left. As a result, the transmitted beam will
(A) converge.
(B) diverge.
(C) form a parallel beam of diameter 1 mm.
(D) form a parallel beam of diameter greater than 1 mm.
(E) form a parallel beam of diameter less than 1 mm.

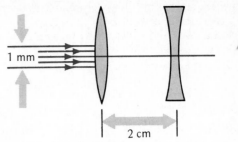

Figure 37-29. Question 37-29.

: **37-30 P** An object 2 mm tall is placed at the origin. A diverging lens of focal length 20 cm is placed on the x axis at $x = 30$ cm. A converging lens with focal length 24 cm is placed on the x axis at $x = 50$ cm. Determine the position and size of the final image formed by these lenses.

: **37-31 P** A telephoto lens consists of a converging lens of focal length $f = +6.0$ cm placed 4 cm in front of a diverging lens of focal length $f = -2.5$ cm. Determine the position of the image (relative to the diverging lens) of a very distant object.

: **37-32 P** Two thin converging lenses, each of 30-cm focal length, are separated by 15 cm. An object is placed 60 cm from the first lens. Where is the final image located?

: **37-33 P** An astronomical telescope is to be constructed with magnification ×120. Its eyepiece lens has a focal length of 2.0 cm. (*a*) What is the required focal length of the objective lens? (*b*) What will be the overall length of the telescope?

: **37-34 P** Assume that a magnifying glass of focal length f (in centimeters) is placed immediately adjacent to the eye when it is viewing an object located close to the principal focal point. The image formed by the magnifying glass is 25 cm distant from the eye, the so-called near point for the most distinct, comfortable vision for most people. Show that the angular magnification of this simple magnifier is given approximately by $25/f + 1$.

: **37-35 P** A microbiologist observes a specimen through a microscope whose tube length is 16 cm. The objective lens has a focal length of 0.4 cm. If the overall magnifying power is ×900, what focal length of the eyepiece is required?

: **37-36 P** An object 2 mm tall is placed at the origin. A diverging lens of focal length 20 cm is placed on the x axis at $x = 30$ cm. A converging lens with $f = 24$ cm is placed on the x axis at $x = 50$ cm. Determine the size and location of the image formed by the lenses.

: **37-37 Q** Is it ever possible to get a real image from a diverging lens?
(A) No, because the rays always diverge from a point behind the lens.
(B) No, because a diverging lens forms no image at all; it

merely spreads out a light beam. Only converging lenses form images.

(C) Yes. A real image is formed whenever the object is placed inside the focal point.

(D) Yes. A real image can be formed if the object distance is negative, that is, if the image from a converging lens is used as the object for the diverging lens.

: 37-38 P Suppose that you have two lenses with focal lengths of 2 cm, two with focal lengths of 20 cm, and two with focal lengths of 150 cm. If you can use any combination of lenses, (*a*) which two would you use for a telescope? (*b*) Which two for a microscope?

: 37-39 P An astronomical telescope's objective lens has a focal length of 50 cm, and the eyepiece has a focal length of 2 cm. (*a*) How far apart should the lenses be placed to form an image at infinity? (*b*) For this case, what is the magnification? (*c*) What should be the separation of the lenses so that an image will be formed 25 cm from the eyepiece?

: 37-40 P Does the order in which two lenses are arranged make any difference in the final image they form? Carry out the following exercise to see. An object is placed at the origin. A converging lens of focal length 20 cm is placed at $x = 30$ cm, and a diverging lens of focal length -20 cm is placed at $x = 100$ cm. (*a*) Determine the position of the image formed by the two lenses. (*b*) Determine the position when the two lenses are interchanged.

Section 37-6   The Lens Maker's Formula

· 37-41 Q The focal length of a thin converging lens is
(A) equal to the radius of curvature of the lens.
(B) the distance at which initially parallel rays come together after passing through the lens.
(C) twice the radius of curvature of the lens.
(D) the distance between the two focal points of the lens.
(E) half the radius of curvature of the lens.

· 37-42 Q A certain thin lens has one convex surface and one concave surface. Both have the same radius of curvature. What is the focal length of the lens?

: 37-43 Q Which of the lenses shown in Figure 37-30 has the smallest positive focal length? (All are made of the same kind of glass.)

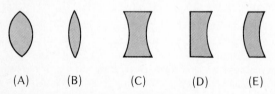

(A)   (B)   (C)   (D)   (E)

Figure 37-30.   Question 37-43.

: 37-44 P You have flat plates of glass with a refractive index of 1.50, and a device for grinding a convex spherical

surface with a radius of curvature of 5.0 cm. What are the focal lengths of the lenses you can produce?

: 37-45 P A certain doubly-convex thin lens made of glass with refractive index 1.60 has a focal length 50 cm. What is (*a*) the type of lens and (*b*) the focal length of this lens when it is immersed in water?

: 37-46 P You wish to design a lens that is to be used underwater with light. The lens is to be converging, its focal length is to be 50 cm, and the two spherical surfaces are to have the same radius of curvature. The lens will consist of very thin transparent material enclosing air. (*a*) Should the surfaces be convex or concave? (*b*) What is their required radius of curvature?

: 37-47 P Measuring the radius of curvature of the convex surface of a lens is not easy without special instruments, but it is relatively easy to measure the thickness of a lens like that in Figure 37-30B at its center and its outer diameter. Suppose that a doubly convex thin lens has equal radii of curvature, a thickness t and an outer diameter d, and its glass has refractive index n. Find the relation for the focal length of the lens in terms of t, d, and n.

Section 37-7   Lens Aberrations

: 37-48 Q If you look closely at an object's image, formed in sunlight by a lens, you will see a fuzzy rainbow around the edge of the image. This effect (called chromatic aberration) originates because
(A) rays that are not close to the lens axis, compared with rays near the axis, are focused at a different position.
(B) different light frequencies travel at different speeds in glass.
(C) some wavelengths interfere constructively and some destructively when they pass through the lens.
(D) the lens creates a diffraction pattern similar to that due to a circular aperture.
(E) light is a transverse wave, as opposed to a longitudinal wave, like sound.

: 37-49 Q Why is it that using a bright light helps you to read fine print?
(A) A bright light increases the number of photons hitting your retina each second.
(B) A bright light produces a brighter image on your retina.
(C) A bright light enables you to use your rods rather than your cones.
(D) A bright light enables you to relax your eye and focus on infinity.
(E) A bright light enables you to use the central part only of the lens of your eye.

Supplementary Problems

37-50 Q A converging lens can be used to focus the light from the sun onto a very small spot. Such a "burning glass" was once widely used to start fires. The temperature one can achieve presumably depends on the temperature of the surface of the sun and on the diameter and focal length of the lens. Would it be possible, if a sufficiently large and powerful lens were used, to achieve, at least in principle, a spot hotter than the sun?

37-51 Q An automobile has just been polished to a very high gloss, and then it rains briefly. Large water drops sit on the polished surfaces as the sun comes out and shines brightly. You later notice that a small spot appears on the waxed surface at the center of each large water drop. Why?

37-52 Q One obvious difference in operation between the eye and a 35-mm single-lens reflex camera is the method for detecting the light. What are other differences?

38 Interference

38-1 Superposition and Interference of Waves

The term *interference,* with which we're stuck because of its long entrenched usage, is an unhappy one. Interference phenomena have to do with the combined effect of waves from two or more sources, and the basic observation is that one wave does not influence (interfere with) the other. The basic proposition governing what happens when two waves arrive at the same location at the same time, the *superposition principle* (Section 17-3), says that the resultant wave disturbance at any location is simply the sum of the instantaneous individual wave disturbances. Thus, the resultant electric field from two electromagnetic waves is given by $E_r = E_1 + E_2$, where the electric fields, E_1 and E_2, of the individual waves are superposed as vectors.*

---

* We could equally well choose to find the resultant magnetic field **B**. It is conventional, however, to choose the electric rather than the magnetic field for an electromagnetic field, especially since the force on charged particles from **E** in a material typically greatly exceeds the force from **B**.

A consequence of the superposition principle (and the best evidence for its validity) is this: if two waves "collide," with both passing through the same region of space at the same time, each wave emerges from the "collision" unchanged in shape. It is just as if the other wave had not been present at all. The progress of one wave is completely independent of the presence of the other. Neither wave affects the other. See Figure 38-1. The superposition principle for electromagnetic waves is illustrated in quite ordinary circumstances. Suppose that you look at some bright object. What you see when emitted or reflected light reaches your eye is totally unaffected—in shape, color, brightness—by any other electromagnetic radiation, light or otherwise, that happens to pass through the line connecting your eye with the sighted object.

As we have seen (Chapter 18), the superposition principle applies also to mechanical waves propagated through a deformable medium to the degree that the relationship between deformation magnitude and deformation force magnitude is strictly linear.

Merging waves are said to interfere. If the resultant wave disturbance is *greater* than the disturbance from either one alone, for example, one wave crest on a second wave crest, we have *constructive interference*. If the resultant wave disturbance is *less* than that from either one separately, for example, one wave's crest on another wave's trough, the waves exhibit *destructive interference*. A point in space at which the component waves always interfere destructively, so that there is always a zero resultant wave disturbance, is called a *node*.

Hereafter in this chapter we shall be concerned primarily with the interference of sinusoidal electromagnetic waves from two sources of the same frequency. We shall concentrate on the electric fields and intensities of radio waves and visible light. The analysis holds, however, for all types of waves. For example, applying the analysis to sound waves requires merely that we replace the electric field (vector) by the pressure difference (scalar).

This is the primary problem to be solved. Suppose that you know how sinusoidal wave sources of the same frequency are placed or arranged. Then, how does the time-average intensity of the combined wave, the average power per transverse area—for light, a measure of brightness—vary with position?

First, why time average? If each source and the waves it generates oscillate sinusoidally with time, so also does the intensity of the resultant wave at any location. We are interested typically not in the instantaneous intensity, but in

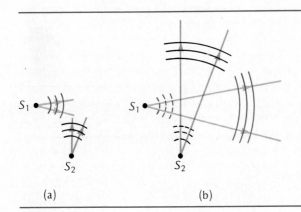

S_1

S_2

(a)

S_1

S_2

(b)

Figure 38-1. When the superposition principle applies, the waves generated by two independent sources do not interact. (a) Before "collision" (b) After.

its average over some period. After all, the human eye does time averaging — we don't see the discrete still frames of a motion picture or a television picture because of the persistence of vision; in visible light, with a frequency of more than 10^{14} Hz, the eye samples a signal over many oscillation cycles; a photographic negative gives an even longer time average of visible light.

As we have seen, the time average of the intensity $\bar{I}$ of a sinusoidal electromagnetic wave at any location is directly proportional to the *square* of the *resultant* electric-field amplitude $\mathbf{E}_r$ at that location:

$$\bar{I} \propto E_r^2 \tag{35-11}$$

For mechanical waves, the intensity is also proportional to the square of the amplitude of the resultant wave disturbance.

What determines $\mathbf{E}_r$? The superposition principle says that

$$\mathbf{E}_r = \mathbf{E}_1 + \mathbf{E}_2$$

where $\mathbf{E}_1$ and $\mathbf{E}_2$ are the electric fields from two sources. The relative magnitudes of $\mathbf{E}_1$ and $\mathbf{E}_2$ depend in turn on the phase difference ϕ between the two waves at the observation point. And this phase difference depends in turn on two factors:

- The *path difference* from the observation point to each of the wave sources.
- Any *source phase difference* between the oscillations at the sites of the two sources.

A simple example in Section 38-2 illustrates these features.

38-2   Interference from Two Point Sources

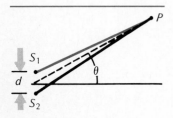

Figure 38-2. *The resultant radiation from two point sources, S_1 and S_2, oscillating in phase, is observed at distant point P.*

The situation, shown in Figure 38-2: two identical sources, S_1 and S_2, oscillating in phase with respect to one another; each source radiates waves of wavelength λ uniformly in all directions (in the plane of Figure 38-2) and the sources are separated vertically by a distance $d = \lambda/2$. (We can think of the sources as electric-dipole radio antennas.) The question: How does the time-average intensity $\bar{I}$ vary with angular position θ (at a large fixed distance from the sources)? Observation point P in Figure 38-2 is taken to be so far from S_1 and S_2 that the lines S_1P and S_2P can be taken as essentially parallel; each line then makes the same angle θ with the perpendicular bisector of the line joining S_1 and S_2.

Consider first point P_1, in Figure 38-3(a). It is far to the right of S_1 and S_2, and here $\theta = 0$. The path lengths from S_1 and S_2 to P are the same. The resultant field E_r at this location is the sum of E_1 and E_2, which are the electric-field amplitudes at P_1 from the individual sources. Fields E_1 and E_2 arrive in phase at P_1, so that we have constructive interference there. It is helpful to represent the sinusoidally oscillating electric fields by phasors (see Section 33-3). The magnitude of a phasor represents the amplitude of the associated simple harmonic oscillation, and the direction of the vector represents the phase of the oscillation. Therefore, phasors $\mathbf{E}_1$ and $\mathbf{E}_2$ are parallel at P_1. These two fields have essentially the same magnitude, so that the resultant field E_r has a magnitude

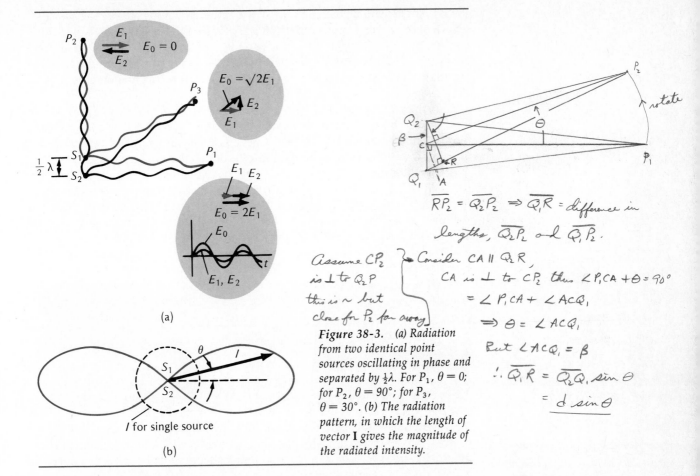

{{handwritten notes beside figure:}}

$\overline{RP_2} = \overline{Q_2 P_2} \Rightarrow \overline{Q_1 R} = $ difference in

lengths, $\overline{Q_2 P_2}$ and $\overline{Q_1 P_2}$.

Assume CP_2
is $\perp$ to $Q_2 P$
this is ~ but
close for P_2 far away

Consider $CA \parallel Q_2 R$,
CA is $\perp$ to CP_2 thus $\angle P_1 CA + \theta = 90°$
$= \angle P_1 CA + \angle ACQ_1$
$\Rightarrow \theta = \angle ACQ_1$

But $\angle ACQ_1 = \beta$

$\therefore \overline{Q_1 R} = \overline{Q_2 Q_1} \sin \theta$
$= d \sin \theta$

(a)

I for single source

(b)

Figure 38-3. (a) Radiation from two identical point sources oscillating in phase and separated by $\frac{1}{2}\lambda$. For P_1, $\theta = 0$; for P_2, $\theta = 90°$; for P_3, $\theta = 30°$. (b) The radiation pattern, in which the length of vector **I** gives the magnitude of the radiated intensity.

twice that of a field from a single source. It follows that at P_1 the intensity, proportional to E_r^2, is four times that for a single source.

Now consider location P_2 ($\theta = 90°$) in Figure 38-3(a). The path lengths to P_2 differ by $\frac{1}{2}\lambda$. The wave from S_2 travels $\frac{1}{2}\lambda$ farther than the wave from S_1 to reach P_2. At P_2, the electric fields from S_1 and S_2 are out of phase by $180°$. Again, the separate fields have essentially the same magnitude, so that we have nearly complete destructive interference at this location. At P_2, the intensity is zero.

Now consider point P_3 in Figure 38-3(a), where now $\theta = 30°$. The situation here is a little more complicated because the interference is neither completely constructive nor completely destructive. From the geometry of the figure, we see that path difference is $d \sin \theta = (\frac{1}{2}\lambda) \sin 30° = \frac{1}{4}\lambda$. The electric fields at P_3 are $90°$ out of phase; this corresponds to the two phasors at right angles. As Figure 38-3(a) shows, the amplitude of the resultant electric field is $\sqrt{2}E_1$, so that the intensity here is proportional to $(\sqrt{2}E_1)^2 = 2E_1^2$. At $\theta = 30°$, we have an intensity twice that from a single oscillator.

We could readily compute the intensity at still other angles to find $\bar{I}$ as a function of θ. The results, displayed in the polar diagram of Figure 38-3(b), show $\bar{I}$ as a polar vector plotted as a function of θ. We see that electromagnetic energy from the two sources is radiated primarily in the directions left and right ($\theta = 0$ and $\theta = 180°$) and none is radiated up or down ($\theta = 90°$ and $\theta =$

$-90°$). The radiation pattern of the two identical antennas separated by a half wavelength is said to consist of two *lobes*. Each antenna by itself radiates a circular pattern. Interference between the two identical sources does not change the total energy radiated (it is twice the energy from one source alone), but interference is responsible for redistributing how this total energy is radiated in various directions.

   Can the radiation pattern be changed without actually moving the sources? Yes. A change in the relative phases of oscillation at the sources will do this. Suppose now that the two identical antennas are again separated vertically by $\frac{1}{2}\lambda$ but now with the sources oscillating $180°$ out of phase. The phase difference at observation point P_1 ($\theta = 0$) is also $180°$ because there is no path difference to P_1. At P_2 ($\theta = 90°$), on the other hand, the $\frac{1}{2}\lambda$ path difference from S_1 and S_2 to P_2 is just compensated by the $180°$ phase difference at the sources, so that the two waves arrive at P_2 *in phase*. Under these circumstances, the intensity is zero at $\theta = 0$ and four times the intensity from one source alone at $\theta = 90°$. (Query: What is the intensity at $\theta = 30°$? Is it still twice that from one antenna? Is the radiation pattern of Figure 38-3(b) merely rotated $90°$?)

   Notice what can be accomplished by changing the relative phase of oscillation of the two antennas: the direction in which the energy is beamed outward in space can be changed without changing the physical location of the antennas. This also means, through the reciprocity principle, that a receiving antenna can be swept through space simply by changing the relative phases of the receivers at the antennas.

   Now that we've seen the basic features of the interference effects from two continuous sinusoidal point sources of waves, let us summarize how one generally finds the time-average intensity as a function of position. Because of the interference between the continuous (coherent) waves from the two sources, the observed intensity pattern is not merely the arithmetic sum of the intensity patterns radiated by the two sources separately. For electromagnetic waves, the time-average intensity $\bar{I}$ at any location P is proportional to the square of the resultant electric-field amplitude E_r at P:

$$\bar{I} \propto E_r^2 \tag{38-1}$$

where the resultant field is merely the vector sum of the electric fields at P from the two sources:

$$\mathbf{E}_r = \mathbf{E}_1 + \mathbf{E}_2 \tag{38-2}$$

If the two sources are approximately the same distance from P, then $\mathbf{E}_1$ and $\mathbf{E}_2$ have nearly the same magnitude. Then the net field at P is determined solely by the relative phase of the two arriving sinusoidal signals. The relative phase difference ϕ at the observation point P is in turn determined by:

   • The *path difference* Δr, the difference between the distances from P to each of the two sources and P.
   • The *source phase difference* Φ between sinusoidal oscillations of the two sources.

   As we know, a path difference of $\Delta r = \frac{1}{2}\lambda$ produces destructive interference and corresponds to a phase difference at the observation point of π radians. The general relation between path difference and the corresponding phase difference is

$$\frac{\text{phase difference}}{2\pi} = \frac{\text{path difference}}{\lambda} \qquad (38\text{-}3)$$

Furthermore, if the two sources are the same distance from P, so that there is no path difference, we can still have destructive interference if the sources oscillate 180° out-of-phase so that $\Phi = \pi$ radians.

From (38-3), the general relation for the relative phase ϕ at the observation point is then

$$\phi = \frac{2\pi}{\lambda} \Delta r + \Phi \qquad (38\text{-}4)$$

If Δr represents the additional path length from source 2 to point P compared with the path length from source 1, then Φ is the phase lag of source 2 behind source 1.

Example 38-1. Three identical omnidirectional radio antennas oscillating in phase and radiating waves of wavelength λ are arranged as shown in Figure 38-4. Source 1 is a distance of $\frac{1}{2}\lambda$ north of source 2, and source 3 is $\frac{1}{2}\lambda$ east of source 2. A receiving antenna, located far to the east of the transmitting antennas, registers absorbed power of 3.0 mW when all three antennas are turned on. (a) Suppose that the source driving antenna 3 is turned off. What power is then registered at the receiving antenna? (b) How must the relative oscillation phases of the three antennas be adjusted to have maximum power delivered to the receiving antenna?

(a) With all three transmitting antennas turned on, waves to the east from sources 2 and 3 destructively interfere because of their $\frac{1}{2}\lambda$ separation; in effect, only the radiation from *one* antenna reaches the receiving antenna. But suppose that antenna 3 is turned off. Then we have constructive interference to the east of sources 1 and 2. The amplitude of the resultant wave arriving at the receiving antenna is doubled. The received power, which is proportional to the intensity at the receiving antenna and therefore proportional to the square of the resultant amplitude of the electric field, is increased by a factor 4. The power received is 4(3.0 mW) = 12.0 mW.

(b) Changing the phase of the sinusoidal signal driving antenna 3 by 180° (switching the two lead wires, for example) will mean that the path difference of $\frac{1}{2}\lambda$ between antennas 2 and 3 is exactly compensated by a source phase difference of $\Phi = \pi$. All three antennas then will constructively interfere for waves to the east. The resultant wave amplitude will be three times the amplitude of one alone, and the power at the receiving antenna is then up by a factor 9 to $3^2(3.0 \text{ mW}) = 27.0$ mW.

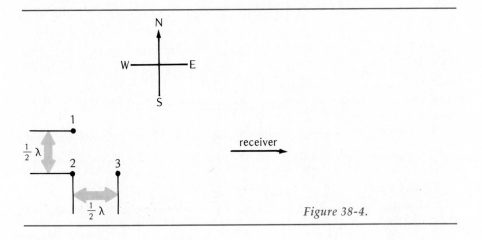

Figure 38-4.

38-3   More on Interference from Two Point Sources

In Section 38-2, we considered the interference pattern for a special case: two identical point oscillators separated by $\frac{1}{2}\lambda$. Here we treat the more general case of two identical point oscillators or sources S_1 and S_2 separated by any distance d (where d typically is large compared with λ). See Figure 38-5. Again, observation point P is so far from S_1 and S_2 (compared with d) that the two rays to P make the same angle θ with the perpendicular bisector of the line joining S_1 and S_2. The geometry of Figure 38-5 then shows that the path difference is

$$\Delta r = d \sin \theta \tag{38-5}$$

Using (38-4), we can express the phase difference ϕ at P in terms of the path difference

$$\phi = \frac{2\pi}{\lambda}\Delta r = \frac{2\pi d}{\lambda}\sin \theta \tag{38-6}$$

Constructive interference corresponds to

$$\text{Path difference:} \quad \Delta r = m\lambda \tag{38-7}$$

or equivalently,

$$\text{Phase difference:} \quad \phi = m(2\pi) \quad \text{where} \quad m = 0, 1, 2, 3, \ldots$$

so that from (38-5), the directions θ for maximum radiated time-average intensity are given by*

$$\text{Max } \bar{I}: \quad \Delta r = m\lambda = d \sin \theta \tag{38-8}$$

where $m = 0, 1, 2, 3, \ldots$. Destructive interference occurs when the path difference is an odd multiple of $\lambda/2$ or the phase difference is an odd multiple of π. Point P in Figure 38-5 corresponds to a path difference of 2λ; the locus of points, all with this same path difference, defines a curve of maximum intensity. The curve is a hyperbola; this follows from the definition of a hyperbola as the locus of points whose distances from two fixed points (here S_1 and S_2) differ by a constant (here 2λ). The asymptotes of the hyperbola make the angle θ relative to the bisector of the line connecting S_1 and S_2. A series of such hyperbolas gives the locations of maximum intensity. A second set of hyperbolas represents the lines of zero intensity, or *nodal lines*. The path difference for these curves is an odd multiple of $\lambda/2$.

The interference pattern from two sinusoidally oscillating point sources can easily be demonstrated with water-surface waves generated by transverse oscillators on the water surface. A photograph of such a ripple-tank interference pattern is shown in Figure 38-6. To observe the interference of visible light from two sources requires special arrangements (to be examined in Section 38-4).

---

* This is the first of what will be a series of equations, in this chapter and the next, that give the conditions for either zero intensity or maximum intensity for various interference or diffraction effects. These equations all look pretty much alike, and they can easily be confused. It's never a good idea to try to learn physics by memorizing specialized formulas, and it is especially risky for interference and diffraction relations. Far better, then, to know the fundamental principles and derive a specialized relation from scratch when it is needed.

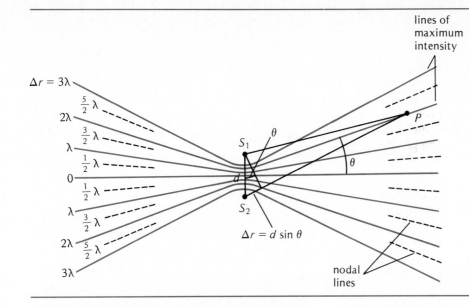

$\Delta r = 3\lambda$

$\frac{5}{2}\lambda$

2λ

$\frac{3}{2}\lambda$

λ

$\frac{1}{2}\lambda$

0

$\frac{1}{2}\lambda$

λ

$\frac{3}{2}\lambda$

2λ $\frac{5}{2}\lambda$

3λ

$\Delta r = d \sin \theta$

lines of
maximum
intensity

nodal
lines

*Figure 38-5.   Lines of
maximum intensity (solid) and
nodal lines (dashed) for two
point sources oscillating in
phase.*

*Figure 38-6.   Photograph of a
water-ripple-tank surface,
showing the interference from
two point sources.*

Example 38-2.   Derive the general expression for the time-average intensity $\bar{I}$ as a function of θ.

Phasors for the electric fields $\mathbf{E}_1$ and $\mathbf{E}_2$ differ in phase by ϕ, as shown in Figure 38-7. Vector $\mathbf{E}_2$ leads $\mathbf{E}_1$ by ϕ. The magnitudes are the same, $E_1 = E_2$. The sum of the horizontal components of the vectors in Figure 38-7 is given by

$$E_1 + E_2 \cos \phi = E_1(1 + \cos \phi) = 2E_1 \cos^2 \frac{\phi}{2}$$

From Figure 38-7, we can see that the horizontal component of resultant field $\mathbf{E}_r$ is $E_r \cos (\phi/2)$. Therefore, equating horizontal components gives

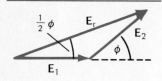

Figure 38-7. Resultant electric field $\mathbf{E}_r$ *from the phasors representing electric fields* $\mathbf{E}_1$ *and* $\mathbf{E}_2$ *out of phase by* ϕ.

$$E_r \cos \frac{\phi}{2} = 2E_1 \cos^2 \frac{\phi}{2}$$

$$E_r = 2E_1 \cos \frac{\phi}{2}$$

But

$$\bar{I} \propto E_r^2$$

so that

$$\bar{I} \propto \cos^2 \frac{\phi}{2}$$

From (38-6), the phase difference ϕ is

$$\phi = \frac{2\pi d}{\lambda} \sin \theta$$

Substituting this result in the relation for I gives

$$\bar{I} = I_0 \cos^2 \left(\frac{\pi d}{\lambda} \sin \theta \right) \tag{38-9}$$

where the maximum value for $\bar{I}$ is written as I_0. For small angles, $\sin \theta \simeq \theta$, so that the intensity varies directly with the square of the cosine, as shown in Figure 38-8.

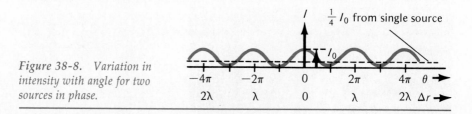

Figure 38-8. Variation in intensity with angle for two sources in phase.

38-4   Young's Interference Experiment

The British physician and scientist Thomas Young showed in an epochal experiment in 1800 that because visible light exhibits interference effects, it consists of waves.

The essential parts of a double-slit interference apparatus are shown in Figure 38-9. Light from a distant ordinary light source S passes first through a narrow single slit.* Wave fronts spreading from the single slit fall on the double slits, S_1 and S_2. The two parallel slits separated by distance d act as wave sources that oscillate in phase and produce circular wave fronts. A pattern of equally spaced interference fringes appears on the screen a large distance D from the slits.† Let y represent the distance on the observation screen from the central maximum. Then, if the interference fringes are closely spaced, and angle θ is small, we can write $\theta \simeq y/D$. The relation (38-8), for the positions of maxima in the interference pattern, then becomes, with $\sin \theta \simeq \theta \simeq y/D$,

---

* The single slit is required for the incoherent light from an ordinary light source. The distinction between coherent and incoherent light is given in Section 38-6; how a single slit diffracts light is given in Section 39-2.

† Young, in his original interference experiment, used pinholes instead of slits.

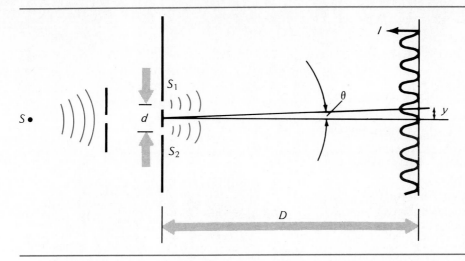

Figure 38-9. *Young's double-slit experiment. Ordinary light first passes through a narrow single slit, then through the parallel slits S_1 and S_2, and finally to the distant observation screen.*

$$\text{Max } \bar{I}: \quad m\lambda = \frac{yd}{D} \qquad (38\text{-}10)$$

where $m = 0, 1, 2, 3, \ldots$

The crucial observation is that the pattern on the screen does not look at all like two parallel narrow openings; the regular alternations in the brightness of light make sense only when light is assumed to consist of waves. In particular, at the location on the screen exactly midway between the two slits, we find a *bright* line, not darkness.

Figure 38-10 is a photograph of the interference fringes from parallel slits. The intensity falls off in both directions from the center because of diffraction effects associated with the finite width of the two slits. (The combined effects of interference, and diffraction by a double slit, are treated in Section 39-6.)

Example 38-3.   The red light from a helium-neon laser falls on two parallel narrow slits separated by 0.50 mm. The interference pattern on a screen 1.00 m away shows that the distance between the dark lines bordering ten interference fringes is 1.27 cm. What is the wavelength of the light?

From (38-10), we have

$$\lambda = \frac{yd}{mD} = \frac{(1.27 \times 10^{-3} \text{ m})(0.50 \times 10^{-3} \text{ m})}{(1)(1.00 \text{ m})} = 6.35 \times 10^{-7} \text{ m} = 635 \text{ nm}$$

It turns out that the single slit (preceding the double slit in Figure 38-9) is not required when the double slit is illuminated by light from a laser. The reasons are given in Section 38-6.

Figure 38-10. *Photograph of the central interference fringes from a double slit.*

Hey, Phenomenal!

They called him "Phenomenal"—literally. That was his nickname in college.

He deciphered the Rosetta stone. This made it possible to read Egyptian hieroglyphics in pyramids and on other ancient Egyptian structures. The discoverer: Thomas Young, M.D. (English, 1773–1829). Young's other big discovery—the experiment and the explanation showing that light consists of waves because of interference effects.

An infant prodigy, Young grew up to be an adult prodigy. He could read at age 2, he had read through the Bible twice by the time he was 6, and by age 20 he had mastered 13 languages on his own.

Although Young was trained as a physician, he had such a wide range of scholarly interests and such a poor bedside manner that he practiced medicine only part-time and with indifference. In his first scientific work, done while he was still in medical school, he showed that the eye can accommodate (focus on objects at various distances) because the eye muscles can change the radius of curvature of the lens of the eye.

The Rosetta stone, now on view in the British Museum, London, has inscriptions dating from about 200 B.C. in two languages and three alphabets. The top is in Egyptian hieroglyphics; the same message is written in the middle in demotic characters, a cursive form of Egyptian hieroglyphics, and at the bottom in Greek. Found in 1799 near the town of Rosetta in Egypt, the Rosetta stone was deciphered by Thomas Young in 1814.

His observations on the interference effects of light, made at age 28, involved first two pinholes and later what has since been known as the Young double-slit experiment. He interpreted the fringes on the basis of

38-5   The Diffraction Grating

The diffraction grating, invented by J. Fraunhofer (1786–1826), is a device for measuring wavelengths of light with high precision. A *transmission* grating consists of a large number N of parallel slits separated from one another by a distance d (greater than wavelength λ). For example, a few thousand closely spaced lines may be scratched over a distance of about 1 cm onto a plate of glass. Another example of a transmission grating, one suitable for use with millimeter-wavelength waves of sound, is a Venetian blind. A *reflection* grating may be produced by a precision ruling machine that scratches parallel grooves with a diamond point on a smooth surface of metal. The device is called a *diffraction* grating because diffraction effects (to be discussed in Section 39-7) enter its more complete description.

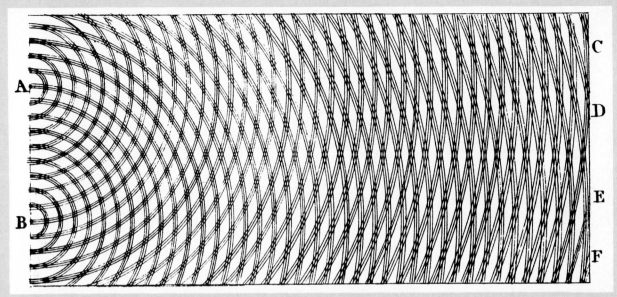

Thomas Young's original diagram showing how waves from two sources produce an interference pattern on a screen.

wave theory, and measured the wavelength of light. But he could not persuade his British contemporaries that the wave theory of light was superior to the particle theory; they stood in awe of Isaac Newton, and could not imagine him to have been wrong in advocating a particle theory of light. What is now known as the superposition principle for waves was first formulated by Young. He was also the first to use the term *energy* in its present meaning.

Young produced a three-color theory for the response by the eye's retina to all colors of light; three colors now serve as the basis for color photography and color television. *Young's modulus* is the name given to the elastic modulus that quantitatively measures a material's susceptibility to being stretched or compressed.

The range of Young's talents and expertise, developed primarily in studying on his own, is staggering. Among other things, he was a consultant on nautical matters to the British Admiralty; he established the formula for mortality tables used by insurance companies; he wrote 60 articles for the fourth edition of the *Encyclopedia Britannica* on topics ranging from bathing to hieroglyphics.

It is easy to see that a diffraction grating produces:

• Intensity maxima at the *same angular locations as a double slit* (with the same separation distance d), namely,

$$\text{Max } I: \quad m\lambda = d \sin \theta \tag{38-8}$$

where $m = 0, 1, 2, 3, \ldots$.
• *Very narrow and intense* interference peaks.

Consider Figure 38-11, which shows N point sources all oscillating in phase at the same frequency arranged along a line with *grating spacing d ($\gg \lambda$)* between adjacent sources (or slits). As before, observation angle θ is measured from the normal to the line of sources, and the observation point P is far from the line of sources. The angle θ in Figure 38-11 is that for the first-order ($m = 1$)

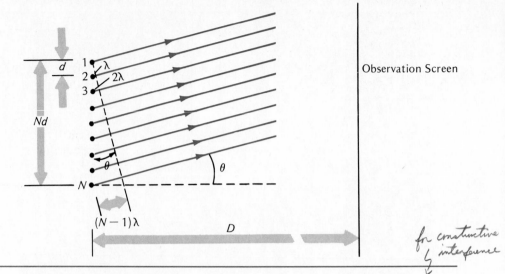

Figure 38-11. Arrangement of sources in a diffraction grating (d > λ).

interference peak, because the path difference from sources 1 and 2 is λ. The path difference is also λ between any other pair of adjacent sources, so that all point sources interfere constructively at distant point P. The condition here for constructive interference is the same as that for two sources.

A simple argument, based on energy conservation, shows that the interference peaks produced by a diffraction grating are much stronger and narrower than peaks from just two slits. When the waves from all N slits interfere constructively at some distant point, the resultant electric field there is N times the electric field from just one slit. Therefore, the intensity is up by a factor N^2. But the total energy radiated per unit time is up by a factor of only N, not N^2. Proportionately more energy (up by factor N) goes to the peak, so that proportionately less energy (down by a factor N) must go in the region between peaks. The inescapable conclusion: the width of interference peaks from a diffraction grating of N slits is smaller by a factor N than the width of peaks from just two slits. The intensity variation for two slits is shown in Figure 38-9; smooth variation (with the square of the cosine) from peak to the adjoining zeros with I dropping to half its peak value midway between the peak and the nearby zero. As more detailed arguments (Section 39-4) show, for a grating the zero is brought closer to the peak by a factor N.

As (38-8) shows, the angle θ at which intensity peaks are radiated depends on the wavelength λ. This means that if the radiation incident upon a diffraction grating consists of many wavelengths, each component wavelength will be deviated differently; the larger wavelengths at the red end of the visible spectrum will be deviated through a larger angle than the shorter wavelengths at the violet end. For m = 0, we have θ = 0 for all wavelengths; the undeviated zero-order "spectrum" consists of all wavelengths, which is white light. But the first-order (m = 1), second-order (m = 2), and still higher-order spectra are increasingly dispersed and may overlap appreciably. The wavelength can readily be computed from the number of lines per unit length on the grating (=1/d) and the measured deviation angle θ. The special advantages of diffraction grating over the prism spectrometer (Section 36-6) for observing a spectrum are:

- The wavelength is directly computed from readily measured quantities.
- The lines are sharp.
- The spectrum can have a large angular spread.

A disadvantage of the diffraction grating is that the incident light is dispersed into several spectra to the left and right of the zero-order central bright white line and thus are not as bright as the single spectrum in the prism spectrometer.

Example 38-4.   A certain diffraction grating has 5000 rulings over 1.0 cm. What is the angular separation in the first order of the extremes in the visible spectrum (from 400 to 700 nm)?

We have $d = \frac{1}{5000}$ cm^{-1}. From (38-8), we have for $m = 1$ and $\lambda = 400$ nm $= 400 \times 10^{-9}$ m,

$$\theta = \sin^{-1}\left(\frac{m\lambda}{d}\right) = \sin^{-1}(4.0 \times 10^{-7}\text{ m})(5.0 \times 10^5\text{ m}^{-1}) = 11.5°$$

Similarly, $\theta = 20.5°$ for $\lambda = 700$ nm, so that the entire visible spectrum lies within the 9 degrees between the violet and red limits.

38-6   Coherent and Incoherent Sources

Suppose that two waves of the same amplitude and wavelength arrive at observation point P in phase. The waves interfere constructively, and location P corresponds to an intensity maximum. Of course, the intensity at P will remain maximum only so long as the phase relationship between the two waves is fixed. If the phase of the source of one wave were to change by π, the interference at P would become destructive and the intensity there would fall to zero. In short, a steady interference pattern is maintained only if the two interfering waves have a fixed phase relationship. Two waves or two sources between which the phase relationship remains *constant* are said to be *coherent* with respect to one another. Examples of coherent sources of waves of the same frequency: two radio antennas driven by the same continuously operating oscillating electric circuit; two loudspeakers driven by the same audio signal generator.

Visible light from an ordinary source is *not* coherent. Ordinary sources consist of many individual atoms emitting short bursts of light at random. An example of an incoherent light source is the heated filament of an incandescent bulb. Here the atoms are thermally excited to upper atomic energy states, from which the atoms make downward transitions discontinuously and randomly. A typical atom emitting light is "on," that is, radiating light, for only about 10^{-8} s. Although this time interval is very short, it is long compared with the oscillation period of about 10^{-15} s for typical visible light. Of course, the radiation from any two emitting atoms *is* coherent over a period of about 10^{-8} s [or over a distance of $(3 \times 10^8\text{ m/s})(10^{-8}\text{ s}) = 3$ m], but since we register the intensity of visible light over a time much longer than 10^{-8} s, the light appears essentially incoherent.

A coherent source of visible light is a *laser*, whose name is an acronym for Light Amplification by the Stimulated Emission of Radiation. A thoroughgoing explanation of laser operation is possible only with the quantum theory;

what follow here are merely some general features. In a laser, the atoms typically are brought to excited states in preparation for radiation, not by thermal excitation, but by *optical pumping,* a process in which light of higher frequency than the light emitted by the atoms is absorbed by the active material, or by an electric discharge. Atoms remain in the excited energy states for much longer than 10^{-8} s. When light of the frequency to be amplified by the laser enters the material, it stimulates the atoms in upper energy states to make transitions to lower energy states. The emitted light is the same frequency as the "stimulating" radiation.

Most important, the waves of light emitted by stimulated atoms are exactly in phase with the waves stimulating the emission. Therefore, the emitted light is coherent with the light stimulating the emission. There is light amplification. In practice, the active material that has been optically pumped to the excited states is held between two parallel reflecting boundaries and the emitted light is made to traverse the region between the boundaries repeatedly through multiple reflections. The intensity of the wave grows because the light first present causes additional atoms to emit light in coherence. The useful light—highly monochromatic, unidirectional, intense, and coherent—leaves the laser by passing through a partially reflecting end mirror. Many technological applications of lasers derive from the fact that they produce, in the visible region, electromagnetic radiation that has coherence properties heretofore available only in radio waves. The laser is treated in more detail in Section 42-6.

38-7   Reflection and Change in Phase

The procedures of classical electromagnetic theory provide a way of working out what happens to the phase of a sinusoidal electromagnetic wave that encounters a boundary between two media. Since these procedures are beyond our immediate scope, what we do here instead is to consider the analogous properties of reflected mechanical waves. This is not a proof, but it does give a useful mnemonic for remembering the results.

First, compare the two traveling wave pulses shown in Figure 38-12. Their shapes are the same, but one is positive and the other negative; the negative pulse is in fact just the positive pulse flipped over (about a horizontal axis). The labels "positive" and "negative" are not useful for a sinusoidal wave. As Figure 38-13 shows, a sine wave flipped over becomes a negative sine wave. Another way of describing the reversed, or flipped, sinusoidal wave is to say that its *phase* has been *shifted* by 180°, or by *one half-wavelength.* In short, the inversion in any wave disturbance can also be described as a 180° phase shift.

In the mechanical-optical analogy, low speed for a mechanical wave corre-

Figure 38-12.   *(a) A wave pulse on a taut string traveling to the right. (b) The same wave pulse inverted (flipped up-down).*

(a)

(b)

(a)

(b)

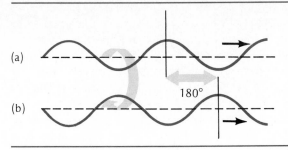

180°

Figure 38-13. (a) A sinusoidal wave traveling to the right. (b) The same wave inverted (flipped up-down). Note that the inverted wave is 180° out of phase with the uninverted wave.

sponds to low speed (or higher refractive index) for an electromagnetic wave. A mechanically hard boundary corresponds to an optical mirror or a conducting surface.

The easily remembered results for a transverse wave pulse encountering a boundary on a stretched string under tension are shown in Figure 38-14 (like Figure 17-8), where the wave speed is lower in the more massive string:

- No phase change in the transmitted wave.
- For slow to faster medium (optically dense to less dense), no phase change in the reflected wave; for fast to slower medium, a 180° phase change in the reflected wave.
- Reflection from a conductor (with no transmitted wave), a 180° phase change in the reflected wave.

The phase change for a reflected wave is illustrated in the interference arrangement known as *Lloyd's mirror*, named for its originator, H. Lloyd, and shown in Figure 38-15. Here we have *two* waves arriving at the observation point P but only *one* source S_1. Waves reach point P both through a direct path from S_1 and through a second path involving reflection in a mirror. The re-

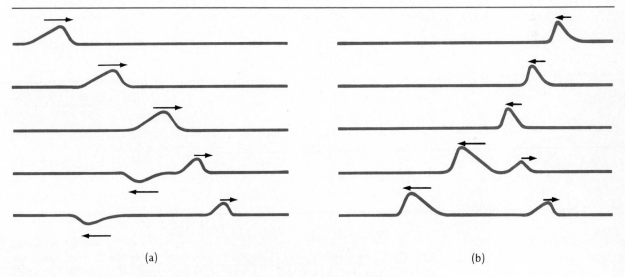

(a)   (b)

Figure 38-14. (a) Wave pulse approaches and is reflected from the boundary layer of a medium of lower wave speed. (b) Wave pulse is incident in low-speed medium and encounters medium of higher wave speed.

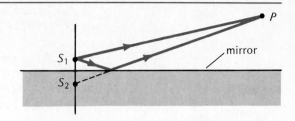

Figure 38-15. Rays from point source S_1 and its mirror image S_2 reach distant point P.

flected ray may be thought to originate from virtual source S_2, located below the mirror surface at the position of S_1's mirror image. The arrangement is just like that for two point sources (Figure 38-5) except that the wave from S_2 undergoes a 180° phase shift; therefore, the intensity pattern is like that for two point sources but with maximum and zero intensity locations interchanged.

Example 38-5. See Figure 38-16. A radar unit (S_1) is located 2.0 m ($\frac{1}{2}d$) above the surface of a smooth lake. It sends out microwave pulses of 10-cm wavelength toward an airplane flying at altitude y above the lake surface and at a horizontal distance of $D = 2.0$ km from the radar transmitter. The radar registers no signal reflected from the airplane. What is the minimum elevation of the airplane above the lake surface?

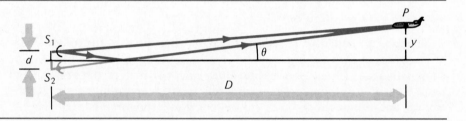

Figure 38-16.

First, if no signal reaches P from S_1, there can be no echo signal back at S_1. Waves from S_1 and S_2 arrive out of phase at P. Because of the 180° phase shift in the reflected waves, the relation (38-8), which gives the intensity maxima for two sources separated by d and oscillating in phase, will now give the locations of the intensity zeros.

$$\text{Zero I:} \quad m\lambda = d \sin \theta = d \frac{y}{D}$$

Minimum elevation corresponds to $m = 1$, so that we have

$$y = \frac{\lambda D}{d} = \frac{(10 \text{ cm})(2 \text{ km})}{(4 \text{ m})} = 50 \text{ m}$$

The lowest altitude is 50 m.

38-8   Interference with Thin Films

Everyone is familiar with the variegated colors one observes from a film of oil on water. The effect comes from interference between the light waves reflected from the top and bottom surfaces of the oil film.

Consider the situation shown in Figure 38-17. Here light waves are incident, for simplicity, nearly along the normal to two parallel interfaces separating media with refractive indices n_1, n_2, and n_3. We are concerned with interference at some distant observation point between the reflected rays from the two interfaces separated by distance d. Whether the interference is constructive or destructive depends on the following circumstances:

- The *film thickness d*. For normal incidence, the path difference for the two rays is $2d$.
- The wavelength λ_m *within the medium* of refractive index n. From (36-6),

$$\lambda_m = \frac{\lambda}{n}$$

where λ is the free-space wavelength.

- The *change in phase*, if any, *of the reflected ray*. Recall (Section 38-7) that the reflected wave undergoes a 180° phase change when light goes from a medium of low refractive index to one of higher (or from a high-speed medium to a low-speed medium), but the reverse situation does not obtain.

The *optical path length* is a useful quantity for light in a refractive medium. It is defined as the geometrical path length multiplied by the refractive index of the medium. As so defined, the optical path length takes into account the shrinking of a wavelength in an optically dense medium and makes the optical path length equivalent to the actual path length in a vacuum. For example, red light with a wavelength of 600 nm in a vacuum becomes, in glass with refractive index 1.5, red light with wavelength of 600 nm/1.5 = 400 nm. A path length in a vacuum of 0.6 mm contains 1000 wavelengths. In glass, the same 1000 wavelengths extend over only 0.4 mm. The optical path length in glass is (0.4 mm)(1.5) = 0.6 mm; this geometrical distance in a vacuum contains the same number of wavelengths.

Consider Figure 38-18(a), in which two flat plates of glass are inclined slightly to produce a wedge-shaped film of air between them. The observed interference pattern, consisting of equally spaced, parallel interference fringes, is shown in Figure 38-18(b). The path difference changes by one wavelength (and the air thickness by $\lambda/2$) as one goes from one fringe to an adjoining one. This arrangement can be used to test for optical flatness (within a fraction of a wavelength, or better than a micron); departure from a perfectly flat plane surface is manifest in the interference pattern as a departure from equally spaced, parallel fringes.

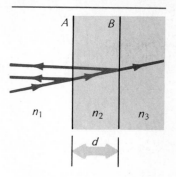

Figure 38-17.   Incident rays are reflected from the boundary between media 1 and 2 and from the boundary between media 2 and 3.

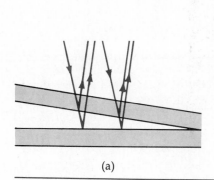

(a)

(b)

Figure 38-18.   (a) Interference from an air wedge. (b) The observed fringes.

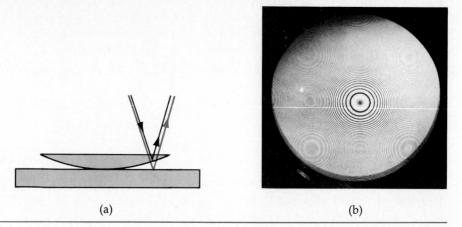

Figure 38-19. *(a) Interference producing Newton's rings. (b) The observed interference pattern.*

(a)   (b)

A modification of this arrangement is shown in Figure 38-19(a). Here a plano-convex lens is in contact with a flat plate of glass. The air wedge has cylindrical symmetry and the interference fringes now appear as concentric circles. The phenomenon, known as *Newton's rings,* was discovered by Robert Hooke (of Hooke's law) and studied, but not explained fully, by Isaac Newton. Figure 38-19(b) is a photograph of light reflected from the flat and spherical surfaces in contact. Note one especially interesting feature; the central region is dark, not bright. At the point of contact, there is no path difference but the two reflected rays are out of phase. At one interface light goes from glass to air; at the other light goes from air to glass.

Example 38-6.   Lenses are often coated with a thin film of transparent magnesium fluoride ($n = 1.38$) to reduce reflection. How thick should the coating be to minimize reflection at the center (550 nm) of the visible spectrum?

When there is minimum reflection from the lens, there must be maximum transmission into the lens. Minimum reflection implies destructive interference between the rays reflected from the two faces of the film. The overall path difference must then be one half-wavelength within the coating material; this means that the coating has a thickness of $\frac{1}{4}\lambda_m$.

What about phase changes in the reflected rays? There is a 180° phase change at the outer surface (air to coating, $n = 1$ to 1.38) and also a phase change at the inner surface (coating to glass, $n = 1.38$ to 1.5). The two phase changes produce no net effect, and the required film thickness then is

$$t = \frac{1}{4}\lambda_m = \frac{\lambda}{4n} = \frac{550 \text{ nm}}{4(1.38)} = 99.6 \text{ nm} \approx 0.1 \text{ } \mu m$$

Coated lenses on cameras and binoculars have a purplish look. The thickness of the coating is chosen to transmit yellow-green light (0.55 μm) in the middle of the visible spectrum. This thickness is not $\frac{1}{4}\lambda_m$ for the red and the violet light at the ends of the spectrum and therefore some of the red and violet light is reflected.

38-9   The Michelson Interferometer

An *interferometer* is an instrument for measuring displacements in terms of wavelength by using the interference between two beams. The Michelson

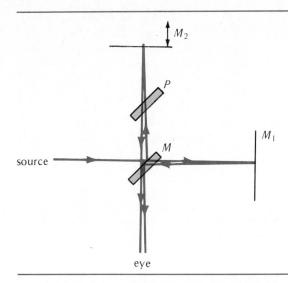

source

eye

Figure 38-20. *Elements of a Michelson interferometer: M, a partially silvered mirror; P, a compensator plate; M_1, a fixed mirror; and M_2, a movable mirror.*

interferometer, invented by the first American Nobel laureate in physics, A. A. Michelson (1852–1931), is commonly used in measuring distances to within a fraction of a wavelength of visible light. Basically, a beam is split into two beams that travel separate paths and then are recombined; a change in the path length of one beam produces a change in the interference pattern of the recombined beams. Specifically, as shown in Figure 38-20, incident light from the source is split by a partially silvered mirror M (half reflected, half transmitted) into two beams that travel separate perpendicular paths to mirrors M_1 and M_2; the reflected beams are then recombined to interfere as they enter the observer's eye. (Compensator plate P, parallel to M and equal to it in thickness, is placed as shown in the figure to ensure that the two optical path lengths, up-down and left-right, are at least approximately equal and the two beams are coherent.) This produces an interference pattern much like that of Figure 38-19(b). Suppose that the central spot in the viewing screen is dark; the two waves arrive there out of phase. Now suppose that M_2 is shifted upward just $\frac{1}{4}\lambda$. The wave reflected from M_2 travels an additional overall path length of $\frac{1}{2}\lambda$, so that the central spot in the viewing screen is now bright. It follows that a shift from one bright fringe to the adjoining bright fringe corresponds to a shift in one mirror of only $\frac{1}{2}\lambda$, so that the Michelson interferometer allows displacements to be measured to within a fraction of a wavelength.

Summary

Definitions

Newton's rings: concentric circular interference fringes produced by interference between waves reflected from flat and spherical surfaces in contact.

Interferometer: device for measuring displacement with high precision by interference effects.

Important Results

The basic principle underlying the interference be-

tween two waves: The time-average intensity $\bar{I}$ at observation point P is proportional to the square of the resultant electric field at P:

$$\bar{I} \propto E_r^2 \qquad (38\text{-}1)$$

where

$$\mathbf{E}_r = \mathbf{E}_1 + \mathbf{E}_2 \qquad (38\text{-}2)$$

and $\mathbf{E}_1$ and $\mathbf{E}_2$ are the separate electric fields from the two sources.

How E_1 and E_2 combine depends on the phase difference ϕ at P, where

$$\phi = \frac{2\pi}{\lambda} \Delta r + \Delta\Phi \qquad (38\text{-}4)$$

Here Δr is the path difference from the two sources to P and Φ is the phase difference between the two sources.

Constructive interference: $\phi =$ integral multiple of 2π.
Destructive interference: $\phi =$ half-integral multiple of 2π.

Two point sources separated by distance d, in phase, and producing waves of wavelength λ (or a double slit with separation d):

$$\text{Max } I: \quad \Delta r = m\lambda = d \sin\theta \qquad (38\text{-}8)$$

where $m = 0, 1, 2, 3, \ldots$.

Angle θ is measured from the perpendicular bisector of the line joining the two point sources.

Diffraction grating, with many slits (or sources) separated by d (where $d > \lambda$): same relation (38-8) for intensity maxima as for the double slit.

Change in phase for reflected electromagnetic beam: *180° phase change* for beam reflected from *interface* leading *to a medium of higher refractive index* (or reflection from a conductor); otherwise, no phase change.

Problems and Questions

Section 38-2   Interference from Two Point Sources

· **38-1 Q** A simple test to determine whether the two speakers in a stereo system are properly wired is this. Compare the bass response with the speakers wired in one sense with that when the wires to one speaker are reversed. The correct wiring corresponds to the louder bass response. Why?

· **38-2 P** The two speakers of a stereo system are in phase and connected to the same 165-Hz source, which produces sound of 2.0-m wavelength. The speakers are separated by 2.0 m. How far back from one speaker must a listener move (along a line perpendicular to the line joining the speakers) until he hears a minimum in sound intensity?

: **38-3 P** Four identical oscillators oscillate in phase, emitting radiation of wavelength λ. They are arranged as shown in Figure 38-21, and two receivers R_1 and R_2 are placed at large (but equal) distances from the sources. (a) Which receiver detects the strongest signal? (b) Which receiver detects the strongest signal if source B is turned off? (c) if source D is turned off? (d) Which receiver can detect which source, B or D, has been turned off?

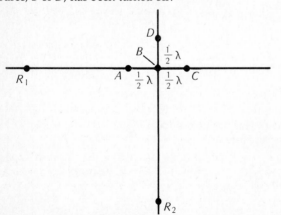

Figure 38-21.   *Problem 38-3.*

: **38-4 P** A plane electromagnetic wave of wavelength 30 cm traveling in a horizontal plane is detected by two vertical electric dipole antennas separated by 10 cm along an east-west line. Measurements show that the signal received by the eastern antenna lags in phase by 60° with respect to the signal picked up in the other antenna. What are the possible directions of propagation of the wave?

: **38-5 P** The radiation pattern of two identical omnidirectional radio oscillators consists of exactly six lobes. What is the separation distance between the oscillators if they radiate at a frequency of 20 MHz?

Section 38-3   More on Interference from Two Point Sources

: **38-6 P** Two omnidirectional microwave transmitting antennas are separated by 7.5 cm and they radiate at a frequency of 10 GHz. What is the angular separation between the lobes of their radiation pattern?

: **38-7 P** Two coherent sources interfere to produce an interference pattern on a distant screen (Figure 38-2). The intensity detected at the screen due to S_1 alone is I_0 and the intensity detected due to S_2 alone is $2I_0$. Sketch the resulting intensity of the interference pattern.

: **38-8 P** Two omnidirectional sources radiate at wavelength λ and in phase. They are placed at the points $(\frac{1}{2}d, 0)$ and $(-\frac{1}{2}d, 0)$ in the xy plane. (a) Show that the condition for maximum intensity at the point (x, y) is

$$\sqrt{(x + \tfrac{1}{2}d)^2 + y^2} - \sqrt{(x - \tfrac{1}{2}d)^2 + y^2} = m\lambda,$$

where $m = 0, 1, 2, \ldots$

(b) Prove that this curve is a hyperbola, and sketch it for the cases $m = 1$, $m = 2$, and $m = 3$.

Section 38-4   Young's Interference Experiment

38-9 Q In Young's double-slit experiment, monochromatic light is incident on two slits and light and dark fringes are observed on a distant screen. If the distance between the two slits is halved, the distance between neighboring dark fringes is

(A) doubled. (D) increased by a factor of 4.
(B) halved. (E) decreased by a factor of 4.
(C) unchanged.

· **38-10 P** Two narrow slits separated by 0.30 mm are illuminated with light of wavelength 496 nm. (*a*) How far are the first three bright fringes from the center of the pattern if observed on a screen 120 cm distant? (*b*) How far are the first three dark fringes from the center of the pattern?

· **38-11 P** Two slits separated by 1.2 mm are placed 1.40 m from a screen. Light of wavelength 480 nm and 640 nm is incident on the slits. What is the separation distance on the screen of intensity maxima of the second order?

: **38-12 Q** White light passes through a double slit and on to a distant screen. For a large interference order, what color will appear on the screen at a node for red light?
(A) Black (that is, no light).
(B) Yellow.
(C) Blue.
(D) Red.
(E) White (that is, a mixture of all colors).

: **38-13 P** In a double-slit experiment performed with light of wavelength 620 nm, an interference pattern is observed on a screen 1.2 m distant from the slits. Adjacent bright fringes near the center of the pattern are separated by 0.5 mm. (*a*) What is the separation of the two slits? (*b*) What is the separation between adjacent fringes if light of wavelength 540 nm is used?

: **38-14 P** Light of wavelength 580 nm is incident on two slits separated by 1.0 mm. An interference pattern is observed on a screen 2.0 m distant. What percentage of error is made in locating the eighth-order maximum if it is not assumed that $D \gg d$ (or equivalently, that $\sin \theta \approx \tan \theta$)?

: **38-15 P** The *Fresnel biprism* produces interference effects analogous to those observed with a double slit. Figure 38-22 shows a point monochromatic source placed a dis-

tance *b* from a small-angle prism. When the source is viewed from the opposite side of the prism, two virtual coherent sources are seen. The light from these two equivalent sources can interfere just as the light from two slits does. Show that the separation of the two sources is $d = 2b\phi(n - 1)$.

: **38-16 P** In a double-slit experiment with monochromatic light, a thin sheet of transparent material of thickness *t* and index of refraction *n* is placed over one of the slits. What now is the condition for the maxima observed on a distant screen?

Section 38-5 The Diffraction Grating

· **38-17 Q** Which wavelengths appear closer to the undeviated beam direction when white light is incident on a diffraction grating?
(A) Short wavelengths.
(B) Long wavelengths.
(C) Either long or short wavelengths, depending on the particular grating spacing used.
(D) Either long or short wavelengths, depending on the angle at which the incident light strikes the grating.

· **38-18 P** A Venetian blind has adjacent slats separated by 4.0 cm. What wavelength of normally incident sound will produce a strong diffraction peak at 30° off the normal to the plane of the blind?

· **38-19 P** Violet light of wavelength 400 nm is incident on a grating with 8000 lines per cm. At what angle will the (*a*) first-order spectrum be located? (*b*) second-order angle spectrum be located? (*c*) third-order spectrum be located?

· **38-20 P** How many lines per centimeter are required in a diffraction grating that will spread the visible spectrum, from 440 nm to 680 nm, through an angle of 30° in first order?

· **38-21 P** A diffraction grating has 5400 lines/cm. It is illuminated by monochromatic light of wavelength 570 nm. At what angle from the normal to the grating will one observe the (*a*) first-order spectrum? (*b*) second-order spectrum? (*c*) third-order spectrum?

: **38-22 P** When monochromatic light is being used with a certain diffraction grating, it is found that the first-order maximum occurs at 8°. What is the highest order of maximum that can be observed?

: **38-23 P** Monochromatic light of wavelength 510 nm is incident on a diffraction grating with 5000 lines/cm. What is the highest-order spectrum that can be observed?

: **38-24 Q** White light is incident on the "black box" apparatus sketched in Figure 38-23. The light striking the screen consists of numerous brightly colored bands. Within the black box there is probably
(A) a prism.
(B) a chromatic disperser.

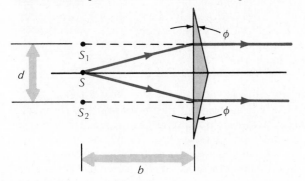

Figure 38-22.   Problem 38-15.

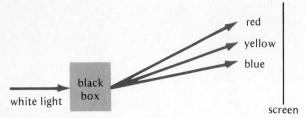

Figure 38-23.   *Question 38-24.*

(C) a diffraction grating.
(D) an interference filter.
(E) a lens.
(F) a polarizer.

∶ 38-25 Q If one wants to disperse light to get a monochromatic light source, why is it better to use a diffraction grating instead of a simple double-slit arrangement?
(A) The slits are closer together in a grating.
(B) The slits can be made narrower in a grating.
(C) The intensity maxima are sharper for a grating.
(D) More orders of interference can be obtained with a grating.
(E) Greater angles of dispersion are obtainable with a grating.

∶ 38-26 Q A light beam consists of two components of light whose wavelengths differ very slightly. An experimenter tries to resolve them with a grating he has available, but can't. How could he improve the resolution, that is, increase the separation between adjacent maxima?
(A) Use a grating with more lines per centimeter.
(B) Use a grating with more total lines and the same number of lines per centimeter.
(C) Use either of the above approaches.
(D) Use a brighter beam.
(E) Move the screen farther from the grating.

∶ 38-27 P Light of wavelength 540 nm falls on a diffraction grating, and a series of lines is observed. When light of unknown wavelength is shone on the same grating, it is found that the sixth-order maximum of the 540-nm light falls at the same place on the screen as the fifth-order maximum for the light of unknown wavelength. What is the wavelength of the unknown?

∶ 38-28 P An experimenter wants to design a grating that will spread the visible spectrum (from 430 nm to 680 nm) through an angular range of 20° in the first order. How many lines per centimeter should the grating have?

∶ 38-29 Q Suppose that a source, a diffraction grating, and a screen were all immersed in water. What effect would this have on the deviation, compared with the normal situation, when the apparatus is used in air? The deviation is the angle through which a given maximum is deviated from the normal to the grating.
(A) The deviation would be increased.

(B) The deviation would be decreased.
(C) The deviation would be unchanged.
(D) The deviation would be increased for some wavelengths and decreased for others.

∶ 38-30 Q Suppose that you are using one diffraction grating 2 cm wide and another 4 cm wide, each with the same number of lines per centimeter. Monochromatic light produces an interference pattern on a distant screen. At a given maximum, the amplitude for the wider grating will be twice as great, since there are twice as many slits. Thus the intensity will be four times as great as for the narrower grating. But only twice as much light enters the wide grating, compared with the narrower one. How do you explain this?
(A) There are only half as many maxima for the wider grating.
(B) Each maximum for the wide grating is sharper than the maxima for the narrow grating.
(C) Each maximum for the wide grating is correspondingly wider than the maxima for the narrow grating.
(D) If a grating is made twice as large, four times as much light will enter it.

∶ 38-31 P Light consisting of two components, of wavelengths λ and $\lambda + \Delta\lambda$, is to be resolved in mth order by a diffraction grating with slit separation d. Show that the angular separation is $\Delta\theta = \Delta\lambda / \sqrt{(d/m)^2 - \lambda^2}$.

∶ 38-32 P Plot intensity as a function of distance from the central maximum for the interference pattern observed on a screen 1 m away, for (*a*) two and (*b*) eight narrow slits of spacing 0.1 mm and wavelength 500 nm.

Section 38-6   Coherent and Incoherent Sources

∶ 38-33 P Explain why two coherent beams of light must travel in parallel (or nearly parallel) paths for complete destructive interference to occur.

∶ 38-34 P Two point sources, S_1 and S_2, are separated by a distance of 1.2×10^{-6} m, with S_2 east of S_1 along an east-west line. Each is radiating light of wavelength 500 nm. An observer at point O, far to the east of the two point sources, detects a light intensity of 3 μW/m² when either of the sources is radiating alone. Calculate (*a*) the intensity from both sources if they are incoherent; (*b*) the intensity from both sources if they are coherent. (*c*) The minimum distance S_2 should be moved toward the observer to get the maximum possible intensity at the observer.

Section 38-7   Reflection and Change in Phase

· 38-35 P A source positioned a distance y above a plane mirror (Lloyd's mirror arrangement) emits light of wavelength 590 nm. See Figure 38-15. An interference pattern is observed on a screen 1.2 m from the source. Fringes 0.8 mm apart are observed on the screen. Determine y.

∶ 38-36 P A microwave transmitter is placed 3.0 cm above a large horizontal conducting sheet. It emits 1.2-cm microwaves, which are detected by a receiver at a distance of 36 cm (measured horizontally). What is the minimum height above the sheet (different from zero) at which no signal will be picked up by the receiver?

Section 38-8   Interference with Thin Films

· **38-37 P** A researcher is studying artificial fibers. She measures the diameter of a fiber by placing it between two flat glass plates 30 cm long, as sketched in Figure 38-24. She illuminates the plates from above with green light of wavelength 546 nm and finds that a lateral distance of 8.2 mm corresponds to a change of 4 orders. What is the diameter of the fiber?

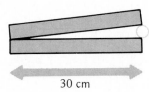

30 cm

Figure 38-24.   Problem 38-37.

: **38-38 P** When you observe the reflected interference colors in a soap film held in a vertical plane, you see a series of brightly colored horizontal bands. These bands keep moving as the liquid in the film gradually sinks downward. Finally the film becomes too thin to support its own weight and it breaks. Just before the film breaks, the colored bands disappear and the film looks black. Which of the following underlies this phenomenon?
(A) Incoherent waves will, on the average, interfere destructively.
(B) Light undergoes a 180° phase shift when reflected from a more dense medium.
(C) As the film thickness goes to zero, all interference effects must vanish.
(D) A film whose thickness is of the order of one wavelength will experience resonance vibrations and absorb all light incident on it.
(E) All the incident light will be transmitted through a very thin film, and none will be reflected.

: **38-39 P** A soap film formed in a loop of wire is, because of its weight, not exactly uniform in thickness, but wedge-shaped in cross section. When the film is illuminated with 546 nm light, one sees that a distance of 2.0 cm separates two bright fringes, four orders apart. By how much does the film differ in thickness over this distance?

: **38-40 P** Light of wavelength 480 nm travels through a liquid of index of refraction 1.42. Two rays initially in phase travel different paths from point A to point B; one path is 2.4 mm longer than the other. What is the phase difference between the two rays when they are brought together at B?

: **38-41 P** Newton's rings are observed when a plano-convex lens is placed on an optical flat plate of glass (Figure

38-19). Show that the area between adjacent bright rings of high order ($m \gg 1$) is approximately constant and equal to $\pi R \lambda$, where R is the radius of curvature of the lens and λ is the wavelength of the light.

: **38-42 P** When a plano-convex lens is placed on a glass optical flat one sees Newton's rings with a dark central spot when viewing from above (Figure 38-19). Show that the radius of the dark interference fringes is given approximately by $r = \sqrt{mR\lambda}$, where $m = 0, 1, 2, \ldots$ and R is the radius of curvature of the lens.

: **38-43 P** An experimental setup using Newton's rings is carried out with two plano-convex lenses, as sketched in Figure 38-25. Show that if the two lenses have radius of curvature R_1 and R_2, then the radius of the maxima of the interference rings is given by

$$r = \sqrt{\lambda\left(m + \frac{1}{2}\right)\frac{R_1 R_2}{R_1 + R_2}}, \qquad \text{where } m = 0, 1, 2, \ldots$$

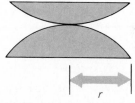

r

Figure 38-25.   Problem 38-43.

: **38-44 P** An antireflection coating is designed to give no reflection at a wavelength of 550 nm, in the middle of the visible spectrum. By what fraction is the intensity diminished at wavelengths of (*a*) 450 nm? (*b*) 700 nm?

Section 38-9   The Michelson Interferometer

· **38-45 P** When one mirror of a Michelson interferometer was moved through a distance d, it was observed that 624 fringes passed through the field of view. Light of 579-nm wavelength was used. How far was the mirror moved?

· **38-46 P** If one mirror in a Michelson interferometer is shifted 0.204 mm, 686 fringes are seen to pass the field of view. What is the wavelength of the light used?

: **38-47 P** An evacuated transparent chamber 9.0 cm long is placed in one arm of a Michelson interferometer. As gas is slowly admitted to the chamber, fringes are observed slowly passing the field of view. When a pressure of 1 atm is reached, 88 fringes of light of wavelength 589 nm have passed. What is the index of refraction of the gas?

: **38-48 P** When a thin film of material of refractive index 1.40 is placed in one beam of a Michelson interferometer, it causes a shift of 38 fringes for light of 540-nm wavelength. What is the thickness of the film?

: **38-49 P** A transparent air-filled chamber 8.0 cm long is placed in one arm of a Michelson interferometer. As the air is evacuated from the chamber, one observes 80 fringes move past the field of view. The light wavelength is 600 nm. What is the index of refraction of air under these conditions?

Supplementary Problems

38-50 P A laser beam is coherent and it can be modulated much as is done with radio waves. One could envision using a light wave as a carrier and modulating it with the range of audio frequencies needed for speech communication (say a bandwidth of 2400 kHz). How many nonoverlapping channels of communication with this bandwidth could be accommodated in the visible spectrum, from 480 nm to 680 nm? Would there be enough such communication bands to assign a specific one to everyone on earth?

38-51 P A microwave oven operates at a frequency of 2.45 GHz. (*a*) What is the purpose of the fan with metal blades that rotates when the microwave oven is turned on? (*b*) Litton Industries, a principal manufacturer of microwave ovens, petitioned the Federal Communications Commission in 1976 to let it operate ovens at a frequency of 10.6 GHz; Litton said that the higher frequency would be much better for browning meat and heating small items. Why should this be? (The FCC denied the request.)

38-52 P A *phased-array radar* antenna is fixed in position, yet it can electronically steer a narrow outgoing beam (or be sensitive to an incoming signal from a minute distant source) because the phase of oscillations to the many individual antenna elements of which it is comprised can be shifted. The PAVE PAWS is the code name for a phased array radar operated by the U.S. Air Force, one in Cape Cod and another in California; it can give early warning for submarine-launched ballistic missiles. The antenna has nearly 3600 individual radiating elements separated from one another by about 0.3 m on each of two 30-m wide faces. The beam direction can be shifted rapidly, within a few microseconds, from one target to another. The PAVE PAWS has a range of about 5000 km; it can detect a target with a frontal area of only 10 m^2; and it can keep track of hundreds of targets. See "Phased-Array Radars," by Eli Brookner, in *Scientific American* **252,** 94 (Jan., 1985).

When all radiating elements of a phased-array antenna are evenly spaced over a plane and oscillate in phase, a strong signal is radiated along the outward normal to the plane. Suppose that the microwave wavelength is 10 cm, and that the beam is to be directed 30° off the normal. What is the required phase shift in the radiating elements?

38-53 Q Ideally, a "stealth" aircraft cannot be detected by reflected radar signals. Assume for simplicity that a stealth aircraft is designed that will foil radar signals of a *single* microwave frequency. How might this be done? Why is designing a broad-band stealth aircraft, one that is immune to radar pulses of microwave wavelengths over a broad band, very difficult?

Diffraction

39

The term *diffraction* is used to describe the distinctive wave phenomena that result from the interference of many (even an infinite number of) waves from point sources oscillating coherently. Diffraction is nothing more than the interference of waves from many sources. The physics is the same as for just two sources; the geometry may be more complicated.

As we shall see, the distinctive diffraction effects of alternating bright and dark bands arise whenever a wave front is impeded by an obstacle or an aperture.

Diffraction effects, like interference effects, may be applied to measure wavelengths with high precision.

39-1   Radiation from a Row of Point Sources

Consider Figure 39-1, where a large number N (here, 12) of identical point oscillators are arranged in a row of total width w. Each source is separated from neighboring sources by the distance d. One might, for example, have 12 equally spaced electric-dipole radio antennas. All point sources oscillate at the

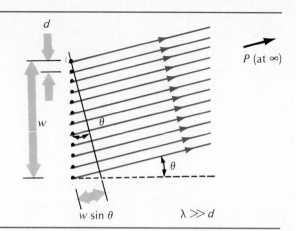

Figure 39-1. *Row of equally spaced point oscillators* ($d \ll \lambda$).

opposite of Fig 38-11

$w \sin \theta$ $\lambda \gg d$

same frequency and are in phase. The sources generate waves of length λ; this wavelength is large compared with the distance d.

We wish to find the radiation pattern of this array of equally spaced oscillators. More specifically, we wish to find the relative intensity, observed at some very distant point P, a fixed distance from the oscillators, as a function of the angle θ between the normal to the line of oscillators and the line joining any of the oscillators to the point P. Any phase difference ϕ at a distant observation point arises solely from the difference Δr in path length between sources. When $\theta = 0$, all rays are drawn horizontally to an infinitely distant point. These effectively horizontal rays all have the same length, so that the sources interfere constructively. Therefore, the resultant electric field at the angle $\theta = 0$ is N (here, 12) times the electric field of any one single oscillator.

At what angle is the intensity first zero? As we shall see, it is much easier to find the angular positions for zero intensity than for maxima in the intensity. The strategy for locating zeros is this: group the oscillators into pairs so that the resultant field at P for every such pair is zero. This means that we must so choose the oscillators and angle that the difference in path length between the pair is $\frac{1}{2}\lambda$ (or an odd multiple of $\frac{1}{2}\lambda$); then each such pair of oscillators will interfere destructively. Suppose, then, that the angle θ is such that the difference in path length between oscillator 1 and oscillator 7 is $\frac{1}{2}\lambda$. At a distant point, the resultant electric field from this pair is zero. But by the geometry of Figure 39-1, we see that oscillator 2 and oscillator 8 then also differ in path length by $\frac{1}{2}\lambda$. Indeed, we can match all the oscillators in pairs — 1 and 7, 2 and 8, 3 and 9, and so on — so that the resultant electric field at point P from every oscillator pair is zero. The path difference between the oscillators at the top and the bottom of the array is, from Figure 39-1, equal to $w \sin \theta$. The path difference between the first and seventh oscillators is $(w/2) \sin \theta$, so that destructive interference between waves from these two implies that $\frac{1}{2}w \sin \theta = \frac{1}{2}\lambda$.

We then have

First-intensity zero: $w \sin \theta = \lambda$

The angle for second-intensity zero is found in similar fashion. Now we divide the array into four *zones*, or groups of oscillators: oscillators 1 through 3, 4 through 6, 7 through 9, and 10 through 12. Angle θ must now be larger, so large, in fact, that the path difference between oscillator 1 and oscillator 4 is $\frac{1}{2}\lambda$.

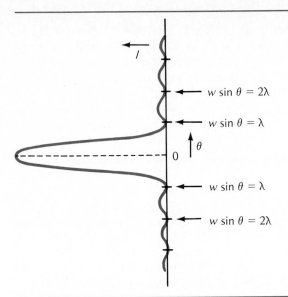

Figure 39-2. Intensity as a function of angle θ.

This pair of sources then interfere destructively at an infinite distance. Similarly, we match oscillators 2 and 5, 3 and 6, and so on, so that again the resultant field of the array is zero. The path difference $w \sin \theta$ between the sources at the extremes of the array is now 2λ, and we have

Second-intensity zero: $w \sin \theta = 2\lambda$

It is apparent that the angles for zero intensity at an infinitely distant observation point are given, in general, by

Intensity zeroes: $w \sin \theta = m\lambda$ (39-1)

where $m = 1, 2, 3, \ldots$ (but not zero). The number of sources N must be large, and they must be spaced closely, less than one wavelength apart.

Figure 39-2 shows the radiated intensity I as a function of θ. (The derivation of the relation for $I(\theta)$ is given in Section 39-6.) This *diffraction pattern* consists of an intense maximum at $\theta = 0$, twice the width of the relatively weaker

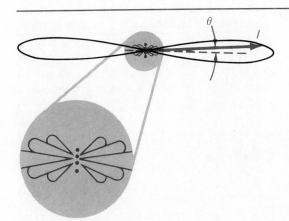

Figure 39-3. Radiation pattern from a row of equally spaced point oscillators. The magnitude of the radius vector from the origin is proportional to the intensity of the radiation in that direction.

secondary maxima that appear symmetrically to its sides. The angular locations of the intensity zeros are given by (39-1). The same information is portrayed in a different graphical form in Figure 39-3, which is the radiation pattern of the array of oscillators; the pattern consists of two strong narrow central lobes, into which most of the radiated energy is directed, together with small side lobes. We see that a linear array of equally spaced antennas (with $d \ll \lambda$) will radiate strongly only in the directions perpendicular to the line of antennas.

39-2   Single-Slit Diffraction

Suppose that monochromatic waves illuminate a single narrow slit with parallel straight sides. What is the intensity pattern of the light falling on a far distant screen? We have already solved this problem! We merely recognize that when a plane wave front of light impinges upon an opening in an otherwise opaque plane, we can imagine each point on the wave front across the slit as a new point source of radiation. All these coherent point sources oscillate in phase. In effect, we have an array of equally spaced point sources spread over the width of the slit. The intensity pattern on the distant screen, therefore, is just what we have already derived for a row of closely spaced oscillators (Section 39-1, Figure 39-2).

An arrangement for observing single-slit diffraction is shown in Figure 39-4. The angular positions of the zeros in the intensity pattern are given as before by

$$\text{Intensity zeros:} \quad w \sin \theta = m\lambda \tag{39-1}$$

where λ is the wavelength, w is now the slit width, and $m = 1, 2, 3, \ldots$ (but not zero). If y is the displacement from the central maximum on the screen and D the distance from slit to screen, we can write for small angular displacements that $\theta \approx y/D$. Equation (39-1) can then be written as

$$w \frac{y}{D} = m\lambda \tag{39-2}$$

where y gives the locations of intensity zeros.

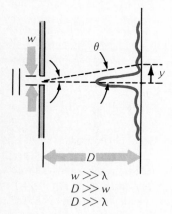

$w \gg \lambda$
$D \gg w$
$D \gg \lambda$

Figure 39-4. *Diffraction from a single slit.*

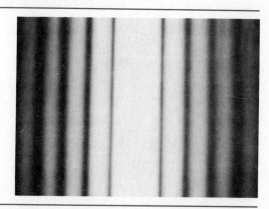

Figure 39-5. *Single-slit Fraunhofer diffraction pattern.*

Figure 39-5 is a photograph of a single-slit diffraction pattern; it corresponds to the intensity variation shown in Figure 39-2.

Note especially what happens if the slit width is reduced: as (39-1) indicates, with the wavelength fixed, the diffraction pattern expands. Light goes far outside the geometrical shadow of the slit edges. Indeed, the diffraction pattern on the distant screen bears no resemblance to the pattern predicted by ray optics: intensity constant over the slit but falling abruptly to zero at the slit's sharp edges.

The wavelengths of visible light are very small compared with the dimensions of ordinary objects. That is why diffraction of light is a fairly subtle effect and you don't ordinarily see diffraction fringes.* Diffraction of much longer wavelengths is commonplace, however. That is why you can often hear a sound source around a corner even though you can't see the source.

The diffraction effects we have described thus far are officially known as examples of *Fraunhofer-type diffraction.* This special type of diffraction applies when both the source and the observation screen are infinitely distant from, or at least a long way from, the diffracting object (here a slit); equivalently, Fraunhofer diffraction implies that wave fronts encountering the diffraction object are plane wave fronts. These special conditions can be met when the source and the screen are actually physically close to a slit. You simply place the light source at the principal focus of a converging lens so that the spherical or cylindrical wave fronts diverging from the source become plane wave fronts as they emerge from the lens and strike the slit. See Figure 39-6. Similarly, suppose that the observation screen is located at the far principal focus of a second converging lens. Then plane wave fronts leaving the slit are brought to a focus on the screen.

The term *Fresnel diffraction* is used to denote the general case in which there are no restrictions on distances or wave fronts. In this chapter we concentrate on the simpler but special case of Fraunhofer diffraction. One important case of Fresnel diffraction is, however, shown in Figure 39-7. Here we see the diffraction pattern produced by a straight edge (one-half of a slit) with the observation screen located very close to the straight edge. We note, first, that the shadow of a sharp straight edge is not perfectly sharp, as expected from ray optics.† Instead, there are variations in intensity—the fringes always characteristic of diffraction—near the edge. Note further that the intensity falls gradually to zero on the shadow side beyond the geometrical edge of the shadow region.

Figure 39-6. *Conditions for Fraunhofer diffraction achieved with source and screen at finite distances from the slit by use of converging lens.*

Example 39-1.   Yellow light from atomic sodium with a wavelength of 589 nm illuminates a single slit. The central dark fringes in the diffraction pattern are found to be separated by 2.2 mm on a screen 1.0 m from the slit. What is the slit width?

From (39-2), we have

$$w \frac{y}{D} = m\lambda$$

---

* To see alternating bright and dark diffraction fringes easily, simply look at a bright source of light through the narrow slit formed when two fingers are pressed together and held close to your eye.
† The first recorded observation of the diffraction of visible light was by F. M. Grimaldi, who noted in 1655 that the shadow of the straight edge was not sharp.

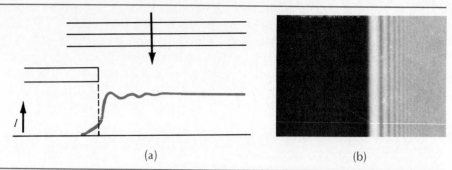

Figure 39-7. *Fresnel diffraction pattern for plane wave diffracted at a straight edge: (a) intensity plot; (b) photo.*

(a)

(b)

Here $2y = 2.2$ mm, and $m = 1$. The slit width is then computed as

$$w = \frac{m\lambda D}{y} = \frac{(1)(589 \times 10^{-9} \text{ m})(1.0 \text{ m})}{1.1 \times 10^{-3} \text{ m}} = 5.4 \times 10^{-4} \text{ m} = 0.54 \text{ mm}$$

39-3   The Double Slit Revisited

The simple interference pattern from two parallel slits, as shown in Figure 38-8 — peaks equally spaced and with *equal*-intensity maxima — applies only if the slits are infinitesimally narrow. Here we consider a pair of real slits each of *finite* width w, with their centers separated by distance d, illuminated by

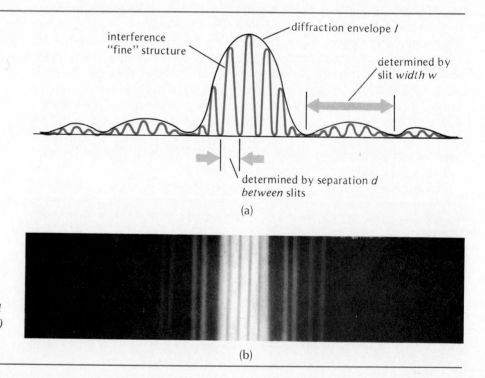

Figure 39-8. *(a) Intensity variation for interference and diffraction by a double slit. (b) Photograph of a double-slit diffraction pattern.*

(a)

(b)

plane waves of wavelength λ. The pattern on a distant screen shows diffraction as well as interference effects.

Imagine first that we cover one slit and leave the other exposed to the incident waves. The intensity pattern on a distant screen is certainly that for single-slit diffraction: a broad, intense central peak flanked by weak, equally spaced diffraction fringes (Figure 39-2). And if we cover the second slit and expose the first, we see the *same* diffraction pattern on the screen (shifted by distance d). The pattern when both slits are exposed is as shown in Figure 39-8:

- *Slow* variations in the *envelope* of the intensity pattern arising from *diffraction* through the slits and controlled by the slit width w. The intensity variation for diffraction is given by (39-7).
- *Rapid variations* in the intensity arising from *double-slit interference* and controlled by the slit separation distance d. The interference intensity variation is given by (38-9).

Another way of describing Figure 39-8 is to say that the interference pattern is "modulated" by the diffraction pattern. The envelope is controlled by diffraction, the fine structure by interference.

39-4 Diffraction and Resolution

Suppose that a point source of light is far distant from an opaque surface with a circular hole of diameter d. Plane wave fronts arrive at the hole, and a diffraction pattern appears on a screen, also far from the aperture. See Figure 39-9(a). The diffraction pattern is as shown in Figure 39-9(b): a bright central, circular spot surrounded by concentric diffraction fringes. The pattern for the circular hole is rather like the diffraction pattern for a single slit (Figure 39-5) but turned, so to speak, in a circle. The first zero in intensity for a single slit of width w is off center by the angle θ, where $\theta \simeq \sin \theta = \lambda/w$. The angular location of the first zero off center for a circular hole is similar. A detailed analysis shows that*

$$\sin \theta \simeq \theta = 1.22 \frac{\lambda}{d} \qquad (39\text{-}3)$$

Here again we see that, as the opening is reduced (d is reduced) for a constant wavelength, the diffraction pattern expands (θ increases). Further, for a hole of fixed size, the diffraction pattern shrinks as the wavelength is reduced.

The essential fact is that a point source of light far from one side of a circular opening is rendered on a distant screen on the other side not as a bright point, but as a smeared bright circle ringed by diffraction fringes. If you know about this diffraction effect, there is no surprise when you see the image of a single point source as a smeared bright circle of light wreathed with fainter rings. Furthermore, two point sources well separated from one another produce a

---

* The number 1.22 is the smallest root of the first-order Bessel function divided by π. The Bessel function enters into the mathematical solutions for arrangements with cylindrical symmetry just as sinusoidal functions appear in arrangements with rectangular symmetry.

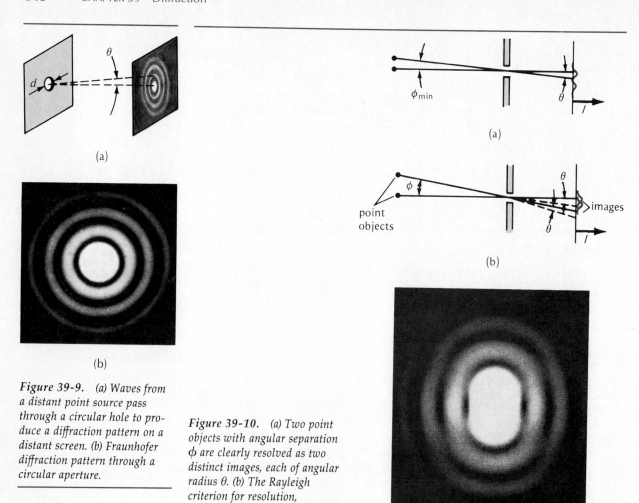

Figure 39-9. *(a) Waves from a distant point source pass through a circular hole to produce a diffraction pattern on a distant screen. (b) Fraunhofer diffraction pattern through a circular aperture.*

Figure 39-10. *(a) Two point objects with angular separation ϕ are clearly resolved as two distinct images, each of angular radius θ. (b) The Rayleigh criterion for resolution, $\phi_{min} = \theta$. (c) Intensity pattern corresponding to part (b).*

nonoverlapping pair of such circular diffraction patterns. But what if two point sources are so close together (in angular position) that their diffraction patterns overlap?

Take a specific example; consider two distant stars with angular separation ϕ. The light from the two point sources goes through an aperture, such as the objective lens of an astronomical telescope.* The images are two smeared diffraction disks of angular "radius" θ, given by (39-3), with their centers separated by ϕ. As shown in Figure 39-10(a), angle ϕ is large enough that the images are clearly resolved. What is the minimum angular separation ϕ_{min} that will permit us to say that there are two stars, not just one? Figures 39-10(b) and (c) show the images for two point sources just barely resolved. The angular separation ϕ here is just equal to the angular radius θ of either image separately. Said somewhat differently, the bright center of one image alone falls

---

* One always views through an aperture, if only that of the eye's pupil.

Figure 39-11. *Photograph of the Very Large Array radio telescope in Socorro, New Mexico. There are 27 giant antennas (diameter, 25 m; mass, 2×10^5 kg), which can be shifted along a Y-shaped track laid out on the desert. Signals from operating antennas are fed to a processing center. The array functions with a resolution like that of a single antenna with a 27 km diameter.*

exactly at the first dark ring of the other. You can just barely tell that there are, in fact, two images. This criterion for resolution, which Lord Rayleigh first proposed, is known as *Rayleigh's criterion for resolution;* it can then be written as

$$\phi_{\min} = \theta = \frac{1.22\lambda}{d} \qquad (39\text{-}4)$$

This resolution criterion applies generally, to any group of sources, not merely to two point sources. After all, any "picture" can be regarded as consisting of point sources.

Resolution is controlled generally by the ratio λ/d, where d is the characteristic dimension of an aperture (the width of a slit, the diameter of a circular hole). To improve resolution, we must use shorter wavelengths or bigger apertures or both. For example, the resolution of a microscope is improved by using short-wavelength, blue light. Telescopes are made big, not to produce more magnification, but to capture more radiation and thereby give a bright picture, and also to reduce diffraction effects and thereby give a sharp picture. Figure 39-11 shows a radio telescope with widely separated receiving "dishes." If the signals from two or more receiving antennas are combined, in a procedure known as long base-line interferometry, the sources can be located with a resolution that is controlled by the greatest distance separating the individual antennas. Although this result is stated without detailed proof, its reasonableness is seen at once when we recall that two separated transmitting antennas (Section 38-2) can concentrate the net radiation into a small angular region.

Example 39-2. What is the size of the smallest detail you can see directly in a hand-held photograph? (Take the diameter of the pupil of a normal eye to be about 4 mm.)

Consider what you do when you examine any object closely. You hold it about 25 cm from your (normal) eyes. At this distance, the image *on the eye's retina* not only is focused but has maximum dimensions. For an object held farther from the eye than 25 cm, the retinal image is smaller; when an object is brought closer than 25 cm, a normal eye cannot bring the image to focus on the eye's retina.

What, then, is the separation distance x between two point sources of light 25 cm away from a circular aperture (the eye's pupil) that will just be resolved as two distinct sources? The angular separation of the two sources measured from the plane of the aperture is $\phi = x/25$ cm. But the minimum value of ϕ for resolution is, from (39-4), $\phi_{\min} = 1.22\lambda/d$. For visible light, diffraction effects will first be evident in the long-wavelength (red) end of the spectrum, $\lambda \simeq 700$ nm. We then have

$$\phi_{\min} = \frac{x}{25 \text{ cm}} = 1.22\frac{\lambda}{d}$$

$$x = \frac{(25 \text{ cm})(1.22)(700 \times 10^{-9} \text{ m})}{4 \text{ mm}} \approx 0.05 \text{ mm} = \frac{1}{20} \text{ mm}$$

Therefore, in a hand-held photograph the eye can just barely discern details only as small as $\frac{1}{20}$ mm. There is really no point in having even finer details in the print. If a photographic negative is to be used to produce a larger print, or a picture is to be projected on a screen, then the resolution requirements may be far more stringent.

Example 39-3.   Show that the smallest separation distance you can "see" with waves of wavelength λ is of the order of λ.

Suppose that we are examining two point objects separated by distance x with a lens of diameter d_1 held a distance d_2 from the objects. See Figure 39-12. (If the lens is used as a simple magnifier, distance d_2 is close to its focal length.) Rays passing through the center of the lens are not deviated, so that rays from the point objects to the center of the lens make angle ϕ, both before entering and after leaving the lens. From the geometry,

$$x = \phi\, d_2$$

and at the limit of resolution

$$\phi = 1.22\frac{\lambda}{d_1} = \frac{x}{d_2}$$

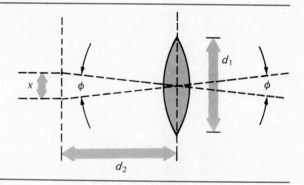

Figure 39-12.

If the lens is to form a bright image and therefore capture a significant fraction of the light from the point source, distances d_1 and d_2 must be comparable. The relation above then yields $x \sim \lambda$.

39-5   X-Ray Diffraction

The wavelengths of x-rays ($\lambda \simeq 0.1$ nm) are of the same order of magnitude as the distance between adjacent atoms in a solid. A crystalline solid is a material in which atoms are arranged in a regular geometric array. It may be used, in

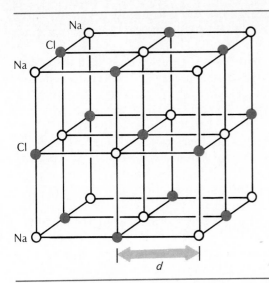

Figure 39-13. *Cubic crystal structure of NaCl.*

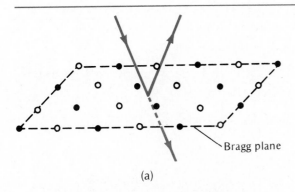

(a)

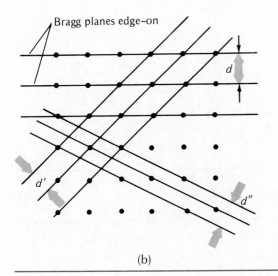

(b)

Figure 39-14. *(a) Atoms in a Bragg plane. (b) Three sets of Bragg planes with different grating spacings.*

effect, as a three-dimensional diffraction grating for measuring x-ray wavelengths. Similarly, x-rays of known wavelengths may be used to deduce the atomic arrangements in crystals.

Sodium chloride has a particularly simple cubic crystalline structure. See Figure 39-13. Sodium ions and chloride ions are located at alternate corners of cubes. Each atom acts as a scattering center for x-rays; strictly, the electrons around the nucleus of each atom are responsible for scattering and diffracting an incoming wave. Strong diffraction involves the cooperative scattering from many atoms, especially those atoms that lie in parallel planes, called Bragg planes, as shown in Figure 39-14(a). The distance between adjacent parallel planes of atoms is the *lattice spacing d.* As Figure 39-14(b) shows, there may be a variety of Bragg planes with differing values of d for a single type of crystalline structure.

Consider now what happens when a wave is incident at angle θ measured with respect to the Bragg planes (not measured with respect to the normal to these planes). Scattering by atoms all lying in a plane is equivalent to the partial reflection of the incident beam from that Bragg plane. Reflected beams from two adjoining parallel planes will interfere constructively when the path difference is an integral multiple of the wavelength. From the geometry of Figure 39-15, we see that the overall path difference (shown with brackets) is $2d \sin \theta$ for two adjoining Bragg planes, so that the condition for strong diffraction is

$$2d \sin \theta = m\lambda \qquad \text{where } m = 1, 2, 3, \ldots \qquad (39\text{-}5)$$

This, the so-called *Bragg relation,* is named after W. L. Bragg, who first derived it.

When polychromatic x-rays illuminate a crystal, only those wavelengths satisfying (39-5) will be strongly diffracted. Or when a single wavelength of

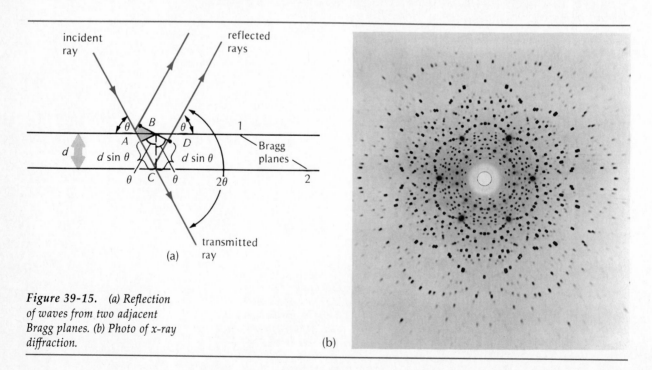

Figure 39-15. (a) Reflection of waves from two adjacent Bragg planes. (b) Photo of x-ray diffraction.

x-rays illuminates a crystal whose orientation in space can be varied, only those particular lattice spacings satisfying the Bragg relation produce strong diffraction.

Such ordinary particles as electrons have wave properties according to the quantum theory; and a crystal exposed to a beam of electrons can also exhibit electron diffraction, an effect that is altogether analogous to x-ray diffraction.

Example 39-4. The lattice spacing for the principal Bragg planes of a sodium chloride crystal is 0.282 nm. For what wavelength x-rays will the first-order diffracted beam be deviated from the incident x-ray beam by 60°?

As Figure 39-15 shows, the angle between the incident and strongly diffracted beams is $2\theta = 60°$, so that the Bragg relation, (39-5), gives

$$\lambda = \frac{2d \sin \theta}{m} = \frac{2(0.282 \text{ nm}) \sin 30°}{1} = 0.282 \text{ nm}$$

39-6 $I(\theta)$ for Single Slit (Optional)

Here we derive the relation for intensity I as a function of observation angle θ for Fraunhofer diffraction through a single slit. We must first find the resultant field $\mathbf{E}_r$ at a distant point; then $I \propto E_r^2$. The resultant field $\mathbf{E}_r$ is computed most simply by representing the continuum of simple harmonic sources spread evenly across the slit of width w by phasor vectors (Section 33-3).

Figure 39-16 shows a number of phasors of equal magnitude E_1 arranged with a constant phase difference ϕ_1 between adjoining vectors. These electric-field vectors represent the contributions from separate oscillators, as in Figure 39-1. The constant phase difference ϕ_1 arises from the constant path difference $\Delta r = d \sin \theta$ between any two adjoining oscillators. The total phase difference ϕ from all N oscillators (with $N - 1$ intervals) is given by

$$\phi = (N - 1)\phi_1$$

$$= (N - 1) \frac{2\pi \Delta r}{\lambda} = \frac{2\pi w}{\lambda} \sin \theta \tag{39-6}$$

As (39-6) shows, the angle ϕ for the observed total phase difference between oscillators at the two extremes of the array increases as the space angle θ increases.

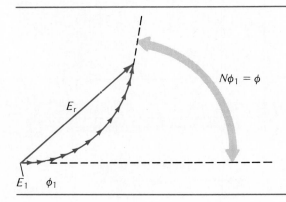

Figure 39-16. *Electric-field vectors (phasors) arranged according to their relative phase differences.*

We wish to find the magnitude of $\mathbf{E_r}$, the resultant electric field at a distant point. Before computing $\mathbf{E_r}$ in detail, let us see qualitatively how the separate vectors must be arranged at the maxima and minima of the intensity pattern of Figure 39-2 and thereby deduce general features of the I versus θ curve.

At the central peak, $\theta = 0$; therefore, $\Delta r = 0$ and $\phi = 0$. The little vectors are all aligned parallel, and one has the maximum possible E_r, namely $E_r = NE_1$. See Figure 39-17. Now, as θ increases, so does ϕ, and the electric fields become progressively out of phase. The vertices of the vectors lie on a circular arc, and the magnitude of $\mathbf{E_r}$ is now less than NE_1. Suppose that the total phase difference ϕ is 2π. The vectors now complete one circle; $\mathbf{E_r}$ is zero, as the intensity is also. Equation (39-6) shows that when $\phi = 2\pi$, then $w \sin \theta = \lambda$, in accord with our earlier finding. Note also that when $\phi = 2\pi$, the vectors for oscillators 1 and 7, and 2 and 8, and so on are antiparallel, corresponding to a $180°$ phase difference, or a $\frac{1}{2}\lambda$ path difference, for each such pair.

The maximum in the first secondary peak occurs very nearly at that space angle θ for which the little vectors make $1\frac{1}{2}$ turns. The magnitude of the resultant $\mathbf{E_r}$ is now much less than NE_1, so that the intensity, $I \propto E_r^2$, at the secondary peak is much less than the intensity of the central maximum. The second zero of intensity corresponds to that space angle θ and phase angle ϕ for which the little electric-field vectors complete two circles.

We can compute the magnitude of $\mathbf{E_r}$ in general by referring to Figure 39-18. The radius of the circle is R, the angle subtended by the chord of length E_r is ϕ, and the length of the corresponding circular arc is NE_1. In either triangle

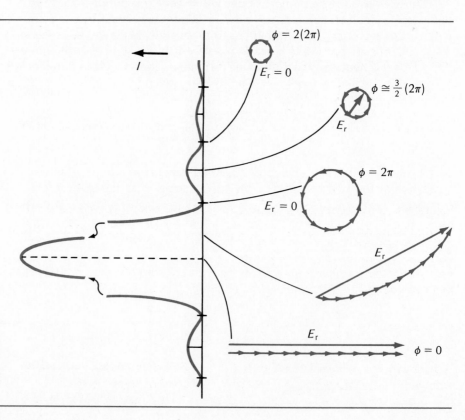

Figure 39-17. *Arrangements of electric-field vectors (phasors) for various locations in the intensity pattern.*

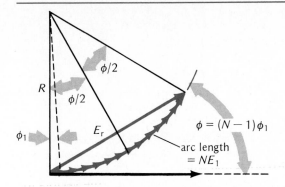

$\phi/2$

R

$\phi/2$

ϕ_1

E_r

$\phi = (N-1)\phi_1$

arc length
$= NE_1$

Figure 39-18.

with angle $\phi/2$, we have

$$\sin \frac{\phi}{2} = \frac{\frac{1}{2}E_r}{R}$$

From the definition of the angle ϕ in radians, we have

$$NE_1 = R\phi$$

This relation holds exactly only if N is *infinite*; the little vectors truly form a circular arc, rather than a polygon, only if their number is infinite.

Eliminating R from the two relations above yields

$$E_r = NE_1 \frac{\sin(\phi/2)}{\phi/2}$$

Therefore, we have for the intensity, $I \propto E_r^2$,

$$I = I_0 \left[\frac{\sin(\phi/2)}{\phi/2} \right]^2 \qquad (39\text{-}7)$$

where I_0 is the intensity with $\phi = 0$ (at $\theta = 0$). Phase angle ϕ is related to the space angle θ through

$$\phi = \frac{2\pi w}{\lambda} \sin \theta$$

Figure 39-2 is a plot of (39-7). We readily verify from this equation that $I = 0$ for $w \sin \theta = m\lambda$, where $m = 1, 2, 3, \ldots$, in agreement with (39-1). Using (39-7), one can also show that the intensities of the secondary peaks relative to the central peak (as 1.00) are 0.045, 0.016, 0.008, ... The central peak is much more intense than the secondary peaks; in fact, more than half of the radiated energy falls within the middle half of the central peak.

39-7   The Diffraction Grating Revisited (Optional)

A grating with many parallel slits is called a *diffraction* grating because the strong and narrow intensity peaks it produces are due to diffraction.

Using an argument like that for a single slit, we show here that the principal

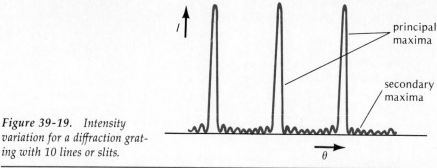

Figure 39-19. Intensity variation for a diffraction grating with 10 lines or slits.

peaks in a diffraction-grating intensity pattern are very narrow. We again consider Figure 38-11, where N sources are separated from one another by distance d (with $d > \lambda$). Angle θ in Figure 38-11 is such that the path difference between adjoining sources is λ. This means that the path difference between the first slit and the Nth slit is $(N - 1)\lambda$. (With N slits, there must be $N - 1$ spaces between adjoining slits.)

Suppose now that angle θ is made just slightly larger, so that the path difference between the first and Nth slit is $N\lambda$, rather than $(N - 1)\lambda$. (The change in angle is very small indeed for a typical value for N, say a few thousand.)

We have already dealt with a situation of this sort in finding the zeros in intensity for an array of oscillators (Section 39-1 and Figure 39-1). Here again we imagine the slits divided into two groups, a top half and a bottom half. At the new, slightly larger angle, the ray from the uppermost slit in the top half will have a path difference of $\frac{1}{2}\lambda$ with respect to the uppermost slit in the lower half. These two rays will destructively interfere. So will every pair of corresponding top-half and bottom-half slits. The principal interference peak falls to zero for a very small change in angle.

Figure 39-19 shows the intensity pattern for a diffraction grating with ten slits. The detailed analysis yielding the curve is like that given in Section 39-6. The principal maxima are more intense at their peaks than the secondary maxima by a factor N. The principal maxima have twice the width of the secondary maxima, and there are $N - 2$ secondary peaks between adjoining principal peaks.

Summary

Definitions

Diffraction: interference from a large number of wave sources.

Fraunhofer-type diffraction: sources and observation screens at infinite distances from diffracting object; equivalently, plane waves incident upon the diffracting object.

Fresnel-type diffraction: sources and observation screens at finite distances from diffracting object.

Rayleigh criterion for the resolution of two adjacent point sources: first zero in the diffraction pattern of one point source alone coincides with the central maximum in the diffraction pattern of the other point source.

Bragg plane: in a crystalline solid, a plane containing many atoms.

Important Results

Single-slit diffraction (Fraunhofer condition) for plane waves of wavelength λ incident upon a slit of width w. The angular locations of the zeros in the diffraction pattern are given by

Zero I: $\sin \theta = \dfrac{m\lambda}{w}$ (39-1)

where $m = 1, 2, 3, \ldots$. This relation also applies to an array of many point sources, equally spaced and oscillating in phase, spread over a distance w.

Circular opening of diameter d. The first zero in the circular diffraction pattern has an angular location given by

First zero I: $\theta = 1.22 \dfrac{\lambda}{d}$ (39-3)

The *resolution of adjacent point objects* as two distinct objects is limited ultimately by diffraction effects. Resolution is improved by reducing the ratio λ/d, where d is a characteristic dimension of an aperture or obstacle.

X-ray diffraction, the Bragg law: strong diffraction from parallel Bragg planes separated by lattice spacing d for waves of wavelength λ incident at angle θ (measured relative to the Bragg planes) corresponds to

$$2d \sin \theta = m\lambda \qquad (39\text{-}5)$$

where $m = 1, 2, 3, \ldots$.

Problems and Questions

Section 39-1   Radiation from a Row of Point Sources

· **39-1 Q** What is the difference between interference and diffraction?
(A) Interference requires coherent light; diffraction does not.
(B) Diffraction describes the interaction of light waves that have passed through some kind of aperture, whereas interference does not.
(C) Interference always occurs between waves of the same amplitude, whereas this is not a requirement for diffraction.
(D) In interference, the waves interacting are either in phase or 180° out of phase, whereas in diffraction the phase difference may have any value.
(E) They are essentially the same thing, with the term *diffraction* used when many sources are involved.

· **39-2 Q** Why are diffraction effects more noticeable for sound waves than for light waves in everyday life?
(A) Humans can detect sound more easily than they can detect light.
(B) Because sound intensities are generally much greater than common light intensities.
(C) Because sound wavelengths are so much greater than light wavelengths.
(D) Because sound wavelengths are so much smaller than light wavelengths.
(E) They are not. Diffraction effects for light are much more noticeable.

: **39-3 P** Consider a straight line of 24 antennas with a separation of 50 cm between adjacent antennas. What is the angular width of the central band in the reception pattern for incoming waves of wavelength (*a*) 5.0 m? (*b*) 2.5 m?

: **39-4 P** Six identical microwave oscillators with frequencies of 10 GHz are aligned along a north-south line with constant spacing of 1.0 cm between adjacent oscillators. Show by a sketch what the radiation pattern is at a great distance from the oscillators (in a horizontal plane).

: **39-5 P** Four identical radio oscillators generate radiation with a 10-m wavelength. The oscillators are aligned along a north-south line. The spacing between adjacent oscillators is 1 m. Sketch the intensity of the radiation at a great distance in a horizontal plane as a function of θ, measured from the east-west line.

Section 39-2   Single-Slit Diffraction

· **39-6 Q** We can understand the origin of the diffraction pattern due to a single slit by considering
(A) the interference of the light passing through one part of the slit with light passing through another part of the same slit.
(B) the slit to be simply a double slit with zero spacing between the slits.
(C) the variation in phase across a given wave front.
(D) the distortion produced when light does not travel along the optic axis of the slit.
(E) the 180° phase shift that occurs when light is diffracted into a more dense medium.

· **39-7 Q** The first maximum away from the central maximum in the diffraction pattern due to a single slit
(A) has approximately the same intensity as the central maximum.
(B) has an intensity about half that of the central maximum.
(C) occurs halfway between the central maximum and the second maximum from the center.
(D) occurs approximately halfway between the first two minima.

39-8 P Monochromatic light of wavelength 589 nm falls on a single slit. The first dark fringes are observed at 20° from the central maximum. What is the slit width?

· **39-9 P** What is the angular full width of the central diffraction maximum for a slit whose width is (*a*) λ? (*b*) 2λ? (*c*) 5λ? (*d*) 10λ?

· **39-10 P** The first minimum of a diffraction pattern of a slit falls at 90° when light of wavelength 580 nm illuminates the slit. What is the slit width?

: **39-11 Q** Indicate for each of the following phenomena whether or not coherence of the light is important. (*a*) Reflection. (*b*) Refraction. (*c*) Interference. (*d*) Diffraction.

: **39-12 P** A light beam consists of two components, one of unknown wavelength and the other of 524-nm wavelength. When this light is incident on a single slit, a diffraction pattern shows that the fifth secondary maximum of the unknown coincides with the sixth secondary minimum of the 524-nm light. What is the unknown wavelength?

: **39-13 P** A slit of width 0.010 mm is illuminated with light of wavelength 550 nm, and Fraunhofer diffraction is observed by using a thin converging lens of 60-cm focal length to project the pattern on a screen. (*a*) What is the angular separation from the central maximum to the first minimum? (*b*) How wide is the central maximum (in centimeters) observed on the screen?

: **39-14 Q** Every particle with momentum $p = mv$ has a wavelength $\lambda = h/p$, where h is the quantum constant, so a beam of particles can exhibit diffraction effects. Suppose that a beam of monoenergetic electrons passes through a narrow slit, and a diffraction pattern is observed on a distant screen. What can be done to narrow the width of the central maximum (that is, to reduce the separation between the two central minima of intensity)?
(A) Increase the electron wavelength.
(B) Reduce the electron speed.
(C) Reduce the potential difference through which the electrons were initially accelerated.
(D) Increase the kinetic energy of the electrons.
(E) Change the width of the slit.

: **39-15 P** A slit of width 0.02 mm is illuminated with light of wavelength 616 nm, and a diffraction pattern is observed on a screen 2 m distant. (*a*) What is the variation in phase for light coming through various segments of the slit at a point on the screen 6 cm from the central maximum? (*b*) What is the ratio of the electric-field amplitude at this point to its value at the central maximum? (*c*) What is the ratio of the light intensity at this point to its value at the central maximum?

Section 39-3   The Double Slit Revisited

· **39-16 P** Light from a helium-neon laser (632.8 nm) is incident on two slits, each 1.6×10^{-6} m wide with centers separated by 0.038 mm. How many bright fringes are contained within the central diffraction maximum?

: **39-17 Q** In our first simple analysis of the two-slit interference pattern formed on a screen, we deduced that all the maxima had the same amplitude. In fact, their intensity decreases rapidly in a direction away from the central maximum. This is because
(A) the slits have finite width, whereas we initially assumed they had negligible width.
(B) the slits have finite spacing, whereas initially we assumed they had negligible separation.

(C) we had not made the approximation $\sin \theta \approx \theta$.
(D) in real experiments, plane waves are not used.
(E) the light used is not actually monochromatic.

: **39-18 P** The interference pattern of two slits is observed and diffraction is significant. More specifically, you observe 13 maxima (arising from interference between the two slits) within the central diffraction maximum. How many interference maxima would you observe within the diffraction maximum adjoining the central maximum?

: **39-19 P** Two slits, each of width 0.1 mm, have their centers separated by 0.8 mm. They are illuminated by light of wavelength 500 nm. On a distant screen, one observes the two-slit interference pattern modulated by the single-slit diffraction pattern, with the result that certain maxima of the interference pattern are "missing" (that is, they have zero intensity). What is the first such missing maximum in this case?

: **39-20 P** A source of light of wavelength 542 nm is placed 1.2 m directly behind a slit, as shown in Figure 39-20. A second slit is located a distance d from the first. The light passing through these slits is detected at a point P on the screen that is equidistant from the two slits. In an inspection of the equipment, the intensity at the screen is measured with slit S_1 covered, with slit S_2 covered, and finally with both slits open. In the last case, the intensity is found to equal the sum of the intensities for each single slit. What is the separation of the two slits?

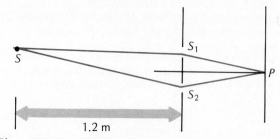

Figure 39-20.   *Problem 39-20.*

Section 39-4   Diffraction and Resolution

· **39-21 Q** Light from a small light source passes through a telescope. The observed intensity pattern is shown in Figure 39-21. It is not certain from the apperance of the spot whether the source is a single oblong source or two closely spaced point sources. One way to decide between these alternatives is to
(A) use light of longer wavelength.
(B) make the hole larger.
(C) make the hole smaller.

Figure 39-21.   *Question 39-21.*

(D) bring the screen closer to the hole.

(E) move the screen farther from the hole.

· **39-22 Q** To resolve the smallest cell structure under a microscope, which of the following combinations would be most effective?

(A) Large-diameter objective lens, red light.

(B) Small-diameter objective lens, blue light.

(C) Large-diameter objective lens, blue light.

(D) Large-diameter objective lens, with as much magnification as possible.

(E) Small-diameter objective lens, with as much magnification as possible.

: **39-23 Q** Diffraction limits the resolving power of a lens. Does this limitation apply to the image formed by a mirror, for example, a telescope with a mirror as objective rather than a lens?

: **39-24 Q** The useful magnification of optical microscopes is limited to about ×400 because

(A) lenses suitable for greater magnification cannot be made.

(B) the length of a microscope with greater magnification would be unwieldy.

(C) the electron microscope is a more suitable instrument for higher magnifications.

(D) of the limit in resolving power due to diffraction.

(E) there is nothing worth looking at with greater magnification.

: **39-25 P** Birds of prey are reputed to have very keen eyesight. Using resolution criteria, estimate the greatest altitude at which an eagle could fly and still see distinctly a rodent 8 cm long. Assume a maximum iris diameter of 10 mm for the eagle's eye and a wavelength of 540 nm.

: **39-26 P** Suppose the surface of the moon is studied with a terrestrial telescope that has an objective diameter of 50 cm. What is the minimum separation of two objects on the moon if they are just barely to be resolved? Assume a wavelength of 500 nm.

: **39-27 P** A surveyor looks at two small objects 200 m away using a transit with a 3.4-cm diameter objective. What is the minimum separation of the objects if they are to be resolved using light of wavelength 550 nm?

: **39-28 P** The world's largest radiotelescope, at Arecibo, Puerto Rico, is a 1000-ft diameter "dish" that fits into a natural mountain basin. (*a*) What is the angular resolution of this telescope when it reflects 10-cm wavelength microwaves? (*b*) What is the angular resolution of the Mt. Palomar reflecting telescope, which has a parabolic mirror with an outer diameter of 200 in., for visible light of 550-nm wavelength?

: **39-29 P** We have examined diffraction effects that result when light passes through a circular aperture in an opaque screen. Discuss qualitatively the nature of the diffraction effects to be expected for the inverse situation, that is, an opaque circular disk placed in a light beam with plane wave fronts in front of a screen.

Section 39-5   X-Ray Diffraction

· **39-30 P** A polychromatic beam of x-rays is incident on a KCl crystal whose lattice spacing is 0.314 nm. What wavelengths will be predominantly diffracted at a scattering angle of 30°?

: **39-31 P** Monochromatic waves have a wavelength equal to the lattice spacing on a crystal. At what angle relative to the direction of the incident beam do diffraction beams occur?

: **39-32 P** A neutron, like any other particle with mass m and velocity v, has a wavelength $\lambda = h/mv$, according to the quantum theory, where h is the basic quantum constant (Planck's constant). A beam of neutrons directed at a crystal can show diffraction effects. The first two diffraction peaks for the Bragg diffraction of thermal neutrons with a wavelength of 0.144 nm are observed at scattering angles (2θ) of 18° and 36° with a crystal of MnO. (*a*) What is the lattice spacing? (*b*) How many additional diffraction peaks are there for this set of planes?

Supplementary Problems

39-33 P What is the right distance to sit away from a 25-in. television picture tube? In the United States the standard television picture consists of 525 horizontal lines. If you sit too close to the picture tube, you see horizontal lines; if you sit too far away from the picture tube, you throw away details your eye cannot see because of the limit of resolution. Take the eye aperture to have a 6-mm diameter; assume light with 550-nm wavelength; a 25-in. picture tube (measured diagonally) has a picture height of about 15 in.

39-34 P A reconnaissance satellite tries to resolve the image of two vehicles separated by 5 m. If diffraction limits the resolution available, what minimum diameter lens must be used when the satellite is at an elevation of 150 km? Assume that the light wavelength is 550 nm.

39-35 Q *Poisson's Spot.* The French physicist, Simeon Denis Poisson (1781–1840) argued that the wave theory of light developed by Augustin Jean Fresnel (1788–1827) could not possibly be right because the following effect was predicted by the wave theory: if a round object is placed in a beam of light, then behind the object at the center of the shadow region you expect to see, not darkness, but a bright spot. The bright spot—often referred to as Poisson's spot—*is* observed. Why?

39-36 Q Diffraction effects show up whenever a regular array of apertures or opaque objects interrupt a beam of light. For example, an ordinary window screen and a distant street lamp produce a readily seen diffraction pattern. What other easily available objects can produce easily observed diffraction effects?

40 Special Relativity

The theory of relativity, primarily the creation in 1905 of Albert Einstein (1879–1955), ranks as one of the two great advances in twentieth-century physics and as one of the greatest triumphs of the human intellect of all time.

Often thought to be esoteric and recondite, the principal features of the theory of special relativity can be set forth using mathematics no more sophisticated than algebra. Relativity theory is no longer conjectural; even its most bizarre predictions have been amply confirmed by experimental test. Many of these predictions conflict with our common sense; indeed, relativity theory shows classical physics to be downright wrong when applied to high-speed phenomena.

40-1 The Constancy of the Speed of Light

All observers measure the speed of light through a vacuum to be the same constant c. This was the starting point for Einstein when he first formulated the theory. (He may not have known of the very experimental evidence that supported this curious postulate.)

The postulate is curious because it claims that all observers will measure the speed of electromagnetic waves through empty space as the same constant value, quite apart from the state of motion of observer or of the source of electromagnetic radiation. In this respect, light differs drastically from other types of waves. Consider sound waves. Their speed through air at room temperature is 340 m/s relative to the medium — air — in which the sound waves propagate. An observer at rest in air measures the pressure disturbance of a sound pulse as advancing a distance of 340 m in 1 s. But if the observer is in motion relative to the air at, say, 40 m/s, in the same direction as that in which the pulse of sound moves, he finds that the speed of the sound pulse is, relative to him, $340 - 40 = 300$ m/s. Only when the observer is at rest in the medium propagating sound does he measure its speed to be 340 m/s.

Physicists used to think that propagating light required a medium. This conjectured medium — the ether — would then constitute the only reference frame in which the speed of light would be c. An observer in motion relative to the ether — for example, an observer attached to the earth as it circled the sun — would then necessarily find the speed of light greater than or less than c by the magnitude of his speed relative to the ether. The speed would be exactly c only if he happened to be at rest in the ether. In short, if electromagnetic waves were like other waves, then their speed would differ from c whenever the observer was in motion relative to the unique reference frame in which the ether was at rest.

Measuring the speed of light is difficult; measuring a small change in c is extraordinarily difficult and requires very subtle experimental procedures. The first significant test was made by A. A. Michelson (1852 – 1931) and E. W. Morley (1838 – 1923) in the famed Michelson-Morley ether-drift experiment of 1887. Its basis is this: If an ether exists and is not rigidly fixed to the earth, then surely at some time during a year the earth will, because of its orbital motion about the sun at a speed of 3×10^4 m/s, be drifting through the ether at this speed, and the speed of light along or against the direction of drift will differ from c by 3×10^4 m/s, or 1 part in 10^4. At the same time, the speed of light when directed at right angles to the drift direction will be unaffected. As a consequence, the round-trip travel time for a pulse of light going "upstream" and then "downstream" (or the reverse) will differ from the round-trip travel time for a pulse of light over the same distance but at right angles to the drift direction. The very minute difference in travel time for the two routes was measured indirectly by Michelson and Morley. Their method involved examining the interference effects between two beams of light that were sent outward at right angles in a Michelson interferometer and then recombined; more specifically, they looked for a shift in the interference pattern arising from a rotation of the instrument and the concomitant interchange of the upstream-and-downstream and right-angle routes.

The Michelson-Morley experiment showed a null effect — no effect attributable to the ether. There is no necessity for assuming that an ether exists. The speed of light is the same for all observers. Other more recent experiments also confirm this result, with higher precision.

40-2 Relativistic Velocity Transformations

A light source at rest in our reference frame directs a beam of light to the right. Some observer moves left at speed $0.2c$ toward the light source. What is the speed of light relative to this observer?

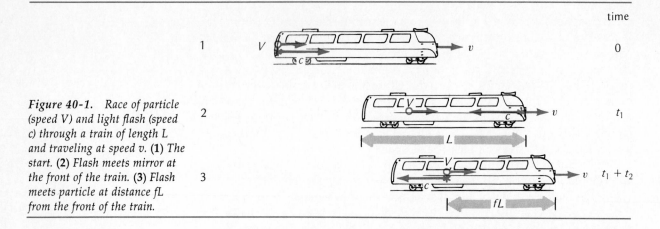

Figure 40-1. *Race of particle (speed V) and light flash (speed c) through a train of length L and traveling at speed v. (1) The start. (2) Flash meets mirror at the front of the train. (3) Flash meets particle at distance fL from the front of the train.*

• *Prerelativity physics* says (using the classical velocity transformation rules, Section 4-6) that the speed should be $c + 0.2c = 1.2c$. (If the observer were to move away from the source, the answer would then be $c - 0.2c = 0.8c$.)

• *Relativity physics* says that there is only one possible value, c.

The classical rules for combining relative velocities are wrong, or at least applicable only for low speeds. Here, we wish to find the relativistic velocity transformations, applicable to all possible speeds.*

Consider this situation, pictured in Figure 40-1. All the following quantities are measured relative to reference frame S, which we can for definiteness imagine to be at rest on earth. A train of length L travels right at speed v. A particle moves to the right at speed V, starting from the back end of the train. A flash of light also travels right at speed c, also starting from the back of the train. This light flash is reflected from a mirror at the front end of the train and then meets the more slowly moving particle. The distance, measured from the front of the train, at which flash meets particle is fL; this means that f is the fractional length of the train at which the two meet.

The time it takes for the light flash to travel from the train's back end to the mirror in front is t_1. The additional time it takes for the flash to travel from the mirror back to the particle is t_2.

The following three statements relate distances, times, and velocities, using their ordinary meanings.

• The total distance traveled by the particle $V(t_1 + t_2)$ from **1** to **3** is just the distance ct_1, traveled by the flash going to the right, less the distance ct_2 the flash travels going left:

$$V(t_1 + t_2) = c(t_1 - t_2) \tag{A}$$

• The distance ct_1 covered by the flash as it goes from the back **1** to the front

---

* The simple derivation of the relativistic-velocity-transformation rule is due to N. David Mermin; it appears in *American Journal of Physics* 51, 1130 (December 1983). How the relativistic-velocity-transformation relations can be derived from the Lorentz coordinate transformation relations (underived) is given in Optional Section 40-7.

2 of the train is just the train's length L plus the distance vt_1 the train advances during time t_1:

$$ct_1 = L + vt_1 \qquad\qquad \textbf{(B)}$$

• We consider the flash during the time it moves left from the train's front end at **2** to the place where it meets the particle at **3**. The distance ct_2 covered by the light flash equals the distance fL from the train's front to the meeting point, reduced by the distance vt_2 by which the train has advanced during time t_2:

$$ct_2 = fL - vt_2 \qquad\qquad \textbf{(C)}$$

We wish to express fraction f in terms of velocities V, v, and c:*

$$f = \frac{(c + v)(c - V)}{(c - v)(c + V)} \qquad\qquad \textbf{(F)}$$

This result applies for the train moving at speed v relative to frame S. But the train need not be moving. Suppose that we now become observers riding with the train in a second reference frame we call S'; in S', the train is at rest. What changes are required?

• The train is observed at rest in S', so that we put $v = 0$.
• The particle's velocity relative to the new reference frame S' is V'.

But the following two items do not change:

• All observers agree on the spot in the train where flash meets particle, so that f, the fractional length measured from the train's front end, is the same as before. (In more formal terms, f is invariant.)
• The speed of light in S' or any other reference frame is still exactly c. Here is where we invoke the basic assumption of special relativity. (The speed of light is an invariant.)

In (F) we put $v = 0$ and replace V by V'. The result is

$$f = \frac{c - V'}{c + V'} \qquad\qquad \textbf{(G)}$$

Finally we equate (F) and (G) and get, after some fairly messy but basically easy algebra, the final result:

$$V = \frac{V' + v}{1 + V'v/c^2} \qquad\qquad \textbf{(40-1)}$$

---

* This is the first of a number of places in this chapter in which intermediate steps in an algebraic development are, in the interest of brevity, relegated to a footnote. To be clear that the derivation works in detail, be sure to check this by filling in the relatively straightforward missing steps.

We wish to eliminate t_1, t_2, and L from (A), (B), (C). First, solving (A) for t_2/t_1 gives

$$\frac{t_2}{t_1} = \frac{c - V}{c + V} \qquad\qquad \textbf{(D)}$$

We get another relation for t_2/t_1 by eliminating L between (B) and (C):

$$\frac{t_2}{t_1} = \frac{f(c - v)}{c + v} \qquad\qquad \textbf{(E)}$$

Equating (D) and (E) yields the result shown in the main text above as (F).

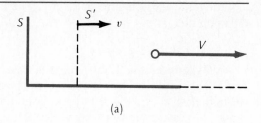

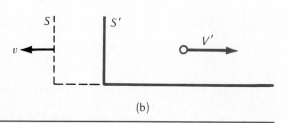

Figure 40-2.   *(a) Relative to S, the particle speed is V and the speed of reference frame S' is v. (b) Relative to S', the particle speed is V' and the speed of reference frame S is −v.*

(a)

(b)

Let us be clear on the meaning of the terms in this equation, which applies not merely for the train in Figure 40-1, but in general. First, we are relating velocities for two reference frames S and S'; their x axes coincide, and all motion is along x.

- v is the velocity of S' relative to S. (By the same token, $-v$ is the velocity of S relative to S'. To switch from S to S', we merely change the sign of their relative velocity v.)
- V is the velocity of some point or particle measured by an observer at rest in S.
- V' is the velocity of the very same point or particle measured by an observer at rest in S'.

See Figure 40-2.

Equation (40-1) gives V in terms of V' and v. To get the inverse relation, we merely interchange V and V' and replace v by $-v$, to get

$$V' = \frac{V - v}{1 - (Vv/c^2)} \tag{40-2}$$

The distinctive relativistic effect comes from the denominator. If any of the speeds is small compared with c, or equivalently, if we can imagine the speed of light to become effectively infinite, then the second term in the denominators of (40-1) and (40-2) is negligible compared with 1, and the relativistic relations reduce, as they must, to the simpler classical relations.*

Another consequence of the velocity rule is that velocities no longer can be combined by the simple rule of vector addition. Instead, the more complicated

---

* The numerators alone in (40-1) and (40-2) are what we get from the classical velocity transformation relations (Section 4-6). For example, with double subscripts for denoting object in motion and reference frame relative to which the velocity is measured, (40-1) becomes, for a low-speed particle p,

$$v_{pS} = V'_{pS'} + v_{S'S}$$

(40-1) and (40-2) relations must be invoked. These relations apply only when the direction of relative motion and the particle in motion are parallel.

Here, as with later relativistic results, we see that classical physics is, in effect, the low-speed limit of relativistic physics.

Example 40-1. A light source is at rest in S; the speed measured in this reference frame is c. What is the speed of light relative to reference frame S', in motion along the direction of the light beam at velocity v?

We use (40-2) with $V = c$:

$$V' = \frac{V - v}{1 - Vv/c^2} = \frac{c - v}{1 - vc/c^2} = \frac{c - v}{c - v}\, c = c$$

Example 40-2. A particle moves east at the speed 24×10^7 m/s $(=0.8c)$ relative to the earth. What is this particle's speed as measured by an observer in a spaceship traveling west relative to the earth at 15×10^7 m/s $(=0.5c)$.

Let S be a reference frame attached to the earth, and S' a reference frame attached to the spaceship. We are then given that

$$V = 0.8c$$

$$v = -0.5c$$

Therefore from (40-1), we have

$$V' = \frac{V - v}{1 - vV/c^2} = \frac{(0.8 + 0.5)c}{1 + (0.5)(0.8)} = \frac{1.3c}{1.4} = 0.93c$$

or

$$V' = 28 \times 10^7 \text{ m/s}$$

The classical velocity combination rule (inapplicable) would have yielded the observed particle speed as $0.8c + 0.5c = 1.3c$, a speed in excess of the speed of light. The relativistic relation ensures that the particle speed not exceed c (here, 93 percent of c).

40-3 Space and Time in Special Relativity

Special relativity reverses the absolute and the relative. In classical physics, the speed of light is relative to a hypothetical medium (the ether); in relativity physics, the speed of light is absolute. In prerelativity physics, time intervals and space intervals are taken as obviously absolute, in agreement with our common sense. Here we see that because the speed of light is absolute, time and space intervals are relative and depend on the state of motion of the observer.*

Time Dilation The speed of a particle means the spatial interval it traverses divided by the corresponding time interval, both measurements made by the same observer. But if light has the same speed for all observers, then space intervals and time intervals may not have unique values for all observers. Said differently, if c is absolute, space intervals and time intervals cannot be absolute; they may depend on the observer's state of motion.

---

* In Optional Section 40-7, it is shown how the space-contraction and time-dilation effects can be derived from the Lorentz coordinate transformation relations, given there without proof.

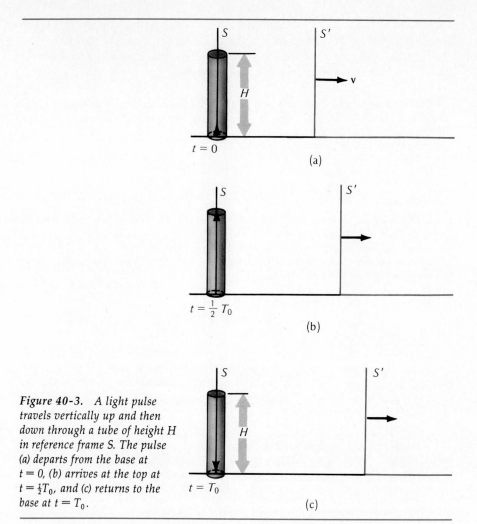

Figure 40-3. *A light pulse travels vertically up and then down through a tube of height H in reference frame S. The pulse (a) departs from the base at $t = 0$, (b) arrives at the top at $t = \frac{1}{2}T_0$, and (c) returns to the base at $t = T_0$.*

In prerelativity physics, the absolute character of space and time is taken as axiomatic and self-evident. If one observer measures the distance between two separated points as, say, 1 m, then other observers will agree that their separation distance is exactly 1 m. Or if one observer clocks a time interval between two events as, say, 1 s, other observers will likewise measure the time interval as precisely 1 s. These seemingly obvious claims, certainly in accord with experience and common sense for all ordinary speeds, are actually fundamentally incorrect.

We wish to find the relationship for time intervals between the same two events as measured in two reference frames, S and S'. Consider the following hypothetical experiment. Observer S has a tube of length H at rest and aligned along the y axis; he sends a pulse of light from the base of the tube along the tube axis until it reaches a mirror at the top end and is then reflected to the base. See Figure 40-3. The departure of the light pulse and its later return to the base take place at the same location in reference frame S. To emphasize that the two events (departure and return of the pulse) are measured by an observer who is at rest with respect to their location, we indicate the time interval S measures as T_0. The pulse of light traverses the total round-trip distance of $2H$ in the time

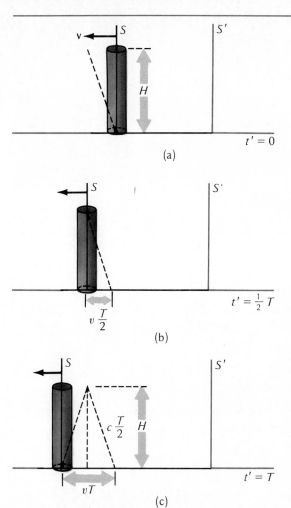

(a)

(b)

(c)

$t' = 0$

$t' = \frac{1}{2}T$

$t' = T$

Figure 40-4. The same events as in Figure 40-3, but now as observed in reference frame S'. The pulse departs from the base at $t' = 0$, (b) arrives at the top at $t' = \frac{1}{2}T$, and (c) returns to the base at $t = T$, all as registered on the clock of observer S'. Over the round trip, the tube advances to the left relative to S' through a distance vT. The speed of the light pulse over the two oblique path segments is also c relative to S'.

interval T_0 at the speed c. So we have

$$c = \frac{2H}{T_0} \tag{40-3}$$

Now consider the same events as seen by observer S', who travels to the right at speed v relative to S. The path of the light pulse is now two oblique lines. See Figure 40-4. As S' sees things, the pulse does not return to its starting location. Overall, the path length S' observes exceeds that $(2H)$ which S observes. But the speed of the light pulse must be the same for both observers. It follows at once that the time interval T recorded by S' between the departure and the return of the pulse to the base of the tube must exceed T_0. As Figure 40-4 shows, the tube moves left a total distance vT in the time interval T, or a distance $vT/2$ in time $T/2$. During the half-time $T/2$, the light pulse covers the oblique distance $cT/2$. From the geometry of the triangle in Figure 40-4, the distances are related by

$$H^2 + \left(\frac{vT}{2}\right)^2 = \left(\frac{cT}{2}\right)^2$$

Using (40-3) to eliminate H from the above equation,* we have

$$T = \frac{T_0}{\sqrt{1 - (v/c)^2}} \qquad (40\text{-}4)$$

This is the fundamental *time-dilation* equation. Keeping straight the meaning of the terms in it is crucial. The time interval T_0 is between two events that occur at the *same location* and are measured on the clock of an observer at rest at this location; T_0 is termed the *proper time* (or rest time). On the other hand, T is the time interval between the *very same two events* but registered on the clock of an observer traveling at speed v relative to the location at which the two events take place (and who therefore sees the two events take place at different locations in his reference frame). Of course, the clocks of the two observers when compared at rest with respect to one another give identical readings. Equation (40-4) shows that in general $T > T_0$. Relative to a moving observer, time intervals are increased, or dilated. The effect is a result of the constancy of the speed of light; it is not attributable to a physical cause, but reflects the relativistic properties of time (and space). The phenomenon is called *time dilation*.

Time-dilation effects are significant only at speeds approaching that of light. Suppose $v = 0.1c$ (a speed about 4000 times greater than that of a satellite orbiting the earth), then

$$T = \frac{T_0}{[1 - (0.1)^2]^{1/2}} = \frac{T_0}{\sqrt{0.99}} = 1.005T_0$$

and T exceeds T_0 by only half of 1 percent. At speeds close to c, the effects are dramatic, however. Suppose now that $v = 0.98c$. Then

$$T = \frac{T_0}{[1 - (0.98)^2]^{1/2}} = \frac{T_0}{\sqrt{0.04}} = 5T_0$$

and T exceeds T_0 by a factor of 5. To see what this means in a particular situation, suppose that the timer for a bomb is set to make the bomb explode in one hour (when the bomb is at rest). But the bomb is set in motion at $0.98c$. So an observer seeing it move by at this speed finds that as read on his clock (and on the clocks of his associates stationed throughout his reference frame), it takes 5 h for the bomb to explode.

No one has observed time-dilation effects with bombs moving at high speeds. But an exactly analogous effect has been observed, and the time-dilation effect confirmed, in experiments with high-speed unstable subatomic particles, such as muons. A muon is created (born) when another unstable particle (a pion) decays; a muon decays (dies) in turn to an electron (together with two uncharged, massless particles called neutrinos). You can't tell when any one muon will be born or die. But the average lifetime of a large number of muons can be given precisely. The half-life of a muon is found to be 1.52×10^{-6} s. This means that if you have a large number of muons at some initial time and these muons remain at rest, one-half will have survived after $1.52 \times$

---

* The vertical distance H, at right angles to the direction of relative motion of the two reference frames, is assumed to be the same for both observers. Although spatial intervals along the direction of relative motion are not the same for all observers, the transverse distances are unchanged.

10^{-6} s has elapsed, while the other half will have decayed. But what if the muons are moving at high speed? Say that 10,000 muons are in motion at $v = 0.98c$. The number of surviving muons is 5000 only after the elapsed time is $T = 5T_0 = 5(1.52 \times 10^{-6}$ s$) = 7.60 \times 10^{-6}$ s. This means that muons in flight at high speed live longer than muons at rest. The increased lifetime arising from time dilation can be observed directly by noting that if a high-speed unstable particle lives longer than the same particle at rest, it will travel a correspondingly greater distance before decaying.*

The increased lifetime of an unstable particle in motion at high speed reflects the properties of time itself (or more properly, space-time), not any physical mechanism. Time dilation applies to any processes taking place with time, including those in biochemical systems. Consider the famous twin paradox, first introduced by Einstein. There are two identical twins. One stays home while the other goes on a round trip in a spaceship to and back from some distant point at such a high speed that time-dilation effects are significant, and is finally reunited with her stay-at-home sister. Time is dilated for the traveling twin, and when the sisters compare their ages on being reunited, they agree that the stay-at-home twin is older than the traveling twin, for whom the time-dilation effect has introduced a shorter period of elapsed time. (The effect is not reciprocal. Whereas the stay-at-home twin has an uneventful history, the traveling twin experiences three profound shocks — the first as the spaceship takes off suddenly; the second as it comes suddenly to rest on arriving at its far destination and immediately reaccelerates to start the homeward portion of the trip; and the third as the spaceship arrives home and is suddenly brought to rest.)

Space Contraction   A time interval between two events depends on the observer's state of motion. Similarly, a spatial interval, or length, may, because of the constancy of c, depend on the state of motion of the observer.

We use the same arrangement as before (Figures 40-3 and 40-4). A tube through which a light pulse travels up and down is at rest in reference frame S. As observed by S', the tube and observer S travel left at speed v, as shown in Figure 40-5. To make the events more picturesque, suppose that observer S' places his meter stick along his x' axis, and that the light pulse makes one burn mark on this meter stick when it leaves the base of the tube and a second burn mark when it returns to the base. The distance between the two burn marks on the meter stick of S' is L_0. The subscript zero emphasizes that this length is measured by observer S', who is at rest relative to his own meter stick.

As indicated earlier, the time interval between the markings of the two spots, as measured by S', is the dilated time T. Reference frame S moves left at speed v over a distance L_0 in time interval T. (All these quantities are measured by S'.) So we have

$$v = \frac{L_0}{T}$$

---

* Significant numbers of high-speed muons, created in the upper portions of the earth's atmosphere when energetic particles from outer space (cosmic radiation) strike the earth, can be observed at sea level because of the time-dilation phenomenon. Without this effect, most muons would not have survived decay long enough, and therefore traveled far enough, to reach the bottom of the atmosphere at sea level.

"The Italian Navigator Has Just Landed."

The moment was truly awesome. Forever after, history would be divided into what happened before or after. At 5:30 A.M. on July 16, 1945, at a remote spot on a desert in New Mexico, the sudden flash was so bright, so blinding, that observers nearly 10 miles from the site of the first nuclear explosion would later struggle for words adequate to describe it.

One of the observing physicists counted seconds from the time of the flash to know when the shock wave, produced by the sudden compression of air at the explosion site, and traveling outward at the speed of sound, would reach him. When it did, he released pieces of paper and watched them being buffeted. He was able to tell at once—because of a computation he had made in advance—that the total "yield" of the plutonium bomb was close to 10^{13} J (the equivalent of about 20 kilotons of TNT).*

Such was the special genius of Enrico Fermi (1901–1954); his penetrating insight could make tough problems look simple.

Fermi mastered analytical geometry at age 10. It was not until he was 14, however, that he first studied physics—he learned it by himself from a sixty-year-old textbook written in Latin. He got his Ph.D. in the very shadow of the place where an earlier Italian physicist had done important physics, at the University of Pisa. World renowned by age 25, Fermi was installed as the occupant of the first chair in physics

at the University of Rome. There he formed what amounted to a school of physics with other Italian colleagues. He did fundamental theoretical work: on quantum statistics, on basic processes in the beta decay of unstable nuclei.

He also began systematic experimental studies on the effect of slow neutrons on various elements. He started with hydrogen and worked his way up the periodic table. When he got to uranium he noticed that peculiar things happened when this element was bombarded with slow neutrons—he thought that transuranic elements were being produced. The correct explanation, that uranium was undergoing nuclear fission, eluded him.

He was given permission to leave Mussolini's fascist Italy to pick up the 1938 Nobel prize in physics at Stockholm. Fermi never returned. Instead, he came to the U.S.A., where his slow-neutron experiments continued. Indeed, the famous letter from Albert Einstein to Franklin D. Roosevelt of August 2, 1939, in which Einstein urged the U.S. President to give prompt governmental attention to the implications of the large energy release in nuclear fission, began with the words "Some recent work by E. Fermi and L. Szilard. . . ."

Fermi was made director of the Manhattan Project at the University of Chicago to produce a controlled self-sustaining nuclear reaction. If energy could be released in a controlled way, it could most likely also be released suddenly. The site of the experiment was the squash courts in the unused football stadium (the University of Chicago had given up intercollegiate football). The big day was December 2, 1942; then

*From *Lawrence and Oppenheimer,* Nuel Pharr Davis (Simon and Schuster, New York, 1968), page 241.

Now consider the distance between the same two burn marks measured by S. Observer S sees S' and his meter stick in motion to the right at speed v, as shown in Figure 40-5. We call L the distance between the two burn marks on the moving meter stick, as measured by observer S. (Observer S, in measuring the length of an object in motion with his own meter stick, must be sure to mark the locations of the two burn marks *simultaneously*.) Relative to S, reference frame S' advances to the right at speed v over a distance L; moreover, this occurs in the *undilated* time interval T_0. Therefore, observer S may write

$$v = \frac{L}{T_0}$$

Eliminating v from the two equations above, we have

the reactor first "went critical," with the uranium generating more energy than it consumed.

The news was telephoned in code by Arthur H. Compton, also of the University of Chicago, to James B. Conant, a leading chemist, president of Harvard University, and another principal figure in the Manhattan Project.

Compton: "Jim, you'll be interested to know that the Italian navigator has just landed in the New World."
Conant: "Were the natives friendly?"
Compton: "Everyone landed safe and happy."

Contemporary physics is replete with references, direct and indirect, to Fermi:

- Fermium is the transuranic element of atomic number 100.
- The fermi is a unit of distance (1 fm = 10^{-15} m that just happens to agree with the official SI unit, femtometer).
- Fermi-Dirac statistics governs the quantum behavior of particles with half-integral spin (called fermions).
- The *Fermi energy*, and the *Fermi surface* show up repeatedly in solid-state physics.
- The Fermi prize (first recipient, Enrico Fermi) is awarded annually by the U.S. Department of Energy for outstanding achievements in nuclear energy.
- Fermilab, near Batavia, Illinois, is a large high-energy research establishment devoted to the study of elementary particles. Protons with a kinetic energy up to 1 TeV (10^{12} eV), the most energetic particles produced by man, are hurled head-on at antiprotons

with the same kinetic energy in the huge accelerating machine known as the Tevatron.

Fermi was a person of prodigious and diverse talents. He did not need to keep an extensive library of books because he found that, without effort, he would pretty much memorize everything he had read. He could recite long stretches of Dante's *Divine Comedy* verbatim. Once, when his automobile broke down in a remote place and it was taken to a small repair shop, *he* fixed the car promptly (and was offered a job as repairman on the spot by the owner). He liked sports requiring physical stamina—especially swimming, mountain climbing, skiing.

Fermi delighted above all in posing interesting problems and solving them. A problem in physics that may at first seem very hard but can be solved—with insight—in just a few deft steps has become known as a "Fermi-type" problem. (Ph.D. candidates in physics at the University of Chicago were once given a test written by Fermi with just one question: "How deep a hole can you dig?")

Fermi also used to pose and solve questions of a more general type. For example, "Within an order of magnitude, how many piano tuners are there in Philadelphia?" How can you possibly do a problem like that? Be bold, but also be reasonable. Use whatever knowledge you have from past experience; if you're not sure, you make a sensible estimate, or still better, a couple of estimates from different, independent approaches. Try it, it works. One of the nice things about this type of question (Philadelphia piano tuners) is that, after you have your final number, you can look up the answer in the back of the book—the Yellow Pages of the Philadelphia Telephone Directory.

$$L = L_0 \left(\frac{T_0}{T} \right)$$

and using the time-dilation relation (40-4), we have finally

$$L = L_0 \sqrt{1 - (v/c)^2} \tag{40-5}$$

This is the basic *space-contraction* relation. The terms mean this: L_0 is the spatial separation, or *proper length,* between two points (lying along the line of relative motion of S and S') and measured by an observer at rest with respect to these points. The contracted length L is the distance between the same two points as measured by an observer traveling at speed v relative to them. We emphasize again that to measure properly the length of an object in motion, the

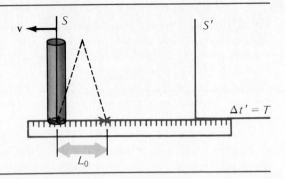

Figure 40-5. *The light pulse of Figure 40-3 makes burn marks on a meter stick attached to reference frame S' as the light flash departs from and returns to the base of the tube. The separation distance is L_0 and the elapsed time interval is T, both relative to S'.*

two ends must be marked *simultaneously.* Lengths are contracted along the direction of relative motion, but not at right angles to this direction.

As (40-5) shows, in general $L < L_0$. Length contraction is significant only for high speeds; for example, with $v = 0.1c$, $L = 0.995L_0$, but for $v = 0.98c$, we have $L = \frac{1}{5}L_0$. At low speeds ($v/c \rightarrow 0$), the relative spatial intervals and time intervals of relativity physics become effectively the absolute space and time intervals of classical physics: $L = L_0$ and $T = T_0$.

The term *space* contraction is used to emphasize that the effect is not due to a physical cause, for example, shrinking because of external pressure or a drop in temperature. The effect reflects instead that space and time intervals are fundamentally changed by the requirement that all observers measure the same speed for light.

Example 40-3. A long, fast spaceship passes an observer stationed at a post fixed to the earth. The spaceship travels at $0.8c$ relative to the earth observer at the post, who notes that the back end of the spaceship passes the post 42 μs after the front end. (a) What is the time interval, relative to observers at rest within the spaceship, elapsing between the instant when the front end aligns with the post and the later instant when the back end is aligned with it? (b) How long is the spaceship, as measured by the crew at rest within the spaceship? (c) What is the length of the spaceship, as measured by observers on earth?

(a) There are basically two events in this problem:

- Front end aligns with post.
- Back end aligns with post.

Relative to the earth observer, the two events take place at the same location, separated by 42 μs. Therefore, this is the *rest* time T_0. The dilated time interval for the same two events relative to spaceship observers is

$$T = \frac{T_0}{\sqrt{1 - (v/c)^2}} = \frac{42\ \mu s}{\sqrt{1 - (0.8)^2}} = 70\ \mu s$$

(b) Relative to spaceship observers, the post travels by at $0.8c$ for 70 μs. The length of the spaceship, clearly at rest relative to them, is

$$L_0 = (0.8c)(70\ \mu s) = (0.8 \times 3.0 \times 10^8\ m/s)(70 \times 10^{-6}\ s) = 16.8\ km$$

(c) The earth observer at the post sees the spaceship in motion, and therefore contracted, with a length of

$$L = L_0 \sqrt{1 - (v/c)^2} = (16.8\ km) \sqrt{1 - (0.8)^2} = 10.1\ km$$

What this means in more detail is this. The observer at the post and a colleague also at rest on earth and separated from him by 10.1 km would find that the two ends of the spaceship were at their respective locations at the same time; that is, when the two earth observers got together later to compare notes, they would find that each of their two watches read the same time when an end of the spaceship was at their location.

40-4 Relativistic Momentum

The constancy of the speed of light fundamentally changes time intervals, space intervals, and the ways in which velocities combine. What about mechanics and such a basic quantity as momentum, which clearly depends on speed? How if at all is it altered for high speeds? What is its appropriate relativistic form?

The starting point is a second postulate of special relativity theory. This postulate, which is also fundamental to classical mechanics, is often unstated or ignored, not because it is so subtle, but because it is so transparently obvious. It is this. *The laws of physics are the same in all inertial frames of reference.* An inertial frame is, of course, a frame of reference in which the law of inertia, or Newton's first law of motion, holds: In an inertial frame an undisturbed object has a constant velocity.* Now the invariance of the laws of physics for all inertial frames means simply this. Such fundamental propositions as the momentum conservation principle, and the mass conservation principle, if they are to be truly laws of physics — propositions that are universally valid — cannot depend on the particular inertial frame in which they are applied. One inertial frame must be just as good as any other.

We consider a very simple collision in two different inertial reference frames. We shall insist that in each of the two reference frames, the total momentum before the collision equals the momentum after and also that mass before equals mass after.

The collision is shown in Figure 40-6, in (a) the center-of-mass frame S' and (b) a laboratory frame S. In more detail we have:

- (a) CM frame. Two identical particles are fired head-on, each at speed V'. They stick together and form a composite particle with mass M_0. This composite object must be at rest if total momentum after the collision is to equal total momentum before. The result also follows simply from the requirements of symmetry; if what happens on the left is mirrored on the right, the composite can show no preference for left or right.
- (b) Lab frame. Here we have the very same collision, but now as seen by an observer in a laboratory reference frame S in which the particle on the right is initially at rest. This means that the composite particle must have velocity V' as measured in the lab.† The mass of the composite is labeled M, *different* from its rest mass M_0; here we anticipate that if we are still to write relativistic momentum as the product of mass and velocity, the mass thus defined may not be independent of speed. In similar fashion, the mass of the single particle at rest is written m_0, and the mass of the single particle in motion is m.

Applying in turn mass and momentum conservation to the collision in Figure 40-6(b) yields:

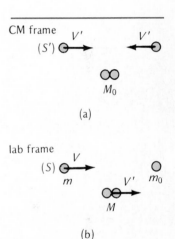

Figure 40-6. *(a) In the center-of-mass reference frame (S'), two identical particles moving initially at speed V' collide head-on and produce the composite particle of mass M_0 at rest. (b) The same collision as viewed from the laboratory reference frame (S), in which the particle on the right is initially at rest with mass m_0. The composite particle with mass M has speed V'. The other particle in motion with mass m has speed V.*

---

* The special theory of relativity is restricted to inertial frames; the general theory of relativity includes accelerated reference frames as well.
† This result follows simply from the fact that if A has velocity V' relative to B, then B has velocity $-V'$ relative to A.

Mass conservation: $m + m_0 = M$ (40-6)

Momentum conservation: $mV + m_0(0) = MV'$ (40-7)

If we were applying classical mechanics, we should say simply that $M = M_0 = 2m_0$ and that $V = 2V'$. But these results do not hold for relativistic speeds. Eliminating M from the two equations above yields

$\frac{V}{V'} = \frac{M}{m} = \frac{m + m_0}{m}$

$$\frac{V}{V'} = 1 + \frac{m_0}{m}$$ (40-8)

Velocities V and V' are also related to one another by the general velocity transformation relation

$$V' = \frac{V - v}{1 - Vv/c^2}$$ (40-2)

In Figure 40-6, V and V' are the velocities in S and S' of the particle on the left. The relative velocity between the two reference frames here is $v = V'$, so that the relation above becomes

$$V' = \frac{V - V'}{1 - VV'/c^2}$$ (40-9)

We find after some manipulation that this equation can be written as*

$$\frac{V}{V'} = 1 + \sqrt{1 - \left(\frac{V}{c}\right)^2}$$ (40-10)

Comparing (40-10) with (40-8), we see that

$$\frac{m_0}{m} = \sqrt{1 - \left(\frac{V}{c}\right)^2}$$

or solving for m,

$$m = \frac{m_0}{\sqrt{1 - (V/c)^2}}$$ (40-11)

Keep in mind what *relativistic* mass m means. It is the quantity by which

---

* Equation (40-9) can be rearranged to become

$$V' - \frac{VV'^2}{c^2} = V - V'$$

Now multiply both sides by c^2/V^3 and collect terms:

$$\left(\frac{V'}{V}\right)^2 + \left(\frac{-2c^2}{V^2}\right)\left(\frac{V'}{V}\right) + \frac{c^2}{V^2} = 0$$

This is a quadratic equation in V'/V, whose solution is

$$\frac{V'}{V} = \frac{1 \pm \sqrt{1 - (v/c)^2}}{(V/c)^2}$$

Taking the reciprocal of this equation and multiplying numerator and denominator on the right by $1 \pm \sqrt{1 - (V/c)^2}$ give finally (40-10). The plus sign was chosen; this ensures that for $V/c \ll 1$, we get the simple classical result $V/V' = 2$.

velocity must be multiplied to yield the quantity, relativistic momentum **p**, that is conserved in every collision.

Therefore,

$$\mathbf{p} = m\mathbf{V} = \frac{m_0 \mathbf{V}}{\sqrt{1 - (V/c)^2}} \qquad (40\text{-}12)$$

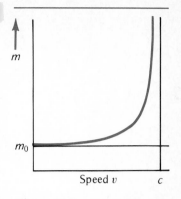

Equation (40-12), which relates the momentum to rest mass m_0, is the fundamental relation; (40-11), which relates m to m_0, is not. How m varies with speed is shown in Figure 40-7. At low speed, $m \simeq m_0$; at speeds close to c, mass m becomes infinite. This means that relativistic momentum increases with velocity at a higher rate than the classical relation $\mathbf{p} = m_0\mathbf{V}$, to which it reduces for low speeds.

An electrically charged particle moves at high speed at right angles to the field lines of a uniform magnetic field. We take the force on the particle to be the time rate of change of momentum, so that

Figure 40-7.   Relativistic mass m of a particle as a function of its speed v.

$$\mathbf{F} = \frac{d(m\mathbf{V})}{dt} = \frac{dm}{dt}\mathbf{V} + m\frac{d\mathbf{V}}{dt}$$

The particle's speed is constant and so is its relativistic mass m, so that $dm/dt = 0$. Then the relation above becomes

$$\mathbf{F} = m\frac{d\mathbf{V}}{dt} = m\mathbf{a}$$

For charge q at speed V in uniform magnetic field **B**, we have

$$F = ma$$

$$qVB = m\frac{V^2}{r} \quad \Rightarrow \quad qrB = mV$$

$$mV = \frac{m_0 V}{\sqrt{1 - (V/c)^2}} = qrB \qquad (40\text{-}13)$$

The relativistic momentum mV is directly proportional in the magnitude of **B** and radius of curvature r. Equation (40-13) provides a simple means of measuring relativistic momentum; simply measure the particle's curvature in a known magnetic field.

40-5   Relativistic Energy

To find the relation for relativistic kinetic energy E_k, we merely do what is done in classical physics: find the work done by a force in bringing a particle from rest to the final speed v.* We write F as $d(mv)/dt$ and write $ds = v\,dt$:

$$E_k = \int_0^s F\,ds = \int_0^s \frac{d}{dt}(mv)\,ds = \int_0^t \frac{d}{dt}(mv)v\,dt$$

$$= \int v\,d(mv) = \int (v^2\,dm + mv\,dv)$$

* Henceforth particle speed is given by a lowercase v.

To integrate the right side, we must recognize that both m and v are variables. The dependence of m on v is given by (40-11). It is simpler to express v in terms of m and then integrate with respect to m as variable. The required relation is*

$$mv\,dv = (c^2 - v^2)\,dm \tag{40-14}$$

Substituting for $mv\,dv$ in the equation for the kinetic energy E_k, we then have

$$E_k = \int_{m_0}^{m} [v^2\,dm + (c^2 - v^2)\,dm] = c^2 \int_{m_0}^{m} dm = mc^2 - m_0 c^2$$

$$E_k = (m - m_0)c^2 \tag{40-15}$$

The relativistic kinetic energy can be regarded as the increase in mass, arising from the particle's motion, multiplied by c^2. At high speeds, the relativistic energy is markedly different from the classical kinetic energy, $\frac{1}{2}m_0 v^2$. The relativistic energy must reduce to the familiar $\frac{1}{2}m_0 v^2$ for $v/c \ll 1$. This is shown in the footnote to Section 9-1. Be careful; relativistic kinetic energy is *not* given by $\frac{1}{2}mv^2$, where m is the relativistic mass. Let us write (40-15) as

$$E_k = E - E_0 = mc^2 - m_0 c^2 \tag{40-16}$$

Here E represents the particle's *total energy*,

$$E \equiv mc^2 = \frac{m_0 c^2}{\sqrt{1 - (v/c)^2}} \tag{40-17}$$

and E_0 is the particle's *rest energy*,

$$E_0 \equiv m_0 c^2 \tag{40-18}$$

(For a system of particles, rest energy E_0 and rest mass m_0 are the system's total energy and mass when its center of mass is at rest.)

Equation (40-17) is the famous Einstein relation. It implies an equivalence of energy and mass, in which mass and energy are interpreted as different manifestations of the same physical entity. A particle at rest has rest mass m_0 and rest energy $m_0 c^2$; in motion, its mass and energy are m and mc^2. Mass and energy need not be regarded as having separate conservation laws of energy and of mass; instead, we consider the two combined into a single, simple law, the conservation law of mass-energy.

Equation (40-17) gives E in terms of v. It is often convenient to express energy E in terms of p. We find the relation by squaring (40-11) and multiplying both sides by $c^4[1 - (v/c)^2]$. We get

$$m^2 c^4 - m^2 v^2 c^2 = m_0^2 c^4$$

---

* Equation (40-11) may be written as

$$1 - \left(\frac{v}{c}\right)^2 = \left(\frac{m_0}{m}\right)^2$$

Taking the differential of this relation gives

$$\frac{-2v\,dv}{c^2} = \frac{-2m_0^2\,dm}{m^3}$$

Combining this relation with the one immediately above it gives (40-14).

This equation can immediately be written more simply as

$$E^2 = (pc)^2 + E_0^2 \qquad (40\text{-}19)$$

Here are the relations in dynamics for the two extreme limits of particle speed:

- *Very low speeds, $v/c \ll 1$ (classical limit):*

$$m \simeq m_0 \qquad E_k = \tfrac{1}{2}m_0 v^2$$
$$p = m_0 v \qquad E_0 \gg E_k$$

- *Very high speeds, $v/c \simeq 1$ (extreme relativistic limit):*

$$m \gg m_0 \qquad E_k \simeq E \simeq pc$$
$$p \simeq E/c \qquad E_0 \ll E_k$$

Here are the most appropriate forms and units for expressing relativistic quantities for particles:

- Speed, relative to the speed of light: v/c.
- Energy in electron volts (or such related units as kilo-, mega-, or giga-electron-volts).
- Mass in unified atomic mass units, where

$$1\ u = \tfrac{1}{12}\ \text{mass of carbon-12 atom}$$
$$= 1.6606 \times 10^{-27}\ \text{kg} = 931.5\ \text{MeV}/c^2$$

- Momentum as energy in eV divided by the speed of light, or eV/c.

Example 40-4.   (a) A proton ($E_0 = 0.938$ GeV) is accelerated from rest across an electric potential difference of 500 V. What is its momentum?

(b) A proton is accelerated in the high-energy accelerator at Fermi National Laboratory, Batavia, Illinois, so that its final kinetic energy is 500 GeV. What is the momentum of such a proton?

(a) The proton's final kinetic energy is

$$E_k = qV = e(500\ \text{V}) = 500\ \text{eV}$$

Since the proton's kinetic energy is much less than its rest energy (500 eV versus 0.938 GeV = 0.938×10^9 eV), we can use the classical kinetic-energy relation, $E_k = p^2/2m_0$:

$$p = \sqrt{2m_0 E_k} = \frac{\sqrt{2(m_0 c^2)E_k}}{c}$$

$$= \frac{\sqrt{2(0.938 \times 10^9\ \text{eV})(500\ \text{eV})}}{c} = 0.97\ \text{MeV}/c$$

(b) The kinetic energy of a 500-GeV proton is much greater than its rest energy, so that the relation $p = E/c$, which applies for very high energies, can be applied here.

$$p = \frac{E}{c} = \frac{E_k}{c} = \frac{500\ \text{GeV}}{c} = 5.0 \times 10^5\ \text{MeV}/c$$

Example 40-5.   As shown in Example 40-4, a 500-eV proton has a momentum of 0.97 MeV/c. What is the radius of curvature of the proton's path when it enters a uniform magnetic field of 0.40 T at right angles to the field lines?

From $p = mv = qr\,B$, (40-13), we get

$$r = \frac{p}{qB} = \frac{(0.97 \times 10^6 \text{ eV}/c)(c/3.0 \times 10^8 \text{ m/s})}{(1.6 \times 10^{-19} \text{ C})(0.40 \text{ T})(1 \text{ eV}/1.6 \times 10^{-19} \text{ J})} = 8.1 \text{ mm}$$

40-6   Mass-Energy Equivalence and Bound Systems

To see the significance of the conservation law of mass-energy, we consider two situations: unbound systems and bound systems.

Unbound Systems   Two particles, each with a rest mass m_0, are projected toward one another, each with speed v relative to an observer in the center-of-mass reference frame. The collision is perfectly inelastic, and the two particles stick together to form a single, composite particle, with rest mass M_0. How is rest mass M_0 of the composite particle related to the rest mass m_0 of each of the separate incident particles? Classical physics would say that M_0 equals $2m_0$ exactly. But in relativity this is not true.

The total energy of the two particles before collision, $2mc^2$, must equal the total energy M_0c^2 of the composite particle after the collision. The total energy of the composite after collision is entirely rest energy, since this object is at rest. The total energy of the particles before collision is, however, their rest energy plus their kinetic energy. Mass-energy conservation then gives

$$M_0c^2 = 2mc^2 = \frac{2m_0c^2}{\sqrt{1 - (v/c)^2}}$$

or

$$M_0 = \frac{2m_0}{\sqrt{1 - (v/c)^2}}$$

The rest mass M_0 of the composite object exceeds the total rest mass $2m_0$ of the incident particles. What has effectively happened is that the kinetic energy of the two particles has become a part of the rest energy of the combined particles after the collision.

Example 40-6.   Two satellites, each with a rest mass of 4000 kg, travel in orbits in opposite directions at a speed of 8.0 km/s with respect to an earth observer. They happen to collide head-on and stick together. What is the change in the total rest mass of the system?

The satellites have equal but opposite momenta; their total momentum is zero. After the collision, the composite object is at rest. The kinetic energy of the incident satellites is converted to rest mass, where

$$\text{Increase in rest mass} = \Delta m = \frac{2E_k}{c^2}$$

where E_k is the initial kinetic energy of each satellite. The speed of each satellite, 8.0 km/s, is much less than the speed of light, and we can properly use the classical expression $E_k = \frac{1}{2}m_0v^2$ for the kinetic energy. Therefore,

$$\Delta m = \frac{2(\frac{1}{2}m_0v^2)}{c^2} = m_0\left(\frac{v}{c}\right)^2 = (4000 \text{ kg})\left(\frac{8.0 \times 10^3 \text{ m/s}}{3.0 \times 10^8 \text{ m/s}}\right)^2 = 2.8 \text{ mg}$$

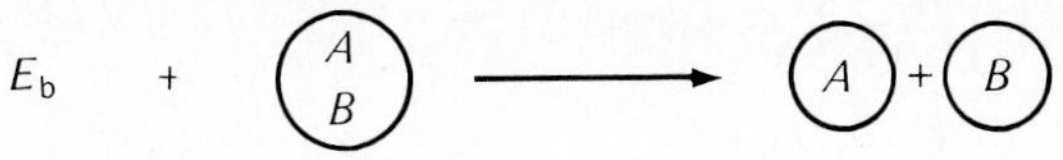

E_b + ⟨A B⟩ ⟶ ⟨A⟩ + ⟨B⟩

Figure 40-8. Symbolic representation of the splitting of two bound particles.

If the two satellites could collide and form a single composite object whose mass could, at least in principle, be measured on a balance, then one would find it to be nearly 3 mg greater than the mass of the two satellites taken separately.

Bound Systems Now consider two particles A and B bound together to form a bound system. To break the composite object into its component parts requires work. We must add energy to the system. The rest mass of the composite system is M_0; the rest masses of the individual particles are m_{0A} and m_{0B}.

The breaking up of the bound system into separated parts is shown symbolically in Figure 40-8. Here E_b is the energy that must be added to the system in order to separate the particles completely. Energy E_b is called the *binding energy*.

Applying mass-energy conservation, we have

$$M_0 + \frac{E_b}{c^2} = m_{0A} + m_{0B} \qquad (40\text{-}20)$$

Since the system is bound and $E_b > 0$, it follows from the equation above that $M_0 < m_{0A} + m_{0B}$. The rest mass of the bound system is *less* than the sum of the rest masses of the individual particles when separated.* The binding energy E_b can be computed simply by knowing the system's rest mass and the rest masses of its constituents. Only for particles bound by the very strong nuclear forces within an atomic nucleus is the binding energy sufficiently great that the mass difference can actually be measured.

Example 40-7. (a) It takes 13.6 eV to ionize a hydrogen atom. By what fraction is the mass changed during the ionization?

(b) For nuclei, the process analogous to ionizing hydrogen is separating a deuteron nucleus into a proton and a neutron. It takes 2.2 MeV. By what fraction is the nuclear mass changed in this process?

(a) If energy is added to a hydrogen atom to separate it into a proton and an electron, then the sum of the masses of proton and electron exceeds the hydrogen mass by $\Delta m = \Delta E/c^2$; here $\Delta E = 13.6$ eV. The rest energy of a proton (and of a hydrogen atom) is about 0.94 GeV. The fractional increase in mass is then

$$\frac{\Delta m}{m_0} = \frac{\Delta E/c^2}{m_0} = \frac{\Delta E}{m_0 c^2} = \frac{13.6\text{ eV}}{0.94\text{ GeV}} = 1.4 \times 10^{-8}$$

The difference is only 1.4 parts in 10^8, a typical value for a chemical process and far too small to be measurable in the masses of the particles.

(b) The rest energy of a proton or of a neutron is approximately 0.94 GeV, and the mass of a deuteron is approximately twice this amount. Strictly, the total mass of the separated proton and neutron exceeds the deuteron mass because energy must be added to the bound system to separate the particles. Here the fractional mass increase is

---

* Compare with the unbound systems above, in which the rest mass of the composite exceeded the rest masses of the separated particles.

$$\frac{\Delta m}{m_0} = \frac{\Delta E}{m_0 c^2} = \frac{2.2 \text{ MeV}}{2(0.94 \text{ GeV})} = 1.2 \times 10^{-3}$$

The mass difference for this nuclear binding of particles, 1.2 parts in a thousand, is so large that it is easily discernible in the masses of the particles.

40-7   The Lorentz Transformations (Optional)

Here, for the record, are the relativistic transformation relations between the three space and one time coordinates (x, y, z, t) in reference frame S and the corresponding coordinates (x', y', z', t') for a second reference frame S' in motion with velocity v relative to S. They are called the Lorentz transformations, after H. E. Lorentz, who first introduced them in 1903 (before Einstein's relativity theory).

Here are the assumptions:

- Corresponding axes are parallel, x with x', y with y', z with z'.
- The two origins coincide at the times when $t = t' = 0$.
- Relative velocity v is of S' along the x axis relative to S. (By the same token, S has velocity $-v$ relative to S'.)
- The transformation relations are so constructed that all observers measure the same speed c for light.

The Lorentz coordinate transformation relations are:

$$x' = \frac{x - vt}{\sqrt{1 - (v/c)^2}} \tag{40-21a}$$

$$y' = y \tag{40-21b}$$

$$z' = z \tag{40-21c}$$

$$t' = \frac{t - (v/c^2)x}{\sqrt{1 - (v/c)^2}} \tag{40-21d}$$

Simply by looking at (40-21) we can see the following.

- Space and time coordinates are not independent: x' depends on t, as well as on x and v. Much more surprising, however, is that t' depends on x, as well as on t, as shown by (40-21d). What S' reads on his clock actually depends, not only on the reading (t) on the clock of S, but also on where (x) S is located.
- For $v \ll c$, the relativistic coordinate transformation relations reduce to the intuitively obvious low-speed relations for space and time: $x' = x - vt$; $y = y'$; $z = z'$; and $t' = t$.
- Equation (40-21) gives x', y', z', t' in terms of x, y, z, t. To get the inverse transformation relations, replace every primed quantity by the corresponding unprimed coordinate (and conversely) while also replacing v by $-v$.

Some important consequences follow immediately.

- The *relativity of simultaneity*. Two spatially separated events that are simultaneous in one reference frame occur in sequence in a second reference frame that is in motion relative to the first. Let two simultaneous events take place in S, one at time $t = 0$ at the origin and the second at the coordinate x. Then from (40-21d), the times of these same two events measured in S' differ by the time interval $(v/c^2)x/\sqrt{1-(v/c)^2}$. Einstein considered this the most profound result in relativity.

- *Time dilation.* A clock is at rest at the origin in S, so that its coordinate is $x = 0$; and we can take $t = T_0$, the rest time. The time interval for the same events as observed by S' is $t' = T$. Equation (40-21d) then gives

$$T = \frac{T_0}{\sqrt{1-(v/c)^2}} \quad (40\text{-}4)$$

- *Space Contraction.* Let a rod be at rest, with one end at the origin of S' and the other at coordinate x'. The rod's rest length is $x' = L_0$. The rod is in motion in S with its two ends observed simultaneously (same t), so that the rod's contracted length is $x = L$. Equation (40-21a) then gives

$$L = L_0 \sqrt{1-(v/c)^2} \quad (40\text{-}5)$$

- *Relativistic velocity transformations.* Let the velocity components of a particle observed in S be V_x, V_y, and V_z. The velocity components of the same particle observed in S' carry primes. By definition $V_x = dx/dt$ and $V' = dx'/dt'$, and so on for the other components. Taking the differential of both sides of (40-21a) and (40-21d) gives

$$dx' = \frac{dx - v\,dt}{\sqrt{1-(v/c)^2}} \quad \text{and} \quad dt' = \frac{dt - (v/c^2)\,dx}{\sqrt{1-(v/c)^2}}$$

Dividing the first relation by the second and applying the definitions yields (40-22a):

$$V'_x = \frac{V_x - v}{1-(v/c^2)V_x} \quad (40\text{-}22\text{a})$$

$$V'_y = \frac{V_y \sqrt{1-(v/c)^2}}{1-(v/c^2)V_x} \quad (40\text{-}22\text{b})$$

$$V'_z = \frac{V_z \sqrt{1-(v/c)^2}}{1-(v/c^2)V_x} \quad (40\text{-}22\text{c})$$

Equation (40-22b) results from taking the differentials of (40-21b) and (40-21d); and similarly for (40-22c).

Summary

Definitions

Rest energy E_0 of a particle with rest mass m_0:

$$E_0 \equiv m_0 c^2 \quad (40\text{-}18)$$

Total relativistic energy E:

$$E \equiv mc^2 \quad (40\text{-}17)$$

where relativistic mass m:

$$m \equiv \frac{m_0}{\sqrt{1-(v/c)^2}} \quad (40\text{-}11)$$

Fundamental Principles

Postulates of the special theory of relativity:

- The speed of light has the same value for all observers, whatever the state of motion of the light source or observer.

● The laws of physics have the same form for all observers in inertial reference frames.

Important Results

Relativistic velocity transformation relation:

$$V' = \frac{V - v}{1 - (v/c^2)V} \qquad (40\text{-}2)$$

where V' is particle velocity relative to S',
   V is particle velocity relative to S, and
   v is velocity of S' relative to S.
All three velocities are along the same line.

Time dilation: $T_0 =$ time interval between two events taking place at the same location in reference frame S;
   $T =$ time interval between same two events as observed in a reference frame with velocity v relative to S:

$$T = \frac{T_0}{\sqrt{1 - (v/c)^2}} \qquad (40\text{-}4)$$

Length contraction: $L_0 =$ length (along direction of relative motion) in which two end points are at rest in S;
   $L =$ length between same two points as measured simultaneously in reference frame with a velocity v relative to S:

$$L = L_0 \sqrt{1 - (v/c)^2} \qquad (40\text{-}5)$$

Relativistic dynamics: The relativistic forms of dynamical quantities assure that the momentum and mass-energy conservation laws are satisfied in all inertial reference frames.

Relativistic momentum **p**:

$$\mathbf{p} = m\mathbf{v} = \frac{m_0\mathbf{v}}{\sqrt{1 - (v/c)^2}} \qquad (40\text{-}12)$$

Relativistic kinetic energy E_k:

$$E_k = mc^2 - m_0c^2 = E - E_0 \qquad (40\text{-}15)$$

Relation between E, E_0, and p:

$$E^2 = E_0^2 + (pc)^2 \qquad (40\text{-}19)$$

Charged particle in magnetic field (with $\mathbf{v} \perp \mathbf{B}$):

$$p = mv = qrB \qquad (40\text{-}13)$$

The relativistic momentum $p = mv$ is proportional to the particle charge q, the radius r of curvature of the path, and the magnitude B of the magnetic field.

To separate a composite particle with rest mass M_0 into its constituent parts with rest masses m_{0A} and m_{0B}, binding energy E_b must be added where

$$M_0 + \frac{E_b}{c^2} = m_{0A} + m_{0B} \qquad (40\text{-}20)$$

Problems and Questions

Section 40-1   The Constancy of the Speed of Light

· **40-1 Q** A fundamental postulate on which the theory of relativity is based is that
(A) light in vacuum always moves with speed c, independent of the velocity of the source or of the observer.
(B) everything is relative.
(C) mass and energy must be equivalent.
(D) the form of the equations of physics depends on the velocity of the reference frame used.
(E) what appears to be vacuum is not really empty space but instead, a substance called ether.

: **40-2 Q** Consider the following thought (*gedanken*) experiment. A flashlamp is placed at the exact center of a boxcar that is traveling at high constant velocity. A man stationed in the car, by using photocells, is able to measure the time at which a light pulse emitted from the lamp strikes each end of the car. A woman at rest on earth can also measure the time at which the light pulses strike the ends of the moving boxcar. By considering such an experiment, Einstein was able to conclude that
(A) the velocity of light is independent of the motion of the source and of the observer.
(B) moving clocks run slow.

(C) a moving boxcar is shortened in its direction of motion.
(D) events that are simultaneous in one inertial frame may not be simultaneous in another.
(E) the laws of physics are the same in all inertial frames.

Section 40-2   Relativistic Velocity Transformations

· **40-3 P** Relative to some observer, A moves east at $0.8c$ and B moves west at $0.8c$. What is the velocity of B relative to A?

: **40-4 P** A particle has a velocity of $0.8c$ to the east relative to a train. The train has a velocity $0.7c$ to the east relative to earth. What is the velocity of the particle relative to the earth?

: **40-5 P** Particle A travels north at $0.1c$. Particle B travels south at $0.1c$. What is the speed of B relative to A computed from (*a*) the relativistic velocity transformation relations and (*b*) the classical velocity transformation relations?

: **40-6 P** Any attempt to "piggy-back" speeds to reach a value over c is frustrated by the relativistic velocity transformation relations. To visualize a specific situation, consider the following arrangement. A ball is thrown at speed $0.9c$ relative to a cart. The cart has speed $0.8c$ relative to a train. The train has speed $0.7c$ relative to the earth. All velocity vectors are in the same direction. What is the velocity of the ball (*a*) relative to the train and (*b*) relative to the earth?

Section 40-3   Space and Time in Special Relativity

· **40-7 Q** You want to travel to a star 100 light-years distant. Assume your maximum life span to be only 90 years. Is it possible for you to reach the star?
(A) No, because nothing can travel faster than light.
(B) No, because your rocket ship will grow shorter and shorter as it approaches the speed of light, so that eventually it will be essentially standing still.
(C) No, because time will pass much faster for you than for a person at rest back on earth.
(D) Yes. Time will pass more slowly for you than for a person back on earth, so as your speed approaches c, the trip could last less than 90 years.
(E) Yes, because speed is only relative. Though you appear to be going at less than the speed of light to an observer on earth, you actually may be moving at speeds much greater than the speed of light.

· **40-8 Q** The proper time is
(A) time measured in any inertial frame.
(B) the shortest time interval between two events.
(C) any time interval measured in seconds, as opposed to time measured in days, months, years, and so on.
(D) greater or less than a time interval measured in another inertial frame moving with respect to the proper frame.

: **40-9 Q** If you were traveling with respect to the distant stars at a speed close to the speed of light, you could detect this by
(A) the increase in your own mass that you would experience.
(B) a change in your pulse rate.
(C) a change in your physical dimensions.
(D) all of the above means.
(E) none of the above means.

· **40-10 P** An astronaut in a satellite circles the earth at a speed of 8200 m/s for the duration of a 14-day mission (as measured by Mission Control in Houston). By how much will his age differ from that of earthbound persons when he returns?

· **40-11 P** A rocket 20 m long when at rest is 16 m long when moving past stationary observers. How fast is the rocket moving?

: **40-12 Q** An important result of special relativity is that we must give up the concept of an "absolute time" independent of the state of motion of the person measuring the time. The principal reason is that
(A) it has not been possible to determine an "origin" of time in the universe.

(B) time is simply a fourth dimension, not unlike spatial dimensions.
(C) energy and mass are equivalent.
(D) the velocity of light is the same for all observers in inertial reference frames, independent of their motion or the motion of the light source.
(E) moving objects are shortened along their direction of motion.

: **40-13 P** How fast would an object have to move so that its length appears to have decreased by a factor of 2?

: **40-14 P** A spacecraft receding from earth with a speed of $0.98c$ emits pulses of radio waves at the rate of 10,000 per second. At what rate are the pulses received on earth?

: **40-15 Q** The timer on a bomb is set to explode the bomb after a time τ (with the bomb at rest). The bomb is set in motion with velocity v. The bomb will then explode, according to a stationary observer, after a time
(A) $\dfrac{\tau}{\sqrt{1 - (v/c)^2}}$
(B) $\dfrac{1}{\tau}\sqrt{1 - \left(\dfrac{v}{c}\right)^2}$
(C) $\tau\sqrt{1 - (v/c)^2}$
(D) τ
(E) $c\tau$

: **40-16 P** How long would a jet plane flying 1000 km/h have to fly before its clocks had lost 1 s with respect to clocks on earth?

: **40-17 P** Pions have a half-life of 2.2×10^{-8} s as measured by an observer at rest with respect to them. An observer in the laboratory sees a beam of pions traveling at $0.995c$. As the beam passes him, he counts 1000 pions. How many pions will be left after the beam travels 10 m farther, as measured in the lab?

: **40-18 P** Electrons in the Stanford Linear Accelerator (SLAC) achieve a final speed that is less than c by only 3 parts in 10^{10}. The accelerator is 2 miles long. (*a*) How long is the accelerator as measured in the rest frame of an electron traveling at the final speed? (*b*) How long does it take, as measured by an observer traveling with the electron, for an electron at the final speed to travel the length of the accelerator? (*c*) How long does it take for an electron with the final speed to travel the length of the accelerator, as measured by an experimenter in the lab?

: **40-19 P** Two satellite space stations are separated by a fixed distance of 1000 km as measured by an observer on one of them. What is the separation of the space stations as measured by an observer in a rocket flying between them with a velocity of $0.8c$?

: **40-20 P** An object in the xy plane moves along the x axis at speed $0.8c$. A stationary observer notes that the moving shape is a square 2.0 cm on a side and that its sides make an angle of 45° with its line of motion. What is the area of the object as measured in its rest frame?

: **40-21 P** A rod of length L_0 is attached to the roof of a car at rest and it makes an angle θ_0 with the horizontal. (a) What is the length of the rod, as determined by a stationary observer on the roadway, when the car is moving with very high speed v? (b) What is the angle between the rod and the horizontal as observed by the stationary observer on the roadway?

Section 40-4   Relativistic Momentum

· **40-22 P** By what factor does the momentum of a particle change when its speed (a) goes from $0.04c$ to $0.08c$ and (b) goes from $0.4c$ to $0.8c$?

· **40-23 P** Calculate the momentum (in MeV/c) of a proton moving with speed (a) $0.01c$; (b) $0.1c$; (c) $0.5c$; (d) $0.9c$.

: **40-24 P** An electron is moving at right angles to a magnetic field of 1.2 T in a path with a radius of 3.0 cm. What is the magnitude of the electron's momentum (expressed in units of MeV/c)?

: **40-25 P** A particle of rest mass m_1 and velocity $v_1 = 0.8c$ collides head-on with a particle of rest mass m_2 moving toward it with speed $v_2 = 0.6c$. The two stick together and are at rest in the laboratory reference frame. Find the ratio of the rest masses, m_1/m_2.

Section 40-5   Relativistic Energy

· **40-26 P** What is the momentum (in units of MeV/c) of an electron of kinetic energy (a) 10 eV and (b) 10 GeV?

· **40-27 P** (a) Through what electric potential difference must an electron be accelerated from rest for its relativistic mass to exceed its rest mass by 10 percent? (b) What would be the electron's final speed?

· **40-28 Q** Confirm that the relation between relativistic momentum p and relativistic energies E_0 and E (40-19) can be represented by a right triangle with sides pc and E_0 and hypotenuse E. Also check that this triangle, which can serve as a useful mnemonic, yields the correct results in the classical and extreme relativistic limits.

: **40-29 P** How much energy is needed to double the momentum of an electron whose velocity is 1.8×10^8 m/s?

: **40-30 P** A particle with speed $0.49c$ has its speed doubled. (a) By what factor does its momentum change? (b) By what factor does its kinetic energy change? Repeat the calculation for a particle with initial speed 10^{-5} $c = 3000$ m/s. (c) By what factor does its momentum change when the speed is doubled? (d) By what factor does the kinetic energy change?

: **40-31 P** (a) What is the momentum of a 100-MeV electron? (b) What result do you get for the momentum if you use the incorrect classical relations $E_k = \frac{1}{2}m_0v^2$ and $p = m_0v$?

: **40-32 P** An electron has speed 2.4×10^8 m/s. Compute the following quantities using both the correct, relativistic expression and the classical approximation. (a) Total energy. (b) Kinetic energy. (c) Momentum.

: **40-33 P** Determine the energy (in MeV) needed to accelerate an electron from (a) $0.10c$ to $0.90c$; and (b) $0.90c$ to $0.99c$.

: **40-34 Q** An electron is accelerated from rest by an electric potential difference. Then the electron enters a uniform magnetic field at right angles to the magnetic-field lines and travels in a circular arc. Sketch a plot of the radius of the arc as a function of the accelerating potential over a range that includes the classical and extreme relativistic regions.

: **40-35 P** An electron is accelerated from rest through a potential difference of 25 kV in a TV tube. (a) What is its final speed? (b) What is the electron's final kinetic energy measured in units of the particle's rest energy?

: **40-36 P** Verify that

$$\frac{1}{\sqrt{1 - v^2/c^2}} = 1 + \frac{E_k}{m_0c^2}$$

: **40-37 P** The most energetic cosmic rays detected are protons with energies of the order of 10^{13} MeV. Our galaxy is about 100,000 light-years across. How long would it take such a particle to traverse the Milky Way as measured in the reference frame of (a) the galaxy? (b) the particle?

: **40-38 Q** If a particle's kinetic energy is equal to its rest energy, E_0, the particle's momentum is
(A) E_0/c
(B) $2E_0/c$
(C) $E_0/2c$
(D) $\sqrt{3}E_0/c$
(E) none of the above.

: **40-39 Q** Indicate whether or not each of the following equations is a valid relativistic relation: (a) $E_k = \frac{1}{2}mv^2$; (b) $p = mv$; (c) $F = ma$; (d) $F = dp/dt$; (e) $E = m_0c^2 + \frac{1}{2}m_0v^2$.

: **40-40 P** An electron is accelerated until its total energy changes from $2m_0c^2$ to $4m_0c^2$. By what factor do each of the following change? (a) Kinetic energy. (b) Speed. (c) Momentum. (d) Mass.

: **40-41 P** In calculating, in terms of m_0c^2, what work must be done to accelerate an electron from rest to the speed $0.9c$, what result do you get (a) using Newtonian mechanics? (b) Using relativistically correct relations?

: **40-42 P** A particle of rest mass M_0 moves with speed V. It decays to two identical particles, each with rest mass m_0 and velocity v. These particles move off symmetrically to the original line of motion, making angles $+\theta$ and $-\theta$ with this direction. (a) Show that

$$m_0 = \frac{M_0(1 - V^2/c^2 \cos^2 \theta)^{1/2}}{2(1 - V^2/c^2)^{1/2}}$$

(b) Determine the maximum allowed value of θ when $V = 0.6c$. (c) Evaluate m_0 when $V = 0.6c$ and $\theta = 45°$.

: **40-43 P** An electron moving with speed $0.8c$ in the x direction enters a region where there is a uniform electric field in the y direction. Show that the x component of velocity of the electron will *decrease*.

Section 40-6 Mass-Energy Equivalence and Bound Systems

· **40-44 Q** Tell whether the mass of the indicated system increases or decreases for each of the following processes: (a) a battery loses its charge; (b) a capacitor is charged; (c) a block of ice melts; (d) an inductor has a current established in it; (e) a spring is stretched; (f) a soldering iron is turned on.

· **40-45 P** How much does the mass of one liter of water increase when it is heated 100 C°?

· **40-46 Q** Suppose a nuclear bomb were exploded in a container strong enough that it could contain the explosion. All the reaction products, including radiation, are trapped within the box. Under these circumstances, the mass of the box and its contents, relative to its value just before the explosion, would
(A) increase.
(B) decrease.
(C) stay the same.
(D) any of the above, depending on the state of motion of the observer relative to the box.

· **40-47 P** Two lumps of clay, each with a rest mass of 0.100 kg and a speed of $0.6c$, collide head-on and stick together. What is the rest mass of the resulting composite lump?

· **40-48 P** What mass of uranium is consumed by a nuclear power plant that produces power at an average rate of 1000 MW for one year?

: **40-49 P** The yield of a nuclear bomb is typically measured in megatons of TNT (that is, in terms of the energy released when 10^6 tons of TNT is detonated). One kilogram of TNT releases about 1 kcal of energy. To what rest mass is one megaton of TNT equivalent?

: **40-50 P** A carbon-12 atom has an atomic mass of exactly 12 u. The $^{12}_6$C nucleus consists of six protons (mass, 1.007 825 u) and six neutrons (mass, 1.008 665 u). (a) How much energy is required to dismember the $^{12}_6$C entirely, into

individual protons and neutrons? (b) How much energy is required to "cut" the $^{12}_6$C nucleus into three helium atoms (mass of $^4_2$He, 4.002 603 u)? (All masses given above include the requisite number of electrons to make the object electrically neutral.)

: **40-51 P** Each fission of a $^{235}_{92}$U nucleus produces 200 MeV. How many uranium nuclei must undergo nuclear fission per second to yield a power of 1 kW?

: **40-52 P** When a $^{235}_{92}$U nucleus undergoes nuclear fission in a nuclear reactor, about one part in 10^3 of its rest mass is converted into kinetic energy. How much fissionable uranium-235 is needed during a one-year period for a nuclear power plant that has an average electrical output of 100 MW? Only about 30 percent of the kinetic energy of the fission particles is converted to electric energy in a nuclear power plant (a fossil-fuel plant has an efficiency of about 40 percent because it can operate at somewhat higher temperatures).

Supplementary Problems

40-53 P In principle, time-dilation effects could be used to make possible interstellar travel that would otherwise be impossible in human life spans. To accomplish this, one is faced with accelerating a spaceship to speeds sufficiently near the speed of light that slowing down the astronauts' biological clocks would allow them to make in 10 years, say, a trip that appears to an earthbound observer to last 10^6 years. Some formidable practical problems will probably prevent realization of this idea for a long time, however. First, humans do not like to accelerate at rates much greater than g, so it would take a very long time to get up to speed. Second, huge amounts of energy would be needed. To confirm this, calculate the energy needed for a ship of mass 10^5 kg for the above trip. In order to gauge adequately the enormous magnitude of the required energy, express your answer in terms of the energy used by the entire world in one year, about 10^{20} J.

40-54 P Serious proposals have been advanced for spaceships propelled by starlight falling on huge mirrored sails. The intensity of light from our star (the sun) near the earth's orbit is 1.4 kW/m². (a) What maximum force could this radiation exert on a sail 1 km × 1 km? (b) What speed could a spaceship of 10,000-kg mass achieve if it were accelerated from rest by this force for one year? (c) A spaceship with a solar sail would be attracted to the sun but repelled by the radiation force on the sail. Show that if the two forces balance at one distance from the sun, the forces will balance at any other distance from the sun if the sail size and orientation are unchanged.

41 Quantum Theory

Relativity physics is one of the two great ideas in twentieth-century physics. Quantum physics is the other. Relativity physics is the physics of the very fast; it changes our conceptions, not only of space and time, but also of mass and energy. Quantum physics is sometimes called the physics of the very small. Certainly atomic and subatomic structure can be comprehended only with quantum theory. But quantum effects also show up in macroscopic phenomena; superconductivity, for example, is basically a quantum effect. Quantum physics revises classical conceptions just as radically as relativity does. As we shall see, the perfect predictability, the clockwork universe, that characterizes Newtonian mechanics is gone in quantum theory. We must deal with probabilities and uncertainties.

41-1 Quantization

One aspect of quantum theory is not entirely new—we find it also in classical physics. It is the idea of *quantization,* whereby quantities come only in certain discrete amounts. Here are examples:

- People. They come only in integers.
- Sides of a coin. We have only heads or tails.
- Electric charge. For an observable particle, always an integral multiple of the basic charge e.
- Frequencies of waves trapped between boundaries. Consider the simplest case, a string attached at both ends. Then the allowed standing-wave patterns are those for which the frequency is exactly an integral multiple of the fundamental frequency (Section 17-6).

One important aspect of quantum theory is that physical quantities that were thought in classical physics to have a continuous range of possible values, are, in fact, quantized.

Quantum theory began in 1900 with Max Planck, who attempted to give a theoretical interpretation of the electromagnetic radiation from a blackbody, a perfect absorber and radiator, and especially of how the intensity of the emitted radiation depended on wavelength. Planck found that only by postulating energy quantization could he produce satisfactory agreement between experiment and theory. A detailed analysis of blackbody radiation is complicated and involves sophisticated arguments. Therefore, we shall introduce quantum concepts through the simpler and in many ways, more compelling arguments that arise in the photoelectric effect.

41-2   Photoelectric Effect

How the photoelectric effect was discovered is an irony of history. Heinrich Hertz in 1887, during the experiments that confirmed Maxwell's theoretical prediction (1864) of the existence of continuous, classical electromagnetic waves, found the following: a charged object loses its charge more readily when it is illuminated by violet light.

The photoelectric effect is this: electromagnetic radiation shines on a clean metal surface and electrons are released from the surface. Conduction electrons in a metal are relatively free to move about the interior but they are bound to the metal as a whole. These electrons may become photoelectrons. The radiation supplies an electron with energy that equals or exceeds the energy that binds the electron to the surface and thereby allows the electron to escape. What matters in releasing an electron from the metal is, in the view of classical electromagnetic theory, simply whether enough energy has reached the initially bound electron. But experiment shows different behavior. The *frequency* of the radiation determines, for any particular kind of emitting surface, whether electrons are released, quite apart from how intense the radiation may be. Unless a certain radiation frequency, characteristic of the material, is exceeded, no electrons are released.

To see how the quantum theory accounts for this circumstance, we put on record the basic properties of the *photon*.

Electromagnetic waves are quantized. They consist of discrete *quanta*, called photons. Each photon has an energy E that depends only on the frequency v (or on the wavelength λ) and is given by

$$E = hv = h\frac{c}{\lambda} \qquad (41\text{-}1)$$

where h is a constant. Indeed h is the fundamental constant of the quantum

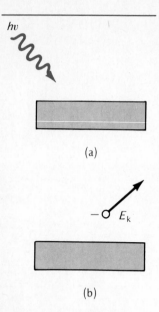

(a)

(b)

Figure 41-1. *The photoelectric effect: (a) A photon with energy hv strikes a photoemitting material; and (b) an electron is released with kinetic energy E_k.*

theory, and it is called *Planck's constant.* Its value was first determined and its significance first appreciated by Planck in 1900. The present value of Planck's constant is

$$h = 6.626\ 176 \times 10^{-34}\ \text{J}\cdot\text{s}$$

According to the quantum theory, a beam of monochromatic light of frequency v consists of particlelike photons. Each has energy hv. A photon travels at the speed of light. It must, on the basis of relativity theory, then have a zero rest mass, and its energy must be entirely kinetic. As long as it exists, a photon moves at speed c. Indeed, the only thing a photon can do is to travel through space at speed c. When it interacts with any object, it ceases to exist. Thus, when a photon strikes an electron bound in a metal, it relinquishes its *entire* energy hv to the single electron it strikes. See Figure 41-1. If the energy the bound electron gains from the photon exceeds the energy binding it to the metal surface, the electron is freed, with the excess energy appearing as kinetic energy of the photoelectron. If the photon has less energy than that binding the electron, it simply cannot be dislodged.

Let ϕ be the energy with which an electron is bound to metal; ϕ is often called the *work function* of the material. The kinetic energy of the released photoelectron is E_k. Then energy conservation yields

$$hv = \phi + E_k \qquad (41\text{-}2)$$

The left side of this equation is the energy initially carried by the incoming photon. The right side tells what happened to it; part of it goes to unbind the electron and the rest appears as the particle's kinetic energy. (Strictly, E_k is the maximum kinetic energy; some energy may be gained by the solid material.)

Clearly, unless $hv > \phi$, no photoelectron can be produced. The *threshold frequency* v_0 for photoemission is then, with $E_k = 0$, from (41-2),

$$hv_0 = \phi \qquad (41\text{-}3)$$

Both v_0 and ϕ are characteristic of the particular photoemitter. Equation (41-2) can also be written as

$$hv = hv_0 + E_k \qquad (41\text{-}4)$$

The basic equation of the photoelectric effect lends itself well to graphical interpretation, as shown in Figure 41-2. A plot of photoelectron kinetic energy, $E_k = hv - \phi$, against photon frequency is a straight line with slope h. The energy intercept gives the work function ϕ; the frequency intercept is the threshold frequency v_0. For a typical photoemitter, ϕ is a few electron volts and v_0 corresponds to violet or ultraviolet light (for example, for potassium, $\phi = 2.30$ eV and $\lambda_0 = c/v_0 = hc/\phi = 539$ nm).

Determining the maximum photoelectron kinetic energy is straightforward; it is measured directly by the stopping electric potential V that brings the most energetic photoelectrons to rest (and therefore the photocurrent to zero). More specifically, $E_k\ (\text{max}) = eV$. The number of photons incident upon a photoemitter is proportional to the intensity of incident radiation; similarly, the number of electrons released per unit time is proportional to the photocurrent (electric current from photoelectrons). Since one photon is extinguished for each photoelectron released, the photocurrent is directly proportional to the incident intensity. This effect is used in practical applications of the photoelectric effect.

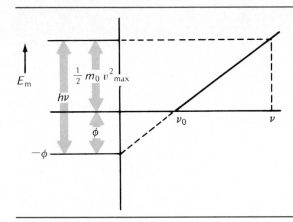

Figure 41-2. *Photon energy plotted as a function of frequency.*

Although the term *photoelectric effect* means, strictly, the release of electrons from a metallic surface, this same effect is seen in more general circumstances, whenever a bound particle is released by the absorption of a photon (see Examples 41-3 and 41-4).

The photoelectric effect provides a fundamentally new insight into the nature of electromagnetic radiation; it is quantized and consists of photons. With the frequency v of the radiation specified, a photon can have but one energy, hv, and the total energy of a monochromatic beam is always precisely an integral multiple of the energy of a single photon. See Figure 41-3.

The granularity of electromagnetic radiation is not conspicuous in ordinary observations; this is simply because the energy of any one photon is very small, and because the number of photons in a light beam of moderate intensity is enormous. The situation is like that found in the molecular theory. The molecules are so small and their numbers so great that the molecular structure of all matter is disclosed only in very subtle observations.

The ideas of wave and particle are apparently mutually incompatible, even contradictory. An ideal particle has vanishing dimensions and is completely localizable. On the other hand, an ideal wave, one with a perfectly defined wavelength and frequency, has infinite extension in space. In the photoelectric effect, light behaves as if it consisted of particles or photons, but this does not mean that we dismiss the incontrovertible experimental evidence of the wave properties of light. Both descriptions must be accepted. How this dilemma is resolved is discussed in Section 41-8, after we have explored more fully other quantum attributes of light.

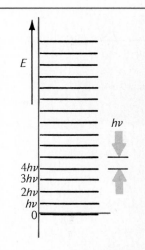

Figure 41-3. *Quantization of monochromatic electromagnetic radiation—the allowed energies are an integral multiple of the energy of a single photon.*

Example 41-1. A radio transmitter broadcasts a continuous power output of 6.0 W at a frequency of 600 kHz. How many radio-frequency photons leave the transmitting antenna each second?

The power output P can be written as the number N of photons emitted per second multiplied by the energy hv per photon. We then have

$$P = Nhv$$

so that

$$N = \frac{P}{hv} = \frac{6.0 \text{ W}}{(6.63 \times 10^{-34} \text{ J} \cdot \text{s})(600 \times 10^3 \text{ s}^{-1})}$$

$$= 1.5 \times 10^{28} \text{ photons/s}$$

Example 41-2.   A small and fairly bright sodium lamp with an output of 1.0 W of yellow light at a wavelength of 590 nm is 1.0 km away from an observer. The observer looks directly at the light source; the diameter of the pupil of his eye is 2.0 mm. How many yellow photons strike the eye's retina per second?

The total number N of photons emitted from the source per second is, as given in Example 41-1, $N = P/h\nu = P\lambda/hc$. The fraction of photons entering the pupil of radius r at a distance R from an effective point source is $\pi r^2/4\pi R^2$, since the surface area of a sphere of radius R is $4\pi R^2$. Therefore, the number n entering the eye per unit time is

$$n = N \frac{r^2}{4R^2}$$

$$= \frac{P\lambda}{(hc)} \frac{r^2}{(4R^2)}$$

$$= \frac{(1.0 \text{ W})(590 \times 10^{-9} \text{ m})(1.0 \times 10^{-3} \text{ m})^2}{(6.6 \times 10^{-34} \text{ J·s})(3.00 \times 10^8 \text{ m/s})(4)(1.0 \times 10^3 \text{ m})^2}$$

$$= 7.4 \times 10^5 \text{ photons/s}$$

Example 41-3.   It takes 13.61 eV to ionize a hydrogen atom. What is (a) the energy and (b) the wavelength of a photon that will ionize hydrogen?

(a) The ionization energy of 13.61 eV for a hydrogen atom is, in effect, its work function,

$$h\nu_0 = \phi = 13.61 \text{ eV}$$

(b) The wavelength is from

$$E = h\nu_0 = h\frac{c}{\lambda_0}$$

given by

$$\lambda_0 = \frac{hc}{E} = \frac{(6.6 \times 10^{-34} \text{ J·s})(3.0 \times 10^8 \text{ m/s})}{(13.61 \text{ eV})(1.60 \times 10^{-19} \text{ J/eV})} = 91.1 \text{ nm}$$

or an ultraviolet photon.

A more general form for the relation between photon energy E in electron volts and wavelength λ in nanometers is

$$\lambda = \frac{hc}{E} = \frac{1239.852 \text{ eV·nm}}{E}$$

Example 41-4.   The masses of a proton, a neutron, and a deuteron (the nucleus of a heavy hydrogen atom ^2_1H) are as follows:

Proton (p):	1.007 825 u
Neutron (n):	1.008 665 u
p + n:	2.016 490 u
Deuteron:	2.014 104 u

(Strictly, the mass of the "proton" includes also the mass of an electron, so that 1.007 825 u is the mass of a neutral hydrogen atom. Similarly, the "deuteron" mass includes an electron, so that 2.014 104 u is actually the mass of the deuterium atom.)

(a) What is the deuteron binding energy? (b) What is the minimum photon energy and maximum wavelength that will separate a deuteron into a proton plus a neutron, a process known as *photodisintegration*?

(a) Finding the deuteron binding energy is strictly an exercise in applying mass-energy conservation. We see above that the total mass of the separated particles exceeds that of the bound system; mass-energy must be added to ^2_1H to yield ^1_1H + n. We have

$$\Delta m = (2.016\ 490 - 2.014\ 104)\ \text{u} = 0.002\ 386\ \text{u}$$

so that the binding energy is*

$$E_b = \Delta m c^2 = (0.002\ 386\ \text{u})c^2\ (931.5\ \text{MeV/u}c^2) = 2.224\ \text{MeV}$$

$1\ u = 931.5\ MeV/_{c^2}$

and the corresponding wavelength is

$$\lambda = \frac{hc}{E} = \frac{1239.852\ \text{eV} \cdot \text{nm}}{2.224\ \text{MeV}} = 5.57 \times 10^{-4}\ \text{nm} = 557\ \text{fm}$$

* Actually, the threshold photon energy for dissociating a deuteron initially at rest into a proton and a neutron exceeds the binding energy slightly. The reason? A photon with an energy of 2.22 MeV has a momentum of 2.22 MeV/c. Therefore, the proton and the neutron cannot be at rest; they must be in motion and also have a total momentum of 2.22 MeV/c. The threshold photon energy is, it can be shown, given by $h\nu = E_b/(1 - E_b/M_d c^2)$, where M_d is the deuteron mass.

41-3   X-Ray Production and Bremsstrahlung

In the photoelectric effect, a photon transfers energy to an electron. The inverse effect is this: an electron loses kinetic energy and creates a photon. The process is most clearly illustrated in the production of x-rays.

When a fast-moving electron comes close to the positively charged nucleus of an atom and is deflected thereby, the electron is accelerated. The accelerated electric charge radiates electromagnetic energy. But in the quantum theory, this radiated electromagnetic energy consists of photons. In short, a deflected electron radiates one or more photons, and the electron leaves the collision site with reduced kinetic energy.

The radiation produced in such a collision is often referred to as *Bremsstrahlung* ("braking radiation" in German). A Bremsstrahlung collision is shown schematically in Figure 41-4. An electron approaches the deflecting atom with a kinetic energy K_1; it recedes with a kinetic energy K_2, having produced a single photon of energy $h\nu$. From energy conservation, we have

$$K_1 - K_2 = h\nu$$

(We can ignore the very small energy of the recoiling atom.)

X-rays were discovered and first investigated in 1895 by Wilhelm Roentgen, who assigned this name because the true nature of the radiation was at first unknown. X-rays, now known to consist of electromagnetic waves, or photons, having wavelengths of about 0.1 nm, pass readily through many materials that are opaque to visible light.

Suppose an electron is accelerated through an electric potential difference V, of several thousand volts, and then strikes a target. The electron acquires a kinetic energy of $K = eV$. We have ignored the electron's kinetic energy as it left the cathode, typically much less than Ve. When the electron strikes the target, it acquires an additional energy, the energy that binds it to the target surface; but this binding energy is also only a few electron volts, and it too can be ignored. When the electron strikes the target, it is brought essentially to rest in a single collision. The most energetic photon that can be produced in a single Bremsstrahlung collision is one whose energy $h\nu_{max}$ is

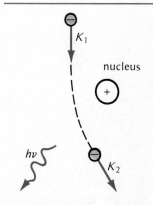

Figure 41-4. *Bremsstrahlung collision of an electron with a nucleus, causing the creation of a photon.*

$$hv_{\text{max}} = eV = K$$

where v_{max} is the maximum frequency of the x-ray photons produced. More typically, an electron loses its energy at the target by heating it or by producing two or more photons, the sum of whose frequencies will then be less than v_{max}. There will be a distribution in photon energies with, however, a well-defined maximum frequence v_{max} or minimum wavelength $\lambda_{\text{min}} = c/v_{\text{max}}$, given by

$$K = hv_{\text{max}} = \frac{hc}{\lambda_{\text{min}}} = eV \tag{41-5}$$

The intensity of x-rays emitted has in fact an abrupt cutoff at the limit v_{max}; this limit is determined solely by the accelerating potential V applied to the x-ray tube, not by the chemical identity of the target material.

41-4 Compton Effect

What happens when a monochromatic electromagnetic wave impinges on a charged particle whose size is much less than the wavelength of the radiation? The charged particle is accelerated principally by the wave's sinusoidally varying electric field. In fact, the particle oscillates in simple harmonic motion at the same frequency as that of the incident radiation. And since the charged particle is accelerated continuously, it radiates electromagnetic radiation of the same frequency. This is what classical theory predicts: scattered radiation with the same frequency as the incident radiation. The charged particle acts as transfer agent; it absorbs energy from the incident beam and reradiates it at the same frequency but scattering it in all directions. Classical scattering theory agrees with experiment for visible light and other, longer-wavelength radiation. A simple example is this: Light reflected from a mirror (a collection of scatterers) undergoes no apparent change in frequency.

To consider scattering from the point of view of quantum theory, we first need the relation for the momentum of a photon. We can get this from relativistic dynamics by regarding the photon as a particle that, because it always travels at speed c, has a zero rest mass and zero rest energy. The general relativistic relation between momentum p and total energy E is

$$E^2 = E_0^2 + (pc)^2 \tag{40-19}$$

and it yields with $E_0 = 0$,

$$p = E/c$$

This result follows also from classical electromagnetic theory (Section 35-6), where it was shown that the linear momentum of a beam of electromagnetic radiation is the energy of the beam divided by c (35-19).

It follows that the momentum of a photon with frequency v and wavelength λ is

$$p = \frac{E}{c} = \frac{hv}{c} = \frac{h}{\lambda} \tag{41-6}$$

The direction of wave propagation is the direction of a photon's momentum.

A photon's momentum increases with frequency. The momentum of a high-frequency (or high-energy) photon, such as a gamma (γ) ray, will exceed by far the momentum of a low-frequency (and low-energy) photon, such as a radio photon. The distinctive feature introduced by the quantum theory is this: electromagnetic momentum occurs not in arbitrary amounts, but only in integral multiples of the momentum h/λ carried by a single photon.

Now consider a photon incident upon a free charged particle at rest. Basically, we have a photon colliding with a particle, and the laws of energy and momentum conservation apply. Figure 41-5 shows the photon and free particle before and after collision. The special advantage in applying the conservation laws is that we need not be concerned with the details of what happens when photon meets electron, but merely with the total energy and momentum going into and coming out of the collision.

As Figure 41-5 shows, if a photon carries momentum and energy into the collision, the struck particle must gain some energy and momentum.

We take the particle to have rest mass m_0 and rest energy $E_0 = m_0 c^2$; it is free and initially at rest. Energy conservation applied to the collision of Figure 41-5 gives

$$hv + E_0 = hv' + E \qquad (41\text{-}7)$$

Here E is the relativistic energy of the recoiling particle after collision. The energies of the incident and scattered photons are hv and hv', respectively. The particle's final energy E (rest energy plus kinetic energy) exceeds initial energy E_0. We immediately see from (41-7) that $hv' < hv$. The scattered photon has *less* energy, a *lower* frequency, and a *longer* wavelength than the incident photon, a result quite different from classical physics. The incident and scattered photons have different frequencies, so that the scattered photon is not to be thought of as merely the incident photon moving in a different direction with less energy. Rather, in the collision the incident photon is annihilated, and the scattered photon is created.

Momentum conservation is implied by the vector triangle of Figure 41-5(c). Here $\mathbf{p} = m\mathbf{v}$ is the relativistic momentum of the recoiling particle. The magnitudes of the momenta of the incident and scattered photons are, respectively, $p_\lambda = hv/c = h/\lambda$ and $p_{\lambda'} = hv'/c = h/\lambda'$. Scattering angle θ is the angle between the directions of $\mathbf{p}_\lambda$ and $\mathbf{p}_{\lambda'}$, the directions of the incident and scattered photons.

The law of cosines applied to the triangle in Figure 41-5(c) yields

$$p_\lambda^2 + p_{\lambda'}^2 - 2p_\lambda p_{\lambda'} \cos\theta = p^2 \qquad (41\text{-}8)$$

We wish to solve (41-7) and (41-8) for the change in wavelength $\lambda' - \lambda = \Delta\lambda$. After some manipulation,* we get

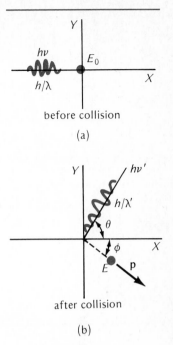

before collision

(a)

after collision

(b)

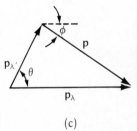

(c)

Figure 41-5. A Compton collision: (a) Before and (b) after the collision. (c) Momentum vectors for the incident and scattered photons and for the electron.

---

* Multiply both sides of (41-8) by c^2 and use $pc = hv$, and we have

$$h^2v^2 + h^2v'^2 - 2h^2vv' \cos\theta = p^2c^2$$

We get a similar relation from (41-7) as follows. Put hv and hv' on one side of the equation and E and E_0 on the other; then square. We get

$$h^2v^2 + h^2v'^2 - 2h^2vv' = E^2 + E_0^2 - 2EE_0$$
$$= 2E_0^2 + p^2c^2 - 2EE_0$$

Now subtract the two equations above, and get

$$h^2vv'(1 - \cos\theta) = E_0(E - E_0) = m_0c^2(hv - hv')$$

Using $v = c/\lambda$ and $v' = c/\lambda'$, we finally get (41-9).

$$\Delta\lambda = \lambda' - \lambda = \frac{h}{m_0 c}(1 - \cos\theta) \qquad (41\text{-}9)$$

This is the basic equation for the Compton effect. It gives the increase $\Delta\lambda$ in the wavelength of the scattered photon over that of the incident photon. Note that $\Delta\lambda$ depends only on the rest mass m_0 of the recoiling particle, Planck's constant h, the speed c of light, and the angle θ of scattering; $\Delta\lambda$ is independent of the incident photon's wavelength λ. The quantity $h/m_0 c$, appearing on the right-hand side of (41-9) and having the dimensions of length, is known as the *Compton wavelength*. Given the scattering angle θ, we can compute the wavelength increase unambiguously, but we cannot predict in advance the angle at which any one photon will emerge.

Suppose the recoiling particle is a free electron, or one that is only very loosely bound to a parent atom. Then $m_0 = 9.11 \times 10^{-31}$ kg, and we compute $h/m_0 c = 2.43$ pm. As (41-9) shows, when $\theta = 90°$, the wavelength change is $\Delta\lambda = h/m_0 c = 2.43$ pm. When θ is 180° and the scattered photon travels in the backward direction, and the recoil electron straight forward, so that the collision is effectively "head-on," the wavelength change is a maximum. Then the electron's kinetic energy is also a maximum.

As (41-9) shows, the increase in wavelength of the scattered photon relative to the incident photon does not depend on the wavelength of the incident photon. *All* photons scattered at $\theta = 90°$ have a wavelength shift of 2.43 pm. For visible light the shift is so small as to be virtually unobservable. An observable shift, one of at least a few percent, can occur for x-rays. For example, an incident x-ray photon with a wavelength of $\lambda = 0.1000$ nm will, when scattered through 90°, produce a scattered photon with a wavelength of 0.1024 nm, or a wavelength increase of 2.4 percent.

The shift in wavelength in the scattering of x-rays was observed first by A. H. Compton in 1922. Figure 41-6 shows schematically the experimental arrangement. At any fixed angle θ, the x-ray detector measures the scattered intensity as a function of wavelength. Figure 41-6 shows two wavelengths for scattered photons. The photon with the expected wavelength *shift* comes from the collision of any incident photon with an essentially *free* electron in the target material. The scattered photons with the *same* wavelength as incident photons come from a photon that has collided with a tightly bound electron. Since a tightly bound electron cannot move without also moving the entire atom, the bound electron's mass is effectively that of the whole atom.

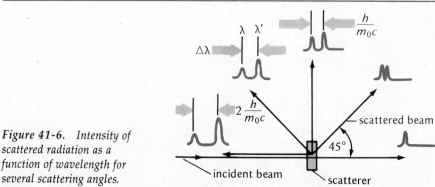

Figure 41-6. *Intensity of scattered radiation as a function of wavelength for several scattering angles.*

41-5   Pair Production and Annihilation

Can a photon's energy be converted into rest mass? The answer is yes, and it is illustrated most directly in the phenomenon of pair production.

Pair Production   What is the minimum energy required to create a single particle? The electron has the smallest nonzero rest mass of all known particles, it requires the least energy for its creation. But a photon has zero electric charge. So the law of electric-charge conservation precludes the creation of a single electron from a photon. But an electron pair, consisting of two particles with opposite electric charges, would be possible. A positively charged particle, called the *positron* and the *antiparticle* of the electron, is in fact observed. The electron and the positron are similar in all ways except in the signs of their charges, $-e$ and $+e$ (and the effects of this difference). Clearly, the minimum energy $h\nu_{min}$ to create an electron-positron pair is

$$h\nu_{min} = 2m_0 c^2 \qquad (41\text{-}10)$$

Since the rest energy $m_0 c^2$ of an electron or a positron is 0.51 MeV, the threshold energy $2m_0 c^2$ for pair production is 1.02 MeV. The photon wavelength corresponding to this threshold is 1.2 pm. Electron pairs can be produced only by γ-ray photons. The general phenomenon in which a particle and its antiparticle are created from electromagnetic radiation is called *pair production*. It is a very emphatic demonstration of the interconvertibility of mass and energy.

If a photon's energy exceeds the threshold energy $2m_0 c^2$, the excess appears as kinetic energy of the created pair.

Pair production cannot occur in empty space. We prove this by showing that energy and momentum cannot simultaneously be conserved in particle-antiparticle production unless the photon is near some massive particle, such as an atomic nucleus. Suppose, for the sake of argument, that a pair has been created in empty space and that we, the observers, are at rest with respect to the center of mass of this two-particle system. Then the total momentum of the pair is zero. But the photon creating the pair would have had some nonzero momentum in this reference frame, since a photon always moves at speed c, whatever the reference frame. Under these imagined circumstances, we should have the momentum of the photon before collision but zero momentum after, clearly a violation of momentum conservation. A photon cannot decay spontaneously to an electron-positron pair in free space; the process can take place only if the photon encounters a massive particle that acquires the requisite (but negligible) momentum.

Figure 41-7 is a schematic drawing of pair production, and Figure 41-8 is a cloud-chamber photograph showing electron-positron pairs. The paths of the charged particles are visible because the charged particles produce ionization effects at which bubbles are formed as the particles travel through the liquid. The oppositely charged particles are deflected into oppositely directed circular arcs by a uniform magnetic field.

Positrons were predicted on theoretical grounds by P. A. M. Dirac in 1928. Four years later, C. D. Anderson observed directly and identified a positron. Electron-positron pairs are now commonly observed whenever high-energy photons interact with matter. Proton-antiproton, neutron-antineutron, and

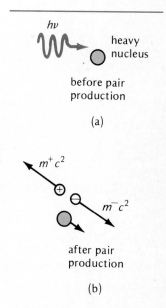

Figure 41-7.   Schematic diagram for pair production.

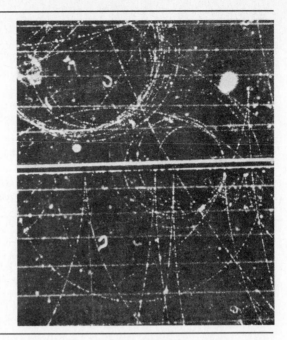

Figure 41-8.   Photograph of pair production (in a cloud chamber).

still other types of pairs can be created. They have higher threshold energies than electron-positron pairs because of their larger rest mass.

Pair Annihilation   A particle-antiparticle pair can annihilate each other and create photons. The process is the inverse of pair production. Suppose an electron and a positron are close together and essentially at rest. Their total linear momentum is initially zero; therefore, a single photon cannot be created. That would violate momentum conservation. Momentum can, however, be conserved if *two* photons, moving in opposite directions with equal momenta, are created. Such a pair of photons would also have equal frequencies and energies. See Figure 41-9. Actually, three or more photons can be created, but with a much smaller probability than for two photons.

Energy conservation implies

$$m_0^+ c^2 + m_0^- c^2 = 2m_0 c^2 = 2h\nu_{min}$$

with an electron and a positron at rest initially. The minimum energy of one photon created by electron-positron annihilation is $h\nu_{min} = m_0 c^2 = 0.51$ MeV.

Annihilation is the ultimate fate of positrons. When a high-energy positron appears, as in pair production, it loses its kinetic energy in collisions as it passes through matter, and finally moves at low speed. Then it combines with an electron and forms a bound system, called a *positronium*; this "atom" decays quickly (10^{-10} s) to two photons of equal energy. Thus, the death of a positron is signaled by the appearance of two annihilation quanta, or photons, of about $\frac{1}{2}$ MeV each. The transitoriness of positrons is due not to an intrinsic instability, but to the high risk of their collision and subsequent annihilation with electrons.

In our part of the universe there is a preponderance of electrons, protons, and neutrons; their antiparticles, when created, quickly combine with them in

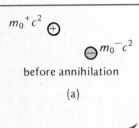

before annihilation

(a)

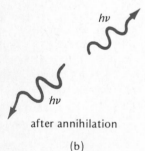

after annihilation

(b)

Figure 41-9.   Schematic diagram showing pair annihilation with the creation of two photons.

annihilation processes. It is conceivable, although at present purely conjectural, that there exists a part of the universe in which positrons, antiprotons, and antineutrons predominate.

Example 41-5.   A highly energetic photon creates an electron-positron pair in a uniform magnetic field of 1.5 T. The two tracks registered in a bubble chamber lie in a plane perpendicular to the magnetic field, and the two radii of curvature are 10 cm and 14 cm. What is the photon energy?

If the electron's kinetic energy far exceeds its rest energy (~ 0.5 MeV), we can write its energy as $E = pc$, where p is the momentum. Likewise for the positron. The momentum of a charged particle in a magnetic field is, from (40-13), given by $p = qrB$, so that

$$h\nu = E^+ + E^- = (p^+ + p^-)c = qBc(r^+ + r^-)$$

$$= \frac{(1.6 \times 10^{-19}\text{C})(1.5\text{ T})(3.0 \times 10^8 \text{ m/s})(10\text{ cm} + 14\text{ cm})}{(1.6 \times 10^{-19} \text{ J/eV})} = 108 \text{ MeV}$$

The assumption initially that the electron rest mass is negligible is justified.

41-6   Matter Waves

As we have seen, a photon of wavelength λ has a momentum given by $p = h/\lambda$. Louis de Broglie in 1924 posed the question, based on the conjectured symmetry of nature, Does every particle with momentum p have associated with it a wavelength λ given by the same relation? The answer of observation is an emphatic yes. Every particle has a wave character, and its wavelength is given by the *de Broglie relation*,

$$\lambda = \frac{h}{p} = \frac{h}{mv} \tag{41-11}$$

where p is the particle's relativistic momentum, m is its relativistic mass, and v is its speed.

What is the wavelength of an ordinary object, say, a pitched baseball ($m = 0.2$ kg, $v = 40$ m/s)? Equation (41-11) yields $\lambda \simeq 10^{-34}$ m. This wavelength is so extraordinarily small (smaller than the size of a proton by a factor of about 10^{19}) that one cannot observe diffraction or interference effects for a baseball. For example, a baseball pitched through an open window — really a single slit — is not diffracted, or deviated off-center. It is, after all, by observing such distinctive wave phenomena as diffraction and interference that we can tell whether energy traveling through space is, in fact, a wave. That is certainly how we confirm that visible light consists of waves. That is also how such short-wavelength electromagnetic radiation as x-rays is confirmed to consist of waves. Recall that x-rays show diffraction effects (Section 39-4) when sent through ordinary crystalline solids; the x-ray wavelength (~ 0.1 nm) is comparable to the distance between adjacent atoms situated in geometrical arrays in a crystal, and diffraction effects are relatively easy to observe.

The key to observing the wave character of ordinary particles is to have the particle wavelength be comparable to the size of a possible diffracting object. What is the kinetic energy of an electron, for example, that yields $\lambda = 0.10$ nm? For an electron of electric charge e accelerated from rest by electric potential difference V, we have, using (41-11),

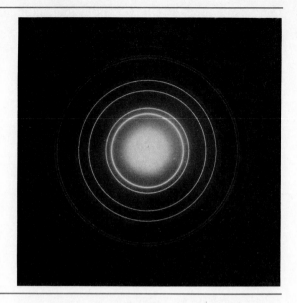

Figure 41-10. *Photograph of electron diffraction.*

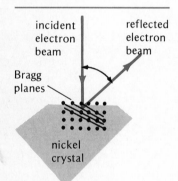

(a)

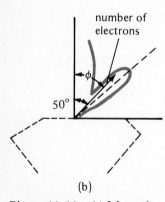

(b)

Figure 41-11. (a) Schematic for the Davisson-Germer experiment. (b) Number of electrons as a function of angle φ.

$$eV = \frac{1}{2}\,mv^2 = \frac{p^2}{2m} = \frac{(h/\lambda)^2}{2m}$$

so that with $\lambda = 1.00 \times 10^{-10}$ m, we have

$$V = \frac{h^2}{2\,me\,\lambda^2} = \frac{(6.63 \times 10^{-34}\ \text{J}\cdot\text{s})^2}{2(9.11 \times 10^{-31}\ \text{kg})(1.60 \times 10^{-19}\ \text{C})(1.00 \times 10^{-10}\ \text{m})^2}$$

$$= 150\ \text{V}$$

A 150-eV electron has a wavelength 0.1 nm. Observing electron diffraction should be and is very much like observing x-ray diffraction. For electron diffraction, the "bright" spots are those locations where many electrons are observed and the "dark" spots are those where few if any are observed. See Figure 41-10. Indeed, essentially all the wave effects that can be demonstrated for visible light — diffraction through a single slit, for example — can also be observed for electrons. The wave properties were first observed in 1927 by C. Davisson and L. H. Germer at AT&T Bell Laboratories. A beam of monoenergetic (and therefore monochromatic) electrons was directed at a single crystal of nickel. The beam was diffracted by reflection from Bragg planes (Section 39-4) within the crystal, and a very pronounced peak was observed, which is explained by electron diffraction. See Figure 41-11.

Not merely electrons show wave effects. The wave character of neutrons, atoms, even molecules is also confirmed in detail. Indeed, all the interference and diffraction effects observed for electromagnetic radiation have been duplicated for particles.

Example 41-6. A nuclear reactor provides a copious supply of neutrons. When a neutron is in thermal equilibrium at temperature T, it has, like a molecule in a gas, an average kinetic energy of $\frac{3}{2}\,kT$ (Section 20-3). A thermal neutron is one with $T = 293$ K, room temperature. Thermal neutrons are especially useful in the phenomenon of neutron diffraction, which is used to study the structure of crystalline materials. Unlike electrons, neutrons are electrically uncharged, so that their behavior is unaffected by electric forces. Show that the wavelength of a thermal neutron is about 0.1 nm.

We have

$$\frac{3}{2} kT = \frac{p^2}{2m} = \frac{(h/\lambda)^2}{2m}$$

$$\lambda = \frac{h}{\sqrt{3mkT}} = \frac{6.6 \times 10^{-34} \text{ J} \cdot \text{s}}{\sqrt{3(1.7 \times 10^{-27} \text{ kg})(1.4 \times 10^{-23} \text{ J/K})(293 \text{ K})}} \approx 0.1 \text{ nm}$$

so that a thermal neutron, a 150-eV electron, and a typical x-ray all have wavelengths of the same order.

41-7   Probability Interpretation of the Wave Function

What is the wave associated with a photon? It is, of course, the oscillating electric field E (or magnetic field B) of an electromagnetic wave.

But what is waving when we associate a wavelength λ with an ordinary material particle? The very bland name *wave function*, typically represented by ψ, is given to the mathematical quantity that represents a particle's wave character. Although ψ cannot be observed directly, the square of the wave function ψ^2 can; it is, as we shall see, proportional directly to the *probability of observing a particle*.*

To see why ψ^2 is proportional to the probability of observing a particle, consider the analogous situation for a photon. A photon's wave function is the electric field E of the associated electromagnetic wave, so that the probability of observing the photon at any location in space would be proportional to E^2. To see that E^2 must indeed represent the probability of observing the photon, recall that the relation between the intensity I of an electromagnetic wave and E^2 is

$$I = \epsilon_0 c E^2 \tag{35-11}$$

so that

$$I \propto E^2$$

Now it is easy to see that in the quantum view of electromagnetic radiation, the intensity, the energy per unit time per unit transverse area, is itself proportional to the probability of observing a photon. Certainly the intensity may be written as

$$I = Nh\nu \tag{41-12}$$

where N is the number of photons, each of energy $h\nu$, passing through a unit transverse area per unit time. But the *number* of photons per unit time per unit area in a beam of radiation must mean really the *average* number. This would certainly be true for a beam of very low intensity, one in which the intensity was so low that on the average, N had the value of, say, 0.5 photon/cm²·s. There is no such thing as half a photon, so that this numerical value of N here

---

* It turns out that the wave function ψ is always a complex mathematical quantity with both a real and an imaginary part. Strictly, ψ^2 represents $\psi * \psi$, where $\psi *$ is the complex conjugate of ψ, and $\psi * \psi$ is then always a real quantity.

Particles, Fields

How can you keep cars from speeding on campus? The problem finally landed on the desk of the Chancellor of Washington University in St. Louis, Missouri. His solution—neat in its simplicity—was the speed bump.* The Chancellor, Arthur H. Compton (1892–1962), was good at analyzing things that bump. He had, after all, been awarded a Nobel prize in physics in 1927 for studying what happens when a photon collides with a particle.

The Compton effect is just one of the basic electron-photon interactions. It, together with the other basic interactions between a photon and a charged particle such as an electron, are shown in the figure. The pictures are those of that branch of contemporary physics known as *quantum electrodynamics* (QED for short), or more generally and simply, *field theory*.

Read each figure with time going forward from left to right. A straight line represents an electron, a wiggly line a photon. A positron corresponds to a straight line with its arrow reversed (a positron regarded as an electron running backward in time).

The interactions are these:

• (a) The primitive photoelectric effect: an electron absorbs a photon and the electron's momentum and energy are changed.

*The story of Compton's design of the speed bump is given in the St. Louis *Globe Democrat* of 7 April 1953.

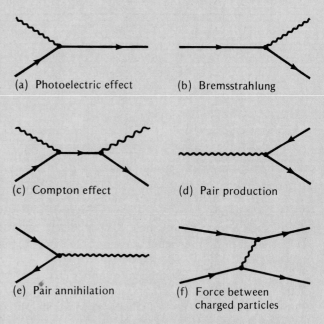

(a) Photoelectric effect

(b) Bremsstrahlung

(c) Compton effect

(d) Pair production

(e) Pair annihilation

(f) Force between charged particles

• (b) Bremsstrahlung: an electron emits a photon and the electron's momentum and energy change, just the reverse of process (a).

• (c) Compton effect: this process takes place as two closely spaced events—first the absorption of the incident photon and then the emission of the scattered photon.

means that over a 2-s interval we observe about one photon on the average crossing a 1-cm² transverse area. In short, N and therefore I and therefore also E^2 are proportional to the probability of locating a photon. In exactly the same way, ψ^2 is proportional to the probability of observing a material particle. See Table 41-1.

We state this important result in more detail. If ψ represents the wave function at the location x, the probability of observing the particle's being between x and $x + dx$ is given by $\psi^2(x)\,dx$:

Table 41-1

	WAVE FUNCTION	PROBABILITY OF OBSERVING ENTITY
Photon	E	E^2
Particle	ψ	ψ^2

• (d) Pair production: a photon becomes an electron and a positron.

• (e) Pair annihilation: an electron-positron pair becomes a photon.

• (f) The electromagnetic force between two charged particles: here one electron emits a photon that is absorbed by the second electron, and thereby each electron has its momentum and energy changed. The force between charged particles is attributed to their trading a photon, the field particle of the electromagnetic interaction.

Look again at the various interactions in the figure, and you see that there is just *one* fundamental process—a vertex, where an incoming and outgoing electron line joins a photon line. It all boils down to this: a photon bumping an electron.

Should we worry that the fundamental laws of momentum and energy conservation seem not to be satisfied in these processes? For example, in (b) we see a free electron blithely coasting at constant velocity and then suddenly emitting a photon. The uncertainty principle says that it's all right. Momentum conservation *may* be violated, but only over a limited region of space. Energy conservation *may*, according to the uncertainty principle, also be violated, but again only over a limited interval of time controlled by Planck's constant.

Take another look at the force between charged particles in (f). The conservation laws of momentum and energy *are* violated when the first electron emits a photon but the momentum and energy debt is repaid when this photon—an unobservable, or *virtual photon*—is absorbed by the second electron. Since the Coulomb force drops off with distance but can extend, even if feebly, to charges separated by an infinite distance, the virtual photon for such a very-long-range interaction must have almost zero energy. Indeed, the infinite range of the electromagnetic force corresponds exactly to the circumstance that the photon as field particle can have zero energy.

This is all very picturesque. But does it work? Of course, the detailed theoretical analysis is far more complicated than the simple diagrams that summarize it. It turns out that these ideas in quantum field theory for electromagnetism produce the very best theory there is in physics. Nothing else shows such nearly perfect agreement between theory and experiment—better than a part in 10^7, by a variety of tests.

Because QED works so well, physicists have been emboldened to use it as the model for *all* interactions between elementary particles. Indeed, the recent discovery of the W and Z particles, described in the panel in chapter 35, is just one example of the validity of this approach.

*To get the latest word on elementary particles and quantum field theory look at recent issues of *Scientific American*. This magazine for the non-specialist interested in science has at least one or two articles each year on this topic. Another source, probably available in your physics library, on latest developments at high-energy accelerating machines, is the *CERN Courier*.

$$\text{Probability of observing a particle in the interval } dx \quad \propto \quad \psi^2 dx$$

Just as the electric field of a photon will generally be a function of both position and time, so too generally will the wave function ψ.

The probability interpretation of waves associated with particles was first given in 1926 by Max Born. That branch of quantum physics that deals with finding the values of ψ is known as *wave mechanics,* or *quantum mechanics.* The two principal originators of the wave mechanics of particles were Erwin Schrödinger (in 1926) and Werner Heisenberg (in 1925), who independently formulated quantum mechanics in different but equivalent mathematical forms.

Maxwell's electromagnetic theory is summarized in the Maxwell's equations; they are the basis for computing values of E. The wave mechanics of matter is governed by the Schrödinger equation; it is the basis for computing values of ψ in any problem in quantum physics. Here the parallel stops,

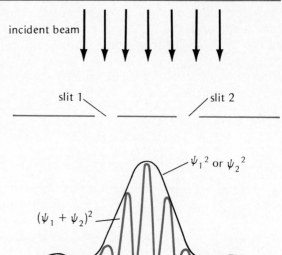

Figure 41-12. *Double-slit diffraction for particles (vertical distances drastically compressed). Wave functions ψ_1 and ψ_2 give the diffraction pattern with either slit 1 or 2 open. The superposed wave function $(\psi_1 + \psi_2)$ gives the pattern when both slits are open.*

however. The electric field has its origin in electric charges, and E gives not only the probability of observing a photon but also the electric force on a unit positive electric charge. But the wave function ψ of the Schrödinger equation is not directly measurable or observable. It does, however, give the most information one can extract concerning any system of objects; and all measurable quantities, such as the energy and momentum, as well as the probability of location, can be derived from it.

Consider the interference of waves that go through two parallel slits. When either of the two slits is closed, the pattern on a distant screen is the typical single-slit diffraction pattern: a broad, central maximum flanked by weaker, secondary maxima (Figure 39-2). When both slits are open, the pattern is as shown in Figure 41-12: interference fine structure within a diffraction envelope. The pattern is not merely two single-slit diffraction patterns superposed; the interference between waves traveling through both of the slits is responsible for the rapid variations in intensity. Here waves (or particles) can take two or more routes from a source to an observation point. We first superpose the wave function from the two separate routes to find the resultant wave function; then we square the resultant wave function to find the probability (or intensity). That is to say, if ψ_1 and ψ_2 represent the wave functions for passage through slits 1 and 2 separately, then $(\psi_1 + \psi_2)^2$, not $\psi_1^2 + \psi_2^2$, gives the probability of observing a particle on the screen. In this view, when a single electron or photon is directed toward a pair of slits, we cannot say through which of the two slits it will pass. When we speak in the language of waves we say, in effect, that the particle passes through *both* slits.

41-8   Complementarity Principle

How can we speak simultaneously of an electron as a particle—a point object—and an electron as a wave—an object spread far over space? Or how can electromagnetic radiation be viewed as a wave phenomenon and also as a collection of particlelike photons? The answer: We can never do the impossible and describe a particular entity *simultaneously* as particle and wave. The two extreme descriptions are mutually incompatible and contradictory.

According to the *principle of complementarity,* enunciated by Niels Bohr in 1928, the *wave and particle aspects* of electromagnetic radiation and of material particles *are complementary.* In any one experiment, we choose either the particle or the wave description. The two aspects are complementary in that our knowledge of the properties of electromagnetic radiation or of particles is partial unless both wave and particle aspects are known. The choice of one description, imposed by the nature of the experiment, precludes the simultaneous choice of the other. The quantities in quantum theory are more complicated than can be comprehended in the simple and extreme notions of wave and particle, notions borrowed from our direct, ordinary experience with large-scale phenomena.

The complementarity principle applied to electrons says this: Electrons in a cathode-ray tube follow well-defined paths and indicate their collisions with a fluorescent screen by very small, bright flashes. A particle model is used to describe electrons in cathode ray-experiments because all the electron energy, momentum, and electric charge is assigned at any one time to a small region of space. The particle nature of electrons is revealed in the cathode-ray experiments; and therefore, by the principle of complementarity, the wave nature of electrons must be suppressed.

The wave nature of electrons shows up in electron diffraction. Here electrons are propagated as waves with an indefinite extension in space, and it is necessarily impossible to specify the location of any one electron. In short, the electron-diffraction experiments exhibit the wave nature of electrons, and by complementarity, the particle nature is necessarily suppressed.

41-9   Uncertainty Principle

Consider this hypothetical experiment. You want to find the location of an electron initially at rest by firing a single photon at it. You can tell whether the photon hits the electron by observing a scattered photon. What kind of photon would be best? Since you can never "see" details smaller than wavelength λ, short-wavelength photons, or gamma rays, will give the highest resolution. But there is a serious complication. Directing a short-wavelength, high-frequency — and therefore high-energy and high-momentum — photon at an electron will deflect the electron, so that you end up learning where the electron *was*, not where it now is. On the other hand, if the electron is to remain nearly undisturbed, you must use long-wavelength light and therefore settle for only a fuzzy idea of its location. This simple illustration of a *gamma-ray microscope,* first discussed in 1927 by Werner Heisenberg, shows that it is fundamentally impossible to specify simultaneously with complete precision certain pairs of physical quantities. More specifically, one important formulation of the Heisenberg *uncertainty principle,* or *principle of indeterminacy,* is this:

$$\Delta p_x \, \Delta x > \frac{\hbar}{2} \qquad\qquad \text{(41-13)}$$

Here Δx is the uncertainty in a particle's location along the x axis and Δp_x is the

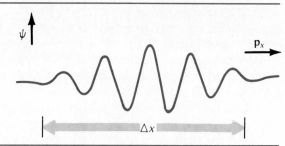

Figure 41-13. *A wave packet.*

uncertainty in the momentum component in that direction. The product of the uncertainties can be no smaller then $\hbar/2$, which is comparable to h, the fundamental constant of quantum theory.* If you know very precisely where a particle is located, the particle's momentum component must be highly uncertain. Or if you know very precisely a particle's momentum (and therefore also its wavelength $\lambda = h/p$), you pay for this knowledge by high uncertainty about where the particle is located. A second formulation of the uncertainty relation is

$$\Delta E \, \Delta t > \hbar/2 \qquad (41\text{-}14)$$

Here ΔE and Δt are the respective uncertainties in energy and time interval. We can see qualitatively the basis for (41-14) as follows. Suppose an oscillation is observed only over a finite time interval Δt. Then the angular frequency of oscillation is uncertain by $\Delta \omega \simeq 1/\Delta t$. But an uncertainty in the frequency of oscillation of a photon also implies an uncertainty in its energy. Again, when either energy or time interval is small, the other quantity must be correspondingly large by an amount controlled by $\hbar$.

Suppose that we want to represent an electron by a wave and yet localize it in space to some degree. It cannot be a single, monochromatic sinusoidal wave; such a wave would necessarily extend to infinity and it certainly would not be localized. We can, however, superpose a number of sinusoidal waves differing in frequency over a range of frequencies $\Delta \nu$ and so have a *wave packet.* The component waves constructively interfere over a limited region of space Δx, identified as the somewhat uncertain location of the "particle," and so yield a resultant wave function ψ of the sort shown in Figure 41-13. Because there is a range of frequency and a range in wavelength, $\Delta \nu$ and $\Delta \lambda$, the associated momentum and energy are necessarily uncertain, and it is impossible to predict precisely where or when the wave packet will reach another point and what the momentum and energy will then be.

Since the uncertainty relation implies an uncertainty in energy ΔE over a time interval Δt, it also implies that the law of energy conservation may actually be violated—by that amount, $\Delta E = \frac{1}{2}\hbar/\Delta t$, but only for the time interval Δt. The greater the amount of energy "borrowed" or "discarded," the shorter the interval over which nonconservation of energy may occur.

To see the uncertainty principle illustrated in a specific situation, consider

---

* The constant on the right side of (41-13), here $\hbar/2$, depends on the precise definitions of Δp_x and Δx. Uncertainties Δp_x and Δx represent the root-mean-square values of several independent measurements. Symbol $\hbar$ (read as "aitch bar") stands for $h/2\pi$. For example, a photon of angular frequency ω has energy $\hbar\omega$.

waves diffracted by a single, parallel-edged slit of width w, as shown in Figure 41-14. The diffraction pattern is formed on a distant screen. The location of the points of zero intensity is given by $\sin \theta = m\lambda/w$, as shown in Section 39-2.

We have not yet specified what sort of wave passes through the slit. If it is electromagnetic radiation, the intensity of the diffraction pattern is proportional to E^2, the square of the electric field at the screen. If on the other hand, the wave consists of a beam of monoenergetic electrons, the intensity is proportional to ψ^2, which is the square of the electron wave function at the screen and gives the probability of finding an electron at any point along the screen. Whatever the wave type, diffraction effects are pronounced only when the wavelength is comparable to the slit width; at the limit of vanishing wavelength, the intensity pattern on the screen corresponds to a geometrical shadow cast by the edges of the slit.

Suppose now that we reduce drastically the amount of incident radiation or the number of electrons, as the case may be. Then, on the screen we no longer

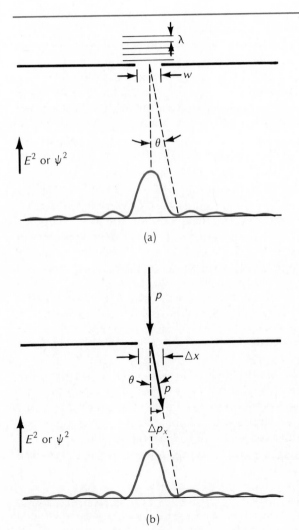

(a)

(b)

Figure 41-14. (a) Monochromatic waves diffracted by a single slit of width w. (b) Uncertainties Δx in position and Δp_x in momentum.

see smooth variations but instead, photons or electrons arriving one by one. The quantity plotted in Figure 41-14 represents the *probability* that a particle will strike a certain spot on the screen. At very low illumination, bright flashes appear over a large area of the screen. As time passes, more and more particles accumulate on the screen, and the distinct bright flashes merge and form the smoothly varying intensity pattern predicted by wave theory.

There is no way of telling in advance where any one electron or photon will fall on the screen. All that wave mechanics permits us to know is the probability of a particle's striking any one point. Before the particles pass through the slit, their momentum is known with complete precision both in magnitude (monochromatic waves) and in direction (vertically down in this case). When they pass through the slit, their location along x, completely uncertain before they reached the slit, is now known with an uncertainty $\Delta x = w$, the slit width. What we don't know, however, is precisely where any one particle will strike the screen. Any particle has approximately a 75 percent chance of falling within the central region. There is an uncertainty in the x component of momentum p_x.

Suppose that Δx is very large. With a wide slit, the uncertainty in position is increased. We are less certain about where an electron is located along x. The uncertainty in the momentum is reduced correspondingly; the diffraction pattern shrinks, and in the limit, essentially all electrons fall within the geometrical shadow. On the contrary, if the slit width is reduced and Δx becomes very small, the diffraction pattern is expanded along the screen. For the increase in our certainty of the electron's position, we must pay by a correspondingly greater uncertainty Δp_x in its momentum.

For a slit width much greater than the wavelength, particles pass through the slit undeviated to fall within the geometrical shadow. This agrees with classical mechanics, where the wave aspect of material particles is ignored. Thus, there is a close parallel in the relationship of wave optics to ray optics, and of wave mechanics to classical mechanics. Ray optics is a good approximation of wave optics whenever the wavelength is much less than the dimensions of obstacles or apertures that the light encounters; similarly, classical mechanics is a good approximation of wave mechanics whenever a particle's wavelength is much less than the dimensions of obstacles or apertures encountered by material particles. Symbolically, we can write

$$\underset{\lambda/w \to 0}{\text{Limit}} \text{ (wave optics)} = \text{ray optics}$$

$$\underset{\lambda/w \to 0}{\text{Limit}} \text{ (wave mechanics)} = \text{classical mechanics}$$

No ingenious subtlety in the design of the diffraction experiment will remove the basic uncertainty. We do not have here, as in the large-scale phenomena encountered in classical physics, a situation in which the disturbances on the measured object can be made indefinitely small by ingenuity and care. The limitation here is rooted in the fundamental quantum nature of electrons and photons; it is intrinsic in their complementary wave and particle aspects.

Example 41-7.   What is the uncertainty in momentum of a 100-eV electron whose position is uncertain by no more than 1.0×10^{-10} m, about the size of an atom. From $\Delta p_x = \hbar/2 \, \Delta x$, we compute $\Delta p_x = 5.3 \times 10^{-25}$ kg·m/s. Now compare this uncertainty with the particle's momentum. The fractional uncertainty is $\Delta p_x / p_x =$ about 10 percent. From these numbers, we see that it is impossible to specify the momentum of an electron confined to atomic dimensions with even moderate precision.

Consider now the uncertainty involved when a 10.0-gm body moves at a speed of 10.0 cm/s; that is, an ordinary-sized object is moving at an ordinary speed. Suppose further that the position of the object is uncertain by no more than 1.0×10^{-3} mm. We are interested again in the fractional uncertainty in momentum. We find $\Delta p_x = 5.3 \times 10^{-29}$ kg·m/s and $p = 1.0 \times 10^{-3}$ kg·m/s; therefore, $\Delta p_x / p_x = 5.3 \times 10^{-26}$! The fractional uncertainty in the momentum of a macroscopic body is so extraordinarily small as to be negligible compared with all possible experimental limitations. The uncertainty principle imposes an important limitation on the certainty of measurements only in the microscopic domain. In the macroscopic domain, the uncertainties are essentially trivial.

Figure 41-15(a) shows the momentum p_x of the electron in our example above,

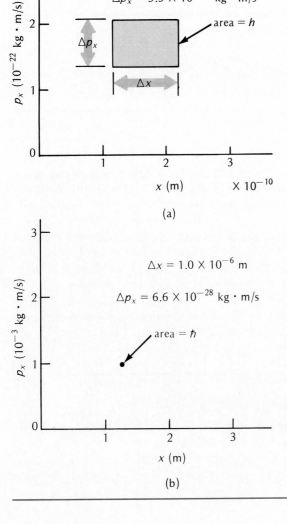

(a)

(b)

Figure 41-15. (a) Uncertainties in position and momentum for an electron confined to atomic dimensions. (b) Plotted to the same scale, the uncertainties in position and momentum for a 10-gm particle. (On the scale of the drawing, the small dot representing $\hbar$ is too big by a factor of about 10^{26}.)

plotted against its position x. The uncertainty principle requires that the shaded area in this figure, the product of the uncertainties in the momentum and the position, be equal in magnitude to $\hbar/2$. If the position is known with high precision, the momentum is rendered highly uncertain; if the momentum is specified with high certainty, the position must necessarily be highly indefinite. It is therefore impossible to predict and follow in detail the future path of an electron confined to essentially atomic dimensions. Newton's laws of motion, which are completely satisfactory for giving the paths of large-scale particles, cannot be applied here. To predict the future course of any particle, one must know not only the forces that act on the particle but also its initial position and momentum. Because *both* position and momentum cannot be known simultaneously without uncertainty, it is not possible to predict the future path of the particle in detail. Instead, wave mechanics must be used to find the probability of locating the particle at any future time.

Now consider again the 10.0-gm body moving at 10.0 cm/s. Figure 41-15(b) shows its momentum and position. Area $\hbar/2$, representing the product of the uncertainties in momentum and position, is so extraordinarily tiny in these macroscopic circumstances that it appears as an infinitesimal point of the figure. Here the classical laws of mechanics may be applied without entailing appreciable uncertainty.

The finite size of Planck's constant is responsible for quantum effects. Quantum effects are subtle because Planck's constant is very small—but not zero. Recall that the relativity effects are subtle because the speed of light is very large—but not infinite. If somehow Planck's constant were zero, the quantum effects would disappear. Thus classical physics may be thought of as the limit of quantum physics as h is imagined to approach zero. Symbolically,

$$\mathop{\mathrm{Limit}}_{h \to 0} (\text{quantum physics}) = \text{classical physics}$$

41-10   The Quantum Description of a Confined Particle

A completely free particle moves in a straight line; it has constant momentum. In wave mechanics, a particle with constant, well-defined momentum must be represented by a monochromatic sinusoidal wave.

Now suppose that such a particle is confined between two infinitely high, hard walls. The particle moves freely back and forth along the x axis, but encounters an infinitely hard wall at $x = 0$ and another at $x = L$; it is, then, confined between these boundaries. The infinitely hard walls correspond to an infinite potential energy V for all values of x less than zero and greater than L. The particle is free between zero and L; therefore its potential energy V in this region is constant. For convenience, we choose the constant potential energy to be zero. The situation we have described is that of a *particle in a one-dimensional box*, or a particle in an infinitely deep potential well. Because the walls are infinitely hard, the particle imparts none of its kinetic energy to them, its total energy remains constant, and it continues to bounce back and forth between the walls unabated.

From the point of view of wave mechanics, we can say that if the particle is confined within the limits stated, then the probability of finding it outside these limits is zero. Therefore, the wave function ψ, whose square represents this probability, must be zero for $x < 0$ and $x > L$. Only those wave functions that satisfy the boundary conditions are allowed. Since the particle has constant momentum magnitude, it is represented by a sinusoidal wave. To satisfy the boundary conditions, only those wavelengths are allowed that permit an integral number of half-wavelengths to be fitted between $x = 0$ and $x = L$. The condition for the existence of stationary, or standing, waves is then

$$L = n \frac{\lambda}{2} \tag{41-15}$$

where λ is the wavelength and n is the *quantum number* having the possible values 1, 2, 3, ...

Figure 41-16 shows the potential well, the wave function ψ, and the probability ψ^2 plotted against x for the first three possible stationary states of the particle in the box. Note that whereas ψ can be negative as well as positive, ψ^2 is always positive. Note also that ψ^2 is always zero at the boundaries. For the first or ground state, $n = 1$, the most probable location of the particle is the point midway between the two walls, at $x = L/2$; for the second state, $n = 2$, however, the least probable location is this point, where, in fact, $\psi = 0$, which is to say that it is impossible for the particle to be located there!

Basically, the boundary conditions on ψ, the fitting of the waves between the walls, restricted the wavelength of the particle to the values given by (41-15). Now if only certain wavelengths are permitted, the magnitude of the momentum also is restricted to certain values, since $p = h/\lambda$. Therefore, the permitted momenta are given by

$$p = \frac{h}{\lambda} = \frac{nh}{2L}$$

Finally, the kinetic energy E_k (and therefore the total energy E of the particle, since the potential energy is zero) is given by

$$E_k = E = \frac{1}{2} mv^2 = \frac{p^2}{2m} = \frac{(nh/2L)^2}{2m}$$

$$E_n = n^2 \frac{h^2}{8mL^2} \tag{41-16}$$

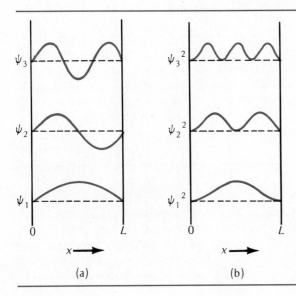

Figure 41-16.   *The first three stationary states for a particle in a one-dimensional box with infinitely high sidewalls: (a) Wave functions. (b) Probability distributions.*

where m is the particle's mass (for nonrelativistic speeds). The subscript n signifies that the possible values of the energy depend only on the quantum number n for fixed values of m and L. The *energy* of the particle in the one-dimensional box is *quantized*. The particle cannot assume just any energy but only those particular energies that satisfy the boundary conditions placed on the wave function.

What are the possible values of the energy if an electron with $m = 9.1 \times 10^{-31}$ kg is constrained to move back and forth within $L = 4 \times 10^{-10}$ m? Setting these values in (41-16) gives for the energy of the first state, $n = 1$, the value $E_1 = 2.3$ eV. Because $E_n = n^2 E_1$, the next possible energies of the particle are $4E_1, 9E_1, 16E_1, \ldots$. The permitted energies of the electron in the atom-sized box are shown in Figure 41-17(a), called an *energy-level diagram*. An electron that is confined to atomic dimensions has possible energies in the range of a few electron volts.

Now consider the allowed energies of a relatively large object confined in a relatively large box. Let us take $m = 9.1$ mg $= 9.1 \times 10^{-6}$ kg, and $L = 4$ cm $= 4 \times 10^{-2}$ m. Equation (41-16) shows that for these values $E_1 = 2.3 \times 10^{-41}$ eV, a fantastically small amount of energy! Figure 41-17(b) is the energy-level diagram for these circumstances [with the energy plotted to the same scale as in Figure 41-17(a)]. The spacing between adjacent energies is so very small that energy is effectively continuous. That is why we never see any obvious manifestation of the quantization of the energy of a macroscopic particle; the quantization is there, but it is too fine to be discerned. This result agrees with the classical requirement that the actually discrete energies of a bound system appear continuous in large-scale phenomena.

The lowest possible energy of a particle in an infinitely deep box is not zero, but E_1. This is in accord with the uncertainty principle. If the particle's energy were zero, with the particle at rest somewhere within the box ($\Delta x = L$), both the momentum p_x and the uncertainty in the momentum Δp_x would be zero. This would violate the uncertainty relation, since the product $\Delta p_x \, \Delta x$ would then be zero.

For an electron confined within atomic dimensions, the energy of the *ground state* is a few electron volts. The electron is never at rest but bounces

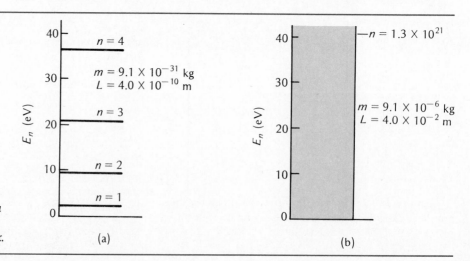

Figure 41-17. *Allowed energies of a particle in a one-dimensional box: (a) Electron in a box of atomic dimensions. (b) A 9.1-mg particle in a 4-cm box.*

back and forth between the confining walls with its lowest possible energy, the so-called *zero-point energy*. This is, of course, true of any confined particle, for example, the one shown in Figure 41-17(b). The minimum speed of the 9.1-mg particle restricted to 4 cm is easily computed to be 9.0×10^{-28} m/s, or a mere 10^{-8} nm per millennium. The particle is effectively at rest.

The problem of the particle in the box is somewhat artificial, but it is important because it reveals the quantization of the energy. Energy quantization occurs basically because only certain discrete values of the wavelength can be fitted between the boundaries.

Summary

Definitions

Photon: A particlelike quantum of electromagnetic radiation.

Work function: The energy required to release an electron from a metallic surface.

Bremsstrahlung: A process whereby an accelerated charged particle radiates photons.

Compton wavelength: For a particle of rest mass m_0, equal to $h/m_0 c$.

Positron: Electron antiparticle with electric charge $+e$.

deBroglie wavelength: Wavelength $\lambda = h/p$ associated with a particle with momentum p.

Wave function ψ: Contains all information concerning a particle or system; ψ^2 is proportional to the probability of observing the particle.

Energy-level diagram: Horizontal lines representing the allowed energies of a quantized system.

Fundamental Principles

Particle properties of electromagnetic waves: Electromagnetic radiation of frequency v and wavelength λ consists of particlelike photons, where a photon has the properties:

Energy: $E = hv = hc/\lambda$ (41-1)

Speed: $v = c$

Rest Energy: $E_0 = 0$

Momentum: $p = E/c = hv/c = h/\lambda$ (41-6)

where Planck's quantum constant is $h \approx 6.63 \times 10^{-34}$ J·s.

Wave properties of particles: According to the quantum theory every particle with momentum p has an associated wavelength λ, where

$$\lambda = h/p = h/mv$$ (41-11)

Uncertainty principle: The uncertainty Δx in the location of a particle along x and the uncertainty Δp_x in the component of the particle's momentum along x are related in quantum theory by

$$\Delta p_x \, \Delta x \geq \hbar/2$$ (41-13)

where $\hbar \equiv h/2\pi$. A similar relation governs uncertainties in energy ΔE and time interval Δt:

$$\Delta E \, \Delta t \geq \hbar/2$$ (41-14)

Important Results

Photoelectric effect: A photon of energy hv dislodges a particle bound with energy ϕ (the work function, $\phi = hv_0$, where v_0 is the threshold frequency) and the particle leaves with maximum kinetic energy E_k.

$$hv = hv_0 + E_k$$ (41-4)

X-ray production, Bremsstrahlung (breaking radiation): A particle emits one or more photons as it is accelerated. If a particle loses all of its kinetic energy E_k in a single collision, the maximum frequency v_{max} and minimum wavelength λ_{min} of the photon are given by

$$E_k = hv_{max} = hc/\lambda_{min}$$ (41-5)

Compton effect: The quantum theory of the scattering of electromagnetic radiation in which a photon collides with a free particle at rest; after the collision the particle acquires kinetic energy and the scattered photon travels at angle θ relative to the direction of the incident photon. The increase in wavelength $\Delta\lambda$ of the scattered photon over the incident photon is related to the scattering angle θ and the rest mass of the particle by

$$\Delta\lambda = (h/m_0 c)(1 - \cos\theta)$$ (41-9)

Pair production: A photon is annihilated and a particle-antiparticle pair (for example, an electron-positron pair) is created. The photon energy for creating particles each with a rest energy $m_0 c^2$ is

$$hv_{min} = 2m_0 c^2$$ (41-10)

Pair annihilation: A particle-antiparticle pair is annihilated and (typically) two photons of equal energy traveling outward in opposite directions are created.

Probability interpretation of the wave function: If the wave function ψ of a particle has the value $\psi(x)$ at location x, the probability of observing the particle between x and $x + dx$ is proportional to $\psi^2 dx$.

Principle of complementarity: The wave and particle descriptions in quantum situations, while they are mutually contradictory, are complementary. If one description is chosen by the experimental arrangements, the other description is suppressed.

Particle in a (one-dimensional) box: The quantization of energies for a system arises basically from the fitting of wave functions between the boundaries. For infinitely high walls at the boundaries of a constant potential, the allowed wave functions are sinusoidal with zeros at the boundaries. The allowed energies of the system with a one-dimensional box of length L are

$$E_n = n^2 h^2 / 8mL^2 \qquad \text{(41-16)}$$

Problems and Questions

Section 41-2   Photoelectric Effect

41-1 Q The photoelectric effect was crucial in the development of modern physics because it demonstrated the
(A) wave nature of the electron.
(B) localization of most of the mass of an atom in the nucleus.
(C) wave nature of light.
(D) particle nature of light.
(E) validity of the special theory of relativity.

41-2 P The visible spectrum covers a range of wavelengths from approximately 450 nm to 680 nm. What are the energies, in electron volts, of photons at the extremes of this range?

· **41-3 P** A 50-kW radio transmitter operates at a frequency of 1.0 MHz. How many photons does it emit each second?

41-4 P What is the number of photons emitted per second in the beam of a 1.0-mW laser with a wavelength of 562 nm.

· **41-5 P** Sodium has a work function of 2.28 eV. What is the longest wavelength of electromagnetic radiation that can release photoelectrons from sodium?

· **41-6 P** Light falls on a metal surface and 2.0×10^{13} photoelectrons are released each second. Suppose that all these electrons are collected at another metal electrode. What is the photocurrent?

: **41-7 P** A radio station transmits 50,000 W with a carrier frequency of 980 kHz. (a) How many photons per second are emitted? (b) A sensitive receiver detects a signal of 6 μW from this station. How many photons per second is this?

: **41-8 P** In a photoelectric experiment the following values for the stopping potential for photoelectrons were found as a function of the wavelength of the incident light:

Wavelength (nm):	250	284	357	382	429	547
Cut-off voltage (V):	3.22	2.45	1.72	1.30	0.95	0.45

(a) Plot the data and determine the corresponding value of h, Planck's constant. (b) Determine the work function of the metal.

: **41-9 P** The work function for cesium, 1.81 eV, is much lower than the work function for other metals. (a) What is the photoelectric cut-off wavelength? (b) What is the maximum electron kinetic energy when cesium is illuminated with light of 512-nm wavelength? (c) What is the speed of such an electron?

: **41-10 Q** Photons of frequency v and wavelength λ release photoelectrons with kinetic energy E_k from a metal. Which statement is not correct?
(A) There is a minimum value of λ below which the electrons will not be emitted.
(B) The number of photoelectrons emitted is proportional to the number of incident photons.
(C) The maximum velocity of the photoelectrons is different for different metals.
(D) For different frequencies, E and v are linearly dependent.
(E) The kinetic energy of a photoelectron will be less than the energy of an incident photon.

: **41-11 P** A mercury arc lamp emits 0.12 W of UV radiation with wavelength 253.7 nm. The arc acts like a point source, and some of the light is incident on a potassium photocathode 1 m from the lamp. The cathode has an effective area of 3.0 cm^2. The work function for potassium is 2.22 eV.

(a) Experimental measurements show that a photocurrent is emitted within less than 10^{-12} s of the time the potassium is illuminated. Suppose, however, that light did not consist of photons. Assume that a potassium atom is a small sphere of radius 0.2 nm that absorbs light. Determine how long it would take for an atom to collect enough energy to emit an electron. Is the result close to the experimental value? (b) What is the energy, in electron volts, of the photons from the lamp? (c) How many photons hit the photocathode each second? (d) What is the saturation current that would be emitted if the photoconversion efficiency (the probability that a given photon will eject an electron) is 4 percent? (e) What is the stopping potential required to prevent any flow of current?

Section 41-3   X-Ray Production and Bremsstrahlung

41-12 P A helium nucleus (an alpha particle) with charge $+2e$ is accelerated from rest through a potential difference of 10 kV. What would be the maximum energy of a photon created when the helium collides with a massive target?

: **41-13 P** The conservation laws of energy and of momentum must hold for every quantum effect. Use this requirement to show that it is impossible for a moving, unbound, single charged particle to slow down and emit a photon. (Hint: View the system in a reference frame in which the particle is initially at rest.)

Section 41-4   Compton Effect

· **41-14 P** An Air Force laser weapon produces a 3-MW pulse of light with a pulse duration of 1 μs. What is the total linear momentum of the pulse of light?

: **41-15 P** A beryllium-8 nucleus in an excited state ($^8_4$Be*) is initially at rest. It then decays to its ground state ($^8_4$Be) with the emission of a 17.6-MeV gamma ray. (a) What is the momentum of the gamma ray? (b) With what momentum does the nucleus (mass ≈ 8 u) recoil?

· **41-16 P** A 12-keV photon collides with a stationary free electron. The scattered photon has a momentum of 10 keV/c. What is the final kinetic energy of the electron?

· **41-17 Q** A photon collides with a stationary electron. After the collision, a scattered photon travels at θ with the direction of motion of the incident photon, and the electron is set in motion. The speed of the scattered photon, relative to an observer at rest with respect to the moving electron, is
(A) c
(D) $(1 + \cos \theta)c$
(B) $(1 - \cos \theta)h/m_0c$
(E) $0.5c$
(C) $(1 - \cos \theta)c$

41-18 Q Which of the following statements concerning the Compton effect is incorrect?
(A) The wavelength of the scattered photon is equal to or larger than the wavelength of the incident photon.
(B) The electron can be given an energy equal to the energy of the incident photon.
(C) The energy of the incident photon equals the kinetic energy of the electron plus the energy of the scattered photon.
(D) The energy the electron acquires is largest when the incident and scattered photons move in opposite directions.
(E) Both energy and momentum are conserved in the process.

: **41-19 P** Assuming a dust particle to have a specific density of 3, determine the minimum size of dust particle that will not be pushed out of the solar system by radiation pressure from the sun. Just outside the earth's atmosphere, the intensity of electromagnetic radiation from the sun is 1.4 kW/m^2.

: **41-20 P** Cesium-137 emits a gamma ray of energy 662 keV. (a) This gamma ray is Compton-scattered through 90°. What is the energy of the scattered photon? (b) What is the minimum energy of the scattered photon, all possible scattering angles considered?

: **41-21 P** In a Compton-scattering experiment, photons with wavelength λ are scattered at 90°. The scattered photons are found to have a wavelength 1.5λ. What is the wavelength of the incident photons?

: **41-22 P** A 4.2-MeV photon is backscattered ($\theta = 180°$) by an electron. (a) What is the wavelength of the scattered photon? (b) What is the kinetic energy of the electron?

: **41-23 P** In the scattering of x-rays from a crystal of NaCl, it is assumed that the scattered x-rays undergo no change in wavelength. Show that this is a reasonable assumption by calculating the order of magnitude of the Compton wavelength for a sodium atom and for a chlorine atom and comparing it with a typical x-ray wavelength of 0.10 nm.

: **41-24 P** What is the wavelength of a photon that can impart a kinetic energy of up to 50 keV to an electron in Compton scattering?

: **41-25 P** In a Compton collision of a photon with an electron, the scattered photon can create an electron-positron pair if it is sufficiently energetic. Show that, no matter how energetic the incident photon, no photon scattered by more than 60° can create an electron-positron pair.

: **41-26 P** The Compton effect occurs for protons as well as for electrons. (a) What is the value of the Compton wavelength for a proton? (b) A 1.0-GeV photon collides with a single proton at rest and the proton recoils in the forward direction. What is the energy of the scattered photon?

Section 41-5   Pair Production and Annihilation

· **41-27 Q** A beam of 0.8-MeV photons is incident on a thin slab of material. Fewer photons emerge from the slab, in the same direction as the incident photons, than enter the slab. A process that can account for this is
(A) Bremsstrahlung.
(B) the Compton effect.
(C) x-ray production.
(D) pair production.
(E) pair annihilation.

· **41-28 Q** Electromagnetic radiation must be assumed to consist of particlelike photons in all of the processes except
(A) x-ray diffraction.
(B) Compton scattering.
(C) the photoelectric effect.
(D) pair production.
(E) pair annihilation.

: **41-29 P** A positron and an electron, each with a kinetic energy of 1.0 MeV, collide head-on and annihilate each other. What is the wavelength of the resulting photons?

: **41-30 P** A 2.0-MeV photon creates an electron-positron pair. If the resulting electron has a kinetic energy of 0.25 MeV, what is the kinetic energy of the positron? (Note that a third heavy particle was present; it acquired some momentum, but not much energy, in this process.)

: **41-31 P** An electron collides head-on with a positron, and two photons are created. The electron and positron each has a speed of 0.8c. What is the energy of one of the photons?

: **41-32 Q** A high-energy photon creates an electron-positron pair when it comes close to a very massive nucleus. The electron and the positron have equal kinetic energies, each equal to the rest energy of an electron. Then the electron emits a single photon when it comes to rest in a collision. If the initial photon creating the electron-positron pair had a wavelength λ_1, the wavelength of the photon produced in the electron collision is

(A) λ_1 (D) $\frac{1}{4}\lambda_1$
(B) $2\lambda_1$ (E) $\frac{1}{2}\lambda_1$
(C) $4\lambda_1$

: **41-33 P** By applying the conservation principles of energy and of linear momentum, show that a single free particle cannot absorb a single photon.

: **41-34 P** An electron and positron at rest mutually annihilate and produce *three* photons. What are (a) the minimum and (b) the maximum photon energies?

: **41-35 P** A positron with kinetic energy of 2.0 MeV collides with an electron at rest. The electron and positron are annihilated and two photons with equal energies are created. What is the angle between the photon momenta?

Section 41-6 Matter Waves

41-36 Q The so-called duality of light refers to the fact that light
(A) is characterized by either a frequency or a wavelength.
(B) may be considered as either electric or magnetic, depending on how it is detected.
(C) behaves in some ways like a wave motion and in some ways like particles.
(D) can create an electron-positron pair, or an electron and a positron can combine to create a photon.
(E) has both energy and linear momentum.

· **41-37 Q** Which of the following is not a feature that photons have in common with electrons?
(A) Ability to transport energy.
(B) Momentum.
(C) Electric charge.
(D) An associated wavelength.
(E) Diffraction.

· **41-38 Q** A photon, an electron, and a helium atom all have the same momentum. The particle(s) with the largest de Broglie wavelength is (are)

$$\lambda = \frac{h}{m\nu} \qquad m\nu = c$$

(A) the photon.
(B) the electron.
(C) the helium atom.
(D) the electron and the helium atom.
(E) All three have the same wavelength.

· **41-39 P** A golf ball with a mass of 49 gm can be given a velocity of 80 m/s with a good drive. What is the de Broglie wavelength of such a ball?

: **41-40 P** Show that if the kinetic energy of a particle is much greater than its rest energy, it has nearly the same de Broglie wavelength as a photon of the same total energy.

: **41-41 P** A photon and a particle have the same wavelengths. How do the following properties of the two compare? (a) their momenta? (b) the particle's total energy and the photon's energy? (c) the particle's kinetic energy and the photon's energy?

: **41-42 P** What is the energy and wavelength of a photon that has the same momentum as a (a) 1.0-MeV electron? (b) 1.0-MeV proton?

: **41-43 P** What is the wavelength of an electron that is accelerated from rest through a potential difference of 25 kV in a color TV set?

: **41-44 P** What is the de Broglie wavelength of a 5-MeV electron? Note that for such an energetic particle Newtonian mechanics does not apply.

: **41-45 P** Show that the de Broglie wavelength of a particle of rest mass m_0 and kinetic energy K can be written

$$\lambda = \frac{hc}{\sqrt{K(K + 2m_0c^2)}}$$

: **41-46 P** In an electron microscope a beam of electrons replaces a beam of light, and electric and magnetic focusing fields replace refracting lenses. The resolving power of a microscope—the smallest distance that can be seen—is approximately equal to the wavelength used in the microscope. A typical electron microscope might use 80-keV electrons. (a) What is the minimum distance that can be resolved with a 80-keV electron microscope? (b) What is the energy of a photon that has the same wavelength as a 80-keV electron? (c) What is the momentum of such a photon? (d) What is the momentum of a 80-keV electron? (e) The photon and electron above have the same wavelengths and thereby would produce the same resolving power. Why is the electron microscope used rather than a photon microscope of the same wavelength?

Section 41-9 Uncertainty Principle

· **41-47 Q** Suppose that we lived in a universe where Planck's constant had the very large value of 1.0 J·s (as opposed to the value of 6×10^{-34} J·s in this universe). A pitching machine fires a stream of baseballs, each of rest mass 0.10 kg, with speed 10 m/s perpendicularly through a window 1 m wide. The balls then strike a distant wall. Under these circumstances,

(A) the kinetic energy of any one ball is appreciably more than 5 J.

(B) all the balls strike the wall within a 1-m-wide strip (that is, within the geometrical shadow of the window).

(C) a significant fraction of the balls will strike the wall outside the 1-m-wide shadow of the window.

(D) the motion of a ball is still well described by Newtonian mechanics.

(E) the momentum of a ball is appreciably larger than 1 kg·m/s.

·41-48 P An electron in an atom of hydrogen is confined to a region of space on the order of 0.1 nm. What is the order of magnitude of the minimum momentum of the electron?

41-49 P A nucleus has a size of the order of 5×10^{-15} m. What is the minimum kinetic energy, consistent with the uncertainty principle, of a proton confined to this region?

: 41-50 P An electron has a velocity of 300 m/s. If this value is accurate to ±0.01 percent, what is the lower limit to the accuracy with which one can locate the position of this electron? Under these circumstances, is it valid to regard the electron as a point object?

: 41-51 P Sodium emits a spectral line with wavelength 589 nm. Because of the uncertainty in the lifetime of the excited state from which the decay takes place, the spectral line has a half-width (that is, width at half-maximum) of 1.16×10^{-5} nm. This is the effective uncertainty in the wavelength. (a) What is the uncertainty in the energy of the photon? (b) What is the mean lifetime of the excited state from which the emitting electron decayed? (c) What is the "size" of the emitted photon (that is, the length of the emitted wave train)?

: 41-52 P The light emitted from an ordinary light source (such as a mercury arc lamp) consists of numerous relatively short wave trains (each perhaps a meter in length). Interference in such a device as the Michelson interferometer can be observed only if the path differences involved do not exceed the length of this wave train (called the coherence length). We can use the uncertainty principle to estimate the coherence length. For the 546-nm line of mercury, the uncertainty in the wavelength is about 0.0005 nm. Estimate the corresponding coherence length.

: 41-53 P Starting with $\Delta x \, \Delta p \approx \hbar/2$, deduce the relation $\Delta E \, \Delta t \approx \hbar/2$ for a free particle with an energy $E = p^2/2m$. The uncertainty in time Δt is related to the uncertainty in position Δx by $\Delta t = \Delta x/v$, where $p = mv$.

: 41-54 P The gamma ray emitted when a Cs-137 nucleus decays has an energy of 662 keV with a line width of 53.0 keV. (a) What is the uncertainty in the wavelength of this photon? (b) What is the uncertainty in the time of emission

of this photon? (c) What is the "size" of this photon, that is, the length of the wave train associated with it?

: 41-55 P Show that if the wavelength of a photon is uncertain by $\Delta\lambda$, the corresponding uncertainties in energy and momentum are

(a) $\Delta E = -\dfrac{hc \, \Delta\lambda}{\lambda^2}$

(b) $\Delta p = -\dfrac{h \, \Delta\lambda}{\lambda^2}$

: 41-56 P The energy of a simple harmonic oscillator can be written

$$E = \frac{p^2}{2m} + \frac{1}{2} kx^2$$

where the natural angular frequency is $\omega = \sqrt{k/m}$. Use the uncertainty principle to estimate the minimum energy of such an oscillator.

Section 41-10   The Quantum Description of a Confined Particle

·41-57 Q A particle of mass m is trapped in a one-dimensional box whose walls are infinitely high and separated by a distance L. The particle's wave function is shown in Figure 41-18. Thus the particle's momentum is

(A) $L/3h$ (D) $3h/L$

(B) $2L/h$ (E) $3h/2L$

(C) $h/2L$

$\lambda = \dfrac{2}{3}L$

$p = \dfrac{h}{\lambda} = \dfrac{3h}{2L}$

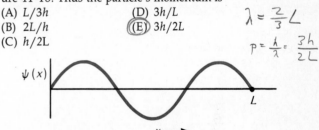

Figure 41-18.   Question 41-57.

41-58 Q A particle of mass m is contained in an infinite one-dimensional square well of width L. Sketched in Figure 41-19 is a plot of the probability of finding the particle at a given value of x when it is in a particular energy state of the system. What is the particle's kinetic energy in this state?

(A) $8mL^2/h^2$ (D) h^2/mL^2

(B) $3h^2/8mL^2$ (E) $9h^2/2mL^2$

(C) $9h^2/8mL^2$

Same ψ as 41-57

thus $\lambda = \dfrac{2}{3}L$

$M = 3$

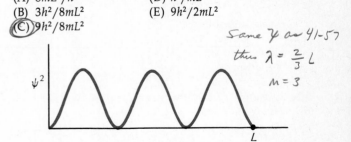

Figure 41-19.   Question 41-58.

: **41-59 Q** If the ground-state energy for a particular particle trapped in a one-dimensional box with infinitely high
side walls is 2.0 eV, what is the next-highest energy the
particle can have?

(A) 3 eV
(B) 4 eV
(C) 6 eV
(D) 8 eV
(E) This energy cannot be determined without knowing
     the width of the box and the mass of the particle.

41-60 P Consider an electron in a box of width 3×10^{-10}
m. What wavelength photon will induce a transition from
the ground state to the $n = 3$ state?

: **41-61 P** What are the first three energy levels (in electron
volts) for a neutron contained in a box of width 10^{-14} m?

: **41-62 P** Plot the first four energy levels for a particle in a
box as a function of the width of the box as the width is
varied from a value L to $4L$.

: **41-63 P** Consider a particle in a box in one dimension.
The width of the box is L. Determine approximately (without integrating) the probability that a particle in the ground
state is within a region of width $0.01L$ centered at (a) $x = 0$;
(b) $x = 0.25L$; (c) $x = 0.50L$; (d) $x = 0.75L$; (e) $x = L$.

: **41-64 P** A small pellet with a mass of 10^{-4} gm is bounc-
ing back and forth in a box of width 1 cm with a speed of 10
m/s. Considering this as a quantum-mechanical problem
of a particle in a box, estimate the value of n for this state of
motion.

Atomic Structure

42

42-1   Nuclear Scattering

An atom as a whole is ordinarily electrically neutral; if we were to remove all the electrons, what would remain would have all the positive electric charge and essentially all the mass. How are the mass and the positive charge distributed? From a variety of experiments, we know that an atom has a "size" (diameter) of the order of 0.1 nm. The positive charge and the mass are confined to at least this small a region, but it is impossible by any direct measurement to see and observe any details of the atomic structure. Indirect measurement must be resorted to. One of the most powerful methods of studying the distribution of matter or of electric charge — in fact, one of the few ways of studying matter of subatomic dimensions — is scattering. It was by the α-particle-scattering experiments of Rutherford that the existence of small, massive atomic nuclei was established.

Here is a simple example of a scattering experiment. We have a large black box; we can't look inside, but we are to determine how the mass is distributed within the box. At the two extremes, the box might be filled completely with some material of relatively low density, such as wood, or be only partly filled with some material of high density. To find out which is the actual distribution, we can use a very simple expedient: shoot bullets into the box and see what happens to them. If all bullets emerge in the forward direction with reduced speeds, then we might infer that the box is filled throughout with material that deflects the bullets only slightly as they pass through. Suppose that on the other hand, we find a few bullets deflected through a large angle from their original paths. Then we might conclude that they collided with small, hard, and massive objects dispersed throughout the box. Notice that we don't aim the bullets; the shots may be fired randomly over the front of the box. This is the essence of the particle-scattering experiments in atomic and nuclear physics.

Ernest Rutherford suggested in 1913 that the positive charge and the mass of an atom are a point charge and a point mass, which compose a *nucleus*. He suggested that his hypothesis be tested by shooting high-speed, positively charged particles (the bullets) through a thin, metallic foil (the black box), and then examining the distribution of the scattered particles. At the time of Rutherford, the only available suitable charged particles were α particles, with energies of several MeV from radioactive materials. An α particle is a doubly ionized helium atom, with a mass several thousand times larger than that of an electron, yet much smaller than the mass of such a heavy atom as gold.

Figure 42-1 shows the essentials of a scattering experiment. A collimated beam of particles strikes a thin foil of scattering material, and a detector counts the number of particles scattered at scattering angle θ. The experiment consists in measuring the relative number of scattered particles as a function of θ.

Now consider qualitatively the paths of the α particles traversing the interior of the scattering foil. Any encounter with an electron *is* inconsequential because the α-particle mass greatly exceeds the mass of an electron; the particle is essentially undeflected in such a collision, and a negligible fraction of its energy is transferred to any one electron. An α particle is appreciably deflected or scattered only by a close encounter with a nucleus. The mass of the nucleus of a gold atom is considerably greater than (50 times) the mass of the α particle; it remains essentially at rest. The α particles and nuclei, both positively charged, repel each other. The only force acting between a nucleus and an α particle, both regarded as point charges, is taken to be the Coulomb electrostatic force. This force varies inversely with the square of the distance; therefore, the force on an α particle, although never zero, is strong only when it is close to a nucleus.

Figure 42-2 shows several paths of α particles as they move through the

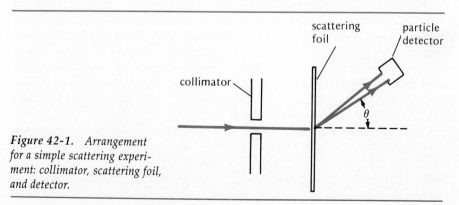

Figure 42-1. Arrangement for a simple scattering experiment: collimator, scattering foil, and detector.

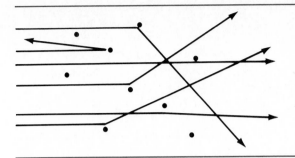

Figure 42-2.   Scattering of α particles by nuclei of a material. (The number of α particles scattered through sizable angles is greatly exaggerated.)

interior of a scattering foil. Most pass virtually undeviated; the chance for a close encounter with a scattering center is fairly remote. But those few particles that barely miss a head-on collision can be deflected at sizable angles. Those extremely rare examples of head-on collisions cause the particle's deflection through 180°; that is, the particle is brought to rest momentarily and then returned along its path of incidence. For point charges, most incident particles are scattered only slightly; but a small but significant number are deflected through large angles. (If the positive charge were distributed uniformly throughout the atom rather than being concentrated in nuclei, virtually no particles would be scattered through large angles.) Rutherford's nuclear hypothesis was confirmed in the experiments of Geiger and Marsden. They found that the measured distribution of the scattered α particles agreed with the distribution predicated on the assumption of scattering through a Coulomb force by point charges; the number of scattered particles was found to vary with scattering angle θ according to $1/\sin^4(\theta/2)$.*

Nuclear scattering reveals that when the constituents of nuclei—positively charged protons and electrically neutral neutrons—are separated from one another by less than ~ 1 fm $= 10^{-15}$ m, a strong, attractive *nuclear force* acts between the nuclear constituents. This force, sometimes referred to simply as *the strong interaction,* is substantially stronger than the repulsive Coulomb force between any pair of protons.

Example 42-1.   An α particle with an initial kinetic energy of 8.0 MeV is fired head-on at a gold nucleus (79 protons). At what distance from the nucleus is the α particle brought to rest?

The problem is solved most easily by applying energy conservation. The α particle loses kinetic energy K and the system acquires electric potential energy U until the α particle comes to rest momentarily at a distance r from the nucleus, where

$$K = U = \frac{k_e q_1 q_2}{r}$$

We have here used (25-8) for the electric potential energy for point charges $q_1 = 2e$ and $q_2 = 79e$ separated by r.

---

* From the standpoint of quantum theory, a beam of monoenergetic particles is, in effect, a beam of monochromatic waves (Section 41-6). The scattering process consists fundamentally of the diffraction of incident waves by scattering centers. It is remarkable that the wave-mechanical treatment of scattering for an inverse-square force yields precisely the same result as that yielded by a strictly classical analysis. For other types of forces, however, the classical and wave-mechanical results differ.

The relation above can be written as

$$r = \frac{k_e(2e)(79e)}{K}$$

$$= \frac{(9.0 \times 10^9 \text{ N} \cdot \text{m}^2/\text{C}^2)(2)(79)(1.60 \times 10^{-19} \text{ C})^2}{(8.0 \times 10^6 \text{ eV})(1.6 \times 10^{-19} \text{ J/eV})}$$

$$= 2.8 \times 10^{-14} \text{ m} = 28 \text{ fm*}$$

* The distance unit used here is the femtometer = fm = 10^{-15} m, also known as a fermi. The fermi unit, used especially for nuclear distances, honors the Italian-American physicist Enrico Fermi.

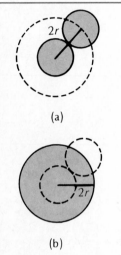

(a)

(b)

Figure 42-3.

Example 42-2.   One thousand spheres, each with a radius 1.0 mm, are dispersed at random and held within a Styrofoam cube 1.0 m on an edge. Then another 1000 identical spherical balls are fired at high speed at random over one face of the cube. How many of the incident spheres are expected to make collisions with spheres held within the block and thereby be scattered from their incident paths?

An incident sphere just misses colliding with a sphere at rest when their centers are separated by a distance of just slightly more than $2r$. See Figure 42-3. Each target sphere then presents an effective target area, or *cross section*, of $\sigma = \pi(2r)^2$ to what are effectively point masses in the incident beam. The target spheres are so widely dispersed that no one sphere is likely to "hide" another. The total target area presented to the incident beam is

$$\text{Target area} = N\sigma = N\pi(4r^2) = 10^3(\pi)4(10^{-3} \text{ m})^2 = 1.3 \times 10^{-2} \text{ m}^2$$

The fractional target area exposed for a 1-m^2 surface is then $1.3 \times 10^{-2} = 1.3$ percent. Therefore, about 1.3 percent of the 1000 incident spheres, or about 13, will be scattered from the forward beam.

42-2   The Hydrogen Spectrum

The simplest atomic system is hydrogen; it consists of only two particles — an electron and a much more massive proton as nucleus — interacting by an attractive Coulomb force. Theoretical description of atomic structure has concentrated first on hydrogen and on its observed behavior, primarily the electromagnetic radiation it emits and absorbs.

To observe the spectrum of isolated hydrogen atoms, one must use gaseous *atomic* hydrogen. In a hydrogen gas, the atoms are so far apart that each one behaves as an isolated system (molecular hydrogen H_2 and solid hydrogen radiate different spectra). One can use a prism spectrometer or a diffraction grating. The hydrogen gas may be excited by an electrical discharge or by extreme heating. The dispersed radiation, separated into its various frequency components, falls on a screen or a photographic plate that gives a record of the frequencies and intensities of the emission spectrum.

The spectrum emitted by atomic hydrogen consists of numerous sharp, discrete, bright lines on a black background. See Figure 42-4. In fact, the spectra of all chemical elements in monatomic gaseous form are composed of such bright lines. The spectrum is known as a *line spectrum*. The *emission spectrum* from atomic hydrogen, then, is a bright-line spectrum characteristic of hydrogen. Each chemical element has its own characteristic line spectrum, so that each spectrum is a characteristic "signature" of the particular element and spectroscopy is a particularly sensitive method of identifying the elements.

Atomic hydrogen at room temperature does not, by itself, emit appreciable electromagnetic radiation, but it can selectively absorb electromagnetic radiation, giving an *absorption spectrum.* The absorption spectrum is observed when a beam of white light (all frequencies present) is passed through atomic hydrogen gas and the spectrum of the transmitted light is examined in a spectrometer. The spectrum consists of a series of dark lines superimposed on the spectrum of white light; this is known as a *dark-line spectrum.* The gas is transparent to waves of all frequencies except those corresponding to the dark lines, for which it is opaque; that is, the atoms absorb only waves of certain discrete, sharp frequencies from the continuum of waves passing through the gas. The absorbed energy is very quickly radiated by the excited atoms, but in all directions, not just in the incident direction. The dark lines in the absorption spectrum of hydrogen are at precisely the same frequencies as the bright lines in the emission spectrum are. Hydrogen is a radiator of electromagnetic radiation only at specific frequencies; it is an absorber of radiation only at the same frequencies.

What holds for atomic hydrogen holds also for other elements—a characteristic set of emitted lines when the atoms radiate, and the same set of frequencies for absorption.

It is easy to see that the observed spectrum of hydrogen cannot be accounted for by classical mechanics and classical electromagnetism. Suppose the hydrogen atoms were merely a miniature solar system with the electron orbiting the nucleus like a planet around the sun. Then, with the electrically charged electron accelerated continuously, the atom would radiate continuously. The frequency of the radiation would be the frequency of the orbiting electron. But if the atom were to lose energy continuously by radiation, its energy would decrease continuously, so that the electron would orbit the nucleus in progressively smaller orbits at progressively higher frequencies. In other words, the atom would collapse rapidly, after having radiated a continuous spectrum.

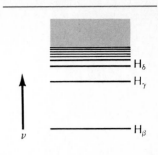

Figure 42-4.   Frequency distribution of radiation from atomic hydrogen in the visible region. This particular group of spectral lines is called the Balmer series.

42-3   Bohr Theory of Hydrogen

The first quantum theory of the hydrogen atom was developed in 1913 by Niels Bohr. The photon nature of electromagnetic radiation had been established at that time, but the wave aspects of material particles were not to be recognized until 1924. The Bohr model was only a first step toward a thoroughgoing wave-mechanical treatment of atomic structure. It retains some classical features but introduces some quantum features. Bohr's theory is therefore transitional between classical mechanics and the wave mechanics developed during the 1920s.

In the Bohr theory, we assume the proton to be at rest and the electron to orbit it in a circle. The force between the electron and the proton is the Coulomb force. An electron of mass m and charge e, is in an orbit of radius r about the nucleus, also of electric charge e.

We want the relation for the total energy—kinetic energy of orbiting electron, and electrostatic energy between electron and proton—of the hydrogen atom, considered a sort of junior solar system. We found (Section 15-6) that when a planet of mass m orbits a gravitational force center of mass M in an orbit of radius r, the total energy E is

$$E = -\frac{GmM}{2r} \tag{15-8}$$

To get the corresponding relation for the electric interaction, we merely replace GmM in the equation by $k_e e^2$, where k_e is the Coulomb force constant and e is the magnitude of the electron's and the proton's electric charge.

The total energy of the hydrogen atom then becomes

$$E = -\frac{k_e e^2}{2r} \tag{42-1}$$

The two-particle system has negative total energy with the two particles bound together. Let the electron radius become infinite so that the atom is dissociated into two separate particles; then from (42-1) we have $E = 0$. The ionization energy of hydrogen is 13.6 eV, so that the total energy of the hydrogen in its normal state must be $E = -13.6$ eV. Substituting this value in (42-1) yields $r = 0.53 \times 10^{-10}$ m, roughly the size of an atom.

The quantum condition used by Bohr to select the allowed atomic energies is this: the angular momentum of the electron as it orbits the nucleus is an integral multiple of Planck's constant h divided by 2π. A particle with linear momentum mv in a circular orbit of radius r has angular momentum mvr (Section 14-1). So the quantization rule is

$$mvr = n\frac{h}{2\pi} = n\hbar \tag{42-2}$$

where $n = 1, 2, 3, \ldots$ and $\hbar$ (read as "aitch bar") represents $h/2\pi$. Only those electron orbits are permitted, according to (42-2), for which the angular momentum is an integral multiple of $\hbar$.*

For a particle of mass m orbiting at speed v in a circle of radius r, Newton's second law requires

$$\Sigma F = ma$$

$$\frac{k_e e^2}{r^2} = m\frac{v^2}{r}$$

---

* Equation (42-2) can be cast in a different form, which leads to a simple interpretation of the allowed electron orbits. We have

$$n\frac{h}{mv} = 2\pi r$$

The quantity h/mv is the de Broglie wavelength λ of the electron, so that the equation above may be written as

$$n\lambda = 2\pi r$$

This relation implies that in going the distance $2\pi r$ around the circumference of the circular electron orbit, an integral number n of electron wavelengths may be fitted; that is, an allowed state is one in which an electron, regarded as a wave wrapped around in a self-completing circle, does not cancel itself out by destructive interference. This latter statement cannot be physically correct, however, because an electron does not orbit the nucleus in an atom as a particle; the electron cannot be regarded as existing only around a sharply defined circular orbit. The electron is a wave that extends in all three dimensions. Therefore, we must regard the relation above as a suggestive mnemonic, not as a rigorous application of wave mechanics.

Aha, That Did It!

They had worked long and hard setting up the experiment, and they were finally ready to take data. Then disaster struck. Almost as if by magic, the apparatus went absolutely haywire, the readings were nonsensical, the whole effort seemed a complete wreck. What did it?

A little later the experimenters learned that at just about the time that things had gone so terribly wrong in the laboratory, Wolfgang Pauli happened to be on a train that passed through their city. Here then was still another example of the dreaded "Pauli effect," a kind of malevolent action-at-a-distance.

Just how did Wolfgang Pauli (1900–1958)—noted of course for his formulation of the exclusion principle that, together with quantum theory, explains much of chemistry—get this reputation? He lived at just the time when relativity theory and quantum theory were being developed, and he himself contributed much to that development. He was invited when he was only 19 to write a comprehensive encyclopedia article on relativity theory. One reader said that it was "mature and grandly conceived" and that it had "sureness of mathematical deduction . . . deep physical insight . . . trustworthiness of the critical faculty." That reader was the founder, Albert Einstein.

Note Einstein's remark on "critical faculty." That was what could make mature, highly accomplished physicists tremble: Pauli's ruthless, searing criticism. He hated fuzzy thinking and half truth. One colleague, Paul Ehrenfest, referred to Pauli as "Scourge of God" (*Gottesgeissel*), using the same phrase as was typical for Atilla, the Hun. We must admit that even when it is rightly understood, quantum theory contains uncomfortable paradoxes; to put it more bluntly, quantum physics is a little crazy. At the time it was first emerging, quantum physics included plenty of half-baked ideas. Pauli was determined to rout them out—the ideas that were not merely crazy, but wrong. No wonder theorists stood in terror of Pauli; if there was a flaw in reasoning, Pauli was sure to spot it and say what was wrong with blunt abruptness.

Pauli could be hard on other people, but he was equally critical of his own work. He was prepared, when the occasion required it, to take the bold step. At one point physicists were toying with the idea that momentum, energy, and angular momentum conservation might not apply to radioactive decay. Pauli's proposal: a massless, chargeless, almost undetectable particle that could carry momentum, energy, and angular momentum, and save the conservation laws. That particle—the neutrino (its name, with an Italian diminutive given by Fermi)—plays a central role in elementary-particle physics and is now recognized to come, in fact, in six distinct varieties.

Like almost everyone else who takes physics seriously, Pauli could become frustrated. He wrote at age 25 to a colleague:

> Physics is very muddled again at the moment; it is much too hard for me anyway, and I wish I were a movie comedian or something like that and had never heard of anything about physics!*

Always the critic, always honest.

*See the article on Pauli in *American Journal of Physics* **43**, 205 (1975). It is a translation into English by Ira M. Freeman of the original article in German by Pascual Jordan.

Solving (42-2) for v and substituting in the equation above give

$$\frac{k_e e^2}{r^2} = m \frac{(n\hbar/mr)^2}{r}$$

$$r_n = n^2 \frac{\hbar^2}{k_e me^2}$$

The smallest allowed radius, the so-called *Bohr radius*, is given by

$$r_1 = \frac{\hbar^2}{k_e me^2} = 5.291\ 77 \times 10^{-11}\ \text{m} \qquad (42\text{-}3)$$

where the values of the known atomic constants have been substituted. The size of the hydrogen atom, ~ 0.05 nm, is in good agreement with experimental values. The allowed radii given above can be written in simpler form as

$$r_n = n^2 r_1 \qquad (42\text{-}4)$$

The radii of the stationary orbits are therefore $r_1,\, 4r_1,\, 9r_1,\, \ldots.$

The allowed values of the atom's energy now result from (42-4) substituted in (42-1):

$$E_n = -\frac{k_e e^2}{2r_n} = -\frac{1}{n^2}\frac{k_e e^2}{2r_1} \qquad (42\text{-}5)$$

We represent the quantity $k_e e^2/2r_1$ by E_I, the hydrogen atom's ionization energy, and the equation above becomes

$$E_n = -\frac{E_I}{n^2} \qquad (42\text{-}6)$$

The only possible energies of the bound electron-proton system that constitutes the hydrogen atom are $-E_I,\, -E_I/4,\, -E_I/9,\, \ldots.$ The permitted energies are discrete; the energy is quantized. The lowest energy (the most negative energy) is that in which the principal quantum number n equals 1; it is called the *ground state*. In the ground state, the energy is $E_1 = -E_I$, and its value from (42-5) and (42-3) is, as computed from fundamental constants,

$$E_1 = -E_I = -\frac{k_e e^2}{2r_1} = -\frac{k_e^2 e^4 m}{2\hbar^2} = -13.605\ 8\ \text{eV} \qquad (42\text{-}7)$$

Figure 42-5 is an *energy-level diagram* for hydrogen. For bound states, E_n is less than zero, and only discrete energies are allowed. As n approaches infinity, the energy difference between adjacent energy levels approaches zero. When n equals infinity, $E_n = 0$, and the hydrogen atom is then dissociated into an electron and a proton; the particles are separated by an infinite distance and both are at rest. In this condition, the atom is ionized, and the energy that must be added to it when it is in its lowest, or ground, state ($n = 1$) to bring its energy

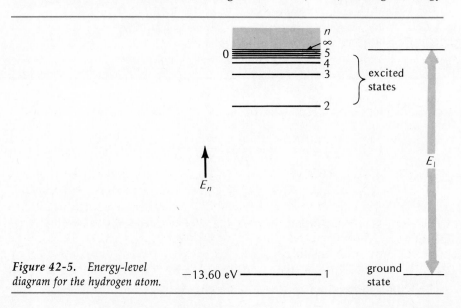

Figure 42-5. *Energy-level diagram for the hydrogen atom.*

up to $E_n = 0$ is just E_I, the ionization energy. When the system's total energy is positive, the electron and proton are unbound; then all possible energies are allowed, and there is a continuum of energy levels.

Each energy shown in Figure 42-5 corresponds to a stationary state in which the atom can exist without radiating. Stationary states above the ground state ($n = 2, 3, 4, \ldots$) are called *excited states*; an atom in one of them tends to make a transition to some lower stationary state. In a downward transition the electron may, very crudely, be imagined to jump suddenly from one orbit to a smaller orbit. It is better to say that the atom as a whole has made a quantum jump. The amount by which the energy of an atom exceeds the energy of the ground state is called the *excitation energy*. The term *binding energy* denotes the energy that must be added to an atom in an allowed state to free the bound particles and thereby make $E_n = 0$.

Now consider the photons emitted when atoms make downward transitions. An atom is initially in an upper, excited state with energy E_u; it makes a transition to a lower state E_1. In the transition, the atom loses energy $E_u - E_1$. Bohr assumed that in such a transition, a single photon having an energy $h\nu$ is created and emitted by the atom. By energy conservation,

$$h\nu = E_u - E_1 \qquad (42\text{-}8)$$

The Bohr theory, as well as more thoroughgoing wave-mechanical treatments, gives no details of the electron's quantum jump nor of the photon's creation.

Now it is easy to compute the wavelengths of the photons that, according to the Bohr model, are radiated by a hydrogen atom. Using (42-8) and (42-6), we have for the frequency

$$\nu = \frac{E_u - E_1}{h} = \left(-\frac{E_I}{n_u^2 h}\right) - \left(-\frac{E_I}{n_1^2 h}\right) = \frac{E_I}{h}\left(\frac{1}{n_1^2} - \frac{1}{n_u^2}\right) \qquad (42\text{-}9)$$

where n_u and n_1 are the quantum numbers for the upper and lower energy states. The wavelengths $\lambda = c/\nu$ of emitted photons may then be expressed as

$$\frac{1}{\lambda} = \frac{E_I}{hc}\left(\frac{1}{n_1^2} - \frac{1}{n_u^2}\right)$$

This equation can be written more simply as

$$\frac{1}{\lambda} = R\left(\frac{1}{n_1^2} - \frac{1}{n_u^2}\right) \qquad (42\text{-}10)$$

where R is the *Rydberg constant*, given by

$$R = \frac{E_I}{hc} = \frac{k_e^2 e^4 m}{4\pi\hbar^3 c} = 1.097\ 37 \times 10^7\ \text{m}^{-1} \qquad (42\text{-}11)$$

Actually, (42-10), which gives the wavelengths in the hydrogen spectrum, was known as an empirical relation before Bohr's quantum theory, with the Rydberg constant chosen to fit the observed wavelengths. The fact that Bohr's quantum theory yielded precisely the value of R, and therefore the wavelengths of light emitted and absorbed by hydrogen atoms, is the most emphatic endorsement of Bohr's theory.

The spectral lines of hydrogen can easily be interpreted in terms of the energy-level diagram, Figure 42-6. Vertical lines represent transitions between

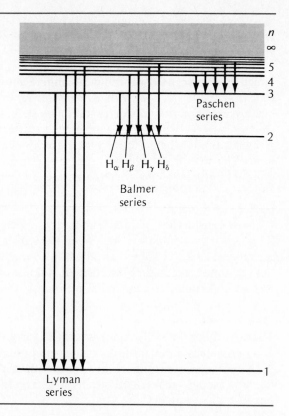

Figure 42-6. *Some possible transitions for atomic hydrogen.*

stationary states; the length of each line is proportional to the respective photon energy and therefore to the frequency. Spectral lines of the *Lyman series* correspond to those photons created when hydrogen atoms in any of the excited states undergo transitions to the ground state ($n_1 = 1$). Transitions from the unbound states ($E > 0$) to the ground state are responsible for the observed continuous spectrum lying beyond the series limit. In a similar way, the *Balmer series* is produced by downward transitions from excited states to the first excited state ($n_1 = 2$). This series of lines, identified by the labels H_α, H_β, H_γ, H_δ, lie in the visible region of the electromagnetic spectrum. Still further emission series involve downward transitions to $n_1 = 3$, $n_1 = 4$, . . . ; these series fall progressively toward longer wavelengths.

Suppose that the entire emission spectrum from an excited hydrogen gas is observed. We then see the simultaneous emission of many photons produced by downward transitions in many atoms from each of the excited states. To observe the entire emission spectrum, we must have very many hydrogen atoms from each of the excited states making downward transitions to all lower states.

Now consider the absorption spectrum. When white light passes through a gas, those particular photons that have energies equal to the energy difference between stationary states can be removed from the beam. These photons are annihilated, as they give their energy to the internal excitation energy of the atoms. The same set of quantized energy levels participates in both emission and absorption; and the frequencies of their emission and absorption lines are identical. (Because atoms remain in an excited state only very briefly, the Lyman series is the only one observed in absorption.)

We have used the fundamental postulates of the Bohr theory implicitly in developing a model of the hydrogen atom. It is useful, however, to isolate them, since they are retained in their essential forms in more complete wave-mechanical treatments of atomic structure:

- A bound atomic system can exist without radiating, but only in certain discrete stationary states.
- The stationary states are those in which the orbital angular momentum mvr of the atom is an integral multiple of $\hbar$.
- When an atom undergoes a transition from an upper energy state E_u to a lower energy state E_1, a photon of energy $h\nu$ is emitted, where $h\nu = E_u - E_1$. If a photon is absorbed, the atom makes a transition from a low energy state to a higher, according to the same relation.

Example 42-3. Some hydrogen atoms are initially in the second excited state. What are the possible energies of photons emitted by these atoms?

The possible transitions, $3 \to 2$, $3 \to 1$, and $2 \to 1$, are shown in Figure 42-7. The photon energies are, from (42-6):

$$3 \to 2: \quad h\nu_{32} = E_3 - E_2 = \left(-\frac{13.61 \text{ eV}}{3^2}\right) - \left(-\frac{13.61 \text{ eV}}{2^2}\right) = 1.89 \text{ eV}$$

$$3 \to 1: \quad h\nu_{31} = (13.61 \text{ eV})\left(1 - \frac{1}{3^2}\right) = 12.09 \text{ eV}$$

$$2 \to 1: \quad h\nu_{21} = (13.61 \text{ eV})\left(1 - \frac{1}{2^2}\right) = 10.21 \text{ eV}$$

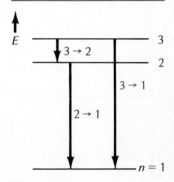

Figure 42-7.

Example 42-4. A hydrogen atom emits a photon corresponding to the Lyman alpha line in the hydrogen spectrum (first excited state to the ground state). (a) With what momentum does the atom recoil on emitting the photon? (b) What is the atom's recoil kinetic energy?

(a) As Example 42-3 shows, the Lyman-alpha transition, $2 \to 1$, produces a 10.21-eV photon. The photon's momentum is $p = E/c = 10.21$ eV/c. The atom, moving opposite to the photon, has the same momentum magnitude.

(b) The hydrogen atom, with a rest energy of $m_0 c^2 = 0.94$ GeV, recoils at relatively low speed with a kinetic energy of

$$K = \frac{1}{2}m_0 v^2 = \frac{p^2}{2m_0} = \frac{(pc)^2}{2m_0 c^2} = \frac{(10.21 \text{ eV})^2}{2(0.94 \times 10^9 \text{ eV})} = 5.5 \times 10^{-8} \text{ eV}$$

The atom's kinetic energy is a very small fraction, only about 5 parts in 10^9, of the photon's energy.

42-4 The Four Quantum Numbers for Atomic Structure

The Bohr quantum theory of hydrogen is only approximately correct. It yields the allowed energies of the hydrogen atom and the hydrogen spectrum fairly satisfactorily. But it does not give the correct quantum relationship for the allowed values of angular momentum. It cannot account for the closely spaced lines (or fine structure) observed in the spectrum. It cannot successfully account for the structure and spectra of atoms with more than one electron. Worst of all, it takes the electron to be a semiclassical particle moving in a

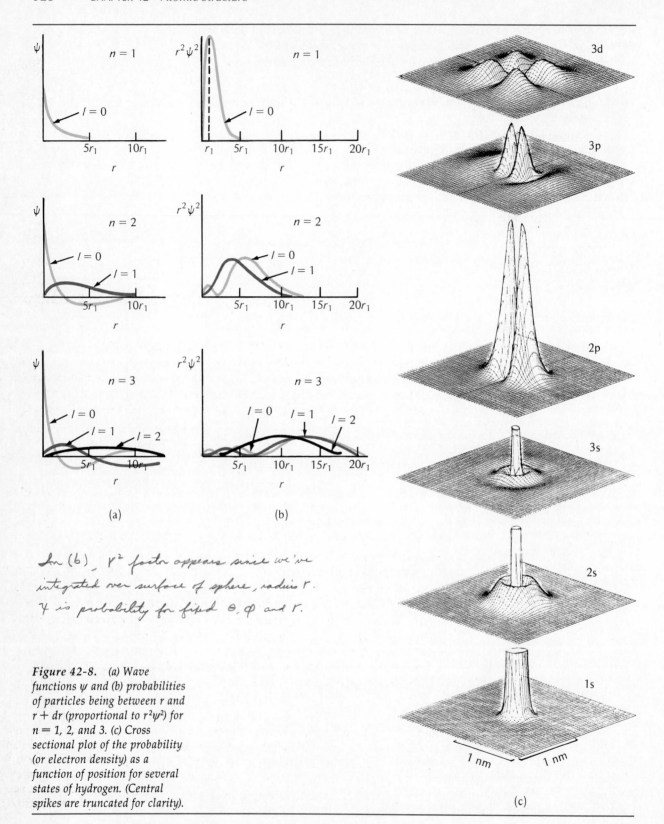

In (b), r^2 factor appears since we've integrated over surface of sphere, radius r. ψ is probability for fixed Θ, φ and r.

Figure 42-8. (a) Wave functions ψ and (b) probabilities of particles being between r and $r + dr$ (proportional to $r^2\psi^2$) for n = 1, 2, and 3. (c) Cross sectional plot of the probability (or electron density) as a function of position for several states of hydrogen. (Central spikes are truncated for clarity).

well-defined classical orbit, not as a wave extending over three-dimensional space.

A thoroughgoing wave-mechanical analysis of the hydrogen atom involves, as a minimum, finding the allowed wave functions and the corresponding energies by solving for the complete solutions of the Schrödinger wave equation for a particle (the electron) subject to an inverse-square attractive Coulomb force from a fixed force center (the nucleus). The analysis is mathematically sophisticated, and we shall not attempt it here. We shall indicate, in an admittedly approximate and qualitative fashion, the principal features that enter, particularly as they are related to the four quantum numbers characterizing atomic structure.

First recall that for the wave-mechanical problem of a particle confined to a one-dimensional potential well (Section 41-10), a single quantum number emerged. This number was, in effect, the number of half-wave segments that could be fitted between the boundaries of the potential well. In this sense, the problem was altogether analogous to that of a wave on a string attached at both ends. The permitted wave patterns, or allowed standing waves, are those for which an integral number of half-wavelengths can be fitted between the reflecting boundaries (Section 17-6).

For a stationary wave in two dimensions — for example, water waves on the surface of a swimming pool — there are two characteristic integers, or quantum numbers, describing the allowed wave patterns. The allowed wave patterns are, of course, those patterns of standing waves that are consistent with the boundary conditions at the edge of the region over which the waves may extend. For a rectangular swimming pool, two quantum numbers give the number of half-waves that can be fitted along the length and width, respectively, of the pool. For a circular pool, we again have two quantum numbers. One quantum number gives the number of half-waves that can be fitted going radially outward from the center, and the other relates to the number of zeros in the wave function as one goes around a circle.

For a three-dimensional potential well — for example, sound waves trapped inside a rectangular parallelepiped and reflecting from the three sets of parallel side walls — we have three characteristic quantum numbers. Each tells in effect the number of half-waves that can be fitted along each of the three dimensions. In general, the number of characteristic quantum numbers for a wave trapped in a potential well equals the number of dimensions of the well.

And so it is for the electron in a hydrogen atom. Here the particle is in a three-dimensional potential well. The potential energy function, $V = -k_e e^2/r$, depends on the electron's distance r from the nucleus. The electron wave function may extend to infinite distances, but it must be zero there. The potential energy depends on distance from the force center, and so too does the electron wavelength. In fact, as r increases and therefore V increases (becomes less negative), the electron wavelength must also increase. The electron wavelength increases as one goes away from the nucleus.

The simplest class of allowed hydrogen wave functions comprises those that depend only on the radial distance r, and are thereby spherically symmetrical. The first three such s wave functions are illustrated in Figure 42-8. Think of the electron in these states as a diffuse spherical ball centered on the nucleus (for $n = 1$) or as a set of concentric spherical shells (for $n > 1$). In classical terms, the electron wave may be thought of as expanding and contracting radially; the electron wave is reflected at both $r = 0$ and $r = \infty$. In general, however, the three-dimensional electron waves need not be spherically symmetrical; such waves require three quantum numbers to tell how the wave function changes with position.

We give below for each quantum number the symbol, the name, the allowed values, the influence of the atom's energy and angular momentum, and some geometrical characteristics of the associated wave functions for hydrogen. The fourth quantum number that enters naturally in the relativistic wave-mechanical treatment of four-dimensional space-time, the so-called spin quantum number, has no exact classical counterpart. Here results of wave mechanisms are given without proof.

Principal Quantum Number _n_   The principal quantum number n enters in the relation giving the approximate total energy of the atom according to the Bohr formula, $E_n = -E_I/n^2$, (42-6), where E_I is a constant. The allowed values for n are 1, 2, 3, ... As n increases, the energy increases and it approaches zero for infinite n. The wave functions extend toward progressively larger values of r as n increases, in the same way that the size of a classical planetary orbit increases with energy.

Orbital Angular-Momentum Quantum Number _l_   The magnitude of an electron's angular momentum is given by

$$L = \sqrt{l(l + 1)}\, \hbar \qquad (42\text{-}12)$$

(Orbital angular momentum classically is that of the orbiting particle around the force center.) Note that the angular-momentum quantization rule is different from $L = n\hbar$, the rule in the Bohr theory.

The orbital angular-momentum quantum number l may take on integral values starting with zero and continuing up to $n - 1$; that is, the l values are restricted to

$$l = 0, 1, 2, 3, 4, \ldots, n - 1$$
$$\text{s, p, d, f, g, } \ldots \qquad (42\text{-}13)$$

(The letter symbols that appear below the numerical values of l are also used to designate the electron state.) For an electron in the state with $n = 1$ (the ground state), the only possible value for l is $l = n - 1 = 1 - 1 = 0$. With $n = 2$, quantum number l may assume the values 0 and 1; the corresponding magnitudes of the orbital angular momentum are, from (42-12), equal to 0 and $\sqrt{2}\hbar$.

Any wave function for which $l = 0$ is spherically symmetrical. The electron has no angular momentum relative to the nucleus (and may be thought of, classically, as passing in an eccentric ellipse through the force center). When $l \neq 0$, the wave functions are not spherically symmetrical but depend on angle. The simplest of these is the p state with $l = 1$. Then the wave function is such that the probability for finding the electron is high at two "knobs" along a line passing through the nucleus, as shown in Figure 42-9. For higher values of l, the wave function shows a more complicated dependence on angle.

When an atom undergoes a transition from one allowed state to another, and a photon is emitted or absorbed, the state and wave functions are changed so that the values of l change by the integer 1, or Δl equals ± 1, for allowed transitions. The permitted energy levels, segregated according to the values of

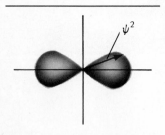

Figure 42-9.   *Polar plot of the variation of ψ^2 with angle for a p ($l = 1$) state. The pattern is symmetrical with respect to rotation about the symmetry axis.*

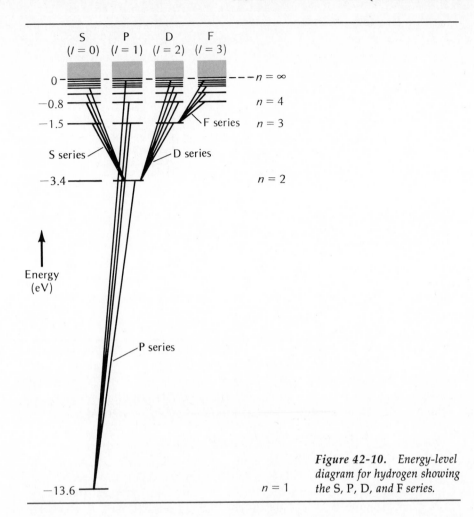

Figure 42-10.   Energy-level diagram for hydrogen showing the S, P, D, and F series.

l, together with the allowed transitions, are shown in Figure 42-10. In hydrogen, the several possible levels for a given n have the same total energy; in atoms with more than one electron, however, the S, P, D, F, . . . levels and the corresponding series of emitted lines differ in energy.

Orbital Magnetic Quantum Number m_l   The orbital magnetic quantum number may, for a given l, assume positive and negative integral values ranging from $+l$ to $-l$; that is,

$$m_l = l, l - 1, l - 2, \ldots, 0, \ldots, -l \qquad (42\text{-}14)$$

For example, for a p wave function, with $l = 1$, the allowed values of the orbital magnetic quantum number are 1, 0, and -1. For a d state ($l = 2$), the possibilities are $m_l = 2, 1, 0, -1,$ or -2.

Orbital angular-momentum number l gives, through (42-12), the *magnitude* of the electron's orbital angular momentum. Orbital magnetic quantum number m_l yields the *component* of the electron's orbital angular momentum L_z along some direction z in space. The angular-momentum component L_z is, in fact,

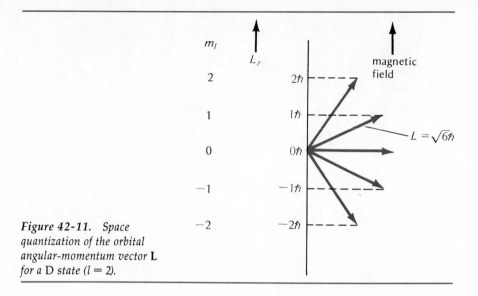

Figure 42-11. *Space quantization of the orbital angular-momentum vector* **L** *for a* D *state* (*l* = 2).

$$L_z = m_l \hbar \qquad (42\text{-}15)$$

For a d state ($l = 2$), for example, the angular-momentum magnitude is $L = \sqrt{l(l + 1)}\,\hbar = \sqrt{2(3)}\,\hbar = \sqrt{6}\,\hbar$, and the allowed projections of this angular-momentum vector along the z axis are $L_z = m_l \hbar = 2\hbar, \hbar, 0, -\hbar$, and $-2\hbar$. To give a more physical interpretation to m_l and the associated angular-momentum component, imagine the angular-momentum vector **L** to be oriented relative to the z in such directions that its components L_z satisfy (42-15). See Figure 42-11.

What dictates the direction of the z axis? An external magnetic field can do so. The phenomenon whereby the component of the angular momentum is restricted to certain discrete values, and therefore the vector **L** is restricted to certain orientations in space relative to z, is sometimes referred to as *space quantization*. The electron is a negatively charged particle, so that when an electron's wave function indicates nonzero angular momentum, the electron also has an associated magnetic moment. Space quantization implies, then, that the electron magnetic moment is restricted to certain discrete orientations in space. Further, since the orientation of a magnet relative to an external magnetic field controls the energy of the magnet, the energy of an atom (and the photons emitted in transitions between allowed states) can exhibit a multiplicity that is referred to as the *Zeeman effect*, after its discoverer.

Spin Magnetic Quantum Number m_s. The fourth quantum number specifying the state and wave function of an electron is the spin magnetic quantum number. It has just two values:

$$m_s = +\tfrac{1}{2} \text{ and } -\tfrac{1}{2} \qquad (42\text{-}16)$$

The idea of electron spin is this. Imagine the electron with its charge smeared over a region of space to be spinning perpetually about an internal axis of rotation. The electron has spin angular momentum that is in addition to and independent of the orbital angular momentum associated with quantum number l. Electron spin actually arises as a necessary consequence of a rela-

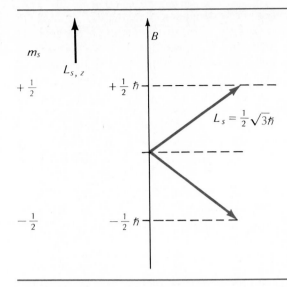

Figure 42-12. Space quantization of electron-spin angular momentum.

tivistic treatment of wave mechanics. The magnitude of the spin angular momentum has just one possible value:

$$L_s = \sqrt{s(s + 1)}\,\hbar = \sqrt{(\tfrac{1}{2})(\tfrac{3}{2})}\,\hbar = \tfrac{1}{2}\sqrt{3}\,\hbar \qquad (42\text{-}17)$$

since the spin quantum number for an electron has the *single* value $s = \tfrac{1}{2}$. Quantum number m_s is related to s as m_l is to l. The component of the spin angular momentum along some direction in space L_{sz} is given by

$$L_{sz} = m_s\hbar = \tfrac{1}{2}\hbar \qquad \text{or} \qquad -\tfrac{1}{2}\hbar \qquad (42\text{-}18)$$

The geometrical interpretation of L_{sz} is shown in Figure 42-12.

A perpetually spinning charged particle constitutes in effect an absolutely permanent magnet, so the energy of an atom differs according to the orientation of the spin vector $\mathbf{L}_s$ relative to a magnetic field.

42-5   Pauli Exclusion Principle and the Periodic Table

Specifying the state of an electron in an atom amounts, in the quantum theory, to specifying the values of each of the four quantum numbers n, l, m_l, and m_s. By the procedures of quantum mechanics, it is possible to compute an atom's energy, its angular momentum, and other of its measurable characteristics for each set of quantum numbers. Indeed, it is possible, at least in principle, to predict all properties of the chemical elements from quantum theory. To calculate chemical properties from quantum theory is difficult, however, because of formidable mathematical difficulties that arise with systems having many component particles. Only the problem of the simplest atom, hydrogen, has been solved completely by relativistic quantum theory. Essentially, experiment and theory agree perfectly.

Quantum theory does provide a wealth of information concerning chemical and physical properties. One of its greatest achievements has been to give a fundamental basis for the periodic table of chemical elements. The key is a principle proposed by W. Pauli in 1924, the *Pauli exclusion principle.* This principle together with the quantum theory predicts and accounts for many of the chemical and physical properties of atoms.

Consider again the energy levels available to the single electron in the hydrogen atom. These energy levels are shown schematically (but not to scale) in Figure 42-13. Here each horizontal line corresponds to a particular possible set of values for quantum numbers n, l, and m_l. For each line there are two possible values of the electron-spin quantum number, $m_s = \pm\frac{1}{2}$. The occupancy of an available state by an electron is indicated here by an arrow, whose direction indicates the electron-spin orientation, up for $m_s = +\frac{1}{2}$ and down for $m_s = -\frac{1}{2}$. For brevity only, the energy levels with principal quantum numbers 1, 2, and 3 are shown. For a given value of n, the states with $l = 0$ turn out to be lowest, states with $l = 1$ next, and so on. For a given value of the orbital angular-momentum quantum number l, the possible values of the orbital magnetic quantum number m_l are shown horizontally arranged. Every one of the states (two for each dash) is available to the electron in the hydrogen atom. Some are *degenerate;* they have the *same total energy,* but are nevertheless, distinguishable when a strong magnetic field or other external influence is applied to the atom.

The rules governing the possible values of the quantum numbers and the number of possible values can be summarized as follows:

For a given n: $l = 0, 1, 2, \ldots , n - 1$ (n possibilities)

For a given l: (42-19)
   $m_l = l, l - 1, 0, \ldots , -(l - 1), -l$ ($2l + 1$ possibilities)

For a given m_l: $m_s = +\frac{1}{2}, -\frac{1}{2}$ (2 possibilities)

Suppose a hydrogen atom is in its lowest, or ground, state. Then the single electron is in the state in which $n = 1$, $l = 0$, $m_l = 0$, and $m_s = -\frac{1}{2}$. If the

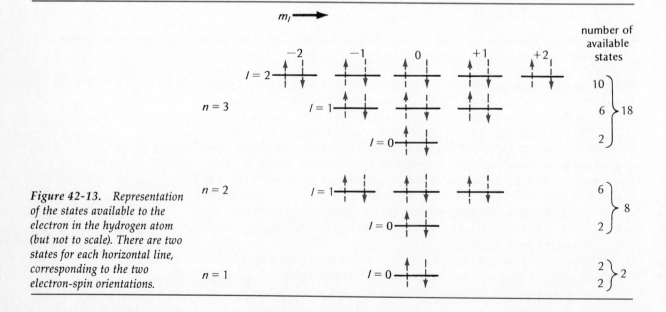

Figure 42-13. Representation of the states available to the electron in the hydrogen atom (but not to scale). There are two states for each horizontal line, corresponding to the two electron-spin orientations.

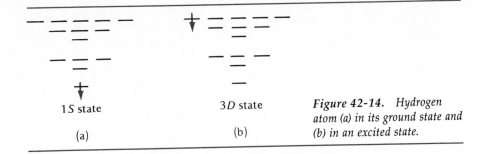

1S state   3D state

(a)   (b)

Figure 42-14.   Hydrogen atom (a) in its ground state and (b) in an excited state.

hydrogen atom gains energy, this may promote the electron to any one of the higher-lying available states. The atom can then decay to the ground state by downward transitions with the emission of one or more photons. Figure 42-14 depicts a hydrogen atom in its ground state and in an excited 3D state ($l = 2$).

Consider next the element lithium, $_3$Li. This atom has three electrons, to be placed in the levels shown in Figure 42-15. If all three electrons were in the lowest level, that with $n = 0$, $l = 0$, and $m_l = 0$, then two electrons would necessarily have to occupy the state with $m_s = +\frac{1}{2}$ and one would occupy the state with $m_s = -\frac{1}{2}$, or the reverse. This amounts to saying that at least two electrons would have the same set of quantum numbers. But all experimental evidence is that in a lithium atom all three electrons are never simultaneously in the state $n = 1$. The lowest-energy configuration, or ground state, for lithium is this; two electrons, one with $m_s = +\frac{1}{2}$, the other with $m_s = -\frac{1}{2}$, are in the $n = 1$ level, while the third electron occupies a state in the $n = 2$ level. See Figure 42-15. We can interpret this behavior as follows. Two of the three electrons in a lithium atom cannot have the same set of four quantum numbers; that is, two electrons cannot exist in the same state.

The evidence from all elements is the same; atoms simply never occur in nature with two electrons occupying the same state. The Pauli exclusion principle formalizes this experimental fact:

> *No two electrons in an atom can have the same set of quantum numbers n, l, m_l, and m_s; or no two electrons in an atom can exist in the same state.*

Exceptions to the exclusion principle, which applies also to systems other than atoms and to particles other than electrons, have never been found. The Pauli principle applies not only to electrons but also to other particles with half-integral spin; the states of particles with integral spin are not limited by it.

Thus, the two electrons in helium in the normal state occupy the two lowest available states indicated in Figure 42-13. No more electrons can be added to the $n = 1$ shell; in helium, $_2$He, the $n = 1$ shell is filled, or closed. The electron spins are then oppositely aligned, and the helium atom has no net angular momentum, either orbital or spin. Furthermore, the two electrons are tightly bound to the nucleus; much energy is required to excite one of them to a higher energy state. That is why helium is chemically inactive.

Suppose that the values of the quantum numbers of each and every electron in an atom are known. Then the electron configuration of the atom is said to be known. A simple procedure is used for specifying an electron configuration. We illustrate it with an example. When a helium atom is in its ground state, each of the two electrons has $n = 1$ and $l = 0$, and their configuration is

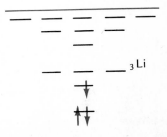

$_3$Li

Figure 42-15.   Electron configuration of lithium in the ground state.

represented by $1s^2$. The leading number specifies the n value, the lowercase letter s designates the orbital quantum number l of individual electrons, and the postsuperscript gives the number of electrons having the particular values of n and l.

The element lithium, with three electrons, has the electron configuration $1s^2 2s^1$, or two electrons in a completely filled $n = 1$ shell and the third in the $n = 2$ shell. Proceeding in this way—adding one electron as the nuclear charge or atomic number increases by one unit, but always with the restriction that no two electrons within the atom can have the same set of quantum numbers—we can confirm the ground-state configurations of the other atoms at the beginning of the periodic table shown in Table 42-1. We see from Figure 42-13 that two electrons can be accommodated in the s subshell of the $n = 2$ shell and six electrons in the p (or $l = 1$) subshell, after which the $n = 2$ is completely occupied and holds its full quota of eight electrons. With the electron configuration $1s^2 2s^2 2p^6$, corresponding to the rare gas element $_{10}$Ne, the electron wave functions are spherically symmetrical and the atom is chemically inert. In general, a filled subshell for any value of l, with electrons occupying states for all positive, zero, and negative values of m_l and m_s, the atom's net orbital and spin angular momentum is zero and the electron distribution is completely spherical. A closed subshell is effectively a spherical shell of charge.

Chemical properties reflect directly the electron configurations. For example, $_1$H, $_3$Li, $_{11}$Na, and $_{19}$K all have one s electron outside a closed subshell; these elements (the alkali metals) readily relinquish this last s electron to become positive ions, or they may contribute the electron in chemical combinations and thereby exhibit a valence of $+1$. On the other hand, the halogen

Table 42-1

ELEMENT	ELECTRON CONFIGURATION FOR THE GROUND STATE				
$_1$H	$1s^1$				
$_2$He	$1s^2$				
$_3$Li	$1s^2$	$2s^1$			
$_4$Be	$1s^2$	$2s^2$			
$_5$B	$1s^2$	$2s^2$	$2p^1$		
$_6$C	$1s^2$	$2s^2$	$2p^2$		
$_7$N	$1s^2$	$2s^2$	$2p^3$		
$_8$O	$1s^2$	$2s^2$	$2p^4$		
$_9$F	$1s^2$	$2s^2$	$2p^5$		
$_{10}$Ne	$1s^2$	$2s^2$	$2p^6$		
$_{11}$Na	$1s^2$	$2s^2$	$2p^6$	$3s^1$	
$_{12}$Mg	$1s^2$	$2s^2$	$2p^6$	$3s^2$	
$_{13}$Al	$1s^2$	$2s^2$	$2p^6$	$3s^2$	$3p^1$
$_{14}$Si	$1s^2$	$2s^2$	$2p^6$	$3s^2$	$3p^2$
$_{15}$P	$1s^2$	$2s^2$	$2p^6$	$3s^2$	$3p^3$
$_{16}$S	$1s^2$	$2s^2$	$2p^6$	$3s^2$	$3p^4$
$_{17}$Cl	$1s^2$	$2s^2$	$2p^6$	$3s^2$	$3p^5$
$_{18}$Ar	$1s^2$	$2s^2$	$2p^6$	$3s^2$	$3p^6$
$_{19}$K	$1s^2$	$2s^2$	$2p^6$	$3s^2$	$3p^6$	$4s^1$
$_{20}$Ca	$1s^2$	$2s^2$	$2p^6$	$3s^2$	$3p^6$	$4s^2$

elements $_9$F and $_{17}$Cl, both with electron configurations of p^5, lack one electron for completing a p shell; such elements readily acquire an additional electron, to become a negative ion or to form chemical compounds, corresponding to a valence of -1.

42-6   The Laser

The term *laser* is an acronym for Light Amplification by the Stimulated Emission of Radiation. Such a device produces unidirectional, monochromatic, intense, and—most important—coherent visible light.

Consider first the processes by which the energy of an atom can change with the emission or absorption of a photon. (See Figure 42-16.)

- *Spontaneous emission.* An atom is initially in an excited state and decays to a lower state as a photon of energy $h\nu = E_2 - E_1$ is emitted. The decay of unstable atoms is governed by an exponential decay law. Typically, an excited atomic state has a lifetime of the order of 10^{-8} s; on the average, the time for an atom in an excited state to decay spontaneously with the emission of a photon is only 10^{-8} s. Some atomic transitions are much slower, however. For such a metastable state, the atomic lifetime may be as long as 10^{-3} s. (The *spontaneous* transition of an atom from a low energy state to a higher is ruled out by energy conservation.)
- *Stimulated absorption.* An incoming photon stimulates, or induces, an atom to make an upward transition, and the photon is thereby absorbed.
- *Stimulated emission.* An incoming photon stimulates an atom initially in an excited state to make a downward transition. As the atom's energy is lowered, the atom emits a photon, which is *in addition* to the photon inducing the transition. One photon approaches the atom in an excited state, and two photons leave. Afterward, the atom is in the lower energy state. Moreover, the two photons both leave in the same direction as that of the incoming photon,

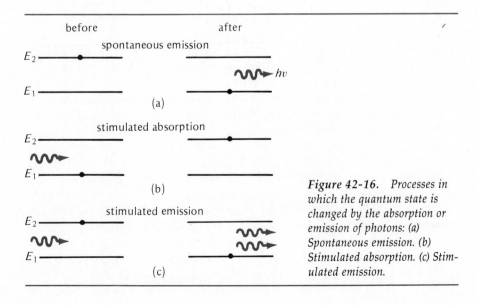

Figure 42-16. *Processes in which the quantum state is changed by the absorption or emission of photons: (a) Spontaneous emission. (b) Stimulated absorption. (c) Stimulated emission.*

and they are *exactly* in phase relative to one another; that is, they are *coherent* (Section 38-6). We can see that stimulated emission produces coherent radiation as follows. Suppose the two photons were out of phase by some amount; then they would at least partially interfere destructively, violating energy conservation. The stimulated emission produces *light amplification*, or photon multiplication. The trick in constructing a laser is to make the stimulated emission dominate competing processes.

The probability of decay by spontaneous emission can be characterized by the mean life of the excited state. Similarly, one can assign probabilities P_a and P_e to the processes stimulated absorption and stimulated emission. Detailed analysis shows that $P_a = P_e$. That is, for a given photon energy and type of atomic system, stimulated emission is just as probable as stimulated absorption. For example, if a certain number of photons directed at a collection of atoms all initially in a low energy state cause, say, a tenth of the atoms to undergo stimulated absorption, then the same number of photons directed at the same collection of atoms in the upper energy state will cause a tenth of the atoms to undergo stimulated emission.

The three processes—spontaneous emission, stimulated absorption, and stimulated emission—apply to free atoms interacting with photons. If a system consisting of many weakly interacting atoms is in thermal equilibrium, still other so-called relaxation processes may operate to change the quantum state of an atom without, however, emission or absorption of photons. An atom in an excited state may, for example, make a *nonradiative transition* to a lower energy state; the excitation energy goes into the thermal energy of the system rather than into creating a photon. On the contrary, an atom may be raised to a higher energy state as the thermal energy of a system decreases.

Now consider a collection of atoms in thermal equilibrium at some temperature T for which $\epsilon_i > kT$, where ϵ_i is the internal energy. The distribution of the atoms among the available energy states is given to a good approximation by the classical Maxwell-Boltzmann distribution. The number $n(\epsilon_i)$ of atoms with energy ϵ_i is $n(\epsilon_i) \propto e^{-\epsilon_i/kT}$. The relative numbers of atoms in the various possible states are controlled by the system's temperature T according to the Boltzmann factor $e^{-\epsilon/kT}$. The numbers of atoms in progressively higher energy states 1, 2, and 3 are n_1, n_2, and n_3, where $n_1 \propto e^{-E_1/kT}$, $n_2 \propto e^{-E_2/kT}$, and $n_3 \propto e^{-E_3/kT}$. Since $E_1 < E_2 < E_3$, it follows that $n_1 > n_2 > n_3$. The ground state is more heavily populated than the first excited state, and the number of atoms occupying higher states is still lower.

Consider first, for simplicity, a collection of atoms that have only two energy states and are in thermal equilibrium. (Such a collection could be atoms with free or nearly free electrons, whose spin direction is aligned or antialigned with an external magnetic field.) The n_1 atoms in the lower energy state exceed the number n_2 of atoms in the upper energy state; see Figure 42-17(a). Suppose further that a beam of photons with energy $h\nu = E_2 - E_1$ illuminates these atoms. We ignore for the moment spontaneous emission and the relaxation processes within the system, and concentrate on stimulated absorption and stimulated emission only. Stimulated absorption depopulates the lower energy state and reduces the number of photons. Stimulated emission depopulates the upper energy state and increases the number of photons. What is the net effect?

The number of photons disappearing by stimulated absorption is proportional to $P_a n_1$ and the number of additional photons created by virtue of

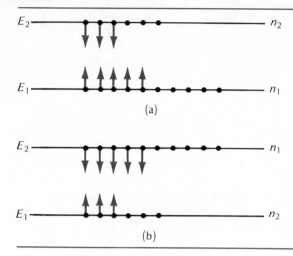

Figure 42-17. *Changes in occupancy of quantized states through stimulated absorption and stimulated emission only. (a) In thermal equilibrium, stimulated absorption dominates stimulated emission, and the number of photons is reduced. (b) For a population inversion, stimulated emission dominates stimulated absorption, and the number of photons is enhanced.*

stimulated emission is proportional to $P_e n_2 = P_a n_2$. But $n_1 > n_2$, so there is net absorption. Absorption dominates emission simply because more atoms occupy the lower energy state than the upper one. Moreover, the net absorption is accompanied by a tendency toward equalization of the populations of the two states.

Now if we were somehow to produce a *population inversion,* in which the atoms occupying the upper energy state outnumber the atoms in the lower state, then emission would dominate absorption. See Figure 42-17(b). With a population inversion, incoming light would be amplified coherently, since the number of additional photons produced through stimulated emission would more than compensate for the number of photons removed through stimulated absorption. Population inversions have been achieved for lasers in a wide variety of materials by several clever procedures, most of which involve a relatively slow relaxation.

A commonly used, continuously operating laser is the helium-neon laser. The population inversion, the essential condition for photon multiplication, is produced through inelastic collisions between excited (asterisked) helium atoms and neon atoms in the ground state. The process can be written

$$\text{He*} + \text{Ne} \rightarrow \text{He} + \text{Ne*}$$

The energy He loses must match the energy Ne gains; that is, the two types of atoms must have excited states with the same or very nearly the same energy. This is indeed the case, as shown in the energy-level diagram of Figure 42-18. The metastable 2s state of helium has an energy of 20.61 eV above the ground state; the 5s state of neon has essentially the same energy, 20.66 eV.* The lifetime of a spontaneous transition of neon from the 5s state to the 3p state is relatively long, but the following transition (3p $\rightarrow$ 3s) is short. This behavior of the neon atom produces a population inversion between the 5s and 3p states, with the lower state less populated. This transition corresponds to a photon

---

* The states for both helium and neon atoms are indicated by the state of a single excited electron. All other electrons in an atom remain in the ground state. For example, the 5s state of Ne has the electron configuration $1s^2 2s^2 2p^5 5s^1$.

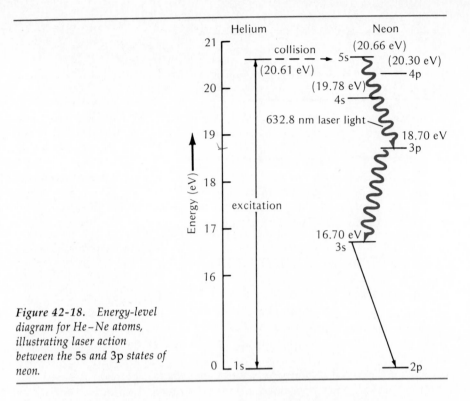

Figure 42-18. *Energy-level diagram for He–Ne atoms, illustrating laser action between the 5s and 3p states of neon.*

wavelength of 632.8 nm, the orange-red color characteristic of ordinary neon tubes.*

A typical helium-neon laser has a mixture of He–Ne gases enclosed in a sealed tube with parallel silvered mirrors (~99 percent reflecting) at its ends. The gas is excited by a dc voltage source that raises some helium atoms to the 2s metastable state. Inelastic collisions transfer energy to neon atoms and increase the population of the neon atoms in the 5s metastable state. Laser action can then occur, since 632.8-nm light reflecting back and forth between the two ends induces more downward transitions than upward transitions. The coherent, monoenergetic, unidirectional beam is therefore amplified, and laser light emerges from the tube end with the smaller reflectivity.

Energy "lost" through the emission of photons in downward transitions from excited atoms is restored through the continuous excitation of atoms by the dc power supply. Only a small fraction of the energy supplied to the excitation of the atoms is converted to the energy of the output coherent laser beam. A typical working efficiency (output laser power to input excitation power) of a He–Ne gas laser is 1×10^{-3} percent.

Besides the gas laser just described, lasers may involve solids, liquids, and semiconducting materials. Lasers operating from the far infrared to the ultraviolet region of the electromagnetic spectrum have been constructed. In every instance, the condition for laser operation is the existence of a pair of quantized energy levels for which a population inversion has been achieved.

Lasers have many technological applications. They are all possible because, with lasers, one can produce in the visible and nearby regions electromagnetic

* Laser action can also take place between other energy states of the neon atom in which population inversion occurs.

radiation that has the coherence properties heretofore available only in radio waves.

Summary

Section 42-4, The Four Quantum Numbers for Atomic Structure, is itself a summary and it is not included here.

Definitions

Alpha particle: helium nucleus.

Rutherford scattering: the scattering of incident charged particles through the Coulomb force by the nucleus as an effective point charge and point mass (in the original experiments suggested by Rutherford, the scattering of alpha particles by gold nuclei).

Cross section: the effective target area presented by a nucleus to a particle.

Types of spectra:

- Bright-line spectrum—collection of sharply defined frequencies in the radiated spectrum, or the emission spectrum.
- Dark-line spectrum—collection of dark lines on the background of the continuous electromagnetic spectrum in the absorption spectrum.

Rydberg constant: the constant R in the empirical relation for the wavelengths λ of the lines in the spectrum of atomic hydrogen

$$\frac{1}{\lambda} = R\left(\frac{1}{n_l^2} - \frac{1}{n_u^2}\right) \qquad (42\text{-}10)$$

where n_u and n_l are integral quantum numbers.

Bohr radius: radius of the electron orbit for hydrogen in its ground state in the semiclassical atomic theory of Bohr.

Ground state: lowest-energy quantized state of a system.

Excited state: quantized energy state above the ground state.

Lyman series (for atomic hydrogen): the group of spectral lines originating from or terminating in the ground state.

Balmer series (in the visible region for atomic hydrogen): the group of spectral lines originating from or terminating in the first excited state of hydrogen.

Spontaneous emission: the process in which an atom in an excited state spontaneously decays to a lower energy state with the creation of a single photon.

Stimulated emission: the process in which one photon induces an atom in an excited state to make a downward transition to a lower energy state with the emission of another photon.

Population inversion: the circumstance in which the population of an excited state exceeds the population in a state of lower energy; the essential condition for the operation of a laser.

Fundamental Principles

Basic postulates of the quantum theory of atomic structure:

- Atoms can exist in stationary states without radiating electromagnetic radiation.
- A quantization principle identifies the stationary states. (In the original Bohr theory of hydrogen, the electron's orbital angular momentum mvr was restricted to integral multiples of $\hbar \equiv h/2\pi$, or $mvr = n\hbar$ (42-2); in modern quantum theory the fitting of wave functions to meet boundary conditions produces the allowed states.)
- The absorption or emission of a photon from an atom corresponds to an equal increase or decrease in the energy of the atom.

$$h\nu = E_u - E_l \qquad (42\text{-}8)$$

Pauli Exclusion Principle: no two electrons in an atom can have the same set of quantum numbers; equivalently, no two electrons can exist in the same state.

Important Results

Bohr quantum theory of the hydrogen atom:

$$\text{Allowed energies} = E_n = -E_I/n^2 \qquad (42\text{-}6)$$

where E_I is the ionization energy and n is the principal quantum number.

$$E_I = k_e^2 e^4 m/2\hbar^2 = 13.6 \text{ eV} \qquad (42\text{-}7)$$

$$\text{Allowed radii} = r_n = n^2 r_1 \qquad (42\text{-}4)$$

$$\text{where Bohr radius} = r_1 = \hbar^2/k_e m e^2$$
$$= 5.29 \times 10^{-11} \text{ m} \qquad (42\text{-}7)$$

Problems and Questions

Section 42-1   Nuclear Scattering

· **42-1 Q** The crucial observation in the Rutherford experiment, in which alpha particles were scattered by gold foil, was that

(A) the alpha particles decayed radioactively before reaching the foil.

(B) appreciable numbers of gold nuclei were dislodged from the foil.

(C) no alpha particles were deflected through large angles.

(D) no alpha particles were observed coming from the region of the foil.

(E) some alpha particles were deflected through large angles and some were deflected through small angles.

· **42-2 P** Which particle would get closer to a nucleus as target if the particle were fired directly at the nucleus, a proton accelerated from rest through an electric potential difference V or an alpha particle accelerated from rest through the same potential difference?

: **42-3 Q** Unless it is aimed directly at a nucleus, an alpha particle can never "strike" a nucleus; the positively charged alpha particle is repelled by the positively charged nucleus. But suppose that an energetic negatively charged particle, such as a negative ion, is fired at a nucleus, but not directly head-on. Will it ever "hit" the nucleus?

: **42-4 P** A Styrofoam cube of edge length L contains N_1 steel balls, each of radius r_1, dispersed randomly throughout the interior of the cube. The balls are small enough that no one ball "hides" another. Some other small balls of radius r_2 are fired at random at the cube. What fraction of the incident balls are removed from the forward beam?

: **42-5 P** (a) A 100-eV proton is fired head-on at an electron at rest. What are the kinetic energies of the two particles afterward? (b) A 100-eV electron is fired head-on at a proton at rest. What are the kinetic energies of the two particles afterward?

: **42-6 P** Alpha particles in the Rutherford scattering experiments can be imagined to move effectively in a straight line until they are a distance much less than an atomic radius from the nucleus. As shown in Example 42-1, an 8.0-MeV alpha particle fired head-on at a gold nucleus comes within a distance of 28 fm of the center of the nucleus. (a) What is the acceleration of the alpha particle at this location? (b) Compute the acceleration of the alpha particle resulting from the Coulomb force of the gold nucleus when the alpha particle is an atomic distance away, say, 1.0 nm from the nuclear center.

: **42-7 P** A *neutron star* is an astronomical object that has undergone such severe gravitational collapse that its electrons, protons, and neutrons have been squeezed together to form a highly dense object consisting solely of neutrons in contact; that is, neutrons separated on the average by a distance of the order of 1 fm. In effect, the density of the neutron star is the same as the density of an atomic nucleus. What is the order of magnitude of this density?

: **42-8 P** Since a neutron has no electric charge, it interacts with a nucleus only through the strong, short-range attractive nuclear force. In effect, a neutron feels a force from a nucleus only when it touches it. The radius R of the nucleus is given approximately by $R = r_0 A^{1/3}$ where A is the mass of the nucleus (equal in atomic mass units, u, to the total number of protons and neutrons in the nucleus) and $r_0 = 1.4$ fm. (a) What is the approximate cross section for neutron absorption by lead-206? Cross sections are usually given with the unit *barn*, where 1 barn $= 10^{-28}$ m^2 = 100 fm^2. (b) A thin foil of lead has an areal density (an indirect measure of thickness) of 100 mg/cm^2. This foil is used as a target for a beam of neutrons. What fraction of the incident neutrons is removed from the forward beam by neutron absorption (density of lead, 11.3×10^3 kg/m^3)?

: **42-9 P** A beam of 5.0-MeV α particles strikes a target of helium atoms in a gas. (a) What is the minimum distance one helium nucleus comes to another helium nucleus? (b) Show that none of the particles can be scattered by an angle greater than 90°.

: **42-10 P** A particle of charge q_1 with initial kinetic energy K is fired head-on at a massive particle of charge q_2 at rest. (a) What is the minimum distance between the two particles? (b) Now suppose that the particle with charge q_1 and initial kinetic energy K is fired so that, if there were no electric force between the particles, it would miss hitting the massive particle by a distance b. What is now the minimum distance between the two charged particles? (*Big hint:* the Coulomb force is a central force.)

Section 42-3   Bohr Theory of Hydrogen

· **42-11 Q** A key experimental observation led Bohr to postulate what has now become the modern model for the atom. This observation was that

(A) for certain metals, blue light would give rise to a photocurrent but red light would not, independent of the intensity of the light.

(B) isolated atoms emit and absorb discrete light wavelengths, as opposed to a continuum of wavelengths.

(C) 6×10^{23} atoms of any element are always found to have a mass equal in grams to the atomic number of the element.

(D) scattering of alpha particles from gold nuclei indicated that the atom's positive charge is fairly uniformly distributed throughout the atom.

(E) the length of a moving object contracts in its direction of motion.

(F) particles such as electrons have a dual nature, acting in some respects like particles and in other respects like waves.

· **42-12 Q** One of Niels Bohr's brilliant insights, which led to our modern understanding of atomic structure, was that

(A) the electron charge is quantized.

(B) an atom's energy can vary continuously from zero to some maximum cutoff value.

(C) the atom's energy is Planck's constant multiplied by the frequency of the orbiting electron.

(D) the electron's angular momentum is quantized.

(E) the atom's energy is an integral multiple of the electron's rest energy.

· **42-13 P** How much energy is required to remove an electron from the first excited state of hydrogen, thereby producing an ion H^+?

· **42-14 Q** For an atom of hydrogen to emit radiation,

(A) it must be in its ground state.

(B) it must make a transition from the ground state.

(C) it must be in an excited state.

(D) it must simultaneously absorb a photon.

(E) it must be fluorescent.

· **42-15 P** The allowed energy levels of a hydrogen atom are characterized by a quantum number n, where $n = 1$ corresponds to the ground state. For which of the transitions listed here would a photon of the shortest wavelength be emitted?

(A) From $n = 100$ to $n = 5$. .0399

(B) From $n = 3$ to $n = 7$.

(C) From $n = 1$ to $n = 5$. $\left(\frac{1}{M_f^2} - \frac{1}{M_i^2} \right)$

(D) From $n = 4$ to $n = 2$.

(E) From $n = 2$ to $n = 1$. .75

· **42-16 P** What is the wavelength of the photon emitted when a hydrogen atom makes a transition from the $n = 5$ state to the $n = 2$ state?

: **42-17 P** Using the Bohr model, determine what energy is required to change a He^+ ion into He^{2+}. The experimental value is 54.4 eV.

: **42-18 Q** Suppose that an atom had only four distinct energy levels. What is the maximum number of spectral lines of different wavelengths that it could possibly emit?

(A) 1.

(B) 2.

(C) 3.

(D) 4.

(E) 5.

(F) 6.

: **42-19 P** A free electron of negligible kinetic energy is captured by a stationary proton to form an excited state of the hydrogen atom. In this process, a photon of energy E is emitted. Shortly thereafter a second photon of energy 10.2 eV is emitted. No further photons are emitted. Deduce the energy E of the first photon emitted under these circumstances.

: **42-20 P** Light of wavelength 409 nm is emitted from a hydrogen arc. What transition produced this emission?

: **42-21 P** What is the wavelength of a photon that will induce a transition from the ground state to the $n = 4$ state in hydrogen?

: **42-22 P** A hydrogen atom can be excited from its ground state to the first excited state when an electron of sufficient kinetic energy hits the atom. What is the minimum electron kinetic energy?

: **42-23 P** Two hydrogen atoms, both initially in the ground state, approach one another with the same initial kinetic energy K_i and collide head-on. As a consequence of the collision, one hydrogen atom is excited to the first excited state while the other hydrogen atom remains in the ground state. What is K_i?

: **42-24 P** The ion Li^{2+} has a nuclear charge of $+3e$. The ion has a single electron, so that the ion is similar to a hydrogen atom. For this ion, calculate (a) the ground-state energy; (b) the wavelength of the photon emitted when the ion makes a transition from the $n = 2$ state to the $n = 1$ state.

: **42-25 P** A free hydrogen atom undergoes a transition from the $n = 3$ state to the $n = 1$ state. A photon is emitted and the atom will necessarily recoil. (a) How much energy, in electron volts, is released in this transition? (b) With what speed does the atom recoil? (c) What is the energy of the photon emitted? (d) What is the kinetic energy of the recoiling atom?

: **42-26 P** When a hydrogen atom is excited to the first excited state, it remains in this state for an average time of about 10^{-8} s before making a downward transition to the ground state. (a) Use the uncertainty principle to find the uncertainty in the energy (in eV) of the excited state. (b) What is the fractional uncertainty in the wavelength of the photon emitted in a transition to the ground state? The finite lifetimes of excited atomic states produce a *natural linewidth* in the spectral lines.

: **42-27 P** A *muonium* atom consists of a muon, an unstable elementary particle with charge $-e$ and a mass 207 times that of an electron, bound to a proton. What are (a) the ionization energy and (b) Bohr radius for muonium?

: **42-28 P** A *positronium* atom consists of an electron and a positron. Each particle may be considered to orbit the atom's center of mass. (a) What is the atom's ionization energy? (b) What is the wavelength of the photon emitted in a transition from the first excited to the ground state?

: **42-29 P** Using the fact that the average kinetic energy of a molecule of a gas at temperature T is $3kT/2$, estimate the minimum temperature of a gas of atomic hydrogen that will produce appreciable ionization of the atoms through collisions that will break up the atoms into protons and electrons (a plasma).

: **42-30 P** Show that the speed of the electron in the first Bohr orbit is, in units of the speed of light, equal to $k_e e^2/\hbar c$. This combination of fundamental atomic constants, usually abbreviated by the symbol α, is called the *fine-*

structure constant (for reasons now of strictly historical interest). The fine-structure constant plays an important role in quantum theory of electromagnetism because it gives a dimensionless relation among the fundamental constants of electromagnetism (k_e and e), quantum theory (h), and relativity (c). ($\alpha \approx 1/137$.)

: **42-31 P** Suppose an "atom" existed with an electron bound to a neutron by the gravitational force. (*a*) From the Bohr model, what would be the energy of the ground state? (*b*) What would be the radius of the first Bohr orbit?

: **42-32 P** In a *positronium* atom, the positively charged proton is replaced by a positively charged positron. For this atom, determine (*a*) the ground-state energy; (*b*) the Bohr radius.

: **42-33 P** Consider a hydrogen atom in a state with very large quantum number n. (*a*) Determine the frequency of a photon emitted in a transition from n to $n - 1$. (*b*) Show that the frequency obtained in (*a*) is approximately equal to the frequency of revolution of the electron in its orbit. This is an illustration of Bohr's correspondence principle, which states that for large n, the results of the quantum theory and classical physics will agree. (*Hint*: $f(n) - f(n - 1) \approx df/dn$ for large n.)

: **42-34 Q** Suppose that an electron makes a transition from the $n = 3$ state to the $n = 2$ state in the Bohr model of the hydrogen atom. Which of the following is then correct?
(A) The electron's kinetic energy decreases, and the potential energy of the atom increases, but its total energy remains the same.
(B) Kinetic energy increases, potential energy decreases, but total energy remains the same.
(C) Kinetic and potential energies both increase and so does total energy.
(D) Kinetic and potential energies both decrease, and so does total energy.
(E) Kinetic energy increases by an amount Δ and potential energy decreases by 2Δ, thereby producing a decrease of Δ in total energy.

: **42-35 P** Consider a hypothetical one-electron atom, in which the series of spectral lines corresponding to transitions that end on the $n = 1$ state have wavelengths 130 nm, 110 nm, 95 nm, 86 nm, . . . , 78 nm. The 78-nm wavelength is the shortest that this atom can emit. (*a*) What is the ionization energy, in electron volts, for the atom? (*b*) What are the energies of the first four energy levels, in electron volts? (*c*) What is the wavelength of the photon emitted in a transition from $n = 4$ to $n = 2$? (*d*) What energy must be supplied to induce a transition from $n = 2$ to $n = 3$?

Section 42-4   The Four Quantum Numbers for Atomic Structure

· **42-36 P** Suppose that a relatively heavy atom such as tungsten (atomic number, 74) has all but one of its electrons

removed. (*a*) How much energy is required to remove this last electron completely if the ionized atom is initially in its ground state? (*b*) What would be the wavelength of the photon emitted if this single remaining electron were to make a transition from the state with $n = 2$ to the state with $n = 1$? In what part of the electromagnetic spectrum would such a photon be found?

· **42-37 P** When an electron with a kinetic energy of at least 4.88 eV collides with a mercury atom in its ground state, the collision is inelastic and the mercury atom emits radiation. The interpretation of this observation (first done in the *Franck-Hertz experiment* of 1914) is that the kinetic energy lost by the electron brings the mercury atom to its first excited state from which the atom then decays back to the ground state with the emission of a photon. What is the wavelength of this photon?

· **42-38 Q** The chemical behavior of an atom is determined by its
(A) mass number.
(B) binding energy.
(C) atomic weight.
(D) atomic number.
(E) number of isotopes.

42-39 Q A hydrogen atom is in a p state. Therefore
(A) the atom has its lowest possible energy.
(B) the atom is ionized.
(C) the atom is in the ground state.
(D) the atom's orbital angular momentum is not zero.
(E) the electron wave function is spherically symmetrical.

: **42-40 P** The characteristic x-rays from any element result from quantum transitions of the innermost electrons in the atom. The x-ray line of shortest wavelength, called the K_α line, is produced by a transition from the shell for $n = 2$ to a vacancy in the shell for $n = 1$. The remaining electron in the innermost shell shields the electric charge of the nucleus; indeed, the electron making the transition "sees" an effective nuclear charge of $Z - 1$, where Z is the atomic number of the element, or equivalently, the total charge in units of e of the nucleus. What is the wavelength of the K_α x-ray line emitted by iron ($Z = 26$)?

: **42-41 P** Sodium atoms strongly radiate two closely spaced yellow lines (called the sodium D lines) with wavelengths of 589.5944 nm and 588.9977 nm. The two lines originate from transitions from two closely spaced excited energy levels to the single ground state of sodium. What is the energy difference between the two upper energy levels?

: **42-42 P** In one fictitious classical model of the electron, it is assumed to be a sphere of radius 2.8×10^{-15} m with electric charge and mass distributed uniformly throughout its volume. (*a*) At what angular speed would the electron have to spin about a diameter to have spin angular momentum with the magnitude $\sqrt{3}\hbar/2$? (b) What would be the tangential speed of a point on the "equator"?

: 42-43 P The total energy of a hydrogen atom may be written as $E = p^2/2m - ke^2/r$, where p is the orbital linear momentum of the electron and r is its distance from the nucleus. If an electron is confined to a distance r, its momentum is uncertain by an amount governed by the uncertainty principle. (a) Show that the atom's energy is a minimum for some distance r_1, where r_1 is in fact the first Bohr radius. (b) Show that for this distance of electron from nucleus, the atom's energy is that of hydrogen in the ground state.

Section 42-5   Pauli Exclusion Principle and the Periodic Table

· 42-44 Q (a) Write down the quantum numbers for the three outermost electrons in magnesium (atomic number 17). (b) Write down the quantum numbers for the lowest-lying excited state in magnesium.

· 42-45 P For each of the following electron configurations, identify the corresponding element. (a) $1s^2 2s^2 2p^6 3s^1$; (b) $1s^2 2s^2 2p^6 3s^2 3p^6$; (c) $1s^2 2s^2 2p^6 3s^2 3p^6 3d^7 4s^2$

· 42-46 P (a) What quantum numbers characterize an electron in the $n = 2$ state in the ion He^+? (b) What is the energy of the He^+ ion in this state?

: 42-47 Q Whether or not a particle obeys the Pauli exclusion principle is determined by the particle's
(A) charge.
(B) energy.
(C) spin angular momentum.
(D) orbital angular momentum.
(E) wavelength.

: 42-48 P Suppose three electrons were placed in a one-dimensional box of width L. What minimum energy would be needed to remove one electron (that is, to ionize the system)?

: 42-49 P Seven identical particles (noninteracting) are placed in a cubical box of side 2.5×10^{-10} m. What is the lowest total energy of the system and what are the quantum numbers of the particles if (a) the particles are elec-

trons? (b) the particles have the same mass as the electron but no spin (that is, they are not subject to the Pauli exclusion principle).

Section 42-6   The Laser

· 42-50 Q A process crucial in operating any laser is
(A) spontaneous absorption of radiation.
(B) stimulated emission of radiation.
(C) conversion of photons into electrons.
(D) conversion of electrons into photons.
(E) splitting a single photon into two photons.

42-51 P A helium-neon laser operates with a 115-V, 2.0-A power supply. The output is 1.0 mW. What is the efficiency for converting electrical energy to coherent light energy?

: 42-52 P A carbon-dioxide pulsed laser produces pulses of 2.0×10^{11} W with a duration of 1.0 ns. The transverse cross section of the laser beam is 0.50 mm². For each pulse, what is (a) the energy, (b) the energy density, and (c) the linear momentum? (d) Suppose that one pulse were absorbed completely by water at 20° C. How much water would be vaporized?

: 42-53 P A helium-neon laser is most commonly operated to produce light of 633-nm wavelength, but lasing action is also possible at other wavelengths. Use Figure 42-1? to identify the transition for producing coherent light of wavelengths (a) 3.4×10^2 nm and (b) 1.15×10^3 nm.

: 42-54 P A helium-neon laser emits light of wavelength 632.8 nm in a transition between two states of neon. For neon atoms in equilibrium at 300 K, what is the ratio of the population of the upper state to the population of the lower state?

: 42-55 P The particles of a certain system have three possible energies — E_1, E_2, and E_3, where $E_1 < E_2 < E_3$. The corresponding number of particles in the three states are $n_1, n_2,$ and n_3, where for thermal equilibrium $n_1 > n_2 > n_3$. Now suppose that the system is irradiated by strong pumping radiation that equalizes the populations in states 1 and 3. Show that a population inversion must exist between one other pair of states, with either $n_3 > n_2$ or $n_2 > n_1$.

Appendixes

A   International System of Units

In the International System of Units (abbreviated SI in all languages) there is one and only one SI unit for each physical quantity, either the appropriate SI base unit itself, defined in the listing below, or the appropriate SI derived unit, formed by multiplication and/or division of two or more SI Base Units, also listed below.

SI Base Units

Meter (m)   The meter is the length equal to the distance traveled in a time interval of $1/299792458$ of a second by plane electromagnetic waves in a vacuum.

Kilogram (kg)   The kilogram is the unit of mass; it is equal to the mass of the international prototype of the kilogram.

Second (s)   The second is the duration of 9192631770 periods of the radiation corresponding to the transition between the two hyperfine levels of the ground state of the caesium-133 atom.

Ampere (A)   The ampere is that constant current which, if maintained in two straight parallel conductors of infinite length, of negligible cross section, and placed 1 meter apart in a vacuum, would produce between these conductors a force equal to 2×10^{-7} newtons per meter of length.

Kelvin (K)   The kelvin, unit of thermodynamic temperature, is the fraction $1/273.16$ of the thermodynamic temperature of the triple point of water.

Candela (cd)   The candela is the luminous intensity, in the perpendicular direction, of a surface of $1/600000$ square meter of a black body at the temperature of freezing platinum under a pressure of 101325 newtons per square meter.

Mole (mol)   The mole is the amount of substance of a system which contains as many elementary entities as there are atoms in 0.012 kilograms of carbon-12. When the mole is used, the elementary entities must be specified and may be atoms, molecules, ions, electrons, other particles, or specified groups of such particles.

SI Derived Units

QUANTITY	UNIT NAME	SYMBOL	BASIC SI UNITS
Frequency	hertz	Hz	s^{-1}
Force	newton	N	$kg \cdot m/s^2$
Pressure	pascal	Pa	N/m^2
Energy	joule	J	$N \cdot m$
Power	watt	W	J/s
Electric charge	coulomb	C	$A \cdot s$
Potential difference	volt	V	W/A
Electric resistance	ohm	Ω	V/A
Conductance	siemens	S	A/V
Capacitance	farad	F	$A \cdot s/V$
Magnetic flux	weber	Wb	$V \cdot s$
Inductance	henry	H	$V \cdot s/A$
Magnetic flux density	tesla	T	Wb/m^2

B   SI Prefixes for Factors of Ten

PREFIX	SYMBOL	POWER
exa	E	10^{18}
peta	P	10^{15}
tera	T	10^{12}
giga	G	10^{9}
mega	M	10^{6}
kilo	k	10^{3}
hecto	h	10^{2}
deka	da	10^{1}
deci	d	10^{-1}
centi	c	10^{-2}
milli	m	10^{-3}
micro	μ	10^{-6}
nano	n	10^{-9}
pico	p	10^{-12}
femto	f	10^{-15}

Examples: MW = megawatt = 10^6 watt

nm = nanometer = 10^{-9} meter

C   Physical Constants

	SYMBOL	VALUE
Acceleration of gravity	g	$9.80665 \ m/s^2$ (standard value)
Standard atmospheric pressure		$1.01325 \times 10^5 \ Pa$
Gravitational constant	G	$6.672 \times 10^{-11} \ N \cdot m^2/kg^2$

	SYMBOL	VALUE
Speed of light	c	2.99792458×10^8 m/s (exact value)
Electron charge	e	1.60219×10^{-19} C
Avogadro's number	N_A	6.0220×10^{23} mol^{-1}
Gas constant	R	8.314 J/mol·K
		8.206 ℓ·atm/mol·K
Boltzmann constant	$k = R/N_A$	1.3807×10^{-23} J/K
		8.617×10^{-5} eV/K
Unified mass unit	u	1.6606×10^{-24} gm (u in gm $= 1/N_A$)
		(1/12)(mass of neutral carbon-12 atom)
Coulomb-law constant	$k_e = 1/4\pi\epsilon_0$	8.98755×10^9 N·m^2/C^2
Permittivity of free space	ϵ_0	8.85419×10^{-12} C^2/N·m^2
Magnetic constant	$k_m = \mu_0/4\pi$	10^{-7} N/A^2 (exact value)
Permeability of free space	μ_0	$4\pi \times 10^{-7}$ N/A^2
Planck's constant	h	6.6262×10^{-34} J·s
		4.1357×10^{-15} eV·s
Mass, electron	m_e	9.1095×10^{-31} kg $= 0.51100$ MeV/c^2
Mass, proton	m_p	1.67265×10^{-27} kg $= 938.26$ MeV/c^2
Mass, neutron	m_n	1.67495×10^{-27} kg $= 939.55$ MeV/c^2

Astronomical Data

Nomenclature: $1.99 \, E \, 30 \equiv 1.99 \times 10^{30}$

OBJECT **Planet** *Satellite*	MASS (kg)	RADIUS (m)	ORBITAL RADIUS (m)	ORBITAL PERIOD
SUN	1.99 *E* 30	6.95 *E* 8	——	——
Mercury	3.28 *E* 23	2.57 *E* 6	5.8 *E* 10	88.0 d
Venus	4.82 *E* 24	6.31 *E* 6	1.08 *E* 11	224.7 d
Earth	5.98 *E* 24	6.38 *E* 6	1.49 *E* 11	1.00 y
Synchronous satellite	——	——	4.15 *E* 7	1.00 d
Moon	7.36 *E* 22	1.74 *E* 6	0.38 *E* 9	27.3 d
Mars	6.34 *E* 23	3.43 *E* 6	2.28 *E* 11	687.0 d
Phobos	2.72 *E* 16	10.4 *E* 3	9. *E* 6	0.318 d
Deimos	1.8 *E* 15	5.0 *E* 3	2.35 *E* 7	1.26 d
Jupiter	1.90 *E* 27	7.18 *E* 7	7.78 *E* 11	11.86 y
1. *Io*	7.87 *E* 22	1.73 *E* 6	4.22 *E* 8	1.77 d
2. *Europa*	4.78 *E* 22	1.49 *E* 6	6.71 *E* 8	3.55 d
3. *Ganymede*	1.54 *E* 23	2.53 *E* 6	1.07 *E* 9	7.15 d
4. *Callisto*	7.35 *E* 22	2.42 *E* 6	1.88 *E* 9	16.6 d
5. *(Smalthea)*	8.3 *E* 18	8.70 *E* 4	1.81 *E* 8	0.498 d
6. *(Hestia)*	3.8 *E* 18	6.70 *E* 4	1.14 *E* 10	251 d
7. *(Hera)*	1.0 *E* 17	2.00 *E* 4	1.16 *E* 10	260 d
8. *(Poseidon)*	7.3 *E* 16	1.80 *E* 4	2.35 *E* 10	2.02 y
9. *(Hades)*	2.2 *E* 16	1.20 *E* 4	2.37 *E* 10	2.07 y
10. *(Demeter)*	9.2 *E* 15	9.0 *E* 3	1.18 *E* 10	0.71 y
11. *(Pan)*	2.2 *E* 16	1.20 *E* 4	2.26 *E* 10	1.91 y
12. *(Adrastea)*	1.3 *E* 16	1.00 *E* 4	2.11 *E* 10	1.71 y
Saturn	5.68 *E* 26	6.03 *E* 7	1.43 *E* 12	29.46 y
Mimas	4 *E* 19	2.72 *E* 5	1.82 *E* 8	0.94 d
Enceladus	7 *E* 19	2.99 *E* 5	2.39 *E* 8	1.37 d

OBJECT **Planet** *Satellite*	MASS (kg)	RADIUS (m)	ORBITAL RADIUS (m)	ORBITAL PERIOD
Tethys	4.9 E 20	5.81 E 5	2.95 E 8	1.89 d
Dione	5.4 E 20	5.98 E 5	3.78 E 8	2.74 d
Rhea	1.8 E 21	8.90 E 5	5.28 E 8	4.52 d
Titan	1.2 E 23	2.38 E 6	1.23 E 9	15.95 d
Hyperion	6.8 E 19	2.01 E 5	1.49 E 9	21.28 d
Iapetus	2.3 E 21	6.47 E 5	3.57 E 9	79.33 d
Phoebe	1.9 E 19	1.32 E 5	1.30 E 10	550.4 d
(Janus)	1.2 E 20	2.41 E 5	1.58 E 8	0.749 d
Uranus	8.68 E 25	2.35 E 7	2.87 E 12	84.02 y
Ariel	5.0 E 20	3.11 E 5	1.92 E 8	2.52 d
Umbriel	1.4 E 20	2.01 E 5	2.67 E 8	4.14 d
Titania	2.1 E 21	5.00 E 5	4.39 E 8	8.70 d
Oberon	1.1 E 21	4.01 E 5	5.86 E 8	13.5 d
Miranda	3.0 E 19	1.21 E 5	1.29 E 8	1.41 d
Neptune	1.03 E 26	2.27 E 7	4.49 E 12	164.8 y
Triton	1.46 E 23	2.01 E 6	3.53 E 8	5.88 d
Nereid	5 E 19	1.36 E 5	5.9 E 9	359.4 d
Pluto	1 E 24	5.7 E 6	5.90 E 12	247.7 y

D   Conversion Factors

Converting units amounts to *multiplying by the factor 1,* and therefore leaving the quantity (but not its units) unchanged.

Example:

$$60 \text{ mi/h} = ? \text{ m/s}$$

We are given that

$$1 \text{ mi/h} = 0.4470 \text{ m/s}$$

so that

$$1 = \frac{(0.4470 \text{ m/s})}{(1 \text{ mi/h})}$$

Multiplying the quantity above by this conversion factor gives

$$60 \text{ mi/h} \frac{(0.4470 \text{ m/s})}{(1 \text{ mi/h})} = 26.82 \text{ m/s}$$

Note the cancellation of the unwanted units, mi/h.

Example:

$$40 \text{ m/s} = ? \text{ mi/h}$$

This time the units m/s are to cancel, so that the conversion factor must have m/s in its denominator. The required conversion factor in this instance is

$$1 = \frac{(1 \text{ mi/h})}{(0.4470 \text{ m/s})}$$

so that

$$40 \text{ m/s } \frac{(1 \text{ mi/h})}{(0.4470 \text{ m/s})} = 89.49 \text{ mi/h}$$

Length
1 inch $= 2.54 \times 10^{-2}$ m
1 ft $= 0.3048$ m
1 mi $= 1.609344$ km
1 Å (Ångstrom unit) $= 10^{-10}$ m $= 0.1$ nm

Mass
1 u $= 1.6606 \times 10^{-27}$ kg
1 lb (avdp) $= 0.4535924$ kg

Energy
1 erg $= 10^{-7}$ J
1 ft-lb $= 1.355818$ J
1 kWh $= 3.6000 \times 10^{6}$ J
1 cal $= 4.183310$ J
1 Btu $= 1054.7$ J
1 eV $= 1.6022 \times 10^{-19}$ J
1 quad $= 10^{15}$ Btu $= 1.0547 \times 10^{18}$ J
1 ton (nuclear equivalent TNT) $= 4.184 \times 10^{9}$ J

Pressure
1 torr $= 133.322$ Pa
1 mmHg $= 133.322$ Pa
1 atm $= 101325$ Pa

Speed
1 mi/h $= 0.4470$ m/s
1 ft/s $= 0.3048$ m/s

E   References

General Textbooks

These introductory physics textbooks are at a somewhat higher level of sophistication and depth than this one:

- Richard P. Feynman, Robert B. Leighton, and Matthew Sands, *The Feynman Lectures on Physics* (3 vol.). Reading, MA: Addison-Wesley Publishing Company, Inc., 1965. A transcription of lectures given over a two-year period to students at CalTech by Nobel prize winner Richard P. Feynman; the insights are frequently brilliant and the going is sometimes very rough.
- Anthony P. French, *Special Relativity* (1968), *Newtonian Relativity* (1971), *Vibration and Waves* (1971), all published in New York by W. W. Norton Company. These are all parts of the M.I.T. introductory physics series.
- Donald G. Ivey (volume 1) and J. N. Patterson Hume (volume 2) *Physics*. New York, N. Y.: John Wiley and Sons, Inc., 1974. All topics are treated rigorously and in depth; the discussion of classical electromagnetism is particularly insightful.

Biographical References

- Isaac Asimov, *Asimov's Biographical Encyclopedia of Science and Technology*. Garden City, New York: Doubleday & Co., Inc., 1964. A one-volume work with entries for all of the principals and many of the secondary contributors, arranged so as to make it easy to spot contemporaries.

• Charles C. Gillispie, ed., *Dictionary of Scientific Biography* (in 16 volumes). New York: Charles Scribner's Sons, 1970. A thorough, scholarly treatment of the subject's scientific contribution, as well as the contributor's life and times. Information for the biographical sketches appearing in panels in this book comes from this source.

General References

• *McGraw-Hill Encyclopedia of Science and Technology* (15 volumes and annual supplements). New York: McGraw-Hill, Inc., 1960. With a distinguished editorial board and consulting editors, this work not only tells about the basic scientific theory but also describes the applications in technology. Articles differ, of course, but most are direct, clear, easy to follow. The annual supplements keep it up to date.

• Robert C. Weast, ed., *Handbook of Chemistry and Physics.* Cleveland, Ohio: The Chemical Rubber Co., published annually. Usually referred to as the "Chemical Handbook," this book is invaluable to any aspiring scientist or engineer. It has an enormous range of specific data, comprehensive tables on physical and chemical properties, and definitions. As technical books go, it is remarkably cheap.

• *Scientific American,* a monthly with clearly written, beautifully illustrated articles of a relatively nontechnical nature on modern developments in science, and prompt notice on all important advances in physics, published by Scientific American, 415 Madison Avenue, New York, N. Y. 10017. Comprehensive sets of "offprints" of individual articles from past issues are available.

• Jearl Walker, *The Flying Circus of Physics.* New York, N. Y.; John Wiley & Sons, 1975. A fun book, with a variety of questions based on more-or-less ordinary observation and whose explication involves more-or-less ordinary basic physics. Get the version that includes answers and detailed references. Some of the questions are easy; some are very tough.

• Weber, Robert L. *A Random Walk in Science* (1973) and *More Random Walks in Science* (1982). London, Institute of Physics. Stories, anecdotes, satire, fun items; this is where I learned about the *N. Y. Times* comment in 1920 on Goddard's work.

16-15 6.19×10^5 N when $\rho = 10^3$ kg/m³; 6.31×10^5 N for salt water with $\rho = 1.02 \times 10^3$ kg/m³

16-17 (a) 1.57×10^5 N (b) 1.96×10^4 N (c) 2.26×10^4 N

16-19 27.2 m

16-21 5.00×10^4 N

16-25 C

16-27 (a) $W_a/(W_a - W_w)$ (b) $(W_a - W_u)/(W_a - W_w)$

16-29 $V_{ice} = 1.00 \times 10^6$ m³ $= (5.00$ m$) \times (2.00 \times 10^5$ m²$)$ if salt water has a density of 1.02×10^3 kg/m³

16-33 4.90×10^8 J

16-37 6.00 mm

16-39 (a) 4 m³/min $= 66.7 \times 10^{-3}$ m³/s (b) 23.6 m/s

16-41 $d = D[1 + (2gx)/(v_0^2)]^{-1/4}$

16-43 E

16-45 A

16-47 83.3 J/m³

16-49 435 m/s

Chapter 17 Mechanical Waves I

17-1 340 m/s

17-3 A

17-5 (a) 62.5×10^{-3} m (b) 72.0° (c) 400 m/s

17-7 1.73 s

17-11 E

17-13 $\sqrt{3}/1$

17-15 (a) $+x$ direction (b) 286 Hz (c) 1.05 m (d) 300 m/s where distances are in meters and time is in seconds.

17-21 (a) 2 cm (b) 50 Hz (c) 1 cm

17-23 (b) ωR

17-25 B

17-27 24 Hz

17-29 (a) Fundamental frequency and wavelength (b) 5/4 (c) The wave velocity, fundamental frequency and the tension change.

17-33 Fundamental = first harmonic = 40.4 Hz

17-35 (a) 3.42 m (b) $y = (0.04$ m$) \sin (4.02 \times 10^3$ s$^{-1})t \sin (3.67$ m$^{-1})x$

Chapter 18 Mechanical Waves II

18-1 A

18-3 F

18-5 34.0 m

18-11 $L_{min} = 40.5$ cm when $c = 340$ m/s

18-13 D

18-15 It will be quadrupled

18-17 2.50×10^{-11} W/m²

18-19 1.41×10^{-4} W

18-21 6.37×10^{-3} W/m² for spherical spreading; double that intensity for hemispherical spreading

18-23 6.91 m

18-25 100 m

18-27 4.0×10^{26} W

18-29 12.6×10^{-6} W

18-33 $v_{car} = 37.8$ m/s; $f = 302$ Hz

18-35 900 m/s $\approx 2.01 \times 10^3$ mph

18-37 $f_0 = 120$ Hz for both (a) and (b)

18-39 No

Chapter 19 Thermal Properties of an Ideal Gas, Macroscopic View

19-1 D

19-3 E

19-5 (a) 296 K, 73.4 °F (b) 77 K, -321 °F (c) -40 °C, 233 K (d) 22.2 °C, 295 K (e) 27 °C, 80.6 °F (f) -196 °C, -321 °F

19-7 -40 °C $= -40$ °F

19-9 B

19-11 B

19-13 119 cm

19-15 $7 \times 10^{-4}(\text{C}°)^{-1}$

19-17 Loses 246 seconds in 1/12 of a 365-day year.

19-19 (a) 7.28×10^{-4} cm² (b) reading low by 15%

19-23 4.00 gm

19-25 0.512 hr

19-27 1.17 kg/m³

19-29 (a) 3.22×10^{16} (b) 1.50×10^{-3} kg/m³

19-31 0.0215%

19-33 26.2 mm³

19-35 (b) 64.2 J

19-37 1.097×10^5 K (high), 3.64×10^4 (low)

19-39 (a) 3/4 atm (b) 1 atm (c) 210 torr

Chapter 20 Thermal Properties of an Ideal Gas, Microscopic View

20-1 8.93×10^{-18} m³ $= 8.93(\mu\text{m})^3$

20-3 3.34×10^{-9} m

20-5 2×10^{27} molecules/m²-s

20-7 1.0137×10^5 Pa

20-13 B

20-15 (a) 6.21×10^{-21} J or 38.8×10^{-3} eV (b) 1.60×10^{-21} J (c) 8.70×10^{-23} J or 0.544×10^{-3} eV

20-17 5.31×10^{11} K

20-19 $(v_{CH_4}/v_{H_2})_{rms} = 0.354$

20-21 (a) 1.01×10^4 K (b) 1.41×10^5 K

20-25 B

20-27 D

20-29 $+8$ cal

20-31 148 K

20-33 2.50 m³·atm

20-35 $C_v = 7.46 \times 10^{-2} \dfrac{\text{kcal}}{\text{kg C}°}$, $C_p = 0.124 \dfrac{\text{kcal}}{\text{kg C}°}$

20-37 3.74×10^4 J

20-39 B

20-41 $C_v = 0.172$ kcal/kg C°

20-43 686 K

20-47 E

20-49 B

Chapter 21   Thermal Properties of Solids and Liquids

21-1 B

21-3 52.0 °C

21-5 0.94 C°

21-7 4.48×10^{-4} C° per pushup

21-9 $T_3 = [C_x T_1 + C_y(T_1 + T_2 - 2T_y) + C_z T_2]/(C_x + C_z)$

21-11 E

21-13 C

21-15 940 W

21-17 6×10^{-4}

21-21 E

21-23 (a) 375 kJ (b) 0.0748 C°

21-25 93.8×10^{-3} C°

21-27 (a) 36 kJ (b) 6.7 C°

21-29 51 C°

21-31 A

21-33 21 kg

21-35 5.8 W

21-37 $\dfrac{H}{L} = \dfrac{2\pi\lambda(T_1 - T_2)}{\ln(1 + a/R_1)}$

21-39 (b) $T_A = \dfrac{(\lambda_2 t_1 + \lambda_1 t_2)T_2 + \lambda_2 t_1 T_1}{2\lambda_2 T_1 + \lambda_1 T_2}$,

$T_B = \dfrac{\lambda_2 t_1 T_2 + (\lambda_2 t_1 + \lambda_1 t_2)T_1}{2\lambda_2 t_1 + \lambda_1 t_2}$

where $T_2 > T_A > T_B > T_1$

21-41 A

21-43 $64.70

21-45 E

21-49 1769 K

21-51 (a) 4.46×10^{26} W (b) 2.02×10^{17} W (c) 10.1×10^6

21-53 (a) 326 W (b) 20.4 s

21-55 (a) 1.2×10^{31} J (b) 4.0×10^{16} s $= 1.2 \times 10^9$ years

Chapter 22   The Second Law of Thermodynamics and Heat Engines

22-1 B

22-3 B

22-5 28.6 hp

22-7 B (assumes start is at T_1)

22-9 (a) 80 J (b) 60 J

22-11 (a) $7/12 = 0.583$ (b) irreversible

22-13 (c) 54.1%

22-17 D

22-19 29.7%

22-21 $e = \dfrac{T_1 - T_3}{T_1}$

22-23 extensive

22-25 C

22-27 $\Delta S = 2R \ln 2$

22-29 $\Delta S = -2R \ln 2$

22-31 $\Delta S = 0.54$ J/K

Chapter 23   Point Electric Charges

23-1 B

23-3 Speeds up

23-5 E

23-7 C

23-9 A

23-11 8.43 N directed at 25.5° above the axis drawn from q_2 to q_1

23-13 14.5 cm left of the origin

23-15 $6.36 \times 10^9 \left(\dfrac{Q}{a}\right)^2$ N, directed downward

23-17 Give each sphere the charge $\frac{1}{2}Q$

23-19 $0.100 \dfrac{kQ^2}{a^2}(\mathbf{i} + \mathbf{j} + \mathbf{k})$ with the particular charge (+ or −) at the origin (0, 0, 0). The nearest neighbor charges are at $(a, 0, 0)$, $(0, a, 0)$, and $(0, 0, a)$; the next nearest charges are at $(a, a, 0)$, $(a, 0, a)$ and $(0, a, a)$; the most remote charge is at (a, a, a).

23-21 At $z = \pm 0.707a$

23-23 6.32×10^5 m/s²

23-25 11.2×10^{-3} g

23-27 C

23-29 1.81×10^3 N

23-31 $\Delta q/q = 9.00 \times 10^{-19}$

23-33 $2.0 \times 10^{-19} \equiv$ fraction of electrons

23-35 6.54 N/C

23-37 C

23-39 9.00 mm from the 4 μC charge and 11.0 mm from the 6 μC charge

23-41 F

23-43 A

23-45 E

23-47 (a) **E** is normal to the side with the two negative charges and directed toward it; $|\mathbf{E}| = 4\sqrt{2}\,\dfrac{kQ}{a^2}$ where $4\sqrt{2}\,k = 5.09 \times 10^{10}$ N·m²/C²

Chapter 24   Continuous Distributions of Electric Charge

24-1 9.0×10^5 N/C

24-3 (a) $\sigma = 3.98 \times 10^{-4}$ C/m² (b) $\sigma = \rho\Delta r = 2.98 \times 10^{-2} \Delta r$

24-5 3.60×10^{10} N/C normal to the sheet and directed toward it

24-7 $\theta = 13.0°$

24-9 $E(z) = \dfrac{kQ}{z^2 - a^2}$ for $z < -a$ and for $z > a$

24-11 $\dfrac{2kQ}{R^2}\left(1 - \dfrac{1}{\sqrt{1 + (R/z)^2}}\right)$ directed away from the disk and along the z axis (the axis of symmetry)

24-13 $E_x(x) = 4k\sigma \tan^{-1}(a/x)$

24-15 8.22×10^{-9} C

24-17 (a) 9.60×10^{-17} N (b) $F_e = 2.30 \times 10^{-18}$ N
(c) $mg = 8.94 \times 10^{-30}$ N
24-19 (a) 56.9×10^{-9} s (b) 5.69 cm
24-21 C
24-23 5.20×10^2 N·m²/C
24-25 $(\phi_E)_{max} = 1.01$ N·m²/C
24-27 $Q/6\epsilon_0$
24-29 D
24-31 B
24-33 D

24-35 $E = \left(\dfrac{\rho}{3\epsilon_0}\right) r = \left(\dfrac{kQ}{R^3}\right) r$ for $r \le R$;

$E = \left(\dfrac{\rho R^3}{3\epsilon_0}\right) \dfrac{1}{r^2} = \dfrac{kQ}{r^2}$ for $r \ge R$

Note: $E = \left(\dfrac{\rho}{3\epsilon_0}\right) r = \dfrac{kQ}{R^2}$ for $r = R$; $Q \equiv \rho(\tfrac{4}{3}\pi R^3)$

24-37 $g(r) = \left(\dfrac{GM}{R^2}\right) \dfrac{r}{R}$ for $r < R$; $g(r) = \dfrac{GM}{r^2}$ for $r < R$;

$g(r = R) = \dfrac{GM}{R^2}$

24-39 $E = \left(\dfrac{\rho}{2\epsilon_0}\right) r = 2\pi k \rho r$ for $r \le R$; $E =$

$\left(\dfrac{\rho R^2}{2\epsilon_0}\right) \dfrac{1}{r} = (2\pi k \rho R^2) \dfrac{1}{r}$ for $r \ge R$. Note: $E(r = R) = 2\pi k \rho R$
24-41 Inside: $E = -4 k\rho_0 t \cos(\pi x/t)$; outside: $E = 4 k\rho_0 t$
24-45 $\sigma = 26.6 \mu C/m^2$
24-47 C
24-49 B
24-51 $Q = 4.53 \times 10^5$ C
24-53 D

Chapter 25   Electric Potential
25-1 E
25-3 8.96×10^4 tons of ice
25-5 (a) -18.0 kV (b) -12.7 kV (c) -6.00 kV
(d) $+1.9$ kV
25-7 (a) For $r < R$: $V = \pi k\rho(R^2 - r^2)$ (b) For $r > R$:
$V = 2\pi k\rho R^2 \ln r/R$ where ρ is the volume charge density.
Note that $\pi k = 2.83 \times 10^{10}$ N · m²/C² and
$2\pi k = 5.65 \times 10^{10}$ N · m²/C²
25-9 A
25-11 $U = -5.40$ J
25-13 D
25-15 D

25-17 (a) $r_a = \dfrac{4}{5} R = \dfrac{32}{15} \dfrac{kZe^2}{mv_0^2}$ (b) $r_b = \dfrac{3}{4} R = 2 \dfrac{kZe^2}{mv_0^2}$

25-19 $U = \dfrac{3}{5} \dfrac{GM_E^2}{R_E} = 2.24 \times 10^{32}$ J where M_E and R_E are
earth's mass and radius, respectively
25-21 E

25-23 $\Delta K = 40.0$ eV $= 6.40 \times 10^{-18}$ J
25-25 12 V
25-27 0.442 mm
25-29 (a) C (b) G (c) B and H
25-31 No.
25-33 E
25-37 A
25-39 B
25-41 B
25-45 (a) $V_L = 2.52V_0$ (b) $\Delta U = U - U_0 = 3.04 V_0^2 r/k$. In
the large droplet, the ratio of charge to radius is reduced.
25-47 (a) 2.22×10^{-4} C (b) 2.22×10^{-5} C
(c) $E_1 = 2.00 \times 10^6$ N/C for 1-m radius; $E_{0.1} = 20.0 \times 10^6$
N/C for 10-cm radius. For a given voltage, a larger
radius of curvature means a smaller value of E.
25-49 $Q_1 = 67 \mu C$, $Q_2 = 133 \mu C$ and $Q_3 = 200 \mu C$,
where the subscripts refer to the sphere's radius in meters.
25-51 (a) $E = 2.09 \times 10^6$ N/C (b) $E = 1.74 \times 10^4$ N/C
(c) $\lambda = 1.16 \times 10^{-8}$ C/m

Chapter 26   Capacitance and Dielectrics
26-1 B
26-5 B
26-7 (a) 4.43×10^{-7} m (b) 16.6 V
26-9 F
26-11 (a) $q_1 = q_2 = q_4 = 68.6 \times 10^{-12}$ C; $V_1 = 68.6$ V,
$V_2 = 34.3$ V, $V_4 = 17.1$ V (b) $q_1 = 120 \times 10^{-12}$ C,
$q_2 = 240 \times 10^{-12}$ C, $q_4 = 480 \times 10^{-12}$ C; $V_1 = V_2 = V_4 =$
120 V where the subscripts refer to the capacitance in pF.
26-17 B
26-19 (a) $q_3 = 120 \mu C$, $V_3 = 40$ V; $q_2 = 40 \mu C$, $V_2 = 20$ V;
$q_4 = 80 \mu C$, $V_4 = 20$ V (b) $q_3 = 180 \mu C$, $V_3 = 60$ V;
$q_2 = 0$, $V_2 = 0$. Note: the subscripts refer to the
capacitance in μF.
26-21 $q_1 = q_2 = 6.55 \mu C$; $q_3 = 29.45 \mu C$ where the
subscripts refer to the capacitance in μF.
26-23 3 in series: $C = C_0/3$; 3 in parallel: $C = 3C_0$; 1 in
series with 2 in parallel: $C = 2C_0/3$; 1 in parallel with 2 in
series: $C = 3C_0/2$
26-25 Still Q_1 on C_1, Q_2 on C_2 and Q_3 on C_3

26-27 $\left(\dfrac{\kappa + 3}{4}\right) C_0$

26-29 C
26-31 (a) remain unchanged (b) decrease
(c) increase (d) increase (e) remain unchanged
(f) increase
26-33 (E)
26-35 C
26-37 5.74 cm³ per cm length. Put a dielectric sheath
around the inner conductor.
26-39 1.96×10^{-9} J
26-41 Subscripts refer to capacitance in μF. Note
$C_{12} = \kappa(4 \mu F) = 12 \mu F$

(a) Before: $V_2 = 10.3$ V, $V_4 = 1.71$ V $= V_8$; after: $V_2 = 10.91$, $V_8 = 1.091$ V $= V_{12}$

(b) Before: $q_2 = 20.6$ μC, $q_4 = 6.86$ μC, $q_8 = 13.7$ μC
After: $q_2 = 21.8$ μC, $q_8 = 8.73$ μC, $q_{12} = 13.1$ μC

(c) Before: $U_2 = 105.8$ μJ, $U_4 = 5.88$ μJ, $U_8 = 11.76$ μJ
After: $U_2 = 119.0$ μJ, $U_8 = 4.76$ μJ, $U_{12} = 7.14$ μJ

26-43 Factor 4

26-47 (a) $\left[1 + (\kappa - 1)\dfrac{x}{L}\right]^{-1} V_0$ (b) $\left[1 + (\kappa - 1)\dfrac{x}{L}\right]^{-1}$
$(\frac{1}{2}C_0 V_0^2)$ where L is the dimension of the plate parallel to x.

26-49 $r = \dfrac{ke^2}{2mc^2} = 1.41 \times 10^{-15}$ m

26-51 (a) $\dfrac{3}{5}\dfrac{kQ^2}{R}$ (b) $\dfrac{1}{2}\dfrac{kQ^2}{R}$ (c) $U_a/U_b = 6/5 = 1.20$

26-53 (a) $U_{\text{Total}} = \left[\dfrac{\kappa}{\kappa - (\kappa - 1)\alpha}\right] U_0$; $U_A/U_{\text{Total}} =$
$\dfrac{\kappa(1 - \alpha)}{\kappa - (\kappa - 1)\alpha}$; $U_D/U_{\text{Total}} = \dfrac{\alpha}{\kappa - (\kappa - 1)\alpha}$ where $U_0 \equiv$
$(\frac{1}{2}C_0 V_0^2)$ with $C_0 \equiv \epsilon_0(A/d)$; note that C_0 is the capacitance when $\alpha = 0$ (i.e., when the dielectric is absent).
(b) $V_A = \left[\dfrac{\kappa(1 - \alpha)}{\kappa - (\kappa - 1)\alpha}\right] V_0$, $V_D = \left[\dfrac{\alpha}{\kappa - (\kappa - 1)\alpha}\right] V_0$
(c) $U_{\text{Total}} = 13.3 \times 10^{-3}$ J; $U_A/U_{\text{Total}} = 2/3$; $U_D/U_{\text{Total}} = 1/3$; $V_A = 66.7$ V; $V_D = 33.3$ V. Note that subscript A refers to the air gap and D refers to the dielectric.

Chapter 27   Electric Current and Resistance

27-1 600 C

27-3 12.4 A/(mm)^2 = 12.4×10^6 A/m^2

27-5 36 mm/s

27-7 (a) 3.93×10^{-6} A (b) 1.96 A/m^2

27-9 30.2 mA

27-11 37.5 s

27-13 C

27-15 C

27-17 (a) 2.22×10^{-6} A (b) 8.88 W (c) 1.37×10^3 years

27-19 (a) 14.1 V (b) 141 mA

27-21 D

27-23 D

27-25 (a) 223 W (b) 1.86 A

27-27 (a) 0.326 Ω (b) 4.75×10^6 A/m^2 (c) 8.15 V
(d) 81.5×10^{-3} V/m (e) 204 W

27-29 2.50 W

27-31 (a) 2.88 Ω (b) 41.7 A (c) 4.30×10^3 kcal (d) 4050 W. The actual power will be greater because the actual resistance will be less than 2.88 Ω

27-33 C

27-35 25.9×10^{-3} Ω

27-37 11.0 Ω

27-39 (a) ~ 14.5 s (b) 2.42 A. It was 2.40 A at 20 °C

27-41 (a) 26.7×10^6 $\Omega \cdot$m (b) 11.9×10^9 pores/m^2
(c) $\sim 9.15 \times 10^{-6}$ m

27-43 (a) 4 s (b) 0.347 RC

27-45 A

27-47 (a) 5.31×10^{-9} F (b) 1.7×10^{10} Ω = 17 kΩ
(c) 62.6 s

Chapter 28   DC Circuits

28-1 0.150 Ω

28-3 11.5 kJ

28-5 D

28-7 7.50 V

28-9 32.0 W

28-11 (a) 0.50 A counterclockwise (b) 1.5 W

28-13 20 V

28-17 C

28-19 1.46 Ω

28-21 A

28-23 4R in series, 37.5 W; 2R in series with 2R in parallel, 60 W; R in series with the combination of R in parallel with 2R in series, 90 W; R in series with 3R in parallel, 112.5 W; 2R in series in parallel with 2R in series, 150 W; R in parallel with 3R in series, 200 W; R in parallel with R in series with 2R in parallel, 250 W; 2R in series in parallel with 2R in parallel, 375 W; 4R in parallel, 600 W; where $R = V^2/P = 96$ Ω = resistance of a single coil

28-25 B

28-27 (a) 10 Ω (b) 30 V

28-29 $\frac{5}{6}R$

28-31 D

28-35 C

28-37 E

28-39 (a) Use 2.50×10^3 ohm series resistor (b) Use 4.80×10^{-3} ohm shunt

28-41 (a) 3.53 V (b) 17.14 V (c) 29.56 V (d) 29.98 V

28-43 (a) $R_a = 50$ Ω (b) $R_b = 11\,250$ Ω
(c) $R_c = 112\,500$ Ω

28-45 E

28-47 5 V bottom, 12 V top

28-49 (a) 1.5 W with three in series; (b) 1.6 W with three in series; (c) 1.6 W with three in parallel.

Chapter 29   The Magnetic Force

29-1 C

29-3 1.89×10^{-4} Wb

29-5 (a) 7.52×10^{-8} Wb (b) 2.74×10^{-8} Wb (c) 0

29-9 (a) Normal to page, directed inward (b) Normal to page, directed outward (c) normal to $\mathbf{v}$, parallel to page, directed downward (d) normal to $\mathbf{v}$, parallel to page, directed to the right (e) force is zero (f) normal to $\mathbf{v}$, parallel to page, directed downward

29-11 C
29-13 C
29-15 (a) 14.4 m (b) 1.52 kHz
29-17 (a) 2.80×10^6 Hz (b) 0.337 m
29-19 45°
29-21 5.08 mm
29-25 F
29-27 A
29-29 33.5 kV with upper plate positive
29-31 0.312 N into the paper
29-33 6.13×10^{-3} N/m
29-35 (a) 2.65×10^8 A/m² (b) Orient along east-west line with conventional current from west to east
(c) 199 kW would be the power needed to suspend a 1-m length of wire with a 1-cm² cross section. Clearly the wire would melt in a very short time.
29-37 IBR
29-39 (a) zero (b) 2.60×10^{-3} N·m
29-41 2.00×10^{-3} N·m
29-43 1.98 T
29-45 $\frac{1}{12}\omega L^2 Q B \sin\theta$

Chapter 30   Sources of the Magnetic Field
30-1 D
30-3 C
30-5 B
30-7 B
30-9 B
30-11 $\lambda = 3.34 \times 10^{-9}$ C/m
30-17 6.28×10^{-6} T

30-19 $B = \dfrac{\mu_0 I}{4R} = 3.14 \times 10^{-7}\,\dfrac{I}{R}$

30-21 $B_z = \dfrac{\mu_0}{2}\,\dfrac{R^2 I}{z^3} = 6.28 \times 10^{-7}\,\dfrac{R^2 I}{z^3}$ where $z \gg R$

30-23 (a) $B_a = 5.66 \times 10^{-7}\,I/a$ (b) $B_b \simeq 5.36 \times 10^{-7}\,I/a$ so $B_a/B_b = 1.055$ (c) No. The approximation is independent of a.
30-25 A
30-27 A

30-29 (a) 0 for $r < R_1$ (b) $\dfrac{\mu_0 I(r^2 - R_1^2)}{2\pi r (R_2^2 - R_1^2)}$ for $R_1 < r < R_2$

where $\mu_0/2\pi = 5.00 \times 10^{-8}$ Wb/A·m (c) $\left(\dfrac{\mu_0}{4\pi}\right)\dfrac{2I}{r}$ for

$r > R_2$ where $\left(\dfrac{\mu_0}{4\pi}\right)(2) = 2 \times 10^{-7}$ Wb/A·m

30-31 $\frac{1}{2}\mu_0 j t$
30-33 C
30-35 15.7×10^{-3} T
30-37 F
30-39 3.20×10^{-3} T

Chapter 31   Electromagnetic Induction
31-1 A, B, C, D
31-3 1.42 Hz
31-5 0.533 V
31-7 B
31-9 1.88×10^3 V
31-11 $|\mathcal{E}| = 126$ V
31-13 0.136 V
31-15 0.711 A

31-17 $|\mathcal{E}| = \mu_0 x\,\dfrac{di}{dt}\,(1 - \sqrt{1 - (R/x)^2})$

31-19 0.815 G

31-21 $\left|\dfrac{d\phi_B}{dt}\right| = \frac{1}{3}\mathcal{E}_0$. **B** either up from page and decreasing or down and increasing.
31-23 $\Delta\phi = 4.80 \times 10^{-2}$ Wb
31-25 G

31-29 $\dfrac{mgR}{(aB)^2}$ (= constant)

31-31 (a) 24.0 m/s² (b) 150 m/s $\simeq$ 336 mph
31-33 R_1/R_2
31-37 B
31-39 B
31-41 D
31-45 (a) zero current (b) $\frac{1}{2}\omega a^2 B$

Chapter 32   Inductance and Electric Oscillations
32-1 1 mH
32-3 Change current at rate of 40.0 A/s
32-5 D
32-7 $\phi = 1.0$ Wb/N where N is the number of turns
32-9 $|\mathcal{E}| = 1.97 \times 10^{-6}\,(R_2^2/R_1)\,di/dt$
32-13 indeterminate
32-19 (a) 6.91 (L/R) (b) 7.60 (L/R)

32-21 (b) $\left(\dfrac{V_0}{6.00 \times 10^5}\right)\tau = (1.67 \times 10^{-6}\,V_0)\,\tau$

32-23 D
32-25 (a) $i_1 = i_2 = 3.53$ A (b) $i_1 = 4.8$ A, $i_2 = 3.0$ A
(c) $i_1 = 0$, $i_2 = -1.8$ A (d) $i_1 = i_2 = 0$
32-27 (a) 43.3 W (b) 45.6 W (c) 2.27 W
32-29 (a) 197 mH (b) 98.7 mJ
32-31 (a) 4.00 mJ (b) 0.750 mJ
32-33 D
32-35 $1.59 \times 10^{-8}(i/r)^2$ in J/m³
32-37 $(\mu_0/8)(Ni/R)^2$
32-39 0.504 T
32-41 (a) 0.989 pF (b) 9.02 pF

32-43 From $\frac{1}{2}f_0$ to $\sqrt{2}f_0$ where $f_0 = \dfrac{1}{2\pi}\dfrac{1}{\sqrt{LC}}$. Note: C is the *minimum* capacitance of a single capacitor and L is the fixed inductance

32-45 (a) Resonant frequency doubled. (b) Total resistance doubled.

32-47 A massless object that is subject to both a spring force ($F = -kx$) and a damping force ($F_D = -kv$).

Chapter 33   AC Circuits
33-1 377 rad/s
33-3 A, B, C
33-5 (a) 2.53 A (c) 6.40×10^{-3} J
33-7 A
33-9 (a) 0.159 pF (b) 1.59×10^{-9} H
33-11 31.7 pF

33-13 (a) $\dfrac{1}{\sqrt{1 + (1/\omega CR)^2}}$

33-15 B
33-17 4 μs
33-19 B
33-21 (a) 20 Ω (b) 286 kΩ (c) 106 kHz
33-25 (a) 16.9 mH (b) 28.1 Ω (c) At 100 Hz, $\phi = 20.7°$; at 500 Hz, $\phi = 62.1°$
33-27 (a) 497 Hz
33-29 E
33-31 (a) 2 A (b) 1.41 A (c) 144 Hz (d) $\theta = 288.375\ \pi$
33-33 D
33-35 (a) 7.07 V (b) 1.77 V (c) 2.89 V
33-37 84.9 W
33-39 (a) $I = 5.00$ A, $\phi = 0°$ (b) $I = 2.73$ A, $\phi = -56.9°$ (c) $I = 3.81$ A, $\phi = +40.4°$
33-41 (a) I_{max} may either increase or decrease; both ω_0 and $\Delta\omega$ decrease. (b) I_{max} may either increase or decrease; ω_0 decreases; $\Delta\omega$ does not change. (c) I_{max} decreases; ω_0 does not change; $\Delta\omega$ increases.
33-43 (a) $R = 24\ \Omega$, $L = 42.4$ mH, $C = 55.3\ \mu$F (b) $P_R = 150$ W, $P_L = 0 = P_C$
33-45 B

Chapter 34   Maxwell's Equations
34-1 D
34-3 C
34-7 (a) $I_R = 75.2$ mA $= I_d$ (b) $I_R = 1.00$ A $= I_d$
34-9 C
34-11 D
34-13 D
34-15 (a) A·m

Chapter 35   Electromagnetic Waves
35-1 C
35-3 D
35-5 C
35-7 A

35-9 (a) 470 nm (b) Blue
35-11 5.26×10^{14} Hz
35-13 E
35-15 3.33×10^{-8} J/m³
35-17 (a) 6.43×10^7 W/m² (b) 2.20×10^5 N/C (c) 7.34×10^{-4} T
35-19 (a) 1.0×10^5 J (b) 3.3×10^{-4} kg · m/s (c) 3.3×10^{-5} N (d) 3.0×10^9 m
35-23 (a) 24 nm (b) 0.20 μm (c) 6.5×10^{-7} J
35-25 0.875 I/c
35-27 (a) 2.41×10^{-23} kg · m/s (b) 7.23×10^{-15} J, 0.0072 μW

Chapter 36   Ray Optics
36-1 610 cm
36-3 90.0 cm
36-7 (b) Doubles the range
36-15 A
36-17 1.63
36-19 2.26×10^8 m/s
36-21 C
36-25 4.26 cm
36-27 40.6°

36-29 $\theta = \cos^{-1}\left(\dfrac{n^2 - 2}{2}\right)$

36-31 A
36-33 1.21 m
36-43 78.5°
36-45 If index $n \geq \sqrt{2}$, then $\theta \leq 45°$ so that $\left(90° - \sin^{-1}\dfrac{1}{n}\right) \geq \theta \geq \sin^{-1}\dfrac{1}{n}$.

36-47 C (If A were true, n could be less than 1.414.)

Chapter 37   Thin Lenses
37-1 E
37-3 5.01 mm
37-7 C
37-11 (a) 1.33 m (b) 3.00 m (c) 0.200 m
37-15 (a) Light intensity cut in half (b) Successive steps of 2
37-17 D
37-21 E
37-23 At $s = 35.7$ cm from her eye
37-25 B
37-27 Far-sighted
37-29 E
37-31 +10.0 cm
37-33 (a) 2.40 m (b) 2.42 m
37-35 1.05 cm
37-37 D
37-39 (a) 52 cm (b) 25 (c) 51.85 cm
37-41 B

37-43 A

37-45 (a) Converging lens (b) 150 cm using $n_{21} = 1.20$

37-47 $f = \left(\dfrac{1}{n-1}\right)\left(\dfrac{d^2 + t^2}{8t}\right)$

37-49 E

Chapter 38   Interference

38-3 (a) R_2 (b) Signal strength the same at R_1 and R_2, (c) R_2, (d) R_2

38-5 22.5 m

38-9 A

38-11 0.37 mm

38-13 (a) 1.49 mm (b) 0.435 mm

38-17 A

38-19 (a) 18.7° (b) 39.8° (c) 73.7°

38-21 (a) 17.9° (b) 38.0° (c) 67.4°

38-23 3rd order (corresponds to $\theta = 49.9°$)

38-25 C

38-27 $\lambda = 648$ mm

38-29 B

38-33 As you move laterally across the area of intersection, the phase relationship between the two beams varies.

38-35 0.443 mm

38-37 40.0 μm

38-39 8.21×10^{-7} m

38-45 0.181 mm

38-47 $n = 1.000\ 288$

38-49 $n = 1.000\ 300$

Chapter 39   Diffraction

39-1 E

39-3 (a) 51.5° (b) 25.1°

39-7 D

39-9 (a) 180° (b) 60° (c) 23° (d) 11.4°

39-11 (a) no (b) no (c) yes (d) no

39-13 (a) $3.15° \simeq 5.50 \times 10^{-2}$ rad (b) 6.60 cm

39-15 (a) $\phi = 6.12$ rad $= 351°$ (b) 2.66×10^{-2} (c) 7.09×10^{-4}

39-17 A

39-19 The maximum for which $\dfrac{d \sin \theta}{\lambda} = m = 8$ at angle $\theta_8 = 5.00 \times 10^{-3}$ rad $= 0.287°$

39-21 B

39-23 Yes. The reflecting surface acts as an aperture in that it reflects (and thereby isolates) only a portion of the incident light.

39-25 1.21 km

39-27 3.95 mm

39-31 60° and 180°

Chapter 40   Special Relativity

40-1 A

40-3 $0.976c$

40-5 (a) $0.198c$, (b) $0.200c$

40-7 D

40-9 E

40-11 $0.6c$

40-13 $0.87c$

40-15 A

40-17 880

40-19 600 km

40-21 (a) $L_0 [\sin^2 \theta_0 + \cos^2 \theta_0 (1 - v^2/c^2)]^{1/2}$ (b) $\tan^{-1} [\tan \theta_0/\sqrt{1 - v^2/c^2}]$

40-23 (a) 9.39 MeV/c (b) 94.5 MeV/c (c) 542 MeV/c (d) 1.93×10^3 MeV/c

40-25 0.563

40-27 (a) 51.1 keV (b) $0.417c$

40-29 4.53×10^{-14} J

40-31 (a) 100 MeV/c (b) 10 MeV/c

40-33 (a) 0.659 MeV (b) 2.45 MeV

40-35 (a) $0.313c$ (b) 4.89 percent

40-37 (a) 100 000 y (b) 297 s

40-39 (a) No (b) Yes (c) No (d) Yes (e) No

40-41 (a) $0.405\ m_0 c^2$ (b) $1.29\ m_0 c^2$

40-45 4.7×10^{-12} kg

40-47 0.25 kg

40-49 42 mg

40-51 3.1×10^{13}

40-53 9×10^6

Chapter 41   Quantum Theory

41-1 D

41-3 7.5×10^{31}

41-5 544 nm

41-7 (a) 7.7×10^{31} (b) 9.2×10^{21}

41-9 (a) 685 nm (b) 0.61 eV (c) 3.47×10^5 m/s

41-11 (a) 5 min (b) 4.88 eV (c) 3.68×10^{-12} s^{-1}

41-15 (a) 17.6 MeV/c (b) 21 keV

41-17 A

41-19 0.04 μm

41-21 4.86 pm

41-23 0.058 fm for Na, 0.048 fm for Cl

41-27 B

41-29 0.82 pm

41-31 0.85 MeV

41-35 71°

41-37 C

41-39 1.7×10^{-34} m

41-41 (a) same (b) particle greater (c) photon greater

41-43 7.8 pm

41-47 C

41-49 0.21 MeV

41-51 (a) 6.65×10^{-27} J (b) 7.9×10^{-9} s (c) 2.38 m

41-57 E

41-59 D
41-61 0.33 pJ, 1.32 pJ, 2.96 pJ
41-63 (a) 0 (b) 10% (c) 20% (d) 10% (e) 0

Chapter 42   Atomic Structure
42-1 E
42-3 No, because of angular momentum conservation
42-5 (a) 0.22 eV (b) 0.22 eV
42-7 10^{17} kg/m³
42-9 (a) 2.3 fm
42-11 B
42-13 10.2 eV
42-15 E
42-17 54.4 eV
42-19 3.4 eV

42-21 12.8 eV
42-23 5.1 eV
42-25 (a) 12.1 eV (b) 3.9×10^4 m/s (c) 12.1 eV
(d) 7.8×10^{-8} eV
42-27 (a) 2.82 keV, (b) 256 fm
42-29 5×10^4 K
42-31 (a) -2.6×10^{-78} eV (b) 1.2×10^{29} m
42-35 (a) 15.9 eV (b) -6.4 eV, -4.6 eV, -2.9 eV,
-1.5 eV (c) 3.5 eV (d) 1.8 eV
42-37 254 nm
42-39 D
42-41 2.1×10^{-3} eV
42-45 (a) $_{11}$Na (b) $_{18}$Ar (c) $_{27}$Co
42-47 C
42-49 (a) 265 eV (b) 42.2 eV
42-51 4.3×10^{-6}
42-53 (a) 4 p $\rightarrow$ 3 s (b) 3 p $\rightarrow$ 4 s

Index

Credits *(Continued from copyright page)*

Researchers (bottom) Courtesy of the National Zoological Park / The Smithsonian Institution. **Figure 4-10:** PSSC Physics, 2nd edition, 1965; D. C. Heath and Company with Educational Development Center, Inc., Newton, MA. **Figure 4-14:** Ira Kirschenbaum / Stock, Boston. **Figure 6-6:** Courtesy of James L. Horton, Robert Marston and Associates, Inc. **Figure 6-13(b):** Museum of Modern Art Film Stills Archive. **Figure 6-38:** Dan Helms / Duomo Photography. **Figure 7-5:** Courtesy of NASA. **Figure 8-9:** Alexander Calder. *Dots and Dashes.* 1959. Pained sheet metal and wire. 60 inches wide. Collection of Howard and Jean Lipman. Photograph by Jerry L. Thompson, N.Y., courtesy of the Whitney Museum of American Art. **Figure 8-16:** Photo by Carlin / The Picture Cube. **Figure 8-17:** Courtesy of Dr. Harold Edgerton, MIT, Cambridge, MA. **Figure 9-5:** Dan Helms / Duomo Photography. **Figure 10-20.** Courtesy of CERN. From *CERN Courier*, Vol. 23, no. 9, November 1983 p. 374. **Figure 14-13:** Courtesy of Sperry Corp. **Figure 14-14:** (left) Paul Sutton / Duomo Photography; (right) Adam J. Stolman / Duomo Photography. **Figure 14-15:** Courtesy of Dr. Harold Edgerton, MIT, Cambridge, MA. **Figure 16-19:** Courtesy of Deutsches Museum. **Figure 20-1:** Princeton Gamma Tech; photo courtesy of Tom Adams. **Figure 20-2:** Courtesy of Field Emission Laboratory, The Pennsylvania State University. **Figure 23-2:** Courtesy of MIT Museum Massachusetts Institute of Technology. **Figure 26-4:** Allyn and Bacon file photo. **Figure 28-2:** Burndy Library. **Figure 29-2:** Fermi National Accelerator Laboratory. **Figure 30-1:** Educational Development Center. **Figure 30-18:** Fermi National Accelerator Laboratory. **Figure 30-22:** After photo by R. W. de Blois in D. Halliday and R. Resnick, *Fundamentals of Physics*. New York: John Wiley, 1981. **Pages 742, 743:** The Institution of Electrical Engineers, London. **Figure 36-10:** Educational Development Center. **Figure 36-19:** Bell Laboratories. **Figure 38-6:** Educational Development Center. **Figure 38-10:** Addison-Wesley Publishing Company. **Page 826:** Ewing Galloway, New York. **Page 827:** Young, Thomas. *A Course of Lectures on Natural Philosophy and the Mechanical Arts.* London: Taylor and Walton, 1845. Courtesy of AIP Niels Bohr Library. **Figure 38-18b:** Bausch and Lomb. **Figures 39-5, 39-7:** Addison-Wesley Publishing Company. **Figure 39-8:** Allyn and Bacon file photo. **Figures 39-9b, 39-10c:** M. Cagnet, M. Francon, J. C.Thrier, *Atlas of Optical Phenomena.* Berlin: Springer, 1962. Courtesy of Springer-Verlag, Heidelberg. **Figure 39-11:** Dan McCoy/Rainbow. **Figure 39-15b:** Eastman Kodak Company. **Figure 41-8:** G. D. Rochester and J. G. Wilson, *Cloud Chamber Photographs of The Cosmic Radiation.* Pergamon Press, 1952. **Figure 41-10:** Courtesy RCA Laboratories. **Figure 42-8c:** John R. Van Wazer and Ilyas Absar, *Electron Densities in Molecules and Molecular Orbitals.* New York: Academic Press, 1975.

Constants

Acceleration due to gravity (g)	9.81 m/s^2
Gravitational constant (G)	6.67×10^{-11} N·m^2/kg^2
Gas constant (R)	8.31 J/mol·K
Avogadro's number (N_A)	6.02×10^{23} mol^{-1}
Boltzmann constant (k)	1.38×10^{-23} J/K
Earth: Mass	5.98×10^{24} kg
Radius (mean)	6.38×10^6 m
Distance (mean) to moon	3.84×10^8 m
Distance (mean) to sun	1.49×10^{11} m
Density of water (20 °C)	1.00×10^3 kg/m^3
Standard atmospheric pressure	1.01×10^5 Pa
Volume of 1 mole ideal gas at STP	22.4 ℓ
Absolute zero temperature	-273 °C
Coulomb-law constant ($k_e = 1/4\pi\epsilon_0$)	9.0×10^9 N·m^2/C^2
Permittivity of free space (ϵ_0)	8.85×10^{-12} C^2/N·m^2
Magnetic interaction constant ($k_m = \mu_0/4\pi$)	10^{-7} T·m/A
Permeability of free space (μ_0)	$4\pi \times 10^{-7}$ N/A^2
Electron charge (e)	1.60×10^{-19} C
Speed of light (vacuum) (c)	3.00×10^8 m/s
Planck's constant (h)	6.63×10^{-34} J·s
Electron mass	9.11×10^{-31} kg $= 0.511$ MeV/c^2
Proton mass	1.67×10^{-27} kg $= 938$ MeV/c^2

Values are to three significant figures. See Appendix C for further information.